Twenty Years Research and Development on Soil Pollution and Remediation in China

Yongming LUO

Science Press
Beijing

Responsible Editor: Dan ZHOU

Published by Science Press
16 Donghuangchenggen North Street
Beijing 100717, P. R. China

Printed in Beijing

ISBN 978-7-03-049898-4

List of Contributors

Xinde CAO	Shanghai Jiao Tong University
Liyuan CHAI	Central South University
Mengfang CHEN	Institute of Soil Science, Chinese Academy of Sciences (CAS)
Shibao CHEN	Chinese Academy of Agricultural Science
Tongbin CHEN	Institute of Geographic Sciences and Natural Resources Research, CAS
Zueng-Sang CHEN	Taiwan University
Xinbin FENG	Institute of Geochemistry, CAS
Shuhai GUO	Institute of Applied Ecology, CAS
Qing HU	South University of Science and Technology of China
Daoyou HUANG	Institute of Subtropical Agriculture, CAS
Shenfa HUANG	Shanghai Academy of Environmental Sciences
Rong JI	Nanjing University
Lin JIANG	Beijing Municipal Research Institute of Environmental Protection
Fasheng LI	Chinese Research Academy of Environmental Sciences
Guanghe LI	Tsinghua University
Huafen LI	China Agricultural University
Shunpeng LI	Nanjing Agricultural University
Shupeng LI	BCEG Environmental Remediation CO., Ltd.
Yongtao LI	South China Agricultural University
Xiaoyong LIAO	Institute of Geographic Sciences and Natural Resources Research, CAS
Yusuo LIN	Nanjing Institute of Environmental Sciences, MEP
Chunling LUO	Guangzhou Institute of Geochemistry, CAS
Yongming LUO	Institute of Soil Science, CAS; Yantai Institute of Coastal Zone Research, CAS
Lena Q. MA	Nanjing University
Zhongren NAN	Lanzhou University
Xiangliang PAN	Zhejiang University of Technology; Xinjiang Institute of Ecology and Geography, CAS
Rongliang QIU	Sun Yat-sen University
Zhenguo SHEN	Nanjing Agricultural University
Jiyan SHI	Zhejiang University

Hongwen SUN	Nankai University
Wenfeng TAN	Huazhong Agricultural University
Chen TU	Yantai Institute of Coastal Zone Research, CAS
Hailong WANG	Zhejiang A & F University; Foshan University
Hongqi WANG	Beijing Normal University
Ligang WANG	Jereh Environment Group
Li WEI	Beijing GeoEnviron Engineering and Technology
Ming-Hung WONG	The Education University of Hong Kong
Longhua WU	Institute of Soil Science, CAS
Qitang WU	South China Agricultural University
Jianmin XU	Zhejiang University
Minggang XU	Chinese Academy of Agricultural Sciences
Yingming XU	Agro-Environmental Protection Institute of Ministry of Agriculture
Xiaoe YANG	Zhejiang University
Fang YU	Chinese Academy for Environmental Planning, MEP
Hongzhen ZHANG	Chinese Academy for Environmental Planning, MEP
Wenhui ZHANG	CECEP DADI Environmental Remediation Co., Ltd.
Yongsheng ZHAO	Jilin University
Qixing ZHOU	Nankai University
Dongqiang ZHU	Nanjing University; Peking University
Lizhong ZHU	Zhejiang University
Yongguan ZHU	Research Center for Eco-Environmental Sciences, CAS; Institute of Urban Environment, CAS

Preface

Soil is an important component of the earth surface system and fundamental to life processes in the terrestrial ecosystems. It is also essential for agricultural production and thus is the material foundation that supports human life. In the past 30 years, with the rapid development of industrialization, urbanization and agricultural intensification in China, large amounts of anthropogenic emissions of different types of pollutants have entered the soil environment through multiple pathways, causing soil pollution. The pressure on soil safety and health is increasing, and the social demand for the control and remediation of polluted soils is extremely urgent. The most important tasks for China's soil environmental protection in this new era are: (i) Strengthening fundamental research, (ii) Developing techniques and equipments, (iii) Establishing a regulatory system, and (iv) developing industrial support on soil pollution and remediation.

In the 1970s, China soil environmental protection focused on preventing the pollution of agricultural soil in sewage irrigated areas. In the 1980s, significant progress was made in the research on pollution control of organochlorine pesticides and heavy metals, including arsenic and chromium. In the early 1990s, the soil environmental background values on a national scale and its regional differentiation were determined. The National Soil Environmental Quality Standard was enacted in 1995 to provide a basis for soil environmental management and pollution prevention. Research on soil remediation and phytoremediation started in the mid 1990s, and the 1st "International Conference on Soil Remediation (SOILREM 2000)" was held in Hangzhou, China in 2000. At the start of the 21st century, the National "973" Program, National Natural Science Foundation of China (NSFC), and the Knowledge Innovation Projects of Chinese Academy of Sciences funded research projects on themes focused on the evolution of rules of soil environmental quality and soil remediation mechanisms. Remediation technologies for polluted soil were also included into the National "863" Program, together with air and water environmental protection technologies. A national soil pollution survey and prevention project was carried out, and the soil pollution status at the national scale was mapped. "Soil Pollution and Remediation" was also included as a discipline for applications in earth sciences by the NSFC. To comply with the practical needs of the soil environmental protection and the developmental needs of soil environmental science and technology, a series of scientific research and thematic projects on soil pollution and remediation were funded nationally. These projects stimulated the nationwide development of scientific research and technology on soil pollution and remediation science, as well as the preliminary establishment of remediation technology systems of heavy metals, persistent organic pollutants, petroleum hydrocarbons and pesticides contaminated soils.

Meanwhile, to strengthen scientific and technical research, personnel training, and innovation on soil environment and remediation, a number of key laboratories and engineering centers have been approved and set up. These include the "Key Laboratory of Soil Environment and Pollution Remediation, Institute of Soil Science, Chinese Academy of Sciences", the "State Environmental Protection Key Laboratory of Soil Environmental Management and Pollution Control" and the national engineering laboratories of soil remediation technologies for farmland and industrial sites. A number of soil remediation enterprises such as BCEG Environmental Remediation Co., Ltd., CECEP DADI Environmental Remediation Co., Ltd., Shanghai Soil Remediation Center, Chongqing Soil Remediation Center, Yonker Environmental Protection Co., Ltd., as well as China Environmental Remediation Industry Alliance have been formed, which effectively expanded the market for soil remediation and promoted industrial development.

A group of academic societies were founded including Soil Environment Committee and Soil Remediation Committee of Soil Science Society of China, as well as Soil and Groundwater Environment Committee of Chinese Society for Environmental Sciences. Several national and international academic conferences and workshops on remediation of polluted soil and groundwater were organized. In particular, the quadrennial "International Conference on Soil Pollution and Remediation (SOILREM)" was launched in 2000 by the Institute of Soil Science, CAS. This has played a leading and motivating role in strengthening research, development, knowledge spreading, international exchange and cooperation on soil pollution and remediation in China. We believe that the "Action Plan for Soil Pollution Prevention and Control" and the future implementation of key science and technology programs on "Soil Pollution Prevention and Control" will further stimulate soil pollution prevention and the remediation industry in China.

The 5th "International Conference on Soil Pollution and Remediation" (SOILREM 2016) and the 13th "International Conference on Phytotechnology" (PHYTOTECH 2016) in September 2016, Hangzhou, China, will showcase the progress and achievements of 20 years research in China to global specialists, scholars, government regulators and industrial delegates who work on soil pollution and remediation. We invited outstanding scientists, engineers and managers who are active in the frontier of soil pollution and remediation research in China, to contribute to a systematic review of the long-term achievements made by the various groups. The positive response to this proposal by many colleagues resulted in 50 review and research articles from all over China, including Hong Kong and Taiwan regions. The authors include scientists, engineers, entrepreneurs and managers from 26 universities, 18 institutes, 4 leading enterprises and 2 government environmental protection departments. The contents cover fundamental research on soil pollution and remediation, technical development, project demonstration, policy and governance. The polluted soil/site types include farmland, industrial sites, mining areas and oilfields, with heavy metals (cadmium, arsenic, copper, chromium, mercury, lead, zinc, nickel, etc.), organic pollutants (PAHs, PCBs, organochlorine pesticides, phthalate esters, halogenated hydrocarbons, etc.), and metal–organic combined pollutants. The remediation techniques mainly include physical and chemical remediation (thermal desorption, soil vapor extraction, *in situ* advanced chemical oxidation, solidification and stabilization), phytoremediation (phytostabilization, phytoextraction by hyperaccumulators, phytoresistence by low accumulation plants), bioremediation (microbial absorption and immobilization, microbial degradation, microbe-enhanced phytoremediation), and the combined remediation by multiple technologies. The governance and policy section mainly consists of laws and regulations, criteria and standards, financial guarantee and industrial market on soil environment and pollution prevention.

These 50 manuscripts make up this book, "Twenty Years Research and Development on Soil Pollution and Remediation in China". We gratefully acknowledge the understanding and support provided by all the contributing authors, peers and colleagues. We hope the publication of this book will be helpful to our national and international peers for the better understanding of the research status and development trends on soil pollution and remediation in China. We also hope that it will promote the collaborative innovation and development of the industry, education, research, and management on soil pollution and remediation in China.

The framework of this book was outlined by Professor Yongming LUO. All the chapters were edited and reviewed by Dr. Chen TU and Professor Yongming LUO. We greatly appreciate all the constructive criticisms and corrections by our peers that made up for our knowledge limitations!

Professor Yongming LUO
Chairman of SOILREM 2016 and PHYTOTECH 2016
August 18, 2016

Contents

List of Contributors
Preface

Part Ⅰ General Reviews

Contamination Processes and Bioremediation of Typical Persistent Toxic Substances in Agricultural Soils in China Yongming LUO, Longhua WU, Ying TENG, Jing SONG, Wuxing LIU, Haibo ZHANG, Chen TU ······ 3
Ecological Restoration of Man-Made Habitats, with Emphasis on Metal Contaminated Sites and Domestic Landfills Ming-Hung WONG ······ 14
Green and Sustainable Remediation Movement in the New Millennium and Its Relevance to China Deyi HOU, Guanghe LI ······ 38
20th Anniversary for Soil Environment Protection in China Yusuo LIN, Guoqing WANG ·· 52
Progress in the Risk Management of Contaminated Sites: Research Activities and Environmental Management of Contaminated Sites at CRAES Qingbao GU, Guanlin GUO, Jin MA, Youya ZHOU, Zengguang YAN, Yunfeng XIE, Bing YANG, Li LIU, Hong HOU, Xiaoming DU, Nandong XUE, Yunzhe CAO, Fujun MA, Ping DU, Liping BAI, Fasheng LI ······ 60
Environmental Damage Assessment Methods for Soil and Groundwater Contamination Fang YU, Dan ZHAO, Ji QI, Jing YUAN, Yanshen ZHANG, Zhihong ZHANG, Hui XIE ······ 81
Ten-Year Review and Prospect of Industrial Contaminated Site Remediation in China Yanli GAO, Shupeng LI, Yanwei WANG, Shaoguo KANG, Yunxiao WEI ······ 96
The Development of China Contaminated Site Remediation Industry: A Decade in Review Qing HU, Yan ZHU, Sijie LIN, Hong WANG, Jingyang GAO ······ 114
The Exploration and Practice on Soil Environmental Protection in the Process of Rapid Urbanization of the Megacity Shanghai Shenfa HUANG, Min WANG, Jian WU, Qingqing LI, Jie YANG, Lin GUO, Jin WANG, Zhihao XU ······ 122

Part Ⅱ Speciation, Bioavailability and Risk Assessment

Copper Speciation and Transformation in Soil-Plant System Jiyan SHI, Lijuan SUN, Cheng PENG, Chen XU, Yuanpeng WANG, Xincai CHEN, Huirong LIN, Jianjun YANG, Tingting LIU, Yingxu CHEN ······ 139
Contribution of Soil Active Components to the Control of Heavy Metal Speciation Wenfeng TAN, Linchuan FANG, Juan XIONG, Hui YIN, Wei ZHAO ······ 152

Application of Oral Bioavailability to Remediation of Contaminated Soils: Method Development for Bioaccessible As, Pb and Cd Hongbo LI, Jie LI, Shiwei LI, Lena Q. MA····· 176

Biochar for Environmental Management: Impacts on the Sorption and Bioavailability of Organic Contaminants in Soil Hongwen SUN, Xinhao REN, Fei WANG, Wen ZHANG, Zunlong ZHOU······ 203

Advance in Health Risk Assessment Methodology of Brownfield Sites in China Lin JIANG, Maosheng ZHONG, Xiaoyang JIA, Tianxiang XIA······ 220

Derivation of Soil Generic Assessment Criteria under the Consideration of Background Exposure and NAPL Mengfang CHEN······ 238

Part Ⅲ Heavy Metal Pollution and Remediation

Principles and Technologies of Phytoremediation for Metal Contaminated Soils-A Review Xiaoe YANG······ 259

Phytoremediation of Cadmium Contaminated Soils Using the Cadmium and Zinc Hyperaccumulator *Sedum plumbizincicola* Longhua WU, Pengjie HU, Zhu LI, Tong ZHOU, Daoxu ZHONG, Yongming LUO······ 308

The Potential of Oilseed Rape and *Thlaspi caerulescens* for Phytoremediation of Cadmium-Contaminated Soil Dechun SU, Rongfeng JIANG, Huafen LI······ 324

Soil Contamination in Arid Region of Northwest China: Status, Mechanism and Mitigation Yahu HU, Zhongren NAN······ 337

Phytoremediation of Heavy Metal Contaminated Soil in Southern China Kengbo DING, Chang LIU, Yetao TANG, Shizhong WANG, Xiange WEI, Yuanqing CHAO, Rongliang QIU······ 347

Advances in Remediation of Acid Agricultural Soils Contaminated by Heavy Metals in South China Qitang WU, Zebin WEI, Xinxian LONG, Chengai JIANG······ 359

Stability and Universal Applicability of Immobilization Effect of Sepiolite on Cadmium in Acid Paddy Soil Yingming XU, Xuefeng LIANG, Lin WANG, Yuebing SUN, Qingqing HUANG······ 367

Systematic Selection and Identification of Vegetable Cultivars with Low Heavy Metal Accumulation and for Food Safety Weitao LIU, Qixing ZHOU, Xuhui LI······ 379

Strategies to Enable the Safe Use of Cadmium-Contaminated Paddy Soils in Southern China Hanhua ZHU, Chao XU, Qihong ZHU, Daoyou HUANG······ 394

Taiwan's Experiences on Soil Amendments, Phytoremediation and Soil Water Managements for the Cadmium-and Arsenic-Contaminated Soils Hung-Yu LAI, Chia-Hsing LEE, Zueng-Sang CHEN······ 403

Arsenic in Soil-Plant System: A Synthesis Guilan DUAN, Yongguan ZHU······ 413

Arsenic Hyperaccumulator *Pteris vittata* L. and Its Application to the Field Tongbin CHEN, Mei LEI, Xiaoming WAN, Jun YANG, Xiaoyong ZHOU······ 425

Microbial Remediation of Heavy Metals and Arsenic Contaminated Environments in the Arid Zone of Northwest China Xiangliang PAN, Varenyam ACHAL, Chenxi ZHAO, Jianying YANG, Deepika KUMARI ······ 437
Cr (Ⅵ)-Reducing Strain and Its Application to the Microbial Remediation of Cr (Ⅵ)-Contaminated Soils Liyuan CHAI, Zhihui YANG, Yan SHI, Qi LIAO, Xiaobo MIN, Qingzhu LI, Chongjian TANG, Lifen LIANG ······ 446
Phytoextraction of Mercury-Contaminated Soil Xinbin FENG, Jianxu WANG ······ 456
Chelant-Enhanced Phytoextraction of Heavy Metal-Contaminated Soils and Its Environmental Risk Assessment Yahua CHEN, Xiangdong LI, Zhenguo SHEN ······ 466
The Immobilization of Heavy Metals in Contaminated Soils Amended with Phosphate-, Carbonate-and Silicate-Based Materials: From Lab to Field Xinde CAO ······ 489
Remediation of Heavy-Metal-Contaminated Soils by Phosphate Fertilizers Minggang XU, Shiwei ZHOU, Shibao CHEN ······ 498
In-Situ Stabilization of Toxic Metals in Polluted Soils Using Different Soil Amendments: Mechanisms and Environmental Implication Shibao CHEN, Bin LIU, Han ZHENG, Nan MENG, Cao CAI, Yongguan ZHU ······ 515

Part Ⅳ Organic Pollution and Remediation

Fate of Several Typical Organic Pollutants in Soil and Impacts of Earthworms and Plants Feifei SUN, Hongyan GUO, Rong JI ······ 527
Adsorption and Reaction of Organic Contaminants on Surfaces of Condensed Carbonaceous Materials Dongqiang ZHU, Heyun FU, Wei CHEN, Mengxing XIE, Linzi ZUO ······ 541
Toxicity, Adsorption and Dissipation of Polycyclic Aromatic Hydrocarbons in Soil Jianming XU, Haizhen WANG, Yan HE, Bin MA ······ 554
Mitigation and Remediation for Organic Contaminated Soils by Surfactants Wenjun ZHOU, Lizhong ZHU ······ 577
Phytoremediation of Polychlorinated Biphenyl Contaminated Soil by Transgenic Alfalfa Associated Bioemulsifier AlnA Hejun REN, Yan WAN, Yongsheng ZHAO ······ 592
Microbial Degradation of Chemical Pesticides and Bioremediation of Pesticide-Contaminated Sites in China Jiandong JIANG, Shunpeng LI ······ 600
Trans-Membrane Transport Theory and Synergistic Remediation Technology for the Bioremediation of Oil Polluted Soil Hongqi WANG, Mengyuan SU, Shuo DIAO, Jie XU, Xiaoxiong WU ······ 615
Site Remediation Technology Research Based on Dynamic Balance Mechanism of Organic Contaminants in Soil/Water/Air Environmental System Wenhui ZHANG ······ 629
Thermal Desorption Technology and Application Ligang WANG, Yong YUE, Pixue WANG ······ 636

Part Ⅴ Metal-Organic Combined Pollution and Remediation

Study on Remediation Technologies of Organic and Heavy Metal Contaminated Soils Shuhai GUO, Fengmei LI, Peijun LI, Sa WANG, Qing ZHAO, Gang LI, Bo WU, Peidong TAI ········ 645

Biodegradable Chelant-Assisted Phytoextraction Chunling LUO, Xiangdong LI, Zhenguo SHEN ········ 665

Speciation, Activity, Transformation and Degradation of Heavy Metals and Organochlorines in Red Soils Yongtao LI, Wenyan LI, Huijuan XU ········ 674

Remediation of Contaminated Soils with Biochar Hailong WANG, Xing YANG, Lizhi HE, Kouping LU, Karin MÜLLER, Kim MCGROUTHER, Song XU, Xiaokai ZHANG, Jianwu LI, Huagang HUANG, Guodong YUAN, Guotao HU, Xingyuan LIU ········ 701

The Research and Development of Technology for Contaminated Site Remediation Xiaoyong LIAO*, Xiulan YAN, Dong MA, Dan ZHAO, Lu SUN, You LI, Yang FEI, Peng LI, Longyong LIN, Huan TAO ········ 720

Application and Case Study of Barrier Technology in Soil and Groundwater Remediation Li WEI, Zhengyong LV, Shucai LI, Guojie FENG, Jingwen LI, Kun SHEN, Zhu MIAO, Hudi ZHU ········ 734

Practice of Green and Sustainable Remediation, Risk-Based Megasite Management in China Hongzhen ZHANG, Jingqi DONG, Ning SUN, Jinnan WANG, Shunze WU ··· 743

Part Ⅰ General Reviews

Contamination Processes and Bioremediation of Typical Persistent Toxic Substances in Agricultural Soils in China

Yongming LUO[1,2], Longhua WU[1], Ying TENG[1], Jing SONG[1], Wuxing LIU[1], Haibo ZHANG[2], Chen TU[2]

[1]Key Laboratory of Soil Environment and Pollution Remediation, Institute of Soil Sciences, Chinese Academy of Sciences

[2]Key Laboratory of Coastal Environmental Processes and Ecological Remediation, Yantai Institute of Coastal Zone Research, Chinese Academy of Sciences

Abstract

With the rapid development of industrialization, urbanization, and intensive agriculture in China, more and more agricultural soils have been contaminated with persistent toxic substances such as cadmium (Cd), polycyclic aromatic hydrocarbons (PAHs) and polychlorinated biphenyls (PCBs), and this has threatened the security of agricultural production, the natural environment, and human health. Therefore, a better understanding of pollution characteristics has always been the scientific frontier in the field of soil environmental science. How to achieve the green, economic, and sustainable remediation of contaminated agricultural soils is still a topic of global concern. The key to solving the problem is to provide insight into the contamination processes and bioremediation mechanisms of pollutants in the soil. Since the last decade the research group has selected typical agricultural soils with low-to-moderate pollution levels representing different regions and has made some breakthroughs in basic scientific problems including source identification and bonding mechanisms, chemical speciation and bioavailability, hyperaccumulation and detoxification mechanisms, and microbe enhanced phytoremediation principles. These studies have provided a theoretical and methodological framework for the green and sustainable remediation of contaminated agricultural soils worldwide.

1. Introduction

Contamination of agricultural soils in China is widespread and serious and threatens the security of agricultural production, the natural environment, and human health. To clarify the contamination processes and mechanisms of typical persistent toxic substances (PTS) in agricultural soils and to elucidate the bioremediation mechanisms of PTS contaminated soils are the important basic scienfic challenges for solving the problem of soil contamination and achieving green and sustainable remediation approaches in China, and have also been a research hotspot and frontier in the field of soil environment and pollution remediation worldwide. Since the last decade the research group has selected typical agricultural soils with low-to-moderate pollution levels representing different regions and has made some innovative achievements on source identification and bonding mechanisms, chemical speciation and bioavailability, hyperaccumulation and detoxification mechanisms, and microbe enhanced phytoremediation principles. These findings have provided a theoretical and methodological framework for the protection of soil environment quality and for the remediation of contaminated agricultural soils in China, and are of great significance in ensuring food security.

2. Soil contamination characterizations and processes

Agricultural soil contamination appears to be the most important soil environmental problem in the world. To discover the contamination sources and characteristics, together with the mechanisms of interactions between soil components and pollutants, and to predict the chemical speciation transformations, bioavailability and risk assessment of toxic substances, have long been fundamental frontier research topics in the field of soil environmental science and the development of phytoremediation technology worldwide. The primary findings on this theme are as follows:

2.1 Sources, bonding mechanism and bioavailability of HMs and POPs pollutants

Soil magnetic properties, stable isotope fractionation and multivariable statistics have been applied in an integrated fashion to identify the different sources of HMs in typical polluted agricultural soils. The results show that dust precipitation, wastewater irrigation and exhaust emissions are the main anthropogenic sources of HMs (Cd, Zn, Cu and Pb) in agricultural soils (Zhang et al., 2008). Most of the HMs in the soils tend to bond to fine particles (mainly to micro/nano metal oxides). Results from X-ray absorption spectroscopy (XAFS) indicate that the surface bonding mechanism correlates strongly with different soil types and HM elements (Hu et al., 2010). Inner-sphere sorption complexation with ionic character dominates the Pb surface speciation in alkaline soil (Vstic Sandic Entisols) together with outer-sphere Pb sorption complexation and ion exchange. Forces of Cd sorption on acid soils (Endogleyic Fe-accumulic Stagnic Anthrosols and Fe-accumulic Gleyic Stagnic Anthrosols) are dipole bond forces and hydrogen bond forces (Fig. 1).

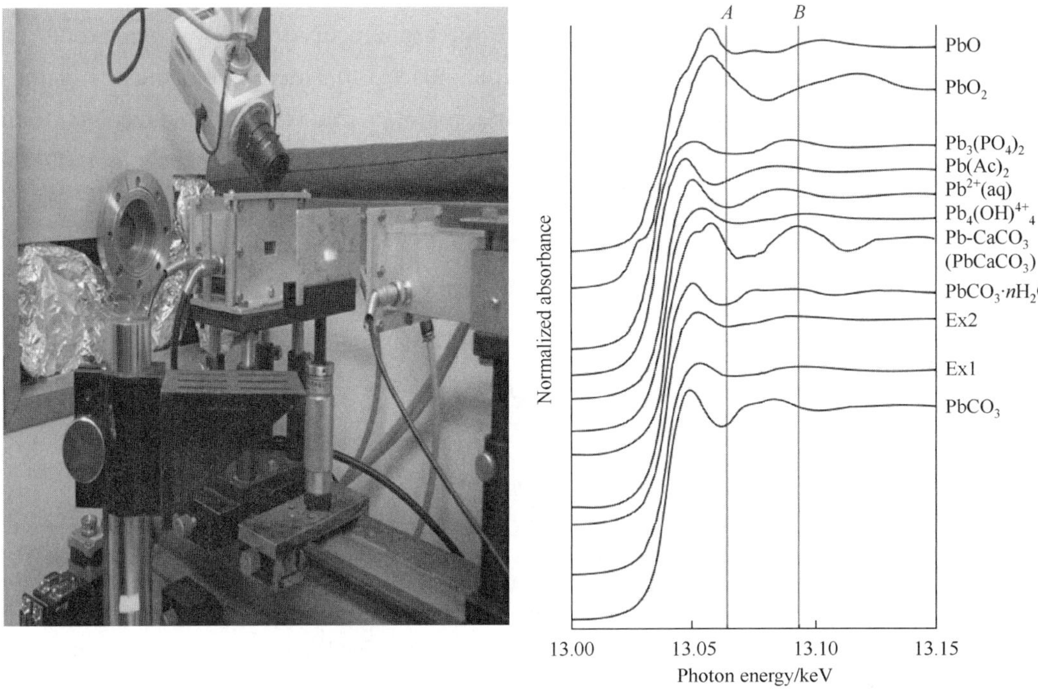

(a) Synchrotron radiation facility

(b) XANES spectra for Pb contaminated soils and reference compounds

Fig. 1 XANES analysis for Pb contaminated soils

Results from a previous regional investigation show that the main components of PAHs in the surface soil from Yangtze River Delta are high-molecular-weight PAHs with four to six rings (Ping et al., 2007). Contamination source identification indicates that PAHs in agricultural soils are derived mainly from the combustion of petroleum, biomass and coal. PAHs in the surface soil can migrate to the subsoil and are distributed unevenly in different sizes of particles. The content of PAHs decreases significantly as particle size is reduced (Ni et al., 2008a). The bioavailability of PAHs is strongly affected by the components and content of soil organic matter (Ni et al., 2008b).

PCB patterns are dominated by 3Cl-CBs, 4Cl-CBs, 5Cl-CBs and 6Cl-CBs on average in topsoil, with the sources from the uncontrolled disassembly of transformers and the incineration of electronic wastes (Gao et al., 2006). PCBs are lipophilic compounds and attach strongly to soil organic matter. The bioavailability, migration and transformation of PCBs in the soil are affected mainly by the soil/water partition coefficient (K_d), octanol-water partition coefficient (K_{ow}), and the quality and quantity of soil organic matter.

2.2 Chemical speciation detection and spatial distribution predicting of HMs in edible crops

HMs can enter soil solutions through bioavailable speciation including ions or soluble metal complexes, diffusion with mass flow to the rhizosphere, absorption by plant roots, and then transport to and accumulation in the shoots. Free ions are considered to be the key factor in the bioavailability of heavy metals. However, this remains a challenge to analyze *in-situ* and represent free ions in the soil solution. Using a modified Donnan Membrane Technique (DMT), the free ionic concentrations of Cd and Hg were measured under different conditions of pH and ion strength (Zhang et al., 2010), and the accumulation of Cd in edible crops was predicted based on the soil Cd concentrations (Fig. 2) (Zhang et al., 2011).

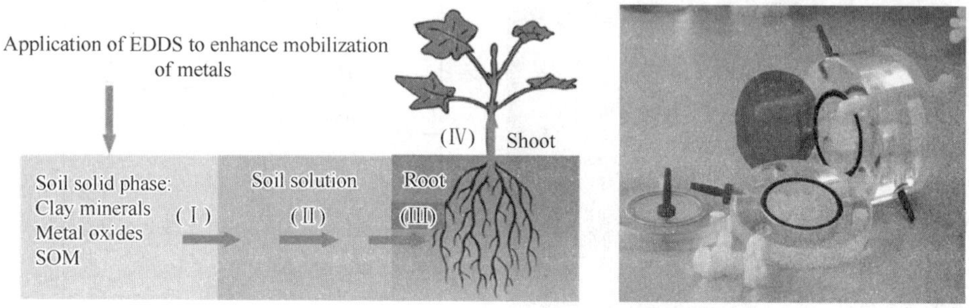

Fig. 2 Predicting of free metal ions in edible crops using a modified Donnan membrane technique

2.3 Soil ecotoxicity and ecological risk assessment on different scales from genes to communities

Chemical analysis cannot accurately or precisely characterize combined pollution and ecological risk for complex soil environments. Developing ecotoxicological diagnosis and risk assessment methods using native organisms in soil ecosystems has been a hotspot in the ecotoxicology field. The project sets up a battery of ecological toxicology diagnosis bioassays by using a series of microbes-insects-earthworms as sensitive receptors. Results from molecular biology, cell biology, biophysiology and biochemistry elucidate the best response times for different sensitive biomarkers such as reproduction and growth of *Folsomia candida*, DNA damage, POD activities and GSH levels of earthworms to the exposure of Cu and PCB contamination (Bu et al., 2011). Based on the different diagnosis scales of genes, cells, individuals, populations and communities, integrated methods of soil ecological toxicity and risk assessment have been established (Fig. 3).

3. Phytoremediation of heavy metals contaminated soils

Screening functional plant species with hyper-or high HM accumulation ability and achieving soil decontamination through the rapid extraction of HMs from the soil have been international hot topics since the last decade. Since the end of the last century the research group has conducted some systematic and innovative constructive field experiments with typical agricultural soils with low-to-moderate HM contamination. The primary findings on this theme are as follows:

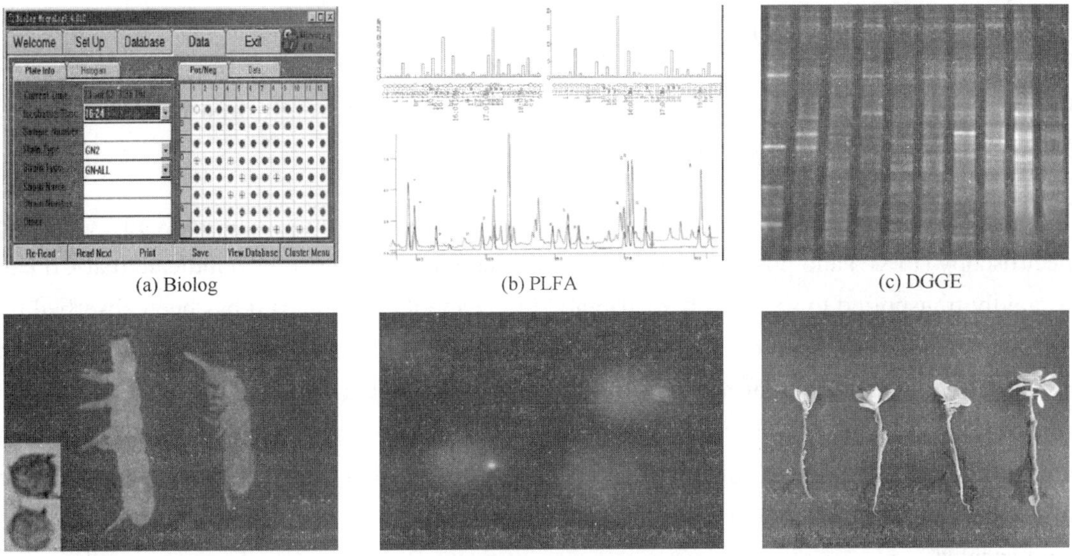

(a) Biolog (b) PLFA (c) DGGE

(d) Soil fauna toxicity indicated by collembola (e) Earthworm cell comet assay (f) Soil vegetal toxicity

Fig. 3 Soil ecotoxicity and ecological risk assessment on different scales from genes to communities

3.1 Sustainable remediation of cadmium/zinc contaminated soils in long-term field trials by a newly discovered cadmium/zinc hyperaccumulator, *Sedum plumbizincicola*

Cd contaminated agricultural soils constitute a problem throughout China. The research group has devoted much effort to the screening of native Cd hyperaccumulators and elucidating their mechanisms in order to develop effective phytoremediation approaches to solve this major challenge. We have firstly reported a newly discovered native Chinese Cd/Zn hyperaccumulator, *Sedum plumbizincicola* (Wu et al., 2013). Phytoremediation with repeated harvests in pot experiments indicates that *S. plumbizincicola* can rapidly decrease the soluble forms, exchangeable forms and total amounts of Cd and Zn in the soil, especially the fractions of Cd and Zn on the surfaces of coarse particles, thus enhancing soil microbial and hydrolase activities, together with the restoration of soil ecological functions and soil environmental quality (Li et al., 2014a, 2014b). The total amounts of HMs in the soil were decreased by 80%-97% (Cd) and 25%-70% (Zn) after nine repeated harvests, levels which are below the National Soil Environment Quality Standard (Level 2). Furthermore, the research group has also pioneered globally the first phytoremediation demonstration sites for multiple HMs-contaminated agricultural soils to further verify the phytoextraction potential of the hyperaccumulators. Under field conditions, increasing the plant density to 440 thousand seedlings per hectare (plant seedling interval 15 cm) had the highest plant biomass and rate of Zn/Cd uptake. After repeated phytoextraction for 2 years, the average Cd removal rates were 12% per year. Moreover, technologies for incineration and re-use of hyperaccumulator biomass have also been developed (Luo et al., 2015b).

3.2　Hyperaccumulation processes and detoxification mechanisms of Cd and Zn in *Sedum plumbizincicola*

Further studies using the advanced technology of micro proton induced X-ray emission spectroscopy (PIXE) and positron emission tracer imaging system (PETIS) indicate that Cd can be rapidly transported to and evenly distributed in the shoots as soon as it has been absorbed by the roots from the soil solution. Zn is largely sequestered in epidermal and parenchyma cells of the vascular bundle. Different nitrogen forms are key factors that regulate Cd uptake from soil solutions by *S. plumbizincicola*. Nitrate can significantly accelerate the absorption and translocation of Cd from the soil solution (Hu et al., 2013, 2015) (Fig. 4). These are considered to be the important mechanisms of HM hyperaccumulation and detoxification for this plant species.

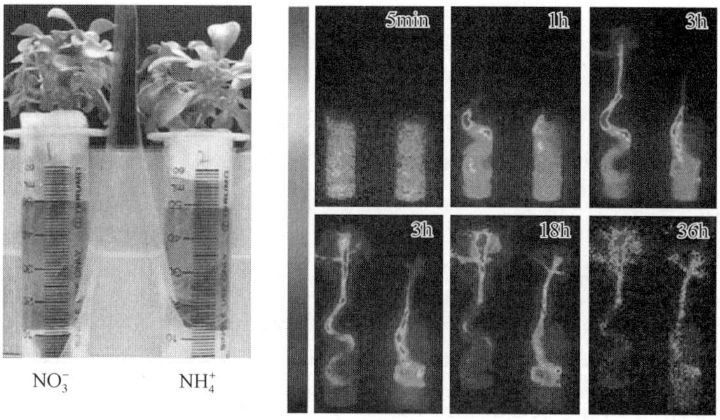

(a) *S. plumbizincicola* fed with NO_3^- or NH_4^+　　(b) Serial images of ^{107}Cd movements in *S. plumbizincicola*

Fig. 4　Serial images of ^{107}Cd movements in *S. plumbizincicola* fed with NO_3^- or NH_4^+

3.3　Mechanisms of organic chelate enhanced phytoremediation of HMs-contaminated soil

The research group has successfully identified an effective ecotype of *Elsholtzia splendens* that produces large biomass and possesses high Cu accumulating ability. Further studies have found that the copper concentration in the roots of *E. splendens* are higher than in the shoots, indicating that Cu cannot easily be transported from the rhizosphere to the shoots within the soil-plant system. Therefore, enhancement of Cu availability in the rhizosphere soil and its mobility within the plant have become key steps in increasing the efficiency of phytoextraction of Cu by *E. splendens*. Here we have proposed for the first time the theory of organic chelate enhanced phytoremediation. Experimental results show that the addition of organic chelates such as EDTA, EDDS and low molecular weight organic acids can significantly enhance soil available HMs and promote HMs transport from root to shoot, and thus accelerate the removal of available HMs from the soil, and also increase the leaching risks of metal elements (Cu, Cd and Fe) (Wu et al., 2003, 2004). Addition of EDDS to the soil can increase Cu concentrations in the vascular tissue of *E. splendens*, and results from synchrotron radiation X-ray fluorescence (SRXRF) studies indicate that Cu is transported in the xylem as the complexed form of Cu-EDDS and stored in the functional organelles, notably leaf epidermal cells, spongy tissue and chloroplasts (Luo et al., 2012).

4. Bioremediation of persistant organic pollutants

PAHs and PCBs are tsypical persistant organic pollutants (POPs) of global concern. Due to their hydrophobicity, PAHs and PCBs attach strongly to soil organic matter and can only be extracted and accumulated by plants with difficulty. Therefore, screening of microbial species with high biodegradation ability for PAHs and PCBs, and revealing their degradation mechanisms have always been a frontier in the soil science and bioremediation field, and have become a key factor to achieve green and sustainable remediation of POPs contaminated soils. The primary findings on this theme are as follows:

4.1 Bioremediation of PAHs contaminated soil with bacterial degradation and fungal laccase oxidation

The research group successfully isolated a bacterial strain named *Paracoccus aminovorans* HPD-2 which can utilize a high-molecular-weight PAH (B[a]P) as the sole carbon and energy source (Fig. 5 (a)). The strain can degrade B[a]P by up to 89.7% in pure culture and the metabolic intermediates of B[a]P were identified as 8-carboxyl-7-hydroxypyrene and dihydroxy phenanthrene and the biodegradation pathway was elucidated. Moreover, the microbial agent can remove 24% of high-molecular-weight PAHs from agricultural soil over a period of 28 days under field conditions (Teng et al., 2010). Furthermore, we have successfully isolated a fungal species, *Monilinia* sp. W5-2, with high laccase activity (Fig. 5 (b)). This fungal strain can

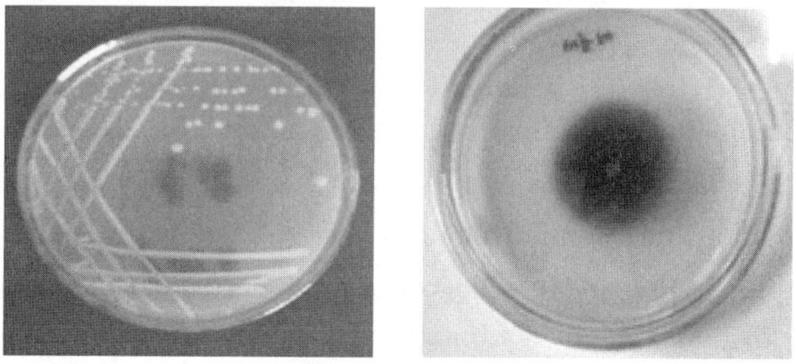

(a) *Paracoccus aminovorans* HPD-2 (b) *Monilinia* sp. W5-2

Fig. 5 Bioremediation of PAHs contaminated soil with bacterial degradation and fungal laccase oxidation transform B[a]P to anthraquinone, an intermediate with lower toxicity, and the biodegradation potential for PAHs was positively related to the ionic potential of PAHs molecules (Wu et al., 2008a, 2008b).

4.2 Reductive dechlorination of PCBs by legume-rhizobium symbiotic nitrogen fixation association

The legume-rhizobium association is possibly the best-known natural symbiosis with the function of N_2-fixation. However, its role in environmental remediation remains unclear. We have found from both pot experiments and field studies conducted over several years that the removal of PCBs from agricultural soil was significantly enhanced by inoculating alfalfa with rhizobia (Teng et al., 2010; Xu et al., 2010). The inoculum can increase plant biomass and enhance the uptake and translocation of PCBs from rhizosphere soil. Furthermore, we have found significantly higher concentrations of PCBs in the root nodules than in the roots or shoots, possibly due to the strong lipophilic nature of PCBs and the high lipid content in root nodules. The most exciting result is that reductive dechlorination of 3Cl-CB was evident from GC chromatograms and a non-invasive ion flux technique, with 2Cl-CB and Cl^- as intermediates, and the reductive dechlorination of PCBs is coupled to nitrogen fixation by alfalfa nodules. Stable isotope probing was used to establish that the bacteroid *Rhizobium meliloti* in the nodule can assimilate ^{13}C-DNA by breaking the benzene ring of ^{13}C-PCB. Moreover, free living *S. meliloti* has a strong PCB degrading potential. The percentage of 2, 4, 4'-CB biodegradation can reach 98.5% with the intermediate HOPDA (Tu et al., 2011a). Therefore, we have elucidated the uptake and transport mechanisms of PCBs by the legume-rhizobium association and have produced a new principle of combined legume-rhizobium remediation for organic chloride polluted soils. The average PCBs removal rates can reach 40% after repeated bioremediation for 2 years in field experiments (Tu et al., 2011b). Based on the above investigations, the research group has proposed the innovative synergistic remediation mechanisms of PCBs reductive dechlorination and symbiotic nitrogen fixation by legume-rhizobium.

4.3 Prepared bed bioremediation of oily sludge in an oilfield by stimulating indigenous microbes

Field-scale bioremediation of oily sludge in prepared beds was studied at Shengli oilfield in northern China. The influence of manure, coarse sand, sawdust, a specialized microbial preparation and greenhouse conditions on the efficiency of removal of oil and grease was evaluated. After bioremediation for 230 d, oil and grease content fell by 32-42 g/kg dry sludge in treated plots, indicating removal of 27% to 46% compared with only 15% in the control plot (Liu et al., 2009). Addition of manure, coarse sand, sawdust and greenhouse conditions significantly increased the amount removed. Moreover, the physico-chemical properties of the sludge in all treated plots improved significantly after bioremediation. Microbial biomass in sludge and community-level physiological profiling examined using BIOLOG microplates were also studied. Total petroleum hydrocarbon degraders and polycyclic aromatic hydrocarbon degraders increased in all treated oily sludge. The activity of sludge microbial communities increased markedly in the treated plots compared with the control. Canonical correspondence analysis showed that differences in substrate utilization patterns were highly correlated with sludge hydrolyzable N, oil and grease content. The biological toxicity of the oily sludge was lower following bioremediation in most of the treated plots as evaluated using *Photobacterium phosphoreum* T3 (Liu et al., 2015).

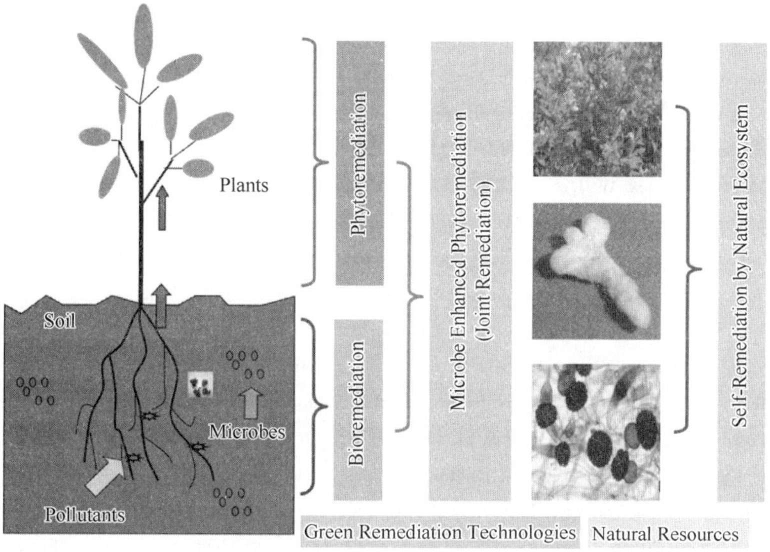

Fig. 6 Green and sustainable self-remediation of polluted soils by natural ecosystem

5. Conclusions

Based on the systematic remediation technologies for contaminated soils including bioremediation, physical-chemical remediation and integrated remediation at home and abroad, six research and development trends in soil remediation are summarized as follows: green and environmentally friendly bioremediation, combined and hybride remediation (Fig. 6), *in situ* remediation, environmentally functional material based remediation, equipment based site remediation, remediation decision supporting system and post-remediation assessment (Luo et al., 2009). In the future, China needs to develop wide use, safe, cost-effective *in situ* bioremediation and physical-chemical stabilization technologies for agricultural farmland soil, to develop safe, land reusable, site-specific physical-chemical and engineering remediation technologies for heavily polluted industrial site, and to develop phytostabilization and eco-engineering remediation technologhies for controlling soil erosion and pollutants transport in mined area. Besides, it also needs to develop national guidelines, standards and management policies for contaminated soil remediation (Luo et al., 2015a, 2015c).

Acknowledgements

This work was supported by the National Natural Science Foundation of China (41230858), and National High Technology Research and Development Program (2012AA06A200).

References

Bu Y Q, Luo Y M, Shan Z J, et al. 2010. Use of the comet assay for detecting DNA damage in earthworms from PCBs-contaminated agricultural soils. Environmental Science and Engineering, 1: 38-45.

Gao J, Luo Y, Li Q B, et al. 2006. Distribution patterns of polychlorinated biphenyls in soils collected from Zhejiang province, east China. Environmental Geochemistry and Health, 28 (1-2): 79-87.

Hu N J, Huang P, Luo Y M, et al. 2010. X-ray absorption fine structure (XAFS) study of the effects of pH on Pb (II) sorption by soil. Spectroscopy and Spectral Analysis, 30 (12): 3425-3429.

Hu P J, Wang Y D, Przybylowicx W J, et al. 2015. Elemental distribution by cryo-micro-PIXE in the zinc and cadmium hyperaccumulator *Sedum plumbizincicola* grown naturally. Plant and Soil, 388 (1-2): 267-282.

Hu P J, Yin Y G, Ishikawa S, et al. 2013. Nitrate facilitates cadmium uptake, transport and accumulation in the hyperaccumulator *Sedum plumbizincicola*. Environmental Science and Pollution Research, 20 (9): 6306-6316.

Ni J Z, Luo Y M, Wei R, et al. 2008a. Distribution of polycyclic aromatic hydrocarbons in particle-size separates and density fractions of typical agricultural soils in the Yangtze River Delta, east China. European Journal of Soil Science, 59 (6): 1020-1026.

Ni J Z, Luo Y M, Wei R, et al. 2008b. Distribution patterns of polycyclic aromatic hydrocarbons among different organic carbon fractions of polluted agricultural soils. Geoderma, 146 (1): 277-282.

Li Z, Wu L H, Hu P J, et al. 2014a. Repeated phytoextraction of four metal-contaminated soils using the cadmium/zinc hyperaccumulator *Sedum plumbizincicola*. Environmental Pollution, 189: 176-183.

Li Z, Wu L H, Luo Y M, et al. 2014b. Dynamics of plant metal uptake and metal changes in whole soil and soil particle fractions during repeated phytoextraction. Plant and Soil, 374 (1-2): 857-869.

Liu W X, Luo Y M. 2015. Soil petroleum pollution and bioremediation. Beijing: Science Press.

Liu W X, Luo Y M, Teng Y, et al. 2009. Prepared bed bioremediation of oily sludge in an oilfield in northern China. Journal of Hazardous Materials, 161 (1): 479-484.

Luo Y M. 2009. Current research and development in soil remediation technologies. Progress in Chemistry, 21 (2/3): 558-565 (in Chinese).

Luo Y M, Li G H, Li F S, et al. 2015a. Research on soil environmental management supporting technologies and frame system of China. Beijing: Science Press (in Chinese).

Luo Y M, Wu L H, Hu P J, et al. 2015b. Research on phytoremediation of Cd/Zn polluted soils by hyperaccumulators. Beijing: Since Press (in Chinese).

Luo Y M, Xia J Q, Zhang H B, et al. 2015c. China's soil environmental quality standard: theories and developing methods. Beijing: Science Press (in Chinese).

Luo Y M, et al. 2012. Phytoremediation of heavy metals polluted soils by elsholtzia plants. Beijing: Science Press (in Chinese).

Ping L F, Luo Y M, Zhang H B, et al. 2007. Distribution of polycyclic aromatic hydrocarbons in thirty typical soil profiles in the Yangtze River Delta region, east China. Environmental Pollution, 147 (2): 358-365.

Teng Y, Luo Y M, Sun M M, et al. 2010. Effect of bioaugmentation by *Paracoccus* sp. strain HPD-2 on the soil microbial community and removal of polycyclic aromatic hydrocarbons from an aged contaminated soil. Bioresource Technology, 101 (10): 3437-3443.

Tu C, Teng Y, Luo Y M, et al. 2011a. Potential for biodegradation of polychlorinated biphenyls (PCBs) by *Sinorhizobium meliloti*. Journal of Hazardous Materials, 186 (2): 1438-1444.

Tu C, Teng Y, Luo Y M, et al. 2011b. PCB removal, soil enzyme activities, and microbial community structures during the phytoremediation by alfalfa in field soils. Journal of Soils and Sediments, 11 (4): 649-656.

Wu L H, Liu Y J, Zhou S B, et al. 2013. *Sedum plumbizincicola X. H.* Guo et S. B. Zhou ex L. H. Wu (Crassulaceae): A new species from Zhejiang Province, China. Plant Systematics and Evolution, 299 (3): 487-498.

Wu L H, Luo Y M, Christie P, et al. 2003. Effects of EDTA and low molecular weight organic acids on soil solution properties of a heavy metal polluted soil. Chemosphere, 50 (6): 819-822.

Wu L H, Luo Y M, Xing X R, et al. 2004. EDTA-enhanced phytoremediation of heavy metal contaminated soil with Indian mustard and associated potential leaching risk. Agriculture, Ecosystems & Environment, 102 (3): 307-318.

Wu Y C, Teng Y, Li Z, et al. 2008a. Potential role of polycyclic aromatic hydrocarbons (PAHs) oxidation by fungal laccase in the remediation of an aged contaminated soil. Soil Biology and Biochemistry, 40 (3): 789-796.

Wu Y C, Luo Y, Zou D, et al. 2008b. Bioremediation of polycyclic aromatic hydrocarbons contaminated soil with *Monilinia* sp.: Degradation and microbial community analysis. Biodegradation, 19 (2): 247-257.

Zhang C B, Wu L H, Luo Y M, et al. 2008. Identifying sources of soil inorganic pollutants on a regional scale using a multivariate statistical approach: Role of pollutant migration and soil physicochemical properties. Environmental Pollution, 151 (3): 470-476.

Zhang H Z, Luo Y M, Song J, et al. 2011. Predicting As, Cd and Pb uptake by rice and vegetables using field data from China. Journal of Environmental Sciences, 23 (1): 70-78.

Zhang H Z, Luo Y M, Weng L P, et al. 2010. Measuring mercury speciation in $CaCl_2$ solutions using donnan membrane technique. Environmental Chemistry, 29 (3): 502-507.

Ecological Restoration of Man-Made Habitats, with Emphasis on Metal Contaminated Sites and Domestic Landfills

Ming-Hung WONG

Consortium on Health, Environment, Education and Research (CHEER), and Department of Science and Environmental Studies, The Education University of Hong Kong

Abstract

Heavy metal contamination of soils and waters has been accompanied with the socio-economic development in China since 1970s. China is rich in mineral resources, and extraction and smelting of metals has generated a large amount of mine tailings, leading to severe soil and water contamination. The subsequent uptakes of toxic metals by crops and fish pose human health risks. In addition, domestic and industrial wastes derived from urban centers and industrial areas are disposed of in landfills, which may also exert harmful effects on environmental and public health, through leaching of contaminants from the embedded wastes. Phytoremediation has been used for restoration of these man-made habitats, which include metal mine tailings, metal contaminated sites (e.g., farm soils), and completed landfills. The present article attempts to review advances made in the past 40 years from our research groups, concerning ecological restoration of these habitats, focusing on the interactions of higher plants-and-soil organisms, including invertebrates such as earthworms, and associated rhizospheric microbes (i.e., fungi and bacteria). Successful ecological restoration of these habitats would depend on the choice of plant species, appropriate combinations with soil organisms, and suitable soil amendments; for enhancing plant growth and ecological succession in the highly stressed conditions

Keywords: phytoremediation, metal tolerant plants, metal hyperaccumulating plants, microbes, earthworms

1. Introduction

China is rich in mineral resources. Extraction and smelting of various metals have generated vast areas of degraded, damaged and contaminated sites, containing elevated concentrations of toxic metals, notably lead (Pb), zinc (Zn), copper (Cu), cadmium (Cd), and arsenic (As), a metalloid. In addition, landfilling of domestic and industrial wastes is a major strategy of solid waste disposal in most countries. As a result, many of these landfills have been generated, emitting landfill gas (consisting of methane, but lacking oxygen), and landfill leachate (rich in nitrogenous compounds and dissolved metals) into the environment.

Ecological restoration refers to the practice of renewing and restoring degraded, damaged, or destroyed ecosystems and habitats. Ecological restoration has been considered as a means of conserving biodiversity and sustainable development worldwide. Therefore, it is necessary to address all the difficulties concerning plant growth and development, in order to accelerate ecological succession of these man-made habitats, which are degraded, damaged or destroyed (Wong, 1986, 2003). The progress in the restoration of degraded land in China (Wong and Bradshaw, 2002), and revegetation of landfill sites (Wong, 1998; Chan and Wong, 2002), have been reviewed, respectively.

The major aim of this paper is to review our major research activities related to ecological restoration of man-made and degraded habitats since the late 1970s, based at the Chinese University of Hong Kong, Hong Kong Baptist University, and then the Education University of Hong Kong; and our collaborators in various institutes of mainland China (notably Sun Yat-Sen University in Guangzhou, and the Institute of Soil Science in Nanjing).

This paper consists of two major parts: ①metal-mined sites, and ②completed landfills.

2. Restoration of metal-contaminated sites

2.1 Use of terrestrial plants for soil remediation

2.1.1 Environmental impacts

Our early investigations focused on properties of the iron ore tailings (Ma On Shan) (Wong and Tam, 1977), and the ecological impacts on the surrounding environment (Wong et al., 1978; Wong, 1981). The tailings generated by mining activities resulted in a large stretch of bare area, without vegetation cover, which imposed adverse effects on the organisms and coastal environment of Tolo Harbour (Hong Kong), through erosion and leaching of toxic metals to the vicinity.

2.1.2 Choice of suitable plant species

It was discovered that populations of grasses growing on sites containing elevated concentrations of metals were tolerant to these metals, through evolution (Brandshaw, 1952). An early study compared the toxicity of several common heavy metals using root elongation of rye grass (*Lolium perenne*) (Wong and Bradshaw, 1982). Toxicity tests, using root elongation, and tolerance tests using plant materials collected from contaminated sites compared with those from normal sites, can be used as standard assays to supplement chemical analysis, in order to obtain a fuller picture of the photo-toxic effects of contaminated sites (Wong et al., 1983; Wong, 1985). The results generated from these simple tests are of value in predicting the toxic effects of heavy metals to plants growing on metal-contaminated sites, and selection of plant species suited for growth. The ultimate goal of revegetation is to stabilize the rather unstable areas, and provide a more pleasant environment.

Early investigations on revegetation of contaminated sites focused on the choice of grass species which could thrive in the stressed environment, i.e., tolerate high concentrations of toxic metals, poor physical and chemical properties, in particular lack of organic matter, nutrients and moisture. Subsequently, it was discovered that *Festuca rubra* (Wong, 1982), and *Paspalum disstichum* and *Cynodon dactylon* (Shu et al., 2002) were capable of simultaneously tolerating Cu, Pb and Zn, metals commonly co-existing in these metal mined sites.

It was revealed that soil seed banks could serve as an important seed source in revegetation of the tailings (Zhang et al., 2001). These plants were found useful in the natural colonization on five different Pb/Zn tailings in China (Shu et al., 2005). Different plant species were collected from these metal-mined sites and propagated by providing them with nutrient solutions, which were subsequently used in land restoration (e.g., Pb/Zn mine tailings, using grasses (Zhang et al., 1996), and legumes (Yang et al., 1997).

It is commonly known that legumes are able to fix nitrogen (N), which is beneficial for plant growth in tailings without sufficient organic matter and N (Wong et al., 2002). Attempts have been made to study the toxic effects of Cd on nodulation and N-fixation of soybean (Chen et al., 2003b), Zn on rhizobia-earleaf Acacia (*Acacia auriculaeformis*) symbiotic association (Zhang et al., 1998), and responses of *Sesbania rostrata* and *S. cannabina* to Pb, Zn, Cu and Cd toxicities (Zheng, 2004).

Chan et al. (2003) further screened four *Sesbania* species (*S. cannabina, S. grandiflora, S. rostrata*, and *S. sesban*), with or without inoculation of rhizobia, for growth on Pb/Zn and Cu mine tailings, based on a 6-month greenhouse pot trial (amended with river sediment or garden soil). Results showed that *S. sesban* and *S. rostrata* had a better growth, in particular *S. rostrata*, as it possessed both root and shoot nodules (while other species only possessed root nodules). It is envisaged that shoot nodules are able to carry out its N-fixing function, even if the root nodules are affected by toxic metals. Being an annual legume, *S. rostrata* can beneficially modify the habitat for the growth of subsequent plant communities.

A field trial has been conducted on the growth and metal accumulation of two *Sesbania species* (*S. rostrata* and *S. sesban*) and vetiver (amended with domestic refuse and/or fertilizer) growing on Pb/Zn mine tailings. It was concluded that intercropping of both species did not show any beneficial effect on individual plant species, in terms of height, biomass, survival rate, and metal accumulation, possibly due to the rather short experimental period of 5 months. In the long run, it is expected that the legumes could provide a stable supply of N, needed by vetiver, to perform its best in soil erosion control (Yang et al., 2003).

In view of the fact that the Chinese Brake Fern (*Pteris vittata*) was able to take up extremely high concentration of As, with a large biomass, Wang et al. (2006) conducted field surveys on As uptake and accumulation in fern species growing at As-contaminated sites of southern China. Wu et al. (2009b) compared the variations in As, Pb and Zn tolerance and accumulation in six populations of the plant grown in As-contaminated and uncontaminated sites in southeast China, based on a hydroponic experiment. It was observed that As tolerance in *P. vittata* resulted from both constitutive and adaptive traits, Pb and Zn tolerances were constitutive properties, and that nonmetallicolous population possessed more effective As hyperaccumulation than metallicolouos populations. Lou et al. (2009) further compared As tolerance, uptake and accumulation between *P. vittata*, the As hyperaccumulator and *P. semipinnata*, non-accumulator. It was revealed that *P. vittata* had a significantly higher uptake of As than *P. semipinnata*. *P. vittata* also had significantly higher accumulation of As in the fronds while *P. ssemipinnata* in the roots. However, high doses of As produced oxidative damages in both plant species. Subsequently, the uptake and accumulation of As by 11 Pteris taxa from southern China were investigated and results revealed a constitutive property of As hyperaccumulation in different populations of *P. cretica* var. *nervosa, P. multifidi, P. oshimensis* and *P. vittata* (Wang et al., 2007).

Woody legumes such as *Acacia auriculiformis* and *Lecucaena leucocephala* would be excellent candidates in revegetating metal-contaminated sites such as lagoon ash (Cheung et al., 2000) and mine tailings (Ma et al., 2003). In addition to being drought tolerant, able to survive in areas contaminated with heavy metals, and lack of organic matter, these species can also fix N. In principle, the toxic metals would be locked up in the woody tissues, and the chance of entering into the food chain would be minimized.

2.1.3 Physiological aspects

The physiological responses of metal tolerance populations of different plant species have been investigated, e.g. Sun et al. (2007) showed that Cd hyperaccumulation leads to an increase of glutathione rather than phytochelatins in the Cd hyperaccumulator *Sedum alfredii*, and Liu et al. (2009) revealed the role of arsenate reductase and superoxide dismutase in As accumulation in four Pteris species.

Pang et al. (2003) further studied the physiological aspects of vetiver for rehabilitation of Pb/Zn mine tailings, and noted that the tailings greatly inhibited the leaf growth, photosynthesis and dry matter production, but stimulated the accumulation of proline and abscisic acid, and enhanced activities of superoxide dismutase, peroxidase and catalase, to detoxify active oxygen species. Sun et al. (2005) observed an increase of glutathione (GSH) in mine population of *Sedum alfreddi* (Zn/Pb hyperaccumlator), accompanying increases in concentrations of Zn and Pb when grown on the mine sites. It was concluded that GSH may be involved in Pb and Zn transport, hyperaccumulation/accumulation and tolerance in mine population of *S. alfredii*.

2.1.4 Organic amendments

The tailings are usually poor in physical and chemical properties, which are not suitable for plant growth (Ye et al., 2000a, 2002b), in particular their acid forming potential (Wong et al., 1988, 1998; Shu et al., 2001). Toxic metals are released into the soil solution under a lower pH, and thereby become more toxic to plants (Ye et al., 2002a). In general, applications of lime, and organic waste, such as manure compost (Ye et al., 1999, 2000b), sewage sludge (Ye et al., 2001b; Chiu et al., 2006), domestic refuse and river sediment (Lan et al., 1998), were effective for revegetating Pb/Zn mine tailings for raising the pH of tailings and provision of organic matter and plant nutrients.

Shu et al. (2002) demonstrated that vetiver (*Vertiveria zizanioides*) performed the best among the four grass species (others were *Paspalum notatium, Cynodon dactylon* and *Imperata cylindrical* var. *major*) tested in the Pb/Zn mine tailings pond, amended with domestic refuse and NPK fertilizer (Lechang Pb/Zn mine, Guangdong Province). This is due to the fact that vetiver is highly tolerant to extreme soil conditions, including heavy metal contamination, and could serve as an excellent candidate to prevent soil erosion, due to its well-developed and extensive root system.

2.1.5 Chelating agents to enhance metal uptake by plants

It is commonly recognized that chelating agents such as citric acid and EDTA are effective in mobilizing toxic metals in soil to enhance their phytoextraction by metal tolerant plants, such as *Thilaspi caerulescens*, *Brassica juncea* and *Hordeum vulgare*. Chen et al (2003c) revealed that addition of citric acid reduced the toxicity of Cd and Pb to radish, and stimulated their transportation from root to shoot. The same effects were noted when EDTA was used for enhancing metal uptake by plants (Cd to Indian mustard, *Brassica juncea* (Jiang et al., 2003); and Cu, Zn, Pb and Cd to *Brassica juncea* (Wu et al., 2003)).

A subsequent experiment compared the uptake of As, Cu and Cd by vetiver (*Vetiveria zizanioides*) and maize (*Zea mays*) for remediation of As-, Zn- and Cu-amended soils, using nine different chelating agents. Results showed that vetiver had a better growth and extraction of various metals than maize, in all treatments. It was also noted that application of NTA in As-amended soil and HEIDA in Cu-amended soil (at the rate of 20 mmol/kg) increased 3-4-fold of As and Cu in shoot of both plants, while application of NTA (20 mmol/kg) increased 37- and 1.5-fold of Zn accumulation in shoot of vetiver and maize, respectively (Chiu et al., 2005).

A study compared the effects of different chelating agents on the solubility and accumulation of As, Cu, Pb and Zn in Chinese Brake Fern (*Pteris vittata*), vetiver (*Vetiveria zizanioides*) and rostrate Sesbania (*Sesbania rostrate*). Results showed that the Chinese Brake Fern contained the highest concentrations of As, Cu, Pb and Zn in the above ground tissue, compared with the other two species, and can be used to simultaneously decontaminate As,

Cu, Pb and Zn from contaminated soils, with the aid of EDTA or HEDTA (Lou et al., 2007). However, the potential environmental risks of metal mobility caused by chelating agents applied for phytoextraction should not be overlooked.

2.1.6 Soil organisms, in particular rhizospheric microbes

Plants interact with other living organisms such as invertebrates and microbes in soil. These organisms could assist plants to thrive in the stressed environment. It is commonly known that earthworms contribute significantly to soil humification processes, and stimulate the mineralization and recycling of nutrients, particular N and phosphorus (P).

2.1.7 Earthworms

It was demonstrated that earthworms (*Pheretima guillelmi* and *Eisena fedita*) and P-solubilizing bacteria had significant effects on microbial growth, enzyme activity, leading to the release of available P (Wan and Wong, 2004). Subsequently, it was noted that earthworm (*Pheretima* sp.) activity decreased the pH of relatively low level of Zn-contaminated soil, with a higher organic matter content, and enhanced growth and metal uptake by metal-hyperaccumulator plants (Cheng and Wong, 2002).

A series of experiments was then conducted on the inoculation of earthworms to improve the physical and chemical properties of Pb/Zn mine tailings, and to provide available nutrients for plant growth. By investigating the effects of earthworm (*Pheretima* sp.) on three sequential ryegrass (*Lolium multiflorum*) harvests for remediating Pb/Zn mine tailings, Cheng and Wong (2008) noted that the presence of earthworms and sequential plant harvesting could be a viable strategy for the remediation of Pb/Zn mine tailings.

Addition of soil in a proportion as little as 20% to the Pb/Zn mine tailings would allow successful inoculation or natural invasion of *Pheretima*, with the increase in bioavailable fraction of Pb and Zn (Ma et al., 2002). Ma et al (2003) investigated the effects of *P. gullelmi* on the growth of a woody legume *Leucaena leucocephala* on Pb/Zn mine tailings and observed that the plant grew well on tailings with a 25% soil amendment, but the earthworm only survived (with active burrowing activity) with a 50% soil amendment. The earthworm enhanced plant yield (by 10%-30%) and plant uptake of P (10%). It also increased bioavailable metal concentrations in the amended spoils, leading to higher plant uptake of both Pb and Zn. It was concluded that the ecological context of phytoremediation should consider the plant-soil-animal interactions that influence metal mobility.

Subsequently, beneficial effects of earthworms (*P. guillelmi*) and arbuscular mycorrhizal fungi (AMF, *Glomus* spp.) on the establishment of the same tree species on Pb/Zn tailings were reported. In general, AMF improved N, P and K uptake by trees, while earthworms had more influence in improving N nutrition. The combined effects of both AMF and earthworms were additive and proved to be beneficial to plant growth, plant nutrition and for protection against uptake of toxic metals (Ma et al., 2006).

2.1.8 Bacteria

Based on a soil column experiment, results showed both positive and negative effects on inoculation of bacteria (*Azotobacter chroococcum, Bacillus megaterium* and *Bacillus musilagionsus*) on metal bioavailability and mobility. Inoculation of bacteria led to an increase of water dissolved organic carbon concentration and a decrease of pH value, which enhanced metal mobility and bioavailability. On the other hand, the inoculation could immobilize metals due to the adsorption by bacterial cell walls and possible sedimentation reactions with P or other anions produced from bacterial metabolism (Wu et al., 2006b). The adsorption kinetics of Pb and Cd by two plant growth promoting rizobacteria: *Azotobacter chroococcum* and *Bacillus megaterium*, were investigated. Results revealed that the former had a better capacity in binding both metals, especially Pb. It was concluded that the presence of bacteria in the rhizosphere may result in the reduction of mobile ions in soil solution (Wu et al., 2009d).

It was further revealed that by inoculating growth promoting rhizobacters (PGPRs) including N-fixing bacteria and P and K solubilizers, into Pb/Zn contaminated soils resulted in enhanced growth of *Brassica juncea*. These microbes also altered metal bioavailability in the soil, thereby protecting the plant from metal toxicity. Although inoculation with rhizobacteria did not influence on the metal concentrations in pant tissues, the much larger above ground biomass produced led to a higher efficiency of phytoextration by the plant (Wu et al., 2006a).

Li et al. (2007) and Li et al. (2010) studied the effects of inoculation of *Burkholderia cepacia*, on Cd/Zn hyperaccumulating plant *Sedum alfredii*, growing in different concentrations of Cd and Zn. The inoculation improved plant growth, P, Cd and Zn uptake, tolerance index, with a better translocation metals from root to shoot significantly. The organic acids secreted by the plant under metal treatments seemed to be a functional metal resistance mechanism that chelates the metal ions extracellularly, thereby reducing their uptake and subsequent impacts on root physiological processes. This has been further confirmed by the column study, investigating the effects of bacteria on metal bioavailability, speciation and mobility in different metal mine soils (Li et al., 2010).

A subsequent study observed that the Cd/Zn hyperacumulating plant (*S. alfredii*) inoculated with rhizosphere bacteria was able to enhance plant metal uptake and removal of phenanthrene simultaneously (Li et al., 2012).

2.1.9 Fungi

It is commonly recognized that arbuscular mycorrhizal fungi (AMF) are excellent in enhancing their hosts to obtain moisture and nutrients (notably P). Interactions between AMF and heavy metals were investigated under a sand culture experiment (Liao et al., 2003), and also in different depths of soil overlying coal fly ash (Bi et al., 2003). A field survey conducted on five mining sites in Chenzhou City, Hunan Province showed that plants thriving in the mine sites adopted different survival strategies, higher colonization rates of AMF enable *Pteris vitatta* (As hyperaccumulator), and *Cynodon dacylon* (metal-tolerant) to accumulate more As in the fronds and roots respectively (Leung et al., 2007).

In addition to possessing a high tolerance to various environmental stresses, such as drought, nutrient deficiency, high levels of toxic metals in soil, vetiver (*V. zizanioides*) is an excellent species for erosion control due to its extensive root system. Both greenhouse (Wong et al., 2007) and field trials (Wu et al., 2010b) confirmed that AMF can further enhance absorption of nutrients and tolerance to toxic metals in the remediation of Pb/Zn mine tailings. Leung et al (2007) and Wu et al (2015b) demonstrated that AMF played a significant role in enhancing As uptake by *Pteris vittata* (As hyperaccumulator) in As-contaminated soils. The positive effects of AMF inoculation on As accumulation and growth of the Chinese brake fern were enhanced by phosphate amendment (Leung et al., 2010a, 2010b; Lou et al., 2010).

By comparing metallicolous and nonmetalliciolous populations of plant species with or without inoculation of AMF, it has been generally revealed that the metallicolous populations of *Sedum alfredii* (Zn/Cd hyperaccumuator) (Deng et al., 2007), *P. vitatta* (As-hyperaccumulator) (Wu et al., 2007), and *Cynodon dactylon* (metal-tolerant plant) (Wu et al., 2010a) had higher accumulations of toxic metals (As, Zn and Cd, respectively) in above ground tissues for hyperaccumulating plants and underground tissues for metal tolerant plants, respectively, when grown on metal-contaminated soils. Another study showed that AMF increased multi-metal tolerance of three leguminous plants by AMF colonization (Lin et al., 2007).

The role of AMF in Zn uptake by red clover growing in a calcareous soil spiked with Zn was studied, using containers with two nylon net partitions (30 μm mesh, which allowed penetration by hyphae but not by roots, so that comparison could be made between mycorrhizal roots, with non-mycorrhizal roots) to separate the central root zone from the two outer hyphal zones. Results obtained confirmed that Zn is taken up and transferred to the host plant via the extra-radical hyphae (Cheng et al., 2003a). However, there are intraspecific differences of AMF in their impacts on metal accumulation, e.g. As by *P. vittata* (Wu et al., 2009c).

A study showed that As uptake by upland rice inoculated with AMF has been lowered, while grain yields were enhanced (Chan et al., 2013; Wu et al., 2015c). In the farm soil moderately contaminated with Cd, AMF inoculation affected the accumulation and portioning of Cd and P in bashfulgrass (*Mimosa pudica*) (Hu et al., 2014a). Further studies are needed to investigate the potential of inoculating AMF into vegetables in order to lower their metal uptake when grown in metal-contaminated farm soils.

2.1.10 Interactions of plants with different soil organisms

Attempts have been made to investigate the interactions of higher plants with two or more soil organisms. These included earthworm-mycorrhiza interaction on Cd uptake and growth of ryegrass (Yu et al., 2005), and the combined effects of AMF and different beneficial bacteria, namely N-fixer, P and K solubilizers with different crops (Wu et al., 2005a). Wu et al. (2009a) noted that dual inoculation with an AMF and rhizobium was able to enhance the growth of alfalfa on coal mine substrates.

The major purpose of intercropping of metal hyperaccumulators and economic crops together is to enhance better growth of economic crops, by lowering the level of toxic metals when grown in metal-contaminated farm soils. The farm soils in South China contained rather high levels of Cd, mainly due to contamination of rock phosphate, and sometime mixing industrial wastes (such as fly ash) with fertilizer. Hu et al. (2013a) showed that AMF induced differential Cd and P phytoavailability via intercropping of upland kangong (*Ipomea aquatic*) with Alfred stonecrop (*S. alfredii*). It was subsequently noted that the combination of AMF and biochar lowered Cd accumulation and raised the yield of upland kangong (Hu et al., 2013b, 2014b).

A study demonstrated that dual inoculation of earthworms and plant growth promoting rhizobacteria (PGRP) (which included N-fixing bacteria, P-solubilizing bacteria and K-solubilizing bacteria) enhanced availability of N and P, and higher activities of urease and acid phosphatase in farm soils. Therefore, dual inoculation of earthworms and PGPR could reduce the amount of chemical fertilizers needed in agriculture (Wu et al., 2012c). There seems to be a need to ensure sufficient supplies of earthworms and their safe storage and transportation for large scale soil restoration projects. However, with regards to producing AMF (on a larger scale) and microbial fertilizer (containing various effective microbes, with nutrients released slowly by these microbes embedded in the waste materials), patents are available (Wu et al., 2005b; Chung et al., 2008b).

2.2 Use of wetland plants for water purification

2.2.1 Purifying nutrients in wastewater

It is commonly known that wetland plants are efficient in removing nutrients (N and P) from wastewater it has been known for a long tiome (as early as 1970s). Eventually, attempts were made to treat heavy metals in wastewater and mine effluents. More recently, constructed wetlands have been employed as low-cost tertiary treatment facilities in purifying nutrients in wastewater. These include the use of constructed wetlands (using cattail and common reed) on N and P removal (Chung et al., 2008a). Wu et al. (2011) indicated that co-culture of both plant species would be more beneficial in the reclamation of wastewater than a monoculture system, as the compositions of organic acids varied between the two plant species.

2.2.2 Purifying heavy metals in wastewater and mine effluents

Lan et al. (1992) used cattails in treating Pb/Zn mine effluent in the 2nd largest Pb/Zn mine in China (Shaoguan, northern Guangdong Province), which achieved higher than 90% removal rates of Pb, Zn and suspended solids. Unlike terrestrial plants, metal tolerance develops through evolution, wetland plants possess constitutional tolerance (innate tolerance, or borne tolerance). Two cosmopolitan species, namely cattail (*Typha latifolia*) and common reed (*Phragmites australis*) have been extensively studied in terms of Zn, Pb, Cd (Ye et al., 1997b, 1997c), and Cu, and Ni (Ye et al., 1997a, 2003a) tolerance, uptake and accumulation. It was observed that the iron plaque formed on root surface is able to sequester toxic metals and thereby minimize its uptake by roots for both species (Ye et al., 1998b, 2001a, 2003b). The results were further confirmed in a later study comparing biomass and metal uptake between two populations of *P. australia* grown in flooded and dry conditions (Ye et al., 1998a).

The study was then extended to other parts of China by investigating 12 common wetland plant species (*Leersia hexandra, Juncus effusus* and *Equisetum ramosisti*, etc.) thriving in metal-contaminated sites (Guangdong, Hunan and Hubei Provinces) in China. It was noted that metal (Pb and Cu) accumulation by wetland plants differed among species, populations and tissues, with some populations accumulating relatively high metal concentrations (even in shoot), above the toxic concentration to common plants. The concentrations of metals taken up by their aboveground and underground tissues were significantly correlated to the concentrations in the soils. These indicated there may be internal detoxification metal tolerance mechanisms (s) (Deng et al., 2004). Further studies showed that populations of 6 wetland plants (including *Alternanthera philoxeroides* and *Beckmannia syzigachne*), collected from both metal contaminated and non-contaminated sites in Guangdong, Hunan, Hebei and Zhejiang Provinces, suggested that innate metal tolerance (Pb and Zn) in wetland plants is related to their special root anatomy, and the reduced root conditions in the rhizosphere, leading to lower metal toxicity (Deng et al., 2006, 2009). Attempts were made to investigate the mechanisms of metal tolerance involved in wetland plants. It was noted that root porosity and radial oxygen loss (ROL) were related to As tolerance and uptake in wetland plants. Those with higher rates of ROL and root porosity tended to thrive better in As-contaminated sites, through forming more Fe/Mn plaque, with higher As concentrations sequestered on root surfaces, and a lower As translocation factor (Deng et al., 2010; Li et al., 2011c).

2.2.3 Uptake of As and Cd by paddy rice

There seems to be an urgent need to reduce As (and also Cd) concentrations in rice grains, a common phenomenon, observed worldwide. Being a wetland plant species, paddy rice behaves similarly to other wetland plant species, thriving in oxygen-depleted environment. Based on results generated from 25 rice cultivars and two pot experiments, it was observed that rice cultivars with high porosities and higher rates of ROL were able to limit the transfer of As to aboveground tissues (including rice grains) (Mei et al., 2009). In fact, elevated As concentrations in As-contaminated soils may induce toxicity to rice plants, with a decrease in ROL, and As methylation. Arsenic species transportation will play a significant role in the uptake and accumulation of As by rice plants (Mei et al., 2012; Wu et al., 2013).

A systematic investigation was then carried out by screening 20 genotypes of rice seedlings, subjecting them to irrigation water containing 0.4 mg As/L. Results showed that different genotypes possessed different root anatomy as reflected by entire root porosity (12.43%-33.21%), were significantly correlated with radial oxygen loss (ROL). The total ROL from entire roots was correlated with metal tolerance (biomass) among the 20 genotypes; while total and inorganic As concentrations in rice grains of the 20 genotypes were negatively correlated with ROL (Wu et al., 2011a). It was further confirmed that genotypic differences in As uptake by rice, and the rice genotypes with higher ROL will result in lower overall As accumulation (Wu et al., 2012a).

The effects of silicon (Si) fertilization in As uptake, and speciation in rice plants with different ROL were investigated, and results indicated that there is a potential to decrease As contamination of rice effectively by combining Si fertilization and the selection of rice genotypes with high ROL (Wu et al., 2015a). Hence, through selecting genotypes to grow in As contaminated soil, and with the addition of Si fertilizer, lower As accumulation in rice grains could be achieved.

Parallel studies on Cd were also conducted as consumption of Cd-contaminated rice is also a public health concern, following the incidence of "Itai itai disease" in Japan. Liu et al. (2006, 2007) showed that variations between rice cultivars in root secretion of organic acids which affected Cd uptake, and low-molecular–weight organic acids (LMWOA) secretion by rice root, seems to be one of the important mechanisms, determining the uptake of Cd by rice plants. Cadmium uptake and tolerance were studied among 20 rice cultivars, based on a field trial and a pot trial, and the rates of ROL were measured. It was observed in both trials that rice cultivars with higher rates of ROL possessed higher capacities for limiting the transfer of Cd to rice grains and straw. Wang et al. (2011, 2013) and Cheng et al. (2014) also demonstrated that ROL and iron plaque formation on roots also altered Cd, Pb and As uptake and distribution in rice plant tissues.

2.2.4 The use of AMF for lowering as uptake in rice grain

Attempts have been made to lower As uptake and accumulation in rice grains, through inoculation of AMF, due to their beneficial effects on plant growth and colonization. A series of experiments has been conducted on the inoculation of AMF into rice plants, including upland and low land rice.

Li et al. (2011a) studied the effects of AMF (*Glomus intraradices* and *G. geosporum)* on cultivars of lowland and upland rice plants, and observed that there was strong functional diversity in AM symbioses, which determined the transfer and uptake of As and P. The lowland rice cultivar inoculated with *G. intraradices* or upland rice cultivar inoculated with *G. geosporum* enhanced As tolerance, grain P content, and grain yield. Li et al. (2011b) further studied the uptake kinetics of different As species (arsenate, arsenite, dimethylarsinic acid, and monomethylarsonic acid) in lowland (paddy) and upland rice colonized with AMF (*Glomus intraradices*). Results demonstrated that the four As species influx exerted significant differences between two rice varieties in low-affinity (0-2.5 mmol/L) uptake systems. In the arsenate treatment, the ratio of arsenate/arsenite was reduced in shoots while increased in roots, due to the presence of *G. intraradices*. Mycorrhizal inoculation also lowered the uptake of arsenite and MMA by both cultivars substantially in both low-and high-affinity (0-0.05 mmol/L) uptake systems. It was further demonstrated that rice varieties with high ROL would favor AM fungal infection, and enhance root ratio of As (Ⅲ) conc./As (Ⅴ) conc. in the presence of AMF. This was possibly caused by the depletion of oxygen by AMF in rhizosphere soil, with As (Ⅴ) transformed to As (Ⅲ) (Li et al., 2013).

The As speciation and uptake by rice, and the effects of ROL and AMF on As uptake by different genotypes of rice, were reviewed and discussed, with the attempt to reduce As in the rice grain (Li et al., 2012).

2.2.5 Wetland plants for purifying other contaminants in wastewater

Studies related to this aspect have been conducted: DDTs and PCBs removal by *P. australis* and *Oryza sativa* (Chu et al., 2006a, 2006b). The physiological and biochemical responses of *O. sativa* to phenantrene and pyrene were investigated (Li et al., 2008).

The combined beneficial effects of AMF and PAH-degrading bacteria were obtained when inoculated to ryegrass for dissipation of phenanthrene and pyrene (Gao et al., 2006, 2010; Yu et al., 2011; Wu et al., 2011b, 2014). It was noted that inoculation of AMF and PAH-degrading bacteria, especially the former, resulted in a positive effect on alleviation of PAH toxicity to ryegrass (Li et al., 2011d), and maize (Wu et al., 2011b). The uptake and transport mechanisms of decabromodiphenul ether (BDE-209) by *O. sativa* were also investigated and the results revealed that wetland plants are able to serve as a tertiary treatment system for purifying this emerging chemical of concern (Chow et al., 2015).

The effects of rapid socio-economic development in South China on the dyke-pond systems have been discussed (Wong et al., 2004). A recent paper reviewed advances made in integrating wetlands with aquaculture (Chen et al., 2016). Constructed wetlands are useful for treating wastewater from small villages, or serving as a tertiary treatment facility to further purify effluent from sewage treatment plants. They also provide a wide range of habitats for wildlife and would be beneficial if aquaculture could be also integrated into the systems.

3. Restoration of completed landfills

Landfilling is a common method for disposing solid waste in most countries. Completed landfills are commonly used for recreational purposes, in which revegetation is highly essential. Most of the early domestic landfills were in fact open dumps, without any measures to contain landfill gas and leachate generated, during anaerobic decomposition of waste materials embedded below, by means of impermeable layers both at the bottom and top of landfills. A cover soil is used to seal the rubbish dumps, and then planted with grasses. Bare patches, in particular in severely eroded areas, are commonly observed, due to the emissions of landfill gas and leachate from these completed landfills.

Early studies focused on the choice of plant species which were able to tolerate high concentrations of landfill gas and leachate. Landfill gas mainly consists of methane and other gases in trace amounts. Due to the lack of oxygen, the appearance of affected soil will exhibit a grey colour, like waterlogged soils. It was recognized that methane itself is inert to plants, but the lack of oxygen is the limiting factor inhibiting plant growth, in particular root growth in areas with high concentrations of landfill gas.

Landfill leachate is generated after heavy rainfall, with rain water entering into the embedded wastes, and seeping out through cracks of the cover soil, without a continuous cover of vegetation. Leachate contains high levels of nitrogenous compounds and occasionally heavy metals such as Cd, Cu, Zn and Pb, which will become more available under reduced conditions, exerting adverse effects on plant growth.

Modern landfills or very often referred to "engineered landfills" possess durable and impermeable liners (made of PVC or bentonite) at the bottom and top (upon closure), with landfill gas and leachate collection systems. The landfill gas is fared on-site or transformed into electricity, while landfill leachate is either treated on-site or transferred to the nearest sewage treatment work. Therefore, the effects of landfill gas and leachate on plant growth are rather limited in these "engineered landfills" .

3.1 Restoration of old landfills ("open dumps")

Landfilling has been adopted as the major method of solid waste disposal in Hong Kong, with all the incinerators phased out since 1990s. There are 13 old completed landfills, and three large-scale engineered landfills under operation in Hong Kong, to deal with the waste generated from a population of over 7 million. Completed landfills will provide additional land for recreational usages, such as golf driving ranges, and foot-ball pitches, with vegetation forming the most important component.

Soil and plant characteristics of landfill sites near Merseyside, England were investigated (Wong et al., 1988). Similar surveys were also conducted in the Gin Drinkers' Bay, Hong Kong, by comparing 13 completed landfills in Hong Kong with a subtropical climate (Wong et al., 1989c, 1989d; Chan et al., 1996). Studying the physical and chemical properties of the contaminated sites could identify the adverse factors inhibiting plant growth and colonization for subsequent restoration. In general, results showed that landfill gas (Chan et al., 1991, 1996; Lan and Wong, 1994) and leachate (Cheung et al., 1993; Chu et al., 1994) were the major factors inhibiting plant growth and establishment on the completed landfill. In addition, the poor physical and chemical properties of landfill cover soil, especially the lack of nutrients and moisture greatly inhibit plant growth (Liang et al., 1999). In completed landfills, the diversity and distribution of soil fauna would indicate stages of ecological succession of completed landfills (Wong et al., 1992; Chan et al., 1997).

Attempts were made to decontaminate landfill leachate by soil filtration (Wong and Leung, 1989b; Wong et al., 1990), and by algal purification (Chu et al., 1996). Landfill leachate was used as irrigation water, which provided nitrogenous compounds for growing trees and vegetables (Wong and Leung, 1989a). It was observed that co-disposing appropriate portions of different waste materials, such as sewage sludge and marine dredging, with municipal refuse, the amount of methane generated could be maximized (Chan et al., 1999a). It was further noted that recirculating leachate to the above mixed waste materials enhanced gas production (Chan et al., 2002).

Screening of suitable tree species for growth on completed landfills was conducted: Seedlings of ten woody plant species (belonging to eight families) were ①planted in two sites of the completed landfill (Gin Drinkers' Bay): one with a high concentration (methane: 1.7%-41.0%, carbon dioxide: 2.0%-39.0%, oxygen: 1.6%-16.3%), while the other one a low concentration (methane: 0%-10%, carbon dioxide: 0%-4.7%, oxygen: 19%-20.3%) of landfill gas; and ②subjected to fumigation using artificial landfill gas (with different portions of the 3 major gases) in the laboratory. It was observed that both the two legumes *Acacia confusa* and *Albizzia lebbek* were more tolerant to landfill gas, among all the species tested, with their roots growing downward, while roots of other trees had their roots growing sideways and even upward (Chan et al., 1991). Legume plant species would be preferred as their associated root nodules, with N-fixing bacteria are able to utilize free N from the air, the most essential nutrient which is always lacking in landfill cover soil (Chan et al., 1998).

It was demonstrated that the high concentration of carbon dioxide and low concentration of oxygen affected growth and nodule activity of another woody legume *Leucaena leucocephala* (Zhang et al., 1995). By providing suitable portions of leachate, growth and N-fixation of two leguminous trees (*A. confusa* and *L. leucocephala*) were enhanced (Chan et al., 1999b).

3.2 Restoration of "engineered landfills"

However, for modern landfills with engineered liners at both the bottom and top of completed landfills, with gas and leachate collection systems; both gas and leachate would not impose significant effects on growth and development of plants on the landfill cover soils.

As part of the restoration programme of one of the engineered landfills namely South–East New Territories (SENT) Landfill, the growth performance of 25 woody plant species subjected to different soil amelioration and seedling planting methods was compared in 1999. Results revealed that the pioneer species, notably *Acacia* species (*A. confusa*, *A. auriculiformis* and *A. mangium*) were much better than native species (such as *Eucalyptus torelliana*, *Schejjlera octophylla* and *Alnus fonnosana*), and soil ameliorations (horticultural soil with fertilizer and horse and pig manure) enhanced seedling survival rate and plant growth (Wong et al., 2016b).

The restoration programme was initiated in 1999, and a long-term monitoring was conducted to evaluate the ecological performance of the restored landfill. The growth performance of planted trees, soil properties, and associated soil fauna of selected landfill sites were monitored and compared with the natural sites near the landfill, twice (summer and winter) yearly, for a period of 12 years (2000-2012). It was shown that the number of plant species at the restored landfill site fluctuated considerably during the past 12 years, compared to the other sites (more natural sites near the landfill). Based on the results of similarity analyses, it was further observed that the plant communities and soil properties (pH, bulk density, moisture content, organic matter, etc.) at the restored landfill site were significantly different, while fauna communities had no significant difference, when compared with the other two sites. Furthermore, soil moisture and stomatal resistance of *A. confusa* were negative correlated showing that the restored site was more subject to drought than the other sites (Chen et al., 2015).

It was further noted that although no soil animal was introduced to the newly restored site in 1999, there were 29, 31 and 44 animal species recorded in 3 different restored sites, 12 years later. Within this monitoring period, 38 plant species were recorded, and 17 of them were self-seeding, with the family Acacia and Mimosaceae dominating at all the restored sites. This reconfirms that being drought and landfill gas tolerant with N-fixing capability, legume woody plants (such as *A. confusa*) are more competitive and suited for growing on landfill cover soils, which lack moisture and N (Wong et al., 2015).

3.3 Biochar as an organic amendment

Biochar is a carbon-rich substance generated by heating biomass (such as leaves and wood chips) at a relatively low temperature (<700 ℃), with a limited supply of oxygen. It is commonly used as soil amendment to improve physical and chemical properties of soil, in particular as a pollutant sorbent for soil and water remediation. In addition to promoting crop growth, biochar can also mitigate odor emission and promote methane oxidation in field soils. In short, biochar can serve as an excellent material for carbon sequestration, promoting higher crop yield and contaminant remediation.

Stabilizing the slope of completed landfills is highly essential in order to avoid soil and water erosion. Chen et al. (2016) reviewed the advances made so far using biochar as a new type of material in the final landfill cover. It was concluded that biochar is an excellent material for enhancing ecological performance and slope stability. It stimulates ecological performance, in terms of productivity (increase biomass), root development (root elongation), and soil microbial activities (better nutrient recycling for plants). It also enhances slope stability, through elevating shear strength (related to water retention characteristics, hydrophobicity, and hydrophilicity), leading to better physical stability of the slope.

An attempt was made to mix biochar with clay (biochar-amended clay [BAC]) as an alternative landfill final cover material, and the results indicated that the gas permeability of the BAC was reduced by increasing the biochar content and degree of compaction. Therefore, BAC can be a potential landfill final cover material in order to minimize the landfill gas generated from waste degradation (Wong et al., 2016a).

4. Conclusions

In this attempt to study the emerging chemicals management issues in development issues in developing countries and countries with economy in transition, "heavy metals" ranked top among these countries (Bouwman et al., 2012). Successful restoration of the man-made habitats derived from mining and smelting of metals will depend on the choice of appropriate plant species, which are able to grow under elevated concentrations of toxic metals. Amendments of inert waste materials (derived from mining activities, such as waste rocks) could serve as an insulation to the toxic metals, and organic waste materials (such as domestic refuse and manure compost) could improve soil properties and provide nutrients for plant growth. Planting woody legumes could lock up a large portion of toxic metals in the woody tissues, and at the same time provide its own N. Intercropping of metal hyperaccumlating plants with crops (with an acceptable uptake of metals) seems to be a strategy to clean up contaminated farm soils, while continuing its economic activities. Nevertheless, plants should be provided with soil organisms (i.e., earthworms and microbes), as much as possible, to assist them to grow in such stressed environments.

Wetland plants could purify nutrients and pollutants effectively with their distinct morphology and physiology. They are able to thrive in a harsh environment. Constructed wetlands have become popular in treating sewage from small villages and serving as a tertiary facility in polishing effluent from sewage treatment plants. In addition to sewage treatment, and provision of a wide range of habitats for biological conservation, it seems a good idea to integrate constructed wetlands with various forms of aquaculture.

Landfilling is the most common method of solid waste disposal in the world. Landfill gas and leachate are the two common problems faced with restoration of these open dumps. Successful revegetation will largely depend on selection of plant species (such as Acacia) which are tolerant to landfill gas (high concentrations of carbon dioxide and methane, and the lack of oxygen). Being tolerant to the landfill gas, drought, and nutrient deficiency, *Acacia* spp. (pioneer species) are excellent candidates to grow on cover soil of these "open dumps" (old landfills). For modern "engineered landfills", landfill gas and leachate would not normally impose adverse effects on plant growth. We could consider planting more native species on these sites. Again, provision of organic amendment and soil organisms will enhance plant growth and development, and accelerate ecological succession.

Acknowledgements

The author would like to express his sincere gratitude to all collaborators and postgraduate students, and financial suport from various funding agents, especially the Hong Kong Research Grants Council and the Croucher Foundation of Hong Kong is duly acknowledged. This is dedicated to Professor A. D. Bradshaw, my fondest mentor and a dear friend.

References

Bi Y L, Li X L, Christie P, et al. 2003. Growth and nutrient uptake of arbuscular mycorrhizal maize in different depths of soil overlying coal fly ash. Chemosphere, 50 (6): 863-869.

Bouwman H, Wong M H, Barra R. 2012. GEF guidance on emerging chemicals management issues in developing countries and countries with economy in transition. UNEP/GEF.

Brandshaw A D. 1952. Populations of Agrostis tenuis resistant to lead and zinc poisoning.

Chan G Y S, Chu L M, Wong M H. 2002. Effects of leachate recirculation on biogas production from landfill co-disposal of municipal solid waste, sewage sludge and marine sediment. Environmental Pollution, 118 (3): 393-399.

Chan G Y S, Wong M H, Whitton B A. 1991. Effects of landfill gas on subtropical woody plants. Environmental Management, 15 (3): 411-431.

Chan G Y S, Wong M H. 2002. Revegetation of landfill sites//Lal R. Encyclopedia of soil science, Marcel Dekker, NY. pp. 6-1161.

Chan G Y S, Zhi H Y, Wong M H. 2003. Comparison of four sesbania species to remediate Pb/Zn and Cu mine tailings. Environmental Management, 32 (2): 246-251.

Chan W F, Li H, Wu F Y, et al. 2013. Arsenic uptake in upland rice inoculated with a combination or single arbuscular mycorrhizal fungi. Journal of Hazardous Materials, 262: 1116-1122.

Chan Y S G, Chu L M, Wong M H. 1997. Influence of landfill factors on plants and soil fauna—an ecological perspective. Environmental Pollution, 97 (1): 39-44.

Chan Y S G, Chu L M, Wong M H. 1999a. Codisposal of municipal refuse, sewage sludge and marine dredgings for methane production. Environmental Pollution, 106 (1): 123-128.

Chan Y S G, Wong M H, Whitton B A. 1996. Effects of landfill factors on tree cover—a field survey at 13 landfill sites in Hong Kong. Land Contamination and Reclamation, 4: 115-128.

Chan Y S G, Wong M H, Whitton B A. 1998. Effects of landfill gas on growth and nitrogen fixation of two leguminous trees (*Acacia confusa*, *Leucaena leucocephala*). Water, Air, and Soil Pollution, 107 (1-4): 409-421.

Chan Y S G, Wong M H, Whitton B A. 1999b. Effects of landfill leachate on growth and nitrogen fixation of two leguminous trees (*Acacia confusa*, *Leucaena leucocephala*). Water, Air, and Soil Pollution, 111 (1-4): 29-40.

Chen B D, Li X L, Tao H Q, et al. 2003a. The role of arbuscular mycorrhiza in zinc uptake by red clover growing in a calcareous soil spiked with various quantities of zinc. Chemosphere, 50 (6): 839-846.

Chen R Z, Wong M H. 2016. Integrated wetlands for food production. Environmental Research, 148: 429-442.

Chen X W, Wong J T F, Mo W Y, et al. 2015. Ecological performance of the restored south east new territories (SENT) landfill in Hong Kong (2000–2012). Land. Degrad. Dev.

Chen X W, Wong J T F, Ng C W W, et al. 2016. Feasibility of biochar application on a landfill final cover—a review on balancing ecology and shallow slope stability. Environmental Science and Pollution Research, 23 (8): 7111-7125.

Chen Y X, He Y F, Yang Y, et al. 2003b. Effect of cadmium on nodulation and N_2-fixation of soybean in contaminated soils. Chemosphere, 50 (6): 781-787.

Chen Y X, Lin Q, Luo Y M, et al. 2003c. The role of citric acid on the phytoremediation of heavy metal contaminated soil. Chemosphere, 50 (6): 807-811.

Cheng H, Wang M, Wong M H, et al. 2014. Does radial oxygen loss and iron plaque formation on roots alter Cd and Pb uptake and distribution in rice plant tissues? Plant and Soil, 375 (1-2): 137-148.

Cheng J, Wong M H. 2002. Effects of earthworms on Zn fractionation in soils. Biology and Fertility of Soils, 36 (1): 72-78.

Cheng J, Wong M H. 2008. Effects of earthworm (*Pheretima* sp.) on three sequential ryegrass harvests for remediating Lead/Zinc mine tailings. International Journal of Phytoremediation, 10 (3): 173-184.

Cheung K C, Chu L M, Wong M H. 1993. Toxic effect of landfill leachate on microalgae. Water, Air, and Soil Pollution, 69 (3-4): 337-349.

Cheung K C, Wong J P K, Zhang Z Q, et al. 2000. Revegetation of lagoon ash using the legume species *Acacia auriculiformis* and *Leucaena leucocephala*. Environmental Pollution, 109 (1): 75-82.

Chiu K K, Ye Z H, Wong M H. 2005. Enhanced uptake of As, Zn, and Cu by *Vetiveria zizanioides* and *Zea mays* using chelating agents. Chemosphere, 60 (10): 1365-1375.

Chiu K K, Ye Z H, Wong M H. 2006. Growth of *Vetiveria zizanioides* and *Phragmities australis* on Pb/Zn and Cu mine tailings amended with manure compost and sewage sludge: a greenhouse study. Bioresource Technology, 97 (1): 158-170.

Chongyu L, Minghung W. 1994. Environmental factors affecting growth of grasses, herbs and woody plants on a sanitary landfill. Journal of Environmental Sciences (China), 6 (4): 504-513.

Chow K L, Man Y B, Tam N F Y, et al. 2015. Uptake and transport mechanisms of decabromodiphenyl ether (BDE-209) by rice (*Oryza sativa*). Chemosphere, 119: 1262-1267.

Chu L M, Cheung K C, Wong M H.1996. Algal purification of pretreated landfill leachate. Toxicol Environ Chem 53: 159-174.

Chu L M, Cheung K C, Wong M H. 1994. Variations in the chemical properties of landfill leachate. Environmental Management, 18 (1): 105-117.

Chu W K, Wong M H, Zhang J. 2006a. Accumulation, distribution and transformation of DDT and PCBs by *Phragmites australis* and *Oryza sativa*. II. Enzyme study. Environmental Geochemistry Health, 28: 169-81.

Chu W K, Wong M H, Zhang J. 2006b. Accumulation, distribution and transformation of DDT and PCBs by *Phragmites australis* and *Oryza sativa* L.: I. Whole plant study. Environmental Geochemistry and Health, 28 (1-2): 159-168.

Chung A K C, Wu Y, Tam N F Y, et al. 2008a. Nitrogen and phosphate mass balance in a sub-surface flow constructed wetland for treating municipal wastewater. Ecological Engineering, 32 (1): 81-89.

Chung M K, Li W C, Wu S C, et al. 2008b. Production of arbuscular mycorrhizal fungi (*Glomus mosseae* and *Acaulospora laevis*) inocula by aeroponic method. Hong Kong Intellectual Property Agent.

Deng D M, Shu W S, Zhang J, et al. 2007. Zinc and cadmium accumulation and tolerance in populations of *Sedum alfredii*. Environmental Pollution, 147 (2): 381-386.

Deng D, Wu S C, Wu F Y, et al. 2010. Effects of root anatomy and Fe plaque on arsenic uptake by rice seedlings grown in solution culture. Environmental Pollution, 158 (8): 2589-2595.

Deng H, Ye Z H, Wong M H. 2004. Accumulation of lead, zinc, copper and cadmium by 12 wetland plant species thriving in metal-contaminated sites in China. Environmental Pollution, 132 (1): 29-40.

Deng H, Ye Z H, Wong M H. 2006. Lead and zinc accumulation and tolerance in populations of six wetland plants. Environmental

Pollution, 141 (1): 69-80.

Deng H, Ye Z H, Wong M H. 2009. Lead, zinc and iron (Fe^{2+}) tolerances in wetland plants and relation to root anatomy and spatial pattern of ROL. Environmental and Experimental Botany, 65 (2): 353-362.

Gao Y, Wu S C, Yu X Z, et al. 2010. Dissipation gradients of phenanthrene and pyrene in the rice rhizosphere. Environmental Pollution, 158 (8): 2596-2603.

Gao Y, Yu X Z, Wu S C, et al. 2006. Interactions of rice (*Oryza sativa* L.) and PAH-degrading bacteria (*Acinetobacter* sp.) on enhanced dissipation of spiked phenanthrene and pyrene in waterlogged soil. Science of the Total Environment, 372 (1): 1-11.

Hu J, Li J, Wu F, et al. 2013a. Arbuscular mycorrhizal fungi induced differential Cd and P phytoavailability via intercropping of upland kangkong (*Ipomoea aquatica* Forsk.) with Alfred stonecrop (*Sedum alfredii* Hance): post-harvest study. Environmental Science and Pollution Research, 20 (12): 8457-8463.

Hu J, Wang H, Wu F, et al. 2014a. Arbuscular mycorrhizal fungi influence the accumulation and partitioning of Cd and P in bashfulgrass (*Mimosa pudica* L.) grown on a moderately Cd-contaminated soil. Applied Soil Ecology, 73: 51-57.

Hu J, Wu F, Wu S, et al. 2013b. Phytoavailability and phytovariety codetermine the bioaccumulation risk of heavy metal from soils, focusing on Cd-contaminated vegetable farms around the Pearl River Delta, China. Ecotoxicology and Environmental safety, 91: 18-24.

Hu J, Wu F, Wu S, et al. 2014b. Biochar and Glomus caledonium influence Cd accumulation of upland kangkong (*Ipomoea aquatica* Forsk.) intercropped with Alfred stonecrop (*Sedum alfredii* Hance). Scientific Reports, 4.

Jiang X J, Luo Y M, Zhao Q G, et al. 2003. Soil Cd availability to Indian mustard and environmental risk following EDTA addition to Cd-contaminated soil. Chemosphere, 50 (6): 813-818.

Lan C Y, Shu W S, Wong M H. 1998. Reclamation of Pb/Zn mine tailings at Shaoguan, Guangdong Province, People' s Republic of China: The role of river sediment and domestic refuse. Bioresource Technology, 65 (1): 117-124.

Lan C, Chen G, Li L, et al. 1992. Use of cattails in treating wastewater from a Pb/Zn mine. Environmental Management, 16 (1): 75-80.

Leung H M, Wu F Y, Cheung K C, et al. 2010a. Synergistic effects of arbuscular mycorrhizal fungi and phosphate rock on heavy metal uptake and accumulation by an arsenic hyperaccumulator. Journal of Hazardous Materials, 181 (1): 497-507.

Leung H M, Wu F Y, Cheung K C, et al. 2010b. The effect of arbuscular mycorrhizal fungi and phosphate amendment on As uptake, accumulation and growth of *Pteris vittata* in As contaminated soil. International Journal of Phytoremediatim 12: 384-403.

Leung H M, Ye Z H, Wong M H. 2007. Survival strategies of plants associated with arbuscular mycorrhizal fungi on toxic mine tailings. Chemosphere, 66 (5): 905-915.

Li H, Man Y B, Ye Z H, et al. 2013. Do arbuscular mycorrhizal fungi affect arsenic accumulation and speciation in rice with different radial oxygen loss? Journal of hazardous materials, 262: 1098-1104.

Li H, Wu C, Wong M H. 2012. Potential strategies for mitigating arsenic contamination in rice grain//Ng J C et al. Understanding the Geological and Medical Interface of Arsenic. CRC Press: London.

Li H, Wu C, Ye Z H, et al. 2011a. Uptake kinetics of different arsenic species in lowland and upland rice colonized with Glomus intraradices. Journal of Hazardous Materials, 194: 414-421.

Li H, Ye Z H, Chan W F, et al. 2011b. Can arbuscular mycorrhizal fungi improve grain yield, As uptake and tolerance of rice grown under aerobic conditions? Environmental Pollution, 159 (10): 2537-2545.

Li H, Ye Z H, Wei Z J, et al. 2011c. Root porosity and radial oxygen loss related to arsenic tolerance and uptake in wetland plants. Environmental Pollution, 159 (1): 30-37.

Li J H, Gao Y, Wu S C, et al. 2008. Physiological and biochemical responses of rice (*Oryza sativa* L.) to phenanthrene and pyrene. International Journal of Phytoremediatim, 10: 1-14.

Li J H, Yu X Z, Wu S C, et al. 2011 d. Responses of bioaugmented ryegrass to PAH soil contamination. International Journal of Phytoremediation, 13 (5): 441-455.

Li W C, Wong M H. 2010. Effects of bacteria on metal bioavailability, speciation, and mobility in different metal mine soils: a column study. Journal of Soils and Sediments, 10 (2): 313-325.

Li W C, Wong M H. 2012. Interaction of Cd/Zn hyperaccumulating plant (*Sedum alfredii*) and rhizosphere bacteria on metal uptake and removal of phenanthrene. Journal of Hazardous Materials, 209: 421-433.

Li W C, Ye Z H, Wong M H. 2007. Effects of bacteria on enhanced metal uptake of the Cd/Zn-hyperaccumulating plant, *Sedum alfredii*. Journal of Experimental Botany, 58 (15-16): 4173-4182.

Li W C, Ye Z H, Wong M H. 2010. Metal mobilization and production of short-chain organic acids by rhizosphere bacteria associated with a Cd/Zn hyperaccumulating plant, *Sedum alfredii*. Plant and Soil, 326 (1-2): 453-467.

Liang J, Zhang J, Chan G Y S, et al. 1999. Can differences in root responses to soil drying and compaction explain differences in performance of trees growing on landfill sites?. Tree Physiology, 19 (9): 619-624.

Liao J P, Lin X G, Cao Z H, et al. 2003. Interactions between arbuscular mycorrhizae and heavy metals under sand culture experiment. Chemosphere, 50 (6): 847-853.

Lin A J, Zhang X H, Wong M H, et al. 2007. Increase of multi-metal tolerance of three leguminous plants by arbuscular mycorrhizal fungi colonization. Environmental Geochemistry and Health, 29 (6): 473-481.

Liu J, Qian M, Cai G, et al. 2007. Variations between rice cultivars in root secretion of organic acids and the relationship with plant cadmium uptake. Environmental Geochemistry and Health, 29 (3): 189-195.

Liu J, Wang D, Xu J, et al. 2006. Variations among rice cultivars on root oxidation and Cd uptake. Journal of Environmental Sciences-Amsterdam, 18 (1): 120-124.

Liu Y, Wang H B, Wong M H, et al. 2009. The role of arsenate reductase and superoxide dismutase in As accumulation in four Pteris species. Environment International, 35 (3): 491-495.

Lou L Q, Ye Z H, Lin A J, et al. 2010. Interaction of arsenic and phosphate on their uptake and accumulation in Chinese brake fern. International Journal of Phytoremediation, 12 (5): 487-502.

Lou L Q, Ye Z H, Wong M H. 2007. Solubility and accumulation of metals in Chinese brake fern, vetiver and rostrate sesbania using chelating agents. International Journal of Phytoremediation, 9 (4): 325-343.

Lou L Q, Ye Z H, Wong M H. 2009. A comparison of arsenic tolerance, uptake and accumulation between arsenic hyperaccumulator, *Pteris vittata* L. and non-accumulator, *P. semipinnata* L.—a hydroponic study. Journal of Hazardous Materials, 171 (1): 436-442.

Ma Y, Dickinson N M, Wong M H. 2003. Interactions between earthworms, trees, soil nutrition and metal mobility in amended Pb/Zn mine tailings from Guangdong, China. Soil Biology and Biochemistry, 35 (10): 1369-1379.

Ma Y, Dickinson N M, Wong M H. 2006. Beneficial effects of earthworms and arbuscular mycorrhizal fungi on establishment of leguminous trees on Pb/Zn mine tailings. Soil Biology and Biochemistry, 38 (6): 1403-1412.

Ma Y, Dickinson N, Wong M. 2002. Toxicity of Pb/Zn mine tailings to the earthworm Pheretima and the effects of burrowing on metal availability. Biology and Fertility of Soils, 36 (1): 79-86.

Mei X Q, Wong M H, Yang Y, et al. 2012, The effects of radial oxygen loss on arsenic tolerance and uptake in rice and on its rhizosphere. Environmental Pollution, 165: 109-117.

Mei X Q, Ye Z H, Wong M H. 2009. The relationship of root porosity and radial oxygen loss on arsenic tolerance and uptake in rice grains and straw. Environmental Pollution, 157 (8): 2550-2557.

Pang J, Chan G S Y, Zhang J, et al. 2003. Physiological aspects of vetiver grass for rehabilitation in abandoned metalliferous mine wastes. Chemosphere, 52 (9): 1559-1570.

Shu W S, Lan C Y, Zhang Z Q, et al. 2000. Use of vetiver and other three grasses for revegetation of Pb/Zn mine tailings at Lechang, Guangdong Province: Field experiment. Internatimal Journal of Phytoremediation, 4 (1): 47-57.

Shu W S, Ye Z H, Lan C Y, et al. 2001. Acidification of lead/zinc mine tailings and its effect on heavy metal mobility. Environment International, 26 (5): 389-394.

Shu W S, Ye Z H, Lan C Y, et al. 2002. Lead, zinc and copper accumulation and tolerance in populations of *Paspalum distichum* and *Cynodon dactylon*. Environmental Pollution, 120 (2): 445-453.

Shu W S, Ye Z H, Zhang Z Q, et al. 2005. Natural colonization of plants on five lead/zinc mine tailings in Southern China. Restoration Ecology, 13 (1): 49-60.

Sun Q, Ye Z H, Wang X R, et al. 2005. Increase of glutathione in mine population of *Sedum alfredii*: a Zn hyperaccumulator and Pb accumulator. Phytochemistry, 66 (21): 2549-2556.

Sun Q, Ye Z H, Wang X R, et al. 2007. Cadmium hyperaccumulation leads to an increase of glutathione rather than phytochelatins in the cadmium hyperaccumulator *Sedum alfredii*. Journal of Plant Physiology, 164 (11): 1489-1498.

Wan J H C, Wong M H. 2004. Effects of earthworm activity and P-solubilizing bacteria on P availability in soil. Journal of Plant Nutrition and Soil Science, 167 (2): 209-213.

Wang H B, Wong M H, Lan C Y, et al. 2007. Uptake and accumulation of arsenic by 11 Pteris taxa from southern China. Environmental Pollution, 145 (1): 225-233.

Wang H B, Ye Z H, Shu W S, et al. 2006. Arsenic uptake and accumulation in fern species growing at arsenic-contaminated sites of southern China: Field surveys. International Journal of Phytoremediation, 8 (1): 1-11.

Wang M Y, Chen A K, Wong M H, et al. 2011. Cadmium accumulation in and tolerance of rice (*Oryza sativa* L.) varieties with different rates of radial oxygen loss. Environmental Pollution, 159 (6): 1730-1736.

Wang X, Yao H, Wong M H, et al. 2013. Dynamic changes in radial oxygen loss and iron plaque formation and their effects on Cd and As accumulation in rice (*Oryza sativa* L.). Environmental Geochemistry and Health, 35 (6): 779-788.

Wong C C, Wu S C, Kuek C, et al. 2007. The role of mycorrhizae associated with vetiver grown in Pb-/Zn-contaminated soils: Greenhouse study. Restoration Ecology, 15 (1): 60-67.

Wong J T F, Chen X W, Mo W Y, et al. 2016. Restoration of plant and animal communities in a sanitary landfill: A 10 - year case study in Hong Kong. Land Degradation & Development, 34 (3): 174-177.

Wong J T F, Chen Z, Ng C W W, et al. 2016a. Gas permeability of biochar-amended clay: Potential alternative landfill final cover material. Environmental Science and Pollution Research, 23 (8): 7126-7131.

Wong J W C, Ip C M, Wong M H. 1998. Acid-forming capacity of lead–zinc mine tailings and its implications for mine rehabilitation. Environmental Geochemistry and Health, 20 (3): 149-155.

Wong M H.1986. Reclamation of wastes contaminated by Cu, Pb and Zn. Environmertal Management 10: 707-13.

Wong M H, Bradshaw A D. 2002. China: Progress in the reclamation of degraded land//Perrow M R, Davy A J. Handbook of Ecological Restoration. Vol. 2, Restoration in Practice. Cambridge University. Press, Cambridge, pp. 89-98.

Wong M H, Bradshaw A D. 1982. A comparison of the toxicity of heavy metals, using root elongation of rye grass, *Lolium perenne*. New Phytologist: 255-261.

Wong M H, Chan K C, Choy C K. 1978. The effect of the iron ore tailings on the coastal environment of Tolo Harbour, Hong Kong. Environmental Research, 15 (3): 342-356.

Wong M H, Chan Y S G, Zhang C, et al. 2016b. Comparison of pioneer and native woodland species growing on top of an engineered landfill, Hong Kong: Restoration programme. Land Degradation & Development, 27 (3): 500-510.

Wong M H, Cheung K C, A Yediler, et al. 2004. The dyke-pond systems in South China: Past, present and future//Wong M H. Wetlands Ecosystems in Asia: Function and Management. Elsevier, Amsterdam, pp. 69-86.

Wong M H, Cheung K C, Lan C Y. 1992. Factors related to the diversity and distribution of soil fauna on Gin Drinkers' Bay landfill, Hong Kong. Waste Management & Research, 10 (5): 423-434.

Wong M H, Kwan S H. 1988. Leachate quality from iron-ore tailings leached with deionized water of different pH. Waste Management & Research, 6 (1): 45-49.

Wong M H, Lau W M, Li S W, et al. 1983. Root growth of two grass species on iron ore tailings at elevated levels of manganese, iron, and copper. Environmental Research, 30 (1): 26-33.

Wong M H, Leung C K. 1989a. Landfill leachate as irrigation water for tree and vegetable crops. Waste management & Research, 7 (1): 311-323.

Wong M H, Leung K C. 1989b. Decontamination of landfill leachate by soil infiltration. Biomedical and Environmental Sciences: BES, 2 (4): 341-357.

Wong M H, Li M M, Leung C K, et al. 1990. Decontamination of landfill leachate by soils with different textures. Biomedical and

Environmental Sciences: BES, 3 (4): 429-442.

Wong M H, Tam F Y. 1977. Soil and vegetation contamination by iron-ore tailings. Environmental Pollution, 14 (4): 241-254.

Wong M H, Yu C T. 1989c. Monitoring of Gin Drinkers' Bay landfill, Hong Kong: I. Landfill gas on top of the landfill. Environmental Management, 13 (6): 743-752.

Wong M H, Yu C T. 1989 d. Monitoring of Gin Drinkers' Bay landfill, Hong Kong: II. Gas contents, soil properties, and vegetation performance on the side slope. Environmental Management, 13 (6): 753-762.

Wong M H, Zhang Z Q, Shu W S, et al. 2002. The role of legumes in the Reclamation of metal mined land in China. The restoration and Management of Derelict Land—Modern Approaches: 286-296.

Wong M H. 1981. Environmental impacts of iron ore tailings—The case of Tolo Harbour, Hong Kong. Environmental Management, 5 (2): 135-145.

Wong M H. 1982. Metal cotolerance to cooper, lead, and zinc in Festuca rubra. Environmental Research, 29 (1): 42-47.

Wong M H. 1985. Toxic effects of iron ore tailings and the response of watercress from tailings at high concentrations of Fe, Zn and Mn. Environmental Pollution Series A, Ecological and Biological, 38 (2): 129-140.

Wong M H. 1988. Soil and plant characteristics of landfill sites near Merseyside, England. Environmental Management, 12 (4): 491-499.

Wong M H. 1998. Landfill; leachates, landfill gas//Alexander D E, Fairbridge R W. Encyclopedia of Environmental Science. Chapman & Hall, NY. pp 356-361.

Wong M H. 2003. Ecological restoration of mine degraded soils, with emphasis on metal contaminated soils. Chemosphere, 50 (6): 775-780.

Wu C, Li H, Ye Z, et al. 2013. Effects of As levels on radial oxygen loss and As speciation in rice. Environmental Science and Pollution Research, 20 (12): 8334-8341.

Wu C, Ye Z, Li H, et al. 2012a. Do radial oxygen loss and external aeration affect iron plaque formation and arsenic accumulation and speciation in rice? Journal of Experimental Botany, 63 (8): 2961-2970.

Wu C, Ye Z, Shu W, et al. 2011a. Arsenic accumulation and speciation in rice are affected by root aeration and variation of genotypes. Journal of Experimental Botany, 62 (8): 2889-2898.

Wu C, Zou Q, Xue S, et al. 2015a. Effects of silicon (Si) on arsenic (As) accumulation and speciation in rice (*Oryza sativa* L.) genotypes with different radial oxygen loss (ROL). Chemosphere, 138: 447-453.

Wu F Y, Bi Y L, Leung H M, et al. 2010a. Accumulation of As, Pb, Zn, Cd and Cu and arbuscular mycorrhizal status in populations of *Cynodon dactylon* grown on metal-contaminated soils. Applied Soil Ecology, 44 (3): 213-218.

Wu F Y, Bi Y L, Wong M H. 2009a. Dual inoculation with an arbuscular mycorrhizal fungus and Rhizobium to facilitate the growth of alfalfa on coal mine substrates. Journal of Plant Nutrition, 32 (5): 755-771.

Wu F Y, Chung A K C, Tam N F Y, et al. 2012b. Root exudates of wetland plants influenced by nutrient status and types of plant cultivation. International Journal of Phytoremediation, 14 (6): 543-553.

Wu F Y, Leung H M, Wu S C, et al. 2009b. Variation in arsenic, lead and zinc tolerance and accumulation in six populations of *Pteris vittata* L. from China. Environmental Pollution, 157 (8): 2394-2404.

Wu F Y, Wan J, Wu S, et al. 2012c. Effects of earthworms and plant growth promoting rhizobacteria (PGPR) availability of nitrogen, phosphorus and potassium in soil. Journal of Plant Nutrition & Soil Science, 44: 2974-86.

Wu F Y, Ye Z H, Wong M H. 2009c. Intraspecific differences of arbuscular mycorrhizal fungi in their impacts on arsenic accumulation by *Pteris vittata* L. Chemosphere, 76 (9): 1258-1264.

Wu F Y, Ye Z H, Wu S C, et al. 2007. Metal accumulation and arbuscular mycorrhizal status in metallicolous and nonmetallicolous populations of *Pteris vittata* L. and *Sedum alfredii* Hance. Planta, 226 (6): 1363-1378.

Wu F Y, Yu X Z, Wu S C, et al. 2011b. Phenanthrene and pyrene uptake by arbuscular mycorrhizal maize and their dissipation in soil. Journal of Hazardous Materials, 187 (1): 341-347.

Wu F, Deng D, Wu S, et al. 2015b. Arsenic tolerance, uptake, and accumulation by nonmetallicolous and metallicolous populations of

Pteris vittata L. Environmental Science and Pollution Research, 22 (12): 8911-8918.

Wu F, Hu J, Wu S, et al. 2015c. Grain yield and arsenic uptake of upland rice inoculated with arbuscular mycorrhizal fungi in As-spiked soils. Environmental Science and Pollution Research, 22 (12): 8919-8926.

Wu F, Yu X, Wu S, et al. 2014. Effects of Inoculation of PAH-degrading bacteria and arbuscular mycorrhizal fungi on responses of ryegrass to phenanthrene and pyrene. International Journal of Phytoremediation, 16 (2): 109-122.

Wu L H, Luo Y M, Christie P, et al. 2003. Effects of EDTA and low molecular weight organic acids on soil solution properties of a heavy metal polluted soil. Chemosphere, 50 (6): 819-822.

Wu S C, Cao Z H, Li Z G, et al. 2005a. Effects of biofertilizer containing N-fixer, P and K solubilizers and AM fungi on maize growth: a greenhouse trial. Geoderma, 125 (1): 155-166.

Wu S C, Cheung K C, Chong K N, et al. 2005b. A new type of bio-organic compound fertilizer. The Chinese Patent Co. Ltd., Hong Kong (File No. 02118921.8, Patent No. ZL02118921.8).

Wu S C, Cheung K C, Luo Y M, et al. 2006a. Effects of inoculation of plant growth-promoting rhizobacteria on metal uptake by *Brassica juncea*. Environmental Pollution, 140 (1): 124-135.

Wu S C, Luo Y M, Cheung K C, et al. 2006b. Influence of bacteria on Pb and Zn speciation, mobility and bioavailability in soil: A laboratory study. Environmental Pollution, 144 (3): 765-773.

Wu S C, Peng X L, Cheung K C, et al. 2009 d. Adsorption kinetics of Pb and Cd by two plant growth promoting rhizobacteria. Bioresource Technology, 100 (20): 4559-4563.

Wu S C, Wong C C, Shu W S, et al. 2010b. Mycorrhizo-remediation of lead/zinc mine tailings using vetiver: A field study. International Journal of Phytoremediation, 13 (1): 61-74.

Yang B, Shu W S, Ye Z H, et al. 2003. Growth and metal accumulation in vetiver and two Sesbania species on lead/zinc mine tailings. Chemosphere, 52 (9): 1593-1600.

Yang Z Y, Yuan J G, Xin G R, et al. 1997. Germination, growth, and nodulation of *Sesbania rostrata* grown in Pb/Zn mine tailings. Environmental Management, 21 (4): 617-622.

Ye Z H, Baker A J M, Wong M H. 2002a. Problems of toxicities//Wong M H, Bradshaw A D. The Restoration and Management of Derelict Land. World Sci. Pub., NY. pp 66-79.

Ye Z H, Baker A J M, Wong M H, et al. 1997a. Copper and nickel uptake, accumulation and tolerance in *Typha latifolia* with and without iron plaque on the root surface. New Phytologist, 136 (3): 481-488.

Ye Z H, Baker A J M, Wong M H, et al. 1997b. Zinc, lead and cadmium tolerance, uptake and accumulation by the Common Reed, *Phragmites australis* (Cav.) Trin. ex Steudel. Annals of Botany, 80 (3): 363-370.

Ye Z H, Baker A J M, Wong M H, et al. 1997c. Zinc, lead and cadmium tolerance, uptake and accumulation by *Typha latifolia*. New Phytologist, 136 (3): 469-480.

Ye Z H, Baker A J M, Wong M H, et al. 2003a. Copper tolerance, uptake and accumulation by *Phragmites australis*. Chemosphere, 50 (6): 795-800.

Ye Z H, Cheung K C, Wong M H. 2001a. Copper uptake in *Typha latifolia* as affected by iron and manganese plaque on the root surface. Canadian Journal of Botany, 79 (3): 314-320.

Ye Z H, Cheung K C, Wong M H. 2003b. Cadmium and nickel adsorption and uptake in cattail as affected by iron and manganese plaque on the root surface. Communications in Soil Science and Plant Analysis, 34 (19-20): 2763-2778.

Ye Z H, Shu W S, Zhang Z Q, et al. 2002b. Evaluation of major constraints to revegetation of lead/zinc mine tailings using bioassay techniques. Chemosphere, 47 (10): 1103-1111.

Ye Z H, Wong J W C, Wong M H, et al. 1999. Lime and pig manure as ameliorants for revegetating lead/zinc mine tailings: A greenhouse study. Bioresource Technology, 69 (1): 35-43.

Ye Z H, Wong J W C, Wong M H, et al. 2000a. Revegetation of Pb/Zn mine tailings, Guangdong Province, China. Restoration Ecology, 8 (1): 87-92.

Ye Z H, Wong J W C, Wong M H. 2000b. Vegetation response to lime and manure compost amendments on acid lead/zinc mine tailings: A greenhouse study. Restoration Ecology, 8 (3): 289-295.

Ye Z H, Wong M H, Baker A J M, et al. 1998a. Comparison of biomass and metal uptake between two populations of *Phragmites australis*grown in flooded and dry conditions. Annals of Botany, 82 (1): 83-87.

Ye Z H, Yang Z Y, Chan G Y S, et al. 2001b. Growth response of *Sesbania rostrata* and *S. cannabina* to sludge-amended lead/zinc mine tailings: A greenhouse study. Environment International, 26 (5): 449-455.

Ye Z, Baker A J M, Wong M H, et al. 1998b. Zinc, lead and cadmium accumulation and tolerance in *Typha latifolia* as affected by iron plaque on the root surface. Aquatic Botany, 61 (1): 55-67.

Yu X Z, Wu S C, Wu F Y, et al. 2011. Enhanced dissipation of PAHs from soil using mycorrhizal ryegrass and PAH-degrading bacteria. Journal of Hazardous Materials, 186 (2): 1206-1217.

Yu X, Cheng J, Wong M H. 2005. Earthworm–mycorrhiza interaction on Cd uptake and growth of ryegrass. Soil Biology & Biochemistry, 37 (2): 195-201.

Zhang J, Liang J, Wong M H. 1995. The effect of high carbon dioxide and low oxygen concentrations in simulated landfill gas on the growth and nodule activity of *Leucaena leucocephala*. Plant Cell Physiology 36: 1431-8.

Zhang Z Q, Lan C Y, Hong M H. 1996. Revegetation of Pb/Zn mine tailings: Germination and seedling establishment of plants. Land Contamination & Reclamation, 4 (4): 269-279.

Zhang Z Q, Shu W S, Lan C Y, et al. 2001. Soil seed bank as an input of seed source in revegetation of lead/zinc mine tailings. Restoration Ecology, 9 (4): 378-385.

Zhang Z Q, Wong M H, Nie X P, et al. 1998. Effects of zinc (zinc sulfate) on rhizobia-earleaf acacia (*Acacia auriculaeformis*) symbiotic association. Bioresource Technology, 64 (2): 97-104.

Zheng Z. 2004. Responses of *Sesbania rostrata* and *S. cannabina* to Pb, Zn, Cu and Cd toxi-cities. Journal of Environmental Sciences, 16 (4): 670-673.

Green and Sustainable Remediation Movement in the New Millennium and Its Relevance to China

Deyi HOU, Guanghe LI

School of Environment, Tsinghua University

1. Introduction

Land is not only a critical component of the earth's life support system, but also a precious resource and an important factor of production in economy. However, historical industrial operations have caused a huge stockpile of contaminated land that is only slowly being remediated. After several decades of clean-up efforts, there are still an estimated 294, 000 contaminated sites in the US and over 300, 000 hectares of potentially contaminated land in the UK (USEPA, 2004; DEFRA, 2006). In China, according to a national survey, soil quality on 16.1% of the nation's land exceeded soil quality standards (MEP, 2014). It is imperative to develop technical solutions as well as socioeconomic and political instruments to achieve sustainable restoration of contaminated land while preventing the further contamination of existing clean lands (Hou, 2011; Hou et al., 2012). The inclusion of sustainability in remediation decision making provides an opportunity to integrate a wide range of considerations: risk control, brownfield regeneration, carbon footprint, water footprint, renewable energy, etc. (Hou and Al-Tabbaa, 2014) On the other hand, the emerging "sustainable remediation" field also presents many new research challenges.

The traditional risk-based decision making process in the remediation field is now increasingly supplemented with sustainability considerations in the on-going "green and sustainable remediation" movement. There are a variety of criteria in determining whether a remediation alternative is sustainable. As discussed in existing publication (Al-Tabbaa et al., 2007), such criteria may include: ①future benefits outweigh cost of remediation; ②environmental impact of the implementation of the remediation process is less than the impact of leaving the land untreated; ③environmental impact of bringing about the remediation process is minimal and measurable; ④the timescale over which the environmental consequences occur, and hence inter-generational risk, is part of the decision making process; ⑤the decision-making process includes an appropriate level of engagement of all stakeholders.

This chapter focuses on exploring sustainability in the remediation field and discuss its relevance to China's remediation industry. China has made a lot of progress in scientific research on soil and groundwater remediation over the past two decades, and a large number of remediation commercial projects have also been carried out in the last decade; however, the concept of green and sustainable remediation is still relatively new for remediation researchers and practitioners in China. On the other hand, researchers in China has conducted extensive research on remediation technologies that meet green and sustainable remediation criteria. We consider it appropriate to summarize such technological development under the umbrella of green and sustainable remediation. The awareness and adoption of green and sustainable remediation by industrial practitioners is also examined in this chapter.

2. The green and sustainable remediation movement

2.1 Sustainable land management

The remediation industry emerged in the 1970s after a few high-profile contaminated sites (e.g. the Love Canal site in the US, and the Lekkerkerk site in the Netherlands) caught politician and the public with surprise, and it was accelerated in the 1980s due to the passage of regulations such as the Comprehensive Environmental Response, Compensation, and Liability Act (CERCLA), commonly known as the Superfund Act in the US, which created funding mechanism and provided broad federal authority to respond to hazardous substance release. The perception that contaminated land was severe but limited in scale led to practices aiming at complete removal of contamination. In the 1990s, European remediation practitioners started to recognize the contaminated land problem as a wide-spread infrastructure problem. This perception transition was partly due to the fast growing number of contaminated site. For instance, in the Netherlands, the number of contaminated sites needing remediation was about 350 sites in the year of 1981, and grew to 300, 000 sites in the year of 1995 (Ferguson, 1999). This perception transition led to the development of a "site specific risk-based contaminated land management" strategy, under the name of "sustainable land management" (CLARINET, 2002a, 2002b; NICOLE, 2002). The risk assessment approach adopted by European practitioners is based on the source-pathway-receptor model, which is also widely used in other countries like the US (e.g. ASTM standard for risk-based corrective action [RBCA]) (ASTM, 2010). In the Risk Based Land Management (RBLM) framework developed by CLARINET, sustainability is a key objective that include the evaluation and optimization of environmental, economic, and social factors (NICOLE, 2005).

There are considerable variability in the adoption of sustainability in remediation practice in various European countries, with countries like the UK where sustainability is widely recognized in regulation and used in practice, as well as countries like Italy where sustainability is not applied in regulation or used by practitioners (Maurer, 2009). The UK plays a leading role in promoting sustainable remediation in Europe. Two organizations, CLAIRE and Surf-UK, are active in advocating sustainable remediation not only in the UK, but Europe wide. In 2010,

Surf-UK, with coordination of CLAIRE and sponsorship from the Home and Communities Agency of the UK government, developed a framework for assessing the sustainability of remediation (Surf-UK, 2010). In addition to risk management, the Surf-UK framework added strong "development" components and "cost saving" components. It has a strong desire for "demonstrating the need not to implement unnecessary or unsustainable remediation measures", which is consistent with a regulation that encourages a "suitable for use" approach in remediation (DEFRA, 2006) and regulatory guidance that promotes sustainable remediation (EA, 2004). A draft European Commission (EC) Soil Framework Directive (SFD) requires that remediation shall be "sustainable actions". The proposed SFD seeks to raise the level of soil protection and contaminated land remediation across the EU. However, the SFD has not been passed due to resistance from Germany, France, Netherlands, Austria, and the United Kingdom (ECJRC, 2012), partly due to concern on the principles of subsidiarity and disproportionate costs (DEFRA, 2012).

In the UK, where the most sustainable remediation discussion occurred in European context, sustainable remediation was largely driven by the demand for sustainable urban renaissance and cost saving. In a 1999 report developed by the Urban Task Force for the former Department of the Environment, Transport, and Region (DETR), it was proposed that all polluted sites be restored by the year 2030 (Rogers, 1999). Driven by a public policy mandating that 60% of new housing development should be built on brownfield land, England had 79% of dwellings built on previously developed land in year 2008 (DCLG, 2009). The sustainable urban development concept and practice led to increasing recognition of secondary environmental impacts associated with remediation operation, as well as economic burden of traditional remediation techniques. In a survey of remediation practitioners conducted in 2004 in the UK (Al-Tabbaa et al., 2007), based on 60 responses from 208 individuals, 40%-50% of the respondents "always" considered environmental sustainability, social sustainability, and long-term sustainability, at least in some aspect of a project in the selection of a remediation technology. And an additional 20% of the respondents "often" considered these sustainability considerations. As for what methodologies were used in determining the best remediation technology for a particular project, the responses showed that practitioners primarily used professional judgement, environmental impact assessment, and cost-benefit analysis. Less than 30% of respondents "always" or "often" used multi-criteria analysis; and less than 10% of respondents "always" or "often" used life cycle analysis.

Various decision support tools have been developed to support sustainable remediation by European practitioners. EURODEMO, a European network for demonstrating innovative remediation technologies, developed a framework for sustainable remediation in 2007 (EURODEMO, 2007). Sustainability evaluation tools based on multi-criteria analysis (Linkov et al., 2005) have been developed by various stakeholders for remediation decision support (Brinkhoff, 2011; Golder, 2012). Such tools may date back to the late 1990s when remediation practitioners in remediation systems based on their scores in risk reduction, environmental merits, and costs (Beinat and Drune, 1997). In the research community, life cycle assessment (LCA) methods have been used in evaluating remediation alternatives due to its comprehensiveness. In a German case study, market-oriented economic appraisal was combined with sustainability evaluation to assess brownfield revitalization options (Schädler et al., 2010). Each of these decision making tools, including sustainability appraisal, is part of many decision factors with their own pros and cons (Bardos et al., 2001).

The sustainable remediation movement in the Europe has encouraged innovative solutions. For instance, the Netherland has for the first time combined groundwater energy (i.e. heat cold storage) with remediation (Slenders et al., 2010). At a former gas manufacturing facility, UK practitioners employed innovative power system to allow for the simultaneous recovery and usage of site contaminants-coal tar. It was demonstrated that such a system can save project cost by 20%-30% and leads to substantial GHG reduction (McLaren et al., 2009). While the long term sustainability, both economically and environmentally, of these systems is still to be monitored and verified, their deployment has demonstrated how the "sustainability" concept can conquest regulatory barriers and promote technological innovation. Another innovative solution proposed by European practitioners is using brownfields to produce biomass, which can be used as feedstock for biofuels (SNOWMAN, 2009). A study in Spain suggests that traditional agricultural practice such as annual liming and application of animal manure may have mitigated soil-plant transfer of heavy metals, raising a potential for sustainable management of such contaminated site (Madejón et al., 2011).

2.2 Green remediation

The sustainable remediation initiatives in the US federal agencies have partly been driven by presidential Executive Orders (EO) 13 423 and 13 514, issued in January 2007 and October 2009 respectively (AFCEE, 2010). Both EOs promote sustainable means in federal agencies' operations. Among the various agencies with heavy involvement in environmental remediation (i.e. as contaminated site owners), the US Air Force developed an excel-based toolkit (i.e. Sustainable Remediation Tool [SRT]) to calculate a range of sustainability metrics (e.g. emissions of CO_2, NO_x, SO_x, PM_{10}, energy consumption, technology cost) for several most commonly used remediation technologies (AFCEE, 2010); the US Army Corps of Engineer published a guidance in 2010 that describes principles and considerations to incorporate green and sustainable practices in remediation (USACE, 2010); the US Navy and its partners developed another excel-based, and building-block type program that calculates sustainability metrics and incorporates footprint reduction analysis (NAVFAC, 2011).

The USEPA is the lead agency responsible for regulatory enforcement as well as who provides technological guidance on environmental remediation, under both CERCRL and RCRA statutes. In 2008, the USEPA published a technology primer on green remediation that incorporates sustainable practices in contaminated site remediation (USEPA, 2008). It stressed the secondary environmental impact from remediation operation, and presented a series of best management practices (BMPs) to enhance the sustainability of remediation projects. The USEPA Office of Solid Waste and Emergency Response (OSWER), a lead federal agency that oversees the Superfund program as well as the Brownfields program, moved further in 2009 by setting its policies to encourages sustainable remediation (USEPA, 2009). Similar green remediation policies and practices have also been adopted by several EPA regions and state governments.

As policies progress, the practice of sustainable remediation is also widely spreading (Petruzzi, 2011; CLU-IN, 2014), concurring with the above government policies and advocacy from various industrial and governmental associations, such as the Sustainable Remediation Forum (SURF) (Ellis and Hadley, 2009; Favara et al., 2011; Holland et al., 2011) and ITRC (ITRC, 2011). Some site owners have also used financial incentives to encourage contractors to employ green practices. Green energy systems, such as solar panel landfill gas converted power system, windmill and geothermal system, have been used at remediation sites as part of green remediation strategies. Bioremediation, phytoremediation, and engineered wetlands are widely used and they employ a wide range of living organism (i.e. bacteria, fungi, pondweed, and trees) and substrates (e.g. biosolids, animal manure). Green chemistry principles are being adopted by researchers who develop new materials and/or processes for environmental treatment (Hoag et al., 2009).

Sustainable practices in the building industry are being used in the remediation industry, including recycling of demolition waste, capture and reuse of rainwater, low maintenance landscape, and treatment building design. These practices follow similar principles as those defined in the United States Green Building Council (USGBC) -Leadership in Energy and Environmental Design (LEED) program (USGBC, 2011). The LEED program becomes more relevant as contaminated site management integrates facility demolition, contamination remediation, and land development. There are private efforts that are integrating LEED practices in facility demolition and remediation planning, partly driven by the potential benefits that may be brought in by LEED certification of future land development (Kats and Alevantis, 2003).

2.3 Green and sustainable remediation methods

In the remediation field, elements of sustainability assessment and thinking are increasingly being incorporated into environmental remediation decision making (NRC, 2011; Hou et al., 2014a). Life cycle assessment (LCA) has been the dominant type of sustainability assessment tool in the published literature (Lemming et al., 2010; Hou et al., 2014b). This may be due to a number of reasons: ①LCA can quantify secondary adverse effects which are often overlooked in traditional decision making; ②LCA evaluates environmental effects across geographical boundaries and along intergenerational time scale to avoid shifting problems from one place to another or from one generation to the next; ③LCA includes a broad range of impact categories; ④LCA is an established method with international standards and professional tools. Other tools like EIA and SEA do not offer all these advantages.

Because of the multi-dimensional nature of sustainability considerations, remediation practitioners have also widely used multi-criteria analysis in sustainability assessment. Many multi-criteria analysis use LCA as a tool in quantifying some effects. For instance, a study in Flanders used LCA to quantify CO_2 footprint of remediation alternatives (Dirk, 2012). The study by Paulus aimed at determining the main contributors of CO_2 emission in order to raise awareness and mitigate environmental emissions. The multi-criteria analysis included three groups: the environmental aspect group, the technical aspect group, and the financial aspect group, with weighting factors 34%, 33%, and 33% respectively. Subsequently, the study ranked the sustainability of the remediation options being evaluated.

There are also multi-criteria analysis studies, especially those addressing sustainable brownfield development, focusing more on social and economic impacts without conducting LCA. Such studies tend to quantify life cycle costs, development area, etc (Schädler et al., 2012). The study by Schädler evaluated various land use alternatives for sustainable brownfield redevelopment. In this framework, "trial and error" iterative planning was used to seek improvement by minimizing remediation expenditure, maximizing site value, and increasing sustainability. The case study automatically integrated spatial data with evaluation tools, thus enabling site designers to include sustainability assessment during early project planning stages.

In addition to the above mentioned specific sustainability assessment case studies, recently the idea of "sustainability linkage" has been proposed to study the effect of sustainability in the context of contaminated land remediation (Bardos, 2012). Bardos considered it important to identify and prioritize sustainability factors with mechanisms linking the changes occurred during remediation with receptors (i.e. those who are affected by such changes). For example, in considering economic sustainability, the construction of an access road can improve access, thus increasing property values in the region of the regeneration project (sustainability gain); in considering environmental sustainability, construction and operation of remediation systems requires the use of materials, thus resulting in the loss of primary resources (sustainability loss); in considering social sustainability, creation of dust from a contaminated site results in wind-blow, which in turn leads to human health risk (sustainability loss). This sustainability linkage approach can be used to combine stakeholder perspectives in a spreadsheet, with benefits such as reduction of complexity, clear rationale for selection criteria and verification performance indicators, and working with an iterative and tiered approach.

3. Green and sustainable remediation in China

3.1 Background

China has a huge environmental burden that is associated with its "world factory" status. One study shows that over 400 million tons of toxic chromium waste has been accumulated and untreated in China, posing a huge risk of land contamination (Gao and Xia, 2011). In the past two decades, many actions have been taken by the Chinese government and industries to address environmental pollution issues, but usually focused on surface water pollution and air pollution. In recent years, as many industrial operations are relocated from downtown areas and brownfields are being re-developed, land contamination issue becomes more stressing, especially after some high profile incidents occurred (e.g. construction workers were poisoned in 2004 at the Songjiazhuang Beijing site, and in 2006 at the Sanjiang Wuhan site) (Xie and Li, 2010). A national soil survey published in 2014 indicated that within the 6.3 million km^2 of land being surveyed, 16.1% of all soils exceeded soil quality standards (MEP, 2014). The *Soil Pollution Prevention and Cleanup Action Plan*, published on May 31, 2016, created a busy schedule for national and local governments: finishing detailed soil investigation for agricultural land by 2018 and for industrial land by 2020, conducting remediation of approximately 0.7 million hectares of seriously contaminated agricultural land by 2020, and utilizing 95% of the nation's contaminated land in a safe manner by 2030.

The environmental remediation industry is expected to grow exponentially in China for a number of reasons: ①China is accelerating its pace of urbanization, which is viewed by the new leadership as one of its four priorities, and many factories in old cities are expected to be relocated to give space for this urbanization process but many of these spaces would be in contaminated states due to historical operations; ②China has historically used industrial wastewater for agricultural irrigation, which has caused large areas of contamination that needs remediation in the wake of public health debate; ③the oil and mining industries have left many contaminations behind in historical operations; ④ waste management practices, include electronic waste recycling activities, may have caused many serious contaminations that need remediation (Hou et al., 2012). Concerns on soil contamination led to a nation-wide investigation on land contamination initiated in 2006 (Zou, 2006). The investigation was completed in 2010, but there are debates whether this information shall be disclosed (Yang, 2010). In addition to soil contamination, groundwater contamination is also a stressing issue. Currently nearly 90% of China's shallow groundwater is polluted, with 37% being too contaminated to be treated for drinking water purpose (Qiu, 2011). Despite of the urgency of the land contamination issue, China lack national regulations and policies in remediation. Several interim technical specifications and guidance, in their draft versions, are typically used by practitioners to monitor contaminated sites (MEP, 2010), perform site investigation (MEP, 2009a), conduct risk assessment (MEP, 2009b), and carry out soil remediation (MEP, 2009c).

Unlike the US and Europe with over three decades of experience on contaminated land remediation, China has little industrial experience and no established regulatory framework in soil and groundwater contamination. Existing legislation demanding remediation has been viewed as a regulatory barrier for applying sustainable remediation (Maurer, 2009). As China's remediation industry is still in its infancy, it provides a great opportunity to incorporate sustainability considerations into policy-making, as well as best practice management in the industry. Among others, China's forthcoming remediation policies can: ①incorporate remediation planning into urban planning processes, for instance, to assign future residential zones to areas where there are none or a smaller number of contaminated sites and where the geochemical and hydrogeological conditions would allow for more cost-effective remediation; ②conduct value engineering by coupling remediation and development, for instance, to save development cost by utilizing the excavation made in remediation; ③stress the importance of site investigation, risk-based management, and "suitable for use" principles; ④encourage green and sustainable remediation practices and reduce secondary impact of remediation operations.

3.2 Research progress in China relevant to GSR

China has conducted a series of research studies aiming at the remediation of contaminated soil. Some of these studies have developed techniques that fit well with the modern GSR concept. Chinese researchers have identified microorganisms which are capable of efficiently degrading pesticides, manufactured treatment reagents consisting of microorganisms, and applied technologies to degrade residual pesticides in crop fields. On the other hand, Chinese researchers have also identified a variety of microorganism species which are capable of degrading petroleum hydrocarbon, and developed bioremediation reactors. In the phytoremediation field, Chinese researchers have also made significant progress. Plant species capable of enriching zinc and cadmium (*Sedum plumbizincicola*) and species capable of accumulating cadmium (*Sedum jinianum*) have been identified and used in large scale to remediate heavy metal contamination. A microorganism species, *Paracoccus* aminovorans have been isolated, which can be used to degrade PAH contamination.

In groundwater remediation, Chinese researchers have developed *in-situ* bioremediation technologies. For semi-volatile petroleum hydrocarbon and heavy-end petroleum hydrocarbon, Chines researchers identified and enriched microorganism species which can degrade these organic compounds with high efficiency. For PAHs and large-molecule petroleum hydrocarbon which are difficult to degrade, Chinese researchers have combined physical separation, chemical oxidation, and biological degradation, to increase removal efficiency and reduce treatment time.

In the China Report on Advances in Environmental Science (2012), Chinese scholars identified four key research directions for groundwater remediation. The second key research direction is entitled "green and environmental friendly remediation technologies". The Chinese scholars consider that the following research directions are critical for future development:

(1) Use green and self-degradable reagents for remediation;

(2) Use solar energy and plant resources for phytoremediation;

(3) Use high-efficiency microorganisms for bioremediation of soil and groundwater

(4) Use various species in the food chain for remediation

(5) Use monitored natural attenuation to manage contaminated sites

All of the above directions align well with the green and sustainable remediation concept. Therefore, it is evident that Chinese researchers have long recognized the importance of green and sustainable remediation. It is expected that more future research will be carried out in this field to strengthen Chinese scholar's understanding and utilization of green and sustainable remediation.

3.3 A survey of remediation practitioners

In order to gauge the awareness and adoption of green and sustainable remediation in China, a questionnaire survey was conducted among remediation practitioners. The survey question was: "how effective is your team in adopting the following 'sustainability' considerations in developing remediation strategies?" The responses were given on a 5-point Likert scale. The anchors were "not at all" ① and "very effective" ⑤. There were 27 sustainability considerations which were rated by each respondent. A link to the online survey was sent to potential participants by emails. The invitation letter included a cover letter explaining the objective of the survey, and assured that the confidentiality of respondent would be guaranteed. To encourage participation, the participants were offered a future summary report of aggregated results to those who explicitly expressed interest. In addition, several reminder emails were sent to all participants who had not responded, or who started, but not finished responding. The target population included all stakeholders involved in environmental remediation decision making and/or practices. The survey participants were mainly contaminated site owners, regulators, and environmental consultants, but also included contractors, technology vendors, environmental groups, etc. While the survey was sent to remediation practitioners worldwide, the study intended to put a primary focus on the US, the UK, and China.

Fig. 1 shows the average score of each sustainability consideration, based on the responses from China, US, and UK. The sustainability considerations with the highest scores are shown on the left, and those with the lowest scores on the right. There is apparent divergence under various socio-economic contexts. China has a lower rate of adoption of most sustainability considerations than the US and the UK. Reducing local community risk and protecting groundwater and surface water are among the top 3 priorities in the US and the UK, but they are ranked the 7th and 8th in China. Instead, the top two sustainability considerations in China are minimizing contaminants left behind and reducing life-cycle cost.

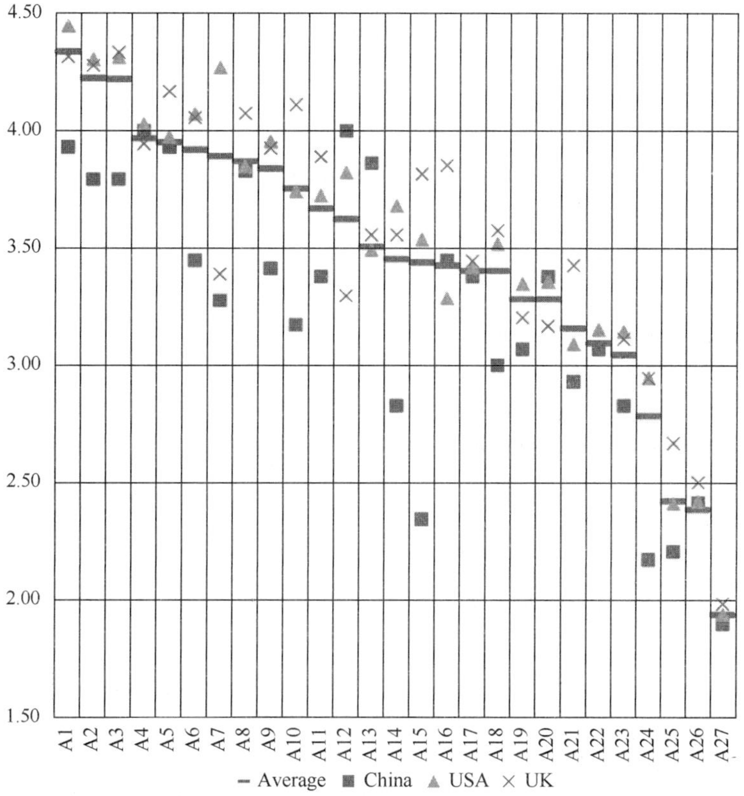

Fig. 1 How effective are 27 sustainability considerations adopted by remediation practitioners in the US, UK, and China (note: list of sustainability considerations are provided in the embedded Table 1)

Based on the questionnaire survey, practitioners in China had much lower acceptance of these sustainability considerations than their US and UK peers: minimizing risk to ecological systems, maximizing area for redevelopment, using monitored natural attenuation rather than active remediation, increasing property value, and enhancing local employment. Based on the qualitative interview, it was evident that the health risk is currently the primary driver of most contaminated site remediation in China. The risk to the environment (i.e. ecological systems) is not at the forefront even though "protection of ecological systems" have been placed a high priority in national environmental protection agenda. It appears that most practitioners in China have not recognized the link between contaminated sites and ecology, likely due to the fact that nearly all contaminated sites undergoing remediation in China are located within well-developed urban systems. According to the questionnaire survey, current remediation practices in China rarely involve monitored natural attenuation. This was probably because most contaminated sites that have been identified in China, tended to be seriously contaminated, and monitored natural attenuation is unlikely to be used alone for their cleanup. In addition, monitored natural attenuation is usually used for groundwater remediation, which is currently not commonly remediated in China, where contaminated site remediation has often focused on soil remediation.

Table 1 List of sustainability considerations

A1	Reducing site worker's risk
A2	Reducing local community risk
A3	Protect groundwater and surface water
A4	Minimizing contaminants leftbehind
A5	Minimizing local scale secondary environmental impacts (e.g. noise, dust, odour, local air quality, traffic, etc.)
A6	Protect habitat and ecosystem
A7	Using in-situ remediation rather than ex-situ remediation
A8	Minimizing waste generation
A9	Minimizing risk to ecological systems
A10	Maximize area for redevelopment
A11	Minimize long-term management (e.g. monitoring) requirement
A12	Reducing life-cycle cost
A13	Using fast-track remediation alternative
A14	Using monitored natural attenuation rather than active remediation
A15	Increase property value
A16	Enhancing reuse and recycling
A17	Conserve natural resources
A18	Encourage public participation and stakeholder involvement
A19	Minimizing water consumption
A20	Using environmental friendly products
A21	Minimizing material use
A22	Minimizing energy use, increasing energy efficiency
A23	Minimizing national to global scale secondary environmental impacts (e.g. greenhouse gas emission, fossil fuel depletion, ozone depletion, etc.)
A24	Enhance local employment
A25	Using sustainable energy
A26	Bring prosperity to disadvantaged community (increase tax revenue, education, security, etc.)
A27	Generating electricity from by-products such as methane gas

It was interesting to find that Chinese practitioners were less effective in maximizing the area for redevelopment. One hypothesis is that, most remediation projects in China would essentially cleanup the entire sites for re-use; therefore the concept of "maximizing area for redevelopment" does not really exist in most cases. It was also interesting that "increasing property value" was rated low by Chinese practitioners. One possible explanation is that the remediation practitioners are often absent from planning processes. Consequently, the practitioners don't view that their decision making has no effect on the property value. Another possible explanation is that the remediation practitioners in China don't believe that different technology options may lead to different post-remediation land values.

4. Conclusions

China has become the world's second largest economy with rapid economic growth over the past three decades, at the cost of serious environmental pollution. The cleanup of contaminated land is an emerging market in China; however, it has been largely driven by land redevelopment rather than regulatory compliance. This development-managerialism may cause issues ranging from under-protection of the environment to over-engineering of remedial systems. Our literature review indicates that green and sustainable remediation is a widely accepted new movement in the remediation industry in western countries. In China, however, the development of green and sustainable remediation is limited to certain research areas such as bioremediation and phytoremediation, and the adoption of green and sustainable remediation among industrial practitioners is very limited. Based on our literature review and survey findings, we consider it imperative to strengthen the adoption of green and sustainable remediation in China. With proper design and cautious implementation, green and sustainable remediation can help to optimize the usage of limited resources, to improve China's environmental quality, enhance public health, curb social inequality, act as a springboard for remediation technology innovation, and help establish a mature remediation market.

References

AFCEE. 2010. Sustainable Remediation Tool User Guide, AFCEE. Air Force Center for Engineering and the Environment, Lackland, Texas.

Al-Tabbaa A, Harbottle M, Evans C. 2007. Robust sustainable technical solutions. Sustainable brownfield regeneration: Liveable places from problem spaces: 203-236.

ASTM. 2010. E2081-00: Standard Guide for Risk-Based Corrective Action. Orriginally approved in 1998. Reapproved in 2010. American Society for Testing and Materials.

Bardos P. 2012. Conceptual site or project models for sustainability assessment//The 2nd International Conference on Sustainable Remediation. Vienna, Austria.

Bardos R P, Mariotti C, Marot F, et al. 2001. Framework for decision support used in contaminated land management in Europe and North America. NATO/CCMS Pilot Study: 9.

Beinat E, Van Drunen M A, Janssen R, et al. 1997. The REC decision support system for comparing soil remediation options. A methodology based on risk reduction, Environmental Merit and Costs. CUR/NOBIS. The Netherlands.

Brinkhoff P. 2011. Multi-criteria analysis for assessing sustainability of remedial actions-Applications in contaminated land development. Chalmers University of Technology.

CLARINET. 2002a. Sustainable Management of Contaminated Land: An Overview, CLARINET. Contaminated Land Rehabilitation Network for Environmental Technologies (succeeded by Common Forum on Contaminated Land in the European Union).

CLARINET. 2002b. An Analysis of National and EU RTD Programmes Related to Sustainable Land and Groundwater Management, CLARINET. Contaminated Land Rehabilitation Network for Environmental Technologies (succeeded by Common Forum on Contaminated Land in the European Union).

CLU-IN. 2014. Green Remediation Focus [WWW Document]. URL http: //www.clu-in.org/greenremediation/

DCLG. 2009. Land Use Change Statistics (England) 2008-provisional estimates (July 2009), DCLG. Department for Communities and Local Government, London, UK.

DEFRA. 2012. Proposed EU Soil Framework Directive [WWW Document]. URL http: //www.defra.gov.uk/food-farm/land-manage/soil/soil-framework-directive/

DEFRA. 2006. Defra Circular 01/2006, Environmental Protection Act 1990: Part 2A, Contaminated Land, Defra. Department for Environment, Food and Rural Affairs, London, UK.

Dirk Paulus. 2012. Improved MultiCriteria Analysis in Remediation Plans in Flanders: Quantifying Sustainability by Introduction of the CO_2-footprint//The 2nd International Conference on Sustainable Remediation. Vienna, Austria.

EA. 2004. Model Procedures for the Management of Land Contamination, Contaminated Land Report 11, EA. Environment Agency, London, UK.

ECJRC. 2012. EU Soil Policy. URL http: //eusoils.jrc.ec.europa.eu/library/jrc_soil/policy/

Ellis D E, Hadley P W. 2009. Sustainable remediation white paper: Integrating sustainable principles, practices, and metrics into remediation projects. Remediation Journal, 19 (3): 5-114.

EURODEMO. 2007. Framework for Sustainable Land Remediation and Management. European Co-ordination Action for Demonstration of Effiient Soil and Groundwater Remediation.

Favara P J, Krieger T M, Boughton B, et al. 2011. Guidance for performing footprint analyses and life-cycle assessments for the remediation industry. Remediation Journal, 21 (3): 39-79.

Ferguson C C. 1999. Assessing risks from contaminated sites: policy and practice in 16 European countries. Land Contamination and Reclamation, 7 (2): 87-108.

Gao Y, Xia J. 2011. Chromium contamination accident in China: viewing environment policy of China. Environmental Science & Technology, 45 (20): 8605-8606.

Golder. 2012. GoldSET: Fast And Reliable Sustainability Decision Support Tool. URL https: //golder.goldset.com/portal/Default.aspx

Hoag G E, Collins J B, Holcomb J L, et al. 2009. Degradation of bromothymol blue by 'greener' nano-scale zero-valent iron synthesized using tea polyphenols. Journal of Materials Chemistry, 19 (45): 8671-8677.

Holland K S, Lewis R E, Tipton K, et al. 2011. Framework for integrating sustainability into remediation projects. Remediation Journal, 21 (3): 7-38.

Hou D. 2011. Vision 2020: more needed in materials reuse and recycling to avoid land contamination. Environmental Science & Technology, 45 (15): 6227-6228.

Hou D, Al-Tabbaa A. 2014. Sustainability: A new imperative in contaminated land remediation. Environmental Science & Policy, 39: 25-34.

Hou D, Al-Tabbaa A, Chen H, et al. 2014. Factor analysis and structural equation modelling of sustainable behaviour in contaminated land remediation. Journal of Cleaner Production, 84: 439-449.

Hou D, Al-Tabbaa A, Guthrie P, et al. 2014. Using a hybrid LCA method to evaluate the sustainability of sediment remediation at the London Olympic Park. Journal of Cleaner Production, 83: 87-95.

Hou D, Al-Tabbaa A, Guthrie P, et al. 2012. Sustainable waste and materials management: national policy and global perspective. Environmental Science & Technology, 46 (5): 2494-2495.

Hou D, Luo J, Al-Tabbaa A. 2012. Shale gas can be a double-edged sword for climate change. Nature Climate Change, 2 (6): 385-387.

ITRC. 2011. Green and Sustainable Remediation: State of the Science and Practice, ITRC. Interstate Technology & Regulatory Council, Washington, DC, USA.

Kats G, Alevantis L. 2003. The costs and financial benefits of green buildings: a report to California' s sustainable building task force. Sustainable Building Task Force.

Lemming G, Hauschild M Z, Bjerg P L. 2010. Life cycle assessment of soil and groundwater remediation technologies: literature review. The International Journal of Life Cycle Assessment, 15 (1): 115-127.

Linkov I, Varghese A, Jamil S, et al. 2004. Multi-criteria decision analysis: a framework for structuring remedial decisions at contaminated sites//Comparative risk assessment and environmental decision making. Springer Netherlands: 15-54.

Madejón P, Barba-Brioso C, Lepp N W, et al. 2011. Traditional agricultural practices enable sustainable remediation of highly polluted soils in Southern Spain for cultivation of food crops. Journal of Environmental Management, 92 (7): 1828-1836.

Maurer O, Workgroup L N S R. 2009. NICOLE' s shared vision on sustainable remediation//Green Remediation Conference, Copenhagen, Denmark.

McLaren. S. Worboys, M., et al. 2009. Ex-situ Thermally Enhanced Coal Tar Recovery-A Low Carbon Option//Green Remediation Conference. Copenhagen, Denmark.

MEP. 2014. National Soil Contamination Survey Report. Ministry of Environmental Protection, Beijing, China.

MEP. 2010. Technical Guidelines for Environmental Monitoring of Sites (Interim Version). Ministry of Environmental Protection, Beijing, China.

MEP. 2009a. The Technical Specification for Environmental Site Investigation (Interim Version). Ministry of Environmental Protection, Beijing, China.

MEP. 2009b. Guidelines for Risk Assessment of Contaminated Sites (Interim Version). Ministry of Environmental Protection, Beijing, China.

MEP. 2009c. Guidelines for Soil Remediation of Contaminated Sites (Interim Version). Ministry of Environmental Protection, Beijing, China.

NAVFAC. 2011. SiteWise Version 2 User Guide, NAVFAC. Naval Facilities Engineering Command, Engineering Service Center, Port Hueneme, CA.

NICOLE. 2005. The Impact of EU Directives on the Management of Contaminated Land, NICOLE. Network for Industrially Contaminated Land in Europe, Utrecht, The Netherlands.

NICOLE. 2002. Need for Sustainable Land Management: Role of a Risk Assessment Based Approach, NICOLE. Network for Industrially Contaminated Land in Europe, Utrecht, The Netherlands.

NRC. 2011. Sustainability and the US EPA.

Petruzzi N M. 2011. A case study on the evaluation and implementation of green and sustainable remediation principles and practices during a RCRA corrective action cleanup. Groundwater Monitoring & Remediation, 31 (2): 63-71.

Rogers Lord. 1999. Towards an urban renaissance: Report by the Urban Task Force, DETR. Department of the Environment, Transport and the Regions, London, UK.

Schädler S, Morio M, Bartke S, et al. 2011. Designing sustainable and economically attractive brownfield revitalization options using an integrated assessment model. Journal of Environmental Management, 92 (3): 827-837.

Schädler S. 2011. Targeted Design and Integrated Evaluation of Land Use Alternatives for Sustainable Brownfield Redevelopment. Universität Tübingen.

Slenders H L A, Dols P, Verburg R, et al. 2010. Sustainable remediation panel: Sustainable synergies for the subsurface: Combining groundwater energy with remediation. Remediation Journal, 20 (2): 143-153.

SNOWMAN. 2009. The Rejuvenate Decision-Making Approach-A Worked Example-Crop Based Systems for Sustainable Risk Based Land Management for Economically Marginal Degraded Land. Sustainable Management of Soil and Groundwater (SNOWMAN).

Surf-UK. 2010. A Framework for Assessing the Sustainability of Soil and Groundwater Remediation, CLAIRE. Contaminated Land: Applications in Real Environments, London, UK.

USACE. 2010. Decision Framework for Incorporation of Green and Sustainable Practices into Environmental Remediation Projects, USACE. United States Army Corps of Engineers, Washington, DC.

USEPA. 2009. Principles for Greener Cleanups. United States Environmental Protection Agency, Office of Solid Waste and Emergency Response.

USEPA. 2008. Green Remediation: Incorporating Sustainable Environmental Practices into Remediation of Contaminated Sites. EPA 542-R-08-002., USEPA. United States Environmental Protection Agency, Washington, DC.

USEPA. 2004. Cleaning Up the Nation's Waste Sites: Markets and Technology Trends, 2004 Edition. United States Environmental Protection Agency, Washington, DC.

USGBC. 2011. LEED 2009 for New Construction and Major Renovations (updated November 2011), USGBC. United States Green Building Council, Washington, DC.

Xie J, Li F. 2010. Overview of the current situation on brownfield remediation and redevelopment in China. The World Bank.

Yang C. 2010. Experts Call for Legislative Proposal and Information Disclosure for Land Contamination. South. Metrop. Dly.

Zou S. 2006. China's First National Survey of Land Contamination. Xinhua News.

20th Anniversary for Soil Environment Protection in China

Yusuo LIN, Guoqing WANG

Nanjing Institute of Environmental Sciences,
Ministry of Environmental Protection of the P. R. China

1. Foreword

It has been nearly 40 years since environmental protection started in China, during which a relatively complete atmospheric, water and solid waste pollution prevention and treatment system has become established; relatively speaking, we have not laid solid foundation for soil pollution prevention, and are still in want of a soil pollution prevention system. In China, however, many efforts have been made in soil pollution protection, and we have conducted a great many conducive explorations and practices, which can be classified into two phrases:

Phrase Ⅰ, from 1980s to 1990s. Some key national scientific and technology projects supported research efforts in agricultural soil background value, national soil and environment background value, and soil environment capacity, during which we have accumulated precious data of soil environment background contents in China. At that time, the soil environment in China was quite favorable on the whole, but Chinese scientists started to pay attention to mining area, sewage irrigation area, and the pollution of arable land soil pollution caused by excessive use of HCH and DDT pesticides. The first Chinese Environmental Quality Standard for Soils was formulated by Chinese scientists, issued in 1995, and enacted in 1996.

Phase Ⅱ, from 2000 to now. The soil pollution problem has been becoming more and more prominent in China, and the issue of soil environment safety has aroused wide social attention. The Chinese Government has attached great importance to soil environment protection, placing soil pollution prevention, together with other important issues like atmosphere and water pollution prevention on top agenda, to enhance our soil pollution prevention efforts in a comprehensive manner, and a soil pollution prevention system with Chinese characteristics has now been forming up, and we have made good progress in soil environmental protection.

2. Establishing complete soil pollution prevention policies and laws

2.1 National policies and laws

To cope with pollution of the three environmental media of atmosphere, water, and soil, China has enacted the Atmospheric Pollution Prevention Law and the Water Pollution Prevention

Law, and now is still in want of dedicated soil pollution prevention legislation. Legal articles related to soil environment protection, however, are present in other relevant laws and provisions, such as the Environmental Protection Act, the Solid Wastes Polluted Environment Prevention and Control Act, the Grassland Law, the Mineral Resources Law, the Land Management Law, Agricultural Law, the Agricultural Product Quality Security Law, as well as the Hazardous Chemicals Control Ordinance, the Provisions on Pesticides Management, the Provisions on Basic Farmland Protection, and the Land the Ordinance on land reclamation. Apparently, this still presents some problems like poor systematicness and pertinence among relevant provisions on soil pollution prevention; making is hard to meet the demand for soil pollution prevention and control. Therefore, it is very pressing for us to make a dedicated law for soil pollution prevention and control. To this end, the Chinese Ministry of Environmental Protection mobilized relevant legal and soil environment specialists to conduct preliminary research on soil environment protection legislation, and in 2013, the Standing Committee of the 12th National People' s Congress listed the Soil Pollution Prevention legislation as the 1st category of legislation plan. Currently, under the leadership of the NPC Environmental and Resources Committee, the Chinese Ministry of Environmental Protection, together with relevant departments, is making joint efforts to prepare the Soil Pollution Prevention and Control Act (draft), and they are expected to submit the draft to the Standing Committee of the People' s Congress for deliberation.

During implementation of the "12th Five Years Plan", the Chinese authorities have issued a series of policy documents to facilitate soil pollution prevention efforts. For example, on February 14, 2011, the Ministry of Environmental Protection upon the approval of the State Council printed and issued the "12th Five Years" Plan for Comprehensive Prevention and Control of Heavy Metals (HUANFA [2011] 17#). On October 1, 2011, the State Council issued the State Council' s Opinions on Facilitating Critical Efforts for Environmental Protection (GUOFA [2011] 35#). On December 15, 2011, the General Office of the State Council issued the "12th FiveYears Plan" for National Environmental Protection (GUOFA [2011] 42#). On January 28, 2013, the General Office of the State Council issued the Circular on Recent Deployment for Soil Environment Protection and Comprehensive Treatment in the Near Future (GUOBANFA [2013] 7#). On October 12, 2013, the 3rd Plenary Session of the 18th Central Committee of the Chinese Communist Party passed the CPC's Decisions on Several Key Problems concerning Further All-rounded Deepening Reform. The State Council also mobilized efforts to prepare the Action Plan for Soil Pollution Prevention and Treatment, which was officially enacted on May 31, 2016 (GUOFA [2016] 31#).

Relevant departments of the State Council also issued some normative documents to enhance soil environment monitoring. For example, on July 7, 2004, the former SEPA (State Environmental Protection Administration) printed and issued the Circular on Solid Efforts for Environmental Pollution Prevention During Corporate Relocation (HUANBAN [2004] 47#).

On July 6, 2008, the Ministry of Environmental Protection printed and issued the Opinions on More Efforts for Soil Pollution Prevention (HUANFA [2008] 48#). On October 27, 2012, the Ministry of Environmental Protection, together with the Ministry of Industry and Information, the Ministry of Land and Resources, and the Ministry of Housing and Urban-Rural Development, altogether printed and issued the Circular on Ensuring Environmental Safety during Redevelopment and Utilization of Previous Industrial and Corporate Premises (HUANFA [2012] 140#). On May 14, 2014, the Ministry of Environmental Protection printed and issued the Opinions on More Efforts for Pollution Shutdown, Relocation and Previous Site Redevelopment and Utilization (HUANFA [2014] 66#).

2.2 Local policies and laws

Recent years, local governments in China have made some conducive exploration in soil pollution prevention legislation, accumulating experiences for legistilation on the national level. In December 2015, Fujian Provincial People' s Government issued Fujian Provincial Measures on Soil Pollution Prevention, which took effect since February 2016. In February 2016, Hubei Provincial People' s Congress issued the Hubei Provincial Provisions on Soil Pollution Prevention, which took effect since October 1, 2016.

Some other local governments have also made some policy explorations on soil pollution prevention. For example, in 2006, Chongqing Municipality Bureau for Environmental Protection printed and issued the Circular on More Efforts for Handling Industrial Remnant Solid Wastes after Corporate Shutdown and Relocation, and Environmental Protection. In 2007, Shenyang Municipal Bureau for Environmental Protection, and Bureau for Planning and Land Resources printed and issued the Administrative Measures for Polluted Site Environmental Treatment and Restoration in Shenyang (trial). In 2008, Chongqing Municipality People' s Government printed and issued the Circular on More Efforts for Treatment and Restoration of Previous Industrial Corporate Premises. In 2011, Zhejiang Provincial People' s Government printed and issued the Zhejiang Provincial Action Plan for Clean Soil, and in 2013, Zhejiang Provincial Department for Environmental Protection, together with Zhejiang Provincial Committee for Economic Information, Finance Department, Department of Land Resources, Department of Housing and Urban-Rural Construction, and Agricultural Department altogether 6 departments printed and issued the Circular on More Supervision and Administration Efforts for Developing and Utilizing Polluted Industrial Corporate Premises. In 2013, Jiangsu Provincial Bureau for Environmental Protection printed and issued the Circular on Standardizing Pollution Prevention and Control Efforts for Industrial Corporate Premises. In 2014, four Shanghai Municipality Bureaus jointly issued the Management Measures for Ensuring Safe Environment for Re-development and Utilization of Industrial Corporate Premises and Municipal Sites (HUHUANB AOFANG [2014] 188#).

3. Establishing sound soil environment protection standards system

Since the first Chinese Environmental Quality Standard for Soils was enacted in 1995, it has been many years, during which more than 50 relevant criteria for soil environment protection, and now we have formed up a preliminary standards system for soil environment protection, including five categories of standards.

3.1 Soil environment quality standards and evaluation standards

These standards mainly include the Soil Environment Quality Standard (GB 15618-1995), the Environment Quality Evaluation Standard for Edible Agricultural Products (Field) Places of Origin (HJ 332-2006), the Environment Quality Evaluation Standard for Greenhouse Vegetable' s Places of Origins (HJ 333-2006), and the Environment Quality Evaluation Standard for the Soil of Exhibition Fair (Provisional) (HJ 350-2007) and so on.

3.2 Standards of technical guidelines

These standards mainly include the Technical Code for Soil Environment Monitoring (HJ/T 166-2004), the Technical Guidelines for Field Environment Survey (HJ 25.1-2014), the Technical Guidelines for Field Environment Monitoring (HJ 25.2-2014), the Technical Guidelines for Polluted Field Risks (HJ 25.3-2014), and the Technical Guidelines for Soil Restoration on Polluted Fields (HJ 25.4-2014) and so on.

3.3 Standards for soil pollutants analysis methods

This category of standards cover analysis methods of As, Hg, Cr, Cu, Zn, Ni, Pb, Cd, Se, Bi, Sb, Be, cyanide and total cyanide, acrylic aldehyde, vinyl cyanide, acetonitrile, volatile organic compounds, volatile aromatic hydrocarbons, volatile halocarbons, phenolic compounds, polychlorinated biphenyl, polycyclic aromatic hydrocarbon, organophosphorus pesticide, benzene hexachloride, DDT and dioxin pollutants in the soil and deposits.

3.4 Standards for soil pollution control

This category of standards include the Standard for Controlling Pollutants in Farming Sludge (GB 4284-1984), the Standard for Controlling Urban Wastes for Agricultural Purpose (GB 8172-1987), the Standard for Controlling Pollutants in Coal Ash for Agricultural Purpose (GB 8173-1987), the Standard for Irrigation Water for Agricultural Purpose (GB5084—1992) (under revision) and so on.

3.5 Basic standards

These standards include the Glossary for Soil Quality (GB/T 18834-2002), the Terminology for Polluted Fields (HJ 682-2014) and so on.

To accommodate the current demand for soil pollution prevention and management in China, the Ministry of Environmental Protection started to revise the Soil Environment Quality Standard in 2006, to address the existing problems among the current soil environment quality standards. This revision will further define the orientation and role of soil environment standards, provide complete soil environment standards system and structure, and promote the soil environment protection standards system in China.

4. Launching soil environment survey and monitoring

After a nation-wide research program on soil environment background value in 1980s, competent Chinese authorities have also mobilized some efforts to conduct fundamental survey on soil environment. In 1999, the Ministry of Land and Resources launched geochemistry multi-purpose regional geochemical survey; by 2014, they had surveyed an area of 1.507 million square km, including 1.386 billion mu of arable land, accounting for 68% of arable land area in China. From 2005 to 2013, the Ministry of Environmental Protection, together with the Ministry of Land and Resources, launched the first nation-wide soil pollution status survey, covering an area of 6.3 million square km. In 2012, the Ministry of Agriculture launched a survey on heavy metals pollution in the soil of agricultural products, covering an area of 1.623 billion mu. Moreover, the environmental and agricultural sectors have also had long-term monitoring survey on the soil environment quality around main wastewater irrigation area, metal ores area, main grain production area, important places of origin of agricultural products, surface water potable water sources and so on.

Different sectors have employed different approaches to measure different items during different time periods, for soil environment survey, and the research precision could not meet the demand for demarcating the soil pollution scope or polluted land parcels; to really get hands on the true picture of soil pollution, and meet the demand for soil pollution risk management and control, and treatment and restoration, the Chinese Government will mobilize some efforts to have detailed survey on soil pollution status. The technological expertise of competent departments will be mobilized, to make full use of the resources of higher institutes and scientific research institutes, and give full play to the social laboratory; and the "five coordinated" requirements including coordinated survey schemes and methods, coordinated evaluation criteria, coordinated laboratory, coordinated quality management, and coordinated time periods for survey will be observed to mobilize implementation. This is another major nation-wide fundamental survey on soil environment, as is also quite rare in the world.

During implementation of the "12th Five Years Plan", the Ministry of Environmental Protection in China has launched some pilot work, engaged in researching into and preparing a scheme for constructing a nation-wide soil environment quality-monitoring network; it is planned to deploy fundamental monitoring and test points, and risk monitoring test points for soil environment quality survey. Meanwhile, we put together relevant soil environment monitoring, agricultural products quality test, pollutant source survey, land utilization data from the environmental protection, land and resources sectors to establish a nation-wide fundamental database for soil environment, and build up a soil environment information management platform to actually share resources.

5. Promoting scientific and technological development and application for soil pollution treatment and restoration

Compared with those developed countries and regions, China has lagged behind, starting up late in making efforts to prevent and treat soil pollution, and in general, our soil pollution treatment and restoration technology research and development and engineering application have lagged behind those of developed countries and regions. It has been 40-50 years since soil restoration started abroad, with specialized and practical soil restoration technological system and completes industrial chain and restoration markets; and now there are proven restoration technology, supporting restoration materials, complete restoration equipment, high-caliber consultant experts, and engineering technicians. Relatively speaking, China has only a short history of 10 years in soil restoration technology research and development, and engineering application, and now still stays in the initial startup stage. Recent years, however, a series of soil restoration techniques out of Chinese independent research and development have started to go to their engineering pilot demo stage, and a series of advanced technological equipment and restoration materials abroad have also been introduced into China. And a batch of arable soil pollution treatment and restoration pilot projects and pollution land parcels restoration engineering projects have started, and the number of consultant agencies for soil pollution treatment and restoration, and professional restoration and supporting services enterprises have increased dramatically, and the soil restoration industry and market are developing rapidly, and have become a growth point for the emerging environmental protection industry and pillar industry.

5.1 Restoring polluted arable soil

Overall, currently the technology and engineering expertise for soil pollution treatment and restoration in China still stays in the phase of exploration, practice, summary, and improvement. In terms of farmland, the polluted arable soil restoration technology research and development level is roughly the same to that of developed countries abroad. Since the ″10th Five Years Plan″, some scientific research programs of the Ministry of Science and Technology have supported our research efforts in plantation restoration, agricultural prevention and control, chemical regulation, agricultural and chemical technology combination, and other control and restoration technology, including heavy metals polluted arable soil plantation restoration techniques, low accumulation species for agricultural prevention and blocking techniques, water and fertilizer regulating and the like arable land soil' s safe utilization techniques, to develop bio-remediation technologies and bio-microorganism combined remediation techniques for organic polluted farmland soils. Recent years, some technologies have started to be widely applied to arable soil restoration. For example, some Chinese scientists have found that ciliate desert-grass-arsenic (As) hyper-accumulator boasts very strong As-toxicity resistance, so research teams have been deployed to Hunan, Henan, Guangxi and other provinces, where they have started their As-polluted arable soil plantation extraction and restoration projects, as represents the largest As-polluted arable soil plantation extraction and restoration project in the world. Some Chinese scientists have also applied a cadmium (Cd) hyper-accumulator *sedum plumbizincicola* to Cd polluted arable soil plantation extraction and restoration projects. Moreover, some Chinese scientists have also developed some bio-carbon blocking and control techniques that can effectively block and control Cd accumulation in the edible parts of rice, making use of industrial wastes-red mud and hydrosulphonyl-rich plant stalk powder as passivators, in combination with the antagonism of Zn against Cd, to reduce the contents of Cd in agricultural products. Some pilot demo project practices have explored the restoration model of restoration, production, and dissemination.

5.2 Treating and restoring polluted soil land parcels

Relatively speaking, technology, construction capacity, and engineering expertise for restoring polluted soil parcels are lagging behind. Initially, cement kiln techniques and land filling techniques were employed, and currently we have developed a complete technological and engineering system featuring thermal desorption, chemical oxidation, soil vapor extraction, and other mainstream technological expertise; this system is applicable to *in-situ* and *ex-situ* treatment. During implementation of the "10th" and "11th" five years plans, the "863" Program of the MOST (Ministry of Science and Technology), the special scientific and technological program for public good and so on were launched, with critical research efforts on generic technology for restoring soil polluted by heavy metals, pesticide, persistent organic contaminants, petroleum and so on. In addition, pilot program for verification has been launched within a small scope. During implementation of the "12th Five Years Plan", the "863" Program of the MOST (Ministry of Science and Technology) has launched a key project for "Polluted Soil Restoration Technology and Demonstration" in the field of environmental technology, speeding up innovations for critical soil restoration technology and equipment. In the future 5-10 years, we shall never stay rested in making efforts to enable our polluted soil (parcels) restoration technological expertise and equipment research and development capacity to reach internationally advanced level, and work out a soil restoration technology system with Chinese characteristics to greatly promote the development of soil restoration industry.

References

General Office of the State Council of the P. R. China. Circular on recent deployment for soil environment protection and comprehensive treatment in the near future (GUOBANFA [2013] 7#). Beijing. October 12, 2013.

Ministry of Environmental Protection of the P. R. China (former SEPA, state environmental protection administration). Circular on solid efforts for environmental pollution prevention during corporate relocation (HUANBAN [2004] 47#). Beijing. July 7, 2004.

Ministry of Environmental Protection, together with the Ministry of Industry and Information, the Ministry of Land and Resources, and the Ministry of Housing and Urban-Rural Development. Circular on ensuring environmental safety during redevelopment and utilization of previous industrial and corporate premises (HUANFA [2012] 140#). Beijing. May 14, 2014.

The State Council of the P. R. China. Action plan for soil pollution prevention & treatment (GUOFA [2016] 31#). Beijing. May 31, 2016.

Progress in the Risk Management of Contaminated Sites: Research Activities and Environmental Management of Contaminated Sites at CRAES

Qingbao GU, Guanlin GUO, Jin MA, Youya ZHOU, Zengguang YAN, Yunfeng XIE, Bing YANG, Li LIU, Hong HOU, Xiaoming DU, Nandong XUE, Yunzhe CAO, Fujun MA, Ping DU, Liping BAI, Fasheng LI*

Department of Soil Pollution Control, Chinese Research Academy of Environmental Sciences
*Corresponding author

1. Introduction

The issue of contaminated sites is receiving increasing public concern in China due to its close relation with industrial activities, disposal of hazardous materials, and environmental accidents. Abandoned chemical plants are considered to be potential sources of contamination and can become brownfield sites if they are not promptly developed. These contaminated sites pose serious problems to human health and the environment in the process of land redevelopment. The most straightforward way to resolve these issues is a risk-based management of the site.

The Chinese government has implemented a series of measures with regard to the protection of the soil environment (Ma et al., 2016). The Action Plan of Soil Pollution Prevention and Control was released on May 31, 2016. Given the current situation, where serious soil pollution exists in many sites, the remediation of polluted soil has become the focus of environmental protection. China plans to launch a comprehensive polluted land remediation and management program in 2016-2020, with 100 polluted farmland sites and 100 polluted construction sites selected as pilot areas to be remediated. The aim is to establish six soil pollution prevention and control demonstration areas, to commence the remediation of 10 million acres of polluted land, and manage the risks of a further 40 million acres of contaminated land.

The Ministry of Environmental Protection (MEP) of China has played a leading role in developing this program. As the most important institution directly affiliated to the MEP, the Chinese Research Academy of Environmental Sciences (CRAES) has conducted many successful studies of soil pollution and remediation. These studies have improved our understanding of soil pollution status and the development of remediation technologies in China, and will aid in the formulation of policies and regulations regarding soil pollution control.

2. Site investigation and risk assessment

2.1 Pollution investigation and source appointment

Contamination from organic chemical plants can cause serious pollution of soil and groundwater ecosystems. The contamination of soil and groundwater was characterized at a typical organic chemical plant site in Chongqing, South China. The concentrations of different contaminants and their three-dimensional (3D) distribution were determined based on the 3D-Krige method (Liu et al., 2015). Organochlorine pesticides (OCPs), specifically dichlorodiphenyltrichloroethane (DDT) and hexachlorocyclohexane (HCH), in the topsoil of a site in Beijing where an OCP manufacturing enterprise was previously established, were quantitatively analyzed to determine the soil pollution status. The results demonstrated that the soil was heavily polluted by HCH and DDT, although the production of these pesticides had been prohibited for more than 20 years (Cong et al., 2008). Petroleum contamination in soil is ubiquitous. The distribution and concentration of polycyclic aromatic hydrocarbons (PAHs) in soil were investigated in an oil-contaminated site, which was used for the disposal of wastewater from a petrochemical plant for about 30 years. The highest concentrations were found for the low molecular weight PAH fraction (Huang et al., 2009). The concentration of heavy metals in a nonferrous mine/smelting site located in South China was investigated. The ecological risks of heavy metals were evaluated using contamination factors and potential ecological risk factors. Multivariate analysis (Pearson's correlation analysis, hierarchical cluster analysis, and principal component analysis) was used to identify heavy metal sources in soil (Zhou et al., 2015, Fig. 1).

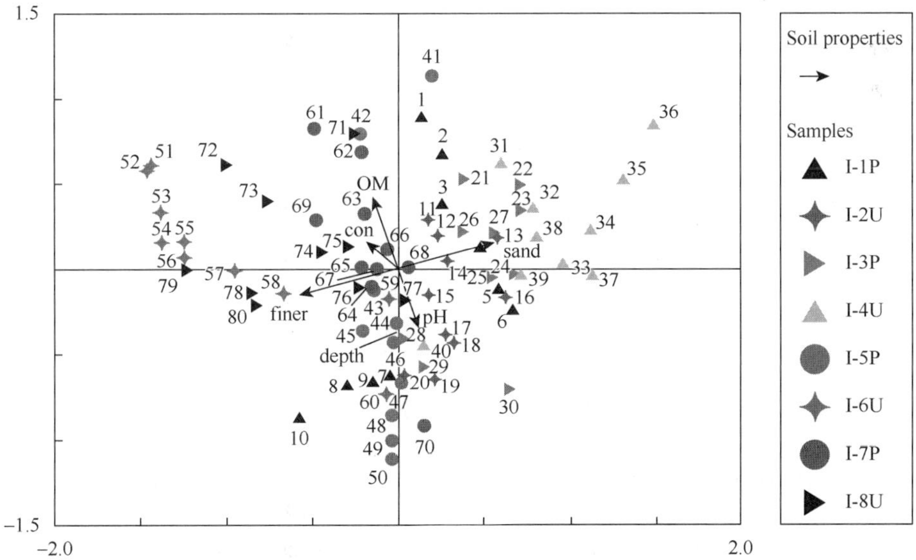

Fig. 1 Biplot of soil properties and soil samples from redundancy analysis (RDA)

2.2 Risk assessment of contaminants and uncertainty analysis

Risk-based cleanup standards provide an increasingly acceptable alternative to regulatory cleanup standards. In a heavy metal-contaminated site, the health risks of arsenic (As), cadmium (Cd), lead (Pb), and zinc (Zn) were assessed independently by the Risk Based Corrective Action (RBCA) model and the Contaminated Land Exposure Assessment (CLEA) models. The two models have different merits, and their application should be selected based on actual site conditions (Shi et al., 2009).

Soil standards for benzo[a]pyrene (B[a]P) were derived by the RBCA and CLEA models. For residential land, the standards determined by the two models were similar and were also comparable to the standards used in other countries. However, for industrial land, a significant difference existed between the standards determined by the two models. The results obtained from the RBCA model were similar to the standards used in other countries, while the results obtained from the CLEA model were significantly higher. The results obtained from these two models differed in two main aspects: one was the difference in the exposure assessment as determined by the different principles and algorithms used in the two models, and the other was the difference in the source and quality of B[a]P toxicology data (Xia et al., 2009).

The risk assessment should acknowledge and quantify the uncertainty in risk predictions. The effects of the spatial heterogeneity of soil organic matter (SOM), soil moisture (Pws), and soil bulk density (ρb) on benzene risk from contaminated soil on the site of a former chemical plant were evaluated using a Monte Carlo simulation (Fig. 2). The sensitivity analysis of the model parameters, especially the physical and chemical parameters of soil, is important in contaminated site risk assessment. The value of the sensitive soil parameters must be determined carefully to ensure a reasonable risk assessment result (Zhang et al., 2012).

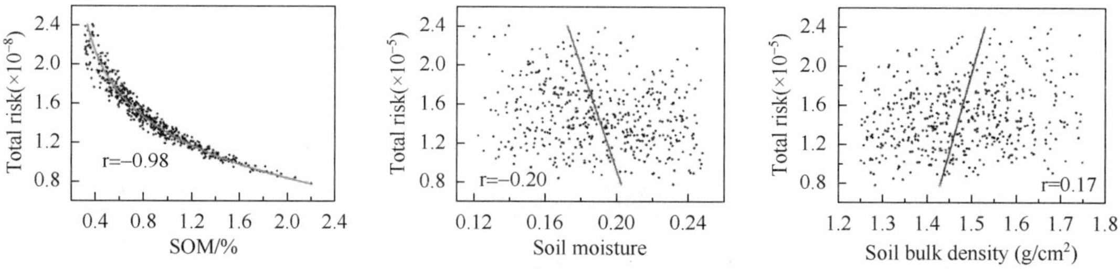

(a) Correlativity between SOM (b) Soil moisture (c) Soil bulk density

Fig. 2 Correlation between soil properties and total carcinogenic risk of benzene

2.3 Spatial pattern and distribution mapping of soil contaminants

The identification of contamination "hotspots" is an important indicator of the degree of contamination in localized areas, and can contribute towards the re-sampling and remedial strategies used in seriously contaminated areas. PAH pollution hotspots at an industrially contaminated site in northern China were analyzed using multivariate statistical and spatial autocorrelation techniques (Fig. 3). The identification of hotspots and the spatial distribution of soil pollutants can play a key role in contaminated site investigation and management (Liu et al., 2013).

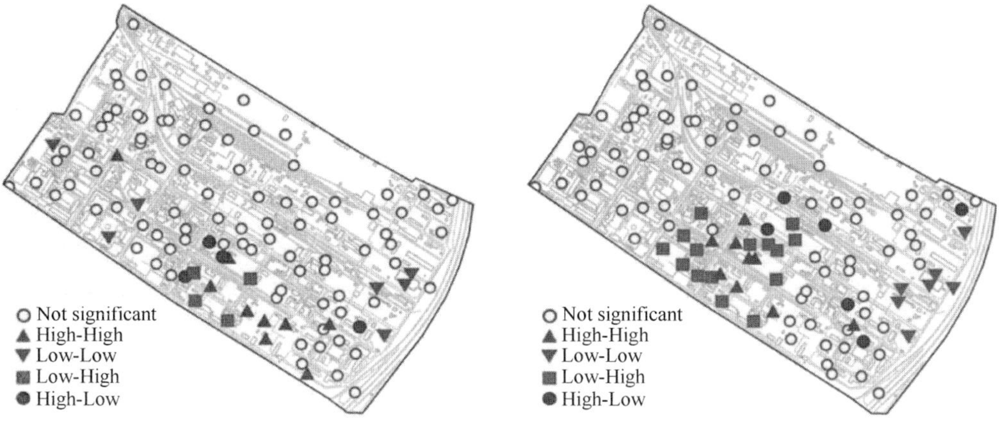

(a) Nap (b) Chr

Fig. 3 Local spatial cluster characteristics for soil PAHs

Mapping the spatial distribution of contaminants in soils is the basis of pollution evaluation and risk control. The distribution of concentration data from contaminated sites is usually severely skewed by the presence of hotspots in contaminated sites. This causes difficulties for accurate geostatistical data transformation. We used an ideal normal transformation method (the Johnson transformation function) prior to the geostatistical analysis of severely skewed data. The method was more accurate than the other models used in the case studies, and its use improved the accuracy of determining remediation boundaries (Liu et al., 2015).

The smoothing effects of widely used spatial interpolation methods lead to underestimates in areas with locally high concentrations of pollutants and overestimates in areas with locally low concentrations. A geostatistical conditional simulation was applied to map the spatial distribution of soil pollutants. A probability map was used to delineate areas that had more potential for contamination and to explicitly assess the uncertainty. The pollution probability-based method facilitates soil pollution control and environmental management decision making (Xie et al., 2015).

3. Technologies used for soil remediation and risk control

3.1 Bioremediation technology

Field bioremediation of oil sludge-contaminated soil was conducted using a landfarming treatment and natural attenuation, in which cotton stalks were incorporated into the surface soil in the Shengli oil field (Fig. 4). The ability of the landfarming treatment to reduce petroleum hydrocarbons and restore soil quality was evaluated. The initial concentration of total petroleum hydrocarbons (TPH) was 12.57 mg/g for oil sludge-contaminated soil. Over a 39-month period of bioremediation, removal efficiencies of 68.84%, 90.04%, and 85.55% for TPH, the saturated fraction, and the aromatic fraction, respectively, were achieved (Fig. 5). The degradation of TPH followed first-order exponential decay kinetics. The soil pH, salinity, alkalinity, nutrient content, organic matter content, and the amount of hydrocarbon degraders were greatly improved (Fig. 6). The results of a Biolog and polymerase chain reaction denaturing gradient gel electrophoresis (PCR-DGGE) analysis revealed an improvement in soil microbial quantity and diversity. The 23 predominant strains that were isolated indicated a shift in soil community structure toward hydrocarbon-degrading species, including *Streptococcus* sp., *Shewanella* sp., *Bacillus* sp., *Pseudomonas* sp., *Marinobactera* sp., and *Thermoanaerobacter* sp. (Wang et al., 2012, Wang et al., 2016).

(a) Mixing the oil sludge and cotton straws (b) Rotary tillage during landfarming treatment

Fig. 4 Field-enhanced bioremediation of oil sludge by landfarming with the addition of straw residues

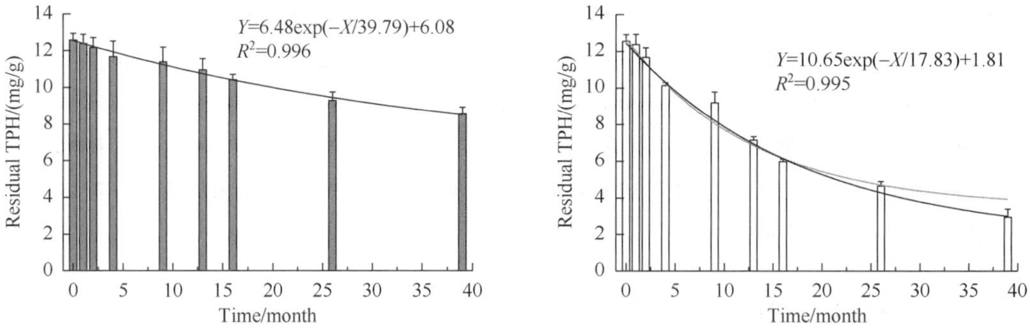

(a) Natural attenuation of oil sludge (b) Landfarming treatment of oil sludge

Fig. 5 Change in residual total petroleum hydrocarbons (TPH) in soil during a 39-month bioremediation involving natural attenuation and landfarming treatment

3.2 Thermodesorption

Thermal treatment was performed in a laboratory-scale rotary kiln with an off-gas treatment system (Fig. 7). The rotary kiln was heated by electricity. The kiln was 2 m long, with an internal diameter of 0.2 m. In each test, the temperature in the kiln was measured by a thermocouple (NiCr-Ni) in the hot zone near the inner wall. The treatment time of soil samples in the hot zone of the kiln (900 mm) was adjusted by changing the speed of the kiln. The inclination angle of the kiln was 3.5°. At the end of the kiln, a collecting container was set up. Off-gas flowed into a condenser and condensed into liquid at 25°C. Off-gas flow through an activated carbon adsorption tank removed uncondensed contaminants.

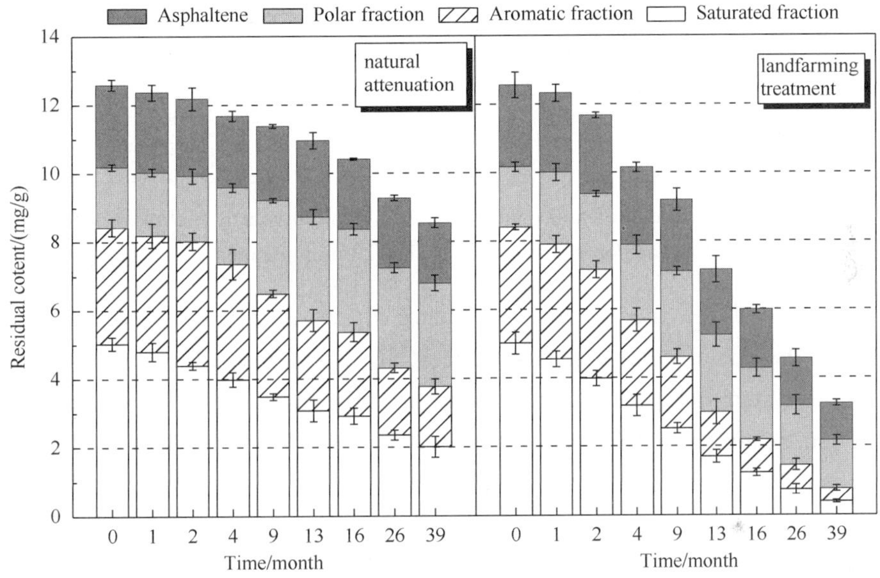

Fig. 6 Temporal changes of the residual concentration of the saturated, aromatic, polar, and asphaltene fractions by natural attenuation and landfarming treatment

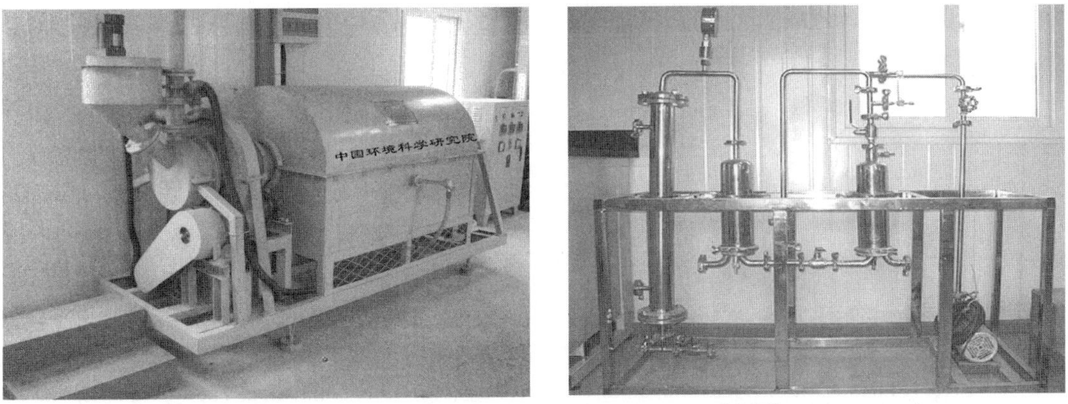

(a) Thermal desorption unit (b) Off-gas treatment unit

Fig. 7 Thermal desorption system

Soil samples were collected from the site of an abandoned pesticide plant in Beijing, China.

The removal efficiency of DDT and HCH was temperature dependent (Fig. 8) (Zhang et al., 2012). After 20 min, more than 97% and 99% of HCH and DDT were removed from soil at 310℃ and 340°C, respectively. The residue met the Standard of Soil Quality Assessment for Exhibition Sites, and the removal rate did not increase thereafter (Fig. 8).

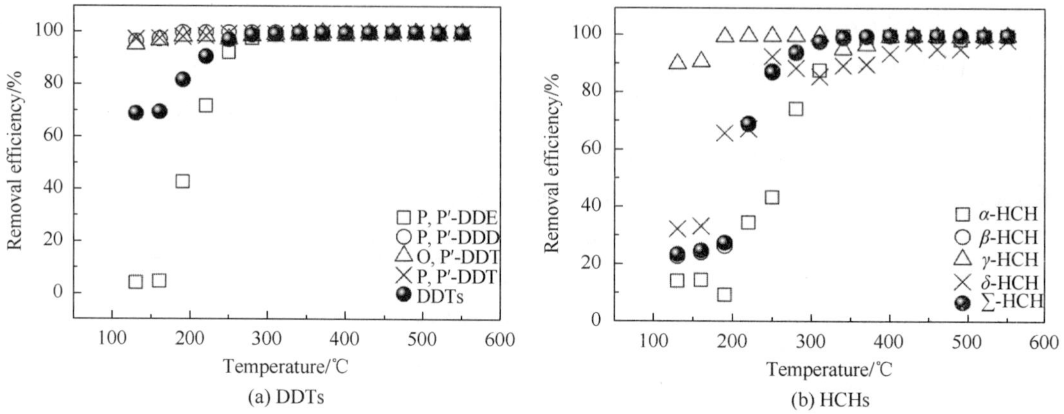

Fig. 8 Removal efficiencies of soils contaminated by dichlorodiphenyltrichloroethane (DDT), hexachlorocyclohexane (HCH) (20-min treatment)

A novel method for the thermal treatment of mercury (Hg)-contaminated soils was developed with the use of citric acid (CA) (Ma et al., 2015). A CA/Hg molar ratio of 15 was adopted as the optimum dosage. Hg concentration in soils was successfully reduced from 134 to 1.1 mg/kg when treated at 400°C for 60 min, and the treated soil retained most of its original soil properties (Fig. 9). To achieve the same cleanup level, the traditional thermal treatment

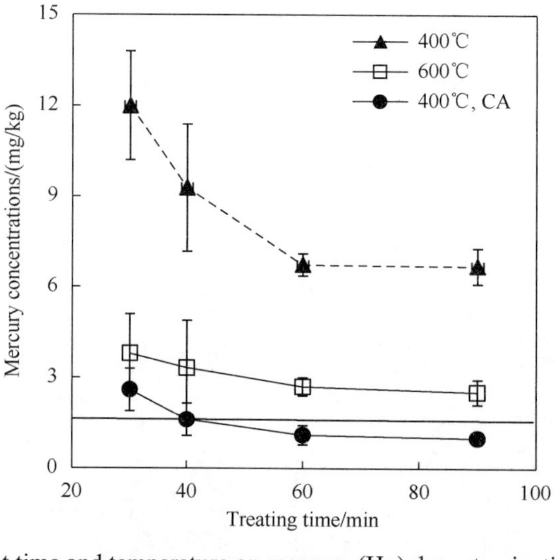

Fig. 9 Effect of treatment time and temperature on mercury (Hg) decontamination. The soil samples were thermally treated with or without citric acid (CA) (CA/Hg: 15). The dashed line represents the maximum Hg concentration in China (1.5 mg/kg, National Standard GB15618-1995)

method requires heating at 700°C for 30 min, which would significantly increase soil pH and sand content, decrease organic carbon content, and change soil color. The proposed method is expected to reduce energy input by 35% compared with the traditional thermal treatments and promote reuse of agricultural soil. Additionally, it should provide greener, more sustainable remediation of Hg-contaminated soil in future engineering applications.

3.3 Mechanical soil aeration

Volatile organic compounds (VOCs) are one of the most abundant chlorinated industrial products.

Mechanical soil aeration is a low-cost, ex-situ remediation technique for volatile organic-contaminated sites. Mechanical soil aeration agitates contaminated soil by tilling or by other means to volatilize contaminants while collecting and disposing of the released gas. This technique is suitable for the volatilization of organic contaminants because it involves simple, low-cost processes (Shi et al., 2012).

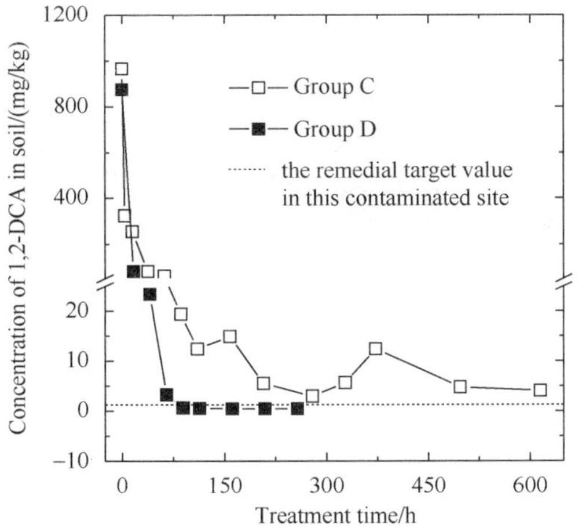

Fig. 10 Effect of soil temperature on the volatilization of 1, 2-DCA. (Group C: in a shed on-site, with an environmental temperature of 2.68 ± 0.87°C and soil temperature of 8.45 ± 1.61°C; Group D: in a greenhouse, with an environmental temperature of 21.47 ± 0.73°C and soil temperature of 14.74 ± 0.28°C)

Traditionally, mechanical soil aeration offers limited efficiency in removing VOCs from soil in the late stage of remediation, especially under low-temperature conditions and for soils with high moisture and clay contents. Tailing is typically described as a behavior for which further removal of contaminants may be achieved only slowly, even if the treatment time is extended, and is associated with the slow, continuous release of contaminants over time. Adding CaO during the late stage of site remediation was found to mitigate tailing (Ma et al., 2015; Fig 10-13).

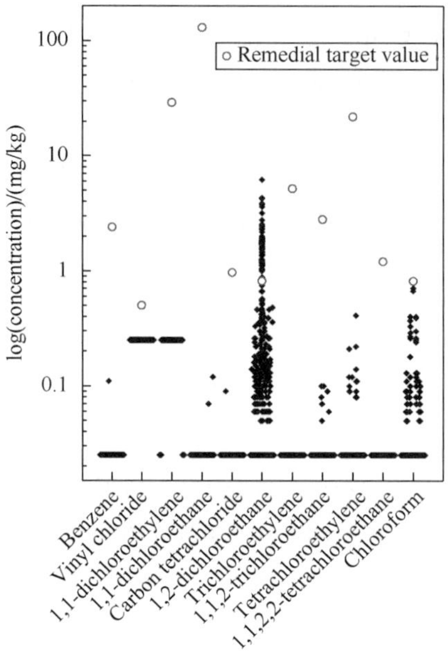

Fig. 11 Soil contamination levels after remediation with mechanical soil aeration

(a) Agitation (b) Air-tight shed (c) Tail gas monitoring system

Fig. 12 Field remediation of chloride hydrocarbons by mechanical aeration

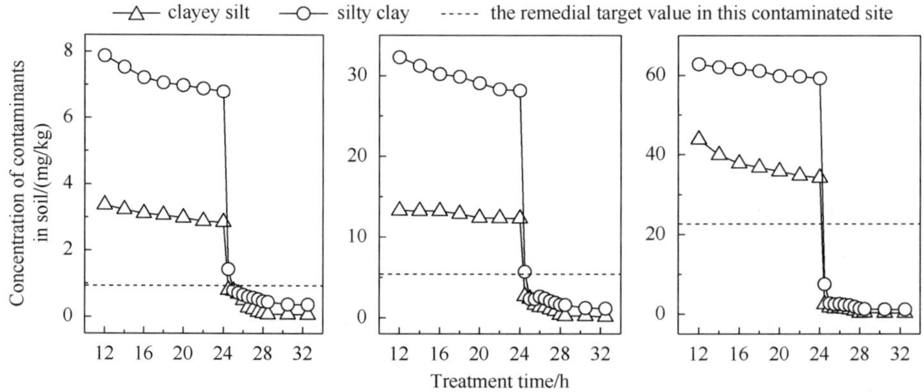

(a) 1,2-DCA (b) TCE (c) PCE

Fig. 13 Rapid changes in soil contaminant concentrations after adding CaO

3.4 Stabilization and solidification

Chemical immobilization by phosphates has been widely used to treat Pb in wastewater and contaminated soils. However, in wastewater and soil, Pb always coexists with other heavy metals, and their competitive reactions with phosphates have not been quantitatively and systematically studied. The immobilization of Pb, Zn, and Cd by mono-, di-, and tripotassium phosphate (KH_2PO_4, K_2HPO_4, and K_3PO_4) was observed in single- and ternary-metal solutions. A competitive reaction occurred in the Pb-Zn-Cd ternary system, with a decrease in immobilization rates following the order of$<$3.6%, $<$78%, and$<$89% for Pb, Zn, and Cd (molar ratios of P: metal$<$1), respectively, compared with a single-metal system (Fig. 14). The reaction of Pb exhibited intense competitiveness, and the phosphates had a stronger affinity for Pb when Cl^- was added. The Pb-phosphate minerals formed by KH_2PO_4 had superior crystalline characteristics and were very stable, with a low leaching rate ($<$0.02%). This study demonstrated that Pb could be immobilized in multi-metal solutions, which will help investigations of the controlling mechanisms of multi-metal pollution systems (Zhang et al., 2015).

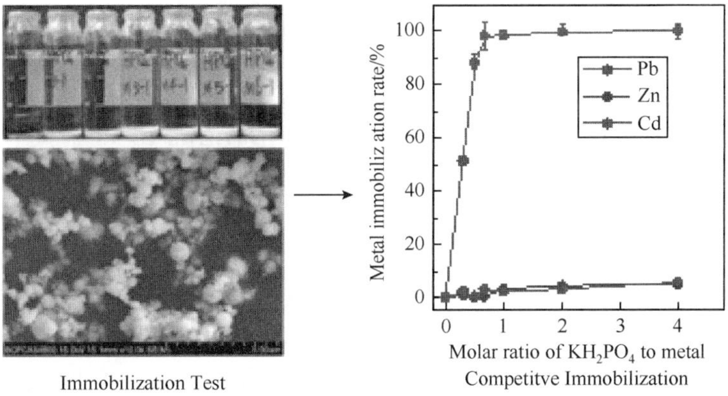

Fig. 14 Immobilization of heavy metals in the contaminated soil of a lead battery production facility

4. The environmental fate and criteria of contaminants

4.1 Vertical transport of PAHs in soils

Studies of the transport of PAHs in soil have mostly been based on colloid-facilitated theories, and have suggested that colloidal matter inherent in the subsurface of soils may provide a favorable phase for contaminant partitioning and, furthermore, may act as agents in contaminant transport. However, transport could be facilitated by changes in the chemical composition of the carrier solution, such as a higher pH, lower ionic strength, and the presence of surfactants or SOM. A simulation experiment with soil columns performed by Zhang (2008) showed that transport of a mixture of PAHs in soil is controlled by both the nature of PAHs (i. e., log*K*oc and molecular weight) and soil properties; thus, soil particle size and SOM play an important role in the vertical transport of PAHs (Fig. 15). Such an approach is of remarkable significance because the transport of hydrophobic organic contaminants in soils, especially in sandy soils, poses a serious threat to groundwater. Furthermore, this may provide insight into the vertical transport of other kinds of hydrophobic organic contaminants and identification of potential groundwater contamination.

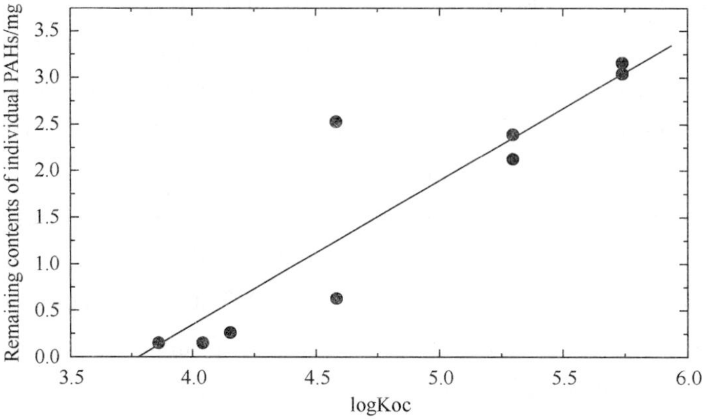

Fig. 15 Linear relationship between the residue levels of individual PAHs (RPAHs) and the partition coefficient (logKoc) for the fine silt column

4.2 Sorption kinetic processes of organic compounds in soils

Sorption processes or partitioning between sorbent (soils or sediments) and sorbate phases play an important role in the transport, deposition, and fate of hydrophobic organic compounds in subsurface systems. Such processes are strongly affected by the chemical conformation and physical structure of SOM. Han (2013) reported the combined effect of the content and the composition of SOM on toluene sorption rates, sorption capacity, and the nonlinear degree of sorption of three typical soils in China (Fig. 16). Such studies have improved our understanding of the sorption mechanism of hydrophobic organic compounds in soils with different SOM fractions. Because the transport of contaminants is often correlated to its sorption behavior, it is possible to predict the residual concentration of contaminants in the soil environment.

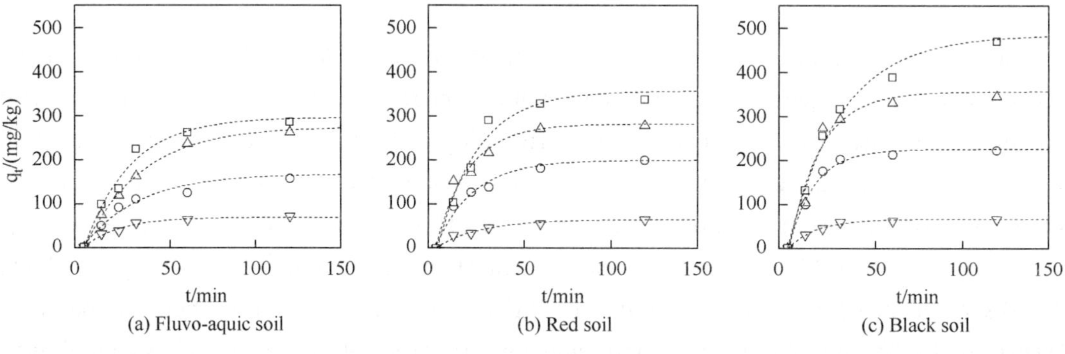

Fig. 16 Experimental and calculated sorption kinetics of toluene in different soil types (a fluvo-aquic soil, b red soil, c black soil) for various initial toluene solute concentrations in the aqueous phase (inverted triangles 14.3 mg/L, open circles 42.8 mg/L, open triangles 71.2 mg/L, open squares 99.5 mg/L). Symbols denote the average qt value of triplicate samples, and the solid line represents fitted results

4.3 Air-soil exchange of organochlorine pesticides

Soils act as the main sink of OCPs, but can also act as a secondary source by revolatilizing the residuals to the atmosphere because the compounds decompose in soils much more slowly than in the atmosphere and water. Yang et al. (2015) used a newly designed passive air sampler to test the vertical concentration profiles of gaseous phase OCPs in a sealed chamber, showing a clear relationship between temperature and the mixture of OCPs that escaped from the soils (Fig. 17). Such findings indicate that revolatilization and re-deposition is at or close to a dynamic pseudo-equilibrium with the overlying air, highlighting that crop management should consider air-soil-plant uptake and the partitioning of persistent organic pollutants (POPs), including OCPs.

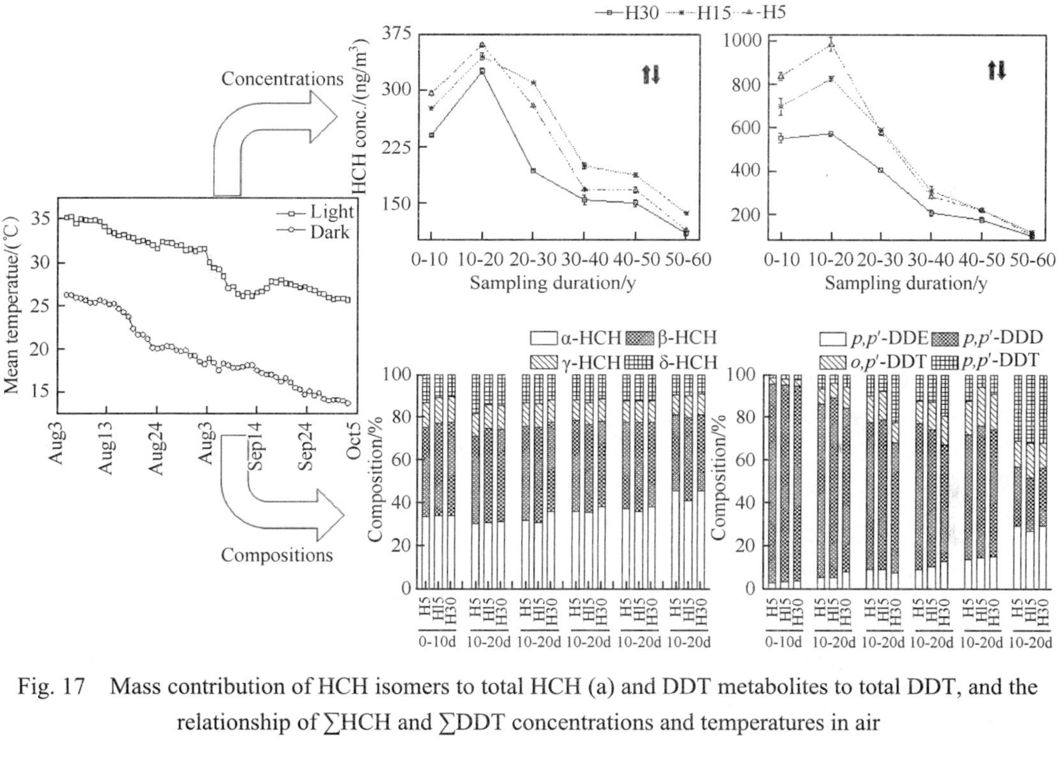

Fig. 17 Mass contribution of HCH isomers to total HCH (a) and DDT metabolites to total DDT, and the relationship of ∑HCH and ∑DDT concentrations and temperatures in air

4.4 Soil environmental criteria for antimony (Sb) based on groundwater protection

Soil environmental criteria values are an important basis for establishing soil quality standards and making soil quality assessments, and are very useful for protecting groundwater. Column experiments were conducted to study the migration of Sb in a primosol soil from Beijing (Fig. 18). Using the concentration of Sb (0.005 mg/L) set in the Standards for Drinking Water Quality of China (GB 5749-2006), the soil environmental criteria value of Sb based on groundwater protection was derived using the Vadose Zone Contaminant Migration Multi-Layered (VZCOMML) model. A value of 0.64 mg/kg was derived, which was much lower than the soil criteria or soil standard values (3.1-41 mg/kg) used in foreign countries and regions. The differences were mainly caused by the small R (or Kd) of Sb in soils, the dilution factor (DF) and the groundwater or drinking water standards (Cgw) of Sb. These studies have enabled the government to establish national standards and control the water resource protection area with regard to groundwater.

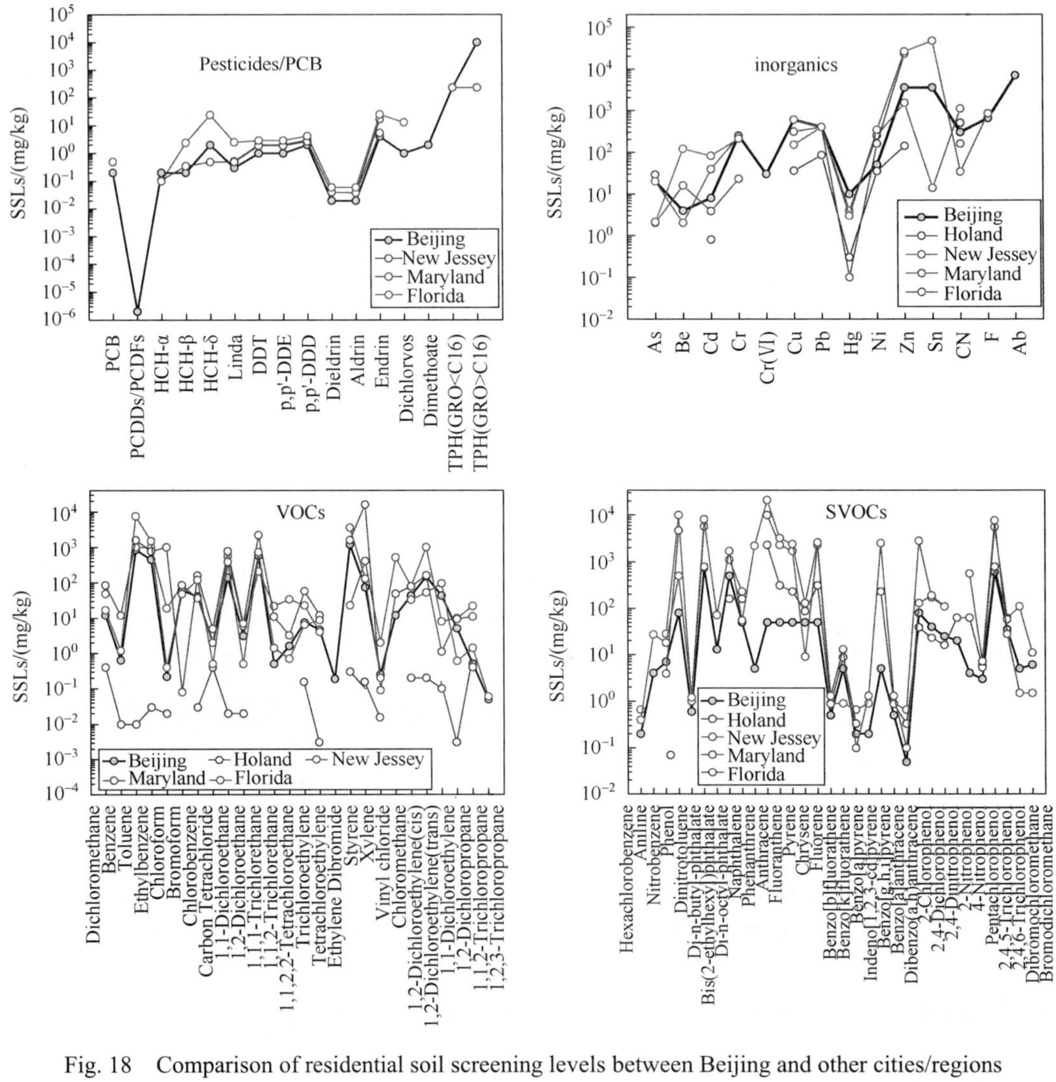

Fig. 18 Comparison of residential soil screening levels between Beijing and other cities/regions

5. Technical support for soil environmental management

Through its affiliation with the MEP of China, CRAES plays an important role in providing technical support for national soil environmental management. A National Soil Survey was initiated in China in 2006, and CRAES was responsible for the development of the Technical Regulations for the Risk Assessment of Regional Contaminated Soils and the Demonstration Project for the Remediation of Sludge-Contaminated Soils in an Oil Field. Furthermore, CRAES took part in many emergency response activities for the management of accidental soil contamination, such as the Emergency Response Program for Explosions in Tianjin (Fig. 19).

A priority action plan for the management of POP-contaminated sites in China was developed, and many activities were launched to aid in the elimination of POPs and the remediation of POP-contaminated soils. Under a collaborative framework with the United Nations Industrial Development Organization (UNIDO), CRAES participated in the development of the Persistent Organic Pollutants: Contaminated Site Investigation and Management Toolkit, and published a book entitled "Vocabulary Handbook of POP-Contaminated Sites", which was distributed to about 500 users working on POP-contaminated sites in UNIDO and other countries, including participants to the Conference of Parties (COP-5), and to countries in Africa, East and Southeast Asia (ESEA), and Gulf Cooperating Countries.

The Soil Screening Levels for Contaminated Sites Assessment in Beijing, the first soil screening levels in China, were developed in 2011; the Technical Guideline for Site Investigation and Risk Assessment in Chongqing was developed in 2016. In addition, an array of management tools for contaminated sites in China has been developed by CRAES. For example, a Remediation Technologies Screening Matrix, as well as a National Framework Inventory of Contaminated Sites was developed as a result of the Sino-Italian Cooperation Project on Contaminated Site and Soil Remediation Management.

(a) Tianjin explosionsite

(b) Changzhou Foreign Language School

Fig. 19 Technical support for the environmental risk assessment of accidental contaminated sites

6. Demonstrations and projects

6.1 Petroleum refinery site

Under the collaborative framework of the Sino-Italian Cooperation Project on Contaminated

Site and Soil Remediation Management, a petroleum refinery contaminated site in Jilin was investigated and the risk to human health was evaluated (Fig. 20). Soil samples from the site

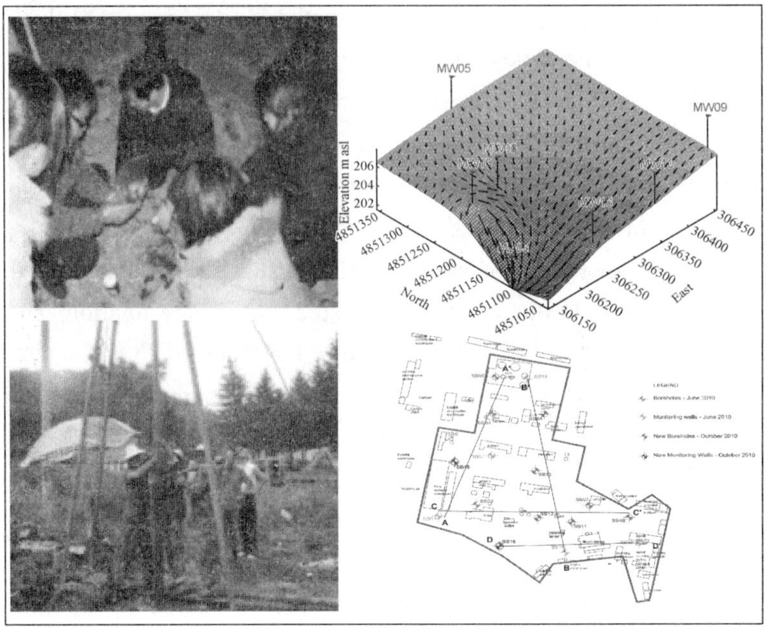

(a) Site investigation

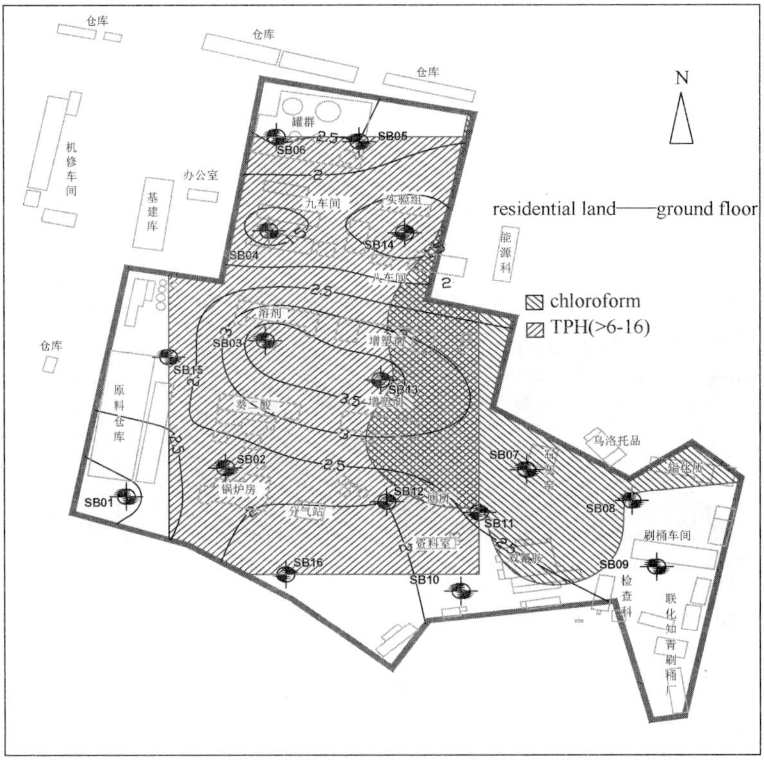

(b) Site risk assessment

Fig. 20 Investigation and risk assessment of contaminated sites in Shijinggou were found to be contaminated by chloroform, benzene, and light petroleum hydrocarbons. Target values for chloroform, benzene, and light petroleum hydrocarbons were calculated, and the boundary of the contaminated area was determined.

6.2 Soil remediation in oil fields

Disposal of oil sludge and remediation of oil-contaminated soils were conducted in Shengli Oil Field. Sludge containing more than 15% oil was excavated and reused, and oil contaminated soils were remediated using phytoremediation, bioremediation, coupled bioremediation and phytoremediation, and biopile technologies (Fig. 21).

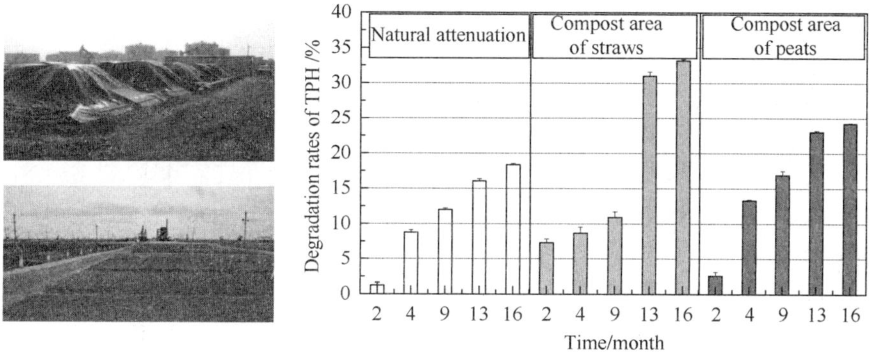

Fig. 21 Remediation of oil-contaminated soils in Shengli Oil Field

6.3 Steel production site

An investigation and risk assessment of the Donghua special steel production site was conducted in 2015. The risk of hydrocarbons, asbestos (As), and hexavalent chromium Cr (VI) contamination in soil was calculated; the amounts exceeded the acceptable risk level (10^{-5}) for human health (Fig. 22). The contaminated soils were treated with incineration, solidification/ stabilization, and bio-degradation technologies.

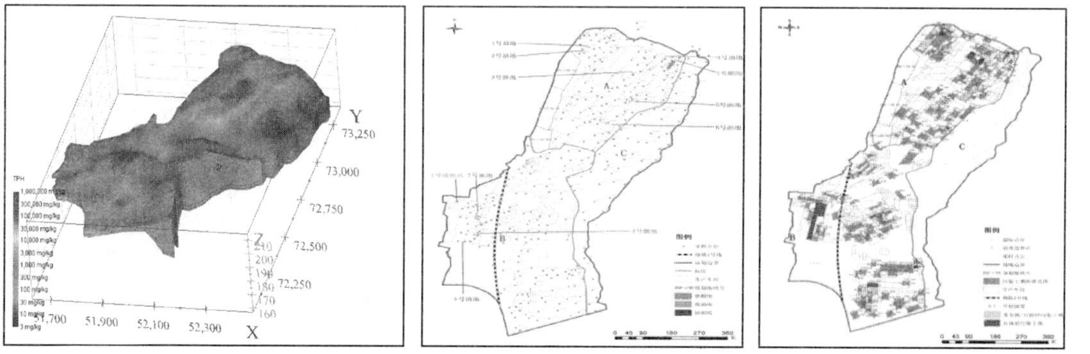

Fig. 22 Investigation and risk assessment of Donghua steel production site

6.4 Medicine production site

An abandoned medicine production site was investigated. A site risk assessment for soil and groundwater was conducted and a remediation plan was proposed. Target values for the remediation of soils contaminated with benzene, chlorobenzene, and petroleum hydrocarbons were calculated, and pilot and bench-scale experiments were conducted to determine the feasibility of thermal desorption for the treatment of contaminated soils. In addition, the monitoring of natural attenuation was proposed for the management of contaminated groundwater (Fig. 23).

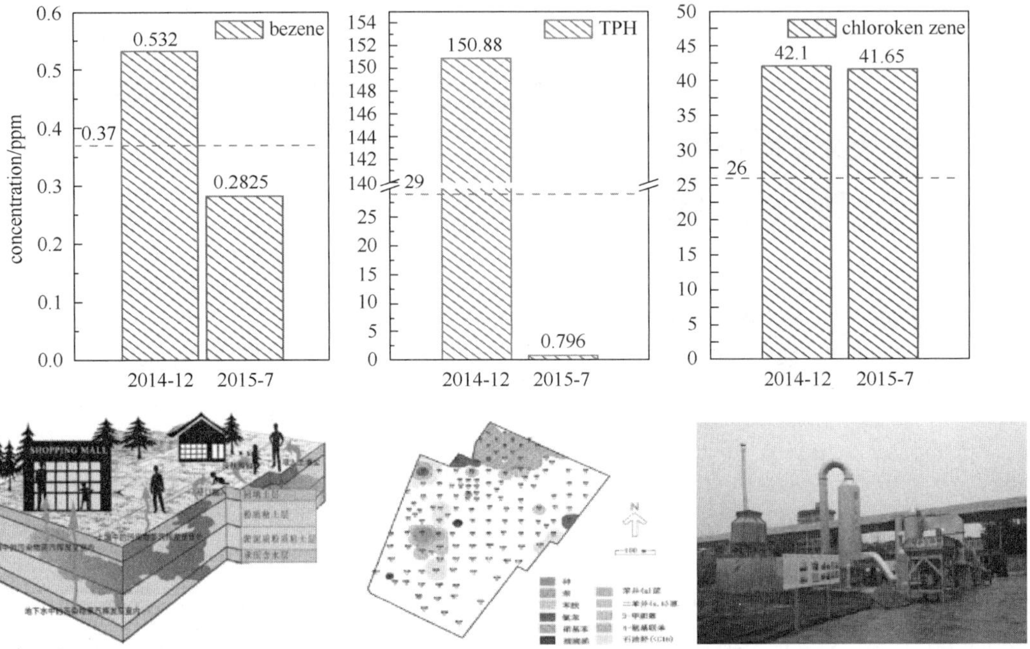

Fig. 23 Risk assessment and remediation plan for an abandoned medicine production site

6.5 Pulp and paper production site

A pulp and paper production site in northeast China was investigated. The soil was contaminated with benzo[a]pyrene. The risk of soil benzo[a]pyrene exceeded the proposed acceptable risk levels. The groundwater in the site was heavily contaminated by wastewater from the pulp and paper production factory. A pump-and-treat technique was used for the treatment of contaminated groundwater (Fig. 24).

6.6 Fertilizer production site

A fertilizer production site in Shijiazhuang, Hebei province was investigated (Fig. 25). The soils were heavily contaminated with ammonia, and strong odors were detected at the site. Groundwater was contaminated by high levels of ammonia. An innovative investigation and risk assessment of the ammonia contaminated site was proposed.

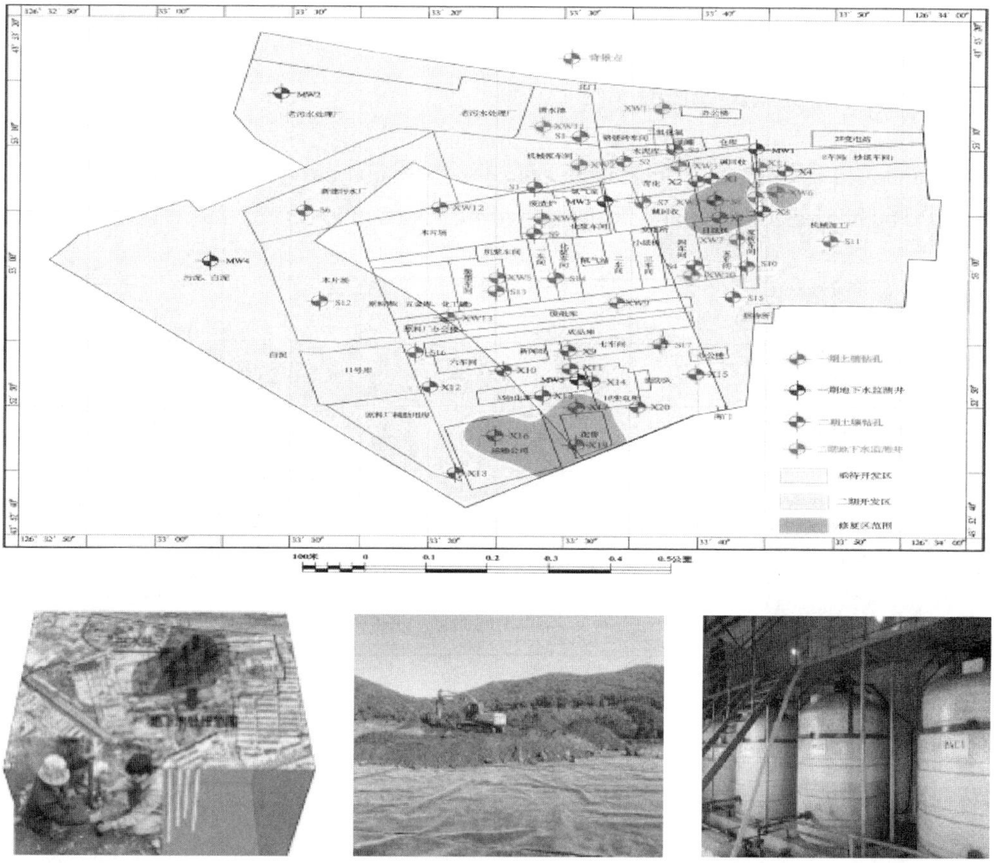

Fig. 24 Investigation, risk assessment, remediation plan, and acceptance of assessment completion for a pulp and paper production site

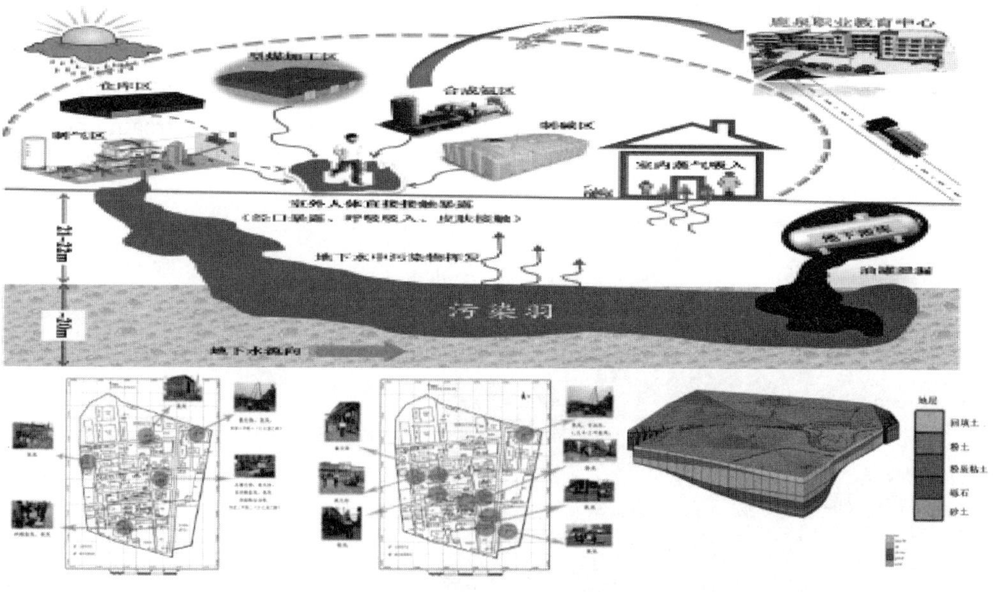

Fig. 25 Preliminary investigation and risk assessment of a fertilizer production site

6.7 Abandoned coking plant site

An abandoned coking plant site was investigated in Shanxi province. Soils were contaminated with As, PAHs, benzene, and total petroleum hydrocarbons (TPH) (Fig. 26).

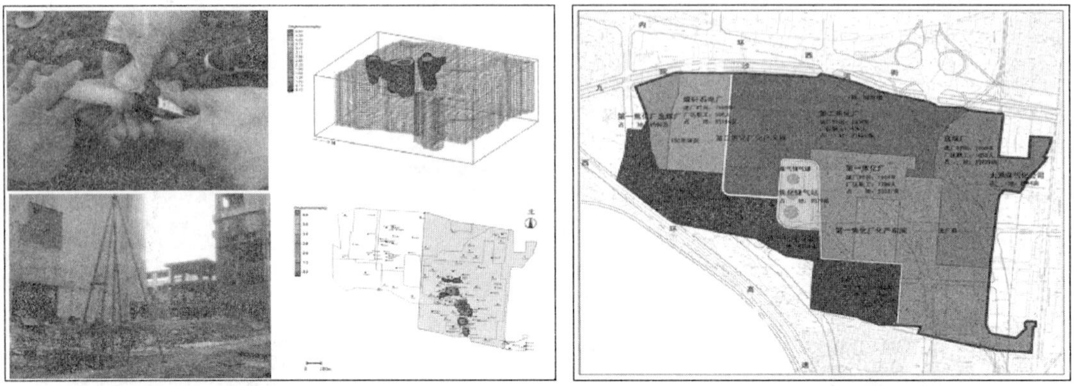

(a) Site investigation　　(b) Different land development

Fig. 26　Investigation and risk assessment of an abandoned coking plant site

6.8 Chloralkali production site

A complete process for site investigation, risk assessment, remediation, and validation was performed at a former chloralkali production site. This is the largest contaminated site that has been remediated in China. Soils at the site were contaminated with heavy metals and VOCs. Groundwater was contaminated with chlorinated hydrocarbons (Fig. 27). An ex-situ soil barrier and landfill technique was used for the disposal of metal-contaminated soils, while ex-situ or on-site ex-situ desorption at room temperature was used for the treatment of VOC-contaminated soils.

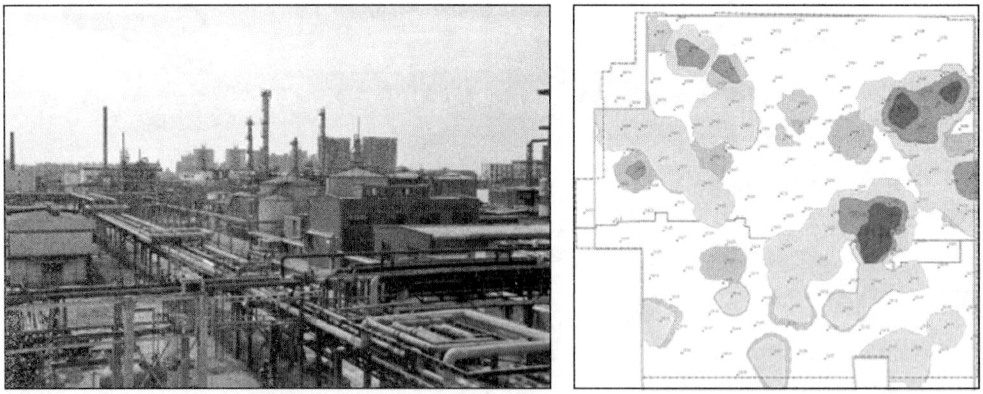

Fig. 27 Investigation, risk assessment, and remediation of a chloralkali production site

6.9 Chromate Plants

Seven heavy metal-contaminated sites in Qinghai Province (mainly chromate plants and the residue piling yards) were investigated (Fig. 28). A remediation plan was developed based on risk assessment results.

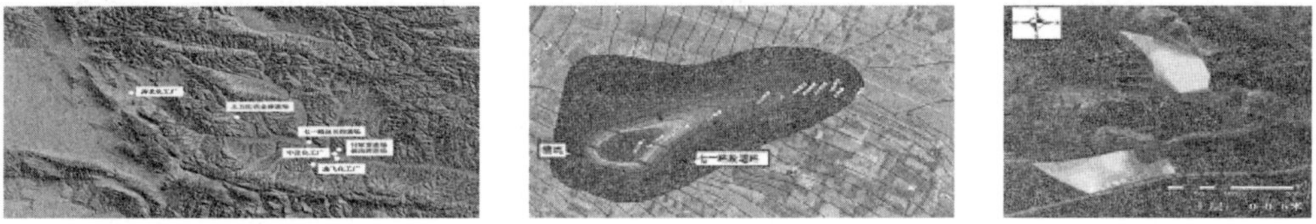

Fig. 28 Hexavalent chromium (Cr (VI)) concentration in soil and groundwater

6.10 Pesticide Plants

A risk assessment of POP-contaminated sites belonging to Shandong Dacheng Pesticide Co., Ltd. was conducted. DDT and Dicofol were investigated at the production sites, and the vertical migration and formation of an accumulation layer were well characterized (Fig. 29).

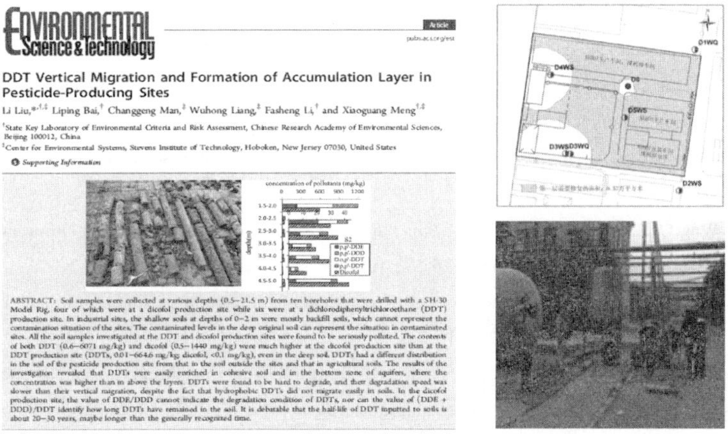

Fig. 29 Risk assessment of POP-contaminated sites belonging to Shandong Dacheng Pesticide Co., Ltd.

References

Cong X, Xue N, Liang G, et al. 2008. Residual characteristics of pollutants in topsoil of a former organocblorine pesticide mann-facturing enterprise. Journal of Agro-environment Science, 27 (3): 850-854.

Xia F Y, Cao Y Z, Li Z Y, et al. 2009. Derivation of soil pollutant standard for benzo[a]pyrene using RBCA and CLEA models. Research of Environmental Sciences, 22 (12): 1445-1451.

Han C, Zhang H, Gu Q, et al. 2013. Toluene sorption behavior on soil organic matter and its composition using three typical soils in China. Environmental Earth Sciences, 68 (3): 741-747.

Shi L Y, Cao Y Z, Zhang J L, et al. 2009. Comparison of application of RBCA and CLEA model for health risk assessment of a heavy metal contaminated site. Research of Environmental Sciences, 22 (2): 241-247.

Liu G, Bi R, Wang S, et al. 2013. The use of spatial autocorrelation analysis to identify PAHs pollution hotspots at an industrially contaminated site. Environmental Monitoring and Assessment, 185 (11): 9549-9558.

Liu G, Niu J, Zhang C, et al. 2015. Accuracy and uncertainty analysis of soil Bbf spatial distribution estimation at a coking plant-contaminated site based on normalization geostatistical technologies. Environmental Science and Pollution Research, 22 (24): 20121-20130.

Liu G, Niu J, Zhang C, et al. 2016. Characterization and assessment of contaminated soil and groundwater at an organic chemical plant site in Chongqing, Southwest China. Environmental Geochemistry and Health, 38 (2): 607-618.

Ma F, Peng C, Hou D, et al. 2015. Citric acid facilitated thermal treatment: An innovative method for the remediation of mercury contaminated soil. Journal of Hazardous Materials, 300: 546-552.

Ma J, Hou H, Zhao L, et al. 2016. China in action to protect soil environment. Science, http: //science. sciencemag. org/content/344 /6185/691.1.e-letters.

Ma Y, Du X, Shi Y, et al. 2015. Low-concentration tailing and subsequent quicklime-enhanced remediation of volatile chlorinated hydrocarbon-contaminated soils by mechanical soil aeration. Chemosphere, 121: 117-123.

Huang Q, Li F S, Gui M Y, et al. 2011. Accumulation and distribution of PAHs in organo-mineral aggregates of oil contaminated site//Computer Distributed Control and Intelligent Environmental Monitoring (CDCIEM), 2011 International Conference on. IEEE, 21-24.

Shi Y, Du X, Li H, et al. 2012. Effects of soil temperature and agitation on the removal of 1, 2-dichloroethane from contaminated soil. Science of the Total Environment, 423: 185-189.

Wang S, Wang X, Zhang C, et al. 2016. Bioremediation of oil sludge contaminated soil by landfarming with added cotton stalks. International Biodeterioration & Biodegradation, 106: 150-156.

Wang X, Wang Q, Wang S, et al. 2012. Effect of biostimulation on community level physiological profiles of microorganisms in field-scale biopiles composed of aged oil sludge. Bioresource Technology, 111: 308-315.

Xie Y F, Du Ping, Chen T B, et al. 2015. Estimating the area of heavy metal contaminated soil using geostatistical conditional simulation. Environmental Pollution and Control, 37 (1): 1-6. (in Chinese)

Yang B, Han B, Xue N, et al. 2015. Air-soil exchange of organochlorine pesticides in a sealed chamber. Journal of Environmental Sciences, 27: 241-250.

Zhang D D, Cao Y Y, Wng Q H, et al. 2012. Effects of Soil Physical-Chemical Properties on Risk Uncertainty in a Contaminated Site. Research of Environmental Sciences, 25 (5): 526-532. (in Chinese)

Zhang J, Dai J, Du X, et al. 2012. Distribution and sources of petroleum-hydrocarbon in soil profiles of the Hunpu wastewater-irrigated area, China's northeast. Geoderma, 173: 215-223.

Zhang X Y, Li F S, Xu D P, et al. 2012. Removal of POPs pesticides from soil by thermal desorption and its effect on physicochemical properties of the soil. Chinese Journal of Environmental Engineering, 6 (4): 1381-1386. (in Chineses)

Zhang Y, Zhu S, Xiao R, et al. 2008. Vertical transport of polycyclic aromatic hydrocarbons in different particle-size fractions of sandy soils. Environmental Geology, 53 (6): 1165-1172.

Zhang Z, Guo G, Teng Y, et al. 2015. Screening and assessment of solidification/stabilization amendments suitable for soils of lead-acid battery contaminated site. Journal of Hazardous Materials, 288: 140-146.

Zhou M, Liao B, Shu W, et al. 2015. Pollution assessment and potential sources of heavy metals in agricultural soils around four Pb/Zn mines of Shaoguan City, China. Soil and Sediment Contamination: An International Journal, 24 (1): 76-89.

Environmental Damage Assessment Methods for Soil and Groundwater Contamination

Fang YU[1], Dan ZHAO[1, 2], Ji QI[1], Jing YUAN[1], Yanshen ZHANG[1], Zhihong ZHANG[1], Hui XIE[1]

[1]Center for Environmental Risk and Damage Assessment, Chinese Academy for Environmental Planning

[2]Institute of Geographic Sciences and Natural Resources Research, Chinese Academy of Sciences

Abstract

Recently, soil and groundwater contamination incidents occur more frequently than before. Emerging environmental incidents call for litigations action on soil and groundwater damage investigation, which pushes forward establishing systematic methods to assess the resulted damage. Our study examined the differences between environmental damage assessment and traditional site investigation and risk assessment, and focused on the methods for some key steps in damage assessment, including general procedure, baseline determination, causation judgment, restoration plan design and loss assessment. Baseline was determined according to background levels, conditions in reference site, criteria, and model outputs. The causation judgment process can be divided into four steps: homology analysis, carrier and medium recognition, migration direction identification, integrity and continuity analysis. Loss of soil and groundwater damage consists of clean-up cost, cost of primary restoration, cost of compensatory restoration, and cost of complementary restoration. Since certain phases and methods in soil and groundwater damage assessment overlaps that of traditional site investigation and risk assessment, it is appropriate to incorporate the latter into the former. It's also critical to develop separate guidelines and manuals for baseline determination, causation judgment and loss assessment, to ensure the incidents investigation and damage assessment is scientific.

Keywords: soil, groundwater, environmental damage assessment, methods

In recent years, soil and groundwater were found out being severely polluted during the site investigation for industrial resettlement in many Chinese cities (Xie and Li, 2010; Luo, 2011; Liao et al., 2011). The nonstandard management of mining activities severely threatened the eco-environment and the adjacent residents (Liu et al., 2009; Yang et al., 2011; Xu et al., 2014). In addition, environmental incidents polluting soil and groundwater occur frequently, including discharge and leakage of dregs and wastewater (Zhao and Li, 2013; Fan and Huang, 2014).

Most of the pollution incidents were solved by the financial support from government, but this is not sustainable. Hence, it is necessary to better implement the "polluter pays" principle, and solve the environmental problems through responsibility investigation and claim. The Central Committee of the Communist Party of China and the State Council paid great attention to the compensation for damage to environment, and the Report to the Eighteenth National Congress of the Communist Party of China required improving the systems of responsibility investigation for ecological environment protection and compensation for environmental damage. In 2015, the *Environmental Protection Law* was revised, requiring that the assessment of environmental damage must be carried out after any unexpected environmental incident occurs, and the assessment reports should be disclosed to the public. In March 2015, the Political Bureau of the CPC Central Committee held a meeting to review and adopt the Opinions on Speeding up the Construction of Ecological Civilization, taking "the compensation system for environmental damage" as a significant system of ecological civilization, and clearly specifying the liabilities of compensation for eco-environmental damage including environmental pollution and ecological destruction. In the *Action Plan for Prevention and Control of Soil Pollution*, it underscored that any institute or individual causing soil pollution should bear the responsibility for restoration according to the "polluter pays" principle, and to perform the cleanup and restoration in an orderly manner. The assessment of environmental damage came into existence to satisfy the needs and provide an important technical support for eco-environmental damage compensation (Lu et al., 2011; Cao and Qi, 2014)

Environmental damage assessment is a process that an assessment agency follows the specific procedure and method, and utilizes scientific and professional knowledge to survey the scope and degree of environmental damage caused by any conducts leading to environmental pollution and ecological damage, so as to recognize the causation of such conducts with environmental damage, determine the measures to restore the environment to the baseline condition and compensate for the interim losses and finally quantify the loss of environmental damage (CAEP, 2014). Starting from the 1970s, the United States had developed laws to investigate environmental incidents. In the subsequent 20 years, it had established an integral system comprising a series of rules, methods and standards for assessment and compensation of environmental damage. Ever since the 1990s, the European Union started assessing environmental damage caused by pollution. In China, the State Oceanic Administration and the Ministry of Agriculture formulated the measures for evaluation of losses related to offshore oil spillage and fishery pollution accidents respectively. It was not until 2011 did the Ministry of Environmental Protection released *Opinions on Assessment of Environmental Pollution Damage* and started studying the rules, methods and standards for assessment of environmental damage systematically (Yu et al., 2013). Nowadays, soil and groundwater damage is a common type of environmental damage at sites where discharge of wastes and wastewater happened, industrial sites, mining areas and large-scale agricultural land, etc.

The assessment of soil and groundwater damage is not only connected intrinsically with, but also clearly distinguished from traditional site investigation and risk assessment. This paper focuses on their differences, including general procedure, determination of baseline, judgment of causation and damage assessment, etc.

1. Procedure of soil and groundwater damage assessment

The United States established two assessment procedures for natural resources damage, namely, DOI (Department of the Interior) rule and NOAA (National Oceanic and Atmospheric Administration) rule. The assessment procedure for natural resources damage, according to DOI rule, consists of four stages, i.e. pre-assessment, assessment planning, assessment and post-assessment. According to NOAA rule, the assessment procedure for natural resource damage has three stages, i.e. assessment, restoration planning and restoration. The damage assessment procedure for environmental resources in the European Union contains five stages, i. e. initial evaluation, determining and quantifying damage (the debit), determining and quantifying the benefits of the remediation (the credit), scaling complementary and compensatory remediation, monitoring and reporting. The DOI rule formulated earlier than the NOAA rule, and stressed the use value of natural resources. The NOAA rule focused on calculating the compensation based on the actual expenses needed to restore the damaged natural resources to the baseline. The European Union further underscored the development of compensatory restoration for the interim losses before natural resources were restored to the baseline.

We learned from the United States and the European Union, and developed procedures for soil and groundwater damage assessment (see Fig. 1). It emphasized on the calculation of losses based on the actual expenses restoring the damaged soil and groundwater resources to the baseline, while considering the compensatory restoration for interim losses before restoring the soil and groundwater resources to the baseline. Hence, the assessment of soil and groundwater damage is conducted based on the survey on the degree and scope of damage to soil and groundwater, the comparison with baseline and the design of restoration plan to estimate the cost for soil and groundwater restoration, so as to assess the losses caused by soil and groundwater pollutions. It mainly consists of damage survey, damage determination, damage quantification and restoration plan design, etc. When restoration is impossible, it is necessary to calculate the value of environmental and ecological services in the region, so as to provide the basis for making a compensation for environmental damage.

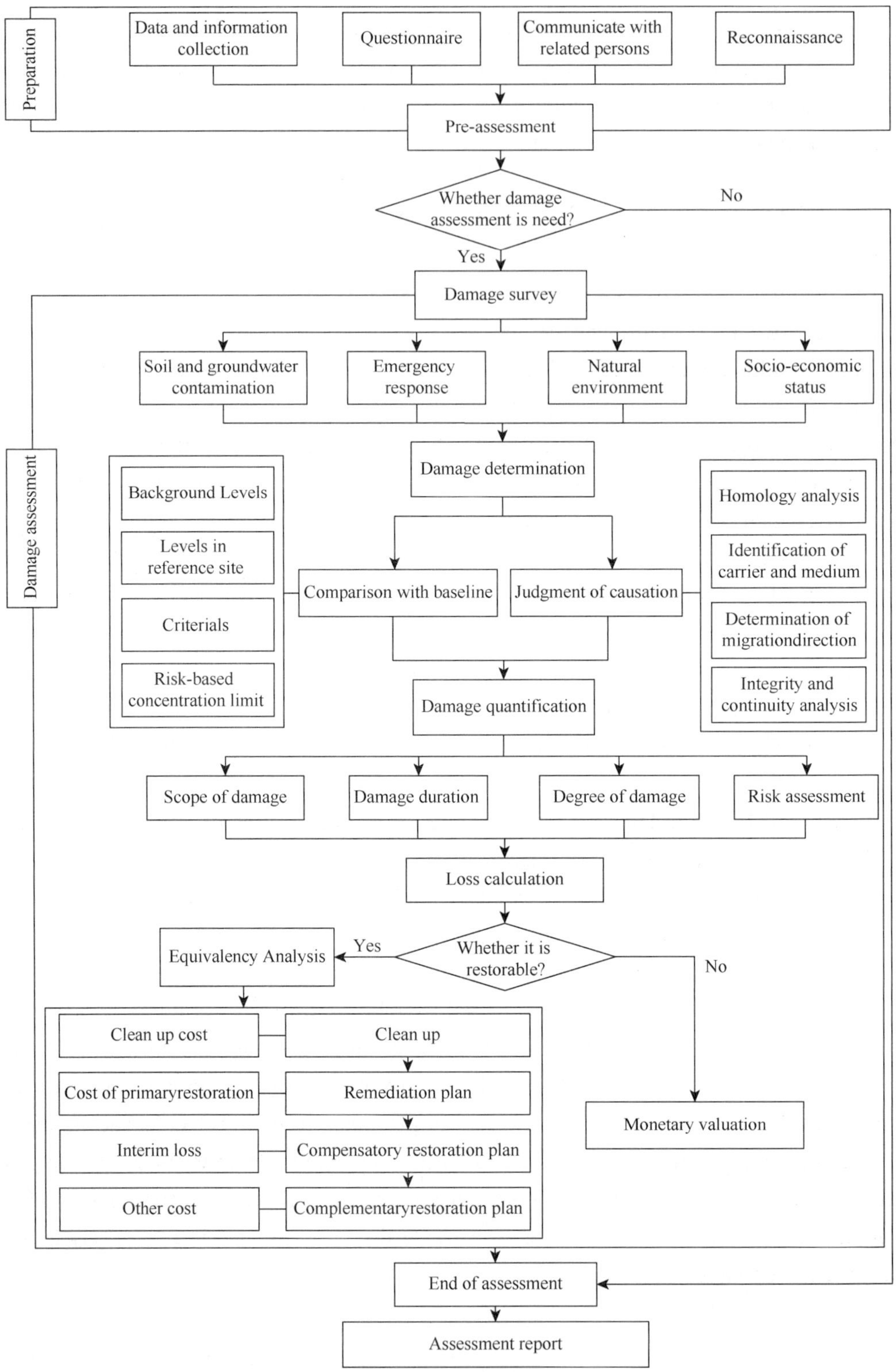

Fig. 1 Procedure for soil and groundwater damage assessment

2. Soil and groundwater baseline level

The degree of soil and groundwater damage is determined by comparing the current level to the baseline level. The baseline level refers to the condition (including the condition of human health, property and ecosystem service) before environmental incident occurred (CAEP, 2014). The baseline level can be determined by the following methodology: ①utilizing historic data; ②surveying reference site; ③referring to existing relevant criteria; ④deriving from risk assessment models outputs. We will discuss each method below.

2.1 Soil and groundwater background levels

The research on background levels of elements in soil, as one of the national key scientific and technologic breakthrough topics in the "Seventh Five-Year Plan" period, identified background levels of more than 60 elements for the main types of soil, and explored regional differentiation and influencing factors. In addition, some scientific research programs were arranged to survey and study the background levels of soil and groundwater for specific indicators in certain regions. The results reveal that China has carried out lots of surveys and studies on the background levels of soil elements in most provinces, among which most surveys focus on heavy metals. The background levels of groundwater are rarely studied, and surveys focus on regular indicators and heavy metals. Hence, in case of environmental damage caused by heavy metals, the background levels in the corresponding region can be used to determine the baseline level. While, due to the limited monitoring indicators and low sampling density, the existing soil and groundwater background level can't fully support the baseline determination.

2.2 Soil and groundwater conditions in reference site

If, by any means, the baseline level are not available, a reference site deem as "before" pollution site should be identified to which data from "after" pollution site are compared. In other words, the data in reference site are taken as the baseline level. The reference site is comparable to the assessed region in terms of demographic characteristics, ecosystem service, etc.

2.3 Soil and groundwater quality standard

Commonly used soil and groundwater quality standard in China are presented in Table 1. They are suggested to be used as baseline value when background data and reference site are not available.

Table 1 Soil and groundwater quality standard

	Medium	Standard	Type
1	Soil	*Environmental Quality Standard for Soils* (GB15618—1995)	National standard
2		*Standard of Soil Quality Assessment for Exhibition Sites* (*Trial*) (HJ/T 350—2007)	Industry standard
3		*Screening Levels for Soil Environmental Risk Assessment of Sites* (DB11T 811—2011)	Local standard
4		*Screening Levels for Soil Environmental Health Risk Assessment of Sites in Shanghai* (*Trial*)	Local standard
5		*Environmental Quality Standard for Soils of Farmland* (*The Third Draft for Opinion*)	To be released
6		*Suggested Screening Levels for Soil Pollution Risks of Soils in Construction Land* (*The Third Draft for Opinion*)	To be released
7	Groundwater	*Quality Standard for Groundwater* (GB/T 14848—93)	National standard
8		*Standards for Drinking Water Quality* (GB5749—2006)	National standard
9		*Groundwater Quality Standard* (DZ/T 0290—2015)	Industry standard

Environmental Quality Standard for Soils (GB15618—1995), the only national standard for soil environmental quality evaluation in China, cannot meet the needs for environmental damage assessment. The reasons are: ①the standard is lack of applicability. This standard is only applicable when evaluating soil quality for farmland, forest and nature reserves, while it is not appropriate for industrial, commercial, residential soil; ②the standard is lack of indicators to fully assess chemical pollution. It only covered eight heavy metals element, that is cadmium, mercury, arsenic, copper, lead, chromium, zinc and nickel, and two pesticide hexachlorobenzene (BHC) and dichlorodiphenyl trichloroethane (DDT) , and it does not include polychlorinated biphenyls (PCBs), polycyclic aromatic hydrocarbons (PAHs), chlorinated hydrocarbons, petroleum, which are commonly seen in oil spill cases and are extremely harmful to human health; ③the range for concentration of certain metal is not feasible, either too stringent or too loose. The first-tier criteria cannot reflect the regional differences objectively since they are uniform provisions nationwide based on the data from the soil background survey in the "Seventh Five-Year Plan" period; and the second and third-tier criteria give too strict screening level for cadmium and too loose screening level for lead, screening level for arsenic are also unreasonable, so the results of soil quality evaluation and agricultural product quality evaluation are inconsistent in many cases. In addition, this standard takes the total amount as the indicator, which can't reflect the actual condition of soils with low content of available heavy metals. *Standard of Soil Quality Assessment for Exhibition Sites* (*Trial*) (HJ/T 350-2007) defines the concentration limit of heavy metals based on the background levels of soil in Shanghai, and screens the concentration limit of organics from the foreign quality standard for soil (Li, 2014), so its applicability is limited as well.

Quality Standard for Groundwater (GB/T 14848-93) covers very few organic pollutants, i.e. BHC and DDT, and does not include halogenated hydrocarbons, aromatic hydrocarbon, and petroleum, etc. In addition, the ranges for some indicators are not acceptable. For instance, the limits of iron, magnesium and total hardness are too strigent. In 2006, the revised *Standards for Drinking Water Quality* (GB5749—2006) added a large number of organic indicators, but it was only applicable to drinkable groundwater, which couldn't always be used as a baseline level.

In all, since *Screening Levels for Soil Environmental Risk Assessment of Sites* (DB11T 811-2011), *Screening Levels for Soil Environmental Health Risk Assessment of Sites in Shanghai* (*Trial)* are developed for soil evaluation in Beijing and Shanghai region, they are useful guidelines to determine the baseline level for these two regions. *Environmental Quality Standard for Soils of Farmland* and *Suggested Screening Levels for Soil Pollution Risks of Soils in Construction Land* are under revision. If completed, they will play an active role in determining the baseline level of soil for farmland and construction land. With regard to groundwater, *Groundwater Quality Standard* (DZ/T 0290-2015) released by the Ministry of Land and Resources in 2015 provides a complete set of indicators, and classifies the criteria for groundwater based on their functions, so it plays a significant role in determining the baseline level of groundwater.

2.4 Determination of baseline based on risk assessment

If the background level can't be determined by methods mentioned above, the outputs of risk assessment model can be used. Risk assessment includes human health risk, ecological risk and environmental risk. Currently, there is no standard method to assess ecological risk and environmental risk in China, but there is the standard for human health risk assessment, *Technical Guidelines for Risk Assessment of Contaminated Sites* (HJ 25.3-2014) and *Guideline for Health Risk Assessment of Groundwater Pollution* (*Trial),* etc., which can be used to determine the baseline level.

3. Judgment of causation for soil and groundwater damage

The judgment of causation is a unique phase of environmental damage assessment, which is different from conventional site investigation. It requires not only logical reasoning but also combination of techniques and methods dependent of environmental media. This section presents the basic principles and useful methods to recognize the causation for soil and groundwater damage.

3.1 Principles of causation judgment

Causation refers to the relationship of cause and effect between objects. The damage to soil and groundwater is through media such as water and air (Bo, 2014). The causal chain of soil and groundwater damage can be visualized as: pollutant①→discharge from contaminated source②→diffusion through medium③→contact with the receptor④→damage occur⑤. Only if each link of the causal chain is proved, causation can be determined (Tang and Zhang, 2012). Now, the mainstream theories of causation, such as, "Possibility Causation", "Epidemiology Causation" and "Indirect Disproof", advocate to recognizing the causation in the principle of presumption. In other words, the prosecutor is not required to provide the complete scientific demonstration for causation, but a certain probability should be satisfied (Qiao, 2012).

In their guidelines for environmental damage assessment, the United States Department of the Interior (USDOI, 1994) and Bureau of Land Management (USDOI, 2008) pointed out that the judgment of causation must determine the migration pathway between the oil or hazardous wastes and the damaged resources. The existence of migration pathway is proved only if it is verified that the resources (water, sediments, soils or plants) under the possible migration pathway contain oils or hazardous wastes of enough concentration; or a model is used to prove that the pollution on possible migration pathway is consistent with oils or hazardous wastes. In their guideline for damage assessment of natural resources (Huguenin et al., 1990), the United States National Oceanic and Atmospheric Administration points out that if the damage is caused by direct exposure to oil leakage, it is necessary to prove that: ①there is a pathway of migration between leakage and concerned natural resource; ②resource is exposed to leakage; ③exposure exerts adverse effect on resources. In the guideline for resource equivalent analysis and environmental damage assessment and quantification (Lipton et al., 2008), European Union points out that the research data, logical analysis, study of specific site, simulation, and deductive reasoning in the references can be used to prove the causation. The evidences of causation are as follows: ①leakage of hazardous chemicals occurs (witnesses, sampling data, photos); ②leaked chemicals migrated to the affected area through air or water (sampling data, witnesses, photos, simulation); ③receptor is exposed to leaked chemicals (witnesses, sampling data, logical assumptions); ④the concentration of chemicals is high enough to cause harmful effect based on the known toxicities of the chemicals (sampling data, simulation); ⑤effect has occurred or may occur (reference data, professional opinions, exposure proved at site or laboratory, simulation). The evaluator is not required to determine the accurate effect of a single incident on the natural resources and services involved, but must verify the causation is reasonable and at least makes a contribution to the effect.

In China, the *Recommended Methods for Environmental Damage Assessment* (2nd Edition) (CAEP, 2014) released in 2014 points out that the causation between environmental pollution and environmental damage are determined in two phases. One is judgment of causation between exposure and environmental damage. The other is establishment and verification of migration pathway from the source to the receptor. The judgment of causation between exposure and environmental damage should comply with the common principles as follows: there is a temperal order between exposure and environmental damage, and the relationship between exposure and environmental damage should be reasonable, consistent and specific. Based on the discharge at the source of pollution, the quality of regional environment and other basic data, the assumptions for source of pollution are brought forward and the exposure pathway is established and verified. In the *Interpretation of the Supreme People's Court of Several Issues on the Application of Law in the Trial of Disputes over Liability for Environmental Torts* promulgated on June 1, 2015, Article 6 points out that, when the aggrieved party requests the compensation, it shall provide the evidences for the following facts: ①the polluter discharges pollutants; ②some damage is caused to the aggrieved party; ③there is correlation between the pollutants discharged by polluter or their secondary pollutants and the damage.

3.2 Methods for causation judgment

There are four phases for judgment process of causation, i.e. homology analysis, carrier and medium identification, determination of migration direction, integrity and continuity analysis.

1) Homology analysis

The first step in the judgment of causation is to perform the homology analysis on the pollution of damaged environment (soil and groundwater) and the pollution at potential contamination source. Homology analysis employs such methods as chemical fingerprinting, multivariate statistical analysis, compound-specific isotope analysis and geographical information technology, etc. Chemical fingerprinting methodology, which was originally applied for analysis of oil spills sources, identifies “oil fingerprints” including N-alkanes, PAHs, isoprenoid hydrocarbon, steranes and terpene, etc. in spilled oils and suspected oil samples (Christensen and Tomasi, 2007; Wang et al., 2013). Afterwards, the concept of “fingerprint” is expanded into other fields, including fingerprint of perfluorinated compounds (PFCs) (Benskin et al., 2010) and fingerprint of wastewater discharged illegally (Wang et al., 2014), etc. Statistical analysis can be employed to study the correlation and compositional features of elements, and identify the source of pollutants based on the indicators of environmental pollution and the characteristics of pollutants in some known sources (Huguenin et al., 1999; Zhang et al., 2012). It mainly involves correlation analysis (Zhou et al., 2015), principal component analysis (PCA) (Davis et al., 2009; Luo et al., 2015), hierarchical cluster analysis (Lee et al., 2006), and enrichment factor analysis (Ye et al., 2011), etc. The physical, chemical, biological and other factors may cause the change of natural abundance of isotope, so that the isotope ratio in the substances from different sources may vary. The stable isotope ratio of element can be utilized to identify the source of element, and track its migration pathway and fate. For instance, the source of pollution and migration pathway can be determined based on the Pb isotope ratio (Cloquet et al., 2006; Luo et al., 2011; Sun et al., 2011), while Cd, Zn, Ni and Hg, etc. can be utilized to track the fractionation caused by human activities (Zhang et al., 2012; Cloquet et al., 2006). In addition, the source of organics can be judged based on the ratio of ^{12}C and ^{13}C (Mansuy et al., 1997; Mazeas and Budzinski, 2002), Cl isotope and H isotope (Sturchio et al., 1998; Pond et al., 2002), etc. Geographical information system (GIS) mainly utilizes computer to generate isopleth maps of contaminant distribution though interpolation, overlaps it with the map containing industries, road and farmland, etc, and analyzes the relationship between abnormal spatial distribution of pollutants and potential source of pollution (Zushi and Masunaga, 2011; Zhang et al., 2012).

2) Identification of carrier and medium

After proving the homology between pollutants in damaged environment (soil and groundwater) and pollutants at pollution source, it is necessary to analyze the migration pathway between source of pollution and damaged environment. The spatial migration of pollutants requires two elements, i.e. medium and carrier. Hence, the analysis on pathway of migration requires medium identification and carrier identification. The identification of carrier and medium can employ the conceptual model of site (CMS), which is established based on geological survey, hydro-geological survey and pollution survey, etc. It centers on "source of pollution—migration pathway—receptor", and covers not only the pollution , but also geologic, hydro-geologic, engineering geologic, environmental background, current status and use planning of site, etc. In the underground environment, the carriers for migration of pollutants normally include infiltrated surface water, groundwater, non-aqueous phase liquids (NAPLs), etc.; and the medium of migration include soil and groundwater, etc. After the infiltrated surface water and NAPLs carrying pollutants enter the soil, the soil becomes a medium. Since pollutants are brought further to the vadose zone and the saturated zone, groundwater becomes a carrier and medium.

3) Determination of migration direction

The third step in the judgment of causation is to determine whether the direction of carrier movement is consistent with the direction of contaminant concentration gradient. Only when the direction of carrier movement is consistent with the direction of contaminant concentration gradient, it is believed that the migration of pollutants is caused by the movement of the carrier, and the migration pathway is reasonable. The movement direction of carrier includes the direction of surface water infiltration, the convection and dispersion direction of NAPLs, and the flow direction of groundwater, etc. The infiltration direction of surface water and the flow direction of groundwater can be determined through the geological and hydro-geological survey, while the flow direction of groundwater can be also measured through some methods including aquifer pumping test, isotopic tracer, charge method, high-density electrical method, etc. The migration direction of NAPLs can be determined through test or model simulation, etc (Agaoglu et al., 2012). The direction of contaminant concentration gradient is determined based on the survey of contamination, and the geological information system software (ArcGIS), environmental visualization system (EVS), Surfer, and Voxler, etc. are utilized to spatial interpolation for the concentration of pollutants in medium or carrier, so as to determine the direction of contaminant concentration gradient.

4) Integrity and continuity analysis

The fourth step in the judgment of causation is to analyze the temporal and spatial continuity of migration pathway. The methods for integrity and continuity analysis include simulation for spatial distribution of pollutants and conversion of pollutants with time. The simulation methods include physical model test, and numerical model simulation, etc. Lots of one-dimension, two-dimension and three-dimension physical models have been used to simulate the migration and conversion of ammonia nitrogen, nitrite nitrogen, nitrate nitrogen, heavy metals or organic pollutants in vadose zone and the saturated zone (Maeda and Bergstrom, 2000; Jacques et al., 2008; Zhang et al., 2010). Based on the current status of pollution, geological conditions and other parameters, numerical simulation model or software is utilized to simulate when the dumped/leaked/discharged pollutant will reach the receptor, and the model is calibrated or verified with the data acquired from laboratory experiments and/or field tests. GMS, MT3D, Visual Modflow and Feflow, etc. are the common numerical softwares for groundwater modeling.

4. Assessment of soil and groundwater damage

The loss of environmental damage is valuated via monetary methods or non-monetary methods. Monetary evaluation include revealed willingness of payment, virtual willingness of payment and stated willingness of payment, etc., while non-monetary methods include resource equalization, service equalization and equivalency analysis, etc. As it is difficult to monetize the value of soil and groundwater resources, it is often to take all the direct and indirect restoration costs and the expenses for compensatory restoration in the process as the loss of damage.

The restoration for damaged soil and groundwater is different from conventional site remediation in several ways:

(1) Targets: The site remediation targets at the acceptable risk-based level, while the restoration of environment targets at the baseline level before environmental damage occurs. When it is unnecessary or infeasible to restore to the baseline level, the environmental quality standard or concentration limit based on risk assessment can be also taken as the target of restoration. However, it is still necessary to calculate the difference between the standard or risk-based concentration limit and the baseline level, so as to provide the basis for compensation;

(2) Tasks: Restoration usually includes environmental remediation and ecological restoration, while site remediation often only refers to environmental remediation;

(3) Procedures: Restoration takes four phases, that is clean-up, primary restoration, compensatory restoration and complementary restoration, while site remediation is referring to clean-up and part of primary restoration;

(4) Methods: Restoration of environment normally involves two ways, i.e. manual restoration and natural restoration, while site remediation often relies on manual activities;

(5) Decision-making: Decision-making on site remediation mainly takes into account technical feasibility, cost, time of restoration proposed by the owner and other requirements. In restoration, the possible interim damage and public acceptance should be taken into account as well, so attention should be paid to the equivalents of cost and effectiveness and the acceptance by the public and potential compensation receivers apart from the aforesaid factors.

Therefore, loss of soil and groundwater damage may be broken down into:

(1) Clean-up cost, i.e. the expenses for the removal of released hazardous substances from the environment, such actions as may be necessary to monitor, assess, and evaluate the threat of release of hazardous substance, the disposal of removed material, or the taking of other actions to prevent, minimize, or mitigate damage to the public health or welfare or environment.

(2) Cost of primary restoration, i.e. the expenses needed to restore the soil and groundwater to baseline level. The primary restoration is sometimes calculated as follows: ①estimate the expenses based on the breakdown of remediation work; ②utilize the models for estimation of remediation cost; ③calculate remediation cost based on the risk-cost function; ④employ remediation cost database for estimation, etc.

(3) Interim Loss, i.e. the compensation for losses of soil and groundwater resources and their services during the restoration period. Equivalency analysis is employed to determine the interim losses based on compensatory restoration (see Fig. 2). The procedure of calculation is as follows: ①identify the damaged soil and groundwater as well as their main services; ② determine the method for compensatory restoration and ensure that the services provided by compensatory resource and damaged resource are comparable and replaceable; ③select a suitable measuring unit as the representation of ecological service level; ④calculate the "debit",

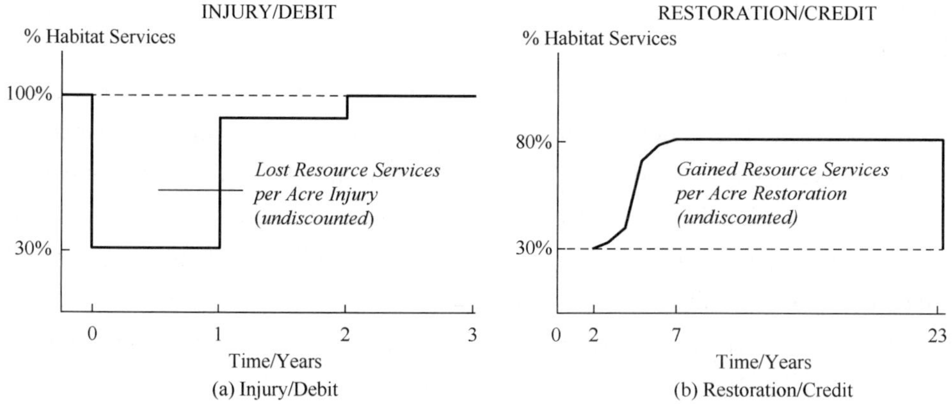

Fig. 2 Principle of equivalency analysis (Hampton and Zafonte, 2002)

i.e. quantification of damaged resources and services; ⑤calculate the "credit", i.e. determination of resources or services provided by the compensatory restoration measure per unit; ⑥establish the equation of "debit" and "credit", and determine the scale of compensatory restoration measure; ⑦calculate the cost of compensatory restoration measure.

(4) Other cost, including cost of complementary restoration, i.e. the cost of any remedial measure taken in relation to soil and groundwater and/or services to compensate for the fact that primary remediation does not result in completely restoring the damaged soil and groundwater and/or services. Complementary restoration should judge whether basic restoration or compensatory restoration achieves the anticipated effect based on post-restoration assessment. If the anticipated effect is not achieved, it is necessary to further determine the complementary restoration plan, and then calculate the cost of restoration.

5. Conclusions

Soil and groundwater damage assessment overlaps the conventional site investigation and risk assessment in various aspects, so it is necessary to better connect them at the beginning of system establishment or criterial formulation, so as to save resources and costs and shorten the assessment period. The specific link of soil and groundwater damage assessment, i.e. determination of baseline, judgment of causation and damage assessment, etc. should be further studied, and corresponding technical standards should be developed, so as to ensure the normative and scientific assessment of soil and groundwater damage.

Acknowledgement

The authors are grateful for the financial support from Scientific and Technological Project of Environmental Benefit *Study of Environmental Damage Assessment Techniques for Acute Water Pollution Accidents* (Project No. 201309060).

References

Agaoglu B, Scheytt T, Copty N K. 2012. Laboratory-scale experiments and numerical modeling of cosolvent flushing of multi-component NAPLs in saturated porous media. Journal of Contaminant Hydrology, 140: 80-94.

Benskin J P, De Silva A O, Martin J W. 2010. Isomer profiling of perfluorinated substances as a tool for source tracking: a review of early findings and future applications//Reviews of Environmental Contamination and Toxicology Volume 208. Springer New York: 111-160.

Bo X B. 2014. Invert and presumption: reflection on the methods of causation judgment in our environmental infringement. Journal of China University of Geosciences (Social Sciences Edition). 14 (6): 68-81.

Cao D, Qi J. 2014. Problems and countermeasures of environmental damage identification and assessment. Environmental Protection. 17: 45-47.

Chinese Academy for Environmental Planning. 2014. Recommended method for environmental damage assessment (the second version).

Christensen J H, Tomasi G. 2007. Practical aspects of chemometrics for oil spill fingerprinting. Journal of Chromatography A, 1169 (1): 1-22.

Cloquet C, Carignan J, Libourel G, et al. 2006. Tracing source pollution in soils using cadmium and lead isotopes. Environmental Science & Technology, 40 (8): 2525-2530.

Davis H T, Aelion C M, McDermott S, et al. 2009. Identifying natural and anthropogenic sources of metals in urban and rural soils using GIS-based data, PCA, and spatial interpolation. Environmental Pollution, 157 (8): 2378-2385.

Department of Interior. 1994. Natural resource damage assessments (43 CFR Part 11). Washington DC: Office of the Secretary of the Interior. 241-301.

Fan L, Huang X. 2014. Criminal responsibility of unauthorized dumping of large quantities of industrial waste water into the river. People's Judicature. 18: 4-5.

Hampton S, Zafonte M. 2002. Calculating compensatory restoration in natural resource damage assessments: recent experience in California//Proceedings of the California and the World Ocean. 2. Huguenin M T, Haury D H, Weiss J C, et al. 1996. Injury assessment: guidance document for natural resource damage assessment under the Oil Pollution Act of 1990. Maryland: National Oceanic and Atmospheric Administration.

Jacques D, Šimůnek J, Mallants D, et al. 2008. Modelling coupled water flow, solute transport and geochemical reactions affecting heavy metal migration in a podzol soil. Geoderma, 145 (3): 449-461.

Lee C S, Li X, Shi W, et al. 2006. Metal contamination in urban, suburban, and country park soils of Hong Kong: a study based on GIS and multivariate statistics. Science of the Total Environment, 356 (1): 45-61.

Liao X Y, Chong Z Y, Yan X L, et al. 2011. Urban industrial contaminated sites: a new issue in the field of environmental remediation in China. Environmental Science, 32 (3): 784-794.

Lipton J, Lejeune K, Calewaert J B, et al. 2008. REMEDE deliverable 13: toolkit for performing resource equivalency analysis to assess and scale environmental damage in the European Union. Belgium: European Union.

Liu B J, Wu F C, Deng Q J, et al. 2009. Pollution characteristics of antimony, arsenic and mercury in human hair at Xikuangshan antimony mining area and Guiyang City, China. Environmental Science, 30 (3): 907-912. Lu J, Zhang H Z, Yu F. 2011. Exploring the mechanism of environmental damage assessment, compensation and remediation. Environmental Protection. 24: 32-34.

Luo X S, Xue Y, Wang Y L, et al. 2015. Source identification and apportionment of heavy metals in urban soil profiles. Chemosphere, 127: 152-157.

Luo Y M. 2011. Contaminated site remediation in China: progresses, problems and prospects. The Administration and Tech. of Enviro. Monit, 23 (3): 1-6.

Maeda M, Bergström L F. 2000. Leaching patterns of heavy metals and nitrogen evaluated with a modified tanks-in-series model. Journal of Contaminant Hydrology, 43 (2): 165-185.

Mansuy L, Philp R P, Allen J. 1997. Source identification of oil spills based on the isotopic composition of individual components in weathered oil samples. Environmental Science & Technology, 31 (12): 3417-3425.

Mazeas L, Budzinski H. 2002. Molecular and stable carbon isotopic source identification of oil residues and oiled bird feathers sampled along the Atlantic coast of France after the Erika oil spill. Environmental Science & Technology, 36 (2): 130-137.

Pan Y Y, Song J, Luo Y M. 2010. Risk management and decision-making process for metallurgical contaminated sites based on human health risk assessment. The Administration and Technique of Environmental Monitoring. 22 (3): 55-61.

Pond K L, Huang Y, Wang Y, et al. 2002. Hydrogen isotopic composition of individual n-alkanes as an intrinsic tracer for bioremediation and source identification of petroleum contamination. Environmental Science & Technology, 36 (4): 724-728. Qiao S M. 2012. Theory and practice of causation presumption for heavy metal pollution: take guangxi cadmium pollution as an example. Research on Rule of Law, 10: 31-38.

Sturchio N C, Clausen J L, Heraty L J, et al. 1998. Chlorine isotope investigation of natural attenuation of trichloroethene in an aerobic aquifer. Environmental Science & Technology, 32 (20): 3037-3042.

Sun G X, Wang X J, Hu Q H. 2011. Using stable lead isotopes to trace heavy metal contamination sources in sediments of Xiangjiang and Lishui Rivers in China. Environmental Pollution, 159 (12): 3406-3410.

United States Department of the Interior. 2008. BLM natural resource damage assessment and restoration handbook. Washington DC: Bureau of Land Management.

Wang C, Chen B, Zhang B, et al. Fingerprint and weathering characteristics of crude oils after Dalian oil spill, China. Marine Pollution Bulletin, 2013, 71 (1): 64-68.

Wang Y, Li C Y, Mo H L, et al. 2014. Evidence analysis for identifying suspect sources of illegally discharged wastewater. Journal of Beijing University of Chemical Technology (Natural Science). 41 (1): 39-45.

XIE J, LI F S. 2012. Remediation and redevelopment of contaminated sites in China. Environmental Protection. 23 (Z1): 15-20.

Xu Y N, Zhang J H, Ke H L, et al. 2014. Human health risk under the condition of farmland soilheavy metals pollution in a gold mining area. Geological Bulletin of China. 33 (8): 1239-1252.

Yang G, Shen F, Zhong G, et al. 2011. Concentration and health risk of heavy metals in crops and soils in a zinc-lead mining area in southwest mountainous regions. Acta Scientiae Circumstantiae, 31 (9): 2014-2021.

Ye C, Li S, Zhang Y, et al. 2011. Assessing soil heavy metal pollution in the water-level-fluctuation zone of the Three Gorges Reservoir, China. Journal of Hazardous Materials, 191 (1): 366-372.

Yu F, Liu Q, Qi J. 2013. Learn from others to improve our environmental damage appraisal and compensation system. Environmental Economy. 11: 38-47.

Yu S, Li X. 2011. Distribution, availability, and sources of trace metals in different particle size fractions of urban soils in Hong Kong: implications for assessing the risk to human health. Environmental Pollution, 159 (5): 1317-1326.

Zhang W, Lin X, Su X. 2010. Transport and fate modeling of nitrobenzene in groundwater after the Songhua River pollution accident. Journal of Environmental Management, 91 (11): 2378-2384.

Zhang Y, Yao Q S, Zhu Y M. 2012. Review of source identification methodologies for heavy metals in solid waste. Chinese Science Bulletin. 57 (33): 3132-3138.

Zhao J T, Li Y K. 2013. Construction of environmental responsibility prevention mechanism in sight of ecological civilization perspective-take the case of land contamination caused by industrial waste waterdumpingas an example. Environmental Protection. 9: 50-52.

Zhong M S, Jiang L, Yao J J, et al. 2013. Case study on health risk assessment based on site-specific conceptual model. Environmental Science, 34 (2): 647-652.

Zhou P, Zhao Y, Zhao Z, et al. 2015. Source mapping and determining of soil contamination by heavy metals using statistical analysis, artificial neural network, and adaptive genetic algorithm. Journal of Environmental Chemical Engineering, 3 (4): 2569-2579.

Zushi Y, Masunaga S. 2011. GIS-based source identification and apportionment of diffuse water pollution: perfluorinated compound pollution in the Tokyo Bay basin. Chemosphere, 85 (8): 1340-1346.

Ten-Year Review and Prospect of Industrial Contaminated Site Remediation in China

Yanli GAO, Shupeng LI*, Yanwei WANG, Shaoguo KANG, Yunxiao WEI

BCEG Environmental Remediation Co., Ltd.

*Corresponding author

Abstract

The industry of contaminated site remediation has been rapidly developed in China in the past ten years. Based on the data of specific remediation projects from January 2007 to May 2016, this paper summarizes and reviews the progress of Chinese remediation industry according to the aspects of project regional distribution, driving force, interannual distribution, capital scale, construction period, engineering quantity, contaminant characterization and technology application overview. On the basis of the reviews, this article prospects the opportunities and challenges for the development of site remediation industry in the future.

Keywords: contaminated site, remediation, organic contaminant, heavy metal, development trend

1. Introduction

The industrialization of China started in the 1950s and rapidly developed thereafter. The industry accounted for more than 61.6% of the gross domestic product in the peak period (NBSC, 2015; Xu and Wang, 2005). According to the data showed in the *China Statistical Yearbook-2015,* the number of industrial enterprise above designated size reached 377 888 in the year of 2014 in China (NBSC, 2015). According to the *China City Statistical Yearbook-2014,* there are 160 795 industrial enterprises above designated size located in municipal district in 658 cities in 2013, accounted for 44.7% of the total (Bureau of urban social and economic research, 2014). In the process of industrialization, the environment has been inevitably polluted (Tang, 2009).

Since the beginning of the new century, the state has successively proposed and implemented policies including "Shifting from the Secondary to the Tertiary Industry","Move into the Industrial Park", and "The Industry Transfer". Almost all medium and large cities in China faced with a large numbers of issues of polluted enterprises closure and relocation. These closed and relocated industrial enterprises left a lot of contaminated sites, where soil, water, wall, equipment and waste pollution were involved (Liu et al., 2013; Luo, 2009a). The contaminated sites are usually located in urban areas, with high values for development and utilization. With more and more urban industrial lands changed into public and residential lands, potential groundwater and soil pollution problems are gradually exposed, affecting the quality of urban life and the safety of human living environment (Luo, 2009b). Many poisoning and polluted incidents are caused by soil and groundwater contamination, for example, the Beijing Songjiazhuang subway construction worker poisoning incident in 2004 (China. com, 2004); the Lanzhou benzene polluted tap water incident in 2014 (People. cn, 2014); and Changzhou foreign language school pollution incident in 2016 (CCTV. com, 2016).

Chinese government has always attached importance to the prevention and control of soil pollution. In early 1980s, the research on agricultural soil background value, domestic soil background value and soil environmental capacity has been initiated, and the *Soil Environmental Quality Standards* was published in 1995. After 2000, the problem of urban soil pollution began to be highlighted in China. However, some of the contaminated sites in cities were directly developed and utilized without risk assessment and remediation, due to the lack of relevant laws, regulations and regulatory policies. The health and environmental risks of contaminated sites obtained high degree of concern after the Beijing Songjiazhuang subway construction worker poisoning incident in 2004, which is considered to be a landmark event. In the same year, the National Bureau of Environmental Protection issued *The Notice on the Effective Prevention of Environmental Pollution in the Process of Enterprise Relocation.* As of June 2016, the proposal draft of the *Soil Pollution Control Act* has been submitted to the Environmental and Resources Protection Committee of the National People's Congress in the legal level. Fujian and Hubei Provinces also have enacted local soil pollution control acts. In contaminated site related policies, over 16 soil pollution control regulations and over 50 soil pollution control and remediation related standard specifications have been released locally and nationally. *The Plan of Soil Pollution Prevention Action* is also released at the end of May 2016.

In the implementation of contaminated site remediation projects, since the BCEG Environmental Remediation Co., Ltd. conducted the first domestic commercial site remediation project in 2007 (European and US enterprises may have individual remediation projects earlier than this period during their investment in China), there has been many accumulations in the research and development and engineering application of contaminated site remediation technologies in China. Independently researched and developed soil remediation technologies such as low temperature thermal desorption, thermal desorption, solidification and stabilization and soil washing, began to enter the engineering demonstration phase or implementation phase. Advanced foreign technologies and remediation materials also started to be introduced into China and applied in projects, such as in-situ thermal desorption, soil vapor extraction, in-situ chemical oxidation, in-situ groundwater bioremediation and remediation reagents. However, published site remediation related papers only presented the current situation, existing problems and industry development trend through summarizing the patent and academic literatures (Luo, 2009a, b; Yu et al., 2010; Liao et al., 2011; Luo, 2011; Huang et al., 2012; Li et al., 2012; Yang et al., 2012; Zhang et al., 2012; Zhao et al., 2013; Gu et al., 2015; Xie et al., 2015; Li et al., 2016; Wang et al., 2016). Papers did not provide interpretation on the development situation of contaminated site remediation industry with specific project data.

This paper gathers information of remediation projects from the national bidding website and official websites of known remediation enterprises. For those projects whose detailed information is not provided publicly, the information were supplemented through investigation, interview and consultation. The total number of collected remediation projects is 400. Excluding assessment and site investigation, mining area remediation, farmland remediation, integrated watershed management, and waste heavy metal and waste water disposal projects, there are 150 industrial pollution remediation projects remaining. On the basis of sorting out and analyzing information in these projects, this paper summarizes characteristics of China's industrial contaminated site in regional distribution, interannual distribution, contaminant type, feature of construction period and application of remediation technologies. This paper also prospects for the development trend of industrial contaminated site remediation in China.

2. The regional distribution feature and driving force of remediation projects

The *National Survey of Soil Pollution* published in 2014 showed that the overall situation of national soil environment is not optimistic. The rate of points that exceeded the standard was 16.1%, and it was 34.9% for industrial waste lands in urban area due to the relocation of highly contaminated industrial enterprises. Therefore, it makes the industrial waste sites the most contaminated area and the highest proportion of land type that exceed the pollution standard (MEP and MLR, 2014).

Based on the project samples collected, the figure for the regional distribution of Chinese industrial contaminated site remediation projects from 2007 to 2016 is shown in Fig. 1. In the past 10 years, the industrial contaminated site remediation projects in China were mainly concentrated in more developed regions such as Beijing, Tianjin, Hebei, Shanghai, Jiangsu, Zhejiang, Hubei, Guangdong and Chongqing. Jiangsu province has implemented the most remediation projects that reached 22. Beijing, Shanghai and Zhejiang also have implemented 21, 18, and 17 projects, respectively. The number of implemented remediation projects in these four provinces and cities accounted for 52% of the national total projects.

In Fig. 1, it also can be clearly seen that heavy metal contaminated sites are dominant in Southwestern and Southern China, while there is little organic contaminated site. This is consistent with the investigation results in the *National Survey of Soil Pollution*, which shows the soil heavy metals in Southwestern and Southern China seriously exceed the regulatory standard due to abundant mineral resources and developed steel and nonferrous metal industries. On the contrary, the majority of contaminated sites in developed regions in Northern and Eastern China are organic sites, because more chemical medicine, chemical fiber, rubber and leather industrial enterprises are concentrated in these regions.

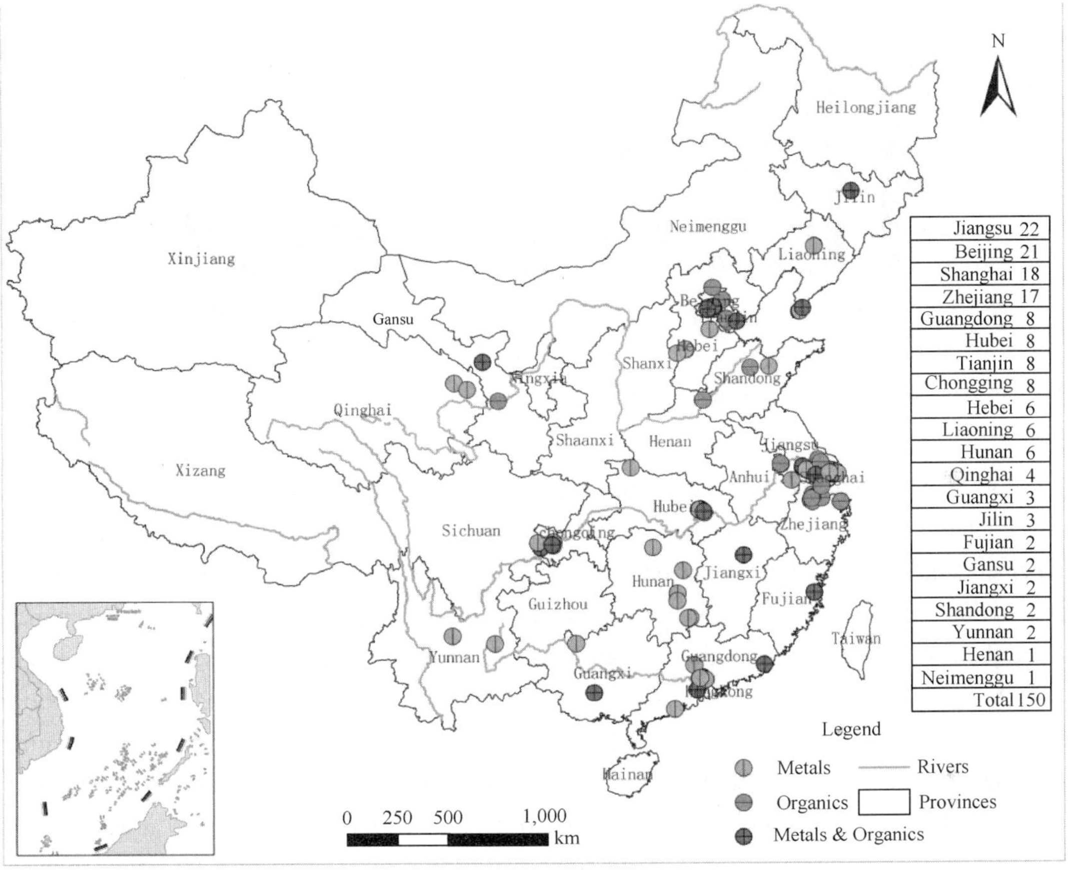

Fig. 1 Regional distribution of industrial contaminated site remediation projects
(Sample number: 150)

The following figure (Fig. 2) shows the relationship between numbers of implemented projects and the house prices in several major cities. The relationship presents a positive correlation and demonstrates that the higher the house prices, the greater the value of land development, and hence the stronger driving force to implement industrial site remediation. For example, the number of implemented remediation projects in Beijing and Shanghai, where the averaged house prices are high (June 2016), are both higher than 18. It indicates the development of house is becoming the internal driving force to the contaminated site remediation in developed regions.

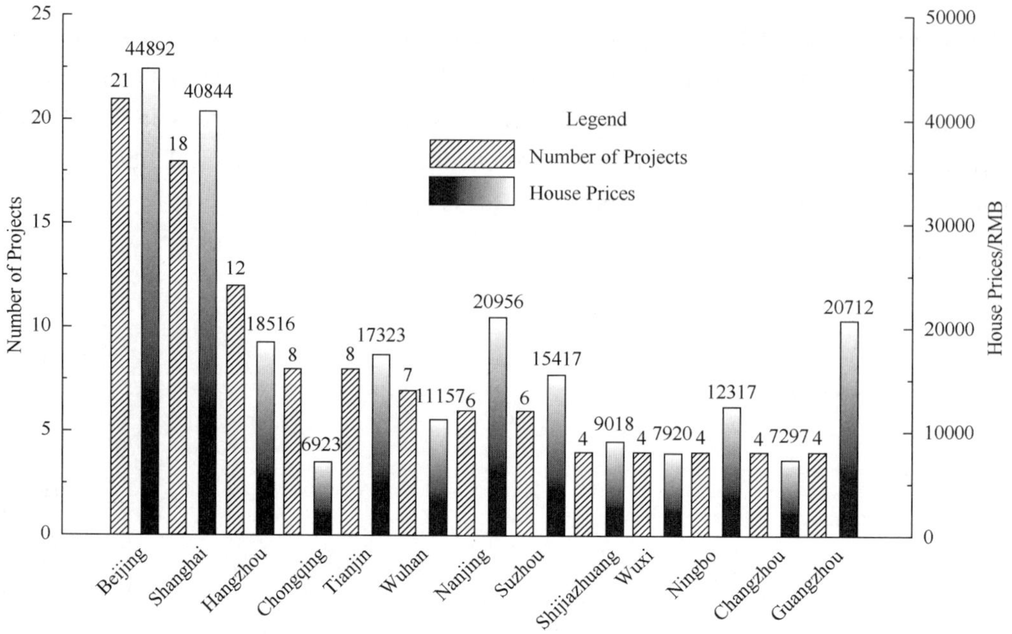

Fig. 2 The comparison for the number of implemented remediation projects and house prices in major cities
(Sample number: 150; data resource: cityhouse. cn, June 2016)

Fig. 2 shows that the averaged house price is relatively low in Chongqing; however, there are 8 contaminated site remediation projects, which is just lower than Beijing, Shanghai, Jiangsu and Zhejiang. This is mainly because Chongqing takes the lead in the area of contaminated site remediation related policies and standards in China. As early as 2004, Chongqing government issued the recommendations on the relocation of key safety hazards enterprises in the main city area. As of 2016, the preliminary statistic shows there are 8 contaminated site remediation related policies and standard documents have been released. Similarly, regions such as Beijing, Shanghai, Zhejiang and Jiangsu that have implemented more remediation projects, also released more remediation policies, guidelines and related documents than other regions at earlier time. In addition, for the implemented remediation projects in Midwestern regions such as Yunnan, Hunan and Guangxi, the sources of funding are mostly from the national special fund for heavy metal pollution control. The above results demonstrate that the national and local policies are another driving force to the contaminated site remediation in China at present.

Table 1 lists top 10 organizations that won the National Natural Science Fund of China from 1995 to 2015, and Fig. 3 provides the ranking of soil remediation related patent application and corresponding remediation project number. Through the comparison on the regional distribution in Table 1 and Fig. 3, it can be concluded that in regions implemented more remediation projects (i.e. Beijing, Zhejiang, Jiangsu and Tianjin), the corresponding regional research institutes obtained more soil remediation related funds and applied more patents. These results indicate technology is one of the driving forces to the soil remediation in China, and advanced technology is the guarantee for the implementation of the remediation projects.

Table 1 Top 10 organizations obtained the soil remediation related National Natural Science Fund (1995-2015; total fund number: 146)

Institute (1995-2015)	Number of NSFC Projects
Institute of Applied Ecology, CAS	11
Research Center for Eco-Environmental Sciences, CAS	6
Nankai University	6
Institute of Geographic Sciences and Natural Resources Research, CAS	6
Huazhong University of Science and Technology	5
Zhejiang University	5
Agro-Environmental Protection Institute, MA	4
Tsinghua University	4
Nanjing University	3
Peking University	3

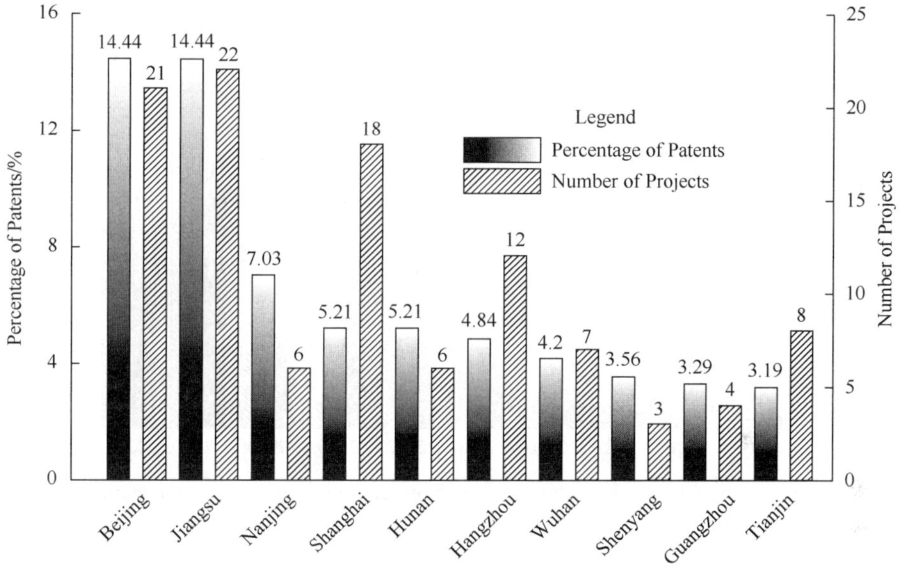

Fig. 3 Percentage of the application of soil remediation related patent in each region

(Total patent number: 1116, only the top 10 regions showed)

3. The interannual distribution feature and capital scale of remediation projects

Since the successful implementation of the first commercial soil remediation project in 2007, the amount of remediation projects and the quantity of capital in China has been increasing year by year (Fig. 4). Before 2011, the annual number of industrial contaminated site remediation projects is less than 10, and the annual fund for remediation is less than 500 million RMB. It is the embryonic stage of the industrial site remediation in China. Starting from 2012, the number of industrial site remediation projects and the amount of funds increased significantly (note: the statistics for the number of projects in 2012 is 11; however, 4 of the project contract values are missing, resulting in a small amount of funds statistics). This is the initial stage of industrial site remediation in China. There are only 2 projects that have contract values over 100 million RMB before 2011, which are Expo Park Contaminated Site Maintenance Project (about 300 million RMB) and Wuhan E-land Comprehensive Treatment Project (232 million RMB). As of May 2016, the statistical number of remediation projects is 19 in the year of 2016. Although the number of projects is half of 2015, there are 6 projects with contract values over 100 million RMB and the half-year remediation capital have exceeded the sum in 2015. With the release of *the Plan of Soil Pollution Prevention Action*, it is reasonable to expect that the amount of contaminated site remediation projects and the quantity of capital will usher in an explosive growth in the next few years.

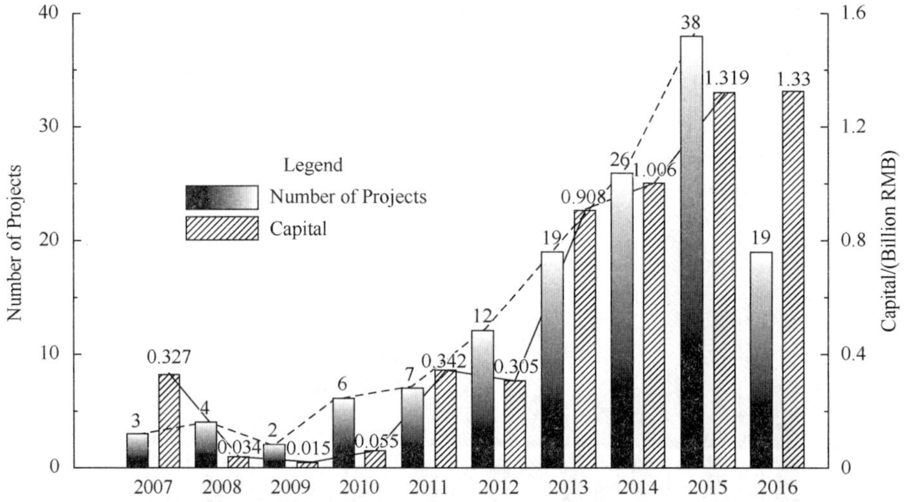

Fig. 4　The interannual change of the number of remediation projects and the capital quantity

(Sample number: 136, partial data is missing; the number of projects with capital quantity: 119; statistics in 2016 is to the end of May)

According to the analysis above, both of the amount of remediation projects and the quantity of capital present rapidly increasing trend since 2011, which could not be achieved without the importance from the state attached to soil remediation. In 2011, the state released the Plan of Heavy Metal Pollution Comprehensive Control in the 12th Five-Year and other 2 soil remediation related documents. Zhejiang and some other provinces also issued local plan for soil cleaning that promote the rapid development of soil remediation industry. Since 2011, the releasing of national and local policies, standards and regulations for contaminated soil remediation are accelerated, which play significant role to promote the development of the industry. Moreover, under the influence of the Luliang chromium residue pollution incident and the Lanzhou benzene polluted tap water incident, the raising of the public awareness on environmental protection and the supervisory opinion of the media on contaminated sites also promote the development of contaminated site remediation.

Fig. 5 presents the distribution of capitals for remediation projects. It shows that the fund scales for the majority of the current remediation projects are at tens of millions level. 80% of the project contract values are lower than 65 million RMB. There are 17 projects over 100 million RMB scale, only accounted for 14.3% of the total.

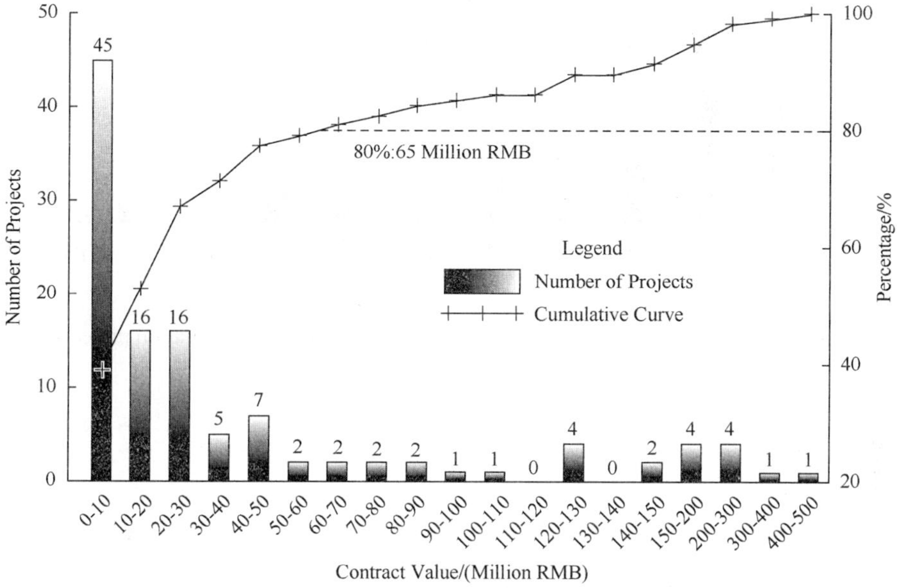

Fig. 5 Classification statistics for the contract value of remediation projects
(Sample number: 115)

4. The feature of construction period and the scale of engineering quantity in remediation projects

Based on the previous analysis, the environmental remediation industry in China has just started and the development of real estate is an important driving force to the contaminated site remediation at this initial stage. Usually, real estate driven projects require urgent construction period to shorten the development cycle. Through the statistical analysis on the collected project samples with certain construction period, the results in Fig. 6 demonstrate that most of the construction periods for remediation projects are tight and 80% of them are shorter than 330 days.

For the scale of engineering quantity in contaminated site remediation projects, Fig. 7 shows that over 1/3 of the remediation projects are lower than 10, 000 m^3 scale and 50% of projects are lower than 30 000 m^3. Only one project has the scale larger than 1 million m^3.

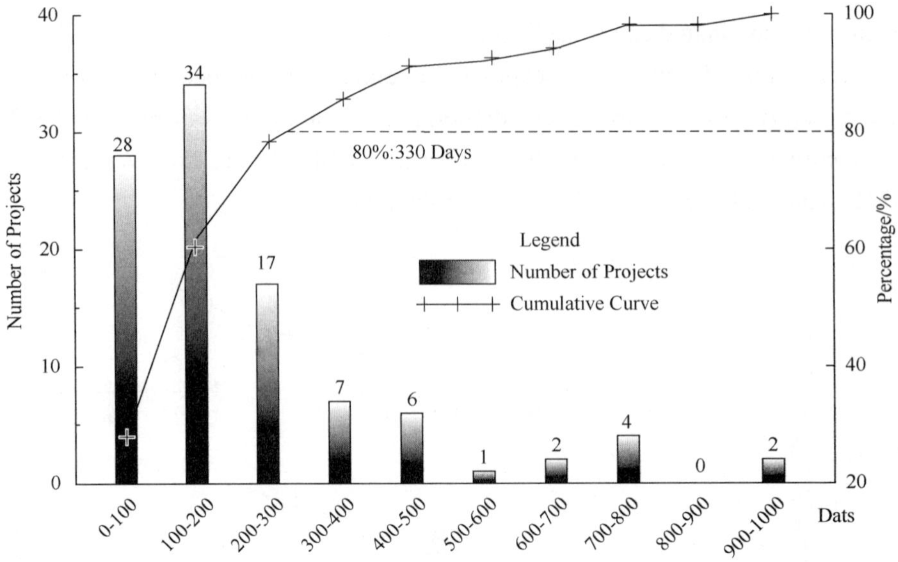

Fig. 6 Classification statistics for the construction period of remediation projects
(Sample number: 101)

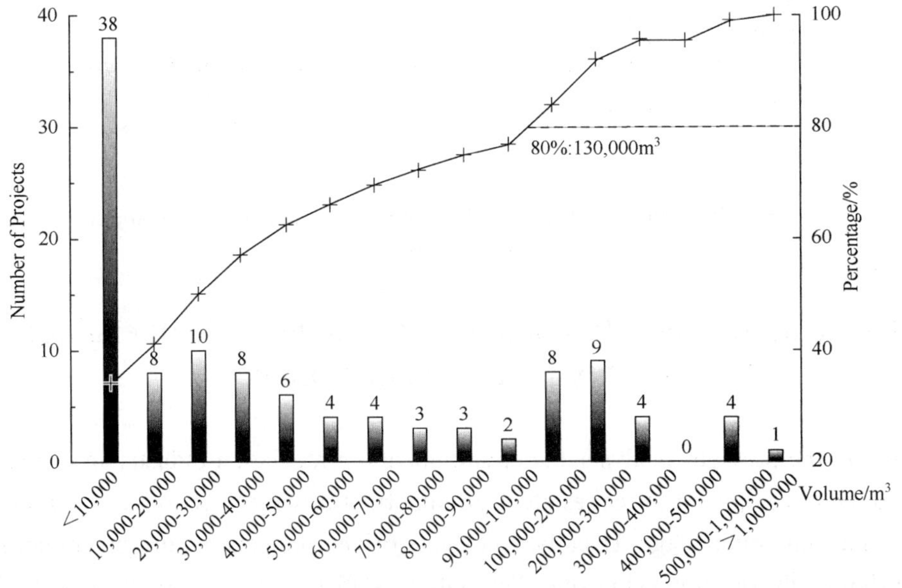

Fig. 7 Classification statistics for the volume of remediation projects
(Sample number: 112)

5. The characteristics of contaminants in remediation projects

An integrated industrial system has formed in China after nearly 70 years of industrialization. Many enterprises experienced lots of product adjustments and site renovations and expansions, which make the type of pollution more complicated. Fig. 8 presents the statistical data for the types of contaminants at industrial contaminated sites in China. The results show that out of 131 sample projects with confirmed pollution types, there are 46 heavy metal sites, 45 combined polluted sites and 40 organic sites. However, results from *the National Survey of Soil Pollution* showed most of the soil pollution type in China is inorganic, while organic pollution comes the second and combined contamination takes smallest proportion (MEP and MLR, 2014). The cause of the difference is probably due to different subjects in these two investigations. The objects of investigation in *the National Survey of Soil Pollution* included large amounts of farmland, forest land, grassland and other agricultural lands, where heavy metal pollutions are heavier than organic pollutions. Statistical project samples in this article are industrial contaminated lands in urban areas, where organic contaminants, especially volatile organic compounds (VOCs) are prone to migrate into ambient environment. Most of VOCs have pungent odor that is easily perceived by people in the neighborhood. Therefore, at the initial development stage of the remediation industry, contaminated sites containing organic matters (especially VOCs) are more likely to attract people's attention, and hence receive priority to be remediated. The statistical results of contaminant types in this paper support the conclusion proposed by Luo et al. that combined or mixed soil pollution were aggravated and complicated in China (Luo, 2009a; Liao et al., 2011; Luo, 2011).

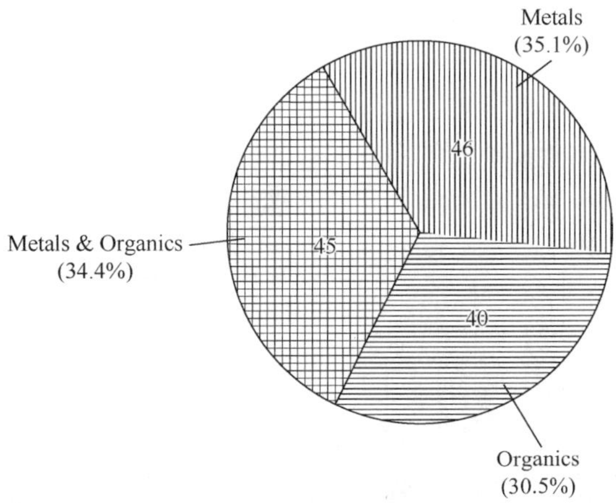

Fig. 8 Types of pollutants at industrial contaminated sites
(Sample number: 131)

Fig. 9 and Fig. 10 present major pollution index of heavy metals and organic matters of industrial contaminated sites in China, compared with the pollution index of U. S. EPA Superfund in 1996. According to the statistics from U. S. EPA (1996), nearly 75% of the contaminated sites in the United States polluted by heavy metals, while heavy metals are not the major contaminants in some of combined polluted sites. The 7 most common heavy metals in the U. S. industrial contaminated sites are lead, chromium, arsenic, zinc, cadmium, copper and mercury, respectively (Evanko and Dzombak, 1997). The 6 most common organic contaminant types are halogenated VOCs, BTEX, SVOCs (others), VOCs (others), PAHs and halogenated SVOCs (USEPA, 2013).

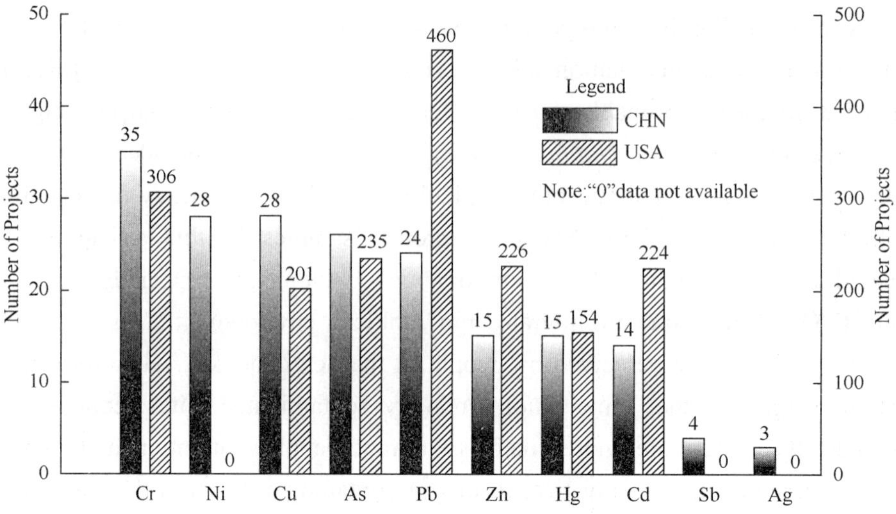

Fig. 9 Comparison of major types of heavy metal pollution in China and U. S.

(Sample number: China, 91)

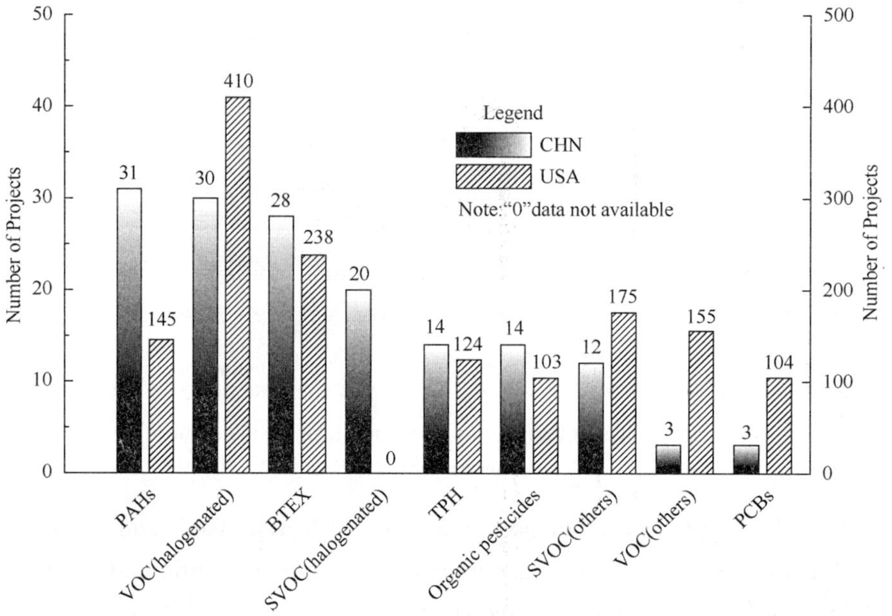

Fig. 10 Comparison of major types of organic pollution in China and U. S.

(Sample number: China, 85; U. S., 977)

Statistical results in Fig. 9 demonstrate that the 7 most common heavy metal contaminants in China are chromium, nickel, copper, arsenic, lead, zinc and mercury, respectively. The largest numbers of sites are contaminated by chromium, probably because China is the largest chromate production country, and pollution of chromium residue obtained great attention (Ji, 2010). There could still be a lot of chromium contaminated sites after the treatment of chromium residue. It is different that lead is the most common contaminant in the United States.

Statistics in Fig. 10 shows the 6 most common organic contaminants in China are PAHs, halogenated VOCs, BTEX, TPHs, halogenated VOCs and Organic pesticides. Halogenated VOCs and BTEX pollutions are more prominent in the U. S. contaminated sites, probably due to large amount of remediation implemented at dry cleaner and gas station sites in the United States. However, these two types of sites are still not received enough attention in China.

6. The feature of technology application in industrial contaminated site remediation

The purposes of contaminated soil remediation are pollution removal, quality restoration, reuse, and safety and health guarantee (Luo, 2009a). Although there are many contaminated soil remediation technologies, some of them are not practicable in industrial contaminated sites due to remediation schedule, site characteristics, environmental risk, or other limitations (Luo, 2009b). The statistical analysis in this study shows that there are over 100 cases of industrial contaminated site remediation projects in China, and most of them are applied with ex-situ technologies. Only 18 projects used in-situ technologies, including 1 in-situ solidification/stabilization, 14 in-situ chemical oxidation/reduction (including 4 in-situ biochemical reduction), and 3 in-situ thermal desorption. They are accounted for 12% of the total (150 projects)

Fig. 11 and Fig. 12 present the application of contaminated site remediation technologies in China in the past ten years. In the five years before 2011, remediation technologies are relatively simple that is mainly consisted of solidification/stabilization (50%, the percentage is calculated as the number of sites applied with specific technology divided by the total amount of sites, same as follows) and incineration (36.3%, including cement kiln co-disposal). In the statistics of this study, there was still no project applied with technologies such as low temperature thermal desorption, thermal desorption, in-situ thermal treatment, chemical reduction and bioventing. In the five years from 2012 to the present, the proportions of solidification/stabilization (40.7%) and incineration (15.8%, including cement kiln co-disposal) have declined. The percentage of the application of incineration/cement kiln technology is significantly reduced, which are mainly due to: few cement plants have the license to treat contaminted soil, so that it usually requires long distance transport and high transport costs; cross-regional transfer permit and remote management problems involved; and difficulties in environmental pollution control during the process of excavation and transportation. In the recent five years, there are successful project cases that are applied with low temperature thermal desorption, thermal desorption, in-situ thermal treatment, chemical reduction and bioventing.

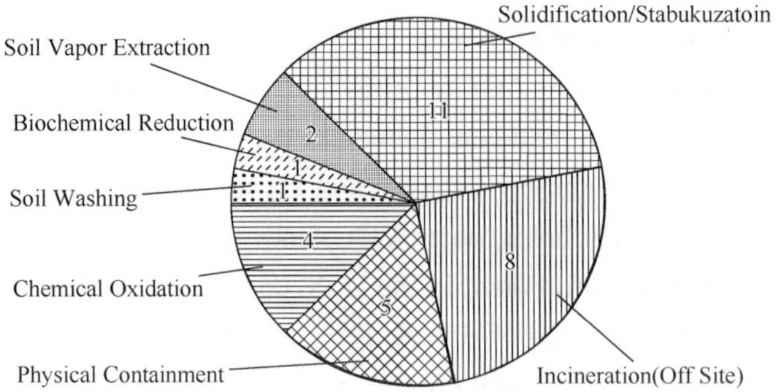

Fig. 11 The application of remediation technologies from 2007 to 2011

(Sample number: 22, Note: some sites used multiple technologies)

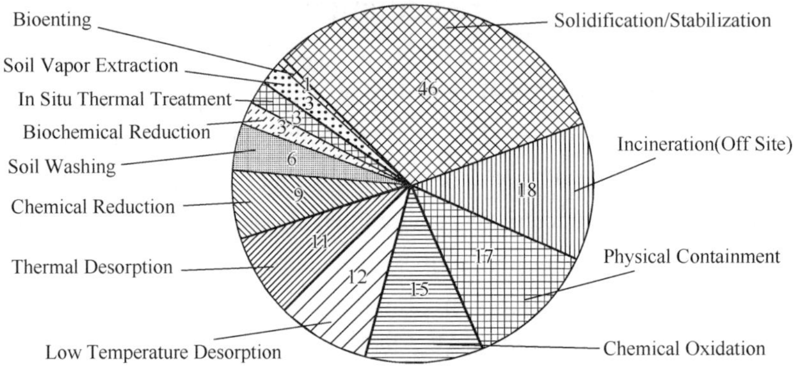

Fig. 12 The application of remediation technologies from 2012 to 2016

(Sample number: 114, Note: Some sites used ncultiple technologies)

The statistics for the application of remediation technologies based on the type of contaminants are illustrated in Fig. 13 and Fig. 14. Results show that there are apparent differences between organic remediation and heavy metal remediation technologies. In the analysis of 46 heavy metal contaminated sites, the number of applications of solidification/stabilization, physical containment and chemical reduction are 34, 12 and 8, respectively, in which the proportion of solidification/stabilization exceeds 70%. However, the most applied technologies in organic contaminated sites are incineration (including cement kiln co-disposal), while the proportion of solidification/stabilization and physical containment are low. The reason is mainly due to the differences in the nature of heavy metal and organic matter and the differences between various remediation technologies.

It is mentioned in the Section 4 that major heavy metal contaminants in China are chromium, nickel, copper, arsenic and lead. The characteristics of these metals are stable, and they only changes in valence and form in the environment. If total removal is the remediation objective, the general cost can be very high. Therefore, main consideration in the selection of heavy metal remediation technology is to change the valence state or form of metals to reduce their migration. Solidification/stabilization and physical containment technologies have obvious advantages in reducing metal migration. In addition, a consensus formed in the long-term management practice is that the remediation objective of heavy metal is mainly to reduce the leaching toxicity and control the migration of contaminants. Moreover, the formations of relevant standard systems also promote the application of these technologies in the remediation of heavy metal contaminated soils.

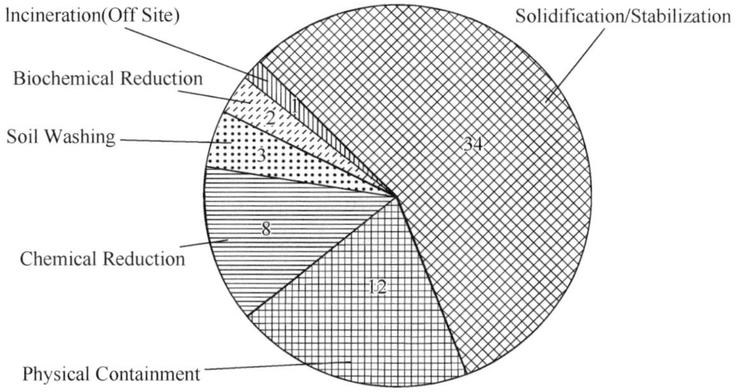

Fig. 13 The application of remediation technologies at heavy metal sites

(Sample number: 46, Note: Some sites used multiple technologies)

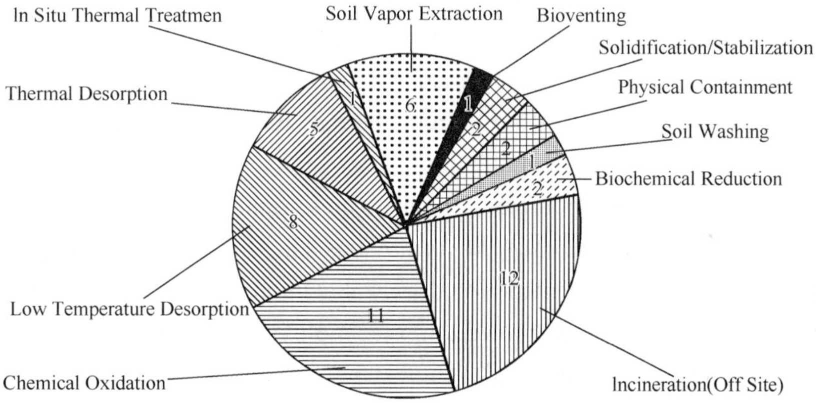

Fig. 14 The application of remediation technologies at organic sites

(Sample number: 40, Note: Some sites used multiple technologies)

Main organic contaminants in China are PAHs, BTEX, halogenated VOCs, TPHs and SVOCs. Most of these contaminants can be easily separated or degraded by thermal treatment, biodegradation and oxidation/reduction. Therefore, most of organic remediation technologies applied in China are thermal treatment (incineration/cement kiln co-disposal, thermal desorption, in-situ thermal desorption), chemical oxidation and low temperature thermal desorption. These technologies are all aiming at reducing the total mass that can totally eliminate the risk of contaminants in the soil and also avoid long-term monitoring and management after remediation. However, solidification/stabilization and physical containment technologies, which are widely used in heavy metal remediation, are seldom applied in organic remediation. On one hand it is because VOCs have strong migration ability that their transports are difficult to control; on the other hand, the cost for long-term monitoring for VOCs is relatively high.

7. The remediation requirement and development trend of industrial contaminated sites

Combined pollutions are common at Chinese industrial sites, where the types of contaminants are complex and the degree and thickness of pollution are significantly different. Soil types of these sites are various, with obvious differences in compositions, characteristics, and spatial distributions. The development of site remediation industry is currently at the initial stage, gradually transferring from growing period to mature period, and from disorder to ordered.

Based on the analysis and summary of the development of industrial contaminated site remediation in the past ten years, there are several requirements as follows:

(1) The remediation management has been developed in western developed countries for 20 to 30 years. A series of laws, regulations, technical guidelines, engineering specifications and action guidelines of contaminated site have formulated as guidance. It has formed a complete monitoring system of contaminated sites and achieved the standardization, informationization and network of contaminated site management (Yu, 2011). China needs to build a nationwide monitoring system for contaminated site remediation and promulgates more detailed and actionable remediation standards and guidelines.

(2) For the remediation technologies, China's core technology, equipment and materials for contaminated site remediation are still at an obvious lag stage and lack of originality. The support of risk control and management are also weak. On one hand, it requires broad-spectrum and high efficiency source-removal remediation technologies that are able to adapt to China's complicated types of contaminants and achieve the remediation goal rapidly; on the other hand, it also requires low-cost and effective source containment technologies and monitored natural attenuation that prevent contaminants from migrating.

(3) The capital demand for contaminated site remediation is huge in China. In the implemented site remediation projects, the majority of them are driven by the real estate, and the minority is supported by national financial funds. Unclear profit and business models have been plagued the development of contaminated site remediation industry. Therefore, it requires to develop more diverse sources of capital investment and to explore different business models.

Based on these requirements, the future development trend of Chinese contaminated site remediation are as follows:

(1) Regulations, standards and management systems of contaminated sites will be gradually

formed. The capacity of environmental management at contaminated sites continues to improve.

In the future, the speed of legislation on the management of contaminated sites in China will be obviously accelerated. The laws and regulatory systems of soil pollution control will be generally established in 2020. The standard system of soil remediation will be initially formed, including the standard for soil environmental quality, assessment, technical guideline, contaminant analysis, pollution control, and other basic standards. The cumulative release of standards will be more than 50 items. The standards for the soil environmental quality of agricultural land and construction land will be released by the end of 2017 (State Council of China, 2016).

In the environmental management, the national monitoring network of soil environmental quality and the basic database of soil environment will be constructed, which builds the platform of informational management for soil environment. Meanwhile, it will provide professional trainings of soil pollution control to national environmental law enforcement officials every 3 years (State Council of China, 2016). With continuous training and the importance attached by all levels of government' s administrative departments of environmental protection, the capability of contaminated site management will be improved rapidly. The contaminated site management will be developed toward the direction of standardization, informationization, networking and transparency.

(2) The rate of originality and localization of technology, equipment and material for contaminated site remediation keeps increasing, and the integrated technical system for green and safety site remediation will be gradually formed.

With more attention paid to the challenges of contaminated sites, China will increase the investment to the research of site remediation technologies. Through the introduction and adsorption of advanced foreign technologies, equipment and remediation materials, it will strengthen the integration progress of domestic production, research and application. In the future, China's technology of contaminated site remediation will be in the transition from ex-situ to in-situ, from simple and individual to complex, coupled and comprehensive, and from technologies applied on single contaminant to integrated technologies utilized on multi-contaminants. Remedial chain technologies, which combine in-situ thermal treatment and in-situ chemical oxidation with other technologies (bioremediation, NMA, etc.), will be more widely used. For the ex-situ remediation, thermal desorption, soil washing and solidification/stabilization will still maintain a large proportion of application. In the selection of remediation technology, equipment and material, it will follow the principles of safety, green and sustainability.

(3) Capital sources and business model tend to be diversified.

Business models of contaminated site remediation will be explored to suit China's condition in the future. It should be fully learned from the experience of foreign contaminated site remediation to establish the responsibility system of "who cause pollution responsible for cleaning up". Innovative investment and financing mechanisms and multi-channel special funds should be created for site remediation, meanwhile encourages social capital to participate in the management of contaminated sites (Cao et al., 2014; Wang et al., 2016). The recently releasing of *the Plan of Soil Pollution Control Action* also explicitly put forward that the leverage function of financial capital should lead more social capital to participate in the soil pollution control and strengthen the cooperation between government and social capital through the Public-Private Partnership model. It should also encourage eligible soil pollution control and remediation enterprises to go public and explore the model to promote the development of soil remediation by issuing bonds (State Council of China, 2016).

References

Bureau of urban social and economic research N. 2014. China city statistical yearbook (2014): Beijing: China Statistics Press.

Cao F, Zhang R, Wu W. 2014. The innovation research on investment and financing mechanism of environmental (soil) remediation industry in the post industrial era. Ecological Economy: 42-45+76.

CCTV. com. 2016. Reporter survey: Schools that are not to be built (changzhou, jiangsu), where closely to three chemical plants contaminated sites. Available: http: //tv.cntv.cn/video/C10616/c8ad1dbeb5e143bdadbe2252c57f9918 [2016-6-10].

China.com. 2004. 3 people poisoning in beijing metro line five subway (songjiazhuang station) construction Available: http: //news. china.com/zh_cn/domestic/945/20040501/11677844.html [2016-6-10].

Evanko C R, Dzombak D A. 1997. Remediation of metals-contaminated soils and groundwater. Available: http: //www.gwrtac.org [2016-6-10].

Gu Q, Hou D, Wu B, et al. 2015. Conception and project practice of green and sustainable site remediation and its implications for China. Chinese Journal of Environmental Engineering, 9 (8): 4061-4068.

Huang D, Zhang S X, et al. 2012. Some opinions of the present condition and countermeasures of polluted site remediation in china. Sichuan Environment: 188-193.

Ji Z. 2010. Development status of china' s chromium salts in recent 50 years. Norganic Chemicals Industry: 1-5+15.

Li L, Zhang X, Li J, Li Y. 2016. Current status of soil pollution and progress of soil reclamation industry. Environmental Science and Management: 45-48.

Li Y, Hu X, Song X, et al. 2012. Advances in soil remediation technologies of urban industrial contaminated sites. Journal of Anhui Agricultural Sciences: 6119-6122.

Liao X Y, Chong Z Y, Yan X L, et al. 2011. Urban industrial contaminated sites: A new issue in the field of environmental remediation in china. Environmental Science: 784-794.

Liu Y, Li Y, Xiao R. 2013. Management experience of industrial contaminated sites in western countries and its implications for china. Ecology and Environmental Sciences: 1438-1443.

Luo Y M. 2011. Contaminated site remediation in china: Progresses, problems and prospects. The Administration and Technique of Environmental Monitoring: 1-6.

Luo Y. 2009a. Trends in soil environmental pollution and the prevention-controlling-remediation strategies in china. Environmental Pollution and Control: 27-31.

Luo Y. 2009b. Current research and development in soil remediation technologies. Progress in Chemistry: 558-565.

MEP, MLR. 2014. National soil pollution survey bulletin. Available: http: //www. zhb.gov.cn/gkml/hbb/qt/201404/W020140417558995804588.pdf [2016-6-10].

NBSC. 2015. China statistical yearbook (2015): Beijing: China Statistics Press.

People. cn. 2014. Benzene pollution in drinking water in lanzhou. Available: http: //society. people.com.cn/n/2014/0412/c1008-2488

7022.html [accessed June 10th 2016].

State Council of China. 2016. The action plan for soil pollution prevention and control. Available: http: //www. gov.cn/zhengce/content/2016-05/31/content_5078377.htm [accessed June 10 th 2016].

Tang D C. 2009. Industrialization, industrystructureandenvironmentpollution— based on manufacturing industry and regional panel model. Soft Science: 6-11.

USEPA. 2013. Superfund remedy report (superfund (fourteenth edition). Available: https: //clu-in.org/download/remed/asr/14/SRR_14th_2013Nov.pdf [accessed June 10th 2016].

Wang J, Zhang F, Ma L. 2016. Research progress of stabilization binder for heavy metals contaminated soil remediation. China Resources Comprehensive Utilization: 49-52.

Wang W, Li S F, Zhu W Y, et al. 2016. Current engineering application and business model of soil remediation industry in China. Environmental Engineering 164-167.

Xie H, Hu Q, Zhang H Q, et al. 2015. A review of the history of contaminated sites remediation in China in the past ten years: problem analysis and experience. Environmental Impact Assessment: 19-23.

Xu K, Wang J. 2005. Course of chinese industrialization: International comparison and complex development strategy. Jianghai Academic Journal: 63-69+238.

Yang Y, He Y M, Luan J L, et al. 2012. Comprehensive analysis on soil remediation technologies of international contaminated sites. Environmental Sciena and Techology, 35 (10): 92-98.

Yu L. 2011. Research on environmental supervision of contaminated sites in the united states and europe. Environment and Sustainable Development: 7-10.

Yu Q, Hou H, Lu L, et al. 2010. Industrial enterprise relocation and their implications to contaminated site management—Beijing and Chongqing case study. Urban Studies: 95-100.

Zhang H, Yu F, Cao D, et al. 2012. Practice of contaminated site remediation technology assessment in developed countries and its enlightenment to china. Environmental Pollution and Control 105-111.

Zhao J Y, Li Y, Li S S, et al. 2013. Polluted soil remediation technology and industry status in China. China Environ. Prot. Ind, 3: 53-57.

The Development of China Contaminated Site Remediation Industry: A Decade in Review

Qing HU, Yan ZHU, Sijie LIN, Hong WANG, Jingyang GAO

Engineering Innovation Center of South University of Science and Technology of China

Abstract

A historical review of the remediation industry of China and USA is summarized, while the common issues between the two countries are discussed. A side-by-side comparison analysis was carried out in terms of site remediation challenges, environmental policies, management guidelines, funding mechanism, and remediation processes. The key finding is that the remediation industry in China has been experiencing a fast growth since 2004 and the international experiences and lessons primarily derived from the United States Superfund program may contribute to China. It is concluded that the solution to address the challenges must be a concerted effort aiming at the formulation of industrial practice code, further emphasis of environmental impact assessment, as well as the establishment of diverse funding mechanisms.

Keywords: site remediation, development review, superfund

As the release of China' s Action Plan for Prevention and Control of Soil Pollution, the topics of soil contamination and remediation are popularly discussed. Moreover, due to the rapid development of national economy and industrial restructuring, there has been a large number of contaminated sites. These sites are experiencing various aspects of conflicts between economic development, pollution control, public events caused by environmental pollution and other issues. Over the past decade, contaminated site remediation industry in China is blooming faster than any other country in the world. In order to break the dilemma of the remediation industry development in China, it is of strategic value to review the history of the remediation industry in the past decade and to learn the comprehensive experiences and lessons derived from the United States remediation industry.

1. The beginning of remediation market in China

Looking back to the developing history, China and the United States shared quite a few common challenges in the environmental protection realm. First, the arising public awareness of environmental protection and a series of environmental public events provided an important opportunity or great start to the remediation industry. In 1962, Rachel Carson's "Silent Spring" directly contributed to the prohibition of pesticides, including DDT and other similar pesticides, which made the public focus on the topic of soil contaminated by pesticides. By 1980, "Love Canal incident" directly promoted the United States Congress passed the "Comprehensive Environmental Responses, Compensation and Liability Act". After that, the Congress ordered Hooker Chemical' s parent company paid $129 million in compensation. As of 2004, after paying off $ 400 million, the Love Canal cleanup project had been completed (Beck, 1979).

Similarly, China contaminated site remediation industry began from the Beijing Metro Line 10 SongJiaZhuang event. After the incident, the relevant environmental authorities launched an onsite monitoring and removed contaminated soil to incineration. The Ministry of Environmental Protection (MEP) issued a series of regulations, which required the relevant parties to take necessary actions after site decommission. SongJiaZhuang event is the start of contaminated site remediation industry in China. This event indicated the competent authority and the public began to be concerned of the remediation and redevelopment of contaminated sites.

At present, China is facing a very severe challenge of soil quality protection. According to the relevant survey, urban industrial site contamination causes more than 5 million mu (a unit of area, 1 mu approximately equal to 0.0667 hectares) of land abandoned every year. The mining contamination lead to over 2 million hectares of tailing zone cannot be reused, of which less than 20% are reclaimed. In addition, informal landfills have contaminated a large amount of soil and groundwater. With the progressing of industry and urbanization, big news and public events caused by soil contamination have gradually attracted government and the public attention, which objectively facilitated the development of contaminated site remediation in recent years.

2. Challenges of remediation industry

Currently, site remediation industry in China is still at the beginning stage. Although it shows a trend of rapid development in general, there are many challenges ahead, such as the small market size, insufficient regulation and policy, lack of contaminated site inventory data, vast number of existed contaminated sites, etc. Coincidentally, the United States remediation industry also have experienced these challenges during its development. In addition, China also has some unique problems, such as the shortage of established remediation technology, limited capital and talent, and chaotic industry.

2.1 Insufficient regulation and policy and lack of regulatory basis of enforcement

In China, national and local environmental protection departments have begun to notice the importance of the management of contaminated sites and released a series of contaminated sites standards in 2014, which include site environment investigation, risk assessment and remediation and so on, to provide the basic technical guidance and support to promote the soil and groundwater pollution prevention and control. However, compared with the water pollution and air pollution that has readily available national law and action plan, such as Water Pollution Prevention and Control Law and Air Pollution Prevention and Control Action Plan, the Action Plan for Prevention and Control of Soil Pollution has just been published in June 2016. Contaminated site remediation industry still lacks the specific law addressing soil pollution prevention and control, directly resulting in the local environmental protection department cannot seek applicable laws or suitable regulations to handle soil remediation issues. In addition, it also causes the competent authority cannot clearly define the main responsibility, but only in a broad sense by referring to the relevant provisions coded in the "Environmental Protection Law".

2.2 Lack of the contaminated site inventory data and vast amount of efforts needed for site remediation

Although the Chinese government took a nationwide survey of soil quality in 2006, the work is only for key industrial enterprises in ten categories. This survey was limited by many factors. For example, the survey covered limited area, which cannot comprehensively reveal the full map of the decommissioned industrial land pollution, is even insufficient to establish a list of contaminated sites similar to the United States's "National Priority List". On the other hand, the survey has been investigated in a limited number of sites, a huge portion of these sites are lack of basic data and information, which make it is almost impossible to create a site-specified document for each contaminated site.

According to relevant survey, heavy metal pollution in China widely existed in 14 provinces. The area of contaminated farmland was estimated about 14.94 million mu, which accounted for 8.3% of the 1.8 billion mu of arable land. The area of contamination industrial land was 12.05 million mu, 48.52% of which were located within the urban area, or about 5.84 million mu. Considering the vast majority of the decommissioned industrial sites without appropriate methods to protect groundwater and soil during its operation period, it is reasonably to assume that all of the existing industrial sites have more or less soil and groundwater pollution.

2.3 Insufficient funding to remediation and shortage of technical ideas and talent

Due to the lack of pollution funding/chasing accountability mechanism, it is hard to define the main responsibility. Statistics show that the government budget support 54.3% of the remediation of contaminated sites as the single funding resources, and government funding combined with remediation enterprise funds accounted for 21%. The remaining, which is less than 1/4 of funds come from the polluter and other channels. On the one hand, the central government provided special fund for remediation of contaminated sites, but the needs were far beyond the budget. On the other hand, the local government paid huge attention to infrastructure construction, thus local governments would be difficult to reserve enough funds to support local remediation projects. It is worth paying attention to Beijing, Shanghai, Nanjing, Hangzhou, Chongqing's local governments and remediation practitioners, who create some new rehabilitation/remediation financing models. More efforts, however, are need to promote these new financing models to be applied in a nationwide scale.

In terms of remediation technology, China is still in the stage of "completely pollution removal, restoration of contaminated sites to the original state". The most popular and mainstream techniques are coming from the United States and Europe. This viewpoint is outdated, and Chinese experts notice that the cost of remediation will be high if they keep this approach. Actually, very little amount of sites can be completely restored in China.

In present China, the site remediation technology is also boorish. The R&D of remediation equipment, remediation reagents, construction management system etc. are still under development. In contrast to the United States as the representative of the developed countries, the remedial objective has been shifted to "contamination containment and pollution prevention to protect human health and environment". By applying contaminated site conceptual model and remediation technology with in-depth understanding, the US site remediation participators are allowed to take site specific remedial solutions based on types of pollution, exposure pathways, modes of impact on human health, which can greatly saves the cost of remediation project and improve the environmental risk management level of government agencies and industry practitioners at the same time. It is reasonable to believe that, with the development of pollution remediation industry and the transformation of the concept, the site remediation of China will also shift to the risk-based remediation.

3. Experience from the US remediation industry

Compared to China's growing remediation industry, the US remediation industry started early, grew quickly, established a comprehensive system, and developed a full industrial chain. It consists of site investigation, risk assessment, remedial design, construction supervision, equipment manufacture, reagent research and application, and so on. In 2012, it was an $8 billion market. The US mature experiences and lessons will help China's remediation industry to grow.

3.1 Improve regulatory framework with a focus on top level design

The rise of the United States contaminated site remediation market, originated in 1980's "Superfund Act" and its supporting regulation of "National Contingency Plan". These two acts not only solved the legal question of "why to remediate the contaminated sites?" in the regulatory frame, but also resolved the question "how to management of contaminated sites? ". In the meanwhile, they also included detailed provisions for Superfund contaminated site project, as shown in Fig. 1. In the Superfund program workflow, special attentions have been paid to site title transaction information and risk assessment. Based on the baseline information of the nation's contaminated sites, combined with site assessment results, a national priority system was implemented, while priorities were given to those sites with sensitive areas or of critical types.

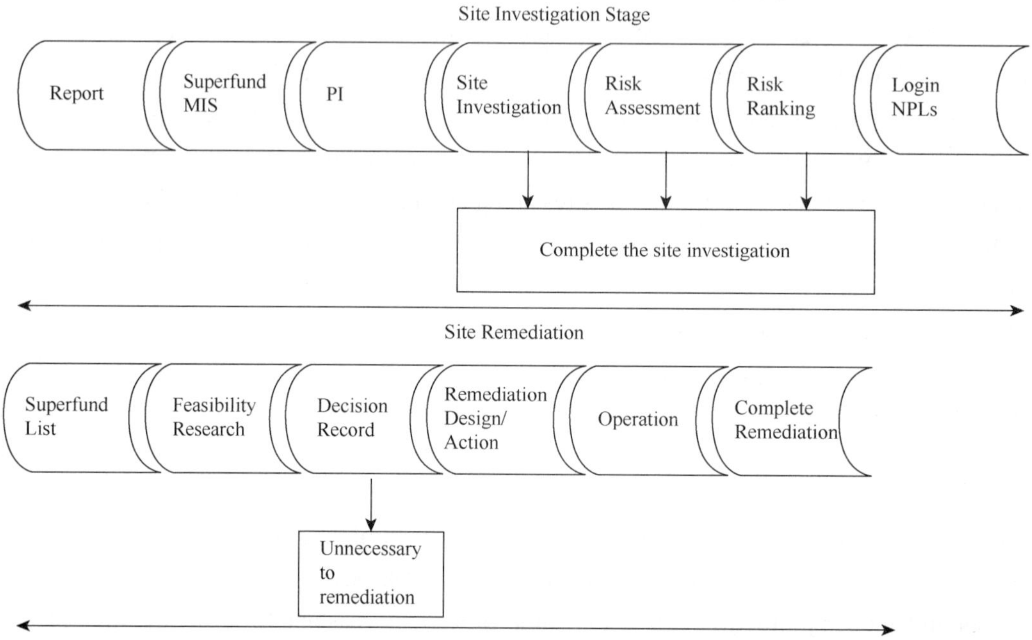

Fig. 1 Working processes of Superfund project

3.2 Detailed market segments, importance on assessment and monitoring

According to Fig. 1, contaminated sites remediation project's life cycle can be divided into four stages: site investigation, risk assessment and feasibility study/remedial design, remediation construction, acceptance and monitoring. In Table 1, the United States remediation market share from 2009 to 2013. In the five year period, remediation market share in the United States is mainly construction, and it increased steadily year by year; this is due to a large number of Superfund projects have progressed from the investigation and design stage into the construction phase, and remedial design market share decreased year by year correspondingly. The market shares of site assessment, risk assessment and feasibility study has been very stable, maintained at 18%, indicating that site remediation is a long-term sustained market, even after more than 30 years of development, there still are a large number of contaminated sites discovered and entered the queue. Similarly, there was a stable market share of about 8% in closure and monitoring. This is because that although a lot of efforts and funds were invested to site cleanup, most contaminated sites could not be cleaned up to the "no residual risk" level. Therefore, after the completion of the remediation construction of the contaminated sites, it is necessary to place multiple institutional controls, such as, planning restrictions, transaction constraint and community protection. Therefore, the assessment of the effectiveness of remediation and long-term monitoring is still an important part of the remediation market.

Table 1 Market share of each part in site remediation industry chain in USA

	2009	2010	2011	2012	2013
Total Revenue/Millions dollars	7530	7670	7850	8070	8320
Remediation Construction/%	62	63	63	64	64
Site Assessments & RI/FS/%	18	18	18	18	18
Remedial Design/%	12	11	11	10	10
Closure & Monitoring/%	8	8	8	8	8

3.3 Dedicated fund

Since contaminated site cleanup projects are often costly, in the initial stages of the remediation industry, to address the cleanup funding problem, "Superfund Act" established a trust fund of $1.6 billion in total for contaminated site cleanup projects. However, single source of funding is not enough to solve the big number of urgent cleanup sites, so another problem the remediation industry was yet to face was the funding source. "Superfund Act" by identifying "potentially responsible party" approach, in accordance with the "polluter pays" principle to solve the funding problem. Compared with the current principles of China's "polluter pays" principle, Superfund Act, in judicial practice, replaces the pollution behavior with ownership identification. Therefore, the potentially responsible party identified by the government is not necessarily the direct party who contaminated the site. Thus, the first line of "potentially responsible parties" in the "Superfund Act" are in fact the current owners of contaminated sites or the current operators. This principle led directly the strong demand from the industrial site owners to clarify the liability of contaminated sites and mitigation mechanism, and ultimately it fostered the formation of a contaminated site management business in both public and private sector.

Diversifying funding sources is another important channel to raise funds for the United States environmental regulators to deal with Superfund sites. Beside the "Superfund" trust fund, the three major group of responsible parties are the federal agencies, the military and private companies. As shown in Table 2, the US Department of Energy is currently the largest responsible party for the cleanup sites. It spent more than $2 billion annually in contaminated site cleanup projects. However, in the past five years, the private sector's portion has been increasing.

Table 2 Funding sources of site remediation in USA/Millions of dollars

	2009	2010	2011	2012	2013
DOE①	2150	2210	2260	2300	2350
DOD②	1340	1370	1400	1420	1440
Superfund (EPA and potential responsible parties)	460	460	470	480	490
State Programs	270	270	270	280	280
Resource Reused and Recycle Act Program	480	490	500	510	530
UST③	450	460	460	470	480
Private Sector	2400	2410	2490	2610	2750
Total	7530	7670	7850	8070	8320

Note: ① Department of Energy; ② Department of Defense; ③ Underground Storage Tank.

4. Suggestion to remediation industry development in China

In the past decade, the development of contaminated site remediation industry in China is a complex process, involving central and local governments and environmental authorities, the responsible parties, the site owners, practitioners. The development of contaminated site remediation industry in China is a gradual process due to the aforementioned problems. It is impossible to solve the problems relying on a single aspect of the industry.

Therefore, on the basis of abroad mature technology and experiences, Chinese environmental authorities should focus on guidance of industrial policies, and proactively carry out top-level design, innovate in financing the contaminated sites cleanup projects, through guidelines, policies regulations and other forms to guide an orderly market competition, healthy development. For practitioners, when making site remediation decisions, should change the mindset from "complete restoration" to "risk-based remediation", emphasis on the guiding role of environmental impact assessment throughout the investigation and cleanup processes. Regarding cleanup technology, we should actively absorb foreign beneficial experiences and advanced technology, change from a single remedy to remedial train, and to improve the cost-effectiveness of remediation while achieving remedial objective and promoting the cleanup technology progress. On the remediation equipment, we should switch from fixed and ex-situ equipment development to in-situ and mobile devices, minimize the environmental impact of contaminated site cleanup processes; on cleanup objectives, we should extend our focus from soil and groundwater to soil vapor and the rehabilitation of the surrounding micro-ecology and so on.

In addition, our current site remediation are mainly focused on remediation construction, and in lack of the necessary pre-cleanup site investigation and post-cleanup monitoring. This defect directly led to the blindly authorization of many remediation projects without remedial action objectives or shortsighted objectives.

With the market further standardized and developed, remediation industry value chain will be further expanded and segmented, upstream and downstream, to a mature market similar to the United States. Facing the aforementioned challenges, only with the holistic efforts of industry, can we achieve an organic development, gradually resolve the existing soil contamination, overcome the lack of funding, and promote the healthy development of this emerging sector.

References

Beck E C. 1979. The love canal tragedy. EPA Journal, 5: 17.

U. S. 1980. House of representatives. The Comprehensive Environmental Response, Compensation, and Liability Act.

The Exploration and Practice on Soil Environmental Protection in the Process of Rapid Urbanization of the Megacity Shanghai

Shenfa HUANG, Min WANG, Jian WU, Qingqing LI,
Jie YANG, Lin GUO, Jin WANG, Zhihao XU

Shanghai Academy of Environmental Sciences
State Environmental Protection Engineering Center for Urban Soil Contamination Control and Remediation

Soil is an important part of the ecological environment. It is not only the repository of biogeochemistry cycle, but also the foundation for human survival and development. As the vital environmental matrix of pedosphere, urban soil is under the intense anthropic disturbance in a long-term. Consequently, the natural, physical and chemical properties of soil have changed profoundly. Unlike natural soil, urban soil forms a kind of unique soil and plays a prominent environmental, social and economic function. It also plays a vital role on urban ecological security and human health.

Urban soil types include construction land soil in urban area such as industry, transportation, commercial, and residential soils etc. Those area includes surrounding farmland soil that is severely disturbed and stressed by the urban development. With the rapid development of urbanization, human activities including industrial emissions, waste, transportation, chemical fertilizers and pesticides have changed the physical, chemical and biological properties of urban soil, which leads to the sharp decline of urban soil quality, and the typical heavy concentrated, heterogeneous and cumulative pollution. Urban soil closely contacts dense urban population and affects not only life health and food safety through food chain, but also the quality of urban environment through water and air (De and Morel, 2000; Scharenbroch et al., 2005; Zhang et al., 2007; Luo et al., 2012). International Soil Society officially established "Urban, Industrial, Mining and Traffic" group in 1998. In 2000, the first "Urban, Industrial, Mining and Traffic Land Soil" international conference was held in Essen, Germany. Since then, researches on urban soil have been rising and deepening in the world, such as distribution, sources and risk assessment of heavy metals, organic matter in urban soil (Li et al., 2001; Manta et al., 2002; Krauss and Wilcke, 2003; Aichner et al., 2007; Morillo et al., 2007; Laidlaw and Filippelli, 2008; Peng et al., 2011; Wu et al., 2016).

As one of the densely populated and fastest urbanized areas, Shanghai, a city with a long history in China, started urban soil environmental research early in 1990s, and carried out a lot of related works. (Lu et al., 1995; Mei et al., 1995; Yang et al., 1997; Nakata et al., 2005; Shi et al., 2008; Jiang et al., 2010; Chen et al., 2012; Wang et al., 2013). The core research team of Shanghai Academy of Environmental Sciences has played an important role in urban soil environmental protection and management. This article focused on the team researched and applications over the past 20 years in the field of urban soil pollution, restoration and management, providing the reference to the peers.

1. Key technology research and development on urban soil pollution analysis, risk assessment and prevention

1.1 Investigation and protection on suburban farmland soil

At present, farmland in Shanghai occupies a large portion of the city area. According to the second national land survey data, it is showed that the city had cultivated land area of 1897.59 square kilometers, accounting for 22.7% of the total area of the land. However, with the high-speed development of social economy in Shanghai and increasingly advancing urbanization, suburban farmland area was shrinking dramatically. The pollution problem derived from it has became a serious issue. Meng et al. (2008), Sun et al. (2010) and Long et al. (2013) measured heavy metal content, analyzed the distribution characteristics and conducted pollution causes in the farmland soil in Shanghai. The results showed that soil quality was affected by a variety of different levels of heavy metals, it exists spatial differentiation and health risk. At the same time, some researches showed that pesticides, polychlorinated biphenyls (PCBs), polycyclic aromatic hydrocarbons (PAHs) and other organic pollutants in the farmland were detected, and they impose a certain risk on human health (Sun et al., 2008; Zhou et al., 2010; Lv et al., 2011; Jiang et al., 2009, 2011). Recently, soil pollution condition investigation also showed that quality of farmland soil in Shanghai is worse than national average level. Major pollutants include PAHs, DDT and heavy metals such as cadmium, nickel and copper etc. Therefore, it is particularly important and urgent to carry out farmland soil related researches.

1) Survey on local soil quality

During the Seventh five-year plan period, the team participated in the Shanghai soil background value investigation. Our team explored different spatial patterns of soil background value of Shanghai, and obtained the soil background values of different natural geographical units of the whole city (Pang et al., 1992; Wang et al., 1992). In 1993, cooperated with Shanghai Academy of Agricultural Sciences, we investigated soil pollution conditions of sewage irrigation areas in Chuansha, Mayi Bang in Pengpu, and zinc-smelting plant in Songjiang. It was found that a huge amount of cadmium was detected in soil, vegetables and rice (Zhang et al., 1993). In 2003, environmental quality investigation of Shanghai Vegetable Base was organized, with total survey area covering 28, 274 hectares, accounting for 50.5% of the total vegetable area. The accumulation situation of heavy metal pollutants in suburb vegetable fields in recent 20 years was determined and the impact and trend of industrial enterprises' influence on farmland soil was analyzed. The results showed that cadmium, mercury, and zinc are the main accumulated heavy metals in vegetable base soil in Shanghai. The accumulation reason not only includes long term suburb industry waste emissions and sewage irrigation in the past, butis also related to the use of calcium superphosphate fertilizer containing cadmium impurities and animal manure organic fertilizer with high content of zinc (Xia et al., 2004; Shen et al., 2006).

2) Analysis and risk control on farmland soil pollution source

With the accelerating of urbanization, the pollution source of farmland soil in Shanghai is not only confined by the agricultural production activity, but also getting more complex. In recent years, the team focused on the present heavy metals, pesticides, polycyclic aromatic hydrocarbons and other key pollutants in agricultural land. We explored the potential sources along with their transition and transformation in farmland soil, to further establish farmland soil quality classification assessment methods and to put forward farmland soil supervision and management system. In addition, other researches such as *the farmland soil ecological risk assessment and prevention, PAHs pollution characteristics, sources and risk prevention and control of the farmland along the traffic route* and so on were launched, to further find out pollution causes and risk characteristics of farmland soil of the city, and provide technical support to protect farmland soil quality and guarantee agricultural product safety.

3) Farmland soil pollution control and remediation

As a basic principle of agricultural soil management, protection and risk control should be a priority. However, for medium and high levels of pollution of farmland soils, for the sake of food security, the landuse characteristics of the farmland soil cannot be changed easily. This is why remediation is needed. Concentration of farmland soil pollution is usually relatively low, while covering large area. General industrial site remediation technologies are not suitable for farmland soil, so the related soil remediation technology or practical engineering case is rare. The team has carried out a beneficial exploration in ecological soil remediation. In 2009, the team carried out the ministry of agriculture public welfare project, which focused on urban agriculture. Based on the retrospective evaluation of agricultural soil pollution, the team combined artificial simulation and natural conditions, integrated and developed soil remediation technology which used to deal with soil salinization and soil heavy metal pollution, and formulated the corresponding applicable technical guide (Huang et al., 2009; Han et al., 2012; Qian et al., 2014).

1.2 Key technology of soil investigation and treatment for construction site

Shanghai is a developed region in China, locating at the junction of the East Coast and the Yangtze River. Due to the multi trans-routes including Yangtze River, Beijing-Shanghai highway, Shanghai-Hangzhou highway in all directions across the city, Shanghai is the national transportation hub. Because of the superior and unique geographical environment, through centuries, Shanghai has the largest comprehensive industrial base in China, yet, those developments also imposed serious industrial pollution on soil environment. The heavy pollution, caused by chemical manufacturing factories, pharmaceutical workshops, electroplate-making factories and textile mills in both urban and rural area, was remained in soil. Our investigation confirmed that the main contaminants in soil include heavy metals, volatile organic compounds (VOCs), semi-volatile organics (SVOCs), total petroleum hydrocarbons (TPH), pesticides and others. A survey found that the contents of the heavy metal and organics in contaminated soil are several times more than those in the reference monitoring point, if sampled in the same period. Even more, these organic solvents formed a free phase in soil underneath. Unfortunately, the contaminated soil leads to secondary contamination in the underground water layer.

In the new century, two-thirds industrial lands have been substituted out for urban infrastructure and tertiary industries because of urban structure and function layout rearrangement. In fact, the original land of the industrial site has been polluted and will impose potential health risks on human life. How to govern and remediate these contaminated soil to meet the needs of the development and utilization has become an urgent problem. (Li et al., 2010; Jiang et al., 2011; Hu et al., 2012)

1) Environmental investigation technology in mega industrial sites

Most of industrial sites in Shanghai have a long history with diversified working process. A great amount of residuals easily formed planar or ribbon plume in soil, presenting a staggered distribution with heavy metal absorption, organic co-solvent, LNAPL and DNAPL. Although traditional investigation techniques are easy to operate (such as spiral and double sleeving machine, etc.), drilling without sealing could lead to cross-contamination and flooded sand, especially in the Yangtze River Delta coastal areas. With the development of investigation techniques, a series of precise and rapid investigation techniques, such as detection of interface structure, X-ray fluorescence, ground-penetrating radar, can instantly and effectively show contaminants distribution in sites and hydrogeological situation. In recent years, our research team has developed and practiced Membrane Interface Probe coupled with a variety of detectors (FID/PID/ECD/EC) applied for monitoring the contaminants in local soil environment. This technique obtained synchronous respond as describing the scope and extent of BTEX, chlorinated hydrocarbons and other polar organic contaminations in chemical sites. Our research found that most free phase of chlorinated hydrocarbons clustered in the underground depth of 6-9 meters with a clustered contamination distribution (Guo et al., 2013) in Shanghai. The coupling technique, Membrane Interface Probe (MIP) coupled to multi-detectors, is a breakthrough of investigation techniques in mega contaminated sites. It is also an effective, rapid and modern technique to detect the contamination situation in groundwater. This technique had been widely applied for exploration in recent super large area-TAOPU SMART CITY. Furthermore, our team has explored the stable carbon isotope technique used in investigation of PAHs contaminated soil and source analysis (Wu et al., 2016); Resistance survey technology (Sha et al., 2016).

2) HHRA technology in super large industrial sites

The risk assessment of soil and underground water has been developed from a single method to diversified assessment methods (Li et al., 2006; Tan et al., 2008). Actually, HHRA has been widely implemented in national environmental management and practice. Ten years ago, our research team adopted risk assessment techniques to build up methodology for soil quality evaluation and remediation issues in super large sites such as Expo site in Shanghai (Geng et al., 2014). Up to now, we have developed local HHRA models on different exposure scenarios (including residential land, green parks, educational land, commercial land, industrial land and municipal land). Local exposure parameters were identified clearly in these models, concerning shanghai regional population characteristics. We found that the modified HHRA model is more precise and suitable on calculating the shanghai population health risk.

American RBCA is the basis of exposure concept modeling on HHRA. However, adjusted soil gas modeling, exposure features in different scenarios, local receptors sign index and on the-spot exposure time are applicable to independent Shanghai HHRA modeling for super-large sites. In recent years, such methodology (Li et al., 2009) has been widely applied in Shanghai International Tourism Resort Disneyland (7.0 km^2), Taopu Science and Technology Industrial Park-SMART CITY (7.9 km^2), Baoshan Nanda area (6.29 km^2) and other redevelopment construction sites and projects. Overall, HHRA technology not only ensures the safety of land, but also saves the cost of construction. Meanwhile, it fulfills the purpose of the practice of low-carbon environmental protection and green governance.

3) Treatment and remediation technology of complex contaminated sites

Since 2004, a series of work, with references of developed countries' experiences, has been carried out to explore the remediation technology of municipal contaminated soil. Research projects sourced from EXPO sites were conducted. National 863 Project named "*Research of Fast blockage, control and remediation technology and equipment development for contaminated soil treatment*" and National Public Welfare Project commissioned by Ministry of Environmental Protection named "*Specific technical specifications of solidification/stabilization and biopile treatments for contaminated soil remediation*" were also conducted in 3 years. By these research, various stabilizing agent, curing agent and bioremediation synergistic agent were developed; technologies of solidification/stabilization and biopile treatment were established systematically; the specific technical specifications of remediation technology were drew up.

From the year of 2010, the remediation research work entered a rapid development period. Remediation engineering of Disney land and several large residential communities were completed successively. Research of solidification/stabilization technology for typical metal contaminated soil including arsenic, chromium, copper, nickel, lead, zinc and beryllium was developed in production (Lu et. al., 2013; Wang et. al., 2013). Moreover, technique research of strengthened soil vapor extraction was practiced to remediate BTEX contamination (Liao, et. al, 2013; Zhu, et. al, 2013). Furthermore, advanced oxidation processes had been researched to remediate the PAHs (Meng et. al., 2013) and oil contaminated soil (Chen et. al., 2016). In addition, a series of advanced oxidation agent (Zhang et al., 2013) and soil synergist (Yu et al., 2016) were developed. The technologies of advanced oxidation processes and soil vapor extraction were also constructed (Luo et al., 2013).

In recent years, collaborated with health risk control and urban planning, the remediation and treatment lead an integrated tendency to deal with water-soil-co-remediation. According to the geo-characteristics and social development trend, our team organized and implemented research projects from Shanghai Committee of Science and Technology named *"The comprehensive assessment of industrial site pollution as well as the research and demonstrated application of remediation integrated technology of co-contaminated soil"*, *"The groundwater pollution survey assessment and the key remediation technique research and demonstration of redevelopment site in massive industrial district"* etc. We studied the remediation technologies at complicated contaminated sites, including heat strengthened soil vapor extraction, chemical oxidation-reduction, the heavy metal ions consolidation and soil leaching process. Moreover, the quick-treatment integrated equipment was developed in view of the Shanghai soil characteristics of high viscosity (Luo et. al., 2013). Overall, in 10 years of site remediation work, the research team has made great effort to build up the soil remediation technology system, including solidification/stabilization, advanced oxidation, biopile treatment, soil vapor extraction and soil leaching technology as the core technologies with self-owned intellectual property right.

2. Urban soil environmental management system based on the full life cycle

2.1 Potential contaminated sites screening, assessment and management

As a historic industrial development megalopolis in China, under development for more than a century, Shanghai urban soil environment has been impacted by the industrial emissions and improper disposal of waste. Urban soil in Shanghai is complex, and it has cumulative characteristics in the long term land use pattern evolution process. It brings obstacles on planning and management of land development and utilization. Since 2004, the team began to explore potential contaminated sites identification and authentication methods in Shanghai, recovering the urban soil potential pollution information for nearly a century of Shanghai urban development. Soil pollution source database was established in seven representative years (1933,

1947, 1956, 1966, 1986, 1996, 2001), reflecting Shanghai urban industry development characteristics (Shen et al., 2006; Huang et al., 2012). In 2014, according to the state and municipality deployment and requirements on soil environmental protection and management, the team launched investigation at industrial enterprises and municipal facilities sites in Shanghai. The potential contaminated site screening techniques were carried out based on key industry pollution feature analysis and typical site recognized environment condition identification.

Meanwhile, potential contaminated site (brownfield) grading evaluation methods were built based on the mechanism of site classification and prioritization. At present, the potential polluted industrial and municipal site directory list has been formed through sites screening. Potentially contaminated sites information collection system based on mobile terminal was established and put into operation. Eventually After all of those potential contaminated sites screening work finish, potential contaminated site management system will be built. By then, information dynamic update and management will be accomplished, providing technical support and foundation platform for full implementation of site environmental management (Huang et al., 2015).

2.2 Policy, management, and guidance in the process of soil environmental supervision

City land resources often change dynamically during the rapid city development. Since the 1990s, the industrial structure has been adjusted in Shanghai. 1/3 of heavy industries were transferred to suburbs or counties; 1/3 of polluted enterprises were shut down; 2/3 industrial land located in the center were replaced for the development of urban infrastructure and the third industry.

In order to protect the environmental safety for redevelopment of the industrial and municipal sites, a document (No.188) was jointly issued by Shanghai EPA, Shanghai City Planning and Land Resources Administration, Shanghai Municipal Commission of Economy and Informatization and Shanghai City Construction and Administration Commission on 30th April, 2014, for risk control and supervision on industrial lands of "12+3" kinds. Under the document, those "12+3" kinds should obey the rule, in the process of dynamic land-use type transformation. No.188 Document was drafted and compiled by our team and other technical specifications (on trial), which provide technical support for the implementation of No.188 Document, have also been prepared.

The technical specifications for local site investigation, monitoring, assessment and remediation were prepared from 2011 to 2015, some of which were issued and tried in succession, including "*Technical specification for environmental site investigation*", "*Technical specification for environmental site monitoring*", "*Technical specification for risk assessment of contaminated site*", "*Technical specification for site soil remediation*", "*Technical specification for site remediation supervision", "Technical specification for site remediation acceptance".* According to the local environmental characteristics and typical contaminated sites, the study of *"Screening Levels for soil environmental risk assessment in Shanghai"* was in consideration of the specific land-use type in Shanghai. This study is one of the important criteria and standards for designing detailed monitoring plan in the further. The technical specifications above have been applied to the investigation, monitoring, assessment and remediation of more than 100 sites and have made benefits for our social and economy. Some other technical specifications for pollution control such as industrial enterprises removal and soil reuse after remediation are still under study.

In addition, in order to provide necessary supplement for establish entire environmental supervision and technical system of the industrial site lifecycle management, the corresponding technical guidelines for industrial periodic investigation and assessment in whole procedure are under study.

3. Typical case analysis of urban soil remediation

With the proceeding of urbanization and the second land exploitation, the demands for site remediation in Shanghai are increasing. Our research group have made full use of the research experience, and integrated the self-developed remediation technique and equipment to solve the problems occurred in the soil remediation project. We have accumulated rich experience in the site remediation, especially for the remediation of large sites with complex conditions, which resulted in the creating and popularizing of the advanced techniques and equipment.

3.1 Investigation and assessment of Expo site and soil remediation

The venues of the 2010 Shanghai World Expo located on both sides of Huangpu River in Shanghai, covering 5.28 square kilometers. Originally, in this area, there were dozens of large industrial enterprises with a long history, such as Jiangnan shipyard, Pusteel Company affiliated to Baosteel group, Nanshi power plant, Shanghai Auxiliaries Factory. In 2004, Shanghai World Expo Coordination launched the project. The project lasted for 6 years and has already totally completed the remediation of more than 200 thousand cubic meters of contaminated soil (heavy metals contaminated soil: 15, 000 m^3, organic contaminated soil: 193, 000 m^3). The project is the first large-scale regional soil remediation program in China. Our group took on the overall project design, organization and implementation, engineering technology consulting and construction supervision.

Based on the relevant science and technology achievements of Ministry of Science and Technology, Ministry of environmental protection, Shanghai Science and Technology Commission, and Shanghai Environmental Protection Bureau, we are the first one introduced risk assessment methods for contaminated sites in the world, proposed soil environmental standard and soil guideline values for the Expo site, and established "*the soil environmental quality evaluation standards for the land use in exhibition (provisional regulation)*" (HJ/T 350-2007), which was the first standard of soil environmental quality assessment of contaminated sites in China. After finishing experiment research and field test, we proposed the technical and engineering schemes of soil remediation in Expo sites. We treated the heavy metal contaminated soils by solidification and stabilization, and remediated the organic soil with TPH and PAHs by composting technology of off-site bioremediation. After remediation, all contaminated soils realized resourceful utilization.

3.2 Investigation and assessment of Disney land and site remediation project

Shanghai Disney Resort covers a total area of 7 square kilometers, with the first-stage construction land covers 3.9 square kilometers. This area locates on the rural-urban continuum in Pu Dong district of Shanghai. Originally, the majority types of land use were agricultural land, residential land and small and medium-sized enterprises. From 2009, our group took the environment investigation, risk assessments, remediation project design and construction supervision. Phased field investigation and technology of health risk assessment were applied to the development and construction of Disney land. We used the method of off-site solidification and stabilization for the heavy metals contaminated soils investigated by first-stage field study. With regard to the volatile and semi-volatile organics contaminated sites investigated by second-stage field study, we combined the techniques of vapor extraction and advanced oxidation treatment to remediate the contaminated soils, to guarantee the safety of land use. This project applied Self-developed craft systems, green environmental functional materials and professional engineering equipment. After remediation, the concentrations of pollutants with high concern all reached preliminary remediation goals in the pre-design, and the project finally gained good economic and social benefits.

3.3 Contaminated soil remediation project in Nanda Area of Baoshan District

Nanda area locates on Baoshan district in Shanghai, which is besides the northwest part of central city in Shanghai. The area covers 6.29 square kilometers, mainly used for industrial and storage land, including about 1000 large and small enterprises. Over time, there were problems in this area such as low industry energy level, poor infrastructure, high potential safety hazard, serious environmental pollution and petition contradictions. From April 2012 to October 2014, our group team started the investigation and identification of pollutants, the preliminary investigation of site environment, detailed survey and health risk assessment. The results of environmental survey and health risk assessment indicated that in parts of this area, the pollution risks of arsenic, chromium, nickel, lead, polycyclic aromatic hydrocarbons, total petroleum hydrocarbons and BTEX exceeded the acceptable level. About 5300 m^3 contaminated soil needed to be remediated to meet the environmental protection standard of further land use. Our group team further worked out the technical proposal of site remediation in Nanda area, and took on the construction project of the total package. Through technical comparison and a pilot-scale test, the contaminated soils were centralized by ex-situ treatment. Soil Washing, gas-phase extraction and advanced oxidation technology were combined to treat the heavy metals, volatile and semi-volatile organics in soil.

3.4 Contaminated site remediation project in Taopu Industrial Zone

Planned Taopu SMART City locates on the northwest of Shanghai, and it covers an area of 7.92 square kilometers. It originated from Taopu industrial zone in the early 1950s, a typical old industrial zone with heavy pollution. It faced the problems of urban regeneration, industrial upgrading and environmental control. From June 2013 to October 2015, our group team took on the overall project design of environmental investigation, assessment and site remediation. With the help of the detailed survey of regional pollution resources and characteristics of soil and groundwater pollution, we, based on the area plan of city transition and upgrade, designed and worked out the technical proposal of contaminated soil and groundwater remediation. For heavy metal contaminated soils, stabilization treatment and soil washing were applied. For semi-volatile organics, advanced oxidation and thermal desorption were proposed. For volatile organics contaminated soils, vapor extraction was used. For contaminated groundwater, in situ chemical oxidation/reduction and groundwater extraction treatment were proposed. Combining site construction conditions, on the basis of groundwater extraction treatment, the remediation was assisted by the isolation barrier technology of cement mixing pile. In the case of redevelopment of large contaminated sites with complicated pollution, the establishment of site remediation proposal in Taopu industrial zone provides a national-wide reference to remediation and treatment in China.

3.5 Typical chlorinated hydrocarbons contaminated groundwater in-situ remediation project in enterprise in production

The foreign invested manufacturing company locates on the Pudong district of Shanghai. Chlorinated hydrocarbons were used to wash metal parts for degreasing treatment in history. In 2007, DNAPLs were self-inspected in the shallow underground water within the site. Afterwards, international consulting expert team was employed to carry out continuous tracing monitoring, risk assessment and remediation experiments. In 2010, our group team was included in the site survey and remediation. Based on related research achievements of national 863 plan, Multiple Phase Extraction (MPE) system was developed and applied to treat the chlorinated hydrocarbons contaminated groundwater, and the free phase of DNAPLs in underground water was recovered effectively. We developed environmental friendly in situ bioremediation materials, and carried out in situ remediation tests to treat underground water by the way of artificially intensified natural attenuation.

4. Urban soil research hotspot and trend in the future

(1) Urban soil suffers from dense and diverse human disturbance, and shows totally different characteristics from natural soil properties. Hence, environmental capacity and the critical threshold of urban soil, pollutant migration and transmission dynamics, aerial interaction effects of water and soil, and the biological effect of pollution need to be studied more deeply and comprehensively. It is helpful to establish a more scientific system on pollution risk assessment.

(2) For the high speed and frequency of land renewal and reuse in the process of urbanization, soil pollution risk is easy to be ignored even spread. How to quickly and easily identify the risk of soil pollution, simplify and strengthen the whole process of soil environmental regulatory process, are the emphases and difficulties in urban soil environmental management in the future.

(3) Environmentally friendly remediation technology with low energy consumption and low emission is the main direction of the future. China has vast land, and different regional soil properties. There are many different kinds of pollutants and complex pollution types. At the same time, the urban land resources are limited, and sites development cycles are short. Hence, it is urgent to develop an efficient, fast and safe remediation technology for the urban site pollution characteristics in our country.

(4) There are shallow buried depth of groundwater and high soil viscosity in the middle and lower area of Yangtze River. Contaminated sites remediation technologies for compound pollution such as heavy metals and organics are in need. This technology includes the eluent research and development of rapid leaching technology for heavy metal polluted soil, in-situ thermal stripping technology and equipment for organic polluted soil, in-situ enhanced bioremediation technology, and soil groundwater joint remediation technology.

(5) Contaminated sites management mode based on the Triad is new, which includes three elements: systematic project planning, dynamic work strategies, and real-time measurement technique. Triad's core technology is based on the systematic project plan to achieve real-time online measurement and formulate dynamic working strategy. With obvious superiority and practicability, Triad contaminated sites management model is the future trend.

References

Aichner B, Glaser B, Zech W. 2007. Polycyclic aromatic hydrocarbons and polychlorinated biphenyls in urban soils from Kathmandu, Nepal. Organic Geochemistry, 38 (4): 700-715.

Chen C C, Li Q Q, Wang J, et al. 2016. Advanced oxidation technology for remediation of petroleum-contaminated tidal flat (in Chinese). Chinese Journal of Environmental Engineering.

Chen Y Y, Wang J, Gao W, et al. 2012. Comprehensive analysis of heavy mentals in soils from Baoshan District, Shanghai: a heavily industrialized area in China. Environmental Earth Sciences, 67 (8): 2331-2343.

De Kimpe C R, Morel J L. 2000. Urban soil management: a growing concern. Soil Science, 165 (1): 31-40.

Fu R B, Liu F, Zhang C B, et al. 2013. Effects of permeable reactive Composite electrodes on hexavalent chromium in the electrokinetic remediation of Contaminated soil. Environmental Engineering Science, 30 (1): 17-22.

Geng C N, Gao Y J, Li D, et al. 2014. Contamination investigation and risk assessment of molybdenum on an industrial site in China. Journal of Geochemical Exploration, 144 (B): 273-281

Guo L, 2013. Natural attenuation mechanism and capability of chlorinated hydrocarbons in shallow groundwater in a study area in Shanghai (in Chinese). Environmental Science and Technology, 26 (3): 9-13.

Hu W X, Ying H M, Zhou J. 2012. Practice of Investigation and Remediation for Contaminated Sites (in Chinese). Beijing: China Environmental Science Press.

Huang S F, Shen G X, Tang H, et al. 2009 The integration and application of restoration techniques for agricultural soil (in Chinese). Shanghai: Shanghai Academy of Environmental Sciences.

Huang S F, Wang M, Wu J, et al. 2015. The investigation of potential contaminated industrial and municipal sites in Shanghai (first period) (in Chinese). Shanghai: Shanghai Academy of Environmental Sciences.

Huang S F, Wang M, Wu J, et al. 2012. Study on the soil environment protection in Shanghai in the 12th Five-Year Plan (in Chinese). Shanghai: Shanghai Academy of Environmental Sciences.

Han S H, Huang S F, Tang H, et al. 2012. A conparative study on the performance of 3 plants for remediation of cadmium contaminated farmland soil (in Chinese). Environmental Pollution and Control, 34 (12): 2-30.

Jiang L, Gong Y Y, et al. 2011. Risk Assessment and Remediation Validation Handbook for Contaminated Site and Facility (in Chinese). Beijing: China Environmental Science Press.

Jiang Y F, Wang X T, Zhu K, et al. 2010. Occurrence, compositional profiles and possible sources of polybrominated diphenyl ethers in urban soils of Shanghai, China. Chemosphere, 80 (2): 131-136.

Jiang Y F, Wang X T, Jia Y, et al. 2009. Occurrence, distribution and possible sources of organochlorine pesticides in agricultural soil of Shanghai, China. Journal of Hazardous Materials, 170 (2-3): 989-997.

Jiang Y F, Wang X T, Wu M H, et al. 2011. Contamination, source identification, and risk assessment of polycyclic aromatic hydrocarbons in agricultural soil of Shanghai, China. Environmental Monitoring and Assessment, 183 (1-4): 139-150.

Tan J, Li Q Q, Loganath A, et al. 2008. Multivariate data analysis of persistent organic pollutants in maternal adipose tissue in Singapore. Environmental Science and Technology, 42 (7): 2681-2687.

Krauss M, Wilcke W. 2003. Polychlorinated naphthalenes in urban soils: analysis, concentrations, and relation to other persistent organic pollutants. Environmental Pollution, 122 (1): 75-89.

Laidlaw M A S, Filippelli G M. 2008. Resuspension of urban soils as a persistent source of lead poisoning in children: a review and new directions. Applied Geochemistry, 23 (8): 2021-2039.

Lu C, Li Q Q, Luo Q S, et al. 2013. Stabilization treatment of available arsenic in contaminated soils and mechanism studies (in Chinese). China Environmental Science, 33 (2): 298-304.

Li X D, Poon C S, Liu P S. 2001. Heavy metal contamination of urban soils and street dusts in Hong Kong. Applied Geochemistry, 16 (11-12): 1361-1368.

Lv J G, Bi C J, Chen Z L, et al. 2011. Characteristics of organochlorine Pesticide residues in agricultural soil of Chongming Island in Shanghai (in Chinese). Environmental Science, 32 (8): 2455-2461.

Long Q, Wang J Y, Da L J. 2013. Assessing the spatial-temporal variations of heavy metals in farmland soil of Shanghai, China. Fresenius Environmental Bulletin, 22 (3A): 928-938.

Li G H, Li F S, Zhang X, et al. 2010. Technologies on Environmental Risk Assessment and Remediation on Contaminated Sites (in Chinese). Beijing: China Environmental Science Press.

Li Q Q. 2010. Procedures and methodology for studying soil remediation objectives based on the human risk assessment-a case study of reuse of PAH-contaminated soil in Shanghai (in Chinese). Journal of Ecology and Rural Environment, 26 (6): 610-615.

Li Q Q, Luo Q S, Zhen W, et al. 2009. Sustainability of soil remediation technologies a case study of in-situ stabilization/ solidification and landfill technology (in Chinese). Soil, 41 (2): 308-314.

Li Q Q, Loganath A, Chong Y S, et al. Levels of persistent organic pollutant residues in human adipose and muscle tissues in Singapore. Journal of Toxicology and Environmental Health, Part A, 2006, 69 (21): 1927-1937.

Luo Q S, Yang J, Zhu Y, et al. 2013. Integrated crushing and mixing equipment (in Chinese): China, CN202655328U.

Luo Q S, Zhu J, Liao Z Q, et al. 2013. Thermally enhanced contaminated-soil vapor extraction advanced oxidation in-situ remediation facility (in Chinese): China, CN202638859U.

Lu Q, Mei Z M, Pan J H. 1995. The evaluation for environment quality of soil and groundwater in an old manufactory (in Chinese). //Symposium of Shanghai Geotechnical Engineering Detecting Centre. Shanghai, 216-220.

Luo X S, Yu S, Zhu Y G, et al. 2012. Trace metal contamination in urban soils of China. Science of the Total Environment, 421: 17-30.

Liao Z Q, Zhu J, Luo Q S, et al. 2013. Thermodesorption of BTEX-contaminated soil (in Chinese). Environmental Chemistry, 32 (4): 646-650.

Manta D S, Angelone M, Bellanca A. 2002. Heavy metals in urban soils: a case study from the city of Palermo (Sicily), Italy. Science of the Total Environment, 300 (1-3): 229-243.

Meng F, Liu M, Shi T G. 2008. Evaluation on environmental quality of heavy metals in agricultural soils of Shanghai (in Chinese). Environmental Science, 29 (2): 428-433.

Meng L. 2013. Regulation of biopile for PAH removal in urban contaminated soil (in Chinese). Environmental Science and Technology, 36 (5): 124-129.

Meng L, Guo L, Li B Z, et al. 2013. Simulation experiment system and method for phytoremediation of PAHs-contaminated soil (in Chinese): China, CN103624070A

Morillo E, Romero A S, Maqueda C, et al. 2007. Soil pollution by PAHs in urban soils: a comparison of three European cities. Journal of Environmental Monitoring, 9 (9): 1001-1008.

Mei Z M, Lu Q, Zhu G. 1995. Investigation for environmental quality of soil and groundwater in the east outer ring road in Shanghai (in Chinese). Symposium of Shanghai Geotechnical Engineering Detecting Centre. Shanghai, 221-224.

Nakata H, Hirakawa Y, Kawazoe, et al. 2005. Concentrations and composition of organochlorine contaminants in sediments, soils, crustaceans, fishes and birds collected from Lake Tai, Hangzhou Bay and Shanghai city region, China, Environmental Pollution, 133 (3): 415-429.

Peng C, Chen W P, Liao X L, et al. 2011. Polycyclic aromatic hydrocarbons in urban soils of Beijing: Status, Source, distribution and potential risk. Environmental Pollution, 159 (3): 802-808.

Pang J H, Wang Y G, Zha J S. 1992. Soil backgrounf values and influential factors of Cu Mn Co As Fe Mo and total rare-earth elements in Shanghai area (in Chinese). Acta Agricultural Shanghai, 8 (2): 65-68.

Qian X Y, Shen G X, Guo C X, et al. 2014. Reclamation of secondary salinized soils in protected vegetable fields using different wastes (in Chinese). Journal of Agro-Environment Science, 33 (4): 737-743.

Scharenbroch B C, Lloyd J E, Johnson-Maynard J L. 2005. Distinguishing urban soils with physical, chemical, and biological properties. Pedobiologia, 49 (4): 283-296.

Sha C Y, Wu J, Wang M, et al. 2016. Stable carbon isotopic composition and distribution of polycyclic aromatic hydrocarbons in soil and source apportionment (in Chinese). Shanghai: Shanghai Academy of Environmental Sciences.

Shen G X, Huang S F, Wu J, et al. 2006. Research on the investigation, disposal and management of unban soil (in Chinese). Shanghai: Shanghai Academy of Environmental Sciences.

Shen G X, Xie Z, Qian X Y, et al. 2006. Investigation and analysis of heavy metal accumulation in the soil of vegetable cropland in Shanghai (in Chinese). Journal of Agro-Environment Science, 25 (S1): 37-40.

Shi G T, Chen Z L, Xu S Y, et al. 2008. Potentially toxic metal contamination of urban soils and roadside dust in Shanghai, China. Environmental Pollution, 156 (2): 251-260.

Sun C, Bi CJ, Chen ZL, et al. 2010. Assessment on environmental quality of heavy metals in agricultural soils of Chongming Island, Shanghai City. Journal of Geographical Sciences, 20 (1): 135-147.

Sun X J, Shi C, Xu S Y, et al. 2008. Converntration and sources of polycyclic aromatic hydrocarbons in surface soil of north suburban Shanghai, China (in Chinese). Research of Environmental Sciences, 21 (4): 140-144.

Wang X T, Miao Y, Zhang Y, et al. 2013. Polycyclic aromatic hydrocarbons

(PAHs) in urban soils of the megacity Shanghai: occurrence, source apportionment and potential human heatlh risk. Science of the Total Environment, 447: 80-89.

Wang J, Luo Q S, Zhang C B, et al. 2013. Stabilization and long-term effect of chromium contaminated soil (in Chinese). Environmental Science, 34 (10): 310-315.

Wu J, Tan J, Wang M, et al. 2016. Analysis for the composition, distribution and sedimental apportionment of a certain oil-polluted tidal marsh source area in terms of TPH and PAH (in Chinese). Journal of Safety and Environment, 16 (1): 282-287.

Wu J, Wang M, Jin Z H, et al. 2016. Review and prospect of research on polycyclic aromatic hydrocarbons in soil environment: a bibliometric analysis based on megadata of Web of Science (in Chinese). Acta Pedologica Sinica (in press).

Wang M, Wu J, Tang H, et al. 2015. Research on pollution source apportionment and Hierarchical control measures of agriculture soils (in Chinese). Shanghai: Shanghai Academy of Environmental Sciences.

Wang Y, Wang Y G, Luo H L, et al. 1992. Background Value of Soil in Shanghai (in Chinese). Beijing: China Environmental Science Press.

Xia D X, Shan Z C, Xie Z, et al. 2004. Environmental quality investigation of "Vegetable Basket" Base in Shanghai (in Chinese). Shanghai: Shanghai Academy of Environmental Sciences.

Yang K, Wang Y, Xu Q X, et al. 1997. Approach on the environmental impact and protection measures on functional replacement of urban industrial land use: a study case of a certain sulfuric acid plant in Shanghai (in Chinese). Pollution Control Technology, 10 (2): 100-102.

Yu Kai, Li Q Q, Zhu Jie, et al. 2016. Method for deep degradation of high concentration chlorinated nitrobenzene in soil by using mixed surfactants and Fenton oxidation (in Chinese): China, CN103624074A.

Zhang C B, Yang J, Luo Q S, et al. 2013. Simulation experiment device and method for the advanced oxidation of contaminated soil by thermal steam activation of activated persulfate (in Chinese): China, CN103624073A.

Zhang G L, Zhao Y G, Yang J L, et al. 2007. Urban soil environment issues and research progresses (in Chinese). Acta Pedologica Sinica, 44 (5): 925-93.

Zhang G L. 2001. Deepening and development of urban soil research (in Chinese). Soil, 2: 111-112.

Zhang C B, Luo Q S, Geng C N, et al. 2010. Stabilization treatment of contaminated soil: a field-scale application in Shanghai, China. Frontiers of Environmental Science and Engineering in China, 4 (4): 395-404.

Zhang R, Zhang W, Liu G, et al, 2016. Changes of lead speciation and microbial toxicity in soil treated with repeated Pb exposure in the presence of BDE209. Environmental Science and Pollution Research, 23 (5): 4621-4628.

Zhou J C, Bi C J, Chen Z L, et al. 2010. Residues of polychlorinated biphenyls in agricultural fields of Chongming Island in Shanghai (in Chinese). China Environmental Science, 30 (1): 116-120.

Zhang J Q, Jiang J H, Zhou Z R, et al. 1993. Research on the Trategy of Rural Environmental Protection in Shanghai (in Chinese). Shanghai: Shanghai Science and Technology Press.

Zhu Jie, Luo Q S, Li X Q. 2013. Thermodesorption of BTEX-contaminated soil using heat conduction (in Chinese). Environmental Chemistry, 32 (8): 1546-1553.

Part Ⅱ Speciation, Bioavailability and Risk Assessment

Copper Speciation and Transformation in Soil-Plant System

Jiyan SHI[1], Lijuan SUN[1], Cheng PENG[1], Chen XU[1], Yuanpeng WANG[1,2], Xincai CHEN[1,3], Huirong LIN[1,4], Jianjun YANG[1,5], Tingting LIU[1,6], Yingxu CHEN[1]

[1]Department of Environmental Engineering, College of Environmental and Resource Sciences, Zhejiang University

[2]Department of Chemical and Biochemical Engineering, College of Chemistry and Chemical Engineering, Xiamen University

[3]Zhejiang HI-TECH Environmental Technology Co.

[4]Institute of Urban Environment, Chinese Academy of Science

[5]Institute of Environmental and Sustainable Development in Agriculture, Chinese Academy of Agricultural Sciences

[6]Institute of Environmental Engineering, Department of Civil Engineering, Tongji Zhejiang College

Abstract

The widespread use of copper (Cu) resulted in anthropogenic inputs to soils through industrial and agricultural practices. Elevated soil Cu concentrations can cause toxicity effects on soil organisms and affect soil ecosystem functioning. The toxicity of Cu to organisms strongly depends on its bioavailability in soil and the sensitivity of the organisms. Molecular-level knowledge about chemical speciation of Cu in soils will advance our ability to predict its reactivities and environmental fates. The speciation and transformation of Cu in soil-plant system were intensively studied during the past decade by combining traditional chemical analysis methods and advanced synchrotron radiation based techniques, including X-ray absorption fine structure (XAFS), X-ray fluorescence (XRF) and scanning transmission X-ray microscopy (STXM). Bioavailability of Cu in soil can be mediated by sulfur and some specific rhizobacterium like Cu-tolerant *Pseudomonas putida* CZ1, which is isolated from rhizosphere of *Elsholtzia splendens*. Excess Cu will cause phytotoxicity to plants, XAFS analysis demonstrated that Cu bound by N/O/S ligands played a key role in Cu detoxification of *E. splendens*. Copper stress leads to different functional protein expression in different *E. splendens* tissues, which includes signal transduction, regulation of transcription and translation, energy metabolism, regulation of redox homeostasis and cell defense. CuO nanoparticles can be absorbed by the roots and transported to the shoots in *E. splendens* and rice (*Oryza sativa* L.), and then transformed into different Cu species, like Cu-citrate, Cu-cysteine. Sulfur and rhizobacterium (such as *Pseudomonas putida* CZ1) enhanced phytoremediation of Cu contaminated soil are considered to be promising, and further practices should be in field conditions.

Keywords: copper, biotransformation, rhizosphere, bioavailability, *Elsholtzia splendens*

1. Introduction

Copper (Cu) is the first element of group Ib of the periodic table of elements, ranking 26th in abundance in the lithosphere (Alloway, 1995). The average abundance of Cu in the earth's crust is 60 mg/kg and Cu concentrations in soil typically vary between 2 and 50 mg/kg (Rankin, 2009). Copper is widely used in human society and it has a long history of development. Copper is mainly used in electrical applications (65%), and construction (25%). Other uses are transport (7%) and coins, sculptures, musical instruments and cookware (Alloway, 1995). The long history and widespread use of Cu has resulted in significant anthropogenic emissions, which have caused increased environmental Cu concentration (Hong et al., 1996).

Copper is an essential trace element and it is vital to the health of all living organisms. It is required for the biological activity of proteins and enzymes as cofactors (Muccifora and Bellani, 2013). It is directly related to photosynthesis, respiration, perception of ethylene and cell wall remodeling, and also plays an indirect role in nitrogen assimilation and abscisic acid biosynthesis (Peñarrubia et al., 2010; Burkhead et al., 2009). Soil with low bioavailable Cu can result in losses in crop yield and deficiency symptoms in livestock, especially in intensive farming systems. Generally, Cu deficiency ($<$5 mg/kg DW in vegetative tissues) in plants will exhibit reduced growth rate and development, with the youngest leaves, the apical meristem and seeds most vulnerable to be affected (Burkhead et al., 2009; Marschner, 2011; Ryan et al., 2013). However, elevated concentration of Cu ($>$20 mg/kg DW in vegetative tissues) can be highly deleterious to plants by inducing the generation of reactive oxygen species (ROS), damaging lipids, proteins or DNA, reducing Fe uptake, even to the point of deficiency (Drążkiewicz et al., 2004; Mehrabi and Wilson, 2007; Burkhead et al., 2009; Marschner, 2011).

The toxicity of Cu in soils to organisms strongly depends on Cu speciation and bioavailability. Synchrotron radiation-based techniques are widely utilized to investigate the speciation of heavy metals in environment and biological systems, providing average oxidation state and local structural information of heavy metals. Understanding the transformation processes of Cu in environment in molecular scale can give clues for the remediation of Cu contaminated soils. This chapter gave a brief introduction of our recent years' research work on the Cu speciation and transformation in soil-plant systems, the potential of chemical and rhizobacterium assisted phytoremediation to Cu-contaminated soils was also discussed.

2. Speciation and transformation of Cu in soils

Factors affecting the ecotoxicity and mobility of Cu include not only the total Cu concentration in soils but also environmental aspects, soil property and Cu bioavailability. Sulfur is an essential element for terrestrial organism. The biogeochemical cycle of sulfur can affect the speciation transformation of Cu in soils. The rhizosphere which is influenced by roots, mycorrhizal fungi, and bacteria has a large specific surface area and an elevated negative charge and could attract metals with strong bonds. The description of the speciation of Cu in rhizosphere soil is of critical importance. During these years, we explored the sulfur and rhizobacterium mediated Cu transformation in soils.

2.1 Speciation characterization of Cu in soils

Molecular-level understanding of soil Cu speciation and distribution assists in management of Cu contamination in mining sites. Multiple synchrotron-based bulk and spatially resolved techniques were applied for the analysis of the speciation and distribution of Cu in a mining site contaminated since 1950s, as well as other related elements (Fe, Ca, Mn, K, Al, and Si) (Yang et al., 2014). Bulk X-ray absorption near-edge structure (XANES) and extended X-ray absorption fine structure (EXAFS) spectroscopy revealed that soil Cu was predominantly associated with Fe oxides instead of soil organic matter. This agreed with the closest association of Cu to Fe by microscopic X-ray fluorescence (μ-XRF) and scanning transmission X-ray microscopy (STXM) nano-analysis, along with the non-occurrence of photoreduction of soil Cu (Ⅱ) by quick Cu $L_{3,2}$-edge XANES spectroscopy (Q-XANES) which often occurs when Cu organic complexes are present. Furthermore, bulk-EXAFS and STXM-coupled Fe $L_{3,2}$-edge nano-XANES analysis revealed soil Cu adsorbed primarily to Fe (III) oxides by inner-sphere complexation. Additionally, Cu K-edge μ-XANES, $L_{3,2}$-edge bulk-XANES, and successive Q-XANES results identified the presence of Cu_2S rather than radiation-damage artifacts dominant in certain microsites of the mining soil. Our study demonstrates the great benefits in use of multiple combined synchrotron-based techniques for comprehensive understanding of Cu speciation in heterogeneous soil matrix, which facilitates our prediction of Cu reactivity and environmental fate in the mining site.

The dynamics and transformation of dissolved and colloidal Cu in the pore water of acontaminated paddy soil after applying ammonium sulphate (AS) and sulfur coated urea (SCU) with various flooding periods (1, 7 and 60 days) were also investigated (Yang et al., 2015). Microscopic X-ray fluorescence (μ-XRF) analysis found a close relationship between Fe and Cu distribution on soil colloids after 60-day flooding, implying the formation of colloidal Fe/Cu sulphide coprecipitates. Copper K-edge XANES directly revealed the transformation of outer-sphere complexed Cu (Ⅱ) species to Cu (Ⅱ) sulphide and reduced Cu_2O in the colloids of S-treated soils after 60-day's flooding. These results demonstrated the great influence of S fertilization on pore-water Cu mobility by forming Cu sulphide under flooding conditions, which facilitated our understanding and control of Cu loss in contaminated paddy soils under S fertilization.

2.2 Effect of sulfur on Cu transformation in rhizosphere soils

Sulfur fertilization in agricultural soils has received much attention in recent years in China as S deficiency dramatically increased due to the increased use of fertilizers with low S content (Wei et al., 2002; Chien et al., 2009). The role of sulfur on the availability of Cu and the bacterial community in rice (*Oryza sativa* L.) rhizosphere was investigated by pot experiments (Shi et al., 2011a). The results showed that with sulfur addition, pH in rhizosphere soil decreased and $Mg(NO_3)_2$ extractable Cu increased significantly. The bacterial community also changed with sulfur addition. Some specific clones having high similarity to *Thiobacillus*, which indicated that sulfur oxidation in the rice rhizosphere could increase the availability of Cu. Sequential extraction and XANES analysis results showed that Cu speciation exhibited some differences between rhizosphere and bulk soil of rice (Lin et al., 2010). In flooded paddy soil, most Cu in the rhizosphere existed as Cu (II), whereas part of Cu transformed to Cu (I) in the bulk soil. Sulfur XANES showed the presence of both oxidized and reduced forms of sulfur in studied soil samples, with more oxidized sulfur in the rhizosphere than in the bulk soil. The speciation change of Cu and sulfur depended on redox conditions. Our findings implied that with higher Eh in rice rhizosphere, transformation of sulfur and organic compounds together contributed to more soluble and exchangeable Cu.

Sulfur fertilization (less than 500 mg/kg) promoted the formation of iron plaque, thus sequestering a large amount of Cu on root surface and decreasing the bioavailability of Cu by inducing transformation of Cu bioavailable fractions (exchangeable, carbonate oxides, or iron and manganese oxide bound to Cu) to Cu bound to organic matter. Copper K-edge XANES revealed that S fertilization increased the percentage of Cu present as Cu_2S and Cu-cysteine in rice rhizosphere soil, thus reducing Cu mobility, leading to reduced Cu uptake by rice roots (Sun et al., 2016).

2.3 Microbe mediated transformation of Cu in rhizosphere soils

Soil microorganisms may play an important role in the behaviors of heavy metals in soils. We studied potential effects of a Cu-accumulator, *Elsholtzia splendens*, and a non-Cu-accumulator plant, *Trifolium repens*, on soil microbial activity and community composition with increasing Cu addition (Wang et al., 2008b). The concentrations of Cu in the shoots of *E. splendens* were 2.1, 2.2 and 2.4 times those of *T. repens* under the treatment of different Cu concentrations. Soil microbial biomass and phosphatase activity in the rhizosphere of *E. splendens* were higher than those of *T. repens*. PCR-denaturing gradient gel electrophoresis (PCR-DGGE) fingerprint analysis revealed that addition of Cu decreased the number of bands in bare soil and soil with *T. repens*. However, there was a significant increase in the number of bands in soil with

E. splendens incorporated with either 200 or 500 mg/kg Cu.

Compared with uncontaminated soil, some species such as *Firmicutes* only existed in the *E. splendens* rhizosphere of contaminated soil, while the very small amount (if any) of some species existed such as *Deinococcus-Thermus*, indicating that the contaminated environment altered the bacterial composition. Moreover, spatial variation of the bacterial community was found among different soil zones. Real-time PCR confirmed the spatial variation via the gene expression of flagellin (*fliC*) and chemotaxis gene (*cheA*). The spatial characteristics of *cheA* expression were consistent with that of DOC and bacterial diversity. DOC and Cu toxicity may affect specific gene expressions such as *fliC* and *cheA*, resulting in bacterial spatial variation in different soil zones (Yuan et al., 2014). Results indicated that *E. splendens,* as a Cu-accumulator, played an important role in governing soil microbial activity and bacterial community composition in the rhizosphere in response to Cu stress.

An extremely heavy-metal-resistant rhizobacterium *Pseudomonas putida* CZ1 (CZ1) was isolated from the rhizosphere of *E. splendens* grown in heavy-metal-contaminated soil. CZ1 exhibited high minimal inhibitory concentration value for Cu with about 3 mmol/L and was capable of removing about 87.2% of Cu in aqueous solutions, with specific biosorption capacities of 24.2 mg/L Cu (Chen et al., 2006). It was found that the optimum pH for Cu (Ⅱ) removal by living and nonliving was 5.0 and 4.5, respectively. At the optimal conditions, Cu^{2+} biosorption increased as the initial Cu concentration increased. The adsorption data with respect to Cu provide an excellent fit to the Langmuir isotherm. The binding capacity of living cells is significantly higher than that of nonliving cells in tested conditions. It demonstrated that about 40%-50% of the Cu was actively taken up by CZ1, with the remainder being passively bound to the bacterium. Moreover, desorption efficiency of Cu (Ⅱ) by living cells was 72.5% under 0.1 mol/L HCl and it was 95.3% by nonliving cells, respectively. It may be due to Cu (Ⅱ) uptake by the living cells enhanced by intracellular accumulation (Chen et al., 2005).

Mechanisms of Cu-binding by CZ1 were ascertained by chemical modifications of the biomass followed by Fourier transform infrared and X-ray absorption spectroscopic analyses of the living or nonliving cells (Chen et al., 2007). A dramatic decrease in Cu (Ⅱ) -binding resulted after acidic methanol esterification of the nonliving cells, indicating that carboxyl functional groups play an important role in the binding of metal to the biomaterial. The XANES analysis showed that the majority of the Cu was bound in both samples as Cu (Ⅱ). The fitting results of Cu K-edge EXAFS showed that N/O ligands dominated in living and nonliving cells. Our results indicate that carboxyl functional groups are the major ligands responsible for the metal binding in CZ1. Adsorption experiments and desorption with 1.0 mol/L CH_3COOK solution indicated that clays (goethite, kaolinite, smectite and manganite) contain more high-affinity copper binding sites than that of CZ1 cells, however, CZ1 cells are involved in more low-affinity Cu-binding sites. Carboxyl group activity is more important at weak-binding sites than at strong-binding sites. These results suggest that CZ1 cells play an important role in regulating the mobility of Cu in the soil environment (Chen et al., 2009).

3. Uptake and transformation of Cu in plants

Copper is an essential element for normal plant metabolism and has also been reported to be among the most toxic metals (Wong and Bradshaw, 2006). Plants can accumulate heavy metals in different plant tissue or cell organelles (Bringezu et al., 1999; Küpper et al., 1999) and also can detoxify heavy metals by chelating them to organic acids or proteins (Lee et al., 1978; Cobbett, 2000). The release of nanoparticles (NPs) to the environment poses an increasing potential threat to biological systems. As one of the most important NPs, CuO NPs are widely used in sensors, catalysts, surfactants, and antimicrobials (Cioffi et al., 2005). The mechanisms of Cu and CuO NPs uptake and biotransformation in plants are systematically demonstrated.

3.1 Copper uptake and transformation in different plant species

Commelina communis is a Cu-tolerant plant that grows in copper mine areas along the middle and lower streams of the Yangtze River (Tang et al., 1999). Our results showed that Cu accumulation in *C. communis* was increased with the drop of Fe concentration in nutrient solution. Cu uptake in *C. communis* was significantly enhanced by Fe deficiency. Cu uptake in Fe-deficient plants were still higher than that of Fe-sufficient plants when P-type ATPase inhibitor Na_3VO_4 was added in the solution, which demonstrated that Fe deficiency induced stimulation of the plasma membrane H^+-ATPase did not play a role in the enhanced Cu uptake. Fe and Cu deficiency could induce root Fe (III) chelate reductase activity in *C. communis*. Both Fe (III) chelate reductase activity and Cu concentration in roots inhibited distinctly by cycloheximide, a kind of protein synthesis inhibitor. In the time course of Fe deficiency, there simultaneously appeared Cu concentration and FCR activity peaks in roots at day 9. All above evidences demonstrated Cu uptake and accumulation in *C. communis* were probably related to the root FCR activity (Chen et al., 2004a). Cu uptake was inhibited by the uncoupler DNP and P-type ATPase inhibitor Na_3VO_4, not by the Ca^{2+} ion channel inhibitor $LaCl_3$, suggesting that Cu could probably be assimilated actively by root and be related with P-type ATPase, but not through Ca^{2+} ion channel. Fe or Zn deficiency could enhance Cu uptake, while 100 μmol/L Cu inhibited Fe, Zn, and Mn accumulation in roots significantly (Shi et al., 2011b).

Elsholtzia splendens is generally considered as a Cu-tolerant and-accumulating plant species, and a candidate for phytoremediation of Cu-contaminated soils. Synchrotron radiation X-ray fluorescence spectroscopy (SRXRF) microprobe was used to study the Cu and other elements distribution in *E. splendens* (Shi et al., 2004). The element (P, S, Cl, K, Ca, Mn,

Fe, Cu, Zn) in the leaf epidermis and cross-sections of the stem and leaf could be checked by SRXRF which was considered a sensitive technique for trace element analysis. The highest Cu levels were measured in the vascular tissues of stem and petiole, while Cu levels in mesophyll were higher than in leaf epidermis. There was a significant correlation between Cu and P, S, Ca in distribution, which suggested P, S, and Ca played an important role in Cu accumulation of *E. splendens*. Based on the significant correlation between Cu and elements Mn, Fe, and Zn in distribution, it seemed that Cu, Mn, Fe, and Zn could be transported by the same transporters with a broad substrate range. XAFS was used to investigate the Cu speciation and biotransformation in *E. splendens* with 300 μmol/L Cu treatment from 10 days to 60 days (Shi et al., 2008). The Cu K-edge XANES revealed that most Cu in roots, stems and leaves exists as divalent Cu. Most Cu in roots, stems and leaves was bound with cell wall and histidine (His)-like ligands, while a minor proportion of the Cu was bound to oxalate and glutathione-like ligands. The fitting results of Cu K-edge EAXFS showed that nitrogen/oxygen (N/O) ligands were dominant in roots, stems and leaves of the plant, while S ligands were rare. All these results suggest that Cu bound by N/O ligands plays a key role in Cu detoxification of *E. splendens*.

Laser ablation coupled with inductively coupled plasma-mass spectrometry (LA-ICP-MS) was used to explore Cu concentration in surface tissue along a longitudinal developmental gradient with meristem, rapidly elongating tissue, and nongrowing tissue in a model system of seedling roots of cucumber (Shi et al., 2009). Copper content of all roots is highest at the apex and falls sharply to lower values by 2 mm from the root tip. Cucumber root growth zones accumulate more of Cu with higher external availability. Copper deposition rates were calculated using a continuity equation with data on local metal content and growth velocity. Deposition rates of Cu is generally highest in the rapidly elongating region, 1.5-3.5 mm, even where Cu concentration is decreasing with position and root age and even when the accumulation is inhibitory to growth.

The localization, biotransformation, and chemical speciation of Cu in root tips of cucumber (*Cucumis sativus*) were investigated using μ-XRF and μ-XANES (Song et al., 2013). The highest content of Cu was found in root cap and meristematic zone whereas low Cu content in elongation and maturation zone. There was a dramatic increase of Cu content in root cap and meristematic zone after treatment with 100 μmol/L $CuSO_4$ for 72 h. The μ-XANES analysis revealed that most of Cu in root tip was bound with alginate, citrate, and cysteine-like ligands whereas rarely deposited in form of CuO. From root cap to maturation zone, the proportion of Cu bound with alginate-like ligands increased whereas that bound with citrate-like ligands decreased. The proportion of Cu bound with cysteine-like ligands increased from root cap to elongation zone but sharply declined in maturation zone. The results suggested that Cu was chelated by S ligands in the cell walls which protect protoplasm against possible damage caused by Cu excess.

3.2 Proteomic response to copper stress in *E. splendens*

To better understand the Cu tolerance/accumulation mechanisms in *E. splendens*, proteomic analysis was performed on *E. splendens* roots and leaves exposed to 100 μmol/L $CuSO_4$ for 3 and 6 days (Li et al., 2009). After 6 days of treatment, Cu accumulation in roots increased much more than that in leaves. SDS–PAGE analysis showed that the proteins changed more intensively in roots than did in leaves upon Cu stress. Two-dimensional gel electrophoresis (2-DE) and image analyses found that 45 protein spots were significantly changed in roots, but only six protein spots in leaves. The abundance of protein spots mostly showed temporal changes. MALDI-TOF MS and LTQ-ESI-MS/MS were used to identify the differently expressed protein spots. The identified root proteins were involved in various cellular processes such as signal transduction, regulation of transcription and translation, energy metabolism, regulation of redox homeostasis and cell defense. The leaf proteins were mainly degraded fragments of RuBisCo and antioxidative protein. The resulting differences in protein expression pattern suggested that redirection of root cellular metabolism and redox homeostasis might be important survival mechanisms of *E. splendens* upon Cu stress. Besides providing insights into Cu stress responses, these Cu-responsive protein data provide a good starting point for further dissection of the Cu tolerance mechanisms in *E. splendens* using genetic and proteomic approaches.

The functional role of the cell wall of *E. splendens* under Cu stress, where several cell wall proteins (CWPs) with possible roles in Cu tolerance/detoxification were identified (Liu et al., 2014). LC-MS/MS approach was performed to analyze the Cu-responsive cell wall proteins and polysaccharides. The majority of the 22 up-regulated proteins were involved in the antioxidant defense pathway, cell wall polysaccharide remodeling, and cell metabolism process. Changes in polysaccharide amount, composition, and distribution could offer more binding sites for Cu ions. The 33 down-regulated proteins were involved in the signal pathway, energy, and protein synthesis. Cu stress conditions can alter the composition (polysaccharide and protein) of the cell wall both qualitatively and quantitatively.

Approximately 40% of the differentially expressed CWPs showed higher abundance in response to Cu stress involved in antioxidant defense, cell wall polysaccharide remodeling, and metabolism process. Up to 60% of the CWPs were in low abundance in response to Cu stress that is involved in signal, energy, and protein synthesis. Polysaccharide analysis confirmed the cell wall remodeling under Cu stress. The amount, composition, and distribution of the cell wall polysaccharides are consequential for plant adaptation to enhance Cu ion levels. Proteome analysis of cell wall confirmed that most proteins were associated with an antioxidant defense response. Hsp 70, small G protein, and RAS-related GTP-binding protein also have essential roles in signal transduction across the cell wall and through the cytoskeleton. Literature provides fundamental information about the role of polysaccharide composition of plant cell wall in metal tolerance and complementary evidence on continuous crosstalk between CWPs and the cytoskeleton. Therefore, knowledge has been expanded on plant stress-related signaling pathways in the cell wall. Cu regulation of these proteins may also not solely respond to abundance changes. Post-translational modification and dynamics are interesting subjects for future investigation.

3.3 CuO nanoparticles transformation and phytotoxicity in plants

The release of nanoparticles (NPs) to the environment poses an increasing potential threat to biological systems. We investigated the phytotoxicity and accumulation of CuO NPs to *E. splendens* under hydroponic conditions (Shi et al., 2014). The 50% effective concentration (EC50) of CuO NPs to *E. splendens* was about 480 mg/L, implying the tolerance of *E. splendens* to CuO NPs. The Cu content in the shoots treated with 1000 mg/L CuO NPs was much higher than those exposed to the comparable 0.5 mg/L soluble Cu and CuO bulk particles. CuO NPs-like deposits were found in the root cells and leaf cells. Cu K-edge X-ray absorption near-edge structure analysis further revealed that the accumulated Cu species existed predominantly as CuO NPs in the plant tissues. All these results suggested that CuO NPs can be absorbed by the roots and transported to the shoots in *E. splendens*.

Metal-based nanoparticles (MNPs) may be translocated and biochemically modified *in vivo*, which may influence the fate of MNPs in the environment. Synchrotron-based techniques were used to investigate the behavior of CuO NPs in rice plants exposed to 100 mg/L CuO NPs for 14 days (Peng et al., 2015a). Micro X-ray fluorescence (μ-XRF) and micro X-ray absorption near edge structure (μ-XANES) analysis revealed that CuO NPs moved into the root epidermis, exodermis, and cortex, and they ultimately reached the endodermis but could not easily pass the Casparian strip; however, the formation of lateral roots provided a potential pathway for MNPs to enter the stele. Moreover, bulk-XANES data showed that CuO NPs were transported from the roots to the leaves, and that Cu (Ⅱ) combined with cysteine, citrate, and phosphate ligands and was even reduced to Cu (Ⅰ). CuO NPs and Cu-citrate were observed in the root cells using STXM.

Natural organic matter (NOM) can interact with engineered nanoparticles (NPs) in the environment and modify their behavior and toxicity to organisms. The phytotoxicity of CuO NPs to rice seedlings in the presence of humic acid as a model NOM was investigated (Peng et al., 2015b). Our results showed that CuO NPs induced the inhibition of root elongation, aberrations in root morphology and ultrastructure, and losses of cell viability and membrane integrity. The adverse effects partly resulted from the generation of reactive oxygen species caused by CuO NPs, which led to lipid peroxidation, mitochondrial dysfunction, and programmed cell death in rice seedlings. However, all the phytotoxicity was alleviated with the addition of humic acid because humic acid coatings on nanoparticle surfaces enhanced electrostatic and steric repulsion between the CuO NPs and the plant cell wall/membrane, reducing contact between NPs and plant and CuO NP-induced oxidative damage to plant cells. Our results shed light on the mechanism underlying NPs phytotoxicity and highlight the influence of NOM on the bioavailability and toxicity of NPs.

4. Chemical and microbe enhanced phytoremediation to Cu contaminated soils

Phytoremediation has recently been promoted as a promising method for the clean-up of polluted soils. For phytoremediation to be possible, the contaminant (s) must be within the plant root zone, be bioavailable and be biologically absorbed. Successful phytoremediation depends mainly on the bioavailability of Cu in the soil, however, the availability of Cu for plants is usually restricted by the complexation of metals within solid soil fractions. Chelate-assisted phytoremediation, the use of synthetic chelators, e.g. ethylenediamine tetraacetate (EDTA), has been used to artificially enhance heavy metal solubility in soil and thus increasing heavy metal phytoavailability.

4.1 The potential of chemical enhanced phytoremediation

Heavy metals in soil are associated with various forms having different bioavailability. The bioavailability of various Cu forms in contaminated soils was investigated using ion-exchange resins, a sequential extraction procedure, and combined with methods including partial dissolution procedure, simulated Cu forms, seedling culture, pot experiment when treated with EDTA, or waste water from monosodium glutamate and citric acid production (Chen et al., 2004b). Results showed that the bioavailability, in decreasing order of different Cu forms to tall fescue is: exchangeable Cu and organic matter bound Cu > Cu bound to carbonate > Fe/Mn oxide bound > residual Cu. Effect of EDTA on the activation of Cu contaminated soil or simulated Cu and the uptake and translocation of tall fescue was better than that of monosodium glutamate waste water (MGW) and citric acid waste water (CAW). EDTA, CAW and MGW all improved the plant availability of different Cu forms in contaminated soil, which should be used in chelate-assisted phytoremediation of heavy metal polluted soil.

Successful phytoremediation depends mainly on the bioavailability of copper (Cu) in the soil. We studied the potential effects of sulfur (S) amendment on mobility of copper and microbial community composition in soil under laboratory conditions (Wang et al., 2008a). The results showed that with S application at 20 g S/kg, soil pH decreased about 3 units and the solubility of the Cu significantly increased after 64 days of incubation. The concentration of Cu in Cu-accumulator *E. splendens* shoots and roots increased with S treatment. Concentration of Cu in the shoots was 156.5 mg/kg under S treatment. It was 2.5 times that of without application of S. PCR-denaturing gradient gel electrophoresis (PCR-DGGE) fingerprint analysis revealed that there were certain groups of acidophilic soil bacteria in the soil after addition of S.

We found specific clones such as 1 (from biofilter-treating hydrogen sulfide and methanol) and 4 (from metal-rich and acidic River Tinto) in the soil with S treatment. The above results indicated that S facilitated the mobility of Cu by soil microorganism and provided a basis for further studies of S-assisted phytoremediation.

4.2 The potential of microbe assisted phytoremediation

Based on the research of *Pseudomonas putida* CZ1 (a heavy-metal-resistant bacterial strain isolated from the rhizosphere of *E. splendens*) on Cu transformation in soils, the effects of CZ1 on the uptake and translocation of Cu in *E. splendens* was explored (Xu et al., 2015). Significant promotion of plant growth coupled with the obvious plant-growth-promoting (PGP) characters of the bacteria suggested that CZ1 would be a plant-growth-promoting rhizobacterium (PGPR) to *E. splendens* under Cu stress condition. The results of inductively coupled plasma optical emission spectrometry (ICP-OES) showed that CZ1 increased the concentration of Cu in the shoots (up to 211.6% compared to non-inoculation treatment) and translocation factor (TF) (from 0.56% to 1.83%) of those exposed to Cu. The distribution of Cu in root cross section measured by synchrotron-based X-ray fluorescence microscopy (SRXRF) indicated that CZ1 promoted the transport of Cu from cortex to xylem in roots, which contributed to the accumulation of Cu in shoots. Furthermore, CZ1 improved the uptake of nutrient elements by plants to oppose to the toxicity of Cu. In summary, *P. putida* CZ1 acted as a PGPR in resistance to Cu and promoted the accumulation and translocation of Cu from root to shoot by element redistribution in plant root; hence, CZ1 is a promising assistance to phytoremediation.

5. Concluding remarks

The toxicity of Cu in soils to organisms strongly depends on Cu speciation and bioavailability. Synchrotron radiation-based techniques are promising method for Cu speciation and distribution characterization in soil-plant system. Sulfur transformation in soils significantly affected the mobility of Cu in soil pore water, the bioavailability of Cu in rhizosphere soil, and the translocation of Cu in plants. Rhizobacterium like *Pseudomonas putida* CZ1 played important role in mediating Cu adsorption and desorption in soil-mineral system, changing the transformation process of Cu in soil-plant systems. The detoxification mechanism of Cu in *E. splendens* was related to the Cu binding to N/O/S ligands. Copper stress leads to different functional protein expression in different *E. splendens* tissues. CuO nanoparticles can be absorbed by the roots and transported to the shoots, and then transformed into different Cu species in plants. Sulfur and rhizobacterium (such as *Pseudomonas putida* CZ1) enhanced phytoremediation of Cu contaminated soil are potential techniques, and further practices should be in field conditions.

References

Alloway B J. 1995. Heavy Metals in Soils, Heavy Metals in Soils. Blackie Academic and Professional, Suffolk, England.

Bringezu K, Lichtenberger O, Leopold I, et al. 1999. Heavy metal tolerance of *Silene vulgaris*. Journal of Plant Physiology, 154 (4): 536-546.

Burkhead J L, Gogolin Reynolds K A, Abdel-Ghany S E, et al. 2009. Copper homeostasis. New Phytologist, 182 (4): 799-816.

Chen X C, Hu S P, Shen C F, et al. 2009. Interaction of *Pseudomonas putida* CZ1 with clays and ability of the composite to immobilize copper and zinc from solution. Bioresource Technology, 100 (1): 330-337.

Chen X C, Shi J Y, Chen Y X, et al. 2007. Determination of copper binding in *Pseudomonas putida* CZ1 by chemical modifications and X-ray absorption spectroscopy. Applied Microbiology and Biotechnology, 74 (4): 881-889.

Chen X C, Shi J Y, Chen Y X, et al. 2006. Tolerance and biosorption of copper and zinc by *Pseudomonas putida* CZ1 isolated from metal-polluted soil. Canadian Journal of Microbiology, 52 (4): 308-316.

Chen X C, Wang Y P, Lin Q, et al. 2005. Biosorption of copper (Ⅱ) and zinc (Ⅱ) from aqueous solution by *Pseudomonas putida* CZ1. Colloids and Surfaces B: Biointerfaces, 46 (2): 101-107.

Chen Y X, Shi J Y, Tian G M, et al. 2004. Fe deficiency induces Cu uptake and accumulation in *Commelina communis*. Plant Science, 166 (5): 1371-1377.

Chen Y X, Shi J Y, Zhang W D, et al. 2004. EDTA and industrial waste water improving the bioavailability of different Cu forms in contaminated soil. Plant and Soil, 261 (1-2): 117-125.

Chien S H, Prochnow L I, Cantarella H. 2009. Recent developments of fertilizer production and use to improve nutrient efficiency and minimize environmental impacts. Advances in Agronomy, 102: 267-322.

Cobbett C S. 2000. Phytochelatins and their roles in heavy metal detoxification. Plant Physiology, 123 (3): 825-832.

Cioffi N, Ditaranto N, Torsi L, et al. 2005. Analytical characterization of bioactive fluoropolymer ultra-thin coatings modified by copper nanoparticles. Analytical and Bioanalytical Chemistry, 381 (3): 607-616.

Drążkiewicz M, Skórzyńska-Polit E, Krupa Z. 2004. Copper-induced oxidative stress and antioxidant defence in *Arabidopsis thaliana*. Biometals, 17 (4): 379-387.

Hong S, Candelone J P, Soutif M, et al. 1996. A reconstruction of changes in copper production and copper emissions to the atmosphere during the past 7000 years. Science of the Total Environment, 188 (2): 183-193.

Küpper H, Zhao F J, McGrath S P. 1999. Cellular compartmentation of zinc in leaves of the hyperaccumulator Thlaspi caerulescens. Plant Physiology, 119 (1): 305-312.

Lee J, Reeves R D, Brooks R R, et al. 1978. The relation between nickel and citric acid in some nickel-accumulating plants. Phytochemistry, 17 (6): 1033-1035.

Li F, Shi J Y, Shen C F, et al. 2009. Proteomic characterization of copper stress response in *Elsholtzia splendens* roots and leaves. Plant Molecular Biology, 71 (3): 251-263.

Lin H R, Shi J Y, Wu B, et al. 2010. Speciation and biochemical transformations of sulfur and copper in rice rhizosphere and bulk soil—XANES evidence of sulfur and copper associations. Journal of Soils and Sediments, 10 (5): 907-914.

Liu T T, Shen C F, Wang Y, et al. 2014. New insights into regulation of proteome and polysaccharide in cell wall of *Elsholtzia splendens* in response to copper stress. PloS One, 9 (10): e109573.

Marschner H. 2011. Marschner's mineral nutrition of higher plants. Academic press. Mehrabi M, Wilson R. 2007. Intercalating gold nanoparticles as universal labels for DNA detection. Small, 3 (9): 1491-1495.

Muccifora S, Bellani L M. 2013. Effects of copper on germination and reserve mobilization in *Vicia sativa* L. seeds. Environmetal Pollution, 179: 68-74.

Peñarrubia L, Andrés-Colás N, Moreno J, et al. 2010. Regulation of copper transport in *Arabidopsis thaliana*: a biochemical oscillator? JBIC Journal of Biological Inorganic Chemistry, 15 (1): 29-36.

Peng C, Duan D C, Xu C, et al. 2015. Translocation and biotransformation of CuO nanoparticles in rice (*Oryza sativa* L.) plants.

Environmental Pollution, 197: 99-107.

Peng C, Zhang H, Fang H X, et al. 2015. Natural organic matter–induced alleviation of the phytotoxicity to rice (*Oryza sativa* L.) caused by copper oxide nanoparticles. Environmental Toxicology and Chemistry, 34 (9): 1996-2003.

Rankin D W H. 2009. CRC handbook of chemistry and physics, edited by David R. Lide.

Ryan B M, Kirby J K, Degryse F, et al. 2013. Copper speciation and isotopic fractionation in plants: uptake and translocation mechanisms. New Phytologist, 199 (2): 367-378.

Shi J Y, Chen Y X, Huang Y Y, et al. 2004. SRXRF microprobe as a technique for studying elements distribution in *Elsholtzia splendens*. Micron, 35 (7): 557-564.

Shi J Y, Gras M A, Silk W K. 2009. Laser ablation ICP-MS reveals patterns of copper differing from zinc in growth zones of cucumber roots. Planta, 229 (4): 945-954.

Shi J Y, Lin H R, Yuan X F, et al. 2011. Enhancement of copper availability and microbial community changes in rice rhizospheres affected by sulfur. Molecules, 16 (2): 1409-1417.

Shi J Y, Peng C, Yang Y Q, et al. 2014. Phytotoxicity and accumulation of copper oxide nanoparticles to the Cu-tolerant plant *Elsholtzia splendens*. Nanotoxicology, 8 (2): 179-188.

Shi J Y, Wu B, Yuan X F, et al. 2008. An X-ray absorption spectroscopy investigation of speciation and biotransformation of copper in *Elsholtzia splendens*. Plant and Soil, 302 (1-2): 163-174.

Shi J Y, Yuan X F, Chen X C, et al. 2011. Copper uptake and its effect on metal distribution in root growth zones of *Commelina communis* revealed by SRXRF. Biological Trace Element Research, 41 (1-3): 294-304.

Song J, Yang Y Q, Zhu S H, et al. 2013. Spatial distribution and speciation of copper in root tips of cucumber revealed by μ-XRF and μ-XANES. Biologia Plantarum, 7 (3): 581-586.

Sun L J, Zheng C Q, Yang J J, et al. 2016. Impact of sulfur (S) fertilization in paddy soils on copper (Cu) accumulation in rice (*Oryza sativa* L.) plants under flooding conditions. Biology and Fertility of Soils, 52 (1): 31-39.

Tang S R, Wilke B M, Huang C. 1999. The uptake of copper by plants dominantly growing on copper mining spoils along the Yangtze River, the People's Republic of China. Plant and Soil, 209 (2): 225-232.

Wang Y P, Li Q B, Hui W, et al. 2008. Effect of sulphur on soil Cu/Zn availability and microbial community composition. Journal of Hazardous Materials, 159 (2): 385-389.

Wang Y P, Li Q B, Shi J Y, et al. 2008. Assessment of microbial activity and bacterial community composition in the rhizosphere of a copper accumulator and a non-accumulator. Soil Biology and Biochemistry, 40 (5): 1167-1177.

Zhou W, Wan M, He P, et al. 2002. Oxidation of elemental sulfur in paddy soils as influenced by flooded condition and plant growth in pot experiment. Biology and Fertility of Soils, 36 (5): 384-389.

Wong M H, Bradshaw A D. 1982. A comparison of the toxicity of heavy metals, using root elongation of rye grass, *Lolium perenne*. New Phytologist: 255-261.

Xu C, Chen X C, Duan D C, et al. 2015. Effect of heavy-metal-resistant bacteria on enhanced metal uptake and translocation of the Cu-tolerant plant, *Elsholtzia splendens*. Environmental Science and Pollution Research, 22 (7): 5070-5081.

Yang J, Liu J, Dynes J J, et al. 2014. Speciation and distribution of copper in a mining soil using multiple synchrotron-based bulk and microscopic techniques. Environmental Science and Pollution Research, 21 (4): 2943-2954.

Yang J J, Zhu S H, Zheng C Q, et al. 2015. Impact of S fertilizers on pore-water Cu dynamics and transformation in a contaminated paddy soil with various flooding periods. Journal of Hazardous Materials, 286: 432-439.

Yuan X F, Luan J, Shi J Y. 2014. Spatial variability of bacteria in the rhizosphere of *Elsholtzia splendens* under Cu contamination. Environmental Science and Pollution Research, 21 (16): 9809-9818.

Contribution of Soil Active Components to the Control of Heavy Metal Speciation

Wenfeng TAN[1], Linchuan FANG[2], Juan XIONG[2], Hui YIN[1], Wei ZHAO[1]

[1]College of Resources and Environment, Huazhong Agricultural University

[2]Institute of Soil and Water Conservation, Chinese Academy of Sciences

1. Introduction

Soil is the central organizer of the terrestrial ecosystem. Mineral, organic components, and microorganisms, which are major solid active components of the soil, profoundly affect the physical, chemical, and biological processes of soils including the behavior, transformation and fate of various nutrients and pollutants (Violante et al., 2002). Heavy metal interaction with soil active components is recognized as being important in controlling heavy metal activities. Colloidal particles of soil organic matter (SOM), clay silicates, metal hydroxides and microorganisms, which have large surface area and are often electrically charged, are considered as important adsorptive surfaces to bind heavy metals.

It is known that the soil has potential ability to immobilize heavy metals by precipitation, adsorption or (bio) transformation and that the soil acts as scavenger and sort of buffer. The buffer capacity of the natural environment strongly influences the impact of toxic metals. Buffering in this sense is described as storage of the heavy metals without a direct effect of these heavy metals on the toxicity experienced at the contaminated soil. Exceedance of this capacity may be harmful to the ecosystem because it implies an increased transport and an increased bio-availability of the toxic metals (Vermeer, 1996). Therefore, heavy metal interaction with those major active components (e.g., minerals, organic components, and microorganisms) of soils has enormous impacts on a series of reactions and processes which are critical to environmental quality and ecosystem health. Fundamental understanding of these reactions and processes at the atomic, molecular, and microscopic levels is essential for remediation of heavy metal pollution in soils, sustaining and enhancing soil health, which includes human health, on a global scale (Huang et al., 2008).

Methodologies, that have been used to quantify the contribution of different sorbent phases to metal sorption in soils, include the following: ①application of sequential extractions, ② development of statistical relationships between the sorption and soil properties, ③observation of similarities in sorption on single sorbents and whole soils, and ④comparison of sorption on natural and treated soils (Weng et al., 2001). However, all these methods have limitations. The contribution of individual sorbents may be difficult or impossible to experimentally identify. In the decision of making process concerning the protection of quality of soil and sediments, it is necessary to make accurate assessments of heavy metal speciation (Van Riemsdijk et al., 2006). This is important for formulating heavy metal standards and choosing appropriate remediation techniques. For purposes such as the risk assessment and land use planning, good prediction of the effects of various changes in land management on the ion speciation is required. Free metal ion activity (or concentration) is the key factor in determining metal bioavailability and toxicity. The dissolved metal pool reflects the metal fraction that could potentially be leached from the soil to ground waters and surface waters (Sigg et al., 2006). Speciation measurements are costly and sometimes very difficult or even not possible. Speciation models can be used and various types of models have been developed to calculate chemical speciation or the distribution of chemicals over all relevant forms (Weng et al., 2006; Xiong et al., 2013).

2. Contribution of Mn oxide in immobilizing heavy metals

Manganese (Mn) oxides are widely distributed in soils, sediments, and ocean manganese nodules; they are characterized by low points of zero charge, high negative charge, large surface areas, and they are actively involved in various chemical reactions. They are considered as important heavy metal adsorbents in the environment (McKenzie, 1980; Post, 1999; O' Reilly and Hochella, 2003). Birnessite is one of the most common Mn oxides. And acid birnessite is highly active in various chemical reactions, which can be achieved by the reduction of potassium permanganate in a strong acidic medium (McKenzie, 1971). It has a hexagonal layer symmetry with layers comprising edge-sharing Mn (Ⅳ) O_6 octahedra, Mn (Ⅲ) O_6 octahedra and vacant Mn octahedral sites (Villalobos et al., 2006). Some Mn^{2+} and Mn^{3+} ions are located above or below vacant Mn octahedral sites in birnessites (Webb et al., 2005).

2.1 Heavy metal adsorption on Mn oxide surface

Employing X-ray diffraction (XRD), X-ray photoelectron spectroscopy (XPS), and fourier transform infrared spectroscopy (FTIR) technologies, the microstructure of birnessites with different Mn average oxidation state (Mn AOS) was characterized. XRD patterns showed Mn AOS in birnessite is negatively correlated to the d (110) spacing (Zhao et al., 2009). XPS analysis indicated that there were two chemistry states of Mn, staturated and undersaturated coordinately Mn, whose relative contents were 83.79%-91.69% and 8.31%-17.21% in the structure, respectively, for the birnessites with different AOS (3.67-3.92). -OH located at vacant Mn octahedral site increased with the increase of AOS in birnessite accounting for the aggregation of crystal grains, which led to the decrease in relative content of undersaturated coordinately Mn. There were three chemistry states of oxygen, lattice oxygen, hydroxide and H_2O, whose relative contents were 50.44%-65.05%, 24.90%-39.27% and 8.07%-12.63% in the structure, respectively. -OH located at vacant Mn octahedral site increased with the increase of AOS in birnessite, which resulted in the increase in relative content of chemistry state of hydroxide in the structure (Zhao et al., 2012a). In FTIR spectroscopy, the band at 899-920/cm was assigned to the bending vibration of -OH located at vacancies. The bands at 1059-1070, 1115-1124 and 1165-1171/cm could be attributed to the vibrations of Mn (III) -OH in MnO_6 layers, and the intensities of these bands increased with decreasing Mn AOS. The bands at 990 and 1023-1027/cm were ascribed to the vibrations of Mn (III) -OH in the interlayers. The band at 564-567/cm was assigned to the vibration of Mn-O located at vacancies (Zhao et al., 2012b).

The heavy metal lead is toxic to animals and vegetations and it can damage the nervous system of humans. Adsorption is one of the important processes that affects the transfer of lead and other heavy metals from the aqueous phase to the solid phase and, thus, influences the distribution, mobility, and bioavailability of heavy metal ions (Xu et al., 2006). Feng et al. (2007) found the adsorption affinity and capacity of birnessite for different heavy metal ions are unlike. Especially, birnessite had the greatest adsorption affinity and amount for Pb^{2+}. The quantity of vacant sites in birnessite largely determines Pb^{2+} adsorption. And the quantity of vacant sites increases with the increase in birnessite AOS values (Zhao et al., 2009). The adsorption of Pb^{2+} on birnessites was investigated by X-ray absorption fine structure (XAFS). The three bonding mechanisms of Pb^{2+} adsorbed on the birnessites with different Mn AOS were proposed: a single-corner-sharing complex on particle edges along the u axis of MnO_6 layers, a double-corner-sharing complex on particle edges along the a or b axis, and a triple-corner-sharing complex in the interlayer above or below vacant sites (Fig. 1). Two Pb-O and two Pb-Mn shells were detected in the birnessites with Pb loading. For the same birnessite at low Pb loading (600 mmol/kg) the two Pb-O shells consisted of 3.1 O atoms at 0.227 nm and 8.2 O atoms at 0.399 nm, respectively. The two Pb-Mn shells consisted of 2.8 Mn atoms at 0.357 nm, and 6.1 Mn atoms at 0.377 nm. At high lead loading (2344 mmol/kg), a decrease in the Pb-O coordination numbers in both the Pb-O shells and in the Pb-Mn coordination

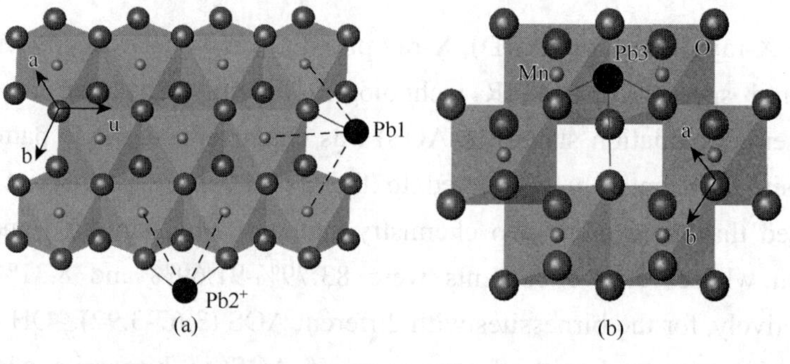

Fig. 1 Illustration of the complex configurations of Pb^{2+} on the birnessite (a) Pb binding to the edge sites; (b) Pb binding to the vacant sites.

Pb1: single-corner-sharing complex;
Pb2: double-corner-sharing complex;
Pb3: triple-corner-sharing complex

numbers in both the Pb—Mn shells resulted from the distortion of the Pb^{2+} coordination environment (Zhao et al., 2011a). FTIR of birnessites with different Mn AOS after Pb^{2+} adsorption showed an increased intensity of the band at 990/cm resulted from Mn (III) in MnO_6 layers partially migrating to interlayers during Pb^{2+} adsorption. The absorption band of vibration of Mn-O located at vacancies could split by coupling of vibrations due to Pb^{2+} and/or Mn^{2+} adsorbed at vacant sites. The large distance between the band at 610-626/cm and that at 638-659/cm might reflect small Mn (III) ions located in Mn (III) -rich rows (Zhao et al., 2012b).

To further confirm the importance of the vacant sites to Pb adsorption, before adsorption experiments, synthetic birnessite with high Mn AOS was treated with Mn^{2+} or Zn^{2+} at different concentrations to occupy the vacant sites (Zhao et al., 2010). After the treatment, Mn AOS and d (110) -interplanar spacings of the birnessites remained almost unchanged as the concentration of the treating Zn^{2+} increased, indicating an unchanged number of vacant Mn octahedral sites, whereas the maximum Pb^{2+} adsorption decreased due to the presence of Zn^{2+} on adsorption sites. The AOS's of the Mn^{2+} -treated birnessites decreased and most of the Mn^{2+} ions added were oxidized to Mn^{3+} ions. A portion of produced Mn^{3+} located above or below vacant sites, which did not affect the number of vacant sites, and the remaining Mn^{3+} migrated to occupy the vacant sites. The d (110) -interplanar spacing of Mn^{2+} -treated birnessites increased, indicative of a decreased vacant Mn octahedral sites. Therefore, the maximum Pb^{2+} adsorption of Mn^{2+} -treated birnessites decreased, and the decrease was greater than that for the corresponding Zn^{2+} -treated birnessites. On the other hand, acid birnessite with low Mn AOS was treated with $Na_4P_2O_7$ at pH 2, 4, 5 respectively, to extract Mn (III) in the structure (Zhao et al., 2011b). After the treatment, Mn (III) located in the layer edge and part of Mn (III) located above or below vacant sites were released to the solution through complexation with $Na_4P_2O_7$. The AOS of Mn of birnessites increased after treatment. While the crystal structure of birnessite did not change after treatment, the amount of Mn (III) located above or below vacant cation sites decreased, and the amount of H^+ located above or below vacant cation sites went up in the structure of birnessites. The amount of vacant cation sites responsible for Pb^{2+} adsorption increased, which led to the increase of the maximum amount of adsorbed Pb^{2+}.

2.2 Heavy metal substitution in Mn oxide structure

The geochemical behaviors of many transition metals (TM), such as Co, Ni, Cu and Fe, are closely related with Mn oxides (especially for birnessites), mainly through substitution of lattice Mn in birnessite layers (Burns, 1993). This substitution is probably governed by the coordination radius (CR), crystal field stabilization energy (CFSE), oxidation state and other properties of the TMs, and environmental conditions, such as pH and the existence of other cations.

By coprecipitation of Co^{2+}, Ni^{2+} or Fe^{3+} with Mn during the crystallization of hexagonal birnessite, it was found that most Co (71%-80%) while only a small fraction of Ni (24%-35%) or Fe (18%-20%) was incorporated into birnessite layers, which could be qualitatively and quantitatively revealed by the TM K-edge extended X-ray absorption fine structure spectroscopy (EXAFS) (Fig. 2). By carefully comparing the electronic structure, CR, CFSE, pairing energy and electronegativity of these TMs in octahedronal crystal field, it was found that layer Mn (III) and Co (III) plausibly adopted low-spin state while Fe (III) located in the birnessite layers exhibited high-spin (HS) configuration which called for further confirmation by other proofs, such as density functional theory (DFT) calculations. The compatibility of birnessite layers with these TMs was negatively correlated with the differences in CR between TMs and the substituted Mn (Ⅳ) or Mn (III) (Yin et al., 2012, 2013). Contrasting to the fact that either coprecipitaion or first sorption and then migration into layers (Manceau et al., 1997) can account for the existence of Co in birnessite layers in natural ferromanganese nodules, only a small part of Ni is incorporated into the manganese layers by coprecipitation with Mn, whereas quite a large part is first adsorbed on the surfaces and then transferred into the layers during large time scales over which ferromanganese nodules accrue (Peacock, 2009). However another interesting phenomenon but still not understandable yet was that the amount of Al^{3+} that can be incorporated into the structure of birnessite was maximized at only 5 at.% (Ogata et al., 2008), taking into consideration the fact that there was a large quantity of Al^{3+} in soil solutions during birnessite crystallization and that there were almost the same CR of Al^{3+} (0.675 Å) and Mn^{4+} (0.670 Å) (Shannon, 1976).

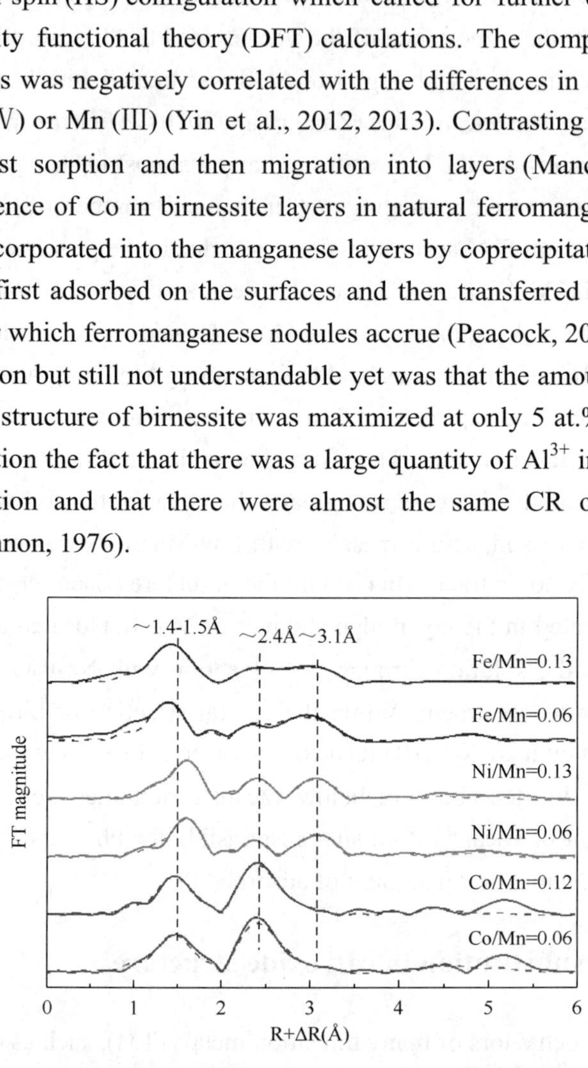

Fig. 2 Fourier transformed TM (Co, Ni, Fe) K-edge EXAFS spectra of doped birnessites (solid lines) overlaid with the best fits of quantification analysis of TM spatial distribution in the mineral structures (dashed lines). The magnitude of the first Me-O pair was normalized for qualitative comparison. Modified from Yin et al. (2013)

Increase in pH can promote the migration of TMs into birnessite layers. With the increase of the solution pH from 4 to 7, Ni that was initially adsorbed above/below Mn octahedral vacancy sites in the phyllomanganate layers was progressively structurally incorporated into the vacancy sites with time, via rearrangement of surface adsorbed Ni with time rather than direct Ni incorporation from solution. And this process is reversible with decreasing pH (Peacock, 2009). Same phenomenon was once observed for Cu^{2+} during EXAFS characterization of sorption complexes on birnessite at pH 4 and 8, in which some Cu was found to become structurally incorporated into the MnO_2 layer by occupying the vacancy sites (Sherman and Peacock, 2010). However first-principles quantum-mechanical geometry optimizations based on DFT eliminated the possibility of Cu insertion into birnessite layers by showing that layer Cu did not exhibit greater stability than Cu adsorbed on vacancies, while convincingly validated by the experimental results for Co, Ni and Zn (Kwon et al., 2013).

In natural environments many TM cations often coexist in the solutions during birnessite crystallizes. The coexistence of one TM cation may promote or suppress the incorporation of the other into the birnessite layers. In a series of Co and Ni co-doped birnessites, EXAFS analysis demonstrated that 74%-79% of the total Co and 23%-39% of the total Ni initially added were inserted into the birnessite layers. By comparing the ratios of content of TM in layers to that of above vacancies of these co-doped birnessites with those single Co or Ni doped samples, it was found that the coexistence of Ni hindered the incorporation of Co into the layers during birnessite crystallization while Co had little effects, neither hindrance nor promotion, on the insertion of Ni into the layers (Fig. 3; Yin et al., 2014).

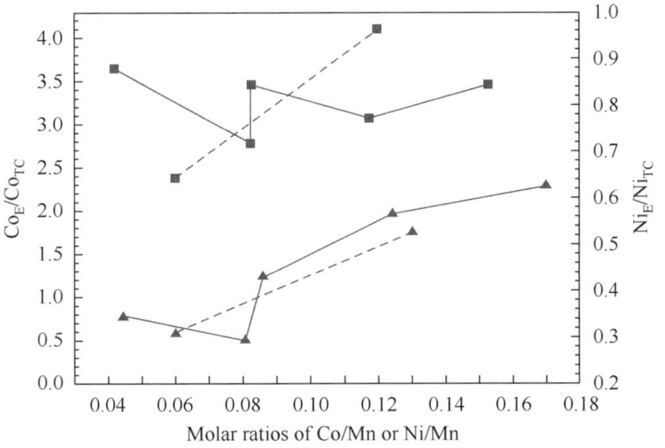

Fig. 3 The ratios of TM (TM=Co, Ni) in birnessite layer (TM_E) to TM adsorbed at vacancies (TM_{TC}) with the increase of the contents of TM in the structures of Co and Ni co-doped (solid lines) and single Co or Ni doped (dashed lines) birnessites (from Yin et al., 2014)

Incorporation of TMs into the layers of birnessite will cause certain changes in the structure and physicochemical properties. Substitution for Mn^{3+} or Mn^{4+}, Co, Ni or Fe doping greatly changed the crystallinity, i. e. the number of layers stacked along the $\mathbf{c}^*$ axis, Mn average oxidation state, specific surface area and thermal stability, the in-plane lattice parameters b and a, and the contents of vacancies in the layers (Zhu et al., 2010; Yin et al., 2012, 2013, 2014). The Mn octahedronal layer symmetry of biogenic birnessite was also sensitive to the coexisting cations during mineral crystallization, and the presence of Ni^{2+} enhanced vacant site formation (Zhu et al., 2010). During the synthesis of birnessite at strong alkaline solutions, coprecipitation of high concentration of Co^{2+} with Mn^{2+} promoted the transition of birnessite layer symmetry from orthogonal to hexagonal by substitution of orderly arranged Mn (III) O_6 octahedron with highly Jahn-Teller distortion by Co (III) (Yin et al., 2015). These changes further affect the behaviors of these birnessites towards the detoxification of heavy metal contaminants. For example, though Ni doping during birnessite crystallization enhanced the formation of vacancies in the layer, Ni-doped hexagonal turbostratic birnessites reduced adsorption capacities for Pb^{2+} and Zn^{2+}, mainly because of the occupation of vacancies and edge sites by a large amount of Ni. These birnessites also exhibited much higher oxidation capability and initial reaction rates towards As (III) oxidation (Yin et al., 2012).

3. Contribution of humic substances in immobilizing heavy metals

Humic substance (HS) is one of the most important adsorbent in soils, sediments and aquatic environments. According to the difference in the solubility of the natural organic particle at different pH, HS can be operationally divided into two class of particles, fulvic acid (FA) and humic acid (HA). Due to the proton dissociation of different types of the acid functional groups, there are a large number of negative charges on the HS which can be involved in the interactions with proton and heavy metal ions. As a consequence, FA and HA were generally believed to play an important role in controlling the mobility, bioavailability, and toxicity of heavy metals which was relate to the species and speciation distribution of heavy metal ions in nature. The binding of the heavy metal ions to HS depends not only on the variable charge properties and chemical heterogeneity of HS, but also on the environmental conditions such as the pH, ionic strength and the presence of the competing ions (Tan et al., 2013). In order to be able to predict the contribution of HS to the heavy metal ions binding accurately, reliable models such as NICA (Non-Ideal Competition Adsorption) -Donnan model (Kinniburgh et al., 1999), Model V/VI (Tipping, 2002) and SHM (Stockholm Humic Model) (Gustafsson and Kleja, 2005) are proposed and developed for heavy metal ions complexation to HS. These models can be operated over a broad range of conditions with the model parameters that are independent of the environmental conditions. Among those models, NICA-Donnan model is the most popular as it is based on relatively simple analytical binding equations and it also takes into account the heterogeneity of HS, the competition of coexisting ions, the stoichiometry of binding reactions and the electrostatic interactions.

In general, the NICA-Donnan model with the material specific parameters is widely applied to describe the heavy metal ions binding to the HS in lab studies. As shown in Fig.

4, the NICA-Donnan model gives an accurate description of the Pb binding to soil HS (the open symbols and solid symbols in Fig. 4) (Xiong et al., 2013). In addition, it also can be used to predict the heavy metal ions binding to the HS in a wide range of the environmental conditions without further measurements in the field and can reveal the binding mechanisms. Based on a series of the experimental data, Milne et al. (2001, 2003) have fitted the generic parameters of NICA-Donnan model to predict the proton and heavy metal ions binding to FA and HA, respectively. When validating the ability of NICA-Donnan model with the generic parameters in predicting heavy metal ions binding in the field, it was found that the values of the model calculation agreed well with those of the experiment measurements over a wide range of environmental conditions for most metal ions such as Cu and Cd, but it failed to predict the Pb binding to the soil HS (Weng et al., 2002). The study of Xiong et al. (2013) on the Pb binding to the soil HS also showed that the Pb binding was overestimated by using the NICA-Donnan model with the generic parameters, especially at low Pb concentrations which are environmentally relevant (black line in Fig. 4). However, a series of sensitivity tests for the parameters showed that for the soil HA the model prediction with the generic parameters could be strongly improved only when the generic n_{Pb1} was replaced by the materials specific value of n_{Pb1} of the soil HA (red line in Fig. 4).

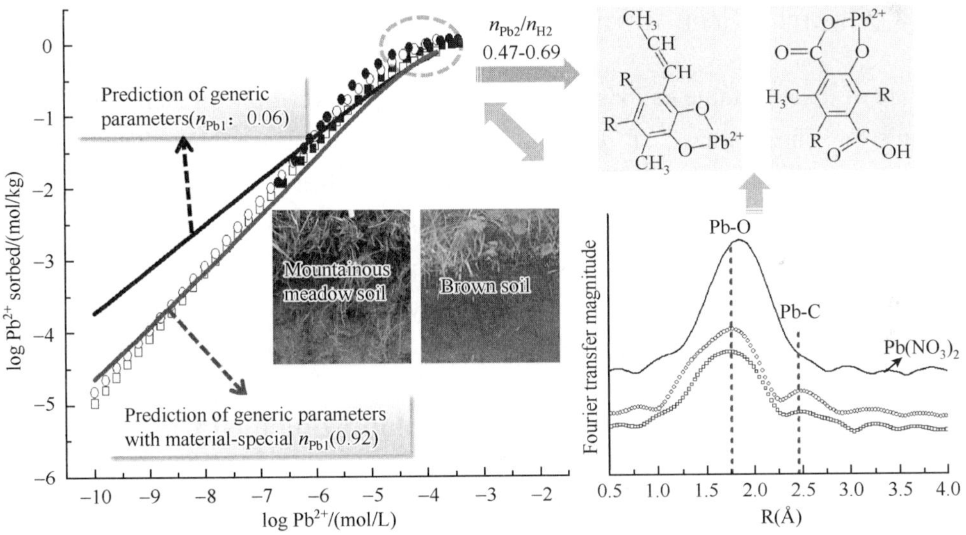

Fig. 4 The NICA-Donnan modeling for the Pb binding to HS with the materials specific parameters and generic parameter; binding mechanism obtained from the NICA-Donnan model parameters and EXAFS spectroscopy (Xiong et al., 2013)

The mechanism of the heavy metal ions binding can be studied by the spectral technique. For example, the XAFS spectroscopy in the Xiong et al. (2013) was directly used to study the coordination environment of Pb binding to soil HS. The vertical dashed lines in Fig. 4 indicated the peak positions for the first and second oscillation, corresponding with, respectively, Pb–O and Pb-C coordination. It suggests that Pb is bound to the HS by the inner-sphere complexation with a C-O-Pb structure. The result of the EXAFS spectra revealed that Pb was coordinated by 3.73-4.73 O atoms at a distance of 2.37-2.40 Å in the first shell, and by 2.49-3.36 C atoms at a distance of 3.17-3.27 Å in the second shell. It is in best agreement with inner-sphere structures that involves bidentate Pb complexes. Taking into account the structure stability, the Pb ions tend to be predominantly complexed by phenolic and/or carboxylic groups located in ortho-position at high Pb concentrations (as the Pb binding structure shown in Fig. 4). Among the NICA-Donnan model parameters, the ratio n_{Pbi}/n_{Hi} was the best suited to obtain the information on the binding stoichiometry. The model interpretation for these ratios was the average number of heavy metal ions bound by an acid functional group. The value 0.47-0.69 for the ratios n_{Pb2}/n_{H2} indicated that the Pb ions were bound by the phenolic groups in a bidentate structure at high Pb concentrations. But the NICA-Donnan model only takes into account the multidentate binding of the heavy metal ions with a few carboxylic groups or with a few phenolic groups; the mixed complexes involving both the carboxylic groups and phenolic groups, which occur in practice, are not included. Therefore, in view of the fact that the options for the bidetate complexes are limited in the NICA-Donnan model, the adsorption mechanism of Pb binding revealed by the model only partially agreed with the results of EXAFS spectroscopy.

4. Contribution of microbe in immobilizing heavy metals

The fate of toxic metallic cations in environment depends largely on their interactions with microorganisms. The biomass of bacteria, fungi, yeasts and algae have been reported for effective and economical removal of a wide variety of toxic heavy metals from wastewater and engineering systems. The bacterial cell wall acts as interface with the environment, and is capable of adsorbing solutes and sensing the presence and abundance of chemical stimulants. Thus, bacterial surfaces are important in mediating many important geological processes, including the transport and cycling of metals and the precipitation and dissolution of minerals.

4.1 Effect of bacterial wall structure on immobilizing heavy metals

Bacterial surfaces contain carboxyl, phosphoryl, hydroxyl, and amino functional groups (Beveridge and Murray, 1980), which gives the bacterial surface a net negative charge above a pH of approximately 2.0. Negatively charged functional group sites are capable of adsorbing large quantities of metals (Fein et al., 1999). Among them carboxyl and phosphoryl groups are considered to be particularly important in metal complexation based on wet chemistry and modeling studies (Ngwenya et al., 2003). It is quite controversial if a universal adsorption behavior exists in different bacterial species in natural environments. Yee and Fein (2001) observed that six Gram-positive and two Gram-negative bacterial species had one common adsorption edge, suggesting that the structures that give rise to metal and proton adsorption onto bacteria are common over a wide range of bacterial species (Fig. 5). Jiang et al. (2004) used attenuated total reflectance FTIR spectroscopy of both gram-positive and gram-negative bacterial species to demonstrate that the infrared spectra of both gram-positive and gram-negative bacteria were similar and exhibit similar variations as a function of changes in pH. The similarity in spectra for this range of bacteria suggests a similarity in binding environment for metals, and this observation supports a universal adsorption behavior that is rooted in similar cell wall functional group chemistries. Although a large number of bacterial species appear to exhibit broadly similar adsorption behaviors, some studies suggest that at least some bacteria have unique adsorptive properties. For example, Borrok et al. (2004) found that bacterial consortia grown from heavily contaminated industrial wastes, groundwater, and soils exhibited different total site densities, demonstrating that bacteria-metal adsorption in contaminated environments cannot be described using average bacteria-metal binding constants. In addition, Kulczycki et al. (2002) demonstrated that a two-site sorption model provided a modestly improved account of Cd^{2+} and Pb^{2+} sorption by Gram-negative *Escherichia coli*, but not Gram-positive *Bacillussubtilis*. Fourier-transform infrared spectroscopy, potentiometric titration, along with sorption experiments using chemically modified bacteria were conducted to compare the behavior of Gram-positive *Bacillus thuringiensis* and Gram-negative *Escherichia coli* as sorbents for Cu^{2+} and Cd^{2+} ions (Fang et al., 2009). The different adsorption behavior suggests that the composition and structure of bacterial cell walls play an important role in binding metals. Therefore, in addition to the work required to better characterize metal-adsorption mechanisms on bacterial cell walls, more research is also required in order to understand the effects exerted on metal speciation by biological structures associated with bacteria in natural environments.

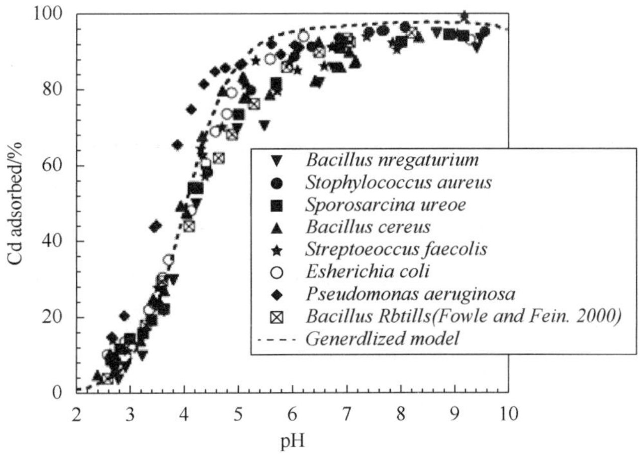

Fig. 5　Cd adsorption onto pure cultures of individual gram-positive and gram-negative bacterial species(from Yee and Fein, 2001)

4.2 Effect of bacterial extracellular polymeric substances (EPS) on immobilizing heavy metals

Among the various reactive components associated with bacterial cell walls, bacterial extracellular polymeric substance (EPS) is of particular importance which affects biofilm formation and cell adhesion to solid substrates (Parikh and Chorover, 2006). EPS can be present in the environment either bound to cell surfaces ("capsular" EPS), released into solution ("free" EPS) or associated with the hydrated matrix of biofilms. These polymers potentially have a significant effect on the acid-base properties and metal-adsorption characteristics of bacterial cells (Toner et al., 2005; Guiné, et al., 2006).

A number of studies have examined metal-EPS binding reactions. Bhaskar and Bhosle (2006) reported that the presence of sodium chloride reduced the concentration of Pb (Ⅱ) and Cu (Ⅱ) binding to *Marinobacter* EPS. Comte et al. (2008) found that the adsorption of Cu (Ⅱ), Pb (Ⅱ), and Cd (Ⅱ) onto sludge EPS increased with increasing pH, and that binding could be described with a single binding constant. Guibaud et al. (2009) reported that Pb (Ⅱ) has a greater affinity for EPS than Cd (Ⅱ), and that these metals have a greater affinity for EPS from sludge than that from pure bacterial strains. Metal-proton exchange is not the only mechanism involved in metal adsorption by EPS and other mechanisms, such as cation exchange, electrostatic interactions, and EPS-induced metal precipitation may also occur (Guibaud et al., 2008). FTIR and XPS indicate that alcohol, carboxyl, amino, and phosphate groups of EPS are responsible for heavy metal binding by EPS (Sun et al., 2009). Sheng et al. (2013) showed that proteins and humic substances in EPS from activated sludge are both strong ligands for Cu (Ⅱ), and that carboxyl groups play a particularly important role in binding Cu (Ⅱ).

Recently a few studies have been performed to investigate the role of tightly bound EPS on metal binding on bacteria (Ueshima et al., 2008; Ha et al., 2010; Kenney and Fein, 2011; Fang et al., 2010b, 2011; Wei et al., 2011). For example, Ueshima et al. (2008) observed that the presence of EPS did not significantly affect the extent of Cd (Ⅱ) removal from solution by *Pseudomonas putida* and it did not appear to strongly affect the Cd (Ⅱ)—binding groups as observed by FTIR. Kenney and Fein (2011) further confirmed that enzymatic removal of EPS from bacteria (*Pseudomonas putida, Shewenellaoneidensis, Rhizobium tropici*, and *Agrobacterium* sp.) does not affect either the proton reactivity or the Cd (Ⅱ) adsorption behavior of the biomass. By contrast, Ha et al. (2010) compared Zn (Ⅱ) and Pb (Ⅱ) adsorption by unmodified *Shewanellaoneidensis* strain (wild-type (WT)) and genetically modified cells with inhibited production of EPS (ΔEPS strain), and found significantly lower Zn (Ⅱ) and Pb (Ⅱ) uptake and binding affinities for the ΔEPS strain. The presence of EPS can significantly enhance Cu (Ⅱ) and Cd (Ⅱ) adsorption capacity on bacteria and the promoting effect was more remarkable on

B. subtilis cells than *P. putida* cells (Fang et al., 2011; Wei et al., 2011). Biosorption mechanisms of Cu (Ⅱ) by EPS from *Bacillus subtilis* were investigated by Fang et al. (2014). The XAFS and ITC results were consistent with inner-sphere binding of Cu (Ⅱ) by carboxyl sites with 2.03 C atoms at distance of 2.96 Å in the second shell, which further suggests that the carboxylic groups are the dominant sites for the Cu (Ⅱ) adsorption by EPS at pH 5.0, and the five-membered chelate rings is the possible structure for Cu (Ⅱ) in EPS (Fig. 6). The molecular binding mechanisms obtained in these studies will add fundamental knowledge of understanding the fate of heavy metals in natural environments.

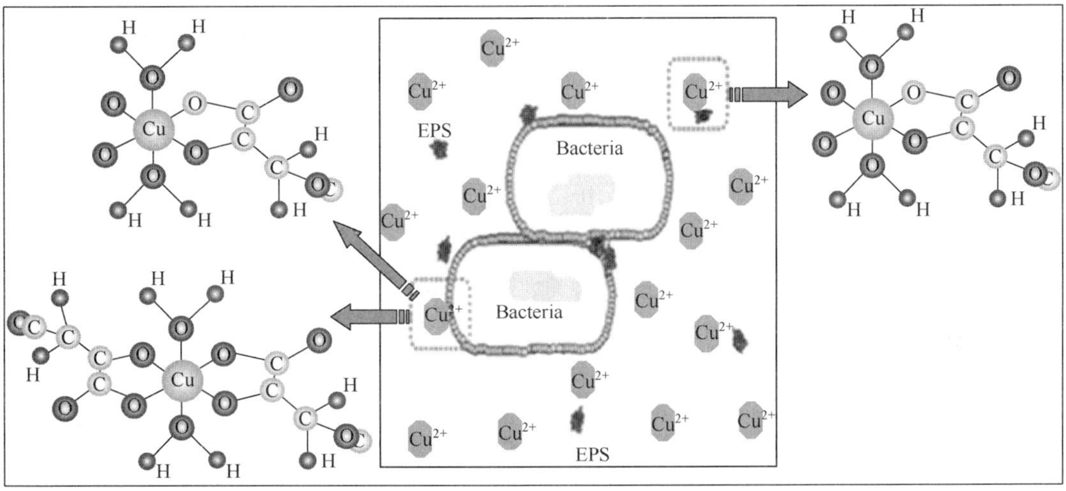

Fig. 6 Cu (Ⅱ) -EPS inner-sphere complexes consist of one five-membered chelate ring (from Fang et al., 2014)

4.3 Molecular binding mechanisms of metals to bacterial surfaces

Metal adsorption onto bacterial surfaces has been studied extensively over the past 30 years. Much of this previous work has been either qualitative or quantified adsorption using a bulk-partitioning approach, making it impossible to estimate the effects of bacteria on mass transport under different or more complex conditions than those studied. Traditional methods of investigating the reactivity of the bacterial surface include acid-base titrations, spectroscopic studies, and bulk adsorption measurements. Recent 10 years, more direct molecular binding mechanisms have been offered by XAFS and isothermal titration calorimetry (ITC) investigations.

Panak et al. (2000, 2002) used time-resolved laser-induced fluorescence spectroscopy (TRLFS), in conjunction with XAFS, to demonstrate that U (Ⅵ) forms inner sphere complexes only with phosphate groups on cell walls of a number of *Bacillus* species at pH 4.5-5.0. Boyanov et al., (2003) reported that Cd binds to the *Bacillus subtilis* predominantly due to phosphoryl binding below pH 4.4, whereas with increasing pH (4.4-6.5), adsorption to carboxyl groups becomes increasingly important. Guiné et al. (2006) reported sulfhydryl ligands were responsible for Zn adsorption to three Gram negative bacterial strains at low loadings of Zn. Fang et al. (2011)

demonstrated that the carboxyl groups play a vital role in the adsorption of Cu (Ⅱ) and Cd (Ⅱ) to the cyanobacterium *Spirulina platensis*. XAFS analysis revealed that Cu (Ⅱ) forms inner-sphere complexes consisting of two five-membered chelate rings on the cyanobacterial surface (Fig. 7). The above molecular-scale investigations demonstrated that sorption of heavy metals onto bacteria is controlled by carboxyl, phosphoryl and perhaps sulfhydryl functional groups on the cell wall of the microorganisms. Recently, the importance of metal-sulfhydryl binding within bacterial cell envelopes was documented, and the nature of the bacterial surface complexes has been characterized by several research groups using EXAFS spectroscopy (Mishra et al., 2010, 2011; Pokrovsky et al., 2012; Song et al., 2012). Yu and Fein (2015) demonstrated that the mechanism of Cd adsorption onto *S. oneidensis* cell envelopes depends strongly on metal loading, due to the presence of sulfhydryl and non-sulfhydryl Cd binding sites that have high and low affinities for Cd, respectively.

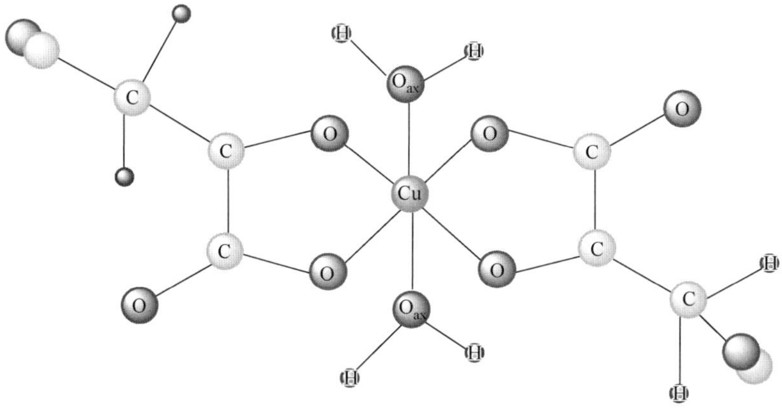

Fig. 7 Proposed model for Cu complexed by two 5-membered chelate rings (Fang et al., 2011)

Calorimetric measurements of metal binding onto bacterial surfaces, interpreted using a surface complexation modeling approach, can be used to infer site-specific enthalpies and entropies of metal adsorption. Weppen and Hornburg (1995) used flow calorimetry to compare the metal sequestration capability of cell wall suspensions harvested from a range of bacterial cells. Their measured bulk enthalpies of metal adsorption (for Cd, Cu, Zn, Pb, Mg, Ca, Sr, and Ba) were all modestly endothermic (+3 to +18 kJ/mol), and in the expected range for divalent cations coordinated with anionic oxygen ligands. Gorman-Lewis et al. (2006) used calorimetric data, in conjunction with surface complexation models of the protonation of the bacterial cell wall, to produce site-specific enthalpies and entropies of Cd adsorption onto the cell wall of *B. subtilis*. The calculated enthalpies of Cd adsorption are typical for Cd complexation with anionic oxygen ligands; the entropies are indicative of inner sphere complexation by multiple ligands; and the calorimetry results also constrain both the stoichiometry of the adsorption reactions and the temperature dependence of the stability constants for the dominant metal-bacterial surface complexes. Cd and Zn complexation onto *Bacillus licheniformis* is entropically driven at low pH and enthalpically driven at circumneutral pH. Zn complexation onto *Bacillus subtilis* is entropically driven, which suggests that *Bacillus licheniformis* has different donor ligands dominating reactivity around circumneutral pH (Gorman-Lewis, 2014).

5. Contribution of Mineral-HS-Microbe composites in immobilizing heavy metals

Soils are complex systems possessing a very high metal-binding capacity, primarily owing to various surface reactive solid particles such as minerals, bacteria and HS. These components interact with each other to form composites that influence the chemical behavior of heavy metals differently compared to the isolated end-member phases. Understanding the speciation and distribution of trace metals in complex systems (clay minerals/bacteria/HS) is crucial in predicting the mobility, bioavailability and toxicity of heavy metals in natural soils and associated environments.

5.1 Mineral-HS composites

Humic substances, such as FA and HA, and the metal (hydr) oxides, such as the iron (hydr) oxides are important reactive colloids in the soil and aquatic environments. On the surface of both HS and metal (hydr) oxides, there are a large number of sites which can interact with the heavy metal ions by the specific adsorption and electrostatic reaction. Those will control the species and speciation distribution of the heavy metal ions, and then to govern its mobility, bioavailability, and toxicity in the nature environment.

It has been reported that mineral-HS complexes have quite different metal adsorption properties from pure minerals. For instance, Wu et al. (2011) find that maximum sorption of Cd on HA-modified Ca-montmorillonite increased by 19.2% compared to adsorption on pure clay end-member. Vermeer et al. (1999) performed Cd^{2+} sorption to PAHA and hematite composites as compared to the end-member systems and discovered that sorption to the composite was non-additive and overall sorption to each sorbent was dependent on the affinity of the metal in solution. For the adsorption of Cu (Ⅱ), Pb (Ⅱ) and HA on natural zeolite tuff binary systems, the complexation structure of Pb-HA/zeolite and Cu-HA/zeolite in three- component system was proposed as S-Pb-HA and S-HA-Cu, respectively (Wang et al., 2008). Lenhart and Honeyman (1999) studied the system of hematite-HA-U (Ⅵ) and found that the additivity rule underestimated the amount adsorbed; they proposed the presence of oxide-uranyl-HA ternary complexes (Type A complexes) to explain the discrepancy. Alcacio et al. (2001) conducted a molecular scale investigation of Cu (Ⅱ) bonding to goethite-HA complexes with EXAFS and XANES measurements; they reported the presence of two types of ternary complexes: Type A complexes with the Cu (Ⅱ) cations bridging the oxide and the HA, and Type B complexes with the cations bonded to the HA, in turn bound of the oxide surface. Arias et al. (2002) measured the competitive adsorption equilibrium isotherms of Cu (Ⅱ) and Cd (Ⅱ) on kaolin in presence and absence of HA. Humic acids were found to enhance the metal adsorption capacity of mineral surfaces.

In order to study the properties of heavy metal ions binding in the lab, usually the chemical speciation model NICA-Donnan model (Kinniburgh et al., 1999) and the CD (Charge Distribution)-MUSIC (Multi Site Complexation) model (Hiemstra and Van Riemsdijk, 1996a) were applied to describe its binding to the HS and iron (hydr) oxides in the binary system, respectively. For the adsorption of heavy metal ions binding to iron (hydr) oxides, the classical fitting methods which not consider the heterogeneity of the surface site was popular for the CD-MUSIC modeling, but it always failed to give correct speciation distribution for the adsorbed heavy metal ions. Therefore, the full site fitting methods was recommended for the CD-MUSIC modeling (Hiemstra et al., 1996b). The results of Xiong et al. (2015) for the CD-MUSIC modeling the adsorption of Pb to goethite indicated that the materials specific parameters obtained from the full sites fitting methods can describe the adsorption edge well. The calculated speciation distribution of the adsorbed Pb ions was significantly different from that of others results and agreed well with the results of the XAFS spectroscopy.

At the environmental conditions, the HS and iron (hydr) oxides are simultaneously present. They can interact with each other due to the oppositely charged surfaces and it will be affected strongly by the co-existed heavy metal ions. The study of Xiong et al. (2015) showed that the amount of soil HA and FA adsorbed to goethite decreased with the increasing pH obviously. The strong adsorption of HS to goethite occurred at low pH; it implies that there is a strong attraction between goethite and HS which decreases gradually with increasing pH. The increasing Pb concentrations also increase the soil HS adsorption to the goethite, especially at the high Pb concentrations. The presence of Pb may affect both the electrostatic attraction between goethite and HS and the electrostatic HS-HS repulsion. In addition, the cation bridging type ternary surface complexes (goethite-Pb-HS) is formed, which promotes the adsorption of soil HS. Compare to soil FA, the amount of soil HA adsorbed by goethite in mg/g was much higher, but the number of HA particles adsorbed per nm^2 of goethite was lower due to the much higher particle mass of HA. The difference between FA and HA adsorption to goethite is largely caused by the difference in the particle size. The calculated particle diameter of the soil HA and FA were 3.6 nm and 1.5 nm, respectively and a well-accepted value of the Stern layer thickness is about 0.8 nm. These indicated that the soil FA adsorbed can be located in the Stern layer with strong conformation change, but the soil HA was adsorbed in the Stern layer and diffuse layer (see Fig. 8).

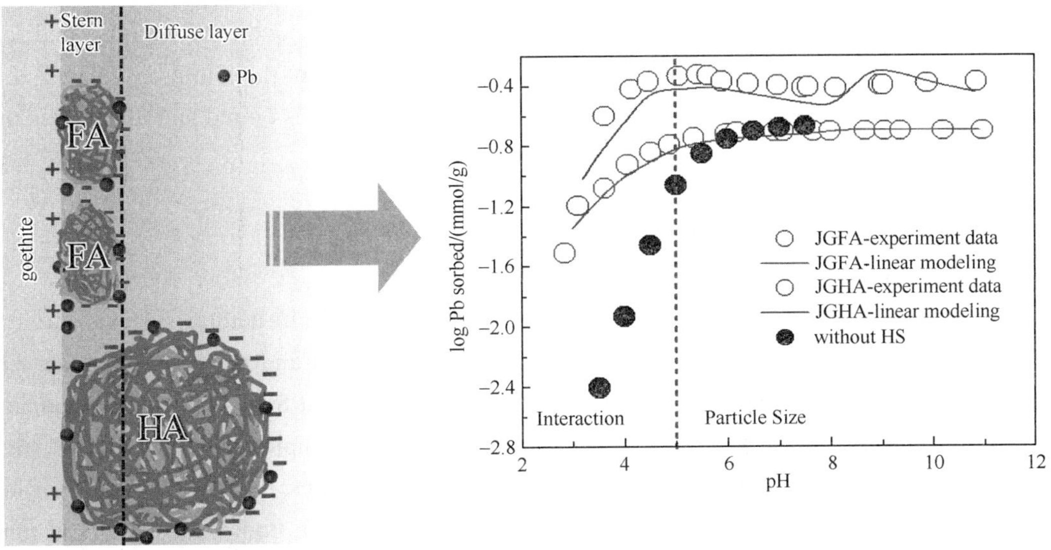

Fig. 8 Schematic presentation for the effect of soil HS on the Pb binding to goethite-soil HS complexes (from Xiong et al., 2015)

The interactions between the HS and iron (hydr) oxides not only determine the binding of HS to the iron (hydr) oxides, they also alter the iron (hydr) oxide surface site availability, charge density, electrostatic potential and the degree of protonation of HS. These changes will affect the amount and speciation of the adsorbed heavy metal ions strongly. For example, in the ternary system (goethite-soil HS-Pb), the Pb binding to goethite can be promoted by the soil HS strongly, especially at low pH (open symbols in Fig. 8). The possible explanations for this behavior are that ①Pb also binds to adsorbed HS, and ②adsorbed HS decreases the electrostatic potentials in the electrical double layer of the goethite-HS-electrolyte interface and this leads to a higher Pb adsorption to goethite. The effect of soil FA on Pb adsorption to goethite is stronger than that of soil HA, due to the fact that the FA particles were primarily adsorbed in the Stern layer, whereas HA particles were present in both the Stern layer and diffuse layer. This caused that the electrostatic potential profile of goethite-FA complex is considerably different from that of goethite-HA complex and affects Pb binding strongly.

When predicting the free heavy metal ions concentrations in the ternary system and soil, the linear additivity rule which is the combination of the NICA-Donnan model and CD-MUSIC model is generally used. In this method, the soil was considered as a set of independent components and it assumed that the iron (hydr) oxides and HS do not interact or the interaction does not affect its behavior in heavy metal ions binding (Weng et al., 2002, 2008). Therefore, the metal ion adsorption behavior obtained from the linear additivity rule may deviates from that in the soil in principle. In the study of Xiong et al. (2015), the linear additivity rule always underestimated the adsorption of Pb to the goethite-soil HS complex at low pH and low initial Pb concentrations; at other conditions the linear additivity rule predicted the experimental results reasonably well. The reason is that at these conditions the adsorbed soil HS strongly reduces the high positive electrostatic potentials in the Stern layer and this promotes Pb binding to goethite. At higher pH values nearly all Pb is bound and this makes the predictions of linear additivity rule relatively easy.

5.2 Mineral-Microbe composites

Microorganisms and minerals affect the speciation of metals in natural systems through complexation reactions with surface sites. Composites of bacteria and minerals can form and affect the mobility and transport of ions in solution. In soils, almost 80%-90% microorganisms are often intimately intermixed with mineral phases to form complex, highly hydrated, high surface-area composites through attractive van der Waals interactions, hydrogen bonding, or ion bridging (Nannipieri et al., 2003). The surface properties of these composite materials may differ dramatically from pure bacterial or mineral surfaces, particularly in terms of surface charge, double-layer properties, numbers of reactive sites, and metal-binding affinity (Templeton et al., 2001; Huang et al., 2005, Fang et al., 2010). Therefore, it is important to make clear the adsorption mechanisms and potentials of heavy metals by bacteria-mineral composites.

Recently, a few studies that had examined ion sorption to bacteria-mineral composites used both natural and synthetic composite materials. Small et al. (1999) observed that the maximum binding capacity of Sr^{2+} (0.034 mmol/g) by the bacteria-amorphous hydrous ferric oxide (HFO) composite was less than the combined sorptive capacity of its components (0.041 mmol/g). At the *Burkholderia cepacia*-metal oxide interfaces, Pb(Ⅱ) was mainly complexed with the high-affinity sites on hematite or corundum at pH 6 and $[Pb]<10^{-6}$ mol/L (Templeton et al., 2001). The EXAFS of *Burkholderia cepacia*-goethite composite revealed that at least 50% of the total sorbed Pb(Ⅱ) was associated with the biofilm at pH<5.5, whereas Pb(Ⅱ) sorbed by the biofilm was considerably decreased due to competition with the goethite surface above pH 6 (Templeton et al., 2003). In the *P. putida* CZ1-goethite composite, the contribution of goethite to the binding of Cu and Zn was smaller than that of bacteria (Chen et al., 2008). These studies demonstrated that the metal retention behavior of the bacteria-mineral composites is quite different from that of the pure bacteria or minerals. Potentiometric titration, ITC, and equilibrium sorption experiments were conducted to investigate the adsorption behavior of Cu(Ⅱ) by *Bacillus thuringiensis*, *Pseudomonas putida* and their composites with minerals (Fang et al., 2010). The interaction of montmorillonite with bacteria increased the reactive sites and resulted in greater adsorption for Cu(Ⅱ) on their composites, while decreased adsorption sites and capacities for Cu(Ⅱ) were observed on goethite-bacteria composites. A gram-positive bacterium *B. thuringiensis* played a more important role than a gram-negative bacterium *P. putida* in determining the properties of the bacteria-minerals interfaces. This result suggests that in natural geologic systems, the similar surface charges on 2:1 phyllosilicate mineral and bacteria may lead to their relatively looser association and formation of new sites, whereas strong electrostatic interactions between the opposite surface charges on hydrous oxide and bacteria may result in a relatively tighter association and the masking of reactive sites (Fig. 9). Recent work by Moon and Peacock (2012, 2013) has examined Cu^{2+} sorption to *Bacillus subtilis*-ferrihydrite composites over a range of pH, Cu^{2+} concentrations, and ferrihydrite/bacteria ratios. In this study the authors found that sorption to the composites was non-additive: the sum of sorption in bacteria and ferrihydrite systems was greater than sorption in the composite system and this deviation increased with the ferrihydrite/bacteria ratio.

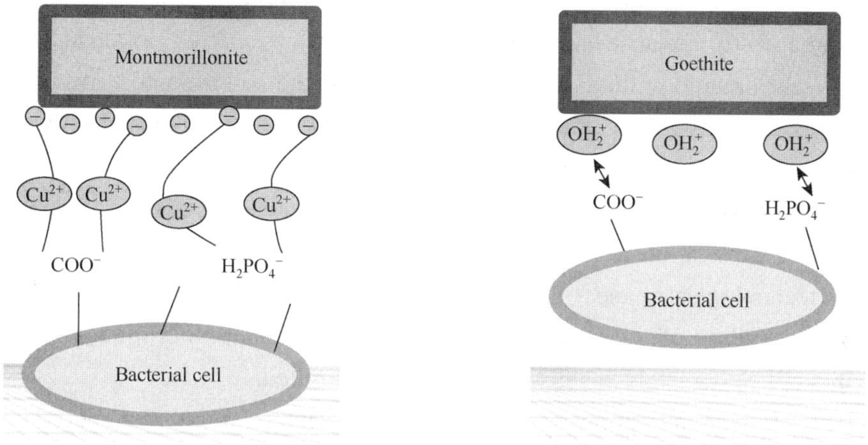

(a) The loose association between bacteria and montmorillonite (b) The tight association between bacteria and goethite

Fig. 9 The loose association between bacteria and montmorillonite; The tight association between bacteria and goethite (from Fang et al., 2010)

5.3 Mineral-HS-Microbe composites

Although a number of studies investigating the adsorption behavior of metal ions on binary mineral-organic or mineral-microbe composites have been conducted, knowledge on the adsorption characteristics of heavy metals to ternary mineral-humic acid-microbe composite is quite limited.

Recently, batch sorption isotherms, ITC, and Cd K-edge EXAFS spectroscopy were applied to investigate the binding characteristics of Cd on montmorillonite (Mont) HA-bacteria composites (Du et al., 2016). Specifically, in the ternary composite, the coexistence of HA and bacteria inhibits Cd adsorption, suggesting a "blocking effect" between humic acid and bacterial cells. Large positive entropies (68.1-114.4 J/mol/K), and linear combination fitting of the EXAFS spectra for Cd adsorbed onto Mont-bacteria and Mont-HA-bacteria composites, demonstrate that Cd is mostly bound to bacterial surface functional groups by forming inner-sphere complexes. These studies also show that natural environments are more complex than the prior laboratory work with one sorbent because non-additive sorption behavior was observed in many studies using natural and synthetic sorbent surfaces. Hence more laboratory work needs to be undertaken using bacteria-mineral composites to be able to accurately predict metal cycling in nature.

6. Summary and Conclusions

In recent years, with the high-speed development of social economy and intensive human activities, the range of soil degradation have been continuously expanded, and soil quality deterioration has been intensified. Particularly with the increase of environmental pollution, the risks and damages to human health caused by the entry of external pollutants such as heavy metals, persistent organic pollutants and antibiotics into the soil attracted great concern of the public. The vast majority of the external pollutants would be adsorbed by soil active components, and are exposed on the soil chemical interface. Soil interfacial interactions have profound effects on the sequestration, degradation and biological metabolism of the pollutants. Therefore, focusing on the chemical behaviors of external pollutants in soil and developing environmental remediation technologies to meet the social and public needs were of prime importance. To date, our understanding of the impacts of soil mineral-organic matter-microbe interactions on the behaviors of heavy metals in soils is still relatively limited.

The development and application of in-situ measurement, micro-examination and element tracing technologies significantly accelerated the development of studies with related to heavy metal interactions with soil active components. Recently the in-situ measurement and the greatly improved understanding of soil microscopic properties based on microscopic spectrum technology at molecular level laid solid foundation for the rapid understanding of immobilizing heavy metals in soil and development of soil remediation. Besides, atomic force microscopy (AFM), confocal laser scanning microscopy (CLSM) and environmental scanning electron microscopy (ESEM) have been widely used to characterize the three dimensional structure, interfacial properties and reaction activities of soil interface. XAS, XPS, nuclear magnetic resonance (NMR), electron paramagnetic resonance and FTIR can be applied to the in-situ analysis of the interactions among soil components with heavy metal and their immobilizing mechanisms. Nanometer-scale secondary ion mass spectrum (NanoSIMS), which combines the ultrahigh-resolution microscope image technique and isotope tracer technique, can demonstrate the spatial distribution characteristics of heavy metals in soil, thus shows great potential in the field of soil pollution and remediation. Besides the application of experimental and instrumental analysis techniques to characterize the biogeochemical structure and reactions of soil, approaches of computational chemistry have also been used in the study of soil components interaction with heavy metals. Quantum chemistry calculation combined with spectral information has greatly enhanced the understanding of these processes at the molecular level. In addition, geochemical models are widely used to predict the speciation and transport of heavy metals that will be important in the control of the solubility, bioavailability and the fate of heavy metals. Future research on this extremely important and exciting area of science should be stimulated to soil pollution and remediation as well as sustain the ecosystem integrity.

References

Alcacio T E, Hesterberg D, Chou J W, et al. 2001. Molecular scale characteristics of Cu (II) bonding in goethite-humate complexes. Geochimica et Cosmochimica Acta, 65 (9): 1355-1366.

Arias M, Barral M T, Mejuto J C. 2002. Enhancement of copper and cadmium adsorption on kaolin by the presence of humic acids. Chemosphere, 48 (10): 1081-1088.

Beveridge T J, Murray R G. 1980. Sites of metal deposition in the cell wall of *Bacillus subtilis*. Journal of Bacteriology, 141 (2): 876-887.

Bhaskar P V, Bhosle N B. 2006. Bacterial extracellular polymeric substance (EPS): a carrier of heavy metals in the marine food-chain. Environment International, 32 (2): 191-198.

Borrok D M, Fein J B, Kulpa C F. 2004. Cd and proton adsorption onto bacterial consortia grown from industrial wastes and contaminated geologic settings. Environmental Science & Technology, 38 (21): 5656-5664.

Boyanov M I, Kelly S D, Kemner K M, et al. 2003. Adsorption of cadmium to *Bacillus subtilis* bacterial cell walls: a pH-dependent X-ray absorption fine structure spectroscopy study. Geochimica et Cosmochimica Acta, 67 (18): 3299-3311.

Burns R G. 1993. Mineralogical applications of crystal field theory. Cambridge University Press.

Chen X C, Chen L T, Shi J Y, et al. 2008. Immobilization of heavy metals by *Pseudomonas putida* CZ1/goethite composites from solution. Colloids and Surfaces B: Biointerfaces, 61 (2): 170-175.

Comte S, Guibaud G, Baudu M. 2008. Biosorption properties of extracellular polymeric substances (EPS) towards Cd, Cu and Pb for different pH values. Journal of Hazardous Materials, 151 (1): 185-193.

Du H, Chen W, Cai P, et al. 2016. Cd (II) Sorption on montmorillonite-humic acid-bacteria composites. Scientific Reports, 6.

Fang L, Cai P, Chen W, et al. 2009. Impact of cell wall structure on the behavior of bacterial cells in the binding of copper and cadmium. Colloids and Surfaces A: Physicochemical and Engineering Aspects, 347 (1): 50-55.

Fang L C, Cai P, Li P X, et al. 2010a. Microcalorimetric and potentiometric titration studies on the adsorption of copper by *P. putida* and *B. thuringiensis* and their composites with minerals. Journal of Hazardous Materials, 181 (1-3): 1031-1038.

Fang L C, Huang Q Y, Wei X, et al. 2010b. Microcalorimetric and potentiometric titration studies on the adsorption of copper by extracellular polymeric substances (EPS), minerals and their composites. Bioresource Technology, 101 (15): 5774-5779.

Fang L, Zhou C, Cai P, et al. 2011. Binding characteristics of copper and cadmium by cyanobacterium *Spirulina platensis*. Journal of Hazardous Materials, 190 (1): 810-815.

Fang L, Yang S, Huang Q, et al. 2014. Biosorption mechanisms of Cu (II) by extracellular polymeric substances from *Bacillus subtilis*. Chemical Geology, 386: 143-151.

Fein J B, Delea D. 1999. Experimental study of the effect of EDTA on Cd adsorption by *Bacillus subtilis*: a test of the chemical equilibrium approach. Chemical Geology, 161 (4): 375-383.

Feng X H, Zhai L M, Tan W F, et al. 2007. Adsorption and redox reactions of heavy metals on synthesized Mn oxide minerals. Environmental Pollution, 147 (2): 366-373.

Gorman-Lewis D. 2014. Enthalpies and entropies of Cd and Zn adsorption onto *Bacillus licheniformis* and enthalpies and entropies of Zn adsorption onto *Bacillus subtilis* from isothermal titration calorimetry and surface complexation modeling. Geomicrobiology Journal, 31 (5): 383-395.

Gorman-Lewis D, Fein J B, Jensen M P. 2006. Enthalpies and entropies of proton and cadmium adsorption onto *Bacillus subtilis* bacterial cells from calorimetric measurements. Geochimica et Cosmochimica Acta, 70 (19): 4862-4873.

Guibaud G, Bordas F, Saaid A, et al. 2008. Effect of pH on cadmium and lead binding by extracellular polymeric substances

(EPS) extracted from environmental bacterial strains. Colloids and Surfaces B: Biointerfaces, 63 (1): 48-54.

Guibaud G, van Hullebusch E, Bordas F, et al. 2009. Sorption of Cd (Ⅱ) and Pb (Ⅱ) by exopolymeric substances (EPS) extracted from activated sludges and pure bacterial strains: Modeling of the metal/ligand ratio effect and role of the mineral fraction. Bioresource Technology, 100 (12): 2959-2968.

Gustafsson J P, Berggren Kleja D. 2005. Modeling salt-dependent proton binding by organic soils with the NICA-Donnan and Stockholm Humic models. Environmental Science & Technology, 39 (14): 5372-5377.

Guiné V, Spadini L, Sarret G, et al. 2006. Zinc sorption to three gram-negative bacteria: combined titration, modeling, and EXAFS study. Environmental Science & Technology, 40 (6): 1806-1813.

Ha J, Gélabert A, Spormann A M, et al. 2010. Role of extracellular polymeric substances in metal ion complexation on *Shewanella oneidensis*: Batch uptake, thermodynamic modeling, ATR-FTIR, and EXAFS study. Geochimica et Cosmochimica Acta, 74 (1): 1-15.

Hiemstra T, Van Riemsdijk W H. 1996. A surface structural approach to ion adsorption: The charge distribution (CD) model. Journal of Colloid and Interface Science, 179 (2): 488-508.

Hiemstra T, Venema P, Van Riemsdijk W H. 1996. Intrinsic proton affinity of reactive surface groups of metal (hydr) oxides: The bond valence principle. Journal of Colloid and Interface Science, 184 (2): 680-692.

Huang Q, Chen W, Xu L. 2005. Adsorption of copper and cadmium by Cu-and Cd-resistant bacteria and their composites with soil colloids and kaolinite. Geomicrobiology Journal, 22 (5): 227-236.

Huang Q, Huang P M, Violante A. 2008. Soil Mineral-Microbe-Organic Interactions. Berlin: Springer.

Jiang W, Saxena A, Song B, et al. 2004. Elucidation of functional groups on gram-positive and gram-negative bacterial surfaces using infrared spectroscopy. Langmuir, 20 (26): 11433-11442.

Kenney J P L, Fein J B. 2011. Importance of extracellular polysaccharides on proton and Cd binding to bacterial biomass: A comparative study. Chemical Geology, 286 (3): 109-117.

Kinniburgh D G, van Riemsdijk W H, Koopal L K, et al. 1999. Ion binding to natural organic matter: Competition, heterogeneity, stoichiometry and thermodynamic consistency. Colloids and Surfaces A: Physicochemical and Engineering Aspects, 151 (1): 147-166.

Kulczycki E, Ferris F G, Fortin D. 2002. Impact of cell wall structure on the behavior of bacterial cells as sorbents of cadmium and lead. Geomicrobiology Journal, 19 (6): 553-565.

Kwon K D, Refson K, Sposito G. 2013. Understanding the trends in transition metal sorption by vacancy sites in birnessite. Geochimica Et Cosmochimica Acta, 101: 222-232.

Lenhart J J, Honeyman B D. 1999. Uranium (Ⅵ) sorption to hematite in the presence of humic acid. Geochimica et Cosmochimica Acta, 63 (19): 2891-2901.

Manceau A, Silvester E, Bartoli C, et al. 1997. Structural mechanism of Co^{2+}oxidation by the phyllomanganate buserite. American Mineralogist, 82 (11-12): 1150-1175.

McKenzie R M. 1971. The synthesis of birnessite, cryptomelane, and some other oxides and hydroxides of manganese. Mineralogical Magazine, 38 (296): 493-502.

McKenzie R M. 1980. The adsorption of lead and other heavy metals on oxides of manganese and iron. Soil Research, 18 (1): 61-73.

Milne C J, Kinniburgh D G , Tipping E. 2001. Generic NICA-Donnan model parameters for proton binding by humic substances. Environmental Science & Technology, 35 (10): 2049-2059.

Milne C J, Kinniburgh D G, Van Riemsdijk W H, et al. 2003. Generic NICA-Donnan model parameters for metal-ion binding by humic substances. Environmental Science & Technology, 37 (5): 958-971.

Mishra B, Boyanov M, Bunker B A, et al. 2010. High-and low-affinity binding sites for Cd on the bacterial cell walls of *Bacillus subtilis* and *Shewanella oneidensis*. Geochimica et Cosmochimica Acta, 74 (15): 4219-4233.

Mishra B, O' Loughlin E J, Boyanov M I, et al. 2011. Binding of HgII to high-affinity sites on bacteria inhibits reduction to Hg0 by Mixed FeII/III phases. Environmental Science & Technology, 45 (22): 9597-9603.

Moon E M, Peacock C L. 2012. Adsorption of Cu (Ⅱ) to ferrihydrite and ferrihydrite-bacteria composites: Importance of the carboxyl group for Cu mobility in natural environments. Geochimica et Cosmochimica Acta, 92: 203-219.

Moon E M, Peacock C L. 2013. Modelling Cu (Ⅱ) adsorption to ferrihydrite and ferrihydrite-bacteria composites: Deviation from additive adsorption in the composite sorption system. Geochimica et Cosmochimica Acta, 104: 148-164.

Nannipieri P, Ascher J, Ceccherini M T, et al. 2003. Microbial diversity and soil functions. European Journal of Soil Science, 54 (4): 655-670.

Ngwenya B T, Sutherland I W, Kennedy L. 2003. Comparison of the acid-base behaviour and metal adsorption characteristics of a gram-negative bacterium with other strains. Applied Geochemistry, 18 (4): 527-538.

Ogata A, Komaba S, Baddour-Hadjean R, et al. 2008. Doping effects on structure and electrode performance of K-birnessite-type manganese dioxides for rechargeable lithium battery. Electrochimica Acta, 53 (7): 3084-3093.

O' Reilly S E, Hochella M F. 2003. Lead sorption efficiencies of natural and synthetic Mn and Fe-oxides. Geochimica et Cosmochimica Acta, 67 (23): 4471-4487.

Panak P, Raff J, Selenska-Pobell S, et al. 2000. Complex formation of U (Ⅵ) with Bacillus-isolates from a uranium mining waste pile. Radiochimica Acta International Journal for Chemical Aspects of Nuclear Science and Technology, 88 (2): 71-76.

Panak P J, Knopp R, Booth C H, et al. 2002. Spectroscopic studies on the interaction of U (Ⅵ) with *Bacillus sphaericus*. Radiochimica Acta, 90 (9-11/2002): 779-783.

Parikh S J, Chorover J. 2006. ATR-FTIR spectroscopy reveals bond formation during bacterial adhesion to iron oxide. Langmuir, 22 (20): 8492-8500.

Peacock C L. 2009. Physiochemical controls on the crystal-chemistry of Ni in birnessite: Genetic implications for ferromanganese precipitates. Geochimica et Cosmochimica Acta, 73 (12): 3568-3578.

Pokrovsky O S, Pokrovski G S, Shirokova L S, et al. 2012. Chemical and structural status of copper associated with oxygenic and anoxygenic phototrophs and heterotrophs: possible evolutionary consequences. Geobiology, 10 (2): 130-149.

Post J E. 1999. Manganese oxide minerals: Crystal structures and economic and environmental significance. Proceedings of the National Academy of Sciences, 96 (7): 3447-3454.

Shannon R D. 1976. Revised effective ionic radii and systematic studies of interatomic distances in halides and chalcogenides. Acta Crystallographica Section A: Crystal Physics, Diffraction, Theoretical and General Crystallography, 32 (5): 751-767.

Sheng G P, Xu J, Luo H W, et al. 2013. Thermodynamic analysis on the binding of heavy metals onto extracellular polymeric substances (EPS) of activated sludge. Water Research, 47 (2): 607-614.

Sherman DM, Peacock CL. 2010. Surface complexation of Cu on birnessite (δ-MnO_2): controls on Cu in the deep ocean. Geochimica et Cosmochimica Acta, 74 (23): 6721-6730.

Sigg L, Black F, Buffle J, et al. 2006. Comparison of analytical techniques for dynamic trace metal speciation in natural freshwaters. Environmental Science & Technology, 40 (6): 1934-1941.

Small T D, Warren L A, Roden E E, et al. 1999. Sorption of strontium by bacteria, Fe (III) oxide, and bacteria-Fe (III) oxide composites. Environmental Science & Technology, 33 (24): 4465-4470.

Song Z, Kenney J P L, Fein J B, et al. 2012. An X-ray absorption fine structure study of Au adsorbed onto the non-metabolizing cells of two soil bacterial species. Geochimica et Cosmochimica Acta, 86: 103-117.

Sun X F, Wang S G, Zhang X M, et al. 2009. Spectroscopic study of Zn^{2+} and Co^{2+} binding to extracellular polymeric substances (EPS) from aerobic granules. Journal of Colloid and Interface Science, 335 (1): 11-17.

Tan W, Xiong J, Li Y, et al. 2013. Proton binding to soil humic and fulvic acids: Experiments and NICA-Donnan modeling. Colloids and Surfaces A: Physicochemical and Engineering Aspects, 436: 1152-1158.

Templeton A S, Trainor T P, Traina S J, et al. 2001. Pb (Ⅱ) distributions at biofilm-metal oxide interfaces. Proceedings of the National Academy of Sciences, 98 (21): 11897-11902.

Templeton A S, Spormann A M, Brown G E. 2003. Speciation of Pb (Ⅱ) sorbed by Burkholderia cepacia/goethite composites. Environmental Science & Technology, 37 (10): 2166-2172.

Tipping E. 2002. Cation binding by humic substances. Cambridge University Press.

Toner B, Manceau A, Marcus M A, et al. 2005. Zinc sorption by a bacterial biofilm. Environmental Science & Technology, 39 (21): 8288-8294.

Ueshima M, Ginn B R, Haack E A, et al. 2008. Cd adsorption onto *Pseudomonas putida* in the presence and absence of extracellular polymeric substances. Geochimica et Cosmochimica Acta, 72 (24): 5885-5895.

Van Riemsdijk W H, Koopal L K, Kinniburgh D G, et al. 2006. Modeling the interactions between humics, ions, and mineral surfaces. Environmental Science & Technology, 40 (24): 7473-7480.

Vermeer R. 1996. Interactions between humic acid and hematite and their effects on metal ion speciation. Landbouwuniversiteit Wageningen (Wageningen Agricultural University).

Vermeer A W P, McCulloch J K, van Riemsdijk W H, et al. 1999. Metal ion adsorption to complexes of humic acid and metal oxides: Deviations from the additivity rule. Environmental Science & Technology, 33 (21): 3892-3897.

Villalobos M, Lanson B, Manceau A, et al. 2006. Structural model for the biogenic Mn oxide produced by *Pseudomonas putida*. American Mineralogist, 91 (4): 489-502.

Violante A, Huang P M, Bollag J M, et al. 2002. Soil mineral-organic matter-microorganism interactions and ecosystem health. Amsterdam: Elsevier Science B. V.

Wang S, Terdkiatburana T, Tadé M O. 2008. Adsorption of Cu (Ⅱ), Pb (Ⅱ) and humic acid on natural zeolite tuff in single and binary systems. Separation and Purification Technology, 62 (1): 64-70.

Webb S M, Tebo B M, Bargar J R. 2005. Structural characterization of biogenic Mn oxides produced in seawater by the marine *Bacillus* sp. strain SG-1. American Mineralogist, 90 (8-9): 1342-1357.

Wei X, Fang L, Cai P, et al. 2011. Influence of extracellular polymeric substances (EPS) on Cd adsorption by bacteria. Environmental Pollution, 159 (5): 1369-1374.

Weng L, Temminghoff E J M, Van Riemsdijk W H. 2001. Contribution of individual sorbents to the control of heavy metal activity in sandy soil. Environmental Science & Technology, 35 (22): 4436-4443.

Weng L, Temminghoff E J M, Lofts S, et al. 2002. Complexation with dissolved organic matter and solubility control of heavy metals in a sandy soil. Environmental Science & Technology, 36 (22): 4804-4810.

Weng L, Van Riemsdijk W H, Koopal L K, et al. 2006. Ligand and Charge Distribution (LCD) model for the description of fulvic acid adsorption to goethite. Journal of Colloid and Interface Science, 302 (2): 442-457.

Weng L, Van Riemsdijk W H, Hiemstra T. 2008. Cu^{2+} and Ca^{2+} adsorption to goethite in the presence of fulvic acids. Geochimica et Cosmochimica Acta, 72 (24): 5857-5870.

Weppen P, Hornburg A. 1995. Calorimetric studies on interactions of divalent cations and microorganisms or microbial envelopes. Thermochimica Acta, 269: 393-404.

Wu P, Zhang Q, Dai Y, et al. 2011. Adsorption of Cu (Ⅱ), Cd (Ⅱ) and Cr (Ⅲ) ions from aqueous solutions on humic acid modified Ca-montmorillonite. Geoderma, 164 (3): 215-219.

Xiong J, Koopal L K, Tan W F, et al. 2013. Lead binding to soil fulvic and humic acids: NICA-Donnan modeling and XAFS spectroscopy. Environmental Science & Technology, 47 (20): 11634-11642.

Xiong J, Koopal L K, Weng L, et al. 2015. Effect of soil fulvic and humic acid on binding of Pb to goethite-water interface: Linear additivity and volume fractions of HS in the Stern layer. Journal of Colloid and Interface Science, 457: 121-130.

Xu Y, Boonfueng T, Axe L, et al. 2006. Surface complexation of Pb (Ⅱ) on amorphous iron oxide and manganese oxide: Spectroscopic and time studies. Journal of Colloid and Interface Science, 299 (1): 28-40.

Yee N, Fein J. 2001. Cd adsorption onto bacterial surfaces: A universal adsorption edge. Geochimica et Cosmochimica Acta, 65 (13): 2037-2042.

Yin H, Tan W, Zheng L, et al. 2012. Characterization of Ni-rich hexagonal birnessite and its geochemical effects on aqueous Pb^{2+}/Zn^{2+} and As (Ⅲ). Geochimica et Cosmochimica Acta, 93: 47-62.

Yin H, Liu F, Feng X, et al. 2013. Effects of Fe doping on the structures and properties of hexagonal birnessites-Comparison with Co and Ni doping. Geochimica et Cosmochimica Acta, 117: 1-15.

Yin H, Li H, Wang Y, et al. 2014. Effects of Co and Ni co-doping on the structure and reactivity of hexagonal birnessite. Chemical Geology, 381: 10-20.

Yin H, Liu Y, Koopal L K, et al. 2015. High Co-doping promotes the transition of birnessite layer symmetry from orthogonal to hexagonal. Chemical Geology, 410: 12-20.

Yu Q, Fein J B. 2015. The effect of metal loading on Cd adsorption onto *Shewanella oneidensis* bacterial cell envelopes: The role of sulfhydryl sites. Geochimica et Cosmochimica Acta, 167: 1-10.

Zhao W, Cui H, Liu F, et al. 2009. Relationship between Pb^{2+} adsorption and average Mn oxidation state in synthetic birnessites. Clays and Clay Minerals, 57 (5): 513-520.

Zhao W, Wang Q Q, Liu F, et al. 2010. Pb^{2+} adsorption on birnessite affected by Zn^{2+} and Mn^{2+} pretreatments. Journal of Soils and Sediments, 10 (5): 870-878.

Zhao W, Tan W, Feng X, et al. 2011. XAFS studies on surface coordination of Pb^{2+} on birnessites with different average oxidation states. Colloids and Surfaces A: Physicochemical and Engineering Aspects, 379 (1): 86-92.

Zhao W, Yin H, Liu F, et al. 2011. Characterization of Pb^{2+} adsorption on the surface of birnessite treatment with $Na_4P_2O_7$ at different pH and the study on the distribution of Mn (III) in the birnessite. Environmental Science, 32 (8): 2477-2484.

Zhao W, Liu F, Feng X H, et al. 2012a. XPS study on birnessites with different average oxidation states. Journal of Central South University (Science and Technology), 43 (2), 776-782. (in Chinese with English abstract)

Zhao W, Liu F, Feng X, et al. 2012. Fourier transform infrared spectroscopy study of acid birnessites before and after Pb^{2+} adsorption. Clay Minerals, 47 (2): 191-204.

Zhu M, Ginder-Vogel M, Parikh S J, et al. 2010. Cation effects on the layer structure of biogenic Mn-oxides. Environmental Science & Technology, 44 (12): 4465-4471.

Application of Oral Bioavailability to Remediation of Contaminated Soils: Method Development for Bioaccessible As, Pb and Cd

Hongbo LI, Jie LI, Shiwei LI, Lena Q. MA

State Key Laboratory of Pollution Control and Resource Reuse,

School of the Environment, Nanjing University

1. Introduction

With rapid urbanization and industrialization over the past three decades, metal contamination in soils has become a serious issue in China (Zhao et al., 2014). According to a joint report issued by the Ministry of Environmental Protection and the Ministry of Land and Resources of China, soils in areas near mining and industry activities have been contaminated, with metal contamination in farmland soils is of particular concern (MEP, 2014). Metals including As, Cd, Co, Cr, Cu, Hg, Ni, Pb, Se, V, and Zn are major soil contaminants, accounting for 82% among metal-contaminated soils. Among metals, Cd ranks the first in contaminated soil samples (7.0%), with As and Pb accounting for 2.7% and 1.5%, highlighting their health risks in contaminated soils.

Soil contamination by As, Pb, and Cd, and their potential impact on human health have been a global issue. Metals in soils derive from both anthropogenic and geogenic origins. Mining and smelting of ores, atmospheric deposition from incineration and burning of fossil fuels, and use of fertilizers and pesticides are the main contributors to As, Pb, and Cd in contaminated sites (Alloway, 1995). Many adverse health effects have been associated with As, Pb, and Cd exposure. As a toxic carcinogen, As is considered a priority pollutant in the USA (ATSDR, 2011; IARC, 2004), leading to various forms of cancer. Lead exposure has been of major public concern due to its well-established adverse neuro-behavioral effects on children (Surkan et al., 2007; Zahran et al., 2009). After phasing out of Pb from gasoline and paint, child blood Pb levels have significantly declined worldwide (Glorennec et al., 2012; Levallois et al., 2014). However, in Pb contaminated sites such as mining and smelting areas, childhood Pb poisoning is still of concern in China (Huo et al., 2007; Yan et al., 2013). Human exposure to Cd can result in obstructive pulmonary disease, emphysema, and kidneys disease (Faroon et al., 2012).

1.1 Metal bioavailability in contaminated soils

Incidental soil ingestion is an important non-dietary pathway for human exposure to As, Pb, and Cd through hand-to-mouth activities, especially for children (Ruby et al., 2012). The estimated incidental daily soil ingestion rate is 23 mg for children (Wilson et al., 2013), but as high as 200 mg has been used (Van Wijnen et al., 1990). Therefore, assessment of metal exposure through soil ingestion has received increasing consideration. However, risk assessment usually uses total metal concentrations in soils, assuming that all As, Pb, and Cd in soils is absorbed into the blood systemic circulation following oral ingestion (i.e., 100% bioavailable). However, only a fraction of soil metal enters the systemic circulation (i.e., bioavailable metal), which is controlled by both metal dissolution from soil in the gastrointestinal (GI) tract and metal absorption across the GI barrier. Metal solubility in soils varies significantly under physiochemical conditions associated with the human GI tract. For example, metals sulfide show low solubility in the acid environment of stomach (Bradham et al., 2011). These differences influence metal exposure via soil ingestion. It is known that soil metal bioavailability varies considerably with soil properties and metal mineral forms (Casteel et al., 2006; Smith et al., 2011, Bradham et al., 2011). Therefore, to accurately assess health risk, it is important to determine metal bioavailability in contaminated soils, i.e., the amount of metals in soils that enters the systemic circulation in humans (Ruby et al., 1996).

Metal bioavailability in contaminated soils refers to the portion of metals that enters the systemic circulation from the GI tract from an administered soil dose. It may be expressed as relative bioavailability (RBA), i.e., the ratio of the the absorbed fraction from metal-contaminated soils to the absorbed fraction from the dosing using soluble references (i.e., sodium arsenate, Pb acetate, or Cd chloride). In vivo animal models including swine and mouse have been developed to measure metal RBA in contaminated soils (Bradham et al., 2011; Smith et al., 2011; Juhasz et al., 2010).

1.2 Metal bioaccessibility in contaminated soils

However, animal assays are time-consuming and costly. As an alternative, in vitro assays have been developed to determine metal bioaccessibility in contaminated soils, i.e., the amount of dissolved metal in simulated gastrointestinal solution, which is potentially available for absorption into the systemic circulation. Common bioaccessibility methods include the in vitro gastrointestinal (IVG; Rodriguez et al., 1999), Solubility/Bioaccessibility Research Consortium (SBRC; Kelly et al., 2002), Deutsches Institut für Normunge method (DIN; DIN 2000), physiologically based extraction test (PBET; Ruby et al., 1996), and unified BARGE method (UBM; Wragg et al., 2011). These assays all include a gastric phase (GP) and intestinal phase (IP), and they vary in gastrointestinal components and analysis parameters. As a result, metal bioaccessibility in contaminated soils varies with in vitro assays (Oomen et al., 2002). For example, based on the gastric phase, SBRC and UBM assays provide higher As bioaccessibility in contaminated soils than IVG, DIN, and PBET assays due to their lower gastric pH (1.5 vs. 1.8-2.5) (Juhasz et al., 2009, 2014a).

Given the variation among in vitro assays, a bioaccessibility method needs to be correlated with in vivo animal data before being used as a surrogate of in vivo assay. Several studies have established in vitro-in vivo correlations (IVIVC) between metal-RBA and metal bioaccessibility using different in vitro assays. For example, Rodriguez et al. (1999) demonstrated a good correlation (r^2=0.67-0.69) between As bioaccessibility by IVG and As-RBA using a swine model in 15 contaminated soils from USA. Similarly, based on a swine model, Juhasz et al. (2009) correlated As bioaccessibility by the SBRC (r^2=0.75) with As-RBA in 12 contaminated soils from Australia. Denys et al. (2012) validated the UBM method to assess As-RBA in 16 contaminated soils from Europe (r^2=0.8). In addition to swine model, a mouse model has also been used to validate in vitro assays. Bradham et al. (2011) showed a strong correlation between As bioaccessibility by the SBRC and As-RBA by a mouse urine model in contaminated soils from USA (r^2=0.88). Using 18 contaminated soils, Schroder et al. (2004) showed a strong correlation (r^2=0.79) between Pb bioaccessibility by the IVG and Pb-RBA based on blood Pb following 15-d soil dosing to juvenile swine. Strong correlations (r^2=0.78-0.90, n=16) were observed between Pb bioaccessibility by UBM and Pb-RBA using a swine model and Pb in the kidney after 14-d soil dosing (Denys et al., 2012). Smith et al. (2011) showed strong correlations (r^2=0.78 and 0.88, n=12) between Pb bioaccessibility by SBRC and Pb-RBA based on a mouse model and a single gavage dose. Schroder et al. (2003) found a good correlation (r^2=0.64) between Cd bioaccessibility based on the IVG assay and Cd bioavailability based on kidney Cd following 15-day dosing to juvenile swine. Cadmium bioaccessibility using the UBM is strongly correlated (r^2=0.77-0.94) with Cd-RBA based on Cd in the liver, kidneys, femur or urine of juvenile swine following 14-day of dosing (Denys et al., 2012). Using 7 Cd-contaminated soils, Juhasz et al. (2010) correlated Cd bioavailability with Cd bioaccessibility based on different assays (SBRC, IVG, PBET, and DIN), showing a linear relationship between Cd-RBA and Cd bioaccessibility by the assays (r^2=0.35-0.84), with the best being the PBET assay (r^2=0.84). These studies demonstrated that metal-RBA in contaminated soils can be predicted using in vitro methods and the established IVIVCs.

However, established IVIVCs are often soil specific, varying with the soils used. For example, when established IVIVCs for SBRC, IVG, DIN, PBET, and UBM assays using Australian soils were compared to those using American soils (Juhasz et al., 2009; Juhasz et al., 2014a), significant differences in the slope and y-intercept of the relationships were observed. The large difference makes the prediction of As-RBA using established IVIVCs less accurate. In addition, established IVIVC have rarely been validated against soils outside those used to establish IVIVCs except for the SBRC by Juhasz et al. (2014b). Therefore, it remains unclear whether established IVIVCs have the ability to predict metal-RBA in contaminated soils outside the models. Furthermore, to date, no study tested the feasibility of in vitro assays to predict metal-RBA in contaminated soils from China.

In this chapter, we review the state of the science in development of in vitro assays to predict metal-RBA in contaminated soils in China. The importance of including metal-RBA for refining metal exposure assessment and modifying soil cleanup standards for contaminated soils will also be discussed.

2. In vivo methods to determine As relative bioavailability in contaminated soils

2.1 Different animal models, biomarkers, and dosing schemes

At the present, different biomarkers (metal concentrations in the blood, kidneys, liver, or urine) have been used to determine As absorption following a single gavaged dose or multiple repeated doses of metal-contaminated soil via diet. At present, the majority of RBA studies utilize swine or mouse bioassay, with either area under the blood metal concentration time curve (AUC) or steady state metal urinary excretion (SSUE) being assessed as the endpoint of metal exposure (Juhasz et al., 2007, 2014a; Bradham et al., 2015). Swine shares many similarities with humans (i.e., As and mineral metabolism, and bone development), which is the preferred animal model to assess metal RBA (Weis et al., 1991; Lunney et al., 2007; Patterson et al., 2008). Compared to swine, mice are cost-effective and easy to handle, and have the potential to be utilized in many labs (El-Masri et al., 2008; Evans et al., 2008; Gentry et al., 2014a, 2014b). In addition, large sample size can be accommodated to ensure the robustness of RBA results. However, these animal models differ in physiological parameters, which may influence metal absorption from contaminated soils, thereby influencing metal RBA measurements. To date, a dearth of information is available comparing metal RBA derived using different animal models, with only one study comparing As-RBA results from mouse and swine models in 12 As-contaminated soils (Bradham et al., 2013).

In addition to different animal models, varying feeding schemes have been employed to measure metal RBA in contaminated soil. A steady state dosing approach has been utilized where animals are exposed to metal-contaminated soil, which is incorporated into feed for up to 10-14 days. In this approach, either metal excreted in urine during the exposure period or the last two days, or metal concentration in the liver or kidneys is used to determine As-RBA (Bradham et al., 2011; Brattin et al., 2013; Denys et al., 2012). This approach provides the advantage of mimicking daily continuous exposure to contaminated soil, however, feed is included in the methodology, which may influence metal absorption (Bradham et al., 2013). For example, inorganic phosphate may compete with As for phosphate transporters in the gastrointestinal barrier, thereby reducing As absorption and RBA. To overcome this issue, alternative approaches have been used to assess metal RBA following administeration of a single gavaged dose of contaminated soil to fasting animals with blood metal AUC as the biomarker (Roberts et al., 2002; Juhasz et al., 2007, 2014a; Li et al., 2015a). With the exclusion of feed and under fasted conditions, this approach may provide a worst case scenario of metal absorption. While different animal models and feeding schemes have been used to assess metal-RBA, comparative studies detailing the influences of these parameters on metal-RBA measurement are lacking.

2.2 Case study of As relative bioavailability in contaminated soils by different in vivo bioassays

A recent study compared As-RBA in 12 contaminated soils (As concentration of 42-1114 mg/kg, Table 1) determined using different animal models (i.e., mouse vs. swine), different feeding schemes (i.e., single gavaged dose vs. multiple diet doses), and different biomarkers (metal concentrations in the blood, urine, liver, or kidneys) to determine whether these operational parameters result in different As-RBA determinations (Li et al., 2016a). In the study, As-RBA in 12 As-contaminated soils with known As-RBA via swine blood AUC model was measured using mouse blood AUC, SSUE, liver and kidneys analyses. Initially, As-RBA was assessed in fasted animals following administeration of a single gavaged dose of sodium arsenate and soil suspension. Blood samples of control and treated mice were collected at 4, 8, 16, 24, and 48 h following gavage to establish the blood As concentration time curve following soil and sodium arsenate exposure. Blood As AUC was calculated for both soil and sodium arsenate administeration. Arsenic-RBA was calculated by dividing the AUC following soil administeration by the AUC following sodium arsenate administeration following dose normalization (Eq. 1):

$$\text{As RBA}\ (\%) = \left(\frac{\text{AUC}_{\text{Oral-soil}}}{\text{AUC}_{\text{Oral-As}}} \times \frac{\text{DR}_{\text{Oral-As}}}{\text{DR}_{\text{Oral-soil}}}\right) \times 100\% \quad (1)$$

where $\text{AUC}_{\text{oral-soil}}$ and $\text{AUC}_{\text{oral-As}}$=AUC for an oral dose of As-contaminated soil and sodium arsenate, respectively; and $\text{DR}_{\text{oral-soil}}$ and $\text{DR}_{\text{oral-As}}$=As dose of orally administered soil and sodium arsenate (mg As/kg bw), respectively.

Table 1 Selected Properties of Soils Used for As-RBA Comparison (Soils from Juhasz et al. (2009) with known As-RBA using Swine Blood AUC Model)

soil source	Sample number	soil properties					As-RBA/%[b]
		Total As	Total Fe	Total Al	Total P	TOC	
		(mg/kg)	(g/kg)[a]	(g/kg)	(mg/kg)	%	
railway corridor	2	267	17.6 (10.6)	22.2	234	0.51	72.2±34.5
	4	42	13.7 (8.5)	18.3	385	1.10	41.6±11.5

Continue table

soil source	Sample number	soil properties					As-RBA/%[b]
		Total As	Total Fe	Total Al	Total P	TOC	
		(mg/kg)	(g/kg)[a]	(g/kg)	(mg/kg)	%	
railway corridor	5	1114	68.3 (16.6)	16.3	874	7.12	20.0±16.5
	10	257	25.8 (9.9)	27.8	242	0.44	10.1±4.3
	16	751	14.5 (11.4)	10.8	422	2.26	22.5±3.8
	18	91	10.0 (3.2)	5.10	130	0.00	80.5±12.0
dip sites	24	713	98.6 (48.6)	94.7	3144	2.60	29.3±15.0
	27	228	17.9 (12.5)	22.4	2941	5.54	43.8±9.7
mine sites	33	807	23.5 (7.1)	10.9	546	0.04	41.7±7.5
	34	577	24.6 (12.5)	17.6	468	0.52	7.0±5.0
gossans	44	190	21.0 (15.6)	8.30	200	0.00	16.4±9.1
	45	88	21.0 (13.2)	9.60	370	1.25	12.1±8.5

a Data in parentheses represent dithionite citrate bicarbonate extractable Fe (Fe oxide); *b* Determined using an in vivo swine model via area under blood As concentration time curve as the biomarker of As exposure by Juhasz et al. (2009).

In addition to mouse blood AUC bioassays, As-RBA in contaminated soils was also assessed using As SSUE and As accumulation in the liver or kidneys following multiple repeated As doses via diet. Prior to in vivo assays, soil was incorporated into mouse basal diet, then the soil-amended feed was suppled to mice for 10 d to collect the urine, liver, and kidneys samples. Urinary As excretion faction (UAEF) was calculated as the ratio of cumulative excreted As in urine to the cumulative dietary intake of As (Eq. 2). Arsenic-RBA was then calculated as the ratio of the UAEF for soil exposure to that for sodium arsenate exposure (Eq. 3).

$$\text{UAEF} = \frac{\text{Cumulative urinary excreted As}}{\text{Cumulative As intake}} \quad \cdots\cdots (2)$$

$$\text{As RBA (\%)} = \frac{\text{UAEF}_{\text{soil}}}{\text{UAEF}_{\text{sodium arsenate}}} \times 100\% \quad \cdots\cdots (3)$$

Results showed that compared to the swine blood AUC model, the mouse blood AUC assay generally yielded lower As-RBA for 11 out of the 12 soils (Fig. 1). However, the differences between the two animal models were insignificant for most samples. A strong linear relationship (r^2=0.83) was observed for As-RBA based on swine and mouse blood AUC models (Fig. 2 (a)). Similarly, strong correlations were observed between mouse blood AUC and mouse SSUE (r^2=0.86, Fig. 2 (b)) and between As accumulations in the urine, liver, and kidneys (r^2=0.75-0.89), suggesting As-RBA values were congruent among different animals, dosing schemes, and biomarkers.

To date, swine is recognized as the preferred animal model to determine As-RBA in contaminated soils and has been widely used to validate in vitro assays to predict As-RBA in contaminated soils by establishing in vivo and in vitro correlations (Denys et al., 2012; Juhasz et al., 2009, 2011). However, compared to swine assay, mouse bioassay offers the advantages of low cost, ease to care, and has the potential to be used in larger sample sizes. The strong linear correlation between the swine and mice As-RBA data suggests that the selection of animal

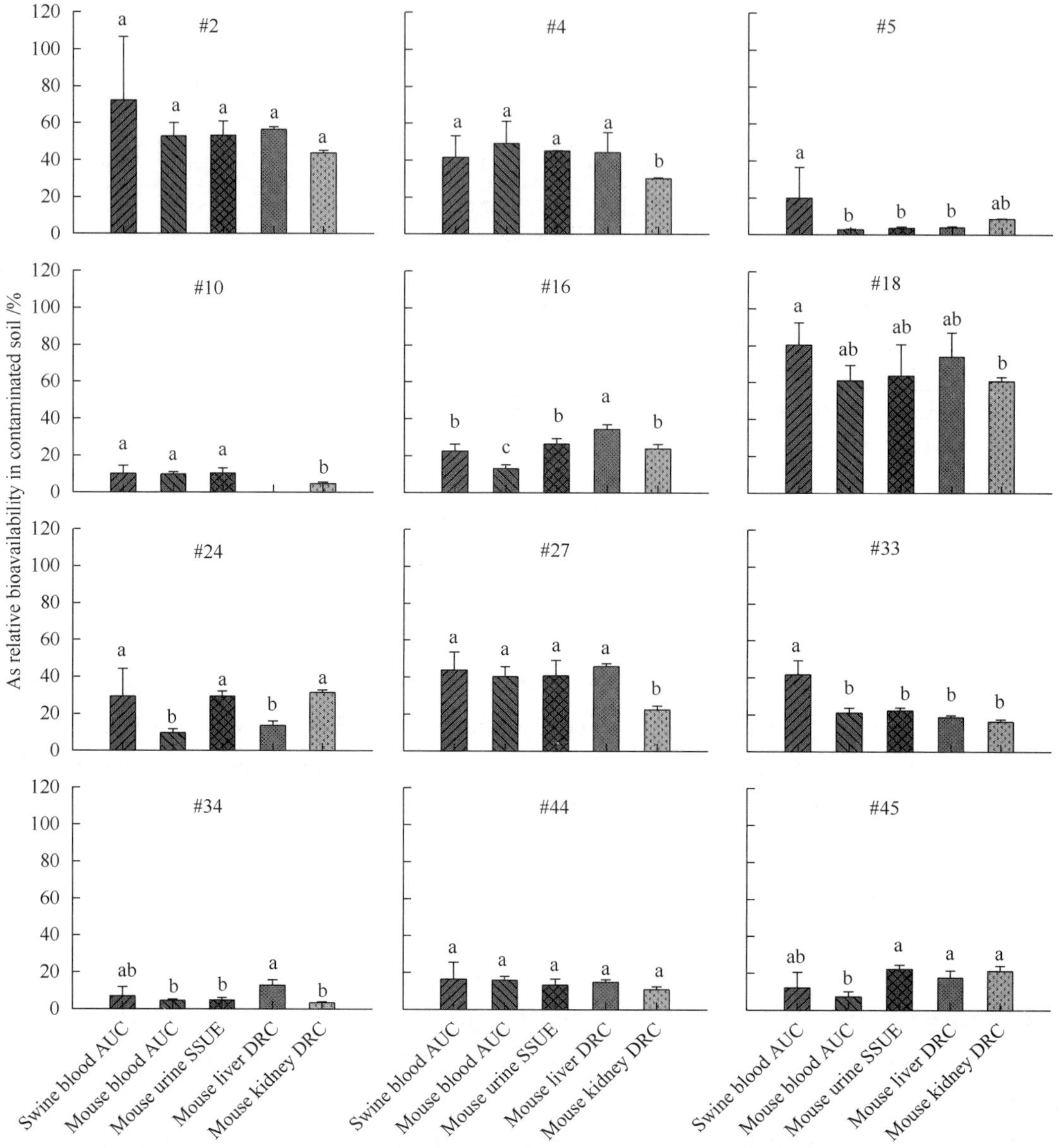

Fig. 1 Arsenic relative bioavailability in 12 contaminated soils measured using swine blood AUC and mouse blood AUC models following a single gavage dose to fasted animals, mouse steady state urinary excretion (SSUE), and mouse liver and kidney dose response curve (DRC) models following multiple doses to soil-amended diets for 10 d. The in vivo mouse bioassays were conducted in triplicates. Arsenic relative bioavailability value could not be calculated for soil #10 using the mouse liver model because the concentration was below the quantification limits (adopted from Li et al., 2016a)

model probably would not significantly influence the role of in vivo assays to validate in vitro assays. Though swine is the preferred animal model for As-RBA determination, mice can be used as a surrogate animal model due to its significant correlation with swine model in addition to its low cost and ease to handle.

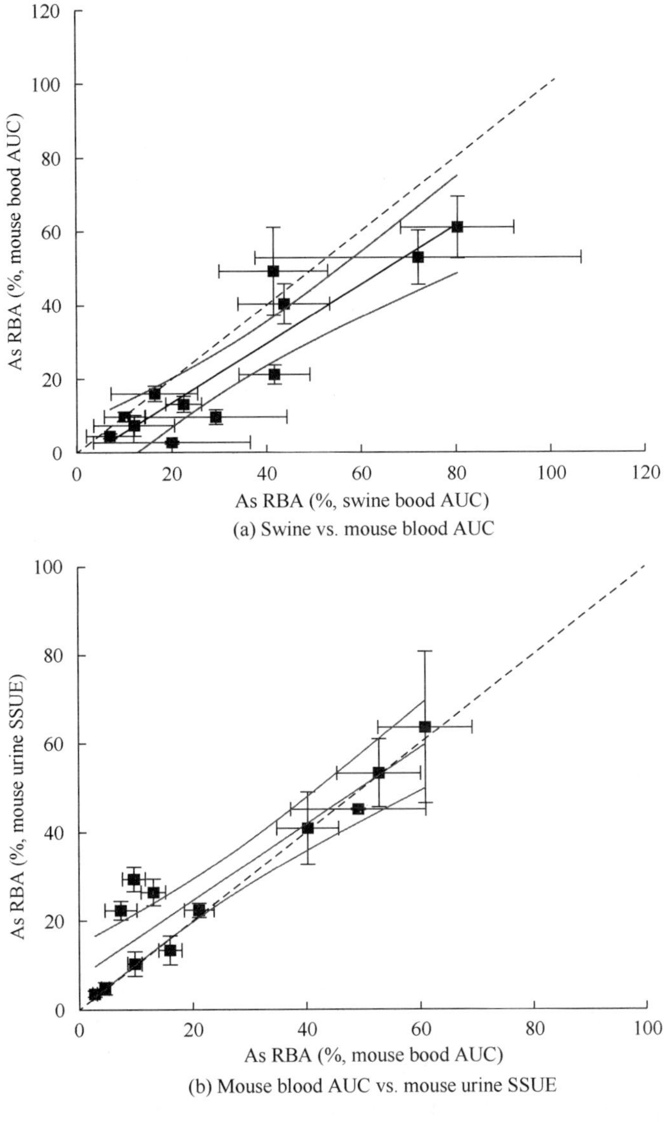

Fig. 2 Comparison of As relative bioavailability (RBA) in 12 contaminated soils determined using swine and mouse models via area under blood As concentration curve (AUC) following a single gavaged administeration to fasted animals (slope=0.80, y-intercept=–2.50, r^2=0.83) (a); and in 12 contaminated soils determined using mouse blood AUC model and mouse SSUE (slope=0.86, y-intercept=7.32, r^2=0.86) (b). The dash line indicated the 1: 1 line between As-RBA measured using two models (adopted from Li et al., 2016a)

Compared to the single gavaged dose, which needs to establish blood As AUC using large animal sizes, the repeated soil dose via diet offers the advantage of small numbers of animals required to determine As-RBA. In addition, the methodology is less stressful on the mice (i.e., consumption of feed vs. gavage of soil slurries) with the steady state method more closely mimicking the status of children's daily exposure to contaminated soils. In term of endpoints, although As urine excretion is a non-invasive approach to assess As-RBA, metabolic cages are required for its precise collection. For multiple dosing schemes, As in the liver or kidneys is a simpler biological endpoint compared to urine. In short, based on the As-RBA comparison study, the mouse liver or kidney assay following repeated doses of soil-amended diet to mice via diet is recommended for future As-RBA studies (Li et al., 2016a). However, limited data are available regarding comparative Pb and Cd RBA studies involving different animal models, different biomarkers, and single gavaged dose vs. steady state models, requiring future studies.

3. Relative bioavailability of As, Pb, and Cd in contaminated soils from China

3.1 Characteristics of 12 metal contaminated soils

Twelve As, Pb, or Cd-contaminated soil samples were collected from various cities in China, representing different geographical locations (farming, mining, and smelting sites) and contamination sources (Tables 2-4).

Total As concentrations in 12 As-contaminated soils range from 22 to 4172 mg/kg, with farming soils having lower As concentrations (36-171 mg/kg) than soils from mining areas (75-1470 mg/kg) and smelting areas (22-4172 mg/kg) (Table 2). Considerable variation in total Fe and Mn, Fe oxides, and TOC is also observed among the 12 soils. Mining impacted soils contain significantly higher total Fe concentrations (1.7-7.9-fold) compared to farming and smelting impacted soils except for sample 11 due to the possible presence of pyrite in mining impacted soils. Similarly, citrate dithionite extractable Fe (reactive Fe) is significantly higher in the mining impacted soils except that two smelting impacted soil samples (10 and 11) also contain high reactive Fe. The high concentration of total and reactive Fe in the two smelting impacted soils might be due to the processing of high Fe minerals in factories nearby the contaminated sites. High total Fe concentrations (120-294 g/kg) have been previously observed in smelting impacted soils from the US (Bradham et al., 2011). Total Mn ranges considerably from 142 to 9807 mg/kg, but its variation among different soil As sources is not obvious.

The 12 Pb-contaminated soils vary greatly in total Pb, Fe, and Mn concentrations and TOC contents (Table 3). Lead concentration in the soils varies by 2 fold, ranging from 215 to 25, 329 mg/kg. Soils 8 and 9 with the highest Pb concentrations (9958 and 25, 329 mg/kg) are from a smelting area. Total Fe concentration is 20.7-219 g/kg, with the highest in soil 9. Soils 11 and 12 from mining areas also have high Pb (1073 and 4164 mg/kg) and Fe concentrations (143 and 115 g/kg). Total Mn concentration is 102-9807 mg/kg, with the highest value in soil 12 from a mining area. Soils 8 and 9 from smelting area also have relatively high Mn concentrations (1086 and 2785 mg/kg). The TOC content in soils range from 0.30% to 14.0%. Generally, mining area soils have relatively lower TOC values (0.30%-1.05%) than smelting area soils (1.03%-13.0%) and farmland soils (1.40%-3.90%).

Cadmium concentration in 12 contaminated soils varies from 3.00 to 296 mg/kg, which is much higher than the Class Ⅱ of Chinese Soil Environmental Quality Standards (0.3 mg/kg for soils with pH<7.5) (Table 4). In addition, soil properties known to influence Cd bioaccessibility also vary among 12 soils including pH (2.8-8.9), total organic carbon (0.64%-5.26%), clay content (2.91–13.4%), and amorphous Al, Fe, and Mn oxides (0.74–5.88 g/kg, 1.76-45.5 g/kg, and 2.58-8463 mg/kg). Elements including Ca (2.22-48.1 g/kg), Fe (21.1-134 g/kg), Zn (50.7-2404 mg/kg), and P (170-1317 mg/kg) may also interact with Cd during its intestinal absorption.

3.2 Metal relative bioavailability in 12 contaminated soils

Tables 2-4 provide a summary of As, Pb, and Cd RBA data following the in vivo assessment of 12 contaminated soils from China. Based on mouse blood AUC bioassays, As-RBA in the 12 soils was 6.38%-73.1%, averaging 29.0% (Table 2). The values are similar to those in contaminated soils measured using swine blood AUC single gavage model (6.98%-80.5%, averaging 33.1%) and mouse steady state urinary excretion multiple dose model (11.1%-52.8%, averaging 32.3%) (Braham et al., 2011; Juhasz et al., 2007). Among the 12 soils, As RBA is significantly lower in the 3 mining soils. In addition, soil 2 (farming soil), and soils 10 and 11 (smelting soils) also contain low As-RBA (Table 1). The low As-RBA for these soils is possibly due to their higher extractable Fe contents (25.3-50.9 g/kg vs. 8.29-16.7 g/kg). Similar association of low As-RBA with Fe oxides has been reported (Bradham et al., 2011; Juhasz et al., 2014a).

Table 2 Basic Properties and As Relative Bioavailability (RBA) in 12 As-Contaminated Soil Samples from China (from Li et al., 2015a)

Soil sample	Contamination source	Sampling site	Total As	Total Fe	Extractable Fe[a]	Total Mn	As RBA
			/(mg/kg)	/(g/kg)	/(g/kg)	/(mg/kg)	/%
As 1	Farming	Jiyuan, Henan	36±1.00	30.4±0.33	11.3±0.55	597±13.7	52.8±10.3
As 2		Hechi, Guangxi	41±7.32	37.5±0.29	25.3±3.36	102±4.04	16.2±3.10
As 3		Jiyuan, Henan	119±1.38	26.5±0.58	8.49±0.17	531±17.3	65.0±6.00
As 4		Huangshi, Hubei	171±1.00	28.8±0.84	16.7±0.39	254±13.3	27.2±8.99

Continue table

Soil sample	Contamination source	Sampling site	Total As	Total Fe	Extractable Fe[a]	Total Mn	As RBA
			/(mg/kg)	/(g/kg)	/(g/kg)	/(mg/kg)	/%
As 5	Mining	Xingren, Guizhou	75±1.14	69.±1.33	29.8±1.78	592±3.51	13.9±5.28
As 6		Gejiu, Yunnan	743±22.0	115±5.11	29.4±3.38	9807±465	6.38±2.08
As 7		Shimen, Hunan	1470±23.9	143±1.31	74.2±1.40	142±5.51	7.29±1.42
As 8	Smelting	Fengxian, Shanxi	22±1.16	30.7±0.58	8.29±0.37	667±5.69	73.1±17.7
As 9		Shuikou, Hunan	87±2.00	22.7±0.55	13.9±0.70	233±4.36	34.1±7.65
As 10		Zhuzhou, Hunan	861±34.0	41.0±0.53	38.1±1.47	1086±10.1	8.31±1.38
As 11		Zhuzhou, Hunan	2556±61.5	219±2.00	50.9±0.65	2785±30.0	14.0±3.69
As 12		Dayu, Jiangxi	4172±62.5	18.2±0.49	10.2±0.50	448±27.2	30.1±5.53

[a] Values represent dithionite–citrate bicarbonate extractable Fe.

Table 3 Selected Properties and Pb Relative Bioavailability (RBA) of 12 Pb-Contaminated Soils from China (from Li et al., 2015b)

Soil sample	Contamination source	Location	Total Pb / (mg/kg)	Total Fe / (g/kg)	Pb RBA /%
Pb 1	Farming	Huangshi, Hubei	215±0.62	28.8±0.84	51.4±16.2
Pb 2		Jiyuan, Henan	734±4.24	30.4±0.33	59.7±19.8
Pb 3		Jiyuan, Henan	1, 306±19.4	26.5±0.58	55.8±16.5
Pb 4		Jiyuan, Henan	1, 543±16.0	27.9±0.48	60.5±20.1
Pb 5	Smelting	Fengxian, Shanxi	250±4.19	30.7±0.58	56.9±18.7
Pb 6		Fenglan, Gansu	515±11.0	37.5±0.29	84.3±12.1
Pb 7		Shuikou, Hunan	1, 174±12.6	22.7±0.55	62.3±25.3
Pb 8		Zhuzhou, Hunan	9, 958±243	41.0±0.53	39.6±5.28
Pb 9		Zhuzhou, Hunan	25, 329±213	219±2.00	30.8±9.39
Pb 10	Mining	Hechi, Guangxi	516±7.03	20.7±0.19	7.00±1.80
Pb 11		Shimen, Hunan	1, 073±4.89	143±1.31	16.4±8.48
Pb 12		Gejiu, Yunnan	4, 164±77.6	115±5.11	26.0±5.35

Table 4 Properties of and Cd Relative Bioavailability (RBA) of 12 Cd-Contaminated Soils from China (from Li et al., 2016b)

Soil	Land uses	Total Cd	Total Ca	Total Fe	Fe_{AM}	TOC	Clay	pH	Cd RBA
sample		/(mg/kg)	/(g/kg)	/(g/kg)	/(g/kg)	/%	/%		
Cd 1	Mining	3.00±0.50	7.37±0.20	134±1.65	45.5±1.56	0.64	13.1±0.24	2.8	83.8±4.07
Cd 2	Mining	9.59±0.23	6.61±0.42	104±6.23	5.10±0.62	1.26	13.4±0.05	7.8	42.7±1.03
Cd 3	Farming	11.1±3.15	14.2±0.32	51.1±0.21	15.4±0.84	2.23	2.91±0.21	7.5	65.4±4.77
Cd 4	Farming	13.6±1.92	42.8±2.26	25.1±1.59	1.95±0.34	4.55	7.64±0.16	8.3	77.0±3.15

Continue table

Soil	Land uses	Total Cd	Total Ca	Total Fe	Fe_{AM}	TOC	Clay	pH	Cd RBA
sample		/(mg/kg)	/(g/kg)	/(g/kg)	/(g/kg)	/%	/%		
Cd 5	Smelting	18.6±1.24	11.1±0.37	51.4±0.79	9.88±0.57	1.47	3.86±0.13	6.5	71.8±3.92
Cd 6	Smelting	25.6±0.88	2.22±0.12	20.7±0.90	6.08±0.52	2.51	9.66±0.18	6.1	75.5±3.95
Cd 7	Farming	36.1±0.71	4.71±0.14	30.2±0.26	12.7±0.12	2.12	7.15±0.46	7.2	51.5±5.86
Cd 8	Farming	59.7±1.32	16.4±0.01	28.4±0.47	1.76±0.06	2.41	8.37±0.03	8.3	56.8±6.49
Cd 9	Residential	130±7.27	6.59±0.40	21.1±1.13	4.51±0.75	3.14	3.44±0.18	7.8	51.5±0.37
Cd 10	Industry	204±12.6	5.59±0.63	34.3±1.78	12.5±0.41	5.26	3.83±0.06	7.8	36.6±1.89
Cd 11	Residential	212±3.57	6.46±0.02	29.9±0.06	8.84±0.13	2.69	3.62±0.06	7.3	68.8±5.09
Cd 12	Smelting	296±20.7	48.1±1.02	29.3±0.87	7.94±0.65	3.16	3.51±0.31	8.9	66.1±2.14

AM: amorphous content of Fe in soil.

For 12 Pb-contaminated soils, the range of Pb-RBA is 7.0%-84%, averaging 46% based on the mouse blood AUC bioassay (Table 3). Like As-RBA, Pb-RBA using the mouse blood model vary with soils, with farming soils being the highest (51%-61%, averaging 57%), followed by smelting soils (31%-84%, averaging 55%) and mining soils (7.0%-26%, averaging 17%) (Table 3). These Pb-RBA values fall within the Pb-RBA range of 1.0%-108% measured in Pb-contaminated soils from various sources (shooting range, mining, and smelting) (Bannon et al., 2007; Schroder et al., 2004). The differences in Pb-RBA among soils can be explained by their different Pb fractionation in soils. Lead in the mining soils is primarily associated with less available forms including F3 (Fe/Mn oxides fraction), O4 (organic fraction), and R5 (residual fraction) (Fig. 3). Particularly, >90% of the Pb in soil 11 is in the residual fraction (R5). The lower Pb in the more bioavailable fractions (exchangeable fraction-E1 and carbonate fraction-C2) in mining soils is consistent with their lower Pb-RBA. Quantifying the contributions of Pb in different fractions to bioavailable Pb shows that Pb-RBA is better correlated with Pb pools in the first and second fractions (E1+C2, r^2=0.52) than Pb in the first fraction E1, first three fractions E1+C2+F3, or the first four fractions E1+C2+F3+O4 (r^2=0.32–0.42). This suggests that bioavailable Pb mainly comes from Pb in the exchangeable and carbonate fractions. To reduce human health risk associated with oral ingestion of contaminated soil, remediation measures are needed to reduce exchangeable and carbonate associated Pb fractions, which contributed the most to bioavailable Pb.

For 12 Cd-contaminated soils, Cd-RBA is 37%-84%, averaging 61% based on the mouse steady state Cd accumulation in mouse liver plus kidneys following multiple soils doses via diet for a 10-d period (Table 4). Compared to As and Pb with average RBA of 29% and 46%, the average Cd-BRA at 61% is significantly higher, suggesting higher bioavailability of Cd in soils compared to As and Pb. The variation in Cd-RBA among soils can also be attributed to Cd speciation in soils. Based on Cd accumulation in the liver and kidneys, Cd-RBA is similar at

34%-90% and 40%-78%, which is consistent with Juhasz et al. (2010). For most soils, no significant difference is found between Cd-RBA based on the liver or kidneys, which is supported by their strong correlation (r^2=0.81). Therefore, combining the two biomarkers is recommended to increase confidence in RBA measurement.

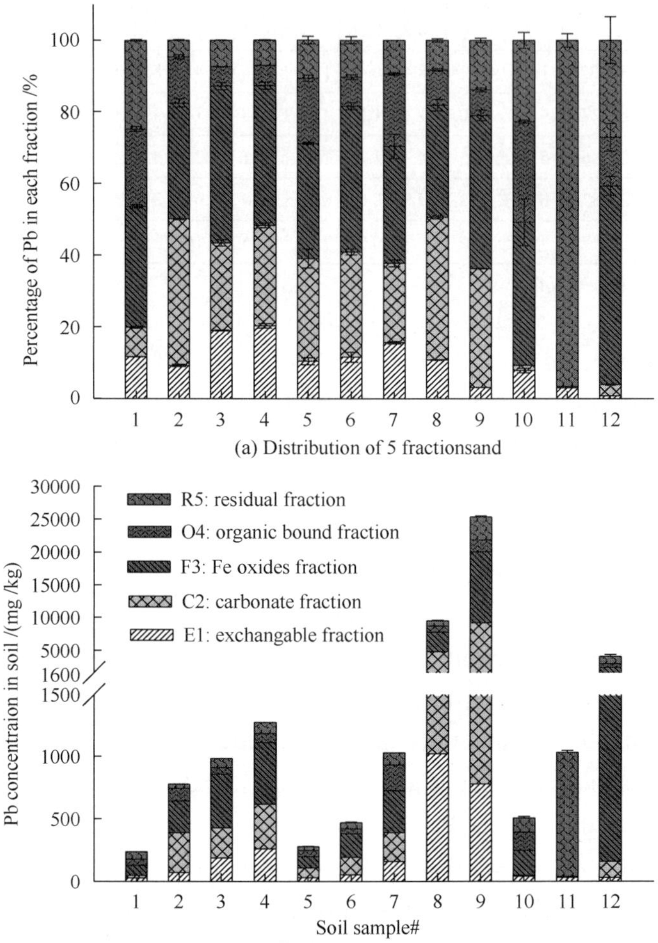

(b) Pb concentration in each fraction

Fig. 3 Distribution of 5 fractions and Pb concentration in each fraction in 12 Pb-contaminated soils. E1=exchangeable fraction; C2=carbonate fraction; F3=Fe/Mn oxides fraction; O4=organic fraction; and R5=residual fraction. Bars represent the mean and standard deviations of triplicates (adopted from Li et al., 2015b)

4. Metal bioaccessibility in contaminated soils from China

Though in vivo animal bioassays can provide an accurate assessment of metal RBA in soils, they are costly and time consuming, therefore not suitable to measure site-specific metal RBA on a large scale. As a result, various in vitro assays that estimate soluble metal fraction from soil in simulated human gastrointestinal fluids (i.e., metal bioaccessibility) have been developed to predict metal RBA. Common in vitro assays including UBM, SBRC, IVG, DIN, and PBET with both gastric (GP) and intestinal phase (IP) have been used to measure bioaccessibility of As, Pb, and Cd in contaminated soils. Table 5 shows compositions and parameters of these in vitro methods. For soils contaminated by As, Pb, or Cd in Tables 2-4, their bioaccessibility has been measured, which varies with assays and soils (Fig. 4-6).

4.1 Arsenic bioaccessibility in contaminated soils

As detailed in Fig. 4, As bioaccessibility in 12 As-contaminated soils varies considerably with assays. Based on gastric phase, As bioaccessibility is 7.6%-59% by UBM (averaging 29%), 2.3%-66% by SBRC (25%), 7.3%-58% by IVG (20%), 0.30%-83% by DIN (22%), and 1.3%-50% by PBET (16%), with the order of UBM>SBRC>IVG>DIN>PBET. The average As bioaccessibility is similar to those by Juhasz et al. (2009) at 33%, 33%, 22%, 15%, and 20%. Higher values by UBM and SBRC assays are attributed to their lower pH at 1.2-1.5 compared to 1.8-2.5 (Juhasz et al., 2009, 2014a). However, in some soil samples, variation of As bioaccessibility does not follow this trend. For examples, for soil 3 and 8, the gastric phase of IVG or PBET assay provides the most conservative results, while for soil 11, UBM is the lowest. This might be related to their differences in soil properties, for example, soil 3 and 8 have the lowest extractable Fe concentrations while soil 11 have the highest (Table 2).

In the intestinal phase, As bioaccessibility in the soils is 5.7%-59% by UBM (averaging 31%), 0.63%-37% by SBRC (15%), 2.3%-46% by IVG (20%), 2.1%-38% (18%) by DIN, and 0.86%-45% by PBET assays (19%), with the SBRC assay generally providing the lowest values (Fig. 4). Compared to gastric phase, As bioaccessibility by the intestinal phase of SBRC is significantly lower for most samples possibly due to precipitation of dissolved Fe in the neutral condition at pH 7 in the intestinal phase and As adsorption onto the newly formed amorphous Fe (Juhasz et al., 2009; 2014a). In contrast to SBRC, As bioaccessibility doesn' t decrease in the intestinal phase for UBM, IVG, DIN, and PBET assays (Fig. 2). The increased As bioaccessibility in the intestinal phase of PBET has been attributed to elevated Fe dissolution in PBET' s intestinal fluid due to organic acids (Juhasz et al., 2014a).

Table 5 Compositions and Parameters of Gastric (GP) and Intestinal (IP) Phases of in vitro Methods SBRC, IVG, PBET, DIN, and UBM for Metal Bioaccessibility Assessment in Soils

Method	Phase	Composition /L	pH	Soil/solution ratio	Extraction time/h	Reference
SBRC	GP	glycine 30.0 g	1.5	1:100	1	Kelley et al. (2002)
	IP	bile 1.75 g, pancreatin 0.50 g	7	1:100	4	
PBET	GP	pepsin 1.25 g, sodium malate 0.50 g, sodiumcitrate 0.50 g, lactic acid 420 μL, and acetic acid 500 μL	2.5	1:100	1	Ruby et al. (1996)
	IP	bile 1.75 g, and pancreatin 0.5 g	7	1:100	4	

Continue table

Method	Phase	Composition /L	pH	Soil/solution ratio	Extraction time/h	Reference
IVG	GP	pepsin10 g, and NaCl 8.77 g	1.8	1:150	1	Rodriguez et al. (1999)
	IP	Bile 3.5g, and pancreatin 0.35g	5.5	1:150	1	
DIN	GP	1 g pepsin, 3 g mucin, 2.9 g NaCl, 0.7 g KCl, 0.27 g KH_2PO_4	2	1:50	2	DIN (2000)
	IP	9.0 bile, 9.0 g pancreatin, 0.3 g trypsin, 0.3 g urea, 0.3 g KCl, 0.5 g $CaCl_2$, 0.2 g $MgCl_2$	7.5	1:100	6	
UBM	Saliva	KCl 0.90 g, NaH_2PO_4 0.89 g, KSCN 0.20 g, Na_2SO_4 0.57 g, NaCl 0.30 g, urea 0.2 g, amylase 0.145 g, mucin 0.05 g, and uric acid 0.015 g	6.5	1:15	10 s	Wragg et al. (2011)
	GP	KCl 0.824 g, NaH_2PO_4 0.266 g, NaCl 2.752 g, $CaCl_2$ 0.4 g, NH_4Cl 0.306 g, urea 0.085 g, glucose 0.65 g, glucuronic acid 0.02 g, glucosaminehydrochloride 0.33 g, bovine serum albumin 1.0 g, mucin 3.0 g, and pepsin 1.0 g	1.2	1:37.5	1	
	IP	KCl 0.94 g, NaCl 12.3 g, $NaHCO_3$ 11.4 g, KH_2PO_4 0.08 g, $MgCl_2$ 0.05 g, urea 0.35 g, $CaCl_2$ 0.42 g, bovine serum albumin 2.8 g, pancreatin 3.0 g, lipase 0.5 g, and bile 6.0 g	6.3	1:97.5	4	

The organic acids such as citrate in the PBET intestinal fluid probably inhibited Fe precipitation in the intestinal phase (Li et al., 2014). However, unlike SBRC, As bioaccessibility by UBM and IVG didn't show a corresponding decrease with decreased soluble Fe (Fig. 4 S1 and S2), which may be attributed to their different gastrointestinal constituents (pepsin and mucin) compared to the SBRC assay (glycine) (Table 5).

4.2 Lead bioaccessibility in contaminated soils

For Pb-contaminated soils, the highest Pb bioaccessibility is obtained by SBRC gastric phase, followed by UBM, IVG, and PBET (Fig. 5). Based on the GP, Pb bioaccessibility in 12 Pb-contaminated soils using UBM, SBRC, IVG, and PBET is 1.1%-84%, 3.0%-99%, 0.46%-71%, and 0.85%-60%, averaging 44%, 69%, 46%, and 28% (Table 3). Lower gastric pH in SBRC (pH 1.5) than IVG and PBET (pH 1.8, and 2.5) partially attributes to the higher Pb bioaccessibility by SBRC. In addition, differences in other assay parameters (e.g., soil:solution ratio) and gastric fluid components (e.g., chyme composition) are probably also responsible. Increased soil:solution ratio of UBM from 1:37.5 to 1:100 significantly increases Pb bioaccessibility from 65%, 57%, and 23% to 88%, 82%, and 30% using soils 2, 7, and 12. In addition, different components in the gastric fluids of the 4 methods (pepsin and mucin for UBM, glycine for SBRC, and pepsin for IVG and PBET, Table S1) may have contributed to the variability in Pb bioaccessibility among assays. Extraction of soil 8 using 10 g /Lof glycine, pepsin, and mucin solution at pH 1.5 shows significantly higher Pb bioaccessibility for glycine (93%) than pepsin (72%) and mucin (77%).

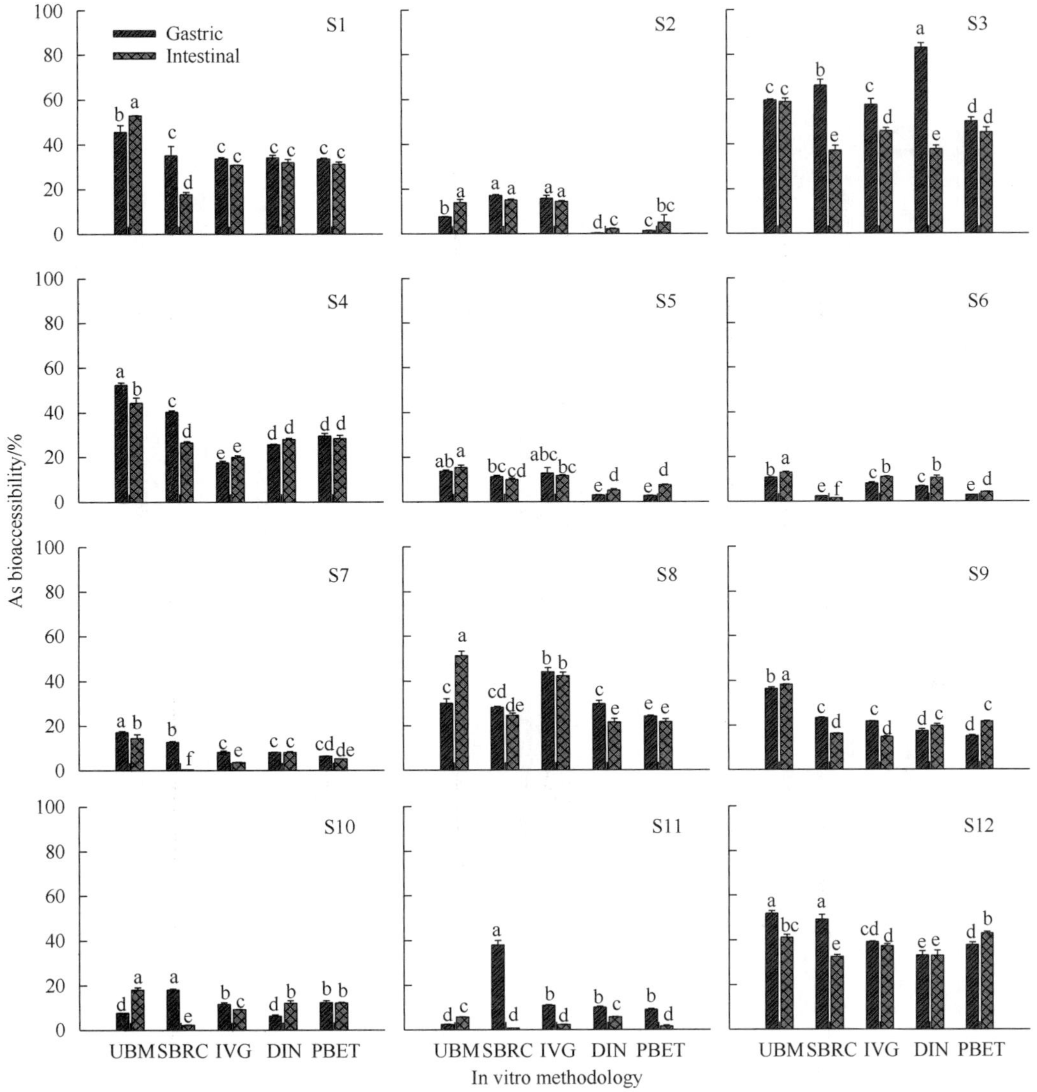

Fig. 4 Arsenic bioaccessibility in 12 As-contaminated soils determined based on the gastric and intestinal phases of UBM, SBRC, IVG, DIN, and PBET methods. Data are shown in mean ± standard deviation of 3 replicate. Means marked with different letters indicate significant ($p<0.05$) differences in As bioaccessibility for each soil (adopted from Li et al., 2015a)

Unlike As bioaccessibility, all in vitro assays show sharply decreased Pb bioaccessibility when they are extended to the neutral intestinal phase (Fig. 5). At neutral intestinal phase, soluble Fe in the gastric phase precipitates as Fe oxides, which adsorbs soluble Pb, decreasing Pb solubility. In addition, Pb may precipitate at the neutral environment and also be absorbed onto solid matrix during intestinal phase extraction (Li et al., 2014).

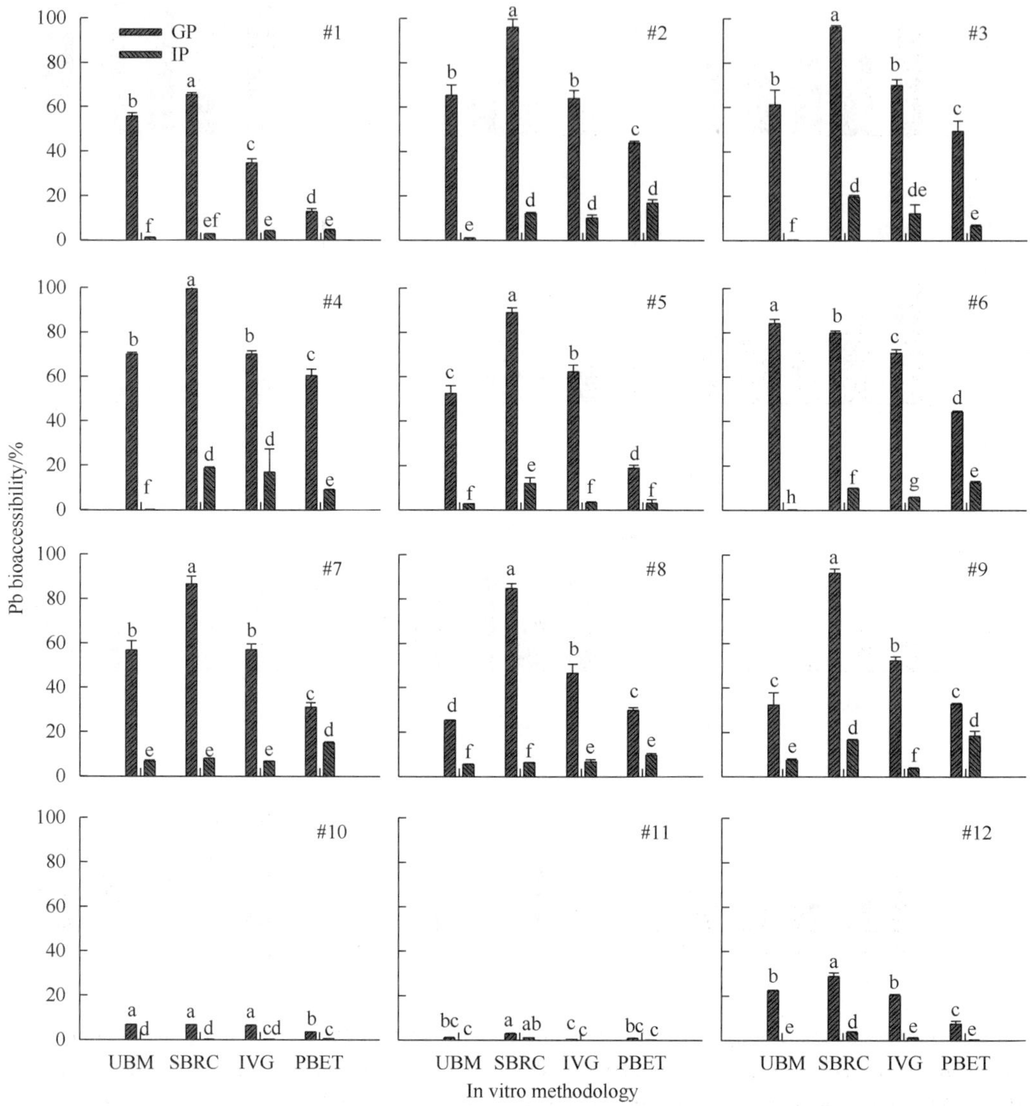

Fig. 5 Lead bioaccessibility in 12 Pb-contaminated soils based on the gastric (GP) and intestinal phases (IP) of UBM, SBRC, IVG, and PBET assays. Bars represent the mean and standard deviations of triplicates. Data with different letters indicate significant ($P<0.05$) differences in Pb bioaccessibility for a given soil (adopted from Li et al., 2015b)

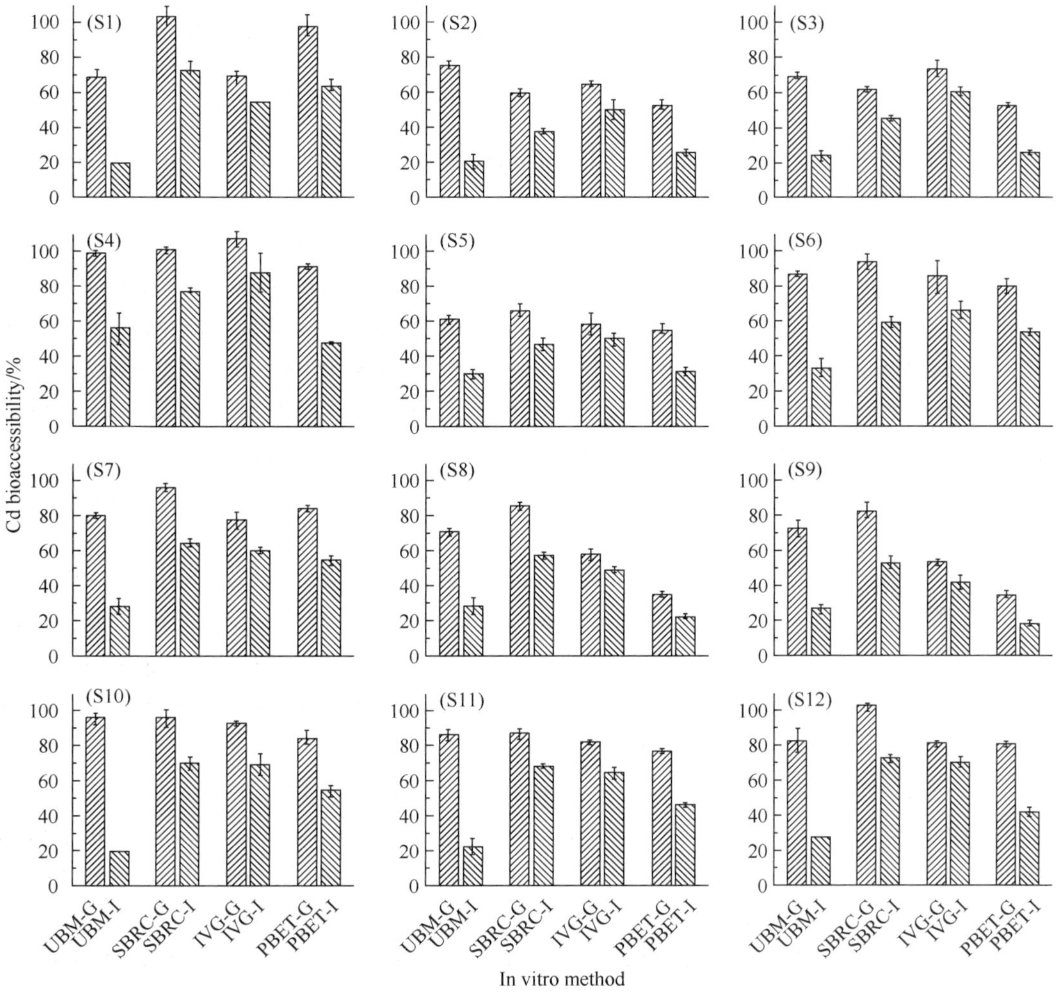

Fig. 6 Comparison of Cd bioaccessibility values based on the gastric (G) and intestinal (I) phases of UBM, SBRC, IVG, and PBET assays in 12 Cd-contaminated soils. Bars represent the mean and standard deviations of triplicates (adopted from Li et al., 2016b)

4.3 Cadmium bioaccessibility in contaminated soils

Cadmium bioaccessibility also varies with in vitro assays (Fig. 6). Generally, at the gastric phase, Cd bioaccessibility based on the SBRC assay is the highest and the PBET the lowest, probably due to pH differences in the assays (Oomen et al., 2002). However, compared to Pb bioaccessibility, variation in Cd bioaccessibility with in vitro assays is smaller. This might be due to significantly higher Cd solubility in gastric fluid in soils compared to Pb, making the difference in Pb bioaccessibility between assays narrower.

In addition, although Cd bioaccessibility decreases from the gastric to intestinal phase for all soils (Fig. 6), the decrease is not as sharp as Pb bioaccessibility (Fig. 5). This might be due to lower absorption of dissolved Cd onto newly formed Fe oxides at the intestinal phase extraction than Pb.

5. Selection of bioaccessibility assay based on correlation with bioavailability data

Due to variation in metal bioaccessibility within and between assays, the suitability of in vitro assays to serve as surrogate for RBA measurements needs to be tested by correlating RBA with bioaccessibility. For 12 As-, Pb-, or Cd-contaminated soils from China (Tables 2-4), the in vivo relative bioavailability-in vitro bioaccessibility correlations (IVIVCs) have been established (Fig. 7-9), showing different bioaccessibility methods for different metals. To test the strength of IVIVC, several criteria, including $r^2>0.6$, slope of 0.8-1.2, and y-intercept close to zero, have been recommended (Wragg et al., 2011).

For As-contaminated soils, all assays show strong linear correlation with As-RBA data (r^2=0.50-0.83), with the IVG assay providing the best estimate of As-RBA (r^2=0.81-0.83) although the IVIVC slope did not meet the criteria of 0.8-1.2 (Fig. 7). This suggests that the IVG assay has the potential to predict As-RBA in contaminated soil from China. In addition, UBM and SBRC assays are also suitable.

For Pb-contaminated soils, the best Pb-RBA predictive in vitro assay is the gastric phase of UBM method with r^2=0.93 (Fig. 8). Compared to the gastric phase, Pb bioaccessibility in the intestinal phase is not correlated with Pb-RBA for all in vitro methods, possibly due to low level of Pb bioaccessibility in the intestinal phase.

For Cd-contaminated soils, the best correlation is observed for the PBET (r^2=0.61-0.70), followed by the IVG assay (0.42-0.52), with poor correlation for the UBM (0.12-0.30) and SBRC (0.14-0.22) and with the gastric phase being better than intestinal phase (Fig. 9). Among the four assays, the PBET is the only one meeting the benchmark of $r^2>0.6$ (Wragg et al., 2011). Our data are consistent with Juhasz et al. (2010) who reported best correlation for the PBET (0.75-0.83) among four assays (SBRC, IVG, PBET, and DIN). Though with large y-intercepts (24 and 33), its stronger linear correlation with Cd-BRA than other in vitro assays.

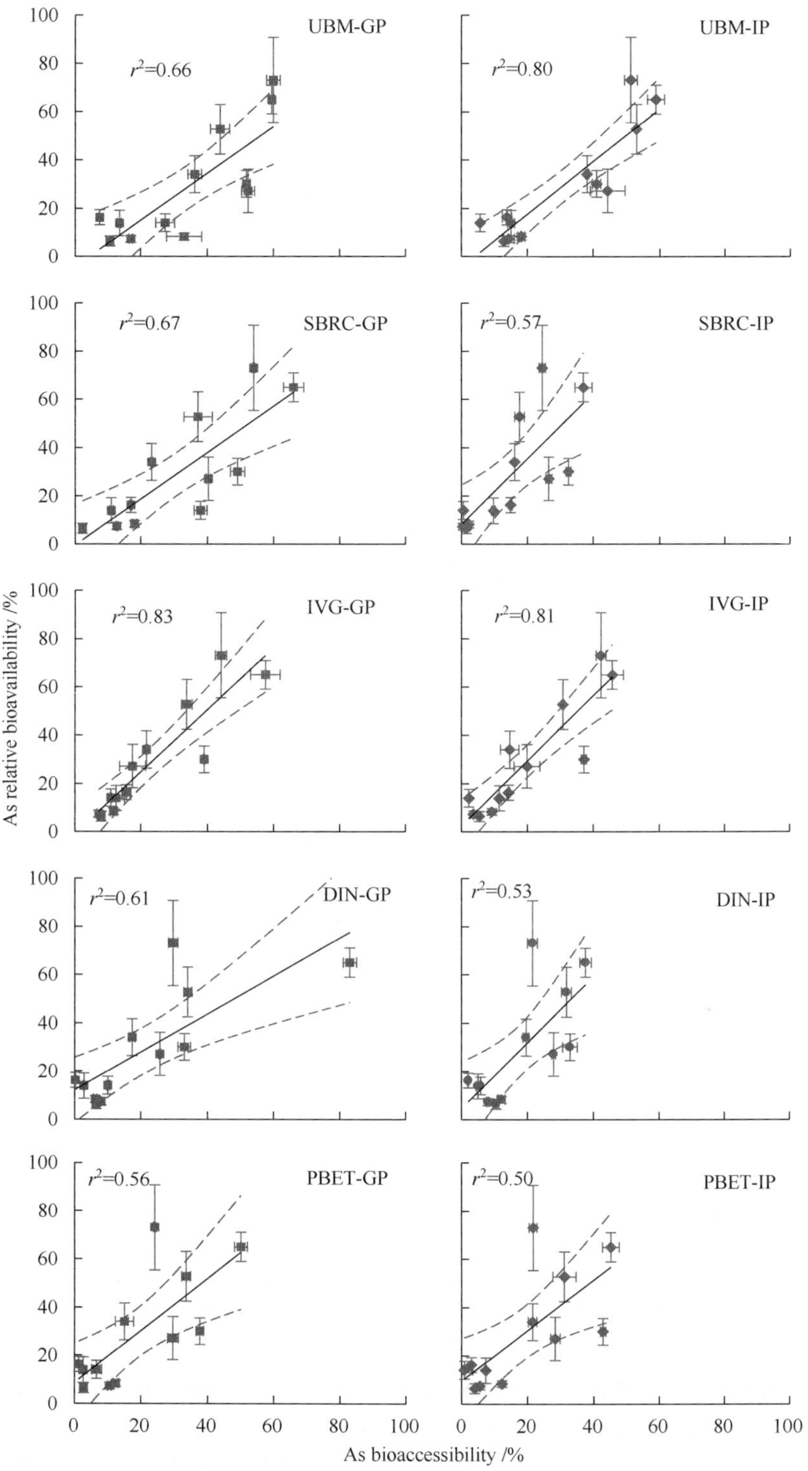

Fig. 7 In vivo–in vitro correlations between As relative bioavailability (RBA) in 12 As-contaminated soils from China using a mouse model and As bioaccessibility determined using gastric (GP) and intestinal phases (IP) of five in vitro assays (UBM, SBRC, IVG, DIN, and PBET) (adopted from Li et al., 2015a)

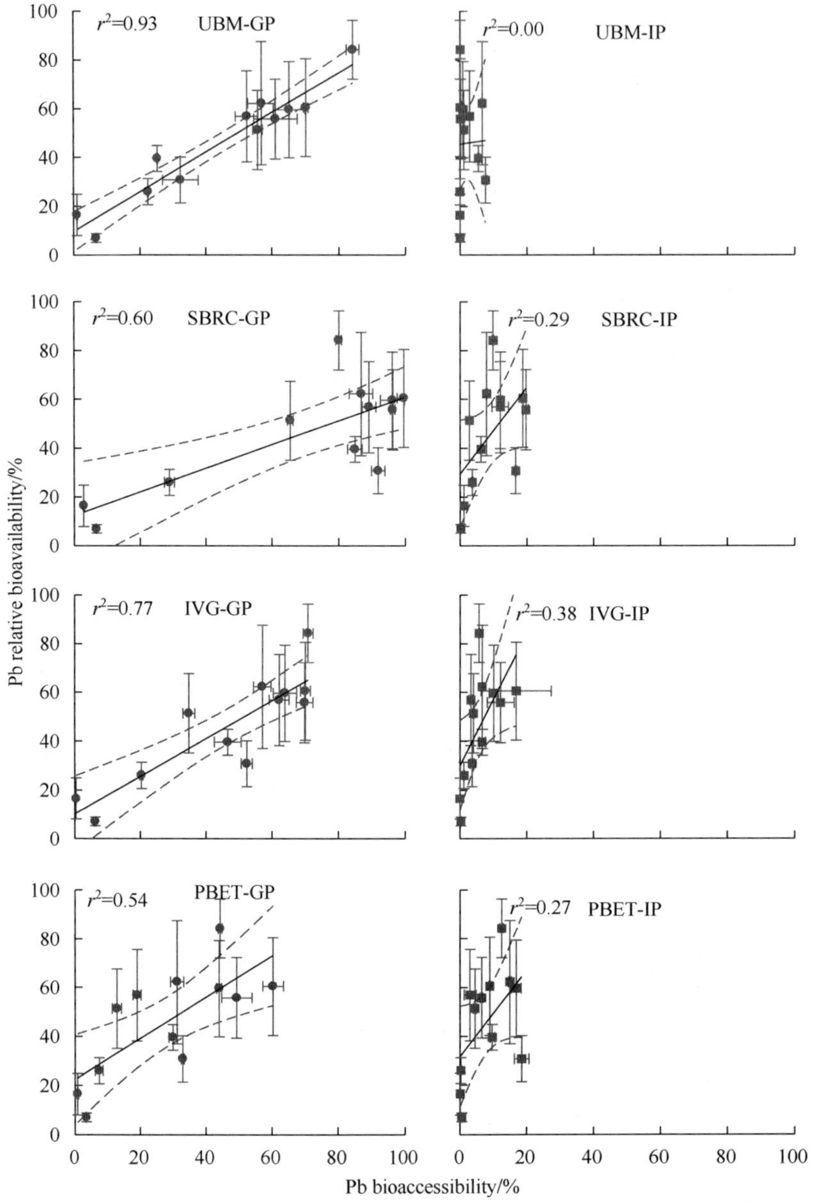

Fig. 8　Comparison of Pb bioaccessibility based on the gastric (GP) and intestinal phases (IP) of UBM, SBRC, IVG, and PBET assays with Pb relative bioavailability using a mouse blood assay in 12 Pb-contaminated soils. The solid lines show the best line of fit, while dotted lines show the 95% confidence intervals (adopted from Li et al., 2015b)

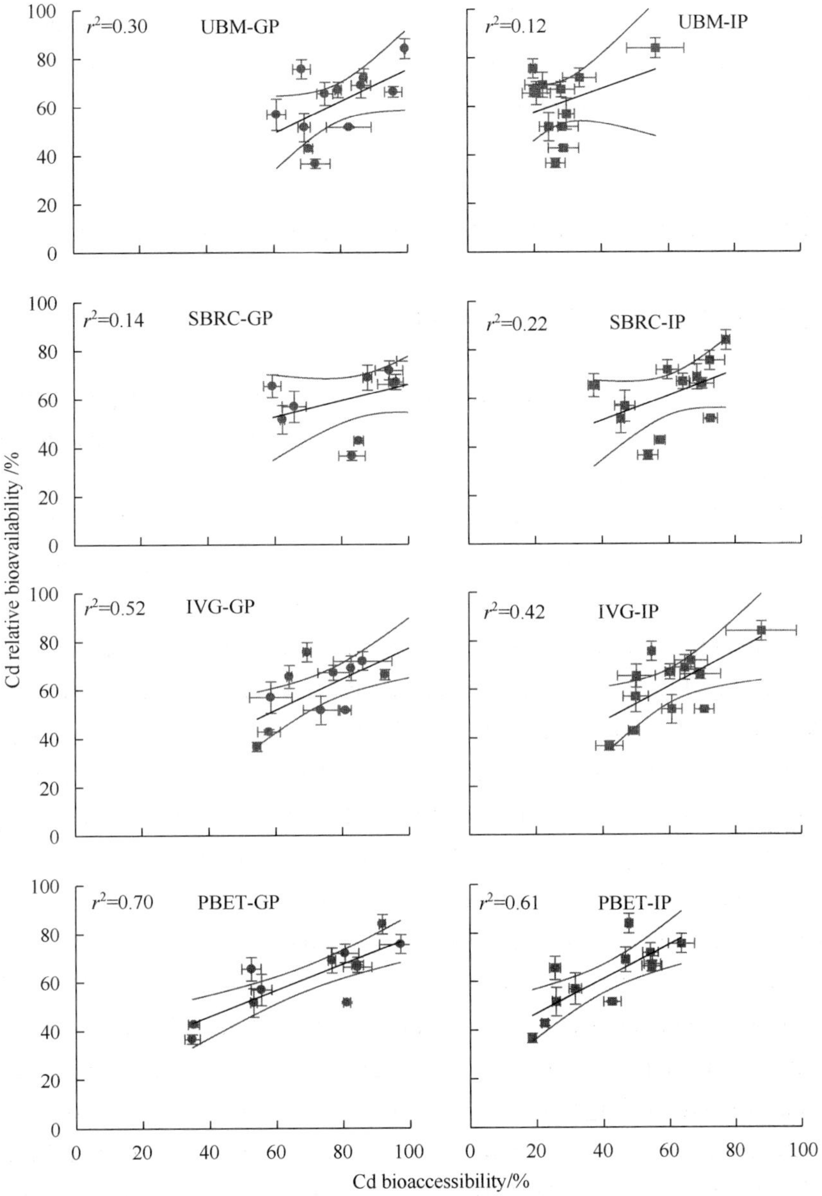

Fig. 9 Comparison of Cd bioaccessibility based on the gastric (GP) and intestinal phases (IP) of UBM, SBRC, IVG, and PBET assays with Cd relative bioavailability using a mouse blood assay in 12 Cd-contaminated soils. The solid lines show the best line of fit, while dotted lines show the 95% confidence intervals (adopted from Li et al., 2016b). assays suggests the PBET is probably suitable to predict Cd-RBA in contaminated soils

Although strong IVIVCs have been established for As, Pb, and Cd contaminated soils for China, future work are still needed to ensure the robustness of in vitro bioaccessibility assays to act as a surrogate of in vivo animal bioassays. For example, collecting more As, Pb, and Cd contaminated soils to represent different contamination sources and soil types as well as comparison of mouse and swing models. In addition, using an independent set of As, Pb, and Cd-contaminated soils to validate the established IVIVCs to predict RBA is also needed.

6. Appling oral bioavailability to remediation of contaminated soils

6.1 Applying to risk assessment of contaminated soils

Following determination of metal RBA using in vivo bioassays or prediction based on in vitro Pb bioaccessibility and the established IVIVCs, metal RBA can be incorporated into human risk assessment via incidental soil ingestion. For soils contaminated by As, Pb, or Cd in China (Tables 2-4), daily As, Pb, and Cd intake was calculated for a child with body weight of 18.6 kg at soil ingestion rate of 60 mg soil per day (Table 6). When metal intake is calculated based on total metal concentrations (i.e., metal concentrations in soils multiplied by daily soil ingestion and divided by body weight), As, Pb, and Cd intake is 0.12-24.0, 0.69-81.7, and 0.01-0.95 μg/ (kg d) (Table 6). However, when metal RBA is considered (i.e., metal concentrations in soils multiplied by daily soil ingestion, metal RBA, and divided by body weight), they are significantly reduced to 0.02-4.05, 0.12-25.2, and 0.01-0.63 μg/ (kg d). The most significant reduction is found for soil with low RBA. For example, for As-contaminated soil 6 with high As concentration of 7430 mg/kg but low As-RBA of 6.38%, its oral ingestion As intake value is 24.0 μg/ (kg d) based on total As concentration; however, when As-RBA is included, it is sharply reduced to 1.53 μg/ (kg d). As a result, As intake is comparable to soil 11 (1.15 μg/ (kg d) even the total As concentration in soil 6 is 3-fold higher than soil 11 (7430 vs. 2556 mg/kg).

To demonstrate the importance of incorporating metal RBA into risk assessment, contribution of metal intake via soil ingestion to provisional tolerable intake values can be calculated. For example, based on total Cd, Cd intake from 12 soils contributed 1.16%–115% (averaging 33%) of the provisional tolerable monthly intake (PTMI at 25 μg/kg bw/month). As expected, the Cd intake contribution to PTMI is much less (0.97%-76%, averaging 19%) based on bioavailable Cd, a reduction of 42%. In addition, incorporating bioavailability into Cd intake calculation results in different influences on Cd intake from soils. For example, for soils 9 and 10 containing 130 and 204 mg/kg Cd, significant difference in their contribution of Cd intake to PTMI is found based on total Cd (50% vs. 79%). However, there is no difference when employing bioavailable Cd (26% vs. 29%). This is due to the higher Cd-RBA in soil 9 than that in soil 10 (52% vs. 37%). Therefore, assessing Cd intake based on total Cd concentration may overestimate Cd exposure in samples with low Cd-RBA, so it is important to consider Cd-RBA in contaminated soils to reliably assess Cd risk associated with incidental soil ingestion.

6.2 Applying to cleanup standard of contaminated soils

The soil cleanup standards are established based on risk assessment assuming all metals in soils are bioavailable. However, we have demonstrated that bioavailable metals vary with metals and soils. Based on our data, the As-RBA in the 12 soils is 6.38%-73.1% (averaging 29%), Pb-RBA in the 12 soils is 7.0%-84% (averaging 46%), and Cd-RBA in the 12 soils is 37%-84% (averaging 61%) based on a mouse model (Tables 2-4). Based on the average RBA, then soil cleanup standards for As-, Pb- and Cd-contaminated soils can be adjusted by dividing 29%, 46% and 61%. Taking Cd for example, based on the Chinese Soil Environmental Quality Standards, the class Ⅱ value is 0.3 mg/kg for soils with pH$<$7.5, which is applicable to agricultural, orchard, and pasture land. By incorporating the Cd-RBA into risk assessment, then the adjusted class Ⅱ values for Cd is 0.49 mg/kg.

Such an approach has been used in the USA. For example, the soil cleanup standard for As-contaminated soils in Florida was 0.8 mg/kg assuming all As in soils is 100% bioavailable. Using a monkey model, the average As-RBA is 38% in Florida soils. As a result, the current soil cleanup standard for As-contaminated soils in Florida has been changed to 2.1 mg/kg. Similar approach can be applied to metal contaminated soils in China. However, it should be noted that the average RBA of As, Pb, and Cd (29%, 46%, and 61%) is based on small size number of metal contaminated soils (n=12, Table 2-4), with considerable variation with contamination sources. For accurately determining soil cleanup standards for different land use, assessment of more soil samples is still needed.

The data suggest that calculation of metal intake based on total metal concentration would significantly overestimate the health risks via soil ingestion, and the overestimation varies with site. In addition, incorporating of metal bioavailability into risk assessment affects the cleanup standards in contaminated soils. Considering metal RBA, some contaminated sites with low metal bioavailability to humans might not need remediation. Thereby, the use of site-specific metal RBA data in risk assessment can significantly lower cleanup costs.

Table 6 Daily As, Pb, or Cd Intake via Soil Ingestion at Ingestion Rate of 60 mg Soil per Day for A Child with A Body Eight of 18.6 kg for Each 12 Soils Contaminated with As, Pb, and Cd from China

daily As, Pb, or Cd intake via soil ingestion based on (μg/(kg d))								
As soil	Total As	As RBA	Pb soil	Total Pb	Pb RBA	Cd soil	Total Cd	Cd RBA
As 1	0.12	0.06	Pb 1	0.69	0.36	Cd 1	0.01	0.01
As 2	0.13	0.02	Pb 2	2.37	1.41	Cd 2	0.03	0.01
As 3	0.38	0.25	Pb 3	4.21	2.35	Cd 3	0.04	0.02
As 4	0.55	0.15	Pb 4	4.98	3.01	Cd 4	0.04	0.03

Continue table

daily As, Pb, or Cd intake via soil ingestion based on (μg/(kg d))								
As soil	Total As	As RBA	Pb soil	Total Pb	Pb RBA	Cd soil	Total Cd	Cd RBA
As 5	0.24	0.03	Pb 5	0.81	0.46	Cd 5	0.06	0.04
As 6	24.0	1.53	Pb 6	1.66	1.40	Cd 6	0.08	0.06
As 7	4.74	0.35	Pb 7	3.79	2.36	Cd 7	0.12	0.06
As 8	0.07	0.05	Pb 8	32.1	12.7	Cd 8	0.19	0.11
As 9	0.28	0.10	Pb 9	81.7	25.2	Cd 9	0.42	0.22
As 10	2.78	0.23	Pb 10	1.66	0.12	Cd 10	0.66	0.24
As 11	8.25	1.15	Pb 11	3.46	0.57	Cd 11	0.68	0.47
As 12	13.5	4.05	Pb 12	13.4	3.49	Cd 12	0.95	0.63

References

Alloway B J. 1995. Heavy metals in soils. Blackie Academic and Professional, London, 368.

ATSDR. 2011. Priority List of Hazardoud Substances. Agency for Toxic Substances and Disease Registry. http: //www.atsdr.cdc.gov/spl/.

Bannon D I, Drexler J W, Fent G M, et al. 2009. Evaluation of small arms range soils for metal contamination and lead bioavailability. Environmental Science & Technology, 43 (24): 9071-9076.

Bradham K D, Diamond G L, Scheckel K G, et al. 2013. Mouse assay for determination of arsenic bioavailability in contaminated soils. Journal of Toxicology and Environmental Health, Part A, 76 (13): 815-826.

Bradham K D, Nelson C, Juhasz A L, et al. 2015. Independent data validation of an in vitro method for the prediction of the relative bioavailability of arsenic in contaminated soils. Environmental Science & Technology, 49 (10): 6312-6318.

Bradham K D, Scheckel K G, Nelson C M, et al. 2011. Relative bioavailability and bioaccessibility and speciation of arsenic in contaminated soils. Environmental Health Perspectives, 119 (11): 1629.

Brattin W, Casteel S. 2013. Measurement of arsenic relative bioavailability in swine. J. Toxicol. Environ. Health, Part A, 76, 449-457.

Casteel S W, Weis C P, Henningsen G M, et al. 2006. Estimation of relative bioavailability of lead in soil and soil-like materials using young swine. Environmental Health Perspectives, 1162-1171.

Denys S, Caboche J, Tack K, et al. 2012. In vivo validation of the unified BARGE method to assess the bioaccessibility of arsenic, antimony, cadmium, and lead in soils. Environmental Science & Technology, 46 (11): 6252-6260.

German Institute for Standardization. 2004. Soil quality-Absorption availability of organic and inorganic pollutants from contaminated soil material. DIN19738-2004.

El-Masri H A, Kenyon E M. 2008. Development of a human physiologically based pharmacokinetic (PBPK) model for inorganic arsenic and its mono-and di-methylated metabolites. Journal of Pharmacokinetics and Pharmacodynamics, 35 (1): 31-68.

Evans M V, Dowd S M, Kenyon E M, et al. 2008. A physiologically based pharmacokinetic model for intravenous and ingested dimethylarsinic acid in mice. Toxicological Sciences, 104 (2): 250-260.

Faroon O, Ashizawa A, Wright S, et al. 2012. Toxicological profile for cadmium.

Gentry P R, Covington T R, Mann S, et al. 2004. Physiologically based pharmacokinetic modeling of arsenic in the mouse. Journal of Toxicology and Environmental Health, Part A, 67 (1): 43-71.

Gentry P R, Haber L T, McDonald T B, et al. 2005. Data for physiologically based pharmacokinetic modeling in neonatal animals: Physiological parameters in mice and Sprague-Dawley rats. Journal of Children' s Health, 2 (3-4): 363-411.

Glorennec P, Lucas J P, Mandin C, et al. 2012. French children' s exposure to metals via ingestion of indoor dust, outdoor playground dust and soil: Contamination data. Environment International, 45: 129-134.

Huo X, Peng L, Xu X, et al. 2007. Elevated blood lead levels of children in Guiyu, an electronic waste recycling town in China. Environmental Health Perspectives, 1113-1117.

IARC. 2004. Some drinking-water disinfectants and contaminants, including arsenic. IARC Monogr Eval Carcinogen Risk Hum, 84, 269-477.

Juhasz A L, Smith E, Nelson C, et al. 2014. Variability associated with As in vivo–in vitro correlations when using different bioaccessibility methodologies. Environmental Science & Technology, 48 (19): 11646-11653.

Juhasz A L, Weber J, Smith E, et al. 2009. Assessment of four commonly employed in vitro arsenic bioaccessibility assays for predicting in vivo relative arsenic bioavailability in contaminated soils. Environmental Science & Technology, 43 (24): 9487-9494.

Juhasz A L, Smith E, Weber J, et al. 2007. Comparison of in vivo and in vitro methodologies for the assessment of arsenic bioavailability in contaminated soils. Chemosphere, 69 (6): 961-966.

Juhasz A L, Weber J, Smith E. 2011. Influence of saliva, gastric and intestinal phases on the prediction of As relative bioavailability using the Unified Bioaccessibility Research Group of Europe Method (UBM). Journal of Hazardous Materials, 197: 161-168.

Juhasz A L, Smith E, Nelson C, et al. 2014a. Variability associated with As in vivo–in vitro correlations when using different bioaccessibility methodologies. Environmental Science & Technology, 48 (19): 11646-11653.

Juhasz A L, Herde P, Herde C, et al. 2014b. Validation of the predictive capabilities of the Sbrc-G in vitro assay for estimating arsenic relative bioavailability in contaminated soils. Environmental Science & Technology, 48 (21): 12962-12969.

Juhasz A L, Weber J, Naidu R, et al. 2010. Determination of cadmium relative bioavailability in contaminated soils and its prediction using in vitro methodologies. Environmental Science & Technology, 44 (13): 5240-5247.

Kelley M E, Brauning S E, Schoof R A, et al. 2002. Assessing oral bioavailability of metals in soil. Battelle Press.

Levallois P, St-Laurent J, Gauvin D, et al. 2014. The impact of drinking water, indoor dust and paint on blood lead levels of children aged 1–5 years in Montréal (Québec, Canada). Journal of Exposure Science and Environmental Epidemiology, 24 (2): 185-191.

Li H B, Cui X Y, Li K, et al. 2014. Assessment of in vitro lead bioaccessibility in house dust and its relationship to in vivo lead relative bioavailability. Environmental Science & Technology, 48 (15): 8548-8555.

Li J, Li C, Sun H J, et al. 2015. Arsenic relative bioavailability in contaminated soils: Comparison of animal models, dosing schemes, and biological end points. Environmental Science & Technology, 50 (1): 453-461.

Li J, Li K, Cui X Y, et al. 2015a. In vitro bioaccessibility and in vivo relative bioavailability in 12 contaminated soils: Method comparison and method development. Science of the Total Environment, 532: 812-820.

Li J, Li K, Cave M, et al. 2015b. Lead bioaccessibility in 12 contaminated soils from China: Correlation to lead relative bioavailability and lead in different fractions. Journal of Hazardous Materials, 295: 55-62.

Li S W, Sun H J, Li H B, et al. 2016b. In vivo validation of different assays to assess cadmium bioaccessibility in contaminated soils. Environment International. (in revision).

Lunney J K. 2007. Advances in swine biomedical model genomics. International Journal of Biological Sciences, 3 (3): 179-184.

Oomen A G, Hack A, Minekus M, et al. 2002. Comparison of five in vitro digestion models to study the bioaccessibility of soil contaminants. Environmental Science & Technology, 36 (15): 3326-3334.

Patterson J K, Lei X G, Miller D D. 2008. The pig as an experimental model for elucidating the mechanisms governing dietary influence on mineral absorption. Experimental Biology and Medicine, 233 (6): 651-664.

Roberts S M, Weimar W R, Vinson J R T, et al. 2002. Measurement of arsenic bioavailability in soil using a primate model. Toxicological Sciences, 67 (2): 303-310.

Rodriguez R R, Basta N T, Casteel S W, et al. 1999. An in vitro gastrointestinal method to estimate bioavailable arsenic in contaminated soils and solid media. Environmental Science & Technology, 33 (4): 642-649.

Ruby M V, Davis A, Schoof R, et al. 1996. Estimation of lead and arsenic bioavailability using a physiologically based extraction test. Environmental Science & Technology, 30 (2): 422-430.

Ruby M V, Lowney Y W. 2012. Selective soil particle adherence to hands: Implications for understanding oral exposure to soil contaminants. Environmental Science & Technology, 46 (23): 12759-12771.

Schroder J L, Basta N T, Casteel S W, et al. 2004. Validation of the in vitro gastrointestinal (IVG) method to estimate relative bioavailable lead in contaminated soils. Journal of Environmental Quality, 33 (2): 513-521.

Schroder J L, Basta N T, Si J, et al. 2003. In vitro gastrointestinal method to estimate relative bioavailable cadmium in contaminated soil. Environmental Science & Technology, 37 (7): 1365-1370.

Smith E, Kempson I M, Juhasz A L, et al. 2011. In vivo–in vitro and XANES spectroscopy assessments of lead bioavailability in contaminated periurban soils. Environmental Science & Technology, 45 (14): 6145-6152.

Surkan P J, Zhang A, Trachtenberg F, et al. 2007. Neuropsychological function in children with blood lead levels$<10\mu g/dL$. Neurotoxicology, 28 (6): 1170-1177.

The Ministry of Environmental Protection. 2014. The Ministry of Land and Resources Report on the national soil contamination survey. http: //www.mep.gov.cn/gkml/hbb/qt/201404/t20140417_270670.htm [2014-8-27]

Van Wijnen J H, Clausing P, Brunekreef B. 1990. Estimated soil ingestion by children. Environmental Research, 51 (2): 147-162.

Weis C P, LaVelle J M. 1991. Characteristics to consider when choosing an animal model for the study of lead bioavailability. Chemical Speciation & Bioavailability, 3 (3-4): 113-119.

Wilson R, Jones-Otazo H, Petrovic S, et al. 2013. Revisiting dust and soil ingestion rates based on hand-to-mouth transfer. Human and Ecological Risk Assessment: An International Journal, 19 (1): 158-188.

Wragg J, Cave M, Basta N, et al. 2011. An inter-laboratory trial of the unified BARGE bioaccessibility method for arsenic, cadmium and lead in soil. Science of the Total Environment, 409 (19): 4016-4030.

Yan C H, Xu J, Shen X M. 2013. Childhood lead poisoning in China: Challenges and opportunities. Environmental Health Perspectives, 121 (10): A294.

Zahran S, Mielke H W, Gonzales C R, et al. 2010. New Orleans before and after Hurricanes Katrina/Rita: A quasi-experiment of the association between soil lead and children’ s blood lead. Environmental Science & Technology, 44 (12): 4433-4440.

Zhao F J, Ma Y, Zhu Y G, et al. 2014. Soil contamination in China: Current status and mitigation strategies. Environmental Science & Technology, 49 (2): 750-759.

Biochar for Environmental Management: Impacts on the Sorption and Bioavailability of Organic Contaminants in Soil

Hongwen SUN, Xinhao REN, Fei WANG, Wen ZHANG, Zunlong ZHOU

MOE Key Laboratory of Pollution Criteria and Environmental Processes, College of Environmental Science and Engineering, Nankai University

Introduction

With the rapid development of industry and agriculture and the acceleration of urbanization, the amount of waste water and waste solid discharged from industrial, agricultural, domestic and commercial sources have increased rapidly. As one of the primary environmental media, soil is a main sink of organic and inorganic contaminants due to its strong sorption capacity, and hence soil is a main media that needs remediation.

Once a contaminant enters soil, it will interact with soil constituents, which will alter its combination (existing) state, i.e. different molecules of the same contaminant may bind to soil sites with different characteristics, such as energy, pore size and flexibility. This exerts great impact on the mobility, bioavailability and chemical reactivity of the contaminant, changing its persistence and risk. On the other hand, the combination state of a contaminant in soil is a main factor affecting the efficiency of a remediation technology.

Biochar is a carbon-rich solid product obtained by pyrolyzing biomass materials, such as wood, manure, and leaves in a closed container with limited air (Manya, 2012). The origin of biochar application can be dated to *Terra Preta de Indio* in the Amazon region, where dark earth was created through the use of slash-and-char technique (Marris, 2006). However, biochar had not received widespread attention until the article "Putting the carbon back: Black is the new green" was published in Nature in 2006 (Marris, 2006). Now, biochar is recognized as an emerging combined means of sequestering carbon, improving soil fertility and remedying contaminated soil (Beesley et al., 2011; Manya, 2012), which has driven a recent surge in biochar research.

1. Biochar production and properties

1.1 Biomass pyrolysis

Biochar is defined as a carbonaceous product of biomass produced by thermochemical conversion technologies in the presence of limited air or no air, which is commonly known as pyrolysis. Pyrolysis is generally divided into fast pyrolysis, flash carbonization and slow pyrolysis, depending on the heating rate and residence time. Fast pyrolysis uses high heating rate (above 200 K/min) with a very short residence time (around 2s), which particularly favors the formation of bio-oil, but not the formation of charcoal (Manya, 2012). The flash carbonization process is an efficient way to convert biomass to charcoal than conventional carbonization or slow pyrolysis. A canister containing a packed bed of a given biomass is placed within a pressure vessel. Air is used to pressurize the vessel to an initial pressure of 1-2 MPa, and a flash fire is ignited at the bottom of the packed bed. After a few minutes, air is delivered to the top of the packed bed and biomass is converted to charcoal (Manya, 2012). Slow pyrolysis or conventional carbonization pyrolysis processes with a residence time of few minutes to several hours and a low heating rate has been used to generate charcoal for many years, and is generally thought to favor for biochar production (Manya, 2012). The conversion process from biomass to biochar is complex and remains unclear. Demirbas (2004) suggested that biochar is produced from biomass through three steps.

1st step: Biomass → Water+Untreated residue

2nd step: Untreated residue → $(\text{Volatile+Gases})_1$+ $(\text{Char})_1$

3rd step: $(\text{Char})_1$ → $(\text{Volatile+Gases})_2$+ $(\text{Char})_2$

In the initial step, moisture and some volatiles lose. In the secondary step, primary biochar is formed. In the third step, the primary biochar produced further decomposes to a carbon-rich residue solid by chemical rearrangement at a very slow rate. Ren (2016) suggested that with the increasing pyrolyzing temperature, there are three obvious pyrolysis processes during the pyrolyzing of pig manure: ① At initial charring stage (60℃-150℃), the native structure of pig manure is preserved with the loss of water and dehydration reactions of the pig manure; ② Within 300℃-400℃, the mass of pig manure decreases dynamically with the pyrolyzing temperature. Some of biopolymers undergo dehydration and depolymerization reactions, forming amorphous centers; ③ Around 700℃, some poorly ordered graphene stacks are generated and embedded in amorphous phases, and more pore structure are formed.

1.2 Biochar properties

Biochar properties are mainly influenced by feedstock properties and pyrolysis conditions. Several variables, such as the highest pyrolysis temperature, pressure, vapor residence time and moisture content have been identified to play critical roles in the slow pyrolysis of biomass (Manya, 2012). Among these, feedstock properties and the highest pyrolysis temperature are considered to be the main factors influencing biochar properties. Biomass materials can be crudely divided into four constituents: lignin, cellulose, hemi-cellulose and inorganic mineral. Since different constituents in biomass are of respective pyrolysis process, the properties of biochar produced from different feedstock types with the same pyrolysis condition are variable. The biochar properties that are most affected by feedstock properties are e.g., yield, total organic carbon, fixed carbon and ash content (Zhao et al., 2013). Many different types of biomass have been used to produce biochars, including plant straw, wood biomass, livestock manure, and solid waste. Biochar produced from these biomass materials could be divided into two groups: plant-derived biochars and animal-derived biochars. In general, the manure-derived biochars contain more ash content and less carbon content than plant-derived biochars (Table 1), due to the presence of minerals such as calcite and quartz in manure feedstocks. Besides, it was found that livestock manure generates a high yield biochar compared to that from plant straw and wood biomass (Zhao et al., 2013). In addition, manure-derived biochars exhibits lower surface areas than those of plant-derived biochars (Table 1). The ash content, elemental compositions and surface area of biochars are also influenced by the biochar pyrolysis temperature. The ash contents of the pig manure-derived biochar were in the range of 49.8%-65.8% and increased as the pyrolyzing temperature increased from 300℃ to 700℃ (Table 1). This could result from the loss of organic matter as the pyrolyzing temperature increases. As the pyrolytic temperature increases, the organic carbon content of biochars increase based on a dry-ash-free basis, while the hydrogen, nitrogen and oxygen contents decrease (Keiluweit et al., 2010). At higher pyrolytic temperatures, some amorphous carbon are converted into aromatic carbon, which results in the higher organic carbon content in biochars pyrolyzed with higher temperature compared to those in biochars with low pyrolytic temperature. Meanwhile, as the charring temperature increases, some volatiles, lignocelluloses and proteins that are present in the biomass decompose (Tsai et al., 2012), which results in the reduction of the O and N contents. Elemental compositions and their calculated molar ratios have been extensively used to reveal the chemistry structure of biochars. The atomic ratios of H/C, O/C, and (O+N) /C are commonly used to characterize the aromaticity and polarity of materials, respectively. The smaller the H/C, the greater the aromaticity. Biochars with larger O/C or (O+N) /C are considered to have greater polarities. Decreased hydrogen, nitrogen and oxygen contents and increased carbon contents with an increasing in pyrolysis temperature result in lower ratios of H/C and (O+N) /C, demonstrating that biochars produced at higher temperatures are heavily carbonized and display highly aromatic structures. While, the higher O/C or (O+N) /C ratio suggests that biochars that are pyrolyzed at lower temperatures have more polar-groups. The surface areas (SA) of the biochars substantially increased with increasing temperature, i.e., 2.89 m^2/g for PBC300 (In this article, BC is abbreviation of biochar and the number after BC refers to pyrolyzing temperature), 7.19 m^2/g for PBC500 and 94.0 m^2/g for PBC700 (Table 1). The increase in the SA with increasing pyrolyzing temperature can be attributed to the transformation of the elemental composition and carbon structure with increasing charring temperature (Keiluweit et al., 2010).

Table 1 Characteristics of biochars produced from different biomass materials with different pyrolytic temperature

Biochar	Feedstock	Pyrolytic tempera-ture/℃	Ash content /%	Elemental composition/%[1]				Atomic ratio/%		Surface area/ (m^2/g)	Aromaticity/%	References
				C	H	O	N	H/C	(O+N) /C			
PBC300	pig manure	300	49.8	62.2	6.53	27.3	4	1.26	0.383	2.89	42	Ren, 2016
PBC500	pig manure	500	62.2	73	4.26	19	3.76	0.699	0.237	7.19	61	Ren, 2016
PBC700	pig manure	700	65.8	75.7	2.75	18.4	3.13	0.436	0.218	94	65	Ren, 2016
MBC350	maize straw	350	15.4	69.6	3.97	25.2	1.18	0.685	0.286	6.71	68	Ren et al., 2016
MBC700	maize straw	700	21.9	55.4	1.86	39.9	1.6	0.403	0.566	265	41	Ren et al, 2016
RBC350	rice straw	350	29.1	62.8	3.79	31.2	2.31	0.725	0.404	9.01	52	Ren et al, 2016
RBC700	rice straw	700	38.2	91.7	3.87	4.22	0.161	0.506	0.036	188	72	Ren et al, 2016
WBC300	pine wood	300	2.25	63.3	4.94	30.9	0.91	0.93	0.38	15.4	72.8	Zhou et al, 2010
WBC500	pine wood	500	3.04	84.1	3.61	11.1	1.07	0.51	0.11	76.5	79.4	Zhou et al, 2010
WBC700	pine wood	700	2.81	92.2	1.96	4.96	8.95	0.26	0.04	376.3	80.7	Zhou et al, 2010

Note: 1 Elemental composition and atomic ratios are on a dry-ash-free basis, H/C is the atomic ratio of hydrogen to carbon, and (O+N) /C is the atomic ratio of the sum of oxygen and nitrogen to carbon.

Table 2 The regression parameters of sorption isotherms of organic pollutants to different types of biochars fitted by Freundlich equation

Sorbent	Feedstock	Pyrolytic tempera-ture/℃	Sorbate	Freundlich			K_d/ (L/kg)			References
				n	K_f	R^2	0.01 S_w	0.1 S_w	0.5 S_w	
Biochar	pine	400	pyrene	0.49±0.01	3.55E5±219	0.997	3.07E+05	9.48E+04	4.17E+04	Zhang, 2011
	pine	700	pyrene	0.18±0.01	4.07E6±126	0.991	3.22E+06	4.87E+05	1.36E+05	Zhang, 2011
	pig manure	300	phenanthrene	0.84±0.03	1.11E4±245	0.994	2.32E+04	1.60E+04	1.24E+04	Ren, 2016
	pig manure	700	phenanthrene	0.52±0.01	1.82E4±326	0.97	1.66E+05	5.50E+04	2.54E+04	Ren, 2016
	pig manure	350	carbaryl	0.51±0.01	689±120	0.991	756	203	112	Ren et al., 2016
	pig manure	700	carbaryl	0.30±0.02	5.24E3±256	0.99	5.99E+03	943	390	Ren et al., 2016
	maize straw	350	carbaryl	0.37±0.07	2.51E3±176	0.993	2.83E+03	524	241	Ren et al., 2016
	maize straw	700	carbaryl	0.22±0.04	1.23E4±285	0.935	1.43E+04	1.77E+03	675	Ren et al., 2016
	pig manure	300	atrazine	0.72±0.02	80.4±4.5	0.998	112	59.1	37.7	Ren, 2016

Continue table

Sorbent	Feedstock	Pyrolytic tempera-ture/℃	Sorbate	Freundlich			K_d/ (L/kg)			References
				n	K_f	R^2	0.01 S_w	0.1 S_w	0.5 S_w	
Biochar	pig manure	700	atrazine	0.51±0.08	2.11E3±337	0.929	3.80E+03	1.23E+03	559	Ren, 2016
	maize straw	200	propranolol	0.99±0.12	0.571±0.052	0.964	0.562	0.549	0.541	Wang, 2016
	maize straw	400	propranolol	0.43±0.04	2.00±0.14	0.978	0.825	0.221	0.0881	Wang, 2016
	maize straw	700	propranolol	0.58±0.07	7.33±0.43	0.949	3.86	1.48	0.761	Wang, 2016

2. Effect of biochar on the sorption of organic contaminants in soil

The environmental fate and uptake by plant or biota of organic contaminants in soil is governed by a variety of physical, chemical and biological processes (Kookana, 2010). These processes, including sorption-desorption, leaching, volatilization, chemical or biological degradation are often complex and dynamic. The transport of organic contaminants from soil media to water, air or food is directly controlled by these processes. Among these processes, sorption-desorption is a pair of processes that regulate the concentration of organic contaminant in soil solution and consequently affect other processes, such as leaching, volatilization, bioavailability and degradation (Kookana, 2010). Due to its porous structure, large surface area, high aromaticity, and surface polar functional groups such as hydroxyl, phenolic hydroxyl and carbonyl, biochars show extraordinary adsorptive affinity for all kinds of contaminants. The application of biochars to soil will influence the occurrence, fate and bioavailability of contaminants.

2.1 Sorption mechanisms of organic contaminants on biochar

The higher organic carbon content and larger surface area of biochar provide large adsorption sites for a series of organic contaminants. Biochar has strong sorption affinities for both apolar and polar contaminants, due to its higher aromaticity and surface polar functional groups (Table 2). As discussed above, biochar properties are mainly influenced by feedstock properties and pyrolysis conditions. The diversity of feedstock types and pyrolytic temperatures determine the variability of biochar properties. Different biochars are of different sorption mechanisms for different organic contaminants, which depends on the properties of both biochar and organic contaminants.

2.1.1 Sorption of apolar organic contaminants on biochar

Biochar exhibits strong affinity for apolar organic contaminants, such as PAHs. There are multiple processes and sorption mechanisms involved in the sorption of apolar organic contaminants. In a previous study, three biochars were produced by heating sawdust at 300℃, 500℃ and 700℃, respectively, and the sorption kinetics and isotherms of phenanthrene (Phen) onto these biochars were examined (Zhou et al., 2010). The results suggested that the sorption process could be divided into very fast, fast and slow stages. The pseudo-second-order model fitted the sorption kinetics data well over the entire sorption duration. The pseudo-first-order and Elovich models did comparatively well for only selected sorption periods. These observations indicated that multi-processes controlled the sorption of Phen on biochars, including diffusion into water film, diffusion on particle surfaces of the biochars, and diffusion in the internal micropores, which resulted in the observed complexity in the sorption kinetics. The differences in the fast sorption on biochars produced at different pyrolytic temperatures were attributed to their different hydrophobicity.

Chen et al. (2011) examined the sorption of PAHs on biochars produced at different temperatures. They found that the sorption isotherms of naphthalene and Phen on BC100 were nearly linear, indicating that partition is the dominant mechanism of BC100. The isotherms of naphthalene and Phen on BC300, BC400 and BC700 exhibited an increasing nonlinearity of concave-downward shape at low-solute concentration, but a practically linear shape at moderate-to-high concentration. In addition, the nonlinearity of isotherms of naphthalene and Phen on biochars increased with the pyrolytic temperature of biochars. The authors suggested that the sorption of PAHs to biochar was a mixed adsorption-partition mechanism. At low solute concentration, adsorption was the main mechanism, but as the solute concentration increased, the surface adsorption sites on biochars reached saturation rapidly. At moderate to high concentration, partition became the dominant mechanism.

To investigate the role of biochar structural characteristics in the sorption of apolar organic contaminants, Zhang (2011) obtained different biochars and modified the biochars by oxidation, oximation and hydrolysis, and studied the sorption of pyrene on these biochars. The results showed that BC700 exhibited stronger nonlinearity (lower n in Freundlich isotherm) and a greater sorption capacity (higher K_f in Freundlich isotherm) as compared to BC400. The greater nonlinearity of the isotherm suggests that the sorption sites of BC700 were more heterogeneously distributed than those of BC400. The greater sorption capacity of BC700 could be explained by its higher aromatic carbon content, surface area, micropore volume, and lower polarity. The elemental composition and structure characteristics of modified biochars were quite different from those of biochars without modification. Chemical modifications resulted in different effects on pyrene sorption nonlinearity for the two biochars. After modifications, the nonlinearity of sorption isotherm decreased for BC700, with n increasing from 0.18 to 0.27, while the nonlinearity increased for BC400, with n decreasing from 0.49 to 0.23. The different effects on sorption nonlinearity for the two BCs may be explained by the change of heterogeneity in sorption sites due to modifications. The BC700 obtained at higher temperature contained more micropores with pore width less than 20 Å. Modifications greatly decreased the number of pores in this range, which led that the pore distribution became more even and the nonlinearity for pyrene sorption decreased. For BC400 obtained at lower temperature, some organic matter residue existed in the original biochar, and the heterogeneity of sorption sites was not high. Modifications increased the heterogeneity of sorption sites and thus the nonlinearity of sorption isotherms increased. Both oxidation and oximation reduced the sorption affinity for pyrene on BC700, and the former caused a greater reduction in K_{oc} (69.1%-73.7% versus 18.7%-33.9%; C_e= 10-100 μg/L). This was probably due to the greater changes in O-containing groups and polarity after oxidation. Hydrolysis did not significantly influence pyrene sorption for BC700. For BC400, oxidation reduced K_{oc} by 2.28%-25.9% and hydrolysis by 29.2%-33.9% (C_e=10-100 μg/L). However, oximation increased sorption affinity of BC400, which was ascribed to the increased accessibility of the interior sorption sites due to the replacement of O by N atoms. However, there were no apparent correlations between sorption strength and polarity, aromatic or aliphatic C content, surface area, or micropore volume for biochar, indicating that the sorption of HOCs by BCs was controlled by multiple factors.

Besides the effects of hydrophobicity, steric effects may affect the sorption of apolar organic contaminants on biochars. Wang (2016) compared the sorption of three PCB analogues on BC700. The n values of different PCBs on BC700 were affected by their planarity: i. e. the sorption isotherm of nonplanar PCB5 (mono-*ortho*-substituted, n=0.408) and PCB4 (di-*ortho*-substituted, n=0.431) was relatively less nonlinear; while the sorption isotherm of planar PCB3 was the most nonlinear (non-*ortho*-substituted, n=0.328). The molecular volume of PCB3 was 583 $Å^3$, being smaller than that of PCB4 and PCB5 (618 $Å^3$ and 612 $Å^3$). The sorption capacity decreased with increasing molecular volume on BC700. The steric effects of the larger nonpolar PCB4 and PCB5 made them more difficult to access to the specific sorption sites on BC700. Hence, the changes in the n with the degree of planarity can be explained by that the specific high-energy sorption sites are easily accessed by the planar PCB3. It is much easier for planar compounds to fit into the pores because the π-cloud overlap and they approach the surface of soot-like material closely. Hence, it is plausible to conclude that the sorption of PCBs on BC700 was primarily controlled by both the surface π-π interaction and pore-filling.

2.1.2 Sorption of polar organic contaminants on biochar

The interaction mechanisms between biochar and polar organic contaminants are different from those between biochar and apolar organic contaminants. Surface O-containing polar functional groups such as, carboxyl and hydroxyl are important characteristics of biochars, especially for biochars acquired at low temperature. Besides, biochars contain electron-deficient and electron-rich moieties. Biochars could interact with organic contaminants, especially for polar organic contaminants through specific forces, including covalent bond, H-bonding and electron donor-acceptor interaction. Ren (2016) found that biochars produced from pig manure could efficiently sorb atrazine. Besides hydrophobic effect and pore-filling, specific forces including π-π electron donor-acceptor interaction and H-bonding interaction between atrazine and biochars were involved in the sorption of atrazine. The lone-pair electrons on the nitrogen atom of atrazine are able to delocalize to the thiazine ring and thus create a tendency to form H-bonding with O-containing polar functionalities in biochars. Another study (Wang, 2016) observed that the sorption of 1-naphthol and 4-nitro-1-naphthol on BC200 derived from corn straw were stronger than naphthalene, because the H-bond and π-π electron donor-acceptor interaction between 1-naphthol and 4-nitro-1-naphthol and the surface of BC200 were higher than those between naphthalene and BC200, due to the hydroxyl- and nitro-substitution.

Biochars produced at lower pyrolytic temperature contain much more surface polar functional groups than those produced at higher temperatures, and may have much higher affinity for polar organic contaminants. Sun et al. (2011) studied the sorption of fluridone on plant-derived biochars produced at different pyrolytic temperatures. The fluridone sorption isotherms on plant-derived biochars displayed an increasing nonlinearity with increasing pyrolytic temperature. However, BC400 had the K_{oc} values (organic carbon-normalized distribution coefficient) at low concentrations much greater than those of biochars produced at higher temperature, which suggested that the binding of fluridone to specific domains in the amorphous carbon phase was the predominant sorption mechanism. Besides, there were significant linear correlations between K_{oc} values and the aromaticity and polarity of biochars. The authors proposed that π-π interactions between fluridone and aromatic components in the biochars and H-bonding between this compound and O-containing moieties played important roles in the sorption. This is consistent with the findings of Yao et al. (2012), which showed a decreasing in the sorption of sulfamethoxazole on biochars with increasing biomass pyrolytic temperature. The authors indicated that the surface functional groups on biochar play a more important role in interacting sulfamethoxazole with biochar than other factors such as surface area or hydrophobicity.

For ionizable organic contaminants, electrostatic attraction or repulsion may exert great effects besides polar interaction. Wang (2016) compared the sorption of propranolol and naphthalene on biochars produced from corn straw at different temperatures (ranging from 200℃ to 700℃). Propranolol is a kind of β-blocker, with pK_a=9.53, and hence propranolol exists mainly as cation under neutral condition. Sorption of propranolol and naphthalene on biochars increased with increasing biomass pyrolysis temperature (200℃ to 700℃) and the values of lg K_{oc} increased from 3.23 to 4.01 for propranolol and increased from 2.78 to 4.83 for naphthalene (C_w=1.5 mg/L). Hydrophobicity played a major effect on naphthalene sorption on biochars produced at low pyrolytic temperature, and the contribution of surface adsorption and pore filling increased with the increase of pyrolysis temperature. There were plentiful polar functional groups on BC200, which favored the electrostatic attraction and other polar interaction, leading to a greater sorption of propranolol compared to naphthalene. The actual adsorption amount calibrated by surface area on BC200 was greater than the value predicted by monolayer adsorption, and hence, it was proposed that oblique attachment or multilayer adsorption may occur on BC200 due to the presence of a large number of polar functional groups. On the other side, the sorption of naphthalene was greater than propranolol on biochars obtained at high pyrolytic temperature, where the pore filling mechanism was enhanced by the relative smaller size and greater hydrophobicity of naphthalene molecules.

The pH is an important factor that affects sorption of organic contaminants on biochars, which is a result of its effects on the surface properties of biochar and the chemical form of contaminants (Wang, 2016). With the increase of pH, the hydrophilicity of biochar increased because the deprotonation of biochar surface groups such as hydroxyl and carboxyl. Hence, the sorption of water molecules on the surface of biochar increases, and the inner hydrophobic sites become less accessible to organic contaminants. That leads to the sorption of apolar organic contaminants on biochar decreased. However, for polar organic contaminants, such as propranolol, when the pH was below its pK_a, the sorption of propranolol increased with the increase of pH. As the pH increased, the proportion of propranolol in neutral molecular form increased, which enhanced the hydrophobicity of propranolol, and the negative charges on the surface of biochar also increased with the pH value, both of which enhanced the sorption of propranolol. As for 1-naphthol, when $7 < pH < pK_a$ of 1-naphthol, the -OH of 1-naphthol dissociates to $-O^-$, the electron-donating strength would be further improved, and the sorption of 1-naphthol increased. With the pH value increasing further, the negative charges on the surface of biochars increased, and electrostatic repulsion between 1-naphthol and biochar increased, which made the sorption of 1-naphthol decreased. The same phenomenon was also found in the sorption of 4-nitro-1-naphthol. As the pH increased, 4-nitro-1-naphthol gradually dissociated, and the hydrophobicity of 4-nitro-1-naphthol decreased, the electrostatic repulsion between 4-nitro-1-naphthol and BC200 increased, which decreased the sorption of 4-nitro-1-naphthol (Wang, 2016).

Humic acid exists extensively in the environment. The effect of humic acid on the sorption affinities of biochars was studied by Wang (2016). Since the aromaticity of humic acid was stronger than BC200, the loading of humic acid increased the adsorption of naphthalene, but humic acid also occupied the adsorption sites on BC200. When the concentration of naphthalene was low, the competition of humic acid was negligible. While, with the increase of naphthalene concentration, the surface adsorption sites on BC200 gradually became saturation, and the competitive effects of humic acid became obvious. The reduced of polarity of BC200 due to humic acid loading weakened the interaction between propranolol and polar functional groups of BC200, hence, the sorption of propranolol on BC200 decreased. As for BC700, humic acid loading also decreased the surface area, pore volume and aromaticity, but increased the polarity.

As a result, the sorption of naphthalene on the surface and the internal pore of BC700 decreased. While, humic acid loading introduced some polar functional groups on the surface of BC700, the sorption of propranolol was promoted, but the sorption of propranolol was suppressed by high concentrations of humic acid because of the reduction in surface area and pore volume. Besides, the simultaneous addition of humic acid with propranolol or naphthalene inhibited the sorption of naphthalene or propranolol on both biochars.

2.1.3 Sorption of co-existing contaminants on biochar

Hazardous organic contaminants always exist in the environment simultaneously, thus the sorption behavior of organic contaminants is often affected by co-existing contaminants, leading to a different behavior from those in single system. Predicted value, only based on the sorption model in single-solute system may deviate from the real environment, which would mis-evaluate their environmental risk. Therefore, understanding the sorption of co-existing organic contaminants on biochar is of great importance. Moreover, study on the co-solute sorption may help us deepen the understanding of the sorption mechanism, and provide us a scientific basis for biochar remediation technology.

Wang (2016) studied the sorption of 4-polychlorinated biphenyls (PCB4) on biochars produced at different temperatures in the presence of propranolol. For BC200, low concentrations of propranolol monomer inhibited the sorption of PCB4 by competing for sorption sites. While, at suitable concentrations, propranolol could form hemimicelle structure on biochar surface and provide a favorable phase for the partitioning of PCB4, leading to an enhanced sorption. For BC700, the hemimicelle could not be formed due to the huge surface area and propranolol generally prohibited PCB4 sorption mainly by competition and pore-blocking mechanism. However, the addition of PCB4 before propranolol abated the prohibition effect, and the sorbed propranolol could further enhanced the subsequent adsorption of PCB4 additionally.

The co-existing effect of propranolol on the sorption of naphthalene and its homologues (1-naphthalenemethanol, 1-naphthol, and 4-nitro-1-naphthol) was also studied by Wang (2016). The sorption mechanisms of naphthalene and its homologue on different biochars were different. When propranolol and naphthalene or its homologues was simultaneously added at a fixed total molar ratio, the sorption of naphthalene and 1-naphthalenemethanol on BC200 was enhanced, because the sorbed propranolol on biochar changed the surface polarity of the biochar and the hydrophobicity of the surface of biochar increased. Since 1-naphthol and propranolol both have polar functional groups, they both could form H-bond and π-π electron donor-acceptor interaction with O-containing polar functionalities in biochars, and hence propranolol competed for the same adsorption sites with 1-naphthol on BC200, which inhibited the sorption of 1-naphthol on BC200. Propranolol could form ion-pair with 4-nitro-1-naphthol and increased the sorption of 4-nitro-1-naphthol on BC200. On BC700, propranolol generally inhibited the sorption of those organic contaminants mainly by competition and pore-blocking mechanism.

Inorganic metal elements existing in soil solution can also influence the sorption behavior of organic contaminants. Wang (2016) examined the sorption behaviors of apolar and polar organic contaminants on biochars in the presence of Cu^{2+} or Ag^{+}. The polar functional groups on BC200 could complex with Cu^{2+} or Ag^{+}, which formed a higher molecular weight and more stable structure on the surface of BC200. In addition, Cu^{2+}can form an inner-sphere type complexation, which drove a fraction of oxygen-containing groups to the interior and hydrophobic sites to the outside. The addition of heavy metal ions enhanced the sorption of naphthalene, but inhibited the adsorption of propranolol on BC200 due to competition for the same adsorption sites. In addition, sorption of heavy metals can weaken the electronegativity of BC200, and reduced the electrostatic repulsion between of 4-nitro-1-naphthol and BC200, therefore, the sorption of 4-nitro-1-naphthol was promoted by the sorption of heavy metals on BC200. However, the sorption of heavy metals on BC700 decreased the sorption of all the organic contaminants by pore-filling effect or occupying the throats of pores. Moreover, heavy metals can also form a hydration shell that directly compete for surface adsorption sites and reduced the accessibility of the inner sites, inhibiting the sorption of organic contaminants.

2.2 Effect of biochar amendment on the sorption behaviors of organic contaminants in soil

Since biochar has a high affinity for organic contaminants, the amendment of biochar to soil could significantly increase the sorption capacity of soil. It has been well documented that sorption coefficient (K_d) and isotherm nonlinearity of organic contaminants in biochar amended soil progressively increase with the content of the biochar applied to soil and the enhancement depends on biochar type and soil properties (Spokas et al., 2009; Zhang et al., 2010; Chen and Yuan, 2011; Dechene et al., 2014). Chen et al. (2011) observed that biochars showed different effects on PAHs sorption in biochar amended soil. Biochar produced at 100℃ applied to soil increased the sorption isotherm linearity, due to its linear-type isotherm. While, biochars produced at higher pyrolytic temperatures demonstrated higher efficiency in increasing the sorption capacity of biochar amended soil, and the sorption isotherms of PAHs in soil amended with those biochars were obvious nonlinearity, which increased with the content of biochar applied to soil. Ren (2016) found that the amendment of biochar to soil led to a significant enhancement in the K_d for atrazine and Phen. While the sorption enhancement varied for biochars. The K_d for atrazine (C_e=0.01-0.5 S_w) in BC300 amended soils were 5.49-2.61 and 3.23-1.80, compared with 4.82-1.81 and 2.01-0.887 in the original black soil and fluvo-aquic soil, respectively. The addition of BC700 led to a higher increase in the K_d value for atrazine (C_e=0.01-0.5 S_w) than BC300 did. While an opposite phenomenon was observed for Phen, the sorption enhancement by the addition of BC300 was greater than that of BC700. This could be attributed to the different sorption mechanisms of the two contaminant and sorption affinities between BC300 and BC700. Atrazine interacts with sorbents by polar interactions such as hydrogen bond, while Phen is sorbed mainly through hydrophobic effect. BC700 has more micropores, which tends to be blocked by soil constituents.

3. Effect of biochar on the bioavailability of organic contaminants in soil

As has been stated above, the amendment of biochar in soil changes the sorption capacity of organic contaminants, which will alter the existing state of the contaminant in soil, and hence have great effects on other environmental processes, such as leaching, volatilization, bioavailability and degradation. Zhang (2011) found that the amendment of biochar and nanotubes (CNT) to soils led to an increased amount of irreversible sorbed pyrene in soils and hence the ratio of irreversible sorbed amount to total sorbed amount increased. Compared with CNT amended soil, the sorption strength and irreversible sorbed amount of biochar amended soil were lower. In addition, biochar and CNT materials reduced the desorption percentage and rapid desorption fraction calculated using the first-order two-compartment model. Besides, biochar and CNT also decreased pyrene bioconcentration in earthworm. The effect of CNT amendment on earthworm BCF was greater than that of biochar, which can be ascribed to its higher sorption strength and irreversible sorbed amount. There was a significant correlation between earthworm BCF and rapid desorption fraction, which could be used as a predicting index for bioavailability of organic contaminants. Yu et al. (2009) investigated the effectiveness of biochar amendment in reducing the bioconcentration of two insecticides (chlorpyrifos and carbofuran) to spring onion. They found that biochar amendment to soil not only significantly decreased the loss of the two pesticides due to the sequestration in soils, but also markedly reduced the plant uptake of the insecticides.

The effect of biochar amendment on the bioavailability of organic contaminants in soil depends on several factors, such as biochar properties and the interaction between biochar and soil. Biochars produced at higher temperatures often have higher sorption affinity for organic contaminants, and were more effective in sequestering the contaminant and reducing its bioavailability than biochars produced at lower temperatures. The interaction between biochar and soil (discussed in the next section) could also influence the effect of biochar on the bioavailability of organic contaminants. Ren (2016) found that the freshly added biochar decreased the desorption rates and percentage of atrazine compared with soils without biochar. The desorption of atrazine was obviously suppressed in BC700 amended soil with low organic matter than those in the same soil amended with BC300. However, due to the aging effect between biochar and soil, the desorption percentage and rapid desorption fraction of atrazine from aged biochar amended soil increased compared with freshly biochar amended soil, indicating that the interaction between biochar and soil increase the bioavailability of atrazine in biochar amended soil.

4. Interaction between soil and biochar

In soils, biochar is likely to undergo a series of chemical and biochemical reactions as well as physical processes that will result in the alteration of its properties with time, such as pore structure, surface properties and aggregation potential, which in turn will affect its sorption affinity for contaminants. However, studies about the impact of the interaction between soil and biochar on biochar properties are rather scarce. To strengthen the study on the interaction between soil and biochar is critical to evaluate the environmental profit of biochar technology, and to guide its healthy development.

Ren (2016) systematically studied the mutual effects between different kinds of soils and biochars, and the impacts and mechanisms on environmental behaviors of two typical organic contaminants, atrazine and Phen. Biochars were produced from pig manure at different pyrolytic temperatures and added to different types of soil. The sorption capacity of BC700 was higher than that of BC300 for both contaminants. Since the biochar samples showed highly sorption affinities for atrazine and Phen, the two soils (black soil: BS; fluvo-auiqc soil: FS) freshly amended with biochars showed dramatically enhancement in the sorption of atrazine (highest by 24.4 times at C_e=0.01 S_w) and Phen (highest by 2.64 times at C_e=0.01 S_w) compared to the bare soils. However, after the biochar was applied to soil, the biochar and soil particles did not exist alone, but interacted with each other. The physicochemical properties of biochar amended soils changed with contact (aging) time, which in turn affected their sorption capacity. After being aged with BC300 for 60 d, the sorption of the two chemicals (atrazine and Phen) by the two soils both increased by different amounts (highest by 1.32 times for atrazine and 1.74 times for Phen at C_e=0.01 S_w), except the reduction in the sorption of Phen on BS; However, after being aged with BC700, the sorption of atrazine on the two soils was markedly decreased as compared to the fresh mixture, which was still higher by up to 1.49 and 5.12 times (at C_e=0.01 S_w) than those on BS and FS, respectively. Whereas, an opposite trend was observed for Phen on the two soils. After being aged with BC700, the sorption of Phen on the two soils increased by some extent as compared to the fresh mixture. These results suggested that besides the physicochemical properties of biochars applied to soil, properties of both soils and chemicals also exert great effects on the aging effects between soil and biochar and their impacts on the sorption of organic contaminants.

Plant roots extensively exist in soil, especially in farmland. Biochar constituents have been reported to be able to be absorbed by plant roots and stimulate root growth (Prendergast et al., 2014). Besides, the porous structure of biochar provides a suitable habitat for many microorganisms by protecting them from predation and desiccation and by providing many of their diverse carbon, energy and mineral nutrient needs (Thies and Rillig, 2009). It is well documented that plant roots release many organic exudates, including low molecular weight organic compounds, polymer degrading enzymes, saccharides and amino acids (Ling et al., 2009; Toyama et al., 2011; Lefevre et al., 2013; Tan et al., 2013). These exudates could not only enhance the activities of microorganisms but also could be adsorbed by biochars, which may affect the properties and sorption capacities of biochars. In order to examine the effects of plant roots on the physical and chemical properties of biochar and its sorption capacity, Ren et al. (2016) cultured biochar produced from pig manure at three different pyrolytic temperatures with wheat roots, and studied the sorption behaviors of atrazine and Phen on fresh biochars and wheat roots aged biochars that had been cropped with wheat for 90 d. The results showed that after being aged with wheat roots for three months, the biochar samples showed significant changes in their physicochemical properties, which depended on biochar types and their distances from the roots. In general, the ash contents of all the root aged biochars decreased, which could be attributed to the joint impact of leaching of nutrient in solution and absorption by plant root. This was especially true for BC300, suggesting that the ash components in biochars obtained at low temperature is more readily released. The elemental compositions of all of the three biochars changed after being aged with wheat roots. The marked changes in the elemental compositions of BC500 and BC700 after aging could be explained by the adsorption of plant root exudates or their metabolites, while the degradation of labile components resulted in the marked changes in BC300. The changes in the biochar properties in turn affected biochar sorption capacities. The sorption of atrazine and Phen by BC300 both increased by different extents after aging, while significantly decreased for BC700, and there were little changes for BC500. The change in atrazine adsorption was greater than the change in Phen adsorption. The complex change patterns were due to the different dominant sorption mechanisms for different biochars and different chemicals.

Clay minerals are the main constituents of soil and can also affect biochar properties. In order to investigate the interaction between biochar and clay minerals, Ren (2016) manually mixed biochars with three different minerals, montmorillonite, kaolin and soil-derived mineral obtained by calcining soil at 750℃ for 4 h. Clay minerals and biochars did not exist alone, but interacted with each other. The interaction between biochars and minerals could be attributed to pore blocking effect or surface adhesion, depending on the properties of minerals and biochars. The interaction of minerals and biochars not only affected their bulk and surface elemental compositions, but also changed their pore structure and surface areas. The aging effect of the same mineral on different biochars was different. After being aged with mineral for 30 d, the surface area of the mixture of BC700 and minerals significantly increased, while the surface area of the mixture of BC300 and minerals decreased compared with the fresh mixture. In addition, the co-existence of dissolved organic matter affected the interaction between biochar and mineral, especially for BC700 and montmorillonite. The surface area of the mixture of

BC700 and montmorillonite after being aged in DOM or root exudate solution increased more obviously than that in $CaCl_2$ solution. The interaction between biochars and minerals also affected the sorption behaviors of atrazine and Phen. The sorption of atrazine and Phen by the mixture of BC300 and mineral decreased by some extent after aging, while significantly increased for the mixture of BC700 and mineral. This could be due to the movement of clay fine particles into the inner moiety of the biochar.

Finally, Ren (2016) conducted a field experiment to investigate the changing in the properties of biochar under the field conditions. A commercial biochar was applied to an agricultural soil and the sorption of atrazine and Phen on biochar amended soils with different aging time ranging from 0 to 2 years were investigated. The fresh amendment of biochar into soil led to an obvious enhancement on the K_d of atrazine (by 3.71 times at C_e=0.01 S_w) and Phen (by 1.33 times at C_e=0.01 S_w) compared with the bare soils without biochar. The biochar and soil particles did not exist alone, but interacted with each other. Along with aging, the properties of biochar changed with surface area increasing first and then decreasing. The aging effects varied with the properties of biochar and soil, which could be ascribed to the degradation of labile moiety of biochar and the interactions between biochar and soil constituents, such as organic matter and inorganic minerals. Correspondingly, the sorption of atrazine and Phen on the biochar amended soils first increased and then markedly decreased with aging time (Fig. 1). Based on the changing trend of the K_d values with aging time, it could be predicted that the sorption capacity of biochar amended soils will decrease to the level of the bare soil after 2.5 years.

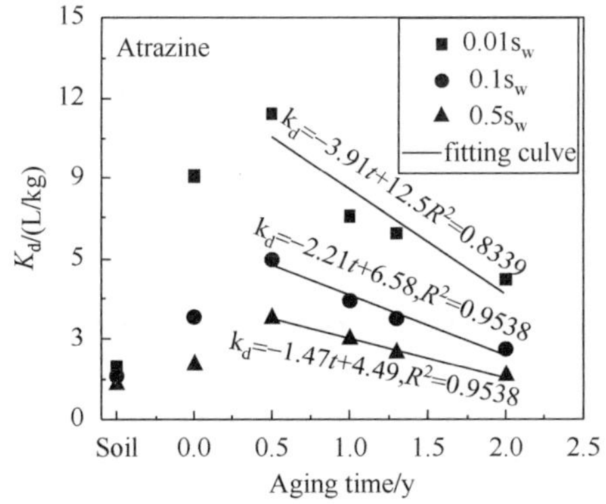

(a) Atrazine

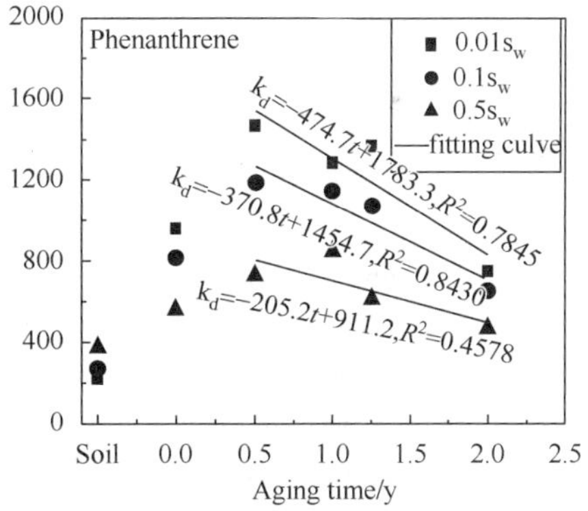

(b) Phenanthrene

Fig. 1 The changing trend of K_d values of atrazine and phenanthrene on biochar amended soils at different aging times

In all, due to its special properties, such as porous structure, larger surface area, highly aromaticity and higher organic matter content, biochar as soil amendment not only could decrease the bioavailability of organic contaminants, but also could proceed a series of geochemical reactions, which could alter its properties and sorption capacity for organic contaminants. There is no doubt that the effectiveness of biochar in sequestration of organic contaminants in soil will change with aging time. However, biochars produced from different biomass materials and at different pyrolytic conditions are likely to be highly heterogeneous in properties, and aging with soil may have different effects on their properties. Therefore, in order to evaluate the environmental profit of biochar technology, and to guide its healthy development, urgent attention of researchers are deserved to further investigate the effect of interaction between biochar and soil on environmental behaviors of organic contaminants in soil.

References

Beesley L, Moreno-Jimenez E, Gmoez-Eyles J, et al. 2011. A review of biochars' potential role in the remediation, revegetation and restoration of contaminated soils. Environmental Pollution, 159: 3269-3282.

Chen B L, Yuan M X. 2011. Enhanced sorption of polycyclic aromatic hydrocarbons by soil amended with biochar. Journal of Soils and Sediments, 11: 62-71.

Chen B L, Zhou D D, Zhu L Z. 2008. Transitional adsorption and partition of nonpolar and polar aromatic contaminants by biochars of pine needles with different pyrolytic temperatures. Environmental Science & Technology, 42: 5137-5143.

Dechene A, Rosendahl I, Laabs V, et al. 2014. Sorption of polar herbicides and herbicide metabolites by biochar-amended soil. Chemosphere, 109: 180-186.

Demirbas A. 2004. Effects of temperature and particle size on biochar yield from pyrolysis of agricultural residues. Journal of Analytical and Applied Pyrolysis, 72: 243-248.

Keiluweit M, Nico P S, Johnson M G, et al. 2010. Dynamic molecular structure of plant biomass-derived black carbon (biochar). Environmental Science & Technology, 44: 1247-1253.

Kookana R S. 2010. The role of biochar in modifying the environmental fate, bioavailability, and efficacy of pesticides in soils: A review. Australian Journal of Soil Research, 48: 627-637.

Lefevre G H, Hozalski R M, Novak P J, et al. 2013. Root exudate enhanced contaminant desorption: An abiotic contribution to the rhizosphere effect. Environmental Science & Technology, 47: 11545-11553.

Ling W T, Ren L L, Gao Y Z, et al. 2009. Impact of low-molecular-weight organic acids on the availability of phenanthrene and pyrene in soil. Soil Biology & Biochemistry, 41: 2187-2195.

Manya J J. 2012. Pyrolysis for biochar purposes: A review to establish current knowledge gaps and research needs. Environmental Science & Technology, 46: 7939-7954.

Marris E. 2006. Putting the carbon back: Black is the new green. Nature, 442: 624-626.

Prendergast-Miller M T, Duvall M, Sohi S P, et al. 2014. Biochar-root interactions are mediated by biochar nutrient content and impacts on soil nutrient availability. European Journal of Soil Science, 65: 173-185.

Ren X H, Zhang P, Zhao L J, et al. 2016. Sorption and degradation of carbaryl in soils amended with biochars: Influence of biochar type and content. Environmental Science and Pollution Research, 23: 2724-2734.

Ren X H, Sun H W, Wang F, et al. 2016. The changes in biochar properties and sorption capacities after beingcultured with wheat for 3 months. Chemsphere, 144: 2257-2263.

Ren X H. 2016. Interactions between biochar and soil and its impacts on sorption and desorption behaviors of atrazine and phenanthrene. Tianjin: Nankai University.

Spokas K A, Koskinen W C, Baker J M, et al. 2009. Impacts of woodchip biochar additions on greenhouse gas production and sorption/degradation of two herbicides in a Minnesota soil. Chemosphere, 77: 574-581.

Sun K, Keiluweit M, Kleber M, et al. 2011. Sorption of fluorinated herbicides to plant biomass-derived biochars as a function of molecular structure. Bioresource Technology, 102: 9897-9903.

Tan S, Yang C, Mei X, et al. 2013. The effect of organic acids from tomato root exudates on rhizosphere colonization of *Bacillus amyloliquefaciens* T-5. Applied Soil Ecology, 64: 15-22.

Toyama T, Furukawa T, Maeda N, et al. 2011. Accelerated biodegradation of pyrene and benzo[a]pyrene in the *Phragmites australis* rhizosphere by bacteria-root exudate interactions. Water Research, 45: 1629-1638.

Tsai W T, Liu S C, Chen H R, et al. 2012. Textural and chemical properties of swine-manure-derived biochar pertinent to its potential use as a soil amendment. Chemosphere, 89: 198-203.

Wang F. 2016. Sorption of polar and apolar organic contaminants in single and multiple systems onto biochars. Tianjin: Nankai University.

Yao Y, Gao B, Chen H, et al. 2012. Adsorption of sulfamethoxazole on biochar and its impact on reclaimed water irrigation. Journal of Hazardous Materials, 209: 408-413.

Yu X Y, Ying G G, Kookana R S. 2009. Reduced plant uptake of pesticides with biochar additions to soil. Chemosphere, 76: 665-671.

Zhang H H, Lin K D, Wang H L, et al. 2010. Effect of *Pinus radiata* derived biochars on soil sorption and desorption of phenanthrene. Environmental Pollution, 158: 2821-2825.

Zhang W, Wang L, Sun H W. 2011. Modifications of black carbons and their influence on pyrene sorption. Chemosphere, 85: 1306-1311.

Zhang W. 2011. The impacts of interactions between carbonaceous materials and soils on sorption and bioavailability of pyrene. Tianjin: Nankai University.

Zhao L, Cao X D, Masek O, et al. 2013. Heterogeneity of biochar properties as a function of feedstock sources and production temperatures. Journal of Hazardous Materials. 256: 1-9.

Zhou Z L, Wu W L, Li Y, et al, 2008. Sorption and desorption behaviors of three PAHs by charcoals. Journal of Agro-Environment Science (in Chinese), 27: 813-819.

Zhou Z L, Lu Y, Sun H W. 2010. Sorption kinetics and isotherms of phenanthrene in charcoals with different properties. Journal of Agro-Environment Science, 29: 476-480.

Advance in Health Risk Assessment Methodology of Brownfield Sites in China

Lin JIANG, Maosheng ZHONG, Xiaoyang JIA, Tianxiang XIA

Beijing Municipal Research Institute of Environmental Protection

National Engineering Research Center for Urban Environmental Pollution Control

Beijing Key Laboratory for Risk Modeling and Remediation of Contaminated Sites

At China's current urbanization rate, it is estimated that 350 million people will be added to its total urban population by 2025 (Woetzel et al., 2009). As cities continue to expand, many older industrial facilities along the edge of, or within, the city boundaries are being relocated or closed, leaving behind derelict, underused and abandoned land contaminated by the former industrial activities, which also are called brownfield sites. These sites can be valuable land for re-development, but require special intervention to bring them back into beneficial use. Since the beginning of this century, attentions have been being paid to the management of such sites, and the first guidance specified on human health risk assessment of brownfield sites in China was published in 2004 by Jiang (Jiang and Wang, 2004). In 2009, the first official technical guideline on investigation and health risk of brownfield sites was issued by Beijing Quality and Technical Supervision Bureau (BQTSB) (BQTSB, 2009), followed by the first official soil screening levels for health risk assessment in 2011 (BQTSB, 2011). The Ministry of Environmental Protection (MEP) started drafting technical guidelines on site investigation and health risk assessment in 2008, which were issued officially in 2014 (MEP, 2014), and the national wide soil screening values has been submitted for approval (MEP, 2016). Besides, some other developed cities, such as Shanghai, Chongqing, Zhejiang, Guangdong, e.g., have been engaged in drafting relevant technical guidelines and soil screening values (ZQTSB, 2013; GQTSB, 2014; SEPB, 2015). In addition to developing regulatory guidelines to support the management of brownfield sites, the methodologies recommended in issued guidelines have been being applied in assessing human health risk of numerous brownfield sites, and results have been used to help regulars make the final management decisions.

The recommended models in the issued guidelines were originated from the methods used in Unite State or other European countries several decades ago, which are all screening level models and the results obtained are always very conservative. The application of such models in practical work has resulted in such conclusions that the derived remediation targets of some carcinogenic contaminants were below the background values, or even below the detection limits of current analytical methods, which are meaningless to the decision makers. For example, the derived screening value or remediation target of arsenic in soil for residential land use scenario is about 0.3 mg/kg, which is an order of magnitude lower than the average background concentration reported for nationwide (Ma, 2010). The typical background concentration of benzo [a] pyrene in soils within the urban area of Beijing is reported to be 0.09 mg/kg (Liu et al., 2010), which is a little higher than the derived acceptable level using the models recommended in the guidelines. The acceptable level of benzene in soil calculated theoretically using the recommended models is about 0.02 mg/kg, even the specific soil physico-chemical properties in Beijing, as well as exposure and building features, are considered.

The problems associated with using the recommended risk assessment models that we are facing now have been confronted by the Americans or Europeans either. To figure out these problems, a tiered approach has been initiated abroad, and more site specific information is recommended to be incorporated in quantifying the health risk. For example, due to ageing effect of hydrophobic organic pollutants and metals in soil, the bioaccessbility or bioavailability of the target contaminants in a specific site can be measured and used to modify the risk assessment. In order to distinguish whether the high concentration of metals is caused by human beings or geochemical process, multi-evidence technology, which is consisted of site-specific background level derivation using geo-statistic assay, spacial distribution analysis and correlation analysis, has been developed. To cope with the conservativeness of vapor intrusion assessment model developed by Johnson and Ettinger in 1991, biodegradation module has been incorporated into the migration model. In addition, concentration of contaminant in soil vapor have been used as screening criteria for vapor inhalation risk assessment in most cities of Unite States and Canada.

The experience abroad can supply us with insights on how to cope with the challenges we are facing in implementing risk assessment on brownfield sites. Herein the advance we have made in China will be reviewed.

1. Incorporating bioaccessiblity in health risk assessment

Conventionally, “total” concentrations analyzed with exhaustive chemical extraction techniques (Reid, 2000; Harmsen, 2007) are employed in risk assessment models as exposure concentrations through oral ingestion, other than the amount that can be desorbed from soil matrix and reach the central blood circulation system where it can bring out adverse health effect (Oomen, 2006; Nathanail, 2007). However, recent researches indicate that hydrophobic organics and heavy metals can be sequestrated or entrapped in soil matrix, and only a fraction can be desorbed from soil matrix to solutions and be available for adsorption or can cross cell membrane of organisms (Semple et al., 2004), resulting in limited bioaccessibility or bioavailability (Tom, 2004). Therefore, concentrations measured by conventional exhaustive analysis assays proposed in regulatory and technical guidelines are too conservative, giving rise to overestimation of health risk and unnecessary cleanups (Reichenberg, 2007). Fortunately, incorporating bioaccessibility into exposure dose assessment is a strategy to cope with the problem, which is accepted by most researchers, regulators and general public.

Theoretically, measuring the concentration of contaminants that can reach the blood using *in-vivo* assay, which is also called bioavailable concentration, is a promising approach to quantify the health risk, but it is time consuming, expensive and unethical. The toxicological dynamic implies that the prerequisite that the contaminants in soil can reach blood circular system is that they can be desorbed from soil matrix in gastrointestinal system, therefore measuring the concentration of contaminants in human gastrointestinal solution using *in-vitro* method mimicking the contaminant dissolution in human gastrointestinal tract, which is called bioaccessible content, has been recommended by researchers as an alternative that can be used as exposure concentration in quantifying health risk through oral ingestion.

Since the proposition of the idea that bioaccessible concentration can be used as surrogate of bioavailable content, much effort have been made in developing *in-vitro* measurement assays, investigating correlation between results of *in-vitro* and *in-vivo* methods, developing predictive models of bioaccessiblity based on soil physicochemical properties as well as constructing models that can predict bioavailability based on bioaccessiblity information. For example, a physiologically based extraction test (PBET), which is a two-step extraction method that simulates the digestive process in the human gastric and intestinal tracts, was developed by Ruby (Ruby et al., 1996) and has been modified as a standardized method as a Pb bioaccessibility test in the stomach because it predicts the bioavailability of Pb very well (Ruby et al., 1996). Medline (1997) developed a simplified method by modifying PBET, which has since been used extensively (Poggio et al., 2009; Luo et al., 2011). In addition, IVG, SBRC, RIVM, DIN, SHIME and TIM have also been applied in metal bioaccessibility studies (Oomen et al., 2002; Marschner et al., 2006; Juhasz et al., 2009b). the average bioaccessibility in agricultural soil from Zhuzhou measured by the SBET method was reported as 16.9% in the stomach (Jining, 2013). The bioaccessibility of As in 17 soil samples from five provinces in China as determined by the PBET method ranged from 2.5% to 65.5% and 1.2% to 31.8% in the gastric and small intestinal phases, respectively (Chen, 2011). The bioaccessibility of As in soil, sediment and slag samples ranged from 1.2% to 27.3% in the stomach and 1.3 to 19.2% in the intestine as indicated by UBM (Wragg et al., 2011).

Jiang and his research group have made progress on bioaccessibility study either (Zhong et al., 2015, 2016), such as investigation on the correlation between bioaccessbility of toxic metals and soil physicochemical properties, development of predictive models of bioaccessibility, evaluation of the economic and environmental benefits originating from incorporating bioaccessiblity of PAHs to health risk assessment through a case study.

For example, Unified Bioaccessibility Model (UBM) was used to measure Arsenic (As) bioaccessibility of 13 soils with different physic-chemical properties and concentrations from

Hunan, Guangxi and Dalian, and the implication of incorporating As bioaccessibility into risk assessment was evaluated. The results revealed that: ① The bioaccessibility of As in stomach was between 3.9%-49.5% with the arithmetic mean being 19.6% while in intestine 1.2%-10.8% with the arithmetic mean being 6.7%, meaning that the bioaccessibility in stomach was 1.2-9.1 times that in intestine, as Fig. 1 shown. ② The most significant factor controlling the bioaccessibility of As in stomach was *w* (TAs) in soil (R^2=0.94, p<0.01, n=13), followed by *w* (TP) (R^2=0.82, p<0.01, n=13) and *w* (TMn) (R^2=0.79, p<0.01, n=13). In intestine, the most significant factor was also *w* (TAs) (R^2=0.83, p<0.01, n=13), followed by *w* (TP) (R^2=0.80, p<0.01, n=13)、As bioaccessible concentration in stomach (R^2=0.76, p<0.01, n=13)、pH (R^2=0.74, p<0.01, n=13)、*w* (TMn) (R^2=0.65, p<0.02, n=13) and *w* (TOM) (R^2=0.59, p<0.04, n=13). ③ The model regressed based on *w* (TAs) and *w* (clay) in soil was tested and able to predict As bioaccessibility in stomach with R^2 (correlation factor), ME (mean erro), RMSE (root mean square error), r_p^2 being 0.97, 0.02, 0.17 and 0.95, respectively. For As bioaccessibility in intestine, the model constructed based only on *w* (TAs) could be used to predict its bioaccessibility precisely, with R^2, ME, RMSE, r_p^2 being 0.90, 0.03, 0.26 and 0.80, respectively. ④ The health risk calculated based on *w* (TAs) was 2.0-15.0 times and 7.3-81.0 times the values when bioaccessibility in stomach and intestine were incorporated, meaning the conservativeness of the current assessment approach assuming the bioaccessibility of contaminants in soil to be 100%.

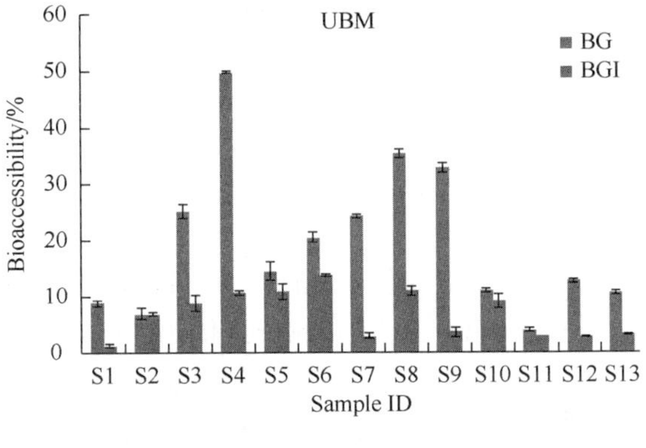

Fig. 1 Bioaccessibility of As measured using UBM

Cadmium (Cd) is also a toxic metal in the soil, and much efforts have been made to control Cd contamination and remediate Cd contaminated soil due to its easy accumulation in rice. Therefore, Unified Bioaccessibility Model (UBM) was used to measure cadmium (Cd) bioaccessibility of 12 soils with different physic-chemical properties and concentrations from Hunan, Guangxi and Dalian. The results revealed that the bioaccessibility of Cd in stomach was 12.24%-81.10% with the average value being 53.60% while in intestine it was 2.01%-43.30% with the average value being 19.74%, as shown in Fig. 2. The bioaccessible concentration in stomach correlated well with total Cd (TCd) (p<0.000, n=12) and total Mn (TMn) (p=0.04, n=12) in soils while in intestine the most significant controlling factor was TCd ($p<0.001$, n=12), followed by bioaccessible concentration in stomach ($p<0.001$, n=12) and TMn (p=0.05, n=12). A model regressed based on TCd and total phosphate (TP) can predict the bioaccessible concentration in stomach very well with R^2 being 0.992, and the bioaccessible concentration in intestine can be predicted precisely using a model regressed based on bioaccessible concentration in stomach and soil pH with R^2 being 0.999. When the bioaccessibility in stomach was considered Were the soil screening values for residential Were and industrial, 1.8 (calculated based on the average bioaccessibility) and 1.2 (calculated based on the maximal bioaccessibility) times the values without considering Cd bioaccessibility while the screening levels were 5.0 (calculated based on the average bioaccessibility) and 2.3 (calculated based on the maximal bioaccessibility) times the values when bioaccessibility in intestine was taken into account.

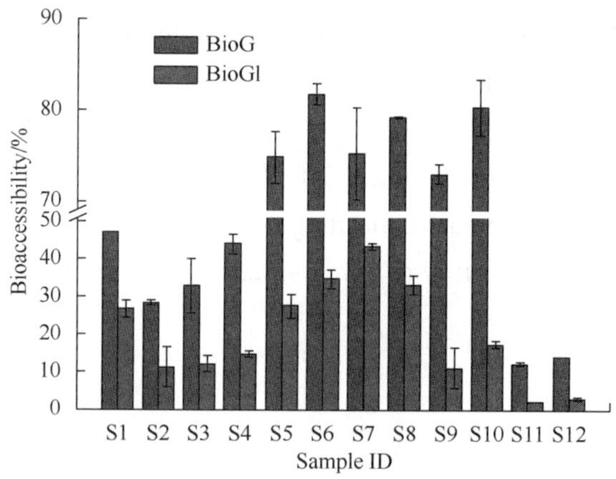

Fig. 2 Bioaccessibility of Cd measured using UBM

Polycyclic aromatic hydrocarbon are common pollutants in coke production plant, and ageing or encapsulation can contribute to the irreversible desorption due to the hydrophobicity. Mild chemical extraction assay was used to measure the bioaccessbility of benzo[a]pyrene (BAP5), indeno[1, 2, 3-cd]pyrene (ICP5) and dibenz[a, h]anthracene (DBA5) in soils from a coke production plant in Beijing, and economic as well as environmental benefits were evaluated when the bioaccessibility data were incorporated into cleanup levels derivation. The results regressed using dual phase desorption model revealed that the fast desorption fractions of BAP5, ICP5 and DBA5 were 0.23 to 0.32, 0.24 to 0.31 and 0.27 to 0.41, respectively, as shown in Fig. 3.

The site-specific cleanup levels derived with bioaccessibility incorporated into the conventional risk assessment model were about one time higher than the generic ones without addressing bioaccessibility, resulting in 16.2% reduction in remediation costs and CO_2 emission volume, as shown in Fig. 4.

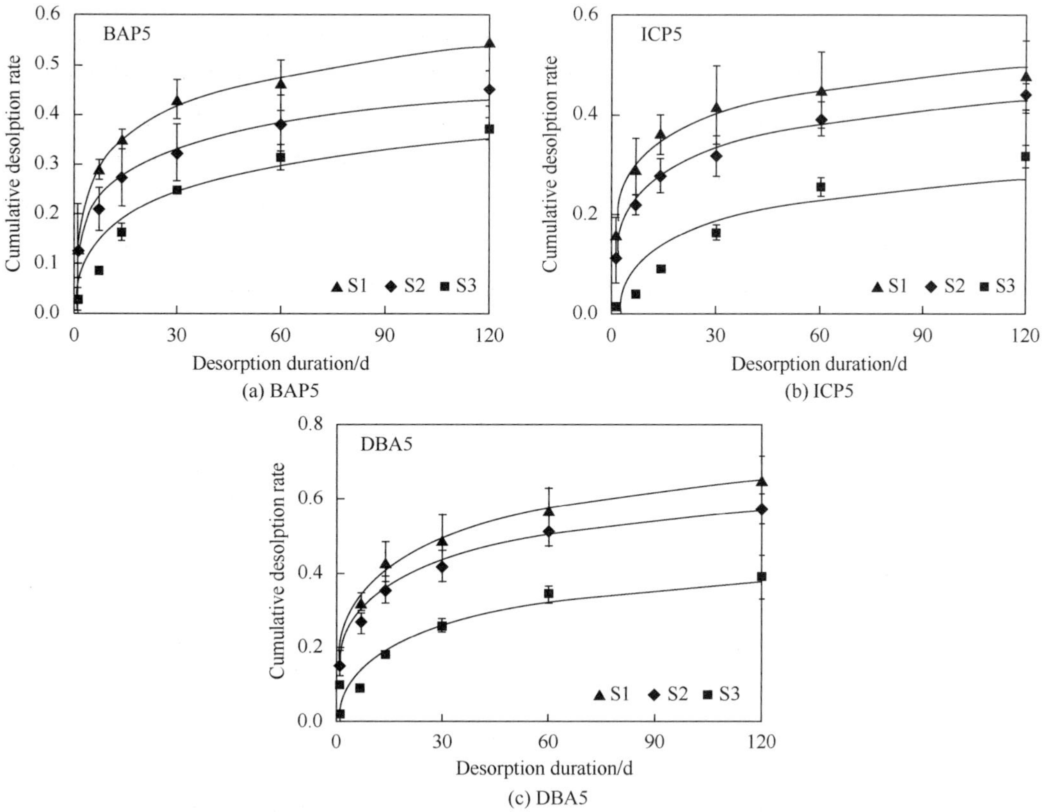

Fig. 3 Desorptiong curves of PAHs in soil

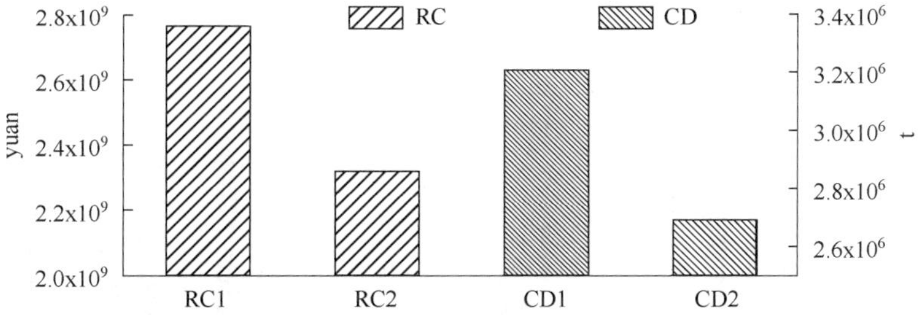

Fig. 4 Comparision of remediation cost and carbon emission

Note: RC1: remediation cost based on generic cleanup level; RC2: remediation cost based on cleanup level incorporating site-specific bioaccessibiltiy; CD1: carbon emission based on generic cleanup level; CD2: carbon emission based on cleanup level incorporating site-specific bioaccessibiltiy

In addition, efforts on developing a bioaccessbility measure assay that can characterize the dissolution of contaminants in gastrointestinal tract of Chinese people have been made. A UBM assay developed by the Bioaccessibility Research Group of Europe was modified based on the assumption that the components of a simulated gastric and intestinal solution listed in the Chinese Pharmacopoeia were representative of the digestive characteristics of Chinese people. The modified UBM was used to assess arsenic bioaccessibility in soils from Zhuzhou, Dalian and Hechi in China. The original UBM method was used for comparison. The results (Fig. 5) revealed that the bioaccessibility of arsenic in the gastric phase measured by the modified UBM (18.1% in average) was consistent with that measured by the unmodified method (19.7%), but in the intestinal phase, the results from the former (23.9%) were much higher than those from the latter (6.9%) due to the absence of bile, a high phosphate concentration and the presence of pepsin in the simulated digestive solution. The soil pH was the predominant factor controlling arsenic dissolution in gastric phase while the bioaccessibility of arsenic in intestine was determined by pH as well as composition in simulated intestinal extract.

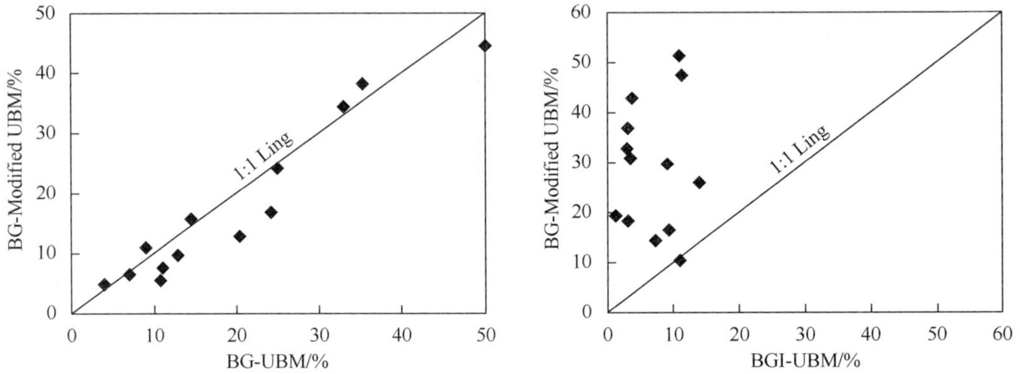

Fig. 5 Comparison of bioaccessibility measured by the unmodified and modified UBM

The addition of pepsin in the gastric phase could enhance the solubility of As, while the effect of trypsin on As dissolution in a simulated intestinal extract was negligible, as determined by the modified UBM method (Fig. 6).

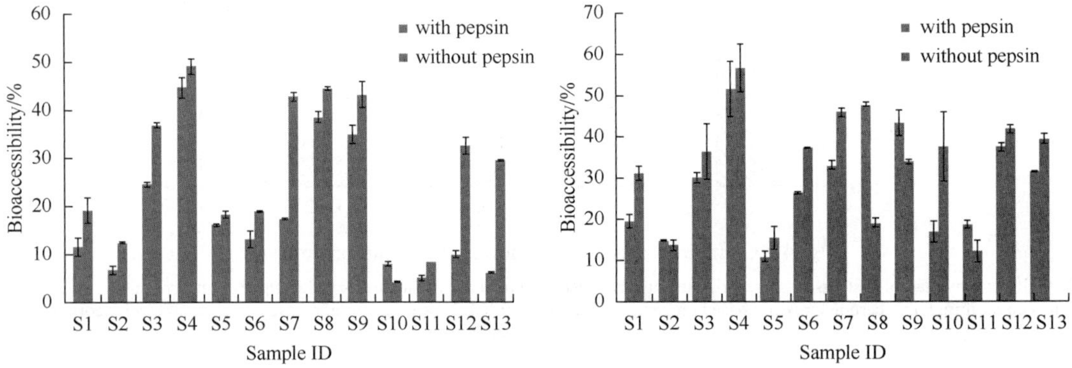

Fig. 6 Effects of pepsin and trypsin on bioaccessibility

2. Vapor intrusion assessment based on measured soil gas concentration

As stated above, inhalation risk from VOCs in contaminated soil is quantified based on the content of contaminants in soil using three-phase partition model in company with one-dimensional diffusion model and box mixing models in China right now. However, the feedback from the practitioners is that the derived remediation targets of carcinogenic pollutants are even below the corresponding reporting limits in laboratories, which are very conservative and can not supply the decision makers with scientific information.

One reason contributed to the above conservativeness is that the vapor concentration in contaminated area in current model is calculated using three-phase partition model theoretically, which can not characterize the real distribution of contaminants among soil particles, soil moisture and soil vapor due to irreversible desorption. For example, the measured concentration of contaminants in soil vapor are one to three orders magnitude lower than that calculated using three-phase partition model (Smith et al., 1990). A comparision study was implemented by Jiang (Jiang et al., 2012) in a coke and chemical plant in Beijing, and the results revealed that the concentration of benzene in soil gas analyzed on site was lower than the one calculated theoretically by ASTM model by one order (Fig. 7 (a)), and benzene decreased sharply within the depth of three meters (Fig. 7 (b)).

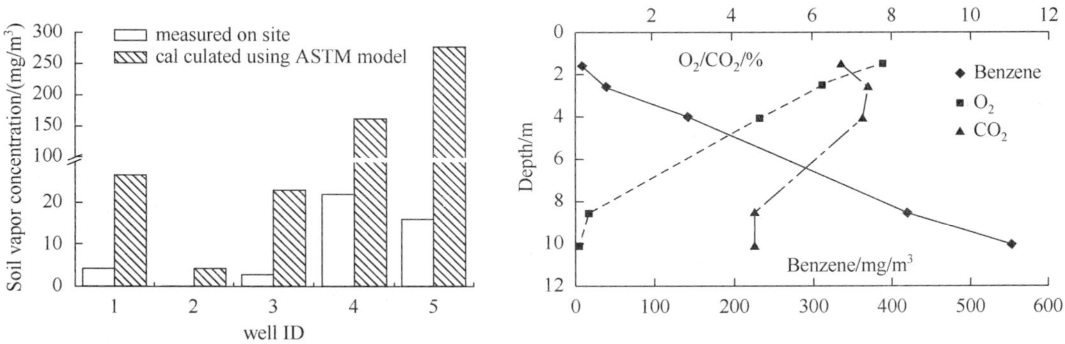

(a) Comparision between measured and calculated concentration

(b) Vapor concentration profile

Fig. 7 Comparision of measured and calculated vapor concentration and distribution profile

Based on the above discovery as well as advance in vapor intrusion assessment abroad, efforts have been made to derive soil vapor screening levels for inhalation risk assessment based on soil physicochemical properties and other specific parameters of China (Zhong et al., 2013). For example, soil gas screening values for benzene, toluene, chloroform and 1, 1-dichloroethylene were derived using the J&E model according to typical vapor intrusion scenarios in China, and comparison with values issued by US EPA and its member states was made as well. The results revealed that the screening values of carcinogenic benzene and chloroform are three to four orders of magnitude lower than the values of non-carcinogenic toluene and 1, 1-dichloroethylene. For the shallow soil gas, the screening values for benzene, chloroform, toluene and 1, 1-dichloroethylene were 9.6×10^2, 2.7×10^2, 1.1×10^7 and 4.0×10^5 μg/m³ under the residential scenario, while were 4.6×10^3, 1.3×10^3, 6.3×10^7 and 2.4×10^6 μg/m³ for the corresponding pollutant under the industrial/commercial scenario. For the deep soil gas, the corresponding values were 1.1×10^3, 3.1×10^2, 1.2×10^7 and 4.5×10^5 μg/m³ according to residential scenario, while were 4.6×10^3, 1.3×10^3, 6.3×10^7 and 2.4×10^6 μg/m^3 based on the industrial/commercial scenario. The maximum allowable indoor air concentrations, soil gas attenuation factors and building parameters, are the main factors in determining the values. Difference between screening values of shallow and deep soil gas is negligible. However, it is significant between the values issued by US EPA and its member states. For shallow soil gas, the difference for benzene, chloroform, toluene and 1, 1-dichloroethylene ranges between 3-309, 0.5-245, 34-2895 and 27-96 000 times, respectively. In addition, for the deep soil gas, the difference ranges between 0.7-35, 13-28, 4.6-323 and 4.5-10 800 times, respectively. Different methods applied to derivate the maximum allowable indoor air concentrations and attenuation factors were the main reasons contributing to this difference. The corresponding average attenuation factors for the shallow and deep soil gas were 2.3×10^{-4} and 2.0×10^{-4}, indicating the difference between them is not significant, but both were lower than the corresponding empiric values, 0.1 for the shallow soil gas and 0.01 for the deep soil gas, applied by US EPA by tow-three orders of magnitude. The attenuation effect contributed by clean soil laid above contaminate source is not significant if adsorption and biodegradation were not considered. Besides, soil vapor screening levels have been derived for Beijing based on the its specific characteristics of soil, building and exposure behavior and have been issued officially in 2016 (BQTSB, 2016).

In addition to derive soil vapor screening values, efforts have also been made on the study of the effect of biodegradation on soil vapor screening levels because more and more researches and field measurements indicate that the concentration of benzene will decrease sharply when it migrate through an oxygen-rich layer (Zhong et al., 2014). For example, effects of biodegradation on attenuation factor ($a_{\text{i-s}}$) and screening level of benzene in soil vapor was studied using the one dimensional model Bio-vapor for sandy soil, and in particular critical influential factors of biodegradation (source intensity c_s, vertical separation distance between building foundation and source LT, depth of aerobic zone in soil L_a, biodegradation rate k_w) were simulated and discussed either. The results (Fig. 8 (a)) indicate that contribution from biodegradation of benzene to $a_{\text{i-s}}$ is negligible when c_s is higher than 5×10^5 mg/m^3. When c_s is no more than 1×10^4 mg/m^3, $a_{\text{i-s}}$ is 1-2 orders of magnitude reduction when the aerobic biodegradation process is included compared to the non-biodegradation case, but the reduction is insensitive to the change of c_s and LT. When c_s falls in between, $a_{\text{i-s}}$ decreases by 2 orders of magnitude as LT increases by 1 order of magnitude (Fig. 8 (b)), and LT is a sensitive parameter. L_a is a critical factor in determining the biodegradation effect, and bio-attenuation decreases by 2 orders of magnitude as L_a just increases from 0.5 m to 1.5 m. Compared with field measured results for L_a, which are always deeper than 1.5 m, the modeled result for sandy soil in the paper is 0.63 m, indicating that Bio-vapor is conservative in predicting the depth of aerobic zone. Therefore determination of L_a through measuring the soil vapor profile in practical project is recommended. Bioattenuation is more sensitive to k_w for light contaminated sites, and $a_{\text{i-s}}$

decreases by 2 orders of magnitude as k_w increases from 0.033 to 2/h when c_s is no more than 5×10^4 mg/m^3. The vapor screening value for benzene is 1-2 orders of magnitude lower when biodegradation is considered for the same conceptual model scenario with no biodegradation.

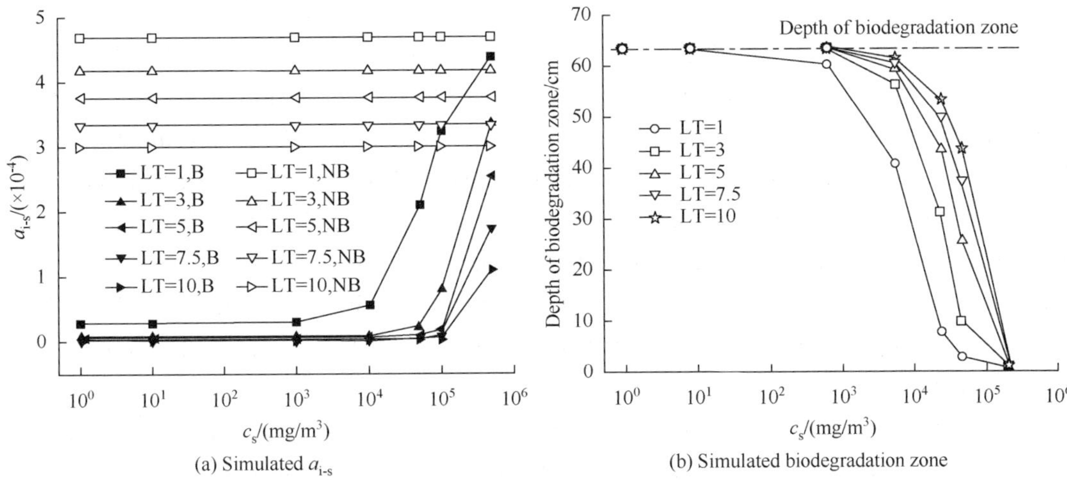

Fig. 8 Impact on a_{i-s} (attenuation factor) by biodegradation at different c_i (source concentration) and LT (source depth)

3. Application of probabilistic analysis in risk assessment

Currently, only a determined value was assigned to each variable used in quantifying the health risk, which is called deterministic risk assessment and very easy for practitioners. However, the value of each variable covers a wide range in reality. For example, soil ingestion rate of child is reported to range from 10 to 1000 mg/d (Day et al., 1975). For safety reason, a very conservative value, usually the upper 95% value, is assigned to each parameter in deterministic risk assessment, contributing to the conservative results obtained using the current method. Besides, information on sensitivity of each parameter in determining the final result cannot be supplied by the current method, resulting in that more efforts cannot be made to investigate more reasonable values for the most sensitive parameters. To cope with the problem, probabilistic risk assessment has been initiated and used a lot in American and European countries for decision making, and technical guideline on how to carry out probabilistic health risk assessment of contaminated site has been issued by US EPA (US EPA, 2001).

Although deterministic risk assessment is the recommended method in the guidelines issued in China, efforts have been made to investigate the implementation of probabilistic risk assessment in contaminated sites management by Jiang (Jiang et al., 2013). For example, based on the previous contamination survey results of a coking plant site in Beijing, the effects of uncertainty of 15 exposure parameters and soil pollutant concentrations on human health risk assessment of benzene and benzo (a) pyrene were investigated with the method of probabilistic risk assessment (PRA). Results showed that the risk values of PRA, which located at the 95th

percentile in the probabilistic range of risk, were lower than that of deterministic risk assessment (DRA) both in surface and subsurface soil of benzene and benzo [a] pyrene, whether single or total exposure pathways. The total risk calculated by PRA ranged from 1.5×10^{-8} to 6.9×10^{-3} (benzene) and 2.3×10^{-9} to 2.2×10^{-3} (benzo [a] pyrene), and the values located at the 95th percentile in the probabilistic range of risk were 3.8×10^{-4} and 1.1×10^{-4}, respectively. The total risk calculated by DRA located at the 96.8^{th} and 99.1^{th} percentiles in the probabilistic range of risk, which were 1.5 and 3.2 times that of PRA, it indicated that DRA overestimated the real risk. The results of parameter sensitivity analysis revealed that the concentration of benzene in subsurface soil $C_{s\text{-}sub}$ (94.63%) and adult exposure duration ED_a (4.12%) contributed the most to the uncertainty in benzene risk assessment, and the concentration of benzo (a) pyrene in surface soil $C_{s\text{-}sur}$ (92.3%), adult exposure duration ED_a (2.40%), soil ingestion rate $IR_{s\text{-}c}$ (2.12%) and child exposure duration ED_c (1.21%) contributed the most to the uncertainty in benzo [a] pyrene risk assessment.

In addition to the above research, a probabilistic risk assessment (PRA) was used to characterize the variability and uncertainty of 17 parameters related to human exposure and 5 related to soil physicochemical properties in deriving the soil cleanup levels of 8 pollutants in a coking plant (Jia et al., 2014). Results showed that the ratios of PRA clean-up level to DRA clean-up level were 1.11-2.49, except for Nap in subsoil (0.9), therefore, the cleanup levels derived by using deterministic risk assessment (DRA) method were in general more conservative than these by using PRA in this case. If contaminants in soil were reduced to the PRA clean-up levels, all the exposure risks would be acceptable but at different levels, and Ben in topsoil was more likely to cause harm to human health than others. The sensitivity analysis revealed that adult exposure duration (ED_a) and child exposure duration (ED_c) both contributed most to the uncertainty (sensitive ratio: 35%-59.8% for ED_a and 6.2%-20.2% for ED_c) in the derivation of all the clean-up levels. Soil physic-chemical parameters, such as soil organic carbon content (f_{oc}), volumetric air content in soil ($\theta_{air,\ vad}$) and volumetric water content in soil ($\theta_{water,\ vad}$), were more sensitive to volatile Ben and Nap and less to semi-volatile BaA-DBA; contrastingly, child soil ingestion rate ($IR_{s\text{-}c}$) were more sensitive to BaA-DBA and less to Ben and Nap.

Application of probabilistic risk assessment (PRA) and deterministic risk assessment (DRA) at a coking plant site was compared by Xia (Xia et al., 2013). By DRA, Hazard Quotient (HQ) following exposure to Naphthalene (Nap) and Incremental Life Cancer Risk (ILCR) following exposure to Benzo [a] pyrene (Bap) were 1.87 and 2.12×10^{-4}. PRA revealed valuable information regarding the possible distribution of risk, and the output showed that Nap HQ ranged from 6.18×10^{-7} to 6.62 and Bap ILCR ranged from 8.67×10^{-10} to 6.89×10^{-4} (Fig. 9).

Risk estimates of DRA located at the 99.59th and 99.76th percentiles in the risk outputs of PRA, indicating that DRA overestimated the risk. Cleanup levels corresponding acceptable HQ level of 1 and ILCR level of 10^{-6} were also calculated for both DRA and PRA. Nap and Bap cleanup levels were 192.85 mg/kg and 0.14 mg/kg by DRA, which would result in only 0.25%

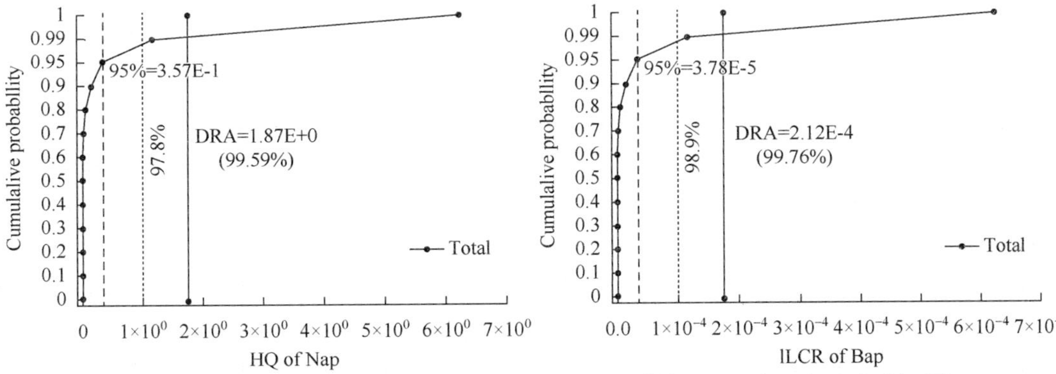

(a) Hazard index of Nap

(b) Incremental carcinogenic risk of Bap

Fig. 9 HQ and ILCR calculated by PRA and DRA approach

and 0.06% of the exposed population to have a risk higher than the acceptable risk, according to the outputs of PRA. The application of PRA on cleanup levels derivation will lift the cleanup levels 1.9 times for Nap and 2.4 times for Bap than which derived by DRA. For this coking plant site, the remediation scale and cost will be reduced in a large portion if the method of PRA is used. Sensitivity analysis was done by calculating a contribution to variance for each exposure parameters ands it was found contaminant concentration in the soil (C_s), exposure duration (ED), total hours per day spent outdoor (ET_{out}), soil ingestion rate (IR_s), the air breathing rate (IR_a) and body weight (BW) were the most important parameters for risk and cleanup levels calculations.

4. Indentifying heavy metal contamination using multi-evidence technology

According to the current assessment assay recommended in issued guidelines, further investigation have to be carried out if concentration of certain heavy metals exceed the health risk screening values. However, site specific background values of heavy metals have not been taken in account for the derived screening values right now, which may result in un-correct judgment whether the high concentration of certain metal is caused by humans or geochemical process. To cope with the problem, a multi-evidence technology, which takes into account the site-specific background value of target metal, the spacial distribution of the metal, the correlation of metal distribution and production facility layout and the correlation of metal distribution and soil physicochemical properties, has been recommended for indentifying whether the high concentration is caused anthropogenically or by geochemical process.

For example, a multiple lines of evidence analysis technology was applied to assess whether soil was contaminated by heavy metals (HMs) in a chemical production site (site 1) and a coke production site (site 2) (Jiang et al., 2013). Firstly, the baseline upper limit concentration for As and Cr^{6+} were derived for the two investigated sites based on the analysis of the relative cumulative probability plots of the two metals and their spatial distribution in the soil. The results indicated that the baseline upper limit concentration for As and Cr^{6+} in site 1 is 29.8 mg/kg, 76.1 mg/kg, respectively, which are much higher than the ones reported for the local area by others. But in site 2, the baseline upper limit concentration for As and Cr^{6+} is 10.6 mg/kg, 33 mg/kg, respectively, which is a little higher than the reported values. Taken the derived baseline concentrations as the assessment criteria, both sites are contaminated by the heavy metals to some degree, which is consistent with the site historical activities review results and element correlation analysis results. In site 1, As concentration in 3.8% samples exceeds the derived baseline concentration, while Cr^{6+} concentration in 6.0% samples exceeds. In site 2, only the concentration of As in 5.2% samples exceeds the derived baseline value. All the above exceeding rates are much lower than the ones based on reported baseline values, 77.7% and 96.7% for As and Cr^{6+} in site 1, respectively and 41.9% for As in site 2. The difference means that for a specific site, it should be keen avoided that the baseline concentrations for heavy metals reported in literatures are applied directly as the criteria to assess whether the site is contaminated or not, which may cause the results unreal and lots of rare resource are allocated to remediate soil that maybe not contaminated.

In addition, probabilistic analysis, also called geostatistic analysis, of Arsenic concentration data in soil samples from 4 brownfield sites within the sub-urban area of Beijing (Chaoyang and Haidian Districts) was implemented with the aim of better understanding the site-specific natural range of variation of Arsenic concentrations in soil and, accordingly, differentiating local background values from"anomalous data" (Marina, 2012). Furthermore, the relationship between site-specific data and the reported wide-scale background, as defined from previous studies for the entire Beijing area, was discussed as well as the different implications on the definition of threshold level concentrations (TLCs), metal soil contamination screening and site assessment studies.

Similar shape of probability plots was observed for concentration data of As in soil from the considered sites. Baseline populations with near normal or low skewness log-normal distributions, encompassing the 88% to 94% of the sampled datasets, were recognized in all cases (Fig. 10). The remaining data (12% to 6% of total) were treated as outliers and associated to either natural subsoil enrichments or possible anthropogenic contribution. The site-specific median values for baseline populations (6 to 8.8 mg/kg) are quite in accordance with the arithmetic or geometric mean of the wide area distribution. Based on the correct understanding of site-specific data variability, an Upper Baseline Concentration (UBC) limit ($+2\sigma$ level), including most of the site-specific background variability, has been selected as significant threshold for site assessment studies in order to discriminate potential contamination. This can avoid a great overestimation (almost one half of the data in this case) of the percentage of "exceeding values". Following appraisal of site-specific Arsenic data variability, UBC values comprised in the range 10.4-12.6 mg/kg were computed for the Chaoyang South District area, where the greatest number of data was available.

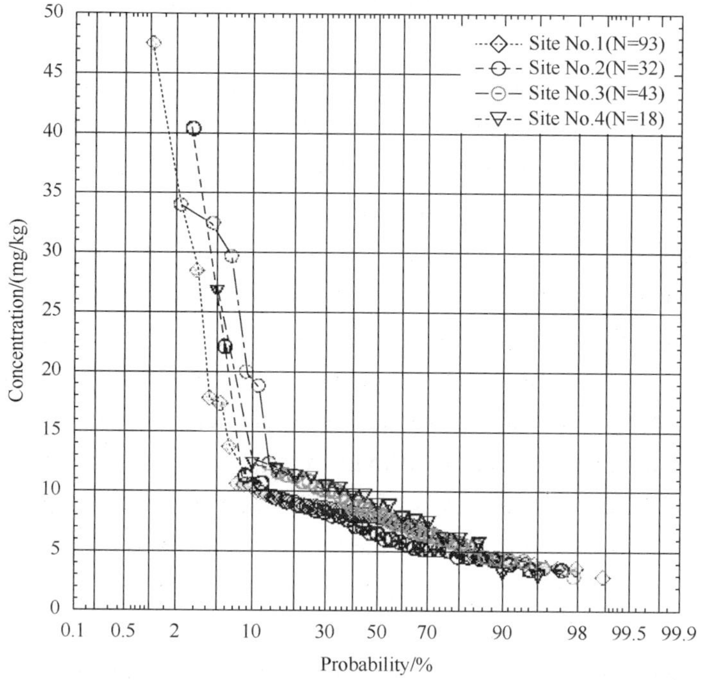

Fig. 10 Normal probability plots for Arsenic in subsoil from the different investigated sites

5. Using tiered approach in assessing groundwater contamination risk

In addition to human health, groundwater contamination attributed to pollutant leaching from vadose soil is also an important issue that should be considered in decision making process. Due to the complexity of pollutant migration in vadose soil and aquifer, the investigation is always expensive and time consuming. Therefore, a tiered approach has been proposed based on the desorption and migration of contaminants from source area to the point of exposure downstream.

The application of the proposed assay was detailed through a case study in which the impact of backfilling treated contaminated soil on groundwater should be assessed (Zhong et al., 2013). The soil was contaminated by 1, 2-dicholorenthane and other 9 organic pollutants and would be remediated before backfilling to meet the pre-set remediation goals based on health risk assessment. The results from tiered Ⅰ assessment indicate that the concentration of 8 contaminants in the leachate within the backfilling soil layer will exceed the assessment standards possible leading to groundwater contamination. However, the results from tiered Ⅱ assessment, in which the adsorption and retardation of vadose zone soil is taken into account and the concentration of pollutants reaching the groundwater table is predicated, reveal that only the concentration of 6 contaminants will exceed the assessment standards. Further, given consideration to the dilution and mixing of the groundwater, tiered Ⅲ assessment was adopted and the results reveal that only 4 contaminants are beyond the standards. Finally, tiered Ⅳ assessment, aiming at predicting the concentration at the target well downstream, was carried out by considering the retardation of contaminants in saturated layer, and the results indicate only 1 pollutant is above the assessment standard. Therefore, it can be seen that the predicted concentration of the target pollutants in advanced assessment levels will be more close to the one at the target drinking water well and the amount of contaminants whose initially-set remediation goals need to be modified will decrease correspondingly, meaning the reduction in pollution prevention cost, although more efforts should be paid and more field data should be collected to implement the advance assessment level.

Besides, generic soil screening values for 27 kinds of VOCs (volatile organic contaminants), 31 kinds of SVOCs (semi-volatile organic contaminants), 11 kinds of pesticides, PCBs and dioxin were derived for groundwater protection using the tiered approach according to different hydro-geological conditions in Beijing (Zhong et al., 2014). The results revealed that the screening values for the downstream areas of the Yongding river alluvial-proluvial fan were the most conservative, followed by the ones for the upstream areas and the ones for the areas in the middle. The screening values, derived using three-phase equilibrium equation combined with groundwater dilution model (method 1), for the areas in the middle were 1.1-1.4 times and 9.9-34.9 times higher than the ones for the upstream and downstream areas, respectively, while the values for the upstream areas were 10.7-24.9 times higher than the ones for the downstream areas. The values for the areas in the middle were 9.8-44.9 times higher than the ones for the downstream areas using SESOIL model combined with groundwater dilution model (method 2). And for the areas in the middle and downstream where continuous aquitard exist, the screening values derived with method 2 were more conservative than the ones derived by method 1. For PAHs, PCBs, Dioxin, most pesticides and esters, whose K_{oc} (organ-water partition coefficients) were high, the results derived by method 2 indicated they would hardly penetrate the clean soil in the vadose zone, meaning less risk of groundwater contamination. For VOCs and phenols, whose K_{oc} were low, the values derived with method 2 were 4.3-18.4 times higher than the ones derived with method 1 for the areas in the middle, while 3.0-24.6 times higher for the downstream areas. Given for the conservative nature of the generic screening levels and its risk screening function, the conservative values among screening levels derived for the upstream areas with method 1 and for the areas in the middle and downstream with methods 2 were recommended as the generic screening values of Beijing to prevent groundwater contamination from soil pollution. Based on the above research, a guideline on assessing groundwater contamination from soil leaching has been issued in Beijing (BQTS, 2016).

6. Conclusions

Since the beginning of this century, contaminated site management has become an important issue in regular environmental management, and risk management strategy has become the core in decision making. Although many problems have been confronted in implementing risk assessment, much efforts have been being made to figure them out, and more research should be carried out to bring the above mentioned scientific insights into the practical work to guide the practitioners how to carry out higher tiered health risk assessment. In addition, the relevant technical guidelines should be drafted as soon as possible.

References

Beijing Quality and Technical Supervision Bureau. 2009. Environmental site assessment guideline (DB11/T 656-2009), Beijing: Beijing Quality and Technical Supervision Bureau.

Beijing Quality and Technical Supervision Bureau. 2011. Screening levels for soil environmental risk assessment of sites (DB11/T 811-2011), Beijing: Beijing Quality and Technical Supervision Bureau.

Beijing Quality and Technical Supervision Bureau. 2016. Technical guideline for investigation and risk assessment of volatile organic compounds in contaminated sites (DB11/T 1280-2015), Beijing.

Beijing Quality and Technical Supervision Bureau. 2016. Environmental assessment guideline for remediated soil reuse from contaminated site (DB11/T 1281-2015), Beijing: Beijing Quality and Technical Supervision Bureau.

Day J P, Hart M, Robinson M S. 1975. Lead in urban street dust.

EPA U S. 2001. Risk assessment guidance for superfund: Volume 3—Process for conducting probabilistic risk assessment chapterl, Part A. Washington, IX.

Guangdong Quality and Technical Supervision Bureau (GQTSB). 2014. Risk screening values for soil heavy metal-The pearl river delta area, Guangdong: Guangdong Quality and Technical Supervision Bureau

Harmsen J. 2007. Measuring bioavailability: From a scientific approach to standard methods. Journal of Environmental Quality, 36 (5): 1420-1428.

Jiang L, Wang Y. 2004. Guidance on site investigation and risk assessment, Beijing: China Environmental Science Press.

Jiang L, Zhong M S, Xia T X, et al.. 2012. Health risk assessment based on benzene concentration detected in soil gas. Research of Environmental Science, 25 (6): 717-723.

Jiang L, Jia X Y, Xia T X, et al. 2013. Research on application of PRA in health risk assessment of soil in a coking plant site, Research of Environmental Science, 26 (2): 220-226.

Jiang L, Zhong M S, Zhu X Y, et al. 2013. Application of multiple lines of evidence analysis technology in the assessment of sites contaminated by heavy metals. Environmental Science, 34 (9): 3641-3647.

Jia X Y, Xia T X, Jiang L, et al. 2014. Application of PRA in deriving soil cleanup level for a coking plant site, China Environmental Science, 34 (1): 2220-2227.

Juhasz A L, Weber J, Smith E, et al. 2009. Assessment of four commonly employed in vitro arsenic bioaccessibility assays for predicting in vivo relative arsenic bioavailability in contaminated soils. Environmental Science & Technology, 43 (24): 9487-9494.

Liu S, Xia X, Yang L, et al. 2010. Polycyclic aromatic hydrocarbons in urban soils of different land uses in Beijing, China: Distribution, sources and their correlation with the city's urbanization history. Journal of Hazardous Materials, 177 (1): 1085-1092.

Li J N, Hou H, Wei Y, et al. 2013. Bioaccessibility and health risk assessment of heavy metals in agricultural soil from Zhuzhou, China. Research of Environmental Sciences, 26: 139-1146.

Medlin E A. 1997. An in vitro method for estimating the relative bioavailability of lead in humans.

Marschner B, Welge P, Hack A, et al. 2006. Comparison of soil Pb in vitro bioaccessibility and in vivo bioavailability with Pb pools from a sequential soil extraction. Environmental Science & Technology, 40 (8): 2812-2818.

Ma W L. 2010. The distribution characteristic of polycyclic aromatic hydrocarbons in soil and air in China and simulation on large scale. Dissertation for the doctoral degree of Engineering, Harbin Institute of Technology.

Ministry of Environmental Protection. 2014. Technical guidelines for environmental site investigation (HJ25.1-2014), Beijing: Ministry of Environmental Protection.

Ministry of Environmental Protection. 2014. Technical guidelines for environmental site monitoring (HJ25.2-2014), Beijing: Ministry of Environmental Protection.

Ministry of Environmental Protection. 2014. Technical guidelines for risk assessment of contaminated site (HJ25.3-2014), Beijing: Ministry of Environmental Protection.

Marina A, Jiang L, Eugenio N, et al. 2015. Arsenic probability distributions in soil from brownfield sites in Beijing (China): baseline evaluation and related implications for site assessment studies, Frontiers of Environmental Science and Engineering, 9 (3): 465-474.

Ministry of Environmental Protection. 2016. Risk screening guideline values for soil contamination of development land (under review), Beijing: Ministry of Environmental Protection.

Nathanail C P, Smith R. 2007. Incorporating bioaccessibility in detailed quantitative human health risk assessments. Journal of Environmental Science and Health Part A, 42 (9): 1193-1202.

Oomen A G, Hack A, Minekus M, et al. 2002. Comparison of five in vitro digestion models to study the bioaccessibility of soil contaminants. Environmental Science & Technology, 36 (15): 3326-3334.

Oomen A G, Brandon E F A, Swartjes F A, et al. 2006. How can information on oral bioavailability improve human health risk assessment for lead-contaminated soils? Implementation and Scientific Basis.

Poggio L, Vrščaj B, Schulin R, et al. 2009. Metals pollution and human bioaccessibility of topsoils in Grugliasco (Italy). Environmental Pollution, 157 (2): 680-689.

Ruby M V, Davis A, Schoof R, et al. 1996. Estimation of lead and arsenic bioavailability using a physiologically based extraction test. Environmental Science & Technology, 30 (2): 422-430.

Reid B J, Stokes J D, Jones K C, et al. 2000. Nonexhaustive cyclodextrin-based extraction technique for the evaluation of PAH bioavailability. Environmental Science & Technology, 34 (15): 3174-3179.

Reichenberg F, Karlson U G, Gustafsson Ö, et al. 2010. Low accessibility and chemical activity of PAHs restrict bioremediation and risk of exposure in a manufactured gas plant soil. Environmental Pollution, 158 (5): 1214-1220.

Smith J A, Chiou C T, Kammer J A, et al. 1990. Effect of soil moisture on the sorption of trichloroethene vapor to vadose-zone soil at Picatinny Arsenal, New Jersey. Environmental Science & Technology, 24 (5): 676-683.

Semple K T, Doick K J, Jones K C, et al. 2004. Peer reviewed: Defining bioavailability and bioaccessibility of contaminated soil and sediment is complicated. Environmental Science & Technology, 38 (12): 228A-231A.

Shanghai Environmental Protection Bureau (SEPB). 2015. Health risk screening level of contaminated soil, Shanghai: Shanghai Environmental Protection Bureau.

Shanghai Environmental Protection Bureau (SEPB). 2015. Technical guidelines for risk assessment of contaminated sites, Shanghai: Shanghai Environmental Protection Bureau.

Wiele V, Richard T, Verstraete W, et al. 2004. Polycyclic aromatic hydrocarbon release from a soil matrix in the in vitro gastrointestinal tract. Journal of Environmental Quality, 33 (4): 1343-1353.

Woetzel J, Mendonca L, Devan J, et al. 2009. Preparing for China's urban billion. McKinsey Global Institute.

Wragg J, Cave M, Basta N, et al. 2011. An inter-laboratory trial of the unified BARGE bioaccessibility method for arsenic, cadmium and lead in soil. Science of the Total Environment, 409 (19): 4016-4030.

Xia T, Jiang L, Jia X, et al. 2014. Application of probabilistic risk assessment at a coking plant site contaminated by polycyclic aromatic hydrocarbons. Frontiers of Environmental Science & Engineering, 8 (3): 441-450.

Yu S, Li X. 2011. Distribution, availability, and sources of trace metals in different particle size fractions of urban soils in Hong Kong: Implications for assessing the risk to human health. Environmental Pollution, 159 (5): 1317-1326.

Zhong M S, Jiang L, Jia X Y, et al. 2013. Deriving soil gas screening value for vapor intrusion scenario, Research of Environmental Science, 26 (9): 979-988.

Zhejiang Quality and Technical Supervision Bureau (ZQTSB). 2013. Guideline for risk assessment of contaminated site, Zhejiang: Zhejiang Quality and Technical Supervision Bureau.

Zhong M S, Jiang L, Yao J J, et al. 2013. Application of tiered approach to assess the impact of backfilling remediated soil on groundwater. Environmental Science, 34 (3): 907-913.

Zhong M S, Jiang L, Yao J J, et al. 2014. Impact of biodegradation on attenuation factor and screening level of benzene in soil vapor, Research of Environmental Science, 27 (2): 178-185.

Zhong M S, Jiang L, Jia X Y, et al. 2014. Derivation of soil screening values for groundwater protection of Beijing, Research of Environmental Science, 27 (8): 80-88.

Zhong M S, Peng C, Jiang L, et al. 2015. Factors controlling As bioaccessibility in aged soils and corresponding health risk, Research of Environmental Science, 28 (2): 267-274.

Zhong MS, Peng C, Jiang L, et al. 2015. Factors controlling Cd bioaccessibility in aged soils and its implication on soil screening values, China Environmental Science, 35 (7): 2217-2224.

Zhong MS, Jiang L, Han D, et al. 2016. Bioaccessibility of As in soils measured using a modified UBM assay based on the gastrointestinal characteristics of Chinese people, Environmental Pollution, under review.

Derivation of Soil Generic Assessment Criteria under the Consideration of Background Exposure and NAPL

Mengfang CHEN*

Key Laboratory of Soil Environment and Pollution Remediation, Institute of Soil Science, Chinese Academy of Sciences

*Corresponding author: mfchen@issas.ac.cn; Tel.: +86 25 86881847

Abstract

Chinese risk assessment guideline for contaminated sites known as C-RAG has been published in 2014 to assess human health risks for contaminated sites and technical algorithm taken is similar to those used in Contaminated Land Exposure Assessment (CLEA) and Risk Based Corrective Action (RBCA) models respectively from the UK and USA. In addition, to comply with C-RAG guidance, Health and Environmental Risk Assessment (HERA) was developed by the Institute of Soil Science, Chinese Academy of Sciences. Common to these guidance and models are the adaptation of a multi-tiered framework and the requirement for developing risk based or generic assessment criteria (GAC) in the earlier tier to allow a generic risk screening to be undertaken prior to a more detailed quantitative risk assessment (DQRA).

Widely used RBCA model employ a simple integration procedure by only partially combining exposure pathways from surface soil with vapour pathways from subsurface soil being excluded in the combination. The minimum from the GAC derived from combined exposure from surface soil and that from vapour pathways from subsurface soil is taken as the integrated Generic Assessment Criteria (GAC) for the purpose of risk screening. However HERA and CLEA models employ an integrated approach to derive GAC by combining multiple exposure pathways. The simplified approach can approximate the integrated GAC only when there is a dominant exposure pathway such as oral ingestion or indoor vapour. But all these models are based on a simple partitioning of a chemical in soil between the sorbed, dissolved and vapour phases without consideration of the presence of Non-Aqueous Phase Liquid (NAPL), and critically fail to consider non-soil background exposure for non-carcinogenic compounds. As a result, the GAC derived may not be considered protective of human health.

This paper reviews numeric integration procedures for the derivation of the GAC under CLEA land use exposure scenarios that consider non-soil background exposure while limiting the GAC for vapour pathways by soil saturation limits. Significance of consideration of soil saturation limits and non-soil background exposure for the derivation of the GAC using an integrated approach is illustrated for a range of organic and inorganic compounds having varied levels of background exposure.

The integration procedure reviewed in this paper is readily adaptable to CLEA and HERA risk assessment models for the derivation of integrated soil GAC.

Keywords: Contaminated Land Exposure Assessment (CLEA), Generic Assessment Criteria (GAC), Average Daily Exposure (ADE), Hazard Index (HI); Soil Saturation Limit (SAL), Tolerable Daily Intakes (TDI), Tolerable Daily Soil Intake (TDSI), Median Daily Intake (MDI).

1. Introduction

Chinese risk assessment guideline for contaminated sites known as C-RAG (MEP, 2014) has been published in 2014 to assess human health risks for contaminated sites and technical algorithm taken is similar to those used in Contaminated Land Exposure Assessment (CLEA) and Risk Based Corrective Action (RBCA) models respectively from the UK and USA (ASTM, 2000; EA, 2009a). In addition, to comply with C-RAG guidance, Health and Environmental Risk Assessment (HERA) was developed by the Institute of Soil Science, Chinese Academy of Sciences and has been widely accepted by environmental scientists and engineer in Chinese contaminated land industry (Chen, 2012). Common to these guidance are the requirement for developing risk based or generic assessment criteria (GAC) in the earlier tier known as generic quantitative risk assessment (GQRA) to allow a generic risk screening to be undertaken prior to a more detailed quantitative risk assessment (DQRA). However, fundamental problems associated with the C-RAG is the exclusion of groundwater exposure pathways in the conceptual exposure model and most significantly the lack of basic research on Chinese specific land use scenarios, soil properties and building parameters.

Calculating the soil GAC requires an integration procedure for combining relevant exposure pathways (Chen, 2010a, 2010b). Internationally accepted RBCA model employs a simple integration procedure by only partially combining exposure pathways from surface soil including soil ingestion, dermal contact and inhalation of dusts and vapours (ASTM, 2000) and cannot be used directly without modifications. The minimum from the GAC derived from combined exposure from surface soil and that from vapour pathways from subsurface soil is taken as the integrated GAC for the purpose of risk screening. The simplified approach can approximate the integrated GAC only when there is a dominant exposure pathway. However, the Hazard Index (HI) calculated from the GAC derived from the simplified approach can often exceed the threshold limit of one, reaching two when there are two equally important exposure pathways. All these models assume a simple partitioning of a chemical in soil between the sorbed, dissolved and vapour phases without consideration of the presence of Non-Aqueous Phase Liquid (NAPL), and critically fail to consider non-soil background exposure for non-carcinogenic compounds (ASTM, 2000; MEP, 2014; EA, 2009a). As a result, the GAC derived under simple integration procedure and without considering background exposure may not be considered sufficiently protective of human health.

The CLEA (Contaminated Land Exposure Assessment) model (EA, 2009a) developed by the Environment Agency in England and Wales is the one of few models which considers background exposure but under the assumption of the linear chemical partitioning. The integration procedures similar to those implemented by CLEA and proposed by Chen (2010a, 2010b) are considered robust as all the applicable exposure pathways are combined during the integration process. These procedures work well if the integrated soil GAC falls below the Soil Saturation Limit (SAL) above which NAPL may be absent in near-surface soil. In case of the GAC exceeding the SAL, the risks from vapour pathways using the linear partitioning approach are likely to be over-estimated particularly when they are significant contributors to the total exposure.

This paper reviews numeric integration procedures for the derivation of the GAC that consider non-soil background exposure while limiting the Average Daily Exposure (ADE) for vapour pathways calculated from a soil concentration at the SAL (Chen, 2010a). Comparison with the simplified integration approach is made and significance of consideration of soil saturation limits and non-soil background exposure for the derivation of the GAC using an integrated approach is illustrated for a range of organic and heavy metal contaminants having varied background levels (Chen, 2010b). The integration procedure presented in this paper can be readily adaptable to HERA and CLEA risk assessment models for the derivation of integrated soil GAC.

2. Theory of exposure assessment

2.1 Estimating the average daily exposure for soil

The methodology for estimating human exposure to soil contamination is generally expressed as Equation 1 (Chen, 2010a, 2010b):

$$\text{ADE}_S = \sum_{j=1}^{t}\sum_{n=1}^{r}\frac{IR_j^n \times EF_j^n \times ED_j^n}{AT_j^n \times BW_j} = C_s \times \sum_{j=1}^{t}\sum_{o=1}^{m} R_o^j + C_s \times \sum_{j=1}^{t}\sum_{pv=1}^{n} R_{pv}^j \quad (1)$$

Where, C_s is the representative soil concentration (mg/kg); *IR*, the chemical intake rate (mg/d); *EF*, the exposure frequency (days/year); *ED*, the exposure duration (years); *AT*, the averaging time (days); *BW*, the body weight (kg); $r=o+pv$, a number of all the exposure pathways, depending on the selection of generic land uses; *j*, a number of age classes to be modeled; *o*, number of oral and dermal pathways; and *pv*, number of vapour and particulate pathways.

For the derivation of the integrated GAC, the ADE is conveniently expressed using unit intake rates: the $\sum R_o$ and $\sum R_{pv}$ terms (μg/kg/d over mg/kg soil) defined as the sum of the ratio of ADE/GAC respectively for oral and dermal, and inhalation exposure pathway.

Except for direct contact pathways, each pathway has analytical fate and transport models to calculate the chemical intake rates (*IR*) or unit intake rates which are proportional to the media transfer factors such as the soil to vapour volatilisation factor, soil to particulate emission factors, and soil to plant concentration factors (ASTM, 2000; GSI, 2008; EA, 2009a). These media transfer factors, combined with generic exposure factors for each standard land use, are then used to estimate the soil ADEs.

Table 1 Summary of Land Uses and Exposure Pathways within the CLEA

Exposure Pathways	Residential with Garden	Residential without Garden	Allotment	Commercial
Direct soil and dust ingestion (Combined)	√	√	√	√
Consumption of site grown vegetables and fruits[a]	√	×	√	×
Indirect soil ingestion attached to vegetable	√	×	√	×
Dermal contact with soil derived dust (Indoor)	√	√	√	√
Dermal Contact with Soil (Outdoor)	√	√	√	√
Inhalation of Soil Derived Dust (Indoor) [b]	√	√	×	√
Inhalation of Soil Derived Dust (Outdoor) [b]	√	√	√	√
Inhalation of Soil Vapour (Indoor) [c]	√	√	×	√
Inhalation of Soil vapour (Outdoor) [d]	√	√	√	√

Note: √, included; ×, excluded; a) multiple algorithms used to calculate soil to plant or fruit concentrations for organic and inorganic compounds (EA, 2009a); b) Particulate emission factors calculated using the USEPA approach with the UK experimental data for air dispersion factor (Q/C) (USEAP, 1996); c) Johnson and Ettinger model (1991) used to calculate soil to indoor volatilization factor; d) ASTM approach for surface soil (ASTM, 2000) used to calculate soil to ambient volatilization factor with UK specific Q/C value. Readers should consult the CLEA Guidance Report on technical algorithms for calculating media transfer factors.

The CLEA model defines four UK standard land uses with a maximum of 10 exposure pathways (Table 1). Plant and fruit uptake pathways are only included in the residential with garden and allotment land uses, and indoor dust and vapour inhalation pathways are excluded in the allotment land use. Generic exposure factors for UK standard land uses are summarized in Table 2. Within the CLEA model, the critical receptors are children up to 6 years old (equivalent to the first 6 age classes) for residential and allotment land uses and working female adults between 16 and 65 years old (age class 17) for commercial land use (Table 2). Default parameters relating to soil, air and building properties used to calculate the medial transfer factors for plant and fruit uptakes and vapour pathways are listed in Table 3.

Table 2 Summary of Generic Exposure Factors within the CLEA

Standards Land Uses	Age Class	Age /years	Body Weight /kg	Body Weight/cm	Exposure Frequency (days/year)	Exposure Duration /year	Averaging Time /days	Childhood Factor (-)		Soil Ingestion Rate /(g/d)	Exposed Skin Surface Area /cm^2	Occupancy Time/hour		Inhalation Rate /(m^3/d)
								Oral	Inhalation			Indoor	Outdoor	
Residential and Allotment	1	0 to 1	5.6	70	180[a] (25) 365[b] (25)	1	2190	0.53	0.51	0.1	366	23	1 (3)	8.5 (10.3)
	2	1 to 2	9.8	80	365 (130)	1		0.66	0.8	0.1	533	23	1 (3)	13.3 (18.8)
	3	2 to 3	12.7	90	365 (130)	1		0.65	0.77	0.1	620	23	1 (3)	12.7 (20.7)
	4	3 to 4	15.1	90	365 (130)	1		0.65	0.74	0.1	742	23	1 (3)	12.2 (19.1)
	5	4 to 5	16.9	100	365 (65)	1		0.74	0.74	0.1	822	19	1 (3)	12.2 (21.3)
	6	5 to 6	19.1	110	365 (65)	1		0.74	0.74	0.1	873	19	1 (3)	12.2 (24.9)
Commercial	17	16 to 65	70	160	170[c] 230[d]	49	17885	1	1	0.05	476	8.3	0.7	14.8

Note: a) Exposure frequencies are for soil ingestion, plant uptake and dermal pathways for residential land use, and plant uptake pathways for allotment land use; b) for Inhalation pathways; c) for outdoor activities; d) for indoor activities including soil ingestion. Values in brackets are for allotment land uses where different from residential land use. Readers should consult CLEA Guidance Report (EA, 2009a) for exposure factors related to consumption of home-grown vegetables and fruits.

Table 3 Generic Soil, Air and Building Parameters within the CLEA Model

a) Building

Land Use	Footprint Area	Living Space Height	Air Exchange Rate	Differential Pressure	Foundation Thickness	Floor Crack Area	Building-Source Separation Thickness	Soil Gas Ingress Rate[2]
	m^2	m	$hour^{-1}$	Pa	m	m^2	m	cm^3/s
Residential[1]	28	4.8	0.5	3.1	0.15	0.04	0.5	25
Commercial[1]	424	9.6	1	4.4	0.15	0.165	0.5	150

Note: 1) default building types are two-storey small terrace house and three-storey pre-1970 office respectively for residential and commercial land uses; 2) Default soil gas ingress rates into the buildings are used to derive Soil Guideline Values within the CLEA model.

b) Soil

Soil Type	Bulk Density	Porosity		Saturated Hydraulic Conductivity	Effective Air Permeability	Fraction of Organic Carbon[1]
	g/cm^3	Air	Water	cm/s	cm^2	—
		cm^3 cm^{-3}				
Sandy Loam	1.21	0.2	0.33	3.56E-03	3.05E-08	0.0348

Note: 1) F_{oc} is based on Soil Organic Matter of 6%.

c) Air

Land Use	Air Dispersion Coefficient (Q/C)[1]	Fraction of Vegetative and Buildings	Transport Factor[3]	Soil Loading Factor[3]
	$g/m^2.s/kg/m^3$	-	g/g	$\mu g/m^3$
Residential	2400	0.75	0.5	50
Allotment	120	0.5	-	-
Commercial	120 (68)[2]	0.8	0.5	100

Note: 1) Q/C default values were taken from Newcastle representing areas of 0.01, 0.5 and 2 ha respectively for residential, allotment and commercial land uses; 2) the value in bracket taken at children's height is used to calculate particulate emission factor for inhalation of indoor dust; 3) parameters used to calculate intake rate for inhalation of indoor dust.

2.2 Estimating the average daily exposure for non-soil background

A Mean Daily Intake (MDI) from non soil background exposure (typically derived from drinking water, air and dietary products) is used to describe adult background intake (EA, 2009b). The MDI is expressed for adults in mg/day or μg/day, and the ADE associated with the MDI for children is allocated using a Childhood Correction Factor (CF) and calculated cumulatively across the age groups using Equation 2 and 3 respectively for oral and inhalation pathways:

$$\mathrm{ADE}^{o}_{\mathrm{MDI}} = MDI^{o} \times \sum_{j=1}^{t} \frac{EF \times ED \times CF^{o}_{j}}{AT \times BW_{j}} \tag{2}$$

$$\mathrm{ADE}^{i}_{\mathrm{MDI}} = MDI^{i} \times \sum_{j=1}^{t} \frac{EF \times ED \times CF^{i}_{j}}{AT \times BW_{j}} \tag{3}$$

Where, $\mathrm{ADE}^{o}_{\mathrm{MDI}}$ and $\mathrm{ADE}^{i}_{\mathrm{MDI}}$ are oral and inhalation background exposure respectively (mg/kg/d); MDI^{o} and MDI^{i}, oral and inhalation MDI (mg/d); CF, Childhood Factor (-).

The CF is the percentage number assigned to receptors and is one for adults as the MDI is derived for adults (Table 2). Exposure frequencies for background exposure occurs daily (i. e 365 days per year) and differ from those associated with generic land uses.

2.3 Toxicological assessment

In the UK, the Health Criteria Value (HCV) is a broad term to describe toxicological parameters. The HCV is a Tolerable Daily Soil Intake (TDSI) for threshold contaminants (i.e non-carcinogen), or an Index Dose (ID) for non-threshold contaminants (i.e carcinogen) representing a minimum acceptable risk level. The TDI is considered as a threshold value below which no appreciable health risk exists. The TDSI is derived from the Tolerable Daily Intake (TDI) taking into account a Mean Daily Intake (MDI) from non soil background exposure.

The TDSI and the ID are expressed in the same units typically the amount of chemical per kilogram body weight (BW) per day (mg/kg BW/d or μg/kg BW/d). The only difference for the derivation of the GAC from a TDI or ID is that the background exposure from non-soil sources is not considered for non-threshold contaminants as is the case for carcinogenic contaminants.

For the purpose of comparison, the oral TDI used for threshold contaminants in the UK corresponds with a Reference Dose (RfD) used in the Integrated Risk Information System (IRIS) database for non-carcinogenic contaminants (USEPA, 2009). The Reference Concentration (RfC) is not directly used in the UK, but rather it is linked to and can be converted to inhalation TDI in mg/kg/d as an input to the CLEA model (Chen, 2009). Toxicological parameters used to describe carcinogenic compounds in the US such as the oral cancer Slope Factor (SF) and inhalation Unit Risk Factor (URF) are not used in the UK. For non-threshold contaminants, setting the oral and inhalation ID (mg/kg/d) requires the user to define a minimum acceptable target risk level in accordance with the principle of As Low As Reasonably Practicable (ALARP). Consequently the CLEA model does not calculate the Incremental Life Cancer Risks (ILCR) from a given soil concentration, but the integrated GAC derived corresponds to the user defined target risk level usually at 1E-5 from which the ID is based (USEPA, 2009).

2.4 Quantification of risks

The ADE_S is then compared with the Health Criteria Values (HCV) which are deemed to be protective of human health. This process is often referred to as a forward calculation as it quantifies the potential risks defined as the sum of the calculated ADE_S based on the soil GAC over the HCV for each exposure route (i. e oral, dermal or inhalation), known as the Hazard Index (HI) from a given soil concentration. The HI less than one indicates that no adverse health effects would occur, whereas the HI greater than one indicates a potential risk that warrants a further investigation.

2.5 Derivation of the Integrated GAC

The process of deriving a soil GAC is known as a reverse calculation. The basic principle is that the HI must be equal to unity for the risk for the contamination to be acceptable.

Under risk based frameworks such as the ASTM Risk Based Corrective Action (RBCA) guidance (ASTM, 2000) and the CLEA guidance (EA, 2009a) it is also generally assumed the TDI or ID for dermal exposure are equal to that for oral exposure due to the lack of primary studies on dermal toxicity. To derive the integrated soil GAC, Equation 4 must be solved.

$$\frac{\mathrm{GAC_{int}} \times \sum_{j=1}^{t}\sum_{o=1}^{m} R_o^j}{TDI^o - \mathrm{ADE_{MDI}^o}} + \frac{\mathrm{GAC_{int}} \times \sum_{j=1}^{t}\sum_{pv=1}^{n} R_{pv}^j}{TDI^i - \mathrm{ADE_{MDI}^i}} = 1 \tag{4}$$

Where $\mathrm{GAC_{int}} \times \sum_{j=1}^{t}\sum_{o=1}^{m} R_o^j = \mathrm{ADE}_S^o$ and $\mathrm{GAC_{int}} \times \sum_{j=1}^{t}\sum_{pv=1}^{n} R_{pv}^j = \mathrm{ADE}_S^i$ are cumulative soil ADE for combined oral and dermal pathways, and combined inhalation pathways respectively across the age groups modeled.

Combined exposure (ADE_{S+MDI}) from soil and non-soil sources is expressed in Equation 5 so that the percentage contributions from soil or background can be calculated given soil concentration at GAC_{int}:

$$\mathrm{ADE}_{S+MDI} = \mathrm{ADE}^{o}_{\mathrm{MDI}} + \mathrm{ADE}^{i}_{\mathrm{MDI}} + \mathrm{GAC}_{\mathrm{int}} \times \left(\sum_{j=1}^{t}\sum_{o=1}^{m} R_{o}^{j} + \sum_{j=1}^{t}\sum_{pv=1}^{n} R_{pv}^{j} \right) \quad (5)$$

3. Critical issues for the derivation of GAC

Although the solution to Equation 4 is simple, three critical factors can complicate the processes for the derivation of the GAC: integration methods, high background exposure and presence of the NAPL.

3.1 Simplified Integration Method

Risk assessment models commonly used in contaminated sites employ a simple integration procedure by only partially combining exposure pathways from surface soils with inhalation of indoor and outdoor vapour pathways from subsurface soils being excluded (*5*). Exposure pathways from surface soil to be combined include soil ingestion, dermal contact, inhalation of particulates and vapours. The GAC is then derived from the minimum of the combined surface soil GAC and the GAC from vapour pathways. The simplified approach can approximate the integrated GAC only when there is a dominant exposure pathway (i.e. indoor vapour for volatile contaminant) (Fig. 1). However, the Hazard Index calculated from the GAC derived from the simplified approach can exceed the threshold limit of one and can be as high as 2 if both the combined and vapour pathways are equally important (Fig. 1).

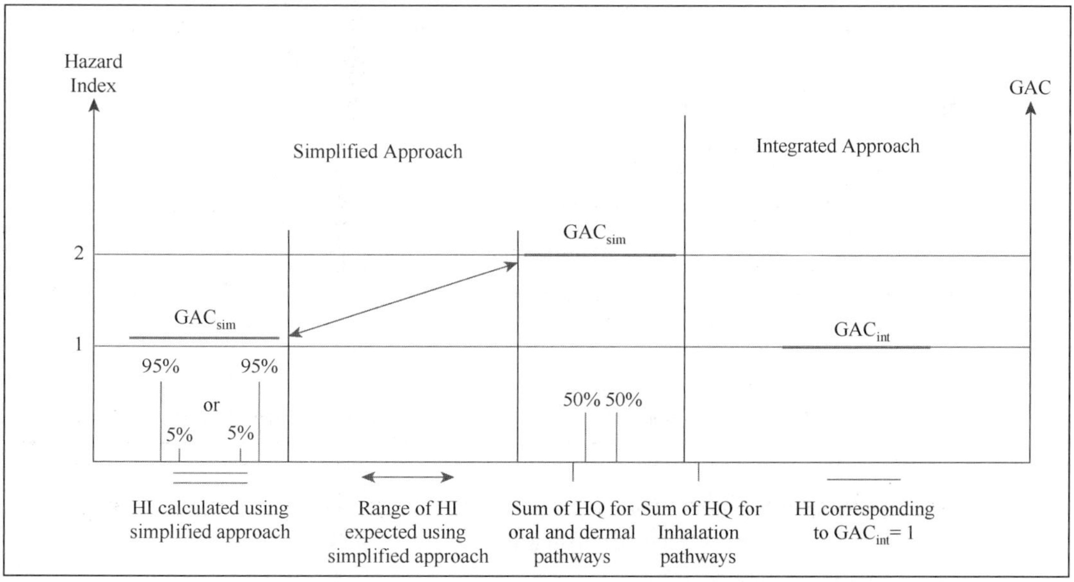

Fig. 1 Comparison of the Simplified and Integrated Approaches for the Derivation of the GAC

3.2 Limiting Background Exposure

Combining exposure pathways is a complex process particularly if it is undertaken under the 50% Rule on background exposure. Where the ADE^{o}_{MDI} or ADE^{i}_{MDI} is smaller than the respective TDI, the solution to Equation 4 is quite straightforward. However problems arise when the ADE^{o}_{MDI} or ADE^{i}_{MDI} is greater than the respective TDI. The difference between the TDI and the ADE_{MDI}, termed as the TDSI, then becomes very small and can even be negative (Fig. 2 (a)), and the resultant GAC would be unrealistically low or it is impossible to solve for soil GAC in the case of negative TDSI (USEPA, 1996). Under these circumstances, for example, the CLEA model applies the so called the "50% Rule" on background exposure.

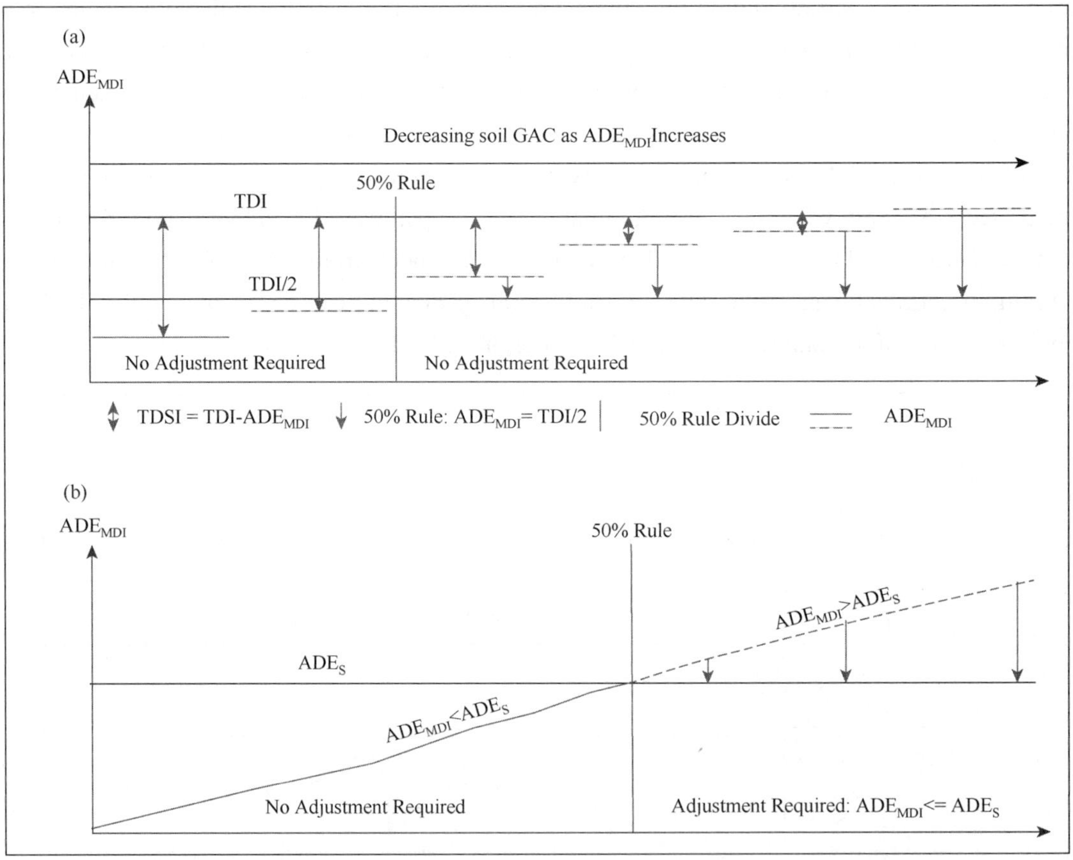

Fig. 2 Effects of the 50% Rule on Background Exposure for (a) Individual and (b) Combined Exposure Pathways

Under the 50% Rule the combined oral and inhalation ADE_{MDI} is limited to a maximum of

50% of the total ADE, to ensure that a minimum of 50% of the total exposure is allocated to soil (EA, 1996, 2009a). Therefore Equation 6 must be satisfied under the 50% Rule (Fig. 2 (b)):

$$\frac{\mathrm{ADE}_{\mathrm{MDI}}}{\mathrm{ADE}_{S}} \leqslant 1 \text{ or } \mathrm{ADE}_{\mathrm{MDI}}^{o} + \mathrm{ADE}_{\mathrm{MDI}}^{i} \leqslant \mathrm{GAC}_{\mathrm{int}} \times \sum_{j=1}^{t}\sum_{o=1}^{m} R_{o}^{j} + \mathrm{GAC}_{\mathrm{int}} \times \sum_{j=1}^{t}\sum_{pv=1}^{n} R_{pv}^{j} \tag{6}$$

When the combined background exposure exceeds the soil ADE, the background exposure will have to be adjusted downward to meet the requirement of the 50% Rule (Equation 6).

3.3 Presence of NAPL

Soil saturation limits for a pure organic compound are usually based on aqueous solubility and saturated vapour concentration and are described in Equations 7 and 8 respectively.

$$C_{\mathrm{satw}} = \frac{S}{\rho} \times \left(\theta_{w} + K_{oc} f_{oc} \rho + H\theta_{a}\right) \tag{7}$$

$$C_{\mathrm{satv}} = \frac{iP_{v}M}{RT} \times \frac{\left(\theta_{w} + K_{oc} f_{oc} \rho + H\theta_{a}\right)}{\rho H} \tag{8}$$

Where C_{satw} is soil saturation limit (mg/kg) based on aqueous solubility; C_{satv}, soil saturation limit (mg/kg) based on saturated vapour concentration; S is the aqueous solubility (mg/L); θ_{w}, volumetric water content in soil pore water (-); θ_{a}, volumetric air content in soil vapour (-); ρ, soil bulk density (g/cm^3); K_{oc}, distribution coefficient (l/kg); f_{oc}, fraction of organic carbon in soil (-); H, Henry Law Constant (-); P_v, the saturated vapour pressure at ambient temperature and pressure (Pa); M, the molecular weight (g/mol); R, the molar gas constant (Pa m^3/mol/K); T, the ambient temperature (K); i, the molar fraction of a compound in the NAPL mixture.

Soil saturation limits are useful indicators for the presence of NAPL (ASTM, 2000; EA, 2009a). Whether the integrated GAC exceeds the SAL depends on not only fate and transport properties, but also toxicological parameters (Fig. 3). When saturation for a pure chemical is reached, the predicted vapour concentrations in indoor and ambient air will no longer increase as soil concentration increases. Therefore the ADE associated with inhalation of indoor and outdoor are constant. Most of risk assessment models such as CLEA uses a linear partitioning between the sorbed, dissolved and vapour phases and do have potential over-estimate the potential risks associated vapour pathways in the presence of NAPL. However, this behavior is not believed to affect exposure pathways such as soil ingestion, consumption of home-grown vegetables and fruits and inhalation of indoor and outdoor dusts (EA, 2009a).

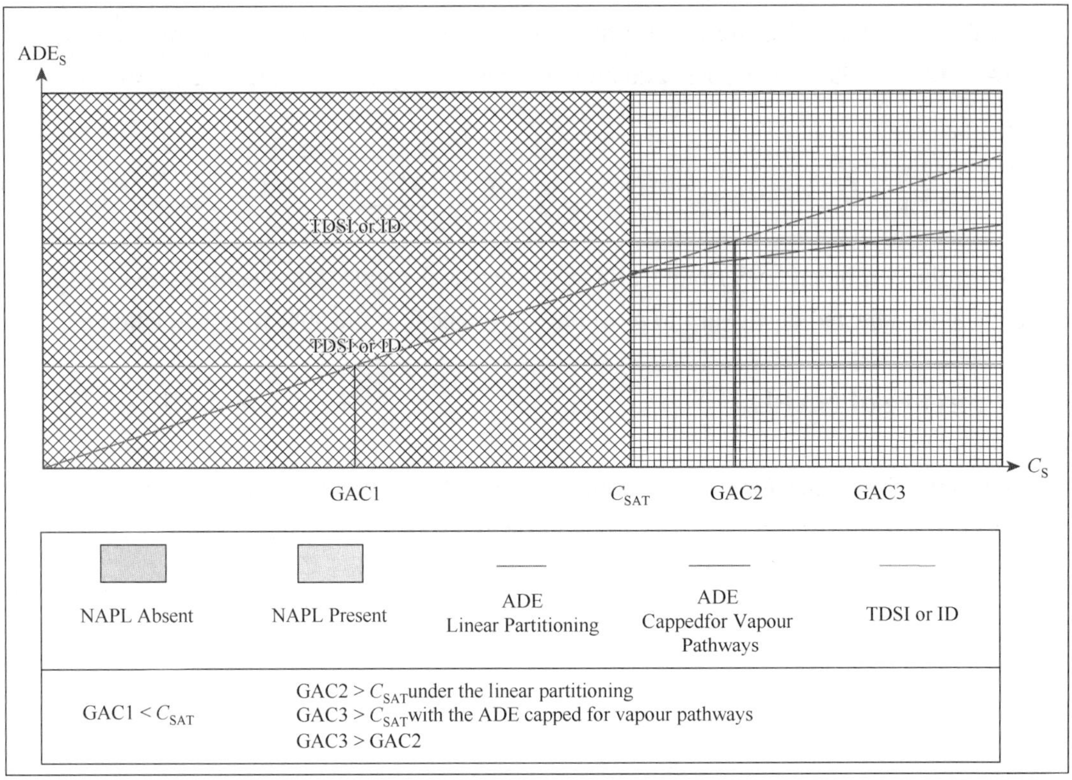

Fig. 3　Relation of the soil ADE Versus Soil Concentration (C_s) and impact on the GAC

The Average Daily Exposure (ADE) normally increases with soil concentration, but should be at a much reduced rate in presence of NAPL as the ADE from vapour pathways is capped (Fig. 3). The integrated GAC derived using the linear partitioning approach is likely to be conservative than that derived by capping the ADE from vapour pathways (Fig. 3).

In practice, the minimum of the soil saturation limits described in Equations 7 and 8 is used to determine if the integrated GAC exceeds the soil saturation limit.

4. Mathematical description and integration procedures under the 50% rule

Chen (2010a) has derived solutions for calculating the integrated GAC in which Equation 4 can be simplified under the 50% Rule on background exposure and the linear partitioning approach. These solutions are summarized in Table 4 and are only valid when the integrated GAC falls below the SAL (Fig. 2). However, the modifications to the solutions derived by Chen (2010a) become necessary if the integrated GAC exceeds soil saturation limits in which case, the ADE contributed by vapour pathways should be adjusted and calculated from the SAL.

Table 4 Mathematical Descriptions of the 50% Rule in Absence of NAPL

Outside the 50% Rule in absence of NAPL[1]	Outside the 50% Rule in presence of NAPL
Equation 9a: $\frac{GAC_{int}\times\sum_{j=1}^{t}\sum_{o=1}^{m}R_o^j}{TDI^o-ADE^o_{MDI}}+\frac{GAC_{int}\times\left(\sum_{j=1}^{t}\sum_{p=1}^{q}R_p^j+\sum_{j=1}^{t}\sum_{v=1}^{s}R_v^j\right)}{TDI^i-ADE^i_{MDI}}=1$	Equation 9b: $\frac{GAC_{int}\times\sum_{j=1}^{t}\sum_{o=1}^{m}R_o^j}{TDI^o-ADE^o_{MDI}}+\frac{GAC_{int}\times\sum_{j=1}^{t}\sum_{p-1}^{q}R_p^j+C_{sat}\times\sum_{j=1}^{t}\sum_{v=1}^{s}R_v^j}{TDI^i-ADE^i_{MDI}}=1$
Under the 50% Rule in absence of NAPL[1]	**Under the 50% Rule in presence of NAPL**
Equation 10a (limiting oral ADE_{MDI}) $\frac{GAC_{int}\times\sum_{j=1}^{t}\sum_{o=1}^{m}R_o^j}{TDI^o-GAC_{int}\times\sum_{j=1}^{t}\sum_{o=1}^{m}R_o^j}+\frac{GAC_{int}\times\left(\sum_{j=1}^{t}\sum_{p=1}^{q}R_p^j+\sum_{j=1}^{t}\sum_{v=1}^{s}R_v^j\right)}{TDI^i-ADE^i_{MDI}}=1$	Equation 10b (limiting oral ADE_{MDI}) $\frac{GAC_{int}\times\sum_{j=1}^{t}\sum_{o=1}^{m}R_o^j}{TDI^o-GAC_{int}\times\sum_{j=1}^{t}\sum_{o=1}^{m}R_o^j}+\frac{GAC_{int}\times\sum_{j=1}^{t}\sum_{p=1}^{q}R_p^j+C_{sat}\times\sum_{j=1}^{t}\sum_{v=1}^{s}R_v^j}{TDI^i-ADE^i_{MDI}}=1$
Equation 11a (limiting inhalation ADE_{MDI}) $\frac{GAC_{int}\times\sum_{j=1}^{t}\sum_{o=1}^{m}R_o^j}{TDI^o-ADE^o_{MDI}}+\frac{GAC_{int}\times\left(\sum_{j=1}^{t}\sum_{p=1}^{q}R_p^j+\sum_{j=1}^{t}\sum_{v=1}^{s}R_v^j\right)}{TDI^i-GAC_{int}\times\left(\sum_{j=1}^{t}\sum_{p=1}^{q}R_p^j+\sum_{j=1}^{t}\sum_{v=1}^{s}R_v^j\right)}=1$	Equation 11b (limiting inhalation ADE_{MDI}) $\frac{GAC_{int}\times\sum_{j=1}^{t}\sum_{o=1}^{m}R_o^j}{TDI^o-ADE^o_{MDI}}+\frac{GAC_{int}\times\sum_{j=1}^{t}\sum_{p=1}^{q}R_p^j+C_{sat}\times\sum_{j=1}^{t}\sum_{v=1}^{s}R_v^j}{TDI^i-\left(GAC_{int}\times\sum_{j=1}^{t}\sum_{p=1}^{q}R_p^j+C_{sat}\times\sum_{j=1}^{t}\sum_{v=1}^{s}R_v^j\right)}=1$
Equation 12a (limiting both oral and inhalation ADE_{MDI}) $\frac{GAC_{int}\times\sum_{j=1}^{t}\sum_{o=1}^{m}R_o^j}{TDI^o-GAC_{int}\times\sum_{j=1}^{t}\sum_{o=1}^{m}R_o^j}+\frac{GAC_{int}\times\left(\sum_{j=1}^{t}\sum_{o=1}^{q}R_p^j+\sum_{j=1}^{t}\sum_{v=1}^{s}R_v^j\right)}{TDI^i-GAC_{int}\times\left(\sum_{j=1}^{t}\sum_{p=1}^{q}R_p^j+\sum_{j=1}^{t}\sum_{v=1}^{s}R_v^j\right)}=1$	Equation 12b (limiting both oral and inhalation ADE_{MDI}) $\frac{GAC_{int}\times\sum_{j=1}^{t}\sum_{o=1}^{m}R_o^j}{TDI^o-GAC_{int}\times\sum_{j=1}^{t}\sum_{o=1}^{m}R_o^j}+\frac{GAC_{int}\times\sum_{j=1}^{t}\sum_{p=1}^{q}R_p^j+C_{sat}\times\sum_{j=1}^{t}\sum_{v=1}^{s}R_v^j}{TDI^i-\left(GAC_{int}\times\sum_{j=1}^{t}\sum_{p=1}^{q}R_p^j+C_{sat}\times\sum_{j=1}^{t}\sum_{v=1}^{s}R_v^j\right)}=1$

Symbol	Unit	Descriptions
$\sum_{j=1}^{t}\sum_{o=1}^{m}R_o^j$	μg/kg/d over mg/kg soil	Sum of the ratios of ADE over GAC from soil source for oral and dermal pathways, where: *m*=3, including soil and dust ingestion, dermal indoor and outdoor for the residential without garden and commercial land uses; *m*=4, representing soil and dust ingestion, dermal outdoor, consumption of site grown vegetables and ingestion of soil attached to vegetables for the allotment land use; *m*=5, representing soil and dust ingestion, dermal indoor, dermal outdoor, consumption of site grown vegetable, ingestion of soil attached to vegetables for the residential with garden land use. *t*=6, for residential and allotment land uses and *t*=1 for commercial land use
$\sum_{j=1}^{t}\sum_{p=1}^{q}R_p^j$	μg/kg/d over mg/kg soil	Sum of the ratios of the ADE over the GAC from soil source for inhalation of dusts, where: q=1, representing inhalation of outdoor dust for allotment land use; q=2, representing inhalation of indoor and outdoor dusts for residential with or without garden and commercial land uses t=6, for residential and allotment land uses and t=1 for commercial land use
$\sum_{j=1}^{t}\sum_{v=1}^{s}R_v^j$	μg/kg/d over mg/kg soil	Sum of the ratios of the ADE over the GAC from soil source for inhalation of vapours, where: s=1, representing inhalation of outdoor vapour for allotment land use; s=2, representing inhalation of indoor and outdoor vapours for residential with or without garden and commercial land uses t=6, for residential and allotment land uses and t=1 for commercial land use
TDI^o	μg/kg/d	Oral TDI
TDI^i	μg/kg/d	Inhalation TDI
ADE^i_{MDI}	μg/kg/d	Inhalation background exposure (Equation 3)
ADE^o_{MDI}	μg/kg/d	Oral background exposure (Equation 2)
C_{sat}	mg/kg	Soil Saturation Limit (minimum of C_{satw} in Equation 7 and C_{satv} in Equation 8)

As illustrated in Table 4, Equation 9a describes the process to calculate the integrated GAC when the background exposure is insignificant (i.e. combined background exposure less than the soil ADE using Equation 9a). Equations 10a to 12a describe the processes to calculate the integrated GAC under the 50% rule on background exposure: Equation 10a to adjust the oral background exposure to equal to the oral ADE; Equation 11a to adjust the inhalation background exposure to equal to the inhalation ADE; and Equation 12a to set the combined background exposure to the combined soil ADE.

The CLEA solves Equations 10a to 12a by using an iterative/repetitive process implemented as a Goal Seek Function using Excel Visual Basic Application (VBA) procedure to find a solution to the soil GAC. It requires numerous iterations to satisfy a pre-determined closure criterion, therefore the time taken to complete a simulation, though not excessive, is considered long for a deterministic model (5).

A starting point to determine when and how the background exposure should be adjusted is to compare the calculated background exposure with the half of the TDI (Fig. 1 (a)) followed by checking if the combined background exposure after the adjustment of either oral or inhalation background exceeds the soil ADE (Fig. 1 (b)). Ultimately Equation 12a would have to be used to set combined background exposure to soil ADE each representing 50% of the total ADE.

There may be cases where the combined background exposure, despite the fact that both oral and inhalation background pass the test of 50% Rule individually, may still exceed 50% of the total exposure if Equation 9a is used. In these cases, the integrated soil GAC may be derived from one of Equations 10a to 12a to be determined by a logical sequence developed by Chen (2010a).

When the integrated GAC calculated from Equations 9a to 12a exceeds the SAL, the soil ADE from vapour pathways should be kept constant to avoid the integrated GAC to be under-estimated (Fig. 2). Equations 9a to 12a are modified into Equation 9b to 12b respectively by capping the ADE from vapour pathways with $C_{sat} \times \sum_{j=1}^{t} \sum_{v=1}^{p} R_v^j$ (Table 4).

Table 5 Summary of Integration Procedures in Absence of NAPL

Solution to Equation 9a (outside the 50% Rule)

$$\mathrm{GAC_{int}} = \frac{1}{\left(\frac{a}{d-g}\right) + \left(\frac{b+c}{e-f}\right)} \tag{13a}$$

Solution to Equation 10a (Limiting oral background exposure)

$$\mathrm{GAC_{int}} = \frac{\left(\frac{2ae-2af+(b+c)d}{a(b+c)}\right) - \sqrt{\left(\frac{2ae-2af+(b+c)d}{a(b+c)}\right)^2 - \frac{4(de-df)}{a(b+c)}}}{2} \tag{14a}$$

Solution to Equation 11a (Limiting inhalation background exposure)

$$\mathrm{GAC_{int}} = \frac{\left(\frac{2d(b+c)-2g(b+c)+ae}{a(b+c)}\right) - \sqrt{\left(\frac{2d(b+c)-2g(b+c)+ae}{a(b+c)}\right)^2 - \frac{4(de-ge)}{a(b+c)}}}{2} \tag{15a}$$

Continued table

Solution to Equation 12a (Limiting both oral and inhalation background exposure)
$$\mathrm{GAC}_{\mathrm{int}}=\frac{\left(\frac{2d(b+c)+2ae}{3a(b+c)}\right)-\sqrt{\left(\frac{2d(b+c)+2ae}{3a(b+c)}\right)^2-\frac{4de}{3a(b+c)}}}{2} \quad (16a)$$

$$a=\sum_{j=1}^{t}\sum_{o=1}^{m}R_o^j \quad b=\sum_{j=1}^{t}\sum_{p=1}^{q}R_p^j \quad c=\sum_{j=1}^{t}\sum_{v=1}^{s}R_v^j \quad d=TDI^{\mathrm{o}} \quad e=TDI^{\mathrm{i}} \quad f=\mathrm{ADE}_{\mathrm{MDI}}^{\mathrm{i}} \quad g=\mathrm{ADE}_{\mathrm{MDI}}^{\mathrm{o}}$$

Note: Alphabetical letters a to g have been used to simplified the above equations and see Table 4 for description of parameters and their units.

Table 6 Summary of Integration Procedures in Presence of NAPL

Solution to Equation 9b (outside the 50% Rule)
$$\mathrm{GAC}_{\mathrm{int}}=\frac{e-f-hc}{b+\frac{a(e-f)}{(d-g)}} \quad (13b)$$
Solution to Equation 10b (Limiting oral background exposure)
$$\mathrm{GAC}_{\mathrm{int}}=\frac{\frac{2ae-2af+db-ahc}{ab}-\sqrt{\left(\frac{2ae-2af+db-ahc}{ab}\right)^2-4\times\frac{2de-2df-dhc}{ab}}}{2} \quad (14b)$$
Solution to Equation 11b (Limiting inhalation background exposure)
$$\mathrm{GAC}_{\mathrm{int}}=\frac{\frac{2db+ae-ahc-2gb}{ab}-\sqrt{\left(\frac{2db+ae-ahc-2gb}{ab}\right)^2-4\times\frac{(d-g)(e-2hc)}{ab}}}{2} \quad (15b)$$
Solution to Equation 12b (Limiting both oral and inhalation background exposure)
$$\mathrm{GAC}_{\mathrm{int}}=\frac{\frac{2db+2ae-3ahc}{3ab}-\sqrt{\left(\frac{2db+2ae-3ahc}{3ab}\right)^2-4\times\frac{(de-2dhc)}{3ab}}}{2} \quad (16b)$$

$$a=\sum_{j=1}^{t}\sum_{o=1}^{m}R_o^j \quad b=\sum_{j=1}^{t}\sum_{p=1}^{q}R_p^j \quad c=\sum_{j=1}^{t}\sum_{v=1}^{s}R_v^j \quad d=TDI^{\mathrm{o}} \quad e=TDI^{\mathrm{i}} \quad f=\mathrm{ADE}_{\mathrm{MDI}}^{\mathrm{i}} \quad g=\mathrm{ADE}_{\mathrm{MDI}}^{\mathrm{o}} \quad h=C_{\mathrm{sat}}$$

Note: Alphabetical letters a to h have been used to simplified the above equations and see Table 4 for description of parameters and their units.

The integration procedure relies on solving Equations 9 to 12 directly. Equations 10 to 12 can be solved by finding one of the roots when rearranged to standard quadratic forms. Equations 13 to 16 are solutions to Equations 9 to 12 and are summarized in Tables 5 and 6 respectively for the absence and presence of NAPL. Other roots have negative GAC and are therefore discounted.

5. Comparison of the GAC from CLEA V1.04 and HERA-Soil

In order to test the integration procedures, the solutions described in Equations 9 to 12 and the logical sequence described by Chen (2010a) have been implemented in our in-house model, HERA-Soil (Human Exposure Risk Assessment model for Soil) (Chen, 2009). Several representative contaminants for metals, semi-volatile and volatile organic compounds known to have a range of background exposure values under standard UK land uses have been selected and the results of the GAC derived have been compared with those derived using the CLEA model.

Toxicological parameters and background exposure parameters are taken from the Environment Agency Toxicological Report (EA, 2009c). Physical and chemical parameters are taken from the Environment Agency Science Report SC050021/SR7 for ethylbenzene, xylene, naphthalene and carbon tetrachloride (EA, 2009b). All other default settings defined in the CLEA guidance report (EA, 2009a) such as soil ingress rates into the buildings, soil to dust transport factors, soil properties, and exposure parameters have been adopted in the testing (Tables 2, 3).

The $\sum R$ terms (unit intake rate) are related to the media transfer factors using technical algorithms recommended for generic quantitative risk assessment (EA, 2009a). These terms are not directly reported in the CLEA model, but the ADE and the GAC are reported in the pathway specific CLEA models. $\sum R$ terms in the CLEA model can therefore be obtained by summing the ADE/GAC for oral and dermal, and inhalation pathways.

Key parameters including ∑R terms used to calculate the integrated soil GAC is summarized in Table 7. The results of the integration and pathway contributions under the 50% Rule are illustrated in Table 8. Table 9 compares the integrated GAC from HERA-Soil and CLEA model.

Table 7 Summary of Key Parameters Used to Derive the GAC_{int}

COC	$\sum_{j=1}^{t}\sum_{o=1}^{m}R_o^j$ a	$\sum_{j=1}^{t}\sum_{p=1}^{q}R_p^j$ b	$\sum_{j=1}^{t}\sum_{v=1}^{s}R_v^j$ c	TDI^o d	TDI^i e	ADE^i_{MDI} F	ADE^o_{MDI} g	C_{sat} h
Ethylbenzene	A: 0.175 B: 0.0112 C: 1.09 D: 0.00052	A: 0.0000235 B: 0.0000235 C: 0.00000081 D: 0.0000029	A: 2.28 B: 2.28 C: 0.00116 D: 0.022	100	170	A: 7.88 B: 7.88 C: 7.88 D: 1.86	A: 0.28 B: 0.28 C: 0.28 D: 0.07	2844
Xylene	A: 0.165 B: 0.0112 C: 1.02 D: 0.00052	A: 0.0000235 B: 0.0000235 C: 0.00000081 D: 0.0000029	A: 1.67 B: 1.67 C: 0.00099 D: 0.0158	179	63	A: 6.79 B: 6.79 C: 6.79 D: 1.60	A: 1.13 B: 1.13 C: 1.13 D: 0.29	3457
Naphthalene	A: 0.136 B: 0.0112 C: 0.832 D: 0.00052	A: 0.0000235 B: 0.0000235 C: 0.00000081 D: 0.0000029	A: 0.074 B: 0.074 C: 0.0021 D: 0.00071	20	0.86	A: 0.17 B: 0.17 C: 0.17 D: 0.17	A: 0.39 B: 0.39 C: 0.39 D: 0.1	432
Carbon Tetrachloride	A: 0.259 B: 0.0112 C: 1.65 D: 0.00052	A: 0.0000235 B: 0.0000235 C: 0.00000081 D: 0.0000029	A: 18.12 B: 18.12 C: 0.0033 D: 0.168	1.42	3.26	A: 3.03 B: 3.03 C: 3.03 D: 0.71	A: 0.0113 B: 0.0113 C: 0.0113 D: 0.0029	7537
Cadmium	A: 0.0084 B: 0.00745 C: 0.0083 D: 0.00045	A: 0.0000235 B: 0.0000235 C: 0.00000081 D: 0.0000029	A: 0 B: 0 C: 0 D: 0	1	0.001	A: 0 B: 0 C: 0 D: 0	A: 0.9 B: 0.9 C: 0.9 D: 9	NR
Nickel	A: 0.0086 B: 0.00742 C: 0.00996 D: 0.00045	A: 0.0000235 B: 0.0000235 C: 0.00000081 D: 0.0000029	A: 0 B: 0 C: 0 D: 0	5	0.006	A: 0.018 B: 0.018 C: 0.018 D: 0.0043	A: 9 B: 9 C: 9 D: 2.29	NR

Note: Unit for TDI and ADE: μg/kg/d; ∑R: μg/kg/d over mg/kg soil; C_{sat}: mg/kg. ∑R is derived under the CLEA default settings; A: Residential with Garden; B, Residential without garden; C: Allotment; D: Commercial; Values in bold and italics indicates the ID for the relevant pathway; See Table 4 for parameter descriptions and Tables 5 or 6 for alphabetical abbreviations; C_{sat}, calculated using foc of 0.0348 for sandy loam soil (Table 3b).

Table 8 Summary of the Integration under the 50% Rule and Pathway Contributions from HERA Model (Chen, 2010b)

COC	ADE^{o}_{MDI}	TDI^{o}	50% Rule?	ADE^{i}_{MDI}	TDI^{i}	50% Rule?	Contribution (%)			
							ADE^{o}_{S}	ADE^{i}_{S}	ADE^{o}_{MDI}	ADE^{i}_{MDI}
Residential and Allotment Land Uses										
Ethylbenzene	0.28	100	NO	7.88	170	NO	A: 6.78 B: 0.47 C: 92.34	A: 88.22 B: 94.73 C: 0.10	A: 0.17 B: 0.17 C: 0.26	A: 4.82 B: 4.64 C: 7.30
Xylene	1.13	179	NO	6.79	63	NO	A: 7.90 B: 0.58 C: 95.64	A: 80.42 B: 87.12 C: 0.09	A: 1.66 B: 1.75 C: 0.61	A: 10.01 B: 10.54 C: 3.66
Naphthalene	0.39	20	NO	0.17	0.86	NO	A: 49.40 B: 7.65 C: 97.16	A: 27.07 B: 50.72 C: 0.02	A: 16.44 B: 29.09 C: 1.97	A: 7.08 B: 12.54 C: 0.85
Carbon Tetrachloride	0.0113	1.42	NO	3.03	3.26	YES	A: 0.71 B: 0.03 C: 98.82	A: 49.47 B: 49.97 C: 0.20	A: 0.34 B: 0.03 C: 0.79	A: 49.47 B: 49.97 C: 0.20
Cadmium	0.9	1	YES	0	0.001	NO	A: 49.93 B: 49.92 C: 50.00	A: 0.14 B: 0.16 C: 0.00	A: 49.93 B: 49.92 C: 50.00	A: 0.00 B: 0.00 C: 0.00
Nickel	9	5	YES	0.0182	0.006	YES	A: 49.86 B: 49.84 C: 50.00	A: 0.14 B: 0.16 C: 0.0041	A: 49.86 B: 49.84 C: 50.00	A: 0.14 B: 0.16 C: 0.0041
Commercial Land Use										
Ethylbenzene	0.0714	100	NO	1.86	170	NO	49.97 (2.35)	48.51 (96.50)	0.06 (0.04)	1.46 (1.11)
Xylene	0.2857	179	NO	1.6	63	NO	25.17 (3.11)	72.35 (93.97)	0.38 (0.44)	2.11 (2.48)
Naphthalene	0.1	20	NO	0.04	0.86	NO	79.08 (38.50)	1.37 (52.30)	10.87 (6.57)	8.69 (2.63)
Carbon Tetrachloride	0.0029	1.42	NO	0.714	3.26	NO	0.24	77.74	0.09	21.93
Cadmium	0.23	1	YES	0	0.001	NO	49.84	0.32	49.84	0.00
Nickel	2.29	5	YES	0.00429	0.006	YES	49.68	0.32	49.68	0.32

Note: Unit for TDI and ADE_{MDI}: μg/kg/d; Values in bold indicate oral or inhalation or combined background exposure are adjusted to comply with the 50% Rule; Values in brackets are the percentage contribution under the linear partitioning approach within the CLEA. See Table 4 for description of parameters.

Table 9 Comparison of the GAC_{int} Derived from the CLEA and HERA model (Chen, 2010b)

COC	Residential with Garden		Residential without Garden		Allotment		Commercial	
	CLEA	HERA	CLEA	HERA	CLEA	HERA	CLEA	HERA
Ethylbenzene	63.32	63.32	70.66	70.66	91.18	91.17	7507.27	120833
Xylene	32.55	32.55	33.49	33.49	173.42	173.37	3833.31	36381
Naphthalene	8.71	8.71	9.22	9.22	23.44	23.41	1117.31	49561
Carbon Tetrachloride	0.089	0.089	0.0899	0.0899	0.85	0.85	15.04	15.04
Cadmium	28.88	28.88	30.16	30.16	58.67	58.68	294.39	294.39

Nickel	110.45	110.49	113.16	113.17	246.65	246.66	989.13	989.17
Note: unit: mg/kg								

As indicated in Table 8, the 50% Rule is applied for cadmium for oral pathways under residential, allotment and commercial land uses, therefore Equation 14a in Table 5 is used for the derivation of the integrated soil GAC. Similarly the 50% Rule is applied for carbon tetrachloride for inhalation pathways under residential and allotment land uses, so that Equation 15a is used instead.

In the case of nickel, Equation 16a is used as the 50% Rule is activated for both oral and inhalation pathways under residential, allotment and the commercial land use.

Equation 13a is applied for ethylbenzene, xylene and naphthalene as the combined background exposure is less than the soil ADE for residential and allotment land uses. However, the integrated GAC for ethylbenzene, xylene and naphthalene under the commercial land use would exceed their respective SAL, therefore Equation 13b in Table 6 instead of Equation 13a in Table 5 is used to derive the integrated GAC which limits the ADE under the vapour pathways (Table 8).

As illustrated in Table 9, the results of the integrated soil GAC from the iterative integration implemented in the CLEA model and the integration procedures as implemented in HERA-Soil (based on the analytical solutions presented in Tables 5 and 6) are almost identical with absolute errors less than 0.0005 except those for ethylebenzene, naphthalene and xylene under the less sensitive commercial land use. The integrated soil GAC for ethylbenzene, xylene and naphthalene are considerably larger in presence of NAPL as the ADE calculated from the SAL for vapour pathways is capped for these contaminants.

6. Summaries and Conclusions

Many risk assessment models used in contaminated sites employ either a simple integration procedure by only partially combining exposure pathways from surface soil for the derivation of integrated soil GAC. This approach though simple to implement the GAC derived can not be considered sufficiently protective of human health in some cases. The CLEA model developed by the Environment Agency in England Wales is the one of few models to consider background exposure which uses an integrated procedure to derive the GAC in absence of NAPL. However for less sensitive commercial land use, the GAC derived is likely to exceed the SAL, the potential risks from vapour pathways for volatile organic compounds are likely to be over-estimated resulting in a conservative GAC for some contaminants such as ethylbenzene, xylene and naphthalene.

This paper demonstrated that the derivation of generic assessment criteria is a complex process particularly when combining oral, dermal and inhalation pathways under the presence of NAPL and 50% Rule on the background exposure. Comparison of the GAC derived using the CLEA and HERA-Soil indicates that the GAC derived from HERA-Soil are identical to those derived by the CLEA model within the SAL, but differs greatly for ethylbenzene, xylene and naphthalene under commercial land use since HERA model limits the ADE for vapour pathways.

The integration procedure presented in this paper are considered robust and can be incorporated into the CLEA or HERA risk assessment models to derive soil GAC to be more protective of human and environmental receptors.

References

ASTM. 2000. Standard guide for risk based corrective action. Report E2081. West Conshohocken, American Society for Testing and Materials.

Chen M. 2009. Technical algorithm for human exposure risk assessment for soil (HERA-Soil). In house model.

Chen M. 2010a. Alternative integration procedures in combining multiple exposure routes for the derivation of generic assessment criteria with the CLEA model. Land Contamination & Reclamation, 18 (2): 135-150.

Chen M. 2010b. Analytical integration procedures for the derivation of risk-based generic assessment criteria for soil. Human and Ecological Risk Assessment, 16 (6): 1295-1317.

Chen M. 2012. Health and Environmental Risk Assessment Software: Technical Manual. Institute of Soil Science, Chinese Academy of Sciences.

EA. 2009a. Updated technical background to the CLEA model. Science Report SC050021/SR3. The Environment Agency in England and Wales.

EA. 2009b. CLEA software (Version 1.04) handbook. Science Report SC050021/SR4. The Environment Agency in England and Wales.

EA. 2009c. Human health toxicological assessment of contaminants in soil. Science Report SC050021/SR2. The Environment Agency in England and Wales.

EA. 2009d. Compilation of data for priority pollutants for derivation of soil guidance values. Science Report SC050021/SR7. The Environment Agency in England and Wales.

GSI. 2008. Modelling and risk assessment software: RBCA tool kit for chemical releases. Version 2. Developed by GSI Environmental Inc.

Johnson P C, Ettinger R A. 1991. Heuristic model for predicting the intrusion rate of contaminant vapors into buildings. Environmental Science & Technology, 1991, 25 (8): 1445-1452.

MEP. 2014. Chinese risk assessment guideline for contaminated sites (C-RAG). HJ 25.3-2014. The Ministry of Environmental Protection of China.

USEPA. 1996. Soil screening guidance: Technical background document. Publication No. EPA/540/R95/128. Office of Solid Waste and Emergence Response. U. S. Environmental Protection Agency.

USEPA. 2009. Integrated Risk Information System. U. S. Environmental Protection Agency.

Part Ⅲ Heavy Metal Pollution and Remediation

Principles and Technologies of Phytoremediation for Metal Contaminated Soils-A Review

Xiaoe YANG

MOE Key Lab of Environmental Remediation and Ecological Health,
Zhejiang University

Abstract

Phytoremediation is attractive and cost effective technology because of its tremendous potential for the treatment of organic and inorganic pollutants. In this mini-review, the progress in 20 years studies of mechanisms and technologies for phytoremediating heavy metal or co-contaminated soils were summarized. Our team has identified several native metal hyperaccumulator and accumulator plants. Plants like *Sedum. alfredii* Hance has been shown to be very effective in phytoremediation of heavy metal and organic pollutant co-contaminated soils. On the other hand, screening metal lower accumulator genotypes of crops like spinach and Pak choi confirmed pollutant safe cultivars for food safety. These pollutant safe cultivars further used with organic soil amendments for the phytoavoidation of heavy metal to make more effective at reducing Cd movement into the food chain than growing alone. Mechanisms of metal hyperaccumulation and accumulation in plants were discussed. The phytoextraction can be largely enhanced by chemical and endophytic microbial assistance. Phytoextraction coupled with phytoavoidation technologies will provide an environmentally friendly alternative to conventional cleanup methods. This is an innovative technology for phytoremediation coupled with crop safe production, and is very useful for providing a new approach for remediating large areas of heavy metal moderately contaminated agricultural soils.

Keywords: Phytoremediation, phytoavoidation, screening, heavy metals, organic pollutants, hyperaccumulator

1. Introduction

Soil contamination with heavy metals has become a worldwide challenge, leading to the losses in agriculture and hazardous health effects as they enter the food chain. This problem may be partially solved by the emerging phytoremediation technology. The cost-effective plant-based approach to remediation takes the advantage of the remarkable ability of plants to concentrate heavy metals in their tissues from the environment. However, the so-called phytoremediation has not achieved a lot partly due to the lack of basic information regarding the fundamental mechanisms employed by metal accumulators. The application of metal accumulators is also obstructed by metal accumulators. The application of metal accumulators is also obstructed by some of their inherent defects, such as small biomass, slow growth rate, and poor adaptability. Hence, before their practical application, the mechanisms responsible for the abnormal ability for metal tolerance and accumulation should be elucidated. Alternatively, efforts should be paid to explore new accumulators with ideal biological traits including high biomass and fast growth rate.

Phytoremediation is a plant-based approach to remediate contaminated soil, including phytoextraction, phytostabilization, phytovolatilization, phytodegradation, phytoavoidation, and phytoremediation coupled with agro-production. It based on the ability of plants to concentrate specific contaminants (mainly heavy motal and organic pollutants) from the environment, and volatilize, degrade or metabolize various molecules in their tissues. Compared with physical and chemical remediation methods, phytoremediation is much more cost-effective and environmentally friendly, and therefore has a promising application prospect (Long et al., 2002; Hu et al., 2014).

It is widely acknowledged that phytoextraction is an effective approach to remove heavy metals from soils (Hu et al., 2014), plants for phytoextraction are defined as hyperaccumulators. There are more than 700 hyperaccumulators found in the world (Li et al., 2011), most of which still have some problems to be settled before applying into practice due to their low biomass and poor adaptability. The technologies of enhancing phytoremediation efficiency, phytoremediation for complexed contaminated soils, phytoremediation coupled with agro-production are required to be developed (Yang et al., 2015). In the last 20 years, we aimed at solving these problems. Screening and identification of metal hyperaccumulators and accumulators like *S. alfredii* Hance, *Elsholtzia splendens*, *E. aryi*, Willows, *Ricinus communis* and *Brassica rape* etc. (He et al., 2003; Huang et al., 2011; Peng and Yang, 2007; Xing et al., 2012; Yang et al., 2015). In addition, we also identified metal low accumulators of rice and vegetables as a supplement for crop safe production (Yang et al., 1998; Diao et al., 2005; Chen et al., 2012; Chen et al., 2010; Chen et al., 2009). Till now, serious of experiments have been conducted to understand the mechanisms of metal hyperaccumulation by plants, and formed a systematic knowledge of phytoremediation in theory and practice.

In contrast to the extensive experimental studies, little attention has been paid to a systematic summary. Thus this paper summarizes the results of comprehensive experimental programs by us, as a guide to the further development of phytoremediation. Topics include: screening and identification of metal hyperaccumulators and low accumulators, mechanisms of metal (Cd, Pb, Cu, Zn) hyperaccumulation and accumulation by plants, microbes-and chemical-assistant phytoremediation, phyto-avoidation of heavy metal contamination in crops, and technologies of phytoremediation coupled with agro-production for complexed contaminated soils.

2. Identification of metal hyperaccumulators/accumulators

2.1 Zn/Cd hyperaccumulator and Pb accumulator*S. alfredii* Hance

Through our survey and screening in different mined areas, we firstly found an extraor-dinary Zn/Cd-hyperaccumulating ecotype of *S. alfredii* Hance in an old Zn/lead mining region in Quzhou, Zhejiang province (Yang et al., 2002) (Fig. 1). This plant has been identified as a Zn/Cd co-hyperacccumulator. It can be healthy grown under Zn supply levels of 240 mg/L (Long et al., 2002), Cd supply level of 400 μmol/L (Yang et al., 2004). Shoot Zn concentration can reach up to 2.9% DW under soil culture (Yang et al., 2005). And shoot Cd concentration can reach to 9000 mg/kg under solution culture (Yang et al., 2004) while up to 989 mg/kg under soil culture (Xiong and Yang, 2005). In addition, we found that *S. alfredii* Hance is also a lead accumulator (He and Yang, 2003), it showed best growth at Pb supply level of 160 mg/L. in solution culture and at Pb supply of 400 mg/kg under soil culture. Shoot Pb concentration in both solution and soil culture conditions could reach over 500 mg/kg (He and Yang, 2003).

Fig. 1 Mined ecotype of *S. alfredii* Hance Zn/Cd hyperaccumulator and Pb accumulator

To date, more than 400 species of hyperaccumulators belonging to 45 families have been identified, of which about 75% are Ni-hyperaccumulators (Reeves and Baker, 2000). So far, about 18 species of Zn-hyperaccumulators and 5 species of Cd hyperaccumulators have been identified. They were found to mainly colonize area with high Zn, Cd, and Pb present in soils due to historical mining activity and the subsequent contamination of top soil with mine spoil rich in heavy metals, and mainly distributed in Europe and Australia (Baker et al., 2000). *S. alfredii* Hance as a Zn/Cd hyperaccumulator is the first observation in China. It could tolerate higher Zn and Cd in growth media, and be able to hyperaccumulate in its shoots as well as *Thlaspi. caerulescens*. In comparison, *S. alfredi* Hance has its characteristics of faster growth rate, larger biomass, asexual reproduction and perennial plant, and the plants can cover nearly 100% of the land surface. Besides, *S. alfredi* Hance could propagate 3 to 4 times a year when the environmental conditions are favorable. According to field survey, shoot dry matter of *S. alfredii* Hance was as high as 4.5 tons/ha. Therefore, there is a great potential in using *S. alfredi* Hance to remediate Cd/Zn/Pb contaminated soils. The identification of *S. alfredii* Hance provides an excellent plant material to study mechanism of plant hyperaccumulation for heavy metals, and a potent new plant species to remediate heavy metals polluted soils via hyperaccumulators.

2.2 Copper and lead high accumulator plants *Elsholtzia*

We firstly discovered Chinese native Cu and Pb-tolerant and accumulating plants, *Elsholtzia splendens* (Haizhou Elsholtzia) and *E. argyi* (Purple Flower Elsholtzia) (Fig. 2), are endemic to Cu and Pb/Zn mine waste deposits respectively (Jiang et al., 2004a, 2004b). *E. splendens* (Haizhou Elsholtzia) has a local nickname of "copper flower" because its growth is confined to highly Cu contaminated soils. It has been identified as a Chinese native Cu-tolerant plant species and copper high accumulator (Yang et al., 1997). Both *Elsholtzia* plants belong to the family of Labiatae.

(a) *E. splendens* (b) *E. splendens* (c) *E. argyi*

Fig. 2 Mined ecotypes of *E. splendens* and *E. argyi*, Cu and Pb accumulators

Since our finding of *E. splendens* and *E. argyi* widely investigation of phytoremediation of Cu contaminated soil using *E. splendens* or *E. argyi* Under solution culture, both *Elsholtzia* plants can tolerate Cu levels of over 300 μmol/L and Pb level over 100 μmol/L. The Cu accumulation in stems can reach over 1000 mg/kg (Jiang et al., 2000; Yang et al., 2000). Ejaz et al. (2008) showed that Pb concentration of *E. argyi* reached to over 300 mg/kg in leaves and 1800 mg/kg in stems grown at 200 μmol/L Pb. *E. splendens* displayed certain tolerance at 100 μmol/L Pb treatment. Lead concentration in stems and leaves of *E. splendens* reached 1600 and 381 mg/kg, respectively (Zhang et al., 2011). Field experiments were conducted in a complexed contaminated soil with metal total levels of 1500 mg/kg Cu, 2000 mg/kg Pb, 1500 mg/kg Zn and 21 mg/kg Cd, respectively. The results showed that both *Elsholtzia* can highly tolerate to complexed metal toxicity, and obtained dry matter yield of 9-12 tons/ha. And *E. argyi* had higher dry matter yields and greater ability of accumulating Pb than *E. splendens*. In Contrast, *E. splendens* had superior in Cu accumulation (Peng and Yang, 2007). Both *E. splendens* and *E. argyi* can effectively clean up lower Cu contaminated water with the former better than the latter (Tian et al., 2008). These two accumulators provide good remediation plant materials for cleaning Cu and Pb from soils. Moreover, plants of *Elsholtzia* are Chinese traditional medicine in common use, and volatile oil which was distilled from their shoots plays important roles in antisepsis and antibiosis, such as Bacillus and Coccus. It has already been reported that *E. splenden*s has remarkedly resistance to epiphyte (Peng et al., 2007). Hence, the medicinal value of the two species should reduce the cost of phytoremediation greatly. Therefore, *E. splendens* and *E. argyi* are potentially good materials for phytoremediation in practice. Considering its high biomass and fast growth, we proposed that *E. splendens* has a great potential for phytoremediation.

2.3 Cadmium high accumulator plants willow clones *Salix*

Willow clones (39 clones) were obtained from the National Willow Germplasm Resources of the Jiangsu Academy of Forestry, Nanjing, China. Systematic studies were conducted to screening and identifying Cd tolerant and accumulating Willows (*Salix* spp.) clones. The results showed large difference in Cd tolerance and accumulation (Fig. 3). Shoot tolerance indexes (TIs) varied between 0.09 and 1.85, and root TIs varied between 0.27 and 1.99 among clones. The large differences in Cd concentration ranged from 64.7 to 663.7 (mg/kg, DW) in leaves, from 118.0 to 308.4 (mg/kg, DW) in stems, and even high in roots among clones (Yang et al., 2015). These results indicated that it is possible to identify Cd high accumulator willow clones for phytoremediation of soils and waters.

2.4 Cadmium accumulator plant *Ricinus communis*

Castor (*R. communis*) species belongs to Euphorbiaceae family, a fast growing C_3 plant, native to tropical Africa. It is an industrial crop because of its oil quality and quantity for plant based industries for making eco-friendly paints and coatings used in chemical industry. Castor attracted attention because of its ability to grow in heavily polluted soil together with its capacity for metal ion accumulation and fast growth rate. In addition, castor is an industrial crop with multiple non-food uses and an excellent rotation and companion crop. It has economic advantage as a cash crop in modern agriculture along with remediation of heavy metal contaminated soils. At soil Cd level of 0.42 mg/kg, *R. communis* varied largely in the uptake and accumulation of Cd, with mean concentrations of 1.22, 2.27 and 37.63 mg/kg DW for Cd in leaf, stem and root. The total uptake of Cd varied from 66.0 to 155.1 μg/pot. *R. communis* has great potential for removing Cd from contaminated soils attributed to its fast growth, high biomass, strong absorption and accumulation for Cd (Huang et al., 2011).

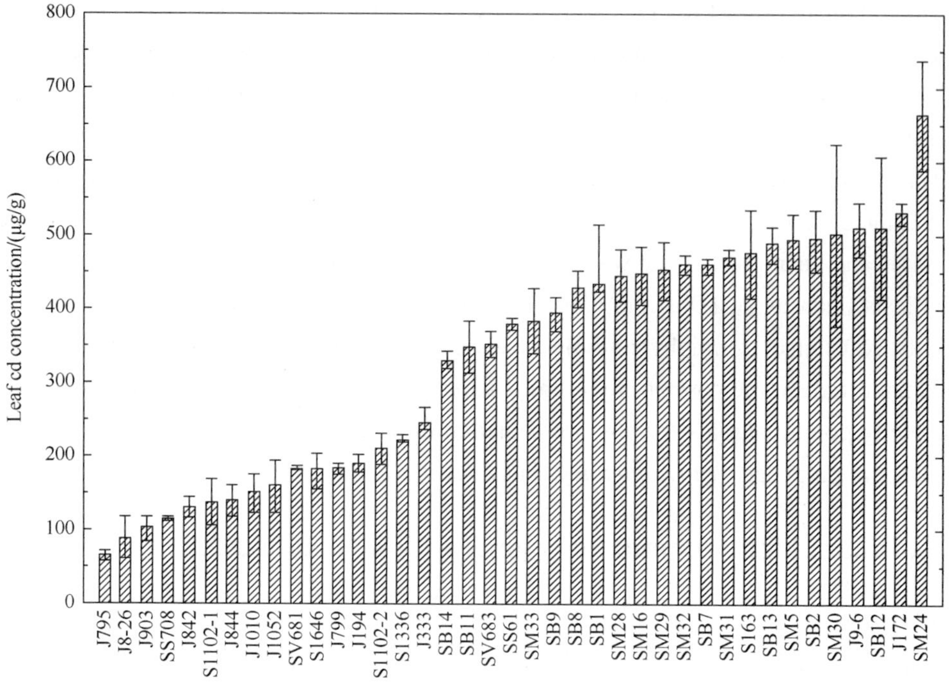

(a)

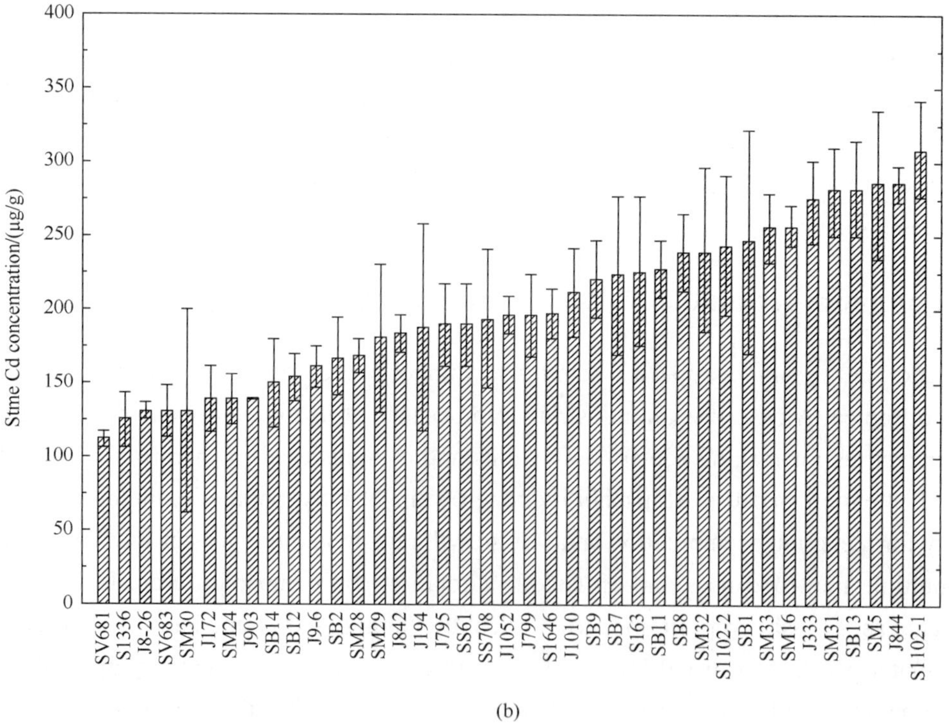

(b)

Fig. 3 Cd accumulation (μg/g DW) in (a) leaves and (b) stems of 39 willow clones exposed to 10 μmol/L Cd for 35 days

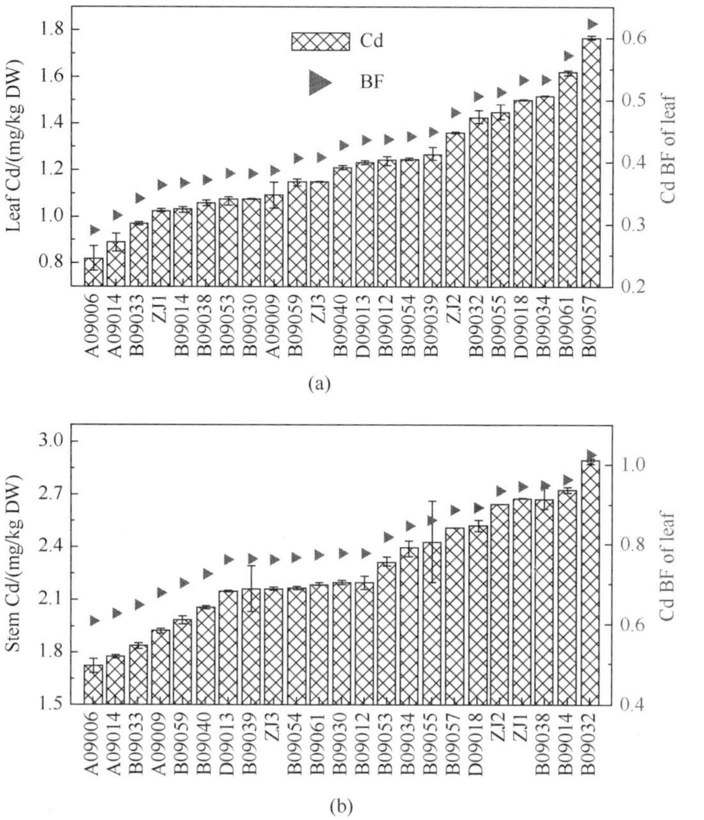

Fig. 4 (a) Cd accumulation in leaf of 23 genotypes of *R. communis*. (b) Cd accumulation in stem of 23 genotypes of *R. communis*

3. Mechanisms of metal hyperaccumulation and/or accumulation

3.1 Overview of Zn and Cd hyperaccumulation Mechanisms

Pioneer researches on *S. alfredii* Hance as a potential zinc (Zn)/cadmium (Cd) hyperaccumulator were initiated in 2002 by Prof. Dr. Xiaoe Yang. In a field survey on an ancient lead (Pb)/Zn mine in Quzhou, Zhejiang Province, China, *S. alfredii* Hance was found prevailing, and grew well in this area. The lab analysis showed that Zn concentration in the shoots ranged from 4134-5000 mg/kg DW (Yang et al., 2002) and Cd concentration in shoots could reach to 9000 mg/kg DW (Yang et al., 2004a), indicating that *S. alfredii* Hance was a promising Zn/Cd hyperaccumulator. Subsequently, a series of greenhouse experiments and field surveys consistently confirmed the remarkably high accumulation, translocation and tolerance abilities of *S. alfredii* Hance under multiple heavy metals conditions (Ni et al., 2004; Yang et al., 2004b, 2006a, 2006b; Long et al., 2006; Zhao et al., 2006).

Past decades researches on hyperaccumulation process were mainly focused on three processes: ①rhizospheric activation and uptake potential; ②translocation and transport component, ③ compartmentation and detoxification. Our past researches develop an explanation theory of "end point control" for metal hyperaccumulation in plant (Fig. 5). It seems that metal hyperaccumulation is mainly driven by shoot accumulation capacity and root uptake potential and regulated by transporters. Shoot accumulation capacity is associated mainly with metal compartmentation, speciation and cell antioxidation; while root uptake potential is mainly related to root tolerance, exudation, absorption, and endophytomicrobe syngestics. Metal transporters play important role in regulating metal sequestration, translocation and cell homeostasis. However, the whole network of metal hyperaccumulation in plant is not fully understood.

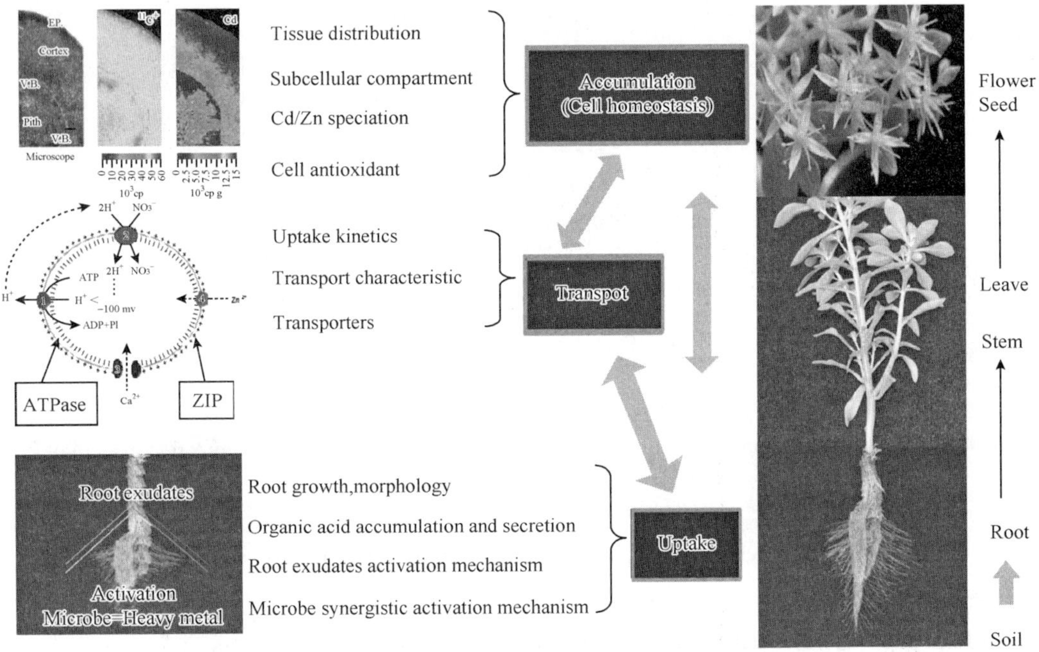

Fig. 5 Mechanisms of Cd/Zn hyperaccumulation in *S. alfredii* Hance

3.1.1 Uptake potential and rhizosperic activation

Rhizosphere is the main interface of plant and environment, and the forefront for plants to take up heavy metals, thus plays a fundamental role in hyperaccumulation process. Soil pH is an important factor controlling metal availability in rhizosphere. Rhizosphere soil pH was reduced by 0.5-0.6 units, as compared to bulk soil after HE *S. alfredii* Hance growth (Li et al., 2011a). The reduced soil pH could significantly increased soil Zn and Cd bioavailability in the rhizosphere and thus facilitated plant Zn and Cd uptake. In contrast, there was no obvious change in rhizosphere pH with NHE *S. alfredii* Hance (Li et al., 2011a, 2011b). HE *S. alfredii* Hance root was reported to be able to exudate more dissolved organic matter (DOM) in the rhizosphere than that of NHE (Li et al., 2012) and HE-DOM had greater ability to form complexes with Zn and Cd than NHE-DOM owing to the higher proportion of hydrophilic fraction (Li et al., 2011a; Li et al., 2013b). The greater DOM in the rhizosphere of HE *S. alfredii* Hance might be the main reason for the reduction of pH. By the addition of citric acid and tartaric acid, short-term (2h) root uptake of ^{109}Cd increased significantly (Lu et al., 2013a). Besides, elevated CO_2 increased the uptake of Cd/Zn of HE *S. alfredii* Hance because of root growth promotion, lower pH and higher photosynthetic carbon uptake rate (Li et al., 2013a; Li et al., 2014a; Li et al., 2015). Therefore, HE *S. alfredii* Hance is able to extract more Zn and Cd from soil pools that are unavailable to nonhyperaccumulator plants.

The hyperaccumulation of *S. alfredii* Hance was also related to its root morphology characteristics. It was shown that Zn concentration in shoot was positively correlated with root length, surface area, and volumes of *S. alfredii* Hance (Li et al., 2005a). With addition of Zn and Cd, the root volume and average diameter in NHE *S. alfredii* Hance were largely reduced, but they were significantly increased in HE *S. alfredii* Hance (Li et al., 2005a, 2005b, 2009). The differential responses of *S. alfredii* Hance root to Zn and Cd between the two populations indicated a higher tolerance of HE *S. alfredii* Hance to heavy metal stress, and accordingly favored its growth and Zn/Cd accumulation. *S. alfredii* Hance roots showed an initial rapid linear Zn absorption followed by a slower linear phase. Compared with NHE, Zn influx was 3-fold higher in HE while Zn transporters were lower in NHE than in HE (Km value of 20.43 and 34.83 μmol/L) (Li et al., 2005a), which supports the enhanced root uptake in *S. alfredii*. Although the Km of ^{109}Cd influx into roots of both ecotypes were similar, the V_{max} was 2-fold higher in the HE compared to NHE *S. alfredii* Hance (Lu et al., 2008). The Cd uptake in HE was 3-4 times higher than the implied Cd uptake calculated from transpiration rate and inhibition of transpiration rate in the HE had no essential effect on Cd accumulation in shoots of the plants (Lu et al., 2009), indicating that there were high affinity transporters for HE to uptake a large amount of Zn and Cd. Several *ZIP* family genes have been cloned and found to respond Zn and Cd stress. But, the detailed identification and analysis are still under way.

3.1.2 Translocation and transport components

After the uptake of Cd and Zn, HE *S. alfredii* Hance can transport the heavy metal to xylem and shoots faster than NHE *S. alfredii* Hance. A $^{65}Zn^{2+}$ efflux experiment revealed 2.7 fold and 3.7 fold more Zn accumulation in root vacuole of NHE than that of HE *S. alfredii* Hance for short-term and long-term, respectively. The half time for Zn efflux from root vacuole was estimated 40% shorter in HE, indicating a 1.8 fold faster efflux from HE root vacuole than that of NHE (Li et al., 2005a). The rate of root-to-shoot translocation of ^{109}Cd in the HE was$>$10 times higher than NHE and HE shoots accumulated dramatically higher ^{109}Cd content (Lu et al., 2008). Cell wall was an important site for Zn and Cd compartmentation in *S. alfredii* Hance root. The root cell walls showed similar Zn adsorption ability in both HE and NHE *S. alfredii* Hance, but they differed in desorption characteristics (Li et al., 2007). Compared to NHE, Zn was bound more loosely to root cell walls in HE whereas Cd was retained more tightly in the NHE due to more free pectic acid residues in the NHE roots cell wall (Li et al.,

2014b). Two *MTP* genes were cloned from HE and NHE *S. alfredii* Hance, and named as *SaMTP1* and *SnMTP1*, respectively. The two cloned *MTP* genes were mainly localized to the tonoplast, and induced under Zn treatment but not influenced by Cd treatment. *SnMTP1* gene was found mainly expressed in root and was induced under high Zn stress, while *SaMTP1* gene expression was induced in shoots but not roots by Zn level (Zhang et al., 2011). The relatively low expression of *SaMTP1* gene in HE root might be attributed to the reduced root sequestration into vacuole and thus maintained Zn more mobile for xylem loading. Recently, we found that *SaHMA3* gene play an important role in Cd translocation and localization into vacuoles (Zhang et al., 2015). However, the Cd sequestration transporters still need to be studied further.

3.1.3 Compartmentation and detoxification

In HE *S. alfredii* Hance, Zn concentration increased in the order of root$<$leaf$<$stem (Li et al., 2006). HE *S. alfredii* Hance had the higher ability to remobilize Zn from mature leaves to mesophyll cells surrounding phloem of new leaves (Lu et al., 2013b), but Zn saturation occurred in HE leaf mesophyll cells and stem vascular bundles relatively earlier, while epidermal layers served as a more important storage sites for Zn accumulation in HE *S. alfredii* Hance (Tian et al., 2009). However, the difference was that mesophyll and vascular cells were more important storage sites for Cd storage of HE *S. alfredii* Hance (Tian et al., 2011a). Besides, the Cd shared the similar storage sites with Ca (Tian et al., 2011a), and Cd uptake and translocation in HE *S. alfredii* Hance was positive associated with Ca pathway (Lu et al., 2010), which might be due to the increasing glutathione (GSH) content after Ca application (Tian et al., 2011b). At the subcellular level, it was observed that Zn was mostly distributed in cell walls and associated with malate in stems and complex with malate in *S. alfredii* Hance leaves (Lu et al., 2014), but aqueous Zn still occupied the highest proportion ($>$55.9%) of Zn speciation (Lu et al., 2013b). However, a larger proportion of Zn was found to be stored in organelles in NHE *S. alfredii* Hance than HE *S. alfredii* Hance (Li et al., 2006), which might partly explain its relatively sensitive to high Zn exposure. Extended X-ray absorption fine structure spectroscopy analysis showed that Cd in the stems and leaves of the HE was mainly associated with oxygen ligands and malic acid (Tian et al., 2011a). Besides, a positive co-relationship of Zn distribution with P or S was revealed in HE *S. alfredii* Hance leaves and stems through SRXRF analysis (Tian et al., 2009). However, this relationship was not evident in NHE *S. alfredii* Hance. Chelation was suggested as a main strategy for heavy metal detoxification or the long distance transportation in plants. Detailed analysis of Zn and Cd speciation might be useful to verify the detoxification mechanism of Zn and Cd.

Over accumulation of hydrogen peroxide (H_2O_2) and superoxide radical (O^{2-}) was proposed as the main reason for phytotoxicity caused by excessive Zn and Cd, and much higher H_2O_2 was induced by Zn treated leaf than Cd treated one (Chao et al., 2008). Research showed that, $>$500 μmol/L Zn in solution caused a marked increase in antioxidant enzyme activities, such as superoxide dismutase (SOD), catalase (CAT), ascorbate peroxidase (APX) and guaiacol peroxidase (GPX).

However, in NHE, >10 μmol/L Zn was sufficient to activate the responses of antioxidant enzyme (Jin et al., 2008b). The different Zn concentration initiated antioxidant responses reflected the higher tolerance ability of HE compared to NHE *S. alfredii* Hance. GSH and ascorbic acid (AsA) contents were found to be increased with Zn or Cd treatment in both HE and NHE *S. alfredii* Hance (Chao et al., 2008; Jin et al., 2008a, 2009), but the total S and GSH contents in shoots were higher under Cd treatment than under Zn treatment (Chao et al., 2008). Furthermore, a dose-dependent decrease in oxidized glutathione and marked increase in reduced glutathione and non-protein thiols were observed in root tips of HE *S. alfredii* Hance, but were not seen in the NHE *S. alfredii* Hance plants after Cd exposure (Tian et al., 2012). Besides, the expression level of *GSH1* in HE *S. alfredii* Hance shoots and roots also increased as Cd level increasing, but decreased in NHE *S. alfredii* Hance (Liang et al. 2014). It was presumed that GSH might directly bind with heavy metals or participate as a signal molecule in HE *S. alfredii* Hance (Jin et al., 2008a, 2008c).

3.2 Mechanisms of Cu and Pb accumulation by plants

3.2.1 Uptake and accumulation

E. splendens, a native Chinese Cu-tolerant and accumulating plant species, has been identified to be tolerant to high Cu concentration and have great potential in remediating contaminated soils (Jiang et al., 2004). *E. splendens* can do hyperaccumualte copper in its shoots with copper concentration up to 1133 and 3417 mg/kg wt when exposed to 500 and 1000 μmol/L Cu in nutrient solution, respectively. Copper uptake efficiency and translocation efficiency were dependent on Cu concentration in nutrient to maintain the concentrations of other essential nutrients, except potassium, within the range considered sufficient for normal growth of higher plants (Yang et al., 2002). *E. splendens* has the greater capacity to absorb Cu in roots from water and translocated Cu from roots to shoots (Tian et al., 2008).

It was found that in the field experiment *E. splendens* can co-tolerate the concentration of 1500 mg/kg Cu in soil (Peng and Yang, 2007). After exposure to 500 μmol/L Cu for 8 d, about 1000 mg/kg Cu were accumulated in the stem and 250 mg/kg Cu in the leaf of *E. splendens* (Peng et al., 2005). And *E. Splendens* can accumulate Cu up to 11.7 mg/plant if grown on the Cu smelter (Jiang et al., 2003).

E. argyi can tolerate external Pb levels up to 200 μmol/L in nutrient solution. The plants can healthy grow in metal complxed contaminated soils with Pb over 3000 mg/kg. Lead concentration in leaf, stem and root increased with lead supply levels and reached to 350, 1600, 4000 mg/kg, respectively when exposed to Pb level of 100 μmol/L (Fig. 6). ME exhibited higher tolerance to excessive levels of Pb in the growth medium, and Pb concentrations in leaves and stem of the ME were 2.6 and 4.5 times respectively, higher than those of the NME (Ejaz et al., 2008). And in *E. splendens*, lead concentration in roots, stems and leaves of could reach 45, 183.6, 1657.6, and 380.9 mg/kg, respectively (Zhang et al., 2011).

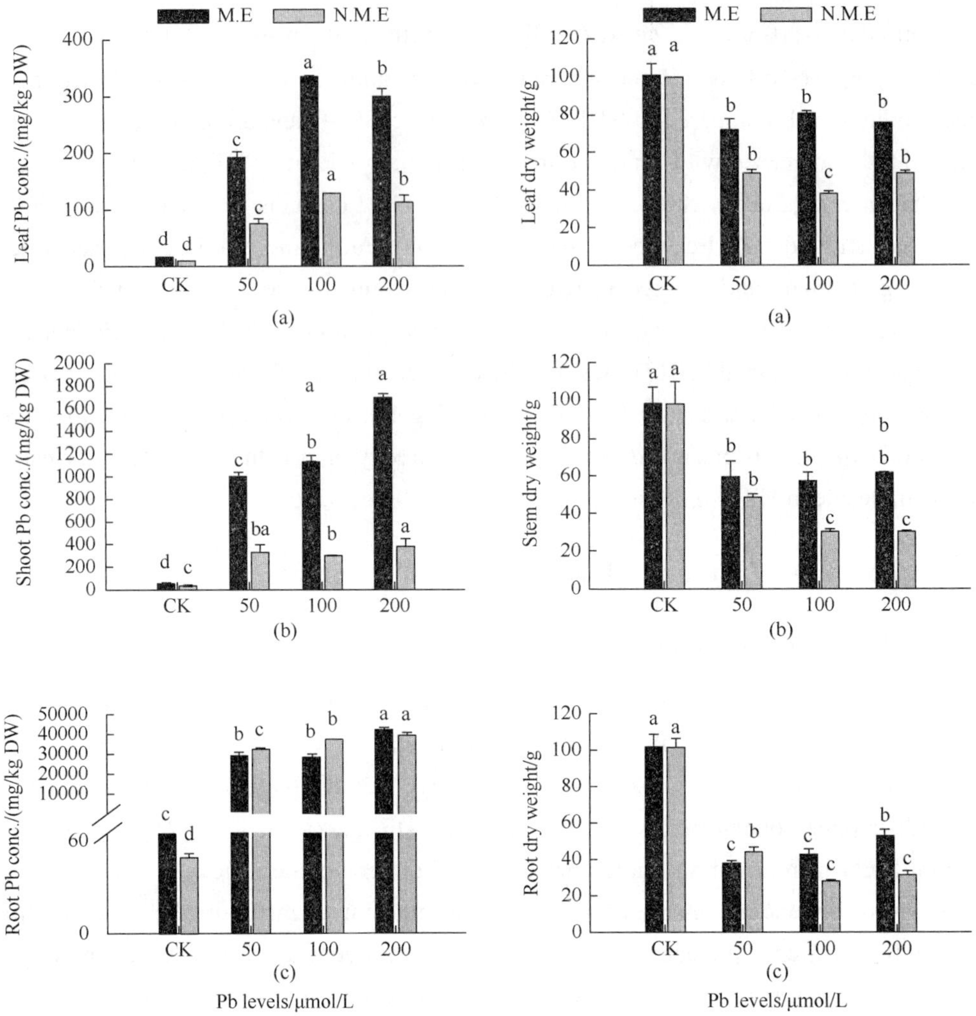

Fig. 6　Growth response and Pb uptake and accumulation in mined ecotype (ME) and non-mined ecotype (NME) of *E. argyi*

Lead concentrations in accumulator and non-accumulator ecotypes of *S. alfredii* Hance. were positively correlated with root length, root surface area, and root volumes (Li et al., 2005). The critical Pb concentration tolerance of *S. alfredii* Hance was 1000 μmol/L, and Pb concentration in different plant parts was in the order of root>stem>leaf (Xiong et al., 2004). Pb concentrations in the shoots of the accumulating ecotype of *S. alfredii* Hance were 1198.1 mg/kg at 0.2 mM treatment level of Pb (Liu et al., 2008). The Pb concentration in roots and shoots of accumulating ecotype (AE) of *S. alfredii* Hance increased with the increase in Pb treatment, at 200 μmol/L Pb, and the maximum Pb concentration after 5 d was 53 775 mg/kg DW in roots and 2506 mg/kg DW in shoots (Gupta et al., 2010). The plantlets of *Pfaffia glomerata* also showed significant lead accumulation in roots (1532 μg/g DW) with a low root to shoot lead translocation (Gupta et al., 2011).

Lead uptake kinetics, xylem loading and the contribution of low root zone temperatures and transpiration streams towards Pb uptake by the ME and NME of *E. argyi* were studied. For xylem loading of Pb both Ae and NAE showed a continuous increment with time and reached the maximum at 12 h; xylem loading of Pb by ME plants was higher than NME. Uptake of Pb was significantly and adversely affected by low transpiration streams (TS), however the overall reduction was much lower in AE than in NAE plants. Pb uptake was less reduced by low root zone temperatures RZT in AE than in NAE plants. It is suggested that apart from the passive uptake of Pb the mined ecotype of *E. argyi* has an active component working simultaneously for the uptake of Pb (Ejaz and Yang, 2008).

Lead uptake characteristics of *S. alfredii* Hance were studied in hydroponics (Liu and Yang, 2008). The results indicated that under low Pb treatment level (10 μmol/L), the concentrations of measured Pb uptake by accumulating ecotype of *S. alfredii* Hance (AE) were higher than the non-accumulating ecotype (NAE) plants. It seems that active Pb-uptake prevailed in AE plants at low external Pb treatment level; in contrast, passive uptake may be dominant in NAE at this treatment level. At Pb level of 50 μmol/L, active uptake may prevail in both ecotypes of *S. alfredii* Hance. After treating with metabolic inhibitor (CCCP), in AE plants, active uptake occupied 33.0%, 18.6% and 20.5% of the total uptake amounts when exposed to Pb levels of 10, 25 and 50 μmol/L, respectively; while in NAE plants these were 18.3%, 14.2% and 9.8%, respectively. In low-temperature (4℃) experiment, similar trend as metabolic inhibitor was noted as active uptake percentages of AE plants were always higher than that of NAE at all treatment level. It seemed that both active and passive Pb-uptake components co-exist in *S. alfredii* Hance. plants, with their relative contribution was higher in AE than in NAE, especially at low Pb levels.

3.2.2 Distribution and compartmentation

In *Elsholtzia* plants, Cu and Pb distribution in different plant organs are: Root>stem>leaf. The root accumulates highest Cu and Pb while stem can hyperaccumulate Pb and cu when exposed to high levels. Leaf contained lowest concentration of Cu and Pb. For *S. alfredii* Hance., Pb distributrion was also root>stem>leaf. The accumulating ecotype (AE) of *S. alfredii* Hance distributed more Pb in stem and leaf than the NAE (Liu et al., 2008). Pb was mostly accumulated in the roots; however only the accumulating ecotype *S. alfredii* Hance showed significant Pb accumulation in shoots (Gupta et al., 2010). And there were also high concentrations up to more than 1000 mg/kg of Pb was transported to the shoots (Fig. 7).

With sequential chemical extraction and enzymatic methods, we found that Cu was mainly bound with cellulose and lignin on the cell wall both in roots and leaves. Water soluble and acid soluble fractions of Cu also accounted for a large proportion of total Cu present in cell, respectively. Pb accumulated mainly on the cell walls of the roots, stems and leaves in *S. alfredii* Hance (He et al., 2003). Micro-XRF analysis of the stem and leaf cross section revealed that Pb was mostly restricted in the vascular bundles and epidermis tissues of both stem and leaf of *E. Splendens* (Zhang et al., 2011). We also found a very low mobility of Pb out of vascular bundles and was largely retained in the cell walls during transportation in plants of *S. alfredii* Hance (Tian et al., 2010), Pb distribution related closely to S distribution (Fig. 7). Further study revealed that the soluble Pb-EDTA complex could be transported and accumulated within the plants of *S. alfredii* Hance, but EDTA does not increase the internal mobility of Pb (Tian et al., 2011).

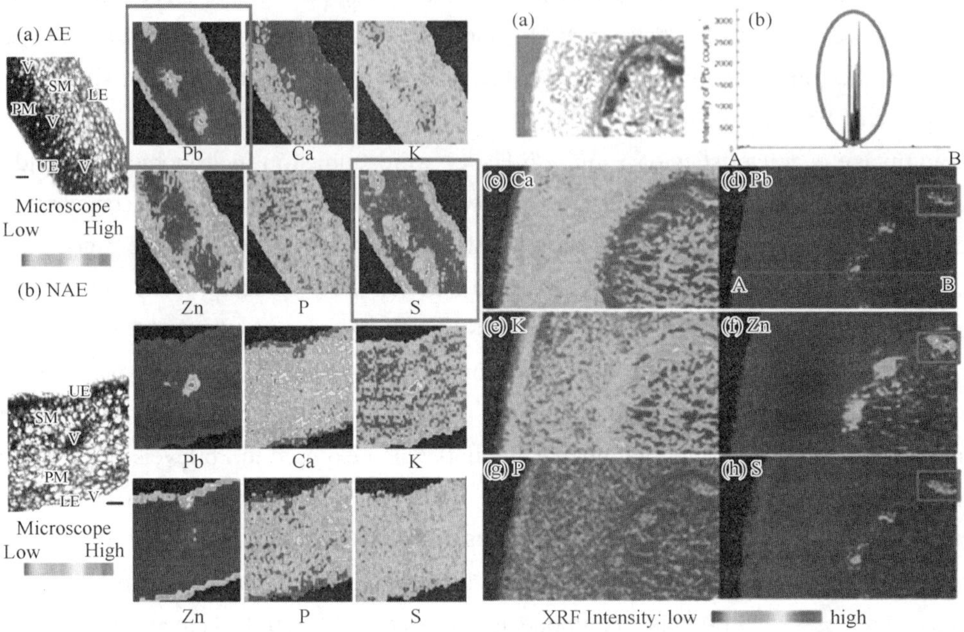

Fig. 7　Lead cellular distribution in leaf (left) and stem (right) sections of *S. alfredii* Hance

3.2.3　Detoxification

Copper localization in cell walls and chloroplasts could mainly account for the high detoxification of Cu in *E. Splendens* (Peng et al., 2005). Also, some physiologically adaptive responses to Cu toxicity and the antioxidant enzyme system were involved in Cu detoxification in leaves of *E. Splendens* observed by Peng (2006). The activities of antioxidative enzymes SOD, XAT, GPX and APX in leaves followed a similar pattern that all of them remained lower in treatments of 50 and 100 μmol/L Cu supply compared to those with Cu treatments lower than 50 μmol/L. the activities of these enzymes dramatically increased when plants grew in nutrient solution with Cu higher than 100 μmol/L. However, SOD and CAT seems to be more important in antioxidative stress initiated by Cu toxicity for their activities were always much higher than those of GPX and APX. The results implied that the oxidative stress was ROS dominantly in the given conditions. MDA content and electrolyte leakage in leaves were lower in treatments of 50 and 100 μmol/L Cu supply compared to those in other Cu treatments. The GSH content and Gr in 50 and 100 μmol/L Cu supply were higher than those of control or that in 500 μmol/L Cu supply. At 500 μmol/L Cu level, Cu detoxification could be attributed to the elevated activities of antioxidant enzymes such as GPX, APX, CAT, SOD, and GR in *E. splendens*; At 50 μmol/L Cu, the relationship between Cu detoxification and the antioxidant system in the plant leaf cell was maintained such that no marked Cu toxicity was exhibited (Peng et al., 2006). The results showed that Cu detoxification capacity of *E. splendens* depends on the balance between the oxidative-defense capacity and excessive Cu-induced oxidative stress.

The roles of free amino acids and organic acids played in Cu tolerance and accumulation were evaluated. Aspartate concentrations in leaves were negatively correlated to Cu concentrations both in the leaves and in solution. This might be due to the metabolic enhancement and the accelerated turnover of Asp in cells under Cu stress. Concentrations of histidine and γ-aminobutyric in correlated to Cu concentrations both in leaves and in solution. Glutamate concentrations in leaves were negatively correlated to GABA concentrations in leaves. Histidine might be involved in Cu uptake and hyperaccumulation in *E. splendens*. γ-aminobutyric might enhance the ability of the plants to tolerate toxic Cu by signaling. The depletion of Glu in lesves was due to the enhanced synthesis of GABA under Cu stress. As far as we know, it was the first time that γ-aminobutyric was found to be involed in heavy metal tolerance and hyperaccumulation (Yang et al., 2005). The role of His played in Cu-hyperaccumulation has not yet been reported in any of reliable documents.

Citric acid concentrations in leaves were negatively correlated to Cu concentrations in leaves. Exterior supply with 250 μmol/L citric acid plus 250 μmol/L Cu significantly decreased Cu concentrations in roots, stems and leaves compared to those supplied with only 250 μmol/L Cu. The plants exposed to Cu plus citric acid grew better than with addition of Cu only in terms of root elongation and shoot height as well as biomass yield. This result implied that citric acid enhanced cu tolerance through declining Cu uptake by *E. splendens*. Citric acid may complex with Cu in nutrient solution and thus decreases the availability of Cu.

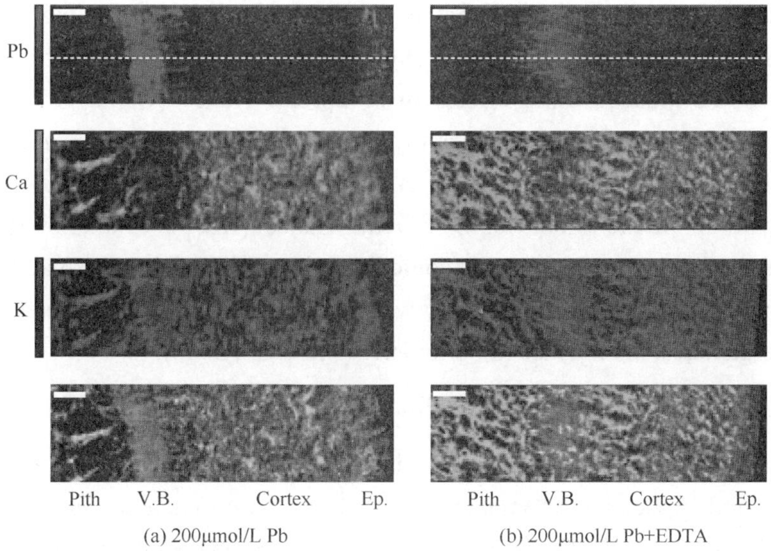

Fig. 8 Effects of (a) Pb and (b) Pb+EDTA application on distribution of Pb across stem of *S. alfredii* Hance (Tian et al., 2010)

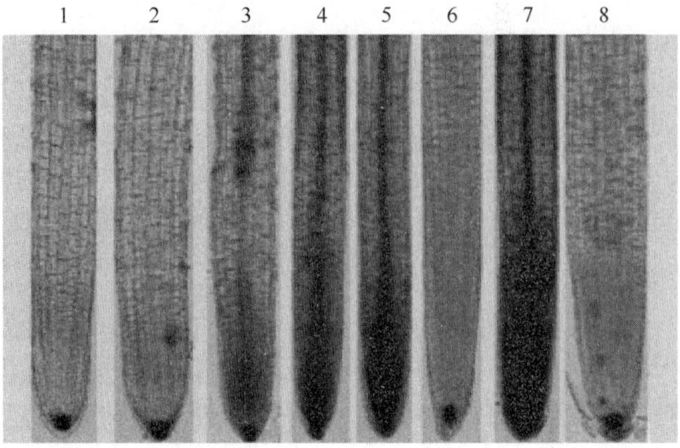

Fig. 9 Effect of lead and EDTA on cell death in accumulating ecotype of *S. alfredii* Hance root. Viability of cells was visualized by Evan's blue staining. The treatments are as follows: Control (1), 25 μmol/L Pb (2), 50 μmol/L Pb (3), 100 μmol/L Pb (4), 200 μmol/L Pb (5), 200 μmol/L Pb+200 μmol/L EDTA (6), 400 μmol/L Pb (7) +200 μmol/L EDTA (8)

The presence of EDTA or Zn alleviated Pb phytotoxicity through changes in *S. alfredii* biomass, root morphology and chlorophyll contents (Liu et al., 2008). The alleviation of Pb toxicity by Zn was mainly due to improved plant defense against oxidative stress by competing with Pb for binding to critical cell constituents in *E. argyi* (Islam et al., 2011). In molecule research, other mechanisms of lead detoxification mentioned are sequestration of cells, the role of antioxidant enzymes and antioxidants. Large amounts of lead deposit in the cell walls of plants, preventing the damage of insoluble substance in the cell (Huang et al., 2009). Lead was predominantly restricted to the vascular bundles of both leaf and stem of the accumulator ecotype (Tian et al., 2010). These results indicated that the distribution of Pb and cell walls of the accumulating ecotype *S. alfredii* Hance could mainly account for the high tolerance to Pb (He et al., 2003) (Fig. 8-9). Enzymatic and non-enzymatic antioxidants play a key role in the detoxification of Pb-induced toxic effects in *S. alfredii* Hance (Gupta et al., 2010). In the absence of PCs, GSH may play an important role of detoxification mechanism in both ecotypes of *S. alfredii* Hance under Pb stress (Gupta et al., 2010). However, GSH played a role in lead detoxification of Pb in the shoots of *S. alfredii* Hance (Huang et al., 2009). Pb tolerance exhibited by accumulating ecotype plants could be mainly attributed to the maintenance of their growth, activity of antioxidant enzymes, as well as an integrity of cell organelles, especially membranous organelles, in response to Pb toxicity (Liu et al., 2008).

3.2.4 Root exudation and metal mobility

Phytoextraction of Cu by *E. splendens* increased with increasing Cu levels in solution, EDTA addition and application of organic manure at an appropriate rate (Jiang and Yang, 2004; Peng et al., 2005). Peng (2005) found that the increased extractability of Cu have mainly attributed to the rhizospheric acidification and chelation by dissolved organic matter (DOM), thus resulting in elevating Cu uptake and accumulation by *E. Splendens* (Peng et al., 2005).

We studied effects of Pb on root exudation of organic acids (OAs) and role of these OAs in soil metal mobility and phytoextraction by *E. argyi* (Ejaz and Yang, 2008). The results showed that the application of different Pb treatments induced the exudation of lactic acid (LA) with comparatively more induction in the AE. The Pb and Cu extractability of root exudates collected from. *E. argyi* under different Pb toxicity levels was tested using two types of soils i. e. Zhuji mined soil (ZMS) and Sanmen mined soil (SMS). The results showed that the overall metal extractability of the root exudates of AE was higher than NAE. The metal mobilization of various organic acids identified in the root exudates was evaluated to be decreased in the order: OA>TA; LA>MA for Pb and OA>TA; MA>LA for Cu, respectively. In SMS, these decreased in the order: OA > LA > MA > OA for Pb, and MA > TA > LA > OA for Cu, respectively. Pot experiment showed that application of OAs brought significant increase in the shoot Pb and Cu concentrations of *Elsholtzia*, especially Pb accumulation in the shoots treated with LA. It is suggested that in *E. argyi*, MA may help the plants to take up exceptionally higher concentration of Pb and Cu from the growth medium but restricts them only up to roots, while LA might be involved in metal mobility in rhizosphere and help root to shoot translocation of these metals. We found that there is a synergistic effect produced by IAA in combination with EDTA on the Pb uptake by *S. alfredii* Hance (Liu et al., 2007). In addition, increasing Zn/Pb levels or citric acid or acetic acid at higher concentrations ($>10^{-3}$ mol/L) could also enhance Pb uptake (Yang et al., 2006, 2010).

4. Microbes-assistant phytoremediation

4.1 Exploitation of plant growth promoting endophytic microbes

4.1.1 Screening and identification of endophytic microbes from hyperaccumulator

Eighty-five metal-resistant bacterial strains were isolated from rhizosphere soils and *S. alfredii* Hance plants grown in the old Pb/Zn mining area using mineral-minimal media containing 1-aminocyclopropane-1-carboxylic acid (ACC) as the sole nitrogen source and high levels of Zn, Cd and/or Pb. Analyses of bacterial 16S ribosomal RNA genes (16S rDNA) revealed that the 85 strains were affiliated to 20 genera of *Proteobacteria* (55%), *Actinobacteria* (27%), *Firmicutes* (11%) and *Bacteroidetes* (7%) and that 45 strains (53%) might represent 15-17 novel species of eight genera. ACC deaminase structure genes (*acdS*) were amplified from 28 strains; ACC deaminase activities were detected from 21 of them in free-living states. The phylogenies of the *acdS* sequences from the other seven strains that did not show ACC deaminase activities were incongruent with those of their 16S rDNA; these *acdS* genes might be evolved through horizontal transfer. All the 85 strains showed differential resistance to high levels of Zn, Cd and/or Pb. The percentages of the obtained bacterial strains with relatively higher metal-resistance were positively correlated to the metal concentrations of the rhizosphere soil, root and shoot tissues. This indicated the nature selection of the high level metals in the soils and plants on the bacterial metal-resistance and adaption. Most of the bacterial strains could produce indole acetic acids and siderophores, or solubilize mineral phosphate, and thus had potentials to promote plant growth and increase metal-solubility from soils.

4.1.2 Colonization of microbes in the root of plant

Inoculating endophytic bacteria was proven as a promising way to enhance phytoremediation.

S. alfredii Hance is a Zn and Cd co-hyperaccumulating plant species found in an old mining area in China. Four bacterial strains, *Burkholderia* sp. *SaZR4*, *Burkholderia* sp. *SaMR10*, *Sphingomonas* sp. *SaMR12* and *Variovorax* sp. *SaNR1*, isolated from surface-sterilized *S. alfredii* Hance plants were used to investigate their endophytic nature and root colonization patterns and effects on phytoextraction of Zn and Cd. Laser scanning confocal microscopy revealed that *gfp*-tagged *SaZR4*, *SaMR12*, and *SaNR1* cells formed biofilms on roots and that *SaZR4* and *SaMR12* cells could invade root tissues (Zhang et al., 2013).

(a) (b)

Fig. 10 (a) Localization of endophytic bacteria in roots and (b) Effects of endophytic bacteria inoculation on growth of *S. alfredii* Hance

Fluorescence in Situ Hybridization (FISH) was also used to investigate *SaMR12* colonization in the root of oilseed and *S. alfredii* Hance. The results showed *SaMR12* could successfully colonize in the inner tissues of root of oilseed and *S. alfredii* Hance. In the space of root cell, the bacteria cell could gather into block and multiply abundantly (Unpublished).

4.2 Effects of endophytic microbe on metal uptake and phytoextraction

4.2.1 Effects of endophytes on growth and metal uptake of *S. alfredii* Hance.

Soil experiments showed that endophytic bacterial strains *SaZR4*, *SaMR12* and *SaNR1* increased the shoot biomass of *S. alfredii* Hance in two harvests and root biomass in the second harvest in the original Pb/Zn mined soil and a multi-metal contaminated paddy soil. The phytoextraction of Zn, Cd, Pb and Cu was significantly enhanced in the two soils. Phytoextraction of Zn was approximately 2-fold after inoculation with *SaZR4* (Zhang et al., 2011). Therefore, continual inoculation of endophytic bacteria to enhance phytoremediation by the perennial *S. alfredii* Hance is applicable.

The inoculation effects of the broad-spectrum endophytic fungus *Piriformospora indica* on hyperaccumulating and non-hyperaccumulating ecotype *S. alfredii* Hance (HE and NHE) plants were studied in a hydroponic system. NHE plants were stressed under treatment of 200 μmol/L Zn or 10 μmol/L Cd and injured under 400 μmol/L Zn or 20 μmol/L Cd whereas HE plants were able to tolerated higher levels of Zn and Cd. *P. indica* colonized in roots of the two ecotypes, promoted their root development, nitrogen and phosphorus uptake, and metal-tolerance (Fig. 11). HE shoots

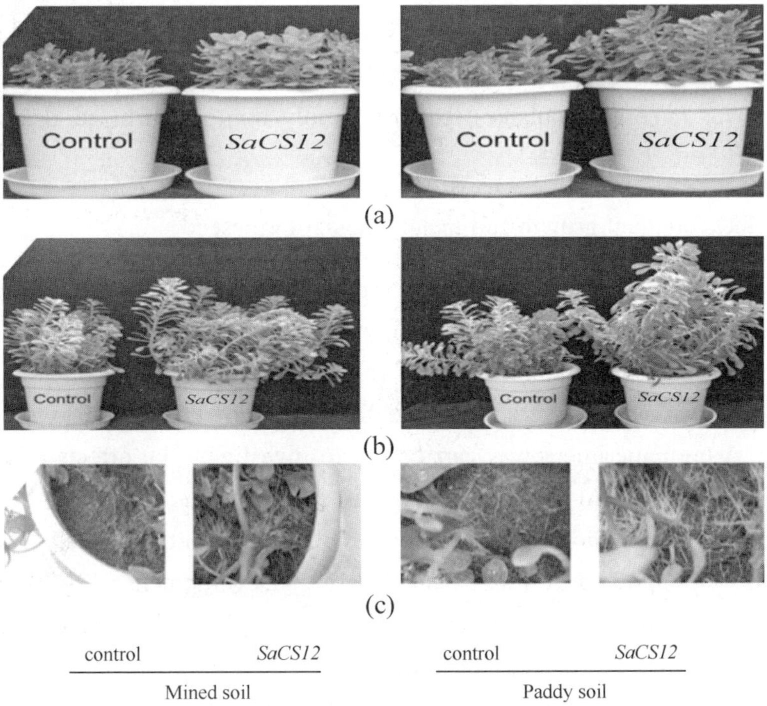

Fig. 11 Effects of *Fusarium oxysporum* on the *S. alfredii* Hance grown (A) after 1 month (B) 5 months (C) in mined soil and paddy soil

accumulated 1.17-1.75 fold Zn or 1.83-2.21 fold Cd compared with the uninoculated controls. *P. indica* reduced Zn or Cd uptake and increased metal-tolerance by NHE plants under 200 μmol/L Zn or 5-10 μmol/L Cd but was not able to reduce the toxic effects under 400 μmol/L Zn or 20 μmol/L Cd. *P. indica* thus has an application potential to enhance phytoextraction by HE *S. alfredii* Hance plants and metal tolerance by non-accumulating plants.

A heavy metal-resistant fungus belonged to the *Fusarium oxysporum* complex was isolated from *S. alfredii* Hance. This *Fusarium* fungus was not pathogenic to plants but promoted host growth. Hydroponic experiments showed that 500 μmol/L Zn or 50 μmol/L Cd combined with the fungus increased root length, branches, and surface areas, enhanced nutrient uptake and chlorophyll synthesis, leading to more vigorous hyperaccumulators with greater root systems. Soil experiments showed that the fungus increased root and shoot biomass and *S. alfredii*-mediated heavy metal availabilities, uptake, translocation or concentrations, and thus increased phytoextraction of Zn (144% and 44%), Cd (139% and 55%), Pb (84% and 85%) and Cu (63% and 77%) from the original Pb/Zn mined soil and a multi-metal contaminated paddy soil. Together, the nonpathogenic *Fusarium* fungus was able to increase *S. alfredii* Hance root systems and function, metal availability and accumulation, plant biomass, and thus phytoextraction efficiency. This study showed an indigenous culturable *Fusarium* fungus other than mycorrhizal fungi to enhance phytoextraction by hyperaccumulators and a new function and application potential for the worldwide distributed soil and plant-associated nonpathogenic *F. oxysproum* fungi. This study also showed a new avenue of microorganism- assisted phytoextraction for hyperaccumulators that are mainly non-mycorrhized Brassicaceae plants (Zhang et al., 2012).

4.2.2 Effects on plant growth and metal uptake of rapeseed

Four plant growth promoting bacteria (PGPB) were used as materials, among them two heavy metal tolerant rhizosphere strains SrN1 (*Arthrobacter* sp.) and SrN9 (*Bacillus altitudinis*) were isolated from rhizosphere soil, while two endophytic strains SaN1 (*Bacillus megatherium*) and SaMR12 (*Sphingomonas*) were identified from roots of the Cd/Zn hyperaccumulator *S. alfredii* Hance. A pot experiment was carried out to investigate the effects of these PGPB on plant growth and Cd accumulation of *Brassica napus* (*B. napus*) plants grown on aged Cd-spiked soil. The results showed that the four PGPB significantly boosted oilseed rape shoot biomass production, improved Soil and Plant Analyzer Development (SPAD) value, enhanced Cd uptake of plant and Cd translocation to the leaves (Fig. 12). The endophytic bacteria performed better in this respect than the rhizosphere bacteria. However, all four PGPB could increase seed Cd accumulation. Due to its potential to enhance Cd uptake by the plant and to restrict Cd accumulation in the seeds, SaMR12 was selected as the most promising microbial partner of *B. napus* when setting up a plant-microbe fortified remediation system (Pan et al., 2016).

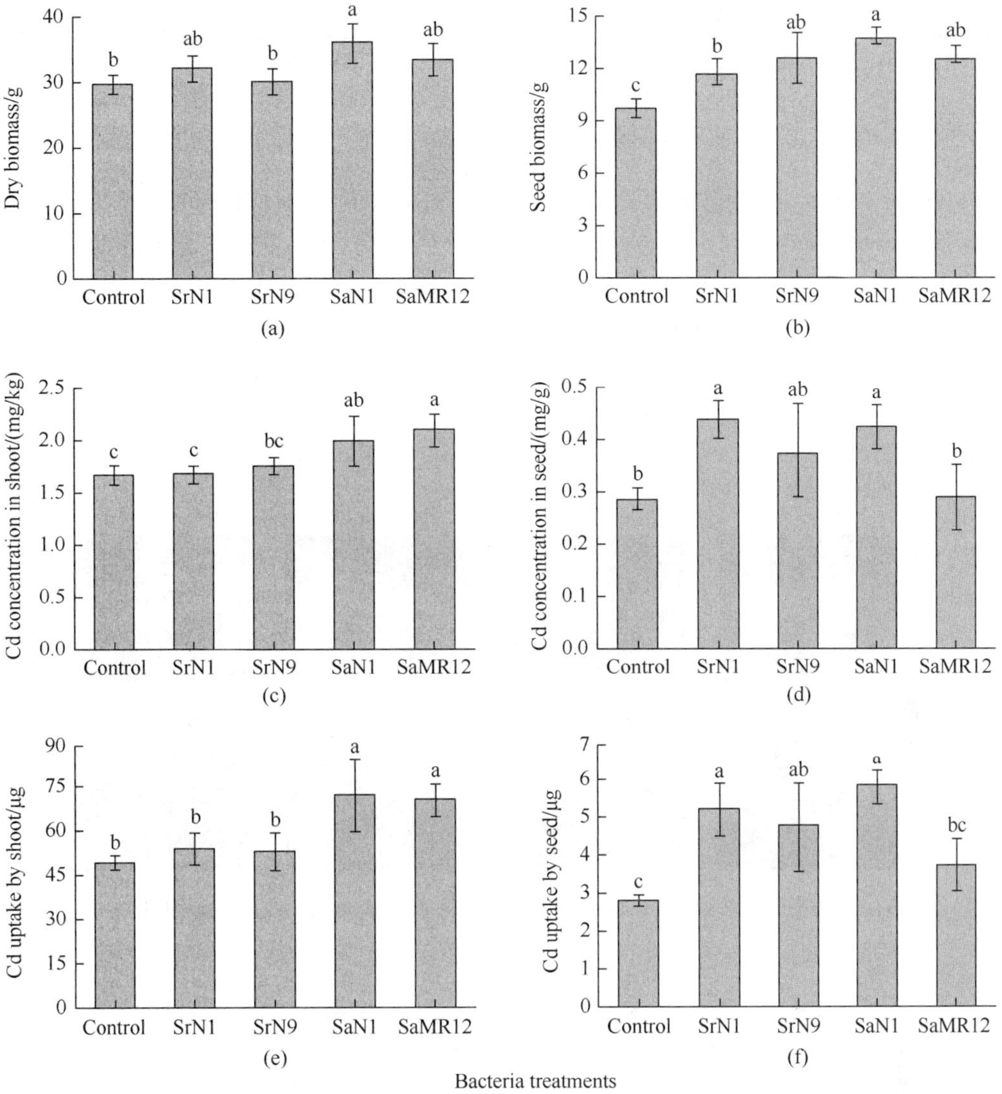

Fig. 12 The influences of PGPB as harvested at mature stage. (a) biomass, (b) seed yield, (c) Cd concentration of shoot, (d) Cd concentration of seed, (e) Cd uptake of shoot, (f) Cd uptake of seed. Error bars represent SD from three individual replicates. The different letters on the error bars indicate significant difference among treatments at $p<0.05$

4.3 Mechanisms of endophytes enhancing metal phytoextraction

4.3.1 Effects on root growth and root hair development

Our serious studies showed that endophytic microbes can considerably enhance root growth and root hair development of the hyperaccumulator. It showed that by inoculating endophytic bacterium *Sphingomonas* SaMR12 to *S. alfredii* Hance. SaMR12 inoculation improved plant biomass, length of roots, number of root tips, and root surface area (Fig. 13).

With The study also showed inoculation with SaMR12 could enhance Zn and Cd accumulation in *S. alfredii* Hance (Chen et al., 2014a, 2014b).

4.3.2 Effects of root exudation of organic acids

Plant growth promoting bacteria have many mechanisms to improve phytoremediation efficiency. Root secretion of oxalic, citric, and succinic acids was also increased after inoculated, which might be alleviated the cadmium toxicity to plant or inhibit the rising trend of oxidative stress of plant. Our study indicated that SaMR12 improved cadmium bioavailability and absorption facility by increasing root-soil contact area and root organic acid secretion; and in the shoot, SaMR12 increased cadmium tolerance by alleviating oxidative stress of plant, so as to enhance the capability of cadmium extraction by plant (Chen et al., 2014b).

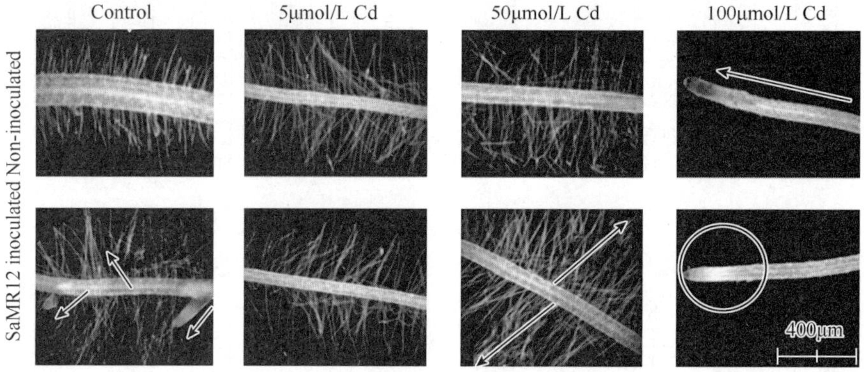

Fig. 13 Effects of endophytic bacteria SaMR12 on root hairs development of *S. alfredii* Hance under different Cd levels

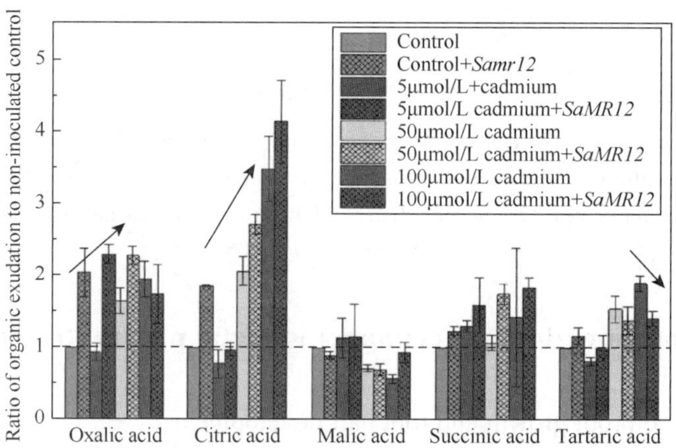

Fig. 14 Organic acid exudation as affected by SaMR12 and Cd treatment. Bars plot mean 6 SD of three replicate experiments. The level of each acid is normalized to the value obtained for the non-inoculated and un-supplemented treatment

4.4.3 Effects on expression of metal tranporters

Compared with control, inoculation of SaMR12 significantly enhanced Cd accumulation of *S. alfredii* Hance under Cd stress. As a number of antioxidant enzymes (SOD, CAT and POD) in *S. alfredii* Hance were not sensitive to the changes in Cd treatment levels even SaMR12 had no significant effects on the antioxidant enzymes activities (Fig. 15). However, Cd exposure

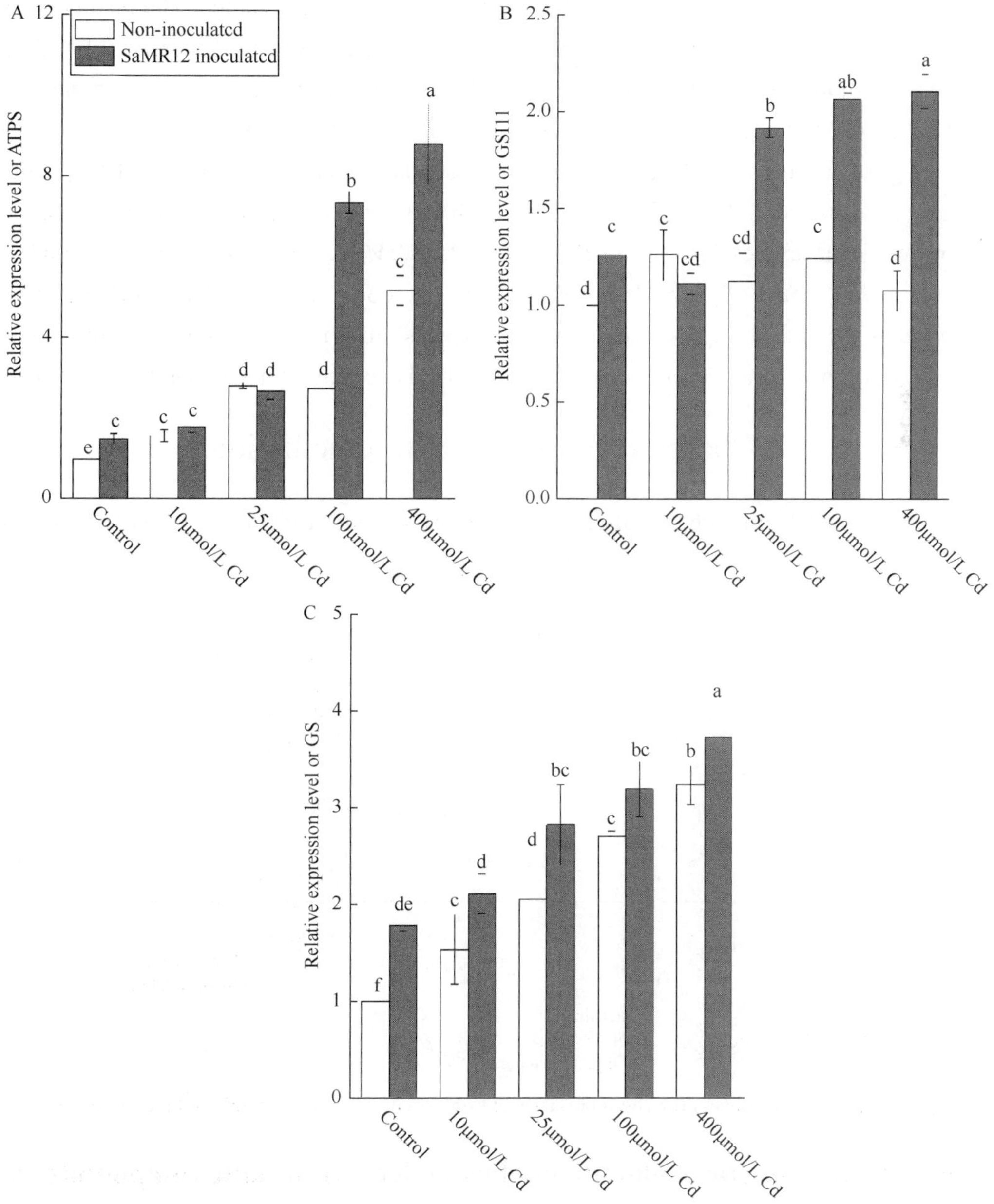

Fig. 15 Effects of Cd and SaMR12 on expression level of ATPS, GSH1 and GS genes. Error bars represent SD from three individual replicates. Different letters indicate significant difference at $p<0.05$

levels and SaMR12 influenced root concentration of H_2O_2, GSH as well as the expression of related genes, such as PER1, ATPS, GS and GSH1. Hence these results indicate that SaMR12 could enhance plant Cd tolerance and accumulation by increasing GSH concentration and related genes expression, and reducing H_2O_2 concentration and root damage (Pan et al., 2016).

We also explored the effects of endophytic bacterial SaMR12 on the expression of three metal transporter families at different Cd treated levels in *S. alfredii* Hance. The results showed that when Cd treated level was within the threshold of toxicity, SaMR12 increased Fe, Zn and Cu content in shoots, elevated expression of *SaIRT1*, *SaHMA2* and *SaNRAMP3* in shoot and *SaHMA2*, *SaNARMP1* in root; suggested that SaMR12 mainly improved the expression of some transporter genes to increase plant acquisition of essential elements and uptake of Cd; while at Cd stress circumstance, SaMR12 increased leave chlorophyll concentration and Fe/Mg content in shoots, elevated expression of *SaIRT1*, *SaZIP3*, *SaHMA2*, *SaHMA3*, and *SaNRAMP6* in shoot and *SaIRT1*, *SaZIP3*, *SaHMA3*, *SaNARMP6* in roots, inferred that SaMR12 mainly increased plant Fe and Mg uptake, regulated the expression of metal transporter genes, so as to improved plant growth and Cd tolerance to enhanced Cd accumulation (Pan and Yang, 2016).

5. Chemical-assistant phytoremediation

We have studied some chemical-assistant phytoremediation technologies since 2002. The basic mechanisms are as showed in Fig. 16.

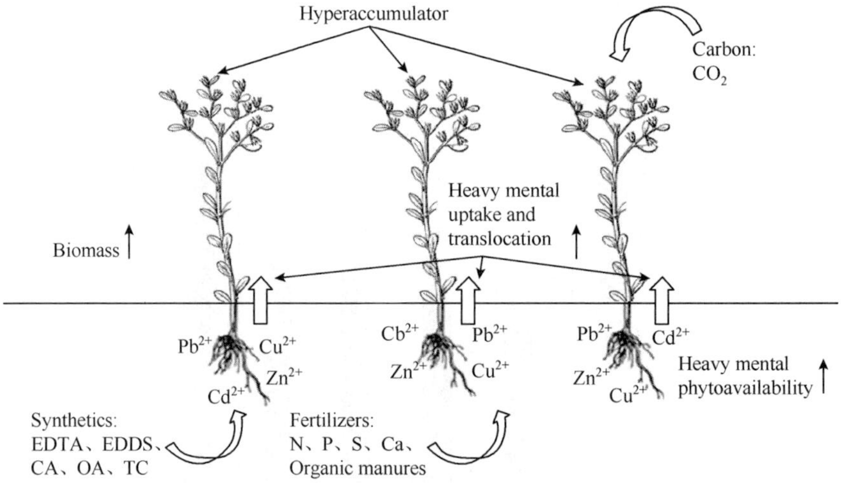

Fig. 16　Basic mechanisms for chemical-assistant phytoremediation of metal contaminated soils

5.1　Effects of synthetic chelators and low molecular organic compounds on enhancing phytoextraction

Some synthetic chelators and organic acids have positive effects on improving the phytoremediation efficiency of hyperaccumulators. Table 1 summarized some research results by us. We reported (Jiang et al., 2004; Yang et al., 2005) that addition of 2.5 mmol/kg EDTA increased the H_2O extractable Cu concentration in soil significantly (from 1.20 mg/kg to 15.78 mg/kg in mined soil and from 0.26 mg/kg to 15.72 mg/kg in paddy soil, in contrast with the control), leading to the shoot Cu concentration of *E. splendens* increased four-fold in mined soil (MS) and eight-fold in paddy soil (PS) compared with the control (Jiang et al., 2004). The efficiency of Cu phytoextraction by *E. splendens* in polluted soils was increased mostly through adding 2.5-5.0 mmol/kg EDTA, while$<$5.0 mmol/kg citric acid (CA) had no remarkable effects (Yang et al., 2005).

The phytoremediation efficiency of *S. alfredii* Hance is also influenced by synthetic chelators and exogenous low molecular weight organic acids. The application of CA or oxalic acid (OA) increased shoot Zn accumulation in non-accumulating ecotype (NHE) *S. alfredii* Hance significantly. But there is no significant effect in shoot Zn accumulation of hyperaccumulating ecotype (HE) *S. alfredii* Hance. However, root Zn accumulation efficiency was prominently increased in both ecotypes of *S. alfredii* Hance (Long et al., 2002). CA might be also involved in the processes of Cd uptake, translocation and tolerance in *S. alfredii* Hance. Therefore, after long-term application of CA or tartaric acid (TA), Cd accumulation was elevated in *S. alfredii* Hance (Lu et al., 2013).

Table 1 Effects of Synthetic chelators and organic acids on phytoremediation effciciency

Hyperaccumulator	Material	Findings	Reference
E. splendens	EDTA	The shoot Cu concentration of E. splendens increased four-fold in mined soil and eight-fold in paddy soil	Jiang et al., 2004
	citric acid	No remarkable effects when citric acid concentration$<$5.0 mmol/kg	Yang et al., 2005
S. alfredii Hance	EDTA	Synergistic effect of IAA/EDTA combination was found on the Pb uptake by *S. alfredii* Hance	Liu et al., 2007
	EDDS	Better Pb accumulation efficiency in low Pb soil	Wang et al., 2009
	oxalic acid	Root Zn accumulation efficiency was prominently increased	Long et al., 2002
	citric acid	Cd accumulation was elevated after long-term application; Root Zn accumulation efficiency was prominently increased	Long et al., 2002; Lu et al., 2013
	tartaric acid	Cd accumulation was elevated after long-term application	Lu et al., 2013

EDTA was found effective for enhancing Pb accumulation in the shoots of *S. alfredii* Hance. And it acted as a chelating agent that assisted Pb to be transported from roots to shoots (Liu et al., 2008). Further study by Liu et al. (2007) showed that there was a synergistic effect of IAA/EDTA combination on the Pb uptake by *S. alfredii* Hance. The treatment of 200 μmol/kg EDTA combined with 10 or 100 μmol/kg IAA increased the Pb accumulation in shoots of *S. alfredii* Hance significantly ($p<0.05$) by 149.2% and 243.7% compared with 200 μmol/kg Pb treatment. Wang et al. (2009) reported that ethylenediaminedisuccinic acid (EDDS) could enhance Pb accumulation in *S. alfredii* Hance as well. And it had a better effect in the low Pb soil (400 mg/kg) than in high Pb soil (1200 mg/kg).

Besides these effects, similar effects are also found in EDDS for Cu and DTPA for Cu/Cd in the shoots of *S. alfredii* Hance (Liu et al., 2008). Though some synthetic chelators could enhance the phytoextraction of heavy metals, the way of reducing potential environmental risks of synthetic chelators still need to be studied.

5.2 Effects of Mineral fertilizers on enhancing phytoremediation

We have conducted a series of hydroponic experiments (Li et al., 2007, 2008; Lin et al., 2007; Zhu et al., 2011), so as to assessed the effects of different nitrogen forms and concentration on *S. alfredii* Hance. The results showed that the application of inorganic nitrogen (urea) significantly ($p<0.05$) promoted the phytoremediation efficiency of Cd compared with organic nitrogen (arginine) in hyperaccumulator *S. alfredii* Hance. While the shoot and root biomass as well as chlorophyll a, b content of inorganic nitrogen treated *S. alfredii* Hance increased markedly ($p<0.05$) (Li et al., 2008). Furthermore, NH_4^+-N has greater ability to promoted Cd accumulation in *S. alfredii* Hance than NO_3^--N (Li et al., 2008). The studies of nitrogen (urea) concentration effects showed that root and shoot growth of *S. alfredii* Hance was promoted consistently when nitrogen level is less than 16. 0 mmol/kg and reached the peak at 16 mmol/kg, while restrained when nitrogen level was more than 32.0 mmol/kg (Li et al., 2007).

Except for N application, applying P appropriately enhanced Zn mobility, absorption, translocation from root to shoot and accumulation in shoots of *S. alfredii* Hance (Sun et al., 2003; Huang et al., 2013). Hydroponic culture experiment showed that Zn contents and accumulations in leaves, stems and roots of *S. alfredii* Hance were significantly increased by increasing P concentrations from 0.5 mmol/kg to 1.0 mmol/kg in nutrient solution (Sun et al., 2003). The pot experiment, conducted by Ni et al. (2004), showed that Zn concentration and accumulation were peaked at 31 mg/kg dose, meanwhile the P concentration in shoots reached 2.09 mg/g DW, where they regarded as a potential index for P application in practice. Recent studies (Huang et al., 2012) indicated that coupled with multiple cuttings, the application of P fertilizers could shorten the time needed for remediation of Zn/Cd contaminated soils. In Table 2, we could conclude that the highest phytoextraction efficiency of Zn and Cd was in the treatments of KH_2PO_4 and $NH_4H_2PO_4$ when P_2O_5 concentration was 352 mg/kg (Huang et al., 2012) indicated that coupled with multiple cuttings, the application of P fertilizers could shorten the time needed for remediation of Zn/Cd contaminated soils (Fig. 17).

Fig. 17 Photographs of *S. alfredii* Hance at the third clipping after 120 days of growth (Huang et al., 2012)

Table 2 Shoot Zn and Cd extraction amounts of different phosphate treat *S. alfredii* Hance (Huang et al., 2012)

P rate (P_2O_5/(mg/kg))	P fertilizer type	Zn extraction amount/(mg/pot)	Cd extraction amount/(μg/pot)
0	KH_2PO_4	5.0±0.64c	119.0±14.43ab
	$Ca(H_2PO_4)_2$	5.0±0.64c	119.0±14.43ab
	NaH_2PO_4	5.0±0.64c	119.0±14.43ab
	$NH_4H_2PO_4$	5.0±0.64c	119.0±14.43ab
22	KH_2PO_4	4.5±0.73cB	114.3±2.85bB
	$Ca(H_2PO_4)_2$	26.4±0.32bA	131.4±9.66aA
	NaH_2PO_4	5.8±0.67bcAB	99.9±5.19bC
	$NH_4H_2PO_4$	6.6±1.26cA	126.9±4.90bcA
88	KH_2PO_4	7.6±0.42bB	128.8±4.03abB
	$Ca(H_2PO_4)_2$	27.1±0.38bB	114.5±9.05aBC
	NaH_2PO_4	7.6±0.61bB	128.9±2.13aB
	$NH_4H_2PO_4$	10.2±1.46bA	147.8±6.71abA
352	KH_2PO_4	12.8±1.12aAB	143.3±6.73aB
	$Ca(H_2PO_4)_2$	212.1±0.85aB	123.4±5.10aC
	NaH_2PO_4	11.1±0.94aB	132.4±4.74aC
	$NH_4H_2PO_4$	13.8±0.69aA	162.6±15.50aA

Different letters indicate statistically different between different treatments ($p < 0.05$), small letters for different phosphorus application rates in the same phosphorus fertilizer treatments and capital letters for different phosphorus fertilizer treatments at the same application rate of P_2O_5.

Sulfur application is another potential method to improve the phytoextraction efficiency of Cd in *S. alfredii* Hance. Cd uptake and translocation efficiency of *S. alfredii* Hance were increased with increasing S (K_2SO_4) supplies in hydroponic culture. And the relationship could be respectively described by logarithmic equation and linear equation. Significant increases were observed in biomass, Cd contents, amounts of Cd accumulated in leaves, stems and roots, total amount s of Cd accumulated in *S. alfredii* Hance when S supplies from 1.5 to 2.25 mmol/kg (Li et al., 2008).

There are other potential mineral fertilizer-assistant phytoremediation methods, such as applying appropriate concentration of exogenous Ca^{2+} to *S. alfredii* Hance to improve its ability of accumulating more Zn (Huang et al., 2008), and so on. All of these findings revealed that using mineral fertilizers to assist phytoremediation has a promising future.

5.3 Effects of carbon and organic manures on enhancing phytoremediation

In recent years, we have examined the effects of elevated CO_2 on improving the phytoremediation efficiency of Cd/Zn hyperaccumulator *S. alfredii* Hance. As Li et al. (2012) reported, after 60d growth, that the total Cd up take per pot and Cd phytoextraction efficiency of *S. alfredii* Hance was significantly ($p < 0.05$) increased in elevated (800 μL/L) CO_2 treatment compared with ambient (350 μL/L) CO_2 treatment, which could be seen from the Table 3. Meanwhile, the shoot and root biomass of elevated CO_2 treated *S. alfredii* Hance was increased by 24.6%-36.7% and 35.0%-52.1%, respectively.

Table 3 Effects of elevated CO_2 treatment on total Cd uptake, bioconcentration factor, phytoextraction efficiency and postharvest available Cd in the soil (Li et al., 2012)

Soil	CO_2 level	Total Cd uptake/(mg/pot)	Bioconcentration factor	Phytoextraction efficiency/%	Post-harvest available Cd/(mg/kg)
CK	Ambient	0.24±0.01	76.6±2.9	15.25±1.07	0.08±0.01
	Elevated CO_2	0.33±0.02*	78.6±1.8	20.47±1.80*	0.06±0.02
Cd5	Ambient	1.92±0.09	61.4±2.1	11.58±1.29	1.43±0.04
	Elevated CO_2	2.86±0.19*	71.9±6.6*	17.18±1.88*	0.99±0.15*
Cd50	Ambient	11.88±1.05	48.0±3.1	9.50±0.89	17.61±1.43
	Elevated CO_2	16.78±1.39*	55.7±3.9*	13.43±1.06*	13.83±2.04*

*Significant differences ($p < 0.05$) between ambient and elevated CO_2 treatments.

Various studies (Li et al., 2013, 2015) showed that elevated CO_2 could change soil microenvironment, increase bioavailability of Cd and Zn through generating more efficient dissolved organic matter (DOM) into rhizosphere (Li et al., 2013) and forming more (8.01% for Cd and 8.47% for Zn, respectively) Cd/Zn-DOM complexes (Li et al., 2014). These changes facilitated metal uptake by the hyperaccumulator *S. alfredii* Hance. What's more, elevated CO_2 increased Pn (105%-149%), Pn_{max} (38.8%-63.0%) and AQY (20.0%-34.8%) of *S. alfredii* Hance significantly ($p<0.05$), and on the contrary, reduced dark respiration and photorespiration, indicating that elevated CO_2 promoted the growth of *S. alfredii* Hance by increasing photosynthetic carbon uptake rate and photosynthetic light-use efficiency (Li et al., 2015). These studies indicated that using elevated CO_2 may be a potential way to improve phytoremediation efficiency of Cd/Zn-contaminated soil by *S. alfredii* Hance.

We also studied the effects of organic manures, and found that the application of pig manure vermicompost (PMVC) enhanced the remediation of Cd and PAHs co-contaminated soil by *S. alfredii* Hance, which inferred to be resulted from increased root exudates by PMVC (Wang et al., 2012). Another pot experiment conducted by Peng et al. (2005) showed that application of organic manure at the rate of 5.0% increased shoot Cu concentration by four times in the *E. splendens* grown on the PS and consequently elevated shoot Cu accumulation by three times compared with the control.

6. Phytoavoidation of heavy metal contamination in crops

6.1 Metal lower accumulation by crops

Phytoavoidation is the reduction of organic and inorganic pollutants uptake into edible parts of crops by using immobilization measures together with low accumulator cultivars. It is very important to select and cultivate pollution safe cultivars (PSC) because most of the heavy metal polluted lands are being used for agricultural purposes. The feasible move to reduce the heavy metal health risk in human through food is the assortment of pollution safe cultivars. The concept of pollution safe cultivars (PSC) came from Yu et al. (2006) who proposed that crop species and their cultivars differ genetically in accumulation of particular heavy metal, the cultivar which accumulates low concentration in their edible parts which is safe for consumption even when grown in contaminated soil. Many pot and field screening experiments has been done on various crops to investigated low and high accumulators of different heavy metals. Therefore, low heavy metal accumulating cultivars in several crop species have been well documented, including rice (Liu et al., 2003; Arao et al., 2003; Cao et al., 2013), Wheat (Zhang et al., 2000; Zhang et al., 2002), barley (Chen et al., 2007), maize (Kurz et al., 1999), peanut (McLaughlin et al., 2000) and Chinese cabbage (Liu et al., 2010a; Liu et al., 2010b; Wang et al., 2014). Whereas, we reported that the Hangzhouyoudonger,

Aijiaoheiye 333, and Zaoshenghuajing cultivars of Pak choi (*Brassica chinensis* L.). Chen et al. (2012) found Cd-PSCs and safe for consumption when grown in low Cd-contaminated soils (≤1.2 mg/kg). Similarly, We (2016) identified low co-accumulator genotypes of Chinese cabbage of Cd and nitrate grown in moderately combined Cd and nitrate contaminated field soil, without posing risk to food safety. In the field culture experiment uptake of Pb was investigated among spinach cultivars to screen out low accumulators of Pb or Pb safe cultivars for food safety (Unpublished).

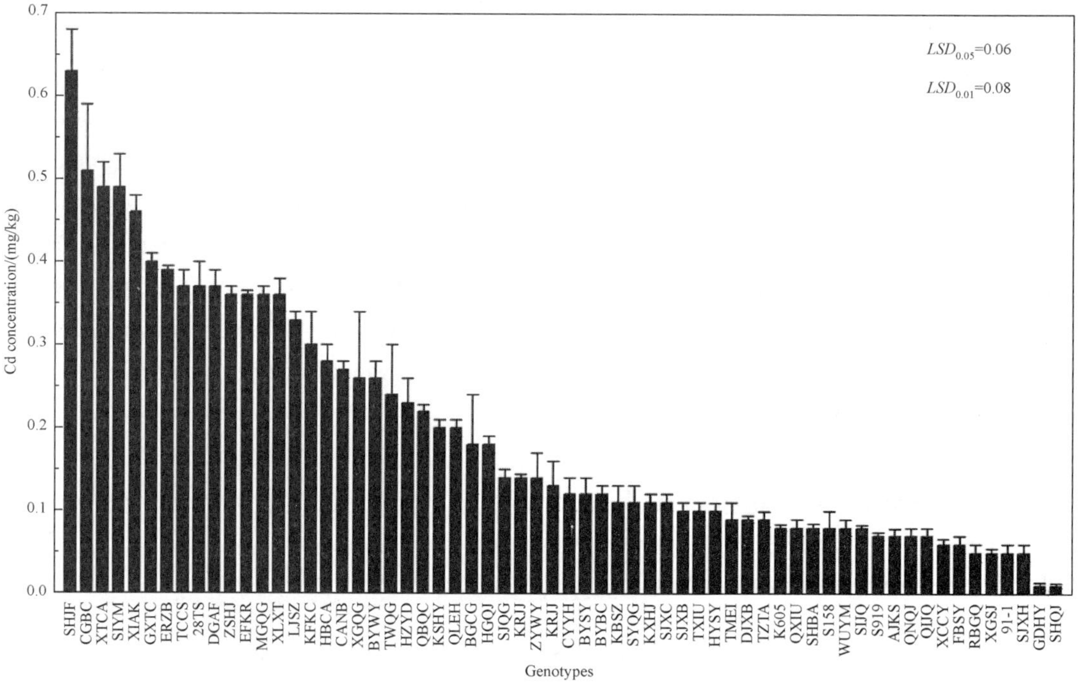

Fig. 18 Cadmium concentrations in 62 Chinese cabbage genotypes grown on the Cd- NO_3^- combined contaminated soil. Plant were harvested after eight weeks growth. All data were means of four replications; bars indicate standard errors of the mean value. The differences among the cultivars are analysed by the *LSD* method (Tang et al., 2016)

The data on the mechanism of low accumulation of heavy metal in different crops is relatively scarce. However, among Pakchoi cultivars in low Cd accumulating cultivars the variations in the shoot Cd concentration were attributed to the differential distribution of Cd within the plant could be the speculated elementarily mechanism of low Cd accumulation of Pak choi (Wang et al., 2014). The diverse tolerance of Cd by *B. chinensis* and *B. pekinensis* might be explained partially by the accumulation of non-protein thiols (NPT) in the shoots of these two plants (Liu et al., 2007). Whereas, Sun et al. (2015) found novel transporter genes HvZIP3 and HvZIP8 in low Cd accumulating cultivar of barley which were identified as being related with low grain cd accumulation. Ueno et al. (2010) reported a gene OsHMA3 from the low Cd-accumulating cultivar of rice which was responsible for restrict translocation of Cd but not other micronutrients from the roots to the above-ground tissues and their over expression caused the low accumulation of Cd by selectively sequestration of Cd into the root vacuoles.

6.2 Amendments for enhancing phytoavoidation

There are different remediation techniques, chemical immobilization is one of them that decrease the concentration of dissolved pollutants by precipitation or sorption (Basta et al., 2001) which supports the phytoavoidation of heavy metals in contaminated soil. For the immobilization of heavy metals in contaminated soils various inorganic (Yang et al., 2003; Park et al., 2011a; He et al., 2013) and organic amendments (Madejón et al., 2006; Park et al., 2011b) have been applied. Inorganic amendments that have been successfully immobilized the heavy metals include rock phosphate, apatite, hydroxyapatite, liming agents, iron and manganese oxides and oxyhydroxides (Keller et al., 2005). In this review we focus on organic amendments as they are cost effective adsorbents and beside immobilization of pollutants also supply plant nutrients and enhance the water retention capacity of soil (Tang et al., 2016). Various studies have assessed the impact of soil organic amendments on immobilization of heavy metals in contaminated soils, but here we summarize the biochar amendments of soils for reducing availability and uptake of heavy metals in soil plant system. Biochar has multiple applications and benefits to use as soil amendment containing C sequestration (Kuzyakov et al., 2009; Woolf et al., 2010), reduction in greenhouse gas emission (Singh et al., 2010), soil as well as water remediation, immobilization of pollutants (Ahmad et al., 2014; Mohan et al., 2014b) and soil fertilization (Chan et al., 2007, 2008).

Numerous studies have been done on land remediation through biochar, but recently researcher's emphasis on immobilization of pollutants especially heavy metals in crops grown in different agricultural soils. For the immobilization of cadmium (Cd) in soil plant system used biochar derived from faecal matter, cow manure, *Prosopis juliflora* and coffee husk (woldetsadik et al., 2016) water hyacinth, rice straw (Yin et al., 2016) compost (Yousaf et al., 2016) nut shield and plum stone magnetic biochar (Trakal et al., 2016). The chicken manure, mushroom soil, corn cobs, corn stover, mixed grass, hop grass, switch grass, pine mulch, pine bark and cow manure derived biochar used for immobilizing mercury (Hg) in soil (Liu et al., 2016). Whereas, for co-contaminated soil with Cd, Pb, Cr, Zn and Cu used pepper stem derived biochar which effectively reduced the concentration of all metals in soil (Park et al., 2016). But there is no published work or data on phytoavoidation of heavy metals in agricultural soils through organic amendments.

Recently, we conducted a pot experiment to evaluate the impact of maize stalk, bamboo and cow manure derived biochar as a soil amendment on cadmium immobilization and uptake by two contrasting Cd accumulating genotypes of *Brassica Chinensis* L. (Pak choi). The maximum reduction of Cd availability in soil was recorded by 54% and accumulation in shoots of low accumulator was 41% with maize straw biochar (Fig. 19), compared to control which did not receive amendment. With bamboo biochar and cow manure derived biochar the availability of soil Cd was reduced by 45% and 39%, while Cd concentration in shoots of low accumulator genotype by 35% and 27%, respectively (Khan et al., Unpublished).

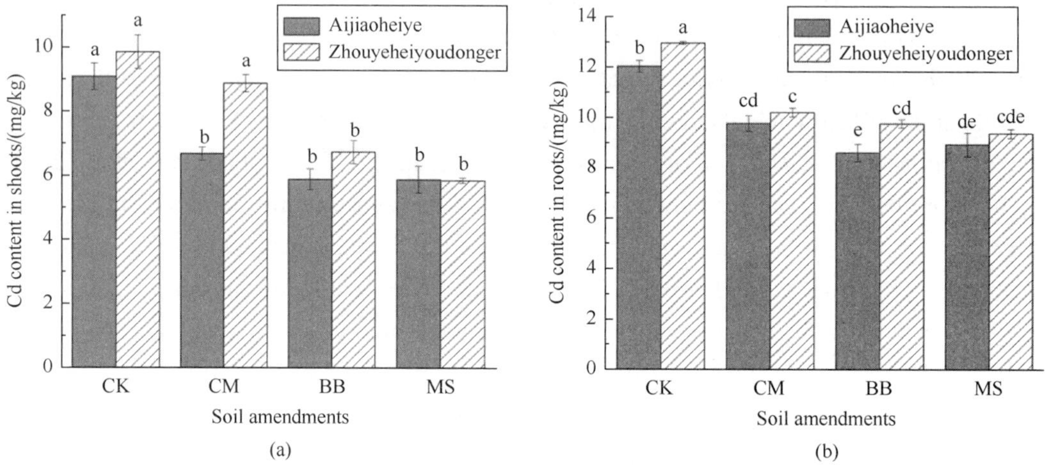

Fig. 19 Effects of different biochar on (a) shoots and (b) roots of *Brassica chinensis* L

Similarly, we conducted a greenhouse experiment in which the cow manure derived biochar was applied as soil amendment at different application rates in Cd contaminated red acidic soil on which contrasting low and high accumulating genotypes of Pak choi were grown. The maximum immobilization of Cd in soil was observed by 70% and its accumulation in shoots by 68% with 6% of biochar application rate, compared to the control which did not receive amendment. Beside immobilization of Cd contents also enhanced shoot dry biomass and availability of essential trace metals (Khan et al., Unpublished). Whereas, the application of humic acid as soil amendment in different Cd contaminated soils also have the potential to reduce the availability of Cd in Haplic-Udic Isohumosol and Udic ferrisol soil by 15%-19% and accumulation in shoots by 25%-34% in both high and low accumulating genotypes of Pak choi compared to the control which did not receive amendment. Experimental findings revealed that application of humic acid increased the bioavailability of Mn and Zn in plant soil system (Khan et al., Unpublished). It has been concluded that organic amendments (biochar and humic acid) positively reduced Cd availability in soil and both low and high Cd accumulator genotypes of *B. chinensis* L. grown in different agricultural soils of China. To make crop safe from health risk with enhancing biomass and essential trace metals in soils and plants as well.

7. Phytoremediation coupled with agro-production (PCA)

Soil contamination in China tends to be severe and complicated. Soil contamination by heavy metals and persistent organic pollutants or secondary salinization has been accelerated in China during the past decade because of rapid urbanization and industrialization. The output of heavy metals and organic pollutants into environment will stay in a comparatively high level for a long period.

7.1 Phytoremediation of Cd-DDTs co-contaminated soils coupled with crop production

Environmental pollution of the soil-crop system is very complex due to the coexistence of old and new pollutants, as well as inorganic and organic compounds. Especially soil pollution caused by heavy metals and persistent organic pollutants (POPs) is widespread concerned. In recent years, the frequent outbreak of environmental pollution accidents due to soil-water pollution puts forward urgent demand for remediation and restoration of contaminated soil.

In this research, the plant-microbe remediation of Cd and DDT was carried out under pot and field experiments to identify bioremediation strategy of Cd-DDs co-contaminated soil and to investigate the mechanism for enhancing bioremediation efficiency. Cadmium (Cd) and dichlorodiphenyltrichloroethane (DDT) or its metabolite residues DDD/DDE (DDT, DDE, and DDD are collectively called DDs) are frequently detected in agricultural soils and agricultural products, posing a threat to human health. The plant-microbe remediation of Cd and DDT was carried out under pot and field experiments to identify bioremediation strategy of Cd-DDs co-contaminated soil and to investigate the mechanism for enhancing bioremediation efficiency.

7.1.1 Plant genotypic differences in co-accumulation of Cd and DDTs

Zhu et al. (2012a) compared the ability of 21 genotypes of *Cucurbita pepo* ssp. in mobilizing and uptake of Cd and DDs (p, p'-DDT, o, p'-DDT, p, p'-DDD and p, p'-DDE) in the co-contaminated soil. The plant genotypes varied greatly in the uptake and accumulation of Cd and DDs, with mean concentrations of 0.26-1.12、0.49-2.25、1.04-4.84 and 1.61-7.72 mg/kg DW for Cd, and 338.4-793.2, 1619-1812, 1273-3548 and 3396-12,811 ng/g DW for DDs in leaf, stem and root, respectively (Fig. 20). The TF and BAF values were 0.79 and 1.68 for Cd, and 0.60 and 1.54 for DDs, respectively. These results indicate that Cucurbita pepo cv. "Tebiexuan mibenwang" has a low ability to absorb and accumulate Cd and DDs from the contaminated soils, but Cucurbita pepo cv. "Riben Hongtianmi" has great potential for accumulating Cd and DDs from moderately co-contaminated soil (Cd≤1.50 mg/kg, DDs≤ 1.00 mg/kg).

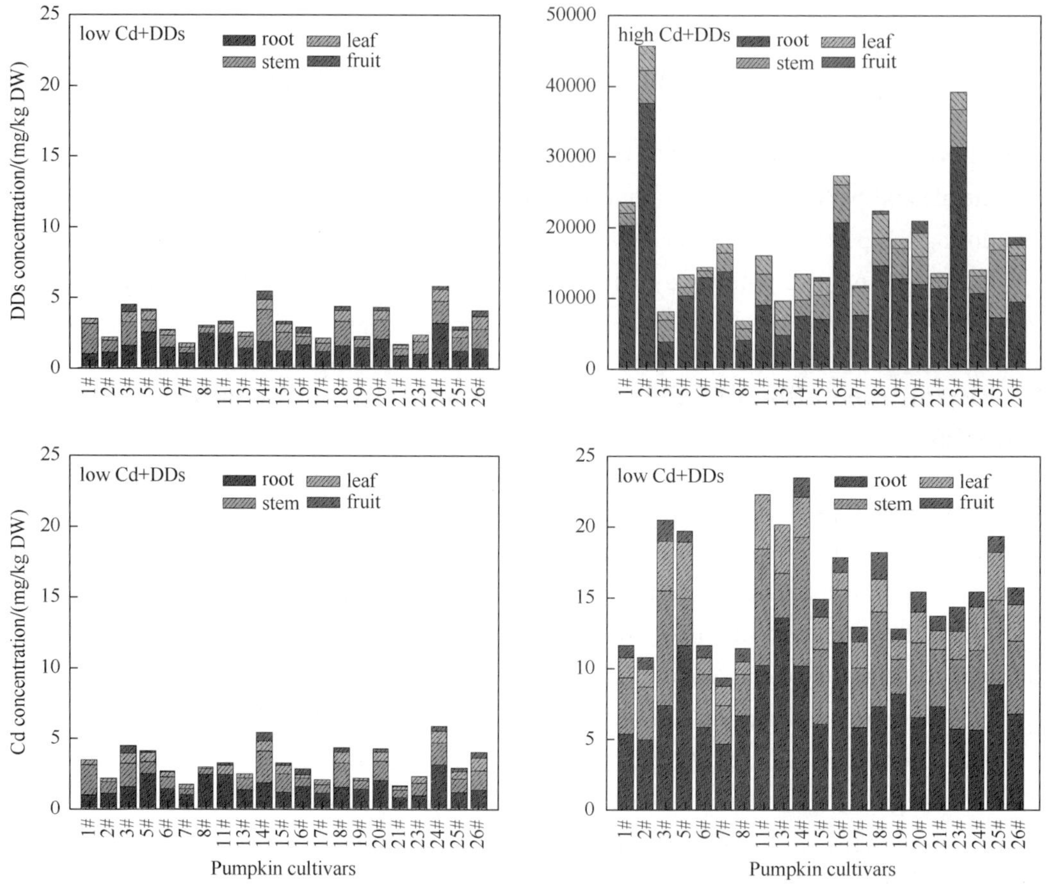

Fig. 20　Distribution of DDs and Cd in different parts of various pumpkin cultivars growing at low (1.35 Cd+ 0.80 DDs mg/kg) and high (3.48 Cd+3.30 DDs mg/kg) levels

The same research was done by Huang et al (2011) on 23 genotypes of *R. communis* by the *R. communis* genotypes varied largely in the uptake and accumulation of DDTs and Cd. These results indicate that *R. communis* has great potential for removing DDTs and Cd from contaminated soils attributed to its fast growth, high biomass, strong absorption and accumulation for both DDTs and Cd.

7.1.2　Phytoremediation of Cd-DDT co-Contaminated Soils

Phytoremediation of Cd-DDTs co-contaminated soils were studied using Cd-hyperaccumulator plant *S. alfredii* Hance (SA) and DDT-degrading microbes (DDT-1) (Fang et al., 2011; Zhu et al., 2012b). Initially, inoculation with DDT-1 was shown to increase *S. alfredii* Hance root biomass in a pot experiment. It was found that when *S. alfredii* Hance was applied together with DDT-1, the levels of Cd and DDs in the co-contaminated soil decreased by 32.1%-40.3% and 33.9%-37.6%, respectively. In the 18 months pot experiment, the levels of Cd and DDs in the co-contaminated soil decreased by 31.1% and 53.6%, respectively (Fig. 21). This result demonstrates that the integrated bioremediation strategy is effective for the remediation of

Cd-DDs co-contaminated soils.

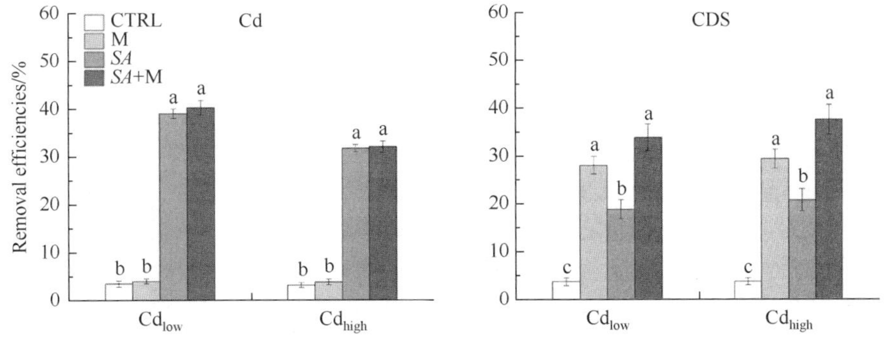

Fig. 21 Effects of *S. alfredii* Hance and DDT-1 strains on removal of Cd and DDS in low and high co contaminated soils in pot experiment (low=1.35 Cd+0.80 DDs mg/kg; high=3.48 Cd+3.30 DDs mg/kg)

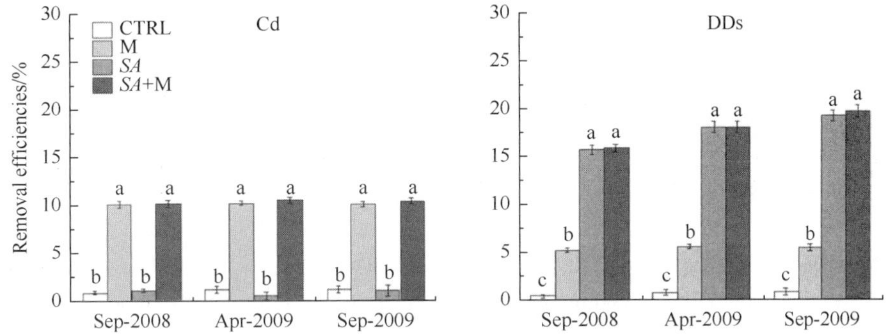

Fig. 22 Removal efficiency of Cd and DDS by sedum and DDT-degrading microbes in field (Cd 0.53+DDTs 0.72 mg/kg)

Rotation or intercropping technology, a widely accepted agronomical practice in China for 2000 years, can increase total crop yields and remediation efficiency through increased resource use efficiency. We found that co-cropping of *S. alfredii* Hance and *Brassica. campestris* ssp. *chinensis* associated with DDT-1 increased the biomass and metal phytoextraction of *S. alfredii* Hance ($p<0.05$), and also enhanced the root growth ($p<0.05$), but had no significant effects on shoot biomass of *B. campestris* ssp. *chinensis*. The combined remediation strategy decreased the Cd concentration by 48.0%-51.5% and 54.7%-71.4% in the roots and shoots of *B. campestris* ssp. *chinensis*, respectively, and also corresponded to a decrease in root DDs concentration by 37%. The removal efficiencies were 30%-46% for Cd and 36.8%-42.7% for DDs. Laser scanning confocal microscopy revealed that gfp-tagged DDT-1 heavily colonized in rhizosphere soil, and on rhizoplane of *S. alfredii* Hance. The soil diversity indices were higher in all of the planted treatments than in the unplanted control. Principal component analysis of bacterial T-RFLP data also revealed strong shifts in bacterial community composition with the planted treatments. The results of this study indicated that the co-cropping of *S. alfredii* Hance and *B. campestris* ssp. *chinensis* associated with DDT-1 appears to be a promising approach for the bioremediation of soils co-contaminated by Cd and DDs, while simultaneously decreasing the pollutant concentration in edible part of *B. campestris* ssp. *Chinensis*, and hence maintaining product safety and reliability of this vegetable (Zhu, 2012a).

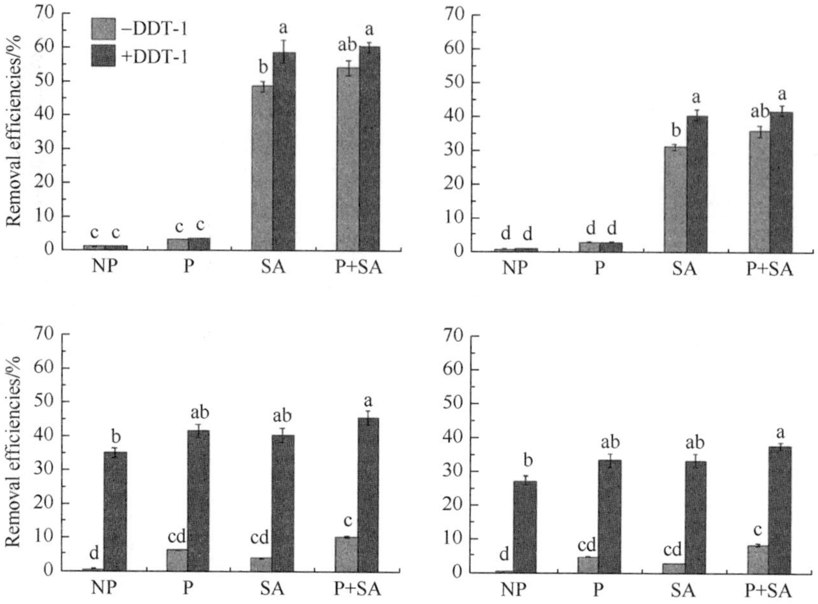

Fig. 23 Removal efficiency of Cd and DDTs by hyperaccumulator and low accumulator co-planting with or without DDT-degrading microbes under successive three crops in field

7.1.3 DDT-degrading microbes under successive three crops in field

Co-cropping *S. alfredii* Hance and *Cucurbita pepo* cv. "Tebiexuan *mibenwang*" associated with DDT-1 increased the root (26.8%) and shoot (21.7%) biomass of cv. T. *mibenwang* ($P<0.05$), but had no significant effects on the root and shoot biomass of *S. alfredii* Hance. The combined remediation strategy did not significantly decrease the Cd concentrations in roots and shoots of both plants. However, for cv. T. *mibenwang*, the bioremediation strategy decreased the shoot DDs concentration by 38.2%-44.5% ($p<0.05$), but had no significant effects on root DDs concentration; for *S. alfredii* Hance, increased and decreased DDs concentration by 34.7%-67.7% and 38.2%-44.5% in roots and shoots, respectively (Fig. 23). The removal efficiencies were 41.9%-60.7% for Cd and 37.5%-45.2% for DDs. For roots of cv. T. *mibenwang*, Cd was largely retained in the cell walls and vessel tissue, DDs was largely retained in central cylinder, and was adsorbed on root hairs, but for shoots, Cd was largely retained in the collenchyma tissue and vacuoles, DDs was largely retained in vessel tissue, vacuoles and glandular trichomes. Laser scanning confocal microscopy revealed that gfp-tagged DDT-1 heavily colonized on rhizoplane of *S. alfredii* Hance and "Tebiexuan *mibenwang*". The soil diversity indices were higher in the treatment than in the unplanted control. The results of this study indicated that the co-cropping of *S. alfredii* Hance and cv. T. *mibenwang* "associated with

DDT-1 appeared to be a promising approach for the bioremediation of soils co-contaminated by Cd and DDs, and decreasing the DDs concentration in shoot of cv. T. *mibenwang*", and ensuring product safety of the pumpkin (Zhu, 2012a).

7.2 Phytoremediation of Cd-PAHS co-contaminated soils coupled with crop production

Cadmium (Cd) and polycyclic aromatic hydrocarbons (PAHs) are of particular concern due to their persistence, potentially carcinogenic, mutagenic, teratogenic properties and their ubiquitous occurrence in the environment. So many regions were co-contaminated by heavy metals and PAHs, a environmental friendly, low-cost, and *in situ* strategies are required.

Li et al. investigated the interaction between Cd and PAHs during the process of soil co-contamination and their effects on phytoremediation of Cd-PAHs co-contaminated soil by a Cd hyperaccumulator *S. alfredii* Hance (Li et al., 2011). Agronomic, chemical, and microbial strategies were also be investigated to improve phytoremediation of Cd-PAHs co-contaminated soils (Wang, 2012a), the same research was also done on other organic pollutants (Fang et al., 2009a, 2009b; Huang et al., 2012; Xiao et al., 2013).

7.2.1 The dissipation of benzo[a]pyrene (B[a]P) in soil

The dissipation of benzo[a]pyrene (B[a]P) in soil was mainly due to microbial degradation (Gu et al., 2010). High Cd concentration (25 mg/kg) significantly inhibited the degradation of B[a]P in the soil, and the addition of pyrene (PYR) with an initial concentration of 250 mg/kg significantly promoted B[a]P degradation. Both desorbing and non-desorbing fractions of B[a]P contributed to B[a]P degradation in soil, however, desorbing fraction contributed more as compared to non-desorbing fraction (Fig. 24) (Wang et al., 2014).

7.2.2 Phytoremediation of Cd-PAHs co-contaminated soil

Wang investigated the growth of *S. alfredii* Hance and the removal effect of contaminants from Cd-PAHs co-contaminated soil via a pot experiment (Wang et al., 2012b). Elevated Cd level (6.38 mg/kg) increased the growth of *S. alfredii* Hance. The presence of PAHs decreased the stimulatory effects of Cd on plant biomass and Cd concentrations in shoots in Cd spiked soil, thus decreasing Cd phytoextraction efficiency. Cadmium removal effect by *S. alfredii* Hance was from 5.8% to 6.7% in the low Cd concentration soil and from 5.7% to 9.6% of the high Cd soil after 60 d, respectively. The elevated Cd inhibited the removal rate of Pyrene and the activity of dehydrogenase in soil. It may attribute to the decrease of the microbial activity in soil. The results demonstrate that the *S. alfredii* Hance could effectively extract Cd with of PHE or PYR from Cd-contaminated soils, but PAHs had negatively affected on phytoextraction of Cd from Cd contaminative soil (Fig. 25).

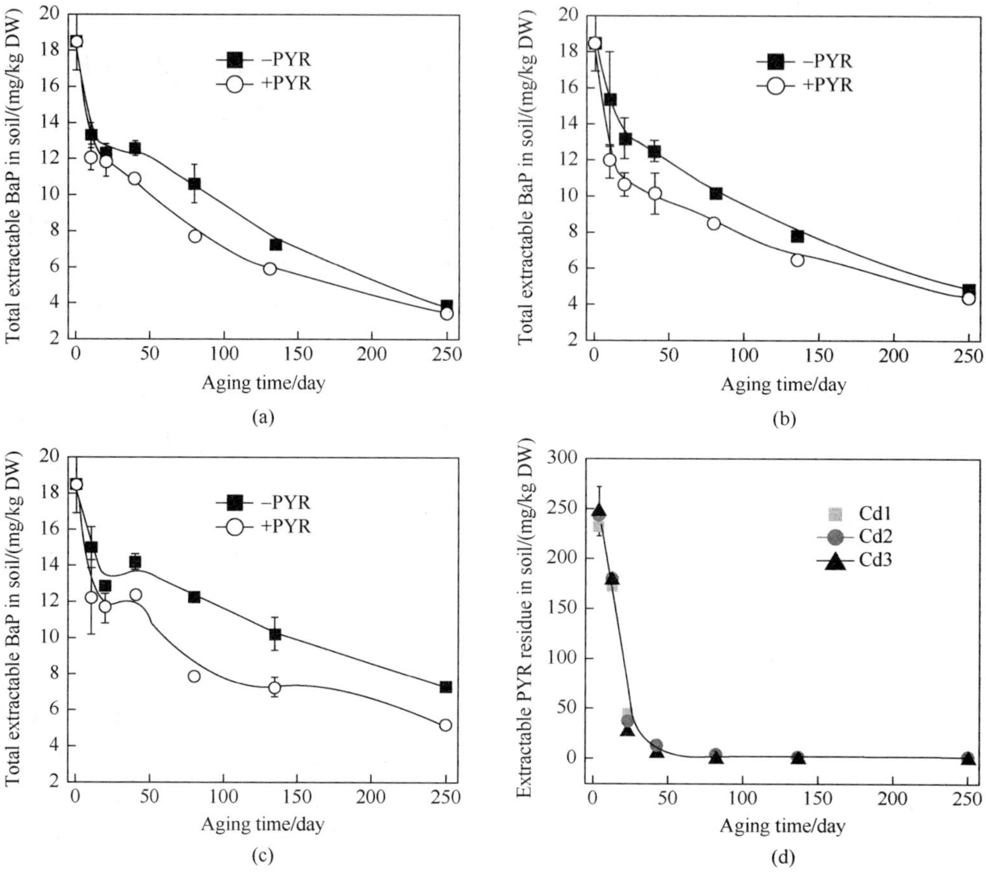

Fig. 24 Effects of pyrene on dynamic changes in total extractable B[a]P concentration in soils at different Cd concentration; dynamic changes in extractable pyrene concentrations in soils at different Cd concentration

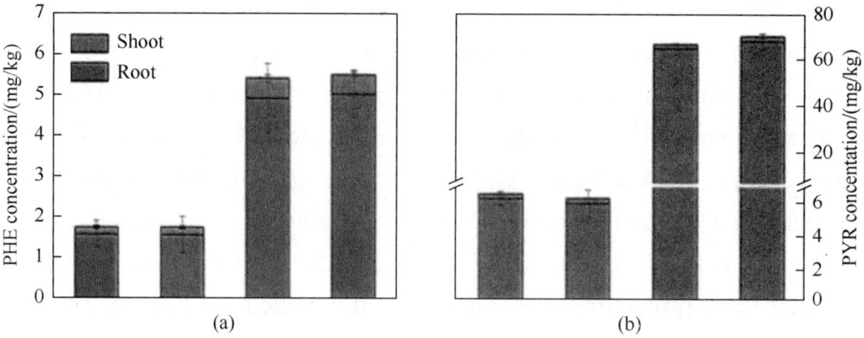

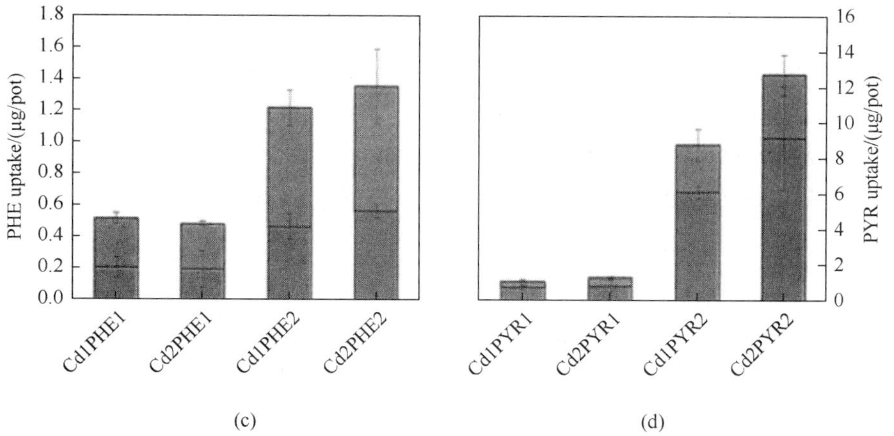

Fig. 25 Concentrations and total uptake of PHE and PYR in shoot and root of plants influenced by Cd and PAH treatments after 60 days of growth. Error bars show standard deviation (n=3)

A pot experiment was carried out to investigate the potential of enhanced phytoextraction of Cd by *S. alfredii* Hance and dissipation of PAHs in co-contaminated soil by the application of pig manure vermicompost (PMVC). Application of PMVC to co-contaminated soil increased the shoot and root dry biomass of *S. alfredii* Hance by 2.27 and 3.93-fold, respectively, and simultaneously increased Cd phytoextraction efficiency by 1.97-fold without inhibiting soil microbial quantity and enzyme activities. The highest dissipation rate of PAHs was observed in Plant+PMVC treatment. The dissipation rates of PHE, PYR and ANT were significantly increased 0.26%, 3.21% and 4.00%, respectively, as compared to the control. However, neither *S. alfredii* Hance nor PMVC enhanced PAHs dissipation when applied separately (Fig. 26). Abundant PAH-degraders in soil was not significantly related to PAHs dissipation rate. Plant+PMVC treatment significantly influenced the bacterial community structure. Enhanced PAHs dissipation in the Plant+PMVC treatment could be due to the improvement of plant root growth, which may result in increased root exudates, and subsequently change bacterial community structure to be favorable for PAHs dissipation, remediation of Cd-PAHs co-contaminated soil by *S. alfredii* Hance can be enhanced by simultaneous application of PMVC (Wang et al., 2012a).

It was investigated by Wang et al. (2012c) and Wang et al. (2013) that the potential for phytoextraction of heavy metals and rhizoremediaiton of PAHs in co-contaminated soil via a pot experiment by co-planting *S. alfredii* Hance with ryegrass (*Lolium perenne*) or castor (*Ricinus communis*). Co-planting with castor decreased the shoot biomass of *S. alfredii* Hance as compared to that in monoculture. Cadmium concentration in the shoot of *S. alfredii* Hance decreased by 64.3% and 47.4% in S-R (*S. alfredii* Hance and ryegrass) and S-C (*S. alfredii* Hance and castor) treatments, respectively, as compared to that in *S. alfredii* Hance monoculture. In contrast, Zn concentration in the *S. alfredii* Hance shoot increased by 19.8% and 25.6% in S-R and S-C treatments, respectively, as compared to that in *S. alfredii* Hance monoculture. Lead concentration in *S. alfredii* Hance shoot in S-R treatment was significantly greater as compared to that in *S. alfredii* Hance monoculture. Total removal of either Cd, Zn or Pb by plants was similar across *S. alfredii* Hance monoculnlre or co-planting with ryegrass or castor, except for the enhanced Pb removal in *S. alfredii* Hance and ryegrass co-planting treatment. Co-planting of *S. alfredii* Hance with ryegrass or castor significantly enhanced the PYR and ANT dissipation as compared to that in the control soil or *S. alfredii* Hance monoculture (Fig. 27).

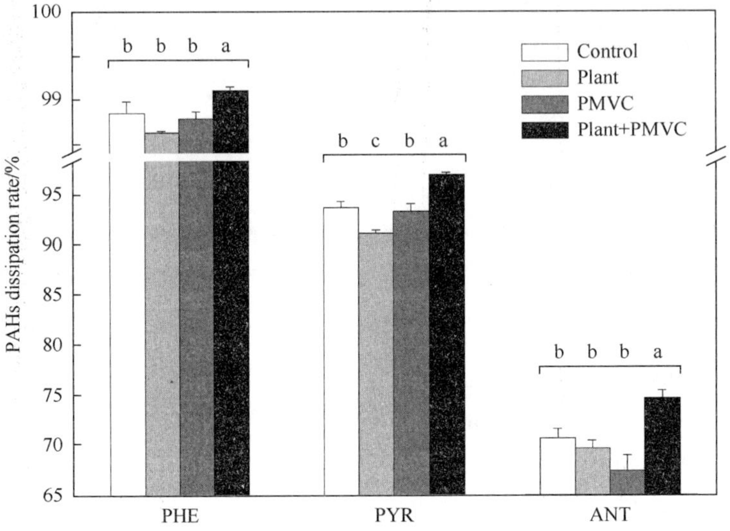

	Plant	*PMVC*	*Plant×PMVC*
PHE%	n.s.	*	**
PYR%	n.s.	*	**
ANT%	*	n.s.	**

Fig. 26 PAHs dissipation rates in different treatments after 90 days. means followed by different letters, with each PAH, are significantly different at $p<0.05$ (based on LSD). Error bars show standard error (n=3). The results of two-way analysis of variance (ANOVA) are shown below the graph, *: $p<0.05$, **: $p<0.01$ and NS: not significant

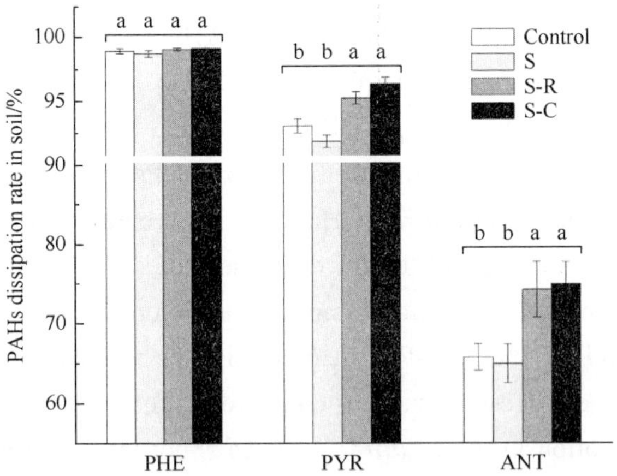

Fig. 27 PAHs dissipation rates in different treatments after 90 days. Means followed by different letters within each PAH are significantly different at $p<0.05$ (base on LSD). Error bars show standard deviation (n=3)

A pot experiment was conducted to investigated the separate and combine effects of non-ionic surfactant (Tween 80) and B[a]P-degrading bacterium on phytoremediation of soils co-contaminated with Cd and high-molecular-weight PAH (B[a]P) by *S. alfredii* Hance. Neither separate nor combined application of Tween 80 and B[a]P-degrading bacterium significantly affected plant growth and Cd uptake and accumulation by plant. Cadmium phytoextraction rate after 120 d of growth varied from 16.8% to 20.8% and from 6.4% to 7.6% in Cd free and add soils, respectively. In general, the application of B[a]P-degrading bacterium can enhance B[a]P dissipation in soil. However, the effects of Tween 80 on B[a]P dissipation were influenced by the interaction between soil Cd concentration and plant. In the soil with high Cd concentration (4.71 mg/kg), the combined application of Tween 80 and B[a]P-degrading bacterium could remove B[a]P most effectively in both planted and unplanted treatments. These results demonstrated that decontamination of Cd and B[a]P can be uptake by *S. alfredii* Hance associated with B[a]P-degrading bacterium. However, the combined application of Tween 80 and B[a]P-degrading bacterium can remove B[a]P more efficiently from soil with relatively higher Cd concentration (4.71 mg/kg). The complicated interactions among Tween 80, B[a]P-degrading bacterium and soil Cd concentration on the competition between B[a]P-degrading bacterium and indigenous microbial community need future investigations (Wang, 2012d).

7.3 Phytoremediation of Cd-nitrate co-contaminated soil coupled with crop production

In modern agriculture, repeated application of chemical fertilizers, especially nitrogen fertilizers, has resulted in excess nitrate in soils and subsequent accumulation of nitrate in crop plants. Nitrate could be stored in the vacuole of plants and enter human body through food chain. Excess nitrate in food represents a threat to human health, as it results in human diseases such as methaemoglobinemia and cancer. Moreover, excessive fertilizer application under protected cultivation of vegetables caused high accumulation of nitrate and cadmium in soil. Phytoremediation for Cd-nitrate co-contaminated soil has been much needed for both vegetable production and safety.

7.3.1 Genotypic difference in Cd-nitrate co accumulation

A field experiment was conducted to screen low co-accumulator genotypes of both Cd and nitrate from 62 Chinese cabbage genotypes grown in a moderately Cd (1.10 mg/kg) and nitrate (235.2 mg/kg) combined contaminated soil. Seven genotypes, i. e., *SIYM*, *TCCS*, *28TS*, *DGAF*, *CANB*, *FBSY* and *SJXH*, were identified as low co-accumulators of Cd and nitrate based on their containing low Cd ($<$0.5 mg/kg DW) and nitrate ($<$3100 mg/kg FW) concentration in the edible parts even when grown in contaminated soils (Table 4). These genotypes were suitable for growing in slightly or moderately contaminated soils without risk to food safety (Tang et al., 2016).

Table 4 Cadmium and nitrate concentrations in the shoots of seven safe Chinese cabbage genotypes.

Genotypes	Cadmium/(mg/kg DW)	Nitrate/(mg/kg FW)
SIYM	0.49±0.04	1071±259.9
TCCS	0.37±0.02	2165±318.7
28TS	0.37±0.03	2143±112.2
DGAF	0.37±0.02	1353±215.3
CANB	0.27±0.01	2179±185.7
FBSY	0.06±0.01	2037±28.7
SJXH	0.05±0.01	2915±228.7

7.3.2 Phytoremediation of Cd-nitrate co-contaminated soils coupled with vegetable safe production

We used a novel phytoremediation system for decreasing Cd and nitrate in soil and vegetable shoots using endophytic microbes (M) colonized Cd-hyperaccumulator plant *S. alfredii* Hance (SA) in rotation with low-accumulator water spinach (*Ipomoea aquatica* Forsk.) and low-accumulator Chinese cabbage (*B. chinensis* L.), and through enhancement of CO_2 under protecred cultivation. The treatments in first rotation were: ①bokchoy (CK), ②CK+CO_2, ③*S. alfredii* Hance; ④*Sedum*+CO_2, ⑤*Sedum*+Microbe (M), ⑥SA+CO_2+M. In second and third rotation, treatment ① and ② planted with conventional water spinach and Chinese cabbage, while treatment ③-⑥ planted with low accumulator water spinach and Chinese cabbage, respectively. After successive rotation of two years, the levels of total Cd, DTPA Cd NO_3^- in the co-contaminated soil decreased by 56.5%, 62.7% and 65.4% respectively. Cd concentration in water spinach and Chinese cabbage shoots were 0.08 mg/kg and 0.01 mg/kg, respectively, NO_3^- concentration in water spinach and Chinese cabbage shoots were 609.24 mg/kg and 419.62 mg/kg, respectively. All the concentration of contaminants in both vegetables were lower than the National Food Safety Standard of China (GB 2762—2012) and safe for consumption (Fig. 28). These result demonstrated that *S. alfredii* Hance was applied together with CO_2 and M was the most effective for the remediation of Cd-NO_3 co-contaminated soils.

(a)

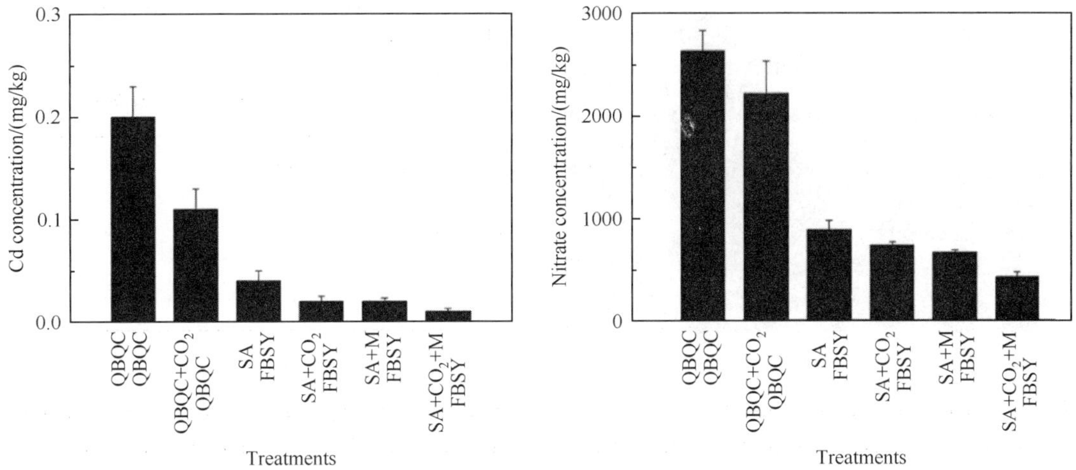

(b) (c)

Fig. 28 Effects of phytoremediation of Cd-nitrate co-contaminated soil on Cd and nitrate accumulation in vegetable shoots after first rotation of phytoremediation. QBQC-conventional cultivar, FBFY-lower accumulator cultivar of Chinese Bokchoy. The treatments in the first rotation were: Bokchoy (CK), CK+CO_2, *S. alfredii* Hance (SA), *Sedum*+CO_2, *Sedum*+Microbe (M), SA+CO_2+M, respectively

7.3.3 Phytoremediation of Cd-Cu co-contaminated soil coupled with crop production

Field experiments were conducted to study the effectiveness of phytoremediation of Cd-Cu co-contaminated soil coupled with vegetable production. We used several rotation and intercropping system, like hyperaccumulator *S. alfredii* Hance intercropping with vegetable crop; accumulator rapeseed rotation with vegetable crops, and accumulator *Elsholtzia splendens* co-cropping with vegetables, etc. After one year rotation or intercropping, the removal of soil Cd and Cu were by 16% and 12%, respectively (Fig. 29. As the tested soil belong to lightly contaminated soil. After one year phytoremediation, the metal concentration in edible parts of vegetables were below National Food Safety Standard of China (GB 2762—2012). These results showed that it is possible and effective for phytoremediation of multi-metal contaminated soil coupled with crop production.

(a) (b)

(c)

Heavy metal in soil /(mg/kg)	Removal of heavy metals/%	
	Elsholtzia	*Sedum*
0.63 (Cd)	0.87	16.40
46.6 (Cd)	12.08	0.05
178 (Zn)	1.29	4.38

(d)

Fig. 29 Phytoremediation of Cd-Cu co-contaminated soil (a) rotation of vegetable crop with Cd accumulating Rapeseed (b) co-cropping of vegetable crops with Cu-accumulating *Elsholtzia* (c) intercropping of *S. alfredii* Hance with crops (d) Concentrations of heavy metal and their removal (%) by *Sedum* and *Elsholtia*

Acknowledgements

This paper were supported by key projects or research plan from Ministry of Science and Technology of China (#2016YFD0800805), from key research plan of Zhejiang Science and Technology Bureau (#2015C02011-1; #2015C03020-1). Author is thankful for all the graduate students and colleagues for great efforts in contributing to these reviews.

References

Chao Y E, Zhang M, Tian S, et al. 2008. Differential generation of hydrogen peroxide upon exposure to zinc and cadmium in the hyperaccumulating plant specie (*Sedum alfredii* Hance). Journal of Zhejiang University Science B, 9 (3): 243-249.

Chen B, Shen J, Zhang X, et al. 2014. The endophytic bacterium, *Sphingomonas* SaMR12, improves the potential for zinc phytoremediation by its host, *Sedum alfredii*. PloS one, 9 (9): e106826.

Chen B, Zhang Y, Rafiq M T, et al. 2014. Improvement of cadmium uptake and accumulation in *Sedum alfredii* by endophytic bacteria Sphingomonas SaMR12: Effects on plant growth and root exudates. Chemosphere, 117: 367-373.

Hua F, Yunlong Y U, Xiaoqiang C H U, et al. 2009. Degradation of chlorpyrifos in laboratory soil and its impact on soil microbial functional diversity. Journal of Environmental Sciences, 21 (3): 380-386.

Fang H, Yu Y L, Wang X G, et al. 2009. Persistence of the herbicide butachlor in soil after repeated applications and its effects on soil microbial functional diversity. Journal of Environmental Science and Health Part B, 44 (2): 123-129.

Fang H, Zhou W, Cao Z, et al. 2012. Combined remediation of DDT congeners and cadmium in soil by *Sphingobacterium* sp. D-6 and *Sedum alfredii* Hance. Journal of Environmental Sciences, 24 (6): 1036-1046.

Gu P, Zhang Q R, Zhou QX, et al. 2010. Isolation of a fungal strain capable of degrading benzo[a]pyrene and its degradation characteristics. Acta Scientiae Circumstantiae, 30 (2): 354-360. (in Chinese)

Gupta D K, Huang H G, Yang X E, et al. 2010. The detoxification of lead in *Sedum alfredii* H. is not related to phytochelatins but the glutathione. Journal of Hazardous Materials, 177 (1): 437-444.

Gupta D K, Nicoloso F T, Schetinger M R, et al. 2011. Lead induced responses of Pfaffia glomerata, an economically important Brazilian medicinal plant, under in vitro culture conditions. Bulletin of Environmental Contamination and Toxicology, 86 (3): 272-277.

He B, Yang X, Ni W, et al. 2003. Pb uptake, accumulation, subcellular distribution in a Pb-accumulating ecotype of *Sedum alfredii*

(Hance). Journal of Zhejiang University SCIENCE A, 4 (4): 474-479.

Huang H, Li T, Gupta D K, et al. 2012. Heavy metal phytoextraction by *Sedum alfredii* is affected by continual clipping and phosphorus fertilization amendment. Journal of Environmental Sciences, 24 (3): 376-386.

Huang H G, Li T Q, Zhu Z Q, et al. 2012. Effects of soluble phosphate fertilizer on Zn/Cd phytoextraction and nutrient accumulation of *Sedum alfredii* H. in co-contaminated soil. Plant Nutrition and Fertilizer Science, 18: 382-389.

Huang H G, Li T Q, Zhu ZQ, et L. 2012. Effects of soluble phosphate fertilizer on Zn/Cd phytoextraction and nutrient accumulation of *Sedum alfredii* H. in co-contaminated soil. Plant Nutrition and Fertilizer Science. 18 (2): 382-389.

Huang H G, Li T X, Zhang X Z, et al. 2008. Effects of exogenous Ca^{2+} on the growth and Zn accumulation of two *Sedum alfredii* Hance ecotypes. Chinese Journal of Applied Ecology, Apr, 19 (4): 831-837.

Huang H, Wang K, Zhu Z, et al. 2013. Moderate phosphorus application enhances Zn mobility and uptake in hyperaccumulator *Sedum alfredii*. Environmental Science and Pollution Research, 20 (5): 2844-2853.

Huang H, Yu N, Wang L, et al. 2011. The phytoremediation potential of bioenergy crop *Ricinus communis* for DDTs and cadmium co-contaminated soil. Bioresource Technology, 102 (23): 11034-11038.

Islam E, Liu D, Li T, et al. 2011. Effect of Pb toxicity on the growth and physiology of two ecotypes of *Elsholtzia argyi* and its alleviation by Zn. Environmental Toxicology, 26 (4): 403-416.

Jiang L, Yang X. 2004. Chelators effect on soil Cu extractability and uptake by *Elsholtzia splendens*. Journal of Zhejiang University Science, 5 (4): 450-456.

Jiang L Y, Yang X E, He Z L. 2004. Growth response and phytoextraction of copper at different levels in soils by *Elsholtzia splendens*. Chemosphere, 55 (9): 1179-1187.

Jiang L Y, Yang X E, Ye Z Q, et al. 2003. Uptake, distribution and accumulation of copper in two ecotypes of *Elsholtzia*. Pedosphere, 13 (4): 359-366.

Jin X F, Yang X E, Islam E, et al. 2008. Ultrastructural changes, zinc hyperaccumulation and its relation with antioxidants in two ecotypes of *Sedum alfredii* Hance. Plant Physiology and Biochemistry, 46 (11): 997-1006.

Jin X, Yang X, Islam E, et al. 2008. Effects of cadmium on ultrastructure and antioxidative defense system in hyperaccumulator and non-hyperaccumulator ecotypes of *Sedum alfredii* Hance. Journal of Hazardous Materials, 156 (1): 387-397.

Jin X, Yang X, Mahmood Q, et al. 2008. Response of antioxidant enzymes, ascorbate and glutathione metabolism towards cadmium in hyperaccumulator and nonhyperaccumulator ecotypes of *Sedum alfredii* H. Environmental Toxicology, 23 (4): 517-529.

Jin X F, Liu D, Islam E, et al. 2009. Effects of Zinc on Root Morphology and Antioxidant Adaptations of Cadmium-Treated *Sedum alfredii* H. Journal of Plant Nutrition, 32 (10): 1642-1656.

Li H H, Hu M H, Li T Q, et al. 2008. Effects of Sulfur on growth, Cadmium Absorption and Accumulation in Hyperaccumulator-*Sedum alfredii* Hance. Journal of Soil and Water Conservation, 22 (6): 71-74.

Li J G, Li T Q, Zhu E, et al. 2007. Effects of nitrogen fertilizer on growth and cadmium accumulation in hyperaccumulator of *Sedum alfredii* Hance. Journal of Soil and Water Conservation, 21 (1): 54-58.

Li J G, Li T Q, Zhu E, et al. 2008. Effects of different nitrogen forms on cadmium accumulation of *Sedum alfredii* Hance. Hance. Journal of Zhejiang University (Agric. & Life), 34 (3): 327-333.

Li J G, Lin G L, Yang C W, et al. 2007. Effects of Organic and Inorganic Nitrogen on Growth and Cadmium accumulation of *Sedum alfredii* Hance. Journal of Agro-Environment Science, 27 (1): 99-104.

Li J G, Zhu E, Li T Q, et al. 2007. Effects of nitrogen fertilizer on biomass, root morphology and Cd accumulation of Cd-stressed *Sedum alfredii* Hance species. Environmental Pollution and Control, 29 (4): 271-275.

Li T, Di Z, Han X, et al. 2012. Elevated CO_2 improves root growth and cadmium accumulation in the hyperaccumulator *Sedum alfredii*. Plant and soil, 354 (1-2): 325-334.

Li T Q, Dong Z S, Jiang H, et al. 2011. Remediation efficiency of Cd-B[a]P combined polluted soil by *Sedum alfredii*. Jour nal o f Zhejiang University (Agric. & Life Sci.), 2011, 37 (4): 465-472.

Li T, Tao Q, Han X, et al. 2013. Effects of elevated CO_2 on rhizosphere characteristics of Cd/Zn hyperaccumulator *Sedum alfredii*.

Science of the Total Environment, 454: 510-516.

Li T, Tao Q, Liang C, et al. 2014. Elevated CO_2 concentration increase the mobility of Cd and Zn in the rhizosphere of hyperaccumulator *Sedum alfredii*. Environmental Science and Pollution Research, 21 (9): 5899-5908.

Li T Q, Tao Q, Di Z Z, et al. 2015. Effect of elevated CO_2 concentration on photosynthetic characteristics of hyperaccumulator *Sedum alfredii* under cadmium stress. Journal of Integrative Plant Biology, 57 (5): 653-660.

Li T, Yang X, Meng F, et al. 2007. Zinc adsorption and desorption characteristics in root cell wall involving zinc hyperaccumulation in *Sedum alfredii* Hance. Journal of Zhejiang University Science B, 8 (2): 111-115.

Li T Q, Yang X E, Yang J Y, et al. 2006. Zn accumulation and subcellular distribution in the Zn hyperaccumulator *Sedum alfredii* Hance. Pedosphere, 16 (5): 616-623.

Li T Q, Yang X E, Jin X F, et al. 2005. Root responses and metal accumulation in two contrasting ecotypes of *Sedum alfredii Hance* under lead and zinc toxic stress. Journal of Environmental Science and Health, Part A, 40 (5): 1081-1096.

LI T Q, YANG X E, HE Z L, et al. 2005. Root morphology and Zn^{2+}uptake kinetics of the Zn hyperaccumulator of *Sedum alfredii* Hance. Journal of Integrative Plant Biology, 47 (8): 927-934.

Li T, Di Z, Islam E, et al. 2011. Rhizosphere characteristics of zinc hyperaccumulator *Sedum alfredii* involved in zinc accumulation. Journal of Hazardous Materials, 185 (2): 818-823.

Li T, Tao Q, Liang C, et al. 2014. Elevated CO_2 concentration increase the mobility of Cd and Zn in the rhizosphere of hyperaccumulator *Sedum alfredii*. Environmental Science and Pollution Research, 21 (9): 5899-5908.

Li T, Tao Q, Liang C, et al. 2013. Complexation with dissolved organic matter and mobility control of heavy metals in the rhizosphere of hyperaccumulator *Sedum alfredii*. Environmental Pollution, 182: 248-255.

Li T, Tao Q, Shohag M J I, et al. 2015. Root cell wall polysaccharides are involved in cadmium hyperaccumulation in *Sedum alfredii*. Plant and Soil, 389 (1-2): 387-399.

Li T, Tao Q, Han X, et al. Effects of elevated CO_2 on rhizosphere characteristics of Cd/Zn hyperaccumulator *Sedum alfredii*. Science of the Total Environment, 2013, 454: 510-516.

Li T, Tao Q, Di Z, et al. 2015. Effect of elevated CO_2 concentration on photosynthetic characteristics of hyperaccumulator *Sedum alfredii* under cadmium stress. Journal of Integrative Plant Biology, 57 (7): 653-660.

Li T, Yang X, Lu L, et al. 2009. Effects of zinc and cadmium interactions on root morphology and metal translocation in a hyperaccumulating species under hydroponic conditions. Journal of Hazardous Materials, 169 (1): 734-741.

Li T, Di Z, Yang X, et al. 2011. Effects of dissolved organic matter from the rhizosphere of the hyperaccumulator *Sedum alfredii* on sorption of zinc and cadmium by different soils. Journal of Hazardous Materials, 192 (3): 1616-1622.

Li T, Xu Z, Han X, et al. 2012. Characterization of dissolved organic matter in the rhizosphere of hyperaccumulator *Sedum alfredii* and its effect on the mobility of zinc. Chemosphere, 88 (5): 570-576.

Li T Q, Yang X E, Jin X F, et al. 2005. Root responses and metal accumulation in two contrasting ecotypes of *Sedum alfredii* Hance under lead and zinc toxic stress. Journal of Environmental Science and Health, Part A, 40 (5): 1081-1096.

Liang J, Shohag M J I, Yang X, et al. 2014. Role of sulfur assimilation pathway in cadmium hyperaccumulation by *Sedum alfredii* Hance. Ecotoxicology and environmental Safety, 100: 159-165.

Liu D, Islam E, Li T, et al. 2008. Comparison of synthetic chelators and low molecular weight organic acids in enhancing phytoextraction of heavy metals by two ecotypes of *Sedum alfredii* Hance. Journal of Hazardous Materials, 153 (1): 114-122.

Liu D, Li T, Yang X, et al. 2007. Enhancement of lead uptake by hyperaccumulator plant species *Sedum alfredii* Hance using EDTA and IAA. Bulletin of environmental Contamination and Toxicology, 78 (3): 280-283.

Liu D, Li T Q, Jin X F, et al. 2008. Lead induced changes in the growth and antioxidant metabolism of the lead accumulating and non-accumulating ecotypes of *Sedum alfredii*. Journal of Integrative Plant Biology, 50 (2): 129-140.

Liu D, Li T Q, Yang X E, et al. 2008. Effect of Pb on leaf antioxidant enzyme activities and ultrastructure of the two ecotypes of *Sedum alfredii* Hance. Russian Journal of Plant Physiology, 55 (1): 68-76.

Long X, Ni W, Ye Z, et al. 2001. Effect of organic acids application on zinc uptake and accumulation by two ecotypes of *Sedum alfredii* Hance. Plant Nutrition and Fertilizer Science, 8 (4): 467-472.

Long X, Ni W, Ye Z, et al. 2006. Zinc tolerance and hyperaccumulation of *Sedum alfredii* Hance: A greenhouse experiment with artificial polluted soils. Bulletin of Environmental Contamination and Toxicology, 76 (2): 264-270.

Lu L, Tian S, Yang X, et al. 2013. Improved cadmium uptake and accumulation in the hyperaccumulator *Sedum alfredii*: the impact of citric acid and tartaric acid. Journal of Zhejiang University Science B, 14 (2): 106-114.

Lu L, Tian S, Zhang J, et al. 2013. Efficient xylem transport and phloem remobilization of Zn in the hyperaccumulator plant species *Sedum alfredii*. New Phytologist, 198 (3): 721-731.

Lu L, Tian S, Yang X, et al. Improved cadmium uptake and accumulation in the hyperaccumulator *Sedum alfredii*: the impact of citric acid and tartaric acid. Journal of Zhejiang University Science B, 2013, 14 (2): 106-114.

Lu L, Tian S, Yang X, et al. 2009. Cadmium uptake and xylem loading are active processes in the hyperaccumulator *Sedum alfredii*. Journal of Plant Physiology, 166 (6): 579-587.

Lu L, Tian S, Yang X, et al. 2008. Enhanced root-to-shoot translocation of cadmium in the hyperaccumulating ecotype of *Sedum alfredii*. Journal of Experimental Botany, 59 (11): 3203-3213.

Lu L, Liao X, Labavitch J, et al. 2014. Speciation and localization of Zn in the hyperaccumulator *Sedum alfredii* by extended X-ray absorption fine structure and micro-X-ray fluorescence. Plant Physiology and Biochemistry, 84: 224-232.

Lu L, Tian S, Zhang M, et al. 2010. The role of Ca pathway in Cd uptake and translocation by the hyperaccumulator *Sedum alfredii*. Journal of Hazardous Materials, 183 (1): 22-28.

Ni W Z, Sun Q, Yang X. 2004. Growth and zinc accumulation of *Sedum alfredii* Hance—a Zn hyperaccumulator as affected by phosphorus application. Bulletin of Environmental Contamination and Toxicology, 72 (4): 756-762.

Ni W, Yang X, Long X. 2004. Comparative studies on zinc tolerance and accumulation between two ecotypes of *Sedum alfredii* Hance in southeastern China. Journal of Plant Nutrition, 27 (4): 627-635.

Pan F, Meng Q, Wang Q, et al. 2016. Endophytic bacterium Sphingomonas SaMR12 promotes cadmium accumulation by increasing glutathione biosynthesis in *Sedum alfredii* Hance. Chemosphere, 154: 358-366.

Peng H Y, Yang X E, Jiang L Y, et al. 2005. Copper Phytoavailability and Uptake by *Elsholtzia Splendens* from Contaminated Soil as Affected by Soil Amendments. Journal of Environmental Science and Health, 40 (4): 839-856.

Peng H Y, Yang X E. 2007. Characteristics of copper and lead uptake and accumulation by two species of *Elsholtzia*. Bulletin of Environmental Contamination and Toxicology, 78 (2): 152-157.

Peng H Y, Yang X E, Jiang L Y, et al. 2005. Copper Phytoavailability and Uptake by *Elsholtzia Splendens* from Contaminated Soil as Affected by Soil Amendments. Journal of Environmental Science and Health, 40 (4): 839-856.

Peng H Y, Yang X E, Tian S K. 2005. Accumulation and ultrastructural distribution of copper in *Elsholtzia splendens*. Journal of Zhejiang University (Science), 6 (5): 311-318.

Peng H Y, Yang X E, Yang M J, et al. 2006. Responses of antioxidant enzyme system to copper toxicity and copper detoxification in the leaves of *Elsholtzia splendens*. Journal of Plant Nutrition, 29 (9): 1619-1635.

Sun Q, Ni W Z, Yang X X, et al. 2003. Effect of phosphorus on the growth, zinc absorption and accumulation in hyperaccumulator–*Sedum alfredii* Hance. Acta Scientiae Circumstantiae, 23 (6): 818-824.

Tang L, Luo W, Tian S, et al. 2016. Genotypic differences in cadmium and nitrate co-accumulation among the Chinese cabbage genotypes under field conditions. Scientia Horticulturae, 201: 92-100.

Tian S, Lu L, Labavitch J, et al. 2011. Cellular sequestration of cadmium in the hyperaccumulator plant species *Sedum alfredii*. Plant Physiology, 157 (4): 1914-1925.

Tian S, Lu L, Zhang J, et al. 2011. Calcium protects roots of *Sedum alfredii* H. against cadmium-induced oxidative stress. Chemosphere, 84 (1): 63-69.

Tian S K, Lu L L, Yang X E, et al. 2012. Root adaptations to cadmium-induced oxidative stress contribute to Cd tolerance in the hyperaccumulator *Sedum alfredii*. Biologia plantarum, 56 (2): 344-350.

Tian S K, Lu L L, Yang X E, et al. 2009. Stem and leaf sequestration of zinc at the cellular level in the hyperaccumulator *Sedum alfredii*. New Phytologist, 182 (1): 116-126.

Tian S, Lu L, Yang X, et al. 2011. The impact of EDTA on lead distribution and speciation in the accumulator *Sedum alfredii* by synchrotron X-ray investigation. Environmental Pollution, 159 (3): 782-788.

Tian S, Lu L, Yang X, et al. 2010. Spatial imaging and speciation of lead in the accumulator plant *Sedum alfredii* by microscopically focused synchrotron X-ray investigation. Environmental Science & Technology, 44 (15): 5920-5926.

Tian S, Peng H, Yang X, et al. 2008. Phytofiltration of copper from contaminated water: growth response, copper uptake and lignin content in *Elsholtzia splendens* and *Elsholtzia argyi*. Bulletin of Environmental Contamination and Toxicology, 81 (1): 85-89.

Wang K, Chen X, Zhu Z, et al. 2014. Dissipation of available benzo[a]pyrene in aging soil co-contaminated with cadmium and pyrene. Environmental Science and Pollution Research, 21 (2): 962-971.

Wang K, Huang H, Zhu Z, et al. 2013. Phytoextraction of metals and rhizoremediation of PAHs in co-contaminated soil by co-planting of *Sedum alfredii* with ryegrass (Lolium perenne) or castor (Ricinus communis). International Journal of phytoremediation, 15 (3): 283-298.

Wang K, Zhang J, Zhu Z, et al. 2012. Pig manure vermicompost (PMVC) can improve phytoremediation of Cd and PAHs co-contaminated soil by *Sedum alfredii*. Journal of Soils and Sediments, 12 (7): 1089-1099.

Wang K, Zhu Z, Huang H, et al. 2012. Interactive effects of Cd and PAHs on contaminants removal from co-contaminated soil planted with hyperaccumulator plant *Sedum alfredii*. Journal of Soils and Sediments, 12 (4): 556-564.

Wang K. 2012d. Enhancement and mechanism of phytoremediation for Cd and PAHs co-contaminated soils. Ph. D. Dissertation.

Wang X, Wang Y, Mahmood Q, et al. 2009. The effect of EDDS addition on the phytoextraction efficiency from Pb contaminated soil by *Sedum alfredii* Hance. Journal of Hazardous Materials, 168 (1): 530-535.

Xiao W, Wang H, Li T, et al. 2013. Bioremediation of Cd and carbendazim co-contaminated soil by Cd-hyperaccumulator *Sedum alfredii a*ssociated with carbendazim-degrading bacterial strains. Environmental Science and Pollution Research, 20 (1): 380-389.

Yang X E, Peng H Y, Jiang L Y, et al. 2005. Phytoextraction of copper from contaminated soil by *Elsholtzia splendens* as affected by EDTA, citric acid, and compost. International Journal of Phytoremediation, 7 (1): 69-83.

Yang J Y, Yang X E, He Z L, et al. 2006. Effects of pH, organic acids, and inorganic ions on lead desorption from soils. Environmental Pollution, 143 (1): 9-15.

Yang X E, Ye H B, Long X X, et al. 2005. Uptake and accumulation of cadmium and zinc by *Sedum alfredii* Hance at different Cd/Zn supply levels. Journal of Plant Nutrition, 27 (11): 1963-1977.

Yang X E, Li T Q, Long X X, et al. 2006. Dynamics of zinc uptake and accumulation in the hyperaccumulating and non-hyperaccumulating ecotypes of *Sedum alfredii* Hance. Plant and Soil, 284 (1-2): 109-119.

Yang X E, Long X X, Ye H B, et al. 2004. Cadmium tolerance and hyperaccumulation in a new Zn-hyperaccumulating plant species (*Sedum alfredii* Hance). Plant and Soil, 259 (1-2): 181-189.

Yang X, Li T, Yang J, et al. 2006. Zinc compartmentation in root, transport into xylem, and absorption into leaf cells in the hyperaccumulating species of *Sedum alfredii* Hance. Planta, 224 (1): 185-195.

Yang X, Long X, Ni W, et al. 2002. *Sedum alfredii* H: a new Zn hyperaccumulating plant first found in China. Chinese Science Bulletin, 47 (19): 1634-1637.

Yang X E, Chao Y E, Ye H B, et al. 2010. Zinc and lead accumulation by two contrasting ecotypes of *Sedum alfredii* Hance at different zinc/lead complex levels. Communications in Soil Science and Plant Analysis, 41 (4): 516-525.

Zhang X, Lin L, Zhu Z, et al. 2013. Colonization and modulation of host growth and metal uptake by endophytic bacteria of *Sedum alfredii*. International Journal of Phytoremediation, 15 (1): 51-64.

Zhang X, Lin L, Chen M, et al. 2012. A nonpathogenic Fusarium oxysporum strain enhances phytoextraction of heavy metals by the hyperaccumulator *Sedum alfredii* Hance. Journal of Hazardous Materials, 229: 361-370.

Zhang J, Tian S, Lu L, et al. 2011. Lead tolerance and cellular distribution in *Elsholtzia splendens* using synchrotron radiation micro-X-ray fluorescence. Journal of Hazardous Materials, 197: 264-271.

Zhang M, Senoura T, Yang X, et al. 2011. Functional analysis of metal tolerance proteins isolated from Zn/Cd hyperaccumulating ecotype and non-hyperaccumulating ecotype of *Sedum alfredii* Hance. FEBS letters, 585 (16): 2604-2609.

Zhang X, Lin L, Zhu Z, et al. 2013. Colonization and modulation of host growth and metal uptake by endophytic bacteria of *Sedum alfredii*. International Journal of Phytoremediation, 15 (1): 51-64.

Zhu E, Liu D, Li J G, et al. 2010. Effect of nitrogen fertilizer on growth and cadmium accumulation in *Sedum alfredii* Hance. Journal of Plant Nutrition, 34 (1): 115-126.

Zhu Z, Yang X, Wang K, et al. 2012b. Bioremediation of Cd-DDT co-contaminated soil using the Cd-hyperaccumulator *Sedum alfredii* and DDT-degrading microbes. Journal of Hazardous Materials, 235: 144-151.

Zhu Z Q. 2012a. Plant-microbe remediation of Cd-DDT co-contaminted soils and its mechanisms. Ph. D. Dissertation.

Phytoremediation of Cadmium Contaminated Soils Using the Cadmium and Zinc Hyperaccumulator *Sedum plumbizincicola*

Longhua WU[1], Pengjie HU[1], Zhu LI[1], Tong ZHOU[1], Daoxu ZHONG[1], Yongming LUO[1,2]

[1]Key Laboratory of Soil Environment and Pollution Remediation, Institute of Soil Science, Chinese Academy of Sciences

[2]Key Laboratory of Coastal Environmental Processes and Ecological Remediation, Yantai Institute of Coastal Zone Research, Chinese Academy of Sciences

Abstract

Sedum plumbizincicola X. H. Guo et S. B. Zhou ex L. H. Wu (Crassulaceae), a new *Sedum* species, was firstly found in 2005 when our research team undertook extensive field investigations for searching metal hyperaccumulator species in Zhejiang province. It was identified as new cadmium (Cd)/zinc (Zn) hyperaccumulator in 2007. During the past decade, great efforts have been taken to understand its metal hyperaccumulating capacities, physiological metal hypertolerance and hyperaccumulation, enhancement of phytoextraction, field application practice of phytoremediation. This paper provides a brief review on the progress on phytoremediation of Cd contaminated soils using this species. Agronomic measures to enhance Cd and Zn phytoextraction efficiency by *S. plumbizincicola* were studied, including cultivation management, intercropping with other plant species, nutrient management, microbial assistant, and so on. Repeated phytoextraction of Cd and Zn contaminated soils using *S. plumbizincicola* was conducted in pot experiments to determine plant metal uptake, the changes of soil metal concentration and speciation and soil microbial properties. So far, several research and demonstration bases have been established in numerous Chinese provinces including Zhejiang, Hu'nan, He'nan, Jiangxi, Jiangsu, Guangdong, and Guizhou. Field assessment results confirmed that phytoextraction using *S. plumbizincicola* is a promising technique for the remediation of slightly Cd-polluted soils without halting normal agricultural production. Safely dispose or treatment of *S. plumbizincicola* biomass is also an important part for practically use of phytoextraction. Incineration and pyrolysis were employed to treat *S. plumbizincicola* biomass.

Keywords: *Sedum plumbizincicola*, Cd, Zn, hyperacumulation, phytoextraction, field application

1. *Sedum plumbizincicola*: a new Cd/Zn hyperaccumulator

1.1 Identification of a new *Sedum* species

Sedum plumbizincicola X. H. Guo et S. B. Zhou ex L. H. Wu (Crassulaceae) was firstly found in 2005 when our research team undertook extensive field investigations in Zhejiang province and collected many specimens in a search for metal hyperaccumulator species (Wu et al., 2006). Some unusual and isolated populations superficially resembling *Sedum alfredii* Hance but producing 4-merous flowers were found in Fuyang and Chun'an counties, together with four other species, namely *S. alfredii*, *Sedum emarginatum* Migo, *Sedum hangzhouense* K. T. Fu and G. Y. Rao and *Sedum bailey* Praeger, that were identified from the same areas. Furthermore, based on geographical distribution, growth habit, phenology, macromorphological characters, stem and leaf anatomical features, seed micromorphology and nrDNA internal transcribed spacers (ITS) sequence data, formal description of the new species *S. plumbizincicola* X. H. Guo et S. B. Zhou ex L. H. Wu and clarification of the affinities between *S. plumbizincicola* and closely related taxa were conducted (Wu et al., 2013a).

S. plumbizincicola, perennial, light green, or yellowish green herb. Roots fibrous. Rhizomes slender and horizontal, yellowish brown or dark brown, to 7 cm long, ca. 4-8 mm diameter. Sterile stems several-branched, erect, 10-25 cm, densely caespitose. Leaves of foliage branches alternate, usually deciduous, crowded distally on stem; leaf blade ovate to obovate-spatulate, 1-5 cm×0.5-1.5 cm, glabrous, apex blunted, base cuneate, pseudopetiolate; adaxial surface bright green, abaxial surface jade green, midrib not convex abaxially. Inflorescence yellow, erect, much-branched, the peduncles up to 35 cm tall, ca. 0.8 cm in diameter. Cyme corymbiform, ca. 8 cm in diameter, many flowered, bracts linear to linear-lanceolate, 5-10 mm× 3-8 mm, apex blunted. Flowers sessile, unequally 4-merous. Sepals 4, narrowly triangulate, ca. 1-2 mm long apex blunted. Petals 4, yellow, lanceolate, 4-6 mm×1-1.5 mm, apex acute. Stamens 8, slightly shorter than the petals, antesepalous ones ca. 3.5-4.5 mm long, antepetalous ones ca. 3 mm long, inserted ca. 1 mm from petal base; filaments greenish, anthers oblong, yellow. Nectar scales inverted trapezia, ca. 0.3-1.0 mm×0.8 mm. Carpels erect, ovoid-lanceolate, ca. 4-5 mm long, connate about one-third at base. Styles ca. 1 mm. Follicles split divergent, tetra-aristiform. Seeds numerous, brown, obovoid-oblong, ca. 0.7-1 mm long, mammillate. Flowering early June-August, fruiting July-September. See Fig. 1.

This new species is known only from the type locality, Zitong town (lead-zinc mining areas) northwest of Hangzhou city, in western Zhejiang province, east China. Annual rainfall ranges from 980 to 2000 mm and occurs mainly in the summer, with a mean annual temperature of 15-18℃. Soils in this area are usually sandy, acidic, highly leached and often shallow. This species has a strong ability to hyperaccumulate Zn and Cd and is a promising taxon for the restoration of metal-polluted soils by phytoremediation. The specific epithet "*plumbizincicola*" refers to the distribution of the species in the lead and zinc mining areas of western Zhejiang province, east China.

(a) Vegetative (b) Flowering states

Fig. 1 *S. plumbizincicola* in its vegetative and flowering states in the original habitat (Wu et al., 2006)

1.2 Cadmium and Zn tolerance and hyperaccumulation in *S. plumbizincicola*

S. plumbizincicola has been reported to be a Cd/Zn hyperaccumulator exhibiting rapid growth, large biomass, asexual reproduction and a perennial habit, with a remarkable potential for phytoextraction of Cd from polluted soils (Liu et al., 2011; Jiang et al., 2010). A field investigation of a mine area found that the shoots of *S. plumbizincicola* hyperaccumulated up to 1470 mg/kg of Cd and 14 600 mg/kg of Zn (dry weight) (Hu et al., 2015). In a hydroponics experiment, Cao et al. (2014) recorded mean Cd and Zn concentrations in the shoots (7010 and 18 400 mg/kg) about seven and five times higher than those in roots (840 and 3000 mg/kg) after exposure to 100 μM $CdSO_4$ and 600 μM $ZnSO_4$, respectively. Cadmium and Zn concentrations in young leaves (4330 and 9820 mg/kg) were about six times and twice as high, respectively, as those in mature leaves (636 and 2620 mg/kg).

1.3 Cellular distribution of elements in *S. plumbizincicola*

Heavy metal sequestration in less metabolically active parts of plants such as the epidermis and trichomes is one of the possible mechanisms for heavy metal detoxification in hyperaccumulators. Cellular distribution of Zn, Cd and other elements in frozen-hydrated sections of *S. plumbizincicola* from a mine area were quantitatively determined using cryo-micro-PIXE (Hu et al., 2015). Results show that Zn was in the roots concentrated in the cortex. In the stems Zn preferentially accumulated in the epidermis (5090 mg/kg wet weight) and in vascular bundles and neighboring parenchyma cells. Interestingly, some vascular bundles of the stem were depleted in Zn. Taking tissue area into account, 22.1%, 46.2% and 31.8% of total Zn and 17.1%, 71.6% and 11.3% of total Cd in the stem were present in the epidermis, cortex and central cylinder, respectively. 81.3% of total Cd and 55.2% of total Zn in the leaves were distributed in the mesophyll. It was concluded that the parenchyma cells, e.g.cortex in stem and mesophyll in leaf, play more important roles in Cd storage and detoxification than Zn in *S. plumbizincicola*. In hydroponically cultured *S. plumbizincicola* supplied with Cd and Zn (Cao et al., 2014), microPIXE analysis showed that Zn was predominantly localized in epidermal cells in both young and mature leaves, but large amounts of Zn occurred in mesophyll cells in young leaves. Leaf tissue fractionation showed that soluble and cell wall fractions were different at the two stages of leaf growth. Young and mature leaves of *S. plumbizincicola* also showed different accumulation and distribution characteristics for Zn and Cd.

2. Agronomic measures to increase cadmium and zinc phytoextraction efficiency by *S. plumbizincicola*

2.1 Cultivation management of *S. plumbizincicola*

S. plumbizincicola is suitable for clone reproduction. The shoot branches are usually picked off for reproduction. A tissue culture method has been established to obtain large amounts of plant materials (Wu et al., 2010). In field application, appropriate plant densities, soil moisture, light and harvest strategy will maximize plant biomass and Cd and Zn uptake and shorten the phytoremediation period. A field plot experiment was conducted to study the effects of planting density on the growth and heavy metal uptake by *S. plumbizincicola* (Liu et al., 2009). The results show that *S. plumbizincicola* had the highest yield and Cd and Zn removal rates when the planting density was 15 cm$\times$15 cm (4.4×10^5 plants/ha). Neither lower nor higher densities increased the performance. Humid dry conditions (70% of water holding capacity) are preferable for plant growth and metal accumulation in extremely dry or flooded conditions (Cui et al., 2009). *S. plumbizincicola* is a shade loving species and its accumulation of Zn and Cd was not reduced when the light intensity decreased to 50% of the control level (Li et al., 2010). Shade tolerance also ensures that it can be employed in intercropping with high straw plants such as maize, sorghum or beans. *S. plumbizincicola* is a perennial and can be harvested two to three times each year, for example before a hot summer or cold winter. When cutting, maintaining a stubble of 3-5 cm will be beneficial for reproduction (Li et al., 2009).

2.2 Intercropping of *S. plumbizincicola* with other plant species

Much effort has been made in the past several years to explore the feasibility of simultaneous grain production and phytoremediation of heavy metal-contaminated soils with *S. plumbizincicola* in the main food crop production areas of China. A pot experiment with a heavy metal-contaminated black soil from Heilongjiang Province, an alluvial soil from Henan

Province, and a paddy soil from Zhejiang Province was conducted to study the effects of intercropping *S. plumbizincicola* during the wheat growing season under wheat (*Triticum aestivum*)-rice (*Oryza sativa*) rotation on the growth of the crops and their uptake of heavy metals (Zhao et al., 2011). The results show that compared with monocultured *T. aestivum*, intercropping with *S. plumbizincicola* increased the soil $NaNO_3$-extractable Zn and Cd significantly, with increments of extractable Zn in test paddy soil, alluvial soil, and black soil of 55%, 32% and 110%, and that of extractable Cd in test paddy soil and black soil of 38 and 110%, respectively. The heavy metal concentration in *T. aestivum* shoots under intercropping *S. plumbizincicola* was 0.1-0.9 times that under monocultured *T. aestivum*, but the intercropping had little effect on the growth of rice or its heavy metal uptake. Although the Cd concentration in rice grain after growth of *S. plumbizincicola* remained>0.2 mg/kg (the food standard Cd limit), it showed a decreasing trend compared with that after growth of *T. aestivum* in monoculture. Intercropping *S. plumbizincicola* in the wheat growth season under the wheat-rice rotation can therefore benefit the phytoremediation of heavy metal-contaminated soil and decrease the food-chain risk of rice grown in rotation.

A field plot experiment was also conducted to study the effect of *S. plumbizincicola* on metal uptake by intercropped *T. aestivum* and a following crop of *Solanum melongena* in the subsequent growing season from heavy metal polluted soil, and to explore potential cropping systems for simultaneous remediation of contaminated soils and production of agricultural crops (Ju et al., 2015). The results show that in the intercropping treatments there was no difference in the biomass of *S. melongena* or *S. plumbizincicola*. Intercropping with *S. plumbizincicola* significantly decreased the Cd concentrations in *T. aestivum* by 52.4% compared to the monoculture treatments and metal concentrations in the following crop of *S. melongena* also decreased markedly. Compared to the control, the soil total Cd concentration in the intercropping treatment declined by 24.3%, showing highly efficient Cd removal. These results suggest that a cropping system consisting of intercropped *S. plumbizincicola* and *T. aestivum* and then rotating with *S. melongena* might realize the combined benefits of phytoremediation of metal polluted soil and the simultaneous safe production of food for human consumption.

2.3 Nutrient management of *S. plumbizincicola* growth

Nutrient management is an important factor that affects plant yield, metal bioavailability and ultimately the metal removal efficiency of hyperaccumulators. Numerous studies have investigated the effects of nutrition status on plant growth and metal accumulation by *S. plumbizincicola* (Shen et al., 2011; Hu et al., 2013; Wang et al., 2014; Arnamwong et al., 2015).

Hu et al. (2013) investigated whether and how the form of nitrogen (nitrate (NO_3^-) versus ammonium (NH_4^+)) influences Cd uptake and translocation and subsequent Cd phytoextraction by *S. plumbizincicola*. Plants were grown hydroponically with N supplied as either NO_3^- or NH_4^+. Short-term (36 h) Cd uptake and translocation were determined quantitatively using a positron-emitting ^{107}Cd tracer and positron-emitting tracer imaging system. The results show that the rates of Cd uptake by roots and transport to the shoots in the NO_3^- treatment were more rapid than in the NH_4^+ treatment. After uptake for 36 h, 5.6% (0.056 μmol/L) and 29.0% (0.290 μmol/L) of total Cd in the solution was non-absorbable in the NO_3^- and NH_4^+ treatments, respectively. The local velocity of Cd transport was approximately 1.5 times higher in roots (3.30 cm/h) and 3.7 times higher in shoots (10.10 cm/h) of NO_3^--fed than NH_4^+-fed plants. Autoradiographic analysis of ^{109}Cd reveals that NO_3^- nutrition enhanced Cd transportation from the main stem to branches and young leaves. Moreover, NO_3^- treatment increased Cd, Ca and K concentrations but inhibited Fe and P in the xylem sap. In 21-day hydroponic culture, shoot biomass and Cd concentration were 1.51 and 2.63 times higher in NO_3^--fed than in NH_4^+-fed plants. It was concluded that compared with NH_4^+, NO_3^- promoted the major steps in the transport route followed by Cd from solution to shoots in *S. plumbizincicola*, namely its uptake by roots, xylem loading, root-to-shoot translocation in the xylem and uploading to the leaves. *S. plumbizincicola* prefers NO_3^- to NH_4^+ nutrition for Cd phytoextraction.

The effects of different N fertilizers, including nitrate, ammonium, a nitrification inhibitor (dicyandiamide, DCD) and a urease inhibitor (n-(n-butyl) thiophosphoric triamide, NBPT) on shoot yield and Cd and Zn accumulation by *S. plumbizincicola* was also conducted under pot experiment conditions (Arnamwong et al., 2015). The soil was contaminated with 0.99 mg Cd/kg and 241 mg Zn/kg. The soil solution pH ranged from 7.30 to 8.25 during plant growth which was little affected by the type of N fertilizer. The $(NH_4)_2SO_4$+DCD treatment produced higher NH_4^+-N concentrations in the soil solution than the $(NH_4)_2SO_4$ and $NaNO_3$ treatment, indicating that DCD addition inhibited the nitrification process. Shoot Cd and Zn concentrations across all treatments showed ranges of 52.9-88.3 mg/kg and 2691-4276 mg/kg, respectively. The $(NH_4)_2SO_4$+DCD treatment did not produce significantly different Cd and Zn concentrations in the xylem sap from the $NaNO_3$ treatment. Plants grown with $NaNO_3$ had higher shoot Cd concentrations than those grown with $(NH_4)_2SO_4$+DCD at 24.0 mg/kg and 15.4 mg/kg, respectively. Nitrogen fertilizer application had no significant effect on shoot dry biomass. Total Cd uptake in the urea+DCD treatment was higher than in the control, urea+NBPT, urea+NBPT+DCD, or urea treatments, by about 17.5%, 23.3%, 10.7%, and 25.1%, respectively.

2.4 Microbial enhancement of *S. plumbizincicola* growth

Functional rhizobacteria have been used in several studies to successfully enhance the phytoextraction efficiency of heavy metal contaminated soils. Microbial enhancement of phytoextraction by *S. plumbizincicola* has been investigated in recent years.

Two plant growth promoting rhizobia (PGPR) strains, *Rhodococcus erythropolis* NSX2 and *Cedecea davisae* LCR1, were isolated (Liu et al., 2015). In a pot experiment inoculation with isolates NSX2 and LCR1 significantly enhanced the growth of *S. plumbizincicola* and its uptake of Cd. 454 pyrosequencing revealed that inoculation with PGPR led to a decrease in microbial community diversity in the rhizopshere during phytoextraction. Specifically, indigenous heavy metal-tolerant PGPR such as *Actinospica*, *Bradyrhizobium*, *Rhizobium*, *Mesorhizobium*, and *Mycobacterium* were selectively enriched in the treatments to which PGPR were added. It is suggested that a unique constitution of microbial communities in inoculated treatments plays a key role in enhancing Cd phytoremediation.

Metal-resistant and 1-aminocyclopropane-1-carboxylate (ACC) utilizing endophytic bacterial strains from tissues of *S. plumbizincicola* were isolated and characterized and their enhancement of the efficiency of phytoextraction of multiple-metal contaminated soils was examined (Ma et al., 2015). Of 42 metal resistant bacterial strains isolated from the tissues of *S. plumbizincicola* grown on Pb/Zn mine tailings, five plant growth promoting endophytic bacterial strains (PGPE) were selected due to their ability to promote plant growth and to utilize ACC as the sole nitrogen source. The five isolates were identified as *Bacillus pumilus* E2S2, *Bacillus* sp. E1S2, *Bacillus* sp. E4S1, *Achromobacter* sp. E4L5 and *Stenotrophomonas* sp. E1L and subsequent testing revealed that they all exhibited traits associated with plant growth promotion such as production of indole-3-acetic acid and siderophores and solubilization of phosphorus. The five strains showed high resistance to heavy metals (Cd, Zn and Pb) and various antibiotics. Further, inoculation of the ACC-utilizing strains significantly increased the concentrations of water-extractable Cd and Zn in the soil. Moreover, a pot experiment was conducted to elucidate the effects of inoculating metal-resistant ACC-utilizing strains on the growth of *S. plumbizincicola* and its uptake of Cd, Zn and Pb in multiple-metal contaminated soils. Of the five strains, *B. pumilus* E2S2 significantly increased root (146%) and shoot (17%) length, fresh (37%) and dry biomass (32%) of *S. plumbizincicola* and plant Cd uptake (43%), whereas *Bacillus* sp. E1S2 significantly enhanced the accumulation of Zn (18%) in plants compared with uninoculated controls. The strains inoculated also showed high levels of colonization in the rhizosphere and plant tissues. Results demonstrate the potential to enhance the phytoextraction of soils contaminated with multiple heavy metals by growing metal hyperaccumulating plants and inoculating with their own selected functional endophytic bacterial strains.

2.5 Other amendments affecting the growth of *S. plumbizincicola*

Organic materials with different functional groups can be used to enhance soil metal bioavailability. Traditional organic materials (rice straw and clover) and ethylenediamine disuccinic acid (EDDS) were applied to enhance metal uptake from polluted soil by *S. plumbizincicola* after repeated phytoextraction (Wu et al., 2012). Changes in pH, dissolved organic carbon (DOC) and metal concentrations were determined in the soil solution after EDDS application. The results show that amendment of the soil with ground rice straw or ground clover resulted in higher concentrations of Cd only (by factors of 1.92 and 1.71 respectively) in *S. plumbizincicola* compared to control soil. Treatment with 3 mmol/kg EDDS increased all the metals studied by factors of 60.4, 1.67, and 0.27 for Cu, Cd, and Zn, respectively. EDDS significantly increased soil solution DOC and pH and increased soil plant-available metals above the amounts that the plants were able to take up, resulting in high soil concentrations of soluble metals and a high risk of groundwater contamination. After repeated phytoremediation of metal contaminated soils the efficiency of metal removal declines as the concentrations of bioavailable metal fractions decline. Traditional organic materials can therefore be much more effective and environmentally friendly amendments than EDDS in enhancing the phytoremediation efficiency of Cd contaminated soil.

A glasshouse pot experiment was also conducted to study the effects of liming on plant growth and Zn and Cd accumulation by *S. plumbizincicola* in a heavy-metal-contaminated acidified paddy soil (Han et al., 2013). Lime application significantly increased the soil pH which reached a maximum of 5.53 after addition of 4.0 g/kg lime to soil, about 1.4 units more than that of the control. *S. plumbizincicola* grew larger after lime application but aboveground biomass did not increase significantly with increasing soil pH. Liming significantly reduced shoot Zn and Cd concentrations and uptake except at the lowest lime application rate (0.5 g/kg soil). This indicates that *S. plumbizincicola* can grow well in acid soil at a soil pH of 4.15, and application of lime did not increase plant extraction of heavy metals. Consequently, it is promising to use this species for Cd and Zn phytoextraction from acid agricultural soils polluted with metals.

Moreover, it was suggested that appropriate application of sulphur fertilizer may promote the growth of *S. plumbizincicola* and the addition of Ca Mg P fertilizer reduced the concentration of the active heavy metals in the soil and the pollution risk of subsequent vegetables (Ren et al., 2013).

3. Repeated phytoextraction of cadmium-and zinc-contaminated soils using the hyperaccumulator *S. plumbizincicola*

The technique of phytoextraction by hyperaccumulators has been shown to be successful in remediating soils polluted with metals. However, the technique requires long periods of remediation effort and successive crops are needed to remove adequate amounts of metals from contaminated soils to achieve safe levels. The available fraction of soil metals decreases over a prolonged phytoextraction time and this may be a major constraint for successful practical phytoextraction of polluted soils. Short-term remediation is not adequate for predicting the dynamics of plant metal uptake and changes in soil metals during the phytoextraction process. It is therefore desirable to study this process over relatively long time periods. Plant growth, plant metal uptake, and changes in soil metals were investigated during repeated phytoextraction of cadmium/zinc contaminated soils with different soil properties using the hyperaccumulator *S. plumbizincicola*.

3.1 Plant metal uptake during repeated phytoextraction

S. plumbizincicola grew well during repeated phytoextration and the plants did not show any visible symptoms of toxicity or growth inhibition but the shoot biomass might decrease because of poor growth conditions such as high temperatures during the growing season (Li et al., 2014b).

Plant metal uptake during repeated phytoextraction depended on which metals were involved, soil pollution levels and soil properties. In the case of soils with different pollution levels but similar soil properties, shoot Cd concentration decreased with harvest time in all soils but shoot Zn declined only in the slightly polluted soils. However, similar shoot Zn concentrations were found in other soils although these soils differed markedly in metal concentrations (Li et al., 2014a). In soils with clear differences in soil properties such as soil pH, plant metal uptake showed different patterns of change during repeated phytoextraction. As the phytoextraction period increased the shoot Zn concentrations in plants growing in acid soils and also shoot Cd concentrations increased in highly Cd-polluted acid soil. However there was no discernible decreasing trend in shoot Zn or Cd during phytoextraction of calcareous soils or in shoot Cd in the acid soil with low total Cd concentration (Li et al., 2014b).

A possible explanation for these results is that plant metal uptake is controlled by both plant active acquisition ability and soil metal availability. Plant active metal acquisition might contribute greatly to metal uptake when there is little soil available metal; conversely, soil metal release may control plant metal uptake when the available fraction is high.

3.2 Effects of repeated phytoextraction by *S. plumbizincicola* on total and extractable metals in contaminated soils

Total soil Cd and Zn decreased greatly in all soils tested after repeated phytoextraction. The extent of decline in metals in soils with different pollution levels were $>$80% for Cd and $>$24% for Zn after the growth of nine crops of *S. plumbizincicola* (Li et al., 2014a). After repeated phytoextraction by seven successive crops the declines in Zn and Cd concentrations in acids soils were $>$37% and 64%, respectively, and in calcareous soils the corresponding values were 12% and 37%, (Li et al., 2014b). Despite these large decreases in total metal concentrations, some soils such as highly polluted and calcareous polluted soils retained metal concentrations above the permission levels. This indicates that long time periods are required for successful phytoextraction or that other measures need to be used to increase the efficiency of metal extraction.

Changes in soil metal availability during phytoremediation have direct effects on removal efficiency and can also illustrate the interactive mechanisms between hyperaccumulators and metal contaminated soils. The changes in metal availability, desorption kinetics and speciation in four metal-contaminated soils during repeated phytoextraction by *S. plumbizincicola* over three years were investigated by chemical extraction and the DGT-induced fluxes in soils (DIFS) model (Li et al., 2016). The available metal fractions (i.e. metal in the soil solution extracted by $CaCl_2$ and by EDTA) decreased greatly by$>$84% after phytoextraction in acid soils and the deceases were dramatic at the initial stages of phytoextraction. However, the decreases in metal extractable by $CaCl_2$ and EDTA in calcareous soils were not significant or quite low. Large decreases in metal desorption rate constants evaluated by DIFS were found in calcareous soils. Sequential extraction indicated that the acid-soluble metal fraction was easily removed by *S. plumbizincicola* from acid soils but not from calcareous soils. Reducible and oxidizable metal fractions showed discernible decreases in acid and calcareous soils, indicating that *S. plumbizincicola* can mobilize non-labile metal for uptake but the residual metal cannot be removed. The results indicate that phytoextraction significantly decreases metal availability by reducing metal pool sizes and/or desorption rates and that *S. plumbizincicola* plays an important role in the mobilization of less active metal fractions during repeated phytoextraction.

3.3 Effects of repeated phytoextraction by *S. plumbizincicola* on soil microbial properties

Heavy metals can lead to decreased soil quality because metals affect the growth, morphology and metabolism of soil microorganisms. Repeated phytoextraction can partly recover soil quality when metals are removed from contaminated land and biological indicators of soil quality can be valid monitoring tools. The influence of repeated phytoextraction over a two-year period by successive crops of *S. plumbizincicola* on soil microbial properties was investigated. Microbial biomass C (C_{mic}), basal respiration and microbial quotient (qM) were significantly and positively correlated and soil metabolic quotient (qCO_2) was negatively correlated with heavy metal concentrations in unplanted soils ($p<0.05$). However, C_{mic}, basal respiration and qM values increased significantly after phytoremediation by five crops over two years compared to unplanted soil. Urease, β-glucosidase, neutral phosphatase and arylsulfatase activities also increased significantly with decreasing heavy metal contents and hydrolase activity was enhanced in planted soil ($p<0.05$) compared to the unplanted control. The results indicate that phytoextraction using *S. plumbizincicola* had beneficial effects on soil microbial and hydrolase activities, with the metal phytoextraction procedure restoring soil quality.

4. Field application of phytoextraction with *S. plumbizincicola*

S. plumbizincicola has been used for field phytoextraction of heavy metal polluted soils from 2006. So far, several research and demonstration areas have been established in numerous Chinese provinces including Zhejiang, Hu'nan, He'nan, Jiangxi, Jiangsu, Guangdong, and Guizhou. This species has always exhibited strong adaption to various climates, soil types and topograpies. It has shown much greater efficiency in the removal of Cd from slightly-to moderately-polluted acid soils than neutral soils.

A large scale (1 ha) field assessment of Cd phytoextraction by *S. plumbizincicola* was started in 2012 in Xiangtan county (Fig. 2), Hu'nan province, with a soil pH of 4.7 and soil total Cd concentrations in the arable layer of 0.49-0.71 mg/kg. Successive phytoextraction with *S. plumbizincicola* was conducted from 2012 to 2015. The plants were transplanted in October and harvested in June the following year. After phytoextraction, different rice cultivars were grown to assess the accumulation of Cd in the grain.

(a) Panorama of remediation base (b) Maturity of field *S. plumbizincicola*

Fig. 2 Demonstration area for Cd phytoextration with *S. plumbizincicola* in polluted acid soils in Xiangtan county, Hu'nan province

In the demonstration area *S. plumbizincicola* grew well with a shoot dry weight of 1.8-5.9 t/ha, Cd concentration of 53.1-94.9 mg/kg, and Cd removal amount of 169-353 g/ha with an average of 244 g/ha. After the first season of phytoextraction the soil Cd concentration (1 ha) decreased from 0.49-0.71 mg/kg to 0.32-0.56 mg/kg, with a removal rate of 11.5%-34.7%, an average of 21.7%. After the second season of phytoextraction the soil Cd concentration in one plot decreased from 0.64 to 0.29 mg/kg, a concentration below the permissible level in China (0.3 mg/kg, GB 15618—1995). The total removal rate of Cd from the arable layer of the soil reached 54.7% after two consecutive periods of phytoextraction. Soil profile analysis shows that the lower concentration of soil Cd occurred mainly in the 2-15 cm depth range, in line with the distribution of *S. plumbizincicola* root systems.

To assess the food safety of phytoextracted soil, several rice cultivars with different Cd accumulation capacities were grown both in phytoextracted soils in which the Cd concentration was 0.3 mg/kg and in non-phytoextracted soils (as controls). Two treatments comprising sepiolite+lime at rates of 0 (control) and 9+0.9 t/ha were applied to the phytoextracted soil as Cd stabilization agents. The results show that the Cd concentration in brown rice of cultivars Milyang 12085 and Milyang 12086 (previously screened as low-Cd-accumulating cultivars) grown in phytoextracted soil were 0.17 mg/kg and 0.13 mg/kg, respectively, values below the permissible level in China (0.2 mg/kg) and regarded as safe. The Cd concentration in brown rice of a hybrid cultivar SXY516 grown in phytoextracted soil was 0.66 mg/kg (exceeding the safety standard) but declined to 0.17 mg/kg after stabilization.

Our field assessment results suggest that repeated phytoextraction of Cd using *S. plumbizincicola* is a promising technique for the remediation of slightly Cd-polluted soils without halting normal agricultural production. For slightly-to moderately-polluted acid soils the time required might be 2-5 years to reduce soil total Cd below the safety level. Low-Cd-accumulating cultivars are recommended for growth on phytoextracted soils and metal stabilization is necessary if ordinary rice cultivars are grown.

5. Treatment of *S. plumbizincicola* biomass

Safely dispose or treatment of *S. plumbizincicola* biomass is also an important part for practically use of phytoextraction. Incineration and pyrolysis were employed to treat *S. plumbizincicola* biomass here. Distribution of incineration and pyrolysis products, characteristics of heavy metal, organic and conventional pollutants emissions, and their migration under incineration and pyrolysis were investigated to optimize the procedure of safely dispose of *S. plumbizincicola* biomass.

5.1 Distribution of incineration and pyrolysis products

S. plumbizincicola sampled from different polluted fields was treated with incineration and pyrolysis processes and significant reductions in the volume of the biomass were observed (Lu et al., 2012; Zhong et al., 2015a, 2005b). The weight losses of *S. plumbizincicola* were 3.56%-9.51%, 49.8%-50.8% and 73.9%-81.2% at three temperature steps of$<$170℃, 170℃-400℃ and 400℃-800℃, respectively, during the pyrolysis process. During incineration the weight losses of *S. plumbizincicola* were 6.17%-10.8%, 49.0%-50.4% and 82.7%-89.4% at three temperature steps of$<$185℃, 185-400℃ and 400-697℃, respectively. By comparison, the weight loss performance of *S. plumbizincicola* was similar in pyrolysis and incineration conditions at low temperatures. At high temperatures the weight loss of sample at the end of the pyrolysis process was slightly lower than in the incineration process. It was best to maintain the temperature$>$700℃ with excess air supply for the thermal disposal of *S. plumbizincicola*. The mass percentage of bottom ash to original biomass of *S. plumbizincicola* after pyrolysis and incineration generally decrease with increasing temperature, and the 13.2%-17.5% reductions after the incineration process were lower than the 31.1%-44.3% reductions after the pyrolysis process (Lu et al., 2012). During pyrolysis of *S. plumbizincicola* the bio-oil yield increased from 22.01 wt% at 450℃ to 31.7 wt% at 650℃ and decreased to 15.9 wt% at 750℃, the tar yield increased to 6.34% at 650℃ and decreased to 3.19 wt% at 750℃, the char yield decreased from 32.43 wt% at 450℃ to 22.49 wt% at 750℃, and the gas yield remained nearly constant up to 650℃ then increased significantly at 750℃. The main components of the pyrolysis oils were acids (at 450℃) which decreased with increasing pyrolysis temperature. Pyrolysis at 650℃ led to the highest yield of alkanes with low-oxygen compounds found in the bio-oil.

5.2 Characteristics of heavy metal emissions and migration under incineration and pyrolysis

Incineration and pyrolysis experiments using *S. plumbizincicola* have been carried out in a laboratory-scale entrained flow tube furnace. The distribution of pollutants in flue gas is an important basis for adaptation of the gas cleaning system for disposal of *S. plumbizincicola*. The pollutant concentration results show that Cd and Zn concentrations in control flue gas were 0.101 and 46.4 mg N/m^3, and the additives Al_2O_3, CaO, kaolin and activated carbon decreased the Cd concentration by 33.0%, 45.2%, 91.2% and 97.6%, respectively, and the Zn concentration by 21.6%, 50.8%, 88.1% and 99.1% in the flue gas during incineration (Wu et al., 2013b). The emission of Cd in the flue gas met the Chinese emission standard after use of the Al_2O_3, CaO, kaolin and activated carbon additives. In addition, the recoveries of Zn and Cd largely decreased with elevated temperatures in the bottom ash and increased in the fly ash, thus following the volatilization behavior of the heavy metals during incineration. Almost 80.0% of Cd was present in the fly ash and 90% of Zn remained in the bottom ash at temperatures of 350-650℃ (Zhong et al., 2015a). When the temperature was above 650℃ most of the Zn was found in the fly ash. The melting point was a key factor determining the recoveries of metals. Elemental sulfur and sulfide increased the retention of Cd and Zn on the bottom ash through the formation of sulfides under the local reducing environment in the furnace. In contrast, the presence of sulfur in the forms of Na_2SO_3 and Na_2SO_4 had little effect on heavy metal volatilization. Chlorine compounds did not increase heavy metal volatilization or partitioning on the fly ash during incineration because most of the chlorine was found to be present as KCl. The Cd and Zn were found to occur as the pure metals, their oxides and carbonates, and silicate compounds were found under low temperature conditions during the incineration process. At 850℃ the heavy metal recovery rates with increasing air flow rate from fly ash were higher than from bottom ash, and this indicates that at higher temperatures both the oxidizing atmosphere and reducing conditions favored the transfer of Cd and Zn to the fly ash phase (Zhong et al., 2015b).

During the pyrolysis of *S. plumbizincicola* kinetic analysis showed that the activation energy changed from 146.4 to 232.4 kJ/mol and the frequency factor changed greatly from 1.34×10^{11} to 8.99×10^{15}/s (Deng et al., 2016). Heavy metals were not detected in the gas. At temperatures of 450-650℃ more than 43.6% of Zn remained in the char and 54.4% was found in the bio-oil at 750℃. More than 87.6% of Cd was vaporized during pyrolysis and was found in the bio-oil. By comparison, mass balances of heavy metals under pyrolysis conditions differed from those under incineration conditions (Lu et al., 2012). In a horizontal tube furnace, incineration favored the volatilization of Cd in contrast to pyrolysis in flue gas. The percentages of heavy metals in bottom ash after pyrolysis were 3.5, 2.7 and 2.3 times higher than those after incineration, especially for Cd, Pb, and Zn, which indicates that the oxidizing atmosphere favors the transfer of these metals to the gaseous phase. However, in an entrained flow tube furnace, the Zn content in flue gas increased with increasing temperature but the Cu and Cd contents fluctuated.

5.3 Characteristics of organic pollutant emissions under incineration processes

The incineration of *S. plumbizincicola* biomass showed high concentrations of total PAHs (35 420 μg/kg) in a laboratory scale entrained flow tube furnace with 86.0% of the PAHs being 3-ring and 13.9% 4-ring (Wu et al., 2013b). Most 3-ring PAHs were present in the gas phase emissions and more ring PAHs were associated with fly ash. The emissions of total PAHs decreased with increasing temperature during incineration of *S. plumbizincicola*, especially in bottom ash. The toxic equivalent quantity (TEQ) of total PAHs showed an increasing trend in flue gas and fly ash and a decreasing trend in bottom ash with increasing temperature. High concentrations of PAHs were detected in flue gas and more than 99% of total PAHs were removed from the flue gas when activated carbon was used as the adsorbent. When silica was used as the bed material the 4-ring PAHs were reduced and 3 ring PAHs increased. The concentrations of PCDD/Fs in flue gas were 131 pg N/m^3 and 107 TEQ pg N/m^3, lower values than the Chinese emission standard (500 TEQ pg N/m^3) but higher than the emission limit of 92.0 pg TEQ N/m^3 set for waste incinerators in many countries. The PCDD/F at the highest concentration was OCDD which was present at 25.4 pg/m^3.

5.4 Characteristics of conventional pollutant emissions under incineration

During the incineration of *S. plumbizincicola* the concentration of CO in flue gas was more than 10 times the Chinese emission standard (GB18484—2001) and the NO_x emission was about 500 mg N/m^3. CO emission decreased but NO_x emission increased with increasing temperature. Additives can change the release of CO and NO_x. The additive CaO significantly increased the emissions of CO and NO_x. When the temperature was over 850℃ the concentrations of CO and NO_x in flue gas were higher than the Chinese emission standard (GB18484—2001). Using kaolin as additive gave the lowest concentration of NO_x, 461 mg N/m^3, which is lower than the 500 mg N/m^3 Chinese emission standard. Using Al_2O_3 also decreased NO_x in flue gas. SO_2 and HCl in flue gas were much lower than their respective emission standards (GB18484—2001).

References

Arnamwong S, Wu L H, Hu P J, et al. 2015. Phytoextraction of cadmium and zinc by *Sedum plumbizincicola* using different nitrogen fertilizers, a nitrification inhibitor and a urease inhibitor. International Journal of Phytoremediation, 17 (4): 382-390.

Cao D, Zhang H Z, Wang Y D, et al. 2014. Accumulation and distribution characteristics of zinc and cadmium in the hyperaccumulator plant *Sedum plumbizincicola*. Bulletin of environmental Contamination and Toxicology, 93 (2): 171-176.

Cui L Q, Wu L H, Li N, et al. 2009. Effects of soil moisture on growth and uptake of heavy metals of *Sedum plumbizincicola*. Soils 41: 572-576

Deng L, Li Z, Wang J, et al. 2016. Long-term field phytoextraction of zinc/cadmium contaminated soil by *Sedum plumbizincicola* under different agronomic strategies. International Journal of Phytoremediation, 18 (2): 134-140.

Han C L, Wu L H, Tan W N, et al. 2013. Bioavailability and Accumulation of Cadmium and Zinc by *Sedum plumbizincicola* after Liming of an Agricultural Soil Subjected to Acid Mine Drainage. Communications in Soil Science and Plant Analysis, 44 (6): 1097-1105.

Hu P J, Wang Y D, Przybyłowicz W J, et al. 2015. Elemental distribution by cryo-micro-PIXE in the zinc and cadmium hyperaccumulator *Sedum plumbizincicola* grown naturally. Plant and Soil, 388 (1-2): 267-282.

Hu P J, Yin Y G, Ishikawa S, et al. 2013. Nitrate facilitates cadmium uptake, transport and accumulation in the hyperaccumulator *Sedum plumbizincicola*. Environmental Science and Pollution Research, 20 (9): 6306-6316.

Jiang J P, Wu L H, Li N, et al. 2010. Effects of multiple heavy metal contamination and repeated phytoextraction by *Sedum plumbizincicola* on soil microbial properties. European Journal of Soil Biology, 46 (1): 18-26.

Ju S Y, Wang J, Mi Y Y, et al. 2015. Phytoremediation of heavy metal contaminated soils by intercropping with *Sedum plumbizincicola* and *Triticum aestivum* and rotation with *Solanum melongena*. Chinese Journal of Ecology 34: 2181-2186.

Li N, Tang M D, Cui L Q, et al. 2010. Effects of light intensity on plant growth and zinc and cadmium uptake by *Sedum plumbizincicola*. Acta Pedologica Sinica, 47 (2): 370-373.

Li N, Wu L H, Luo Y M, et al. 2009. Effects of harvesting way of *Sedum plumbizincicola* on its zinc and cadmium uptake in a mixed heavy metal contaminated soil. Acta Pedologica Sinica, 46: 725-728.

Li Z, Jia M Y, Wu L H, et al. 2016. Changes in metal availability, desorption kinetics and speciation in contaminated soils during repeated phytoextraction with the Zn/Cd hyperaccumulator *Sedum plumbizincicola*. Environmental Pollution, 209: 123-131.

Li Z, Wu L H, Hu P J, et al. 2014. Repeated phytoextraction of four metal-contaminated soils using the cadmium/zinc hyperaccumulator *Sedum plumbizincicola*. Environmental Pollution, 189: 176-183.

Li Z, Wu L H, Luo Y M, et al. 2014. Dynamics of plant metal uptake and metal changes in whole soil and soil particle fractions during repeated phytoextraction. Plant and Soil, 374 (1-2): 857-869.

Liu L, Wu L H, Li N, et al. 2009. Effect of planting densities on yields and zinc and cadmium uptake by *Sedum plumbizincicola*. Environmental Science, 30 (11): 3422-3426.

Liu L, Wu L, Li N, et al. 2011. Rhizosphere concentrations of zinc and cadmium in a metal contaminated soil after repeated phytoextraction by *Sedum plumbizincicola*. International Journal of Phytoremediation, 13 (8): 750-764.

Liu W X, Wang Q L, Wang B B, et al. 2015. Plant growth-promoting rhizobacteria enhance the growth and Cd uptake of *Sedum plumbizincicola* in a Cd-contaminated soil. Journal of Soils and Sediments, 15 (5): 1191-1199.

Lu S Y, Du Y Z, Zhong D X, et al. 2012. Comparison of trace element emissions from thermal treatments of heavy metal hyperaccumulators. Environmental Science & Technology, 46 (9): 5025-5031.

Ma Y, Oliveira R S, Nai F, et al. 2015. The hyperaccumulator *Sedum plumbizincicola* harbors metal-resistant endophytic bacteria that improve its phytoextraction capacity in multi-metal contaminated soil. Journal of Environmental Management, 156: 62-69.

Ren J, Wu L H, Liu H Y, Luo YM. 2013. Effect of amendments on phytoextraction efficiency and metal uptake of following vegetable in heavy metal contaminated soil. Soils 45: 1233-1238.

Shen L B, Wu L H, Han X R, et al. 2011. Effects of nutrient regulation and control on plant growth and Zn/Cd uptake by hyperaccumulator *Sedum plumbizincicola*. Soils, 43 (2): 221-225.

Wang J, Shen L B, Li Z, et al. 2014. Effects of nitrogen forms on growth and Zn/Cd uptake of *Sedum plumbizincicola*. Journal of Agro-Environment Science 33: 2118-2124.

Wu L H, Ben A L, Tan W N, et al. 2010. A tissue culture method for Zn/Cd hyperaccumulator *Sedum plumbizincicola*. China Invention Patent, CN101869077A.

Wu L H, Li Z, Akahane I, et al. 2012. Effects of organic amendments on Cd, Zn and Cu bioavailability in soil with repeated phytoremediation by *Sedum plumbizincicola*. International Journal of Phytoremediation, 14 (10): 1024-1038.

Wu L H, Liu Y J, Zhou S B, et al. 2013. *Sedum plumbizincicola* XH Guo et SB Zhou ex LH Wu (Crassulaceae): a new species from Zhejiang Province, China. Plant Systematics and Evolution, 299 (3): 487-498.

Wu L H, Zhong D X, Du Y Z, et al. 2013. Emission and control characteristics for incineration of *Sedum plumbizincicola* biomass in a laboratory-scale entrained flow tube furnace. International Journal of Phytoremediation, 15 (3): 219-231.

Wu L H, Zhou S B, Bi D, et al. 2006. *Sedum plumbizincicola*, a new species of the Crassulaceae from Zhejiang, China. Soils, 38 (5): 632-633.

Zhao B, Shen L B, Cheng M M, et al. 2011. Effects of intercropping *Sedum plumbizincicola* in wheat growth season under wheat-rice rotation on the crops growth and their heavy metals uptake from different soil types. Chinese Journal of Ecology, 22 (10): 2725-2731.

Zhong D X, Zhong Z P, Wu L H, et al. 2015. Thermal characteristics and fate of heavy metals during thermal treatment of *Sedum plumbizincicola*, a zinc and cadmium hyperaccumulator. Fuel Processing Technology, 131: 125-132.

Zhong D X, Zhong Z P, Wu L H, et al. 2015. Thermal characteristics of hyperaccumulator and fate of heavy metals during thermal treatment of *Sedum plumbizincicola*. International Journal of Phytoremediation, 17 (8): 766-776.

The Potential of Oilseed Rape and *Thlaspi caerulescens* for Phytoremediation of Cadmium-Contaminated Soil

Dechun SU, Rongfeng JIANG, Huafen LI

College of Resources and Environmental Sciences, China Agricultural University

The presence of heavy metals in agricultural ecosystems has raised concerns not only for crop quality but also for human health. The recent join report on the current status of soil contamination in China issued by the Ministry of Environmental Protection (MEP) and the Ministry of Land and Resources (MLR) of the People's Republic of China shows, among the heavy metals and metalloids, Cd ranks the first in the percentage of soil samples (7.0%) exceeding the MEP limit (the Ministry of Environmental Protection and the Ministry of Land and Resources, 2014). Special consideration should be paid to heavy metal pollution in soil-plant systems, especially cadmium (Cd), because of its high mobility and relatively lower toxic concentrations in organisms. Phytoremediation has emerged as a potential cost-effective and environmental friendly alternative to engineering-based technologies.

1. Phytoremediation of mustard oilseed rape

Although metal hyperaccumulation is not a common phenomenon, some plant families, such as the Brassicaceae, are particularly well represented. *Thlaspi caerulescens* CT. *caerulescens* is best known accumulator of zinc (Zn) and cadmium (Baker, 1987; Baker et al., 1994). Mustard oilseed rape (*Brassica juncea* L.) is one of the staple crops in China and also a member of the Brassicaceae. China has a wide variety of oilseed rape varieties, among which some genotypes may be Cd accumulators.

1.1 Selection and characteristics Cd uptake by mustard oilseed rape

Nineteen varieties of mustard oilseed rape (*Brassica juncea* L.) obtained from different provinces of China and Indian mustard seeds obtained from the National Herbarium of the USA were used in the experiments. Among the 19 varieties, only 5 genotypes of oilseed rape had higher Cd uptake than Indian mustard (Fig. 1). There were no significant differences in amounts of shoot Cd uptake among these five varieties of the oilseed rape and Indian mustard. Number 13 (Xikou Huazi) and number 17 (Zhucang Huazi) had relatively higher shoot Cd uptake compared to Indian mustard (number 1) (Su and Wong, 2004).

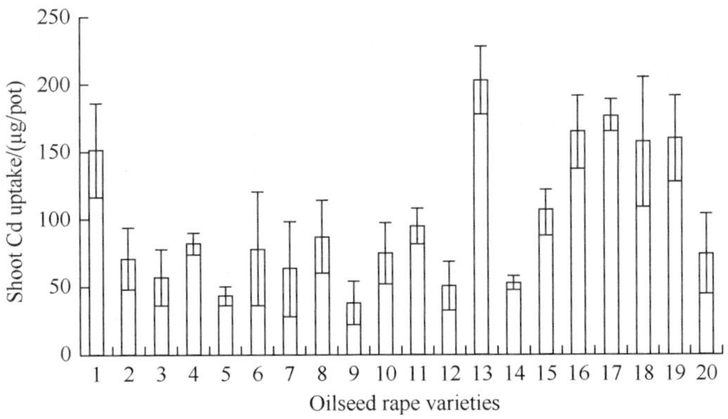

Fig. 1 Shoot Cd uptake by 19 different varieties of mustard oilseed rape (*Brassica juncea* L.) (numbers 2 to 20) and Indian mustard (number 1) grown in soil artificially spiked with 20 mg/kg Cd (Su and Wong, 2004)

With an increase in Cd concentration in the soil from 0 to 40 mg/kg, the shoot Cd concentration increased with a linear correspondence (Table 1). The Cd concentration in shoot of the two oilseed rapes and Indian mustard did not differ significantly at each initial Cd concentration. Comparing with the hyperaccumulation model species-Indian mustard, oilseed rape of Xikou huazi had the same or even better ability of extraction Cd from Cd contaminated soil. The shoot Cd removal percentage of Xikou huazi was 3.5% to 3.9% at a soil Cd concentration range of 5 to 20 mg/kg with a growth period of 42 days, while Indian mustard and Zhcang huazi were lower than 2.7% (Su and Wong, 2004).

Table 1 Cd uptake and shoot removal rate by selected oilseed rape cultivars grown in a loamy soil spiked with different concentration of Cd (Su and Wong, 2004)

Soil Cd/(mg/kg)	Varieties of *Brassica juncea*	Shoot dry weight /(g/pot)	Shoot Cd conc./(mg/kg)	Shoot Cd uptake in total plant/%	Soil Cd removal rate by shoot/%
0	Indian mustard	3.00±0.50	0.83±0.29	92	--
	Xikou huazi	2.98±0.70	0.83±0.26	95	--
	Zhucang huazi	3.35±0.69	0.67±0.29	96	--
5	Indian mustard	3.23±0.49	33.0±4.2	87	2.7
	Xikou huazi	4.16±0.16	31.8±1.8	92	3.5
	Zhucang huazi	2.42±0.27	45.5±8.5	94	2.6
10	Indian mustard	3.06±0.51	64.2±6.0	89	2.6
	Xikou huazi	3.95±0.74	67.2±12.0	92	3.5
	Zhucang huazi	3.36±0.76	76.3±14.2	94	3.3
20	Indian mustard	2.23±0.63	182±23*	88	2.6
	Xikou huazi	4.58±0.29**	127±12*	88	3.9
	Zhucang huazi	1.94±0.81	198±18*	94	2.5
40	Indian mustard	0.71±0.01*	315±25*	90	0.7
	Xikou huazi	0.34±0.11*	306±43*	94	0.4
	Zhucang huazi	0.44±0.12*	352±39*	95	0.5

*Significantly different from the other Cd concentration treatments at $p<0.05$ according to LSD-test. Comparisons are among the same varieties.

**Significantly different from Indian mustard and Zhucang Huazi at $p<0.05$ according to LSD-test. Comparisons are among the same soil Cd concentration.

The relationship between shoot Cd concentration of two oilseed rape and Indian mustard and different soil Cd concentration could be simulated by mathematic quadratic curve (Fig. 2). At soil Cd concentration of 0-100 mg/kg, with increase in the soil Cd concentrations, the shoot Cd concentration of oilseed rapes and Indian mustard ascended in parabola trend. At soil Cd concentration of 100 mg/kg, the shoot Cd concentration of oilseed rape Zhongyou Za-1 reached 160.1 mg/kg, Xikou Huazi and Indian mustard reached 127.4 mg/kg and 108.9 mg/kg, respectively (Fig. 2). The accumulation coefficient of Zhongyou Za-1, Xikou Huazi and Indian mustard was 1.6, 1.3 and 1.1, respectively (Ru et al., 2004).

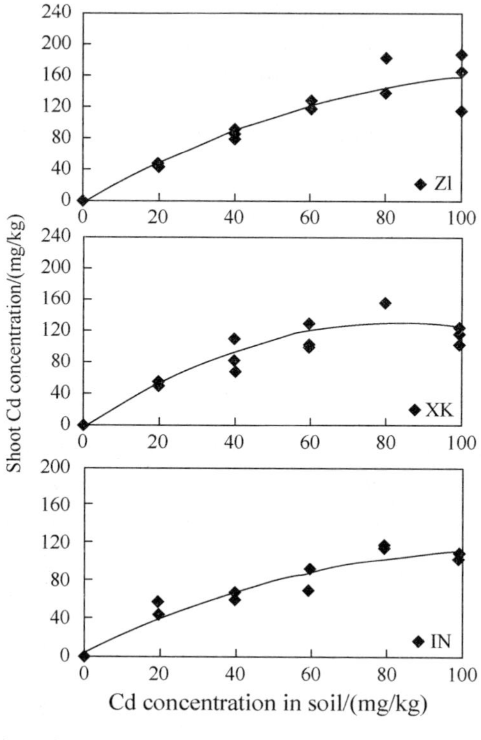

Fig. 2 Shoot Cd concentration of two oilseed rapes (Z1, Zhongyou Za-1; XK, Xikou Huazi) and Indian m ustard (IN, Indian mustard) grown in different Cd concentrations of soil (Ru et al., 2004)

1.2 Cd speciation and tolerance mechanisms of oilseed rape

Pot experiments were conducted under greenhouse conditions to investigate rhizosphere Cd speciation and mechanisms of Cd tolerance in different oilseed rape species (*Brassica juncea* L.) The chemical forms of Cd in a plant is closely related to its toxicity. The concentrations of ethanol-extractable, HCl-extractable, and residual Cd in the oilseed rapes Xikou Huazi, Zhongyou 821, and Indian mustard were very low. No significant differences were observed among oilseed rape varieties (Fig. 3). Water-extractable, NaCl-extractable, and acetic acid-extractable Cd predominated in Cd-tolerant oilseed rape Xikou Huazi, non-tolerant oilseed rape Zhongyou 821, and Indian mustard. Percentage of water-extractable Cd in non-tolerant oilseed rape Zhongyou 821 was 30.2%, while it was only 25.2% and 21.4% in Cd-tolerant oilseed rape Xikou Huazi and Indian mustard, respectively (Ru et al., 2006).

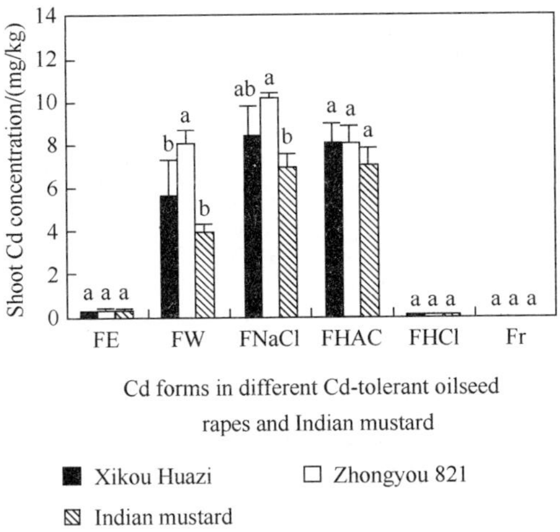

Fig. 3 Cd concentration of different extracted forms in oilseed rapes and Indian mustard tissue grown in soil with 80 mg/kgCd. Different letters atop of the bars indicated significant difference of varieties

FE: ethanol-extractable Cd; FW: water-extractable Cd; FNaCl: NaCl-extractable Cd; FHAC: acetic acid-extractable Cd; FHCl: HCl-extractable Cd; Fr: residual Cd (Ru et al., 2006)

The exchangeable Cd concentration in the rhizosphere was significantly lower than in non-rhizosphere soil at soil Cd concentrations of 30 and 60 mg/kg due to the absorption of plant (Fig. 4). Compared with non-rhizosphere soil of oilseed rape Xikou Huazi, Zhongyou 821, and Indian mustard, exchangeable Cd concentration in the rhizosphere decreased 29.3%, 28.0%, and 21.9%, respectively, at a soil Cd concentration of 30 mg/kg. At a soil Cd concentration of 30 mg/kg, no differences in carbonate-bound, Fe-Mn oxide-bound and organically-bound Cd were found between the rhizosphere and non-rhizosphere and among different oilseed rapes and Indian mustard. The rhizosphere organically bound Cd concentration was significantly higher than the non-rhizosphere soil of oilseed rape Xikou Huazi at a soil Cd concentration of 60 mg/kg, enhanced by 34% (Fig. 4) (Ru et al., 2006).

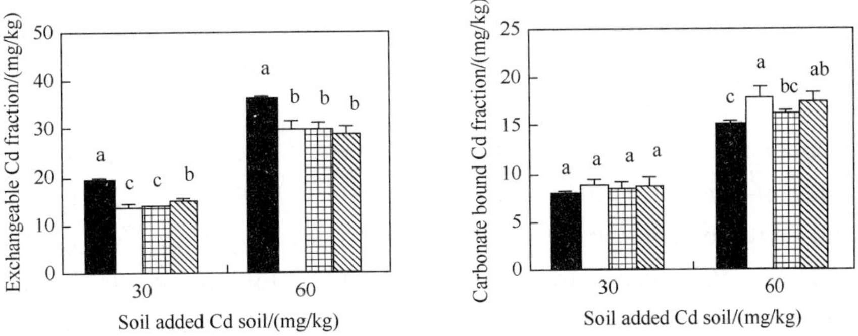

(a) Exchangeable fraction

(b) Carbonate bound fraction

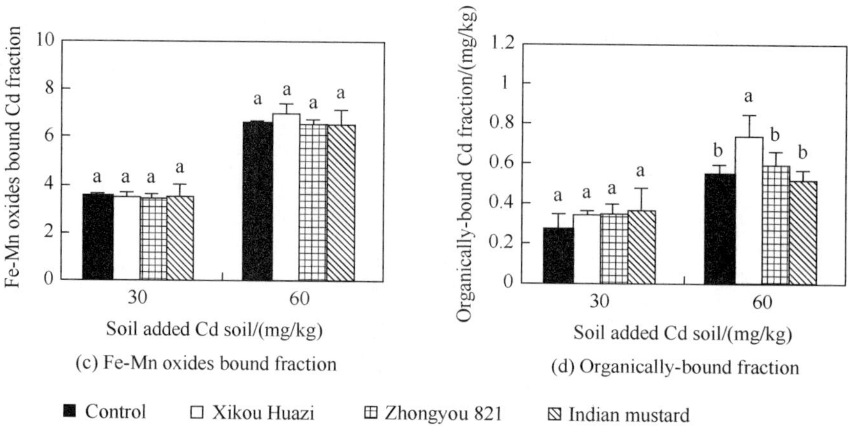

(c) Fe-Mn oxides bound fraction

(d) Organically-bound fraction

Fig. 4 Distribution of Cd speciation in the rhizosphere and non-rhizosphere of oilseed rapes and Indian mustard grown in different Cd concentration soils. Different letters atop of the bars indicated significant difference in rhizosphere and non-rhizosphere of all varieties (Ru et al., 2006)

2. Phytoremediation of *T. caerulescens*

T. caerulescens is one of the best known examples of hyperaccumulators, being able to hyperaccumulate Zn, as well as Cd in some accessions. Hyperaccumulation of metals involves at least three key processes: efficient absorption by roots, efficient root-to-shoot translocation and hypertolerance through internal detoxification (Baker et al., 2000; Pollard et al., 2002; Assunção et al., 2003; McGrath and Zhao, 2003).

2.1 Variation in uptake, translocation and compartmentation

After seedlings were exposed to radio labelled Cd and Zn (5 μM each) for 24 h, uptake of ^{109}Cd and ^{65}Zn, expressed on the basis of root FW, varied 5.0-fold and 6.7-fold, respectively, among the 11 *Thlaspi accessions* (Fig. 5 (a), 5 (b)). The Ga accession of *T. caerulescens* had the highest Cd uptake, while the Vi accession had the highest Zn uptake. The *T. praecox* accession from northern Slovenia also had a high Cd uptake. Within 24 h, 12%-59% and 15%-66% of ^{109}Cd and ^{65}Zn, respectively, had been transported to shoots (Fig. 5 (c), 5 (d)). The variation in translocation efficiency was 4.9-fold and 4.4-fold for Cd and Zn, respectively, among the 11 accessions (Xing et al., 2008).

Cd compartmentation in the root cells of the Ga and Pr accessions was investigated using the method of short-term ^{109}Cd efflux. The time-dependent kinetics of ^{109}Cd efflux from Ga and Pr roots could be resolved into three linear phases. After exposure to ^{109}Cd for 1 d, Ga roots accumulated more ^{109}Cd than Pr roots, with 1.7-fold, 3.3-fold and 4.4-fold larger amounts of ^{109}Cd in the root cell walls, cytoplasm and vacuoles, respectively, than those of Pr roots. Proportionally more ^{109}Cd was in the vacuoles of Ga roots (74%) than in those of Pr roots (57%)

(Table 2) (Xing et al., 2008).

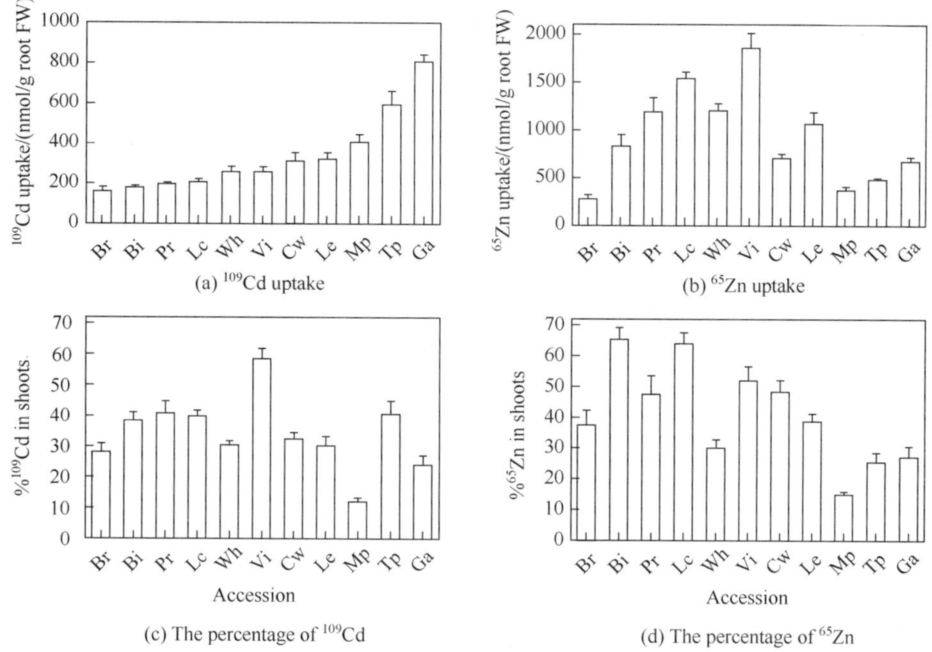

Fig. 5 Variation in the uptake of (a) ^{109}Cd; (b) ^{65}Zn and the percentage of (c) ^{109}Cd; (d) ^{65}Zn translocated to shoots among 11 accessions of *T. caerulescens* and *T. praecox*. *T. caerulescens* (Bizkaia, Bi; Black Rock, Br; Clough Wood, Cw; Ganges, Ga; La Calamine, Lc; Lellingen, Le; Monte Prinzera, Mp; Prayon, Pr; Viviez, Vi; Whitesike, Wh); *T. praecox* (Žerjav, Tp) (Xing et al., 2008)

Table 2 ^{109}Cd compartmentation and half-life ($t_{1/2}$) for ^{109}Cd efflux from different root cell compartments of the Ga and Pr accessions of *T. caerulescens* (Xing et al., 2008)

Accession		Cell wall	Cytoplasm	Vacuole
Ga	Amount of ^{109}Cd/(nmol/g FW)	145.5±31.0	87.1±14.6	659.8±34.1
	Percentage distribution of ^{109}Cd	16.2±3.2	9.6±1.2	74.2±3.1
	Efflux half-life $t_{1/2}$/min	2.4±0.4	25.9±3.8	2306±231.0
Pr	Amount of ^{109}Cd/(nmol/g FW)	83.8±2.2	26.6±2.8	149.0±10.8
	Percentage distribution of ^{109}Cd	32.6±1.1	10.3±1.0	57.1±1.6
	Efflux half-life $t_{1/2}$/min	3.3±0.6	27.5±2.1	2332±360.9

2.2 Effect of nitrogen form on Cd uptake

The NO_3^- treatment produced higher shoot and root biomass than the NH_4^+ treatment (Fig. 6 (a), 6 (b)). In the absence of the MES buffer, both shoot and root Cd concentrations were markedly higher (2-2.2 fold, $p<0.001$) in the NO_3^- treatment than in the NH_4^+ treatment (Fig. 6

(c)). When MES was used to buffer solution pH, the difference in root Cd concentration between the two N treatments became insignificant, but the shoot Cd concentration was 2.3 fold higher in the NO_3^- treatment than in the NH_4^+ treatment ($p<0.001$, Fig. 6 (d)) (Xie et al., 2009).

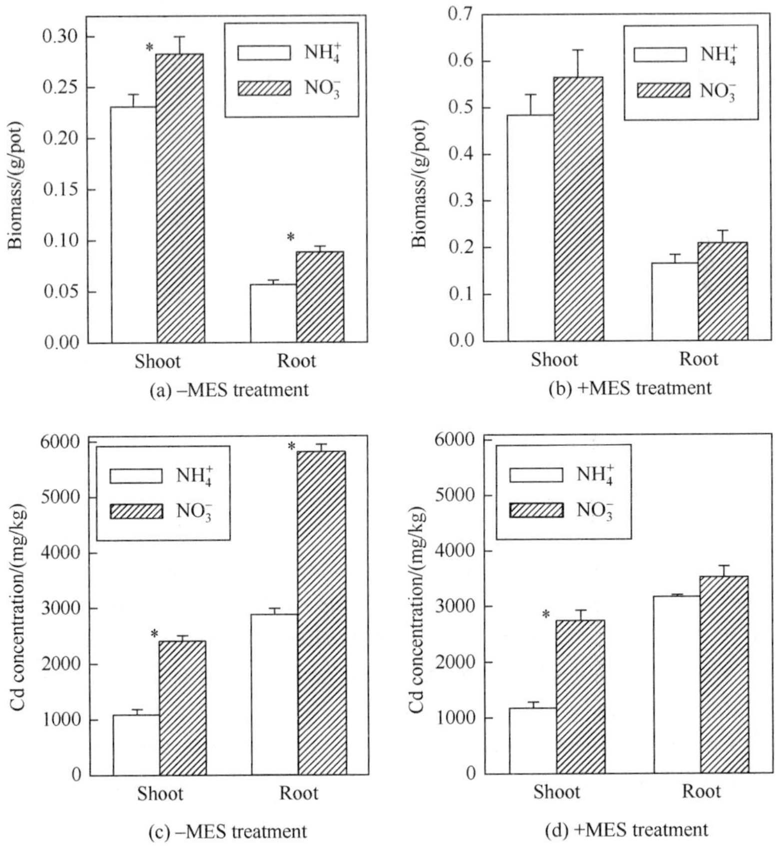

Fig. 6 Effects of N forms on the biomass of shoots and roots of *T. caerulescens* (a, b) and concentrations of Cd ((c), (d)) in shoots and roots in the pH unbuffered ((a), (c)) and pH buffered ((b), (d)) hydroponic experiments.*indicates a significant difference between the NH_4^+ and NO_3^- treatment ($p<0.05$) (Xie et al., 2009)

The addition of NH_4^+ or NO_3^- increased shoot biomass by 72% and 141%, respectively, over the control in the soil rhizobox experiment (Fig. 7 (a)). Shoot Cd concentration varied from 67 to 555 mg/kg dry weight (Fig. 7 (b)). The bioconcentration factor for Cd (i. e. the ratio of shoot Cd to soil Cd concentrations of 2.75 mg/kg) ranged from 25 to 202. Shoot Cd concentration was significantly ($p<0.001$) affected by the additions of different N forms, and followed the order of Control> NO_3^- > NH_4^+ > NH_4^++DCD (Fig. 4 (b)). The total amount of Cd accumulated in the shoots accounted for 1%-7.8% of the total Cd in the soil and followed the order of NO_3^- > Control> NH_4^+ > NH_4^++DCD (Xie et al., 2009).

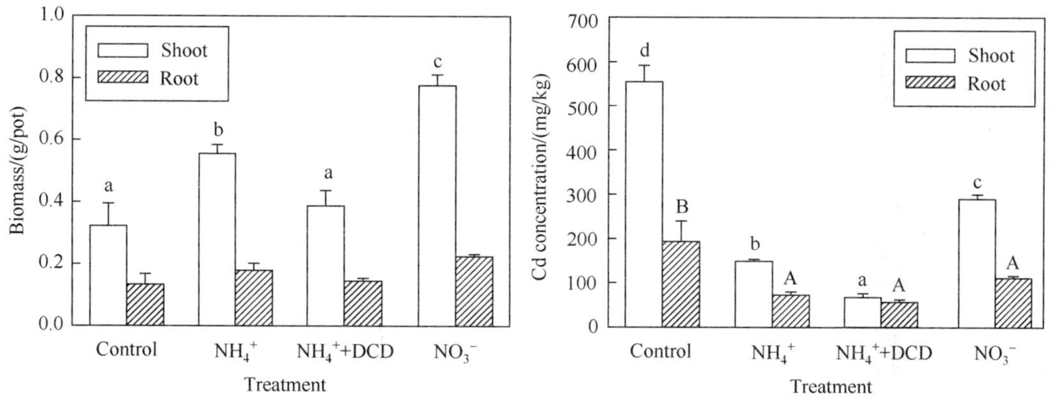

(a) *Thlaspi caerulescens* (b) Concentrations of Cd

Fig. 7 Effects of N forms on the biomass and Cd concentration in shoots and roots in the rhizobox experiment. Different letters indicate significant difference at $p<0.05$ (Xie et al., 2009)

2.3 Extraction efficiency

The selected soil was moderately contaminated with Cd (total Cd 2.87 mg/kg) due to past irrigation with sewage effluent in the water. The Cd concentrations in *T. caerulescens* and pakchoi and their phytoextraction efficiency were presented in table 3. Compared with pakchoi, *T. caerulescens* had a higher Cd concentration and phytoextraction efficiency. The Cd concentration in plants and phytoextraction efficiency increased with the prolonged growth duration.

Table 3 Extraction efficiency of *T. caerulescens*

Growth period (d)	Cd concentration/(mg/kg)		Phytoextraction efficiency/%	
	Pakchoi	*T. caerulescens*	Pakchoi	*T. caerulescens*
25	28.6±2.1	113±18	1.39±0.27	4.28±0.58
50	46.9±7.9	158±15	1.82±0.33	14.74±1.53

3. Cd accumulation in the rotation system

Intercropping and crop rotation allows full use of light, heat, water and soil resources which is the essence of China traditional agriculture; these techniques have made significant contributions to sustainable crop production. Murakami et al. (2008, 2009) revealed that phytoextraction by a high-Cd-accumulating rice cultivar could reduce the seed level of Cd found in successive soybeans and rice grown on paddy soils contaminated with moderate levels of Cd.

3.1 Rotation of rapeseed and Chinese cabbage

The pot experiment was conducted by using two kinds of soils: natural Cd-contaminated soil (soil A) and artificial Cd-contaminated soil (soil B). Soil A was collected from a vegetable field with total Cd 3.40 mg/kg, and soil B was added with Cd at 3 mg/kg. The Cd removal by shoot for the rapeseed cultivar Zhucang Huazi ranged from 0.24% to 1.0% in natural Cd-contaminated soil (soil A) and from 0.63% to 1.2% in artificial Cd-contaminated soil (soil B) during the growth period of 4-7 weeks; whereas that for the rapeseed cultivar Chuanyou Ⅱ-93 ranged from 0.25% to 0.61% and from 0.38% to 0.90%, respectively (Table 4). Chinese cabbage grown in the pots with soil A for five weeks after harvesting 2 rapeseed cultivars exhibited a significantly lower Cd concentration compared to that grown in the control pots (not planted with rapeseed). However, the rotation of rapeseed did not lower the Cd concentration of Chinese cabbage on soil B (Table 5) (Su et al., 2010).

Table 4 The uptake and removal of Cd by the shoot of two rapeseed cultivars on the natural and artificial Cd-contaminated soils (Soils A and B) at different growth time (Su et al., 2010)

Soil	Rapeseed cultivar	Cd uptake parameter	Growth time/weeks			
			4	5	6	7
A	ZhucangHuazi	Shoot mass/(g/pot)	0.97±0.10	1.81±0.50	2.95±0.32	4.04±0.13
		Cd concentration/(mg/kg)	4.52±0.78	5.35±0.54	5.24±1.43	4.17±1.02
		Cd uptake/(μg/pot)	4.12	9.68	15.5	16.9
		Cd removal by shoot/%	0.24	0.57	0.91	1.0
	Chuan You Ⅱ-93	Shoot mass/(g/pot)	1.14±0.07	2.17±0.16	3.63±0.43	4.96±0.27
		Cd concentration/(mg/kg)	3.75±1.30	4.39±0.74	3.10±1.27	2.08±0.98
		Cd uptake/(μg/pot)	4.27	9.53	11.2	10.3
		Cd removal by shoot/%	0.25	0.56	0.66	0.61
B	Zhucang Huazi	Shoot mass/(g/pot)	0.94±0.14	1.64±0.17	2.11±0.18	2.51±0.11
		Cd concentration/(mg/kg)	10.1±1.7	11.1±1.3	8.13±1.51	7.32±0.95
		Cd uptake/(μg/pot)	9.47	18.2	17.2	18.4
		Cd removal by shoot/%	0.63	1.2	1.2	1.2
	Chuan You Ⅱ-93	Shoot mass/(g/pot)	0.93±0.19	1.68±0.10	2.41±0.32	2.90±0.25
		Cd concentration/(mg/kg)	6.19±0.62	7.04±0.71	6.35±1.70	4.56±0.80
		Cd uptake/(μg/pot)	5.75	11.8	15.3	13.2
		Cd removal by shoot/%	0.38	0.79	1.0	0.90

Table 5 Cadmium concentrations in the shoot of Chinese cabbage grown on natural Cd-contaminated soil (soil A) and artificial Cd-contaminated soil (soil B) after cropping rapeseed cultivars Chuanyou Ⅱ-93 and Zhucang Huazi (Su et al., 2010)

Soil	Growth time of rapeseed	Shoot Cd of Chinese cabbage	
		After Chuanyou Ⅱ-93	After Zhucang Huazi
	weeks	mg/kg	
A	0 (Control)	6.5a*	6.5a
	4	6.1a	5.8ab
	5	6.0ab	5.2b
	6	5.6ab	4.7b
	7	5.5b	4.7b
B	0 (Control)	19.3a	19.3a
	4	22.4a	19.8a
	5	20.6a	18.9a
	6	20.1a	19.2a
	7	20.3a	18.3a

*Values followed by the same letter (s) within the same soil in the same column are not significantly different at $p<0.05$ level.

3.2 Rotation of oilseed rape and rice in Cd-contaminated agricultural land

This field experiment analyzed the phytoremediation effects of oilseed rape in moderately Cd-contaminated farmland and the food safety of successive rice in an oilseed rape-rice rotation system. The accumulation of Cd differed significantly in the two oilseed rape cultivars (Table 6). The Zhucang Huazi cultivar absorbed more Cd from the soil than Chuanyou Ⅱ-93. The total Cd uptake by cultivar of Zhucang Huazi were 1.58 to 1.76 times that of Chuanyou Ⅱ-93 in 2009 and 2010, respectively. The phytoextraction efficiencies of Zhucang Huazi and Chuanyou Ⅱ-93 ranged from 0.09 to 0.13% and 0.06 to 0.08% in 2009 and 2010, respectively, with that of Zhucang Huazi were 50 and 63% higher than Chuanyou Ⅱ-93, respectively (Yu et al., 2014).

Table 6 Shoot biomass, Cd uptake and Cd phytoextraction efficiency at harvest of oilseed rape (Yu et al., 2014)

Year	Cultivar	Dry matter weight (t/ha)		Cd uptake (g/ha)		Phytoextraction efficiency*/%
		Straw	Rapeseed	Straw	Rapeseed	
2009	Zhucang Huazi	3.12±0.17a	2.23±0.08a	1.55±0.09a	0.77±0.02a	0.09
	Chuanyou Ⅱ-93	2.84±0.01a	2.21±0.19a	1.16±0.04b	0.31±0.05b	0.06
2010	Zhucang Huazi	3.31±0.19a	1.72±0.14a	2.95±0.16a	0.35±0.02a	0.13
	Chuanyou Ⅱ-93	3.20±0.14a	1.66±0.24a	1.73±0.11b	0.15±0.04b	0.08

*Cd phytoextraction efficiency=total Cd uptake/(soil Cd con.×soil weight) ×100%. Soil weight of 2.6×10^6 kg/ha (0-20 cm) was used in this calculation. Different letters within the same year indicate the significant differences between the two cultivars ($P<0.05$).

Analysis of the Cd concentrations in rapeseed and straw revealed significant differences between the two cultivars of oilseed rape at harvest in 2009 and 2010 (Fig. 8). Zhucang Huazi had a stronger ability to accumulate Cd than Chuanyou Ⅱ-93. The Cd concentrations in rapeseed and straw of Zhucang Huazi were significantly higher than that of Chuanyou Ⅱ-93 in both years. The rapeseed Cd concentration in the Zhucang Huazi was 45% and 5% higher in 2009 and 2010, respectively, than the maximum level of Cd (0.2 mg/kg) permitted by the Chinese standard (MH, 2012). However, the rapeseed Cd concentrations in the Chuanyou Ⅱ-93 were all below the maximum allowed level in both years (Yu et al., 2014).

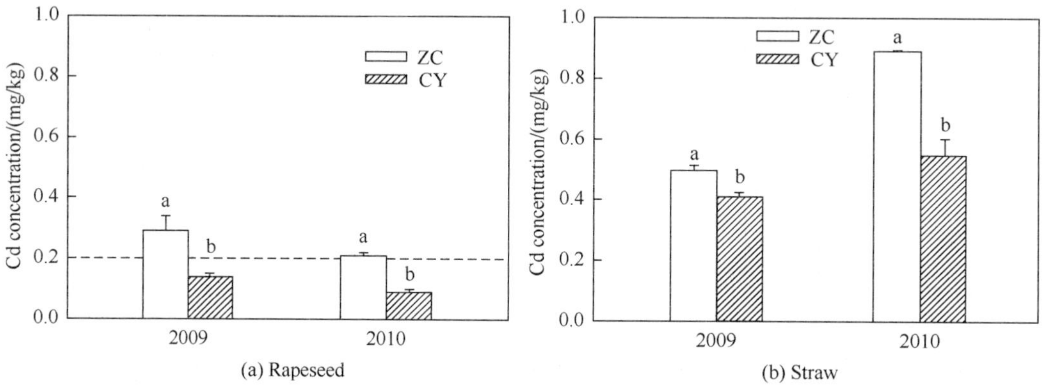

Fig. 8 Cd concentrations in rapeseed and straw at maturity. Different letters within the same year indicate the significant differences between the two cultivars ($p<0.05$). Vertical bars represent means±SE of three replicates. The dashed line represents the maximum level of Cd permitted in foods (0.2 mg/kg) by the Chinese standard (GB2762—2012). ZC and CY denote Zhucang Huazi and Chuanyou Ⅱ-93, respectively (Yu et al., 2014)

The previous planting of oilseed rape significantly influenced the concentrations of Cd in rice. The previous planting of the Zhucang Huazi and Chuanyou Ⅱ-93 cultivars of oilseed rape increased the Cd concentration in rice grain (Xiushui 113) by 102 and 63%, respectively, compared with the previous fallow treatment in 2009. In 2010, a different rice cultivar (Zhejing 22) was used; similar effects were also observed. However, the Cd concentration in Zhejing 22 rice grain in all treatments was lower than that of Xiushui 113. In contrast to the grain, the Cd concentrations in straw and husk of Zhejing 22 were higher than that of Xiushui 113 (Fig. 9) (Yu et al., 2014).

The rapeseed cultivar Zhucang Huazi exhibited high Cd accumulation rates, higher than the legal limit for human consumption (0.2 mg/kg); Cd concentrations in the cultivar Chuanyou Ⅱ-93 were all below the maximum allowed level. Planting oilseed rape increased the uptake of Cd by the successive rice crop compared with a previous fallow treatment. Most Cd concentrations of brown rice were below the maximum allowed level. The phytoextraction efficiency was lower in the moderately Cd-contaminated soil in field experiments. The results suggest screening rice cultivars with lower Cd accumulation can assure the food safety; the mobilization of heavy metals by roots of different plant species should be considered during crop rotation to assure food safety.

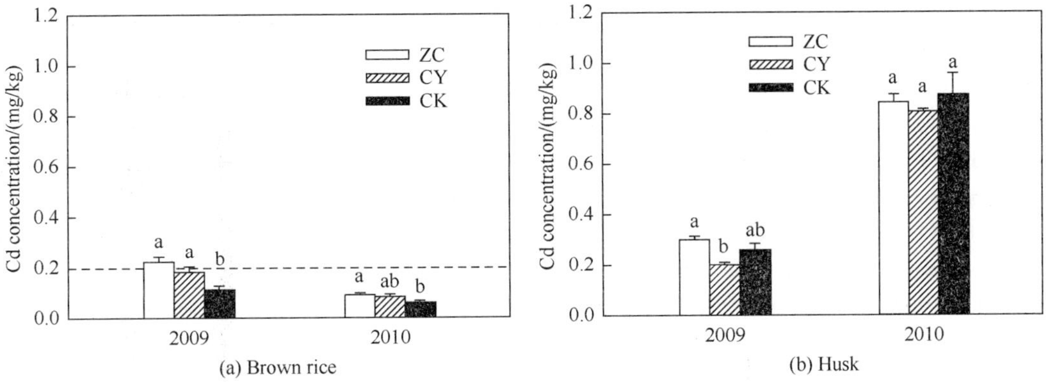

Fig. 9 Effects of previous cultivation on the total Cd concentrations in brown rice (a) and husk (b) at maturity
Different letters within the same year indicate the significant differences among the treatments ($p<0.05$). Vertical bars represent means ± SE of three replicates. The dashed line represents the maximum level of Cd in foods (0.2 mg/kg) permitted by the Chinese standard (GB2762—2012). ZC, CY and CK denote Zhucang Huazi, Chuanyou Ⅱ-93, and the control (no planting of oilseed rape), respectively (Yu et al., 2014)

References

Assunção A G L, Schat H, Aarts M G M. 2003. *Thlaspi caerulescens*, an attractive model species to study heavy metal hyperaccumulation in plants. New Phytologist, 159 (2): 351-360.

Baker A J M, McGrath S P, Reeves R D, et al. 2000. A review of the ecology and physiology of a biological resource for phytoremediation of metal-polluted soils. Phytoremediation of Contaminated Soil and Water. Metal Hyperaccumulator Plants: 85-107.

Baker A J M, Reeves R D, Hajar A S M. 1994. Heavy metal accumulation and tolerance in British populations of the metallophyte *Thlaspi caerulescens* J. & C. Presl (*Brassicaceae*). New Phytologist, 127 (1): 61-68.

Baker A J M. 1981. Accumulators and excluders-strategies in the response of plants to heavy metals. Journal of Plant Nutrition, 3 (1-4): 643-654.

McGrath S P, Zhao F J. 2003. Phytoextraction of metals and metalloids from contaminated soils. Current Opinion in Biotechnology, 14 (3): 277-282.

MH. 2012. Maximum Levels of Contaminants in Foods. Ministry of Health of the People's Republic of China, GB2762—2012.

Murakami M, Ae N, Ishikawa S, et al. 2008. Phytoextraction by a high-Cd-accumulating rice: Reduction of Cd content of soybean seeds. Environmental Science & Technology, 42: 6167-6172.

Murakami M, Nakagawa F, Ae N, et al. 2009. Phytoextraction by rice capable of accumulating Cd at high levels: reduction of Cd content of rice grain. Environmental Science & Technology, 43 (15): 5878-5883.

Pollard A J, Powell K D, Harper F A, et al. 2002. The genetic basis of metal hyperaccumulation in plants. Critical Reviews in Plant Sciences, 21 (6): 539-566.

Ru S H, Wang J Q, Su D C. 2003. Characteristics of Cd uptake and accumulation in two Cd accumulator oilseed rape species. Journal of Environmental Sciences (China), 16 (4): 594-598.

Ru S, Xing J, Su D. 2006. Rhizosphere cadmium speciation and mechanisms of cadmium tolerance in different oilseed rape species.

Journal of Plant Nutrition, 29 (5): 921-932.

Su D C, Wong J W C. 2004. Selection of mustard oilseed rape (*Brassica juncea* L.) for phytoremediation of cadmium contaminated soil. Bulletin of Environmental Contamination and Toxicology, 72 (5): 991-998.

Su D C, Wei-Ping J, Man Z, et al. 2010. Can cadmium uptake by Chinese cabbage be reduced after growing Cd-accumulating rapeseed?Pedosphere, 20 (1): 90-95.

The Ministry of Environmental Protection; the Ministry of Land and Resources, 2014. Report on the national soil contamination survey. http://www.zhb.gov.cn/gkml/hbb/qt/201404/t20140417270670.htm[2014-4-17]

Xie H L, Jiang R F, Zhang F S, et al. 2009. Effect of nitrogen form on the rhizosphere dynamics and uptake of cadmium and zinc by the hyperaccumulator *Thlaspi caerulescens*. Plant and Soil, 318 (1-2): 205-215.

Xing J P, Jiang R F, Ueno D, et al. 2008. Variation in root-to-shoot translocation of cadmium and zinc among different accessions of the hyperaccumulators *Thlaspi caerulescens* and *Thlaspi praecox*. New Phytologist, 178 (2): 315-325.

Yu L, Zhu J, Huang Q, et al. 2014. Application of a rotation system to oilseed rape and rice fields in Cd-contaminated agricultural land to ensure food safety. Ecotoxicology and Environmental Safety, 108: 287-293.

Soil Contamination in Arid Region of Northwest China: Status Mechanism and Mitigation

Yahu HU, Zhongren NAN*

Key Laboratory of Western China's Environmental Systems (Ministry of Education) and Gansu Key Laboratory for Environmental Pollution Prediction and Control, College of Earth and Environmental Sciences, Lanzhou University

*Corresponding author

1. Introduction

The status of soil contamination in China has attracted much public attention both domestically and internationally (Zhao et al., 2015). According to the latest figures from the Ministry of Environmental Protection (MEP) and the Ministry of Land and Resources (MLR) of the People's Republic of China, 16.1% of China's surveyed land is contaminated by heavy metals and metalloids such as arsenic (As), cadmium (Cd), mercury (Hg), and lead (Pb). This is because the discharge of large quantities of heavy metals over last three decades coinciding with the rapid industrialization and urbanization in China. In addition, 19.4% of surveyed arable land has levels of pollution higher than the soil environmental quality standard set by the MEP. Cadmium ranks the first in the percentage of soil samples (7.0%) exceeding the MEP limit, because of the high input rate, averaging 0.004 mg Cd/(kg yr) in the plough layer (Luo et al., 2009).

Northwest China is located in the innermost center of the Eurasia continent, covers approximately 1/3 of China's land area (Shi et al., 2007). Due to the arid region of northwest China is rich in mineral resources, large areas of agricultural lands near mining and smelting sites have been polluted heavily by heavy metals because of deposition of particulate matter, dumping of industrial wastes and tailings, and, especially, irrigation of untreated industrial wastewater (Mendez and Maier, 2008; Hu et al., 2013). This is due to the ore being only a small fraction of the total volume of the mined material, and some additional metals occur naturally within the raw ore becoming inevitable by-products, for example, zinc (Zn) and Pb smelters release large amounts of Cd into the environment (Adriano, 2001). Taking Baiyin region in Gansu Province as an example, this region has been a major non-ferrous metal mining and smelting base in northwest China since the 1950s. Results from field investigations have demonstrated that soils in this region are heavily contaminated by large quantities of heavy metals and metalloids such as As, Cd, and Pb (Nan et al., 1999; Nan and Zhao, 2000; Liu, 2003).

Dongdagou is an escape canal in Baiyin region that collected both domestic sewage and industrial wastewater discharged by non-ferrous metal mining and smelting plants and several other factories located in the mid to upper reaches of the canal. Due to the shortage of irrigated water, agricultural lands along Dongdagou were irrigated with wastewater for 40 years from the 1960s. It is estimated that more than 1330 ha agricultural land in this region has been contaminated with various degrees of heavy metals; one third of the farmland has been heavily contaminated according to the standard set by the MEP (Hu et al., 2014b). The concentrations of Cd in agricultural soils along this canal range from 3.4 to 77.9 mg/kg (Liu, 2003). The Cd concentrations measured in forage range from 5.4 to 32.7 mg/kg in corn, and in wheat grains from 1.98 to 4.12 mg/kg; in the kidneys of sheep and cows concentrations were 25.39 and 27.2 mg/kg, respectively (Liu, 2003). The Cd concentrations measured in human hair and urine in this region were several times higher than in unpolluted regions (Gansu Environmental Monitoring Station, 1984). Recently, at first, local residents of both adults and children were losing their teeth. Soon even goats that ate the grass near the river began to lose teeth. Finally, after 2000, most of the agricultural land along this canal was abandoned because of the severe soil degradation induced by heavy metal contamination.

Heavy meals can accumulate in the surface soil and persist for a long time due to their nondegradable nature (Adriano, 2001). Then, soils containing high concentrations of heavy metals can become a new pollution source to groundwater through leaching, to air through wind erosion, and to low-lying areas through water erosion (Hu et al., 2014b). Heavy metals can eventually pose risks and threats to food production, human and animal health, and soil quality (Kayser et al., 2000). Contamination of agricultural lands by heavy metals in arid region of northwest China has received considerable attention and been regarded as a great environmental concern (Nan et al., 2011). Thus, urgent measures are needed to remediate arable lands in arid region contaminated with heavy metals.

2. Soil-environmental interfacial processes of heavy metals

The interfacial processes of heavy metals in soil depend on many factors, including soil type, land use type, pH, cation exchange capacity (CEC), redox potential (Eh), soil organic matter (SOM), oxides, and agricultural activity (Wang et al., 2010; Zhao et al., 2015). Results from batch experiments showed that the adsorption capacity of heavy metals in arid soil under the condition of single contamination was Pb>Cu>Zn (Hu et al., 2009). Similarly, under the condition of combined contamination with multi-metals, the adsorption capacity was in the order of Pb>Cu>Cd>Zn>Ni. The adsorption model of heavy metals in arid soil is well fitted with Freundlich isotherm equation, regardless of soil contamination regimes.

In the case of the chemical species, results from the single extraction method showed that extractable Cu concentrations in arid soil generally decreased as EDTA>Citric acid>NH_4Ac>

NH_4NO_3>$MgCl_2$≈Tartaric acid>Acetic acid>Oxalic acid. The extraction capacity of Ni was in the order of Citric acid>EDTA>NH_4Ac>tartaric acid>NH_4NO_3≈Acetic acid≈Oxalic acid>$MgCl_2$ (Wang et al., 2009). EDTA and citric acid showed better extractability of both Cu and Ni compared to other extractants, respectively. This is probably because the formation of stable metal-chelant complexes, showing that the main sorption mechanism of both Cu and Ni in arid soil is specific sorption or precipitation, other than ion exchange (Wang et al., 2010). In addition, using the sequential extraction procedures under the condition of multi-metal contamination, Cu was distributed mainly in the organic fraction at low levels of contamination; however, most of Cu would be bounded with carbonate at relatively high levels of contamination (Wang and Nan, 2009). Zinc and Pb were distributed more in the carbonate fraction and less in Fe-Mn oxides fraction (Liao et al., 2010). In contrast, Cd sorbed was distributed mainly in both the carbonate bound fraction and exchangeable fraction (Yang et al., 2011). That is, Cd has a relatively high leaching and migration potential as compared with other heavy metals. In general, carbonate bound fraction is the major solid-phase component for many heavy metals in arid soil (Wang et al., 2009). Accordingly, heavy metals would accumulate in the upper layer of the arid soil because of the adsorption by $CaCO_3$ or precipitation as carbonate (Hu et al., 2014a, 2014b). This is demonstrated by the distribution pattern of heavy metals in soil profile in Dongdagou irrigated land as well (Fig. 1).

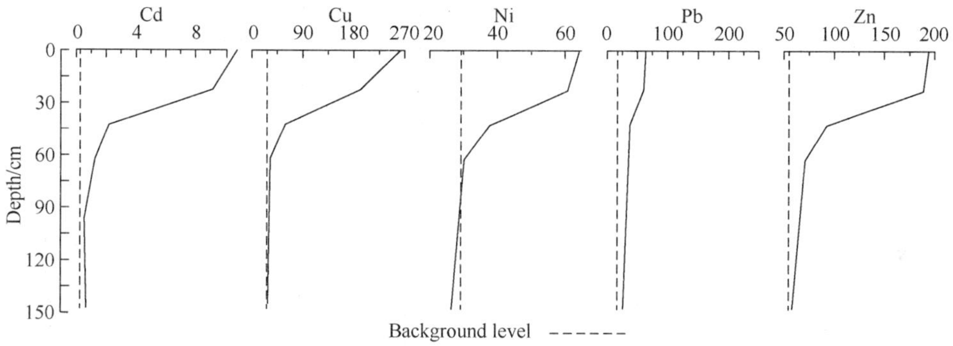

Fig. 1 Concentrations of heavy metals in soil profile in Dongdagou irrigated land/(mg/kg)

3. Accumulation mechanisms of heavy metals in soil-crop systems

It is speculated that heavy metals have a low mobility and bioavailability in arid soil than in acid soil with equal soil metal concentrations (Adriano, 2001). However, crops grown on arid soils are demonstrated that they can take up and accumulate large quantities of metals into their body as well (Nan et al., 1999; Nan and Zhao, 2000). This is largely because the high levels of soil contamination (Wang et al., 2015; Zhao et al., 2015), interactions between metals (Nan et al., 2002a; Zhao et al., 2014), and changes of soil properties like pH, Eh, and liable organic carbon (Nan and Chen, 2001b; Nan et al., 2002b, 2011).

Results from pot culture experiments showed that concentrations of heavy metals (Cd, Cu, Ni, Pb, and Zn), without regard to metal type, in tissues of both wheat (*Triticum aestivum* L.) and corn (*Zea mays* L.) tended to decrease in the order of roots > stems > grains, respectively (Wang et al., 2011b, 2015; Yang et al., 2014). This trend is in agreement with the finds observed from field investigations by Nan and Chen (2001a, 2001b) and Wang et al. (2009). Considering levels of soil contamination, the bioaccumulation factor (BCF) values of Cd ranged from 3.93-7.66 for roots, 0.72-2.41 for stems, 0.46-0.91 for grains in wheat, respectively (Yang et al., 2014). The BCF values of Ni in tissues of wheat ranged from 0.70-1.61 for roots, 0.09-0.15 for stems, and 0.04-0.09 for grains, respectively (Wang et al., 2015); whilst the BCF values of Pb ranged from 0.25-0.36 for roots, 0.01-0.04 for stems, and in grains were less than 0.01, respectively (Wang et al., 2011b). It is noted that Cd is more available to crops than other metals, probably posing a great threat to human health. Especially, Cd uptake of crops could be enhanced with the interaction between metals. For example, Cd concentrations in grains of wheat and corn increased as an increase in levels of soil Zn, since the synergistic effect of Cd and Zn (Nan et al., 2002a). Coexistence of Pb in high levels accelerated Cd uptake as well (Zhao et al., 2014), possibly because Pb competes more effectively for exchange sites on the colloidal surfaces than Cd, releasing more Cd into the soil solution (Hu et al., 2013).

Table 1 Concentrations of heavy metals in grains of crops/(mg/kg)

Metal	Experiment type	Crop type	Metal concentration		MLs*	References
			Soil	Grain		
Cd	Pot culture	Wheat	0.35-7	0.23-2.61	0.1	Yang et al., 2014
	Pot culture	Maize	0.35-7	0.01-0.08	0.2	Yang et al., 2014
	Field survey	Wheat	10.36	0.61	0.1	Nan et al., 2002a
	Field survey	Maize	10.11	0.58	0.2	Nan et al., 2002a
Ni	Pot culture	Wheat	50-1000	3.7-39	1**	Wang et al., 2015
	Field survey	Wheat	139-1099	1.79-5.47	1	Wang et al., 2009
Pb	Pot culture	Wheat	75-1500	0.07-0.56	0.2	Wang et al., 2011b
	Field survey	Wheat	145-413	0.12-4.03	0.2	Nan and Chen, 2001a

*Maximum levels of contaminants in foods (Ministry of Health of the People's Republic of China, 2012). **No standard for grains, the standard for oils and fats is used here.

As compared with crops, vegetables grown on heavy metal-contaminated arid soils have a high risk to human health, since vegetables have a strong ability to take up metals from soils and then translocate them to their aboveground edible parts (Yang et al., 2011; Wang et al., 2012). Among the common vegetables, cole (*Brassica campestris* L.) is a hyperaccumulator for Cd; its root and foliar BCF values for Cd ranged from 3.66-6.74 and 8.03-16.30, respectively (Wang et al., 2011a). In general, crops in arid region grown on soils contaminated by heavy metals may have a low risk to human health; nonetheless, much attention has to be paid on consumptions of grains produced in heavily contaminated land (Table 1). In reality, chemical extraction procedures are suitable to rapidly evaluate bioavailability of metals and predict their risks. In the case of Cd and Zn, the exchangeable fractions made the most contribution on the uptake of Cd and Zn (Yang et al., 2011). Citric acid and NH_4Ac extraction procedures provided a good indication of Cu bioavailability (Wang et al., 2009).

4. Remediation studies in arid regions of northwest China

Soil washing is a process that uses physical and/or chemical techniques to separate contaminants from soil and sediments. Contaminants are concentrated into a much smaller volume of contaminated residue, which is either recycled or disposed (ITRC, 1997). Soils contaminated with heavy metals can be treated with soil washing systems on-site or off-site. Fig. 2 shows a government-run soil pollution clean-up project using on-site soil washing technique along Dongdagou in Gansu Province, northwest China. The total metal concentrations in the surface soil across this site ranged from 20 to 39 mg/kg for Cd, 183-497 mg/kg for Cu, 202-568 mg/kg for Pb, and 417-1, 148 mg/kg for Zn (Hu et al., 2013). According to Chinese soil environmental quality standards (CEPAS, 1995), concentrations of Cd and Zn in soils being disposed were higher than the class III values ($pH>6.5$) for Cd 1 mg/kg and Zn 500 mg/kg, respectively. After treatment, levels of soil Cd and Zn decreased to be lower than the standards set by the MEP, respectively. Although soil washing systems with washwater containing chelating agents can effectively remove the target metals from soils, soil degradation because of losses of nutrients and damage of soil structure is accompanying with the clean-up process. On the other hand, the area of the contaminated arable land in this site is only 3.7 ha, but the cost was as high as 1.68 million dollars. Accordingly, soil washing is not suitable to be used to remediate arid arable soils contaminated with heavy metals in large areas.

Phytoremediation is considered cost-effective and environmentally friendly compared with the conventional technologies used so far, which include excavation, burial, and soil washing (Raskin et al., 1997). Phytoextraction and phytostabilization are the two basic strategies for phytoremediation treating arable soils contaminated with heavy metals. Phytoextraction has been proposed as a possible alternative to the destructive and costly techniques currently used to clean up soils contaminated with heavy metals. Three groups of plants are possible candidates for phytoextraction of heavy metal-contaminated soil: hyperaccumulators, crop plants, and bioenergy plants (Keller, 2006). Due to concerns regarding the use of hyperaccumulators and crop plants (Pulford and Watson, 2003), a great deal of attention has been recently focused on determining the potential of bioenergy plants, such as trees, for phytoextraction (Robinson et al., 2000; Mertens et al., 2004, 2006).

The major advantages of using trees are their relatively large shoot biomass and fast growth rate (Pulford and Watson, 2003). Among potential trees, poplars have reportedly shown high tolerance to and uptake of Cd and Zn (Vandecasteele et al., 2003; Laureysens et al., 2004; Mertens et al., 2004; Wu et al., 2010; Lettens et al., 2011).

Fig. 2　A government-run soil pollution clean-up project using soil washing technique along Dongdagou in Gansu Province, northwest China

Phytostabilization, aimed at the fixation of heavy metals in the soil, seems to be the most suitable technology for remediating sites contaminated with multi-metals (Van Nevel et al., 2007), since few plants can extract high concentrations of more than one element in their tissues (Ernst, 2005; Van Nevel et al., 2007). In particular, in the case of highly metal-contaminated soils that are too expensive to remediate by conventional means or are too long to remediate by phytoextraction, phytostabilization by revegetation may be the most relevant technology in order to stabilize the metals in the soil (Ernst, 2005). Trees are well suited to phytostabilization due to their deep root systems, high transpiration rate, high metal tolerance, and ability to grow on nutrient-poor soils (Pulford and Watson, 2003; Van Nevel et al., 2007; Mendez and Maier, 2008). Trees can stabilize metals in the soil by physically preventing migration via wind or water erosion, leaching, and soil dispersion; alternatively, they can immobilize the metals through uptake and accumulation by the roots into the plant, absorption on the root, or precipitation in the root zone (Pulford and Watson, 2003; Mertens et al., 2004; Mendez and Maier, 2008).

Fig. 3 A field phytoremediation trial using poplar along Dongdagou in Gansu Province, northwest China

Whether phytoextraction or phytostabilization is the goal, the plants used need to be suited to the local climate and soil characteristics. In the case of the arid soils, the ideal plants must be both drought and salt-tolerant (Keller et al., 2003; Mendez and Maier, 2008). Fig. 3 shows a field remediation trial using poplar (*Populus alba* L. var. *pyramidalis* Bunge) along Dongdagou in Gansu Province, northwest China. *Populus pyramidalis* is a variety of *P. alba* of which only male specimens have been identified. It is found widely in desert and arid regions of northwest China and is used for lumber production and ecological protection in terms of its fast growth rate, high biomass production, and elevated tolerance to the stresses of drought, salt, and insects (Hu et al., 2013). Results from pot culture and field sampling experiments have shown that *P. pyramidalis* has a strong ability to accumulate Cd and Zn in its aboveground tissues (Hu et al., 2013, 2014a). However, *P. pyramidalis* has been demonstrated to be an excluder to Cu and Pb, since concentrations of Cu and Pb were always within the normal ranges and were never higher than 8 and 5 mg/kg, respectively (Hu et al., 2013). Thus, *P. pyramidalis* is suitable for phytoextraction of Cd-contaminated arid arable soils (Hu et al., 2016b), whilst it is efficient to phytostabilization of Cu and Pb in combination with grasses of *Peganum harmala* L. and shrubs of *Nitraria praevisa* Bobr. (Hu et al., 2016a). In the field trial, chelants such as EGTA can be applied to increase the phytoextraction efficiency of *P. pyramidalis* for Cd (Hu et al., 2014b). Meanwhile, collection of the litter fall would be necessary due to the relatively high foliar concentrations of Cd and Zn.

5. Conclusions and future perspectives

Drylands, including arid, semi-arid and dry subhumid areas, cover 41% of Earth's terrestrial surface and are inhabited by>38% of the total global human population. However, many of the mineral resources usually distribute in drylands, leading to universal contamination of agricultural lands along the mining and smelting sites. In particular, the pollution area is expanding because of erosion by wind and rain induced by changes of precipitation pattern under the condition of global warming. Unfortunately, most of the *in situ* and *ex situ* remedial methods are performed better in wetter regions than in drier regions. Therefore, in the future, more efforts are needed to pay on research and development of new and proper mitigation measures in arid regions, taking account of soil and metal types, soil contamination level and type, climate, plants, and the future land use type.

Acknowledgements

This study was financially supported by the National Natural Science Foundation of China (NSFC 41501337 and 41401334), and the Fundamental Research Funds for the Central Universities at Lanzhou University (lzujbky-2015-138).

References

Adriano D C. 2001. Trace elements in terrestrial environments: biogeochemistry, bioavailability and risks of metals (2nd edition). New York: Springer-Verlag: 276.

CEPAS (Chinese Environmental Protection Agency of the States). 1995. National Soil-Environmental Quality Standard of China (GB 15618—1995).

Ernst W H O. 2005. Phytoextraction of mine wastes-Options and impossibilities. Chemie der Erde-Geochemie der Erde, 65: 29-42.

Gansu Environmental Monitoring Station, Gansu Environmental Research Institute. 1984. The influence of cadmium pollutant upon the human being's health in Baiyin city. Environmental Research, 1: 15-21.

Hu X N, Nan Z R, Wang S L, et al. 2009. Sorption and desorption of copper, zinc and lead in the irrigated desert soil from the oasis in the arid regions, northwest China. Ecology and Environmental Sciences, 18 (6): 2183-2188.

Hu Y, Nan Z, Jin C, et al. 2014a. Phytoextraction potential of poplar (*Populus alba* L. var. *pyramidalis* Bunge) from calcareous agricultural soils contaminated by cadmium. International Journal of Phytoremediation, 16 (5): 482-495.

Hu Y, Nan Z, Su J, et al. 2013. Heavy metal accumulation by poplar in calcareous soil with various degrees of multi-metal contamination: implications for phytoextraction and phytostabilization. Environmental Science and Pollution Research, 20 (10): 7194-7203.

Hu Y, Nan Z, Su J, et al. 2014b. Chelant-assisted uptake and accumulation of Cd by poplar from calcareous arable soils around Baiyin nonferrous metal smelters, northern China. Arid Land Research and Management, 28 (3): 340-354.

Hu Y, Su J, Nan Z, et al. 2016a. Method for remedying calcareous soil polluted by heavy metals of copper and lead through phytostabilization. Patent CN 104126405.

Hu Y, Su J, Nan Z, et al. 2016b. Phytoextraction restoring method of heavy metal cadmium-polluted calcareous farmland soil. Patent CN 104084419.

Interstate Technology and Regulatory Council (ITRC). 1997. Technical and regulatory guidelines for soil washing. Metals in Soils Work Team, Washington, DC.

Kayser A, Wenger K, Keller A, et al. 2000. Enhancement of phytoextraction of Zn, Cd, and Cu from calcareous soil: the use of NTA and sulfur amendments. Environmental Science & Technology, 34 (9): 1778-1783.

Keller C. 2006. Factors limiting efficiency of phytoextraction at multi-metal contaminated sites//Phytoremediation of metal-contaminated soils. Springer Netherlands: 241-266.

Keller C, Hammer D, Kayser A, et al. 2003. Root development and heavy metal phytoextraction efficiency: comparison of different plant species in the field. Plant and Soil, 249 (1): 67-81.

Laureysens I, Blust R, De Temmerman L, et al. 2004. Clonal variation in heavy metal accumulation and biomass production in a poplar coppice culture: I. Seasonal variation in leaf, wood and bark concentrations. Environmental Pollution, 131 (3): 485-494.

Lettens S, Vandecasteele B, De Vos B, et al. 2011. Intra-and inter-annual variation of Cd, Zn, Mn and Cu in foliage of poplars on contaminated soil. Science of the Total Environment, 409 (11): 2306-2316.

Liao Q, Wang S L, Nan Z R, et al. 2010. Experiments on accumulation and chemical forms of Cd, Pb, Zn and Ni in arid oasis soils. Agricultural Research in the Arid Areas, 28 (5): 108-114.

Liu Z P. 2003. Lead poisoning combined with cadmium in sheep and horses in the vicinity of non-ferrous metal smelters. Science of the Total Environment, 309 (1): 117-126.

Luo L, Ma Y, Zhang S, et al. 2009. An inventory of trace element inputs to agricultural soils in China. Journal of Environmental Management, 90 (8): 2524-2530.

Mendez M O, Maier R M. 2008. Phytostabilization of mine tailings in arid and semiarid environments-an emerging remediation technology. Environmental Health Perspectives, 116 (3): 278-283.

Mertens J, Vervaeke P, De Schrijver A, et al. 2004. Metal uptake by young trees from dredged brackish sediment: limitations and possibilities for phytoextraction and phytostabilisation. Science of the Total Environment, 326 (1): 209-215.

Mertens J, Vervaeke P, Meers E, et al. 2006. Seasonal changes of metals in willow (*Salix* sp.) stands for phytoremediation on dredged sediment. Environmental Science & Technology, 40 (6): 1962-1968.

Ministry of Health of the People's Republic of China. 2012. National standards for food safety and maximum levels of contaminants in foods (GB 2762—2012).

Nan Z, Cheng G. 2001a. Accumulation of Cd and Pb in spring wheat (*Triticum aestivum* L.) grown in calcareous soil irrigated with wastewater. Bulletin of Environmental Contamination and Toxicology, 66 (6): 748-754.

Nan Z, Cheng G. 2001b. Copper and zinc uptake by spring wheat (*Triticum aestivum* L.) and corn (*Zea Mays* L.) grown in Baiyin region. Bulletin of Environmental Contamination and Toxicology, 67 (1): 83-90.

Nan Z, Li J, Zhang J, et al. 2002. Cadmium and zinc interactions and their transfer in soil-crop system under actual field conditions. Science of the Total Environment, 285 (1): 187-195.

Nan Z, Liu X, Zhao Z, et al. 2011. Chemical behavior and ecological risk assessment of heavy metals in arid oasis soil-crop systems. China Environmental Science Press.

Nan Z, Zhao C. 2000. Heavy metal concentrations in gray calcareous soils of Baiyin region, Gansu province, PR China. Water, Air, and Soil Pollution, 118 (1-2): 131-142.

Nan Z R, Zhao C Y, Li J J, et al. 1999. Field survey of Cd and Pb contents in spring wheat (*Triticum aestivum* L.) grain grown in Baiyin City, Gansu province, People's Republic of China. Bulletin of Environmental Contamination and Toxicology, 63 (4): 546-552.

Nan Z, Zhao C, Li J, et al. 2002b. Relations between soil properties and selected heavy metal concentrations in spring wheat (*Triticum aestivum* L.) grown in contaminated soils. Water, Air, and Soil Pollution, 133 (1-4): 205-213.

Pulford I D, Watson C. 2003. Phytoremediation of heavy metal-contaminated land by trees—a review. Environment International, 29 (4): 529-540.

Raskin I, Smith R D, Salt D E. 1997. Phytoremediation of metals: using plants to remove pollutants from the environment. Current Opinion in Biotechnology, 8 (2): 221-226.

Robinson B H, Mills T M, Petit D, et al. 2000. Natural and induced cadmium-accumulation in poplar and willow: Implications for phytoremediation. Plant and Soil, 227 (1-2): 301-306.

Shi Y, Shen Y, Kang E, et al. 2007. Recent and future climate change in northwest China. Climatic Change, 80 (3-4): 379-393.

Vandecasteele B, Lauriks R, De Vos B, et al. 2003. Cd and Zn concentration in hybrid poplar foliage and leaf beetles grown on polluted sediment-derived soils. Environmental Monitoring and Assessment, 89 (3): 263-283.

Van Nevel L, Mertens J, Oorts K, et al. 2007. Phytoextraction of metals from soils: how far from practice?. Environmental

Pollution, 150 (1): 34-40.

Wang B, Zhou Q, Nan Z, et al. 2011a. Accumulation and migration of Cd and Pb in cole grown in the arid oasis soil. Acta Agriculturae Boreali-Occidentalis Sinica, 20 (3): 62-68, 80.

Wang S, Nan Z. 2009. Copper sorption behavior of selected soils of the oasis in the middle reaches of Heihe River Basin, China. Soil & Sediment Contamination, 18 (1): 74-86.

Wang S, Nan Z, Liu X, et al. 2009. Accumulation and bioavailability of copper and nickel in wheat plants grown in contaminated soils from the oasis, northwest China. Geoderma, 152 (3): 290-295.

Wang S, Nan Z, Liu X, et al. 2010. Availability and speciation of Cu, Zn, and Pb added to irrigated desert soil. Polish Journal of Environmental Studies, 19 (4): 865-869.

Wang X, Nan Z, Ding W, et al. 2012. Chemical fraction of heavy metals in an oasis soil and their bioavailability to cole crops. Arid Land Research and Management, 26 (2): 166-180.

Wang Y, Wang S, Nan Z, et al. 2015. Effects of Ni stress on the uptake and translocation of Ni and other mineral nutrition elements in mature wheat grown in sierozems from northwest of China. Environmental Science and Pollution Research, 22 (24): 19756-19763.

Wang Z W, Nan Z R, Wang S L, et al. 2011b. Accumulation and distribution of cadmium and lead in wheat (*Triticum aestivum* L.) grown in contaminated soils from the oasis, north–west China. Journal of the Science of Food and Agriculture, 91 (2): 377-384.

Wu F, Yang W, Zhang J, et al. 2010. Cadmium accumulation and growth responses of a poplar (*Populus deltoids* ×*Populus nigra*) in cadmium contaminated purple soil and alluvial soil. Journal of Hazardous Materials, 177 (1): 268-273.

Zhao F J, Ma Y, Zhu Y G, et al. 2015. Soil contamination in China: current status and mitigation strategies. Environmental Science & Technology, 49 (2): 750-759.

Zhao Z J, Nan Z R, Wang Z W, et al. 2014. Interaction between Cd and Pb in the soil-plant system: a case study of an arid oasis soil-cole system. Journal of Arid Land, 6 (1): 59-68.

Yang Y, Nan Z, Zhao Z. 2014. Bioaccumulation and translocation of cadmium in wheat (*Triticum aestivum* L.) and maize (*Zea mays* L.) from the polluted oasis soil of Northwestern China. Chemical Speciation & Bioavailability, 26 (1): 43-51.

Yang Y, Nan Z, Zhao Z, et al. 2011. Chemical fractionations and bioavailability of cadmium and zinc to cole (*Brassica campestris* L.) grown in the multi-metals contaminated oasis soil, northwest of China. Journal of Environmental Sciences, 23 (2): 275-281.

Phytoremediation of Heavy Metal Contaminated Soil in Southern China

Kengbo DING[1], Chang LIU[1], Yetao TANG[1,2], Shizhong WANG[1,2], Xiange WEI[1,2], Yuanqing CHAO[1,2], Rongliang QIU[1,*]

[1]School of Environmental Science and Engineering, Sun Yat-sen University

[2]Guangdong Provincial Key Laboratory of Environmental Pollution Control and Remediation Technology

[*]Corresponding author

Soil heavy metal contamination was found much more serious in Southern China than the North. In Southern China, contaminated soils are difficult to be remediated or reclaimed due to its characteristics of high acidity, low nutrients and high heavy metals contents. For decades, we have worked on soil remediation and made progresses on ①discovery of heavy metals tolerant plants and hyperaccumulators, ② screening for better soil amendments, and ③ suitable remediation systems for contaminated soils of different degrees.

1. Phytostabilization of highly-contaminated mining soils

1.1 Remediation of acid mine drainage at the source

Acid mine drainage (AMD) resulting from the oxidation of pyrite and other metal sulfides has caused significant environmental problems, including acidification of rivers and streams as well as leaching of toxic metals. With the goal of controlling AMD at the source, we evaluated the potential of tetraethylorthosilicate (TEOS) and n-propyltrimethoxysilane (NPS) coatings to suppress pyrite oxidation. Results showed that TEOS-and NPS-based coatings reduced chemical oxidation (H_2O_2) of pyrite by as much as 59% and 96% (based on Fe release), respectively, while biological oxidation (iron-oxidizing bacteria *Acidithiobacillus ferrooxidans*) of pyrite was reduced by 69% and 95%, respectively (Diao et al., 2013). These results were attributed to the formation of a dense network of Fe-O-Si and Si-O-Si bonds on the pyrite surface that limited permeation of oxygen, water, and bacteria (Fig. 1) (Diao et al., 2013).

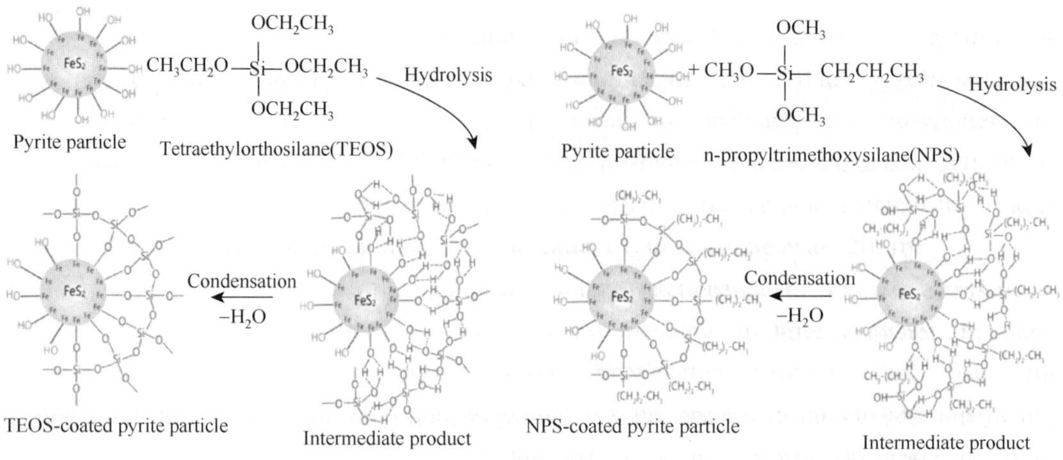

(a) Mechanism of pyrite oxidation suppressionby TEOS-based coating

(b) Mechanism of pyrite oxidation suppressionby NPS-based coating

Fig. 1 Mechanisms of pyrite oxidation suppression by silane-based coatings (Diao et al., 2013)

A strain of sulfate-reducing bacteria, designated strain "DBM", also had potential to remediate AMD. Strain DBM reduced 10 mmol/L of sulfate completely to sulfide within 7 d, and it recovered its sulfate reduction ability after 7 d of aerobic growth (Qiu et al., 2009). Furthermore, strain DBM effectively precipitated 0.40 mmol/L copper during its growth (Qiu et al., 2009). Elemental composition of the resulting microbial precipitate was studied using electro-dispersive X-ray spectroscopy, and it was found that the ratio of S:Cu was 1.07, which was consistent with the formation of copper sulfide (Qiu et al., 2009).

1.2 Screening heavy metal tolerant plants for phytostabilization

Greenhouse pot experiments were conducted to determine the growth response, metal tolerance, and phytostabilization potential of *Jatropha curcas* L. The higher tolerance index (> 90%) with no phytotoxic symptoms and growth reduction in moderately contaminated soils showed that this plant has the ability to tolerate polymetallic acid mine tailings (Wu et al., 2011). The physiological response to Cd and its accumulation characteristics of *J. curcas* at different Cd treatments were analyzed to explore its metal defense mechanism. With the Cd concentration increased, the antioxidant enzymes activities-Superoxide Dismutase, peroxidase was first increased and then decreased, while the Catalase activity was clearly increased than the control at Cd=200 mg/kg (Li et al., 2012). Contents of Malondialdehyde, proline, soluble protein, soluble sugar, acid thiol and glutathione in the leaves was increased with increasing of Cd concentration, which reduced the Cd toxicity to the plant in certain extent (Li et al., 2012). These results showed that *J. curcas* had a great potential in phytoremediation of heavy metal contaminated soils. Addition to *J. curcas*, other tolerant plants, e.g., kenaf (*Hibiscus cannabinus*) (Yang et al., 2013) and ramie (*Boehmeria nivea*) (Yang et al., 2010), are also candidates for phytostabilization.

1.3 Screening of soil amendments

Limestone, dolomite, fly ash and organic fertilizer were applied to enhance phytostabilization in a five-year field trial at the surrounding area of Dabao Mountain Mine. Among the amendments, dolomite and fly ash had better effects than limestone and organic fertilizer. With the application of dolomite and fly ash, the aboveground dry mass production of kenaf reached 14-15 t/hm^2, which was similar to that on normal soils, and the heavy metal concentrations in the bast fiber and stem of kenaf decreased significantly, as compared with the control (Yang et al., 2013). Furthermore, the mass of the bast fiber accounted for 32%-38% of the shoot production, and the extractable heavy metal concentrations in the bast fiber could meet the standard of technical specifications of ecological textiles in China, suggesting that the bast fiber had potential economic value (Yang et al., 2013).

The growth performance of *J. curcas* was also improved with the addition of limestone, due to the increase in soil pH and decrease in phytoavailable soil Al, Zn, and Cu (Wu et al., 2011). In the soils with lower levels of pollution, high bio-availabilities of Cu and Pb appeared to be the primary factors inhibiting the growth of *J. curcas*, while in highly polluted soils, the growth inhibition of *J. curcas* mainly resulted from the high bio-availability of Cd, Cu and Zn, as well as Al toxicity caused by the strong acidity (Li et al., 2009). Results of the dose-response effects of lime stone addition showed that the optimum dosage of limestone was 0.25% for soils with lower pollution levels and 0.5% for soils with higher pollution levels (Li et al., 2009). Additionally, the combination of dolomite with *J. curcas* significantly improves soil remediation in terms of stimulating root exudates and other metabolites which could be utilized by soil microbial communities to promote growth and inhibit the uptake of metals in *J. curcas* (Du et al., 2009).

Biochar, another soil amendment, also has great potential in phytostabilization. It was found that the biochar derived from pyrolysis treatment of sewage sludge could effectively remove Pb^{2+} from acidic solution with the capacities of 16.11, 20.11, 24.80, and 30.88 mg/g at initial pH 2, 3, 4 and 5, respectively (Lu et al., 2012). Our results showed that Pb sorption primarily involved the coordination with organic hydroxyl and carboxyl functional groups, which was 38.2%-42.3% of the total sorbed Pb varying with pH, as well as the coprecipitation or complex on mineral surfaces, which accounted for 57.7%-61.8% and led to a bulk of Ca^{2+} and Mg^{2+} release during sorption process (Fig. 2) (Lu et al. 2012). Factors related to the properties of biochar, e.g., feedstocks (Lu et al., 2013), pyrolysis temperatures (Lu et al., 2013, Zhang et al., 2013), coexisting humic acids (Zhou et al., 2015), aging time (Fang et al., 2016), and solution chemistry (Zhang et al., 2015) were also studied for future application.

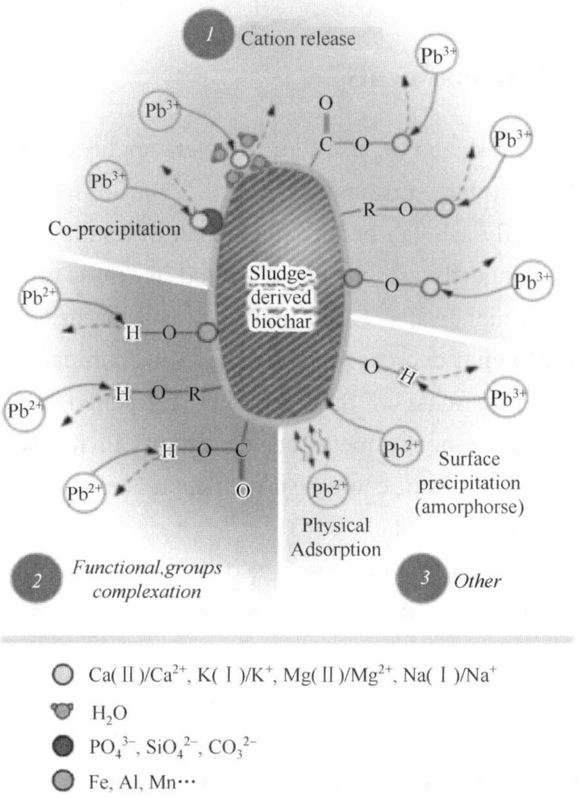

Fig. 2 Conceptual illustration of Pb adsorption mechanism on SDBC (sludge-derived biochar), including: ①metal exchange with Ca^{2+} and Mg^{2+}, attributing to co-precipitation and innersphere complexation with complexed humic matter and mineral oxides of SDBC; ②surface complexation with free carboxyl and hydroxyl functional groups, and innersphere complexation with the free hydroxyl of mineral oxides and other surface precipitation; ③others, innersphere complexation with the free hydroxyl of mineral oxides and other surface precipitation (Lu et al., 2012)

1.4 Enhancing phytostabilization with PGPR

Plant growth promoting rhizobectrium (PGPR) could further enhance phytoremediation efficiency (Wang et al., 2013). Mechanisms of soil Pb immobilization by *Bacillus subtilis* DBM, a bacterial strain isolated from a heavy-metal-contaminated soil, were investigated. Adsorption and desorption experiments with living bacterial cells as well as dead cells revealed that both extracellular adsorption and intracellular accumulation were involved in the Pb^{2+} removal from the liquid phase (Bai et al. 2014). Of the sequestered Pb (Ⅱ), 8.5% was held by physical entrapment within the cell wall, 43.3% was held by ion-exchange, 9.7% was complexed with cell surface functional groups or precipitated on the cell surface, and 38.5% was intracellularly accumulated (Bai et al., 2014). Another PGPB, *Enterobacter* sp. strain EG16, could also tolerates high external Cd concentration and is able to produce siderophores and the plant hormone indole-3-acetic acid (IAA), both of which contribute to plant growth promotion. Cd exposure resulted in global regulation at the transcriptomic level, with the bacterium switching to an energy-conserving mode by inhibiting energy-consuming processes while increasing production of stress-related proteins (Chen et al., 2015). The stress response system included increased import of sulfur and iron, which become deficient under Cd stress, and redirection of sulfur metabolism to maintain intracellular glutathione levels in response to Cd toxicity (Chen et al., 2015). Increased production of siderophores, responding to Cd-induced Fe deficiency, was not only involved in the Cd stress response systems of EG16, but may also play an important role in promoting plant growth as well as alleviating the Cd-induced inhibition of IAA production (Chen et al., 2015). In a pot experiment, inoculation of a Pb resistant PGPR DBM1 (*Arthrobacter* sp.) also effectively increased biomass and chlorophyll contents of kenaf, with more effective at higher Pb stress (Chen et al., 2013).

2. Phytoextraction of moderately-contaminated soils

2.1 Multi-Heavy metal enriched hyperaccumulators in Southern China

Over the past 20 years, a lot of native hyperaccumulators which can enrich multi-heavy metals have been discovered in Southern China. The field survey showed that the wild population of *Arabis paniculata*, the first found Pb/Zn/Cd hyperaccumulator, was hyper-tolerant to extremely high concentrations of Pb, Zn and Cd, and could accumulate in shoots an average level of 2300 mg/kg dry weight (DW) Pb, 20800 mg/kg Zn and 434 mg/kg Cd, with their translocation factors (TFs) all above one (Tang et al., 2009a). Under natural conditions, the average Zn and Cd concentrations in shoot of *Picris divaricate* were 5911 and 246 mg/kg (DW), respectively (Tang et al., 2009b), and the average Zn and Cd concentrations in shoot of *Potentilla griffithii* were 6250 and 178 mg/kg (DW), respectively (Qiu et al., 2006; Hu et al., 2007).

2.2 Metal enrichment mechanisms in plants

The mechanisms of heavy metal uptake, accumulation, distribution and detoxification have been studied in a wide range of hyperaccumulators. Vacuolar sequestration and cell wall binding play a major role in the detoxification process. The predominant sequestration of Zn and Cd in cell walls of roots and in vacuoles of epidermis and bundle sheath of leavesmay play amajor role in strong tolerance and hyperaccumulation of Zn and Cd in *P. griffithii* (Hu et al., 2009). Further experiments described that up to 94% and 70% of the total Zn and Cd in the leaves of *P. griffithii* were present in the protoplasts, and more than 90% Cd and Zn in the protoplasts were localized in the vacuoles (Qiu et al., 2011). Complexation is also involved in the detoxification process of hyperaccumulators. Cd exposure increased the concentrations of glutathione (GSH) and phytochelatins (PCs) in the roots, and the molar ratios of PCs:Cd in the roots were close to the value expected to PC-mediated Cd sequestration in plants, which implies the potential role of GSH and PCs in Cd accumulation and Cd complexation in the roots of *A. paniculata* (Zeng et al., 2009).

While, cysteine had significant positive correlation with Zn concentration in plants, suggesting that cysteine may be involved in Zn tolerance of *A. paniculata* (Zeng et al., 2011a). *A. paniculata* was against to Zn stress mainly by enhancement of energy metabolism including auxin biosynthesis and protein metabolism to maintain plant growth and correct misfolded proteins (Zeng et al., 2011b). In the case of Cd, plants adopted antioxidative/xenobiotic defense and cellular metabolism to keep cellular redox homeostasis and metal-transportation under Cd stress (Zeng et al., 2011b).

Lately, Tang (2012) found that the significant isotope fractionation in *Noccaea caerulescens* indicated of high-and/or low-affinity transport systems operating in three ecotypes. In *Silene vulgaris*, however, there was no significant isotope fractionation between whole plant and rhizosphere soil and between root and shoot, suggesting that this species appears to have a particular Zn homeostasis (Tang et al., 2012). Isotope fractionation of Ni and Zn was also observed during plant uptake and translocation processes of *N. caerulescens* (Deng et al., 2014). This result revealed that Zn could compete with Ni during the uptake process in high Zn treatment, which reduced Ni concentration in plants and decreased the extent of Ni isotope fractionation, indicating that plants might take up Ni through a low-affinity transport system of Zn (Deng et al., 2014). A most recent work by Deng (2016) found that phloem sap of *N. caerulescens* is enriched in Ni and malate, the majority of which moves upward to young tissues (Fig. 3). Phloem translocation may play an important role for Ni accumulation in young leaves of *N. caerulescens* (Deng et al., 2016).

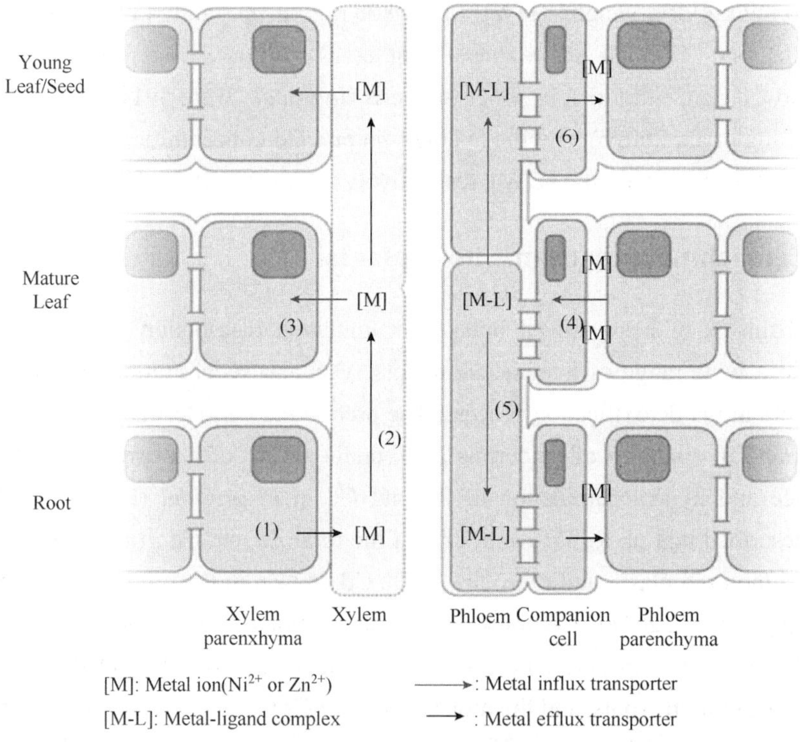

Fig. 3 Conceptual model for long distance transport of heavy metals in *Noccaea caerulescens* (Deng et al., 2016)

(1) Xylem loading; (2) Xylem transport; (3) Xylem unloading; (4) Phloem loading; (5) Phloem translocation; (6) Phloem unloading

2.3 Physiological functions of toxic metals in plants

Cd is generally considered as a toxic element; however, evidences showed that Cd had certain physiological functions in hyperaccumulators. Suitable levels of Pb, Zn and Cd could stimulate the biomass production and chlorophyll concentrations of *A. paniculata* (Fig. 4 (g)) (Tang

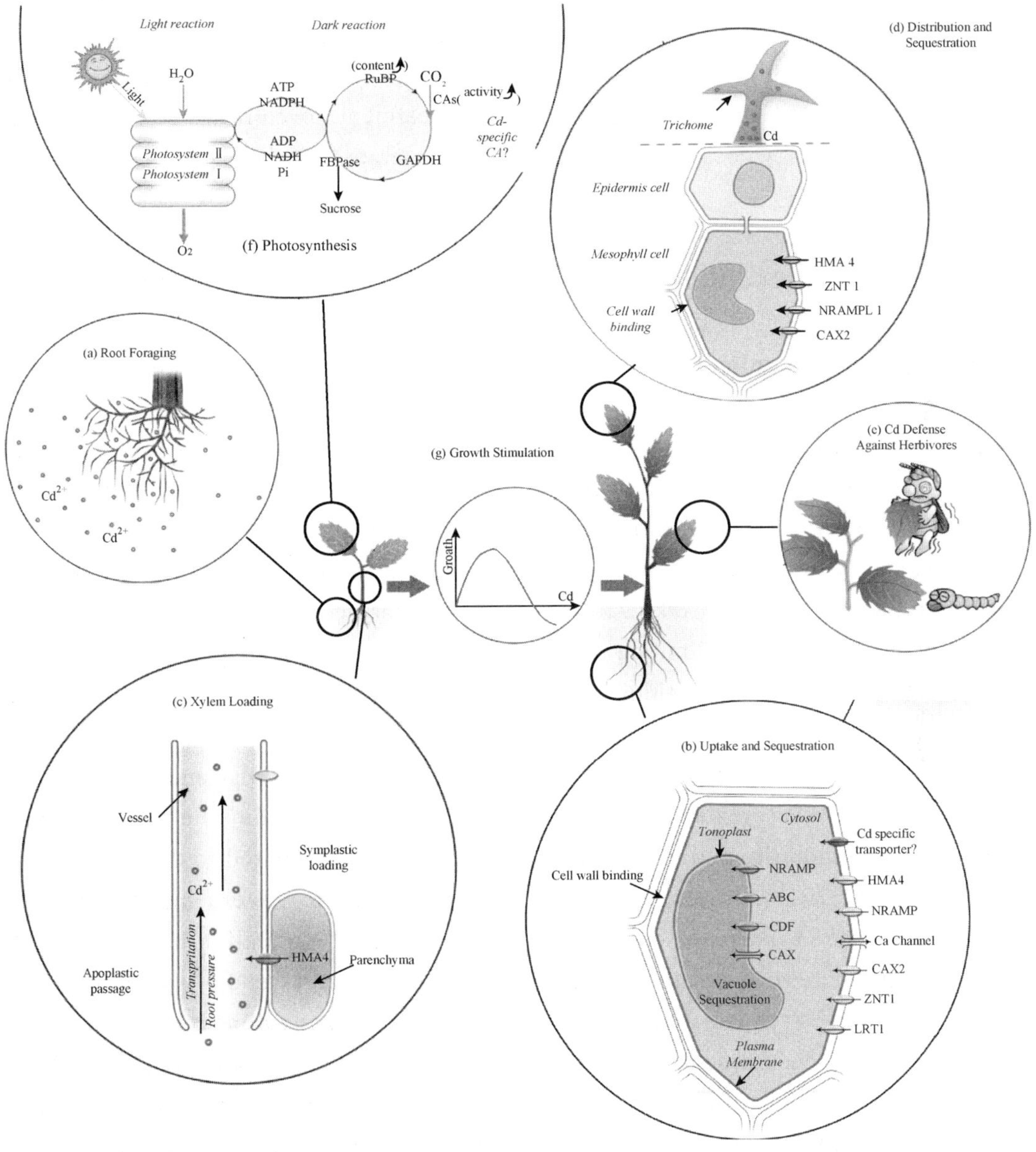

Fig. 4 Molecular and physiological mechanisms involved in Cd hyperaccumulation and detoxification (a)-(d), and their evolutionary significance for Cd hyperaccumulators (e)-(g) (Qiu et al., 2012)

et al., 2009a). As for *Sedum alfredii*, plant exhibited strong root foraging responses for Zn and Cd, allocating 90% of its root biomass to Zn and Cd rich patches when grown on a substrate with heterogeneous distribution of supplied Zn and Cd. This root foraging is responsible for maintaining high levels of Zn and Cd accumulation in *S. alfredii* (Fig. 4 (a)) (Liu et al., 2010). In *P. griffithii*, Zn supplement significantly reduced the Cd concentration in roots and further enhances the Cd translocation in shoots, especially petioles. This indicates strong competition between Cd and Zn uptake in root level and efficiently translocated these metals into the shoots via a common transport system (Qiu et al., 2011). Further study also found that although the mesophyll cell size was reduced, chloroplast ultrastructure remained intact at the highest Cd treatment, revealing that *P. divaricata* chloroplast and the enzymes of carbon assimilation tolerate high levels of Cd (Fig. 4 (f)) (Ying et al., 2010).

3. Phytoexclusion of slightly-contaminated croplands

3.1 Screening rice cultivars for low heavy metal or metalloid accumulation

Understanding the complex biotic and abiotic interactions invoked by the rice (*Oryza sativa* L.) root system in oxygen-depleted soil is an important step in screening genotypes for low toxic metal or metalloid accumulation. Cadmium (Cd) uptake and tolerance were investigated among 20 rice cultivars based on a field experiment (1.2 mg Cd/kg in soil) and a soil pot trial (control, 100 mg Cd/kg). Results indicated that rice cultivars with higher rates of ROL had higher capacities for limiting the transfer of Cd to rice and straw (Wang et al., 2011). Similar results were found in screening genotypes for As (Mei et al., 2009) and Pb (Cheng et al., 2014).

Rice genotypes with higher rates of radial oxygen loss (ROL) and at the bolting stage, tended to have greater effects on rhizosphere Eh, pH, Fe^{3+}/Fe^{2+} quotients, As fractionation and mobility and also on Fe plaque formation compared to those with lower ROL (Mei et al., 2012). In aerated treatments, rice tended to have a higher Fe plaque formation than in a stagnant solution, with the greatest formation at the root tip decreasing with increasing distances away, in accordance with a trend of spatial ROL (Wu et al., 2012). Plaques sequestered As on rice roots, with arsenate almost double that with arsenite, leading to decreased As accumulation in both roots and shoots (Wu et al., 2012). Increased Fe plaque was formed on rice roots at the ear emergence stage due to the increased ROL and, thus, this stage could therefore be an important period to limit the transfer and distribution of As in rice plants (Wang et al., 2013). In turn, elevated soil As may induce As toxicity towards rice plants, leading to the decrease of ROL (Wu et al., 2013).

3.2 External exclusion by soil amendments

Chemical stabilization with alkaline amendments could be an effective and stable soil remediation strategy for attenuating metal bioavailability and reducing rice metal uptake. The results from a pot experiment indicated that the application of fly ash (20 and 40 g/kg) and steel slag (3 and 6 g/kg) increased soil pH from 4.0 to 5.0-6.4, decreased the phytoavailability of heavy metals by at least 60%, and further suppressed metal uptake by rice (Gu et al., 2011). Significant correlations were found between C_{DGT} (the concentrations of metals measured with DGT) and the concentration of a combination of metal fractions (exchangeable, bound to carbonates, and bound to Fe/Mn oxides), unraveling the labile species that participate in the flux of metal resupply (Qiu et al., 2012). The capability of metal resupply, as reflected by the R (ratio of C_{DGT} to pore water metal concentration) values, significantly decreased in the amended soils (Qiu et al., 2012). X-ray diffraction analysis indicated the mobile metals were mainly deposited as their silicates, phosphates and hydroxides in amended treatments (Gu et al., 2011). Another pot experiment showed that soil pH and available silicon content increased with the increase of the amount of slag application and along with the decrease of the slag particle size (Deng et al., 2011).

3.3 Internal exclusion in rice

Field trial showed that steel slag application could dramatically increase rice production and decrease the heavy metal concentration in polished rice. Heavy metal concentration in shoot of rice plant decreased significantly with the slag application and was far lower than that in root (Deng et al., 2011). Heavy metal concentration in the root with the slag application sharply decreased in compared to CK (Deng et al., 2011). Moreover, it was found metal translocation from stem to leaf was dramatically restrained by adding amendments, which might be due to the increase of silicon concentration and co-precipitation with heavy metals in stem (Gu et al., 2011).

In a hydroponic experiment, we examined the uptake, xylem loading and localization of Zn in rice seedlings with the additional silicate supply. The silicate addition significantly increased the seedling biomass, and decreased Zn concentration in both root and shoot of seedlings and in xylem sap flow (Gu et al., 2012). Zinpyr-1 fluorescence test and Energy-dispersive X-ray spectroscopy analysis showed the concentration of biologically active Zn^{2+} decreased, and Zn and Si co-localized in the cell wall of metabolically less active tissues, especially in sclerenchyma of root (Gu et al., 2012). The fractionation analysis further supported silicate supply increased about 10% the cell wall bound fraction of Zn (Gu et al., 2012). Silicon was confirmed to enhance plant resistance to toxic elements, and its beneficial role was mainly based on external and internal plant mechanisms.

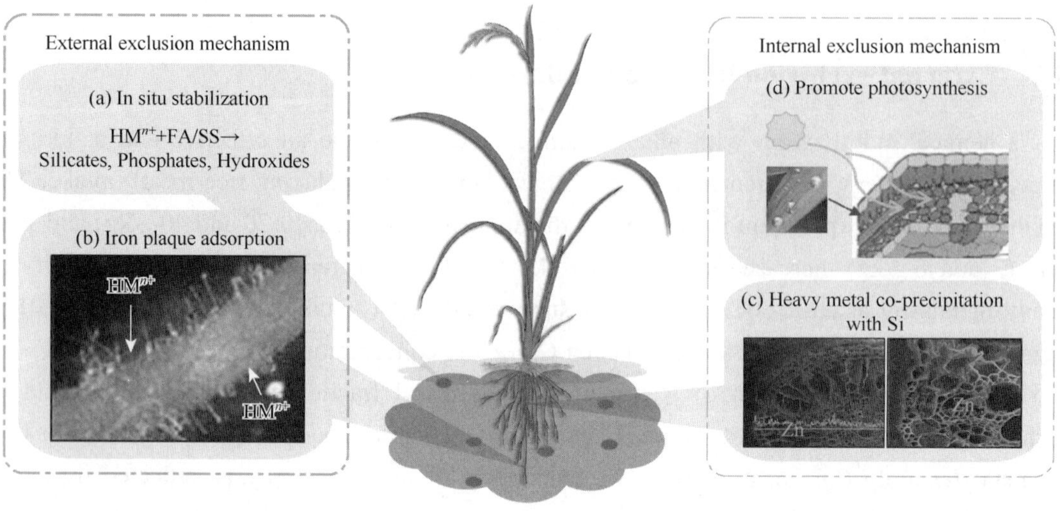

Fig. 5 Phytoexclusion mechanism of rice cultivation with fly ash/steel slag addition

4. Summary and perspective

Our works demonstrated that phytostabilization, phytoextraction and phytoexclusion are effective in remediation of heavy metal contaminated soils of different degrees (Fig. 6). However, future studies are still required to explore the interaction of soil, bacteria and plant, as well as its effects on heavy metals. Pedogenesis, local climate and other natural factors should be also included in soil phytoremediation. Furthermore, phytomining and agromining are not available for most metals, remaining a great challenge for us.

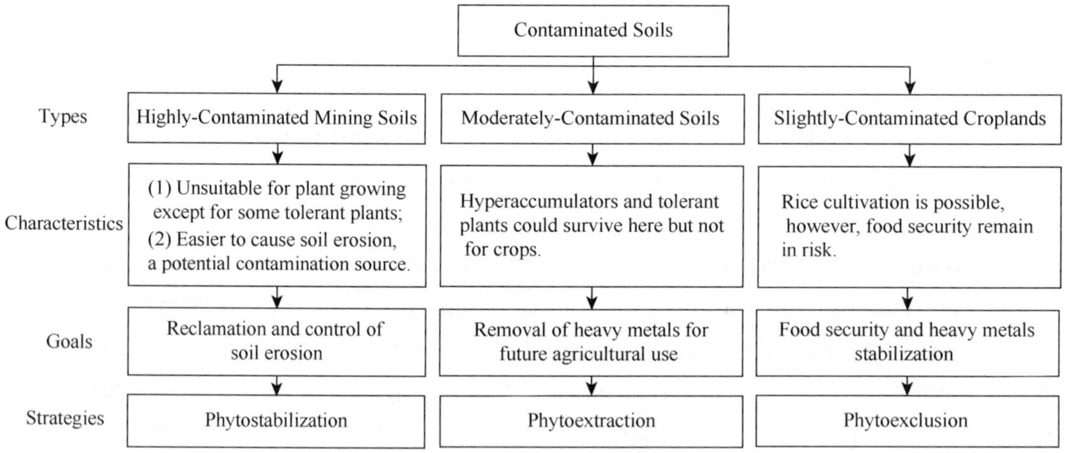

Fig. 6 Classified management of different contaminated soils

References

Bai J, Yang X H, Du R Y, et al. 2014. Biosorption mechanisms involved in immobilization of soil Pb by *Bacillus subtilis* DBM in a multi-metal-contaminated soil. Journal of Environmental Sciences, 26: 2056-2064.

Chen Y M, Bai J, Yang Y X, et al. 2013. Phytoremediation of Pb polluted soil by kenaf with assistance of plant growth promoting rhizobectrium (PGPR). Journal of Agro-Environment Science, 32 (11): 2159-2167.

Chen Y, Chao Y, Li Y, et al. 2016. Survival strategies of the plant-associated bacterium Enterobacter sp. strain EG16 under cadmium stress. Applied and environmental microbiology, 82 (6): 1734-1744. Cheng H, Wang M Y, Wong M H, et al. 2014. Does radial oxygen loss and iron plaque formation on roots alter Cd and Pb uptake and distribution in rice plant tissues?Plant and Soil, 375 (1): 137-148.

Deng T H B, Cloquet C, Tang Y T, et al. 2014. Nickel and zinc isotope fractionation in hyperaccumulating and nonaccumulating plants. Environmental Science and Technology, 48 (20): 11926-11933.

Deng T H B, Gu H H, Qiu R L. 2011. Ameliorative effects of steel slag application on multi-metal contaminated soil and heavy metal uptake of rice. Journal of Agro-Environment Science, 30 (3): 455-460.

Deng T H B, Tang Y T, Ent A V D, et al. 2016. Nickel translocation via the phloem in the hyperaccumulator *Noccaea caerulescens* (Brassicaceae). Plant and Soil, on line, 1-11.

Diao Z H, Shi T H, Wang S Z, et al. 2013. Silane-based coatings on the pyrite for remediation of acid mine drainage. Water Research, 47 (13): 4391-4402.

Du R Y, Bai J, Wang S Z, et al. 2011. Response of soil microbial community function to chemical aided remediation of multi-metal contaminated soils using *Jatropha curcas*. Acta Scientiae Circumstantiae, 31 (3): 575-582.

Fang S E, Tsang D C W, Zhou F S, et al. 2016. Stabilization of cationic and anionic metal species in contaminated soils using sludge-derived biochar. Chemosphere, 149: 263-271.

Gu H H, Qiu H, Tian T, et al. 2011. Mitigation effects of silicon rich amendments on heavy metal accumulation in rice (*Oryza sativa* L.) planted on multi-metal contaminated acidic soil. Chemosphere, 83 (9): 1234-1240.

Gu H H, Zhan S S, Wang S Z, et al. 2012. Silicon-mediated amelioration of zinc toxicity in rice (*Oryza sativa* L.) seedlings. Plant and Soil, 350 (1-2): 193-204.

Hu P J, Qiu R L, Senthilkumar P, et al. 2009. Tolerance, accumulation and distribution of zinc and cadmium in hyperaccumulator *Potentilla griffithii*. Environmental and Experimental Botany, 66 (2): 317-325.

Hu P J, Zhou X Y, Qiu R L, et al. 2007. Cadmium tolerance and accumulation features of Zn-hyperaccmulator *Potentilla griffithii* var. velutina. Journal of Agro-Environment Science, 26 (6): 2221-2224.

Li Q F, Qiu R L. 2012. Cadmium physiological tolerance and accumulation characteristics of *Jatropha curcas* L. Journal of Agro-Environment Science, 31 (1): 42-47.

Li Q F, Qiu R L, Shi N, et al. 2009. Remediation of strongly acidic mine soils contaminated by multiple metals by plant reclamation with *Jatropha curcas* L. and addition of limestone. Acta Scientiae Circumstantiae, 29 (8): 1733-1739.

Liu F J, Tang Y T, Du R J, et al. 2010 Root foraging for zinc and cadmium requirement in the Zn/Cd hyperaccumulator plant *Sedum alfredii*. Plant and Soil, 327 (1): 365-375.

Lu H L, Zhang W H, Wang S Z, et al. 2013. Characterization of sewage sludge-derived biochars from different feedstocks and pyrolysis temperatures. Journal of Analytical and Applied Pyrolysis, 102: 137-143.

Lu H L, Zhang W H, Yang Y X, et al. 2012. Relative distribution of Pb^{2+}sorption mechanism by sludge-derived biochar. Water Research, 46 (3): 854-862.

Mei X Q, Wong M H, Yang Y, et al. 2012. The effects of radial oxygen loss on arsenic tolerance and uptake in rice and on its rhizosphere. Environmental Pollution, 165: 109-117.

Mei X Q, Ye Z H, Wong M H. 2009. The relationship of root porosity and radial oxygen loss on arsenic tolerance and uptake in rice grains and straw. Environmental Pollution, 157 (8-9): 2550-2557.

Qiu H, Gu H H, He E K, et al. 2012. Attenuation of metal availability in polluted soil after treating with fly ash and steel slag as related to the stability of remediation. Pedosphere, 22 (4): 544-553.

Qiu R L, Fang X H, Tang Y T, et al. 2006. Zinc Hyperaccumulation and Uptake by *Potentilla Griffithii* Hook. International Journal of

Phytoremediation, 8 (4): 299-310.

Qiu R L, Tang Y T, Zeng X W, et al. 2012. Mechanisms of Cd Hyperaccumulation and Detoxification in Heavy Metal Hyperaccumulators: How Plants Cope with Cd//Progress in Botany 73. Springer Berlin Heidelberg, 2012: 127-159.

Qiu R L, Thangavel P, Hu P J, et al. 2011. Interaction of cadmium and zinc on accumulation and sub-cellular distribution in leaves of hyperaccumulator *Potentilla griffithii*. Journal of Hazardous Materials, 186 (2-3): 1425-1430.

Qiu R L, Zhao B L, Li Q F, et al. 2009. Sulfate reduction and copper precipitation by a *Citrobacter* sp. isolated from a mining area. Journal of Hazardous Materials, 164 (2-3): 1310-1315.

Tang Y T, Cloquet C, Sterckeman T, et al. 2012. Fractionation of stable zinc isotopes in the field-grown zinc hyperaccumulator *Noccaea caerulescens* and the zinc-tolerant plant *Silene vulgaris*. Environtal Science and Technolgy, 46 (18): 9972-9979.

Tang Y T, Qiu R L, Zeng X W, et al. 2009a. Lead, zinc, cadmium hyperaccumulation and growth stimulation in *Arabis paniculata* Franch. Environmental and Experimental Botany, 66 (2): 126-134.

Tang Y T, Qiu R L, Zeng X W, et al. 2009b. Zn and Cd hyperaccumulating characteristics of *Picris divaricata* Vant. International Journal of Environment and Pollution, 38 (1-2): 26-38.

Wang M Y, Chen A K, Wong M H, et al. 2011. Cadmium accumulation in and tolerance of rice (*Oryza sativa* L.) varieties with different rates of radial oxygen loss. Environmental Pollution, 159 (6): 1730-1736.

Wang X, Yao HX, Wong MH, Ye ZH. 2013. Dynamic changes in radial oxygen loss and iron plaque formation and their effects on Cd and As accumulation in rice (*Oryza sativa* L.). Environmental Geochemistry and Health, 35 (6): 779-788.

Wang Y L, Lin Q Q, Li Y, et al. 2013. Application potential of siderophore-producing rhizobacteria in phytoremediation of heavy metals-contaminated soils: A review. Chinese Journal of Applied Ecology, 24 (7): 2081-2088.

Wu C, Li H, Ye Z H, et al. 2013. Effects of As levels on radial oxygen loss and As speciation in rice. Environmental Science and Pollution Research, 20 (12): 8334-8341.

Wu C, Ye Z H, Li H, et al. 2012. Do radial oxygen loss and external aeration affect iron plaque formation and arsenic accumulation and speciation in rice. Journal of Experimental Botany, 63: 2961-2970

Wu Q H, Wang S Z, Thangavel P, et al. 2011. Phytostabilization potential of *Jatropha curcas* L. in polymetallic acid mine tailings. International Journal of Phytoremediation, 13 (8): 788-804.

Yang B, Zhou M, Shu W S, et al. 2010. Constitutional tolerance to heavy metals of a fiber crop, ramie (*Boehmeria nivea*), and its potential usage. Environmental Pollution, 158 (2): 551-558.

Yang Y X, Lu H L, Zhan S S, et al. 2013. Using kenaf (*Hibiscus cannabinus*) to reclaim multi-metal contaminated acidic soil. Chinese Journal of Applied Ecology, 24 (3): 832-838.

Ying R R, Qiu R L, Tang Y T, et al. 2010. Cadmium tolerance of carbon assimilation enzymes and chloroplast in Zn/Cd hyperaccumulator *Picris divaricata*. Journal of Plant Physiology, 167 (2): 81-87.

Zeng X W, Ma LQ, Qiu RL, et al. 2009. Response of non-protein thiols to Cd exposure in Cd hyperaccumulator *Arabis paniculata* Franch. Environtal Experimental Botany, 66 (2): 242-248.

Zeng X W, Ma L Q, Qiu R L, et al. 2011a. Effects of Zn on plant tolerance and non-protein thiol accumulation in Zn hyperaccumulator *Arabis paniculata* Franch. Environmental and Experimental Botany, 70 (2): 227-232.

Zeng X W, Qiu R L, Ying R R, et al. 2011b. The differentially-expressed proteome in Zn/Cd hyperaccumulator *Arabis paniculata* Franch. in response to Zn and Cd. Chemosphere, 82 (3): 321-328.

Zhang W H, Mao S Y, Chen H, et al. 2013. Pb (Ⅱ) and Cr (VI) sorption by biochars pyrolyzed from the municipal wastewater sludge under different heating conditions. Bioresource Technology, 147: 545-552.

Zhang W H, Zheng J, Zheng P P, et al. 2015. Sludge-derived biochar for arsenic (III) immobilization: effects of solution chemistry on sorption behavior. Journal of Environmental Quality, 44: 1119-1126.

Zhou F S, Wang H, Fang S E, et al. 2015. Pb (Ⅱ), Cr (VI) and atrazine sorption behavior on sludge-derived biochar: Role of humic acids. Environmental Science and Pollution Research, 22 (20): 16031-16039.

Advances in Remediation of Acid Agricultural Soils Contaminated by Heavy Metals in South China

Qitang WU, Zebin WEI, Xinxian LONG, Chengai JIANG

College of Natural Resources and Environment, Key Laboratory on Soil Environment and Waste Reuse in Agriculture of Guangdong Higher Education Institutes, South China Agricultural University

1. Introduction

Heavy metal contamination of agricultural lands is widespread, particularly in China, and cadmium (Cd) is the major pollutant among heavy metals based on contaminated land area (MEPC, 2014). Bioaccumulation of Cd in the food chain through soil-plant systems can be very harmful to human health, may cause Itai-itai disease which occurred in Japan in 1970s (Adriano, 1986). Hence, studies on the accumulation of Cd in plants have attracted considerable attention (Salt et al., 1995).

In South China there are many metal co-mines and thus soils contaminated by acid mine drainage (AMD) and multi-metals are frequently encountered (Wang et al., 2001). However, the need for agricultural production is increasing because of population expanding. Therefore, it is urgently needed to remediate the contaminated soils and to improve food safety. Technologies in low-cost and permeating non-stop of agricultural productions are especially welcome. Our group has undertaken studies according to these needs on reducing the heavy metals in crops and in soils.

2. Fertilizers and amendments for reducing heavy metal uptake by crops

Changing some agricultural management such as fertilizers and cultivars may reduce the contamination of crops at a low cost. Thus, pot and field trials were carried out to study the effects of different chemical fertilizers, industrial and agricultural wastes with the potential of agricultural application, on the uptake and accumulation of heavy metals by plants. Results showed that choosing appropriate fertilizer sources significantly reduced the uptake of heavy metals by plants (Wu et al., 1989, 1994). The chemical fertilizer sources influenced particularly the soil pH, water soluble and exchangeable metal content. It was recommended to use $Ca(NO_3)_2$ or NH_4HCO_3, Ca-Mg-phosphate and K_2SO_4 instead of commonly used chemical fertilizers: $CO(NH_2)_2$, super-phosphate and KCl for Cd-contaminated soils.

Pot experiment indicated that, among 8 industrial and agricultural waste amendments, the application of alkaline waste from sugar-cane industry of $CaCO_3$-treatment process was effective and economical for treating acid Cd/Pb-contaminated soils (Fu and Wu, 1995). Field experiment showed that the waste-$CaCO_3$ was also useful to reduce the bioavailability and the uptake of Hg by a leaf-vegetable (Table 1).

Table 1 Effects of waste-$CaCO_3$ on Hg uptake by *Brassica parachinensis*

Waste-$CaCO_3$ addition/(kg/m^2)	0	0.075	0.15
Soil pH	4.73	5.62	6.53
0.1 M $CaCl_2$ extractable Hg/(μg/kg soil)	3.47	2.75	1.65
Plant yield/(kg/plot)	5.12 a*	5.23 a	5.15 a
Hg content in plant/(mg/kg FW)	0.011 a	0.009ab	0.005b

*According to Duncan's test, the means followed by the same letter are not significantly different (p>0.05).

3. Low-accumulating cultivars/varieties

Low Cd cultivars/varieties of rice (*Oriza Sativa*), Cantonese cabbage (*Brassica parachinensis*) and corn were selected (Wu et al., 1994, 1999; Xu et al., 2002; Guo et al., 2010). Generally, low Cd cultivar/variety has been named if Cd concentration in the edible part of the cultivar was only the half of the average concentration of the studied cultivars (at least 5 cultivars/varieties). This means that using this cultivar could reduce the metal concentration in crop by about 50% than the average level using normal cultivars.

Among 24 rice cultivars grown on the same contaminated soil, the hybrids such as Shanyou-63 harvested a higher yield than others, but Cd content in husked rice was also higher (Wu et al., 1999). The high consuming quality cultivars such as Yeaosimiao, Zengchengsimiao and Heinuo resulted in a lower Cd content in husked rice. The normal cultivars varied more greatly in yield and Cd content. Cd content in husked rice for high accumulating cultivars was more than 2-fold higher than that of the low accumulating cultivar. Under the same hydroponic cultivation conditions, the low Cd cultivar, Yeaosimiao, showed significantly lower Cd translocation rate to grains (16.2%) than the high accumulating cultivar (Shanyou-63, 38.4%). This is maybe the most important factor causing the differences among the rice cultivars in Cd accumulation.

Low-Cd cultivars of *B. parachinensis* have lower translocation of Cd to areal parts (Xu et al., 2002) and lower free and simple complex of Cd but higher amono-acid bound Cd in stem juice than high-Cd cultivars (Wu et al., 2007a). Lvbao-701, a low-accumulating and high-tolerance cultivar of *B. parachinensis* to Cd, was compared with a high-accumulating cultivar, Chixin-4, to elucidate the mechanisms involved in low Cd accumulation in the crop (Sun et al., 2016). Root cell walls were separated by dissolving the cytoplasm with organic solvent (Hart et al., 1992), and the Cd concentration and phytochelatins (PCs) in plant roots were measured. Furthermore, the PC synthase gene fragment was cloned and expressed under

Cd stress (Sun et al., 2016). The results showed that Lvbao-701 had a significantly higher content and proportion of Cd combined with cell walls in roots (68%-77%) than Chixin-4 (35%-54%) under Cd stress. The proportions of Cd bound to cell walls obtained with dissolving method were consistently higher than those obtained with conventional grinding method. Cd stress induced PC production and resulted in a higher PC content in Lvbao-701 than in Chixin-4. Cloning and expression of a cDNA fragment of the PC synthase gene from *B. parachinensis* indicated that, in both Lvbao-701 and Chixin-4, PC synthase gene was expressed mainly in roots, and was induced by Cd. PC synthase gene expression in the roots of Lvbao-701 was higher than in Chixin-4. Therefore, keeping a large amount of Cd in root cell walls and producing more PCs in roots are probably the main reasons Lvbao-701 exhibits lower Cd translocation to shoots and stronger tolerance to Cd than other cultivars.

4. Combination of low-Cd cultivars and amendments

A field experiment was conducted to investigate the effectiveness of soil amendments (lime, Nano-Si foliar solution and diatomite) on the growth of and metal uptake by 3 maize (*Zea mays* L.) cultivars grown in an acidic contaminated soil that was irrigated with river water contaminated by Pb and Zn mining wastewaters (Guo et al., 2011). The addition of lime significantly increased the corn grain yields and decreased the concentrations of Zn and Cd in the grain and shoots of maize compared with the control. Among the three maize cultivars, the proposed low-Cd cultivar, Yunshi-5, accumulated lowest amount of Cd and Zn in grains. The concentrations of Cd in the grains of Yunshi-5 treated with lime were conform to the old Chinese feed standards (Cd$<$0.2 mg/kg) but exceeded the new Chinese feed standards (Cd$<$0.1 mg/kg) in this soil with a Cd content of about 1 mg/kg and a pH about 4.6. It is suggested that a combination of low-Cd corn and chemical fixation could effectively reduce Cd to enter the food chain but might not be enough if the soil contains too high available Cd.

At the same location, a field with lower Cd concentration (about 0.8 mg/kg soil) was also tested for 2 years with 3 treatments: control without lime, limed every year and limed only first year (residual effect of lime) (Shao et al., 2010). Two crops of maize were planted each year. Results showed that the yield of the plant increased significantly for the treatments of continuous application and residual effect of lime, but the better treatments was continuous application of lime. The treatments of continuous application of lime lowered the heavy metals (Cd, Pb, Zn, Cu) content significantly in grains as compared with the control and the Cd and Pb contents conformed to the new Chinese feed standards (Cd$<$0.1 mg/kg; Pb$<$0.2 mg/kg). However, residual effect of lime did not decrease significantly the heavy metals (Cd, Pb, Zn, Cu) contents in corn grains in the fourth crop as compared with the control. It shows that the persistence of liming effect is only one and a half years.

5. Co-crop/intercrop of low-Cd cultivar with hyperaccumulator

Low accumulating cultivars or varieties combining with appropriate fertilizers or amendments may produce safe foods or feeds on contaminated lands, but the heavy metals remained in the contaminated soil and it requires special monitoring and management of these agricultural lands for decades. A co-crop/intercrop of low-Cd cultivar with metal hyperaccumulator was proposed to produce safe food/feed and meanwhile to reduce the metal content in soil (Wei et al., 2005).

A co-planting system studied in field experiment was comprised of a Zn-and Cd hyperaccumulator (*Sedum alfredii*) and a low-Cd crop (*Z. mays*, cultivar Yunshi-5) (Wu et al., 2007b). Results showed that yields of *S. alfredii* were significantly increased by co-planting with *Z. mays* in spring-summer seasons. Our study further revealed that concentrations of Zn, Pb, and Cd in the corn grains of *Z. mays* conform to the Chinese hygiene standards for animal feeds and in the other parts of maize conform to the Chinese organic fertilizer standards. The uptake of Zn by *S. alfredii* was significantly enhanced by co-planting with *Z. mays*. When the MC was applied to the co-planting system in the soil contaminated with Zn, Cd, and Pb, the highest phytoextraction rates were observed. This study suggested that the use of the hyperaccumulator *S. alfredii* and the lowaccumulating crop *Z. mays* in the co-planting system with the addition of the MC was a more promising approach than the use of a single hyperaccumulator with the assistance of EDTA. This approach not only enhances the phytoextraction rates of the heavy metals but also simultaneously allows agricultural practices with safe feed products in the metal-contaminated soils.

In order to verify the effectiveness of long-term application of the technology of inter-cropping remedying heavy metal contaminated soils, a 3-year field experiment was conducted to have three treatments and 5 croppings, Treatment IC (inter-cropping of *S. alfredii* and corn), Treatment Mono-cropping (*S. alfredii*) and Treatment IC+MC (mixture of chelators consisting of citric acid and EDTA) (Zhou et al., 2014). Results show that the total Cd and Zn in the soil decreased continuously with the progress of the experiment (Fig. 1). After the 5 croppings, the soils in Treatments IC and IC+MC were remedied to meet the criteria of the National Standard for Soil Environmental Quality (GB15618—1995). Soil Cd decreased from 1.21-1.27 mg/kg to 0.29-0.30 mg/kg and soil Zn from 280-311 mg/kg to 196-199 mg/kg. But no significant change was observed in soil Pb. MC did not enhance much phytoextraction of heavy metals from the soil. Heavy metal content in the plant of *S. alfredii* and heavy metal phytoextraction rate of the plant was not decreased with the year. Mono-cropping of *S. alfredii* was higher in biomass production than the plant in intercropping in all the four rounds and so in total phytoextraction of heavy metals. However, in Treatment IC, the corn turned out to meet the standard for feed and even the standard for food in the fourth cropping. Calculation based on harvested *S. alfredii* shows that the plant contributed 32.5%-36.5% to the decrease in soil Cd, while corn did only 0.47%-0.60% and the remaining 63.0%-66.9% was attributed to leaching loss and other factors. Soil Zn was in a similar case. These findings indicate that in the studied acid soil (pH 4.7) downward leaching and *S. alfredii* phytoextraction both play important roles in reducing soil Cd and Zn.

However, in the previous experiment, corn yield decreased continuously due to probably only providing chemical fertilizers to crops and lack of organic manure to prevent fixation of metals happening which reduces the phyto-extraction. In order to select appropriate organic amendments for cropping hyperaccumulator and normal plants on contaminated soils and establish the relationship between Cd adsorption characteristics of soil amendments and their capacity to reduce Cd uptake by plants, batch adsorption experiments with 11 different clay minerals and organic matters and a pot experiment with the same amendments were carried out (Sun et al., 2014). The pot experiment was conducted with *S. alfredii* and maize (*Z. mays*) in a co-cropping system. The results showed there was a significant negative correlation between the n value of Freundlich equation and Cd uptake by plants, and between the logarithm of the stability constant K of the Langmuir equation and plant uptake, but not between the highest adsorption amount of the Langmuir equation and plant uptake. Humic acids (HAs) and mushroom manure increased Cd uptake by *S. alfredii*, but not maize, thus they are suitable as soil amendments for the co-cropping *S. alfredii* and maize. The stability constant K in these cases was 0.14-0.16 L/mg and n values were 1.51-2.19. The alkaline zeolite and mica had the best fixation abilities and significantly decreased Cd uptake by the both plants, with $K \geqslant 1.49$ and $n \geqslant 3.59$.

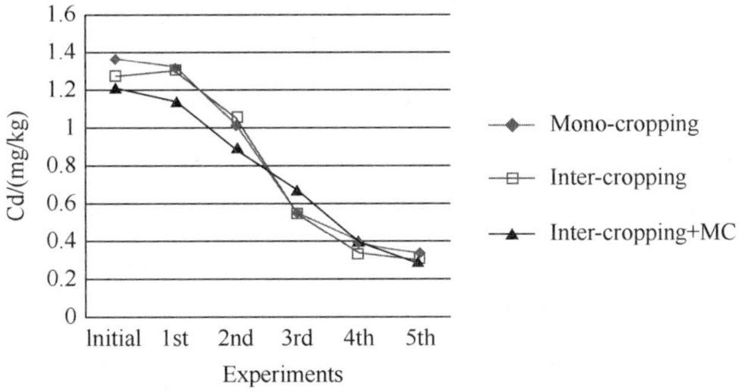

Fig. 1 Changes in content of Cd in soil after 5 croppings in the field experiment (From Zhou et al., 2014)

The value of each point is the mean of 4 replicates. Mono-cropping: only *Sedum alfredii*, Inter-cropping: inter-planting *S. alfredii* and maize, MC: Mixture of Chelators (Citric acid+EDTA, applied at 5 mmol/kg soil).

6. Chemical washing, chemical enhanced phytoremediation and control of underground water contamination by fixation in deep layers of the soil profile

Chemical washing is necessary for efficacious remediation of multi-metals contaminated soils, for example, Pb is difficult to be phytoextracted due to the lack of real hyperaccumulator for this metal. In the literature, ethylenediaminetetraacetic acid (EDTA) is the most frequently cited chelating agent for potentially toxic trace metals extracted from soils because EDTA has strong chelating ability for different heavy metals (Lestan et al., 2008). Other washing chemicals include salt and high-concentration chloride solution, such as calcium chloride and iron (III) chloride (Makino et al., 2007). However, people pay more attention to the leaching risks of heavy metals in soil washing technology (Wu et al., 2004; Luo et al., 2006). Therefore, a horizontal permeable barriers to reduce the leaching risk were tested in pot experiments (Kos and Lestan, 2003).

However, after EDTA-enhanced soil washing, a significant part of metal-EDTA complexes are retained in the soil by the adsorption with the mineral surface such as soil iron oxides (Nowack, 2002). The potential phyto-availability of toxic metals left in soil after soil washing may not be adequately reduced and therefore requires further investigation. Soil washing with acidic chelators followed by liming is under consideration for remediating multi-metals-contaminated soils (Guo et al., 2013). This technique was tested by pot experiments with a mixture of chelators (MC) as a washing agent. After washing and liming treatments, corn (*Zea mays*) was planted to assess bioavailability of heavy metals. The results showed that the concentrations of Cd, Pb and Zn in the corn grain with the washing treatment were still higher than the food standard limits in China and liming the washed soil did not effectively decrease Cd and Pb concentrations in the corn grain. Liming the washed soil significantly increased heavy metals concentrations in leachate. A phytoextraction with *Sedum alfredii* was additionally performed after corn harvest to further decrease heavy metals in the soil. The concentrations of Cd, Pb, Zn and Cu in *S. alfredii* after washing were increased by 42.8%, 51.2%, 7.3% and 25.8%, respectively, compared to non-washed soils. This study suggests that the bioavailability of heavy metals could still be high after washing, the phytoextraction by *S. alfredii* after soil washing is an effective approach to deplete the mobile Cd activated by soil washing. Liming the chelator washed soil may release the adsorbed metal-chelates, significantly increased the release of metal-EDTA complexes into the soil solution during corn cropping.

In a field experiment on multi-metal contaminated soil, we investigated the efficiency of Cd, Pb, Zn, and Cu removal by only mixture of chelators (MC) or combining with $FeCl_3$ (Guo et al., 2016). After washing treatment, a co-cropping system was performed for heavy metals to be extracted by *S. alfredii* and to produce safe food from *Z. mays*. The concentrations of heavy metals in groundwater were analyzed to evaluate the leashing risk of soil washing with $FeCl_3$ and MC. Results showed that addition of $FeCl_3$ was favorable to the removal of heavy metals in the topsoil. Metal leaching occurred mainly in rain season during the first co-cropping. The removal rates of Cd, Zn, Pb, and Cu in topsoil were 28%, 53%, 41%, and 21% with washing by $FeCl_3$+MC after first harvest. The application of $FeCl_3$ reduced the yield of *S. alfredii* and increased the metals concentration of *Z. mays* in first harvest. However, after amending soil with lime and organic fertilizer, the metals concentration of *Z. mays* in $FeCl_3$+MC treatment were similar to that only washing by MC. The grains and shoots of *Z. mays* were safe for use in feed production. Soil washing did not worsen groundwater contamination during the study period, but the concentration of Cd in groundwater was higher than the limit value of Standard concentrations for Groundwater IV. This study suggests that soil washing using $FeCl_3$ and MC for the remediation of multi-metal contaminated soil is potential feasibility.

Wei et al. (2011) reported no adverse impact to leached water with three MC applications (total application amount, 15 mmol MC/kg soil) and three crops in an acid ferralsol. The leached Cd and Pb accumulated in the subsoil (40-60 cm) for MC treated soil compared to the initial subsoil concentrations. The enhancement of deep layer fixation of heavy metals was proposed with addition of lime and $Fe^{3+}$to subsoil (Wu et al., 2011). Results suggested that the leaching risk of heavy metals is minimal in the acid ferralsol regions such as South China and can be further reduced by amending the soil of deep layers (Wei, 2015).

References

Adriano, D C. 1986. Trace Elements in the Terrestrial Environment. New York: Springer-Verlag: 106-155.

Fu X H, Wu Q. T. 1995. Effect of some solid waste applications on the uptake of Cd and Pb by leaf vegetables. Journal of Agro-Environment Science, 14: 145-149.

Guo X F, Wei Z B, Qiu J R, et al. 2010. Differences between corn cultivars in accumulation and translocation of heavy metals. Journal of Ecology & Rural Environment, 26: 367-371.

Guo X, Wei Z, Wu Q T, et al. 2011. Cadmium and zinc accumulation in maize grain as affected by cultivars and chemical fixation amendments. Pedosphere, 21 (5): 650-656.

Guo X, Wei Z, Penn C J, et al. 2013. Effect of soil washing and liming on bioavailability of heavy metals in acid contaminated soil. Soil Science Society of America Journal, 77 (2): 432-441.

Guo X, Wei Z, Wu Q T, et al. 2016. Effect of soil washing with only chelators or combining with ferric chloride on soil heavy metal removal and phytoavailability: Field experiments. Chemosphere, 147: 412-419.

Hart J J, Di Tomaso J M, Linscott D L, et al. 1992. Characterization of the transport and cellular compartmentation of paraquat in roots of intact maize seedlings. Pesticide Biochemistry and Physiology, 43 (3): 212-222.

Kos B, Leštan D. 2003. Induced phytoextraction/soil washing of lead using biodegradable chelate and permeable barriers. Environmental Science & Technology, 37 (3): 624-629.

Leštan D, Luo C, Li X. 2008. The use of chelating agents in the remediation of metal-contaminated soils: a review. Environmental Pollution, 153 (1): 3-13.

Luo C, Shen Z, Lou L, et al. 2006. EDDS and EDTA-enhanced phytoextraction of metals from artificially contaminated soil and residual effects of chelant compounds. Environmental Pollution, 144 (3): 862-871.

Makino T, Sugahara K, Sakurai Y, et al. 2006. Remediation of cadmium contamination in paddy soils by washing with chemicals: selection of washing chemicals. Environmental Pollution, 144 (1): 2-10.

MEPC (Ministry of Environmental Protection of China). 2014. Communique on the investigation of soil pollution in China (in Chinese). http://www.zhb.gov.cn/gkml/hbb/qt/201404/t20140417_270670.htm. [2015-7-31].

Nowack B. 2002. Environmental chemistry of aminopolycarboxylate chelating agents. Environmental Science & Technology, 36 (19): 4009-4016.

Shao L, Guo X F, Shi X F, et al. 2010. Effect of Lime on Heavy Metals Uptake by Zea mays and the Persistence of the Liming Effect.

Journal of Agro-Environment Science, 29 (10): 1986-1991.

Sun Y, Wu Q T, Lee C C C, et al. 2014. Cadmium sorption characteristics of soil amendments and its relationship with the cadmium uptake by hyperaccumulator and normal plants in amended soils. International Journal of Phytoremediation, 16 (5): 496-508.

Sun Y, Ye H., Wei Z B, et al. 2016. Root cell wall and phytochelatins in low cadmium cultivar of *Brassica parachinensis*. Pedosphere (in press).

Salt D E, Prince R C, Pickering I J, et al. 1995. Mechanisms of cadmium mobility and accumulation in Indian mustard. Plant Physiology, 109 (4): 1427-1433.

Dong W Q Y, Cui Y, Liu X. 2001. Instances of soil and crop heavy metal contamination in China. Soil and Sediment Contamination, 10 (5): 497-510.

Wei Z B. Wu Q T, Long X X. 2005. Phytoremediation of heavy metal contaminated soil with mixed chelators in co—crop system. Journal of Agro-Environment Science, 24 (6): 1262-1263.

Wei Z B, Guo X F, Wu Q T, et al. 2011. Phytoextraction of heavy metals from contaminated soil by co-cropping with chelator application and assessment of associated leaching risk. International Journal of Phytoremediation, 13 (7): 717-729.

Wei Z B. 2015. Remediation of Cd/Pb Contaminated Soil by Phytoremediation Coupling with Chemical Treatments. PhD Thesis. South China Agricultural University, Guangzhou.

Wu L H, Luo Y M, Xing X R, et al. 2004. EDTA-enhanced phytoremediation of heavy metal contaminated soil with Indian mustard and associated potential leaching risk. Agriculture, Ecosystems & Environment, 102 (3): 307-318.

Wu Q T, Morel J. L., Guckert A. 1989. Effect of nitrogen source on Cd uptake by plants. Compte Rendu de L'Academie des Sciences, Serie Ⅲ, 309: 215-220.

Wu Q T, Chen L, Wang G S. 994. Effect of chemical fertilizers and crop cultivars on Cd accumulation by plants. Journal of South China Agricultural University, 15 (4): 1-6.

Wu Q. T., Chen L, Wang G. S. 1999 Accumulation of Cd by different rice cultivars and its mechanisms. Acta Ecol. Sinica, 19 (1): 104-107.

Wu Q T, Xu Z L, Ye H, et al. 2007. Chemical composition of root and stem saps in relation to cadmium resistance and accumulation in Brassica parachinensis. Pedosphere, 17 (3): 352-359.

Wu Q T, Wei Z B, Ouyang Y. 2007. Phytoextraction of metal-contaminated soil by *Sedum alfredii* H: Effects of chelator and co-planting. Water, Air, and Soil Pollution, 180 (1-4): 131-139.

Wu Q T, Wei Z B, Qiu J R. 2011. A chemical washing and deep-layer fixation combined technology to remediate heavy-metals-contaminated soils. China patent No: ZL200910040403.8.

Xu Z L, Wu Q T, Yi Y L. (2002) Cd resistance of different cultivars of *Brassica parachinensis*. Acta Ecologica Sinica, 22 (4): 571-576.

Zhou J L, Shao L, Zhu H R, et al. 2014. Phytoremediation of inter-cropping with chemical enhancement of heavy-metal-contaminated acid Soil: A long-term field experiment. Acta Pedologica Sinica, 51 (5): 143-152.

Stability and Universal Applicability of Immobilization Effect of Sepiolite on Cadmium in Acid Paddy Soil

Yingming XU, Xuefeng LIANG, Lin WANG, Yuebing SUN, Qingqing HUANG

Key Laboratory of Original Environmental Quality of MOA

Agro-Environmental Protection Institute of Ministry of Agriculture

Recent nationwide surveys show that 16% of the soil samples, 19% for the agricultural soils, are contaminated based on Chinese soil environmental quality limits, mainly with heavy metals and metalloids (Zhao et al., 2015). 82.8% of the contaminants existed in agricultural fields were inorganic contaminants such as Cd, Hg, As, Cu and Pb. Agricultural production on heavy metal polluted soils is concerned because of the potential exposure to heavy metals via dietary intake. Immobilization relies on the addition of immobilization agents to polluted soils to decrease the bioavailability of heavy metals, which was cost-effective and environmentally compatible (Lee et al., 2009; Lombi et al., 2002).

At present immobilization agents can be categorized into pH regulating agents and sorption agents (Kim et al., 2012). The pH regulating agents such as calcite and lime can induce the precipitation of heavy metals in the forms of hydroxides or carbonate, or affect the surface charge of soil particles to enhance sorption. Sorption agents include biochar and clay minerals, and so on. Biochar application to a contaminated soil can reduce the plant uptake of Zn, Pb and Cd (Puga et al., 2015) and bioavailability of soil Cd and Pb (Bian et al., 2014), and suppress the accumulation of cadmium in japonica rice (Zhang et al., 2014).

Rice (*Oryza sativa* L.) was considered as a particular crop that had high Cd uptake in its grain. The acidic soils of Cd-contaminated areas can contribute to a higher dietary intake of Cd (Aziz et al., 2015). How to reduce or diminish the Cd content of brown rice to ensure the food safety is an urgent task.

The objective of this chapter is to introduce the long-term stability and universal applicability of immobilization effects of sepiolite on Cd contaminant in acid paddy soil through large area *in-situ* demonstration in different fields with various cultivars.

1. Stability of remediation effect of sepiolite

1.1 Experimental design

Sepiolite as natural clay from the Hebei province was utilized as immobilization agent for cadmium contaminants in paddy soil. The composition of sepiolite determined by XRF was

41.7% CaO, 16.8% MgO, 7.4% Al_2O_3, and 32.5% SiO_2.

The stability of remediation effect of sepiolite on Cd contaminants in paddy soil were verified by *in-situ* immobilization demonstration of consecutive years in Hunan province. The paddy soil with total Cd content 1.68 mg/kg was severely polluted by Cd due to historical smelting and mining. The pH value of paddy soil was about 5.5 in average. Four treatments were designed as follows: sepiolite at dosage 0.75 kg/m^2 (T1), 1.50 kg/m^2 (T2), 2.25 kg/m^2 (T3) and the control (CK). The data of the first year was labeled as 2013 (1). In the second year, the data for the groups without additional sepiolite, marked as 2013 (1) + 2014 (0), and additional sepiolite recorded as 2013 (1) +2014 (1). There were two rice cultivars Zhonglianyou 950 (ZLY-950 in short) and Fengyou 9 (FY-9 in short) of hybrid *O. sativa L. subsp. hsien* Ting selected.

1.2 Dynamics of grain yield and Cd contents of brown rice

There were no statistical differences among all the treatments for the grain yields of FY-9 and ZLY-950, respectively. Sepiolite was natural mineral without any excess nutrients or heavy metals, so that it had no obvious promotion or inhibition effect on grain yields.

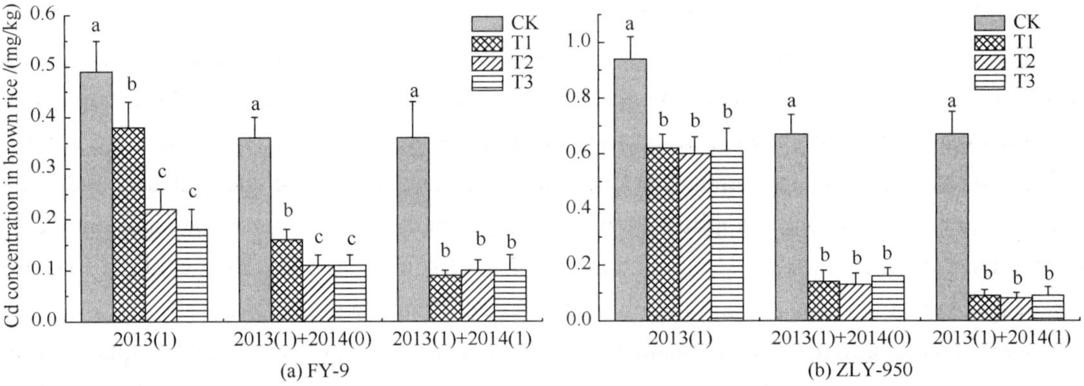

Fig. 1 Dynamics of Cd contents of brown rice in two consecutive years (Liang et al., 2016)

In the first year, the immobilization effect of sepiolite on Cd contaminants in acid paddy soil was significant. Cd contents of brown rice were 0.49 mg/kg and 0.75 mg/kg for FY-9 and ZLY—950, respectively, which exceeded the maximum limited level of GB2762—2012 and CAC 153-1995. Sepiolite can remarkably reduce Cd contents of brown rice. In the second year without additional sepiolite, the immobilization effect was remained as shown in Fig. 1. Meanwhile for the group of consecutive addition of sepiolite, the similar reduction contents were found and there were no obvious statistical differences between 2013 (1) +2014 (0) and 2013 (1) +2014 (1). The immobilization effect of sepiolite on Cd contaminant can last two years at least, and the additional sepiolite in the second year was unnecessary in the aspect of uptake of Cd by rice. For the other cultivar ZLY-950, the similar results were observed.

1.3 Changes of pH and extractable Cd contents in consecutive years

The pH changes of paddy soil in the consecutive years were shown in Fig. 2. Sepiolite could increase the pH values of soil by different contents, which was attributed to the reaction between $H^+$and $CaCO_3$ existed in the natural sepiolite. In the second year without any additional sepiolite, the pH regulating effect was also remained, indicating the stability of pH lifting effect of sepiolite. In the second year, the additional sepiolite had no further impact on pH values of paddy soil, which revealed that the pH regulating effect of sepiolite could last two years at least.

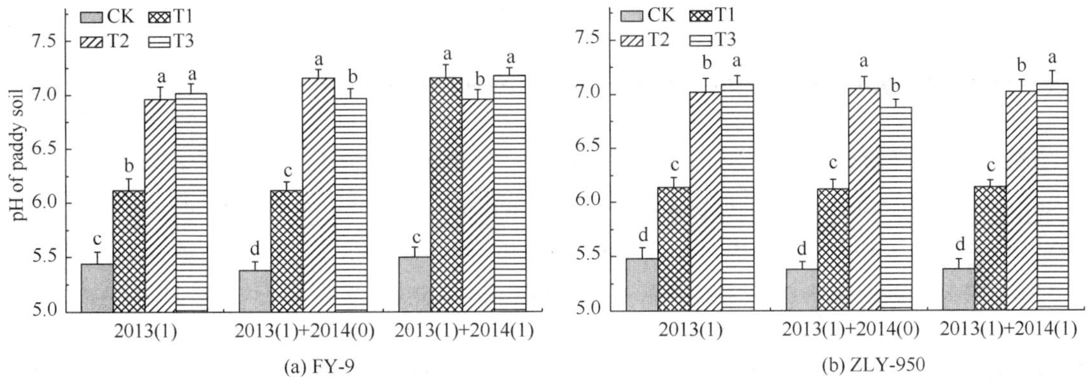

Fig. 2 Dynamics of pH of paddy soils in two consecutive years (Liang et al., 2016)

The point of zero charge (PZC) of paddy soil in CK was 5.75. After the application of sepiolite, the surface of soil particle transferred to negative charge, which was benefit for the sorption of metal cation. The tendency of pH changes of paddy soil indicated that once the pH value increased after addition of sepiolite in the first year, it would maintain in the second year.

Sepiolite had no effects on the total Cd concentration in soil, but had significant effect on the bioavailability of Cd contaminants in paddy soil. 0.025 M HClextractable Cd contents of paddy soil was selected as the index of bioavailable Cd content and showed better correlation for fitting Cd concentrations in brown rice for both rice cultivars. The extractable Cd contents in paddy field planted FY-9 decreased obviously after the addition of sepiolite in the first year as shown in Fig. 3. And the second year without any additional sepiolite, the extractable Cd contents maintained as before, especially for T3, the extractable Cd content decreased further compared to that in the first year. The phenomena was also found in another cultivar ZLY-950. The data revealed that the interaction between sepiolite and Cd contaminant in soil was a long-term complex process. The dynamic changes of pH and extractable Cd content of paddy soil in two consecutive years demonstrated the stability of immobilization effect of sepiolite on Cd contaminants in paddy soil.

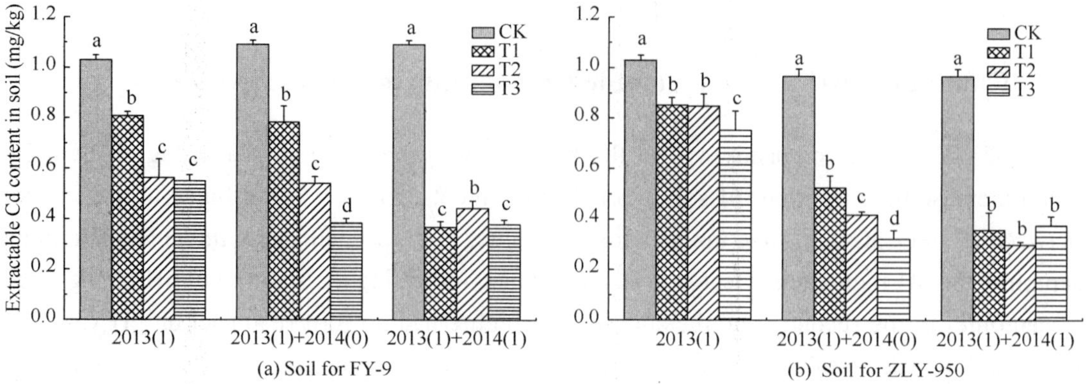

Fig. 3　Extractable Cd contents in paddy soil in consecutive years (Liang et al., 2016)

1.4　Distribution of species fraction in consecutive years

Sequential extraction was used to assess the fractions of Cd species. The data in Fig. 4 clearly showed that in the first year, Exc-Cd fraction was dominant fraction existed in soil environment. Sepiolite acted to strongly modify the fraction of Cd from phytoavailable to less phytoavailable phases, which was consistent with the results of 0.025 mol/L HCl extractable Cd content of paddy soil. Exc-Cd fraction reduced and Res-Cd fraction increased obviously. In the second year, Exc-Cd fraction in CK had no obvious changes compared to that in last year and still exceeded 40% of the total Cd content. Meanwhile in the group without additional amendment, Exc-Cd fraction in sepiolite treatments was similar to that in 2013, which indicated that the immobilization effect in the aspect of Cd fraction could last at least two years.

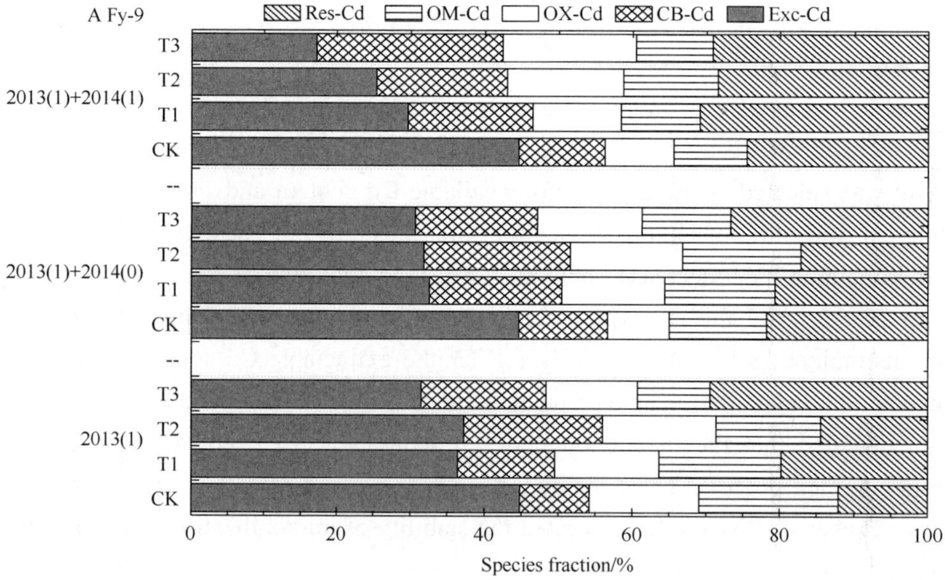

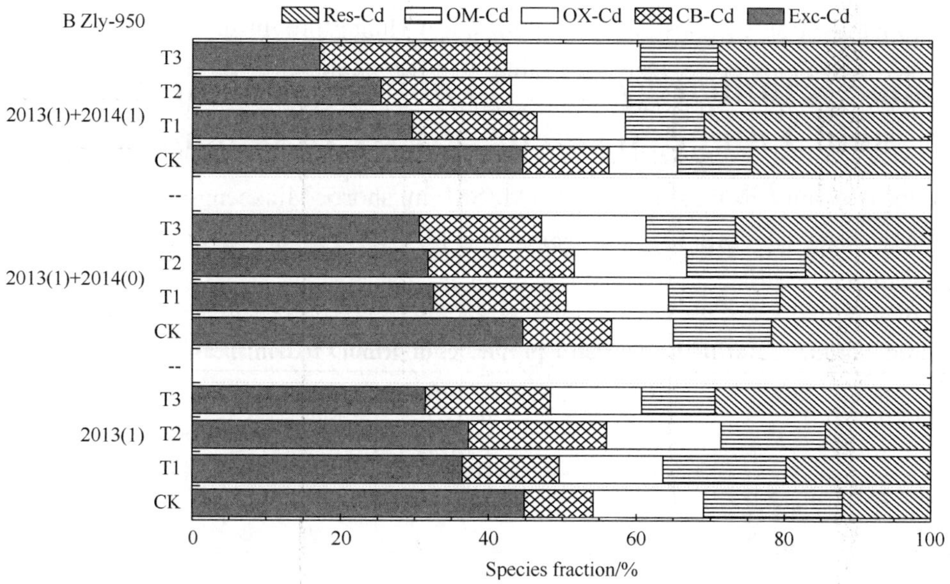

Fig. 4 Dynamics of species fraction of Cd in paddy soil in two consecutive years (Liang et al., 2016)

The additional sepiolite in 2013 (1) +2014 (1) enhanced the immobilization effect and Exc-Cd fraction decreased further compared to that in 2013 (1) +2014 (0), especially at high application dosage T3. However, additional sepiolite increased the remediation cost. Compared to the remediation effect of the second year, the additional sepiolite was not necessary.

The distribution of species fraction revealed that the reaction between soil contaminants and sepiolite added into soil was a long-term complex process. Because soils are heterogeneous, heavy metals in soils can be involved in a series of complex chemical and biological interactions including oxidation-reduction, precipitation and dissolution, volatilization, and surface and solution phase complexation (Selim and Michael, 2001).

1.5 Summary

The stability of immobilization effect of sepiolite on Cd contaminants in paddy soil was confirmed by two consecutive years filed scale in situ demonstration. The Cd content of brown rice, 0.025 mol/L HCl extractable Cd content and exchangeable Cd content of paddy soil decreased remarkably even in the second year without any additional sepiolite. The results indicated the interaction of sepiolite and cadmium was a long-term process and the immobilization effect could last at least two years.

2. Universal applicability of remediation effect of sepiolite

2.1 Demonstration design

In 2015, universal applicability of sepiolite on Cd pollutants in paddy soil were studied by *in-situ* immobilization demonstrations conducted in Hunan province. The basic parameters of the selected six fields were listed in Table 1. Five application dosages were designed as follows: 0.5 kg/m^2 (T1), 1.0 kg/m^2 (T2) and 2.0 kg/m^2 (T3), 3.0 kg/m^2 (T4) and the control. Six different rice cultivars were selected and all of them were main cultivars with large planting area in local county. Mingzhu Simiao (MZSM in short), Minsheng-6 (MS-6), Minsheng-7 (MS-7), Minsheng-10 (MS-10) and Minsheng-13 (MS-13) are hybrid rice cultivars and Huanghuazhan (HHZ) is conventional rice cultivar.

Table 1 The basic properties of demonstration areas

No.	Total Cd/(mg/kg)	pH	Available N/(mg/kg)	Olsen-P /(mg/kg)	Available K/(mg/kg)	OM /(mg/kg)	CEC/(cmol/kg)
A	4.25	5.93	148.9	20.0	63.5	41.15	14.10
B	2.95	5.31	211.0	26.4	65.7	59.96	16.95
C	3.31	5.96	172.2	16.8	43.1	81.07	17.60
D	3.52	5.53	200.9	22.6	48.1	87.90	16.78
E	2.50	6.40	178.0	17.8	43.6	90.89	16.67
F	2.42	5.26	149.7	23.7	64.1	62.51	18.25

2.2 Effects on grain yield and Cd accumulation of brown rice

The grain yields of brown rice of different cultivars in field A to F were in the range 6836-7764 kg/hm^2. Rice production had been more or less stressed due to long-term metal contamination. There were no statistical differences among all the treatments. The data indicated that sepiolite had no adverse impact on the grain yields. Pearson correlation analyses revealed that the grain yield had no obvious relationship with the Cd content of brown rice and the extractable Cd contents of paddy soil.

As shown in Tables 1-2, the Cd contents of brown rice in CK of the six fields were in the range of 0.35-2.45 mg/kg. Sepiolite could reduce Cd content of brown rice by contents of 36.06%-70.56% for MS-13 in field A and 56.24%-60.51% for HHZ in field E. And the minimum Cd content after sepiolite application was 0.12 mg/kg for MS-13 and 0.11 mg/kg for HHZ, which could meet the guideline of Chinese national standard GB 2762—2012. Meanwhile, the minimum Cd content of brown rice were 0.24 mg/kg for MS-7 in field C and 0.29 mg/kg for MS-10 in field D, which were less than the guideline of CAC 153—1995. Although the minimum Cd contents of brown rice were 1.18 mg/kg for MZSM in field B and 0.47 mg/kg for MS-6 in field F, respectively, the maximum reduction contens were up to 50.31% and 34.63%, respectively. The maximum reduction contents varied from each other due to the bioavailability of Cd contaminates, the dosage of sepiolite and the selected rice cultivars. The combined remediation methods taking the advantage of the immobilization of sepiolite and certain low-accumulating rice cultivars were recommended for food safe and sustainable agriculture in the future (Bolan et al., 2013).

In the dosage range of 0.5-3.0 kg/m^2, the Cd contents of brown rice decreased with the increase of application dosages. For most demonstrations, sepiolite can decrease Cd content of brown rice obviously even at the minimum dosage 0.5 kg/m^2. For MS-13 in field A, MZSM in field B, HHZ in field E and MS-6 in field F, there were no obvious statistical differences between the huge application dosage 2.0 kg/m^2 and 3.0 kg/m^2. Based on the remediation effect and the cost of sepiolite, the application dosage 1.0-2.0 kg/m^2 was recommended as optional dosage for the remediation of Cd polluted acid paddy soil.

2.3 Effects on soil pH and bioavailability of Cd

The addition of sepiolite led to a remarkable increase of soil pH. It was $CaCO_3$ composite in natural sepiolite that resulted in the increase of soil pH value. Soil pH plays a significant role in determining the equilibrium of heavy metals at soil particle and solution interfaces (Gu et al., 2011). Increases in soil pH affect metal species directly and favor the formation of insoluble hydroxides, metal-carbonate precipitates, metal-organic complexes. Increased pH in soil results in the generation of more negatively charged binding sites on the surface of soil colloids and therefore consequential decreases in the bioavailability of heavy metals (Pehlivan et al., 2009).

Table 2 The Cd contents of brown rice, pH and extractable Cd contents of soil

		Cd content of brown rice/(mg/kg)	pH of paddy soil	Extractable Cd content of soil/(mg/kg)
A	CK	0.42 ± 0.06^a	6.87 ± 0.05^c	0.52 ± 0.06^a
	T1	0.27 ± 0.03^b	6.95 ± 0.06^c	0.46 ± 0.07^a
	T2	0.23 ± 0.03^b	7.39 ± 0.08^b	0.26 ± 0.04^b
	T3	0.13 ± 0.03^c	7.69 ± 0.11^a	0.25 ± 0.05^b
	T4	0.12 ± 0.04^c	7.71 ± 0.12^a	0.22 ± 0.04^b
B	CK	2.38 ± 0.15^a	5.31 ± 0.07^d	0.71 ± 0.08^a
	T1	2.08 ± 0.16^b	6.38 ± 0.08^c	0.38 ± 0.04^b
	T2	1.63 ± 0.11^c	7.00 ± 0.12^b	0.23 ± 0.04^c
	T3	1.18 ± 0.08^d	6.92 ± 0.11^b	0.16 ± 0.04^c
	T4	1.34 ± 0.09^d	7.39 ± 0.14^a	0.16 ± 0.03^c
C	CK	1.16 ± 0.12^a	5.96 ± 0.06^c	1.77 ± 0.15^a
	T1	1.03 ± 0.08^a	6.13 ± 0.10^b	1.25 ± 0.1^b
	T2	0.59 ± 0.06^b	7.32 ± 0.08^a	0.75 ± 0.08^c
	T3	0.49 ± 0.07^b	7.37 ± 0.12^a	0.65 ± 0.06^c
	T4	0.29 ± 0.04^c	7.48 ± 0.09^a	0.48 ± 0.03^d

Continue table

		Cd content of brown rice/(mg/kg)	pH of paddy soil	Extractable Cd content of soil/(mg/kg)
D	CK	1.35 ± 0.10^{a}	5.53 ± 0.09^{e}	2.37 ± 0.11^{a}
	T1	1.11 ± 0.05^{b}	5.93 ± 0.06^{d}	1.96 ± 0.09^{b}
	T2	0.68 ± 0.08^{c}	6.36 ± 0.11^{c}	1.50 ± 0.08^{c}
	T3	0.49 ± 0.07^{d}	7.02 ± 0.08^{b}	0.49 ± 0.05^{d}
	T4	0.24 ± 0.06^{e}	7.46 ± 0.12^{a}	0.33 ± 0.08^{e}
E	CK	0.35 ± 0.09^{a}	6.40 ± 0.07^{c}	1.35 ± 0.08^{a}
	T1	0.13 ± 0.07^{b}	6.95 ± 0.06^{b}	0.81 ± 0.04^{b}
	T2	0.13 ± 0.08^{b}	6.93 ± 0.11^{b}	0.84 ± 0.07^{b}
	T3	0.11 ± 0.04^{b}	6.96 ± 0.04^{b}	0.83 ± 0.09^{b}
	T4	0.15 ± 0.05^{b}	7.65 ± 0.07^{a}	0.71 ± 0.06^{b}
F	CK	0.73 ± 0.08^{a}	5.26 ± 0.05^{e}	1.68 ± 0.13^{a}
	T1	0.71 ± 0.06^{a}	6.38 ± 0.07^{d}	1.12 ± 0.11^{b}
	T2	0.61 ± 0.09^{ab}	6.81 ± 0.09^{c}	0.96 ± 0.08^{bc}
	T3	0.55 ± 0.05^{ab}	7.02 ± 0.11^{b}	0.84 ± 0.08^{cd}
	T4	0.471 ± 0.09^{b}	7.51 ± 0.08^{a}	0.73 ± 0.06^{d}

The phytoavailable Cd contents in paddy soil determined by 0.025 M HCl decreased by addition of sepiolite. In the dosage range of 0.5-2.0 kg/m^2, 0.025 M HCl extractable Cd contents decreased with the increase of application dosage. For Field A, B, E and F, there were no obvious statistical difference between the huge dosage 2.0 kg/m^2 and 3.0 kg/m^2. Based on the analyses of Cd contents of brown rice, the application dosage 1.0-2.0 kg/m^2 was also recommended.

The uptake of Cd from paddy soil by the rice plants varies from each other, which depended on the special cultivars. For example, the accumulation of Cd in brown rice of HHZ was low even in the condition of high extractable Cd content of paddy soil, but MZSM can be regarded as hyper-accumulating cultivar, which can accumulate much Cd in brown rice even at low extractable Cd concentration in paddy soil. At present an approach for safe crop production was the cultivation of low metal accumulating crop species (Cao et al., 2014). Natural variations in the concentrations of Cd among cultivars have been well documented in staple crops including rice. It is physiological and genetic mechanisms that determine the Cd transport in rice for low Cd-accumulating cultivars. Cadmium-safe cultivars or low-accumulating cultivars were recommended to combine with immobilization to ensure the crop safety (Hongjiang et al., 2014).

Table 3 Pearson correlation of the relative parameters of brown rice and paddy soil

Index		a[Cd]brown rice	grain yield	pH	b[Cd]HCl	Exc-Cd	CB-Cd	Ox-Cd	OM-Cd	Res-Cd
[Cd]brown rice	A	1	–0.756	–0.925*	0.914*	0.951*	–0.950*	–0.896*	–0.725	–0.851
	B	1	–0.596	–0.889*	0.931*	0.935*	–0.939*	–0.622	–0.828	–0.784
	C	1	0.093	–0.974**	0.970**	0.988**	–0.936*	–0.926*	–0.747	–0.811
	D	1	–0.182	–0.983**	0.966**	0.955*	–0.869	–0.788	–0.791	0.013
	E	1	–0.464	–0.873	0.941*	0.941*	–0.711	0.114	0.638	–0.995**
	F	1	0.305	–0.897*	0.940*	0.964**	–0.781	0.016	0.361	–0.706
pH	A	–0.925*	0.738	1	–0.966**	–0.982**	0.960**	0.992**	0.798	0.662
	B	–0.889*	0.434	1	–0.978**	–0.878	0.912*	0.785	0.773	0.619
	C	–0.974**	–0.095	1	–0.957*	–0.939*	0.840	0.936*	0.812	–0.803
	D	–0.983**	0.164	1	–0.990**	–0.977**	0.942*	0.874	0.808	–0.122
	E	–0.873	0.396	1	–0.983**	–0.930*	0.782	0.362	–0.840	0.821
	F	–0.897*	–0.140	1	–0.990**	–0.961**	0.717	–0.313	–0.431	0.842
[Cd]HCl	A	0.914*	–0.880*	–0.966**	1	0.992**	–0.984**	–0.945*	–0.842	–0.767
	B	0.931*	–0.529	–0.978**	1	0.856	–0.899*	–0.659	–0.859	–0.617
	C	0.970**	0.012	–0.957*	1	0.972**	–0.912*	–0.987**	–0.861	0.109
	D	0.966**	–0.264	–0.990**	1	0.953*	–0.941*	–0.893*	–0.855	0.187
	E	0.941*	–0.460	–0.983**	1	0.946*	–0.734	–0.221	0.817	–0.902*
	F	0.940*	0.140	–0.990**	1	0.969**	–0.705	0.259	0.488	–0.860
Exc–Cd	A	0.951*	–0.833	–0.982**	0.992**	1	–0.992**	–0.965**	–0.811	–0.779
	B	0.935*	–0.567	–0.878	0.856	1	–0.919*	–0.762	–0.728	–0.897*
	C	0.988**	0.142	–0.939*	0.972**	1	–0.952*	–0.923*	–0.759	0.112
	D	0.955*	–0.179	–0.977**	0.953*	1	–0.908*	–0.807	–0.668	–0.029
	E	0.941*	–0.220	–0.930*	0.946*	1	–0.859	–0.104	0.602	–0.919*
	F	0.964**	0.233	–0.961**	0.969**	1	–0.856	0.198	0.323	–0.741
CB–Cd	A	–0.950*	0.830	0.960**	–0.984**	–0.992**	1	0.950*	0.747	0.829
	B	–0.939*	0.307	0.912*	–0.899*	–0.919*	1	0.830	0.626	0.668
	C	–0.936*	0.072	0.840	–0.912*	–0.952*	1	0.857	0.592	–0.823
	D	–0.869	0.025	0.942*	–0.941*	–0.908*	1	0.978**	0.819	–0.387
	E	–0.711	0.067	0.782	–0.734	–0.859	1	0.105	–0.436	0.686
	F	–0.781	–0.321	0.717	–0.705	–0.856	1	–0.127	0.085	0.361

*. Correlation is significant at the 0.05 level (2-tailed).**. Correlation is significant at the 0.01 level (2-tailed).

a $[Cd]_{brown\ rice}$: Cd contents of brown rice; *b* $[Cd]_{HCl}$: 0.025 M HCl extractable Cd content of paddy soil

Sequential extraction was used to investigate the distribution of Cd species before and after the application of sepiolite. Sepiolite shifted Cd distribution from Exc-Cd fraction to CB-Cd and Res-Cd fractions.

The Pearson correlation of the relative parameters such as Cd contents of brown rice, pH, HCl extractable Cd content, and tessier extraction contents were listed in Table 3. Cd contents of brown rice were negatively correlated with pH and positively correlated with 0.025 mol/L HCl extractable Cd contents and Exc-Cd fraction of paddy soil. 0.025 mol/L HCl extractable Cd contents were negatively correlated with CB-Cd fraction and positively correlated with Exc-Cd fraction of paddy soil. Exchangeable Cd contents were negatively correlated with pH and CB-Cd fraction of paddy soil. Carbonate binding Cd contents were positively correlated with pH and negatively correlated with 0.025 mol/L HCl extractable Cd fraction of paddy soil.

2.4 Summary

In-situ large area demonstrations in different paddy fields have certified the universal applicability of immobilization effect of sepiolite on cadmium pollution in acid paddy soil. Sepiolite can significantly decrease Cd contents of brown rice of various cultivars, and reduce the bioavailability of Cd in paddy soil. The recommended application dosage was 1.0-2.0 kg/m^2 based on the immobilization effect and total cost. The combined remediation methods taking the advantage of the immobilization of sepiolite and certain low-accumulating rice cultivars were recommended for food safe and sustainable agriculture as shown in Fig. 5.

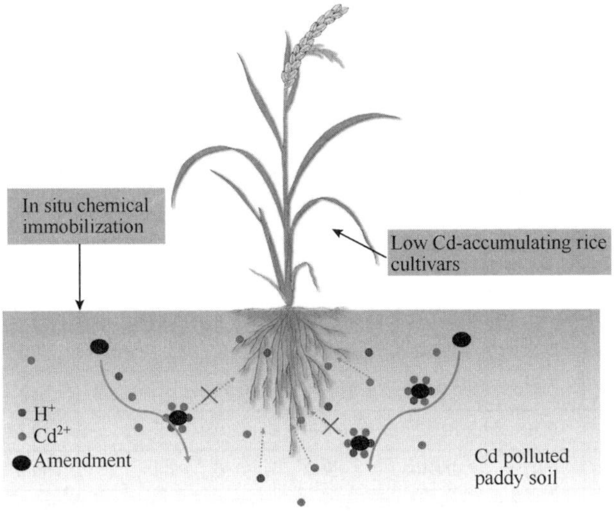

Fig. 5 Schematic of the combination of chemical immobilization and low Cd-accumulating rice cultivar (Liang et al., 2014)

3. Immobilization mechanism

Some mechanisms for the decrease of phytoavailability of heavy metal in amended soils include chemical and physiological processes. The chemical processes mainly involve chemical immobilization due to pH increases (Zhu et al., 2012) and sorption effects. In order to identify the chemical interaction between Cd^{2+} and sepiolite, sorption experiments of Cd^{2+} were carried out. Sepiolite sample was dispersed in $Cd(NO_3)_2$ solutions. The sorption product SEP-Cd was characterized by X-ray diffraction (XRD) and X-ray photoelectron spectroscopy (XPS).

The XRD patterns of sepiolite confirmed that the composition of natural sepiolite was sepiolite, calcite, augite, talc and SiO_2. For SEP-Cd, the structures of sepiolite were maintained but new phases for $Cd(OH)_2$ and $CdCO_3$ were found, indicating the precipitation mechanism.

Cd $3d_{5/2}$ peak could be divided into two separate peaks. The peak at 405.3 eV corresponds to either $Cd(OH)_2$ (Moulder and Chastain, 1995) or $CdCO_3$ (Hammond et al., 1975). Three characteristic peaks were separated for C1s spectrum in the SEP-Cd sample. The peak at 289.6 eV represents the peak value of $CdCO_3$ (Hammond et al., 1975). Sepiolite, containing $CaCO_3$ as a pH regulating immobilization agent, has the ability to immobilize Cd in soil by precipitation in the form of $Cd(OH)_2$ or $CdCO_3$. The specific adsorption mechanism of Cd on sepiolite is due to the complex composition of natural sepiolite. It can also stabilize heavy metals by surface complexation with surface functional groups.

4. Highlights

Large area demonstrations have certified the universal applicability of immobilization effect of sepiolite on Cd pollution in acid paddy soil. Sepiolite significantly reduced HCl extractable Cd contents of paddy soil, resulting in a notable decrease in the Cd concentration in brown rice. The immobilization effect can last at least two years. The recommended application dosage was 1.0-2.0 kg/m^2 based on the immobilization effect and total cost.

The combined remediation methods taking the advantage of the immobilization of sepiolite and certain low-accumulating rice cultivars were recommended for food safe and sustainable agriculture.

References

Aziz R, Rafiq M T, Li T, et al. 2015. Uptake of Cadmium by Rice Grown on Contaminated Soils and Its Bioavailability/Toxicity in Human Cell Lines (Caco-2/HL-7702). Journal of Agricultural and Food Chemistry, 63 (13): 3599-3608.

Bian R, Joseph S, Cui L, et al. 2014. A three-year experiment confirms continuous immobilization of cadmium and lead in contaminated paddy field with biochar amendment. Journal of Hazardous Materials, 272: 121-128.

Bolan N S, Makino T, Kunhikrishnan A, et al. 2013. Cadmium Contamination and Its Risk Management in Rice Ecosystems.//Donald LS. Advances in Agronomy. 119. Amsterdam: Academic Press: 183-273.

Cao F, Wang R, Cheng W, et al. 2014. Genotypic and environmental variation in cadmium, chromium, lead and copper in rice and approaches for reducing the accumulation. Science of The Total Environment, 496: 275-281.

Gu H H, Qiu H, Tian T, et al. 2011. Mitigation effects of silicon rich amendments on heavy metal accumulation in rice (*Oryza sativa* L.) planted on multi-metal contaminated acidic soil. Chemosphere, 83 (9): 1234-1240.

Hammond J S, Gaarenstroom SW, Winograd N. 1975. X-ray photoelectron spectroscopic studies of cadmium-and silver-oxygen surfaces. Analytical Chemistry, 47 (13): 2193-2199.

Hongjiang Z, Xizhou Z, Tingxuan L, et al. 2014. Variation of cadmium uptake, translocation among rice lines and detecting for potential cadmium-safe cultivars. Environmental Earth Sciences, 71 (1): 277-286.

Kim K R, Kim J G, Park J S, et al. 2012. Immobilizer-assisted management of metal-contaminated agricultural soils for safer food production. Journal of Environmental Management, 102: 88-95.

Lee S H, Lee J S, Jeong Choi Y, et al. 2009. In situ stabilization of cadmium-, lead-, and zinc-contaminated soil using various amendments. Chemosphere, 77 (8): 1069-1075.

Liang X, Han J, Xu Y, et al. 2014. In situ field-scale remediation of Cd polluted paddy soil using sepiolite and palygorskite. Geoderma, 235-236: 9-18.

Liang X, Xu Y, Xu Y, et al. 2016. Two-year stability of immobilization effect of sepiolite on Cd contaminants in paddy soil. Environmental Science and Pollution Research, doi: 10.1007/s11356-016-6466-y.

Lombi E, Zhao F J, Zhang G, et al. 2002. In situ fixation of metals in soils using bauxite residue: chemical assessment. Environmental Pollution, 118 (3): 435-443.

Moulder J F, Chastain J. 1995. Handbook of X-Ray Photoelectron Spectroscopy. Waltham: Physical Electronics Division, Perkin-Elmer Corp.

Pehlivan E, Özkan A M, Dinç S, et al. 2009. Adsorption of $Cu^{2+}$and Pb^{2+}ion on dolomite powder. Journal of Hazardous Materials, 167 (1-3): 1044-1049.

Puga A P, Abreu C A, Melo L C A, et al. 2015. Biochar application to a contaminated soil reduces the availability and plant uptake of zinc, lead and cadmium. Journal of Environmental Management, 159: 86-93.

Selim H M, Michael C A. 2001. Sorption and Release of Heavy Metals in SoilsHeavy Metals Release in Soils. Boca Raton: CRC Press: 1-29.

Zhang Z Y, Meng J, Dang S, et al. 2014. Effect of Biochar on Relieving Cadmium Stress and Reducing Accumulation in Super japonica Rice. Journal of Integrative Agriculture, 13 (3): 547-553.

Zhao F J, Ma Y, Zhu Y G, et al. 2015. Soil Contamination in China: Current Status and Mitigation Strategies. Environmental Science & Technology, 49 (2): 750-759.

Zhu Q H, Huang D Y, Liu S L, et al. 2012. Flooding-enhanced immobilization effect of sepiolite on cadmium in paddy soil. Journal of Soils and Sediments, 12 (2): 169-177.

Systematic Selection and Identification of Vegetable Cultivars with Low Heavy Metal Accumulation and for Food Safety

Weitao LIU[1], Qixing ZHOU[1,*], Xuhui LI[2]

[1]Key Laboratory of Pollution Processes and Environmental Criteria, Ministry of Education/Tianjin Key Laboratory of Urban Ecology Environmental Remediation and Pollution Control, College of Environmental Science and Engineering, Nankai University

[2]Key Laboratory of Environment Change & Water-land Pollution Control, College of Environment and Planning, Henan University

[*]Corresponding author

Abstract

It becomes increasingly concern in how to produce safe vegetable crops as human foods in the slightly and moderately contaminated agriculture soils, especially in China, a country with the largest population in the world. In recent years, the selection and breeding of low accumulating cultivars (LACs), i.e. cultivars accumulate specific pollutants at a level low enough for safe consumption even when grown in contaminated soil, has attracted widespread attention and been believed as an alternative choice for remediation of heavy metal contaminated soils to ensure food safety. Thus, we summarized our studies on selection and identification of heavy metal LACs among vegetables during the last decade so as to provide possible reference value for the future studies. Moreover, the further research directions and limitations are also discussed in the chapter.

Keywords: heavy metal pollution, low accumulating cultivars (LACs), food safety, selection and identification

1. Introduction

The term heavy metal refers to a series of metals and metalloids with atomic mass over 20 and specific gravity above 5, which can be toxic to both plants and animals even at very low concentrations (Rascio and Navari-Izzo, 2011). Activities such as mining and smelting of metal ores, industrial emissions and applications of insecticides and fertilizers have all contributed to elevated levels of heavy metals in the environment (Yoon et al., 2006). With the rapid development of industry and extensive application of agrochemicals in recent years, heavy metal pollution in soils becomes an increasingly serious global environmental issue (Krämer, 2005; Liu et al., 2013).

According to the Bulletin of National Soil Pollution Survey issued in April 2014, 16.10% of the soil samples and 19.40% for the agricultural soils are contaminated based on the Grade II limits of the Environmental Quality Standard for Soils (GB 15618—1995), mainly with heavy metals and metalloids (Zhang et al., 2015; Zhao et al., 2015). Among the heavy metals and metalloids, cadmium (Cd) ranks the first in the percentage of soil samples (7.0%) exceeding the limit and 1.5% of the soil samples is contaminated by lead (Pb). Cd and Pb are two of the most toxic metals which can be accumulated by agricultural crops, inhibiting plant growth and mineral nutrition, and can cause diseases in animals and human beings (Liu et al., 2010a, 2010b).

Soil is an important source for heavy metals in crops and vegetables since the plants' roots can absorb these pollutants from soil, and transfer these pollutants to their edible parts, which poses a potential risk to human health via food chain (Liu et al., 2011). There are several pathways available to reduce heavy metal risks to human health. One is to remediate and/or to restore heavy metal-contaminated soils, and hence to ensure food safety (Yu et al., 2006). Conventional methods include *ex situ* excavation and landfill of the top contaminated soils, which are highly effective and clear-cutting but is too expensive (Zhou and Song, 2004). In recent ten years, phytoremediation has been widely considered as a novel low-cost technology to remediate metal-or metalloid-contaminated soils (Chaney et al., 1997). However, in general, its application is still very limited due to low biomass of hyperaccumulators, unavailability of the suitable plant species and unprofitable production during remediation (Pilon-Smits and Freeman, 2006; Liu et al., 2010b).

The other way is to select and breed low accumulating cultivars (LACs), i.e. cultivars accumulate specific pollutants at a level low enough for safe consumption even when grown in contaminated soil. The concept of LACs is grounded on the basis of prior literatures, which have shown that significant difference occurs in the uptake and distribution of trace elements among plant species and among cultivars within species (Liu et al., 2007; Zhu et al., 2007; Grant et al., 2008). As an alternative choice, selection and breeding of heavy metal LACs has attracted much more attentions and interests (Grant et al., 2008). It has been reported that the selection and breeding of Cd LACs had been successfully undertaken in sunflower and durum wheat (Penner et al., 1995; Li et al., 1997). For the last decade, we pay much more attention to screen out LACs among different vegetable cultivars and have successfully identify some LACs of Cd and Pb, which are suitable to be planted in low Cd or Pb polluted soils without harm to human health (Li et al., 2009, 2010a, 2010b, 2012, 2016).

2. Selection of LACs from vegetables

2.1 Selection of Cd and Pb LACs from Chinese cabbage

Chinese cabbage, an important leafy vegetable with 1500-year history in China, is now widely cultivated in the world. In China, the cultivated area of Chinese cabbage in 2005 alone was up to 2.67 million ha, with 1.38 billion (10^9) ton of production, which accounted for 15% and 19% of the total cultivated area and total yield of all vegetable crops, respectively (Chinese Ministry of Agriculture, 2007). It has been estimated that vegetables contribute to≥70% of Cd uptake by humans (Ryan et al., 1982). Therefore, the screening of LACs from vegetables, especially from Chinese cabbage, is an important health issue.

It is prerequisite to develop an operable methodology for evaluating and identifying LACs, thus providing genetic materials for breeding (Zeng et al., 2008). However, most of the previous studies reported on the selection of LACs were conducted in hydroponic culture or by the addition of heavy metal to non-polluted soils. The screening and evaluation of LACs in one soil or hydroponic culture for breeding purposes may not be sufficient, because the uptake of heavy metal in plants is considerably affected by various soil factors, e.g., pH, CEC, clay content, total carbon content, concentrations of heavy metals including Cd, etc. (Li and Chaney, 1995). In addition, Cd in soils occurs in complicated forms because of its association with a number of physicochemical forms that in turn influence its bioavailability. Therefore, our systemic screening methodology i.e. pot-culture screening combined with field-culture screening may provide more accurate and reliable information.

2.1.1 Site description and Safety standard

A pot-culture experiment was carried out under open field conditions in Shenyang Station of Experimental Ecology, Chinese Academy of Sciences (41°31'N and 123°41'E), which is located at the south of Shenyang City, Liaoning Province, China. The average annual temperature of the site is about 5℃-9℃ and the annual precipitation is 650-700 mm. The frostless period lasts 127-164 days per year. The tested soil was collected from an agricultural area in the station and classified as meadow burozem. After the pot-culture experiment, a field-culture experiment was carried out on a farm (42°38'N and 123°55'E) in the Dongling District, Shenyang City, Liaoning Province, China. The meteorological conditions in the field experimental site are basically similar to those in the pot-culture site.

The National Food Hygiene Standard of China (NFHSC) was employed to measure the safety of consuming the shoot of Chinese cabbage grown in heavy metal contaminated soils. The maximum permissible concentration (MPC) of Cd and Pb (GB15201—94) in vegetables for safe consumption is 0.05 mg/kg and 0.20 mg/kg fresh weight (FW), respectively. Since cabbage shoots contain 90% water (the average value in this study), shoot Cd and Pb concentration with less than 0.50 and 2.0 mg/kg DW, respectively, may be considered as a standard for identifying LACs among different cabbage cultivars.

2.1.2 Cd and Pb accumulation in the pot-culture experiment

Significant differences ($p<0.05$) in shoot Cd concentrations were observed among 40 Chinese cabbage cultivars, which were tested under the three Cd treatments (1.0, 2.5 and 5.0 mg/kg) (Fig. 1). The Cd concentrations in vegetable shoots ranged from 0.22 to 2.46, from 0.90 to 14.1 and from 2.14 to 22.2, with average values 0.88, 4.45 and 7.76 mg/kg, respectively. Under treatment T_1 (Cd=1.0 mg/kg), 60% (24/40) cabbage samples examined in this study exceeded 0.50 mg/kg DW.

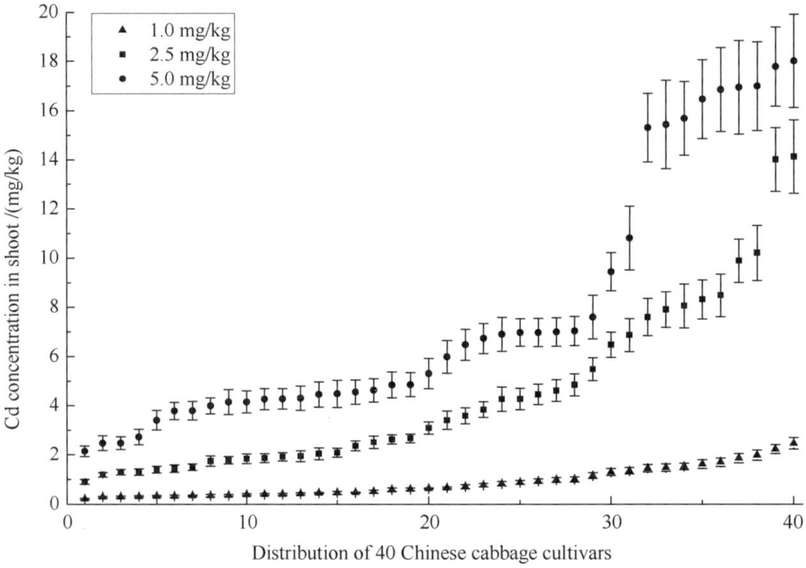

Fig. 1 Cd concentration in the shoots of 40 Chinese cabbage cultivars under three Cd treatments

Significant ($p<0.05$) differences in shoot Pb concentration were observed among the 30 tested cultivars under Pb treatments 500 and 1500 mg/kg (Fig. 2), ranging from 0.52 to 8.68 and from 1.86 to 16.2, with the mean of 3.01 and 6.87 mg/kg DW, respectively. The results indicated that the Pb concentrations in shoots of the Chinese cabbage cultivars increased with the increased Pb concentrations in the soils. Under Pb treatments 500 and 1500 mg/kg, Pb concentrations in 60% (18/30) and 86.7% (26/30) of the tested cabbage cultivars exceeded 2.0 mg/kg DW, respectively

2.1.3 Enrichment factors of different cabbage cultivars in the pot-culture experiment

Enrichment factors (EFs) can be used to evaluate the ability of plants to accumulate heavy metals. It is calculated as the ratio of the Cd concentration in shoots to the Cd concentration in soils (Chen et al., 2004). Overall, the average EFs of various cabbage cultivars increased with the increased Cd concentration in soils (Fig. 3). The lowest EFs under three Cd treatments

(1.0, 2.5 and 5.0 mg/kg) were found in New Beijing 3, only 0.19, 0.34 and 0.61, respectively. The highest EFs under three treatments were found in Suancaiwang and/or Beijingxiaoza 56, ranging from 3.46 to 3.85. Under the three Cd treatments, the EFs in 9 cabbage cultivars were lower than 1.0, including New Beijing 3, Saixin 5, Fengyuanxin 3, Shuishiying 91-12, Liaodaqiukang, Qiukangduobao 3, Qiuguan, Beijing-12 and Beijingdabaoxin (Fig. 3).

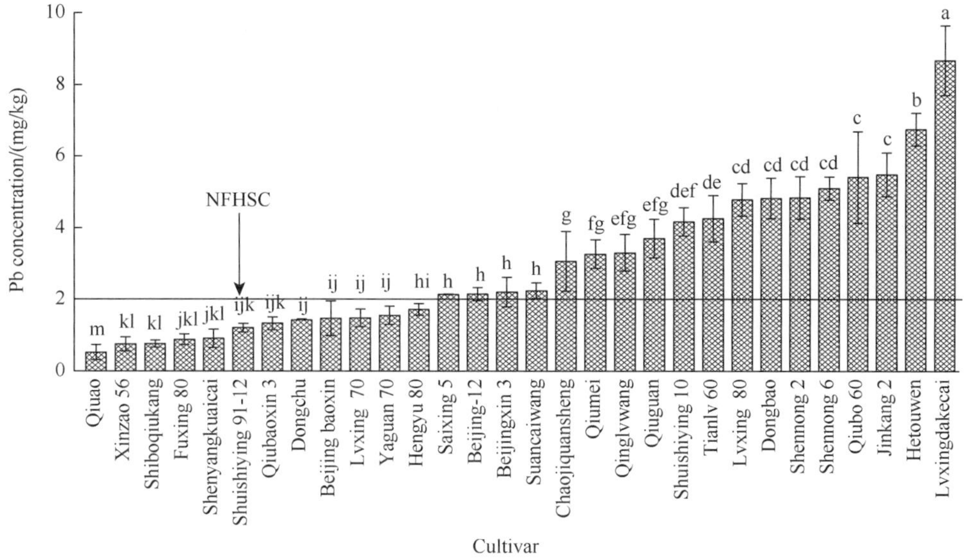

(a) Pb treatments 500

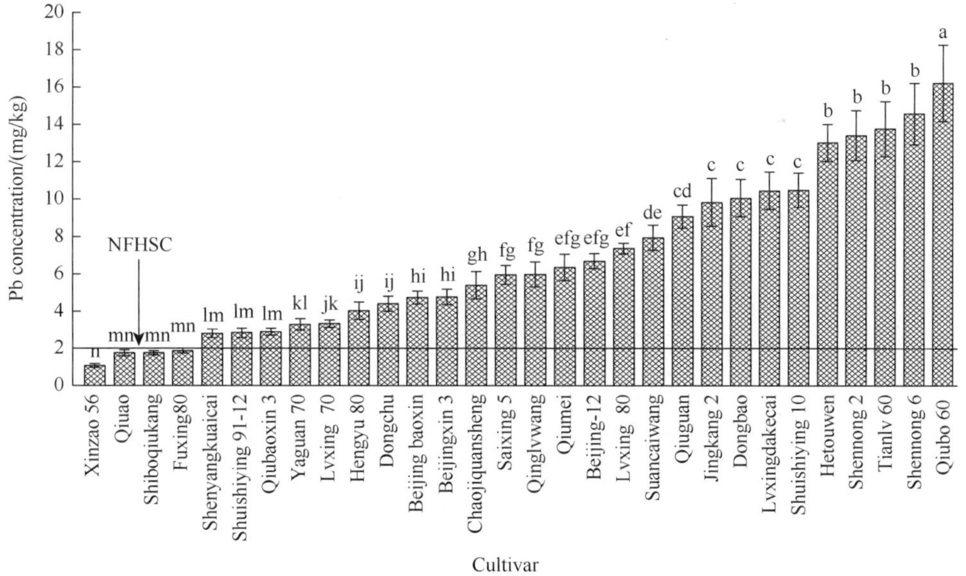

(b) Pb treatments 500 mg/kg

Fig. 2 Shoot Pb concentrations under Pb treatments 500 and 1500 mg/kg in the pot-culture experiment Note: NFSHC, The maximum limitation of Pb according to the National Food Hygiene Standard of China (NFHSC); Different letters on the top of the histogram indicate significant difference at $p<0.05$ level

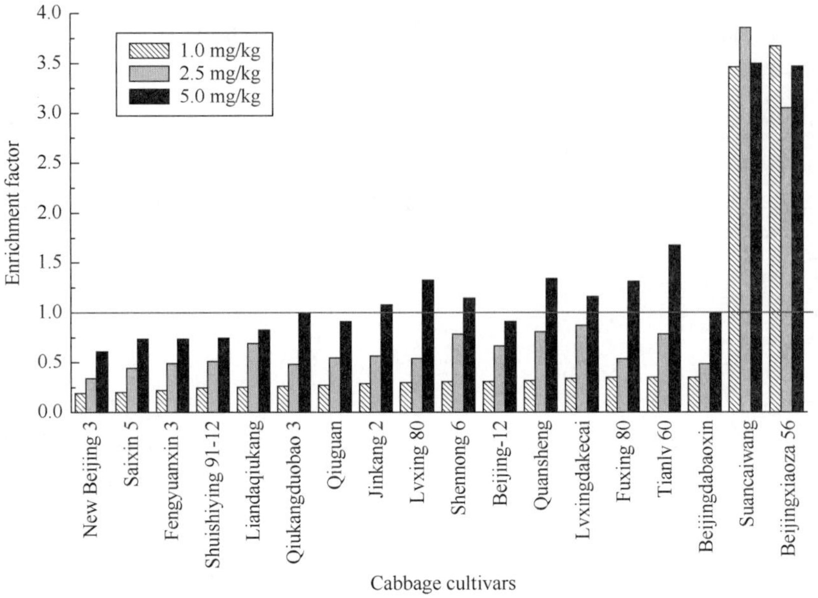

Fig. 3 Average enrichment factor of 18 Chinese cabbage cultivars under three Cd treatments

There is significant ($p<0.05$) difference in the EFs of the selected cabbage cultivars among different cultivars under the Pb treatments 500 and under 1500 mg/kg, ranging from 0.01 to 0.17 and from 0.01 to 0.11, respectively. However, there were not significant differences of EFs between different Pb treatments (500 and 1500 mg/kg). Generally, the EFs decreased with the increased Pb concentration in soils (Table 1).

Table 1 Enrichment factors and translocation factors in 30 Chinese cabbage cultivars

Cultivars	Enrichment factors		Translocation factors	
	500 mg/kg	1500 mg/kg	500 mg/kg	1500 mg/kg
Qiuao	0.001	0.001	0.026	0.012
Xinzao 56	0.002	0.001	0.033	0.015
Shiboqiukang	0.002	0.001	0.034	0.026
Fuxing80	0.002	0.001	0.035	0.027
Shenyangkuaicai	0.002	0.002	0.057	0.029
Shuishiying 91-12	0.002	0.002	0.057	0.029
Qiubaoxin 3	0.003	0.002	0.061	0.039
Dongchu	0.003	0.003	0.061	0.040
Beijing baoxin	0.003	0.003	0.063	0.046
Lvxing 70	0.003	0.002	0.064	0.049
Yaguan 70	0.003	0.002	0.066	0.051
Hengyu 80	0.003	0.003	0.068	0.055
Saixing 5	0.004	0.004	0.074	0.056
Beijing-12	0.004	0.004	0.085	0.063

Continue table

Cultivars	Enrichment factors		Translocation factors	
	500 mg/kg	1500 mg/kg	500 mg/kg	1500 mg/kg
Beijingxin 3	0.004	0.003	0.086	0.074
Suancaiwang	0.004	0.005	0.096	0.076
Chaojiquansheng	0.006	0.004	0.100	0.101
Qiumei	0.007	0.004	0.108	0.105
Qinglvwang	0.007	0.004	0.119	0.140
Qiuguan	0.007	0.006	0.169	0.155
Shuishiying 10	0.008	0.007	0.207	0.163
Tianlv 60	0.009	0.009	0.211	0.175
Lvxing 80	0.010	0.005	0.226	0.180
Dongbao	0.010	0.007	0.260	0.193
Shennong 2	0.010	0.009	0.282	0.196
Shennong 6	0.010	0.011	0.283	0.201
Qiubo 60	0.011	0.010	0.305	0.236
Jingkang 2	0.011	0.007	0.373	0.275
Hetouwen	0.014	0.009	0.422	0.292
Lvxingdakecai	0.017	0.007	0.424	0.316

2.1.4 Translocation factors of different cabbage cultivars in the pot-culture experiment

Translocation factors (TFs) of heavy metals can be used to evaluate the capacity of plants to translocate heavy metals from roots to shoots. It is calculated as the ratio of the Cd concentration in plant shoots to the Cd concentration in plant roots (Liu et al., 2011). Fig. 3 shows the average TFs of the 18 cabbage cultivars under three Cd treatments (1.0, 2.5 and 5.0 mg/kg), ranging from 0.24 to 2.62, from 0.57 to 1.87 and from 0.46 to 1.99, respectively. The lowest TFs under three treatments were found in Shuishiying 91-12 and/or New Beijing 3. Similar to the EFs, the highest FTs were also detected in Suancaiwang and/or Beijingxiaoza 56. The TFs of the 5 cabbage cultivars were lower than 1.0, including New Beijing 3, Saixin 5, Fengyuanxin 3, Shuishiying 91-12 and Liaodaqiukang under all Cd treatments (Fig. 4).

There is significant ($p<0.05$) difference in the TFs of the selected cabbage cultivars among different cultivars under Pb treatments 500 and under 1500 mg/kg, ranging from 0.026 to 0.424 and from 0.012 to 0.316, respectively. It indicated that Pb uptake in Chinese cabbage cultivars was sequestered in their roots, and only a small portion of Pb could be translocated to the aerial parts. The results also showed that the TFs in the selected cultivars decreased with an increase in the Pb concentrations in soils (Table 1). More than 60% of Pb in the selected cultivars distributed in the roots under the Pb treatment 500 mg/kg and there was no significant ($p > 0.05$) difference among 30 cultivars. However, under the Pb treatment 1500 mg/kg, more Pb was accumulated in the shoot compared with that under the Pb treatment 500 mg/kg. There was significant ($p < 0.05$) difference in Pb distribution among the selected cultivars. Under the treatment 1500 mg/kg, about 70% of the total Pb uptake was distributed in the shoot of Qinglvwang, Qiuguan and Dongchu. The phenomenon may indicate that the accumulation of Pb in shoot will increase to some extent with increasing Pb concentration in soils.

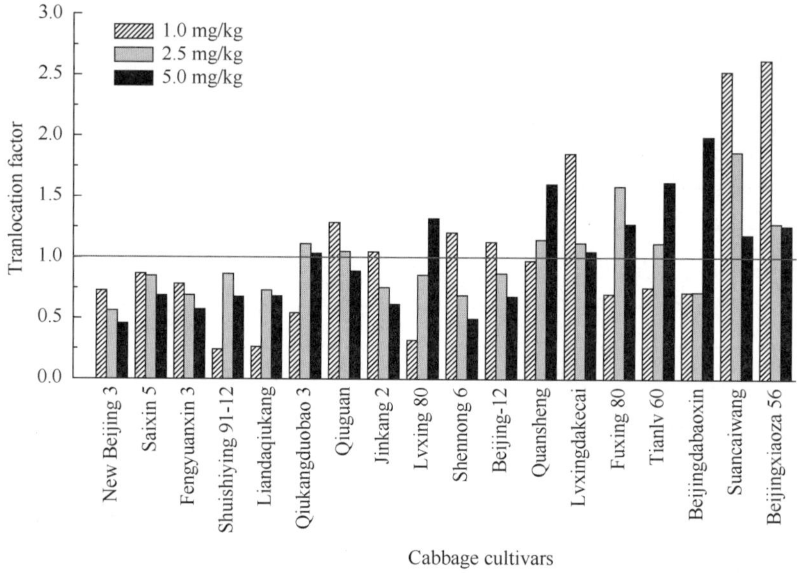

Fig. 4 Average translocation factor of 18 Chinese cabbage cultivars under three Cd treatments

2.1.5 Cd and Pb tolerance in different cabbage cultivars in the pot-culture experiment

Heavy metal tolerance of plants can be measured by their variation in shoot biomass response to heavy metal toxicity (compared with control) (Sun et al., 2009, Wei and Zhou, 2008). Generally, Chinese cabbage cultivars had tolerance to Cd toxicity to some extent. The shoot biomass of 13 cultivars under treatments (1.0, 2.5 and 5.0 mg/kg) did not decrease significantly ($p > 0.05$) compared with that under control (Fig. 5). However, the shoot biomass of Saixin 5, Qiuguan and Qiubo 60 decreased significantly ($p < 0.05$) compared with the control indicating its poor tolerance to Cd stress. On the contrary, Liaodaqiukang, Suancaiwang and Beijingxiaoza 56 had higher tolerance to Cd toxicity.

Similarly, Chinese cabbage cultivars had tolerance to Pb toxicity to some extent. The shoot biomass of 26 cultivars under treatments (500 and 1500 mg/kg) did not decrease significantly ($p > 0.05$) compared with that under control (Fig. 6). However, the shoot biomass of Qinglvwang, Beijingxin 3, Suancaiwang and Xinzao 56 decreased significantly ($p < 0.05$) compared with the control, indicating its poor tolerance to Pb stress.

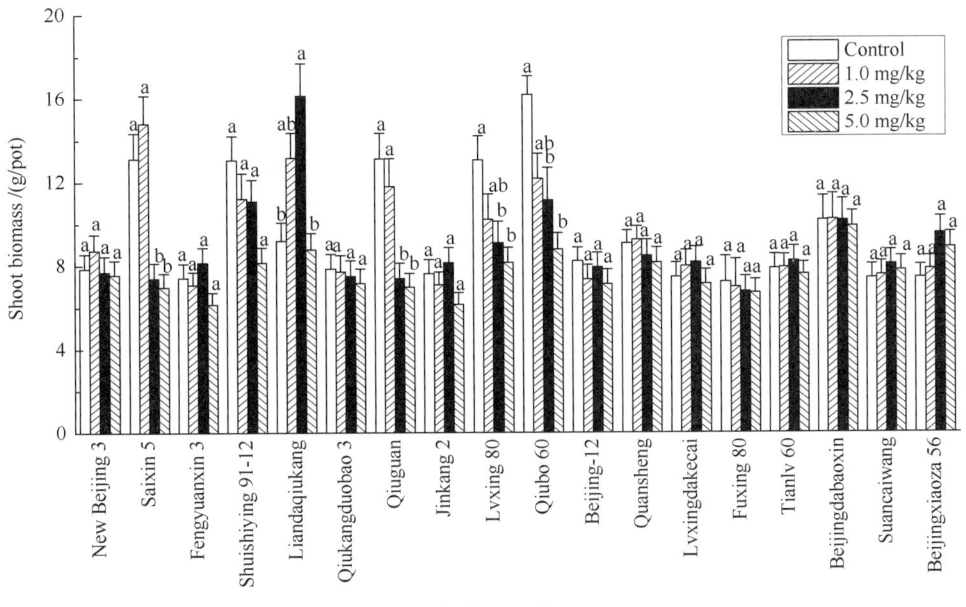

Fig. 5 The shoot biomass of 18 cultivars under different Cd treatments. The same letter on the top of the columns means not significant ($p>0.05$) different between different treatments

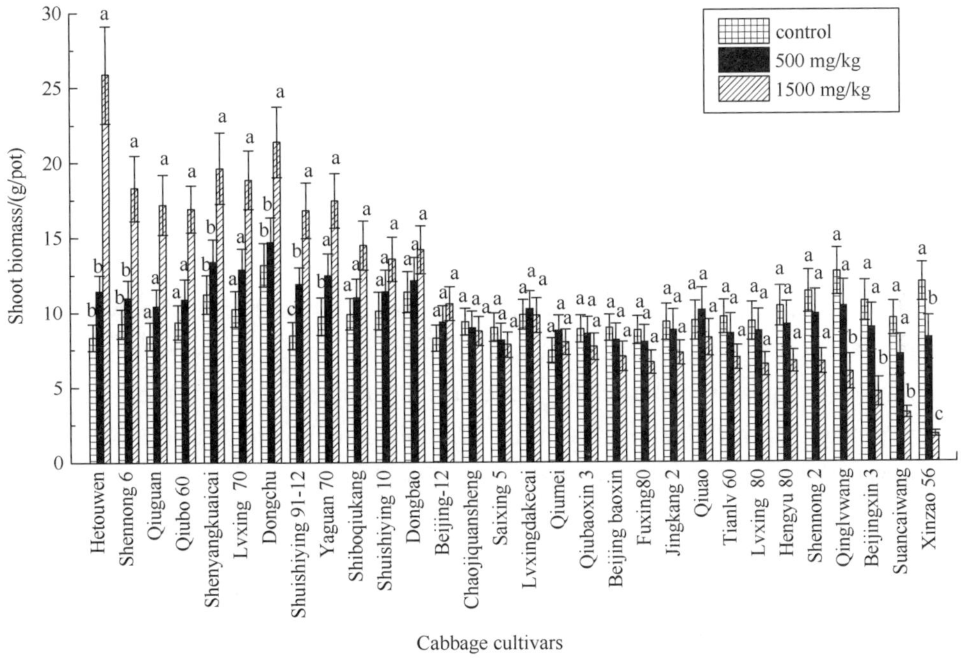

Fig. 6 The shoot biomass of 30 cultivars under three Pb treatments (control, 500 and 1500 mg/kg). The same letter on the top of the columns means not significant ($p>0.05$) different between different Pb treatments

2.1.6 Cd and Pb accumulating characteristics in the field-culture experiment

7 cabbage cultivars were further tested in the field-culture experiment to confirm the

low-Cd accumulation properties in these cultivars. Their Cd concentrations, EFs, and TFs were shown in Fig. 7. The Cd concentrations in the 7 cabbage cultivars ranged from 0.32 to 1.56 mg/kg, with a mean of 0.80 mg/kg. The Cd concentrations in New Beijing 3, Saixin 5 and Fengyuanxin 3 were lower than 0.5 mg/kg, which is similar to the results from the pot-culture experiment. However, Cd concentrations in Shuishiying 91-12 and Liaodaqiukang were 0.66 and 0.84 mg/kg and were significantly different from the results in the pot-culture experiment.

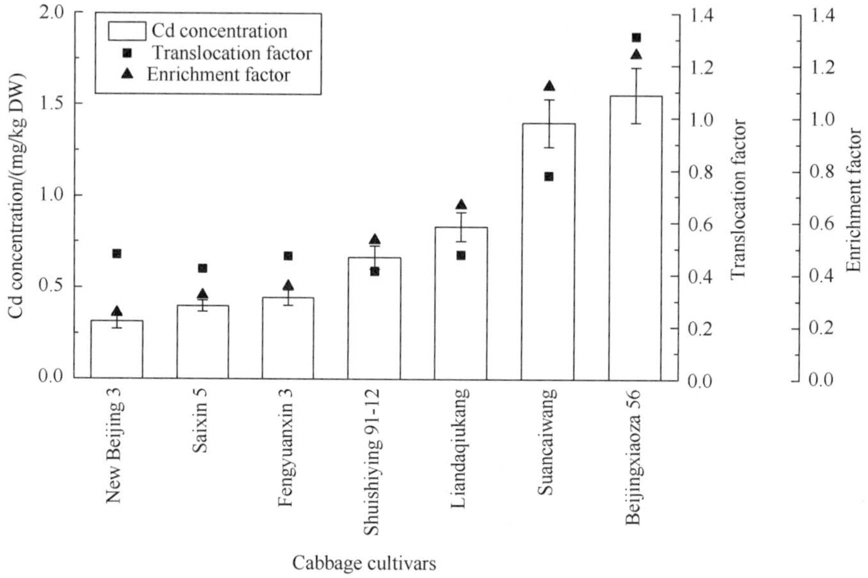

Fig. 7 Cd concentration, translocation factor and enrichment factor in 7 Chinese cabbage cultivars in the field-culture experiment

The Pb concentrations in 12 cultivars ranged from 0.49 to 2.15 mg/kg, with a mean of 0.96 mg/kg. Except for Hengyu 80, Pb concentrations in other cultivars were lower than 2.0 mg/kg, which is similar to the results under the Pb treatment 500 mg mg/kg in the pot-culture experiment. Meanwhile, the TFs in the 12 cabbage cultivars were all lower than 0.15, relatively lower than that in pot-culture experiment (Fig. 8).

2.1.7 Selection of LACs among Chinese cabbage cultivars

Based on the former literatures and our research, we suggest that the perfect LACs of Cd should have these indispensable traits: ①heavy metal contents in their edible parts should be lower than MPC; ②EF$<$1.0; ③TF$<$1.0; and ④they can tolerate Cd when growing in contaminated soils as measured by shoots biomass (Liu et al., 2009). When the four standards were used, New Beijing 3 and Fengyuanxin 3 could be finally confirmed as Cd LACs. Qiuao, Shiboqiukang and Fuxing 80 were selected as Pb LACs, which were safe to consumers if they were cultivated in low-Pb ($<$382.25 mg/kg) contaminated soils.

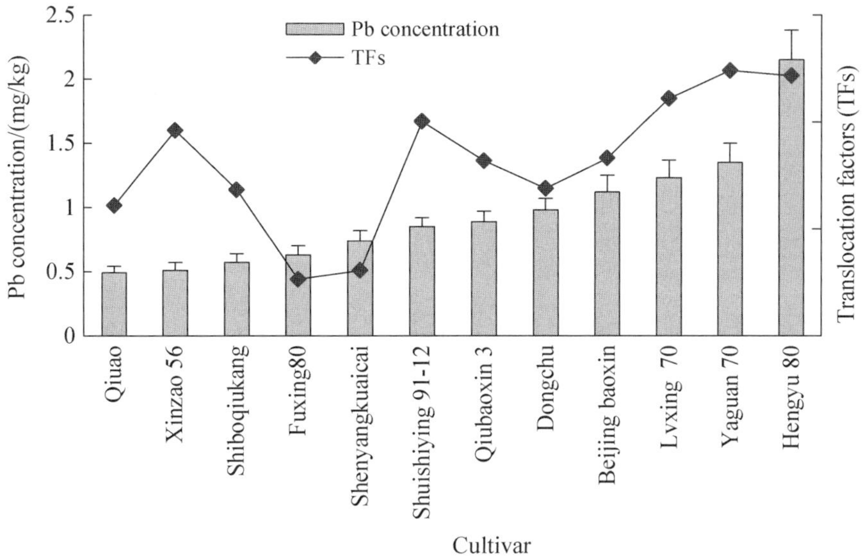

Fig. 8 Pb concentration and Translocation factors in 12 cultivars in the field-culture experiment

2.2 Selection of Cd LACs from welsh onion

With a cultivation history of approximately 2500 years, welsh onion (*Allium fistulosum* L.) is one of the most important vegetables in China and other countries of Northeast Asia, Europe, and Northern America. Its cultivation area and production accounted for approximately 3% of the total vegetables in China (Li et al., 2012). For welsh onion, the MPC of Cd is 0.10 mg/kg FW (fresh weight) according to both the Food Safety Standardization in China (GB 2762—2005) and the Commission Regulation (No. 629/2008).

A change in the concentration of Cd in the pseudostem among various cultivars under different treatments was depicted in Fig. 9. There was a significant difference ($p<0.05$) in the concentration of Cd in the pseudostem observed among 25 tested cultivars. The accumulation of Cd in the pseudostem ranged from 0.08 mg/kg to 0.20 mg/kg, and with an average value of 0.14 mg/kg. All the cultivars tested exceeded the maximum permission concentration (MPC) value under T1 treatment except for No. 7 (Ribentiegancongwang) and No. 16 (Wuyeqi), of which the Cd concentration were 0.08 and 0.10 mg/kg, respectively. Under T2 or T3 treatment, the concentrations of Cd in pseudostems varied from 0.18 to 0.41 mg/kg and from 0.26 to 0.61 mg/kg, with mean value of 0.29 and 0.44 mg/kg, respectively. The Cd concentration in pseudostems of all cultivars exceeded the MPC value under T2 or T3 treatment, which suggested that identification for Cd-excluding cultivars would not ensure the food safety when growing in serious contaminated soils.

The criterion on yield and the quality was also very important for vegetables. The yield of No. 7 cultivar was 45.65±3.40 (mean value±S.D.), 42.76±0.64, 35.73±3.96 and 39.56±2.81 g/pot, under CK, T1, T2 and T3 treatments, respectively. The yield decreased 4.4%-11.6%

than controls when growing in contaminated soils, but no significant difference (p=0.227-0.850, LSD test) with the value of CK was observed. For No. 16 cultivar, the yield was 43.95±4.21, 55.37±3.82, 44.58±5.30 and 47.18±3.65 g/pot under four treatments, respectively. The yield in Cd contaminated soils increased 7.3%-26.0% compared with that in the control group. One way ANOVA test showed that T1 treatment increased the yeild significantly (p=0.046, LSD test), but there was no significant (p=0.542-0.905, LSD test) difference among CK, T2 and T3 treatment.

In the plot experiment carried out in 2010, the Cd concentration in pseudostem of No. 7 cultivar were 0.041±0.003 and 0.054±0.001 mg/kg FW under two different Cd treatments, respectively. For No. 16 cultivar, these data were 0.046±0.002 and 0.066±0.011 mg/kg FW, respectively. In summary, No. 7 and No. 16 cultivars met all the screening criteria of Cd LACs except for some trophic indexes of certain trace elements. Thus, the two cultivars could be identified as Cd LACs.

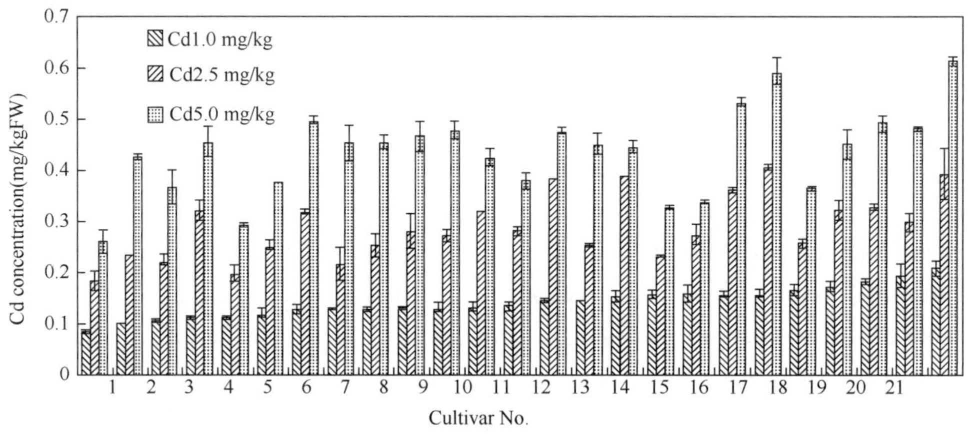

Fig. 9 Cd accumulation in edible parts of different cultivars (mean±S.D.)

3. Discussion

It seemed that the selected Chinese cabbage cultivars had higher tolerance to soil Cd and Pb toxicity. It was observed that the biomass of Chinese cabbage cultivars did not decrease significantly under high Cd and Pb treatment and even increased under low Cd and Pb treatment (Fig. 5, Fig. 6). Similar positive and neutral responses of biomass to Cd stress have also been found in a wide range of species (Yu et al., 2006). An impossible mechanism for the stimulatory effect of Cd may be that metal ions can serve as activators of enzyme (s) in cytokinin metabolism, which accelerates the growth of plants. It was reported that cytokinins can delay senescence while maintaining the chlorophyll and photosynthetic activity of leaf tissues and can take part in the remediation of the effect of some low-dose heavy metals (Nyitrai et al., 2004).

The cultivar variation in the heavy metal accumulation may be associated with several physiological aspects, such as different acquisition capacities of root for heavy metal, variation in the ability to take up heavy metal and translocation and accumulation capacity of heavy metal (Zhu et al., 2007). Herren and Feller (1997) observed that the redistribution of Cd within cereal shoots is highly relevant for the accumulation of Cd in the grain (Herren and Feller, 1997). Similarly, Dunbar et al. (2003) suggested that internal distribution rather than uptake led to the low Cd accumulation in edible parts of plants (Dunbar et al., 2003). But Harris et al. (1981) reported that low uptake rather than internal distribution may affect the genotypic difference (Harris et al., 1981). In our study, we think that low uptake and internal distribution may affect the genotypic difference. However, the factors governing processes of Cd uptake, translocation and accumulation in plants are not yet understood and needed further studies to gain a deeper insight.

Moreover, soil amelioration adding, e.g.organic matter, lime, phosphates, has been used for many years with the aim to improve plant growth and reduce soil phytotoxicity to crops by decreasing mobility and bioavailability of toxic elements (Liu et al., 2010a). The combing of LACs with use of soil amendments may be more effective at reducing heavy metal movement into the food chain than growing LACs alone.

Although cultivar selection is proved to be effective to reduce potential heavy metal accumulation in plants, there are still some constrains in the selection and breeding of LACs. Heavy metal concentrations in plant tissues varied with pollution level in soils, plant genotypes and environment factors (Liu et al., 2009). Moreover, the selection and breeding of LACs is time consuming, expensive in chemical analysis and influenced by both soil and management practices (Grant et al., 2008). Whether the differences in lead uptake among cultivars could be maintained over several crop rotations was another unresolved subject. Therefore, further studies should be carried out to resolve these problems for secure agricultural production.

Acknowledgements

The work was jointly supported by the Ministry of Science and Technology, People's Republic of China as a key project of the hi-tech research and development program (863) of China (grant No. 2006AA06Z386). The authors would also like to thank to the National Natural Science Foundation of China (41471411), the Tianjin Research Program of Application Foundation and Advanced Technology (15JCYBJC22700) and the Program for Changjiang Scholars and Innovative Research Team in University (IRT13024).

References

Chaney R L, Malik M, Li Y M, et al. 1997. Phytoremediation of soil metals. Current Opinion in Biotechnology, 8 (3): 279-284.

Chen Y, Shen Z, Li X. 2004. The use of vetiver grass (*Vetiveria zizanioides*) in the phytoremediation of soils contaminated with heavy metals. Applied Geochemistry, 19 (10): 1553-1565.

Chinese Ministry of Agriculture, 2007. Cultivated area and yield of vegetable crops in China in 2005. Chinese Vegetable 1, 40-41 (in Chinese).

Dunbar K R, McLaughlin M J, Reid R J. 2003. The uptake and partitioning of cadmium in two cultivars of potato (*Solanum tuberosum* L.). Journal of Experimental Botany, 54 (381): 349-354.

Grant C A, Clarke J M, Duguid S, et al. 2008. Selection and breeding of plant cultivars to minimize cadmium accumulation. Science of the Total Environment, 390 (2-3): 301-310.

Harris M R, Harrison S J, Wilson N J, et al. 1981. Varietal differences in trace metal partitioning by six potato cultivars grown on contaminated soil. Proceedings of the international conference on heavy metals in the environment, Edinburgh: CEP Consultants.

Herren T, Feller U. 1997. Transport of cadmium via xylem and phloem in maturing wheat shoots: Comparison with the translocation of zinc, strontium and rubidium. Annals of Botany, 80 (5): 623-628.

Krämer U. 2005. Phytoremediation: novel approaches to cleaning up polluted soils. Current Opinion in Biotechnology, 16 (2): 133-141.

Li X H, Zhou Q X, Sun X Y, et al. 2016. Effects of cadmium on uptake and translocation of nutrient elements in different welsh onion (*Allium fistulosum* L.) cultivars. Food Chemistry, 194: 101-110.

Li X H, Zhou Q X, Wei S H, et al. 2012. Identification of cadmium-excluding welsh onion (*Allium fistulosum* L.) cultivars and their mechanisms of low cadmium accumulation. Environmental Science and Pollution Research, 19 (5): 1773-1780.

Li Y M, Chaney R, Schneiter A, et al. 1997. Screening for low grain cadmium phenotypes in sunflower, durum wheat and flax. Euphytica, 94 (1): 23-30.

Li, Y M, Chaney, R L. 1995. Genotypic variation in kernel cadmium concentration in sunflower germplasm under varying soil conditions. Crop Science Society of America 35, 137-141.

Liu J, Qian M, Cai G, et al. 2007. Uptake and translocation of Cd in different rice cultivars and the relation with Cd accumulation in rice grain. Journal of Hazardous Materials, 143 (1-2): 443-447.

Liu W T, Ni J C and Zhou Q X. 2013. Uptake of heavy metals by trees: Prospects for phytoremediation. Materials Science Forum, 743-744: 768-781.

Liu W T, Zhou Q X, An J, et al. 2010a. Variations in cadmium accumulation among Chinese cabbage cultivars and screening for Cd-safe cultivars. Journal of Hazardous Materials, 173 (1-3): 737-743.

Liu W T, Zhou Q X, Sun Y B, et al. 2009. Identification of Chinese cabbage genotypes with low cadmium accumulation for food safety. Environmental Pollution, 157 (6): 1961-1967.

Liu W T, Zhou Q X, Zhang Y L, et al. 2010b. Lead accumulation in different Chinese cabbage cultivars and screening for pollution-safe cultivars. Journal of Environmental Management, 91 (3): 781-788.

Liu W T, Zhou Q X, Zhang Z N, et al. 2011. Evaluation of cadmium phytoremediation potential in Chinese cabbage cultivars. Journal of Agricultural and Food Chemistry, 59 (15): 8324-8330.

Nyitrai P, Bóka K, Gáspár L, et al. 2004. Rejuvenation of ageing bean leaves under the effect of low-dose stressors. Plant Biology, 6 (6): 708-714.

Penner G A, Bezte L J, Leisle D, et al. 1995. Identification of RAPD markers linked to a gene governing cadmium uptake in durum wheat. Genome, 38 (3): 543-547.

Pilon-Smits E A H, Freeman J L. 2006. Environmental cleanup using plants: biotechnological advances and ecological considerations. Frontiers in Ecology and the Environment, 4 (4): 203-210.

Rascio N, Navari-Izzo F. 2011. Heavy metal hyperaccumulating plants: How and why do they do it?And what makes them so interesting?Plant Science, 180 (2): 169-181.

Ryan J A, Pahren H R, Lucas J B. 1982. Controlling cadmium in the human chain: A Review and rationale based on health effects. Environmental Research, 28: 251-302.

Sun Y B, Zhou Q X, Wang L, et al. 2009. Cadmium tolerance and accumulation characteristics of *Bidens pilosa* L. as a potential Cd-hyperaccumulator. Journal of Hazardous Materials, 161 (2-3): 808-814.

Wei S H, Zhou Q X. 2008. Screen of Chinese weed species for cadmium tolerance and accumulation characteristics. International Journal of Phytoremediation, 10 (6): 584-597.

Yoon J, Cao X D, Zhou Q X, et al. 2006. Accumulation of Pb, Cu, and Zn in native plants growing on a contaminated Florida site. Science of the Total Environment, 368 (2-3): 456-464.

Yu H, Wang J L, Fang W, et al. 2006. Cadmium accumulation in different rice cultivars and screening for pollution-safe cultivars of rice. Science of the Total Environment, 370 (2-3): 302-309.

Zeng F R, Mao Y, Cheng W D, et al. 2008. Genotypic and environmental variation in chromium, cadmium and lead concentrations in rice. Environmental Pollution, 153 (2): 309-314.

Zhang X, Zhong T, Liu L, et al. 2015. Impact of Soil Heavy Metal Pollution on Food Safety in China. Plos One, 10 (8): e0135182.

Zhao F J, Ma Y, Zhu Y G, et al. 2015. Soil Contamination in China: Current Status and Mitigation Strategies. Environmental Science and Technology, 49 (2): 750-759.

Zhou Q X, Song Y F. 2004. Remediation of contaminated soils: principles and methods. Beijing, China: Science Press (in Chinese).

Zhu Y, Yu H, Wang J L, et al. 2007. Heavy metal accumulations of 24 asparagus bean cultivars grown in soil contaminated with Cd alone and with multiple metals (Cd, Pb, and Zn). Journal of Agricultural and Food Chemistry, 55 (3): 1045-1052.

Strategies to Enable the Safe Use of Cadmium-Contaminated Paddy Soils in Southern China

Hanhua ZHU, Chao XU, Qihong ZHU, Daoyou HUANG*

Key Laboratory of Agro-ecological Processes in Subtropical Region,
Institute of Subtropical Agriculture, Chinese Academy of Sciences

*Corresponding author

1. Introduction

With the rapid progression of industrialization, urbanization, and modern agriculture, large amounts of pollutants, especially heavy metals, are entering the soil in China's agricultural areas. According to the survey by the Ministry of Environmental Protection and the Ministry of Land Resources, 19.4% for the agricultural soils are contaminated based on China's soil environmental quality limits, cadmium being the primary contaminant. The total area of agricultural soils contaminated by cadmium is 10 million ha (7% of China's cultivated land), and is mostly distributed in the southern rice region (Ministry of Environmental Protection P. R. C. and Ministry of Land and Resources P. R. C. 2014). Cadmium in the soil might enter the food chain relatively easily by absorption into agricultural products and therefore poses a considerable threat to human health (Toppi and Gabbrielli, 1999; Chen et al., 2016). An investigation from the Ministry of Agriculture indicated that over 10% of brown rice are Cd-contaminated (Li and Xu, 2015).

More recently, as a result of the ever-increasing problem of agricultural Cd contamination, researchers in China and abroad have been studying the effectiveness of techniques such as agronomic regulation, *in-situ* immobilization, soil washing and flushing, and electrokinetics (Li and Xu, 2015; Zhu et al., 2012). The total area polluted by Cd is large and widely distributed, with varying degrees of pollution. In order to guarantee the security of the food supply and the quality of agricultural products of China, while also ensuring that the agricultural production on 1.8 billion ha of cultivated land does not simultaneously and abruptly stop, the only feasible way is to treat the issue of Cd contamination while maintaining production. In response to the seriousness of the problem of Cd contamination in China's southern rice paddies, our research team proposed a methodological philosophy of "clarify the problem, treat locally, strategize by zone, and proceed with caution". With this philosophy in mind, a system of techniques developed to treat the Cd-contaminated paddy soils while maintaining agricultural production. Thus, the aim of this study was to develop a model for the safe production of rice crops in slightly Cd-contaminated paddy soils in China.

To achieve this aim, our study employed and investigated several methods in slightly contaminated paddies, such as the use of low-cadmium rice varieties, moisture management in the paddies, and *in-situ* immobilization. The aim of these methods was to ensure the quality and safety of agricultural products without disrupting planting models.

2. Cultivation of low-cadmium rice varieties

Previous research has demonstrated significant differences in Cd accumulation among different rice varieties (Chen et al., 2016). Shi et al. (2009) planted 110 varieties of hybrid and super hybrid rice in slightly a Cd-contaminated paddy soil. Their research showed that Cd accumulation can differ by a factor of 14 among different rice varieties. Cao et al. (2014) studied Cd accumulation in 158 newly-bred rice varieties. Their results showed that hybrid rice varieties are much more capable than conventional rice in terms of cadmium accumulation. However, there is a lack of data regarding which low-cadmium rice varieties are suitable for large-scale planting in the Cd-contaminated paddy soils of southern China. The present study collected a total of 473 rice varieties commonly grown in southern China (Table 1). All varieties were planted in a soil with a Cd concentration of 1.78 mg/kg and a pH of 4.72 to compare and analyze their Cd accumulation abilities.

Table 1 Sources of rice varieties used to test Cd accumulation

Province	Number of varieties
Hunan	212
Fujian	21
Jiangxi	29
Yunnan, Guangxi	96
Zhejiang	67
Sichuan, Chongqing	48
Total	473

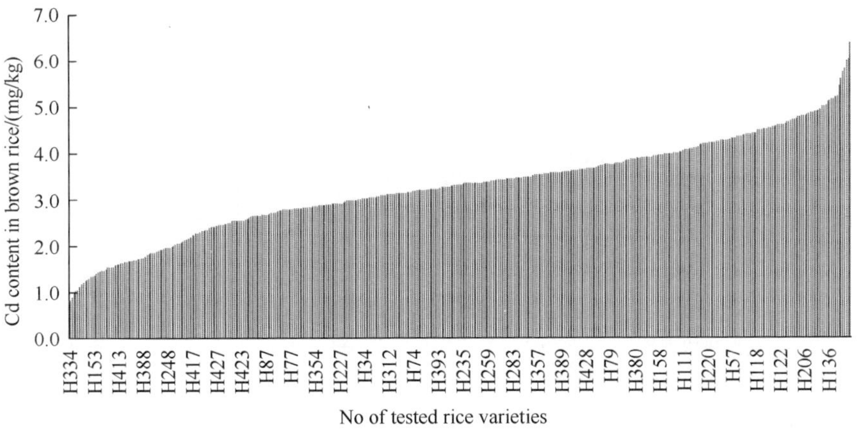

Fig. 1 Abilities of rice varieties from southern China to accumulate Cd

Our results indicated that there was a large variation in Cd accumulation among the 473 popular rice varieties investigated in this study (Fig. 1). Brown rice variety LIANG YOU 616 had the highest cadmium content, reaching 6.39 mg/kg, 7.6 times higher than variety JIN YOU 268, which had the lowest cadmium content among the varieties tested. And rice varieties with the low Cd accumulation were JIN YOU 268, XIANG ZAO XIAN number 32, ZHU LIANG YOU 189, ZHU LIANG YOU number 1, CHUAN NONG YOU 528, SHUO FENG number 2, JIN YOU 433, Y LIANG YOU 1998, YOU I651, and ZHU LIANG YOU 06. The Cd contents of the brown rice in these varieties were in the range of 0.84-1.29 mg/kg. These varieties could be used to mitigate the threat of Cd contamination in the paddy soils of southern China.

3. Water management in paddies

Moisture conditions in paddy soils affect the soil's redox potential. Moisture affects the pH of the soil, as well as the presence of iron and manganese oxides, thereby affecting the phytoavailability of the Cd in the soil. (Zhu et al., 2012; Li and Xu, 2015). Therefore, in-paddy testing was carried out in soil with Cd contents of 0.37 mg/kg, 0.63 mg/kg, and 0.88 mg/kg, with pH levels ranging from 5.3 to 5.5. Our results showed that, compared to conventional water management, the use of flooding during the entire growth period could, to a certain extent, reduce the cadmium content of the brown rice (Fig. 2). The Cd content of early rice was reduced by 16.9%-30.4%, while the cadmium content of later rice was reduced by 13.9%-23.3%. In soil with a Cd content of 0.37 mg/kg, this technique basically guaranteed that the Cd content of early and late rice could be kept to below the national contaminant limits in food crops of 0.2 mg/kg (GB2762—2012). In soil with a Cd content of 0.63 mg/kg, the Cd content of early rice fell within the standard of 0.2 mg/kg. However, when the Cd content of the soil was 0.88 mg/kg, the Cd contents of both early and late rice failed to meet this standard.

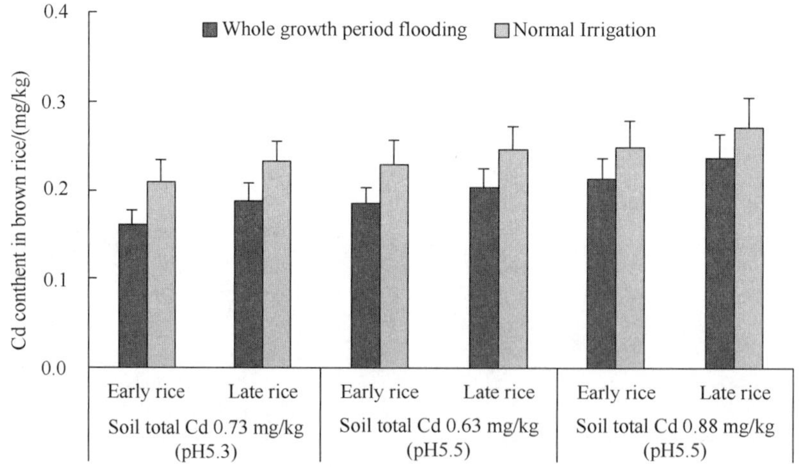

Fig. 2 Effect of water management on Cd content of rice

Water management with controlled flooding during the entire growth period is an effective technique to reduce Cd content in rice grown in Cd-contaminated paddy soils. However, this method requires a source of abundant, clean water for irrigation, as there is a risk of side effects such as gleization of the soil, which can lead to increased arsenic content in the rice. The method must be combined with appropriate sun-drying of the paddy.

4. *In-situ* immobilization techniques

In recent years, the *in-situ* immobilization of high-Cd soil has been gaining attention as an economical and effective remediation measure (Feng et al., 2013). Commonly used amendments can essentially be divided into lime substances, phosphate substances, biochar, substances containing iron and manganese, and organic substances. By increasing the pH of the soil, thereby causing Cd to co-precipitate or to be absorbed or chelated, the mobility and phytoavailability of Cd in the soil is reduced, thus reducing the content of Cd in brown rice. In order to ensure the effectiveness of this technique, and to keep the price of amendment low, we mainly selected agricultural waste products as raw materials for amendments. Continuous, *in-situ* immobilization experiments were developed in high-Cd soils over the course of several growing seasons. The soil used in the experiment was reddish-yellow mud from red quaternary clay. See Table 2 and Table 3 for the basic properties of the soil and the heavy metals content of the amendments. Eight different treatment were established: control (CK) with no amendments, lime (L) (90 g/m^2 of quicklime), slag (S) (1125 g/m^2), bagasse (B) (1125 g/m^2), lime+slag (LS), lime+bagasse (LB), slag+bagasse (SB), and lime+slag+bagasse (LSB). All amendment materials were raked into the fields by hand 7 days prior to transplanting in the first season (2013 late rice). The rice was managed using conventional methods. Rice shoot and soil surface samples were taken at harvest time.

Table 2 Select physical and chemical properties of tested soil

Properties	Values
pH	5.72
Organic matter content (g/kg)	34.2
Total N (g/kg)	1.84
Available P (m-g/kg)	26.9
Available K (m-g/kg)	80
Cation exchange capacity (CEC, cmol/kg)	10.1
Clay content (%)	48.9
Total Cd (m-g/kg)	1.2
$CaCl_2$-extractable Cd (m-g/kg)	0.053

Table 3 Heavy metals contents and pH of amendments

Amendments	pH	Heavy metals (mg/kg)					
		Cd	Cu	Cr	Pb	Zn	Ni
Lime	12.75	0.24	1.17	1.00	0.98	5.82	1.62
Bagasse	10.14	0.76	21.87	2.81	4.42	47.68	1.06
Slag	9.08	1.20	76.12	4.19	2.03	322.51	9.19

Table 4 Effects of amendments on soil pH and $CaCl_2$-extractable Cd

Treatments	2013-late rice		2014-early rice		2014-late rice	
	pH	$CaCl_2$-extractable Cd mg/kg	pH	$CaCl_2$-extractable Cd mg/kg	pH	$CaCl_2$-extractable Cd mg/kg
CK	5.72±0.05 d	0.053±0.003 a	5.93±0.03 d	0.050±0.02 a	6.17±0.04 d	0.042±0.007 a
L	6.25±0.16 c	0.026±0.007 b	6.21±0.06 cd	0.032±0.003 b	6.47±0.11 cd	0.026±0.005 bc
S	6.56±0.17 bc	0.022±0.003 b	6.62±0.07 ab	0.016±0.003 cd	6.63±0.14 bc	0.016±0.003 cd
B	6.20±0.10 c	0.028±0.003 b	6.22±0.09 cd	0.033±0.005 b	6.38±0.06 cd	0.031±0.005 ab
LS	6.93±0.20 ab	0.005±0.001 c	6.81±0.11 a	0.015±0.001 cd	7.01±0.10 a	0.008±0.002 d
LB	6.56±0.12 bc	0.020±0.005 b	6.48±0.13 bc	0.025±0.005 bc	6.63±0.08 bc	0.017±0.003 cd
SB	7.09±0.10 a	0.004±0.001 c	6.88±0.16 a	0.015±0.003 cd	6.92±0.06 ab	0.011±0.000 d
LSB	7.03±0.18 a	0.007±0.003 c	6.80±0.10 a	0.013±0.002 d	7.01±0.18 a	0.010±0.005 d

Data represent the mean±SE, n=3.

Effects of amendments on soil pH and $CaCl_2$-extractable Cd

A three-way analysis of variance (ANOVA) indicated that the application of lime, slag, and bagasse significantly increased soil pH ($p<0.05$), and that the effects were sustained for at least three harvest seasons (Table 4). In the late rice season of 2013 (first season), the soil pH of the control treatment (CK) was 5.72, which increased to 5.93 and 6.17 in the early (second season) and late (third season) harvest seasons of 2014, respectively. After the application of the different amendments, the soil pH significantly increased by 0.48 to 1.37 units ($p<0.05$) in the first season. While no significant difference in soil pH were observed for the three amendments applied as single treatments ($p>0.05$), the pH of the soils with amendments applied as mixtures were higher than in soils where the three amendments were applied alone. Similar increases in soil pH resulting from the different amendments were also observed in both 2014 harvest seasons.

The $CaCl_2$-extractable Cd in the CK soil slightly decreased from 0.053 mg/kg to 0.042 mg/kg during the three harvest seasons (Table 4). A three-way ANOVA demonstrated that treatments L, S, and B significantly decreased the $CaCl_2$-extractable Cd, while this interaction was only observed in S×B in the first rice season and in L×S and L×B in the second rice season ($p<0.05$). Compared with the control treatment, the application of lime alone (L) significantly reduced the $CaCl_2$-extractable Cd by 51.2%, 36.1%, and 38.2% in the first, second, and third rice seasons, respectively. Similar to treatment L, the application of slag (S) and bagasse (B) alone significantly reduced the $CaCl_2$-extractable Cd 59.1%-67.0% and 26.4%-47.4%, respectively, during the three rice seasons. Moreover, the application of LS, LB, SB, and LSB significantly reduced the $CaCl_2$-extractable Cd in soil by 69.8%-89.9%, 49.4%-61.4%, 69.2%-93.3%, and 73.3%-86.6%, respectively, during the three rice seasons. Generally, the three amendments successfully reduced the $CaCl_2$-extractable Cd in soil in the following order: S>L≈B, where the mixtures were more effective than the three amendments alone.

Effects of amendments on Cd uptake

A three-way ANOVA indicated that L significantly affected Cd accumulation in brown rice in the first and third rice seasons, S significantly affected Cd accumulation in the second and third rice seasons, and B significantly affected the Cd accumulation in only the first rice season ($p<0.05$) (Table 4). In the CK, the Cd contents of brown rice were 1.42, 0.50, and 1.44 mg/kg for the late rice season in 2013 and the early and late rice seasons in 2014, respectively (Fig. 3). Compared with the CK, the application of L, S, and B alone decreased the Cd contents of brown rice by 42.3%, 15.8%, and 40.8%, respectively, in the first rice season. Similarly, the Cd contents of brown rice decreased by 49.8%, 18.9%, and 10.1% after the application of L, S, and B alone, respectively, in the third season. However, the application of L and B alone increased the Cd contents of brown rice by 8.1% and 18.1%, respectively compared to the CK in the second rice season. Correspondingly, a steady decrease in the Cd content of brown rice was observed with the application of the mixtures, with the exception of the SB treatment. During the three rice seasons, the application of LS, SB and LSB reduced the Cd contents of brown rice 38.3%-58.3%, 55.6%-65.0%, and 60.0%-70.9%, respectively.

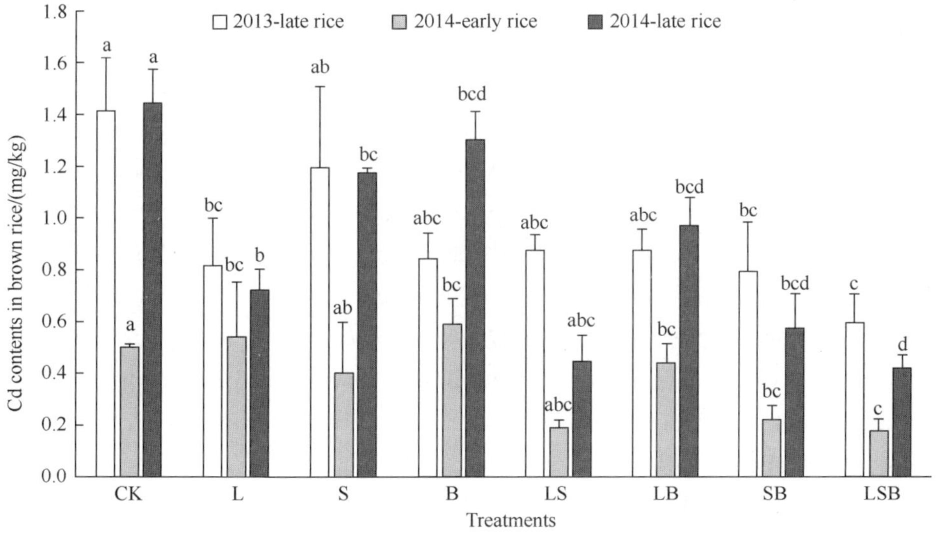

Fig. 3 Effect of amendments on the Cd content of brown rice

5. Techniques for the safe use of slightly Cd-contaminated paddy soils: model construction and application

Although the techniques described above can effectively decrease the accumulation of Cd in rice, they have a very limited range of appropriate uses. We have established that the use of these techniques effectively guarantees safe production in slightly Cd-contaminated paddy soils. Next, the study combined the use of low-cadmium rice varieties (V), flood irrigation (I), lime to increase soil pH (P), and the use of amendments or foliar application of trace elements (n) to create a model of VIP and VIP+*n* techniques. Experiments and demonstrations were developed in Changsha, Zhuzhou, and Xiangtan of Hunan Province in China.

A total of 48 test sites were selected for two growing seasons in slightly (Soil total Cd contents≤0.6 mg/kg), moderately (0.6 mg/kg≤Soil total Cd content≤1.0 mg/kg), and heavily (1.0 mg/kg≤Soil total Cd content) Cd-contaminated paddy soils. There were 32 slightly contaminated sites, 7 moderately contaminated sites, and 9 heavily contaminated sites. There was one control group at each site using local rice varieties and conventional irrigation. VIP and VIP+*n* treated test and demonstration sites were at least 5 ha in area.

Low Cd-accumulation rice varieties (V) comprised the following five early rice varieties: XIANG ZAO XIAN number 32, XIANG ZAO XIAN number 45, ZHONG JIA ZAO 17, ZHU LIANG YOU 819, and ZHU LIANG YOU 189. There were also three low Cd-accumulation varieties of late rice: XIANG WAN XIAN number 12, XIANG WAN XIAN number 13, and JIN YOU 59.

Water management included conventional irrigation (i.e., sun-drying the paddies during tilling and ripening stages) and the flood irrigation technique (I) (i.e., maintaining a layer of water throughout the development period). Lime was used at 750～1800 kg/ha depending on soil quality and pH value (P). There were five soil conditioners used: XIANG YU FENG, SHI LI KANG, ZHONG KE, XING WANG HONG, and MATA. Five organic fertilizers were used, such as JIN KUI ZI microbial agents. In addition, 6 foliar resistance control agents were used, such as WO NONG and ZHONG KE (+*n*). Methods and quantities used were in accordance with manufacturers' instructions.

Results from demonstration experiments in slightly and moderately Cd-contaminated paddy soils with over two seasons showed that, when using none of these techniques and growing local rice varieties, the pH values of average early rice soil were 5.70 and 5.28 respectively. Additionally, the average available Cd content in the soils was 0.37 mg/kg and 0.52 mg/kg respectively, the average Cd content in rice brown was 0.26 mg/kg and 0.53 mg/kg, and the proportion of brown rice that met national contaminant limits in food crops was 23.3% and 16.0%. The available Cd content in soils for the late rice season was 0.56 mg/kg and 0.42 mg/kg, the average soil pH was 5.86 and 5.43, and the Cd content in rice brown was 0.43 mg/kg and 0.57 mg/kg. The proportion of rice that met national contaminant limits in food crops was 30.7% and 10.3% (Table 5, Fig. 4).

In slightly and moderately contaminated fields, the use of VIP techniques increased the pH of soil for early rice season by 0.05 and 0.03 units, respectively, compared with the control group, while the average available Cd content of the soild decreased by 4.6% and 34.9%. Additionally, the average Cd content of brown rice decreased by 34.3% and 47.1% in slightly and moderately contaminated fields, respectively, and rice that met the national contaminant limits in food crops increased to 74.7% and 47.8%. The average pH of soil for late rice season increased slightly, the average available Cd content of the soil decreased, and the average Cd content of brown rice decreased by 37.1% and 50.7%. In addition, the proportion of rice that met the national contaminant limits in food crops increased to 70.0% and 43.3%. Using VIP+*n* techniques in slightly and moderately contaminated soils, the average soil pH increased by 0.15 and 0.10 units, respectively. The average effective cadmium content in the soil decreased by 13.5% and 19.4%, the average Cd content of brown rice decreased by 48.4% and 50.2%, and the proportion of brown rice that met the national contaminant limits in food crops was 80.0% and 52.7%. The average pH of soil for late rice season increased by 0.26 and 0.33 units. The average available Cd content of the soil decreased, the average cadmium content of brown rice decreased by 55.6% and 60.5%, while the proportion of rice that met the national contaminant limits in food crops was 78.1% and 50.0%. In heavily contaminated paddy soils, the changes in soil pH and the available Cd content of the soil were similar to slightly and moderately contaminated paddy soils. The Cd content in brown rice decreased by 23.3% to 51.4%. However, less than 12% of the brown rice met the national contaminant limits in food crops. In conclusion, the VIP and VIP+*n* models guarantee the production of rice products that meet the national contaminant limits in food crops in slightly Cd-contaminated paddy soils. In moderately Cd-contaminated paddy soil, these techniques are useful in reducing Cd content; however, the proportion of rice that meets the national contaminant limits in food crops should be further improved for increased safety. These methods are not very effective in heavily Cd-contaminated paddy soils.

Table 5 Effect of VIP and VIP+*n* on soil pH, available Cd and brown rice Cd contents

Contamination Level	Treatment	Early rice						Late rice					
		DTPA-Cd in soil		Soil pH		Cd in brown rice		DTPA-Cd in soil		Soil pH		Cd in brown rice	
		Mean mg/kg	Reduction %	Mean	Increase	Mean mg/kg	Reduction %	Mean mg/kg	Reduction %	Mean	Increase	Mean mg/kg	Reduction %
slightly (*n*=32)	CK	0.37	–	5.7	–	0.26	–	0.56	–	5.86	–	0.43	–
	VIP	0.36	4.6	5.74	0.05	0.17	34.3	0.55	–	5.89	0.03	0.27	37.1
	VIP+*n*	0.32	13.5	5.85	0.15	0.13	48.4	0.56	–	6.12	0.26	0.19	55.6
moderately (*n*=7)	CK	0.52	–	5.28	–	0.53	–	0.42	–	5.43	–	0.57	–
	VIP	0.34	34.9	5.31	0.03	0.28	47.1	0.42	–	5.52	0.09	0.28	50.7
	VIP+*n*	0.42	19.4	5.39	0.1	0.26	50.2	0.42	–	5.76	0.33	0.23	60.5
heavily (*n*=9)	CK	1.01	–	5.48	–	0.87	–	0.64	–	5.56	–	0.94	–
	VIP	0.91	9.9	5.51	0.04	0.67	23.3	0.62	–	5.77	0.21	0.56	40.4
	VIP+*n*	0.76	24.9	5.62	0.15	0.62	28.9	0.54	15.4	5.82	0.26	0.46	51.4

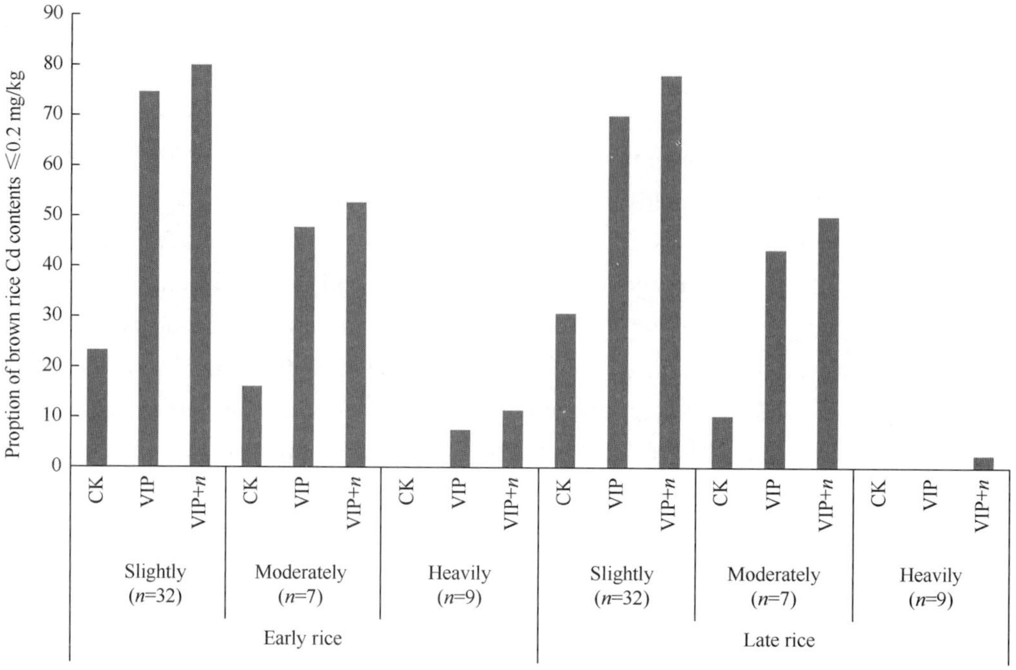

Fig. 4 Proportions of rice crops meeting the national contaminant limits in food crops used the VIP and VIP+ n techniques

References

Cao F, Wang R, Cheng W, et al. 2014. Genotypic and environmental variation in cadmium, chromium, lead and copper in rice and approaches for reducing the accumulation. Science of the Total Environment, 496: 275-281.

Chen D, Guo H, Li R, et al. 2016. Low uptake affinity cultivars with biochar to tackle Cd-tainted rice—a field study over four rice seasons in Hunan, China. Science of the Total Environment, 541: 1489-1498.

Di Toppi LS, Gabbrielli R.1999. Response to cadmium in higher plants. Environmental and Experimental Botany, 41: 105-130.

Feng R, Qiu W, Lian F, et al. 2013. Field evaluation of in situ remediation of Cd-contaminated soil using four additives, two foliar fertilisers and two varieties of pakchoi. Journal of Environmental Management, 124: 17-24.

Li J, Xu Y. 2015. Immobilization of Cd in a paddy soil using moisture management and amendment. Chemosphere, 122: 131-136.

Ministry of Environmental Protection P. R. C., Ministry of Land and Resources P. R. C. 2014. Report on soil pollution in China. http://www.sdpc.gov.cn/fzgggz/ncjj/zhdt/201404/t20140418-607888.html.[2014-4-18]

Shen X, Huang D Y, Ren X F, et al. 2016. Phytoavailability of Cd and Pb in crop straw biochar-amended soil is related to the heavy metal content of both biochar and soil. Journal of Environmental Management, 168: 245-251.

Shi J, Li L, Pan G. 2009. Variation of grain Cd and Zn concentrations of 110 hybrid rice cultivars grown in a low-Cd paddy soil. Journal of Environmental Sciences, 21 (2): 168-172.

Zhu Q H, Huang D Y, Liu S L, et al. 2012. Flooding-enhanced immobilization effect of sepiolite on cadmium in paddy soil. Journal of Soils and Sediments, 12 (2): 169-177.

Taiwan's Experiences on Soil Amendments, Phytoremediation and Soil Water Managements for the Cadmium- and Arsenic-Contaminated Soils

Hung-Yu LAI[1], Chia-Hsing LEE[2], Zueng-Sang CHEN[2]

[1]Department of Post-Modern Agriculture, MingDao University

[2]Department of Agricultural Chemistry, Taiwan University

1. Heavy metal contamination in Taiwan

Soil contamination with heavy metals (HMs), mainly resulting from mining and smelting, is a serious global environmental problem. In Taiwan, HMs-contaminated paddy soils were mainly produced by the illegal discharges of surrounding electroplating plants. According to previous survey of Environmental Protection Administration (EPA) of Taiwan, there are approximately 470 ha of crop lands have been contaminated with HMs till the end of 2008. Most of these lands were located in northern and central Taiwan and contaminated mainly with copper (Cu) and nickel (Ni), followed by zinc (Zn) and cadmium (Cd). These contaminated lands should be conducted further soil remediation to decrease the concentrations of aqua regia soluble HMs to below the regulation announced by the Soil and Groundwater Pollution Remediation Act (SGWPR Act) in Taiwan. Many soil remediation techniques were used in Taiwan to treat these HMs-contaminated soils include acid washing and soil turnover or attenuation. However, more HMs-contaminated were discovered recently because of illegal discharges into irrigation water system still existed.

The soil turnover/attenuation method was the most popular technique by the policy makers to be used in Taiwan because it was cheaper and more acceptable for farmers compared with other techniques. This method mixed the contaminated soils of the surface layer with appropriate amounts of subsurface layer to reduce the total concentration of HMs to meet the soil regulation. While the total amount of HMs in soil remains unchanged and soil characteristics manifest changed. Further application of fertilizers or organic materials, however, is necessary to recover the soil fertility or soil functions for framer's sustainable agriculture (Lai et al., 2007).

2. Phytoremediation experiences in Taiwan

2.1 Selection of hyperaccumulators

Phytoremediation was recommended as one of the remediation options besides engineering and chemical remediation methods for these HMs-contaminated cropped fields. Many researches were carried out to study the phytoextraction efficiency of indigenous plants. Besides pot experiments there were at least four times of field experiments had been *in-situ* conducted in the HMs-contaminated sites of northern and central Taiwan (Chen and Lee, 1997; Lai and Chen, 2009; Lai, et al., 2010). The removal efficiencies of soil HMs of more than forty species of herbal and woody indigenous plants were investigated and some of them were regarded as Cd hyperaccumulators. These plants include rainbow pink (*Dianthus chinensis*), Indian mustards (*Brassica juncea*), French marigold (*Tagetes patula*), and impatiens (*Impatiens walleriana*). The capacities of accumulation and tolerance were further investigated from 2004 till now.

Nevertheless, the differences between pot experiments and field studies make it difficult to evaluate the removal capacity of these plants because the soil Cd concentration and soil characteristics were not the same. We thus summarized the study results in the last three decades and used BCF factor (Bioconcentration factor=ratio of shoot concentration to soil concentration) as an indicator to compare the phytoextraction capacities of these plants (Lai and Chen, 2011). Experimental results showed that the same plant species *in-situ* grown in Cd-contaminated sites had higher BCF compared with those of the pot experiments (Fig. 1). Plants grown in Cd-contaminated field sites can remove more amounts of Cd from the soils because they have more root space and better growth condition if they have similar biomass. The BCF also exponentially decreased with increasing the soil Cd concentration.

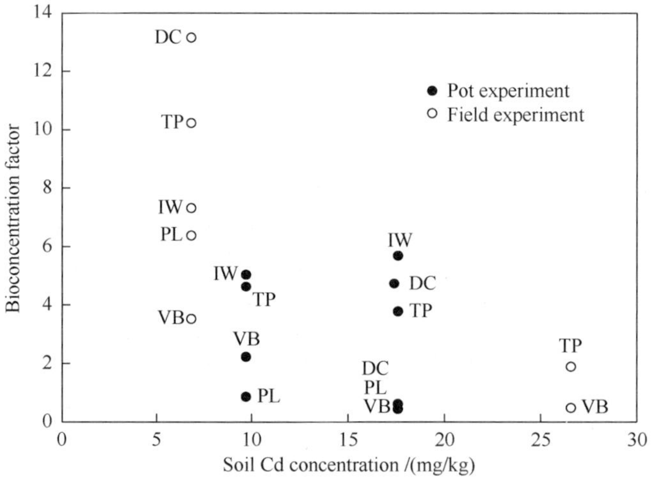

Fig. 1 Bioconcentration factor of different plant species grown in artificially Cd-contaminated soils or *in-situ* grown in Cd-contaminated sites for 35 to 50 days (Modified from Lai and Chen, 2011) (TP—French marigold; VB—Garden verbena; IW—Impatiens; PL—Star cluster; SS—Scarlet sage; BJ-A—India mustard Afghanistan; BJ- Ⅰ —India mustard India; BJ-P—India mustard Pakistan, and DC—Rainbow pink)

2.2 Chemical-enhanced phytoremediation

Rainbow pink's hyperaccumulation capacity was firstly confirmed in Cd-contaminated sites of northern Taiwan (Chen and Lee, 1997). After growing for five weeks, the Cd concentration of the shoots increased from 1.56 to 115 mg/kg. Lai and Chen (2004, 2005, 2006) further grown its seedlings in the artificially Cd-, Zn-, and lead-(Pb-) contaminated soil with total concentration of 19.0, 503, and 931 mg/kg, respectively. Two concentrations of EDTA ($C_{10}H_{14}N_2O_8Na_2$, 5 or 10 mmol/kg) were applied to increase the soil availability of these HMs. Experimental results showed that EDTA significantly increased the concentrations of these HMs in the soil solutions, especially for Pb. The concentrations of Cd and Pb in the shoots also significantly increased.

Soil samples were artificially spiked with single or combined HMs to investigate the phytoextraction efficiency of rainbow pink and also the interactions between Cd, Zn, and Pb (Lai and Chen, 2005). Experimental results showed that the application of 5 mmol EDTA/kg significantly increased the availability of Cd, Zn, or Pb concentrations in various HMs-contaminated soils and also the accumulated HMs concentrations in the shoots of rainbow pink. The application of 2 mmol EDTA/kg might too low to enhance the phytoextraction effect for the study soils. The accumulation of HMs of rainbow pink was affected by the interactions between Cd, Zn, and Pb (Lai and Chen, 2006). The existence of Pb of the Cd-contaminated soil enhanced the uptake of Cd in rainbow pink in the treatments of 2 mmol EDTA/kg. Cadmium inhibited the concentration of Zn without applying EDTA. In the treatment of 5 mmol EDTA/kg, Cd or Zn in the Pb-contaminated soil inhibited the uptake of Pb in rainbow pink.

Nevertheless, EDTA has low biodegradability and is persistent in the soils for extended periods (Tandy, et al., 2006). Many studies also reported the negative influences of applying chelating agents on the activity of microorganisms (Römkens et al., 2002; Ultra et al., 2005) and enzymes (Epelde et al., 2008). The influences of applying various chelating agents (DTPA $C_{14}H_{23}N_3O_{10}$; EDDS $C_{10}H_{13}N_2O_8Na_3$; EDTA $C_{10}H_{14}N_2O_8Na_2$) on soil quality of five soil series in Taiwan was also evaluated (Lai, 2015a). Experimental results showed that the application of 5 mmol/kg of above chelating agents had negative effects on the soil quality. This was especially true for the high electrical conductivity (EC) of soil when DTPA was used as the chelating agent. This result revealed that chemical-enhanced phytoremediation was not a suitable practice from the viewpoints of costs and soil and environmental quality.

2.3 Detoxification and Tolerance Mechanisms

Impatiens (*Impatiens walleriana*) was a herbal plant commonly found in Taiwan and it was good plant to be as a potential Cd hyperaccumulator from our previous studies. The Cd accumulated in the shoots grown in the Cd-contaminated soils (Cd-10 and Cd-20 mg/kg) were more than 100 mg/kg and the BCF and TF (Transfer factor=ratio of shoot concentration to root concentration) were both more than unity (Table 1) (Lin et al., 2010). In the soils treated with higher Cd concentrations (Cd-20, Cd-40, and Cd-80 mg/kg), the maximum accumulated Cd concentration was more than 1, 100 mg/kg (Table 1) (Wei et al., 2012). Nevertheless, higher concentrations of soil Cd had advised effect on the biomass of impatiens.

Table 1　The Cd concentration, BCF, and TF of impatiens grown in the Cd-contaminated soils

Treatments	Concentration/(mg/kg)		BCF	TF
	Root	Shoot		
Lin et al. (2010)				
Cd-10	29.5±9.6	48.9±11.7	5.0	1.7
Cd-20	99.0±8.4	100±11	5.7	1.0
Wei et al. (2012)				
Cd-20	128±21	282±34	14.0	2.2
Cd-40	196±67	342±103	9.1	1.7
Cd-80	462±169	1168±140	14.7	2.5

In order to understand the detoxification and tolerance mechanisms of impatiens under Cd stress, the subcellular distribution and chemical forms of Cd in the root and shoot of impatiens were studied. Experimental results of subcellular distribution showed that Cd was mainly compartmentalized in the soluble fraction and the cell wall fraction of the roots and leaves, respectively. The leaf area and transpiration rate had linear relationships with the Cd accumulated in the shoots (Lai, 2015b). Results of a hydroponic experiment showed that the chemical forms of Cd in the various organs of impatiens were quite different (Lai and Cai, 2016). The Cd accumulated in the roots was primarily compartmentalized in the ethanol and deionized water extractable chemical forms with high migration abilities, but in the shoots was mainly associated in the cell wall or with pectate and protein. Similar results were observed when impatiens was grown in the soils contaminated with Cd (10-120 mg/kg). Cadmium was mainly compartmentalized in the soluble fraction in the roots, which has a high migration capacity. After further translocate to the shoots, the Cd was mainly compartmentalized in the cell wall fraction as a mechanism of tolerance. Most of the Cd in the various organs of impatiens was mainly in the forms of pectate and protein-integrated (Lai, 2015c).

3. Techniques for decreasing HMs uptake by crops

3.1 Soil amendments for inhibiting Pb and Cd uptake by wheat

Seeds of wheat were grown in the Cd-and Pb-contaminated soils collected from contaminated sites in northern Taiwan and soil amended with various amendment materials, such as compost, zinc oxide, calcium carbonate, calcium carbonate mixed with zinc oxide, and calcium carbonate mixed with compost. Soil amendments of calcium carbonate or calcium carbonate mixed with zinc oxide (or compost) significantly decreased the availability of Cd and the Cd accumulated in the grains, leaf and stem, and husk also significantly decreased accordantly (Lee et al., 2004).

3.2 Soil amendments for inhibiting arsenic uptake by rice

Arsenic ion (As) can pose significant human health risks as a contaminant of drinking ground water, which has been found in many countries, including Taiwan (Chen et al., 1995). Generally, As mobility is high in soils with low redox conditions (Xu et al., 2008), high pH values (Quazi et al., 2011), low content of clay (Adriano, 2001) and low content of free iron oxides (Heikens et al., 2007). Particularly, amorphous iron oxides may have capacity for As sorption four-times higher than that of well-crystalline iron oxides (Linge and Oldham, 2002). Arsenite (As (III)) ion is more labile and more toxic than other As speciation and unfortunately, the proportion of As (III) ion to total As concentration of soil solutions in paddy fields is high (Roberts et al., 2011). Paddy rice is a staple foodstuff in Asia and becomes a main route of As intake because it accumulates As more efficiently compared with other cereals (Su et al., 2010). In addition, rice growth or yields in paddy fields may decrease in soils with high levels of As content because of As phytotoxicity (Talukder et al., 2014).

Arsenate (As (V)) ion is a chemical analog of phosphate (P_i) and shares the uptake pathway of rice with the same transporters (Shin et al., 2004). Previous works have lent credible support to this assumption through hydroponic experiments (Geng et al., 2005). Similarly, application of silicates (Si) was assumed to decrease As accumulation in rice plant trough competitive uptake against arsenite (Seyfferth and Fendorf, 2012). However, soil experiments in Taiwan have revealed that As accumulation in rice remained or even increased after application of silicate at reasonable application rates, predominately because of competitive adsorption in soils which leads to elevation of As levels in soil solutions and consequently, the Si/As ratios can't largely increase as shown in Fig. 2 (Lee et al., 2014). Syu et al. (2015) tested the foliar application of Si to rice plant accordingly, but As accumulation in rice seedlings was not affected. Moreover, reasonable phosphorous applications into soils were also found insufficient to inhibit As uptake by rice (Lee et al., 2016).

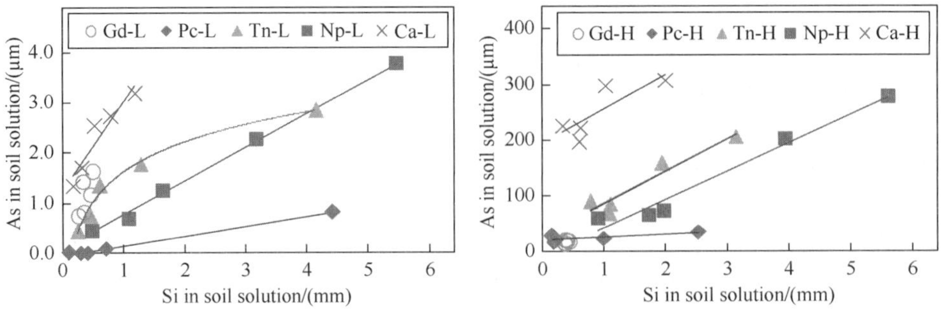

Fig. 2 The relationships between Si and As concentration in solutions of five disparate soils treated with 0-2 folds of recommended Si amount for rice. Values are average values during incubation (Modified from Lee et al., 2014)

3.3 Soil water management for arsenic immobilization in Taiwan

Soil water management is a practically easier technique for decreasing As phytoavailability. During rice cultivation, temporary drainage could result in the (As (III)) transformation to (As (Ⅴ)) , decrease the mobility and phytotoxicity of As to rice plant and consequently, inhibit the As accumulation in brown rice (Somenahally et al., 2011). The effects of drainage occasion and soil types on the efficiency of aerobic treatments have been tested in Taiwan. Chang and Chen (2012) revealed that soil water drainage can decrease the As level in the soil solution but not in the brown rice. Huang and Chen (2013, 2014) further indicated that As concentration of brown rice can be decreased by soil water drainage regardless of drainage occasion in well-drained soils, while the inhibition effect slightly revealed only for drainage treatment before flowering stage and also for saturated treatment for poor-drained soils (Fig. 3). Accordingly, the efficiency of soil water management for reducing As mobility or phyto-availability is intuitively limited by poor-drained soils (e.g. sticky texture and dispersive particulates). Extending the drainage period can be applied to reduce the As concentration of brown rice, but rice growth and productivity may be inhibited.

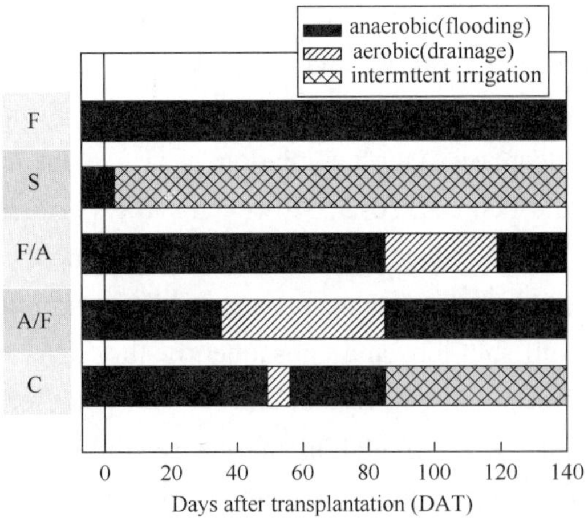

	Poor drained soil		Well drained soil	
	Yield	Grain As conc.	Yield	Grain As conc.
F	-	-	-	-
S	NS	Lower	Higher	Lower
F/A	NS	NS	NS	Lower
A/F	NS	Lower	Higher	Lower
C	NS	NS	Higher	Lower

F: flooding; S: saturated; A: aerobic drainage; C: conventional; NS: not significant

Fig. 3 The effects of soil water treatments and soil drainage conditions on rice yields and As concentrations of brown rice (Modified from Huang and Chen, 2013, 2014)

Integrated remediation techniques are evaluated in Taiwan. Since phosphorous competitive for rice uptake against As (V) but not the predominate arsenite ion (As (III)) in paddy soils, Chin and Chen (2015) tested the effects of different phosphorous application rates under saturated water management and found that there are no competitive effect between phosphorous and (As (V)) for rice uptake. In the other hand, husk is less decomposable and more consistent in shape and structure than other green wastes, which can improve soil structure, increase soil porosity and consequently, enhance soil drainage condition for fine-textured or clayey soils (Kaboosi et al., 2012). A recent research of Penido et al. (2016) also concluded that either fresh husk or husk ash are attractive amendments for improve rice nutrition and decreasing As uptake through the beneficial impacts of husk-derived Si. However, application of organic matters into soils may increase the bioavailability and mobility of As (Wang and Mulligan, 2006). Since both benefit and advertise effects may occur, the overall effects of husk application into soils on As concentration of brown rice were evaluated in Taiwan. Chen et al. (2015) revealed an overall effect of husk aggravated As mobility and phyto-availability in poor-drained soils even though drainage were improved, which were caused by husk decomposition and subsequent As release in the reducing condition.

3.4 Arsenic speciation of brown rice and their health risk implications

Arsenite (As (III)) ion accumulated in brown rice is probably to have a maximum level at 0.2 mg As (III)/kg rice, regardless of water managements, amendments, and soil types, as the same value as the limit announced by Codex in 2015. In comparison, dimethyl arsenic (DMA) is likely to be the predominate As species of excess total As concentration in brown rice. Several studies had shown this phenomenon for rice with low As levels (<1 mg/kg). Related Taiwan's researches (Chen et al., 2015; Chin and Chen, 2015; Huang and Chen, 2013) obtained the brown rice with high As levels up to 4.5 mg/kg and even corroborated the same As distribution pattern (Fig. 4). In an in-vivo assessment, the oral bioavailability of monomethyl arsenic acid (MMA), DMA, and (As (V)) was about 16%, 33% and 90% of As (III), respectively (Juhasz et al., 2006). Moreover, DMA is most likely have no carcinogenic effects and was found to have LC50 values 133-557 times higher than arsenite (As (III)) for human hepatocytes under several in-vitro assessments (Petrick et al., 2000). The Joint Food and Agriculture Organization of the United Nations (FAO)/WHO Expert Committee on Food Additives (JECFA) also highly concerned and adapted the maximum level of 0.2 mg/kg for inorganic As (As (III)) in polished rice, but not for organic As (DMA) (WHO, 2010, 2014). In these studies, although high As concentrations in brown rice were found, only few samples had higher inorganic As levels. Therefore, the human health risks of brown rice production in reviewed studies still under control and needs further investigation on this hot issue.

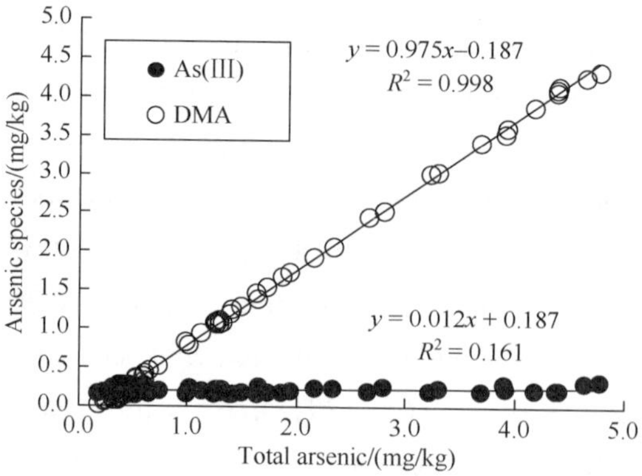

Fig. 4 The distribution of As (III) and DMA in brown rice cultivated in two poor-drained soils with disparate properties and As levels (10-24 mg As/kg) under different soil water managements (flooding and drainage) with husk application rates (0 and 2.5%) (Modified from Chen et al., 2015)

It should be addressed that the DMA accumulation in brown rice grown in the acidic, clayey and high iron oxide soils which provided with low As mobility in soil was even much higher than that of neutral, moderate soil textured and low iron oxide soils which had higher levels of As concentrations in soil solutions (Chen et al., 2015; Chin and Chen, 2015; Huang and Chen, 2013). Since arsenic methylation is most likely to be occurred in soil but not in plant (Lomax et al., 2012), the mentioned acidic soils probably have either higher population of arsenic methylation microorganisms or better environmental conditions for methylation. Furthermore, organic As speciation translocated from rice shoot to grain could be much faster than inorganic As speciation (Carey et al., 2011), leading to the higher DMA levels in rice grain in these kinds of soils. Therefore, soil properties may control the As speciation accumulation by rice more strongly than soil water treatments and soil amendments.

References

Adriano D C. 2001. Trace elements in terrestrial environments: biogeochemistry, bioavailability, and risks of metals. New York: Springer.

Carey A M, Norton G J, Deacon C, et al. 2011. Phloem transport of arsenic species from flag leaf to grain during grain filling. New Phytologist, 192 (1): 87-98.

Chang H Y, Chen Z S. 2013. The effect of different water managements on rice arsenic content in two arsenic-spiked soils//E3S Web of Conferences. EDP Sciences, 1.

Chen Z S, Lee D Y. 1997. Evaluation of remediation techniques on two cadmium polluted soils in Taiwan. Remediation of Soils Contaminated with Metals: 209-223.

Chen P H, Lee C H, Chen Z S. 2015. The effects of rice husk and water management on arsenic concentration and arsenic speciation of brown rice grown in arsenic-contaminated soils.

Chen S L, Yeh S J, Yang M H. et al. 1995. Trace element concentration and arsenic speciation in the well water of a Taiwan area with endemic blackfoot disease. Biological Trace Element Research, 48 (3): 263-274.

Chin P L, Chen Z S. 2015. Effects of water management and phosphorous application on arsenic concentration of pore water in paddy soils with rice cultivation.

Epelde L, Hernández-Allica J, Becerril J M, et al. 2008. Effects of chelates on plants and soil microbial community: comparison of

EDTA and EDDS for lead phytoextraction. Science of the Total Environment, 401 (1): 21-28.

Geng C N, Zhu Y G, Liu W J, et al. 2005. Arsenate uptake and translocation in seedlings of two genotypes of rice is affected by external phosphate concentrations. Aquatic Botany, 83 (4): 321-331.

Heikens A, Panaullah G M, Meharg A A. 2007. Arsenic behaviour from groundwater and soil to crops: impacts on agriculture and food safety//Reviews of Environmental Contamination and Toxicology. New York: Springer: 43-87.

Huang T H, Chen Z S. 2013. Water management and soil characteristics affect the As concentration in the brown rice of *Oryza sativa*.

Huang T H, Chen Z S. 2014. Soil water management effect on As concentration of brown rice grown in two different soil properties of arsenic-contaminated soils. 20th World Congress of Soil Science: 508.

Juhasz A L, Smith E, Weber J, et al. 2006. In vivo assessment of arsenic bioavailability in rice and its significance for human health risk assessment. Environmental Health Perspectives, 114: 1826-1831.

Kaboosi K, Liaghat A, Hosseini S H. 2012. The feasibility of rice husk application as envelopematerial in subsurface drainage systems. Irrigation and Drainage, 61 (4): 490-496.

Lai H Y. 2015. Negative effects of chelants on soil qualities of five soil series. International Journal of Phytoremediation, 17 (3): 228-234.

Lai H Y. 2015. Effects of leaf area and transpiration rate on accumulation and compartmentalization of cadmium in *impatiens walleriana*. Water, Air, & Soil Pollution, 226: 2246.

Lai H Y. 2015. Subcellular distribution and chemical forms of cadmium in *Impatiens walleriana* in relation to its phytoextraction potential. Chemosphere, 138: 370-376.

Lai H Y, Cai M C. 2016. Effects of extended growth periods on subcellular distribution, chemical forms, and the translocation of cadmium in *Impatiens walleriana*. International Journal of Phytoremediation, 18 (3): 228-234.

Lai H Y, Chen Z S. 2004. Effects of EDTA on solubility of cadmium, zinc, and lead and their uptake by rainbow pink and vetiver grass. Chemosphere, 55 (3): 421-430.

Lai H Y, Chen Z S. 2005. The EDTA effect on phytoextraction of single and combined metals-contaminated soils using rainbow pink (*Dianthus chinensis*). Chemosphere, 60 (8): 1062-1071.

Lai H Y, Chen Z S. 2006. The influence of EDTA application on the interactions of cadmium, zinc, and lead and their uptake of rainbow pink (*Dianthus chinensis*). Journal of Hazardous Materials, 137 (3): 1710-1718.

Lai H Y, Chen Z S. 2009. *In-situ* selection of suitable plants for the phytoremediation of multi-metals–contaminated sites in central Taiwan. International Journal of Phytoremediation, 11 (3): 235-250.

Lai HY, Chen ZS. 2011. Prior to a successful phytoextraction: pot experiments and field scale studies on the total removal capacity by garden flowers grown in cadmium-contaminated soils in Taiwan. pp. 737-750.

Lai H Y, Juang K W, Chen Z S. Large-area experiment on uptake of metals by twelve plants growing in soils contaminated with multiple metals. 2010. International Journal of Phytoremediation, 12 (8): 785-797.

Lai H Y, Su S W, Lin C C, et al. 2007. A study on the recovery of soil fertility of two metal-contaminated soils after treating with soil acid washing. Journal of Science and Technology (National Yunlin University of Science and Technology), 16: 39-46.

Lee C H, Wu C H, Syu C H, et al. 2016. Effects of phosphorous application on arsenic toxicity to and uptake by rice seedlings in As-contaminated paddy soils. Geoderma, 270: 60-67.

Lee C H, Huang H H, Syu C H, et al. 2014. Increase of As release and phytotoxicity to rice seedlings in As-contaminated paddy soils by Si fertilizer application. Journal of Hazardous Materials, 276: 253-261.

Lee T M, Lai H Y, Chen Z S. 2004. Effect of chemical amendments on the concentration of cadmium and lead in long-term contaminated soils. Chemosphere, 57 (10): 1459-1471.

Lin C C, Lai H Y, Chen Z S. 2010. Bioavailability assessment and accumulation by five garden flower species grown in artificially cadmium-contaminated soils. International Journal of Phytoremediation, 12 (5): 454-467.

Linge K L, Oldham C E. 2002. Arsenic remobilization in a shallow lake. Journal of Environmental Quality, 31 (3): 822-828.

Lomax C, Liu W J, Wu L, et al. 2012. Methylated arsenic species in plants originate from soil microorganisms. New Phytologist,

193 (3): 665-672.

Penido E S, Bennett A J, Hanson T E, et al. 2016. Biogeochemical impacts of silicon-rich rice residue incorporation into flooded soils: Implications for rice nutrition and cycling of arsenic. Plant and Soil, 399 (1-2): 75-87.

Petrick J S, Ayala-Fierro F, Cullen W R, et al. 2000. Monomethylarsonous acid (MMA III) is more toxic than arsenite in Chang human hepatocytes. Toxicology and Applied Pharmacology, 163 (2): 203-207.

Quazi S, Datta R, Sarkar D. 2011. Effects of soil types and forms of arsenical pesticide on rice growth and development. International Journal of Environmental Science & Technology, 8 (3): 445-460.

Roberts L C, Hug S J, Voegelin A, et al. 2010. Arsenic dynamics in porewater of an intermittently irrigated paddy field in Bangladesh. Environmental Science & Technology, 45 (3): 971-976.

Römkens P, Bouwman L, Japenga J, et al. 2002. Potentials and drawbacks of chelate-enhanced phytoremediation of soils. Environmental Pollution, 116 (1): 109-121.

Seyfferth A L, Fendorf S. 2012. Silicate mineral impacts on the uptake and storage of arsenic and plant nutrients in rice (Oryza sativa L.) [J]. Environmental Science & Technology, 46 (24): 13176-13183.

Shin H, Shin H S, Dewbre G R, et al. 2004. Phosphate transport in Arabidopsis: Pht1; 1 and Pht1; 4 play a major role in phosphate acquisition from both low-and high-phosphate environments. The Plant Journal, 39 (4): 629-642.

Somenahally A C, Hollister E B, Yan W, et al. 2011. Water management impacts on arsenic speciation and iron-reducing bacteria in contrasting rice-rhizosphere compartments. Environmental Science & Technology, 45 (19): 8328-8335.

Su Y H, McGrath S P, Zhao F J. 2010. Rice is more efficient in arsenite uptake and translocation than wheat and barley. Plant and Soil, 328 (1-2): 27-34.

Syu C H, Huang C C, Jiang P Y, et al. 2015. Effects of foliar and soil application of sodium silicate on arsenic toxicity and accumulation in rice (*Oryza sativa* L.) seedlings grown in As-contaminated paddy soils. Soil Science and Plant Nutrition: 1-10.

Talukder A, Meisner C A, Sarkar M A R, et al. 2014. Effects of water management, arsenic and phosphorus levels on rice yield in high-arsenic soil-water system. Rice Science, 21 (2): 99-107.

Tandy S, Ammann A, Schulin R, et al. 2006. Biodegradation and speciation of residual SS-ethylenediaminedisuccinic acid (EDDS) in soil solution left after soil washing. Environmental Pollution, 142 (2): 191-199.

Ultra Jr V U, Yano A, Iwasaki K, et al. 2005. Influence of chelating agent addition on copper distribution and microbial activity in soil and copper uptake by brown mustard (*Brassica juncea*). Soil Science & Plant Nutrition, 51 (2): 193-202.

Wang S, Mulligan C N. 2006. Effect of natural organic matter on arsenic release from soils and sediments into groundwater. Environmental Geochemistry and Health, 28 (3): 197-214.

Wei J L, Lai H Y, Chen Z S. 2012. Chelator effects on bioconcentration and translocation of cadmium by hyperaccumulators, *Tagetes patula* and *Impatiens walleriana*. Ecotoxicology and Environmental Safety, 84: 173-178.

World Health Organization. 2010. Exposure to arsenic: a major public health concern. Geneva: Public Health and Environment: 1-5.

WHO.2014. "Report of the thirty-seventh session of Codex Alimentarius Commission (REP14/CF)." Joint FAO/WHO Food Standards Programme, FAO, Rome.

Xu X Y, McGrath S P, Meharg A A, et al. 2008. Growing rice aerobically markedly decreases arsenic accumulation. Environmental Science & Technology, 42 (15): 5574-5579.

Arsenic in Soil-Plant System: A Synthesis

Guilan DUAN[1], Yongguan ZHU[1, 2,*]

[1]State Key Laboratory of Urban and Regional Ecology, Research Centre for Eco-Environmental Sciences, Chinese Academy of Sciences

[2]Key Laboratory of Urban Environment and Health, Institute of Urban Environment, Chinese Academy of Science

*Corresponding author

Arsenic (As) is a global contaminant, and it can enter humans through both drinking water and food. Upon my return from Australia, As biogeochemistry has been the key topic in our research group in the last 15 years or so. We have published more than 100 papers in international journal, including Nature Plants, Plant Physiology, New Phytologist and many in Environmental Science & Technology etc. We have also published several review papers, such as in Annual Review of Earth and Planetary Science (Zhu et al., 2014), Trends in Plant Science (Ye et al., 2012), Current Opinion in Biotechnology (Zhu and Rosen, 2009). Many of our studies have featured in journals, such as Science, Nature and Trends in Plant Science. This paper is a synthesis of the research findings from our lab in the 10 years or so.

1. Arsenic in rice

In areas with no significant contamination of As in water, food represents the major pathway of human exposure to this topic element (Zhu et al., 2014, Fig. 1). We have analyzed the data from the China National Nutrition and Health Survey (CNNHS) and collecting reported values of iAs in major food groups, we developed a framework to calculate average iAs daily intake for different regions of China, and to quantify the cancer risks from iAs in food deterministically and probabilistically (Li et al., 2011). Based on this framework, both per individual and for total population estimates were estimated. For the total population, daily iAs intake is 42 μg/day, and rice food contuibuted the largest total iAs intake accounting for about 60%. Cancer risk from food iAs intake is 106 per 100, 000 and 177 per 100, 000 for adult individuals and the median population, respectively. Population in the regions of South China has a higher cancer risk than that in the Northern region and the total population. Cancer slope factor, ingestion rates of rice, aquatic products and iAs concentration in rice have higher contribution to variance of the incremental lifetime cancer risk. From this study, we concluded that rice should be the largest contributor of iAs through food route for Chinese people. The population from the South regions of China has greater cancer risk than those from the North and the whole population in China.

Fig. 1 Human exposure to groundwater arsenic through irrigation of rice paddy and drinking water (Zhu et al., 2014)

In relation to As exposure analysis from food for Chinese population, we have also conducted national field survey on As levels in market rice (Zhu et al., 2008). To do this, two approaches were undertaken to characterize the As content of Chinese rice. one was a national market basket survey (n=240) in provincial capitals, sourcing grain from China's premier rice production areas. Another one was to reflect rural diets, paddy rice (n=195) directly from farmer fields were collected from three regions in Hunan, two of the sites were within mining and smeltery districts, and the third was devoid of large-scale metal processing industries. Total As concentrations were determined in all the samples while a subset (n=33) were characterized for As species. The vast majority (85%) of the market rice possessed total As levels$<$150 ng/g. However, The rice collected from mine-impacted regions were found to be highly enriched in As, with highest As concentrations up to 624 ng/g. Inorganic As (A*si*) was the predominant species detected in all of the samples, with A*si* levels in some samples exceeding 300 ng/g. The ratio of A*si* concentration to total As levels in polished and unpolished Chinese rice was successfully predicted. Based on this survey, the mean baseline concentrations for A*si* in Chinese market rice were estimated to be 96 ng/g, while levels in mine-impacted areas were higher, with ca. 50% of the rice in one region predicted to fail the national standard.

2. Arsenic biotransformation in soils

Biotransformation of As includes biological reduction, oxidation, methylation, and conversion to some more complex organic arsenicals. As speciation determines its toxicity and mobility in soil-plant system and beyond. The transformation of different As species in soil is facilitated by enzymes encoded by genes from environmental microbes. Over the last few years, we have made significant contribution to the understanding of As biotransformation, biomethylation in particular (Zhu et al., 2014, Fig. 2).

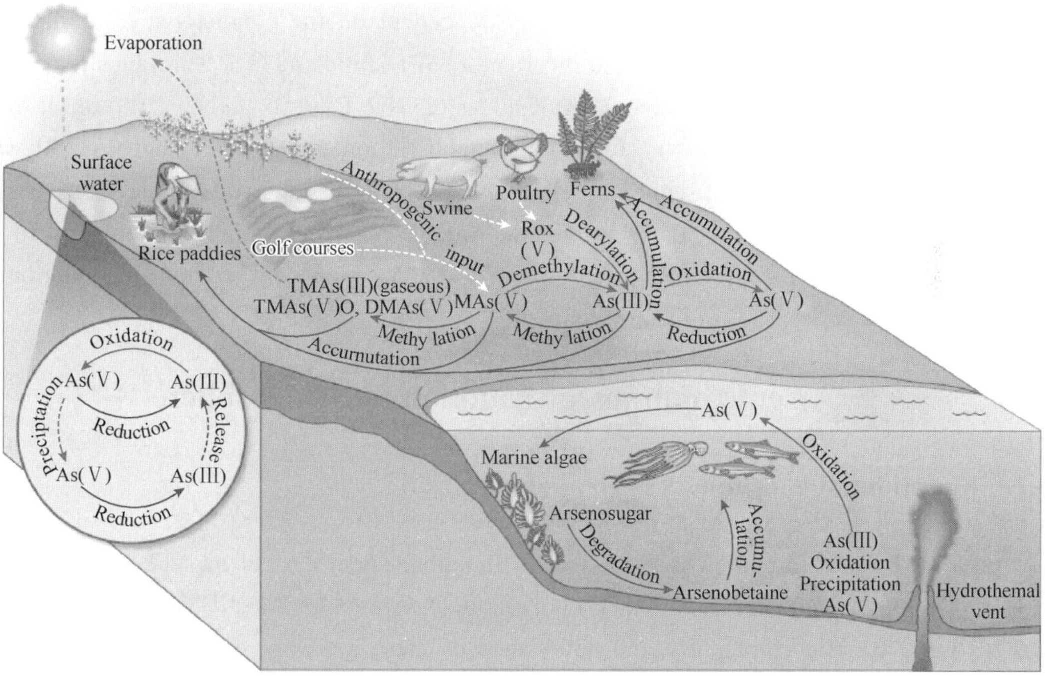

Fig. 2 Arsenic biogeochemical cycles (Zhu et al., 2014)

Cyanobacteria are ubiquitous in soils, aquatic systems, and wetlands. We investigated biotransformation and volatilization of As by three photosynthetic cyanobacteria (Microcystis sp. PCC7806, Nostoc sp. PCC7120, and Synechocystis sp. PCC6803) (Yin et al., 2011). Cyanobacteria is efficient in accumulating As, up to 0. 38-45 gk/g dry weight dry weight when treated with 100 mmol/L sodium As (III) for 14 d. In Cyanobacteria, inorganic arsenate [As (V)] and As (III) were the predominant species, and As (V) making up>80% of total As. Organic arsenicals were detected after exposure to higher As concentrations, and when treated with As (V) for 6 weeks, volatile arsenicals were produced. The genes encoding the arsenite S-adenosylmethionine methyltransferase (*ArsM*) were cloned from these three cyanobacteria. When expressing *ArsM* in an As-hypersensitive strain of *Escherichia coli*, each conferred resistance to As (III). Two purified *ArsM* homologs (*SsArsM* from *Synechocystis* sp. PCC6803 and *NsArsM* from *Nostoc* sp. PCC7120) were shown to methylate As (III) to trimethylarsine in vitro. Given that *ArsM* homologs are widespread in *cyanobacteria*, suggesting that *cyanobacteria* play an important role in As biogeochemistry.

In addition to cynaobacteria, we have identified the molecular mechanisms of As (III) methylation in different organisms. Wang et al. (2014) indentified and characterized an As (III) methyltransferase (*ArsM*) from an archaeon Methanosarcina acetivorans C2A. expressing MaArsM in As-sensitive strain of E. coli resulted in As (III) resistance through As methylation and subsequent volatilization. Purified MaArsM protein showed the function in catalyzing the formation of various methylated products from As (III) in vitro. We further found that methylation of As (III) by MaArsM is highly dependent on the characteristics of the thiol cofactors, homocysteine, and dithiothreitol were more efficient than GSH. Through site-directed mutagenesis, three conserved cysteine (Cys) residues (Cys62, Cys150, and Cys200) in MaArsM were demonstrated to be necessary for As (III) methylation, and Cys150 and Cys200 were required for the methylation of monomethylarsenic. These results showed a molecular pathway for As methylation in archaea and highlighted the role of archaea in As biogeochemistry. To our knowledge, this is the first study focused on the mechanism of arsenic methylation in archaea, which is widespread in the environment.

3. Arsenic biotransformation in rice rhizosphere

3.1 Functional genes for As biotransformation

Arsenic biotransformation is mediated by functional genes, such *aroA*-like genes for As (III) oxidation, *arrA* and *arsC* for As (V) reduction, and *ArsM* for As (III) methylation (Zhu et al., 2014, Fig. 3). The community and abundance of these genes is significantly affected by rhizosphere condition, and affect rice As uptake in turn. However, little is known about interactions between root and rhizosphere microbes involved in As biotransformation. Jia et al. (2014) used two rice cultivars with different radial oxygen loss (ROL) ability to investigate the impact of microbially mediated As redox changes on As tansformation. Results showed that the higher ROL cultivar (Yangdao) had lower As uptake than lower ROL cultivar (Nongken). The abundance of the As (III) oxidase gene (*aroA*-like) was more significantly enhanced by straw application than that of As (V) reductase gene (*arsC*) and As (V) respiratory reductase gene (*arrA*), resulting in more As oxidation and sequestration in the rhizosphere. The community of As (III)-oxidizing bacteria in the rhizosphere was dominated by α-Proteobacteria and β-Proteobacteria and was influenced by rice straw application, growth stage, and cultivar. Rice straw incorporation into the soil increased As release from soil and accumulation in rice plants. These results highlighted that uptake of As by rice roots is influenced by microbial processes in the rhizosphere, and root ROL and organic matter application can affect these processes significantly.

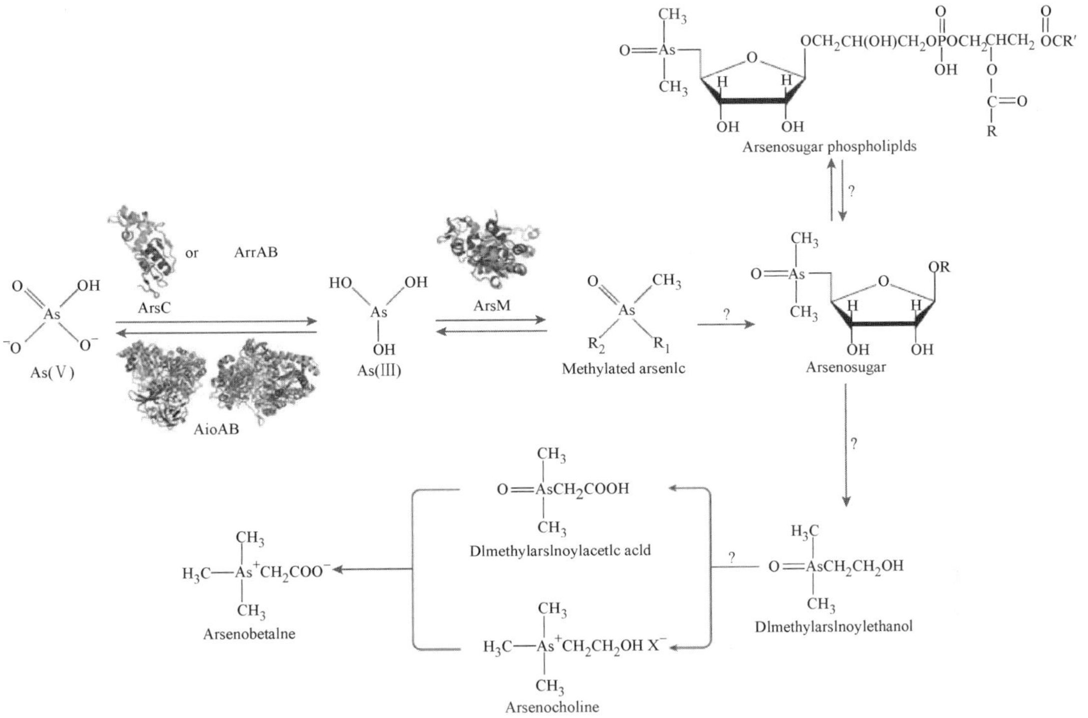

Fig. 3 Arsenic species and their biotransformations (Zhu et al., 2014)

Not only rhizosphere, soil organic matter amendment can also affect As biotransformation. Huang et al. (2012) investigated how amended OM influences As speciation and microbial processes in soil and soil solution. They conducted microcosm experiments with predried and fresh As contaminated paddy soils to investigate microbial mediated As transoformation and volatilization by amending different OM. The arsine flux from the microcosm amended with 10% clover and dried distillers grain were significantly higher than the control. Trimethylarsine (TMAs) was the dominant As species derived from clover treatment, whereas the primary As species from dried distillers grain was TMAs and arsine (AsH_3), followed by monomethylarsine ($MeAsH_2$). The predominant As species in the soil solutions of clover and dried distillers grain treatments were dimethylarsinic acid (DMAA) and As (Ⅴ), respectively. OM addition did not increase or even decrease the abundance of As (III) oxidizers, but increased the activities of As (III) oxidizing bacteria (harboring *aroA*-like genes). In contrast, OM amendment increased the abundance of As (Ⅴ) reducers (carrying the *arsC* gene), and enhanced the activity of As (Ⅴ) reducers when amended with clover. Through this study, we found that OM addition significantly increased As methylation and volatilization from paddy soil. In addition, the results suggested that physiologically active bacteria responsible for As oxidization, reduction, and methylation are coexisted in paddy soil, and regulating As speciation in soil environment.

3.2 Distribution of As biotransformation functional genes in paddy soils

Aarsenic biotransformation in paddy soils is mainly driven by microbes. However, the distribution, diversity and abundance of genes involved in As metabolism in paddy soils have not been well characterized. Through PCR-based methods, we revealed that As (III) oxidase genes (*aioA*), respiratory As (V) reductase genes (*ArrA*), As (V) reductase genes (*arsC*), and As (III) Sadenosylmethionine methyltransferase genes (*arsM*) exhibited wide distribution, high diversity, and abundance of in 13 paddy soils collected across Southern China (Zhang et al., 2015). These As biotransformation genes are mainly from rice rhizosphere bacteria, such as *Proteobacteria, Gemmatimonadales,* and *Firmicutes.* We found that there was a significant correlation of gene abundance between *arsC* and *ArsM*, this suggests that these two genes coexist well in the microbial As resistance system. Through, redundancy analysis (RDA), soil pH, EC, total C, N, As, and Fe, C/N ratio, SO_4^{2-}–S, NO_3^-–N, and NH_4^+–N were found to be the key factors driving the diversity of microbial community. To our knowledge, this is the first study provides an overall picture of microbial communities involved in As biotransformation in paddy soils, it also provides insights into the critical role of paddy fields on As biogeochemical cycle.

4. Iron plaque affects the micro-interface processes of As in rice rhizosphere

Iron plaque is commonly formed on the surfaces of roots of aquatic plants, including *Oryza sativa.* Iron plaque is mainly formed by the oxidation of ferrous to ferric iron and the precipitation iron oxide on the root surface. It has been demonstrated that iron plaque had very strong binding affinity to As (V) and played an important role on the As uptake by rice roots and transport within rice plants. Liu et al. (2004a) demonstrated that phosphorus deficiency can induce the iron plaque forming, and iron plaque can sequestrate As, and consequently reduce the translocation of As from roots to shoots of rice. They conducted hydroponic experiment to investigate the effect of phosphorus (P) nutrition and iron plaque on As (V) uptake by rice roots, and translocation within seedling. They found that rice roots appeared reddish after 24 h in -P solution (without P), while no color change on the +P treated roots, indicating -P induced the formation of iron plaque. As concentrations in iron plaque were significantly higher in -P plants than in +P plants, while As concentrations in shoots were significantly lower in -P plants than in +P plants. In addition, in -P rice plants, the total uptake of As was mainly concentrated on the surface of rice roots, whereas most As in +P plants was accumulated in the roots, which is ready for translocation to shoots. This indicated that iron plaque might sequestrate As, and consequently reduced the translocation of As from roots to shoots.

To further investigate the effects of iron plaque and genotypes on As (Ⅴ) uptake and translocation to shoots, Liu et al. (2004b) induced iron plaque by different concentration of iron on the root surface of different rice genotypes. They found that the amount of iron plaque was positively correlated with Fe^{2+} concentrations in solutions. However, the amount of iron plaque on root surface did not exhibited significantly positive correlation with As concentrations in the roots, and there were no significant differences in As concentrations in the roots between the three genotypes. However, As concentrations in shoots differed significantly between genotypes, and were positively correlated with iron concentrations in the shoots. The results suggest that iron plaque on rice root surface may act as a "buffer" for As in the rhizosphere. As sequestration in iron plaque can also significantly affect As accumulation and speciation in rice grains. Liu et al. (2006) used a compartmented soil-glass bead culture system investigated relationship between iron plaque and As accumulation and speciation in mature rice plants with different capacities of forming iron plaque. They characterized the mineralogical properties of iron plaque and As sequestration in plaque on the rice roots by using X-ray absorption near-edge structure spectra and extended X-ray absorption fine structure. Iron plaque was dominated by (oxyhydr) oxides, which with 81%-100% were ferrihydrite, and a minor amount of goethite (19%). Arsenic in iron plaque was mainly sequestered with amorphous and crystalline iron (oxyhydr) oxides, and As (Ⅴ) was the predominant species. Between genotypes, there was significant variation in iron plaque formation. For all genotypes, the distribution of As in different components of mature rice plants followed the following order: iron plaque>root>straw>husk>grain. Between genotypes, As accumulation in grain differed significantly. For all genotypes, inorganic As and dimethylarsinic acid (DMA) were the main As species in rice grain, and there were large genotypic differences in the ratio of DMA to inorganic As in grains.

Except iron plaque on rice root surface, there is also manganese plaque on rice root s surface (Liu et al., 2005). After 12 h induction, both Fe and Mn plaques were clearly visible as reddish or brown coatings on the rice root surface, and Fe plaque was much more apparent than Mn plaque. When As (Ⅴ) was supplied, As concentrations in Fe plaque were significantly higher than those in no plaque (control) and Mn plaque treatments, and much higher than those supplied with As (Ⅲ). This study showed that Fe plaque had priority to form than Mn plaque, and they had higher affinity to As (Ⅴ) than to As (Ⅲ).

Iron can also associate with nitrogen to affect As metabolism in the rice rhizosphere. Chen et al. (2008) investigate the effect of different forms of N fertilizers on As uptake by rice by conducting a pot microcosm experiment. They found that the nitrate addition decreased Fe plaque formation on the root surface, and decreased dissolved Fe (Ⅱ) concentrations in the rhizosphere soil solution, compared to the non-amended control. Nitrate addition also reduced As uptake by the rice plant. These results suggest that nitrate may inhibit Fe (Ⅲ) reduction and/or stimulate nitrate-dependent Fe (Ⅱ) oxidation, resulting higher Fe (Ⅲ) minerals in the soil. Higher Fe (III) minerals content in the soil could cause As coprecipitation with, or adsorption to, Fe (III) minerals in the soil. Therefore, they found that nitrate addition also reduced As uptake by the rice plant. Compared to a non-treated control, addition of ammonium enhanced Fe (III) reduction, probably coupled to NH_4^+ oxidation in the nonrhizosphere, which would thus increase As uptake by the rice plant.

5. Arsenic translocation and transformation within plants

5.1 Arsenate reduction by arsenate reductase

Enzymatic reduction of As (Ⅴ) to As (III) is the first step in As (Ⅴ) metabolism in all organisms studied. Duan et al. (2005) demonstrated that efficient As (Ⅴ) reduction in roots is one of the main mechanisms of As hyperaccumulating by Chinese brake fern (*Pteris vittata* L.). Root extracts from *P. vittata* can reduce As (Ⅴ) to As (III) and the reaction mechanism was similar to the previously reported Acr2p, an AR from yeast (Saccharomyces cerevisiae), using glutathione as the electron donor. Substrate specificity and sensitivity to inhibitors of fern AR was also similar to Acr2p. Fern AR had a Michaelis constant value of 2. 33 mM for As (V), which was 15-fold lower than the purified Acr2p. The AR-specific activity of the fern roots was about 7 times higher than those of roots and shoots of non tolerate plants. No AR activity was detected in the Acr2 gene knockout mutant of Arabidopsis (Arabidopsis thaliana). In shoots of *P. vittata*, we could not detect AR activity either. this study demonstrated that As (III) in fronts of *P. vittata* may come mainly from the reduction of As (Ⅴ) in roots, and AR in roots plays an important role in As detoxification *P. vittata* fern.

ACR2 is the As (Ⅴ) reductase of yeast (*Saccharomyces cerevisiae*). The rice genome contains two ACR2-like genes, *OsACR2. 1* and *OsACR2. 2*, which may be involved in regulating As metabolism in rice. Duan et al. (2007) cloned both *OsACR2* genes and expressed them in *Escherichia coli* and yeast (*Saccharomyces cerevisiae*) strains in which the arsenate reducatse genes was disrupted. *OsACR2. 1* complemented the As (Ⅴ) hypersensitive phenotype of both *E. coli* and yeast, while *OsACR2. 2* showed much less ability to complement. The purified gene products were demonstrated to reduce As (Ⅴ) to As (III) *in vitro*, and both exhibited phosphatase activity. In agreement with the results of complementation, *OsACR2. 1* exhibited higher arsenate reductase activity than *OsACR2. 2*. Mutagenesis of cysteine residues in HC (X) 5 R motif led to completely loss activities of phosphatase and As (Ⅴ) reductase. In *planta*, the expression of *OsACR2. 1* was much stronger than that of *OsACR2. 2,* and increased dramatically after exposure to As (Ⅴ), while the expression of *OsACR2. 2* was observed only in roots following As (Ⅴ) exposure.

5.2 Arsenic biotransformation and volatilization in transgenic rice

Inorganic As can be biotransformed to organic As, As (III) S-adenosylmethyltransferase enzymes catalyze As (III) methylation to mono-, di-, and trimethylated As species, some of which are less toxic than As (III). However, no methyltransferase gene has been identified in plants. By expressing an *ArsM* gene from the soil bacterium Rhodopseudomonas palustris in Japonica rice (*Oryza sativa*) cv *Nipponbare*, we generated a transgenic rice which produced methylated As species (Meng et al. 2010). They found that both monomethylarsenate (MAs (Ⅴ)) and dimethylarsenate (DMAs (Ⅴ)) were detected in the roots and shoots of transgenic rice, and volatile arsenicals were detected from the transgenic rice after 12 d exposure to As (III). This study demonstrated As methylation and volatilization could be achieved by expression of an *ArsM* gene in rice induces, theoretically providing a potential strategy for phytoremediation of As contaminated soil and water.

5.3 Spatial distribution and temporal concentration in rice

Although As uptake by root, translocation and transformation have studied thoroughly, there is little information about spatial distribution of As and the temporal variation of As concentration in whole rice plants at different growth stages. In order to gain insights into the transport and distribution of As in intact rice plants and its unloading into the rice grain, the spatial distribution of As and the temporal variation of As concentration in whole rice plants at different growth stages were investigated. In this study, Zheng et al. (2011) investigated in situ As distribution in the leaf, internode, node and grain by using synchrotron-based X-ray fluorescence (SXRF). They found that total As concentrations in vegetative tissues increased during the 2 wk after flowering. In the caryopsis, the concentration of dimethylarsinic acid (DMA) decreased progressively with its development, but inorganic As concentration remained stable. The ratios of As concentration between neighboring leaves or between neighboring internodes were c.0.6. During the caryopsis developing, As accumulated in the center at its early stage and then in the ovular vascular trace. DMA accumulated mainly in the caryopsis before flowering, whereas inorganic As was mainly transported into the caryopsis during grain filling. These results indicated that there are different pathways for the unloading of inorganic As and DMA to grains. Moreover, nodes were demonstreated as check-point in As distribution in rice shoots. This study should be the first time description of the dynamic As distribution and accumulation within rice plants from seedling to mature.

5.4 Arsenic transporting to plant phloem

Phloem transport has been shown to be central for arsenic translocation to the grains. In rice grains, approximately 90% of the As (III) was transported via the phloem. In the recent decades, membrane transporters that catalyse As uptake by roots, and translocation through the xylem, have been characterized in many plants, including rice. However, the transporters responsible for As loading into the phloem and on into the seeds have not yet identified. Duan et al. (2016) demonstrated that inositol transporters in the phloem in Arabidopsis also transport As (III). Expression *AtINT2* or *AtINT4* in *Saccharomyces cerevisiae* led to increased As accumulation and high sensitivity to As (III). Expressing *AtINT2* in *Xenopus laevis* oocytes induced As (III) import activity. Disruption of *AtINT2* or *AtINT4* in *Arabidopsis thaliana* resulted in reduction of As concentrations in phloem, silique and seed in plants fed with As (III) through the roots, comparing with wild-type plants. As concentrations in silique and seed of the mutants were significantly lower than that of wild-type plants when fed with As (III) through the leaves. This study demonstrated that in Arabidopsis, inositol transporters, *AtINT2* and *AtINT4*, are responsible for As (III) loading into the phloem, the key step of As loading to seeds. This study is the first identification of transporters regulating As accumulation in plant seeds. If these findings are applicable to rice, then inositol transporters could be the candidates for future genetic modification for reducing As content in rice grain without affecting yield production.

6. Genetic control of As accumulation and speciation in rice grain

An understanding of genetic variation and the effect of environments on rice As accumulation is needed for breeding rice cultivars with low grain As. To this aim, an international cooperative study compared a range of common cultivars grown in contaminated sites across Bangladesh, China and India (Norton et al., 2009). In this study, 13 cultivars were grown at 2 field sites each in Bangladesh, India and China. They found that total grain As was a significant affected by site, genotype, and site by genotype interaction. Correlations were observed between sites in Bangladesh and India, but not observed between countries or within the two sites in China. Among these 13 cultivars, grains of 7 cultivars were analyzed As speciation. Results revealed that percentage As*i* was significantly affected by site, genotype, and site by genotype interaction. Although a number of genotypes that in general had low grain As across the multiple field sites have been identified, the variations across different environments would pose difficulty for breeding consistently low grain As (and low As*i*) over multiple environments. This variation rises the limitation of traditional breeding approaches and requirement genetic approaches.

The genetic control of As accumulation in rice plants was further illustrated by identification the quantitative trait loci (QTLs) associated with As accumulation in rice. Zhang et al. (2008) mapped the QTLs associated with As accumulation in rice by using a doubled haploid population, which was established by anther culture of F1 plants from a cross Japonica cultivar CJ06 and Indica cultivar TN1 (*Oryza sativa*). They detected four QTLs for As concentrations in the map. At the seedling stage, two QTLs was mapped, one was on chromosome 2 for As concentrations in shoots, and the other one on chromosome 3 for As concentrations in roots. At maturity, two QTLs were found for As concentrations in grains. They were on chromosomes 6 and 8, and contributed to 26.3% and 35.2% phenotypic variance, respectively. Among As concentrations in roots, shoots and grains, there was no common loci were detected. However, the QTL on chromosome 8 was found to be colocated for As concentrations in grain at maturity and phosphorus (P) concentrations in shoot at seedling stage. This study provided an insight into the genetic control of As uptake and accumulation in rice, and will be useful to develop biomarkers for breeding or selecting rice culturs with low grain As accumulation.

7. Summary

Understanding the way in which As is transported and transformed within soil-plant system is critical for mitigation As in rice, from soil to grain and to humans. During the recent decades, significant progresses have been made in this field, including As biotransformations in rice rhizosphere, metabolism within rice plants, and accumulation in rice grains. Zhu et al. (2014) reviewed the biotransformation of As from the perspective of the formation of earth and described the evolution of life, and the connection between As geochemistry and biology. This paper is an overview of the research findings from our lab in recent years, including molecular mechanisms of As redox and methylation cycles as well as other As biotransformations, and also the prospective implications of As biotransformation in environmental remediation and food safety, with particular emphasis on groundwater As contamination and As accumulation in rice.

References

Chen X P, Zhu Y G, Hong M N, et al. 2008. Effects of different forms of nitrogen fertilizers on arsenic uptake by rice plants. Environmental Toxicology and Chemistry, 27 (4): 881-887.

Duan G L, Hu Y, Schneider S, et al. 2016. Inositol transporters *AtINT2* and *AtINT4* regulate arsenic accumulation in arabidopsis seeds. Nature Plants, 2: 15202.

Duan G L, Zhou Y, Tong Y P, et al. 2007. A CDC25 homologue from rice functions as an arsenate reductase. New Phytologist, 174 (2): 311-321.

Duan G L, Zhu Y G, Tong Y P, et al. 2005. Characterization of arsenate reductase in the extract of roots and fronds of Chinese brake fern, an arsenic hyperaccumulator. Plant Physiology, 138 (1): 461-469.

Huang H, Jia Y, Sun G X, et al. 2012. Arsenic speciation and volatilization from flooded paddy soils amended with different organic matters. Environmental Science & Technology, 46 (4): 2163-2168.

Jia Y, Huang H, Chen Z, et al. 2014. Arsenic uptake by rice is influenced by microbe-mediated arsenic redox changes in the rhizosphere. Environmental Science & Technology, 48 (2): 1001-1007.

Li G, Sun G X, Williams P N, et al. 2011. Inorganic arsenic in Chinese food and its cancer risk. Environment International, 37 (7):

1219-1225.

Liu W J, Zhu Y G, Hu Y, et al. 2006. Arsenic sequestration in iron plaque, its accumulation and speciation in mature rice plants (Oryza sativa L.). Environmental Science & Technology, 40 (18): 5730-5736.

Liu W J, Zhu Y G, Smith F A, et al. 2004b. Do iron plaque and genotypes affect arsenate uptake and translocation by rice seedlings (Oryza sativa L.) grown in solution culture. Journal of Experimental Botany, 55 (403): 1707-1713.

Liu W J, Zhu Y G, Smith F A, et al. 2004. Do phosphorus nutrition and iron plaque alter arsenate (As) uptake by rice seedlings in hydroponic culture. New Phytologist, 162 (2): 481-488.

Liu W J, Zhu Y G, Smith F A. 2005. Effects of iron and manganese plaques on arsenic uptake by rice seedlings (Oryza sativa L.) grown in solution culture supplied with arsenate and arsenite. Plant and Soil, 277 (1-2): 127-138.

Meng X Y, Qin J, Wang L H, et al. 2011. Arsenic biotransformation and volatilization in transgenic rice. New Phytol., 191 (1): 49-56.

Norton G J, Duan G L, Dasgupta T, et al. 2009. Environmental and Genetic Control of Arsenic Accumulation and Speciation in Rice Grain: Comparing a Range of Common Cultivars Grown in Contaminated Sites Across Bangladesh, China, and India. Environmental Science & Technology, 43 (21), 8381-8386.

Wang P P, Sun G X, Zhu Y G. 2014. Identification and characterization of arsenite methyltransferase from an archaeon, Methanosarcina acetivorans C2A. Environmental Science & Technology, 48 (21): 12706-12713.

Ye J, Rensing C, Rosen B P, et al. 2012. Arsenic biomethylation by photosynthetic organisms. Trends in Plant Science, 17 (3): 155-162.

Yin X X, Chen J, Qin J, et al. 2011. Biotransformation and volatilization of arsenic by three photosynthetic cyanobacteria. Plant Physiol, 156 (3): 1631-1638.

Zhang J, Zhu Y G, Zeng D L, et al. 2008. Mapping quantitative trait loci associated with arsenic accumulation in rice (Oryza sativa). New Phytologist, 177 (2): 350-356.

Zhang S Y, Zhao F J, Sun G X, et al. 2015 Diversity and abundance of arsenic biotransformation genes in paddy soils from southern China. Environmental Science & Technology 49 (7): 4138-4146.

Zheng M Z, Cai C, Hu Y, et al. 2011. Spatial distribution of arsenic and temporal variation of its concentration in rice. New Phytol. 189 (1): 200-9.

Zhu Y G, Rosen B P. 2009. Perspectives for genetic engineering for the phytoremediation of arsenic-contaminated environments: from imagination to reality. Current Opinion in Biotechnology, 20 (2): 220-224.

Zhu Y G, Sun G X, Lei M, et al. 2008. High percentage inorganic arsenic content of mining impacted and nonimpacted Chinese rice. Environmental Science & Technology, 42 (13): 5008-5013.

Zhu Y G, Yoshinaga M, Zhao F J, et al. 2014. Earth abides arsenic biotransformations. Annual Review of Earth and Planetary Sciences, 42: 443.

Arsenic Hyperaccumulator *Pteris vittata* L. and Its Application to the Field

Tongbin CHEN, Mei LEI, Xiaoming WAN, Jun YANG, Xiaoyong ZHOU

Institute of Geographic Sciences and Natural Resources Research, Chinese Academy of Sciences

Abstract

Arsenic contaminated soil is a serious problem in China, which seriously threatens the safety of drinking water and food safety. Phytoremediation, an emerging, plant-based technology for the removal of toxic contaminants from soil, has been receiving renewed attention. The discovery of the As hyperaccumulator *Pteris vittata* L. opens a door for the phytoremediation of arsenic contaminated soils. The mechanisms of arsenic uptake, translocation, and transformation by this plant have keen studied. The hyperaccumulation and hypertolerance characteristics and mechanisms have keen disclosed. Antioxidative system, compartmentation and chelation may contribute to the tolerance of hyperaccumulators. The reduction of As (Ⅴ) to As (Ⅲ) (>50% of As in aboveground parts existed as As (Ⅲ)), the high concentration of phosphor in root (48%-53% of P in *P. vittata* distributed in roots), and the efficient translocation of As from roots to aboveground parts (percentage of aboveground As accounted for 73%-77% of total As) were suggested as essential processes for the hyperaccumulation of As in *P. vittata*. Studies on population differences and multi-metals accumulating theories were also summarized. *In situ* phytoremediation projects using *P. vittata* have been established, with high As removal rates achieved. The internationally first phytoremediation project was established in Chenzhou, Hunan Province. Afterwards, more phytoremediation projects were established in the Guangxi Zhuang Autonomous Region, Yunnan Province, Henan Province, Hebei Province, and Beijing City. During the phytoremediation practices, some problems than cannot be confronted in the lab were faced; and through both lab and field investigation, some key cultivation and management issues were studied, including the reproduction, optimized fertilizer, intercropping technology, harvesting, incineration and cost calculation. Based on these previous study, future research directions were pointed out.

Keywords: arsenic, field practice, hyperaccumulation, population, *Pteris vittata* L.

1. Introduction

Heavy metal contamination in soil has become an environmental concern worldwide. In

March 2015, the Chinese Ministry of Environmental Protection and the Ministry of Land and Resources released the first-ever results of a nationwide soil pollution survey, which revealed startling statistics such as one-fifth of arable land is polluted and contaminated with inorganic chemicals like cadmium (Cd), nickel (Ni) and arsenic (As). Serious soil contamination with As has been reported at numerous sites worldwide. Arsenic in soils can enter the food chain through drinking water or agricultural products, posing risk to human health and environmental safety.

The concept of using hyperaccumulators to remove heavy metals from contaminated soils was proposed first by Rufus Chaney from U. S. Department of Agriculture (Chaney et al., 1997). By planting hyperaccumulators in contaminated soils, heavy metals can be removed through harvesting the above ground parts. The phytoremediation of metal-contaminated soils offers a low-cost method for soil remediation and some extracted metals may be recycled for value. Phytoremediation is a cost-effective, engineering-economical and environmental friendly technique (Kramer, 2005).

An arsenic hyperaccumulator *Pteris vittata* L. was first discovered in China by means of field survey and greenhouse cultivation (Ma et al., 2001; Chen et al., 2002). Both field survey and pot trials showed that Chinese brake had large accumulating capacity to arsenic. The bioaccumulation coefficients of the above ground parts and rhizoids of *P. vittata* were as high as 71 and 80 respectively. The aboveground As concentration of 5070 mg/kg on a dry matter basis was noticed. Not only has *P. vittata* extraordinary tolerance and accumulation to arsenic, but it grew rapidly with great biomass, wide distribution and easy adaptation to different environmental conditions as well. Therefore, it has great potential in the future remediation of arsenic contamination.

Since the discovery of the hyperaccumulating characteristic of *P. vittata*, a large number of studies have been conducted on hyperaccumulating theory and the field practice. *P. vittata* has become an essential plant material for the investigation of arsenic physiology and biochemistry such as arsenic absorption, translocation and detoxification mechanisms (Rascio and Navari, 2011). At the same time, *in situ* phytoremediation practices using *P. vittata* have been established on both farmlands and residential areas, with high removal rates of As achieved (Kertulis et al., 2006; Chen et al., 2007; Ebbs et al., 2010). This is a review of recent developments in basic and applied research relevant for the *P. vittata*-based clean-up of As contaminated soils.

2. Hyperaccumulating theories

2.1 Hyperaccumulation characteristics

P. vittata showed super tolerance to As. Under field circumstance, *P. vittata* can grow on soils with As concentration being 50-4030 mg/kg, even on mine tailings with As concentration being as high as 23, 400 mg/kg (Chen et al., 2002). *P. vittata* also showed super As accumulating ability. In the field, the As concentration in aboveground parts can be as high as 1540 mg/kg, with a bioaccumulating factor of 7.43. In the pot trial, the aboveground As concentration can be as high as 4383 mg/kg, with a bioaccumulating factor of 77.6. *P. vittata* growing on soil with As concentration of 6 mg/kg showed an aboveground As concentration of 755 mg/kg, with a bioaccumulating factor of 126 (Ma et al., 2001). Different from normal plants, the large amount of As accumulated in *P. vittata* did not lead to an decrease in biomass; in contrary, a certain amount of As can increase the biomass of As. According to the field survey, the height of *P. vittata* can reach 2 m, with the biomass being 36 t/ha. Besides, *P. vittata* can quickly translocate As from roots to aboveground parts (Kitajima et al., 2008). Therefore, due to its super tolerance to As, hyperaccumulation of As, efficient absorption of As and high biomass, *P. vittata* has huge potential to be utilized in phytoextraction.

The distribution of As in *P. vittata* was in the order of pinnae＞stem＞root. Arsenic in the pinnae of *P. vittata* was significantly higher than the rest tissue (Chen et al., 2002). The micro-distribution of As was analyzed using synchrotron radiation X-ray fluorescence (SRXRF) and Environmental scanning electron microscope fitted with an energy dispersive X-ray micro analyzer. It was shown that As was relatively high in the midrib than in its surrounding tissues, and the difference between midrib and the surrounding tissues was the most distinct in the tip of pinna. That is to say that *P. vittata* had strong ability to transport As across mesophyll cell membrane (Chen et al., 2005a). The relative weight of As in the pinnate trichome, which contained the highest concentration of As among all tissues of the plant, was 2.4 and 3.9 times as much as that in the epidermal and mesophyllous cells, respectively, indicating the essential role of trichome in As hyperaccumulation (Li et al., 2005b).

P. vittata can accumulate As from different As compounds, including NaMMA, CaMMA, K_2HAsO_4, Na_2HAsO_4 and $Ca_3(AsO_4)_2$ (Tu and Ma, 2002). Arsenite (AsIII) treated *P. vittata* can translocate As at a higher rate than arsenate (AsⅤ) treated *P. vittata* (Su et al., 2008). Arsenic in *P. vittata* was mainly in form of inorganic AsⅤ or AsIII, with a small amount of $FeAsO_4$, $AlAsO_4$ and $Ca_3(AsO_4)_2$ (Tu and Ma, 2002; Zhang, 2002). Further research has found that As existed in root mainly as AsⅤ, while mainly in form of AsIII in pinnae (Huang et al., 2004). A reduction reaction from AsⅤ to AsIII was found in root, and was further located at the endodermis of rhizoid (Huang et al., 2004; Lei et al., 2012a). This reduction reaction was evidenced by the study indicating that the extracts of the root of *P. vittata* was found to be able to reduce AsⅤ to AsIII (Duan et al., 2005), and the research indicating that As in the xylem was mainly in the form of As (III) (Su et al., 2008). Thiols were further found to play a role of reducing agent in this reaction (Lei et al., 2013a). Using Synchrotron Radiation Extended X-ray Absorption Fine Structure (SR-EXAFS), it has been further indicated that As in the pinnae of *P. vittata* existed mainly as As (III)-O, whereas in the root, As (Ⅴ)-O predominated (Huang et al.,

2004). A limited amount of As existed as As-SH. In another As hyperaccumulator *P. cretica*, similar chelating model was detected (Huang, 2004).

2.2 Hyperaccumulation mechanism

For ordinary plants, As is a toxic element, whereas *P. vittata* can hyperaccumulate As and maintain normal physiological activity. The potential reason for the hypertolerance and hyperaccumulation is a research focus. It has been suggested that antioxidative system, compartmentation and chelation may contribute to the tolerance of hyperaccumulators.

Antioxidative system. It has been found that both enzymatic and nonenzymatic antioxidants played significant roles in As detoxification and hyperaccumulation in *P. vittata*, and the former is more important at low As exposure (less than or equal to 20 mg/kg), whereas the latter is more critical at high As exposure (50-200 mg/kg) (Cao et al., 2004). A comparison study of *P. vittata* and a non-hyperaccumulator *P. ensifortnis* indicated that upon exposure to 133 μmol/L As, concentrations of chlorophyll, protein and carotenoids increased in *P. vittata* whereas they decreased in non-hyperaccumulator *P. ensifortnis* (Singh et al., 2006). Similar comparison study found that arsenate reductase (AR) activities of roots, superoxide dismutase (SOD) activities and As concentrations in both roots and fronds of the four *Pteris* plants increased when exposed to As-contaminated soils, indicating that AR and SOD may play important roles to accumulate and detoxify As in the four *Pteris* species (Liu et al., 2009).

Compartmentation. Compartmentation was proposed as an essential mechanism for the hyperaccumulators to tolerate heavy metals, which was confirmed by the investigation on the sub-cellular distribution of As in *P. vittata* (Chen et al., 2005b). Trichome and vacuoles were suggested to be compartments where *P. vittata* store excess As in pinnae (Li et al., 2005b; Lombi et al., 2002). When *P. vittata* was grown in a nutrient solution without additional As, most of the accumulated As was isolated to the cell wall. However, in plants growing in a nutrient solution containing 0.1 or 0.2 mmol/L As, approximately 78% of the total As accumulated within the pinna. The proportions of As accumulation in the cytoplasmic supernatant fraction were 78% of that in the pinna and 61% of that in the plant. Results suggest that As accumulates in the pinna where it is primarily distributed in the cytoplasmic supernatant fraction. The role of As compartmentalization may be intricately linked with As detoxification in *P. vittata*.

Chelation. Phytochelatins (PCs) was suggested to be an important chelator for As in *P. vittata*, which can decrease the amount of As in form of free ion, therefore lower the hurt of As to plant tissue. With the exposed As concentration increasing from 150 mg/kg to 300 mg/kg, thiols and glutathione in *P. vittata* significantly increased (Srivastava et al., 2009). Similarly, PCs synthesis was induced upon exposure to AsV in *P. vittata*, And the As concentration correlated significantly with PC2 concentration in both roots and shoots, but not with GSH

(Zhao et al., 2003). However, the molar ratio of PC-SH to As was 0.09 and 0.03 for shoots and roots, respectively, suggesting that only a small proportion (1%-3%) of the As in *P. vittata* can be complexed with PCs. Therefore, it is suggested that PCs play a limited role in the hypertolerance of As in *P. vittata* (Zhao et al., 2003). This suggestion was supported by theoretical calculation, which found that for the detoxification of As by PCs, the molar ratio of SH to As should be more than 3, whereas the molar ratio of SH to As in *P. vittata* was much less than 3 (Shoji et al., 2008). Through *in vivo* regulation of sulfur metabolism, it has been found that thiols appeared able to act as both a reductant and a chelator of As in *P. vittata* and the ratio of-SH to As may have determined the specific role of-SH in *P. vitatta* (Lei et al., 2013b).

Besides the hypertolerance of *P. vittata* to As, the processes contributing to the hyperaccumulation of As in *P. vittata* was another research focus. The reduction of As (Ⅴ) to As (Ⅲ) (>50% of As in aboveground parts existed as As (Ⅲ)), the high concentration of phosphor in root (48%-53% of P in *P. vittata* distributed in roots), and the efficient translocation of As from roots to aboveground parts (percentage of aboveground As accounted for 73%-77% of total As) were suggested as essential processes (Luongo and Ma, 2005).

Reduction of As (Ⅴ) to As (Ⅲ) in root. Although the reduction of As (Ⅴ) to As (Ⅲ) was observed in both hyperaccumulators and non-hyperaccumulators, the aboveground As (Ⅲ) percentage in *P. vittata* (83%-84%) was significantly higher than non-hyperaccumulators (Fayiga et al., 2008). After 24 h exposure to AsⅤ, more than 90% of As in pinnae was in form of As (Ⅲ) (Tu et al., 2004). Through the comparison of *P. vittata* and ramie, it has been found that the chelation of As with GSH prevented the upwards translocation of As in ramie, whereas in *P. vittata*, As was mainly transported in form of free As (Ⅲ) ion, which favored the efficient upwards translocation of As (Huang et al., 2008). Further study indicated that arsenite was the predominant species in the xylem sap, accounting for 93%-98% of the total As, and more As was translocated to fronds in the arsenite-treated than in the arsenate-treated plants (Su et al., 2008). Therefore, the efficient reduction of As (Ⅴ)to As (Ⅲ) may be one of the key mechanisms for *P. vittata* to hyperaccumulate As.

Relationship between P and As. Due to the similarity between phosphate and arsenate, it has been proposed that the uptake of As in *P. vittata* occurred via a high-affinity P translocation system. The addition of P has been reported to inhibit As uptake (Huang et al., 2007; Tu et al., 2004) and the amount of As has correlated positively with the amount of P in plant tissues and cells (Chen et al., 2002). Through the comparison between As hyperaccumulator and non-hyperaccumulator, it was disclosed that the hyperaccumulation of As by *Pteris* plant may be contributed by the transporter with high affinity to As (Ⅴ), which was also the transporter of P, therefore, there was a competitive absorption between As and P (Poynton et al., 2004). To further investigate the interaction of As and P in *P. vittata*, a synchrotron X-ray microprobe was utilized to study the micro-distribution and micro-speciation of As and P (Lei et al., 2012b). It was found that As (V) was reduced to As (III) at the endodermis of *P. vittata* rhizoids. As and P displayed great similarity during the transportation process from epidermis to cortex of *P. vittata*, with the addition or absence of As or P, implied the co-transport of arsenate and phosphate in rhizoids. From the endodermis to the vascular bundle, arsenite were not co-transported with P, detected from the influence of P deficiency and energy inhibition on the transport to two elements.

Efficient translocation of As from roots to aboveground parts. Transpiration was found to be an important process contributing to the efficient translocation of As from roots to aboveground parts (Wan et al., 2015). The As accumulation of *P. vittata* increased proportionally with an increase in the As exposure concentration. Lowering the transpiration rate by 28%-67% decreased the shoot As concentration by 19%-56%. Comparison of As distribution under normal treatment and shade treatment indicated that transpiration determines the distribution pattern of As in pinnae. In terms of the ecotypic difference, the *P. vittata* ecotype from moister and warmer habitat had 40% higher transpiration and correspondingly 40% higher shoot As concentration than the ecotype from drier and cooler habitat. Transpiration is the main driver for *P. vittata* to accumulate and re-distribute As in pinnae.

2.3 Population differences

P. vittata has a wide distribution in China. Through field investigation and lab experiments, it has been found that the variation in As accumulation among different *P. vittata* populations can be-5.4 times (Wu et al., 2009). Both differences in plant biomass and As accumulating ability were observed (Cai et al., 2007). There were dramatic difference in plant height, biomass and frond number. The plant height ranged from 2916 to 6812 cm, frond number from 1810 to 6010 per plant, shoot fresh weight from 150 to 540 g per plant and root fresh weight from 2013 to 9419 g per plant. The distinct difference among the genotypes was also found in shoot and root As accumulation, ranging from 643.10 to 3009.03 mg/kg and from 26.34 to 112.38 mg/kg, respectively. There was a positive association between As accumulation and some growth characters, including plant height, frond and bud numbers, and shoot fresh weight.

Such difference among populations were related to rhizoids' As uptake and As species conversion through a comparison of 4 *P. vittata* populations collected from 4 sites located in one mining area with a gradient of soil As concentrations (from 108 to 7527 mg/kg) (Wan et al., 2013). Populations from low-As habitats had 80% higher shoot As concentrations than those from high-As habitats. In contrast, populations from high-As habitats had-5 times higher biomass when exposed to the same concentration of As. Further study indicates that the rhizoids' As uptake and As species conversion are important physiological activities bridging soil As concentration and As accumulation or tolerance of *P. vittata*. Depending on the target soil As concentration, selecting different *P. vittata* populations can result in-8 folds difference in remediation efficiency. Results provide essential information for phytoremediation theory and practice.

2.4 Multi-metals accumulating theories

In the real cases, multi-metals often co-exist, whereas hyperaccumulators normally hyperaccumulate only one element. This contradiction inspires the investigation on potential multi-metals hyperaccumulator (Wan et al., 2016b). Different populations of *P. vittata* that can hyperaccumulate As and cadmium (Cd), As and zinc (Zn), and As and lead (Pb) were identified.

The results showed that *P. vittata* could survive in pot soils spiked with 80 mg/kg of Cd and tolerated as great as 301 mg/kg of total Cd and 26.8 mg/kg of diethyltriaminepenta acetic acid (DTPA)-extractable Cd under field conditions (Xiao et al., 2008). The highest concentration of Cd in fronds was 186 mg/kg under a total soil concentration of 920 mg As/kg and 98.6 mg Cd/kg in the field. The Cd-tolerant ecotype of *P. vittata* extracted effectively As and Cd from the site co-contaminated with Cd and As, and might be used to remediate and revegetate this type of site.

Similarly, the potential of As hyperaccumulator *P. vittata* to remediate sites co-contaminated with Zn and As was investigated (An et al., 2006). *P. vittata* had a very high tolerance to Zn and grew normally at sites with high Zn concentrations. In addition, *P. vittata* could effectively take up Zn into its fronds, with a maximum of 737 mg/kg under field conditions. The high Zn tolerance, relatively high ability to accumulate Zn, and great capacity to accumulate As under conditions of suppression by high Zn suggest that *P. vittata* could be useful for the remediation of sites co-contaminated with Zn and As.

The *P. vittata* population isolated from a Pb-Zn mine in Yunnan province, China is a potential extractor of As and Pb co-contamination (Wan et al., 2014; Xie et al., 2010). Hydroponic experiment found that the highest frond As and Pb concentrations in mining population of *P. vittata* reached 12219 mg/kg and 992 mg/kg, respectively. The interaction between As and Pb in *P. vittata* was further more disclosed. Pb (2 mg/L) improved the frond As concentration by 60% to 150% in mining populations of *P. vittata*.

3. Field experiences

In situ phytoremediation projects using *P. vittata* have been established on both farmlands in China and residential areas in America, with high As removal rates achieved (Chen et al., 2007; Ebbs et al., 2010). The internationally first phytoremediation project was established in Chenzhou, Hunan Province. Afterwards, more phytoremediation projects were established in the Guangxi Zhuang Autonomous Region, Yunnan Province, Henan Province, Hebei Province, and Beijing City (Table 1).

Table 1 Phytoremediation projects of arsenic contaminated soil in China

Place	Contaminant	Technology	Remediation efficiency
Chenzhou, Hunan Province	As	Phytoextraction	● Through 3-5 year remediation, soil As content decreased from 50 mg/kg to<30 mg/kg, reaching the second national soil quality standard (GB15618—1995)
Shimen, Hunan Province	As	Phytoextraction and intercropping	● Phytoextration technology removed 13% As from soil each year ● In the intercropped system, agricultural products from intercropped cash crops meet the national standard
Huanjiang, Guangxi Zhuang Autonomous Region	As, Cd and Pb	Phytoextraction, intercropping and phytobarrier	● Phytoextration technology removed 10.5% Cd and 28.6% As from soil after 2-year remediation. ● In the intercropped system, the yield of maize, rice and sugarcane increased by 154%, 29.6% and 105%, respectively; and the As, Cd and Pb concentration in corn kernel decreased by 39%, 4.1% and 4.9%, respectively ● In the phytobarrier system, the over standard rate of heavy metals in agricultural products was less than 5%
Gejiu, Yunnan Province	As and Pb	Phytoextraction	● Phytoextration technology removed 18% As and 14% Pb from soil each year
Huize, Yunnan Province	As	Phytoextraction	● Phytoextration technology removed 12% As from soil each year
Jiyuan, Henan Province	As, Cd and Pb	Phytoextraction	● Phytoextration technology removed 13.9% As, 0.5% Pb and 16.1% Cd from soil each year ● After 2-year remediation, 338.5g As, 36.2 g Cd and 104.5 g Pb were removed from each Mu soil
Fangshan, Beijing City	As	Phytoextraction	● Phytoextration technology removed 17.2% As from soil each year
Dabaoshan, Guangdong Province	As, Cd, Cu, Pb and Zn	Intercropping and phytobarrier	● Products of cash crops met the national standards
Baoding, Hebei Province	AS	Phytoextraction	● Phytoextraction removed 8% As from soil each year

During these phytoremediation practices, some problems than cannot be confronted in the lab were faced, and through both lab and field investigation, some key cultivation and management issues were studied.

Reproduction. The supply of enough sporelings is the first step for a successful application of *P. vittata* to an As contaminated site. The germination and growth conditions of *P. vittata* was optimized. The effects of soil pH, soil calcium (Ca) concentration, and temperature on the sexual propagation of *P. vittata* was investigated. The best temperature was 25℃, and the germination was both accelerated and increased by high soil pH and Ca concentration (Wan et al., 2010).

Fertilizer. The optimum nitrogen (N) and P fertilizers were selected. The total As accumulation in the plants grown in As-supplied soil, under different forms of N fertilizer, decreased as $NH_4HCO_3 > (NH_4)_2SO_4 >$ urea $>$ Ca $(NO_3)_2 > KNO_3 >$ CK. The plants treated with N and As accumulated up to 5.3-7.97 mg As/pot and removed 3.7%-5.5% As from the soils, compared to approximately 2.3% of As removal in the control. NH_4^+-N was apparently more effective than other N fertilizers in stimulating As removal when soil was supplied with As at initiation (Liao et al., 2007).

Field experiment showed that the yields of *P. vittata* were enhanced with increasing P addition, however there was no further increase when rate of P addition was more than 200 kg/hm^2. Application of P fertilizer could enhance As concentration, and As concentration in the aboveground of *P. vittata* was depressed because of excessive P addition. Theoretically, As concentration of plant reaches a maximum of 1622 mg/kg when rate of P addition is 340 kg/hm^2. Moreover, the results also showed that P application was helpful to maintain a balance of available As between before transplanting and after harvest (Liao et al., 2004).

Intercropping. Phytoremediation of heavy metal contaminated soil often costs a long period. The agricultural production was interrupted during the remediation, which makes it difficult to promote the engineering applications of phytoremediation. Intercropping of hyperaccumulators vegetation and economical crops could get economic benefits during the remediation. Therefore, intercropping system which can remediate contaminated soil and produce safe products at the same time was established (Wang, 2012). A series of cash crops with low As accumulating ability were screened out. The key parameters, such as the planting distance, the population difference, were established.

Harvesting. *P. vittata* belongs to perennial plants, indicating that it can be harvested for several times during growth period. The effects of harvesting on As accumulation and phytoextraction efficiency of *P. vittata* was studied by pot experiment. The As uptake rates by *P. vittata* in the first harvesting ranged from 20 to 35 μg/(plant·d), which was significantly lower than those of the second and third harvesting. Therefore, more harvesting would increase the As accumulation and phytoextraction efficiency of As of *P. vittata*, indicating harvesting was an economical and convenient way for the phytoremediation when utilized *P. vittata* to improve the phytoextraction efficiency of As (Li et al., 2005a). Through field practice, 16 kg As, 8.5 kg Pb and 9.2 kg Zn can be removed from one hectare soil each year by harvesting 2 times per year (Xie et al., 2010).

Incineration of hyperaccumulator biomass. Disposal of hyperaccumulators with extremely high concentration of heavy metals is an important step for the phytoextraction practice. Incineration results reveal that 24% of total As accumulated by *P. vittata* (H-As) containing high As content (1170 mg/kg) is emitted at 800℃, of which 62.5% of the total emitted As is volatilized below 400℃. EXAFS and thermo gravimetric experiments showed that carbon originating from biomass incineration might catalyze As (Ⅴ) reduction (Yan et al., 2008).

Cost and benefit. Soil remediation is one of the most difficult management issues for municipal and state agencies because of its high cost. A two-year phytoremediation project for soil contaminated with As, Cd and Pb was implemented to determine the essential parameters for soil remediation (Wan et al., 2016a). Results showed highly efficient heavy metal removal. Costs and benefits of this project were calculated. The total cost of phytoremediation was

US\$75 375.2/hm^2 or US\$37.7/m^3, with initial capital and operational costs accounting for 46.02% and 53.98%, respectively. The costs of infrastructures (i.e., roads, bridges, and culverts) and fertilizer were the highest, mainly because of slow economic development and serious contamination. The cost of phytoremediation was lower than the reported values of other remediation technologies. Improving the mechanization level of phytoremediation and accurately predicting or preventing unforeseen situations were suggested for further cost reduction. Considering the loss caused by environmental pollution, the benefits of phytoremediation will offset the project costs in less than seven years.

4. Future prospects

The requirement of remediating a large area of heavy metal contaminated farmland in China is a unique problem in the world. Through decades of lab and field investigation, the basic structure of phytoremediation technology for As contaminated farmland has been proposed. Phytoremediation is confirmed to be an economical, easy and green technology, which well fit for the current farmland contamination status. However, there are still some issues waiting for further study.

(1) The potential value of the harvested hyperaccumulator biomass. In France, the nickel (Ni) hyperaccumulator has been reused for the phytomining of Ni. However, due to the low economical value of As, the potential reuse approach for *P. vittata* still needs further investigation.

(2) The control of weed. In the practice, it has been found that weed is a serious problem, which significantly increase the management cost. The present herbicide can also hurt *P. vittata*. Therefore, the development of an environment-safely herbicide for *P. vittata* is one of the next research targets.

(3) The strengthening measures of phytoextration. Compared to chemical or physical technologies, phytoextraction requires a relatively long time. Measures to promote the extraction of As from soil by *P. vittata* still requires further study.

References

An Z Z, Huang Z C, Lei M., et al. 2006. Zinc tolerance and accumulation in *Pteris vittata* L. and its potential for phytoremediation of Zn-and As-contaminated soil. Chemosphere 62: 796-802.

Cai B, Zhang G, Chen T. 2007. Genotypic difference in growth and As accumulation in *Pteris vittata*. Journal of Zhejiang University 33: 473-478.

Cao X D, Ma L Q, Tu C. 2004. Antioxidative responses to arsenic in the arsenic-hyperaccumulator Chinese brake fern (*Pteris vittata* L.). Environmental Pollution: 317-325.

Chaney R L, Malik M, Li Y M, et al.1997. Phytoremediation of soil metals. Current Opinion in Biotechnology, 8: 279-284.

Chen T B, Huang Z C, Huang Y Y, et al. 2005a. Distributions of arsenic and essential elements in pinna of arsenic hyperaccumulator *Pteris vittata* L. Science in China Series C-Life Sciences, 48: 18-24.

Chen T B, Liao X Y, Huang Z C, et al. 2007. Phytoremediation of arsenic-contaminated soil in China, in: Willey, N.

(Ed.), Phytoremediation. Humana Press: 393-404.

Chen T B, Wei C Y, Huang Z C, et al. 2002. Arsenic hyperaccumulator *Pteris vittata* L. and its arsenic accumulation. Chinese Science Bulletin, 47: 902-905.

Chen T B, Yan X L, Liao X Y, et al. 2005b. Subcellular distribution and compartmentalization of arsenic in *Pteris vittata* L. Chinese Science Bulletin: 2843-2849.

Duan G L, Zhu Y G, Tong Y P, et al. 2005. Characterization of arsenate reductase in the extract of roots and fronds of Chinese brake fern, an arsenic hyperaccumulator. Plant Physiology, 138: 461-469.

Ebbs S, Hatfield S, Nagarajan V, et al. 2010. A Comparison of the dietary arsenic exposures from ingestion of contaminated soil and hyperaccumulating *pteris* ferns used in a residential phytoremediation project. International Journal of. Phytoremediation, 12: 121-132.

Fayiga AO, Ma LQ, Rathinasabapathi B. 2008. Effects of nutrients on arsenic accumulation by arsenic hyperaccumulator *Pteris vittata* L. Environmental and Experimental Botany, 62: 231-237.

Huang Z C, An Z Z, Chen T B, et al. 2007. Arsenic uptake and transport of *Pteris vittata* L. as influenced by phosphate and inorganic arsenic species under sand culture. Journal of Environmental Sciences-China, 19: 714-718.

Huang Z C, Chen T B, Lei M, et al. 2008. Difference of toxicity and accumulation of methylated and inorganic arsenic in arsenic-hyperaccumulating and-hypertolerant plants. Environmental Science & Technology, 42: 5106-5111.

Huang Z, 2004. EXAFS study on arsenic species and transformation in arsenic hyperaccumulator. Science in China Series C, 47: 124.

Huang Z C, Chen T B, Lei M, et al. 2004. Direct determination of arsenic species in arsenic hyperaccumulator *Pteris vittata* by EXAFS. Acta Botanica Sinica, 46: 46-50.

Kertulis Tartar G M, Ma L Q, Tu C, et al. 2006. Phytoremediation of an arsenic-contaminated site using *Pteris vitrata* L.: A two-year study. International Journal of Phytoremediation, 8: 311-322.

Kitajima N, Kashiwabara T, Fukuda N, et al. 2008. Observation of arsenic transfer in leaf tissue of hyperaccumulator fern by utilizing synchrotron radiation micro-XRF imaging. Chemistry Letters, 37: 32-33.

Kramer U, 2005. Phytoremediation: novel approaches to cleaning up polluted soils. Current Opinion in Biotechnology, 16: 133-141.

Lei M, Wan X M, Huang Z C, et al. 2012. First evidence on different transportation modes of arsenic and phosphorus in arsenic hyperaccumulator *Pteris vittata*. Environmental Pollution, 161: 1-7.

Lei M, Wan X M, Li X W, et al. 2013. Impacts of sulfur regulation *in vivo* on arsenic accumulation and tolerance of hyperaccumulator *Pteris vittata*. Environmental and Experimental Botany, 85: 1-6.

Li W, Chen T, Liu Y, 2005a. Effects of harvesting on As accumulation and removal efficiency of As by Chinese brake (*Pteris vittata* L.) (in Chinese, and abstract in English). Acta Ecologica Sinica, 25: 538-542.

Li W X, Chen T B, Chen Y, et al. 2005b. Role of trichome of *Pteris vittata* L. in arsenic hyperaccumulation. Science in China Series C-Life Sciences, 48: 148-154.

Liao X Y, Chen T B, Xiao X Y, et al. 2007. Selecting appropriate forms of nitrogen fertilizer to enhance soil arsenic removal by *Pteris vittata*: A new approach in phytoremediation. International Journal of Phytoremediation, 9: 269-280.

Liao X, Chen T, Xie H, et al. 2004. Effect of application of P fertilizer on efficiency of As removal form Ascontanimated soil using phytoremediation: Field study (in Chinese, and abstract in English). Acta Scientiae Circumstantiae, 24: 455-462.

Liu Y., Wang, H B, Wong, M. H., et al. 2009. The role of arsenate reductase and superoxide dismutase in As accumulation in four *Pteris* species. Environment International, 35: 491-495.

Lombi E, Zhao F J, Fuhrmann M, et al. 2002. Arsenic distribution and speciation in the fronds of the hyperaccumulator *Pteris vittata*. New Phytologist, 156: 195-203.

Luongo T, Ma L Q. 2005. Characteristics of arsenic accumulation by *Pteris* and non-*Pteris* ferns. Plant and Soil, 277: 117-126.

Ma L Q, Komar K M, Tu C, et al. 2001. A fern that hyperaccumulates arsenic-A hardy, versatile, fast-growing plant helps to remove arsenic from contaminated soils. Nature, 409: 579-579.

Poynton C Y, Huang J W, Blaylock M J, et al. 2004. Mechanisms of arsenic hyperaccumulation in *Pteris* species: root As influx and translocation. Planta, 219: 1080-1088.

Rascio N, Navari Izzo F. 2011. Heavy metal hyperaccumulating plants: How and why do they do it?And what makes them so interesting?Plant Science, 180: 169-181.

Shoji R, Yajima R, Yano Y. 2008. Arsenic speciation for the phytoremediation by the Chinese brake fern, *Pteris vittata*. Journal of Environmental Sciences-China, 20: 1463-1468.

Singh N, Ma L Q, Srivastava M, Rathinasabapathi B. 2006. Metabolic adaptations to arsenic-induced oxidative stress in *Pteris vittata* L and *Pteris ensiformis* L. Plant Science, 170: 274-282.

Srivastava M, Ma LQ, Rathinasabapathi B, et al. 2009. Effects of selenium on arsenic uptake in arsenic hyperaccumulator *Pteris vittata* L. Bioresource Technology, 100: 1115-1121.

Su Y H, McGrath S P, Zhu Y G, et al. 2008. Highly efficient xylem transport of arsenite in the arsenic hyperaccumulator *Pteris vittata*. New Phytologist, 180: 434-441.

Tu C, Ma L Q. 2002. Effects of arsenic concentrations and forms on arsenic uptake by the hyperaccumulator Ladder brake. Journal of Environmental Quality, 31: 641-647.

Tu S, Ma L Q, MacDonald G E, et al. 2004. Effects of arsenic species and phosphorus on arsenic absorption, arsenate reduction and thiol formation in excised parts of *Pteris vittata* L. Environmental and Experimental Botany, 51: 121-131.

Wan X M, Lei M, Liu Y R, et al. 2013. A comparison of arsenic accumulation and tolerance among four populations of *Pteris vittata* from habitats with a gradient of arsenic concentration. Science of the Total Environment, 442: 143-151.

Wan X, Lei M, Chen T. 2016a. Cost–benefit calculation of phytoremediation technology for heavy-metal-contaminated soil. Science of The Total Environment, 563: 796-802.

Wan X, Lei M, Yang J. 2016b. Two potential multi-metal hyperaccumulators found in four mining sites in Hunan Province, China. CATENA. doi: 10.1016/i.catena.2016.02.005.

Wan X M, Lei M, Chen T B, et al. 2015. Role of transpiration in arsenic accumulation of hyperaccumulator *Pteris vittata* L. Environmental Science and Pollution Research, 22: 16631-16639.

Wan X M, Lei M, Chen T B, et al. 2014. Phytoremediation potential of *Pteris vittata* L. under the combined contamination of As and Pb: beneficial interaction between As and Pb. Environmental Science and Pollution Research, 21: 325-336.

Wan X M, Lei M, Huang Z C, et al. 2010. Sexual propagation of *Pteris vittata* L. influenced by pH, calcium, and temperature. International Journal of Phytoremediation, 12: 85-95.

Wang X. 2012. *Pteris vittata* L.-economic crops intercropping technique model for remediation of heavy metal contaminated soil (in Chinese, abstract in English), Institute of Geographic Sciences and Natural Resources Research, Chinese Academy of Sciences. Graduate University of Chinese Academy of Sciences, Beijing.

Wu F Y, Leung H M, Wu S C, et al. 2009. Variation in arsenic, lead and zinc tolerance and accumulation in six populations of *Pteris vittata* L. from China. Environmental Pollution, 157: 2394-2404.

Xiao X Y, Chen T B, An Z Z, et al.2008. Potential of *Pteris vittata* L. for phytoremediation of sites co-contaminated with cadmium and arsenic: The tolerance and accumulation. Journal of Environmental Sciences, 20: 62-67.

Xie J, Lei M, Chen T, et al. 2010. Phytoremediation of soil co-contaminated with arsenic, lead, zinc and copper using *Pteris vittata* L.: A field study (in Chinese, abstract in English). Acta Scientiae Circumstantiae, 30: 165-171.

Yan X L, Chen T B, Liao X Y, et al. 2008. Arsenic transformation and volatilization during incineration of the hyperaccumulator *Pteris vittata* L. Environmental Science & Technology, 42: 1479-1484.

Zhang W, Cai Y, Tu C, et al. 2002. Arsenic speciation and distribution in an arsenic hyperaccumulating plant. Science of the Total Environment, 300 (1-3): 167-177.

Zhao F J, Wang J R, Barker J H A, et al. 2003. The role of phytochelatins in arsenic tolerance in the hyperaccumulator *Pteris vittata*. New Phytologist, 159 (2): 403-410.

Microbial Remediation of Heavy Metals and Arsenic Contaminated Environments in the Arid Zone of Northwest China

Xiangliang PAN[1, 2], Varenyam ACHAL[2], Chenxi ZHAO[2], Jianying YANG[3], Deepika KUMARI[2]

[1]College of Environment, Zhejiang University of Technology

[2]Xinjiang Key Laboratory of Environmental Pollution and Bioremediation, Xinjiang Institute of Ecology and Geography, Chinese Academy of Sciences

[3]College of Resources and Environmental Sciences, Xinjiang University

1. Heavy metal and arsenic pollution in northwest China

The northwest arid zone accounts for nearly 25% of the total territory area of China. It has a wealth of oil, natural gas and mineral resources. For example, Xinjiang, one of the provinces in the northwest China, possesses 2.19 trillion tons of coal, more than 40% of the total coal deposits in China. Apart from coal, metal mineral deposits, including chromium, gold, iron, vanadium, copper, nickel, tungsten, molybdenum, lead and zinc, are also very ample in Xinjiang and are on the top list of mineral resources of China. Deposits of magnesite, fluorite, sulphur, kyanite, salt, kaolin, asbestos, vermiculite, gypsum, graphite, perlite and zeolite are also abundant in Xijiang (Pirajnoa et al., 2011). Over three thousand mines have been being exploited. Huge amounts of wastewaters and slags containing toxic metal elements have been produced during extraction and smelting. Soils, ground waters and surface waters in and around mines are seriously polluted by heavy metals and arsenic. In some sites near gold mines, mercury and arsenic concentrations in soils can be up to hundreds of ppm. Crops in some downstream areas are also polluted by heavy metals such as cadmium and mercury. In addition, large areas of soil in Xinjiang, Inner Mongolia and Ningxia is polluted with high levels of arsenic due to geological reasons. For example, arsenic pollution of soil and groundwater was found from the Aibi Lake to the east of Manasi River, which covers 250 km in length (Sun, 2004). Heavy metal and arsenic pollution poses a great risk to human being and ecosystem in northwest China.

2. Challenges for bioremediation of heavy metal pollution in the arid zone

The northwest China is the typical arid zone in the world. Its climate is extremely dry. In most areas, annual precipitation is below 400 mm while the evaporation is around 2500 mm.

Therefore, biological remediation of contaminated environments is limited by water shortage, highly salinization of soils and low temperature during six month winter. Phytoremediation, which performs well in the south China, cannot work well in the arid zone because the plants cannot survive or grow well. Similarly, some microbial remediation technologies that work well in the warm and humid areas cannot be applied to the arid zone because of the harsh environments.

In addition to the unfavorable natural constraining factors, some conventional physicochemical remediation technologies are restricted due to their prohibitively high cost because soils in northwest China are always polluted in large areas. For example, at least 10,000 km^2 of soils are contaminated by arsenic in Kuitun, Xinjiang. To remediate soils using chemical agents or materials may also cause serious secondary pollution. In comparison with the physicochemical methods, microbial remediation of heavy metals and arsenic using extremophiles is a cost-effective and ecologically friendly option. Abundant microbial diversities in desert, highly salinized soil and water and permafrost in the arid zone provide a new solution to decontamination of environments. Some of the bacteria, fungi and cyanobacteria that live in these extreme habitats can be isolated and used to immobilize heavy metals and arsenic.

Unlike organic pollutants, which can be abiotically or biotically decomposed into small less toxic or innocuous molecules, heavy metal ions themselves are at atomic level and cannot be decomposed into smaller fractions any more. Therefore, there are only two technological choices for decontamination of heavy metals and arsenic contaminated environments. One is to thoroughly remove them from the environments using some chemical leaching reagents. The other is to immobilize them to reduce their mobility and bioavailability. In the arid zone, the latter is more feasible than the former because of the scarce precipitation.

3. Microbial carbonate precipitation based remediation of heavy metals and arsenic

It is well known that bacteria, both heterotrophic bacteria and autotrophic bacteria, can induce carbonate precipitation (MICP). The heterotrophic bacteria can trigger carbonate precipitation both actively and passively. The former is a ubiquitous process and is attributed to the ionic exchanges through the cell membrane. The latter pathway is owing to some special metabolic processes that change the chemical conditions of the microdomain around the cell to accelerate carbonate precipitation. The typical passive pathways include ureohydrolysis and photosynthesis. Bacterial ureohydrolysis is omnipresent in various environments. When urea is hydrolyzed by ureolytic bacteria into NH_4^+ and HCO_3^- ambient pH consequently increases and this induces precipitation of calcite. The autotrophic bacteria such as cyanobacteria induce carbonate precipitation because of their consumption of the ambient CO_2 around the cell during photosynthesis, which increases water pH of the microenvironment near the cell. Furthermore,

EPS on the surface of cell usually has strong binding capacity for Ca^{2+} (Pan, 2010). The supersaturation of Ca^{2+} and carbonate ions in the microdomain around the cell formed as both pH and Ca^{2+} concentration increase and consequently calcium carbonate minerals precipitated. Metal ions with ion radius close to Ca^{2+}, such as Cd^{2+} and Cu^{2+}, may be incorporated into the calcite crystal by their replacement of Ca^{2+} in the lattice during precipitation of calcite (Pan, 2009). This makes possible for remediation of heavy metals and arsenic based on MICP (Kumari et al., 2016).

The feasibility of immobilization of heavy metals and arsenic in soils and water through calcite precipitation induced by ureolytic bacteria and cyanobacteria has been assessed in our laboratory and the field. Biocementation of mine tailings based on MICP was also investigated.

3.1 Bioimmobilization of heavy metals and arsenic in soils

3.1.1 Copper

An indigenous calcifying bacterial strain CR1, identified as *Kocuria flava*, was isolated from soil of mining area, Urumqi, China. An extensive copper bioremediation capacity of this isolate was studied based on MICP. *K. flava* CR1 removed 97% of copper when initial Cu concentration was 1000 mg/L. The isolate produced significant amount of urease (472 U/mL), an enzyme that lead to calcite precipitation. The isolate removed 95% of copper from contaminated soil. MICP process in bioremediation was further confirmed by FTIR and XRD analyses. FTIR analysis showed two different forms of calcium carbonate i.e., calcite and aragonite and the results were well supported by XRD. For the first time, ability of *K. flava* has been documented in bioremediation of polluted soil. This study showed that MICP based bioremediation by *K. flava* is a viable, environmental friendly technology for cleaning-up the copper contaminated site (Yang et al., 2016).

3.1.2 Lead

Lead (Pb), one of the most emerging toxic heavy metals, is persistently discharged from electric battery manufacturing, lead smelting and mining activities (Cho et al., 2004). High levels of Pb has been reported in soils worldwide (Harrison et al., 1981). Soil lead pollution is a serious environmental issue in both urban and mine areas in northwest China.

Pb bioremediation in soils were investigated using calcite precipitating bacterium, *Kocuria flava*. Soil spiked with 100 mg/kg Pb was incubated with bacterial cells and nutrient broth media containing 2% urea and 25 mM $CaCl_2$. After 7 days of bioremediation with *K. flava* CR1 the soluble-exchangeable Pb content decreased from 85.4 mg/kg in control soil to 14.2 mg/kg whereas the concentration of carbonate-bound Pb in remediated soil increased significantly to 35.1 mg/kg from 5.9 mg/kg in control soil. Compared with the control soils, carbonate-bound Pb, organic-bound Pb and residual Pb are the dominant fractions in the bioremediated soil with the lowest concentration of soluble exchangeable Pb, indicating the lowest mobility and bioavailability of Pb. XRD analysis and X-ray diffraction patterns analysis showed that the major lead species found in contaminated soils include lead carbonate ($PbCO_3$), lead oxides (PbO/PbO_2), and hydrocerussite ($Pb_3(CO_3)_2(OH)_2$). MICP based bioremediation lead to transformation of Pb-minerals into geochemically stable calcite species as calcite beside vaterite. Pb oxyanions may substitute for the carbonate group in the calcite structure (Achal et al., 2012c).

3.1.3 Strontium

Strontium is a naturally occurring alkaline earth metal that is common in marine systems and crustal materials (Kasimtseva, 2010). Its radioactive form (^{90}Sr) is highly toxic and may pose a great risk to the population. Microbially induced calcite precipitation (MICP) was evaluated for its potential to remediate strontium from aquifer quartz sand. A Sr resistant urease producing *Halomonas* sp. was characterized for its potential role in bioremediation. *Halomonas* sp. SR4 was highly effective in immobilization of soluble-exchangeable Sr in aquifer quartz sand, indicated by the decreasing of the soluble-exchangeable Sr content in control of 70.2 mg/kg to 14.2 mg/kg after bioremediation. The amount of carbonate-bound Sr in remediated aquifer quartz was about four times higher than in control sand. The bacterial strain removed 80% of Sr from soluble-exchangeable fraction of aquifer quartz sand. X-ray diffraction detected calcite, vaterite and aragonite along with calcite-strontianite ($SrCO_3$) solid solution in bioremediated sample with indications that Sr was incorporated into the calcite. Scanning electron micrography coupled with energy-dispersive X-ray further confirmed MICP process in remediation. The study showed that MICP sequesters soluble strontium as biominerals and could play an important role in strontium bioremediation from both ecological and greener point of view (Achal et al., 2012b).

3.1.4 Chromium

The hexavalent form of chromium, Cr (Ⅵ), is a highly toxic metal. Cr (Ⅵ) is highly soluble and mobile in soils and groundwater. Cr (Ⅵ) mainly comes from industrial processes such as electroplating, metal finishing, leather, mining and by human activities as a combustion product (Dennis et al., 2003). Exposure to Cr (Ⅵ) may cause kidney failure, anaemia, dermal allergy and respiratory effects, including lung cancer (Atsdr, 2002).

The current mainstream bioremediation of Cr (Ⅵ) is to reduce Cr (Ⅵ) to the less toxic and insoluble Cr (Ⅲ) by various Cr (Ⅵ) reducing bacteria (Kathiravan et al., 2011; Desai et al., 2008). The problem with the bioreduction remediation technology lies in re-release as Cr (Ⅵ) after potential oxidation of Cr (Ⅲ). Therefore, remediation of Cr (Ⅵ) based on MICP process was also evaluated. Artificially contaminated soil with Cr (Ⅵ) as high as 100 mg/kg were treated based on MICP. The urease producing *Bacillus cereus* YR5 was employed for the treatment of soil, which resulted in remarkably lower Cr (Ⅵ) in the bio-available soil fraction in one month. In order to evaluate the bioremediation efficiency of Cr (Ⅵ) contaminated soil, it is very important to measure a level of metal toxicity in edible plant tissues to make sure about their consumption by animals or humans. After the bioremediation, *Pisum sativum* seeds were planted in the soil and harvested after 60 days. Compared to Cr (Ⅵ) content of 7.5-18 mg/kg in the dry weight of different tissues of *P. sativum* grown in contaminated soil, bioremediation resulted in Cr (Ⅵ) content of only 0.34-0.65 mg/kg in its root, stem and leaf. Further, EDS analysis revealed sharp peak of Cr in stems of *P. sativum* grown in contaminated soil, which were not remarkable when grown in treated soil. The results revealed MICP as highly efficient technology for heavy metal remediation and plants grown in such treated soil will be safe for consumption purposes (Kumari et al., 2014a).

3.1.5 Cadmium

The ureolytic *Exiguobacterium undae* for immobilize cadmium in contaminated soil at low temperature (10℃) was evaluated. More than 90% of cadmium in the tested soil was converted from the soluble-exchangeable fraction to carbonate-bound fraction in 14 days. Surface and structural analyses revealed that cadmium was co-precipitated in calcite crystals in the form of $CdCO_3$ as well precipitated as $CdCO_3$. Urease and dehydrogenase were enhanced in activities, which were not affected by the testing temperatures, in the tested soil as a consequence of MICP. MICP with *E. undae* is an effectively biological technology to immobilize cadmium in soils of cold regimes (Kumari et al., 2014b).

The other case of Cd remediation is based on MICP of photosynthetic bacteria. Some cyanobacteria species that can rapidly induce calcium carbonate precipitation during photosynthesis were screened for their removal of heavy metals. In one of our studies, cadmium (Cd)-containing water was treated by three cyanobacterium *Nostoc calcicola* reactors. *N. calcicola* can significantly increase aquatic phase pH in the reactor and decrease dissolved inorganic carbon content in the effluent. During the two-month operation of the bioreactors, high Cd removal efficiency was obtained in the range of 98.16 ($\pm$1.49)% to 98.70 ($\pm$1.83)%. Cd was mainly in organic-bound fraction, followed by a small amount of carbonate and exchangeable fractions in the *N. calcicola* cell. A crystalline compound of Ca, C, and O, as well as a small amount of Cd on the surface of *N. calcicola* cell, was observed, indicating that the crystallization of $CaCO_3$ induced by *N. calcicola* contributes to the removal of Cd in the reactors. Although the results revealed that calcifying cyanobacterium reactor is a promising way to remove Cd from water, assessment of its long-term performance of heavy metal removal is needed (Zhao et al., 2014).

3.1.6 Arsenic

Arsenic (As) is the most commonly found highly toxic metalloid element in the northwest

China. The first endemic arsenic area was found in Kuitun reclamation area of Xinjiang, China in 1980s. Millions of rural residents in northwest China are under exposure to high levels of arsenic. Arsenic occurs in soils mainly in the trivalent form arsenite [As (III)] and pentavalent form arsenate [As (V)]. Long term exposure to arsenic at high levels via drinking water and foods can exert various harmful effects on human health, including increase of the risk of skin cancer and tumors of the bladder, kidneys, liver and lungs (NRC, 1999).

Highly salinized soil in the northwest China makes a great challenge toward the traditional technologies. On one hand, arsenic is highly mobile in salinized soil. Unlike other heavy metals such as Pb which may be precipitated at high soil pH, arsenic shows higher mobility when the salinity and pH increase. On the other hand, high soil salinity and pH, together with the dry climate, limit the application of phytoremediation in northwest China. In the warm and humid south China, *Pteris vittata* (Chinese brake fern) is widely used as a cost effective method for removal of arsenic from soils (Fayiga et al., 2004; Xiao et al., 2008).

The ability of a novel indigenous bacterium *Sporosarcina ginsengisoli* has been investigated in the remediation of As (III) contaminated soil. The As (III) tolerant bacterium *S. ginsengisoli* CR5 was isolated from As contaminated soil of Urumqi, China. We investigated the role of microbial calcite precipitated by this bacterium to remediate soil contaminated with As (III). The bacterium was able to grow at high As (III) concentration of 50 mm. In order to obtain arsenic distribution pattern, five stage soil sequential extraction was carried out. Arsenic mobility was found to significantly decrease in the exchangeable fraction of soil and subsequently the arsenic concentration was markedly increased in carbonated fraction after bioremediation. Microbially induced calcite precipitation (MICP) process in bioremediation was further confirmed by ATR-FTIR and XRD analyses. XRD spectra showed presence of various biomineralization products such as calcite, gwihabaite, aragonite and vaterite in bioremediated soil samples. The results from this study have implications that MICP based bioremediation by *S. ginsengisoli* is a viable, environmental friendly technology for remediation of the arsenic contaminated sites (Achal et al., 2012a).

3.2 Biocementation of mine tailings based on MICP

Mine tailings are important sources for heavy metals contamination in the northwest China due to long term exploitation of mineral resources. Metallic mine tailings usually contain high concentrations of metals such as Cu, Pb, Zn, Hg, Cd, Cr and As. Mine tailings are usually piled up in open air without any preventive measures against leaching and wind erosion in northwest China. The toxic microparticles of mine tailings can be easily diffused to surrounding environments with wind and heavy metals can be readily leached from mine tailings to groundwater (Concas et al., 2006).

3.2.1 Acidic copper mine tailings

Acidic copper mine tailings, one of the common types of metal mine tailings in Xinjiang, is characterized with high content of sulfide minerals. It shows high acidity because oxidation of sulfides to sulfate. Its leachate always contains high levels of heavy metals which are released from sulfides or other minerals during their oxidation. The acidic mine drainage is the most challenged environmental pollution issue associated with mining activities. However, the current bioremediation techniques using sulfate reducing bacteria are generally not applicable to the acidic mine tailings because of the sensitivity of sulfides to dynamic change of redox potential. Phytoremediation is also impossible due to very low pH and high toxicity of the mine tailings. There is an urgent need to take measures to immobilize heavy metals in mine tailings and prevent their leaching out to environment.

An indigenous calcifying urease producing bacterial strain was isolated from copper mine tailing soil and used in the bioimmobilization of copper, lead and cadmium from mine tailing soils. Phylogenetic analysis of the 16S rRNA gene sequence of the strain identified it as *Bacillus firmus* XP8. The efficiency of bioimmobilization was based on microbially induced calcite precipitation (MICP). Five stage soil sequential extractions were carried out to obtain distribution patterns of different toxic metals. The mobility of toxic metals was found to be significantly reduced in their exchangeable fraction in mine tailing soil and their concentrations were markedly increased in carbonated fraction after bioremediation. Cu^{2+}, Pb^{2+} and Cd^{2+}concentrations in the exchangeable fraction in the bioremediated tailings were only 1.3, 3.6 and 1.4 mg/kg, respectively. In the untreated tailings, the soil contained 16.7, 12.5 and 11.5 mg/kg exchangeable copper, lead and cadmium, respectively. The fractionation of the untreated mine tailing soil followed the order of the residual > the soluble-exchangeable > the carbonated > the Fe-Mn oxides > the organic matter. However, the metals in the bioremediated mine tailings were dominated by the carbonate bound fraction (45%-55%). The carbonated Cu-, Pb-and Cd-fractions in bioremediated mine tailing increased to 55%, 45% and 48%, respectively with respect to the control. The species of the toxic metals associated with organic matter and the residual fractions did not change significantly in the control and the bioremediated samples. Scanning electron micrography showed the precipitation of calcite by bacterial cells. Calcium carbonate minerals such as alcite, gwihabaite and aragonite were identified in the microbial precipitates in the biormediated tailings soils using XRD spectroscopy spectra. MICP processes hold great promise in the remediation of mine tailing soils and efficient removal of toxic metal ions such as copper, lead and cadmium (Yang et al., 2016).

3.2.2 Cr Slag

There is huge amount of Cr slag in northwest China (e.g., Turpan) and Cr slag has become a serious problem as a contaminant in soil or water. Serious Cr slag contamination accidents that cause death of livestock and illness of human beings are frequently reported in China. High concentrations of chromate are found in Hunan, Tianjin, Jinzhou and Yunnan ranging from 250-7060 mg/kg Cr (Ⅵ) in soil (Wang et al., 2011; Xu et al., 2011). There is an urgent need to seek cost-effective methods to appropriately treat Cr slag.

Here we demonstrate a calcifying ureolytic bacterium *Bacillus* sp. CS8 for the bioremediation of chromate (Cr (Ⅵ)) from chromium slag based on microbially induced calcite precipitation (MICP). A consolidated structure like bricks was prepared from chromium slags using bacterial cells, and five stage Cr (Ⅵ) sequential extraction was carried out to know their distribution pattern. Cr (Ⅵ) mobility was found to significantly be decreased in the exchangeable fraction of Cr slag and subsequently, the Cr (Ⅵ) concentration was markedly increased in carbonated fraction after bioremediation. One-month bioremediation reduced Cr (Ⅵ) in exchange fraction of 124.8 mg/kg Cr (Ⅵ) to 2.6 mg/kg in bacterial specimens, associated with a significant increase in the carbonate bound Cr (Ⅵ) fraction. Furthermore, the biocemented Cr slag bricks show high compressive strength with low permeability. The leaching column tests show remarkable decreases in Cr (Ⅵ) concentration after bioremediation. SEM-EDS confirmed the incorporation of Cr (Ⅵ) into the calcite surface forms a strong complex and this leads to obstruction in Cr (Ⅵ) release into the environment (Achal et al., 2013).

4. Summary

Some bacteria including cyanobacteria show excellent MICP capacity and efficient removal of heavy metal ions from soil solution or water. Heavy metal ions are incorporated into the lattice of $CaCO_3$ by substituting for Ca^{2+} or may also coprecipitated with $CaCO_3$. Because precipitation is scarce while evaporation is very high in the arid lands, soils always have higher pH and high salinity, MICP based bioremediation offers a promising option for decontamination of heavy metals and arsenic in arid lands.

References

Achal V, Pan X, Zhang D. 2011. Remediation of copper-contaminated soil by Kocuria flava CR1, based on microbially induced calcite precipitation. Ecological Engineering, 37 (10): 1601-1605.

Achal V, Pan X, Fu Q, et al. 2012a. Biomineralization based remediation of As (Ⅲ) contaminated soil by Sporosarcina ginsengisoli. Journal of Hazardous Materials, 201: 178-184.

Achal V, Pan X, Lee D J, et al. 2013. Remediation of Cr (Ⅵ) from chromium slag by biocementation. Chemosphere, 93 (7): 1352-1358.

Achal V, Pan X, Zhang D. 2012b. Bioremediation of strontium (Sr) contaminated aquifer quartz sand based on carbonate precipitation induced by Sr resistant *Halomonas* sp. Chemosphere, 89 (6): 764-768. V

Achal V, Pan X, Zhang D, et al. 2012. Bioremediation of Pb-contaminated soil based on microbially induced calcite precipitation. J Microbiol Biotechnol, 22 (2): 244-247.

ATSDR. 2002. Draft Toxicological Profile for Several Trace Elements U. S. Department of Health and Human Services.

Atlanta, GA: Agency for Toxic Substances and Disease Registry.

Cho DaeH, Yoo ManH, Kim EuiY. 2004. Biosorption of lead (Pb^{2+}) from aqueous solution by Rhodotorula aurantiaca. Journal of Microbiology and Biotechnology, 14 (2): 250-255.

Concas A, Ardau C, Cristini A, et al. 2006. Mobility of heavy metals from tailings to stream waters in a mining activity contaminated site. Chemosphere, 63 (2): 244-253.

Paustenbach D, Finley B, Mowat F, et al. 2003. Human health risk and exposure assessment of chromium (Ⅵ) in tap water. Journal of Toxicology and Environmental Health Part A, 66 (17): 1295-1339.

Fayiga A O, Ma L Q, Cao X, et al. 2004. Effects of heavy metals on growth and arsenic accumulation in the arsenic hyperaccumulator *Pteris vittata* L. Environmental Pollution, 132 (2): 289-296.

Harrison R M, Laxen D P H, Wilson S J. 1981. Chemical associations of lead, cadmium, copper, and zinc in street dusts and roadside soils. Environmental Science & Technology, 15 (11): 1378-1383.

Kasimtseva N Ⅴ. 2010. Removal of arsenic and strontium from aqueous solution using ironoxide coated zeolitized tuff.

Kumari D, Li M, Pan X, et al. 2014. Effect of bacterial treatment on Cr (Ⅵ) remediation from soil and subsequent plantation of Pisum sativum. Ecological Engineering, 73: 404-408.

Kumari D, Pan X, Lee D J, et al. 2014. Immobilization of cadmium in soil by microbially induced carbonate precipitation with Exiguobacterium undae at low temperature. International Biodeterioration & Biodegradation, 94: 98-102.

Kumari D, Qian X Y, Pan X L, et al. Microbially-induced Carbonate Precipitation for Immobilization of Toxic Metals. Advances in Applied Microbiology, ISSN 0065-2164, Volume 94, chapter 2,

NRC. 1999. Arsenic in Drinking Water. National Academy of Sciences, Washington DC.

Pan X L. 2009. Micrologically induced carbonate precipitation as a promising way to in situ immobilize heavy metals in groundwater and sediment. Research Journal of Chemistry & Environment, 13(4): 3-4.

Pan X L. 2010. Microbial extracellular polymeric substances: The ignored but crucial bio-interface affecting mobility of heavy metals in environment. Research Journal of Biotechnology, 5(3): 3-4.

Pirajno F, Seltmann R, Yang Y. 2011. A review of mineral systems and associated tectonic settings of northern Xinjiang, NW China. Geoscience Frontiers, 2 (2): 157-185.

Sun G. 2004. Arsenic contamination and arsenicosis in China. Toxicology and Applied Pharmacology, 198 (3): 268-271.

Wang Z, Chen J, Chai L, et al. 2011. Environmental impact and site-specific human health risks of chromium in the vicinity of a ferro-alloy manufactory, China. Journal of Hazardous Materials, 190 (1): 980-985.

Xiyuan X, Tongbin C, Zhizhuang A N, et al. 2008. Potential of *Pteris vittata* L. for phytoremediation of sites co-contaminated with cadmium and arsenic: the tolerance and accumulation. Journal of Environmental Sciences, 20 (1): 62-67.

Xu Y, Yang Z, Xiang R. 2011. Reduction of chromium contaminated soils by microorganism. Environmental Chemistry, 30: 555-560 (in Chinese).

Yang J, Pan X, Zhao C, et al. 2016. Bioimmobilization of heavy metals in acidic copper mine tailings soil. Geomicrobiology Journal, 33 (3-4): 261-266.

Zhao C, Fu Q, Song W, et al. 2015. Calcifying cyanobacterium (Nostoc calcicola) reactor as a promising way to remove cadmium from water. Ecological Engineering, 81: 107-114.

Cr (Ⅵ)-Reducing Strain and Its Application to the Microbial Remediation of Cr (Ⅵ)-Contaminated Soils

Liyuan CHAI, Zhihui YANG, Yan SHI, Qi LIAO, Xiaobo MIN, Qingzhu LI, Chongjian TANG, Lifen LIANG

School of Metallurgy and Environment, Central South University
National Engineering Research Center for Heavy Metals Pollution Control and Treatment

1. Introduction

Chromium is an important heavy metal that is widely used in industrial processes such as ore refining, electroplating, production of steel and alloys, metal plating, tannery, wood preservation, and pigmentation etc (Cheung et al., 2007). Cr (Ⅵ)-containing wastewater and untreated Cr (Ⅵ)-containing slag were released into the environment directly and caused serious pollution. In the natural environment, Cr (Ⅲ) is thermodynamically stable and less toxic (Viti et al., 2003) and it is easily formed as precipitate Cr $(OH)_3$ or Cr_2O_3 (Puzon et al., 2005). Consequently, Cr (Ⅵ) is a toxic element and the reduction of Cr (Ⅵ) to stable Cr (Ⅲ) is considered as an efficient way to remove Cr (Ⅵ) pollution in soils.

For the remediation of Cr (Ⅵ)-contaminated soil, physicochemical treatment technologies include ion exchange, chemical reduction, adsorption, precipitation and electrokinetic remediation (Krishna and Philip, 2005). However, these methods consume high amounts of energy and large quantities of chemical reagents which is not economically feasible. Furthermore, the resultant metal-containing chemical sludge can be a potential source of metal pollution (Pei et al., 2009). Alternatively, biological processes for treating chromium contaminated sites are becoming very promising because of its high efficiency, low operating cost, short operation time and eco-friendly (Srividya and Mohanty, 2009; Jeyasingh and Philip, 2005; Srivastava and Thakur, 2006). Therefore, bioremediation of Cr (Ⅵ) by microorganisms has been suggested as a potential alternative to the existing physicochemical technologies.

2. Cr (Ⅵ) reducing bacterial strain

2.1 Isolation of Cr (Ⅵ)-reducing bacterial strain

A Cr (Ⅵ)-reducing bacterial strain was isolated from the heaping site of chromium-containing slag. It was identified as *Pannonibacter phragmitetus*, and named as BB, based on physiological, biochemical characteristics and 16S rRNA gene sequence analysis.

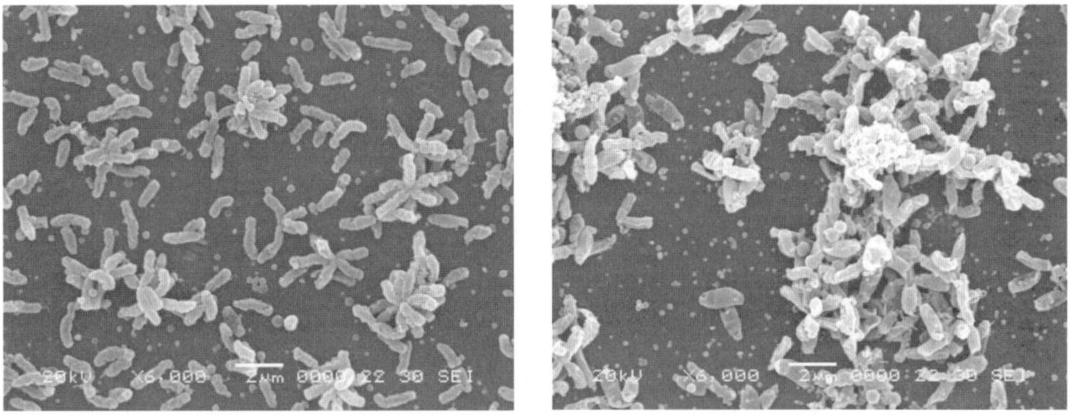

(a) Initial stage of reduction (b) End stage of reduction

Fig. 1 Morphology of *Pannonibacter phragmitetus* BB cells reduced 100 mg/L of Cr (Ⅵ) indicated by scanning electron microscopy

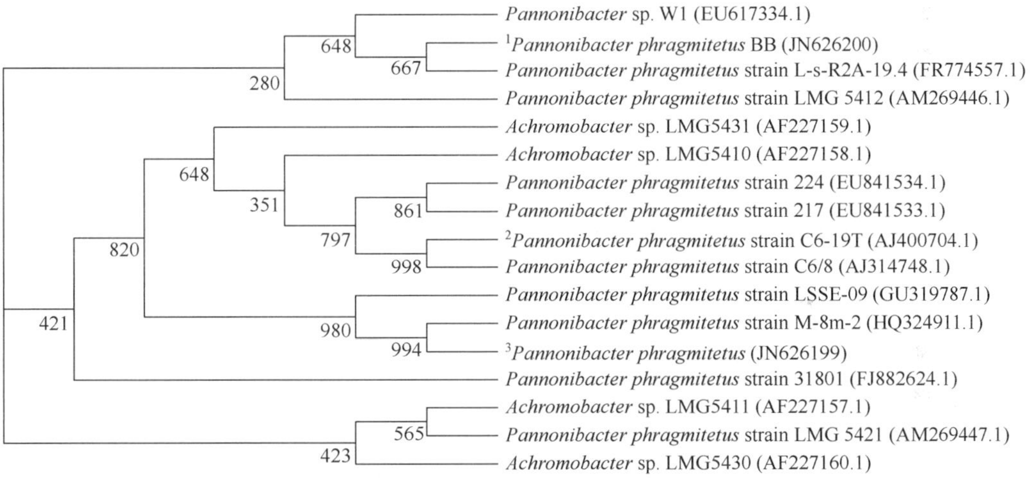

(random number generatorseed = 57, trails = 1,000)

Fig. 2 Phylogenetic tree obtained from 16S rRNA sequence comparisons 1355 bp bases

2.2 Identification of Cr (Ⅵ)-reducing genes of *Pannonibacter phragmitetus* BB

The genome of *Pannonibacter phragmitetus* BB were sequenced by the Illumina technology. The genome size reached to 5.1 Mb. The GC content, scaffold number and contig number were 63.52%, 25 and 366, respectively. Based on the genome annotation, it was found that Cr (Ⅵ)-resistant genes were > Scaffold2_gene_831and > Scaffold7_gene_96. The Cr (Ⅵ)-reduction genes were > Scaffold5_gene_106 and > Scaffold1_gene_445, relating to the nitroreductase and nitrite reductase.

Table 1　Genes related with the metabolization of chromium in the genome

Gene number	Similar sequence	Gene length/bp
>Scaffold2_gene_831	Chromate transport protein ChrA [Pelagibacterium halotolerans B2]	1368
>Scaffold7_gene_96	Chromate transport protein ChrA [Pelagibacterium halotolerans B2]	1368
>Scaffold5_gene_106	Nitrite reductase, copper-containing protein [Starkeya novella DSM 506]	1128
>Scaffold1_gene_445	Nitroreductase [Rhizobium sp. PDO1-076]	633
>Scaffold13_gene_46	Nitroreductase family protein [Polymorphum gilvum SL003B-26A1]	654
>Scaffold17_gene_27	Nitroreductase family protein [Polymorphum gilvum SL003B-26A1]	594
>Scaffold9_gene_43	Putative NAD (P) H quinone oxidoreductase, PIG3 family [Polymorphum gilvum SL003B-26A1]	1023
>Scaffold15_gene_42	Outer membrane nitrite reductase, putative [Burkholderia thailandensis TXDOH]	480
>Scaffold11_gene_63	Putative NADH-flavin oxidoreductase [Stappia aggregata IAM 12614]	1101

2.3　Expression of Cr (Ⅵ)-reducing gene

Cr (Ⅵ)-reduction gene (gene 106) in the genome of *Pannonibacter phragmitetus* BB was expressed in *E*. coil competent cells. Gene 106 was successfully linked to pET-22b (+) and transformed into *E*. coil competent cells. The formation of inclusion bodies was reduced by changing the temperature and concentration of IPTG. Protein purification kit was used to purify the expressed protein and western-blot technique was used to verify the target protein. With the presence of sufficient amount of NADPH, the enzyme activities of the purified target protein were tested at different temperature and pH. The optimum temperature and pH of the target protein were 35℃ and pH 7.0, respectively. Enzyme kinetic of Cr (Ⅵ) reduction process was fitted well with the anti-function equation, the K_m and V_{max} values were 14.55 nmol/L and 34.46 nmol/(min· mg), respectively.

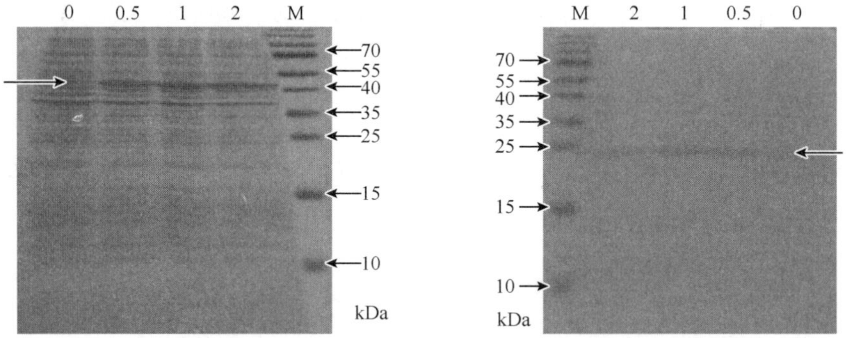

(a) Gene 106　　(b) Gene 445

Fig. 3　Expression Cr (Ⅵ)-reducing genes of under different concentrations of IPTG

3. Microbial reduction of Cr (Ⅵ)

3.1 Cr (Ⅵ) reduction capacity of *Pannonibacter phragmitetus* BB

This bacterium displayed high Cr (Ⅵ) reduction capability that 1917 mg/L Cr (Ⅵ) could be completely reduced in 24 h, and the maximum reduction rate was 562.8 mg/(L·h). *Pannonibacter phragmitetus* BB was able to use several carbon sources such as lactose, fructose, glucose, pyruvate, citrate, formate, lactate, NADPH and NADH as electron donors, among which lactate had the greatest ability to promote the reduction process.

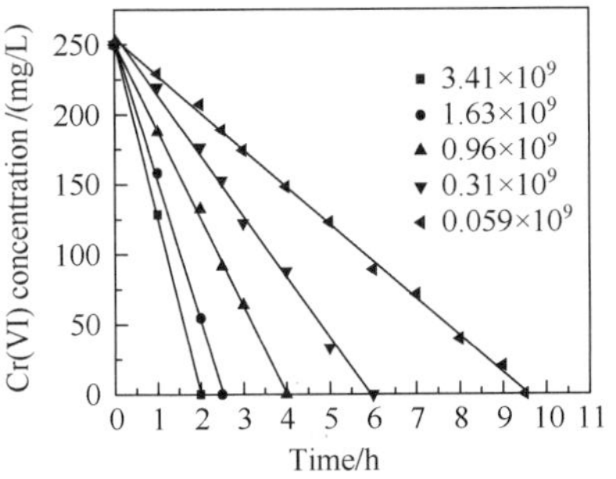

(a) Initial cell concentration

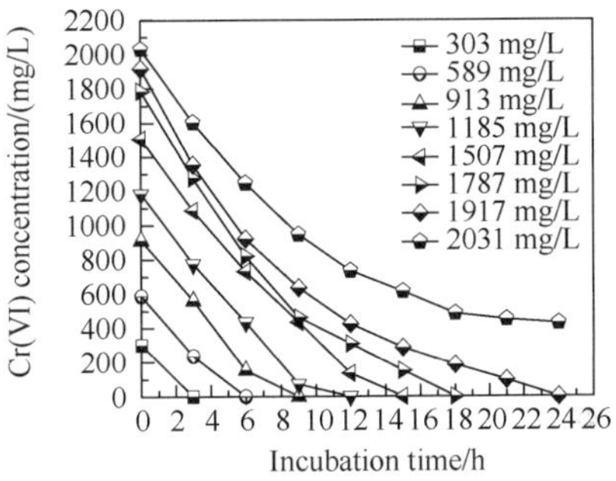

(b) Initial Cr(Ⅵ) concentration

Fig. 4 Cr (Ⅵ) reduction by *Pannonibacter phragmitetus* BB with different initial cell concentrations and Cr (Ⅵ) concentration at pH 9 and 30℃

3.2 Cr (Ⅵ) reduction dynamics

During the procedure of Cr (Ⅵ) microbial reduction in soil, the concentration of NO_3^-, Mn^{2+}, SO_4^{2-}, $Fe^{2+}$and Cr (Ⅵ) were monitored. It was found that Cr (Ⅵ) reduction occurred after NO_3^-, $Mn^{2+}$and Fe^{2+}reduction, but before SO_4^{2-}reduction. The reduction sequence was NO_3^--N_2>Mn^{4+}-Mn^{2+}>Fe^{3+}-Fe^{2+}>Cr (Ⅵ)-Cr (Ⅲ) >SO_4^{2-}-H_2S. The Cr (Ⅵ) reduction was proved to be a two-step process (Fig. 5): Cr (Ⅵ) concentration in soils exponentially declined within 36 h and the kinetic equation of Cr (Ⅵ) reduction was: C=–5.56×exp (t/7.33) + 15,338.88 (R^2=0.995). Thereafter, Cr (Ⅵ) concentration in soils linearly decreased and the kinetic equation of Cr (Ⅵ) reduction was: C=1157.65–14.43t (R^2=0.999).

3.3 Identification of Cr (Ⅵ) reduction product

As shown in Fig. 6, the peaks of Cr 2p3/2 and Cr 2p1/2 in the X-ray photoelectron spectra were located at 577.1±0.1 and 586.9±0.1 eV, respectively (Fig. 6 (b)). The Cr 2p3/2 binding energies of Cr (Ⅲ) hydroxides have been reported to be in the range of 576.5-576.9 eV and spin–orbit splitting for Cr (Ⅲ) between the Cr 2p3/2 and Cr 2p1/2 peaks is around 9.9 eV, while the 2p3/2 peak for Cr (Ⅵ) has been shown between 579.0 and 579.8 eV and spin-orbit splitting for Cr (Ⅵ) is in the range of 8.7-9.4 eV (Asami and Hashimoto, 1977). The Cr 2p3/2 binding energies of Cr $(OH)_3{\cdot}4H_2O$ was located at 577.0 eV and 577.1 eV respectively (Jung et al., 2007). In Fig. 6 (b), the peaks and their spin–orbit splittings demonstrated that Cr (Ⅵ) was converted into Cr (Ⅲ) as the form of Cr (Ⅲ) hydroxides.

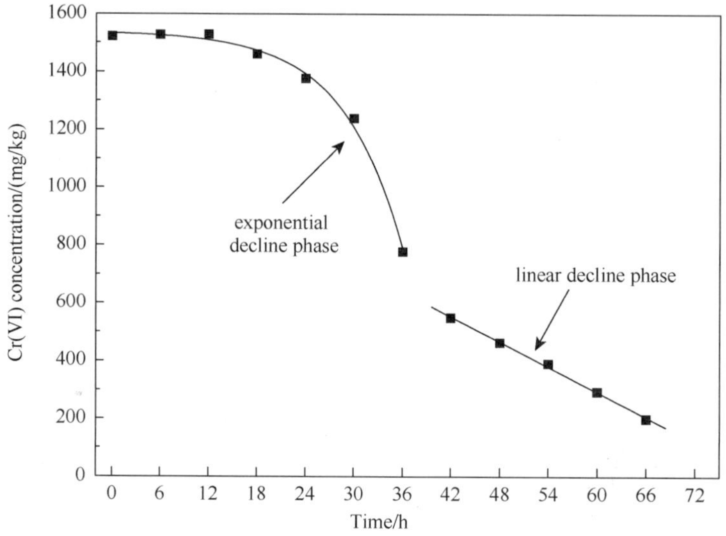

Fig. 5 Change concentration of Cr (Ⅵ) with time

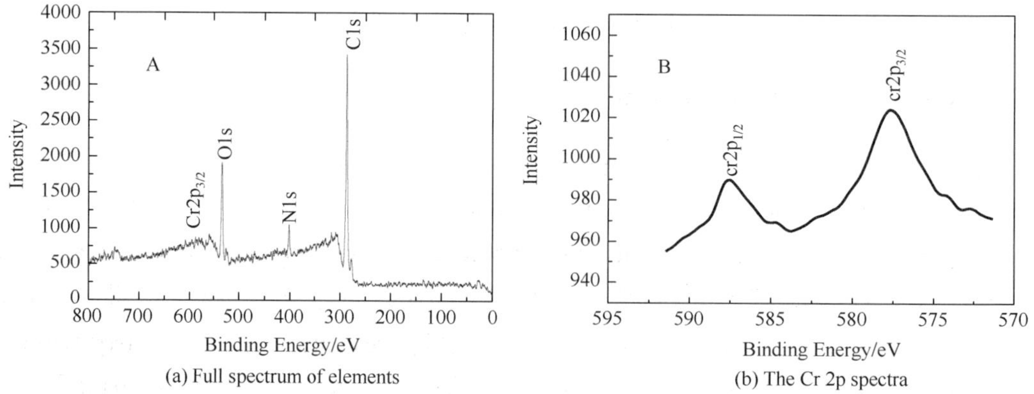

Fig. 6 X-ray photoelectron spectra of Cr (Ⅵ) reduction production

4. Remediation technique of Cr (Ⅵ)-contaminated soils

4.1 Process route

In order to promote the uniform distribution of *Pannonibacter phragmitetus* BB in soil and enhance Cr (Ⅵ) leaching, a recycling bioleaching route was proposal. In process route, soil cores were crushed and pass through a 2-3 cm sieve, and heaped in a tank. The bacterial suspension was pumped to spray the soil heap through spargers. The leachate was collected at the bottom of tank, transferred into a biological reduction tank where the Cr (Ⅵ) reduction occurred. Thereafter, bacteria supernatant in the biological reduction tank was recycled to spray the soil heap till Cr (Ⅵ) in the leachate was not detected.

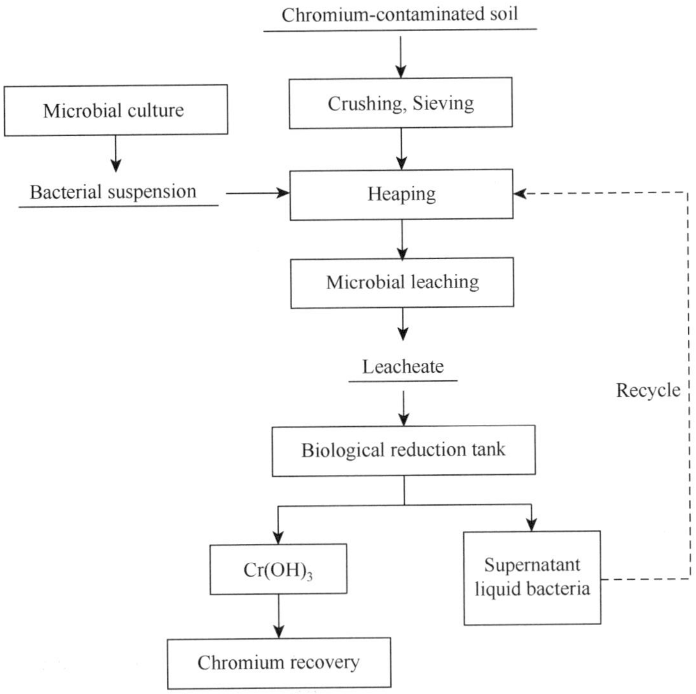

Fig. 7 Process route of microbial remediation of Cr (Ⅵ)-contaminated soil

4.2 Optimum parameters

The removal of Cr (Ⅵ) in the contaminated soils was affected by soil particle size, spray intensity, circulation mode and soil depths. The optimum soil particle size was less than 2 cm, and the spray intensity was in the range of 29.6-59.2 mL/min. The tap water spray followed by continuous spray with bacterial suspension revealed better efficiency of Cr (Ⅵ) remediation than continuous spray with bacterial suspension. In the treated soil, the concentrations

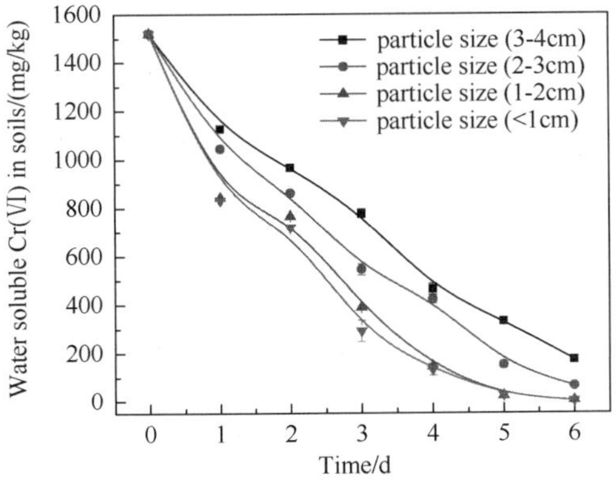

(a) Particle size

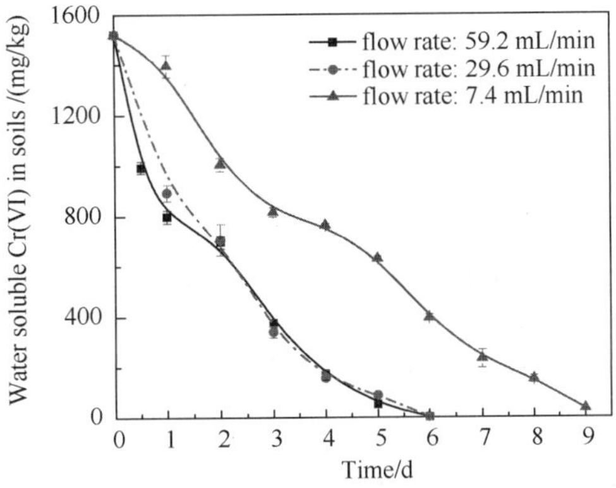

(b) Spray intensity

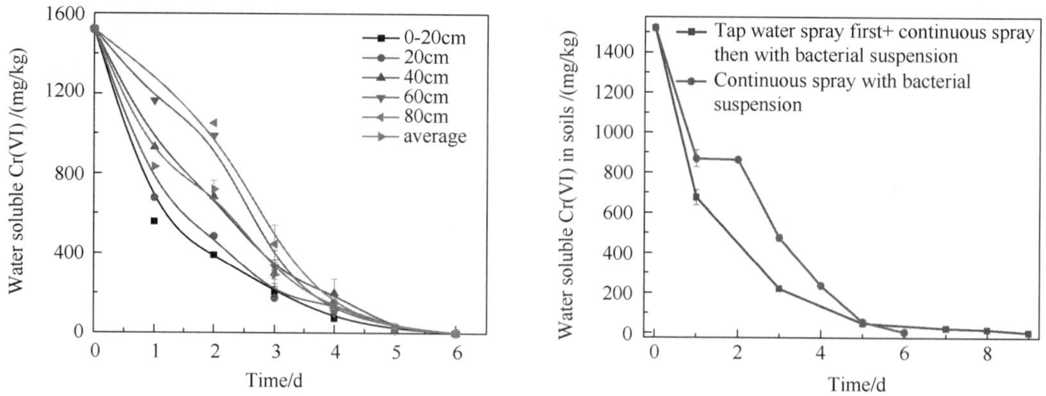

(c) Soil depths (d) Circulation mode

Fig. 8 Cr (Ⅵ) removal in soils under different parameters

of water soluble, exchangeable, carbonates-bonded Cr (Ⅵ) fractions declined from 1520.54 mg/kg to 0.68 mg/kg, from 34.83 mg/kg to 0.01 mg/kg and from 13.55 mg/kg to 0.68 mg/kg, respectively. Meanwhile, a corresponding increase in residual, Fe and Mn oxides-bonded and organics-bonded Cr (Ⅲ) were obtained.

4.3 Pilot remediation project

A pilot experiment with a scale of 25 tons/batch was conducted under the above optimum parameters to demonstrate the remediation feasibility of Cr (Ⅵ) contaminated soil by using *Pannonibacter phragmitetus* BB. Cr (Ⅵ)-contaminated soils at two levels (48 and 488 mg/kg, and represent soil 1 and soil 2, respectively) were treated in pilot experiment. Cr (Ⅵ) in soils was completely removed after running 7-10 days. The Cr (Ⅵ) leaching concentration of soil 2 declined from 53.8 mg/L of the initial concentration to 0.41 mg/L (HJ/T 299—2007) at 10 days, which was lower than the Cr (Ⅵ) limited value of chromium-containing slag used as roadbed material and concrete aggregate according to the Environmental Protection Technical Specifications for Pollution Treatment of the Chromium Residue (HJ/T 301—2007).

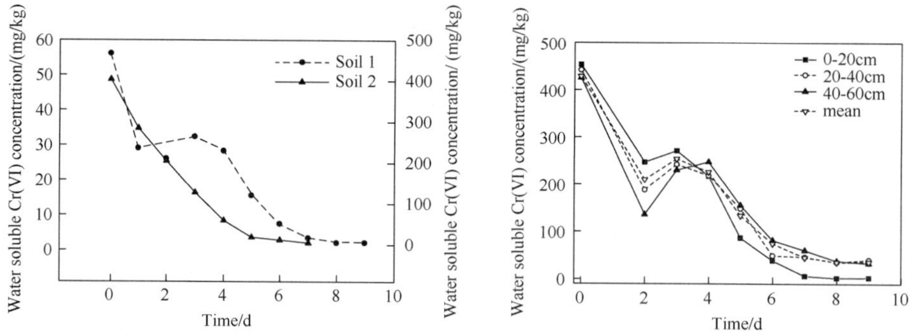

(a) different soils (b) different soil depths

Fig. 9 The remediation efficiency of chromium-contaminated soils

4.4 Changes of *Pannonibacter phragmitetus* BB population in the remediation process

During the bioremediation process, the bacterial population was monitored. Restriction fragment length polymorphism (RFLP) revealed that the relative abundance of *Pannonibacter phragmitetus* BB ranged from 35.5% to 74.8% during the bioremediation process, which was dominant in the soil bacterial community. Real-time quantitative PCR revealed that the population of *Pannonibacter phragmitetus* BB in soil maintained relatively stable at the beginning 48 hours (about 2.0×10^9 cells per gram soil), and the concentration of *Pannonibacter phragmitetus* BB increased dramatically after the beginning 48 hours, suggesting that *Pannonibacter phragmitetus* BB can play a key role and offer competitive advantages in the remediation of Cr (Ⅵ) contaminated soil.

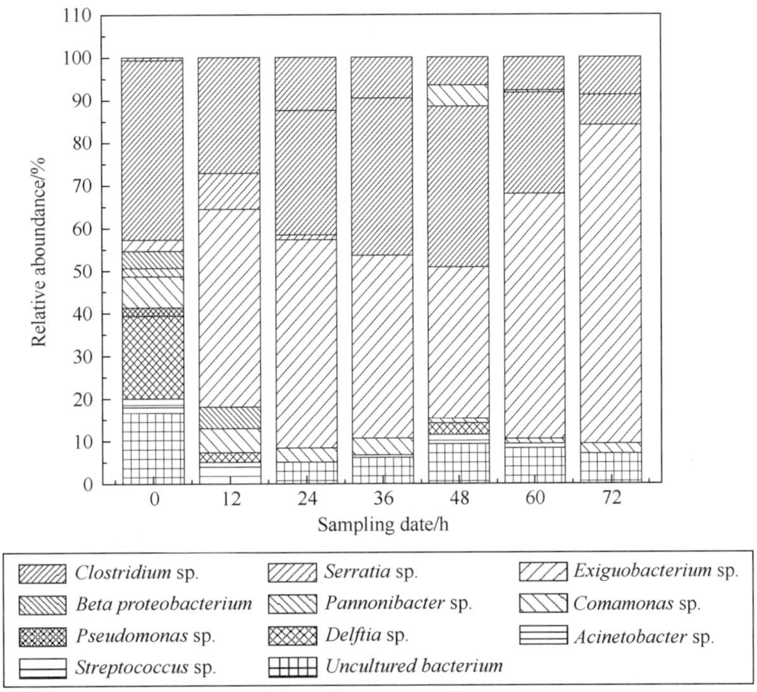

Fig. 10 Changes of soil microbial communities in the bioremediation process

4.5 Stability of chromium in treated soils

Pannonibacter phragmitetus BB was utilized to remedy the Cr (Ⅵ)-contaminated soils at two levels (S1 and S2) by the leaching experiment, and the chemical remediation with ferrous sulfate was used as a control. Under leaching with the simulated acid rain, the release amounts of total chromium and Cr (Ⅵ) from the remedied soil by *Pannonibacter phragmitetus* BB more efficiently deceased as compared with that by ferrous sulfate remediation (Fig. 12). Carbonate-bounded, exchangeable, and organics-bonded chromium were the major chromium releasing sources under the simulated rain leaching.

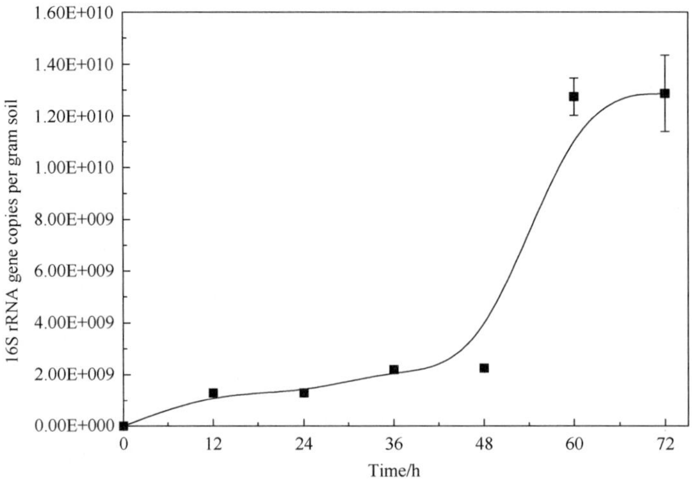

Fig. 11　Variation of *Pannonibacter phragmitetus* BB in the bioremediation process

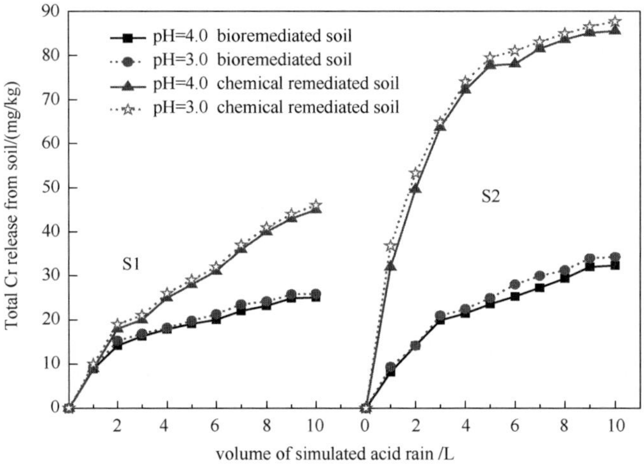

(a) Total Cr

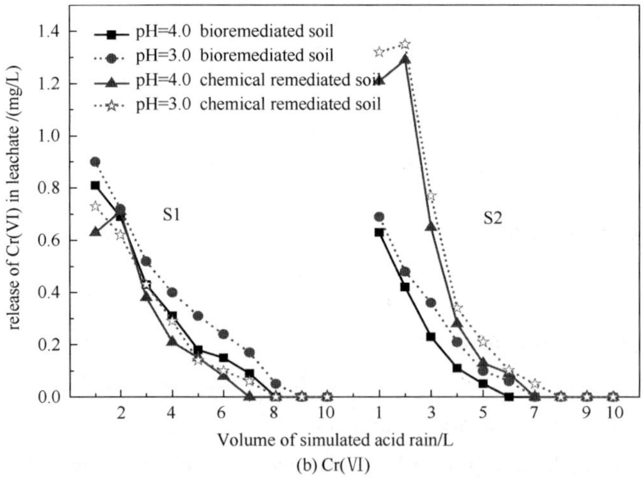

(b) Cr(Ⅵ)

Fig. 12　Cr release in soils with prolonging leaching time

5. Conclusion

A bacterial strain identified as *Pannonibacter phragmitetus* BB was proved to be able to effectively reduce Cr (Ⅵ). Two reduction-related genes (>Scaffold5_gene_106 and >Scaffold1_gene_445), relating to the nitroreductase and nitrite reductase were involved in Cr (Ⅵ) reduction. The Cr (Ⅵ) reduction contained two-step processes including a exponential decline within 36 h and a linear decrease process subsequently. When *Pannonibacter phragmitetus* BB was used for the remediation of Cr (Ⅵ) contaminated soil, the preferable result was obtained at the soil particle size less than 2 cm and the bacterial spray intensity of 29.6-59.2 mL/min with tap water spray followed by continuous spray with bacterial suspension. Moreover, microbial remediation can effectively reduce chromium release form soils. The results demonstrate that *Pannonibacter phragmitetus* BB has advantages and potential application to bioremediation of Cr (Ⅵ) contaminated soil.

References

Asami K, Hashimoto K.1977. X-ray photoelectron-spectra of several oxides of iron and chromium. Corrosion Science, 17: 559-570.

Cheung K H, Gu J D. 2007. Mechanism of hexavalent chromium detoxification by microorganisms and bioremediation application potential: a review. International Biodeterioration and Biodegradation, 59 (1): 8-15.

Jeyasingh J, Philip L. 2005. Bioremediation of chromium contaminated soil: optimization of operating parameters under laboratory conditions. Journal of Hazardous Materials, 118 (1-3): 113-120.

Jung Y, Choi J, Lee W. 2007. Spectroscopic investigation of magnetite surface for the reduction of hexavalent chromium. Chemosphere, 68 (10): 1968-1975.

Krishna R K, Philip L. 2005. Bioremediation of Cr (Ⅵ) in contaminated soils. Journal of Hazardous Materials, 20: 109-117.

Pei Q H, Shahir S, Santhana Raj A S, et al. 2009. Chromium (Ⅵ) resistance and removal by *Acinetobacter haemolytic*us. World Journal of Microbiology and Biotechnology, 25: 1085-1093.

Puzon G J, Roberts A G, Kramer D M, et al. 2005. Formation of soluble organo-chromium (Ⅲ) complexes after chromate reduction in the presence of cellular organics. Environmental Science and Technology, 39: 2811-2817.

Srivastava S, Thakur I S. 2006. Evaluation of bioremediation and detoxification potentiality of *Aspergillus niger* for removal of hexavalent chromium in soil microcosm. Soil Biology and Biochemistry, 38 (7): 904-1911.

Srividya K, Mohanty K. 2009. Biosorption of hexavalent chromium from aqueous solutions by Catla catla scales: equilibrium and kinetics studies. Chemical Engineering Journal, 155: 666-673.

Viti C, Pace A, Giovannetti L. 2003. Characterization of Cr (Ⅵ)-resistant bacteria isolated from chromium-contaminated soil by tannery activity. Current Microbiology, 46: 1-5.

Phytoextraction of Mercury-Contaminated Soil

Xinbin FENG, Jianxu WANG

State Key Laboratory of Environmental Geochemistry, Institute of Geochemistry, Chinese Academy of Sciences

Abstract

Mercury (Hg) is considered to be a "priority hazardous substance" due to its toxicity, mobility, long-range transport in the atmosphere, and methylation processes. The contamination of soil with Hg has attracted public concern as it can be easily incorporated into the food chain. The development of appropriate technologies to manage the risk of Hg in soil is particularly important. Chemically assisted phytoextraction has been proposed as a viable technology to remediate heavy metal-contaminated soil. Thiosulphate can greatly enhance Hg accumulation and transportation in plants. Thiosulphate ($S_2O_3^{2-}$) can combine with Hg^{2+} to form Hg $(S_2O_3)^{2-}$ complexes, which can easily pass through the epidermis and exodermis of root and subsequently load into the stem though the xylem. With the assistance of thiosulphate, the *Brassica juncea var. DPDH* has shown the ability to accumulate Hg from soil. A series of field plots were set in Qinzheng city where the soils were contaminated with Hg due to past industrial activities. The soil was characterized as having a low concentration of bioavailable Hg, but a high concentration of potentially bioavailable Hg (organically bound Hg). The bioavailable Hg may be replenished by the potentially bioavailable pool due to changes in soil biogeochemical factors during the crop growing period. Thiosulphate-assisted *B juncea var. DPDH* has been used to remediate Hg-contaminated soil at Qinzheng. There was a significant Hg fractionation in soil before and after remediation. The potentially bioavailable Hg was transformed into residual fractions after remediation. We propose that thiosulphate-assisted phytoextraction can be used to manage the Hg risks in Qinzheng soil by decreasing the pool of potentially bioavailable Hg.

Keywords: bioavailable Hg, field trial, mechanism, phytoextraction, thiosulphate

1. Introduction

Mercury (Hg) is regarded as a "priority hazardous substance" by the United States Environmental Protection Agency (USEPA) due to its toxicity, mobility, long-range transport in the atmosphere, and methylation processes (Wang et al., 2012a; Malarbi et al., 2015). The Minamata convention on Hg was adopted in 2013, with the aim of protecting the environment and human health from the adverse effects of Hg. Mercury is also listed as one of the key heavy metals in the "12th Five-Year" plan for the comprehensive prevention and control of heavy metal pollution, issued by the Ministry of Environmental Protection of China (MEPC). China is the largest Hg producer, user, and emission source in the world (Streets et al., 2005). The major Hg emission sources in China include non-ferrous metal smelting, coal combustion, cement production, battery and fluorescent lamp production, and biomass burning (Zhang et al., 2015). Inefficient recovery systems and the inadequate management of Hg-containing wastes has led to a large area of land becoming contaminated with Hg. In Hg mining areas, the long history of Hg mining and retorting activities had led to serious Hg contamination of soil (Feng and Qiu, 2008). Under anaerobic conditions, such as in paddy rice soil, Hg can be microbially transformed into methylmercury (MeHg), which is more toxic than inorganic Hg (Zhang et al., 2010). Recent studies have shown that the seeds of rice plants can accumulate an abnormally high concentration of MeHg from the soil (Meng et al., 2014). The consumption of Hg-contaminated rice is a major pathway by which people are exposed to MeHg (Zhang, et al., 2010). To protect humans from Hg poisoning through food ingestion, the development of appropriate technologies to remediate Hg-contaminated soil is very important.

2. Phytoextraction of Hg-contaminated soil

Phytoextraction, the use of natural plants to remove heavy metals from soils, is recognized as a cost-effective and green technology (Kumar et al., 1995). For phytoextraction to be worthwhile, the candidate plants should accumulate substantially high concentrations of heavy metal in their aboveground tissues (Sheoran et al., 2011). No plants have been identified as Hg hyperaccumulators, with most plants that have been investigated as having low accumulation factors (Lomonte et al., 2010). This can be attributed to the limited bioavailability of Hg in soil (Poulin et al., 2016). Some scientists have used a sequential extraction method to analyze Hg speciation in soil, and it has been found that the majority of Hg is associated with the less bioavailable fractions (residual fractions) (Wang et al., 2011). In general, Hg-dissolved organic matter (DOM) complexes are the dominant Hg speciation in DOM-rich soil. Nagy et al. (2011) found that the binding affinity of Hg to reduced DOM (S_{red}) increased with the S_{red}/Hg ratio, and the binding of the highest to lowest affinity of Hg corresponds in order to the Hg_4S_x cluster, linear Hg $(SR)_2$, thiolated aromatic, and five-membered chelate ring sites. The soil sample usually contains approximately10 wt % organic matter, with the reduced sulfur content typically ranging from about 0.1-2 wt %, which is sufficient to bind even unusually high concentrations of Hg (Nagy et al., 2011). In soil samples collected from the Wanshan mercury mine in China, Yin et al. (2016) reported that over 80% of the Hg in soils were present as both α-HgS and β-HgS. It is very challenging for researchers to use natural phytoextraction processes to remediate DOM-rich soil where Hg has been stabilized by DOM or is present as cinnabar. In this context, an important trait is that not all of the Hg in the soil can transfer into the food chain, where it can cause harm to humans. The bioavailable fraction, which only accounts for a minor fraction of the total Hg, presents the highest environmental risk in soil (Paustenbach et al., 1997; Chojnacka et al., 2005). The remediation of the bioavailable fraction may therefore be a feasible way to manage the risk of Hg in soil (Robinson et al., 2015; Wang et al., 2016).

Chemically assisted phytoextraction has a higher remedial efficiency than natural phytoextraction due to the chelant-enhanced bioavailability of heavy metals (Meers et al., 2008). The screening of the optimum chemical ligands is a crucial step for successful phytoextraction. The optimum chelating ligands can not only enhance heavy metal accumulation in plant tissues, but they also have little impact on heavy metal leaching (Wang et al., 2016). Wang and Gorge (2006) found that potassium iodide (KI) could induce Hg accumulation in the root of willow, but the transportation of Hg from root to the aboveground tissue was limited. Moreno et al. (2005) reported that thiosulphate enhanced Hg accumulation in plants grown on Hg-contaminated tailings.

2.1 The screening of optimal chelating ligands to enhance Hg accumulation from Hg-contaminated soils

To determine the optimal chelating ligand for Hg phytoextraction, a series of pot experiments were conducted to investigate the ability of seven chelating ligands: ammonium thiosulphate, sodium thiosulphate, ammonium sulfate, ammonium chloride, sodium nitrate, ethylenediaminetetraacetic acid (EDTA), and sodium sulfite, to enhance plant uptake of Hg from Hg-contaminated soil (Wang et al., 2016). Wang et al. (2016) found that the Hg concentration in the roots of Indian mustard were significantly higher than in non-treated control plants when ammonium thiosulphate, sodium thiosulphate, ammonium sulfate, and sodium sulfite were added to the soil (Fig. 1). Only the ammonium thiosulphate and sodium sulfite treated plants had a significantly higher Hg concentration in their stems and leaves compared with non-treated control plants. This suggests that ammonium thiosulphate and sodium sulfite were more efficient at inducing Hg accumulation in Indian mustard than the other chelating ligands. After harvesting the plants, the bioavailable Hg concentration in the soil was analyzed. We found that ammonium thiosulphate-treated soil had a significantly lower bioavailable Hg concentration than that of the other treatments (Wang et al., 2016). Therefore, ammonium thiosulphate may be the optimal chelating ligand for phytoextraction, due to its large potential for enhancing Hg accumulation in plants, while decreasing the bioavailable Hg concentration in the soil. The effect of different amounts of ammonium thiosulphate on Hg accumulation by *Chenopodium glaucum* was investigated. When thiosulphate was applied at a rate of 8 g per kg of soil (dry weight), the Hg concentration in the plant tissues was significantly higher than in the other treatments (Wang, 2012).

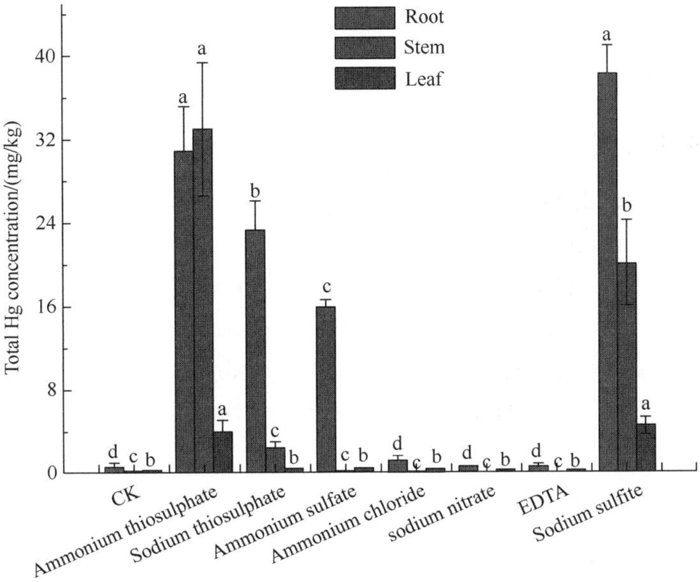

Fig. 1 Effect of different chelating ligands on mercury (Hg) accumulation in the roots, stems, and leaves of Indian mustard. The different letters among the different treatments indicate a significant difference at $p<0.05$

2.2 The mechanism of thiosulphate-induced Hg accumulation in plants

According to the "hard and soft" acid–base principle, Hg is a 'soft metal' and will preferentially form soluble chemical complexes with sulfur-containing ligands (Wang et al., 2011). The physiological mechanism of plant uptake of Hg remains unknown. Lomonte et al. (2014) have reported the use of micro-proton-induced X-ray emission (PIXE) spectrometry to study the spatial distribution of Hg in vetiver grass; they found that thiosulphate induced Hg transportation from the epidermis to the vascular bundle of root. Wang et al. (2012b) used X-ray absorption near edge structure (XANES) to examine Hg speciation in thiosulphate-treated *Brassica Juncea*; their results indicated that the majority of Hg in *B. Juncea* was in a form similar to β-HgS, suggesting that Hg was associated with sulfur-containing ligands in plant tissues. The thiosulphate ion ($S_2O_3^{2-}$) has a high binding constant to Hg^{2+} (lg *K*, 29.27). We propose that the $S_2O_3^{2-}$ could readily combine with Hg (Ⅱ) to form soluble Hg $(S_2O_3)^{2-}$ in soil. The plant may preferentially uptake the Hg ($S_2O_3^{2-}$) complex over other soluble Hg complexes. The epidermis and exodermis of roots have abundant cysteine-containing proteins, such as aquaporins. The $Hg^{2+}$is generally retained in the epidermis and exodermis of roots due to the strong binding ability between this metal and thiol-containing groups. However, when $Hg^{2+}$is chelated with $S_2O_3^{2-}$, it is less likely to bind with cysteine-containing proteins (Lomonte et al., 2014). Wang et al. (2012a) also reported that the sulfate transporters embedded in the plasma membrane might be involved in Hg $(S_2O_3)^{2-}$ transportation in plants. Althongh the significant progress have been achieved in the last decade, the mechanism of thiosulphate- induced Hg accumulation in plants should be further investigated.

2.3 Ammonium thiosulphate enhanced phytoextraction of Hg from historically Hg-contaminated soil

2.3.1 The ability of six plant species to accumulate Hg from historically Hg-contaminated soils

A field trial was conducted to investigate the ability of six plant species to accumulate Hg from historically Hg-contaminated soils at Wanshan mercury mine in China. *Brassica juncea var. LDZY*, *B. juncea var. ASKYC*, *B. juncea var. YDJC*, *B. juncea var. DPDH*, *B. juncea var. CHBD* and *C. glaucum* were used as candidates for thiosulphate-assisted phytoextraction (Wang et al., 2012). Both non-thiosulphate-treated (control) and thiosulphate-treated plots were designed for each plant species. The total Hg concentration in plants and bioaccumulation factors were analyzed. The total Hg concentration in the soil samples could reach 510 mg/kg, which was nearly two orders of magnitudes higher than the maximum allowable total Hg concentration (1.5 mg/kg) in soils set by the Chinese government. Fig. 2 shows the total Hg concentration in the six plants growing in control soils. Without thiosulphate treatment, Hg concentrations in the roots and aboveground tissues of the six plants were below 1.2 mg/kg, which was too low to be used for phytoextraction. The application of thiosulphate substantially increased the Hg concentration in both the roots and aboveground tissues of the six plants (Fig. 3). In both *B. juncea var. DPDH* and *B. juncea var. CHBD*, the highest Hg concentration was found in their aboveground tissues. Although the differences in the Hg concentration in the roots among the six plants were statistically insignificant (except *C. glaucum*), the average Hg concentration in the roots of *B. juncea var. DPDH* were higher than in the roots of the other plant species.

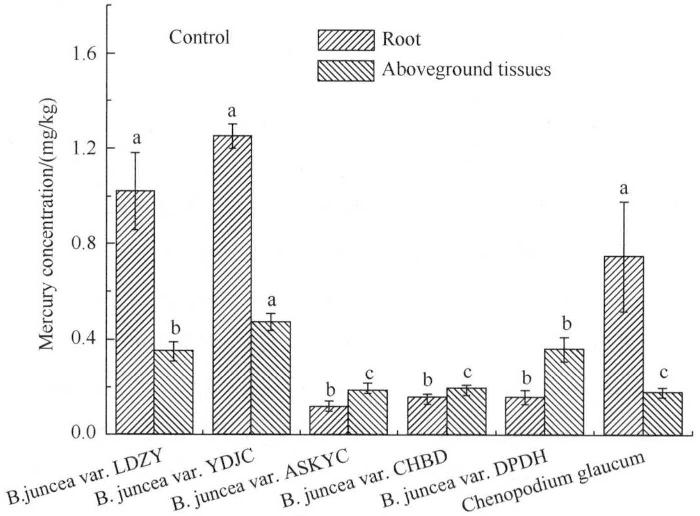

Fig. 2 Mercury concentration in the roots and aboveground tissues of the control plants. The different letters among the different plants indicate a significant difference at $p<0.05$

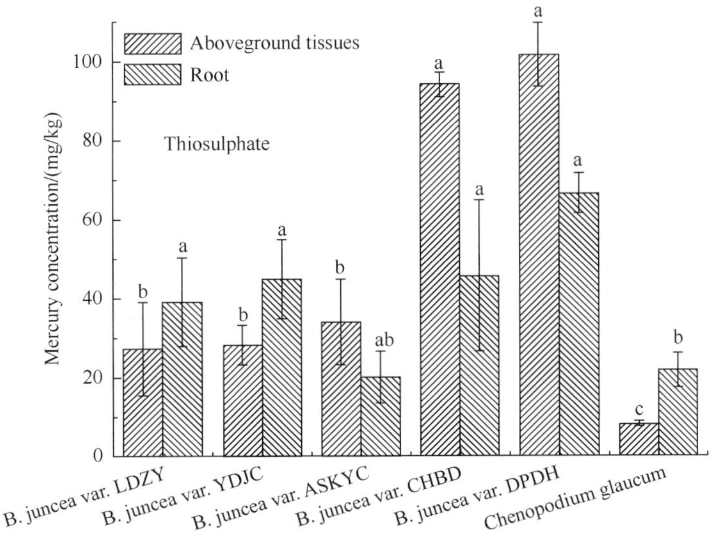

Fig. 3 Mercury concentration in the roots and aboveground tissues of the treated plants. The different letters among the different plants indicate a significant difference at $p<0.05$

Table 1 shows the bioaccumulation factors (BAF) for both the control and thiosulphate treatments. All of the plants had a very low BAF in the control plots compared with those in the thiosulphate treatments. The roots and shoots of *B. juncea var. DPDH* had a higher BAF than the other plants. Overall, *B. juncea var. DPDH* had a greater ability to accumulate Hg than the other plant species, and it therefore may be an ideal candidate plant for the phytoextraction of Hg.

Table 1 Comparison of the bioaccumulation factor (BAF) for both the control and thiosulphate treatments (mean ± sd, *n*=3)

	Control		Thiosulphate	
	BAF_{shoot}	BAF_{root}	BAF_{shoot}	BAF_{root}
B. juncea var. LDZY	<0.001	0.002±nd*	0.06±0.03	0.09±0.03
B. juncea var. DPDH	<0.001	<0.001	0.24±0.02	0.16±0.01
B. juncea var. CHBD	<0.001	<0.001	0.25±0.04	0.12±0.04
B. juncea var. ASKYC	<0.001	<0.001	0.1±0.03	0.05±0.005
B. juncea var. YDJC	<0.001	0.003±nd*	0.09±0.01	0.13±0.03
Chenopodium glaucum	<0.001	0.005±nd*	0.04±0.01	0.15±0.03

nd*: no data; sd: standard deviation.

2.3.2 Thiosulphate-assisted phytoextraction—A case study at a Hg-contaminated industrial site in Guizhou province

The performance of thiosulphate-assisted phytoextraction under field condition was investigated. The Hg-contaminated farmland was located close to the Guizhou Organic Chemical Plant (GOCP), which used Hg as a catalyst for the production of acetaldehyde in the past (Feng and Qiu, 2008). The Hg-containing wastewater released from the GOCP through river discharges have resulted in the serious contamination of farmlands, which depend on the river as an irrigation source. The total Hg concentration in the experimental farmland soil was 14 mg/kg (Wang et al., 2014c). Fig. 4 shows photographs of the field plot. The *B. juncea var. DPDH* was used for phytoextraction. The total Hg concentrations in soil before and after remediation (both thiosulphate-treated and non-thiosulphate-treated) were statistically insignificant ($p>0.05$), suggesting that thiosulphate-assisted phytoextraction was less efficient in removing total Hg from soil (data not shown). However, there was a significant fractionation of Hg in the soil, the concentration of organically bound Hg in thiosulphate-treated soil decreased, while the concentration of residual Hg increased in comparison to both the initial soil (before remediation) and non-thiosulphate-treated soil (after remediation-control) (Fig. 5).

(a) Before planting (b) 40-days after planting

Fig. 4 Photographs of the field plot

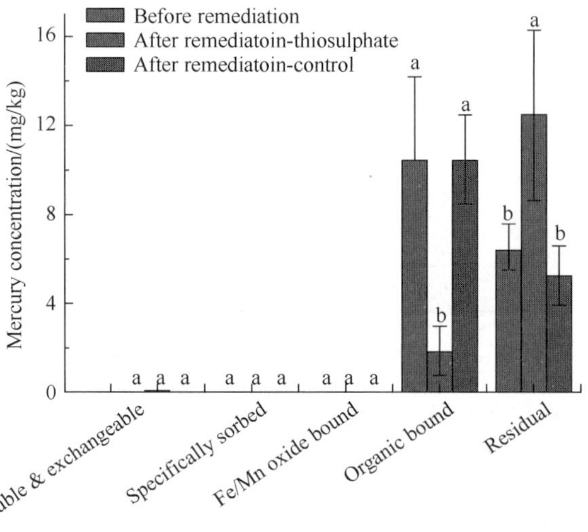

Fig. 5 Distribution of Hg fractions in soil before and after remediation. Different letters among the different treatments indicate a significant difference at $p<0.05$

In general, the soluble and exchangeable, and specifically sorbed fractions are the bioavailable fractions that can be easily taken up by the plant and enter the food chain (Wang et al., 2011). The Fe/Mn oxide bound and organically bound fractions are potentially bioavailable fractions that can be solubilized when the biogeochemical factors of soil change; e.g., pH, Eh, and DOM content (Wang et al., 2014c). The residual fraction mainly contains cinnabar and mineral lattice-associated Hg, which have limited bioavailability in soil. In Qinzheng soil, the concentration of soluble and exchangeable, specifically sorbed, and Fe/Mn oxide-bound Hg fractions only account for 0.2% of the total Hg concentration, with the majority of Hg associated with the organically bound and residual fractions. The concentration of bioavailable Hg fractions was approximately 30 ng/g, which was similar to or even lower than that in background soil. In the non-contaminated soil samples collected from the southwest of India, Navya et al. (2015) found that total Hg and bioavailable Hg concentrations ranged from 2 to 683 and from 0.1 to 230 ng/g, respectively. It is a paradox that crops have high Hg concentrations in their tissues, but the corresponding soil contains very little bioavailable Hg. It is more likely that the bioavailable Hg in soil may be replenished during the crop-growing period. Wang et al. (2014a) reported that the organically bound Hg in soil could be solubilized by the roots and subsequently taken up by plants. We propose that organically bound Hg may be the main source of bioavailable Hg in Qinzheng soil, and the removal of just this fraction may be effective in managing the risk of Hg migration in soil. In this case study, thiosulphate-assisted phytoextraction induced the transformation of Hg speciation in soil, with the organically bound fraction transformed into a residual fraction. This suggests the use of phytoextraction to manage the risk of Hg in Qinzheng soil by decreasing the concentration of organically bound Hg. The application of chelating ligands may induce leaching of Hg into underground water. Wang et al. (2014a) studied soil profiles from a thiosulphate-treated plot to evaluate the distribution of bioavailable Hg at different soil depths; the results showed that the addition of thiosulphate had little impact on the vertical transportation of bioavailable Hg.

Acknowledgements

This study was financed by the Natural Science Foundation of China (41303068, 41573082), the Science and Technology Foundation of Guizhou Province (No. [2013]2218 and [2014]2169), the Youth Innovation Promotion Association of the Chinese Academy of Sciences, and Opening Fund of the State Key Laboratory of Environmental Geochemistry (SKLEG2015903).

References

Chojnacka K, Chojnacki A, Gorecka H, et al. 2005. Bioavailability of heavy metals from polluted soils to plants. Science of the Total Environment, 337 (1-3): 175-182.

Feng X B, Qiu G L. 2008. Mercury pollution in Guizhou, Southwestern China —An overview. Science of the Total

Environment, 400 (1-3): 227-237.

Kumar P B A N, Dushenkov V, Motto H, et al. 1995. Phytoextraction-the use of plants to remove heavy-metals from soils. Environmental Science and Technology, 29 (5): 1232-1238.

Lomonte C, Doronila A I, Gregory D, et al. 2010. Phytotoxicity of biosolids and screening of selected plant species with potential for mercury phytoextraction. Journal of Hazardous Materials, 173 (1-3): 494-501.

Lomonte C, Wang Y D, Doronila A, et al. 2014. Study of the spatial distribution of mercury in roots of vetiver grass (*Chrysopogon Zizanioides*) by micro-pixe spectrometry. International Journal of Phytoremediation, 16 (11): 1170-1182.

Malarbi D, Falciglia P P, Vagliasindi F G A. 2015. Removal of Hg from real polluted sediments using enhanced-EK decontamination: verification of experimental methods and batch-test preliminary results. Journal of Chemistry, 1-9.

Meers E, Tack F M G, van Slycken S, et al. 2008. Chemically assisted phytoextraction: a review of potential soil amendments for increasing plant uptake of heavy metals. International Journal of Phytoremediation, 10 (5): 390-414.

Meng B, Feng X, Qiu G, et al. 2014. Localization and speciation of mercury in brown rice with implications for Pan-Asian public health. Environmental Science and Technology, 48 (14): 7974-7981.

Moreno F N, Anderson C W N, Stewart R B, et al. 2005. Mercury volatilisation and phytoextraction from base-metal mine tailings. Environmental Pollution, 136 (2): 341-352.

Nagy K L, Manceau A, Gasper J D, et al. 2011. Metallothionein-like multinuclear clusters of mercury (Ⅱ) and sulfur in peat. Environmental Science and Technology, 45 (17): 7298-7306.

Navya C, Gopikrishna V G, Arunbabu V, et al. 2015. Distribution and fractionation of mercury in the soils of a unique tropical agricultural wetland ecosystem, southwest coast of India. Environmental Monitoring and Assessment, 187 (12): 1-11.

Paustenbach D J, Bruce G M, Chrostowski P. 1997. Current views on the oral bioavailability of inorganic mercury in soil: implications for health risk assessments. Risk analysis, 17 (5): 533-544.

Poulin B A, Aiken G R, Nagy K L, et al. 2016. Mercury transformation and release differs with depth and time in a contaminated riparian soil during simulated flooding. Geochimica et Cosmochimica Acta, 176: 118-138.

Robinson B H, Anderson C W N, Dickinson N M. 2015. Phytoextraction: where's the action?. Journal of Geochemical Exploration, 151: 34-40.

Sheoran V, Sheoran A S, Poonia P. 2010. Role of hyperaccumulators in phytoextraction of metals from contaminated mining sites: a review. Critical Reviews in Environmental Science and Technology, 41 (2): 168-214.

Streets D G, Hao J, Wu Y, et al. 2005. Anthropogenic mercury emissions in China. Atmospheric Environment, 39 (40): 7789-7806.

Wang J, Feng X, Anderson C W N, et al. 2011. Ammonium thiosulphate enhanced phytoextraction from mercury contaminated soil-results from a greenhouse study. Journal of Hazardous Materials, 186 (1): 119-127.

Wang J, Feng X, Anderson C W N, et al. 2012a. Remediation of mercury contaminated sites—a review. Journal of Hazardous Materials, 221: 1-18.

Wang J, Feng X, Anderson C W N, et al. 2012b. Implications of mercury speciation in thiosulfate treated plants. Environmental Science and Technology, 46 (10): 5361-5368.

Wang J, Feng X, Anderson C W N, et al. 2014a. Thiosulphate-induced mercury accumulation by plants: metal uptake and transformation of mercury fractionation in soil-results from a field study. Plant and Soil, 375 (1-2): 21-33.

Wang J, Feng X, Anderson C W N, et al. 2014b. Effect of cropping systems on heavy metal distribution and mercury fractionation in the Wanshan mining district, China: Implications for environmental management. Environmental Toxicology and Chemistry, 33 (9): 2147-2155.

Wang J, Xia J, Feng X. 2016. Screening of chelating ligands to enhance mercury accumulation from historically mercury-contaminated soils for phytoextraction. Journal of Environmental Management, doi: 10.1016/j. jenvman.2016.05.031.

Wang Y, Greger M. 2006. Use of iodide to enhance the phytoextraction of mercury-contaminated soil[J]. Science of the Total Environment, 368 (1): 30-39.

Wang J X. 2012. The establishment and mechanism study of phytoextraction of mercury contaminated soil at mercury mining district.

Guiyang.

Wang J X, Zhang J F, Feng X B, et al. 2014c. Effect of ammonium thiosulphate addition on mercury accumulation by *Brassica juncea* var. HPDH—a field study. Asia Journal of Ecotoxicology, 9 (5): 992-997.

Yin R, Gu C, Feng X, et al. 2016. Distribution and geochemical speciation of soil mercury in Wanshan Hg mine: Effects of cultivation. Geoderma, 272: 32-38.

Zhang H, Feng X, Larssen T, et al. 2010. In inland China, rice, rather than fish, is the major pathway for methylmercury exposure. Environmental Health Perspectives, 118 (9): 1183.

Zhang L, Wang S, Wang L, et al. 2015. Updated emission inventories for speciated atmospheric mercury from anthropogenic sources in China. Environmental Science & Technology, 49 (5): 3185-3194.

Chelant-Enhanced Phytoextraction of Heavy Metal-Contaminated Soils and Its Environmental Risk Assessment

Yahua CHEN[1], Xiangdong LI[2], Zhenguo SHEN[1]

[1]College of Life Sciences, Nanjing Agricultural University,
[2]Department of Civil & Environmental Engineering, The Hong Kong Polytechnic University

Abstract

The *in situ* phytoremediation, using plants to restore the deteriorated soils, is a promising technology in cleanup of polluted sites. The potential use of high biomass plants was investigated for the EDTA or EDDS-enhanced phytoextraction of metals from contaminated soils. The approaches for enhancing the efficiency of chelant-assisted phytoextraction and the environmental risk (metals leaching) of applying EDTA and EDDS were also investigated. The results showed that, compared to the monocotyledon species (e.g., barley, wheat, sorghum, corn, vetiver grass, etc.), the dicotyledon species (i.e., mung bean, buckwheat, pea, oil sunflower, mustard, etc.) had a higher sensitivity to the EDTA or EDDS treatments, which could be reflected by the severer toxicity of seedlings and the higher concentrations of heavy metals in the shoots. Considering the three factors (the metal level of the shoot, the biomass of plants, and the growing seasons of plants), mustard and pea were more suitable used in the cool seasons, while corn, sunflower, mung bean and buckwheat could be selected in the warm seasons. Vetiver grass might be suitable used in the phytostabilization of heavy metals in soils. There was a new approach that the increase of soil temperature could improve the efficiency of heavy metals accumulation of plants from soils by ten-forty times in comparison with the normal chelator application. This was the highest efficiency of chelators application reported in the literature up to now. Soil column leaching tests and a field study were carried out to investigate the leaching behavior of heavy metals in chelant-enhanced phytoextraction. Results showed that, the soluble heavy metals in the contaminated soils after EDTA application could be persistent in the soils for a long time (at least 32 weeks), which could be easily leached down to the deep soils under the large rainfall percolation. On the contrary, EDDS application had a little effect on the movement of heavy metals in the soils. The results showed that, the potential leaching associated with EDDS addition may be controlled under field conditions. The use of biodegradable chelators for enhancing phytoextraction of heavy metal-contaminated soils is generally regarded as an environmentally acceptable approach.

Keywords: phytoremediation, phytoextraction, EDTA, EDDS, heavy metal-contaminated soils

1. Introduction

Heavy metal contamination of soils is one of major environmental problems in the world. The *in situ* phytoremediation, using plants to restore the deteriorated soils, is a promising technology in cleanup of polluted sites due to the characters of less destructive, low cost and environmentally friendly nature. Among various types of phytoremediation, phytoextraction- removing metals from soils through harvesting the plant biomass which accumulate metals, is regarded as a complete metal clean-up strategy, and becomes more and more appealing in recent decades. Two approaches have been proposed for phytoextraction of heavy metals: ①The continuous phytoextraction using of natural hyperaccumulator plants; ②The chemically enhanced phytoextraction: based on the use of high-biomass plants that are induced to take up large amounts of metals of which the mobility in soil is enhanced by chemical treatments (Salt et al., 1998). As to the chemically enhanced phytoextraction, we mainly focused on the selection of chelators and plant species, the optimization methods of chelators application, the adoption of some measures (such as heating treatment) for increasing the efficiency of phytoextraction and decreasing the leaching of heavy metals, as well as the evaluation of the safety in the process of phytoextraction in the past 20 years.

2. Synergism of plant-chelator interactions for cleanup of metalliferous soils

2.1 EDTA-enhanced phytoextraction of Pb from contaminated soils using ten higher biomass plant species

In a pot experiment, the potential use of ten plant species, including six dicotyledon species and four monocotyledon species, was investigated for the EDTA-enhanced phytoextraction of Pb from contaminated soil. The tested soil samples were artificially amended with Pb in the form of $PbCO_3$ (lead carbonate), and the total Pb in soil was 861 mg/kg.

Ten crop species, including six dicotyledon and four monocotyledon, were used: corn (*Zea mays* L.), sorghum (*Sorghum bicolor* L.), mung beans [*Vigna radiat* (L.) R. Wilczek var. *radiata*], oil sunflowers (*Helianthus annuus* L.), common buckwheat (*Fagopyrum esculentum* Moench.), cabbage [*Brassica rapa* L. subsp. *chinensis* (L.) Hanelt], Chinese mustard (*Brassica juncea* L. Czern. et Coss.), pea (*Pisum sativum* L.), wheat (*Triticum aestivum* L.), and barley (*Hordeum vulgare* L.). The seeds of all of the crops were sown in pots. After germination, the plants were thinned to ten plants per pot for all species.

On the 16th day after sowing, EDTA was applied to the surface of the soil at rates of 0 (control), 2.5 and 5.0 mmol EDTA/kg of soil as a 50 mmol/L Na_2-EDTA solution for each pot. Plant shoots were harvested 2 days before, and 2 and 7 days after the EDTA treatment, respectively. The shoot materials were wet-digested with a mixture of $HNO_3/HClO_4$ for elemental analysis using ICP-AES.

Results showed that, the application of EDTA affected the growth of the plants and the production of shoot biomass (Table 1). The dry weights of the shoots of all plants decreased as the rate at which EDTA was applied increased. Monocotyledon species showed relatively less response to the addition of EDTA, exhibiting little chlorosis, with small reductions of shoot biomass. However, six dicotyledon species tested were highly sensitive to the EDTA treatment. The leaves of these plants exhibited visual symptoms (such as curling, chlorosis, necrosis and stunting) of metal or EDTA toxicity 1-2 days after the application of EDTA, followed by a rapid senescence and drying of the plant shoots. The toxic effect was most prominent on mung bean and buckwheat.

Table 1 Dry weights of the shoots of ten plant species 2 and 7 days after the application of EDTA. Error bars represent ± SE (*n* =3)

Crop species	EDTA /(mmol/kg soil)	Shoot dry wt/(mg/plant)	
		2d	7d
Dicotyledon			
Mung bean	0	71.1±7.6 (N) *	---
	2.5	50.7±4.8 (III)	Nd**
	5.0	48.3±5.4 (IV)	Nd
Buckwheat	0	41.7±9.3 (N)	78.7±11.3 (N)
	2.5	28.0±3.2 (II)	25.6±7.3 (IV)
	5.0	24.1±5.4 (IV)	Nd
Sunflower	0	280±41 (N)	533±39 (N)
	2.5	209±31 (II)	241±37 (IV)
	5.0	182±25 (II)	164±30 (IV)
Cabbage	0	230±27 (N)	303±48 (N)
	2.5	210±38 (II)	190±28 (II)
	5.0	168±36 (III)	185±20 (III)
Pea	0	95.0±7.2 (N)	167±29 (N)
	2.5	75.0±7.0 (II)	103±15 (IV)
	5.0	80.0±6.6 (III)	Nd
Mustard	0	35.6±6.5 (N)	132±28 (N)
	2.5	22.2±3.4 (II)	78.9±10 (III)
	5.0	22.9±4.2 (III)	35.0±5.3 (IV)

Continue table

Crop species	EDTA /(mmol/kg soil)	Shoot dry wt/(mg/plant)	
		2d	7d
Monocotyledon			
Maize	0	367±29 (N)	520±61 (N)
	2.5	263±24 (I)	430±29 (I)
	5.0	245±40 (I)	399±49 (II)
Sorghum	0	75.0±5.8 (N)	133±30 (N)
	2.5	65.0±5.7 (I)	128±23 (I)
	5.0	63.8±10.5 (I)	93.3±17 (II)
Wheat	0	62.2±5.7 (N)	105±12 (N)
	2.5	60.0±2.8 (N)	76.3±9.2 (N)
	5.0	58.8±5.7 (N)	76.3±5.8 (I)
Barley	0	61.3±3.5 (N)	124±11 (N)
	2.5	56.3±8.6 (N)	102±4.8 (N)
	5.0	51.3±4.6 (N)	106±11 (I)

Note: *The Figures in the brackets stand for the degrees of harm to plant seedlings as judged by observing those the pot experiment. N: stands for no visual harm in leaf; I: stands for the area of visible necro-macula in leaves was less than 5% ; II: the area of necro-macula in leaves was about 20% ; III: the area of necro-macula in leaves was about 50% ; and IV: stands for the oncoming death of seedlings.

**Nd denotes no data because these species of plants did not survive.

The application of EDTA at a rate of 2.5 or 5.0 mmol/kg dramatically increased the concentrations of metals in the shoots of plants (Table 2). In the 2.5 and 5.0 mmol/kg EDTA treatments, the Pb concentrations in the shoots of the six dicotyledon species ranged from 1000 to 3000 mg/kg of dry matter, which were higher than those of the monocotyledon species ranged from 330 to 830 mg/kg of dry matter. The highest amount of phytoextrated Pb (2.9 mg Pb/pot) was achieved in sunflowers, due to the high concentration of Pb in their shoots and large biomass, followed by corns (1.8 mg Pb/pot) and peas (1.1 mg Pb/pot).

Table 2 Amounts of Pb accumulation in the shoots of ten species of plants 2 and 7 days after the application of EDTA. Error bars represent±SE (*n*=3)

Crop Species	EDTA /(mmol/kg soil)	Pb concentration/(μg/g DW)		Amount of Pb accumulation/(μg/plant)	
		2 d	7 d	2 d	7 d
Dicotyledon					
Mung bean	0	56.9±7.3	—	4.1±0.9	—
	2.5	1430±143	Nd	72.3±6.3	Nd
	5.0	1490±182	Nd	72.5±14.4	Nd
Buckwheat	0	25.6±6.2	40.3±2.9	1.0±0.2	3.2±0.7
	2.5	1660±118	2500±168	46.2±2.1	63.4±13.8

Continue table

Crop Species	EDTA /(mmol/kg soil)	Pb concentration/(μg/g DW)		Amount of Pb accumulation/(μg/plant)	
		2 d	7 d	2 d	7 d
	5.0	1820±162	Nd	43.3±5.8	Nd
Sunflower	0	53.5±8.9	56.9±7.3	14.9±2.8	30.0±1.9
	2.5	389±55	1160±68	81.2±11.4	278±39
	5.0	391±68	1800±175	69.5±8.5	286±30
Cabbage	0	24.6±4.3	31.7±4.9	5.7±1.5	9.7±2.3
	2.5	262±39	475±47	54.3±5.5	89.4±3.7
	5.0	452±51	459±42	76.3±19.1	84.7±9.8
Pea	0	9.97±1.50	13.7±3.2	0.9±0.1	2.3±0.9
	2.5	1040±52	1110±152	78.5±11.2	113±13
	5.0	1390±106	Nd	112±14	Nd
Mustard	0	28.7±4.4	30.2±5.3	1.0±0.2	3.9±0.8
	2.5	461±46	1260±99	10.2±1.7	99.1±12.2
	5.0	761±101	2900±164	17.3±2.7	101±9.9
	Monocotyledon				
Maize	0	42.3±3.8	36.7±5.9	15.8±2.6	18.9±2.5
	2.5	332±45	295±30	86.7±6.5	128±13
	5.0	649±63	445±48	159±26	179±34
Sorghum	0	24.1±4.2	18.7±1.8	1.8±0.4	2.5±0.3
	2.5	302±19	271±40	19.8±3.0	34.5±6.5
	5.0	572±36	833±63	36.2±4.2	76.9±8.7
Wheat	0	12.3±1.8	14.2±2.9	0.8±0.2	1.5±0.4
	2.5	271±18	233±35	16.3±1.2	17.7±2.9
	5.0	387±25	329±17	22.7±2.3	25.1±2.9
Barley	0	16.5±2.0	18.2±3.7	1.0±0.1	2.3±0.5
	2.5	125±20	223±22	6.96±0.87	22.8±2.8
	5.0	94.3±14	360±56	4.80±0.46	38.1±5.0

Note: *Nd denotes no data because these species of plants did not survive.

Considering the three factors (the shoot level of metals, the biomass of plants, and the growing seasons of plants), mustard and pea were more suitable used in the cool seasons. On the contrary, corn, sunflower, mung bean and buckwheat could be selected in the warm seasons in field phytoextraction practice.

3. Approaches for enhancing phytoremediation of heavy metals-root damage and heating treatment

3.1 Effect of root damage on phytoextraction of heavy metals-solution culture experiment

In hydroponics, roots of corn (*Zea mays* L.) were pretreated with cutting, flapping or heating stress, and then exposed to the 250 μmol/L Pb solutions with or without 250 μmol/L EDTA addition. The results showed that the mechanical damage of roots by cutting or flapping had no effect on increasing shoot Pb level, however, the pre-heating treatment (50℃ for 0.5 h to 12 h) significantly facilitated the Pb transportation from roots to shoots (Fig. 1). Compared to the Pb treated alone, the addition of EDTA to the Pb solutions alleviated the phytotoxicity of seedlings and decreased Pb concentrations in roots, but increased Pb levels in shoots (Fig. 2).

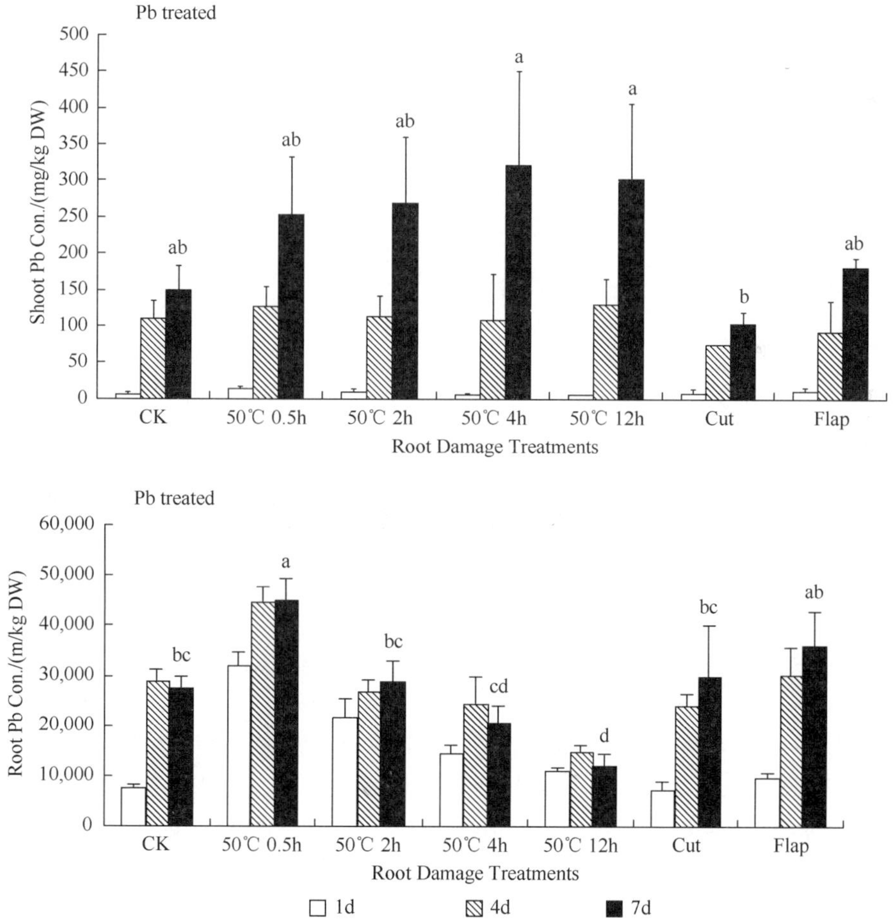

Fig. 1 Effects of root damage on Pb concentrations in the shoots and roots of corn seedlings

Roots of the 2-leaf-old corn seedlings were pretreated with high temperature (dipped into 50℃ water) for 0 (CK), 0.5, 2, 4, 12h. Other roots were cut out 0-2 cm root tips (Cut), or flapped with rules (Flap). Afterwards, seedlings were exposed at the 250 μmol/L $Pb(NO_3)_2$ solution for 1, 4, or 7 days in the solution culture experiment. Bars represent SEs. Different letters above the columns of 7d indicate a significant difference at $p<0.05$ according to the Duncan tests.

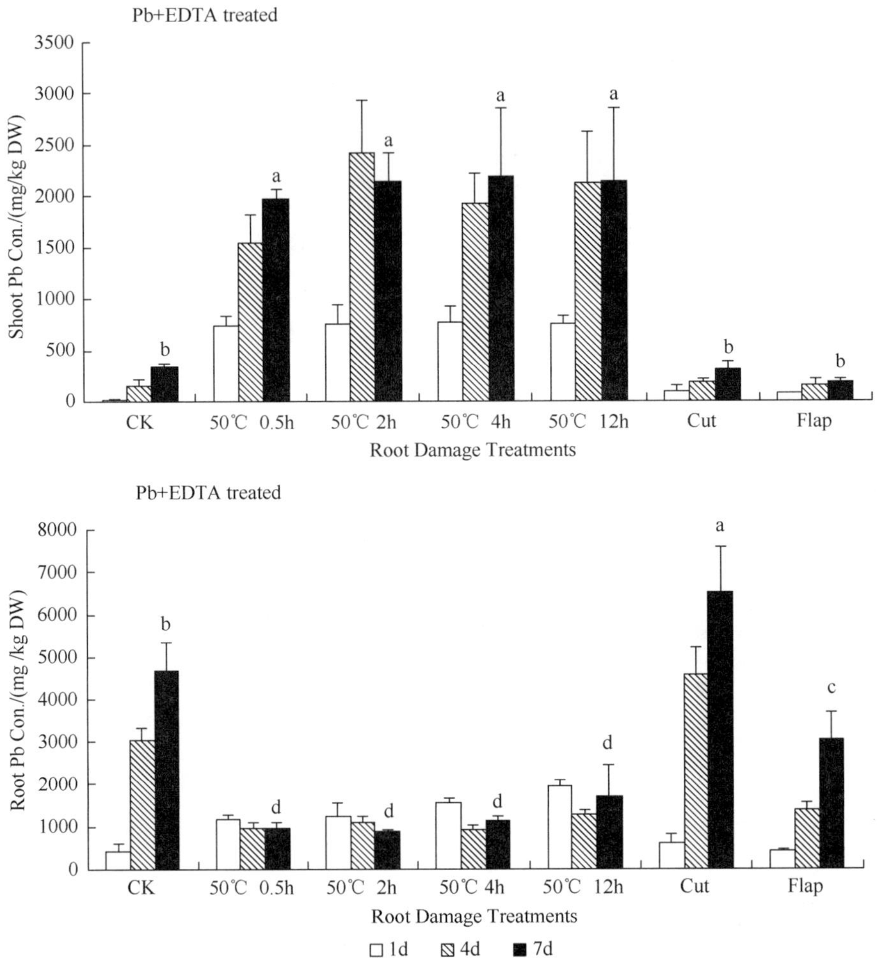

Fig. 2 Effects of root damage on Pb concentrations in the shoots and roots of corn seedlings

Roots of the 2-leaf-old corn seedlings were pretreated with high temperature (dipped into 50℃ water) for 0 (CK), 0.5, 2, 4, 12h. Other roots were cut out 0-2 cm root tips (Cut), or flapped with rules (Flap). Afterwards, seedlings were exposed at the 250 μmol/L $Pb(NO_3)_2$+250 μmol/L EDTA solutions for 1, 4, or 7 days in the water culture experiment. Bars represent SEs. Different letters above the columns of 7d indicate a significant difference at $p<0.05$ according to the Duncan tests.

3.2 Effect of hot EDTA solution on the accumulation of heavy metals by shoots-pot experiment

In pot experiment, the effect of hot EDTA solution on the accumulation of heavy metals by shoots of corn (*Zea mays* L.) and pea (*Pisum sativum* L.) were examined. Compared to normal EDTA (25℃) treatment, the application of 95℃ EDTA solution to the soil surface enhanced shoot metals uptake. The total removal of Pb by corn and pea increased by 8-fold and 12-fold, respectively, in a highly Pb-contaminated soil (Fig. 3); and the total removal of Cu by corn and pea increased by 6-and 9-fold, respectively, in a highly Cu-contaminated soil (Fig. 4). These results suggested that the heating treatment enhanced the efficiency of phytoextraction heavy metals. This new approach could help to minimize the amount of chelate applied in the field and reduce the potential risk of heavy metals' leaching to deep soils and groundwater.

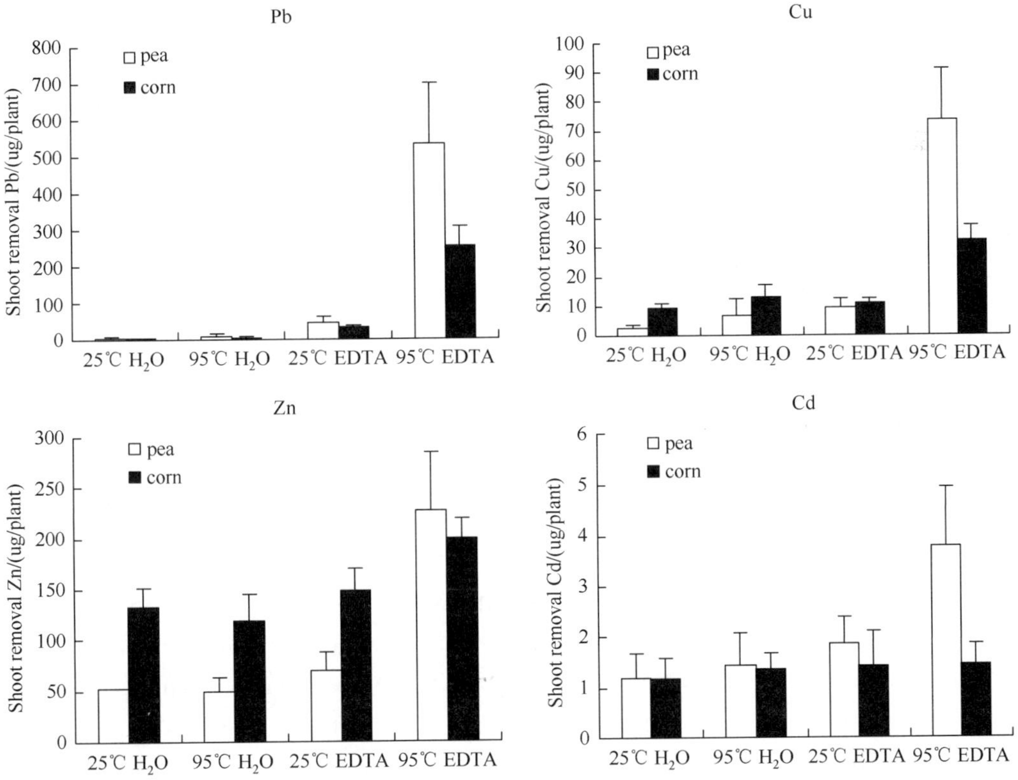

Fig. 3 Total removal of Pb, Cu, Zn and Cd by shoots of pea and corn seedlings 7 days after soil amendments treatment

The seedlings grew in the artificially heavy metal contaminated soils. 340 mL EDTA solution (at 25℃ or 95℃) were applied to the soil surface at a rate of 1.0 mmol EDTA/kg soil, 25℃ or 95℃ H_2O applied as the controls, respectively. (*n*=4; error bars; standard deviation).

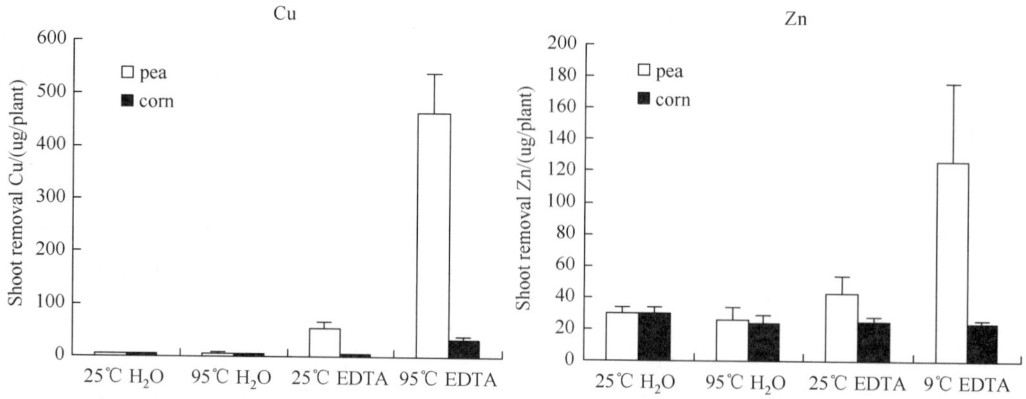

Fig. 4 Total removal of Cu and Zn by shoots of pea and corn seedlings 7 days after soil amendments treatment The seedlings grew in the field heavy metal contaminated soils. 340 mL EDTA solution (at 25℃ or 95℃) were applied to the soil surface at a rate of 1.0 mmol EDTA/kg soil, 25℃ or 95℃ H_2O applied as the controls, respectively. (*n*=4; error bars; standard deviation)

3.3 Increasing soil temperature for enhancing phytoextraction-pot experiments

Pot experiments were carried out to investigate the effects of increased soil temperature through different treatment schemes on the shoot uptake of metals by mung bean and corn seedlings grown in an artificially multi-metal contaminated soil and a naturally Cu-contaminated soil amended with EDTA or EDDS.

3.3.1 Effects of heating treatment stage on the concentrations of heavy metals in the shoots of mung bean

The artificially multi-metal contaminated soil was used in the pot experiment. The concentrations of Pb, Cu, Zn, and Cd in soils were 1940, 205, 515, and 2.12 mg/kg soil, respectively. The seeds of mung bean [*Vigna radiat* (L.) R. Wilczek var. *radiata* cv. VC2768A] were sown in pots. On the 14th day after sowing (2 days before chelators addition, named-2 d), the pots with seedlings were put into a thermostat water-bath tank at 25℃, 50℃ or 80℃ for 3 h, respectively. The waterline outside of pots was kept at 5 cm below the soil surface. On the 16th day after sowing (the day of chelators addition, 0 d), 50 mL of chelator (as 5.0 mmol/L Na_3-EDDS or Na_2-EDTA solution) was applied to the soil surface with the final application rate of 1.0 mmol/kg soil. The control pots received 50 mL H_2O. Followed by the chelator addition, the pots were treated at 25℃, 50℃ or 80℃ for 3 h as the above method. On the 18th day after sowing (2 days after chelator addition, 2 d), some pots with the addition of EDTA (without heating treated) were imposed on 25℃, 50℃ or 80℃ treatments for 3 h with the method described above. Plant shoots were cut from the cotyledonary node 7 days after the chelator addition. Shoot samples were used for elemental analysis using AAS.

Fig. 5 shows the concentrations of Pb, Cu, Zn and Cd in the shoots of mung bean 7 days after the application of chelators (EDDS or EDTA) with different temperature (25℃, 50℃, 80℃) treatment. Results showed that the 3 h high temperature (50℃ or 80℃) treatment of soils after the application of chelators significantly increased the concentrations of heavy metals in plant shoots (Fig. 5).

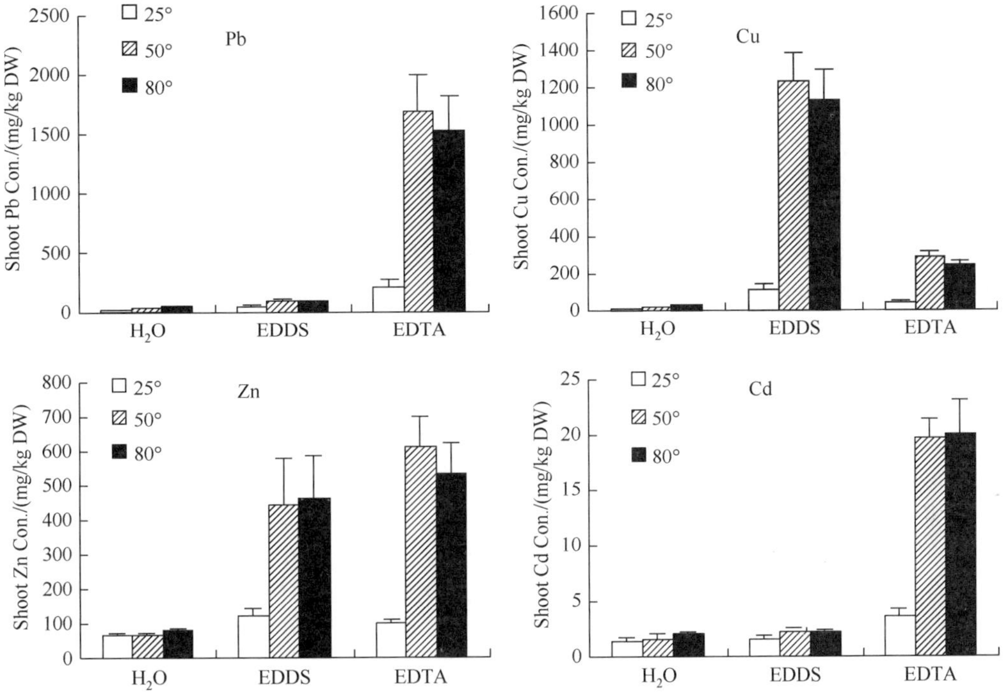

Fig. 5 The concentrations of Pb, Cu, Zn, and Cd in shoots (mg/kg DW) of mung bean 7 days after chelators (EDDS and EDTA) addition at a rate of 1.0 mmol/kg soil. Seedlings grew on the artificially heavy metal contaminated soils and the roots were treated with different temperature (25℃, 50℃, 80℃) for 3 hours followed by the chelators addition in the Experiment 1 (n=4, error bars-standard deviation).

Fig. 6 shows the concentrations of Pb, Cu, Zn and Cd in the shoots of mung bean 7 days after EDTA addition with different heating treatments. Under the same temperature for 3 h, heating treatment 2 days after the application of EDTA (2 d) produced the highest concentrations of Pb, Cu, Zn and Cd in the shoots, followed by the heating treatment on the day when EDTA was applied (0 d), then the heating treatment 2 days before the application of EDTA (-2 d).

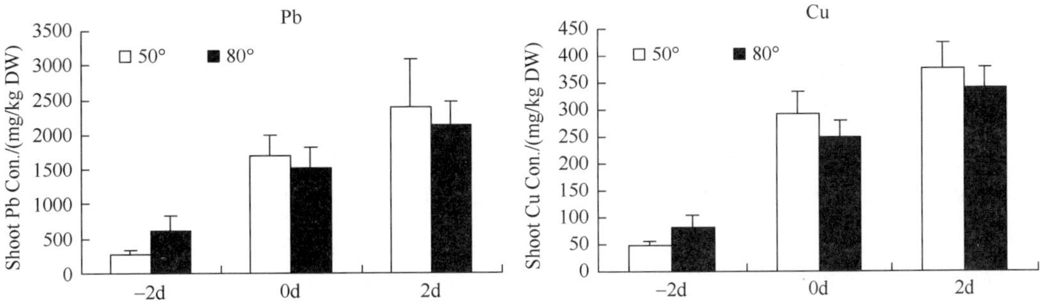

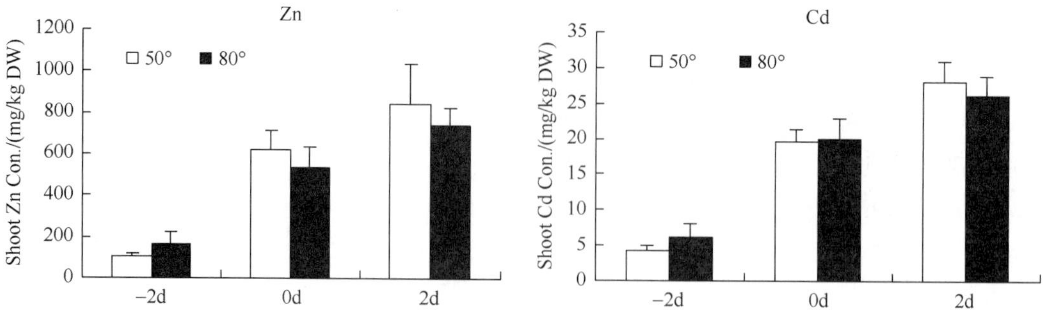

Fig. 6 The concentrations of Pb, Cu, Zn and Cd in shoots (mg/kg DW) of mung bean 7 days after EDTA (1.0 mmol/kg soil) addition

Seedlings grew on the artificially heavy metal contaminated soils and the roots were treated with 50℃ or 80℃ for 3 hours at 2 days before the EDTA addition (-2 d), the same day of EDTA addition (0 d) and 2 days after EDTA addition (2 d) in the Experiment 1 (n=4, error bars-standard deviation).

3.3.2 Effects of underground PVC tubes circulated with hot water on the shoot uptake of Cu by mung bean and corn from Cu-contaminated soils

The naturally Cu-contaminated soils (2500 g/pot) were placed in plastic pots (18.0 cm diameter × 15.0 cm height). Within each pot, a polyvinyl chloride (PVC) tube (1.5 m in length, 18 mm in inner diameter) was curved along the inner wall of the pot and embedded into soils.

Soil surface of each pot was divided into 8 sections. Seeds of corn and mung bean were sterilized, soaked, and then inter-sown at 4 sections. After germination, seedlings of two crops were thinned to 10 to 12 plants per pot.

On the 10th day after sowing, 250 mL of EDDS or EDTA was applied to the soil surface at a rate of 1.0 mmol/kg soil as 10 mM Na_3-EDDS or Na_2-EDTA solution. Pots treated with 250 mL H_2O was used as the control group. On the 12th day after sowing, that was 2 days after chelators addition, PVC tubes were circulated with 50℃ water for 4 hours at a flow rate of 6 L/min. On the 13th day after sowing, some pots were heating-treated again with the above method. On the 17th day after sowing, that was 5 days after the chelator addition, shoots of corn and mung bean were sampled, cleaned, acid digested and analyzed for metal concentrations. Plant samples were digested with a mixture of HNO_3 and $HClO_4$ (87:13, by volume) and diluted with 5% HNO_3 for elemental analysis using AAS.

Results showed that, without chelator addition, increasing soil temperature by underground tubes circulated with 50℃ water increased the shoot Cu concentrations about 2-fold in the two crops. With chelators (EDDS or EDTA) application, increasing soil temperature could enhance the concentrations of Cu in the shoots 17-fold to 20-fold for corn, and 5-fold to 6-fold for mung bean, respectively, compared to the normal temperature treatments (Table 3).

Total Cu phytoextraction by the shoots of two crops is also shown in Table 3. With chelator assistance, although the shoot dry matter yields of corn and mung bean in the heating treatments decreased about to 50%-80% of the normal chelator treatments (data not shown), the total phytoextraction of Cu by shoots of two crops increased about 4-fold to 14-fold in the heating treatments, in comparison with the normal chelator treatments (Table 3). The increased shoot Cu uptake could be reflected directly by the toxicity symptoms in the leaves of two crops.

Compared with the heating treatment one time (50℃), heating treatments conducted two times (50℃+50℃) did not further improve the uptake of Cu in the shoots of two crops, no matter with or without chelator addition (see Table 3).

Table 3 Effects of heating treatment by underground tubes circulated with hot water on the concentration of Cu and the total uptake of Cu in the shoots of corn and mung bean seedlings. Values are means±SD (*n*=4)

Chelators	Heating Treatments*	Shoot Cu Con. /(mg/kg DW)		Shoot Uptake Cu/(μg/plant)	
		Corn	Mung bean	Corn	Mung bean
Ck	25℃	23.8±1.1	7.8±1.3	9.67±2.50	1.47±0.32
	50℃	40.5±2.9	17.2±4.5	17.2±2.5	2.91±0.84
	50℃+50℃	39.6±3.2	19.3±5.2	15.3±2.8	3.11±0.92
EDDS	25℃	60.8±5.2	331±70.0	22.1±2.8	45.5±5.2
	50℃	1060±243	1770±250	219±39	199±22
	50℃+50℃	992±214	1630±281	187±59	182±41
EDTA	25℃	34.7±4.6	274±33.2	13.4±3.5	38.4±5.6
	50℃	697±186	1660±750	180±91	191±81
	50℃+50℃	722±247	1230±418	167±73	125±42

*. 12-day-old corn or mung bean seedlings grown in the Cu-contaminated soils were treated with or without 1.0 mmol soil/kg chelators (EDDS or EDTA). Two days after the chelators addition, the soils were heated by underground tubes circulated with 50℃ water for 4 h (50℃). Some pots were treated again in the following day (50℃+50℃). Shoots were harvested 5 days after the chelators treatment.

4. The leaching behavior of heavy metals in chelant-enhanced phytoextraction and the environmental risk assessment

4.1 Short soil-leaching columns planting sunflower

The leaching behavior of heavy metals as a result of applying EDTA to the surface of heavy metal-contaminated soils was investigated using short soil-leaching columns (9.0 cm diameter, 20 cm height) by the percolation of artificial rainfall.

Soil samples were artificially contaminated with Pb (2500 mg/kg soil) in the forms of $Pb_3(OH)_2(CO_3)_2$ (lead hydroxide carbonate) and PbS (lead sulfide) at a ratio of 1:1; with Cu (500 mg/kg soil) in the form of $CuCO_3$ (copper carbonate); with Zn (1000 mg/kg soil) in the forms of $ZnCO_3$ (zinc carbonate) and ZnS (zinc sulfide) at a ratio of 1:1; and with Cd (15 mg/kg soil) in the form of Cd $(NO_3)_2 \cdot 4H_2O$ (cadmium nitrate). The seeds of oil-sunflower (*Helianthus annuus* L. cv. S61) were initially sown in four soil columns (9.0 cm inner diameter, 20 cm height). After germination, one single seedling with a similar biomass was allowed to grow in the soil columns.

On the 50th day after the sunflower seeds were sown, 5.0 mmol of EDTA/kg soil in the form of a Na_2-EDTA solution was applied to three soil columns without plant and to three soil columns with sunflowers. The soil columns were irrigated to field capacity with de-ionized water three days after the application of EDTA. Artificial rainwater was then applied gradually to the top of each soil column. The volumes of the leachate were then measured. The application of rainwater was conducted for 16 h per day for three days. The leachate solutions were analysed for heavy metals using ICP-AES (Perkin-Elmer Optima 3300 DV).

After the leaching experiment, sunflower shoots and roots were gently removed from the soil and carefully washed with tap water. The roots and shoots were further separated with scissors and washed with de-ionized water, and subsequently dried for metal analyze.

Results showed that: in the leachates from the soil columns where EDTA had not been applied, the concentrations of Pb, Cu, Zn and Cd in the leachates were very low, indicating that only a very small amount of heavy metals had been removed by the rainwater. The cumulative amounts of Pb, Cu, Zn and Cd were less than 1% of the initial concentrations of metal in the soil.

After the EDTA treatment, the concentrations of Pb, Cu, Zn and Cd in the leachates increased rapidly to the highest levels at leachate volumes of 300 mL and 400 mL, respectively, from the column with sunflowers and without plant. Subsequently, the concentrations of metals declined and reached an equilibrium of about 800 mL in leachate volume. The majority of the leached heavy metals occurred in the first 800 mL of leachates from the soil columns.

Table 4 shows the total amounts of heavy metals taken up by sunflower seedlings grown in the soil columns after the application of EDTA. The total uptake of Pb, Cu, Zn and Cd by sunflower seedlings was 0.62, 0.42, 2.23 and 0.12 mg, respectively. On the other hand, the total amount of Pb, Cu, Zn and Cd leached from the soil columns planted with sunflowers was 116, 136, 237 and 3.25 mg, respectively. About 4% of the initial total Pb, 16% of Cu, 13% of Zn and 23% of Cd in soils were leached from the soil columns. The total amount of heavy metals absorbed by the roots and shoots of sunflowers only accounted for a very small proportion ($<$1%) of the total leached heavy metals. No significant differences were found between the amounts of Pb, Cu, Zn and Cd leached from the columns with and without sunflower plants (Table 4). Therefore, the growth of sunflower plants did not change the leaching patterns of heavy metals in columns of soil under the present experimental conditions.

Table 4 Total amounts of metals in the leachates after the application of 1000 mL of rainwater and in the sunflower seedlings grown in the soil columns

Metals	Total initial content of metals in soil columns/mg	Leachates/mg*			Sunflower/mg
		Without EDTA With plants	With EDTA No plants	With EDTA and plants	With the application of EDTA
Pb	3280	1.58	126±10	116±8	0.62
	(100)**	(0.05)	(3.84)	(3.54)	(0.02)
Cu	860	3.19	133±5	136±6	0.42
	(100)	(0.37)	(15.5)	(15.8)	(0.05)
Zn	1730	6.47	225±5	237±3	2.23
	(100)	(0.38)	(13.1)	(13.7)	(0.13)
Cd	15.8	0.129	3.64±0.29	3.25±0.15	0.12
	(100)	(0.82)	(23.0)	(20.6)	(0.73)
Ca	8880	373	485±29	482±40	34.2
	(100)	(4.20)	(5.46)	(5.43)	(0.39)
Fe	9350	0.479	44.0±0.2	33.0±4.5	0.78
	(100)	(0.005)	(0.47)	(0.35)	(0.01)
K	2120	29.7	25.3±0.9	26.5±2.5	41.8
	(100)	(1.40)	(1.19)	(1.25)	(1.97)
Mg	671	45.8	57.9±2.3	57.6±5.8	7.6
	(100)	(6.83)	(8.64)	(8.58)	(1.13)

*. After the application of 1, 000 mL of rainwater (the amount of rainwater applied was nearly equivalent to 158 mm of rainfall precipitation within 2 days).

**. The data in the brackets are the percentage (%) of the total initial content of metals in soil columns.

4.2 Long soil leaching experiment–Vetiver grass on heavy metal retention in soil columns

Vetiver grass (*Vetiveria zizanioides*) is a tall (1-2 m), fast-growing, perennial tussock grass. It has a long (3-4 m), massive and complex root system, which can penetrate to the deeper layers of the soil. Owing to its unique morphological, physiological and ecological characteristics such as its massive and deep root system, and its tolerance to a wide range of adverse climatic and edaphic conditions, including elevated levels of heavy metals, the interest in this grass is increasing in recent years.

Each of the five polyethylene leaching columns (9.0 cm inner diameter, 60 cm height) was packed in layers, as follows: (from bottom to top): ①2 cm layer of acid-washed quartz sand (< 2 cm); ②a circle of fiberglass (2 mm), ③4900 g of air-dried soil (8.7% water content) equivalent to a layer of 56 cm in the column. Three seedlings of vetiver grass (the root length was about 20 cm) were transplanted into three of the soil columns (Column No. 1, No. 2 and No. 3). Another two columns (Column No. 4 and No. 5) without vetiver grass were used as the control group.

On the sixtieth day after the vetiver grass was transplanted, the first 800 mL of heavy metal leachate solution (Solution 1 obtained from 5.1) were applied to the four soil columns (No. 1, No. 2, No. 4 and No. 5), which was equal to 31.4 mm of rainfall per column. Subsequently, other heavy metal leachate solutions (obtained from 5.1) were added to these columns at two-day intervals. As for soil column No. 3, artificial rainwater was added instead at the same volume. After two days, the artificial rainwater was applied to each column at 55 mm precipitation every day for eleven days (equal to 350 mL of rainwater applied to each column per day).

The leachate solutions were collected at the bottom of the soil columns and measured for volume on a daily basis. A sub-sample of the leachate solutions was filtered through Whatman No. 42 filter paper, digested with concentrated HNO_3, then diluted with 5% HNO_3 and analysed for heavy metals by ICP-AES.

During the leaching experiment, the vetiver grass shoots (in columns No. 1, 2 and 3) were sampled and analysed for element concentration by ICP-AES, and the vetiver roots were collected and analysed at a later stage, as previously described.

The results of long-soil column experiment showed that soil matrix with planted vetiver, could re-adsorb 98%, 54%, 41%, and 88% of the initially applied Pb, Cu, Zn, and Cd, which may reduce the risk of heavy metals flowing downwards and entering the groundwater.

Table 5 The leachate solution volume collected under the five soil columns

Treatment times	Added solution volume/mL	Leachate solution volume/mL				
		No.1*	No.2	No.3	No.4	No.5
Day 1	200 (31.4) #	0	0	0	23	38
Day 3	200 (31.4)	0	0	0	164	171
Day 5	200 (31.4)	0	0	0	153	159
Day 7	200 (31.4)	0	0	0	168	169
Day 9	350 (55.0)	0	27	24	304	310
Day 10	350 (55.0)	197	174	202	318	321
Day 11	350 (55.0)	219	268	244	339	346
Day 12	350 (55.0)	212	282	216	317	321
Day 13	350 (55.0)	222	294	266	325	331
Day 14	350 (55.0)	158	207	165	303	312
Day 15	350 (55.0)	299	284	288	333	327
Day 16	350 (55.0)	224	246	245	308	328
Day 17	350 (55.0)	283	279	304	322	335
Day 18	350 (55.0)	282	255	241	312	320
Day 19	350 (55.0)	256	275	269	329	338
Sum	4650 (730.6)	2352	2591	2466	4020	4126

*. The different treatment of soil columns (No.1 and No.2: planted vetiver grass and applied a heavy metals leachate solution; No.3: planted vetiver grass without heavy metal leachate solution application; No.4 and No.5: without vetiver plant and with heavy metal leachate solution application).

#. Data in the brackets stand for the estimated rainwater precipitation/mm.

Table 6 The retention of heavy metals in the five long soil columns

Heavy metals	The amount of metal/mg	The amount of leached metals/mg				
		No.1*	No.2	No.3	No.4	No.5
Pb	122	2.19 (1.79%) #	2.41 (1.97%)	0.008	44.8 (36.6%)	49.3 (40.3%)
Cu	134	59.8 (44.6%)	63.3 (47.3%)	0.049	99.4 (74.3%)	109 (81.3%)
Zn	247	138 (55.7%)	155 (62.6%)	0.106	191 (77.2%)	208 (84.3%)
Cd	3.51	0.35 (9.9%)	0.52 (14.9%)	0.002	2.15 (61.3%)	2.12 (60.5%)

*. The different treatment of soil columns (No.1 and No.2: planted vetiver grass and applied a heavy metals leachate solution; No.3: planted vetiver grass without heavy metal leachate solution application; No.4 and No.5 without vetiver plant and with heavy metal leachate solution application).

#. Data in the brackets stand for the percentage of leached heavy metals.

4.3 The leaching behavior of heavy metals in contaminated soils during the process of EDDS/EDTA-enhanced phytoremediation

Using soil column leaching test, the leaching behavior of heavy metals in soils was studied in the process of EDDS or EDTA-enhanced phytoextraction, and the roles of high-biomass plants (corn and vetiver grass) on the decrease of heavy metals leaching were also examined.

Each of the 24 polyethylene leaching columns (115 cm height) was packed in layers, as follows: (from bottom to top): ①2 cm layer of acid-washed quartz sand (<2 cm); ②a circle of fiberglass (2 mm), ③deep-soils: heavy metal-contaminated soils equivalent to a layer of 80 cm in the column, ④up-soils: un-contaminated soils equivalent to a layer of 25 cm in the column (Table 7). Nine seedlings of vetiver grass (the root length was about 20 cm) were transplanted into nine of the soil columns. Another nine columns were sown corn seeds. Six columns without plants were used as the control group.

Table 7 Selected physicochemical properties of tested soils

Soils in different layers	Texture/%			Con. of heavy metals in soils/(mg/kg)			
	①	②	③	Pb	Cu	Zn	Cd
Up-soil (0-25 cm)	35.8	42.8	21.4	1480	203	388	2.39
Deep-soil (25-105 cm)	39.3	44.7	17.0	44.6	28.4	97	0.87

Texture: ①>0.05 mm ②0.05-0.001 mm ③<0.001 mm.

When the seedlings (corn and vetiver grass) height was about 1 m (with higher biomass), 2.5 mmol of EDTA or EDDS/kg soil was applied to 18 soil columns (6 without plants, to 6 soil columns with corn seedlings, and to 6 soil columns with vetiver grass). Another 6 columns (3 with corn seedlings and 3 with vetiver grass) were applied with water (as control).

From the 1st to 20th day after EDTA or EDDS application, 0, 0, 0, 0, 27, 0, 2, 4, 11.5, 25, 8, 0, 0, 36, 34, 1, 0, 0, 36 and 30 mm of artificial rainfall were applied into the columns. From 21st to 25th day after chelator application, 50.9 mm of rainfall was added each day.

The leachate were collected at the bottom of the soil columns and measured for the volume on a daily basis. The sub-samples of the leachate were analysed for heavy metals by ICP-AES.

Owing to the compactness of soils after water irrigation, up-soils (0-20 cm) and deep-soils (20-40, 40-60, 60-80, and 80-100 cm) were sampled, and used for soluble and total heavy metals analysis by ICP-AES. Plant shoots were also sampled.

Results showed that, application of EDDS into soil surface did not lead to the leaching of heavy metals from the soil columns (80 cm in height) after 480 mm of rainfall percolation within 25 days. However, EDTA application led to the migration of heavy metals from up-soil to deep-soil in the columns, and the migration of metals tightly related to the amount of added rainwater with or without planting (Table 8-11).

Table 8 Total volume of leachates and the total amount of leached heavy metals

Chelator treatment	Plant species	Total leached volume/mL	Total amount of leached heavy metals/mg			
			Pb	Cu	Zn	Cd
EDTA	No-plant	3 602±199	15.2±3.4	31.7±6.5	23.9±3.9	0.80±0.04
	Corn	1 530±266	3.6±0.9	2.4±2.4	2.0±0.7	0.08±0.14
	Vetiver	1 018±117	0.9±1.2	2.3±0.1	1.7±0.1	0.04±0.00
EDDS	No-plant	6 505±572	—	—	—	—
	Corn	2 648±525	—	—	—	—
	Vetiver	1 992±215	—	—	—	—

—. Not Detected.

Table 9 Content of water soluble-heavy metals in the different layer of soils

Chelator treated (Soil columns)	Plant species	Heavy metals	Water soluble heavy metal in the different layer of soils/(mg/kg)			
			0-20 cm	20-40 cm	40-60 cm	60-80 cm
EDTA	No plant	Pb	1.29±0.11	0.31±0.09	0.11±0.04	0.11±0.06
	Corn		4.24±2.99	6.89±11.4	0.03±0.02	0.05±0.05
	Vetiver		62.3±8.4	5.09±4.70	0.11±0.12	0.01±0.01
	No plant	Cu	0.06±0.01	0.33±0.28	0.50±0.53	1.57±1.82
	Corn		1.68±1.39	6.77±9.47	0.35±0.30	0.48±0.78
	Vetiver		20.25±3.05	5.70±3.47	1.31±0.97	0.19±0.21
EDDS	No plant	Pb	0.64±0.22	—	—	—
	Corn		0.60±0.19	—	—	—
	Vetiver		0.82±0.31	—	—	—
	No plant	Cu	0.51±0.15	—	—	—
	Corn		0.52±0.05	—	—	—
	Vetiver		0.47±0.29	—	—	—

Table 10 Total Pb concentration of soils in the different layers under EDTA-treated

EDTA-treated columns	Total Pb concentrations/(mg/kg)			
	No-EDTA	EDTA-corn	EDTA-vetiver	EDTA-no plant
0-20	1 492±56a	1 435±58a	1 476±29a	1 392±21b
20-40	65.5±14.4a	187±88b	135±38b	205±125b
40-60	45.3±5.8a	68.1±12b	58.0±7.2ab	83.8±5.6c
60-80	43.8±3.1a	50.5±4.3ab	46.5±2.8ab	60.3±13.4b

Table 11 Total Cu and Pb concentration of soils in different layers under EDDS-treated

Heavy metals	EDDS treated column	Total heavy metal concentrations/(mg/kg)			
		No-EDDS	EDDS-corn	EDDS-vetiver	EDDS-no plant
Cu	0-20	195±17a	188±15a	178±28a	193±19a
	20-40	29.6±2.0a	31.2±2.7a	36.2±3.0ab	36.9±1.7b
	40-60	27.1±1.1a	29.0±2.2a	29.6±2.0a	28.5±1.8a
	60-80	27.1±1.6a	29.1±0.7a	29.8±2.1a	28.4±0.3a
Pb	0-20	1 506±48a	1 488±56a	1 498±43a	1 458±75a
	20-40	59.1±5.6a	73.7±2.6b	73.4±9.3b	82.0±10.1b
	40-60	55.8±5.2a	52.9±6.1a	48.6±3.1a	53.9±7.8a
	60-80	46.8±2.3a	48.3±3.2a	44.7±4.5a	46.0±3.4a

Table 12 Pb and Cu concentrations in the shoots of vetiver and corn 25 days after EDTA or EDDS treatment

chelators	Shoot Pb Concentrations/(mg/kg)		Shoot Cu Concentrations/(mg/kg)	
	vetiver	corn	vetiver	corn
−EDTA	1.32	5.78	12.52	7.48
+EDTA	11.49±1.86	20.35±1.57	12.42±0.81	9.39±1.44
-EDDS	1.17	4.46	13.75	8.29
+EDDS	5.26±0.41	6.75±0.69	15.33±0.56	10.92±1.01

Table 12 showed the Pb and Cu concentrations in the shoots of vetiver and corn 25 days after EDTA or EDDS treatment. The seedlings of corn or vetiver grass could be effectively delay the movement of heavy metals downward, and vetiver grass showed more effective than corn (Table 8-11). This implies that, combined planting with some high-biomass plants, EDDS can be regarded as a good chelator candidate for the environmentally safe phytoextraction heavy metals from contaminated soils. It should be careful to use of EDTA in the field phytoremediation, especially in some heavily raining areas.

4.4 Metal leaching along soil profiles after the EDDS application—A field study

The field experiment was carried out during June 1-September 19 of 2009 on a farmland in the vicinity of a past copper mine (Wu et al., 2011), located at the Tangshan Town of Nanjing City, Jiangsu Province, Eastern China. The air temperature during this experiment period was in the range of 19-37℃ with an average of 27.3℃, and the precipitation was recorded with detailed rainfall events shown in Fig. 1. The soil is sandy loam soil (U. S. Texture Triangle), and the sites were previously planted by crops, such as maize (*Zea mays* L.), Kang Kong (*Pomoea aquatica* Forsskal), lettuce (*Ctuca sativa*), and amaranth (*Maranthus mangostanus* L.).

An experimental plot of 9 m×3 m was delineated in the center of the test field. Maize seeds were directly planted in a hole-seeding way on June 1. On June 18, when the maize seedlings were about 15 cm tall, seedlings were thinned to a density of 3.3×10^5/ha.

On August 14, during the flowering period of maize plants, 6 plots (1.5 m×1.5 m) with uniform growth conditions were selected, and a subplot with a size of 60 cm×80 cm was chosen in the center of each plot. Two different treatments were conducted on the subplots: the control, and the treatment with EDDS followed by irrigation with hot water (90℃) 2 days later (Chen et al., 2008). EDDS was applied with 4.8 L of 78 mmol/L Na_3EDDS solution (Fluka Chemie GmbH, UK) in a single application to the surface of subplots which equaled to a dosage of 3 mmol/kg soil (calculated based on a soil depth of 20 cm and a soil bulk density of 1.3 g/cm^3). To make the EDDS solution uniformly irrigated to the soil surface, each subplot was divided into 48 grids with a size of 10 cm×10 cm, and each grid was irrigated with 100 mL EDDS solution or water (the control group). Two days later, EDDS-treated subplots were irrigated with 4.8 L 90℃ hot water using the same irrigating method, by which treatment could facilitate metal uptake by roots and transport upward to the shoots (Chen et al., 2008). For the control group, the same dosage of water at normal ambient temperature was irrigated.

On August 24, 84 days after the maize was sowed (10 d after the EDDS application), the above ground parts of maize plants were harvested, and metals in plant tissues were analyzed after acid digestion.

Soil samples along a 60 cm depth soil profile were collected in each subplot with a stainless soil probe on 1, 8, 15, 22, 29, 36 d after the application of EDDS. The top 6 samples were collected at every 5 cm along 0-30 cm depth, and the other 3 samples were collected in every 10 cm along 30-60 cm depth. Five equally sized, and discrete subsamples were taken per subplot for every layer from evenly spaced, adjacent sampling points to form composite samples. Samples were immediately taken to laboratory and stored in refrigerator. Total metals in the soil samples were determined with ICP-AES after strong acid digestion. In addition, the soil samples were extracted with DIW (pH 6.8), and the soil solution was used for the analysis of major elements (including Al, Ca, Mn and Fe) and trace metals (e.g., Cu, Zn and Pb).

In the present study, soluble metals were analyzed during a period of 36 d after EDDS application along the 60 cm depth soil profiles under field conditions. The results showed the application of EDDS to the soils could significantly solubilize trace metals from the soil matrix, and the soluble Cu and Zn reached 123 and 11 mg/kg at the 0-5 cm of the soil 1 d after the EDDS application, which was 412 and 32 times, respectively, of the control (Fig. 7). The increase of soluble metals was mainly confined to the top 20 cm of the soil profiles, and a sharp decreasing trend was observed from the soil surface to 25 cm depth. Below 25 cm depth, no significant increase of soluble metals was found.

Total Cu content in the soil varied widely at different depth with an average about 494 mg/kg in the top 15 cm, and decreased to 235 mg/kg at 15-30 cm, and then drop to 197 mg/kg below 30-60 cm. No significant difference ($p<0.05$) in the soluble metals was observed between various depths in the control soil plots (no EDDS treatment) although the total metals varied from 200 to 600 mg/kg. Therefore, the heterogeneity of soil was not the major reason for the decreasing trend of soluble metals when EDDS was applied.

The highest extractable metals were recorded 1 d after EDDS application (August 15), and soluble metals in the soils decreased quickly with time (Fig. 7). Seven days later (August 22), the concentrations of soluble Cu were decreased by 28%, 68%, 74% and 72% at 0-5, 5-10, 10-15 and 15-20 cm of soil profile, respectively, compared with those values obtained 1 d after EDDS application. The concentrations of soluble Pb and Zn also dramatically decreased in the upper layers of soil. Below 25 cm depth, no significant increase of soluble Cu, Pb and Zn was observed. During a period of 8 d after EDDS application, no rainfall event occurred. Thus, the observed decreases of soluble metals were probably due to the biodegradation of these metal-EDDS complexes. As shown in Fig. 8, detectable EDDS was only limited to the top 20 cm, and the amount of EDDS decreased very quickly with time. The degradation of EDDS could be influenced by the property of the soil, such as pH, metal compositions and concentrations, and the dose of applied EDDS. Vandevivere et al. (2001) observed that the Ca-, Mg-, Cd-, Fe (III)-, Al-, Pb-, and Cr (III)-EDDS complexes biodegraded readily at an average rate of 0.3 mmol/d. The calculated half-life of EDDS in sludge-amended soil was 2.5 days (Jaworska et al., 1999). Meers et al. (2005) estimated that the effective half-life of EDDS ranged between 3.8 and 7.5 days when the dose that was applied ranged from 0.8 to 4 mmol/kg.

After 22 d of EDDS application (September 5), no significant difference was observed in the concentrations of soluble Cu, Pb and Zn between EDDS-treated soils and untreated control along the whole profile (Fig. 7). It is interesting to note that two rainfall events occurred (with no runoff forming) on the 8th days (August 22) and 16-18th days (August 30-September 1) after EDDS application. However, no significant difference was observed in the concentrations of soluble Cu, Pb and Zn at a 25-60 cm depth between soils collected before (August 22) and after these rainfall events (August 29 and September 5). It indicated that the rainfall did not result in an increase in the concentration of soluble metals in deeper soil layers.

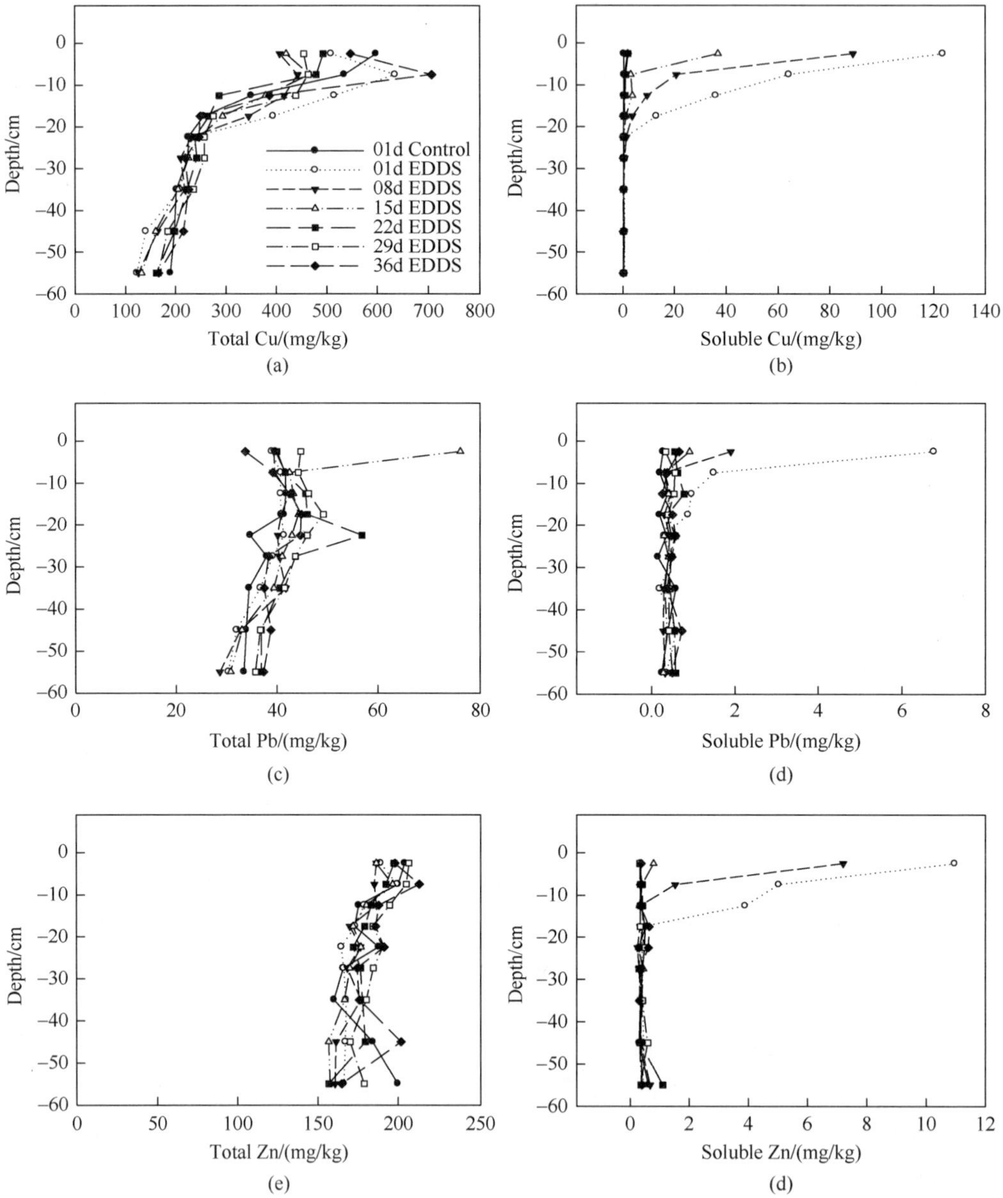

Fig. 7 The variation of soluble and total Cu (a) and (b), Pb (c) and (d), and Zn (e) and (f) along the 0-60 cm soil profile

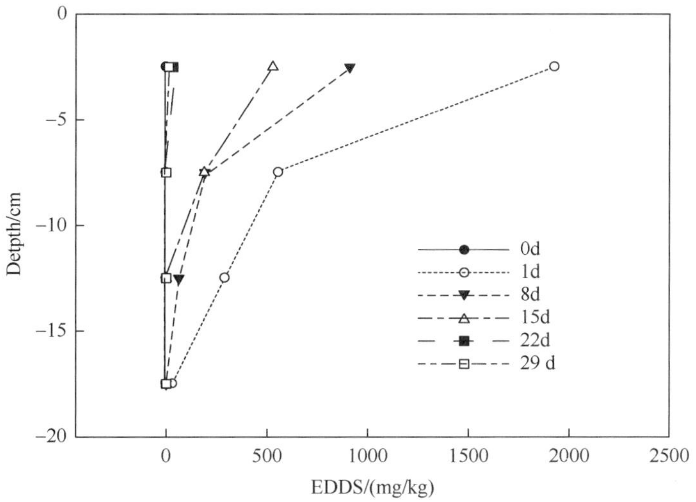

Fig. 8 The variation of EDDS along the soil profile with the time

In the present study, EDDS was found to be fully degraded during the first 22 d after EDDS was applied to soil (Fig. 8). The rapid biodegradation of EDDS made it more attractive over EDTA for significantly low leaching of associated metals to the surrounding environment after chelant application. When EDTA was applied to the soil, high soluble metals could remain several months even in the deeper soil (Neugschwandtner et al., 2008). Meers et al. (2008) observed a full degradation period of 54 d after EDDS applied. The shorter phase of EDDS degradation may be attributed to the lower application dose of EDDS to contaminated soil in our study.

References

Chen Y H, Li X D, Shen Z G. 2004. Leaching and uptake of heavy metals by ten different species of plants during an EDTA-assisted phytoextraction process. Chemosphere, 57 (3): 187-196.

Chen Y, Liu L, Wang G, et al. 2007. The leaching behavior of heavy metals in contaminated soils during the process of chelant-enhanced phytoremediation. Journal of Nanjing Agricultural University, 30 (4): 46.

Chen Y H, Mao Y, He S B, et al. 2007. Heat stress increases the efficiency of EDTA in phytoextraction of heavy metals. Chemosphere, 67 (8): 1511-1517.

Chen Y H, Shen Z G, Li X D. 2004. The use of vetiver grass (*Vetiveria zizanioides*) in the phytoremediation of soils contaminated with heavy metals. Applied Geochemistry, 19 (10): 1553-1565.

Chen Y H, Wang C C, Wang G P, et al. 2008. Heating treatment schemes for enhancing chelant-assisted phytoextraction of heavy metals from contaminated soils. Environmental Toxicology and Chemistry, 27: 888-896.

Jaworska J S, Schowanek D, Feijtel T C J. 1999. Environmental risk assessment for trisodium [S, S]-ethylene diamine disuccinate, a biodegradable chelator used in detergent applications. Chemosphere, 38 (15): 3597-3625.

Meers E, Ruttens A, Hopgood M J, et al. 2005. Comparison of EDTA and EDDS as potential soil amendments for enhanced phytoextraction of heavy metals. Chemosphere, 58 (8): 1011-1022.

Meers E, Tack F M G, Verloo M G. 2008. Degradability of ethylenediaminedisuccinic acid (EDDS) in metal contaminated soils: Implications for its use soil remediation. Chemosphere, 70 (3): 358-363.

Neugschwandtner R W, Tlustoš P, Komárek M, et al. 2008. Phytoextraction of Pb and Cd from a contaminated agricultural soil using different EDTA application regimes: laboratory versus field scale measures of efficiency. Geoderma, 144 (3): 446-454.

Salt D E, Smith R D, Raskin I. 1998. Phytoremediation. Annual Review of Plant Biology, 49 (1): 643-668.

Vandevivere P C, Saveyn H, Verstraete W, et al. 2001. Biodegradation of metal-[S, S]-EDDS complexes. Environmental Science & Technology, 35 (9): 1765-1770.

Wang A, Luo C, Yang R, et al. 2012. Metal leaching along soil profiles after the EDDS application—a field study. Environmental Pollution, 164: 204-210.

Wu F, Liu Y, Xia Y, et al. 2011. Copper contamination of soils and vegetables in the vicinity of Jiuhuashan copper mine, China. Environmental Earth Sciences, 64 (3): 761-769.

The Immobilization of Heavy Metals in Contaminated Soils Amended Amended with Phosphate-, Carbonate- and Silicate-Based Materisls: From Lab to Field

Xinde CAO

School of Environmental Science and Engineering, Shanghai Jiao Tong University

Abstract

The amendment-induced immobilization of heavy metals has been considered as a cost-effective and easily-operational method for remediation of contaminated soils. This method relies on the amendments to convert heavy metals into in the stable forms, by sorption, precipitation, complexation, ion exchange or redox process, thereby decreasing mobility and bioavailability of heavy metals. Among the amendments, phosphate-, carbonate-, and silicate-based materials was most-commonly used. In this paper, a review was made on the development of our research on such an immobilization treatment with more attention paid to the molecular mechanism and field application. The impact factors, long-term effects, possible secondary pollution, and effects on the soil property related to the treatment were also discussed.

Keywords: amendment, contaminated soil, field demonstration, heavy metals, immobilization, lab test

1. Introduction

Soil contamination with heavy metals is of great concern over the world. The primary sources of Pb contamination include industrial activities such as mining, smelting of metals and the use of Pb-containing products such as paint, gasoline, and pesticides (Hettiarachchi et al., 2001). Use of Pb bullets/shots as ammunition at shooting ranges is under increasing scrutiny as a potentially significant source of Pb pollution (USGS, 2002). Battery recycling sites are often contaminated with Pb as well as other metals (Zn, Cu, and As) (USEPA, 1991). Mining and smelting of Cu and Zn metal ores are important sources of Cu and Zn environmental degradation to soil and water sources (Dudka and Adriano, 1997). Although Cu and Zn are not a human health concern, their phytotoxic levels can result in soil erosion by wind and water, thereby increasing human exposure to other metal contaminants (Pb and Cd) (Basta et al., 2001).

Human exposure to contaminated soils includes drinking of the water with heavy metals released from the soil and consumption of edible plants grown in the contaminated soil (Basta et al., 2001, Madrid et al., 2006). In addition, accidental hand-to-mouth ingestion by humans, especially children, is of health concern (Chaney et al., 1994). Therefore, implementing soil remediation practices to reduce metal availability in soils is necessary to protect human health. In situ chemical immobilization is a cost-effective remediation approach for the reduction of metal mobility and bioavailability in contaminated soils (Cao et al., 2011a). Amendments added to the soil immobilize a contaminant and reduce leachable concentrations to an acceptable level (Basta et al., 2001). Among the amendments, phosphate-, carbonate, and silicate-based materials was most-commonly used (Liang et al., 2012).

In this paper, a review was made on the development of our research on such an immobilization treatment with more attention paid to the molecular mechanism and field application. The impact factors, long-term effects, possible secondary pollution, and effects on the soil property related to the treatment were also discussed.

2. Reactions of heavy metals with amendments

The reaction of heavy metals with amendments is crucial for the immobilization effectiveness of heavy metals. It was shown that phosphate rock (PR) has the highest affinity for Pb, followed by Cu and Zn, with sorption capacities of 138, 114, and 83.2 mmol/kg, respectively (Cao et al., 2004). The greatest stability of Pb retention by PR can be attributed to the formation of insoluble fluoropyromorphite ($Pb_{10}(PO_4)_6F_2$) (Fig. 1 (a)) , which was primarily responsible for Pb immobilization (up to 78.3%), with less contribution from the surface adsorption or complexation (21.7%), compared to 74.5% for Cu and 95.7% for Zn. Flow calorimetry indicated that Pb and Cu adsorption onto PR was exothermic, while Zn sorption was endothermic (Fig. 1 (b)). Our research indicated that PR can effectively immobilize Pb, Cu, and Zn with the most effective for Pb.

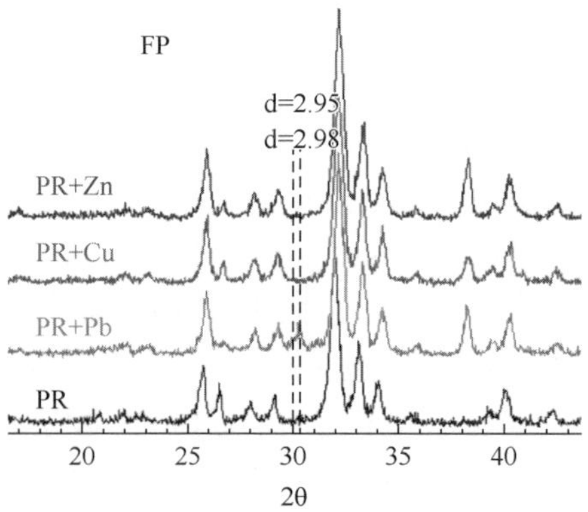

(a) X-ray diffraction patterns

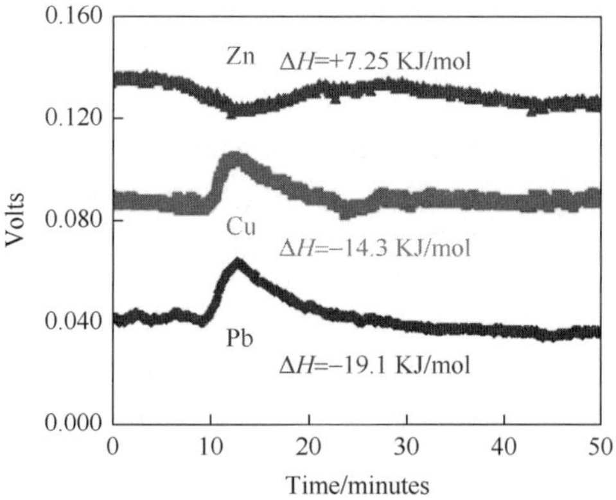

(b) Flow calorimetr

Fig. 1 X-ray diffraction patterns of the residual solids from phosphate rock reaction with 0.4 mM Pb, Cu, and Zn, respectively (a); and flow calorimetric analysis of phosphate rock reaction with 0.4 mM Pb, Cu, and Zn, respectively (b). FP, fluoropyromorphite [$Pb_{10}(PO_4)_6F_2$]; PR, phosphate rock (Cao et al., 2004)

Combination of PR with $FeSO_4$ was proven to be effective in simultaneously immobilizing Zn and Cr (Ⅵ) within the tested solution pH range of 5.5-8.5 (Zhao et al., 2016). Over 95% Zn and Cr (Ⅵ) was removed by PR with $FeSO_4$. Zn was stabilized via formation of insoluble minerals such as $Zn_3(PO_4)_2$ or CaZn $(CO_3)_2$ while Fe^{2+}induced reduction of Cr (Ⅵ) into stable Cr $(OH)_3$ or $Cr_xFe_{(1-x)}(OH)_3$ which was responsible for Cr (Ⅵ) immobilization.

A carbonaceous material, biochar (BC) pyrolyzed from biomass waste, is increasingly being recognized as a green, cost-effective amendment for environmental remediation. A study (Cao et al., 2009a) showed that the dairy manure-derived biochar can sorb Pb as much as 680 mmol/kg. Sorption of Pb by biochar followed a dual Langmuir-Langmuir model, attributing to Pb precipitation (84%-87%) and surface sorption (13%-16%). Chemical speciation Modeling, X-ray diffraction, and infrared spectroscopy indicated that Pb was precipitated as β-$Pb_9(PO_4)_6$ or as $Pb_3(CO_3)_2(OH)_2$. A further study (Xu et al., 2013) indicated that dairy manure-derived biochar was also effective for sorbing Cu, Zn, and Cd from aqueous solutions, with the highest sorption for Cu. Retention of the three metals by the biochar was mainly attributed to their precipitation with PO_4^{3-} and CO_3^{2-}, with less from surface adsorption through complexation of heavy metal by phenonic-OH site or delocalized π electrons (Fig. 2). The biochar derived from rice husk was effective in removing all the four heavy metals, with the removal capacities of 65.5-140 mmol/kg. The sorption was mostly due to the comlexation with ionized hydroxyl-O^- groups (Xu et al., 2014). Biochars produced from bagasse and hickory chip showed high sorption abilities of Hg (Ⅱ) (Xu et al., 2016). The sorption of Hg (Ⅱ) by bagasse biochar was mainly attributed to the formation of $(-COO)_2Hg^{II}$ and $(-O)_2Hg^{II}$. The Hg removal by hickory chip biochar was mainly resulted from the π electrons of C=C and C=O induced Hg−π binding. Further X-ray photoelectron spectroscopy analysis indicated the possibility of reduction of the Hg (Ⅱ) to Hg (Ⅰ) by phenol groups or π electrons during the removal of Hg (Ⅱ) by both biochars. Overall, the waste biomass can be converted into value-added biochar as an

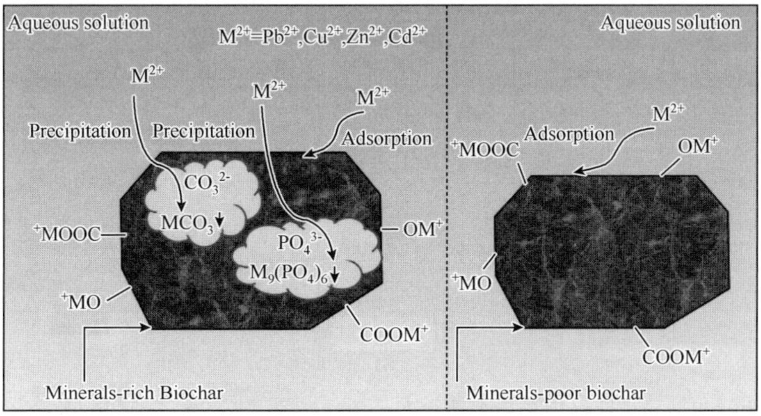

Fig. 2 Schematic diagram of heavy metals sorption by biochar with difference contents of mineral components (Xu et al., 2013, 2014)

effective sorbent for various heavy metals; the sorption effectiveness varied with the different metals and is affected by the biochar production temperature; and the mineral components (e.g., PO_4^{3-}, CO_3^{2-}) originated in the biochar serve as additional sorption sites, contributing to the biochar's high sorption capacity (Xu et al., 2015) (Fig. 2). Turning dairy manure into biochar as a sorbent has many environmental implications, e.g., reuse of solid waste and immobilization of heavy metals.

3. Field demonstration

To assess the efficiency of amendment-induced immobilization of heavy metals in soils, a series of field demonstration were conducted at contaminated sites. Phosphate was applied in a contaminated soil at an abandoned battery recycling site at a P/Pb molar ratio of 4.0 with three treatments: T1: 100% of P from H_3PO_4(PA), T2: 50% of P from PA+50% of P from Ca $(H_2PO_4)_2$(PC), and T3: 50% of P from PA+5% PR in the soil (Cao et al., 2002, 2003a, 2009b; Chen et al., 2003; Melamed et al., 2003). All P amendments effectively transformed soil Pb from the non-residual (sum of exchangeable, carbonate, Fe/Mn, and organic) to the residual fraction, with residual Pb increase by 11%-55%. After 4-month incubation, TCLP extractable Pb in the P-treated soils was reduced from 82 mg/L to below EPA's regulatory level of 5 mg/L in the surface soil. Lead immobilization was attributed to the P-induced formation of chloropyromorphite (Fig. 3 and Fig. 4).

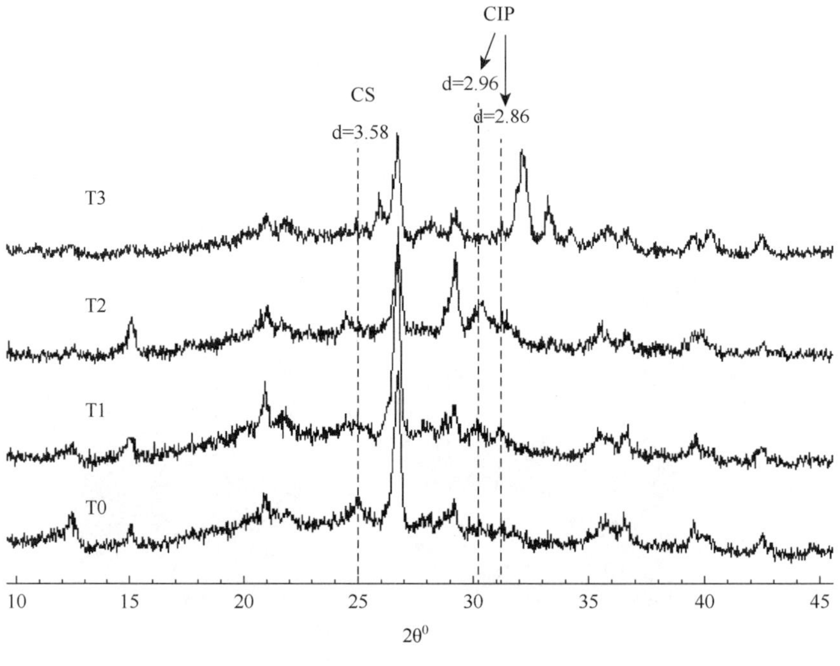

Fig. 3 X-ray diffraction patterns of the surface soils (0-10 cm) in the control and P-treated plots on 220 days after first P application. T0, the control; T1, H_3PO_4 alone; T2, H_3PO_4+Ca $(H_2PO_4)_2$, and T3, H_3PO_4+phosphate rock. CS, cerussite [$PbCO_3$]; ClP, chloropyromorphite [$Pb_{10}(PO_4)_6Cl_2$]

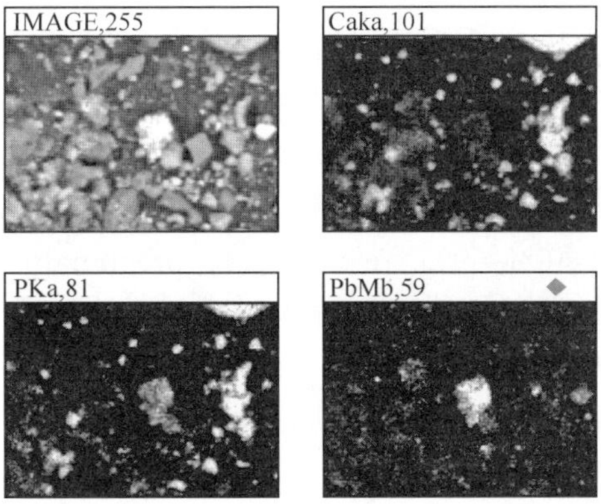

Fig. 4 Scanning electron microscopy elemental maps of a surface (0-10 cm) soil clay sample collected from the T2 plot treated with H_3PO_4+Ca $(H_2PO_4)_2$

Occurrence of chloropyromorphite was evident at 220 days after P addition for PA and PC treatments and at 330 days for PA+PR treatment (Cao et al., 2002). Lead immobilization shows a long-term stability. Formation of an insoluble chloropyromorphite mineral was also found in the subsurface soil (Fig. 5). Although P amendment enhanced metal uptake in the roots of St. Augustine grass (*Stenotaphrum secundatum*), it significantly reduced metal translocation from root to shoot, especially Pb via formation of a pyromorphite-like mineral on the membrane surface of the root.

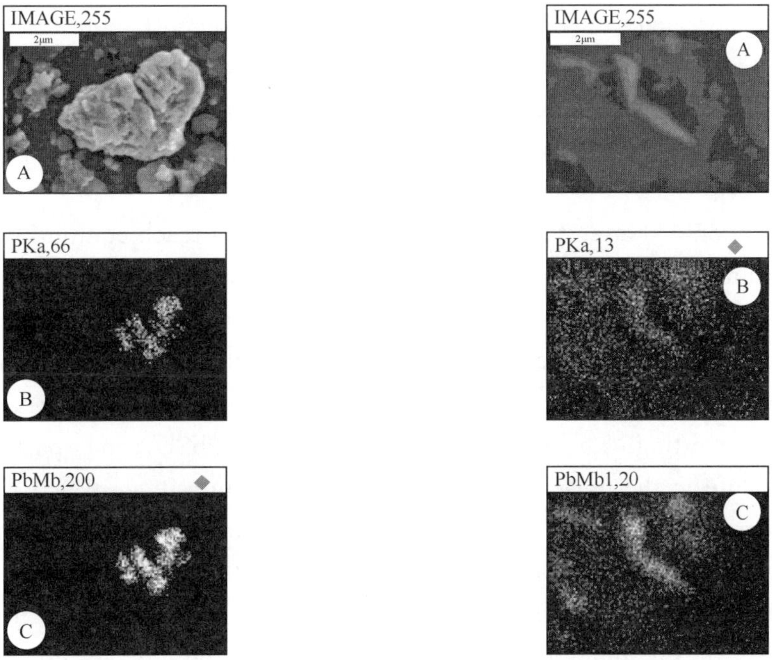

(a)a subsurface soil clay samples collected from the depth of 30-40 cm

(b)a rhizospheric soil collected from around the root of St. Augustine grass

Fig. 5 Scanning electron microscopy elemental maps of a subsurface soil clay samples at depth of 30-40 cm (a) and a rhizospheric clay (b) collected from the T2 plot treated with H_3PO_4+Ca $(H_2PO_4)_2$. (A) chloropyromorphite crystal, (B) Ca, (C) P, (D) Pb, Scale bar=5 μm

The further test confirmed that all three treatments were also effective in transforming soil Zn and Cu from the non-residual to the residual phase with the Zn and Cu in the residual phase increased by up to 15% and 13%, respectively, compared to the untreated soil (Cao et al., 2009b). Correspondingly, water soluble Cu and Zn were reduced by 31%-80% and 40%-69%, respectively, presumably due to their sorption on minerals (e.g., calcite and phosphate phases) following CaO addition. However, P had little effect on the Cu and Zn phytoavailability, while the bioaccessibility of Cu and Zn assessed by SBET were even elevated by up to 48% and 40%, respectively, in the PA and PR+PA treatments.

An experiment was conducted to determine the ability of biochar to immobilize Pb in a contaminated soil collected from a shooting range site (Cao et al., 2011b). The soil treated with the dairy manure biochar showed more than 57% reduction in Pb in 0.01 M $CaCl_2$ extraction. Lead concentrations in TCLP solutions were reduced by 70%-89%. Uptake of Pb by earthworms (*Eisenia fetida*) was reduced by up to 79%. Phosphorus originally contained in biochar induced conversion of soil $PbCO_3$ to insoluble hydroxypyromorphite $Pb_5(PO_4)_3(OH)$, which was presumably responsible for soil Pb immobilization.

After 120-d incubation of the PR with $FeSO_4$ in an abandoned electroplating contaminated soil, the TCLP extractable Zn and Cr (Ⅵ) were by 40% and 94%, respectively, compared to the untreated soil (Zhao et al., 2016). Biosolid compost application significantly reduced plant As uptake from the soil at an abandoned chromated copper arsenate (CCA)-treated wood site by 79%-86%, relative to the untreated soils. This suppression is possibly because of As adsorbed by biosolid organic matter, which reduced As phytoavailability (Cao et al., 2003b).

Phosphorus could shorten the time for Pb dissolution to reach equilibrium from $>$180 min to $<$60 min and reduce soluble Pb via transformation of Pb-solid phases into geochemically stable chloropyromorphite. The intensity and magnitude of chloropyromorphite were higher at the low pHs ($<$5) due to enhanced dissolution of Pb minerals, allowing more Pb to form chloropyromorphite (Cao et al., 2008a). Our results indicate that P was effective in converting Pb from PbO, $PbCO_3$ and $PbSO_4$ to sparingly soluble chloropyromorphite in the presence of soils. Acidic condition (e.g., pH=5) was favorable for this transformation. Among the three Pb minerals tested, P was most effective in reducing Pb solubility from $PbSO_4$, followed by $PbCO_3$ and PbO (Fig. 6). This study clearly demonstrates the importance of the form of Pb contamination and soil pH in determining the effectiveness of Pb immobilization. This information will be useful to optimize P treatment application rates as well as design parameters for a more effective in situ remediation technology. Total Pb concentration in contaminated soils is not adequate in optimizing treatment application rates and immobilization effectiveness because various forms of Pb exhibit different solubility(Cao et al., 2008a).

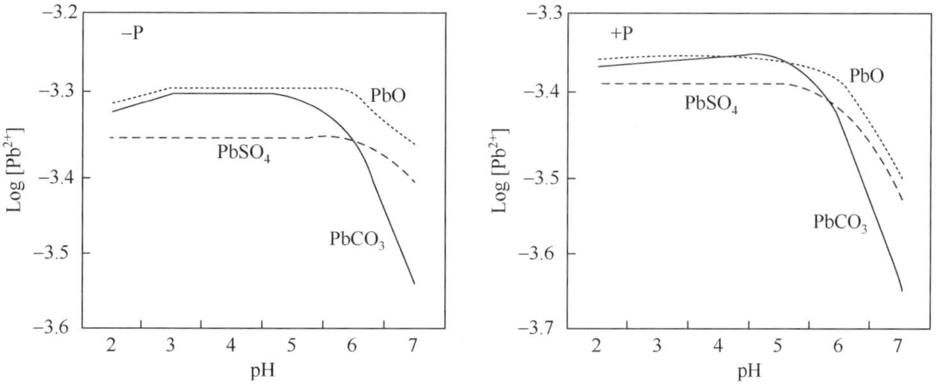

(a) Absence of P (b) Presence of P

Fig. 6 Predicted solubility of various lead minerals in a 0.01 mol/L KNO_3 aqueous solution as a function of pH in the absence and presence of P, with the following assumptions: $[Pb^{2+}]=1.0\times10^{-3}$ mol/L, $[SO_4^{2-}]=1.0\times10^{-3}$ mol/L, $[CO_2]=3.0\times10^{-3}$ mol/L, $[Cl^-]=5.0\times10^{-3}$ mol/L, $[Ca^{2+}]=5.0\times10^{-3}$ mol/L, and $[H_2PO_4^-]=3.0\times10^{-3}$ mol/L

4. Future study

Although extensive work has demonstrated the feasibility of the technology for heavy metal immobilization, negative impact resulting from this technology was also observed which needs further demonstration. A study (Cao and Ma, 2004) indicated that addition of PR to the CCA soil increased As accumulation by 4.56-9.3 times for carrot and 2.45-10.1 times for lettuce due to increased soil water-soluble As via a possible replacement of arsenate by phosphate in soil. The observation means that PR even increased As bioavailability. Caution should be taken for phosphate application since it enhances As accumulation in plants. A mixture of PA and PR was effective in metal immobilization, with less soil pH reduction and less soluble P, however, use of PA alone should be limited to minimize soil pH reduction and potential eutrophication risk (Cao and Harris, 2008, Cao et al., 2008b, 2013). Caution should be also exercised when PA is amended to the soil co-contaminated with Cu and Zn since the acidic condition of SBET could increase Cu and Zn bioaccessibility though their water solubility could be reduced (Cao et al., 2009b).

Planting could affect leachability of metals in the P-amended soils, specially leaching of colloid fraction. The conventional assessment on leaching risks of heavy metals by determining dissolved metals (filtered through 0.45 μm pore size membrane) in leachates could be underestimated since colloid fraction may also contribute to the leaching. Planting enhanced the leaching of Pb and Cu in all soils, especially the colloid fraction probably due to colloid formation induced by the rhizosphere effect. In contrast, a dramatic decrease in Zn leaching occurred due to planting. It seems that planting is not beneficial for Pb and Cu immobilization. Thus, cautions should be exerted for growing plants in the P-amended contaminated soil, especially in the soil contaminated with multiple Pb, Cu, and Zn metals (Fang et al., 2012).

The amendments used for the heavy metal immobilization has am impact on the soil pH.

For example, soluble phosphate compounds could reduce soil pH which may enhance mobility of the untargeted metals, while carbonate or silicate minerals could increase soil pH which may affect soil property and availability of nutrients.

Acknowledgements

The author is grateful for the financial support from Ministry of Environmental Protection of China (No. 201537023), Natural Science Foundation of China (No. 21537002, 20877056, 21077072, 21377081, 21428702), and Key Laboratory of Bio-organic fertilizer Creation of Ministry of Agriculture (No. BOFC2015KA04).

References

Basta N T, Gradwohl R, Snethen K L, et al. 2001. Chemical immobilization of lead, zinc, and cadmium in smelter-contaminated soils using biosolids and rock phosphate. Journal of Environmental Quality, 30 (4): 1222-1230.

Cao X D, Harris W. 2008. Magnesium and carbonate interactive effects on calcium phosphate precipitation. Environmental Science & Technology 42: 436-442.

Cao X D, Dermatas D, Xu X, et al. 2008b. Immobilization of lead in shooting range soils by means of cement, quicklime, and phosphate amendments. Environmental Science and Pollution Research, 15 (2): 120-127.

Cao X D, Liang Y, Zhao L, et al. 2013. Mobility of Pb, Cu, and Zn in the phosphorus-amended contaminated soils under simulated landfill and rainfall conditions. Environmental Science and Pollution Research, 20 (9): 5913-5921.

Cao X D, Ma L Q, Chen M, et al. 2002. Impacts of phosphate amendments on lead biogeochemistry at a contaminated site. Environmental Science & Technology, 36 (24): 5296-5304.

Cao X D, Ma L Q, Chen M, et al. 2003a. Phosphate-induced metal immobilization in a contaminated site. Environmental Pollution, 122 (1): 19-28.

Cao X D, Ma L Q, Rhue D R, et al. 2004. Mechanisms of lead, copper, and zinc retention by phosphate rock. Environmental Pollution, 131 (3): 435-444.

Cao X D, Ma L Q, Shiralipour A. 2003b. Effects of compost and phosphate amendments on arsenic mobility in soils and arsenic uptake by the hyperaccumulator, *Pteris vittata* L. Environmental Pollution, 126 (2): 157-167.

Cao X D, Ma L Q, Singh S P, et al. 2008a. Phosphate-induced lead immobilization from different lead minerals in soils under varying pH conditions. Environmental Pollution, 152 (1): 184-192.

Cao X D, Ma L Q. 2004. Effects of compost and phosphate on plant arsenic accumulation from soils near pressure-treated wood. Environmental Pollution, 132 (3): 435-442.

Cao X D, Ma L, Gao B, et al. 2009a. Dairy-manure derived biochar effectively sorbs lead and atrazine. Environmental Science & Technology, 43 (9): 3285-3291.

Cao X D, Ma L, Liang Y, et al. 2011b. Simultaneous immobilization of lead and atrazine in contaminated soils using dairy-manure biochar. Environmental Science & Technology, 45 (11): 4884-4889.

Cao X D, Wahbi A, Ma L, et al. 2009b. Immobilization of Zn, Cu, and Pb in contaminated soils using phosphate rock and phosphoric acid. Journal of Hazardous Materials, 164 (2): 555-564.

Cao X D, Wei X X, Dai G L. 2011a. Combined pollution of multiple heavy metals and their chemical immobilization in contaminated soils: A review. Chinese Journal of Environmental Engineering, 5 (7): 1441-1453.

Chaney R L, Ryan J A, Umweltschutz F. 1994. Risk based standards for arsenic, lead and cadmium in urban soils. Frankfurt, Germany: Dechema.

Chen M, Ma L Q, Singh S P, et al. 2003. Field demonstration of in situ immobilization of soil Pb using P amendments. Advances in

Environmental Research, 8 (1): 93-102.

Dudka S, Adriano D C. 1997. Environmental impacts of metal ore mining and processing: A review. Journal of environmental quality, 26 (3): 590-602.

Fang Y, Cao X D, Zhao L. 2012. Effects of phosphorus amendments and plant growth on the mobility of Pb, Cu, and Zn in a multi-metal-contaminated soil. Environmental Science and Pollution Research, 19 (5): 1659-1667.

Hettiarachchi G M, Pierzynski G M, Ransom M D. 2001. In situ stabilization of soil lead using phosphorus. Journal of Environmental Quality, 30 (4): 1214-1221.

Liang Y, Cao X D, Zhao L, et al. 2014. Biochar-and phosphate-induced immobilization of heavy metals in contaminated soil and water: Implication on simultaneous remediation of contaminated soil and groundwater. Environmental Science and Pollution Research, 21 (6): 4665-4674.

Liang Y, Wang X C, Cao X D. 2012 Immobilization techniques for remediation of heavy metal contaminated soils using phosphate, carbonate, and silicate materials: A review. Environmental Chemistry (in Chinese), 31: 1-11

Madrid F, Romero A S, Madrid L, et al. 2006. Reduction of availability of trace metals in urban soils using inorganic amendments. Environmental Geochemistry and Health, 28 (4): 365-373.

Melamed R, Cao X D, Chen M, et al. 2003. Field assessment of lead immobilization in a contaminated soil after phosphate application. Science of the Total Environment, 305 (1): 117-127.

US Environmental Pollution Agency (USEPA). 1991. Engineering Bulletin: Selection of Control Technologies for Remediation of Lead Battery Recycling Sites, U. S. Environmental Protection Agency Office of Research and Development, Cincinnati, OHIO.

US Geological Survey (USGS). 2002. Lead in December 2001 Mineral Industry Surveys. U. S. Geological Survey Minerals Information Publications Service 984 National Center, Reston, VA.

Xu X Y, Cao X D, Zhao L. 2013. Comparison of rice husk-and dairy manure-derived biochars for simultaneously removing heavy metals from aqueous solutions: Role of mineral components in biochars. Chemosphere, 92 (8): 955-961.

Xu X Y, Schierz A, Xu N, et al. 2016. Comparison of the characteristics and mechanisms of Hg (II) sorption by biochars and activated carbon. Journal of Colloid and Interface Science, 463: 55-60.

Yoon J K, Cao X D, Ma L Q. 2007. Application methods affect phosphorus-induced lead immobilization from a contaminated soil. Journal of Environmental Quality, 36 (2): 373-378.

Remediation of Heavy-Metal-Contaminated Soils by Phosphate Fertilizers

Minggang XU, Shiwei ZHOU, Shibao CHEN

National Engineering Laboratory for Improving Quality of Arable Land, Institute of Agricultural Resources and Regional Planning, Chinese Academy of Agricultural Sciences

1. Introduction

During the last three decades, heavy metal pollution in China's agricultural soils has become a serious situation due to highly intensive cultivation and rapid industrialization and urbanization. As a result, about 10 million ha of cultivated land (8.2% of the country's arable land) has been polluted (China Geological Survey, 2015), bringing serious threats to the ecosystem and public health. Recently, frequent occurrence of Cd toxicity in rice caused widespread public concern and anxiety, highlighting China's dependence on food derived from contaminated soils (Bi et al., 2013; Liu et al., 2013) .

Disposal of metal-contaminated soils is very time-consuming and costly. Since Gauglitz et al. (1992) showed that hydroxyapatite (HA) has the capacity to immobilize heavy metals, the use of phosphate materials as amendments has been a promising way to remediate heavy-metal-contaminated soils (Miretzky and Fernandez-Cirelli, 2008; Selim, 2015) . Because high-quality agricultural soil resources are insufficient to supply food for a very large population, many crops are still planted in contaminated soils in China, giving rise to two issues that need to be considered: ①the ability of the soils to retain lower metal solubility and mobility for reducing the risk of heavy metal transport to water and plant, and ②the ability to provide adequate nutrients for plant growth and grain production. Phosphate materials that can perform as both fertilizers and amendments address each of these issues.

Herein, we presented a detailed summary of more than ten years of research on remediation of heavy-metal-soils by phosphate fertilizers, with special focus on the chemical speciation and the ionic interaction between phosphate and heavy metal.

2. Application of phosphate materials to contaminated soils promotes crop biomass and reduces toxic metal uptake by plant

Shoot and root biomass of crops could be reduced greatly by heavy metals in contaminated soils, and there were significantly negative relationship between crop growth and soil pollution degree. However, the addition of phosphate materials such as HA, tri-superphosphate (TSP), di-ammonium phosphate (DAP) and phosphate rock (PR) could offset the inhibition induced by heavy metals in red soil contaminated by Cd, lead (Pb), copper (Cu) and zinc (Zn), where the growth of rape (*Brassica campestris*, L.) was the same or even better than that on non-contaminated soil; and HA exhibited the largest efficiency in promoting the biomass (Chen et al., 2007) . Generally, shoot growth is more sensitive than root growth to phosphate addition, and could be promoted markedly. Similar results were obtained in mustard (*Brassica oleracea*, L.) grown on Pb-contaminated yellow earth, where the shoot and root biomass increased significantly under application of phosphate materials, with the largest effect of HA (Chen and Zhu, 2004) .

Fig. 1 also showed that after application of phosphate fertilizers to Cd-contaminated soils rape shoot biomass increased remarkably, comparable to and even better than that of controls (non-contaminated soils); whereas there was no significant difference with and without application of nitrogen and potassium fertilizers, suggesting that not nitrogen and potassium but phosphorus played the important role in promoting crop growth in contaminated soils. Furthermore, it could be observed from Fig. 1 that the effectiveness to promote rape growth on acidic and alkaline contaminated soils followed the same order of $Ca(H_2PO_4)_2 > KH_2PO_4 > NH_4H_2PO_4$, with a more obvious effect in alkaline soil, so that rape biomass in heavily contaminated cinnamon soil (1.0 mg Cd/kg) increased by 991% under addition of $Ca(H_2PO_4)_2$.

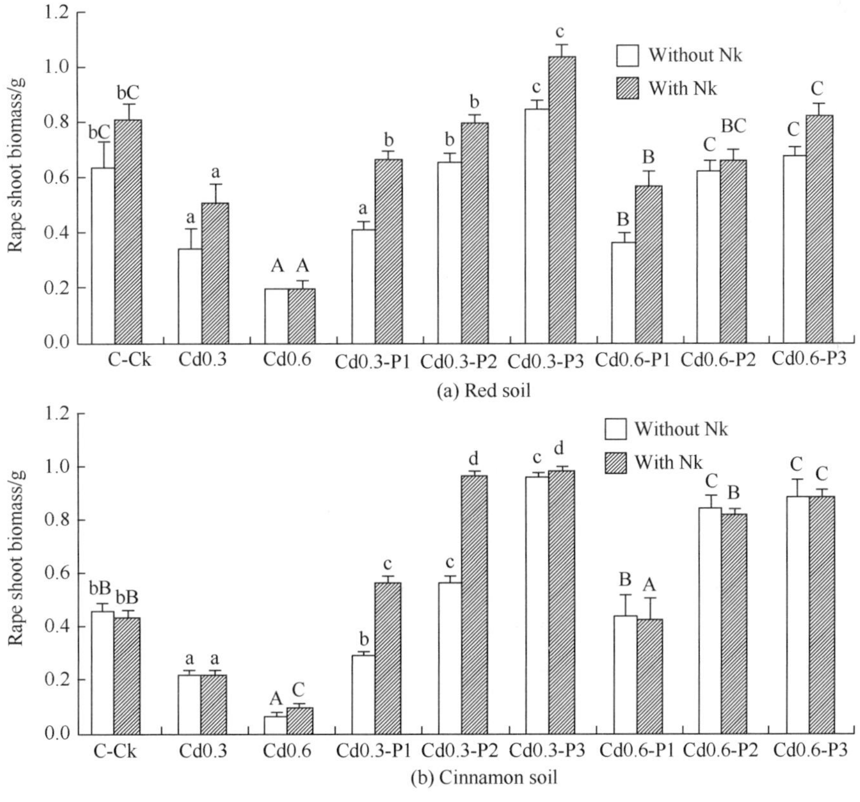

Fig. 1 Rape shoot biomass in Cd-contaminated soils under application of different phosphate fertilizers (P1: $NH_4H_2PO_4$, P2: KH_2PO_4, P3: $Ca(H_2PO_4)_2$, and molar ratio of N: P_2O_5: K_2O=0.2: 0.92: 0.12 for treatments with NK) (Chen, 2009)

Application of phosphate fertilizers not only promoted crop growth, as mentioned above, but also decreased toxic metal content in plant. Fig. 2 showed that under application of phosphates to heavy-metal-contaminated red soil, the concentration of Pb in rape shoot decreased by 32.6%-57.9%, whereas Cd decreased by only 9.8%-39.1%. The effectiveness of phosphate reducing Pb uptake by rape is the largest, with the decrease order of Pb>Zn, Cu>Cd; HA had the largest effect in reducing metal uptake, following the order: HA > PR > TSP > DAP (Fig. 2), which is consistent with the trend observed in crop growth in metal-contaminated soils with phosphate amendments (Chen and Zhu, 2004; Chen et al., 2007).

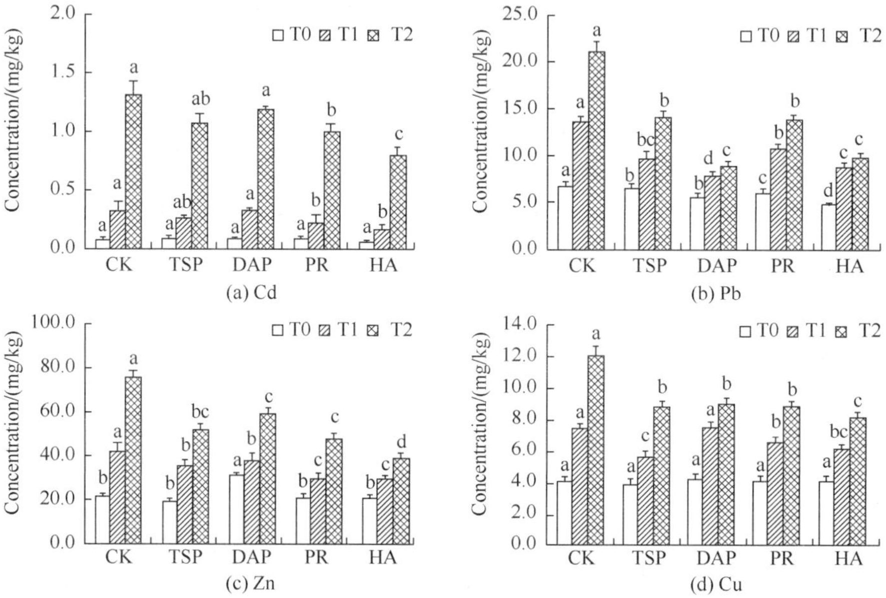

Fig. 2 Rape shoot metal concentration in contaminated red soil under application of different phosphate materials (T0: non-contaminated soil; T1: slightly contaminated soil with the content of Cd, Pb, Zn and Cu of 0.6, 100, 66 and 30 mg/kg, respectively; T2: heavily contaminated soil with the content of Cd, Pb, Zn and Cu of 1.5, 300, 200 and 100 mg/kg, respectively) (Chen et al., 2007)

However, application of different phosphate fertilizers to different soils with a different pollution degree would lead to different efficiency in reducing toxic metal uptake by crops, due to their different solubility, complexing capacity towards heavy metals, interaction between P and metal, and soil environment. Chen (2009) observed that after application of phosphate fertilizers ($Ca(H_2PO_4)_2$, KH_2PO_4 and $NH_4H_2PO_4$) to heavily contaminated red soil (0.6 mg Cd/kg), Cd uptake by rape significantly decreased (by 23%-58%), whereas it significantly increased (by 22%-71%) in heavily contaminated cinnamon soil. But, the trend was just the opposite of that found in the two soils with a slight pollution degree. Moreover, the effect of $Ca(H_2PO_4)_2$, whether increasing or decreasing Cd uptake by rape, was always the largest among three phosphate fertilizers applied to acidic and alkaline soils (Chen, 2009) .

In a word, plant growth could be promoted greatly and metal uptake was reduced markedly by phosphate, so in situ chemical immobilization of heavy metals especially Pb in soils using phosphate materials as a widely accepted technology has been successfully applied to many contaminated soils (Miretzky and Fernandez-Cirelli, 2008; Selim, 2015) . However, the complex changes occurred for Cu, Zn and Cd, causing increase or decrease in toxic metal content in plant grown on phosphate-amended contaminated soils, which requires us to focus on soil type and soil pollution degree, soil heavy metal speciation and crop species, phosphate sources and amount, soil nutrients, etc., during remediation of heavy metals in soils using phosphate fertilizers. Our study showed that for acidic contaminated soils the most effective phosphate amendment was HA or $Ca(H_2PO_4)_2$, which was consistent with the results of some researchers (Zhu et al., 2004; Mignardi et al., 2013) . But many other studies similar in design showed that either PR or DAP or TSP was the optimized amendment (Basta and McGowen, 2004; Abou Zied, 2007; Thawornchaisit and Polprasert, 2009; Fang et al., 2012) , indicating a complex interaction between P and heavy metals depending on the properties of phosphate materials, heavy metals and soils.

It is well known that soluble phosphate materials could provide an abundance of solution phosphorus and increase the efficiency of metal phosphate precipitate/mineral formation, but their environmental risk such as eutrophication associated with P leaching also increased. In contrast, application of insoluble phosphate materials such as HA and PR to metal-contaminated soils might reduce this risk. Additionally, HA and PR could also act as liming materials due to containing some $CaCO_3$ and thus neutralize soil acidity in dissolution process. Obviously, the immobilization efficiency of insoluble phosphates was relatively low because of their dissolution limit. Therefore, in order to minimize environmental risk and maximize immobilization effectiveness, a more realistic approach may be to combine soluble phosphates with insoluble phosphates for field application to heavy-metal-contaminated soils so that soluble phosphates could quickly reduce the concentration of heavy metals to below an acceptable level, while insoluble phosphates would supply constant P sources in soils to react with heavy metals for long-term immobilization, as suggested by Cao et al. (2003) . Besides the sources of phosphate, we must also pay attention to the application rate of phosphate. Based on the stoichiometric P to Pb ratio of pyromorphites, a P/M_{total} molar ratio of 3/5, where M_{total} is the sum of the molar content of heavy metals in soils, has been suggested by many researchers. However, a higher P/M_{total} molar ratio was often used to ensure sufficient solution P and thus high immobilization efficiency of heavy metals in soils, such as 2.0, 3.0, 4.0, 5.0 and even more than 10. To sum up, the optimum application rate for the immobilization of heavy metals in soils has not been specifically studied regardless of phosphate sources.

3. Immobilization mechanisms of heavy metals in phosphate-amended soils

As mentioned above, numerous phosphate fertilizers such as HA, PR, $Ca(H_2PO_4)_2$ and

DAP have been used to immobilize heavy metals in soils, where there was different efficiency due to different solubility and complexing capacity towards different heavy metals, resulting in different interaction mechanism between P and metal. Here, the following three main processes were involved in the interaction: ① phosphate-induced adsorption of metal ions, ② the formation of metal phosphate precipitates or minerals, and ③ surface adsorption or complexation of metals (Zhou and Xu, 2007; Selim, 2015) .

3.1 Phosphate-induced adsorption of metal ions

Specific adsorption of phosphate on hydrous oxides of iron, aluminum, and manganese by ligand exchange with surface hydroxyl could result in the increase in negative charge and/or pH, and thus inducing the increase in metal retention by soils. Some studies have even reported the formation of stable mineral-phosphate-metal ternary surface complexes (Pérez-Novo et al., 2011; Elzinga and Kretzschmar, 2013; Tiberg et al., 2013) .

After insoluble phosphate materials were added, soil pH usually increased to a certain degree, where it seems that the increase in alkaline soil was more obvious than that in acidic soil. For example, after application of low amount of HA (2.5 g P/kg) to Pb-contaminated yellow earth (pH 7.13), soil pH increased by 0.2 units ($p<0.05$) ; after application of high amount of HA (5 g P/kg), the value of soil pH increased by 0.4 units. However, after application of HA and PR to contaminated red soil (pH 6.4), soil pH increased slightly ($p>0.05$) (Chen and Zhu, 2004; Chen et al., 2007) . On the contrary, after soluble phosphate materials were added to strongly acidic contaminated soil, the pH was increased markedly, especially under combined application of nitrogen with potassium, e.g., almost 1 unit of pH increase was observed in those heavily Cd contaminated soil treatments with NK (Fig. 3) . Additionally, it was also found from Fig. 3 that the effect order of phosphate on soil pH was $Ca(H_2PO_4)_2>KH_2PO_4>NH_4H_2PO_4$, but there was no significant difference among them. Based on above view, contaminated soil pH was promoted greatly because of the specific adsorption of P added, which would directly reduce metal solubility and increase its sorption by soils.

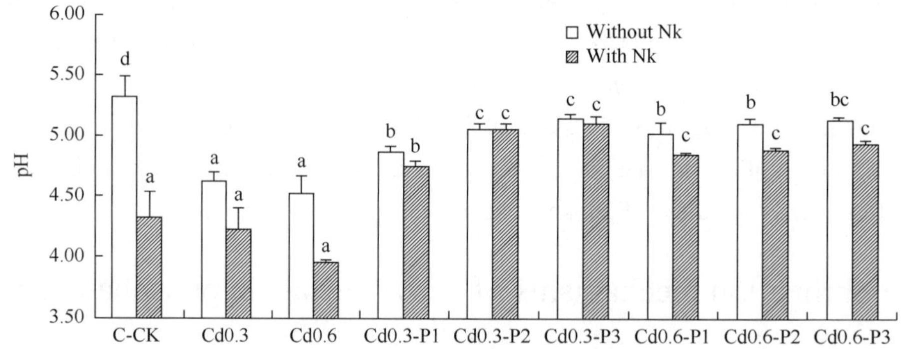

Fig. 3 Cd-contaminated red soil pH under application of different phosphate fertilizers (P1: $NH_4H_2PO_4$, P2: KH_2PO_4, P3: $Ca(H_2PO_4)_2$, and molar ratio of N: P_2O_5: K_2O=0.2: 0.92: 0.12 for treatments with NK) (Chen, 2009)

In fact metal adsorption on soils under the presence of phosphate materials was very complex, depending on not only phosphate ion, but on its accompanying cations. Fig. 4 exhibited Cd adsorption isotherms on red soil under presence of different type and amount of phosphate fertilizers. Cd adsorption on red soil received phosphate treatment was lower than those that received no phosphate treatment, and the difference was more obvious with increasing P application rate, in particular under presence of $Ca(H_2PO_4)_2$, with the order of $Ca(H_2PO_4)_2 > KH_2PO_4 > NH_4H_2PO_4$. As compared to no phosphate application, Cd maximum adsorption with maximum addition concentration of 20 mg Cd/L on soil received $Ca(H_2PO_4)_2$ at rate of 200, 400, 800 and 1600 mg P/kg decreased by 2.5%, 4.7%, 7.6% and 7.9%, respectively. For Cd adsorption on cinnamon soil, the effect trend of phosphate was same as that on red soil (Chen, 2009) . This means that accompanying cations of phosphate fertilizers apparently compete with Cd for adsorption sites and thus inhibit Cd adsorption on soils. The inhibition of bivalent Ca^{2+} showed significant difference when compared with univalent K^+ or NH_4^+.

Cd desorption on red soil under presence of phosphate fertilizers was also investigated and showed that the desorption amount decreased significantly with increasing P application rate, with the same effect order of $Ca(H_2PO_4)_2 > KH_2PO_4 > NH_4H_2PO_4$ (Chen, 2009) . For example, after application of 1.5 g P/kg of $Ca(H_2PO_4)_2$, Cd desorption amount decreased from 42 mg/kg to 32.9 mg/kg (by 9.1 mg/kg), compared to no P addition. It could be deduced from these results that phosphate materials may have two side effects: ①They may strengthen the adsorption of metal ions by soils due to ligand exchange adsorption of phosphate anions, which would reduce the metal desorption rate. ②They may also weaken the adsorption of metal ions by soils, in view of the strong competition of accompanying cations in phosphate fertilizers, in particular calcium phosphate fertilizers. Therefore, we must focus on not only the property and amount of phosphate anions but also the type and concentration of accompanying cations during researching on phosphate-induced adsorption mechanism of heavy metal ions on soils.

3.2 The formation of metal phosphate precipitates or minerals

Phosphate addition does not generally change the content but does affect the existing species of heavy metals in soils, accordingly impacting their solubility/mobility and availability. Application of phosphate amendments caused heavy metals in soils to shift from forms with high availability to the most strongly bound fractions such as residual forms (precipitates and/or minerals of metal phosphate), which is the main mechanism responsible for the immobilization of heavy metals in contaminated soils. Zhang (2007) found that after application of $Ca(H_2PO_4)_2$ to heavily contaminated red soil and paddy soil (200 mg Cu/kg and 400 mg Zn/kg), available Cu and Zn were decreased significantly by 8.3%-11.6% and 3%-4.6%, respectively; and the effect of $Ca(H_2PO_4)_2$ was higher in acidic red soil.

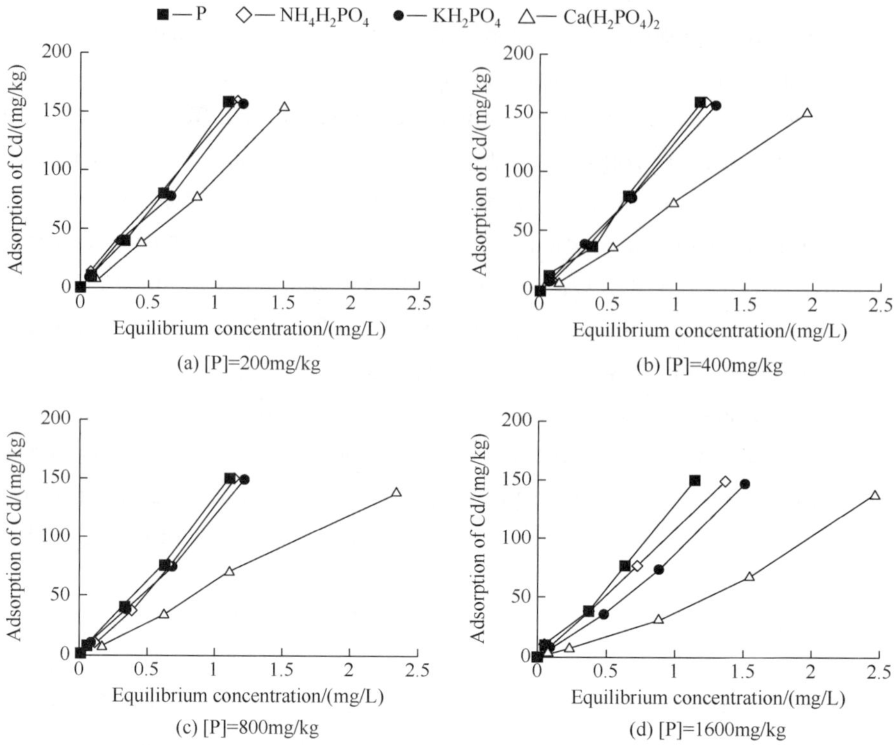

Fig. 4 Cd adsorption isotherms on red soil under presence of phosphate fertilizers (Chen, 2009)

Furthermore, using modified Tessier's method, we investigated the speciation changes of heavy metals in red soil under absence and presence of phosphate materials, as shown in Fig. 5. It could be observed that all P treatments substantially modified the partitioning of heavy metals from the potentially available phase to less available phase, where addition of phosphate significantly decreased metals in the first four fractions (exchangeable, EX; organic bound, OC; carbonate bound, CB; amorphous Fe and Al oxides-bound, OX) and significantly increased the residual fraction (RES), in particular for Pb. RES%-Pb increased by 35.6%-51.9%, whereas RES%-Zn and RES%-Cd increased by 8.4%-37.6% and 22.4%-42.5%, respectively. P amendment was therefore less effective for immobilization of Zn and Cd than for Pb. This was because Pb-phosphate precipitates (pyromorphites) were easier to form in view of their extremely low solubility ($\log K_{sp}$=−71.63-83.7, compared to −35.3 for $Zn_3(PO_4)_2$ and −38.1 for $Cd_3(PO_4)_2$) (Selim, 2015).

Scanning electron microscope (SEM) analysis on the rape root surface and contaminated red soil received different phosphate application evidenced that pyromorphites could form in both soil and root surface membrane (Fig. 6 and Fig. 7). The element dispersive spectrometer (EDS) analysis also showed that Pb particles contained a significant amount of PbO, P_2O_5, and CaO, indicating the occurrence of pyromorphites (Chen et al., 2007). However, Cd and Zn were less associated with P compared with Pb in heavy-metal-contaminated red soil under application of phosphate materials.

Numerous studies have proved that the RES-Pb increased markedly after application of different phosphates to Pb-contaminated soils, resulting from the formation of pyromorphites. And among pyromorphites (fluoro-, chloro-, bromo- and hydroxyl-pyromorphite) chloropyromorphite was expected to be the dominant form due to its less solubility and the ubiquity of chloride in nature (Ryan et al., 2001) . According to solubility constants of metal-phosphate precipitates, the metal affinity for phosphate ions usually followed the sequence: Pb$\gg$Zn$>$Cu$>$Cd, so other metal-phosphate mineral formation except pyromorphites may not take place in a wide range of soils, as confirmed by some studies (Cao et al., 2003, 2004; Raicevic et al., 2005; Chen et al., 2007) . This suggested that although they are efficient in immobilizing heavy metals such as Pb, Zn, Cu and Cd, phosphate fertilizers should be preferentially used as amendments in the remediation of Pb-contaminated soils, because of more efficient stabilization of Pb by the in situ formation of insoluble pyromorphite-like minerals.

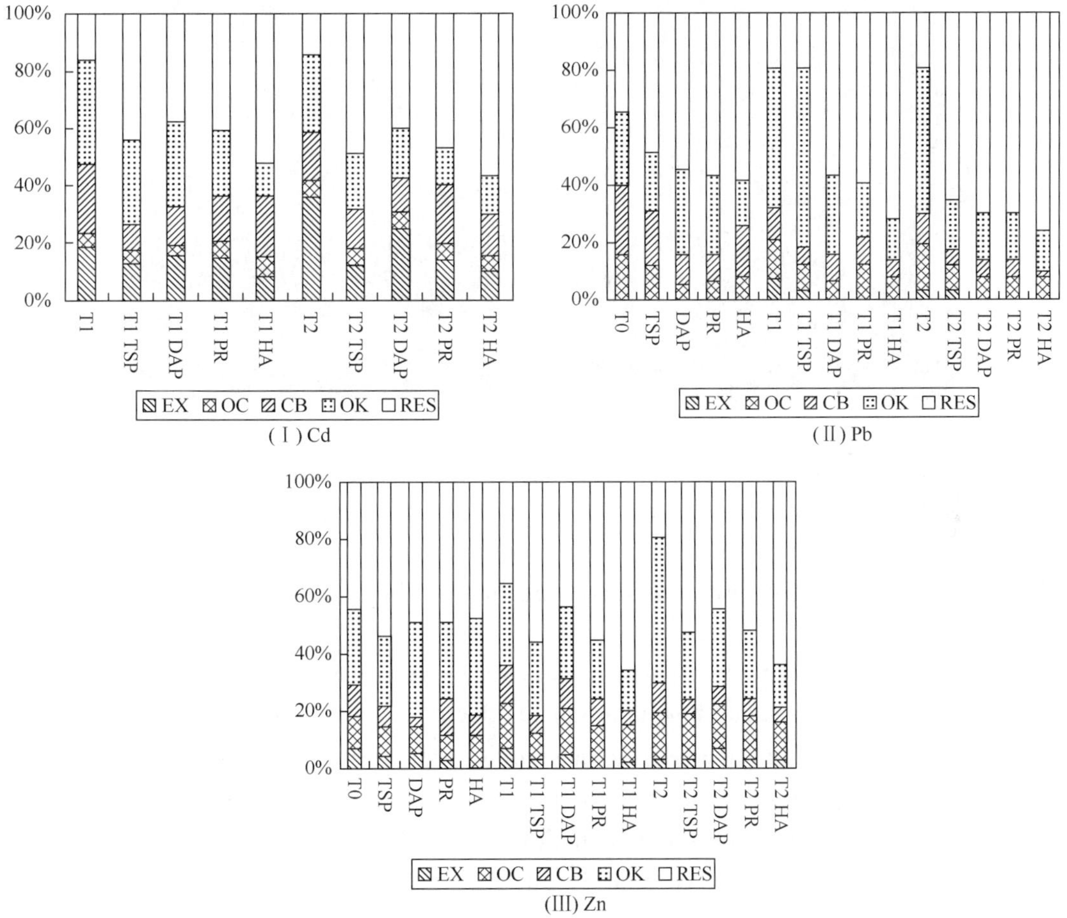

Fig. 5 Relative percentages of heavy metals (Ⅰ Cd, Ⅱ Pb and Ⅲ Zn) in each fraction of red soil from control (no phosphate addition) and P-treatment plots (the concentration of Cd in T0 was below the limit, so the data was omitted) (T0: no pollution; T1: Cd, Pb and Zn content is 0.6, 100 and 66 mg/kg, respectively; T2: Cd, Pb and Zn content is 1.5, 300 and 200 mg/kg, respectively) (Chen et al., 2007)

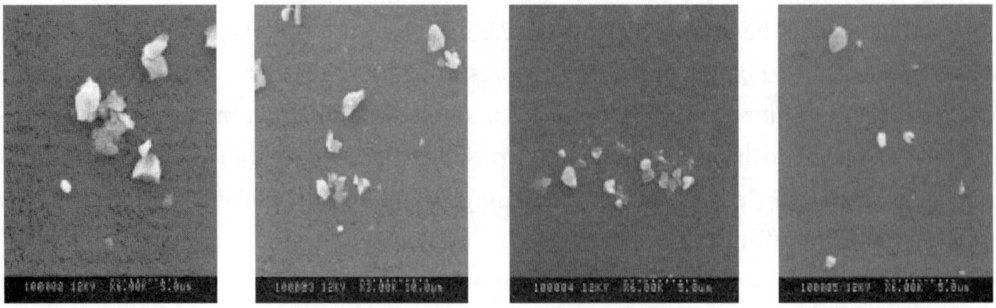

(a) T2-HA　(b) T2-PR　(c) T2-DAP　(d) T2-TSP

Fig. 6　Scanning electron microscopy maps of pyromorphite crystal in different phosphate treated red soils (a: T2-HA; b: T2-PR; c: T2-DAP, d: T2-TSP) (T2: Cd, Pb and Zn content in red soil is 1.5, 300 and 200 mg/kg, respectively) (Chen et al., 2007)

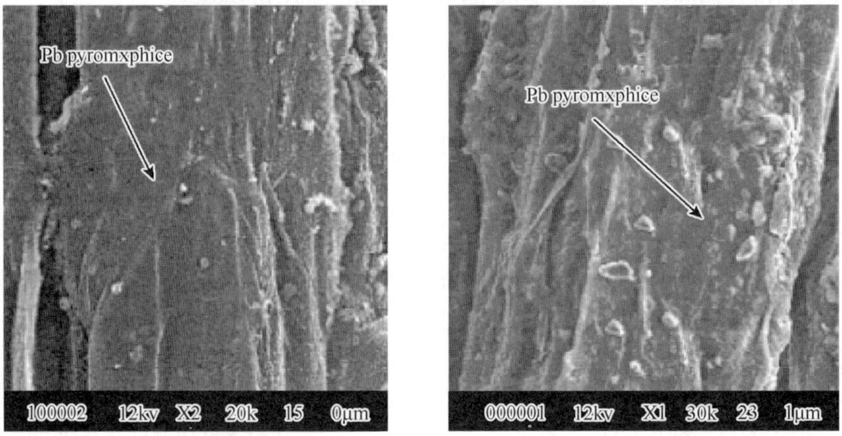

(a) T2-HA　(b) T2-PR

Fig. 7　Scanning electron microscopy maps of rape root grown in contaminated red soil treated with T2-HA and T2-PR (T2: Cd, Pb and Zn content is 1.5, 300 and 200 mg/kg, respectively) (Chen et al., 2007)

3.3　Surface adsorption or complexation of metals

The surface of insoluble phosphate materials such as HA and PR can directly adsorb or complex with heavy metals, and thus immobilize effectively these contaminants. But, there are different contributions from the surface adsorption/complexation for different metals. For example, Cao et al. (2004) showed that for the immobilization of heavy metals by PR, the contribution from the surface adsorption/complexation of Pb (21.7%) was much lower than that of Cu (74.5%) and Zn (95.7%) . Xu et al. (1994) also revealed that surface complexation and co-precipitation were the most important mechanisms for the immobilization of Zn and Cd by HA, with more significance for Cd co-precipitation.

Compared with adsorption/complexation of heavy metals on HA or PR surface, the co-precipitation of heavy metal with dissolved phosphate ions played an more important role, resulting in the substitution of calcium by other cations (M) such as Cd and Zn in HA crystal structure ($M_xCa_{10-x}(PO_4)_6(OH)_2$) (Xu et al., 1994) . Cations with larger ionic radii than Ca^{2+} (r=0.099 nm) is easier to be incorporated in the apatite structure, thus Zn^{2+} (r=0.069 nm) may become much more difficult to co-precipitate with Ca^{2+} into apatite structure than Cd^{2+} (r=0.097 nm) ; and Pb^{2+} (r=0.119 nm) co-precipitation should be the most obvious. Mavropoulos et al. (2002, 2004) demonstrated that Pb was firstly incorporated into HA structure, with the formation of Pb-Ca HA ($Pb_xCa_{10-x}(PO_4)_6(OH)_2$), but Pb-Ca HA was evolved continuously where its Ca/Pb molar ratio decreased continuously until it reached a more stable structure of a pure hydroxyl-pyromorphite ($Pb_{10}(PO_4)_6(OH)_2$) .

The immobilization mechanisms of heavy metals in phosphate-amended soils are very complex and unclear, because of complex interactions among phosphates, heavy metals and soils, especially the direct reaction between phosphate anions and heavy metals, and the competition between accompanying cations and heavy metals. In fact, long-term application of phosphate fertilizers, soil properties such as pH, organic matter, and cation exchange capacity were changed deeply, which would also influence and control the bioavailability of heavy metals in soils, besides those interactions mentioned above. Maybe the important role of soil pH should be paid special attention to elucidate the immobilization mechanisms of heavy metals by phosphates, for it affected not only the solubility and reaction process of heavy metals in soils, but also the solubility and function of P-amendments. Zhang and Ryan (1998) found that the formation of pyromorphites depended on soil solution pH, where the precipitation (chloro-pyromorphite) dominated at pH 2.0-5.0, whereas surface adsorption of Pb on HA became dominant at pH 6.0-7.0.

4. Long-term environmental risks for the immobilization of heavy metals by phosphate fertilizers

As mentioned above, to ensure sufficient solution P and high immobilization efficiency, higher rate of phosphate fertilizers was usually applied to heavy-metal-contaminated soils. Then some environmental and ecological risks might occur, including second-pollution of heavy metals from P-amendments, and eutrophication of water body deriving from P leaching.

4.1 Second-pollution of heavy metals from P-amendments

Although it generally contains a concentration of most heavy metals such as Cd (Table 1), under normal cropping practice, the impact of phosphate fertilizer application on trace metal accumulation in receiving soils is not as serious as it might appear, in particular in China in view of its very low Cd content in either phosphate rocks or phosphate fertilizers. Jiao et al. (2012) evaluated that the effect of long-term application of phosphate fertilizers on the accumulation of heavy metals in soils has been limited and localized. Based on this estimate of

Lu et al. (1992), every year about 8 million tons of P_2O_5 need be applied to China's croplands, resulting in 37.3 tons of Cd was input into soils, which is equivalent to 0.497-0.746 g Cd/ha. On the other hand, the maximum environmental capacity for Cd in China's croplands ranged from 0.73 to 6.36 kg/ha (more than 1000 times higher) (Lu et al., 1992) . This means that long-term application of domestic phosphate fertilizers will not cause environmental pollution derived from its heavy metal input. In Red Soil Experimental Station of Qiyang, Hunan province, annual application of 120 kg P_2O_5/ha (superphosphate with 0.06 mg Cd/kg) (Liu et al., 2009) would retain 7.2 mg/ha of Cd input, then soil Cd does not exceed the critical value (0.6 mg/kg, viz.1.35 kg/ha) until more than 180 thousand years, ignoring its uptake by crops and leaching. After more than 25 years of continued application of phosphate fertilizers, the contents of heavy metals in China's agricultural soils in 42 long-term experimental sites are all similar to the control (without fertilization), greatly lower than soil environmental quality criteria (unpublished data) .

Table 1 Cadmium contents in phosphate rocks and fertilizers in China and some other countries (Lu et al., 1992)

Country	Phosphate deposit	Range	Average	Country	Content in P-fertilizers / (mg/kg)
		/ (mg/kg)			
China	All	0.1-571	15.3	China	0.1-2.93
	Most*	0.1-4.4	0.98	USA	7.4-15.6
Russia	Kola	-	0.3	Canada	2.1-9.3
USA	Florida	3-12	7	Australia	18-91
	N. C.	-	36	Sweden	2-3
Togo	Togo	38-60	53		
Morocco	Khouribga	1-17	12		
	Joussoufia	-	4		
Algeria	Algier	-	23		
Tunisia	Gafsa	55-57	56		

*Deducting a small portion of phosphate rocks from Guangxi province.

These results all showed that soil Cd pollution did not appear under long-term application of phosphate fertilizers for normal crop growth and grain production. But, some imported phosphate fertilizers (e.g. those from Australia and USA) containing high Cd concentrations should be used carefully. In addition, the use of calcium-magnesium phosphate (CMP), in particular in acidic soils, was encouraged, because it usually contained lower Cd and higher pH than superphosphate (Lu et al., 1992) .

However, regarding the immobilization of heavy metals in soils by phosphate addition, we must focus on the potential pollution from high levels of addition of phosphate materials.

According to the common application rate (P/M_{total} molar ratio=3/5), we calculate the application rates of phosphates (average 14.7% P_2O_5 or 6.42% P) to be 39, 184 kg/ha in a contaminated paddy soil in Zhejiang, China (Pb: 388 mg/kg, Zn: 2286 mg/kg, Cd: 1.6 mg/kg, Cu: 1481 mg/kg) (Zhang et al., 2005) ; and thus Cd accumulation rate in the paddy soil was evaluated to be 23.9 g/ha, based on its average content in phosphate fertilizers (0.61 mg/kg) (Lu et al., 1992), which was 50 times as high as that accumulated in non-polluted soils. For heavier smelter-contaminated soil (Cd: 2810 mg/kg; Cu: 455 mg/kg; Pb: 2770 mg/kg; Zn: 45400 mg/kg) (Wei et al., 2015), the application rate up to 481.8 t/ha would be needed, and 293.9 g/ha of Cd would be input, which is almost equivalent to one-third of the maximum environmental capacity of China's croplands (Lu et al., 1992) . Obviously, the accumulation of contaminants from phosphate fertilizers would be serious when they were used to immobilize heavy metals in mining/smelter soils, where the risk of second-pollution of heavy metals would increase significantly.

4.2 Eutrophication from long-term application of phosphate fertilizers

Eutrophication is a wide spread and growing problem with sharply negative effects on water quality. The accelerated eutrophication of most freshwaters is limited by P inputs, which is mainly runoff from land use of phosphate fertilizers, although phosphorus added to soils has very low mobility because it is adsorbed strongly to soil minerals and also precipitated readily with calcium, iron, or aluminum. That is, there is strong relationship between agricultural phosphorus leaching and water eutrophication (Daniel et al., 1998) .

Addition of excessive amounts of phosphate materials to heavy-metal-contaminated soils might increase the risk of P leaching and eutrophication to water body. As shown in Table 2, it could be concluded that application of soluble phosphate materials such as phosphorous acid (PA), TSP, DAP and calcium phosphate posed a greater risk of P loss than insoluble phosphate materials such as HA and PR. Moreover, P leaching increased remarkably with the increase of amount of soluble phosphates, in particular in acidic soils (Dermatas et al., 2008) . Therefore, it should be noted especially that soluble phosphate sources caused the risk of P leaching into water body during immobilizing heavy metals in acidic soils.

Besides mentioned above, there are also some other issues, such as nutrition imbalance, need to be addressed. Long-term overuse of phosphate fertilizers may inhibit the uptake of some necessary elements like manganese (Mn), calcium (Ca) etc. by plants (Boisson et al., 1999; Theodoratos et al., 2002) . Thus, the use of phosphate amendments should be comprehensively evaluated with respect to potential side effects as well as the immobilization effectiveness of contaminants and the cost. More importantly, the type and rate of phosphate materials and also application method should be carefully studied to maximize the immobilization effectiveness and minimize the side effects.

Table 2 Assessment of phosphorus leaching in contaminated soils under application of different phosphates

Phosphate*	Added rate	P leaching	Reference
PA, HA (total 4710 mg P/kg)	0	0.04 mg/L	Zupančič et al., 2012
	n (PA): *n* (HA) =0: 1	0.26 mg/L (0.045% of added)	
	n (PA): *n* (HA) =0.33: 1	1.15 mg/L	
	n (PA): *n* (HA) =0.75: 1	1.76 mg/L	
	n (PA): *n* (HA) =1.28: 1	7.17 mg/L	
	n (PA): *n* (HA) =2: 1	13.0 mg/L	
	n (PA): *n* (HA) =3: 1	14.3 mg/L (2.88% of added)	
DAP	0	Not detected	McGowen et al., 2001
	460 mg P/kg	0.32 mg/column	
	920 mg P/kg	0.78 mg/column	
	2300 mg P/kg	2.31 mg/column	
PR	0	Not detected	Basta and McGowen, 2004
	60 g/kg	Not detected	
	180 g/kg (layered)	Not detected	
	180 g/kg (mixed)	Not detected	
Limestone	170 g/kg	Not detected	
DAP	10 g/kg (=2300 mg P/kg)	2.31 mg/column (<1% of added)	
	90 g/kg	335 mg/column (10.5% of added)	
MCP in PATF soil (pH 8.7)	25 g/kg	11 mg/L	Dermatas et al., 2008
	125 kg/kg	307 mg/L	
MCP in FDR26 soil (pH 5.2)	6 g/kg	18 mg/L	
	45 g/kg	331 mg/L	
100% PA	Molar ratio of P/Pb=4 (6935 mg P/kg)	6-20% of added	Cao et al., 2002
50% PA+50% Ca $(H_2PO_4)_2$		3-16% of added	
50% PA+50% PR		1-7% of added	
PA	0	0.29 mg/L	Yang et al., 2002
	10 g P/kg	335 mg/L	
PR	0	0.04	Wei, 2010
	18.11 g/kg	0.06	
PR+TSP	9.05 g/kg+6.34 g/kg	0.20	
TSP	12.68 g/kg	0.22	

*PA: H_3PO_4; MCP: monobasic calcium phosphate.

5. Conclusions

Phosphate materials could supply nutrients to plants as well as immobilize heavy metals, thereby reducing their bioavailability in soils, so addition of phosphate is considered to be a successful and economical remediation strategy for those slightly contaminated soils used for growing crops in China.

After application of phosphate fertilizers to both acidic and alkaline soils contaminated by heavy metals, crop growth was promoted where the shoot biomass was even increased by 991%, and toxic metal uptake by crops was inhibited where metal concentration in shoot was even decreased by 58.4%. This indicated that phosphate materials as amendments could remediate effectively heavy-metal-contaminated soils. Furthermore, we found that HA and $Ca(H_2PO_4)_2$ seem to have the best effect. However, long-term overuse of phosphate fertilizers may pose environmental risk, such as second-pollution of heavy metals; eutrophication to water body from P leaching, etc. Thus, the type and rate of phosphate source and also its application method should be studied comprehensively, and phosphate materials should be used carefully, to maximize the immobilization effectiveness and minimize the side effects. Generally, we should prefer to use P-amendments for slightly contaminated soils, in particular, the immobilization by phosphate should be avoided in heavier arsenic (As) -contaminated soils; in addition, soluble and insoluble phosphates should be combined together in actual remediation of contaminated soils.

There are different and complex interaction among phosphate fertilizer, heavy metal and soil, leading to different immobilization efficiency, in which three main mechanisms were involved, including ①phosphate-induced adsorption of metal ions, ②the formation of metal phosphate precipitates or minerals, and ③phosphate surface adsorption or complexation of metals. Generally speaking, Pb immobilization proceeded dominantly through a pyromorphite precipitation, whereas phosphate-induced adsorption and/or co-precipitation were responsible for Cd immobilization.

It is important to focus on the long-term stability of metal immobilization in phosphate-amended soils, because the addition of phosphate did not change total metal concentrations. With time, continued P consumption by plants may cause the decomposition or activation of metal-phosphate precipitates/minerals formed previously; biogeochemical changes may also affect the stability of metal immobilization. Phosphate fertilizers as chemical amendments could introduce elements capable of inducing or affecting the activities of certain lithotrophic microbes that could influence vital geochemical processes such as mineral dissolution and formation, weathering, and organic matter mineralization (Udeigwe et al., 2011), and thereby affecting the stability of immobilized heavy metals. But, very little research has been done and little is known about the long-term stability of phosphate-induced immobilization of heavy metals in soils. Other issues should be also focused on in the future, for example, the transformation and transport of heavy metals in soil-plant systems after phosphate addition in order to distinguish the interaction mechanisms between P and metal; environmental and ecological risks from long-term application of phosphate to metal-contaminated soils in order to evaluate the remediation potential of phosphate application to actual field-contaminated soils.

Acknowledgements

We would like to gratefully acknowledge the financial support from the Ministry of Science and Technology of China (National Basic Research and Development Program-2002CB410809) and Natural Science Foundation of China (41271254) .

References

Abou Zied M M A. 2007. Reducing bioavailability of some heavy metals in a contaminated soil using phosphate amendments. Egyptian Journal of Soil Science, 47: 9-22.

Basta N T, McGowen S L. 2004. Evaluation of chemical immobilization treatments for reducing heavy metal transport in a smelter-contaminated soil. Environmental Pollution, 127 (1): 73-82.

Bi X, Pan X, Zhou S. 2013. Soil security is alarming in China's main grain producing areas. Environmental Science & Technology, 47 (14): 7593-7594.

Boisson J, Ruttens A, Mench M, et al. 1999. Evaluation of hydroxyapatite as a metal immobilizing soil additive for the remediation of polluted soils. Part 1. Influence of hydroxyapatite on metal exchangeability in soil, plant growth and plant metal accumulation. Environmental Pollution, 104 (2): 225-233.

Cao X, Ma L Q, Chen M, et al. 2002. Impacts of phosphate amendments on lead biogeochemistry at a contaminated site. Environmental Science & Technology, 36 (24): 5296-5304.

Cao R X, Ma L Q, Chen M, et al. 2003. Phosphate-induced metal immobilization in a contaminated site. Environmental Pollution, 122 (1): 19-28.

Cao X, Ma L Q, Rhue D R, et al. 2004. Mechanisms of lead, copper, and zinc retention by phosphate rock. Environmental Pollution, 131 (3): 435-444.

Chen M. 2009. Different Effects and Mechanisms of Phosphate on the Adsorption and Desorption of Cadmium in Red Soil and Cinnamon Soil. Master's thesis, Agricultural University of Hebei, Baoding.

Chen S, Xu M, Ma Y, et al. 2007. Evaluation of different phosphate amendments on availability of metals in contaminated soil. Ecotoxicology and Environmental Safety, 67 (2): 278-285.

Chen S, Zhu Y. 2004. Effects of different phosphorus-compounds on Pb uptake by *Brassica Oleracea*. Acta Scientiae Circumstantiae, 24 (4): 707-712.

China Geological Survey. 2015. Geochemical Survey Report on the Cultivated Lands in China. http: //www.cgs.gov.cn/xwtzgg/jrgengxin/123194.htm

Daniel T C, Sharpley A N, Lemunyon J L. 1998. Agricultural phosphorus and eutrophication: A symposium overview. Journal of Environmental Quality, 27 (2): 251-257.

Dermatas D, Chrysochoou M, Grubb D G, et al. 2008. Phosphate treatment of firing range soils: Lead fixation or phosphorus release? Journal of Environmental Quality, 37 (1): 47-56.

Elzinga E J, Kretzschmar R. 2013. In situ ATR-FTIR spectroscopic analysis of the co-adsorption of orthophosphate and Cd (II) onto hematite. Geochimica et Cosmochimica Acta, 117: 53-64.

Fang Y, Cao X, Zhao L. 2012. Effects of phosphorus amendments and plant growth on the mobility of Pb, Cu, and Zn in a multi-metal-contaminated soil. Environmental Science and Pollution Research, 19 (5): 1659-1667.

Gauglitz R, Holterdorf M, Franke W, et al. 1992. Immobilization of heavy metals by hydroxylapatite. Radiochimica Acta, 58-59 (2): 253-257.

Jiao W, Chen W, Chang A C, et al. 2012. Environmental risks of trace elements associated with long-term phosphate fertilizers applications: A review. Environmental Pollution, 168: 44-53.

Liu J, Lv J, Xu M, et al. 2009. Effect of long-term fertilization on content and activity index of Cu and Cd in red soil. Ecology and Environmental Sciences, 18 (3): 914-919.

Liu Y, Wen C, Liu X. 2013. China's food security soiled by contamination. Science, 339 (6126): 1382-1383.

Lu R, Shi Z, Xiong L. 1992. Cadmium contents of rock phosphates and phosphate fertilizers of China and their effects on ecological environment. Acta Pedologica Sinica, 29 (2): 150-157.

Mavropoulos E, Rossi A M, Costa A M, et al. 2002. Studies on the mechanisms of lead immobilization by hydroxyapatite. Environmental Science & Technology, 36 (7): 1625-1629.

Mavropoulos E, Rocha N C C, Moreira J C, et al. 2004. Characterization of phase evolution during lead immobilization by synthetic hydroxyapatite. Materials Characterization, 53 (1): 71-78.

McGowen S L, Basta N T, Brown G O. 2001. Use of diammonium phosphate to reduce heavy metal solubility and transport in smelter-contaminated soil. Journal of Environmental Quality, 30 (2): 493-500.

Mignardi S, Corami A, Ferrini V. 2013. Immobilization of Co and Ni in mining-impacted soils using phosphate amendments. Water, Air, & Soil Pollution, 224 (2): 1447-1456.

Miretzky P, Fernandez-Cirelli A. 2008. Phosphates for Pb immobilization in soils: A review. Environmental Chemistry Letters, 6 (3): 121-133.

Pérez-Novo C, Bermúdez-Couso A, López-Periago E, et al. 2011. Zinc adsorption in acid soils: Influence of phosphate. Geoderma, 162 (3-4): 358-364.

Raicevic S, Kaludjerovic-Radoicic T, Zouboulis A I. 2005. In situ stabilization of toxic metals in polluted soils using phosphates: Theoretical prediction and experimental verification. Journal of Hazardous Materials, 117 (1): 41-53.

Ryan J A, Zhang P, Hesterberg D, et al. 2001. Formation of chloropyromorphite in a lead-contaminated soil amended with hydroxyapatite. Environmental Science & Technology, 35 (18): 3798-3803.

Selim H M. 2015. Phosphate in soils: Interaction with micronutrients, radionuclides and heavy metals. CRC Press, Boca Raton, FL.

Thawornchaisit U, Polprasert C. 2009. Evaluation of phosphate fertilizers for the stabilization of cadmium in highly contaminated soils. Journal of Hazardous Materials, 165 (1-3): 1109-1113.

Theodoratos P, Papassiopi N, Xenidis A. 2002. Evaluation of monobasic calcium phosphate for the immobilization of heavy metals in contaminated soils from Lavrion. Journal of Hazardous Materials, 94 (2): 135-146.

Tiberg C, Sjöstedt C, Persson I, et al. 2013. Phosphate effects on copper (II) and lead (II) sorption to ferrihydrite. Geochimica et Cosmochimica Acta, 120: 140-157.

Udeigwe T K, Eze P N, Teboh J M, et al. 2011. Application, chemistry, and environmental implications of contaminant-immobilization amendments on agricultural soil and water quality. Environment International, 37 (1): 258-267.

Wei J, Chen M, Song J, et al. 2015. Assessment of human health risk for an area impacted by a large-scale metallurgical refinery complex in Hunan, China. Human and Ecological Risk Assessment: An International Journal, 21 (4): 863-881.

Wei X. 2010. Phosphate-Induced immobilization of heavy metals in multi-metal contaminated soils. Master's thesis, Xi'an University of Science and Technology, Xi'an.

Xu Y, Schwartz F W, Traina S J. 1994. Sorption of Zn^{2+} and Cd^{2+} on hydroxyapatite surfaces. Environmental Science & Technology, 28 (8): 1472-1480.

Yang J, Mosby D E, Casteel S W, et al. 2002. In vitro lead bioaccessibility and phosphate leaching as affected by surface application of phosphoric acid in lead-contaminated soil. Archives of Environmental Contamination and Toxicology, 43 (4): 399-405.

Zhang Q. 2007. Effects of phosphate and lime on stability of Cu and Zn in contaminated soils and their influencing factors. Master's thesis, Graduate College of Chinese Academy of Agricultural Sciences, Beijing.

Zhang P, Ryan J A. 1998. Formation of pyromorphite in anglesite-hydroxyapatite suspensions under varying pH conditions. Environmental Science & Technology, 32 (21): 3318-3324.

Zhang X, Zhu Y, Chen B, et al. 2005. Arbuscular mycorrhizal fungi contribute to resistance of upland rice to combined metal contamination of soil. Journal of Plant Nutrition, 28 (12): 2065-2077.

Zhou S, Xu M. 2007. The progress in phosphate remediation of heavy metal-contaminated soils. Acta Ecologica Sinica, 27 (7): 3043-3050.

Zhu Y, Chen S, Yang J. 2004. Effects of soil amendments on lead uptake by two vegetable crops from a lead-contaminated soil from Anhui, China. Environment International, 30 (3): 351-356.

Zupančič M, Lavrič S, Bukovec P. 2012. Metal immobilization and phosphorus leaching after stabilization of pyrite ash contaminated soil by phosphate amendments. Journal of Environmental Monitoring, 14 (2): 704-710.

In-Situ Stabilization of Toxic Metals in Polluted Soils Using Different Soil Amendments: Mechanisms and Environmental Implication

Shibao CHEN[1], Bin LIU[1], Han ZHENG[1], Nan MENG[1], Cao CAI[2], Yongguan ZHU[2]

[1]Institute of Agricultural Resources and Regional Planning, Chinese Academy of Agricultural Science

[2] Institute of Urban Environment, Chinese Academy of Sciences

1. Introduction

The environmental safety of field soil has become severe in China with the boost of industrialization and urbanization. In the past decades, remediation of the polluted field soils has been paid more efforts in China; various remediation technologies of soil contaminated by heavy metals have been studied, including physical remediation, *in-situ* chemical fixation, biological remediation and agronomic remediation et al. Disposal of metal-contaminated soils is very time-consuming and costly. Over the past two decades, *in-situ* chemical fixation has been put forth for consideration as an alternative to the remediation of contaminated field soils because of its low cost and high efficiency. Since Gauglitz et al. (1992) showed that hydroxyapatite (HA) has the capacity to immobilize heavy metals, the use of phosphate materials as amendments has been a promising way to remediate heavy-metal-contaminated soils. High-quality agricultural soil resources are insufficient in China to supply food for a very large population, many crops are still planted in contaminated field soils, which giving rise to two issues that need to be considered: ①the ability of the field soils to retain lower metal solubility and thus mobility for reducing the risk of heavy metal transport to waters and crops, and ②the ability to provide adequate nutrients for plant growth. The phosphate materials, bone char, clay mineral and the mixture of these materials, are often used as soil additives in *in-situ* fixation remediation practices, can perform as both fixation amendments and fertilizers address each of these issues.

2. Fixation mechanisms of metals by phosphate materials in amended soils

For the past two decades, numerous natural and synthetic phosphate materials have been used to immobilize the heavy metals in contaminated soils, including soluble phosphate fertilizers such as triple superphosphate, diammonium phosphate, monocalcium phosphate, potassium dihydrogen phosphate and phosphoric acid, and insoluble phosphate such as hydroxyapatite, apatite and phosphate rock et al. Varied effects of immobilization of metals were observed for different phosphate sources due to their solubility and complexing capacity to different heavy metals in soils, and the interaction of phosphate with lead (Pb) was usually much stronger than that with other heavy metals such as cadmium (Cd), zinc (Zn) and copper (Cu) et al. most probably because of different mechanisms of metal immobilization that were very complex and were suggested to be attributed to the following three main processes: phosphate-induced adsorption of metal ions, the formation of metal phosphate precipitates or minerals, and surface adsorption or complexation of metals as described in Equations (1) - (4) .

$$Ca_{10}(PO_4)_6(OH)_2 + xM^{2+} \rightarrow (M_xCa_{10+x}^-)(PO_4)_6(OH)_2 + xCa^{2+} \quad (1)$$

$$Ca_{10}(PO_4)_6(OH)_2(s) + 14H^+(aq) \rightarrow 10Ca^{2+}(aq) + 6H_2PO_4^-(aq) + 2H_2O \quad (2)$$

$$10M^{2+}(aq) + 6H_2PO_4^-(aq) + 2H_2O \rightarrow M_{10}(PO_4)_6(OH)_2(s)^- \quad (3)$$

$$Ca_{10}(PO_4)_6(OH)_2(s) + xM^{2+} \rightarrow (Ca_{10+x}^-M_x)(PO_4)_6(OH)_2(s) + xCa^{2+} \quad (4)$$

The formation of precipitates and/or minerals is the main mechanism responsible for the immobilization of heavy metal in soils, especially in mining/smelting soils, using phosphate addition. It should be noted that according to the solubility of metal phosphates, which is presented in Table 1, Pb phosphates, especially pyromorphites ($Pb_5(PO_4)_3X$, where X=fluorine (F) , chlorine (Cl), bromine (Br), OH, representing fluoro-, chloro-, bromo-and hydroxypyromorphite, respectively), are the most stable compounds, and hence, the interaction of Pb and P through the formation of pyromorphites is an important immobilization mechanism controlling the mobility and availability of heavy metals in soils. Moreover, chloropyromorphite is several orders of magnitude less soluble than hydroxyl-, bromo-, and fluoropyromorphite (Table 1), and due to the ubiquity of chloride in nature, chloropyromorphite was expected to be the dominant form in the immobilization of Pb in phosphate-amended soils. Numerous studies have proved that the residual fraction of Pb increased after application of different phosphates resulting from the formation of pyromorphites, and the identification of pyromorphites formed in phosphate-amended soils has also been carried out by different techniques, especially XRD, XAFS, and SEM-EDX.

3. Effects of different phosphorus compounds on Pb uptake by crops

Effects of different phosphorus compounds (hydroxylapatite, HA; phosphate rock, PR; and single superphosphate, SSP) on the phytoavailibility and speciation of Pb in a contaminated soil were examined. Results showed that the Pb concentrations in both shoots and roots of the vegetable plants (*Brassica Oleracea*, BO) decreased with increasing quantities of added P compounds, and the HA treatment had the best effect at the level of 5000 mg (P) /kg compared with other treatments, in which Pb concentrations in shoots and roots of BO plants decreased by 52.2% and 73.1% respectively compared with the control treatment. At the addition level of 5000 mg (P) /kg the effect of treatments followed this order of HA>PR>HA+SSP>SSP. Sequential extraction results indicated that the addition of different phosphate compounds transformed soil Pb from non-residual fractions to the residual fraction substantially. The effect of treatments followed this order at the equivalent level of added phosphate: HA>PR>HA+SSP>SSP. The results suggested that HA amendment could lower the bioavailability and increase the geochemical stability of soil Pb, it therefore has the potential for in-*situ* remediation of Pb-contaminated soils. Sequential extraction results indicated that the addition of soil amendments transform soil Pb from non-residual fractions to residual fraction substantially. The effect of treatments followed this order at the equivalent P addition: HA>PR>HA+SSP>SSP.

4. The effect of grain size of rock phosphate amendment on metal immobilization in contaminated soils

When rock phosphates (RP) are used to remediate Pb-contaminated soils, their effectiveness is likely affected by their grain size. In this study, the effect of grain size of rock phosphate on the effectiveness of heavy metal immobilization in two contaminated soils was measured in pot experiment. Rock phosphate was used with four different grain sizes: <35, 35-72, 72-133 and 133-266 μm. The application rate of rock phosphate in two soils was determined based on P-metals (Pb, Zn, Cu and Cd) molar ratio of 5.0 in the soils. The results showed that rock phosphate of the smallest grain size (<35 μm) was superior to all of other grain sizes more than 35 μm for reducing uptake in plant (*Brassica oleracea* L.) shoots for Cd (19.6%-50.0%), Pb (21.9%-51.4%) and Zn (22.4%-34.6%), respectively, as compared with the soil without application of rock phosphate. Sequential extraction analysis indicated that rock phosphate was most effective for soil Pb to induced transformation from non-residual fractions to a residual fraction than that for Zn and Cd. Such transformation was probably through dissolution of Pb associated with exchangeable (EX), organic fraction (OC), acidic fraction (AC) and amorphous Fe and Al oxides-bound (OX) fraction and precipitation of pyromorphite-like minerals. Results suggested that the rock phosphate with small grain size was superior to that with large grain size for in situ remediation technology.

5. Comparison of Pb (Ⅱ) immobilized by bone char meal and phosphate rock: characterization and kinetic study

Adsorption isotherms and kinetics of aqueous Pb (Ⅱ) by bone char meal (BM) and phosphate rock (PR) were investigated using batch experiments. Pb removal was characterized and effects of pH and reaction time on Pb removal efficiency by BM/PR were also investigated. Results indicated that Pb removal efficiency by BM and PR is mainly controlled by dissolution of phosphatic components associated with apatite, followed by subsequent precipitation of geochemically stable pyromorphite [Pb_{10} $(PO_4)_6$ $(OH, Cl)_2$]. Sorption kinetics results indicated that Pb sorption onto BM or PR was generally biphasic, with initially fast reactions followed by slow and continuous reactions. Pb removal efficiency by BM and PR increased with increasing pH except at a pH higher than 6.15. Sorption isotherms showed that BM has a much higher Pb removal rate than PR in solution; the greater capability of BM than of PR to remove aqueous Pb indicates its potential as another promising way to remediate Pb-contaminated media.

6. Pb immobilization by phosphorus in soil and its mechanisms with XRD, SEM and TEM evidence

The mechanisms of immobilization of Pb in contaminated soil by different phosphates (P) amendments were studied by using X-ray diffractometry (XRD), scanning electron microscope (SEM) and transmission electron microscope (TEM) etc. The X-ray diffraction patterns of the treated soils shown that the formation of the prominent peaks of chloropyro-morphite [Pb_{10} $(PO_4)_6Cl_2$] at 1296 nm, Ca-pyromorphite at 1288 nm and 1321 nm respectively, indicating that less soluble Pb-mineral were formed in-situ in the treated soils. However, the peak at 1296 nm was higher in SSP_5, HA_5 and HA+SSP_2 soils than in PR_5, suggesting less chloropyromorphite formed in PR_5 due to its less solubility. Formation of chloropyromorphite in the treated soils and the plant tissue after P addition were further evidenced by a SEM and TEM microscopy, which showed that Pb was deposited on the cell wall of the plant roots. Combination of SEM, TEM, and XRD results permitted the conclusion that lead ions were immobilized by precipitating $Pb_{(9x)}$ Ca_x $(PO_4)_6X_2$ (X=Cl, OH) compounds in the treated soils and the plant tissues. Based on these results, P-amendments were effective in reducing Pb mobility via in-*situ* formation of insoluble chloropyromorphite etc. in the soils.

7. Effects of phosphate amendments on Pb extractability and movement of phosphorus in contaminated soil

To investigate the effects of different phosphates amendments on the bioavailability of Pb and the movement of phosphorus in soil profile during the remediation of Pb-contaminated soils with single superphosphate (SSP), phosphate rock (PR) and hydroxyapatite (HA) application at two rates of P addition (2500-5000 mg/kg) were studied by using soil column experiment. The results showed that the extractable Pb concentrations by 0.01 mol/L $CaCl_2$ were significantly reduced in surface soils (0-10 cm) with the SSP, PR and HA, and the concentrations of extractable Pb decreased with increasing the P addition. At the P addition level of 5000 mg/kg, the reduction of the concentrations of extractable Pb with SSP, HA and PR reached 86.6%, 81.1% and 89.7%, respectively compared with control. Total phosphorus and plant available P (Olsen-P) in the subsurface soil (0-10 cm) was significantly high after the 270 days' addition of the amendments. However, the total and plant available P (Olsen-P) decreased sharply with the soil depth expended. All the P-treatments, except for SSP_5 treatment, had little marginal increment of total phosphorus and plant available P (Olsen-P) in the depth of 50 cm soil profile. When the depth reached more than 70 cm, no significant difference was found with all the treatments in the total phosphorus and plant available P (Olsen-P) . In other words, no significant fertilizer P moved below 70 mm depth including the soluble SSP fertilizer. This reason may be the strong P adsorption by the surfaces of soil minerals.

8. Effect of bone char application on Pb bioavailability in Pb-contaminated soil

The effects of bone char (BC) application on the bioavailability of Pb in a polluted soil from Hunan Province, China were examined. The Pb-contaminated soil was treated with two types of bone char, one from the UK and the other from China. The bioavailability of Pb was determined in terms of the uptake by Chinese cabbage (*Brassica chinensis* L.), sequential extraction and X-ray diffraction analysis. The results indicate that the Pb concentrations in both shoots and roots decreased with increasing quantities of added bone char, and the application of BC from the UK at the rate of 1.6% (w:w) had the largest effect. Lead Pb concentrations in the shoots and roots decreased by 56.0% and 75.9%, respectively, whereas the application of BC from Zhejiang Province, China at the rate of 1.6% (w:w) reduced Pb concentrations in the shoots and roots to 2.04 mg/kg and 8.42 mg/kg, respectively, only 45.8% and 30.2% compared to the control treatment. Sequential extraction results indicate that the addition of bone char, as a metal-immobilizing agent, substantially transforms soil Pb from non-residual fractions to the residual fraction. The transformation was further confirmed using X-ray diffraction studies.

9. Alleviation of cadmium-induced root growth inhibition in crop seedlings by nanoparticles amendments

The short-term effects of six types of nanoparticles (NPs) (Kaolin, montmorillonite, hydroxyapatite, Fe_3O_4, α-Fe_2O_3 and β-Fe_2O_3) on the EC_{50}s (Cd) for root growth of four plant species (i.e. tomato, cucumber, carrot and lettuce) were investigated using standard toxicity testing. NPs and Cd influencing on growth of the plant were as well as tested, respectively. Scanning electron microscopy (SEM) equipped with the element dispersive spectrometer (EDS) was used to observe the interaction of NPs prepared with EC_{50}s (Cd) as the solvents with the root surface and identify the mechanisms of Cd toxicity reduction to the root growth induced by NPs additives. The results showed that the seedling growth was negatively related to the exposure concentration of Cd, among the tested plants, the sensitive endpoint appeared in the order of tomato>carrot>lettuce>cucumber according to the EC_x measured. The root growth was not significantly inhibited by the presence of NPs except for HAP on tomato, but was noticeably promoted by particular NPs suspensions prepared with EC_{50}s (Cd) as the solvents at higher test concentrations compared with the controls (Cd, EC_{50}s) with one exception for Kaolin. Microscopy images showed roots of tested plants exposed to Cd exhibited a decrease in root diameter and root wilt, and the disintegration of the root epidermis, the clutter root surface showed the evident stress under Cd solution, after the addition of NPs, many root hairs and no disintegration on the surfaces of the root system can be observed, NPs crystal also occurred on the plants root surface. The element dispersive spectrometer (EDS) analysis showed that the precipitation mainly contributed to phytotoxicity reduction by the NPs.

10. Removal of Cd, Pb and Cu from solution using thiol and humic acid functionalized Fe_2O_3 nanoparticles

Humic acid (HA) and 3-Mercaptopropyltriethoxysilane (MPTES) were successfully coated onto the surface of Fe_2O_3 (α and γ) nano-particles as characterized using IR and BET-N_2 analysis, the potential use of the naked and functionalized nano-Fe_2O_3 particles as novel nano-sorbents for removal of Cd, Pb and Cu ions in solution was investigated in this study. The result indicated that the sorption of Cd, Pb and Cu ions by the nano-particles can be fitted well using Langmuir isotherm; all the adsorbents exhibited definitely adsorption ability to Cd, Pb and Cu ions in solution. The sorption maximum and sorption affinity on the nano-particles for Pb (aq) were always higher than Cu (aq) and Cd (aq), the sorption maxima for the Pb, Cd and Cu followed the order Pb>Cu>Cd. Among the nano-sorbents, the Fe_2O_3 (α and γ) nano-particles coated with HA exhibited higher sorption ability to metal ions than the naked and thiolated Fe_2O_3 nano-particles, the sorption maxima of α-Fe_2O_3/HA for Pb (aq) reached 151.5 mg/g, which was significantly higher than the values of 116.3 and 84.0 mg/g observed for the α-Fe_2O_3 and α-Fe_2O_3/MPTES particles. However, no increased sorption maxima was observed for the thiolated Fe_2O_3 nano-particles (Fe_2O_3/MPTES) for the metal ions compared with the naked Fe_2O_3 nano-particles in this study. The greater capability of Fe_2O_3/HA to adsorb Cd (aq), Pb (aq), Cu (aq) indicates its potential use as another promising way to remediate metals-contaminated water.

11. Effects of nanoscale amendments on uptake and transfer factors of Cd by carrot (*Daucus carota*) in polluted soils

A pot experiment was conducted to study the effects of different nanoscale amendments (hydroxyapatite (HAP), red mud (RM), Fe_3O_4, humic acid (HA) -Fe_3O_4) on the uptake and transfer factors of Cd by carrot in two kinds of Cd-polluted soils. The results showed that nanoparticles (NPs) mentioned above could increase plant biomass and tolerance index (TI) of carrot significantly. The application of NPs could decrease the concentration of Cd of different carrot parts, the Cd concentrations in both shoots and roots of carrot decreased with increasing quantities of added NPs. The Cd concentrations in shoots of carrot decreased 78.8%, and in roots 67.8% with the most scope respectively, as compared with the control treatment. The application of nanoscale amendments in polluted soils could decrease transfer factors (TF) and biological concentration factors (BCF) of Cd by carrot in different extent, the mechanism may probably due to the transformation of soil Cd from non-residual fractions to residual fraction substantially after addition of the nano-amendments. In general, the efficiency of different amendments on the Cd bioavailability reduction followed this order at the equivalent additions: RM-HAP > HA-Fe_3O_4 > Fe_3O_4.

References

Abbaspour A, Golchin A. 2011. Immobilization of heavy metals in a contaminated soil in Iran using di-ammonium phosphate, vermicompost and zeolite. Environmental Earth Sciences, 63 (5): 935-943.

Aide M, Whitener K, Westhoff E, et al. 2008. Effectiveness of triple superphosphate amendments in alleviating soil lead accumulation in Missouri Alfisols. Soil & Sediment Contamination, 17 (6): 630-642.

Basta N T, McGowen S L. 2004. Evaluation of chemical immobilization treatments for reducing heavy metal transport in a smelter-contaminated soil. Environmental pollution, 127 (1): 73-82.

Betts A R, Chen N, Hamilton J G, et al. 2013. Rates and mechanisms of Zn^{2+} adsorption on a meat and bonemeal biochar. Environmental Science & Technology, 47 (24): 14350-14357.

Bi X, Pan X, Zhou S. 2013. Soil security is alarming in China' s main grain producing areas. Environmental Science & Technology, 47 (14): 7593-7594.

Bolan N S, Adriano D C, Naidu R. 2003. Role of phosphorus in (im) mobilization and bioavailability of heavy metals in the soil-plant system. Reviews of environmental contamination and toxicology. Springer New York: 177: 1-44.

Brown S, Chaney R, Hallfrisch J, et al. 2004. In situ soil treatments to reduce the phyto-and bioavailability of lead, zinc, and cadmium. Journal of Environmental Quality, 33 (2): 522-531.

Cao X, Liang Y, Zhao L, et al. 2013. Mobility of Pb, Cu, and Zn in the phosphorus-amended contaminated soils under simulated landfill and rainfall conditions. Environmental Science and Pollution Research, 20 (9): 5913-5921.

Cao X, Ma L Q, Singh S P, et al. 2008. Phosphate-induced lead immobilization from different lead minerals in soils under varying pH conditions. Environmental Pollution, 152 (1): 184-192.

Cao X, Wahbi A, Ma L, et al. 2009. Immobilization of Zn, Cu, and Pb in contaminated soils using phosphate rock and phosphoric acid. Journal of Hazardous Materials, 164 (2): 555-564.

Chen S B, Zhu Y G, Ma Y B. 2006. The effect of grain size of rock phosphate amendment on metal immobilization in contaminated soils. Journal of Hazardous Materials, 134 (1): 74-79.

Chen S, Xu M, Ma Y, et al. 2007. Evaluation of different phosphate amendments on availability of metals in contaminated soil. Ecotoxicology and Environmental Safety: 67 (2): 278-285.

Cotter-Howells J D, Champness P E, Charnock J M. 1999. Mineralogy of Pb-P grains in the roots of *Agrostis capillaris* L-by ATEM and EXAFS. Mineralogical Magazine, 63, 777-789.

Elouear Z, Bouhamed F, Bouzid J. 2014. Evaluation of different amendments to stabilize cadmium, zinc, and copper in a contaminated soil: Influence on metal leaching and phytoavailability. Soil and Sediment Contamination: An International Journal, 23 (6): 628-640.

Fang Y, Cao X, Zhao L. 2012. Effects of phosphorus amendments and plant growth on the mobility of Pb, Cu, and Zn in a multi-metal-contaminated soil. Environmental Science and Pollution Research, 19 (5): 1659-1667.

Fayiga A O, Ma L Q. 2006. Using phosphate rock to immobilize metals in soil and increase arsenic uptake by hyperaccumulator *Pteris vittata*. Science of the Total Environment, 359 (1): 17-25.

Gauglitz R, Holterdorf M, Franke W, et al. 1992. Immobilization of heavy metals by hydroxylapatite. Radiochimica Acta, 58 (2): 253-258.

Hamon R E, McLaughlin M J, Cozens G. 2002. Mechanisms of attenuation of metal availability in in situ remediation treatments. Environmental Science & Technology, 36 (18): 3991-3996.

Hashimoto Y, Takaoka M, Shiota K. 2011. Enhanced transformation of lead speciation in rhizosphere soils using phosphorus amendments and phytostabilization: An X-ray absorption fine structure spectroscopy investigation. Journal of Environmental Quality, 40 (3): 696-703.

Hettiarachchi G M, Pierzynski G M, Ransom M D. 2000. In situ stabilization of soil lead using phosphorus and manganese oxide. Environmental Science & Technology, 34 (21): 4614-4619.

Hodson M E, Valsami-Jones E, Cotter-Howells J D, et al. 2001. Effect of bone meal (calcium phosphate) amendments on metal release from contaminated soils—a leaching column study. Environmental Pollution, 112 (2): 233-243.

Jiao W, Chen W, Chang A C, et al. 2012. Environmental risks of trace elements associated with long-term phosphate fertilizers applications: A review. Environmental Pollution, 168: 44-53.

Kamiishi E, Utsunomiya S. 2013. Nano-scale reaction processes at the interface between apatite and aqueous lead. Chemical Geology, 340: 121-130.

Kassir L N, Darwish T, Shaban A, et al. 2012. Mobility and bioavailability of selected trace elements in Mediterranean red soil amended with phosphate fertilizers: Experimental study. Geoderma, 189: 357-368.

Knox A S, Kaplan D I, Paller M H. 2006. Phosphate sources and their suitability for remediation of contaminated soils. Science of the Total Environment, 357 (1): 271-279.

Kumpiene J, Lagerkvist A, Maurice C. 2008. Stabilization of As, Cr, Cu, Pb and Zn in soil using amendments-a review. Waste Management, 28 (1): 215-225.

Liu R, Zhao D. 2013. Synthesis and characterization of a new class of stabilized apatite nanoparticles and applying the particles to in situ Pb immobilization in a fire-range soil. Chemosphere, 91 (5): 594-601.

Luo L, Ma Y, Zhang S, et al. 2009. An inventory of trace element inputs to agricultural soils in China. Journal of Environmental Management, 90 (8): 2524-2530.

Mavropoulos E, Rossi A M, Costa A M, et al. 2002. Studies on the mechanisms of lead immobilization by hydroxyapatite. Environmental Science & Technology, 36 (7): 1625-1629.

Melamed R, Cao X, Chen M, et al. 2003. Field assessment of lead immobilization in a contaminated soil after phosphate application. Science of the Total Environment, 305 (1): 117-127.

Mignardi S, Corami A, Ferrini V. 2012. Evaluation of the effectiveness of phosphate treatment for the remediation of mine waste soils contaminated with Cd, Cu, Pb, and Zn. Chemosphere, 86 (4): 354-360.

Misra V, Pandey S D. 2005. Immobilization of heavy metals in contaminated soil using nonhumus-humus soil and hydroxyapatite. Bulletin of Environmental Contamination and Toxicology, 74 (4): 725-731.

Padmavathiamma P K, Li L Y. 2010. Effect of amendments on phytoavailability and fractionation of copper and zinc in a contaminated soil. International Journal of Phytoremediation, 12 (7): 697-715.

Park J H, Bolan N, Megharaj M, et al. 2012. Relative value of phosphate compounds in reducing the bioavailability and toxicity of lead in contaminated soils. Water, Air & Soil Pollution, 223 (2): 599-608.

Raicevic S, Kaludjerovic-Radoicic T, Zouboulis A I. 2005. In situ stabilization of toxic metals in polluted soils using phosphates: Theoretical prediction and experimental verification. Journal of Hazardous Materials, 117 (1): 41-53.

Sanderson P, Naidu R, Bolan N. 2014. Ecotoxicity of chemically stabilised metal (loid) s in shooting range soils. Ecotoxicology and Environmental Safety, 100: 201-208.

Singh S P, Ma L Q, Harris W G. 2001. Heavy metal interactions with phosphatic clay. Journal of Environmental Quality, 30 (6): 1961-1968.

Spuller C, Weigand H, Marb C. 2007. Trace metal stabilisation in a shooting range soil: Mobility and phytotoxicity. Journal of Hazardous Materials, 141 (2): 378-387.

Thawornchaisit U, Polprasert C. 2009. Evaluation of phosphate fertilizers for the stabilization of cadmium in highly contaminated soils. Journal of Hazardous Materials, 165 (1): 1109-1113.

Udeigwe T K, Eze P N, Teboh J M, et al. 2011. Application, chemistry, and environmental implications of contaminant-immobilization amendments on agricultural soil and water quality. Environment International, 37 (1): 258-267.

Zhu Y G, Chen S B, Yang J C. 2004. Effects of soil amendments on lead uptake by two vegetable crops from a lead-contaminated soil from Anhui, China. Environment International, 30 (3): 351-356.

Part Ⅳ Organic Pollution and Remediation

Fate of Several Typical Organic Pollutants in Soil and Impacts of Earthworms and Plants

Feifei SUN, Hongyan GUO, Rong JI

State Key Laboratory of Pollution Control and Resource Reuse, School of the Environment, Nanjing University

1. Introduction

Soil is a major sink of many organic pollutants in the environment such as polycyclic aromatic hydrocarbons (PAHs), pesticides, and endocrine disrupting chemicals. Most hydrophobic organic contaminants tend to accumulate to soil matrix, which causes more risks to soil animals and plants, furthermore endangering the human health. When entering the soil, micropollutants undergo processes such as mineralization, degradation, binding to soil matrixes, transformation to various metabolites, and bioaccumulation in organisms. These processes usually occur contemporary indicating a complicated transformation of the micropollutants in soil environment. Radioactive ^{14}C as a tracer is an efficient approach to detect the fate of organic pollutants in soil and allow doing mass balance. When combined to gas chromatography-mass spectrometry (GC-MS) or liquid chromatography-mass spectrometry (LC-MS), ^{14}C tracer can facilitate the identification of metabolites in soil and organisms.

In the last ten years, the scientists in the School of the Environment at Nanjing University investigated the environmental fate of some typical emerging micropollutants and traditional organic pollutants of wide concerns in soil. The impacts of earthworms and plants on the fate of these pollutants in the soil, and accumulation of parent and metabolites of the pollutants in soil, earthworms and plants were also studied. In most of the studies ^{14}C-tracer were applied. We summarized the results related to several aromatic compounds, i.e., nonylphenol, tetrabromobisphenol A, polycyclic aromatic hydrocarbons, and chlorophenols. These studies were mainly financed by National Natural Science Foundation of China (NSFC), Chinese Ministry of Science and Technology (MOST), Sino-Swiss Science and Technology Cooperation (SSSTC) .

1.1 Nonylphenol (NP)

As a typical endocrine-disrupting micropollutant, the major source of 4-nonylphenol (4-NP) is anaerobic breakdown of the nonionic surfactant nonylphenol polyethoxylates that is widely used in the world, thus 4-NP is ubiquitous in the environment. 4-NP is more persistent, lipophilic, and toxic than the parent polyethoxylates. Due to its high hydrophobicity (log K_{ow}=4.48), 4-NP can easily accumulate in organisms and may interfere with the secretion of cytokines in human placenta at environmental concentration levels (from pM to nM) . Thus 4-NP has been listed as a priority substance in the Water Framework Directive since the late 1970s.

In treated sludge of wastewater treatment plants (i.e., biosolids), 4-NP tends to adsorb on the sludge at concentrations varying from a few mg NP/kg sludge up to several thousand mg NP/kg sludge. With increasing land application of the biosolids as fertilizer, large amounts of 4-NP were released into the soil, and these 4-NP tends to adsorb to soil or accumulates easily in organisms. Actually, high concentrations of 4-NP have been detected in soil that can reach to several hundred mg/kg dry weight.

Consequently, much attention has been paid to degradation of NPs in both oxic and anoxic soils, and most of these studies were carried out using technical NP (*t*-NP) mixtures or the isomer 4-NP_1. 4-NP_1 does not exist in technical 4-NP mixtures, and the latter is a mixture of *ortho*-and *para*-substituted NP isomers. *para*-substituted NP isomers are the predominant components comprising 86%-94% of technical NP (*t*-NP) mixtures, thus the fate of these isomers in soil should be paid more attention (Liu et al., 2008; Shan et al., 2010b; Shan et al., 2011) .

1.2 Tetrabromobisphenol A (TBBPA)

Tetrabromobisphenol A [4, 4′-isopropylidenebis (2, 6-dibromophenol), TBBPA] is the most commonly used brominated flame retardant (BFR) and accounts for approximately 60% of all applied brominated flame retardants. Its primary applications are in printed circuit boards as a reactive flame retardant and in plastic polymers as an additive in acrylonitrile-butadiene-styrene (ABS) resin and high impact polystyrene.

The undesirable effects of TBBPA include its activity as a potential endocrine disrupter and as a source of oxidative stress in a wide variety of organisms. Both in vivo and in vitro studies have shown negative physiological effects of TBBPA. A growing body of studies has shown that TBBPA has thyroid hormonal activity and leads to immunological effects and inhibition of synaptic neurotransmitter uptake. At environmentally relevant concentrations, TBBPA is particularly toxic to fish and other aquatic organisms. TBBPA has been shown to exhibit significant thyroid- and estrogen-hormone activity as well as hepatotoxicity, immunotoxicity, nephrotoxicity, and neurotoxicity.

Due to the extensive use, TBBPA can be released into the environment during the production, use, and disposal of flame-retardant-containing products. TBBPA has become a ubiquitous environmental contaminant, and has been detected in air, dust, water, sewage sludge, sediment, soil, human and animal tissues. TBBPA concentrations have been reported to be as high as 450 mg/kg in contaminated soil. Given its widespread use, TBBPA has become one of the worldwide environmental contaminants (Liu et al., 2013; Li et al., 2014; Li et al., 2015a; Li et al., 2015b) .

1.3 Chlorophenols (CPs)

Chlorophenols are world widely used as synthetic intermediates in industry and biocides in agriculture or for wood preservation. As a pesticide, herbicide, and antiseptic, pentachlorophenol (PCP) was once used worldwide and can enter soils through waste products or as a by-product on industrial sites. In the last two decades, most developed countries have restricted the use of PCP for domestic and agricultural applications, as a result of concerns about its risk to human health. However, PCP is continually used for treatment of utility poles, wharf pilings, and railroad ties thus still represents an environmental hazard. In China, PCP and its salts were extensively used as pesticides for fighting against schistosomiasis in the last few decades and its production in China amounted to about 20% of its global production. PCP is considered to be a major environmental pollutant because of its toxicity and recalcitrance, and is regulated as one of the priority pollutants by the U. S. Environmental Protection Agency (Shan et al., 2011; Sun et al., 2013) .

1.4 Polycyclic aromatic hydrocarbons (PAHs)

PAHs are of concern due to their ubiquitous distribution in the environment and potential toxicity to organisms. The contamination of PAHs in soil has been a hot topic in environmental studies. Recently, increased concentrations of PAHs are often detected in various environmental media, particularly, more than 90% of PAHs environmental burden has been found in surface soil. In the latest 100-150 years, concentrations of PAHs in soil have increased in wide-ranging latitude of the world, especially in urban areas. Crops can take up PAHs. Thus, food chain contamination is an important pathway through which these toxic pollutants enter the human body: 88%-98% of human exposure to PAHs is connected with food.

PAHs can be retained in soils and subjected to many processes and reactions, including adsorption and desorption, volatilization, degradation, plant uptake, and leaching. Consequently, the fate of PAHs in soil has attracted the attention of the scientific community (Du et al., 2011) .

2. Fate of selected organic pollutants in soil

2.1 Mineralization

Mineralization of pollutants in soil is a process to transform organic chemicals to inorganic substance i.e. CO_2, representing a complete degradation. The rate of the mineralization is usually in agreement with the degradation rate of the pollutants in soil and has a high relationship to the soil type, organic matter content, redox condition, and the microbial activity.

The mineralization of 4-NP was low in various soils. ^{14}C-radioactive tracer showed that mineralization of 4-NP_{111} in the rice paddy soil was about 3.8% of the initially applied radioactivity within 49 days, while the level of radioactive CO_2 in silt loam paddy soil was higher and this was agreement with a better degradation of NP in the soil derived from silty loam deposit with higher organic matter (Shan et al., 2011). In a red soil, 3.84±0.02% of 4-NP_{111} was mineralized at the end of the oxic incubation for 49 days. The addition of ammonium decreased the mineralization of 4-NP_{111} in the soil to 0.1%-2.0% of the total radioactivity (Liu et al., 2014).

In anoxic soil, only minor dissipation of 4-NP_{111} was observed. The amount of $^{14}CO_2$ produced during the anoxic incubation was less than 0.1% (Liu et al., 2014). Therefore, we may draw a conclusion that 4-NP_{111} is persistent under anoxic conditions in paddy soils. However, amendment of nitrate to the soil may increase the mineralization of 4-NP_{111} (Liu et al., 2008).

Unlike 4-NP, TBBPA can be degraded in both oxic and anoxic soils. In an oxic soil slurry, only 1.1±0.3% TBBPA was mineralized after 20 days of incubation, and amendment with the bacterium *Sphingomonas* sp. strain TTNP3 significantly increased the mineralization to 2.0±0.2% (Li et al., 2014). While in a submerged soil, the mineralization of TBBPA was significantly higher than in oxic and anoxic soils, about 11.5±0.4% of TBBPA was mineralized, likely reflecting its higher rate of aerobic degradation in the oxic layer of the top soil (Sun et al., 2014).

Wetland plants such as rice and reed may restrain the mineralization of TBBPA in submerged soil, only 5% of the TBBPA was mineralized in the submerged soil with the growth of rice and reed (Sun et al., 2014). This reduced mineralization in planted submerged soils could be attributed to ①plant-mediated alteration of TBBPA metabolism in the soil and ②a reduction in the amount of mineralizable labeled residues (parent and metabolites) because of their substantial uptake by the plants. The similar phenomena were also observed about the phenanthrene mineralization in lysimeter system that rice and wheat were planted (Du et al., 2011). The presence of wheat plants in soil resulted in reductions of the gaseous losses of phenanthrene from soil surfaces (Du et al., 2011). This is mainly due to that VOCs and CO_2 emitted from the soil could be taken up by plant surfaces in the soil/plant chamber.

2.2 Degradation

The degradation rate of persistent organic pollutants is usually highly dependent on the microbial species and activity in the soil, thus for many pollutants different redox conditions can cause large difference in degradation degree and metabolism.

2.2.1 Degradation in oxic soil

1) NPs

The degradation of 4-NPs in soil was highly isomer specific (Shan et al., 2011). A 58 days' incubation study showed that the most rapid dissipation was observed for the linear isomer 4-NP_1 with a $t_{1/2}$ only 1.4 days in a clay paddy soil, while the branched 4-NP_{111}, the main component of *t*-NP mixtures, was degraded considerably more slowly ($t_{1/2}$=10.3 days). Considering that 4-NP_1 does not exist in *t*-NP mixtures, rapid degradation of 4-NP_1 may not represent the real 4-NP persistence in oxic soils. The observed higher recalcitrance of the branched 4-NP isomers than 4-NP_1 can be attributed to the alkyl chain structure at the benzene ring. The branched isomers have a quaternary α-C on the alkyl chain, and this structure is regarded resistant to ω-and β-oxidation. The length of the side chain at α-C seems to be the most important factor for their degradation. The three isomers with an ethyl side chain at α-C (4-NP_{111}, 4-NP_{112}, 4-NP_{65}) showed a longer $t_{1/2}$ than 4-NP_{38} with two methyl side chains at α-C. For branched 4-NP isomers, the branch number of the alkyl chain was a factor affecting isomer degradation. The isomers with an alkyl side chain branched at two positions (such as 4-NP_{111} at α-C and γ-C and 4-NP_{112} at α-C and δ-C) exhibited a longer $t_{1/2}$ than the isomer 4-NP_{65} with only one branch at α-C. 4-NP_{111}, the main component of *t*-NP mixtures, was degraded considerably more slowly ($t_{1/2}$=10.3 days) (Shan et al., 2011).

In a gleyic hydragric anthrosol soil, the formation of the metabolite of 4-NP_{111} was rapid, after 5 days of incubation, the amounts of the metabolite in the extract accounted already for 24.2% of the initial radioactivity and appeared to be stable during incubation, whereas the relative amounts of the metabolite in the extractable residues increased and the extractable free 4-NP_{111} decreased, indicating the progressive transformation of 4-NP_{111} into the metabolite by soil microorganisms (Shan et al., 2011).

The half-life of 4-NP_{111} in the red soil (16.1±1.6 days) was higher than that in the oxic clay paddy soil (10.3±3.1 days), and the amendment with ammonium significantly ($p<0.05$) inhibited the degradation of 4-NP_{111} in the red soil and increased its half-life from 16.1±1.6 days to 68.0±7.7 days (Liu et al., 2014). The negative effects on degradation and mineralization of 4-NP_{111} in soil may due to ①competition as nitrogen nutrient source between nitrogen-containing pollutants (e.g., atrazine) and the added mineral nitrogen. ②competition for available nutrients other than nitrogen between pollutant degraders and soil organic matter degraders; ③stimulation of the primary growth of fungi in soil by nitrogen, which may alter microbial processes; ④large amounts of nitrogen inhibition of microbes, that are adapted to oligotrophic soil environment; and ⑤oxidation of high concentrations of ammonium may result in an increase of soil acidity and accumulation of nitrite, a toxic intermediate of nitrification (Liu et al., 2014).

2) TBBPA

TBBPA can degrade in both oxic soil and anoxic soil conditions, while more persistent was observed in oxic soil. In a sandy soil slurry, the half-life of TBBPA was 40.8±10.4 days, while the presence of strain TTNP3 (half-life $t_{1/2}$=26.7±5.9 days) significantly ($p<0.05$) accelerated the removal of TBBPA in soil slurry (Li et al., 2014).

Under oxic conditions, TBBPA can be *O*-methylated to form its mono- and dimethyl ethers (MeO-TBBPA and diMeO-TBBPA, respectively) by microorganisms (Li et al., 2014, 2015) . In the oxic soil slurry with nutrient amendment, TBBPA can be degraded to a variety of products, including polar single-ring compounds, MeO-TBBPA, and diMeO-TBBPA. Four interconnected pathways were proposed for TBBPA degradation in the native soil slurry and the slurry amended with strain TTNP3: one major pathway, ①oxidative skeletal rearrangements accompanied by reduction and *O*-methylation, and three minor pathways, ②*O*-methylation, ③ *ipso*-substitution, and ④reductive debromination. In the presence of strain TTNP3, *O*-methylation was stimulated (Li et al., 2014, 2015) .

2.2.2 Degradation in anoxic and submerged soils

1) Anoxic soil

Anaerobic degradation of 4-NP_{111} in the paddy red soil was negligible within 49 days incubation, resulting in accumulation of 4-NP_{111} in the soil (Liu et al., 2014), while the brominated flame retardants TBBPA may be reductively debrominated to less brominated bisphenol A (BPA) and finally to BPA under anoxic conditions (Liu et al., 2013) . In a paddy rice soil, TBBPA dissipated completely under anoxic incubations within 195 days with a half-life of 36 ± 3 days. The debromination end product BPA was detected after 30 days and increased to about 54% of the initially applied TBBPA at the end of the anoxic incubation (Liu et al., 2013) . When alter the incubation condition to oxic soil, the anaerobically accumulated and stable BPA degraded rapidly with a half-life of 11 ± 3 days. The results demonstrated that sequential anoxic-oxic treatment is benefit to the removal of TBBPA in soil (Liu et al., 2013) .

2) Submerged soil

Submerged soil such as wetland soil and paddy fields that are kept mostly water-flooded with 5-10 cm in the whole growing season represents an important ecosystem in environment. One of the main distinguishing features of submerged soil is the presence of oxic-anoxic interface, thus the degradability of pollutants in submerged soil is of great importance since sequential anaerobic and aerobic processes could affect the fate of organic contaminants, especially for halogenated compounds that can be dehalogenated in anoxic soil layer and then further degrade to other metabolites.

The study of Liu et al. (2014) shows that accelerated removal of 4-$NP_{111}$was not observed in submerged soil. The degradation percentage in the Wushan paddy soil without nitrate addition was only 8%, while the addition of nitrate stimulated the degradation of 4-NP_{111} in both soils, especially in the Wushan paddy soil, which had higher content of organic matter than the red soil (Liu et al., 2008) . Approximately 30% of the initial radioactive 4-NP_{111} was recovered as radioactive metabolites from the combined extracts of the Wushan paddy soil amended with nitrate after 75 days of incubation (Liu et al., 2014) .

As mentioned above, TBBPA can be completed degraded by sequential anoxic-oxic incubation (Liu et al., 2013) . In a submerged soil with an oxic-anoxic interface, more than 90% of labeled TBBPA was removed after 66 days of incubation and the half-life ($t_{1/2}$) was 20.8±0.1 days, which was faster than the reported dissipation of TBBPA under anoxic conditions in a rice paddy soil ($t_{1/2}$ 36 days), an estuarine sediment ($t_{1/2}$ 30-40 days), salt marshes ($t_{1/2}$ 70 to>130 days) and a heavy clay soil ($t_{1/2}$ 430 days) (Sun et al., 2014) . Loss of TBBPA in the submerged soil was also faster than the dissipation of TBBPA under oxic conditions in a freshwater sediment ($t_{1/2}$ 40 days), a sandy soil slurry ($t_{1/2}$ 41 days), and a heavy clay soil ($t_{1/2}$ 65-93 days), suggesting that the presence of an oxic-anoxic interface, such as in wetland soils and paddy rice fields, provides a suitable environment for the rapid removal of TBBPA (Sun et al., 2014) .

After 35 days of incubation, about 29.5%±6.0% of the initial TBBPA was transformed to extractable metabolites in the submerged soil (Sun et al., 2014) . In the unplanted soil, four pathways likely participate in the transformation of TBBPA: ①oxidative skeletal cleavage, ②*O*-methylation, ③type II *ipso*-substitution, and ④reductive debromination, similar to degradation of TBBPA under oxic conditions (Li et al., 2014) . Anaerobic and aerobic transformation of TBBPA took place simultaneously in these soils, with further transformation of the metabolites by either route, and reflected the complexity of TBBPA metabolism in flooded soil (Sun et al., 2014) .

2.3 Formation of bound residues

Bound-residue formation is considered as a xenobiotic detoxification process in soil, as they are regarded as stable and recalcitrant toward microbial attack and transport. Several mechanisms are responsible for formation of bound residues such as sorption, diffusion into micropores, and interaction with humic substances. Microbial activity plays an important role in aging processes, e.g., soil microbes would increase formation of bound residues of organic pollutants by incorporating the pollutants or their metabolites into soil organic matter.

Bound-residue formation is a major dissipation process of most organic xenobiotics in soil. We found that almost 30% of phenanthrene was bound to soil after 100 days of incubation when the soil was planted with wheat or rice. Even though with the growth of wheat and rice, the proportion of the bound residues kept stably during 300 days of incubation (Du et al., 2011) .

In oxic soil, bound-residue formation was the major mechanism for the initial dissipation of 4-NP_{111} in the soil. In the red paddy soil, after 49 days of incubation 60.9%±1.7% of the initially 4-NP_{111} was bound to the soil (Liu et al., 2014) . The amount of bound residues of 4-NP_{111} in soils depended on the clay and organic matter contents. The Wushan paddy soil, with higher content of organic matter, contained relatively more amounts of NP residues than the red paddy soil (Liu et al., 2008) .

Amendment of ammonium inhibited the formation of bound residues in the soil (Liu et al.,

2014), which accounted for 23.7%±0.2% of the radioactivity after 49 days. In a clay paddy soil, the formation of bound residues was also rapid and represented the main fate of 4-NP_{111} (54% after 54 days of incubation) (Shan et al., 2011) . The amount of bound residues in the sterilized soil was negligible (Shan et al., 2011; Liu et al., 2014), which confirmed the importance of microorganisms in the bound residue formation and revealed the connection of bound-residue formation to biodegradation of 4-NP_{111}. In the submerged soil, the bound residues of 4-NP_{111} increased when nitrate was present (Liu et al., 2008) .

Pollutants which are covalently bound to soil organic matter are considered as an integral portion of soil organic matter and have little or no risks to the environment, the residues that are bound to soil via non-covalent bondings may be easily released again into soil and therefore pose risks. A silylation procedure was used to distinguish the chemically bound residues from that via physicochemical enclosure (Shan et al., 2011) .

Chemically bound residues of 4-NP_{111} in the humin fraction resulted from covalent binding of 4-NP_{111} or its metabolites, such as hydroquinone or short chain organic acids, to soil organic matter through ester or ether bonds. Oxidative coupling of phenolic compounds to soil organic matter, which may be mediated by enzymes (such as laccase) or abiotic catalyst (such as manganese dioxide), could also contribute to the binding of 4-NP_{111} residues to the humin. The increase in residues chemically bound to humin in the active soil during incubation suggested that 4-NP_{111} was transformed continuously into stable residues in soil by binding to the soil matrix (Shan et al., 2011) .

TBBPA also tended to form bound residues in different soils. In oxic soil slurry, the portion of the bound residues of TBBPA was around 20% of the initial amount after 20 days of incubation, but significantly ($p<0.05$) higher in the slurry amended with strain TTNP3 (Li et al., 2014) .

In the anoxic paddy rice soil, a large amount (about 17%) of bound residues of TBBPA was observed already in the sterilized anoxic soil, suggesting the role of abiotic aging processes in the bound-residue formation during the anoxic incubation, such as adsorption on soil matrixes, diffusion into micropores of soil organic and inorganic aggregates. In the active soil, the bound residues increased to 30% within 155 days of anoxic incubation, underlining the role of microbial activity in the bound-residue formation. During the subsequent oxic incubation, the bound residues in the active soils increased rapidly from 30% to 54% within 15 days, while in the sterilized soil the bound residues increased only slightly, highlighting again the role of microbial activity in the bound-residue formation. There is an interesting phenomena: TBBPA and the less brominated intermediates (mono-, di-, and tri-BBPA), which were detected with low amount (TBBPA) or not detectable (the intermediates) in the anoxic soils before changing to the oxic state, were observed at considerable amounts (1.9%-11.1%) of the initial equivalent and 42.1% of the initially applied TBBPA and the anaerobically formed bound residues, respectively, were released as the lower brominated BPAs (mono-, di-, and tri-BBPA) during the subsequent 70 days of oxic incubation (Liu et al., 2013) .

In the submerged soil, the bound residues of TBBPA increased over time after 66 days, accounting for 60.8%±3.0% of initial TBBPA (Sun et al., 2014) . Since the formation of TBBPA-derived bound residues in soil was low (<35%) under anoxic conditions but increased under oxic conditions (Liu et al., 2013), the large amount of bound residues in the submerged soil can be attributed to the oxic layer in the upper soil (Sun et al., 2014) .

3. Impacts of earthworms and plants on dissipation of selected organic pollutants in soil

3.1 Earthworms

In most terrestrial ecosystems, earthworms are a kind of predominant soil fauna that play crucial roles in the cycling of carbon and nutrient and strongly affect microbial activities in the soil. The surface soil in their habitat may be completely transformed by the earthworms into casts after several years. Thus the activities of earthworms can intensively affect the biotransformation of organic contaminants in the soil.

Actually, a number of studies demonstrated that earthworms accelerated the degradation of various organic pollutants, including pesticides, PAH, PCBs, herbicides, 2, 4, 6-trinitrotoluene (TNT) and nonylphenol. The biodegradation of organic pollutants in soil can be promoted by the improvement of soil aeration and nutritional status caused by earthworms. Earthworms could also lessen the binding of pollutants to soil and remobilize previously soil-bound pollutants. Organic pollutants can accumulate in earthworm body as free or bound-residues. Here we summarize the effects of the geophagous earthworm *Metaphire guillelmi*, a Chinese anecic species, on the fate of organic pollutants in soil.

3.1.1 Sorption

The activity of earthworms can affect many behaviors of the organic pollutions in soil, one of them is the sorption. Chlorophenols are world widely used biocides and their sorption to soil matrix was impacted by the chlorination degree and molecules hydrophobicity. The sorption of the chlorophenols on the soil and earthworm casts were in the order of pentachlorophenol>2, 4-dichlorophenol>2, 4, 6-trichlorophenol, while casts can increase the sorption of these three chemicals largely with a higher sorption of 11%, 24% and 14%, respectively, being mainly attributed to the higher contents of the fine soil particles and possible modification of the organic matter through the gut passage (Shan et al., 2011) .

The same phenomenon was also observed for sorption of 4-NP_{111} to casts. The sorption of 4-NP_{111} on the cast (K_d=1564) was significantly higher than on the parent soil (K_d=1474) (Shan et al., 2011). The higher sorption capacity of the fresh cast was probably attributed to the higher content of fine particles (silt and clay) in the fresh cast through the earthworm gut passage. A desorption hysteresis of 4-NP_{111} on the soil and fresh cast was observed. Although the fresh cast had a higher sorption capacity for 4-NP_{111}, the desorption hysteresis of 4-NP_{111} on the fresh cast was significantly lower than that of the parent soil, which indicates that the affinity of 4-NP_{111} on the fresh cast was weaker than that on the parent soil (Shan et al., 2011) .

Besides higher contents of the fine soil particles, there were several potential factors affecting the sorption capacity of the casts such as the changes of the nature of organic matter, e.g. polarity, hydrophobicity, and molecular configuration, besides, geophagous soil fauna (including earthworms) may lead to an accumulation of aromatic components in the ingested soil by selective digestion of the proteinaceous components of soil humic substances (Shan et al., 2010a) . Such selective digestion behavior would inevitably modify the composition and surface properties of soil organic matter and increase the possibility of exposure of hydrophobic compounds, and therefore increase the sorption capacity of the soil organic matter by increasing partitioning or diffusion of hydrophobic compounds into the soil organic matter. In addition, feeding activity of the earthworms, by grinding soil aggregates in their gut, may increase the accessibility of organic matter in the ingested soil, which may also increase the sorption of the chlorophenols on the casts (Shan et al., 2011) .

3.1.2 Mineralization

Mineralization of 4-NP_{111} in the soil was significantly ($p<0.05$) decreased by the presence of the earthworm *M. guillelmi* (Shan et al., 2010b) . Compared with the control soil, the decrease in the mineralization by *M. guillelmi* was 35%. The reasons might be that organic pollutants may be enclosed in aggregates of cast after their passage through the earthworm gut, resulting in the protection of pollutants against further microbial degradation and might also be related to the higher sorption of 4-NP_{111} on the fresh cast. The high contents of ammonium and nitrate in the earthworm casts might also suppress the mineralization of 4-NP_{111} in the fresh cast because a higher mineral nitrogen content may inhibit the enzyme activity of specific resident decomposers that are capable of degrading the recalcitrant aromatic structures of pollutants and transformation of pollutants.

3.1.3 Bioaccumulation and bound-residue formation

M. guillelmi accumulated large amounts of 4-NP_{111} residues, including free 4-NP_{111} and conjugated residues, and most of the accumulated 4-NP_{111} residues (77% of the total accumulation) were present as the bound, non-extractable forms in the earthworm tissues (Shan et al., 2010b) .

The uptake of 4-NP_{111} by *M. guillelmi* was fast within the first 4 days of the exposure and then slowly reached a steady-state phase. The epidermis-adsorbed radioactivity accounted only for 0.7% of the total radioactivity in the earthworm, which indicated that most of the accumulated 4-NP_{111} residues were present in the intestinal tract of the earthworm (Shan et al., 2010b) .

Bound-residue formation in the earthworms was also rapid, accounting for 60% of the total accumulated radioactivity already at the beginning, the bioaccumulation kinetics of 4-NP_{111} in the earthworms fitted well to the two-compartment accumulation model (Shan et al., 2010b) . In contrast to the bioaccumulation, the elimination of the residues of 4-NP_{111} from the earthworms was slow. After living in rice paddy soil for 16 days, the earthworms still contained 51% of the initially accumulated total residues and 57% of the initial bound residues. The bound residues were the predominant radioactive form in the earthworms, accounting for 77% of the total radioactivity.

Large amounts of the less-polar metabolite (36%, 2-nitro-4-nonylphenol) and the polar metabolites of 4-NP_{111} were found in the extract of the earthworms, whereas only$<$9% of the free form of 4-NP_{111} was detected, which suggested that 4-NP_{111} and its metabolites were largely transformed in the earthworm body (Shan et al., 2010b) . Glucuronide conjugates probably played a less important role in the transformation of 4-NP_{111} and its metabolites in the earthworms. Sequestration of the free form and metabolites of 4-NP_{111} in the chloragocyte tissues maybe one possible pathway for the bound-residue formation in the earthworm bodies. Another mechanism for the formation of the bound residues might be an oxidative coupling, that is, covalent binding of 4-NP_{111} to the cellular components of the earthworms.

3.2 Plants

Phytoremediation has emerged as a promising strategy because it appears to be economically valuable, environmentally friendly, and aesthetically attractive. Over the last decade, there have been extensive reports on the phytoremediation of contaminated soils using various plants.

3.2.1 Degradation of TBBPA in rhizosphere of wetland plants

Rhizosphere provides a suitable environment for removing micropollutants from soils. In the rhizosphere, root exudates simultaneously stimulate the microbial growth, co-metabolism and supply surfactants increasing the bioavailability of pollutants. Submerged soil containing an anoxic-oxic interfaces can accelerate many biogeochemical processes, and the microsites containing soil-water-plants have been proven to accelerate many biotransformation processes of pollutants. Wetland plants transport O_2 to their roots from aboveground, thereby forming more oxic zones within anoxic soil compartments to stimulate the activity of aerobic microorganisms. At the same time the root exudates supply carbon and nutrients for microbial growth. Thus, the rhizosphere of wetland plants can accelerate both aerobic and anaerobic degradation of halogenated compounds.

Using paddy rice (*Oryza sativa*) and common reed (*Phragmites australis*), the fate of TBBPA in submerged soil-plant system was investigated in laboratory scale with flow-through system (Sun et al., 2014). The study showed that comparing to the control ($t_{1/2}$ 20.8±0.1 days), the appearance of reed strongly accelerated the dissipation of TBBPA in submerged soil with a half-live only 11.4 days. In the soils with plant growth, bound residues accounted for 50.1%±4.0% and 40.0%±5.8% of the radioactivity in rice and reed seedlings, respectively. The lower amount of bound residues of TBBPA in the planted than the unplanted soil likely reflects either the accumulation of TBBPA residues in the plants or the release of bound residues from the soil by the plant roots.

Both rice and reed stimulation altered the amounts of the metabolites as well as the time course of their occurrence in the soil. The presence of the reed strongly stimulated the transformation of TBBPA in the soil, and more polar metabolites such as unknown polar metabolites, single-ring metabolites, BPA were found in the rhizosphere of rice and reed (especially reed) (Sun et al., 2014). The presence of the reed and rice plants apparently stimulated anaerobic reduction processes, monobrominated BPA (monoBBPA) and BPA were detected in the rhizosphere but not observed in the unplanted soil, thus in the two planted soils a complete debromination of TBBPA to BPA was observed in a stepwise process, via tribrominated BPA (triBBPA), dibrominated BPA (diBBPA), and monoBBPA. In addition, in the reed soil aerobic *O*-methylation occurred more rapidly than anaerobic debromination. While both plants enhanced the formation of MeO-TBBPA, rice seedlings stimulated further *O*-methylation and debromination of MeO-TBBPA to diMeO-TBBPA and MeO-triBBPA, respectively. The plants stimulated the anaerobic debromination and the aerobic *O*-methylation of TBBPA (Sun et al., 2014).

3.2.2 Phytoremediation of soil contaminated by pentachlorophenol

Among the trees studied, willow displays many characteristics suitable for phytoremediation, such as fast growth, easy propagation, deep rooting, adaption to a wide range of climatic conditions, and tolerance to temporarily waterlogged environments.

We assessed the ability of willow, earthworms, and horseradish alone or in combinations (willow/earthworms and willow/horseradish) to remove PCP from soil in field trials (Sun et al., 2013). The result showed that willow had no effect on PCP removal from soil after 15 days, but after 30 days willow significantly ($p<0.05$) increased the removal of PCP. After 45 days, 88.3% of PCP remained in the soil without willow growth, while 47.2% in soil with willow growth. Horseradish significantly decreased the concentrations of PCP in the soil and PCP removal increased with increasing concentration of horseradish.

The combination of willow/horseradish and willow/earthworms significantly increased the amount of PCP removed ($p<0.05$). On all dates sampled, the efficiency of PCP removal by willow/earthworm was greater than that of willow/horseradish, which in turn was greater than that of willow alone. On the day 45, 23.1% of PCP remained in the soil with willow/horseradish, and only 2.2% remained with willow/earthworms (Sun et al., 2013) .

3.3.2 Accumulation of pollutant residues in plants

Plants can take up organic pollutants from soil to aboveground parts or accumulate them on root parts up to a high concentration, resulting in risks to food safety if the pollutants accumulate in eatable parts. Using ^{14}C tracer, we can quantify organic pollutants and their residues in the plants.

In the soil planted with a rice-wheat rotation in field, ^{14}C-labeled phenanthrene was accumulated in different organs of wheat or rice at the end of grain harvest. For both wheat and rice, ^{14}C radioactivity in five organs of the plants decreased in the following order: roots > leaves > shells > stems > grains. The observation of radioactivity in grains suggests that pollutants in soil can be incorporated into field-grown food, threatening its security (Du et al., 2011) .

Uptake of polybrominated diphenyl ethers (PBDEs) from soil by rice-wheat rotation was studied over 4 years in outdoor lysimeters. BDE-209 and seven lower-brominated PBDEs (BDE-28, -47, -99, -100, -153, -154, and -183) were detected in crops. Relatively high concentrations of BDE-209 (1.8-101.1 μg/kg) were detected in all organs of plants grown in spiked soil, which indicated that BDE-209 was taken up, translocated, and accumulated in both wheat and rice grains. It is noteworthy that the total proportion of lower-brominated PBDEs in ΣPBDE in plants was much higher than in surface soil (0-20 cm) samples, especially BDE-100 was not detected in soil but was detected in plants, along with BDE-154, whose concentration in plants was higher than in soil. These results suggested that other lower-brominated congeners in soil were more readily taken up by plants or that BDE-209 might be further debrominated inside plants (Du et al., 2013) .

Both rice and reed seedlings showed a high potential to accumulate TBBPA and its metabolites from the soil (Sun et al., 2014) . Reed seedlings accumulated significantly ($p < 0.05$) more radioactivity than rice seedlings. During the incubation, most of the radioactivity that accumulated in the plants was localized to their roots. The radioactivity translocation factor (TF) was 0.21 on day 35 in the rice seedlings, but 0.067 and 0.18 on day 66 in the rice and reed seedlings, respectively, indicating: ① a decreasing translocation factor (TF) with increasing plant growth, attributable to the formation of the hydrophobic *O*-methylated bromoBPAs in the soil during the incubation and the tendency of these compounds to adsorb onto plant roots, and ② the greater accumulation of TBBPA and its metabolites in the above-ground parts of reed rather than rice seedlings (Sun et al., 2014) .

References

Du W, Ji R, Sun Y, et al. 2013. Fate and ecological effects of decabromodiphenyl ether in a field lysimeter. Environmental Science &

Technology, 47 (16): 9167-9174.

Du W, Sun Y, Cao L, et al. 2011. Environmental fate of phenanthrene in lysimeter planted with wheat and rice in rotation. Journal of Hazardous Materials, 188 (1): 408-413.

Li F, Jiang B, Nastold P, et al. 2015. Enhanced transformation of tetrabromobisphenol A by nitrifiers in nitrifying activated sludge. Environmental Science & Technology, 49 (7): 4283-4292.

Li F, Wang J, Jiang B, et al. 2015. Fate of tetrabromobisphenol A (TBBPA) and formation of ester- and ether-linked bound residues in an oxic sandy soil. Environmental Science & Technology, 49 (21): 12758-12765.

Li F, Wang J, Nastold P, et al. 2014. Fate and metabolism of tetrabromobisphenol A in soil slurries without and with the amendment with the alkylphenol degrading bacterium *Sphingomonas* sp. strain TTNP3. Environmental Pollution, 193: 181-188.

Liu J, Shan J, Jiang B, et al. 2014. Degradation and bound-residue formation of nonylphenol in red soil and the effects of ammonium. Environmental Pollution, 186: 83-89.

Liu J, Wang Y, Jiang B, et al. 2013. Degradation, metabolism, and bound-residue formation and release of tetrabromobisphenol A in soil during sequential anoxic-oxic incubation. Environmental Science & Technology, 47 (15): 8348-8354.

Liu Q, Ji R, Hommes G, et al. 2008. Fate of a branched nonylphenol isomer in submerged paddy soils amended with nitrate. Water Research, 42 (19): 4802-4808.

Shan J, Brune A, Ji R. 2010. Selective digestion of the proteinaceous component of humic substances by the geophagous earthworms *Metaphire guillelmi* and *Amynthas corrugatus*. Soil Biology & Biochemistry, 42 (9): 1455-1462.

Shan J, Jiang B, Yu B, et al. 2011. Isomer-specific degradation of branched and linear 4-nonylphenol isomers in an oxic soil. Environmental Science & Technology, 45 (19): 8283-8289.

Shan J, Wang T, Li C, et al. 2010. Bioaccumulation and bound-residue formation of a branched 4-nonylphenol isomer in the geophagous earthworm *Metaphire guillelmi* in a rice paddy soil. Environmental Science & Technology, 44 (12): 4558-4563.

Shan J, Wang Y, Wang L, et al. 2014. Effects of the geophagous earthworm *Metaphire guillelmi* on sorption, mineralization, and bound-residue formation of 4-nonylphenol in an agricultural soil. Environmental Pollution, 189 (12): 202-207.

Shan J, Xu J, Zhou W, et al. 2011. Enhancement of chlorophenol sorption on soil by geophagous earthworms (*Metaphire guillelmi*) . Chemosphere, 82 (2): 156-162.

Sun F, Kolvenbach B A, Nastold P, et al. 2014. Degradation and metabolism of tetrabromobisphenol A (TBBPA) in submerged soil and soil-plant systems. Environmental Science & Technology, 48 (24): 14291-14299.

Sun Y, Ren L, Li J, et al. 2013. The Co-application of willow and earthworms/horseradish for removal of pentachlorophenol from contaminated soils. Soil and Sediment Contamination: An International Journal, 22 (5): 498-509.

Adsorption and Reaction of Organic Contaminants on Surfaces of Condensed Carbonaceous Materials

Dongqiang ZHU[1, 2], Heyun FU[1], Wei CHEN[3], Mengxing XIE[1], Linzi ZUO[1]

[1]School of the Environment, Nanjing University

[2]School of Urban and Environmental Sciences, Peking University

[3]College of Environmental Science and Engineering, Nankai University

1. Introduction

Condensed carbonaceous materials (CCMs) discussed herein refer to the inert and aromatic-rich materials (e.g., coal, kerogen, coke, char, charcoal, etc.) that are derived from either pyrolysis or diagenesis of biomass. CCMs are ubiquitous in natural soil and sediment environments (Skjemstad et al., 1996; Schmidt et al., 1999; Song et al., 2002; Cornelissen et al., 2005). For example, black carbon (BC) was estimated to account for 2%-13% of soil total organic carbon (TOC), and 5%-18% of sediment TOC (Cornelissen et al., 2005). In some fire-impacted soils, the BC content can be as high as 30%-45% of the TOC (Skjemstad et al., 1996; Schmidt et al., 1999). Due to the high surface hydrophobicity and relatively large specific surface area, CCMs can strongly adsorb apolar hydrophobic organic contaminants such as polycyclic aromatic hydrocarbons (PAHs) and polychlorinated biphenyls (PCBs), particularly at environmentally relevant low solute concentrations, therefore heavily influencing the environmental fate of contaminants (Bucheli and Gustafsson, 2000; Accardi-Dey and Gschwend, 2003; Cornelissen et al., 2005; Liu et al., 2008). Owing to the unique physicochemical properties, CCMs are also able to strongly mediate the transformation (e.g., redox and hydrolysis) reactions of organic contaminants adsorbed on the carbon surfaces (Oh and Chiu, 2009; Xu et al., 2010; Tang et al., 2011; Fu and Zhu, 2013; Fu et al., 2014;Chen et al., 2014a, 2014b).

In recent years, engineered synthetic CCMs (e.g., biochar and carbon nanomaterials) have attracted tremendous interest in many environmental applications including water and wastewater treatment and soil remediation (Mauter and Elimelech, 2008; Beesley et al., 2011; Hale et al., 2011). Thus, understanding the adsorption behavior and the reactivity of organic contaminants on CCMs is very important not only for predicting the environmental fate and risk of contaminants, but also for exploring the environmental applications of these materials. Hence, in the last two decades a large number of researchers around the world including our research group have devoted to illustrating the underlying mechanisms and key factors controlling the adsorption and reaction processes of organic contaminants on CCMs. In the present chapter, we will review the relevant research papers with a focus on those published by our research group. Special attention will be paid on how the electronic and structural properties of sorbates, as well as the surface chemistry and pore structure of sorbents affect sorption. We will also summarize our recent findings on the mechanisms and key factors controlling the reactivity of contaminants on CCMs.

2. Adsorption mechanisms of organic contaminants on CCMs

The graphitized carbon surfaces of CCMs provide highly hydrophobic adsorption sites for hydrophobic organic contaminants. However, the hydrophobic effect alone cannot fully explain the strong adsorption of organic contaminants on these materials. For example, we found that the adsorption affinity of cyclohexane to carbon nanotubes (CNTs) was only similar to that of benzene and toluene, but was up to two orders of magnitude weaker than that of nitroaromatics, in spite of the much higher or comparable hydrophobicity of cyclohexane (Fig. 1 (a)) (Chen et al., 2007b). The stronger adsorption of aromatic compounds on CNTs can be attributed to the π-π stacking/coupling interaction with the highly polarizable graphitic surfaces (Chen et al., 2007b; Wang et al., 2014). Additionally, due to the surface defects and the polarization of the edges, the graphitic surfaces of CCMs contain electron-rich and electron-depleted sites (Fig. 1 (b)), which can induce strong π-π electron-donor-acceptor (EDA) interaction correspondingly with π-electron-acceptor compounds (such as nitroaromatics, tetracyclines and sulfonamides) and π-electron-donor compounds (such as aromatic amines and phenolics), resulting in an extra sorption driving force and in turn enhanced adsorption relative to those compounds that are neither π donors nor π acceptors (such as benzene and chlorinated benzenes) (Zhu and Pignatello, 2005; Chen et al., 2007b; Ji et al., 2009a, 2009b; Zuo et al., 2016b). The strength of the π-π EDA interaction was found to be dependent on the tendency of the substituent group to accept or donate electrons and its substitution number. For charged and/or highly polar organic contaminants (such as phenolics, tetracyclines and sulfonamides), the adsorption on CCMs can also be affected by electrostatic force and polar interaction (such as H-bonding and Lewis acid-base interaction) (Ji et al., 2011; Wang et al., 2012; Xie et al., 2014).

The adsorption of organic contaminants is also closely related to the surface chemistry of CCMs. Depending on the type of adsorbent, the oxygen-containing functional groups may suppress the adsorption of both polar and nonpolar compounds on charcoal due to the competitive effect caused by water molecules associated to these groups (Zhu et al., 2005) or facilitate the adsorption of polar compounds on CNTs by increasing the polar interaction (Piao et al., 2008). The contrasting results observed between charcoal and CNTs might be reconciled by the interplay between sorbate-sorbent interaction, water-sorbent interaction and sorbent porosity, in that the competitive effect of water molecules associated with the surface oxygen functional groups would be more pronounced if the adsorption of organic contaminants occurs primarily in restricted micropores rather than on external surfaces. Recently, we found that nitrogen doping on the graphitic surface of CNTs could markedly affect the adsorption affinity of organic contaminants (Zuo et al., 2016b). Compared with the nondoped CNTs, the nitrogen-doped CNTs exhibited much stronger adsorption (2-10 folds) for the π-donor compounds (2-naphthol and 1-naphthalmine), but similar adsorption for the weak π-donor compound (naphthalene) and slightly lower adsorption for the π-acceptor compound (1, 3-dinitrobenzene). Due to the high electron negativity of nitrogen atoms, the nitrogen doping increased the π-electron accepting ability of CNTs, and thus facilitated the adsorption of π-donor compounds via the abovementioned π-π EDA mechanism.

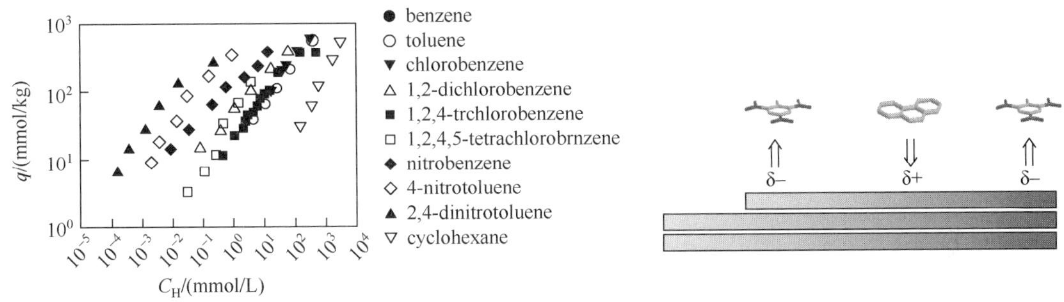

(a) Adsorption isotherms (b) π-π surface of carbonaceous materials

Fig. 1 Adsorption of different organic contaminants on synthetic carbon nanotubes (From Zhu and Pignatello, 2005; Chen et al., 2007b)

In addition to surface chemistry, pore structure (shape and size) of CCMs is also a key factor influencing adsorption of organic contaminants. For microporous CCMs (such as charcoal and activated carbon/AC), adsorption of low-sized adsorbate molecules could be facilitated by the micropore-filling effect when the adsorbate molecular size is close to the adsorbent pore size; alternatively, adsorption of large-sized adsorbate molecules (such as humic substances and bulky pharmaceuticals) could be suppressed due to the size-exclusion effect (Zhu and Pignatello, 2005; Ji et al., 2010b; Fu et al., 2011; Liu et al., 2011; Zuo et al., 2016a). Consistently, we found that after normalization of adsorbent surface area, microporous AC exhibited enhanced adsorption for low-sized phenol and nitrobenzene, but suppressed adsorption for large-sized tetracycline and tylosin, as compared with nonporous graphite powder (Fu et al., 2011). Owing to the much higher assessable surface area, the adsorption of bulky tetracycline on CNTs was one to two orders of magnitude stronger than the adsorption on commercial microporous AC (Ji et al., 2009). Nonetheless, carbon nanomaterials are not necessarily always superior adsorbents to ACs especially when the adsorbate molecules are aliphatic and cannot induce π-π electron coupling interaction. For example, compared with CNTs and graphene nanosheets, microporous ACs exhibited more profound micropore-filling effect and therefore higher adsorption for the aliphatic *trans*-1, 2-dichlorocyclohexane molecule, attributed to the good match between the cubic molecular geometry of the adsorbate and the irregular-shaped micropores of ACs (Wang et al., 2014). Additionally, as individual CNTs aggregate into bundles in aqueous solution to form flexible porous interstices, the adsorption kinetics of flexible bulky molecules such as tylosin to CNTs could be pronouncedly impeded due to a combination of restricted diffusion of adsorbate molecules and dynamic configuration rearrangement of adsorbent (Ji et al., 2010b). In order to optimize the adsorption of flexible bulky molecules, we further exploited the adsorption properties of template-synthesized micro- and mesoporous carbons toward tetracycline, tylosin and humic acid (Ji et al., 2010b; Liu et al., 2011). The surface area-normalized adsorption of tetracycline and tylosin on the synthesized carbons was very close to that on nonporous graphite, reflecting complete accessibility of the adsorbent surface area (Fig. 2 (a)). Moreover, resulting from the regular-shaped, fixed and open three-dimensional pore structure, the synthesized carbons showed much faster adsorption kinetics as compared with ACs and CNTs (Fig. 2 (b)). Notably, the adsorption capacity of CNTs and ACs for pharmaceutical chemicals could be dramatically enhanced by KOH activation because the treatment created substantially more surface areas and pore volumes of adsorbents (Ji et al., 2010c ; Fu et al., 2011).

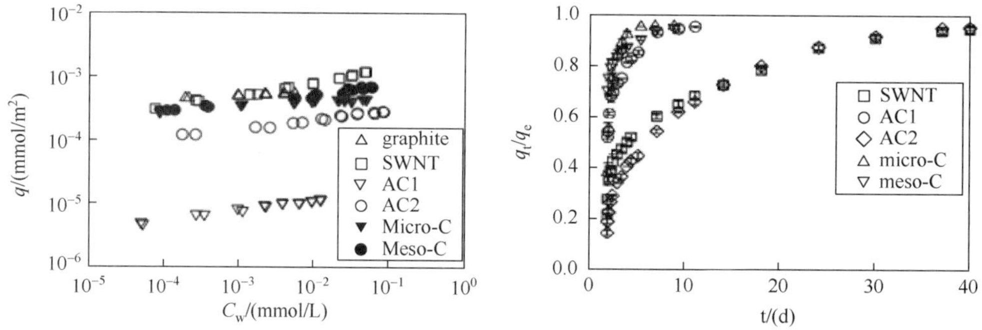

(a) Adsorption isotherms (b) Adsorption kinetics

Fig. 2 Adsorption isotherms and adsorption kinetics of tylosin on different carbonaceous materials: graphite, single-walled carbon nanotubes (SWNT), activated carbons (AC1 and AC2), and template-synthesized microporous carbon (Micro-C) and mesoporous carbon (Meso-C) (From Ji et al., 2010b)

3. Effects of solution chemistry on adsorption of organic contaminants on CCMs

At contaminated sites, organic contaminants often coexist with a complex suite of dissolved organic matter, heavy metals and many other ionic species, which may influence the adsorption of organic contaminants on CCMs. Thus, we systematically investigated the effects of various aqueous solution parameters (e.g., pH, ionic strength, coexisting heavy metals and dissolved organic matter) on adsorption of organic contaminants to CCMs. Changing pH is expected to affect sorption on CCMs—the pH can significantly affect the electronic properties and charge distribution of both adsorbate molecules and CCMs as their functional groups undergo the protonation-deprotonation transition. We found that the pH-dependent adsorption patterns of sulfonamide antibiotics (sulfapyridine and sulfamethoxazole) on CNTs and wood- derived charcoal were very similar to that on nonporous, oxygen functionality-free graphite (Ji et al., 2009a; Xie et al., 2014). This implied that the adsorption was predominated by the graphitic surfaces of CCMs, and the observed pH dependency of adsorption was mainly due to the protonation-deprotonation transition of sulfonamides. Increasing pH facilitated the deprotonation of sulfonamide group and hence decreased the hydrophobicity as well as the π-electron-acceptor ability, resulting in suppressive effects on adsorption. Interestingly, the adsorption of nitroaromatics on CNTs was found to increase with increasing pH (Chen et al., 2008a). The increase of pH apparently facilitated the deprotonation of the acidic functional groups (e.g., carboxyl and phenolic groups) on CNTs, and thus promoted the π-electron-donor ability of the graphitic surface, leading to enhanced π-π EDA interaction of nitroaromatics.

Studies on the effect of coexisting metal ions on adsorption of organic contaminants to CCMs are limited in the literature. Cations (such as Na^+ and Ca^{2+}) can electrostatically screen the surface charge on CCMs, and hence may affect the adsorption of ionic contaminants by interfering with the electrostatic interaction. For example, we found that the adsorption of tetracycline on CNTs decreased (up to 4.5 times) as the ionic strength (Na^+or Ca^{2+}) increased from 0.01 mM to 0.1 mM (Ji et al., 2010a). This was because the electrostatic screening of the oppositely charged groups on tetracycline molecules and CNT surfaces mitigated their mutual electrostatic attraction. In contrast to the base cations $Na^+$and Ca^{2+}, heavy metal ions (e.g., Cu^{2+}) can strongly coordinate to the Lewis-base functional groups of both CNTs and tetracycline, and therefore enhance the adsorption by serving as a bridging agent (Fig. 3 (a)) (Ji et al., 2010a). For nonionic compounds (such as naphthalene), negligible effects of Cu^{2+} were observed on the adsorption to CNTs (Fig. 3 (b)) (Chen et al., 2008a). In contrasting to CNTs, consistent suppressive effects by Cu^{2+} were shown on adsorption of organic contaminants to microporous CCMs (e.g., charcoal and biochar) (Chen et al., 2007a, Xie et al., 2014). The Cu^{2+}-induced adsorption suppression was attributed to surface complexation of Cu^{2+} to form hydration shells of dense water that directly compete with organics for adsorption surface area. This mechanism is expected to be more effective when the adsorption of organic contaminants occurs primarily in restricted micropores of the adsorbent.

It has been well-documented that humic substances can strongly adsorb on CCMs via van der Waals interaction and other specific mechanism (such as π-π interaction, H-bonding and electrostatic forces) (Newcombe and Drikas, 1997; Daifullah et al., 2004; Wang et al., 2009; Yang and Xing, 2009). Previous studies have shown that coadsorption of humic substances on conventional porous CCMs (e.g., ACs) significantly suppressed adsorption of organic contaminants due to combined effects of molecular sieving, pore blockage and competitive adsorption (Kilduff and Wigton, 1999; Kwon and Pignatello, 2005; Pignatello et al., 2006). We explored these mechanisms to explain the influence of dissolved humic substances on the adsorption properties of several porous and nonporous CCMs (e.g., CNTs, charcoal and biochar) (Chen et al., 2008a; Ji et al., 2009a, 2010a, 2011; Xie et al., 2014). The presence of dissolved humic acid consistently led to slightly to moderately reduced adsorption (20%-50%) of the test adsorbates on CCMs (Fig. 3).

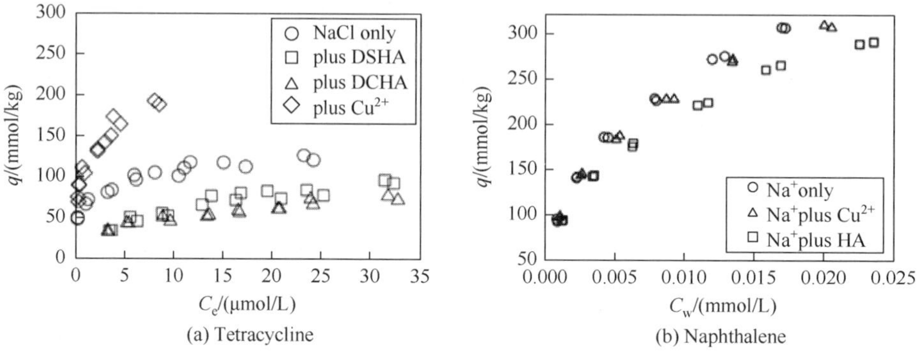

Fig. 3 Effect of Cu^{2+} (initially at 50 mg/L) and dissolved soil humic acid (DSHA, initially at 50 mg/L) on adsorption of tetracycline and naphthalene on carbon nanotubes (From Ji et al., 2010a; Chen et al., 2008a)

4. Contribution of CCMs to soil sorption and nonideal sorption behavior of CCMs

Naturally occurring soils often contain fairly high content of CCMs (e.g., kerogen, coal and BC), which exhibit much stronger adsorption affinity for apolar hydrophobic organic contaminants (such as PAHs and PCBs) as compared with other soil components (e.g., amorphous organic matters and minerals) (Accardi-Dey and Gschwend, 2003; Bucheli and Gustafsson, 2000; Cornelissen et al., 2005). Thus, CCMs may dominate the sorption process and hence the fate of these contaminants in soil environments. We compared the sorption isotherms of nine nonionic aromatic compounds varying in polarity and electronic properties to four surface soils of eastern China (Liu et al., 2008), and found the key role of BC in sorption of apolar sorbates, especially at low solute concentrations. For example, in spite of the low content (ranging from 0.05% to 0.34%), BC accounted for 21%-95% of the overall sorption of phenanthrene to the test soils. The observed high adsorption affinity of phenanthrene to BC can be attributed to a combination of large specific surface area of BC, strong hydrophobic effect and specific π-π EDA interaction as discussed above. The contribution of CCMs to soil sorption of ionic organic contaminants (pharmaceuticals) was also evaluated using simulated soil BCs (e.g., plant biomass-derived biochars) (Ji et al., 2011; Xie et al., 2014). It was found that the contribution of BC to the overall sorption to bulk soil was important for sulfamethoxazole, but negligible for tetracycline (Ji et al., 2011). Owing to the strong surface complexation and/or cation exchange reaction, sorption of tetracycline to clay minerals and soil humic acids was remarkably stronger than sulfamethoxazole; as a result, for tetracycline the importance of BC in the overall sorption to bulk soil was substantially decreased.

For an accurate assessment of the sorption contribution of CCMs, it is prerequisite to know their structural characteristics and content in soil. Prior to the characterization and quantification, CCMs are often treated by thermal or chemical oxidation to remove the amorphous organic carbon (Cornelissen et al., 2005). The oxidation treatment, however, could change the content and properties of CCMs contained in soils, particularly those less-graphitized CCMs, and consequently impact sorption of organic contaminants. We found that the organic carbon content of coals (lignite and anthracite) was markedly lowered after the thermal oxidation treatment (Xie et al., 2013). According to the analysis of solid-state ^{13}C nuclear magnetic resonance (^{13}C NMR), the heat treatment introduced significant amounts of oxygen-containing functional groups (such as ketones/aldehydes, COO/N-C$=$O and aromatic C-O groups) to lignite (approximately 11.3%) and to anthracite (approximately 22.3%) (Li et al., 2012). Hence, the heating-induced oxidation lowered the surface hydrophobicity of coals, resulting in pronouncedly decreased sorption affinity (up to one order of magnitude) for aromatic compounds (Xie et al., 2013). Thus, we came to the conclusion that when the heating treatment was applied, one might significantly underestimate the sorption contribution of CCMs due to the combined effect of reduced organic carbon content and decreased hydrophobicity of the less graphitized CCMs (such as kerogens and coals).

In addition to the predominant role in affecting sorption affinity, CCMs are also the key component responsible for the nonideal sorption (such as multiphasic and slow sorption kinetics and irreversible sorption/hysteresis) observed on natural soils and sediments. Previous studies have attributed slow sorption kinetics to slow relaxation and slow pore diffusion of CCMs (Pignatello and Xing, 1996; Huang and Weber, 1997), and attributed hysteresis to irreversible sorbent conformation alteration (e.g., pore deformation) induced by sorbate (Lu and Pignatello, 2002; Braida et al., 2003; Yang et al., 2004). Our recent study showed that sorbent structural heterogeneity might be a critical factor causing nonideal adsorption of organic contaminants on CCMs (Shi et al., 2011). The removal of amorphous humic substances and/or mineral fraction from a lignite coal significantly accelerated the sorption kinetics and meanwhile alleviated the sorption hysteresis of 1, 3-dinitrobenzene (Fig. 4), which likely resulted from the enhanced exposure of the high-energy adsorption sites consisting primarily of highly graphitized carbons. This finding implies that the strong sequestration of organic contaminants by CCMs might be greatly attenuated by natural biogeochemical and physicochemical processes such as diagenesis of humic substances and weathering of mineral compositions.

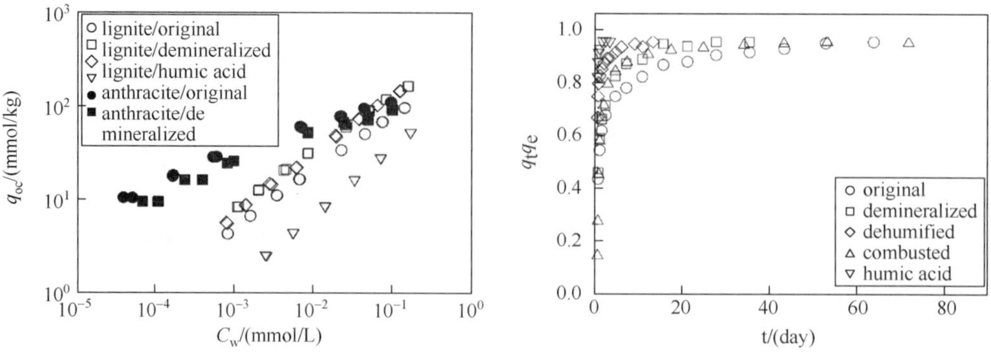

(a) Isotherms to lignite and anthracite (b) Kinetics to lignite

Fig. 4 Effect of removal of mineral compositions or humic substances on 1, 3-dinitrobenzene sorption isotherms to lignite and anthracite and kinetics to lignite (From Shi et al., 2011)

5. Reaction of organic contaminants on CCMs

In addition to providing extraordinarily strong adsorption sites, CCMs may also act as an effective reactor and/or mediator in the transformation reaction of the adsorbed organic contaminants. It is well recognized that the graphitic surfaces of CCMs have very high electric conductivity, which may enable them to accelerate redox reactions of organic contaminants by enhancing electron transfer. Furthermore, CCMs often contain large amounts of functional groups, including carboxyl, hydroxyl and amino groups, with the type and content depending on the carbon source and formation conditions, many of which possess acidity or basicity. Thus, CCMs may also mediate some acid-catalyzed and/or base-catalyzed reactions (e.g., nucleophilic and hydrolytic reactions) of organic contaminants. Understanding whether and how such reactions of organic contaminants may occur on CCMs is essential not only to assessing the fate of the contaminants but also to designing CCMs-based adsorption-reaction coupled systems for effective environmental remediation. For example, we observed efficient reductive dechlorination of AC-adsorbed trichloroethylene (TCE) by nano-scale zero valent iron, even though the TCE molecules could not make direct contact with the iron particles (Tang et al., 2011). It was proposed that the AC served as the conductor for the transfer of electrons or hydrogen atom from the surface of zero valent iron to the adsorbed TCE molecules (Fig. 5 (a)). More recently, we found that graphene oxide (GO) and CNTs had a strong mediation effect on the reduction of nitrobenzene and hexachloroethane by sulfide (Fu and Zhu, 2013; Fu et al., 2014). In particular, the observed pseudo-first-order rate constant (k_{obs}) for nitrobenzene reduction was raised by nearly two orders of magnitude in the presence of only 5 mg/L GO. Besides enhanced electron transfer, we proposed a mechanism of activation of the adsorbed molecules by carbon atoms at the zigzag edges to explain the strong mediation capacity of carbon nanomaterials (Fig. 5 (b)). In parallel, Chen et al. reported that functionalized (e.g., hydroxylated, carboxylated and aminated) CNTs could promote the dehydrochlorination of 1, 1, 2, 2-tetrachloroethane (TeCA), wherein the functional groups served as bases to catalyze the reaction (Chen et al., 2014a). Notably, the mediation efficiencies of carbon nanomaterials in the abovementioned reactions were all markedly higher than that of dissolved humic acids, which are probably the most common and widely distributed natural mediator in aquatic environments. However, the strong adsorption on CCMs may also cause reaction suppression of organic contaminants. It was shown that the apparent reaction kinetic constants for dehydrochlorination of AC-adsorbed TeCA were only 13%-70% of the respective kinetic constants of dissolved TeCA, in spite of the base catalytic effect of the oxygen-containing functional groups on AC (Chen et al., 2014b). This was ascribed to the sequestration of TeCA in the micropores of AC, impeding the hydroxide ion transfer to the TeCA molecules.

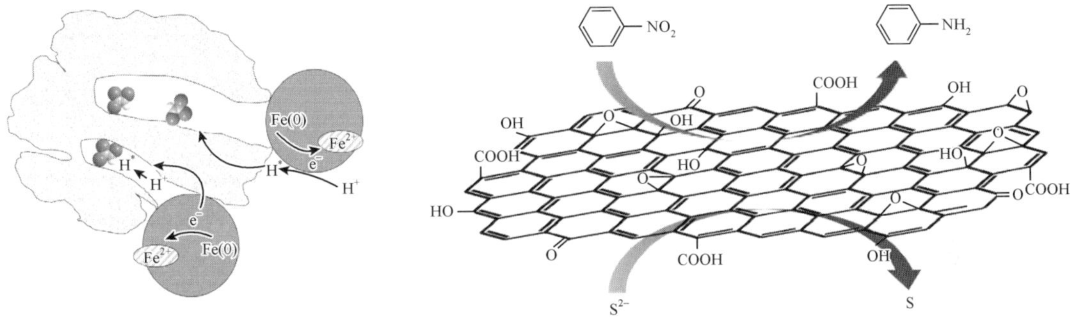

(a) Microporous activated carbon on trichloroethylene reduction by nano-scale zero valent iron

(b) Graphene oxides on nitrobenzenereduction by sulfide

Fig. 5 Schematic description of mediation effect of microporous activated carbon on trichloroethylene reduction by nano-scale zero valent iron and graphene oxides on nitrobenzene reduction by sulfide (From Tang et al., 2011; Fu and Zhu, 2013)

The influences of surface chemistry of CCMs as well as solution chemistry conditions on the CCM-mediated reactions were also examined in the studies discussed above. It was shown that the mediation efficiency of CNTs on hexachloroethane reduction was proportional to the specific surface area of CNTs (an index of defect sites), in comparison with overdosed nonporous graphite that had even higher total surface area but much less surface defects (Fu et al., 2014). Chen et al. compared the dehydrochlorination kinetics of TeCA between 17 different ACs varying in origin and physicochemical properties (Chen et al., 2014a). They found that the second-order rate constant (k_e) decreased with the surface area of ACs, but linearly increased with the AC oxygen content serving as the catalytic sites (Fig. 6). As discussed earlier, humic substances can be strongly adsorbed on CCMs via hydrophobic effect and specific interaction, and thus may block the active sites on CCMs responsible for the catalytic effect. Consistently, it was reported that the presence of dissolved humic acid suppressed the CNT- and GO-mediated dechlorination reaction of hexachloroethane (Fu et al., 2014). In contrast, Chen et al. found that the mediation effect of CNTs on TeCA dehydrochlorination was unaffected by the presence of dissolved humic acid (Chen et al.,

2014b). The overall effect of dissolved humic acid was proposed to be controlled by the interplay of two factors: dissolved humic acid could cover the mediation sites by adsorption on CNTs; humic acid adsorption could also facilitate the dispersion of CNTs, causing more mediation sites available.

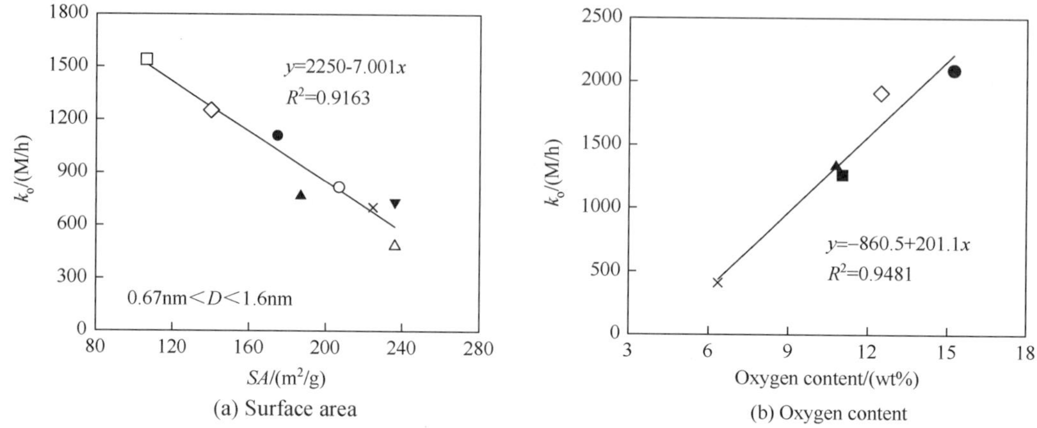

Fig. 6 Effect of surface area and oxygen content of activated carbon on dehydrochlorination kinetics of 1, 1, 2, 2-tetrachloroethane (From Chen et al., 2014a)

6. Perspectives

The adsorption behavior and the reactivity of organic contaminants on CCMs need to be fully understood for the sake of better prediction of the fate of contaminants in soils as well as better application of these materials in soil remediation. Extensive investigation has been devoted to figure out the key factors and the underlying mechanisms affecting these processes. Nevertheless, many issues remain unresolved and warrant further study. For example, to better understand the adsorption mechanisms on natural CCMs and their contributions to the overall soil sorption, nondestructive and comprehensive methods that can adequately characterize these materials without altering their contents and structures are in a critical need. The adsorption on CCMs is controlled by various interplaying mechanisms, which would be affected by the environmental conditions in different ways. Therefore, further investigations are still needed to better understand the relative importance of individual mechanisms in complex environmental conditions. The mechanisms for adsorption hysteresis on CCMs are not well understood either. A new quantitative method that is able to link irreversible sorption with sorbent characteristics is highly demanded. Compared with the sorption properties, the mechanisms for the transformation reaction of organic contaminants on CCMs, particularly the structure-activity relationship between the chemical reactivity and the CCM surface and pore characteristics, are much less understood. From a practical standpoint, a promising research direction for soil remediation is the development of multi-functional contaminant removal reagents. For example, AC or biochar might be combined with natural and synthetic redox reagents (such as reduced iron and sulfur species and potassium permanganate) to make a soil amendment so that adsorption and degradation can be achieved simultaneously. This can be particularly interesting for the removal of low-level contaminants, for which conventional technologies can be costly and highly inefficient.

References

Accardi-Dey A M, Gschwend P M. 2003. Reinterpreting literature sorption data considering both absorption into organic carbon and adsorption onto black carbon. Environmental Science & Technology, 37 (1): 99-106.

Beesley L, Moreno-Jiménez E, Gomez-Eyles J L, et al. 2011. A review of biochars' potential role in the remediation, revegetation and restoration of contaminated soils. Environmental Pollution, 159 (12): 3269-3282.

Braida W J, Pignatello J J, Lu Y, et al. 2003. Sorption hysteresis of benzene in charcoal particles. Environmental Science & Technology, 37 (2): 409-417.

Bucheli T D, Gustafsson Ö. 2000. Quantification of the soot-water distribution coefficient of PAHs provides mechanistic basis for enhanced sorption observations. Environmental Science & Technology, 34 (24): 5144-5151.

Chen J, Chen W, Zhu D. 2008a. Adsorption of nonionic aromatic compounds to single-walled carbon nanotubes: effects of aqueous solution chemistry. Environmental Science & Technology, 42 (19): 7225-7230.

Chen J, Zhu D, Sun C. 2007a. Effect of heavy metals on the sorption of hydrophobic organic compounds to wood charcoal. Environmental Science & Technology, 41 (7): 2536-2541.

Chen W, Duan L, Wang L, et al. 2008b. Adsorption of hydroxyl-and amino-substituted aromatics to carbon nanotubes. Environmental Science & Technology, 42 (18): 6862-6868.

Chen W, Duan L, Zhu D. 2007b. Adsorption of polar and nonpolar organic chemicals to carbon nanotubes. Environmental Science & Technology, 41 (24): 8295-8300.

Chen W, Li Y, Zhu D, et al. 2014a. Dehydrochlorination of activated carbon-bound 1, 1, 2, 2-tetrachloroethane: Implications for carbonaceous material-based soil/sediment remediation. Carbon, 78: 578-588.

Chen W, Zhu D, Zheng S, et al. 2014b. Catalytic effects of functionalized carbon nanotubes on dehydrochlorination of 1, 1, 2, 2-tetrachloroethane. Environmental Science & Technology, 48 (7): 3856-3863.

Cornelissen G, Gustafsson Ö, Bucheli T D, et al. 2005. Extensive sorption of organic compounds to black carbon, coal, and kerogen in sediments and soils: mechanisms and consequences for distribution, bioaccumulation, and biodegradation. Environmental Science & Technology, 39 (18): 6881-6895.

Daifullah A A M, Girgis B S, Gad H M H. 2004. A study of the factors affecting the removal of humic acid by activated carbon prepared from biomass material. Colloids and Surfaces A: Physicochemical and Engineering Aspects, 235 (1): 1-10.

Fu H, Yang L, Wan Y, et al. 2011. Adsorption of pharmaceuticals to microporous activated carbon treated with potassium hydroxide, carbon dioxide, and steam. Journal of Environmental Quality, 40 (6): 1886-1894.

Fu H, Zhu D. 2013. Graphene oxide-facilitated reduction of nitrobenzene in sulfide-containing aqueous solutions. Environmental Science & Technology, 47 (9): 4204-4210.

Fu H, Guo Y, Chen W, et al. 2014. Reductive dechlorination of hexachloroethane by sulfide in aqueous solutions mediated by graphene oxide and carbon nanotubes. Carbon, 72: 74-81.

Hale S E, Hanley K, Lehmann J, et al. 2011. Effects of chemical, biological, and physical aging as well as soil addition on the sorption of pyrene to activated carbon and biochar. Environmental Science & Technology, 45 (24): 10445-10453.

Huang W, Weber W J. 1998. A distributed reactivity model for sorption by soils and sediments. 11. Slow concentration-dependent sorption rates. Environmental Science & Technology, 32 (22): 3549-3555.

Ji L, Chen W, Zheng S, et al. 2009. Adsorption of sulfonamide antibiotics to multiwalled carbon nanotubes. Langmuir,

25 (19): 11608-11613.

Ji L, Chen W, Duan L, et al. 2009b. Mechanisms for strong adsorption of tetracycline to carbon nanotubes: a comparative study using activated carbon and graphite as adsorbents. Environmental Science & Technology, 43 (7): 2322-2327.

Ji L, Chen W, Bi J, et al. 2010a. Adsorption of tetracycline on single - walled and multi - walled carbon nanotubes as affected by aqueous solution chemistry. Environmental Toxicology and Chemistry, 29 (12): 2713-2719.

Ji L, Liu F, Xu Z, et al. 2010b. Adsorption of pharmaceutical antibiotics on template-synthesized ordered micro-and mesoporous carbons. Environmental Science & Technology, 44 (8): 3116-3122.

Ji L, Shao Y, Xu Z, et al. 2010c. Adsorption of monoaromatic compounds and pharmaceutical antibiotics on carbon nanotubes activated by KOH etching. Environmental Science & Technology, 44 (16): 6429-6436.

Ji L, Wan Y, Zheng S, et al. 2011. Adsorption of tetracycline and sulfamethoxazole on crop residue-derived ashes: implication for the relative importance of black carbon to soil sorption. Environmental Science & Technology, 45 (13): 5580-5586.

Kilduff J E, Wigton A. 1999. Sorption of TCE by humic-preloaded activated carbon: Incorporating size-exclusion and pore blockage phenomena in a competitive adsorption model. Environmental Science & Technology, 33 (2): 250-256.

Kwon S, Pignatello J J. 2005. Effect of natural organic substances on the surface and adsorptive properties of environmental black carbon (char): pseudo pore blockage by model lipid components and its implications for N_2-probed surface properties of natural sorbents. Environmental Science & Technology, 39 (20): 7932-7939.

Li Y, Cao X, Zhu D, et al. 2012. Characterization of coals and their laboratory-prepared black carbon using advanced solid-state ^{13}C nuclear magnetic resonance spectroscopy. Fuel Processing Technology, 96: 56-64.

Liu F, Xu Z, Wan H, et al. 2011. Enhanced adsorption of humic acids on ordered mesoporous carbon compared with microporous activated carbon. Environmental Toxicology and Chemistry, 30 (4): 793-800.

Liu P, Zhu D, Zhang H, et al. 2008. Sorption of polar and nonpolar aromatic compounds to four surface soils of eastern China. Environmental Pollution, 156 (3): 1053-1060.

Lu Y, Pignatello J J. 2002. Demonstration of the "conditioning effect" in soil organic matter in support of a pore deformation mechanism for sorption hysteresis. Environmental Science & Technology, 36 (21): 4553-4561.

Mauter M S, Elimelech M. 2008. Environmental applications of carbon-based nanomaterials. Environmental Science & Technology, 42 (16): 5843-5859.

Newcombe G, Drikas M. 1997. Adsorption of NOM onto activated carbon: electrostatic and non-electrostatic effects. Carbon, 35 (9): 1239-1250.

Oh S Y, Chiu P C. 2009. Graphite-and soot-mediated reduction of 2, 4-dinitrotoluene and hexahydro-1, 3, 5-trinitro-1, 3, 5-triazine. Environmental Science & Technology, 43 (18): 6983-6988.

Piao L, Liu Q, Li Y, et al. 2008. Adsorption of L-phenylalanine on single-walled carbon nanotubes. The Journal of Physical Chemistry C, 112 (8): 2857-2863.

Pignatello J J, Kwon S, Lu Y. 2006. Effect of natural organic substances on the surface and adsorptive properties of environmental black carbon (char): attenuation of surface activity by humic and fulvic acids. Environmental Science & Technology, 40 (24): 7757-7763.

Pignatello J J, Xing B. 1995. Mechanisms of slow sorption of organic chemicals to natural particles. Environmental Science & Technology, 30 (1): 1-11.

Schmidt M W I, Skjemstad J O, Gehrt E, et al. 1999. Charred organic carbon in German chernozemic soils. European Journal of Soil Science, 50 (2): 351-365.

Shi X, Fu H, Li Y, et al. 2011. Impact of coal structural heterogeneity on the nonideal sorption of organic contaminants. Environmental Toxicology and Chemistry, 30 (6): 1310-1319.

Skjemstad J O, Clarke P, Taylor J A, et al. 1996. The chemistry and nature of protected carbon in soil. Soil Research, 34 (2): 251-271.

Song J, Peng P, Huang W. 2002. Black carbon and kerogen in soils and sediments. 1. Quantification and characterization. Environmental Science & Technology, 36 (18): 3960-3967.

Tang H, Zhu D, Li T, et al. 2011. Reductive dechlorination of activated carbon-adsorbed trichloroethylene by zero-valent iron: carbon as electron shuttle. Journal of Environmental Quality, 40 (6): 1878-1885.

Wang B, Chen W, Fu H, et al. 2014. Comparison of adsorption isotherms of single-ringed compounds between carbon nanomaterials and porous carbonaceous materials over six-order-of-magnitude concentration range. Carbon, 79: 203-212.

Wang F, Zhu D, Chen W. 2012. Effect of copper ion on adsorption of chlorinated phenols and 1-naphthylamine to surface-modified carbon nanotubes. Environmental Toxicology and Chemistry, 31 (1): 100-107.

Wang X, Tao S, Xing B. 2009. Sorption and competition of aromatic compounds and humic acid on multiwalled carbon nanotubes. Environmental Science & Technology, 43 (16): 6214-6219.

Xie M, Lv D, Shi X, et al. 2013. Sorption of monoaromatic compounds to heated and unheated coals, humic acid, and biochar: Implication for using combustion method to quantify sorption contribution of carbonaceous geosorbents in soil. Applied Geochemistry, 35: 289-296.

Xie M, Chen W, Xu Z, et al. 2014. Adsorption of sulfonamides to demineralized pine wood biochars prepared under different thermochemical conditions. Environmental Pollution, 186: 187-194.

Xu W, Dana K E, Mitch W A. 2010. Black carbon-mediated destruction of nitroglycerin and RDX by hydrogen sulfide. Environmental Science & Technology, 44 (16): 6409-6415.

Yang C, Huang W, Xiao B, et al. 2004. Intercorrelations among degree of geochemical alterations, physicochemical properties, and organic sorption equilibria of kerogen. Environmental Science & Technology, 38 (16): 4396-4408.

Yang K, Xing B. 2009. Adsorption of fulvic acid by carbon nanotubes from water. Environmental Pollution, 157 (4): 1095-1100.

Zhu D, Pignatello J J. 2005. Characterization of aromatic compound sorptive interactions with black carbon (charcoal) assisted by graphite as a model. Environmental Science & Technology, 39 (7): 2033-2041.

Zhu D, Kwon S, Pignatello J J. 2005. Adsorption of single-ring organic compounds to wood charcoals prepared under different thermochemical conditions. Environmental Science & Technology, 39 (11): 3990-3998.

Zuo L, Ai J, Fu H, et al. 2016. Enhanced removal of sulfonamide antibiotics by KOH-activated anthracite coal: Batch and fixed-bed studies. Environmental Pollution, 211: 425-434.

Zuo L, Guo Y, Li X, et al. 2015. Enhanced adsorption of hydroxyl- and amino-substituted aromatic chemicals to nitrogen-doped multiwall carbon nanotubes: A combined batch and theoretical calculation study. Environmental Science & Technology, 50 (2): 899-905.

Toxicity, Adsorption and Dissipation of Polycyclic Aromatic Hydrocarbons in Soil

Jianming XU, Haizhen WANG, Yan HE, Bin MA

Institute of Soil and Water Resources and Environmental Science, Zhejiang Provincial Key Laboratory of Agricultural Resources and Environment, Zhejiang University

1. Introduction

Polycyclic aromatic hydrocarbons (PAHs) are typical persistent organic pollutants (POPs) that mostly generated from the incomplete combustion of fossil fuels, waste incineration, forest and prairie fires, and industrial processes (Bamforth and Singleton, 2005; Zeng et al., 2010a). They are widespread distributed in environments, such as soil, air, water, sediment, etc. (Johnson et al., 2005). Because of their potential bioaccumulation and high toxicity, great attention has been paid to study the behavior of PAHs in soil and to develop effective practices to remediate PAHs contaminated soil in the past decades. In this chapter, some work about the toxicity (Ma et al., 2010a), adsorption (Li et al., 2013; Zeng et al., 2014; He et al., 2015a; Ma et al., 2011, 2016) and dissipation of PAHs (Ma et al., 2010b, 2010c; He et al., 2015b) in soil conducted in our group these years are reviewed.

2. Evaluation of toxicity risk of PAHs in crop rhizosphere of contaminated field using sequential extraction

PAHs have been classified as priority monitored and controlled pollutants by the United States Environmental Protection Agency and other national government agencies due to widespread distribution and persistence in the environment and their toxic, mutagenic and carcinogenic effects to humans (Bojes and Pope, 2007). Normally, the toxic equivalency factors (TEFs) method is a numerical procedure which can provide an interim approach until more data become available to assess the risk of mixtures containing structurally related compound (Delistraty, 1997). In order to evaluate the toxicity risk of soils polluted by PAHs, TEF were employed for converting the PAHs concentrations to toxicity concentrations based on relatively carcinogenic potential of other PAHs to benzo[a]pyrene (B[a]P). However, PAHs persisting in soil would exhibit declining extractability and bioavailability to organisms with increasing contact time (Northcott and Jones, 2001). Rhizosphere effect, including promoting microbial activities and releasing secretions, dominates the phytoremediation through interactions with microorganisms and absorption of root directly (Mackova et al., 2006). The fate of PAHs in rhizosphere is expected distinct with bulk soil as the bioavailability of PAHs would be significantly different. The discrepancy has been masked by the exhaustible extraction procedures that used single organic solvent as an extracting (Oleszczuk, 2009). Solid-phasemicro-extraction (Hynes et al., 2004), XAD-2 resin (Parrish et al., 2005) and passive sampling devices (Allan et al., 2009) were all valuable for evaluating the bioavailability of PAHs in soils for plants and microorganisms, but the contribution of rhizosphere to degradation of PAHs would also obscure by the formation of non-extractable PAHs in soils. Hence, in this section, the role of rhizosphere effect for toxicity risk evaluation in long-term PAHs polluted soils by using sequential extraction approach was discussed (Ma et al., 2010a).

2.1 Sequential extracted concentrations of PAHs in rhizosphere and bulk soil

The scheme of sequential extraction included methanol/water extractable PAHs, butanol extractable PAHs, DCM extractable PAHs, humic acid-bound PAHs, crude humin-bound PAHs, and organic-C enriched humin-bound PAHs. The sum of six sequentially extracted fractionations in rhizosphere soils of all four plant-species was lower than that in bulk soils, especially in oat rhizosphere. The primary fractionation was DCM extractable fraction among three extractable fractions and was crude humin-bound PAHs among three non-extractable fractions. The profiles of sequential extraction in bulk soils varied with plant species, particularly the proportions of n-butanol extractable and DCM extractable PAHs (Ma et al., 2010a).

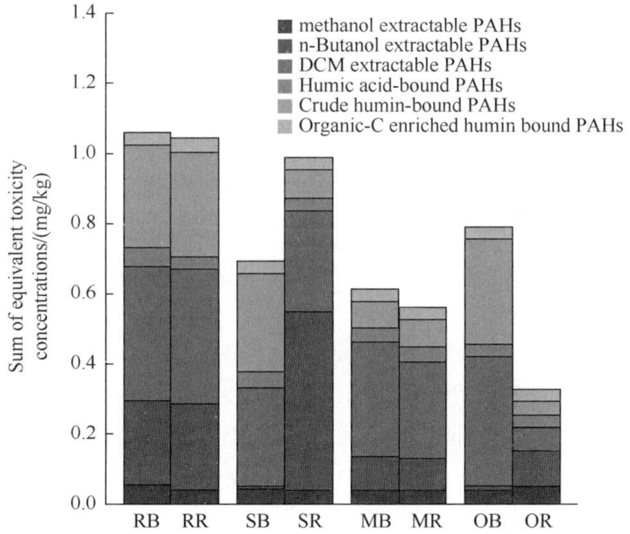

Fig. 1 The toxicity equivalency concentration of PAHs in rhizosphere and bulk soils based on sequential extraction RR, rhizosphere soil of rape; RB, bulk soil of rape; SR, rhizosphere soil of soybean; SB, bulksoil of soybean; MR, rhizosphere soil of maize; MB, bulk soil of maize; OR, rhizosphere soil of oat; OB, bulk soil of oat (Ma et al., 2010a)

After substituting the equivalency toxicity concentration for the concentration of individual PAHs, the profiles of sequential extraction were remarkably distinct (Fig. 1). The total toxicity level and equivalency concentration of six sequentially extracted fractions from soybean, maize, and oat were significantly lower than those under rape. In rhizosphere of soybean and oat, the sum of equivalency toxicity concentrations was significantly lower than that in corresponding bulk soils because the proportion of equivalency toxicity concentrations of crude humin-bound PAHs was sharply decreased in rhizosphere. The proportion of equivalency toxicity concentrations for methanol extractable PAHs was significantly lower than that of methanol extractable PAHs in total PAHs concentrations. The proportion of equivalency toxicity concentrations for n-butanol extractable PAHs was dramatically increased in soils under rape and was remarkably decreased in both rhizosphere and bulk soils under soybean and bulk soil of oat. The discrepancy of n-butanol extractable PAHs was much more remarkable among plant species for equivalency toxicity concentration, which was as low as 0.013 mg/g under soybean and reached to 0.49 mg/g under rape (Ma et al., 2010a).

2.2 PAHs in plant tissues

Both the total PAHs concentration and equivalency toxicity concentration of B[a]P were low comparing with the concentrations in soils. The total PAHs concentration in root was significantly different among plant species. However, the total PAHs concentration in shoot was similar with each other, except the lower PAHs concentration in soybean. When the equivalency toxicity factor was proposed, total equivalency toxicity concentration in root was significantly higher than that in shoot except soybean. Total PAHs concentration in plant tissues was only approximately 5% of that in soils. Similarly, the contribution of root uptake for PAHs was only 2.4% in soil spiked with naphthalene, phenanthrene, and pyrene (Su and Zhu, 2008). The PAHs in shoot were considered to accumulate from atmosphere directly other than transport from root (Su and Zhu, 2008; Keyte et al., 2009). Therefore, the discrepancy between total PAHs concentration and toxicity equivalency concentration of shoot and root revealed the different profile of PAHs in shoot and root.

Further, the results of partial least square regression showed that PAHs in plant tissues could be effectively predicted by the sequentially extracted concentrations of PAHs in soils as all coefficients of multiple determinations exceed 0.6. In soybean matrix, the second component had the highest weight for root and the third component for shoot. The sequentially extracted fractions with the greatest weight values in corresponding components were crude humin-bound PAHs and DCM extractable PAHs for root and shoot, respectively. Result of partial least squares regression indicated that the concentration of PAHs in shoot and root was correlated with different fraction of solvent extractable PAHs in soils. However, this correlation did not make a point that PAHs in shoot or root were taken from such fractions, but indicated similar partition coefficient of plant tissues and correlated PAHs fractions.

In conclusion, not only the concentration but also the toxicity equivalency concentration of PAHs in rhizosphere of crops sampled in a field polluted with PAHs was investigated. Results showed that the discrepancy of toxicity equivalency concentrations between rhizosphere and bulk soils was much more significant than that of total PAHs concentrations. The non-extractable fractions of PAHs in field soils were significantly greater than in spiked soils. Moreover, extractable PAHs fractions in field soils were also more strongly associated with soil components than in spiked soils. It has also shown that PAHs concentration in plant tissues was generally correlated with non-extractable fractions and strongly associated fractions. The findings suggested that the toxicity equivalency factors should be considered when evaluating the potential role of rhizosphere processes in PAHs cleanup in soils (Ma et al., 2010a).

3. Adsorption of PAHs in soil

Adsorption of PAHs to soil influences their transport, bioavailability, and fate in natural environments (Weber et al., 2002; Krauss and Wilcke, 2005). Understanding of the adsorption and fate of PAHs in soil environment is fundamental to food security and remediation strategies for soil contamination (Gao and Zhu, 2005). Although many studies have explored the adsorption of PAHs in soil, understanding regarding partition and distribution of PAHs among water, particles and colloidal matter is still limited due to the difficulty of differentiating and quantifying the concentrations of PAHs between dissolved and colloid-bound phases (King et al., 2004). In this section, the adsorption of PAHs by soil nanoparticles will be introduced (Li et al., 2013; Zeng et al., 2014; He et al., 2015a). Further, the importance of fungal cell wall components in adsorption of PAHs, and consequently the role of fungi in PAHs bioremediation will be presented (Ma et al., 2011, 2016).

3.1 Adsorption of PAHs by natural soil nanoparticles

With the rapid development of nanotechnology, engineered nanoparticles (NPs) have been widely used in various fields (Zhang et al., 2008). However, previous toxicological studies indicate that engineered NPs may enter the body and be toxic to animals (Tang et al., 2008), plants (Stampoulis et al., 2009), and bacteria (Choi and Hu, 2008). Due to the perceived environmental risks of engineered NPs, recent researches have increasingly focused on natural NPs (Tsao et al., 2011). Natural NPs in soils may participate in essential ecological services, since they have special characteristics arising from their nanoscale sizes and large surface areas. Therefore, experiments about the adsorption of PAHs on soil nanoparticles and natural minerals (hematite, montmorillonite and kaolinite) were conducted to understand the role of natural NPs in the adsorption of PAHs.

Aqueous solubility enhancement experiments were conducted to derive the partition coefficients of phenanthrene (PHE) between water and six natural soil NPs. The coefficients were approximately exponentially reduced with increasing concentrations of NPs, low concentrations of NPs (50 mg/L) had significant high adsorption capacities for phenanthrene. Further experiments based on dynamic light scattering technique and column tests were performed to examine the aggregation and mobility of soil NPs and how they influence PHE mobility in porous media (Li et al., 2013). NPs have high and reversible adsorption on surfaces of porous media with aggregation taking place during their transport and they largely control the mobility of PHE in sand columns (Fig. 2).

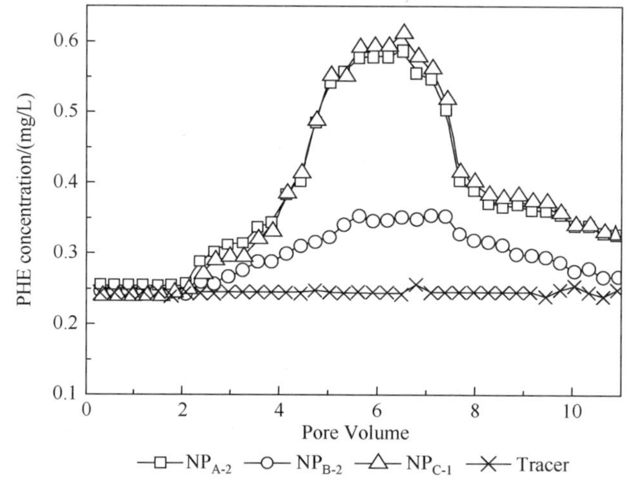

Fig. 2 Breakthrough curves of the facilitated transport of phenanthrene in porous medium (Li et al., 2013)

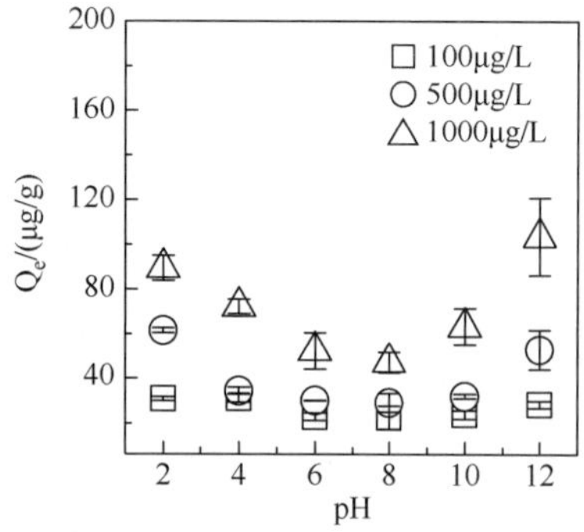

Fig. 3 Sorption of PHE onto hematite NPs versus the electrolytic pH (Zeng et al., 2014)

Meanwhile, the interaction between mineral nanoparticles (hematite, montmorillonite and kaolinite) and phenanthrene (PHE) and factors during the PHE sorption were investigated. In terms of hematite NPs, the PHE sorption capacity was higher at relatively high or low pHs (e.g. below 4.0 and above 10.0), possibly due to the larger available surface area of the hematite NPs, caused by the higher values of netcharges and charge density (Fig. 3). Changes in pH might thus affect the sorption of PHE by hematite NPs, through modification of the surface characteristics of the sorbent and the electronic properties of the sorbate molecules (Zeng et al., 2014). However, sorption of PHE by montmorillonite and kaolinite NPs increased within the pH range of 4.0 to 10.0 (Fig. 4). The influence of different ionic strengths indicated that the amounts of PHE sorbed by these three mineral NPs all decreased as a concentration function of different types of ions (e.g. Na^{+}, K^{+}, $Mg^{2+}$and Ca^{2+}), with the underlying mechanism possibly being due to four interactions i.e. hydrogen-bonding, competitive sorption by ions in the ambient solution, screening effects and aggregation effects. The results confirmed that the surface chemistry of natural NPs, the chemical properties of PHE, and solution chemistry (e.g. pH and ionic strength) of the electrolyte all played an important role in PHE sorption by natural NPs. By comparison of both sorption capacity and ecologic advantages, our results suggested that natural mineral NPs would be more competitive and efficient for PHE sorption than engineered NPs. This finding increases our knowledge regarding the environmental function of natural NPs (such as hematite, montmorillonite and kaolinite) for organic pollutants remediation through manipulating their interfacial behavior (He et al., 2015a).

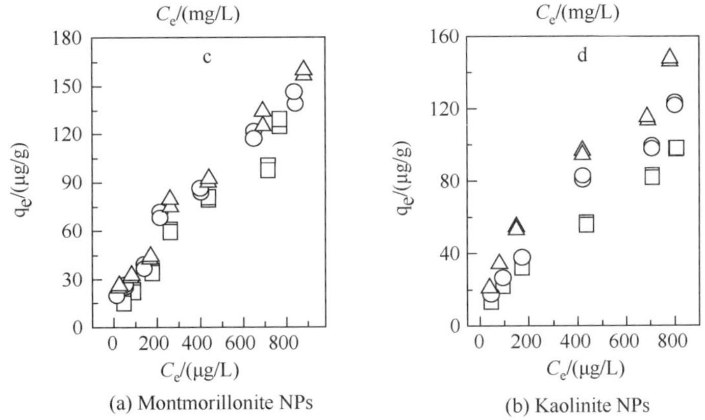

Fig. 4 Sorption of PHE on montmorillonite NPs and kaolinite NPs respectively at different pH (He et al., 2015a) pH: 4 (□), 6 (○), 10 (△)

3.2 Assessing adsorption of PAHs on *Rhizopus oryzae* cell wall components with water-methanol cosolvent model

Microbial treatments of contaminated soils are an efficient, financially affordable, and adaptable way because of potential advantages: the complete degradation of the pollutants, lower treatment cost, greater safety, and less soil disturbance (Habe and Omori, 2003). Plenty of fungi (filamentous fungi, basidiomycetes, white-rot fungi, and deuteromycetes) have been found to remove PAHs efficiently, even for five-ring PAHs which is difficult to be degraded (Peng et al., 2008). Therefore, an understanding of the fungi uptake and translocation mechanisms of PAHs is crucial for guidance of the field application of microbial remediation with fungi. Therefore, we chose *Rhizopus oryzaeas* a typical fungus, observed the sorption of naphthalene, fluorene, phenanthrene, and pyrene as model PAHs on *Rhizopus oryzae* cell wall and fitted results with water-methanol cosolvent models.

The cell wall-cosolvent partition coefficients (K_m) of PAHs were determined for *Rhizopus oryzae* cell walls by controlling the volume fraction of methanol ranging from 0.1 to 0.5. Five cosolvent models were employed for extrapolating the cell wall-water partition coefficients K_w) in pure water. The extrapolated K_w values of four PAHs on *R. oryzae* cell walls were ranged from 2.9 to 5.1. Comparison of various K_w values of pyrene generated from extrapolation and the QSPR model, together with predicted different (PD), mean percentage deviations (MPD), nd root mean square errors (RSE), revealed that the performance of the LL and Bayesian models were the best among all five tested cosolvent models (Fig. 5). This study suggests that *R. oryzae* cell walls play an important role in the partitioning of PAHs during bioremediation because of the high K_w of fungal cell walls (Ma et al. 2011).

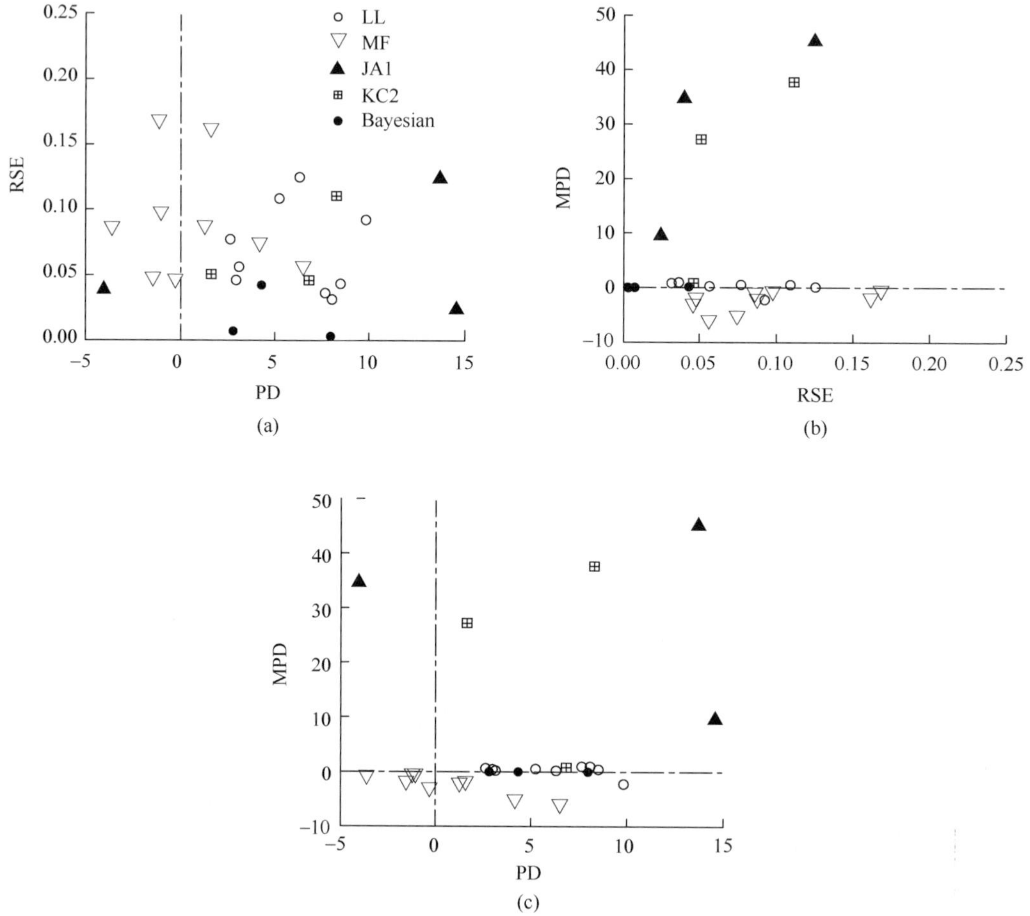

Fig. 5 Plots of predicted difference (PD) verse root mean square error (RSE), RSE verse mean percentage deviation (MPD), and PD verse MPS for various cosolvent models. (a), PD verse RSE, (b), RSE verse MPD, (c), PD verse MPS (Ma et al., 2015)

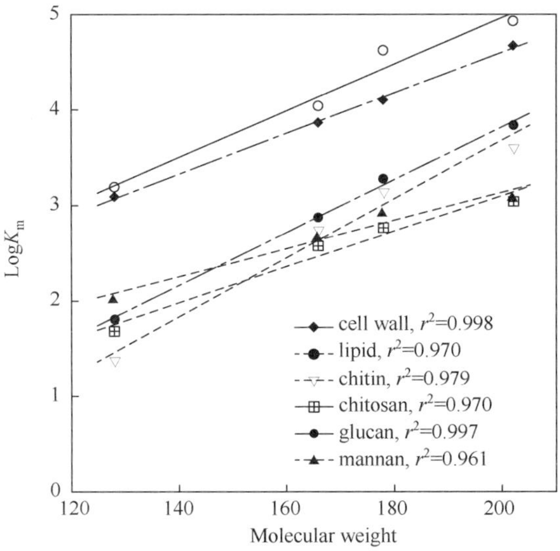

Fig. 6 Logarithm of the cosolvent water partition coefficients of all cell wall components as a function of molecular weights of PAHs (Ma et al., 2016)

The contribution of different fungal cell wall components in adsorption of PAHs is still unclear. We isolated *Rhizopus oryzae* cell walls components with sequential extraction, characterized functional groups with NEXAFS spectra, and determined partition coefficients of PAHs on cell walls and cell wall components with cosolvent model. Spectra of NEXAFS indicated that isolated cell walls components were featured with peaks at 532.7 and 534.5 eV energy. The lipid cosolvent partition coefficients were approximately one order of magnitude higher than the corresponding carbohydrate cosolvent partition coefficients. Comparing with the relationship between K_{ow} and K_m, the correlation coefficients between K_m and molecular weight were all increased except the K_{mchi} values. It indicated that molecular weight was a better descriptor of K_m than K_{ow} for PAHs (Fig. 6). The partition coefficients for four tested carbohydrates varied at approximate 0.5 logarithmic units. Partition coefficients between biosorbents and water calculated based cosolvent models ranged from 0.8 to 4.2. The present study proved the importance of fungal cell wall components in adsorption of PAHs, and consequently the role of fungi in PAHs bioremediation (Ma et al., 2016).

4. Dissipation of PAHs in rhizosphere soil

The elimination of PAHs from contaminated soil is based on various remediation approaches geared toward relevant cleanup standards. Given a wide range of catabolic reactions mediated by microbial enzymes, it is not surprising that initial PAHs bioremediation research focused on the biodegradation by microorganisms (Mackova et al., 2006). Rhizoremediation is a significant form of bioremediation of PAHs in soil. The rhizosphere could increase the microbial population, elevate quantities of readily available carbons derived from root exudation, and change soil redox status relative to bulk soils (Anderson et al., 1993; He et al., 2005). In this section, meta-analysis on dissipation of PAHs in the rhizosphere would be presented to explore the efficiency in accelerating PAHs dissipation varied widely among soil habitat and PAHs types, as well as plants species (Ma et al., 2010b). The removal process of PAHs at the aerobic-anaerobic root-soil interface in paddy field would be also discussed (He et al., 2015b).

4.1 Dissipation of PAHs in rhizosphere: synthesis through meta-analysis

Many studies focused on rhizosphere effects on PAHs dissipation for the contribution of plants on remediation (Mueller and Shann, 2007; Su and Zhu, 2008a). However, a synthetic analysis of the reasons underlying differential PAHs dissipation in the soil-plant system is lacking. Based on the large body of published literature, meta-analysis was used to explore the interaction between plant growth and dissipation of PAHs in soil. The results showed that plants had significant promotional influence on the dissipation of PAHs, representing a negative effect size in the full model (Fig. 7). There was a non-significant difference in the effect sizes between soils with mixed and single PAHs, but a significant difference ($p<0.0001$) between freshly spiked and aged field-contaminated soils. The Hedges' d (d) valve was used to adjust the degree of the effect size. By convention, a value of d greater than or equal to 0.8 is a large effect, d equal to 0.5 is a moderate effect and d equal to 0.2 is a small effect (Gurevitch et al., 1992). Based on these criteria, plants had a large effect on PAHs dissipation in all groups (all the effect size estimates were significantly different from zero), especially in the freshly-spiked soils, in which case plants had an effect size larger than 3 on the dissipation of PAHs.

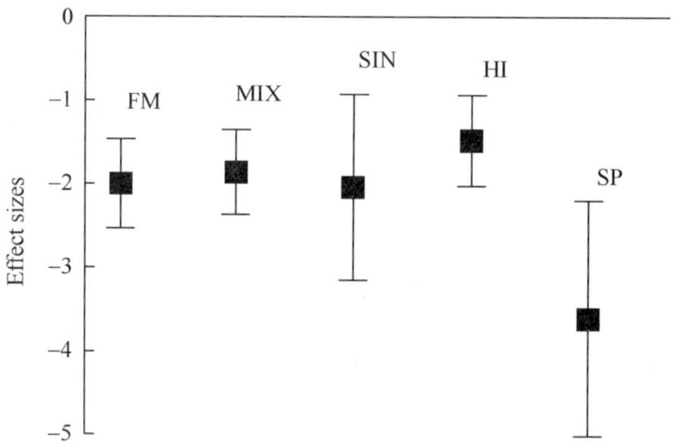

Fig. 7 Plant effect on PAH dissipation. The mean and the 95% confidence interval (vertical bars) are shown for each analysis. No confidence interval overlaps zero; hence, all means are significantly different from zero. FM, full model; MIX, soils with mixed PAHs; SIN, soils with single PAHs; HI, field-contaminated soils; SP, spiked soils (Me et al., 2010b)

Plants had also a positive effect on the number of cultivable PAHs degraders in soils based on the most probable number (MPN), as shown in Fig. 8. These indicated a positive effect of the number of microbial populations capable of degrading PAHs in therhizosphere. These findings represents the first quantitative evidence of the promoting impact of plants on PAH dissipation and the abundance of PAHs decomposers, as well as negative influence of PAHs on plant growth (Ma et al., 2010a).

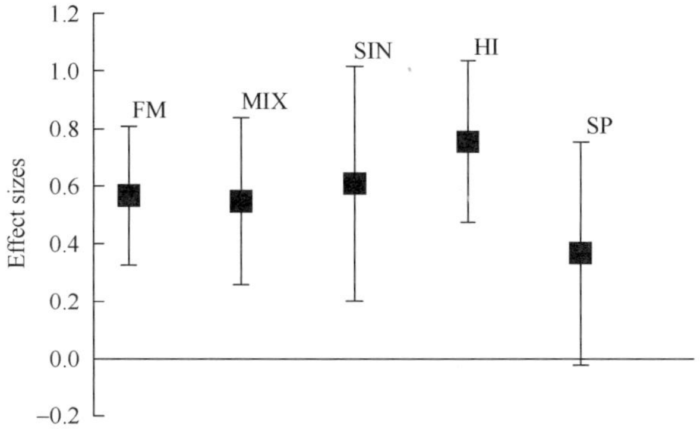

Fig. 8 Plant effect on the number of PAH decomposers. The mean and the 95% confidence interval (vertical bars) are shown for each analysis. No confidence interval overlaps zero; hence, all means are significantly different from zero. FM, full model; MIX, soils with mixed PAHs; SIN, soils with single PAHs; HI, field-contaminated soils; SP, spiked soils (Me et al., 2010b)

Overall, the first quantitative evidence of the promoting impact of plants on PAHs dissipation and the abundance of PAH decomposers, as well as negative influence of PAHs on plant growth were presented. Traditional reviews of phytoremediation emphasize the variation among species, habitats, and contamination types and doses, whereas our analyses not only captured this variation through the distribution of effect sizes but also revealeda significant promoting effect of the rhizosphere despite all this variation. The dynamics of rhizosphere effects and the resultant PAHs dissipation in the soil matrix are complex. The rhizosphere effects are influenced by many factors, including root exudates, microbial communities, soil structure, root architecture, water and nutrient conditions, and pollutant stress. These factors may vary widely among different studies, species, and developmental stages; however, analyses presented here emphasize the generality ofa promoting effect of the rhizosphere. The most important result of these analyses is the consistency of the mean effect sizes regardless of other influencing variables in each study (Ma et al., 2010b).

Further, since few studies paid attention to the rhizoremediation of PAHs compounds other than 16-priority control PAHs of USEPA, we also examined the role of molecular structure in determining the rhizosphere effect on PAHs dissipation (Ma et al., 2010c). Effect size in meta-analysis was employed as activity dataset for building quantitative structure activity relationship (QSAR) models and accumulative effect sizes of 16 PAHs were used for validation of these models. By using the QSAR model, the activities of untested chemicals can be predicted by only requiring information of the chemical structure (Zhao et al., 2008). Based on the genetic algorithm combined with partial least square regression, models for comprehensive dataset, Poaceae dataset, and Fabaceae dataset were built. The results showed that information indices, calculated as information content of molecules based on the calculation of equivalence classes from the molecular graph, were the most important molecular structural indices for QSAR models of rhizosphere effect on PAHs dissipation. The QSAR model, based on the molecular structure indices and effect size, has potential to be used in studying and predicting the rhizosphere effect of PAHs dissipation (Ma et al., 2010c).

4.2 Dissipation of PAHs in rhizosphere at the aerobic-anaerobic soil interface

Rice (*Oryza sativa* L.) is the most common and important cereal crop in Asia, especially China. It can change the redox status of waterlogged paddy soil and establishes an aerobic-anaerobic root-soil interface that may further induce changes in rhizosphere microbial community (i.e., some specific aerobic microbes) (Begg et al., 1994; Pedersen et al., 1995), and thereby the microbial-induced degradation of xenobiotics (Hayat et al., 2011). Microbial degradation is considered to be the major dissipation process of PAHs in environment (Haritash and Kaushik, 2009), aerobic oxidation catabolism has been regarded as the most important for soil removal of PAHs. As a consequence, the dissipation processes of PAHs involved in paddy field would be more complex than in other ecosystems. Thus, a pot experiment was conducted to reveal the removal of two PAHs (phenanthrene and pyrene) in rhizosphere at the aerobic-anaerobic interface of paddy soil during rice growth (He et al., 2015b).

The results demonstrated that in flooded paddy soils cultivated with rice, the dissipation of PAHs in rhizosphere would be more complex relative to those in rhizosphere of dryland soils, because of the special aerobic-anaerobic heterogeneous microbial habitats possibly developed by oxygen secretion of rice root. Besides the interactive influence of soil properties, PAHs species, and cultivation time, the differences in dissipation of PAHs would even be attributed to differences in genotypes within a species. The spiked phenanthrene and pyrene were removed quickly following rice cultivation, with the larger dissipation rates achieved by PAHs-sensitive cultivar in both soils. Therefore, when dealing with in situ phytoremediation through rice plantation, one should be aware that PAHs tolerant cultivar does not inevitably lead to the higher dissipation of PAHs relative to the sensitive one. Irrespective of difference in genotypes, our results also suggested that rhizosphere effect of rice for facilitating dissipation of PAHs was larger for pyrenethan for phenanthrene. This indicates that phytoremediation through rice plantation would be more efficient for PAH species of high molecular weight than those of low molecular weight, likely due to the more significant enhancement of bioavailability of SOM-bound PAH fraction induced by root-released labile C under flooded condition. The microbial-mediated mechanism that underpins the facilitated dissipation of pyrene following rice plantation might be the selective enrichment of aerobic functional bacteria capable of catalyzing oxidative degradation of pyrene by producing dioxygenase in rice rhizosphere (He et al., 2015b).

For an improved understanding, further effort is necessary toward more direct evidence that is able to well illustrate the aerobic-anaerobic heterogeneous microbial habitats using more advanced in situ technologies. Especially, DNA-based molecular sequencing could be also used to identify the specific functional aerobic groups as well as their aerobic catabolizing genes causing the differentiations in PAHs dissipation among different treatments (soil type, rice cultivars) and different soil microbial habitats (rhizosphere and non-rhizosphere) during rice plantation in paddy field (He et al., 2015b).

5. Microbial functional profiling during PAHs degradation in soil

The phytoremediation potential of the rice ecosystem for organic pollutants has received close attention recently (Su and Zhu, 2008b; Ma et al., 2010b). The most important factors that influence PAHs degradation, microbial communities, will vary spatially along the redox gradient (Aaen et al., 2011). However, most studies have compared the communities in the rhizosphere with those in the bulk soil, but the heterogeneity in the rhizosphere has been ignored (Sipila et al., 2008), and the response of microbial metagenome to PAHs degradation in the rice rhizosphere remains poorly understood. In this section, changes in microbial community structure affected by the amendment of black carbon in the soil will be introduced (Wang et al., 2012). Further, the spatial and temporal variations of microbial communities and predicted metagenome along the rice rhizosphere gradient during PAHs degradation will be discussed (Ma et al., 2015).

5.1 Microbial communities analysis affected by black carbon in PAHs spiked soil

During past decades, enrichment of black carbon (BC) in the environment has drastically increased. In the presence of BC, PAHs may apparently be a decreased environmental risk subsequently leading to inaccurate estimations and prediction of safe ecological limits and remediation end points. Currently, relations between sorption of BC and biological processes have mainly focused on their uptake by benthic organisms and much less on their microbiology and bioremediation in soils and sediments. Soil-BC-microbe interactions should be more fully evaluated as their interactions need a better understanding in order to gain improved knowledge about the effects of organic contaminants on soil microbial activity and community structure. Thus, the influence of BC amendment over time on the microbial community in PAHs spiked soil was evaluated using PLFAs profiles. The results showed that the amendment with BC caused a marked decrease in the dissipation (ascribed to mainly degradation and/or sequestration) of phenanthrene residues from soil. Extracted phenanthrene in black soil with 1% BC were higher, oppositely in red soil, 0.5% BC amendments were higher. Further, changes in microbial community structure in the presence of BC with time were detected by PLFAs, and the increasing time of incubation was a more effective factor in changing concentrations than different BC amendments. However, BC had little effect on the microbial community structure of phenanthrene-spiked soils, as indicated by principal component analysis of the PLFA signatures. Meanwhile, results suggested that Gram-positive bacteria were more adversely affected than Gram-negative bacteria, fungi, and actinobacteria, and the latter two groups appeared to be mainly responsible for phenanthrene degradation in soil (Wang et al., 2012).

5.2 Reconstructed metagenomes reveal changes of microbial functional profiling during PAHs degradation in soil

Metagenome analysis provides insight into the physiological potential of the microbial community and enables the identification of the different metabolic pathways in an ecosystem under different environmental conditions (Guazzaroni et al., 2013; van de Vossenberg et al., 2013). However, the metagenome related to PAHs degradation in the rice rhizosphere remains poorly understood. Therefore, bacteria and archaea 16S rRNA genes in the rice rhizosphere soils spiked with different PAHs concentrations were amplified and separated by denaturing gradient gel electrophoresis. The dominant bands in the DGGE gel were excised and sequenced to identify major microbial species in the microbial community. These 16S rRNA gene sequences were mapped to fully sequenced genomes to reconstruct a predicted metagenome, which can used for predicting the functional profiling changes in rice rhizosphere soils during PAHs degradation.

5.2.1 The response of the microbial community to PAHs degradation in the rice rhizosphere

The structure of the bacterial community in the rice rhizosphere soils with different PAHs concentration, distances from the root surface and culture time were firstly obtained by the DGGE analysis. Then the effects of constrained environmental factors, including PAHs concentration, distance from the root surface, and culture time, on the bacterial and archaeal communities in the rhizosphere during PAHs degradation were analyzed with RDA. These constrained environmental factors were only responsible for 10.7% of the variation of the prokaryotic communities in the rice rhizosphere during PAHs degradation, with 4.5% for PAHs concentration, 4.0% for distance from root surface, and 2.2% for culture time. The results for fitting environmental factors onto an ordination showed that PAHs concentration ($r^2=0.57$, $p=0.001$) and distance from the root surface ($r^2=0.44$, $p=0.001$) influenced the microbial communities, but the effect of culture time ($r^2=0.05$, $p=0.54$) could be ignored (Ma et al., 2015).

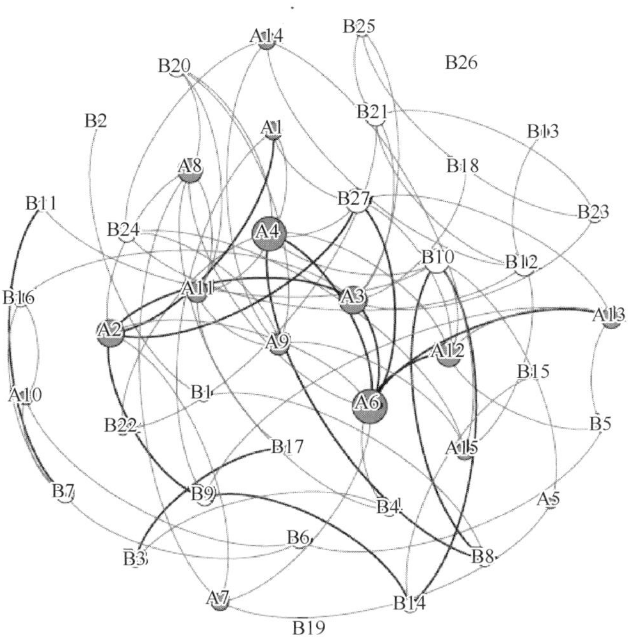

Fig. 9 Correlation network plot showing the correlations between archaea and bacteria from communities in rice rhizosphere during PAHs degradation. Black and white nodes labeled with A and B denotesarchaeal and bacteria, respectively. The following numbers denote the band numbers. The black and grey lines indicate significant negative and positive correlations ($P<0.05$) between two nodes, respectively. The node size denotes the number of connecting line of nodes (Ma et al., 2015)

The relationship of different species in communities is given in the correlation network (Fig. 9). All bacterial and archaeal species were involved in the correlation network, except bacterial band 19 related to *A. imshenetskii* and band 26 related to *I. album*. Hub nodes bacterial bands 10 and 27, and archaeal bands 2-4 and 6, which were the nodes with the maximum number of edges in the network, play an important role in rhizosphere communities. Bacterial band 10 and archaeal bands 2, 3, and 6 were species constantly present in the rhizosphere. Bacterial band 27 was an important component in lower concentration PAHs rhizosphere soils at 2 to 3 mm from the root surface and in H-PAHs (higher concentration of PAHs) rhizosphere soils at 1 mm from the root surface. Archaeal band 4 was an important component in L-PAHs (lower concentration of PAHs) rhizosphere soils at 1 mm from the root surface. The negative correlation in the network suggested competition between species. Most of the negative correlation in the present network was related to the nodes with high edge numbers. This implies that these hub node species competed with each other, and occupied similar ecological niches (Ma et al., 2015).

5.2.2 Predicted metagenome in the rice rhizosphere

A predicted metagenome was reconstructed based on the distance between 16S rRNA genes of fully sequenced genomes and the sequence of sliced bands from DGGE gel.

Abundances of functional gene classes, including defense mechanism, replication and recombination and repair, and hypothesis protein, were high in H-PAHs rhizosphere soils at 1 mm from the root surface and L-PAHs rhizosphere soils at a distance of 2 mm from the root surface (Fig. 10). The opposite trend in abundance occurred with the gene classes for RNA processing and modification, inorganic ion transport and metabolism, energy production and conversion, and intracellular trafficking, secretion, and vesicular transport.

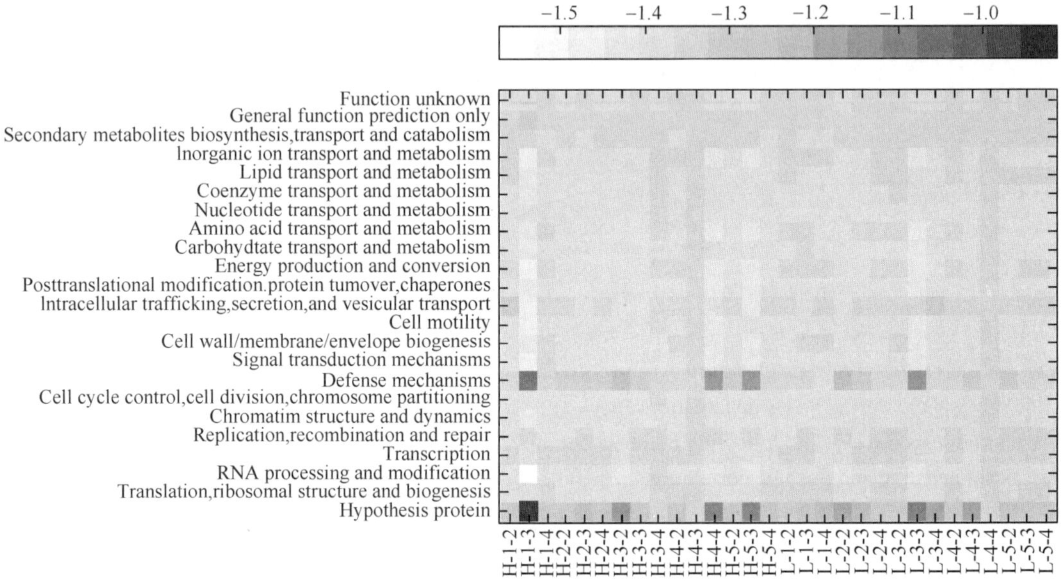

Fig. 10 Heat map indicating difference in relative abundances of gene classes among predicted metagenomes from rice rhizosphere soils during PAHs degradation. The color indicated the logarithmic percentage values for each gene class. The metagenomes labelled with L and H denotes the communities in rhizosphere soils spiked with low and high PAHs concentrations, respectively. The first number after L and H ranging from 1 to 5 denote the distance from root surface. The second number after L and H ranging from 2 to 4 denotes the sampling time at 15, 30 and 45 days after planting, respectively. The horizon and vertical side bar denotes different clusters in samples and genes, respectively (Ma et al., 2015)

Thirty metagenomes were hierarchically clustered based on fifty dominant functional genes, each contributing more than 0.1% to the reconstructed predicted metagenomes. These thirty metagenomes in the rhizosphere soils were clustered into two groups, which were dominated with metagenomes in the rhizosphere soils with high and low PAHs concentrations, respectively. However, some metagenomes in the L-PAHs rhizosphere soils were clustered with metagenomes in soil with high PAHs concentrations, including the metagenomes in L-PAHs rhizosphere soils at 2 mm from the root after 15 days cultivation and at 1, 2 and 4 mm from the root after 30 days cultivation. Reconstructed metagenomes in the H-PAHs rhizosphere soils at 3 to 5 mm from the root after 45 days of cultivation were clustered with the metagenome in soil with low PAHs concentrations. Our previous study showed that the

PAHs degradation rate in the former cluster was greater than in the latter (Ma et al., 2012). The highest degradation rate was detected in L-PAHs rhizosphere soil at 2 mm and in the H PAHs soils at 1 mm from the root. Therefore, the high abundance functional genes in these metagenomes were involved. In metagenomes of low rate PAHs in rhizosphere soil at 2 mm from root surface, the abundance of the functional genes for diaminopimelate epimerase, D-glucarate permease, and oxalate/formate antiproter protein was greater than in other metagenomes. In metagenomes of high rate PAHs at 1 mm from the root surface, the abundance of functional genes for diaminopimelate epimerase, chemotaxis protein methyltransfersase, 4-oxalocrotonate decarboxylase, phospholipase, PP-loop superfamily ATPase, biopolymer transport ExbD-related transmembrane protein, nuclease subunit of the excinucleasecompex protein, and P pilus assembly protein/porin PapC protein were clearly greater than in other metagenomes.

The biodegradation of PAHs catalyzed by microbial enzymes is the most important approach for the remediation of PAHs contamination in the rhizosphere. The pathway of PAHs degradation included three important enzyme categories, dioxygenase, dehydrogenase, andaldolase. The abundance of the dioxygenase gene was greater in both H- and L-PAHs rhizosphere soils at 1 mm from the root surface (Fig. 11). The abundance of the aldolase gene was great in both H- and L-PAHs rhizosphere soils at 1 to 2 mm from the root surface, and also in the H-PAHs rhizosphere soils at 4 mm and 5 mm from the root surface. The abundance of the dehydrogenase gene was similar in all rhizosphere soils.

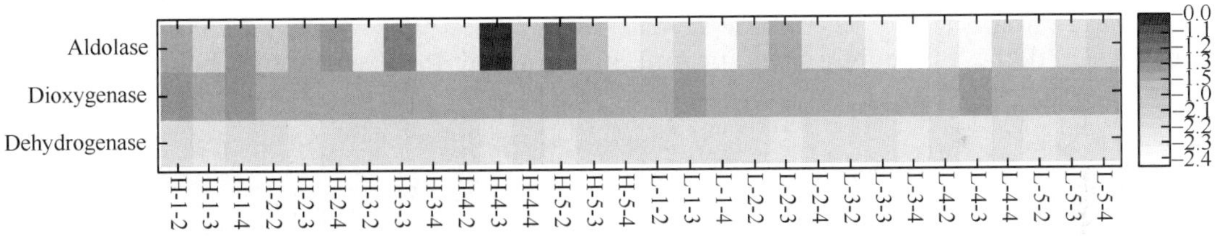

Fig. 11 Heat map indicating difference in relative abundances of aldolase, dioxygenase and dehydrogenase from rice rhizosphere soils during PAHs degradation. The color indicated the logarithmic percentage values for each gene. The metagenomes labeled with L and H denotes the communities in rhizosphere soils spiked with low and high PAHs concentrations, respectively. The first number after L and H ranging from 1 to 5 denotes the distance from root surface. The second number after L and H ranging from 2 to 4 denotes the sampling time at 15, 30 and 45 days after planting, respectively (Ma et al.,2015)

In conclusion, distance to root surface and PAH concentrations affect the microbial communities and metagenomes in rice rhizosphere. The abundance of critical functional genes relating to PAHs degradation in metagenomes mirrored the PAH degradation potential in rice rhizosphere. This finding suggest that the predicted metagenomes reconstructed from 16S rRNA marker gene sequences provide further insights into the spatial variation and dynamics of microbial functioning that occur during bioremediation of organic pollutants in soil (Ma et al., 2015).

6. Biodegradation of PAHs by *Massilia* sp. isolated from soil

Reducing PAHs in the environment is critical to ecosystem and human health, and biodegradation by microorganisms has been generally considered to be one of the primary means for the removal of PAHs from environment (Zeng et al., 2010a; Xu et al., 2013a). The high efficiency PAHs-degrading strain and the relevant biodegradation characteristics were still of great value for improving PAHs bioremediation in environment (Gu et al., 2016). Following the first three PHE-degrading bacteria strains isolated and reported in 1928 (Evans et al., 1965), many other PAHs-degrading bacteria have been isolated and characterized by investigators thereafter. However, yet biodegradation ability of the isolated *Massilia* sp. on OPs remains to be tested. Besides, there exist limited number of reports on isolating and characterizing the *Massilia* sp. in biodegradation of PAHs.

6.1 Isolation of novel strains *Massilia* sp.

The isolation and collection of novel phenanthrene (PHE) -degrading strains *Massilia* sp. were based on the methods of Tian et al. (2008) and Zeng et al. (2010). *Massilia* sp. WF1 and *Massilia* sp. WG5 were isolated from a PAHs-contaminated soil in Wuxi, Jiangsu Province, China (31.604°N, 120.476°E). Scanning electron microscope (SEM) observations indicated that the cells of both WF1 and WG5 were rod shaped and cell size was about 0.3-0.4 μm in width and 2-4 μm in length (Fig. 12). According to the 16S rRNA gene sequences and the phylogeny tree information, both of the two strains belong to genus *Massilia* (Wang et al., 2016).

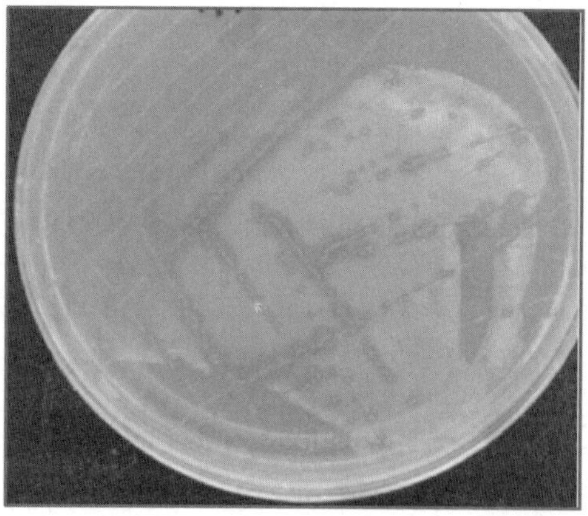

(a) WF1 on solid MSM plate with PHE

(b) The scanning electronic microscope photographs of *Massilia* sp. WF1

Fig. 12 The growth of strain *Massilia* sp. WF1 on solid MSM plate with PHE, the scanning electronic microscope photographs of *Massilia* sp. WF1

6.2 Biodegradation and its trans-membrane transport by *Massilia* sp.

Massilia sp. WF1 was capable of degrading PHE by using it as the sole carbon source and energy (Wang et al., 2016). The PHE of 100 mg/L was almost completely degraded by *Massilia* sp. WF1 over a 2-d period at 28℃ and pH 6.0 (Fig. 13). The kinetics of PHE biodegradation by *Massilia* sp. WF1 was well represented by the Gompertz model. Comparing with the earlier findings, *Massilia* sp. WF1 exhibited a highly effective PHE-degrading ability, and thus it has a greater potential in bioremediation of PHE contaminated environment. The *Massilia* sp. WG5 presented the similar ability for effective PHE degradation when compared with WF1 (the data were not shown). Further, the pH values, culture temperatures and initial PHE concentrations can affect PHE biodegradation. The pH 6.0, temperature 28℃ and initial PHE concentration of 100 mg/L were proved as the optimal PHE degradation conditions of *Massilia* sp. WF1 in a range of pH (5.0-8.0), temperatures (20℃-35℃), and PHE concentrations (25-400 mg/L).

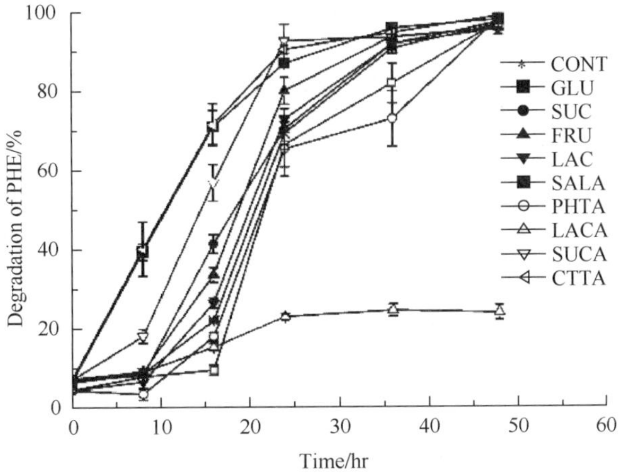

Fig. 13 Dynamics of phenanthrene (PHE) biodegradation by *Massilia* sp. WF1 in different carbon sources treatments. CONT, control treatment (PHE as the sole carbon source) ; GLU, glucose; SUC, sucrose; FRU, fructose; LAC, lactose; SALA, salicylic acid; PHTA, phthalic acid; CITA, citric acid; SUCA, succinic acid; LACA, lactic acid. Error bar represents the standard deviation of triplicate independent measurements

Besides, carbon sources can also affect the efficiency of PHE biodegradation by *Massilia* sp. WF1, such as glucose, citric acid, and succinic acid displaying distinct promotion, while lactic acid producing a strong inhibition effect on PHE biodegradation (Fig. 13). Hence, the synergistic effect on PHE biodegradation along with glucose, citric acid, or succinic acid as co-substrate would then be beneficial for successful bioremediation of PHE contaminated environments by *Massilia* sp. WF1. Salicylic acid (SALA) and phthalic acid (PHTA) were not utilized by *Massilia* sp. WF1 and had no obvious effect on PHE biodegradation (Fig. 13). Only two metabolites, 1-hydroxy-2-naphthoic acid (1H2N) and PHTA, were identified in

PHE biodegradation process. Quantitatively, nearly 27.7% of PHE was converted to 1H2N and 30.3% of 1H2N was further metabolized to PHTA. However, the PHTA pathway was broken and the SALA pathway was ruled out in PHE biodegradation process by *Massilia* sp. WF1 , as shown in Fig. 14 (Wang et al., 2016).

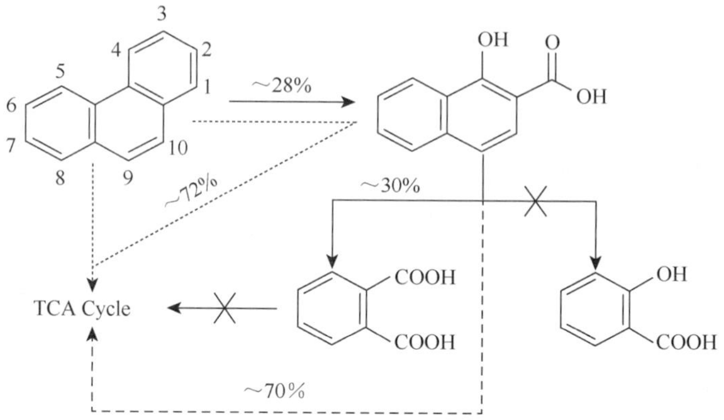

Fig. 14 Proposed biodegradation pathways of phenanthrene by *Massilia* sp. WF1. Percentage represents the conversion rate which deduced from metabolites concentrations. Solid line: the documented pathway; solid line with cross: the pathway is non-operative; dash line: tentative pathway from 1-hydroxy-2-naphthoic acid except converted to phthalic acid; dot line: tentative pathway directly from phenanthrene (Wang et al., 2016)

It was observed that for most microorganisms, the PAHs biodegradation process occurred intracellularly (Xu et al., 2013) and the trans-membrane transport of PAHs was the first step in biodegradation (Li et al., 2014; Fayeulle et al., 2014). Nevertheless, the mechanisms of PAHs entering into microorganisms and the fate of PAHs in cells have received considerably less attention (Xu et al., 2013; Fayeulle et al., 2014). Hence, explorations of the trans-membrane transport mechanism combined with biodegradation of PAHs are of great value in investigating the bioremediation of PAHs. Based on the measured values and regression, the intracellular and extracellular dissipation rate constants of PHE (1.0 mg/L) by *Massilia* sp. WF1 were almost equivalent. The transport rate constant of PHE from outside solution into cells for *Massilia* sp. WF1 were higher than the efflux rate constant of PHE from cells to outside solution. These phenomena combined with the high biodegradation capacity of PHE by *Massilia*sp. WF1 indicated that the amount of PHE that transported into cells decreased mainly due to biodegradation. The inhibition treatments of 4℃ and the presence of sodium azide, colchicine and cytochalasin B did not prevent PHE entering into the cell of *Massilia* sp. WF1, thus, the passive trans-membrane transport mechanism was involved in PHE entering into the cells of *Massilia* sp. WF1 (Gu et al., 2016).

6.3 Complete genome sequence of *Massilia* sp.

In order to further investigate the pathway of PHE degradation in *Massilia* sp. WF1 and

WG5, the complete genome sequence of both strains were performed. The data showed that the two strains have the similar complete genome sequences. The complete genome of *Massilia* sp. WG5 contains a circular chromosome of 5831847 bp with a 65.71% G+C content, and two circular plasmids, one of 203106 bp with 60.68% G+C content and the other of 14623 bp with 63.11% G+C content. Of the 5559 predicted genes, 5140 were protein-coding genes, and 111 were RNA genes. A total of 4922 genes were assigned a putative function. The genes were classified into 22 COGs functional categories. The metabolic networks were found according to KEGG pathway analysis, there are 201 genes involved in the pathway of xenobiotics biodegradation and metabolism. However, in these 201 genes, only 2 genes are related to PHE degradation and only 2 other genes are found in the pathway of PAHs degradation (Lou et al., 2016).

Acknowledgements

This work was supported by the Major Program of National Natural Science Foundation of China (41130532), and the Fundamental Research Funds for the Central Universities.

References

Aaen K N, Holm P E, Priemé A, et al. 2011. Comparison of aerobic and anaerobic [^{3}H] leucine incorporation assays for determining pollution-induced bacterial community tolerance in copper-polluted, irrigated soils. Environmental Toxicology and Chemistry, 30 (3): 588-595.

Allan I J, Booij K, Paschke A, et al. 2009. Field performance of seven passive sampling devices for monitoring of hydrophobic substances. Environmental Science & Technology, 43 (14): 5383-5390.

Anderson T A, Guthrie E A, Walton B T. 1993. Bioremediation in the rhizosphere. Environmental Science & Technology, 27 (13): 2630-2636.

Bamforth S M, Singleton I. 2005. Bioremediation of polycyclic aromatic hydrocarbons: current knowledge and future directions. Journal of Chemical Technology and Biotechnology, 80 (7): 723-736.

Begg C B M, Kirk G J D, Mackenzie A F, et al. 1994. Root-induced iron oxidation and pH changes in the lowland rice rhizosphere. New Phytologist, 128 (3): 469-477.

Bojes H K, Pope P G. 2007. Characterization of EPA' s 16 priority pollutant polycyclic aromatic hydrocarbons (PAHs) in tank bottom solids and associated contaminated soils at oil exploration and production sites in Texas. Regulatory Toxicology and Pharmacology, 47 (3): 288-295.

Choi O, Hu Z. 2008. Size dependent and reactive oxygen species related nanosilver toxicity to nitrifying bacteria. Environmental Science & Technology, 42 (12): 4583-4588.

Delistraty D. 1997. Toxic equivalency factor approach for risk assessment of polycyclic aromatic hydrocarbons. Toxicological & Environmental Chemistry, 64 (1-4): 81-108.

Evans W C, Fernley H N, Griffiths E. 1965. Oxidative metabolism of phenanthrene and anthracene by soil pseudomonads. The ring-fission mechanism. Biochemical Journal, 95 (3): 819.

Fayeulle A, Veignie E, Slomianny C, et al. 2014a. Energy-dependent uptake of benzo [a] pyrene and its cytoskeleton-dependent intracellular transport by the telluric fungus Fusarium solani. Environmental Science and Pollution Research, 21 (5): 3515-3523.

Gao Y, Zhu L. 2005. Phytoremediation for phenanthrene and pyrene contaminated soils. Journal of Environmental Sciences-

Amsterdam, 17 (1): 14-18.

Gu H, Lou J, Wang H, et al. 2016. Biodegradation, biosorption of phenanthrene and its trans-membrane transport by *Massilia* sp. WF1 and phanerochaete chrysosporium. Frontiers in Microbiology, 7.

Guazzaroni M E, Morgante V, Mirete S, et al. 2013. Novel acid resistance genes from the metagenome of the Tinto River, an extremely acidic environment. Environmental Microbiology, 15 (4): 1088-1102.

Gurevitch J, Morrow L L, Wallace A, et al. 1992. A meta-analysis of competition in field experiments. American Naturalist: 539-572.

Habe H, Omori T. 2003. Genetics of polycyclic aromatic hydrocarbon metabolism in diverse aerobic bacteria. Bioscience, Biotechnology, and Biochemistry, 67 (2): 225-243.

Haritash A K, Kaushik C P. 2009. Biodegradation aspects of polycyclic aromatic hydrocarbons (PAHs): A review. Journal of Hazardous Materials, 169 (1): 1-15.

Hayat T, Ding N, Ma B, et al. 2011. Dissipation of pentachlorophenol in the aerobic-anaerobic interfaces established by the rhizosphere of rice (*Oryza sativa* L.) root. Journal of Environmental Quality, 40 (6): 1722-1729.

He Y, Xu J, Tang C, et al. 2005. Facilitation of pentachlorophenol degradation in the rhizosphere of ryegrass (*Lolium perenne L.*). Soil Biology & Biochemistry, 37 (11): 2017-2024.

He Y, Zeng F, Lian Z, et al. 2015a. Natural soil mineral nanoparticles are novel sorbents for pentachlorophenol and phenanthrene removal. Environmental Pollution, 205: 43-51.

He Y, Xia W, Li X, et al. 2015b. Dissipation of phenanthrene and pyrene at the aerobic-anaerobic soil interface: differentiation induced by the rhizosphere of PAH-tolerant and PAH-sensitive rice (*Oryza sativa* L.) cultivars. Environmental Science and Pollution Research, 22 (5): 3908-3919.

Hynes R K, Farrell R E, Germida J J. 2004. Plant-assisted degradation of phenanthrene as assessed by solid-phase microextraction (SPME). International Journal of Phytoremediation, 6 (3): 253-268.

Johnson D L, Anderson D R, McGrath S P. 2005. Soil microbial response during the phytoremediation of a PAH contaminated soil. Soil Biology & Biochemistry, 37 (12): 2334-2336.

Keyte I, Wild E, Dent J, et al. 2009. Investigating the foliar uptake and within-leaf migration of phenanthrene by moss (*Hypnum cupressiforme*) using two-photon excitation microscopy with autofluorescence. Environmental Science & Technology, 43 (15): 5755-5761.

King A J, Readman J W, Zhou J L. 2004. Determination of polycyclic aromatic hydrocarbons in water by solid-phase microextraction-gas chromatography-mass spectrometry. Analytica Chimica Acta, 523 (2): 259-267.

Krauss M, Wilcke W. 2005. Persistent organic pollutants in soil density fractions: distribution and sorption strength. Chemosphere, 59 (10): 1507-1515.

Li W, Zhu X, He Y, et al. 2013. Enhancement of water solubility and mobility of phenanthrene by natural soil nanoparticles. Environmental Pollution, 176: 228-233.

Li Y, Wang H, Hua F, et al. 2014. Trans-membrane transport of fluoranthene by *Rhodococcus* sp. BaP-1 and optimization of uptake process. Bioresource Technology, 155: 213-219.

Lou J, Gu H, Wang H, et al. 2016. Complete genome sequence of *Massilia* sp. WG5, an efficient phenanthrene-degrading bacterium from soil. Journal of Biotechnology, 218: 49-50.

Ma B, Chen H, He Y, et al. 2010a. Evaluation of toxicity risk of polycyclic aromatic hydrocarbons (PAHs) in crops rhizosphere of contaminated field with sequential extraction. Journal of Soils and Sediments, 10 (5): 955-963.

Ma B, He Y, Chen H, et al. 2010b. Dissipation of polycyclic aromatic hydrocarbons (PAHs) in the rhizosphere: Synthesis through meta-analysis. Environmental Pollution, 158 (3): 855-861.

Ma B, Chen H, Xu M, et al. 2010c. Quantitative structure-activity relationship (QSAR) models for polycyclic aromatic hydrocarbons (PAHs) dissipation in rhizosphere based on molecular structure and effect size. Environmental Pollution, 158 (8): 2773-2777.

Ma B, Xu M, Wang J, et al. 2011. Adsorption of polycyclic aromatic hydrocarbons (PAHs) on *Rhizopus oryzae* cell walls: Application of cosolvent models for validating the cell wall-water partition coefficient. Bioresource Technology, 102 (22): 10542-10547.

Ma B, Wang J, Xu M, et al. 2012. Evaluation of dissipation gradients of polycyclic aromatic hydrocarbons in rice rhizosphere utilizing a sequential extraction procedure. Environmental Pollution, 162: 413-421.

Ma B, Lyu X F, Zha T, et al. 2015. Reconstructed metagenomes reveal changes of microbial functional profiling during PAHs degradation along a rice (*Oryza sativa*) rhizosphere gradient. Journal of Applied Microbiology, 118 (4): 890-900.

Ma B, Lv X, He Y, et al. 2016. Assessing adsorption of polycyclic aromatic hydrocarbons on *Rhizopus oryzae* cell wall components with water-methanol cosolvent model. Ecotoxicology and Environmental Safety, 125: 55-60.

Mackova M, Dowling D, Macek T. 2006. Phytoremediation-Rhizoremediation. //Mackova M, Dowling D, Macek T. *Focus on Biotechnology*, 1-3, 5-21, 23-31, 33-55, 57-67, 69-167, 169-225, 227-257, 259-300.

Mueller K E, Shann J R. 2007. Effects of tree root-derived substrates and inorganic nutrients on pyrene mineralization in rhizosphere and bulk soil. Journal of Environmental Quality, 36 (1): 120-127.

Northcott G L, Jones K C. 2001. Partitioning, extractability, and formation of nonextractable PAH residues in soil. 1. Compound differences in aging and sequestration. Environmental Science & Technology, 35 (6): 1103-1110.

Oleszczuk P. 2009. Application of three methods used for the evaluation of polycyclic aromatic hydrocarbons (PAHs) bioaccessibility for sewage sludge composting. Bioresource Technology, 100 (1): 413-420.

Parrish Z D, Banks M K, Schwab A P. 2005. Assessment of contaminant lability during phytoremediation of polycyclic aromatic hydrocarbon impacted soil. Environmental Pollution, 137 (2): 187-197.

Pedersen O, Sand-Jensen K, Revsbech N P.1995. Diel pulses of O_2 and CO_2 in sandy lake sediments inhabited by *Lobelia dortmanna*. Ecology, 76 (5): 1536-1545.

Peng R H, Xiong A S, Xue Y, et al. 2008. Microbial biodegradation of polyaromatic hydrocarbons. FEMS Microbiology Reviews, 32 (6): 927-955.

Sipilä T P, Keskinen A K, Åkerman M L, et al. 2008. High aromatic ring-cleavage diversity in birch rhizosphere: PAH treatment-specific changes of IE 3 group extradiol dioxygenases and 16S rRNA bacterial communities in soil. The ISME Journal, 2 (9): 968-981.

Stampoulis D, Sinha S K, White J C. 2009. Assay-dependent phytotoxicity of nanoparticles to plants. Environmental Science & Technology, 43 (24): 9473-9479.

Su Y H, Zhu Y G. 2008. Uptake of selected PAHs from contaminated soils by rice seedlings (*Oryza sativa*) and influence of rhizosphere on PAH distribution. Environmental Pollution, 155 (2): 359-365.

Tang J, Xiong L, Wang S, et al. 2008. Influence of silver nanoparticles on neurons and blood-brain barrier via subcutaneous injection in rats. Applied Surface Science, 255 (2): 502-504.

Tian Y, Liu H J, Zheng T L, et al. 2008. PAHs contamination and bacterial communities in mangrove surface sediments of the Jiulong River Estuary, China. Marine Pollution Bulletin, 57 (6): 707-715.

Tsao T M, Chen Y M, Wang M K, et al. 2011. Structural transformation and physicochemical properties of environmental nanoparticles by comparison of various particle-size fractions. Soil Science Society of America Journal, 75 (2): 533-541.

van de Vossenberg J, Woebken D, Maalcke W J, et al. 2013. The metagenome of the marine anammox bacterium 'Candidatus Scalindua profunda' illustrates the versatility of this globally important nitrogen cycle bacterium. Environmental Microbiology, 15 (5): 1275-1289.

Wang P, Wang H, Wu L, et al. 2012. Influence of black carbon addition on phenanthrene dissipation and microbial community structure in soil. Environmental Pollution, 161: 121-127.

Wang H, Lou J, Gu H, et al. 2016. Efficient biodegradation of phenanthrene by a novel strain *Massilia* sp. WF1 isolated from a PAH-contaminated soil. Environmental Science and Pollution Research: 1-11.

Weber W J, Kim S H, Johnson M D. 2002. Distributed reactivity model for sorption by soils and sediments. 15. High-concentration co-contaminant effects on phenanthrene sorption and desorption. Environmental Science & Technology, 36 (16): 3625-3634.

Xu N, Bao M, Sun P, et al. 2013. Study on bioadsorption and biodegradation of petroleum hydrocarbons by a microbial consortium. Bioresource Technology, 149: 22-30.

Zeng F, He Y, Lian Z, et al. 2014. The impact of solution chemistry of electrolyte on the sorption of pentachlorophenol and phenanthrene by natural hematite nanoparticles. Science of The Total Environment, 466: 577-585.

Zeng J, Lin X, Zhang J, et al. 2010. Isolation of polycyclic aromatic hydrocarbons (PAHs) -degrading *Mycobacterium* spp. and the degradation in soil. Journal of Hazardous Materials, 183 (1): 718-723.

Zhang Y, Chen Y, Westerhoff P, et al. 2008. Stability of commercial metal oxide nanoparticles in water. Water Research, 42 (8): 2204-2212.

Zhao C, Boriani E, Chana A, et al. 2008. A new hybrid system of QSAR models for predicting bioconcentration factors (BCF). Chemosphere, 73 (11): 1701-1707.

Mitigation and Remediation for Organic Contaminated Soils by Surfactants

Wenjun ZHOU, Lizhong ZHU

Department of Environmental Science, Zhejiang University

The contamination of soils by organic pollutants has become a widespread environmental problem. The mitigation and remediation of organic contaminated soils has become a major serious concern as some organic contaminants are toxic, mutagenic, or carcinogenic and have potential threat on human health. Soil contamination by the organic compounds is a complex process and difficult to treat due to many reasons like the tendency of adsorption of contaminants onto the soil matrix, low water solubility, limited rate of mass transfer for biodegradation and so on (Paria, 2008). Various physical, chemical, biological technologies and their combinations have been attempted to remediate organic contaminated soils. Among them, the surfactant-enhanced technologies have been suggested as promising technologies for the mitigation and remediation of organic contaminated soils (Zhu et al., 2010). Surfactants can enhance the desorption and removal of hydrophobic organic compounds (HOCs) from soil solid into aqueous phases by increase the aqueous-phase solubilities of HOCs via micelle solubilization (Zhu et al., 2010), which would improve the bioavailability of HOCs for microbial remediation and phytoremediation (Zhu et al., 2010). On the other hand, cationic surfactants can enhance the sorption and retention of soils for organic contaminants and then reduce their concentrations and mobility in soil solution, their bioavailability and risk for plants because cationic surfactants can be strongly sorbed by soils and the sorbed surfactant acts as an effective sorbent for HOCs (Zhu et al., 2003). Therefore, surfactants can effectively enhance the mitigation and remediation of organic contaminated soils (Zhu et al., 2010).

1. Mitigation of organic contaminated soils by cationic surfactants

For the slightly contaminated agricultural soils, remediation is not the best choice because of the relative high-cost, especially for areas that are short of cultivated land. How to reduce the bioavailability of organic contaminants and restrain their transposition from soil to plant is an interesting problem. Plants uptake organic contaminants via several pathways: root uptake from the soil, foliar absorption from the atmosphere, and particle-phase deposition onto the waxy cuticle of the leaves (Zhu et al., 2010). Thus, decreasing the effective concentration of organic contaminants in the soil solution and restraining the volatilization of soil contaminants may be the key to reduce the uptake of plants for soil contaminants.

1.1 Interaction of organic contaminants with soil-adsorbed surfactants

Basic studies have found that the sorption of cationic surfactant onto the soil can form a new sorption phase for the organic contaminant and then result in a significant increase of the apparent partition capability of the organic contaminant, which can significantly enhance the sorption of HOCs onto soils and subsequently reduce their concentrations as well as their mobility in the soil solution (Zhu et al., 2010). In the soil-water system with a cationic surfactant, the organic contaminants should be sorbed onto both the natural organic matters and the soil-adsorbed surfactants. The apparent solute distribution coefficient (K_d^*) in solid-water mixtures with a coexisting surfactant may be expressed in the form of (Zhu et al., 2003, 2010)

$$K_d^* = \frac{K_{oc} f_{oc} + K_{ss} f_{soc}}{1 + K_{mn} X_{mn} + K_{mc} X_{mc}} \qquad (1)$$

where f_{oc} and f_{soc} are the natural organic-carbon fraction and the surfactant organic-carbon fraction in the solid, respectively. K_{oc} and K_{ss} are the carbon-normalized solute distribution coefficients with the natural organic matter and the soil-sorbed surfactant in solids, respectively.

Contaminant sorption coefficients with soil-adsorbed surfactants, K_{ss}, were always higher than respective partition coefficients with surfactant micelles (K_{mc}) and natural organic matter (K_{oc}). At low sorbed-surfactant levels, the resulting soil-adsorbed surfactant via the cation-exchange process appears to form a thin organic film, which effectively "adsorbs" the contaminants, resulting in very high K_{ss} values. At high surfactant levels, the sorbed surfactant appears to form a bulk like medium that behaves essentially as a partition phase (rather than an adsorptive surface), with the resulting K_{ss} being significantly decreased and less dependent on the Cationic surfactant loading (Zhu et al., 2003).

1.2 Surfactant enhanced soil retention for organic contaminants

The sorption capacities of soils for cationic surfactant and the subsequent partition of HOC into the soil-sorbed surfactant are the key factors in enhancing soil retention of HOCs. Due to the complexity of soil composition, the sorption capacities of different soil components for cationic surfactant and the soil-sorbed surfactant onto different components for HOCs were of great difference (Lu and Zhu, 2012a). Aside from the clay minerals, soil organic matter such as humic acid (HA) also is a very important adsorbent for the cationic surfactant. The saturated sorption capacity of the HA for dodecylpyridinium bromide (DDPB) is similar to that of montmorillonite, and is about one or two orders of magnitude higher than illite or kaolinite. The sorbed surfactant obviously enhances the soil sorption of PAHs. However, the partition capacity of the sorbed-surfactant on different soil components for HOCs is extremely different due to the different configurations. For each pure soil component, the K_{ss} value for the partition of HOCs onto the soil-sorbed surfactant is approximately linearly related with the surfactant sorption amount in low loading levels.

Cationic-nonionic mixed surfactants also showed great efficiency in enhancing soil retention capability for organic contaminants (Jin et al., 2013). The cationic-nonionic mixed surfactants, DDPB and Triton X-100 (TX100), significantly enhanced soil retention of phenanthrene. The maximum sorption coefficient K_d^* of phenanthrene for contaminated soils treated by mixed surfactants was about 24.5 times that of soils without surfactant (K_d) and higher than the combined effects of DDPB and TX100 individually, which was about 16.7 and 1.5 times K_d, respectively.

The cationic surfactants sorbed on soils also have great influence on the volatilization of PAHs from a paddy soil (Lu and Zhu, 2012b). The cationic surfactant, DDPB, affected both liquid-vapor and solid-vapor processes of PAHs in soil. At DDPB concentrations below the critical micelle concentration (CMC), movement of PAHs from the bulk solution to the gas-liquid interface appeared to be facilitated by interaction between PAHs and the surfactant monomers adsorbed at the gas-liquid interface, promoting the volatilization of PAHs from solution. However, when DDPB was greater than the CMC, volatilization was inhibited due to the solubilization of PAHs by micelles. On the other hand, the sorption of soil-sorbed surfactant significantly inhibited the solid-vapor volatilization of PAHs. The overall effect of the two simultaneous effects of DDPB on liquid-vapor and solid-vapor processes apparently inhibited the volatilization loss of PAHs from soil.

1.3 Reducing plant uptake of PAHs by cationic surfactant-enhanced soil retention

Reducing the transfer of contaminants from soils to plants is a promising approach to produce safe agricultural products grown on contaminated soils. The performance of cationic surfactant-enhanced soil retention in reducing plant uptake of polycyclic aromatic hydrocarbons (PAHs) was investigated (Lu and Zhu, 2009). Two cationic surfactants, CTMAB and DDPB, were separately utilized to enhance the soil retention of PAHs then reduce the available concentrations for three kinds of leafy vegetables. The addition of cationic surfactants significantly retarded desorption of PAHs from soil and the aqueous concentration of PAHs sharply decreased with the increasing dose of cationic surfactants, which were apparently caused by the sorption of surfactants onto soils and the functionality of soil-sorbed surfactant as a sorptive phase for PAHs. The addition of CTMAB or DDPB to contaminated soils also significantly reduced the uptake of PAHs by vegetables (Fig. 1). The maximum decreases of phenanthrene and pyrene under CTMAB treatment were about 66% and 51% for chrysanthemum, 62% and 71% for cabbage, and 34% and 53% for lettuce, respectively. The shoot concentration factors (SCFs) of phenanthrene and pyrene significantly decreased with increasing CTMAB dose. CTMAB was more appropriate to employ than DDPB due to the toxicity of DDPB on plant growth. The most effective dose of CTMAB was 100-200 mg/kg.

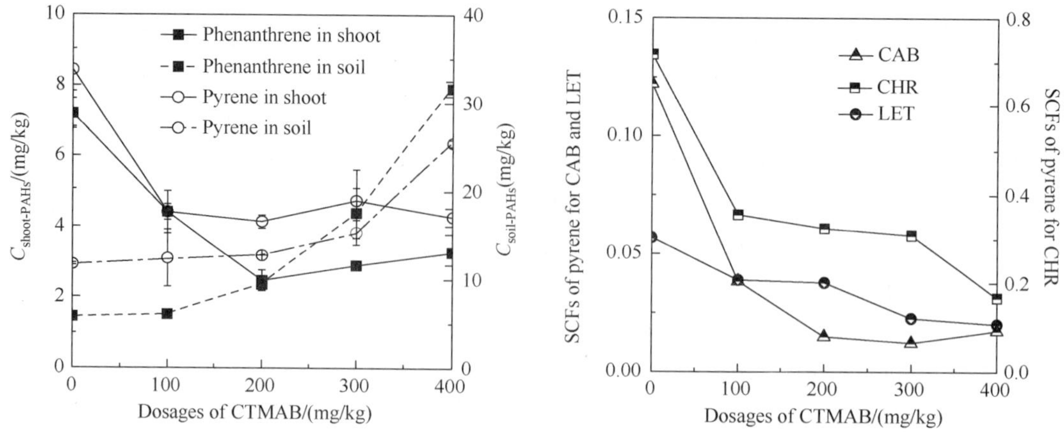

(a) Concentrations of phenanthrene and pyrene in soils and chrysanthemum shoots

(b) Shoot concentration factors pyrene as a function of CTMABdosages.

Fig. 1 Effect of CTMAB dose on PAH uptake in vegetables

2. Surfactant enhanced soil washing of hydrophobic organic compounds (HOCs)

For most heavily contaminated sites, the removal of HOCs from contaminated soils is becoming a major concern as their water-insoluble characteristics may retain those compounds in solid phase and then pose a long-term threat to ecological safety and human health. Surfactant enhanced soil washing have been suggested as a promising technology for the desorption and removal of HOCs from contaminated soils because surfactants can increase the aqueous-phase solubilities of HOCs via micelle solubilization and improve the mass transfer of HOCs from solid into aqueous phases by accumulating the HOCs in the micelles and by decreasing the interfacial tension. However, surfactants can also be adsorbed by the solid matrix and thereby lead to HOCs partitioning into immobile soil-adsorbed surfactants, which would enhance the sorption of HOCs onto soils (Zhou and Zhu, 2005, 2008). Thus, a quantitative evaluation for any potential surfactant remediation approach must consider the sorption of surfactants onto soils and subsequent HOC partitioning to each phase to maximize efficiency and minimize remediation costs.

2.1 Effect of surfactants sorption on the removal of organic contaminants from contaminated soils

With the presence of surfactants in soil-water system, two additional competitive processes affect the distribution of HOCs between the aqueous phase and the solid phase: ①the partition of HOCs into aqueous surfactant micelles in phase and ②the partitionn of HOCs into the soil-sorbed surfactant (Zhou and Zhu, 2005, 2008). Thus, the apparent solute distribution coefficient (K_d^*) defines the solute mass balance of HOCs in solid-water-surfactant system and can be expressed as (Zhou and Zhu, 2005, 2008)

$$K_d^* = \frac{K_d + K_s Q_s}{1 + C_{mic} K_{mc}} \tag{2}$$

where K_d is the intrinsic distribution coefficient of solute between natural organic matter and water without surfactant; Q_s is the sorbed-amount of surfactant on solid; K_s and K_{mc} are the solute distribution coefficient with the solid-sorbed surfactant and the micelle-water partition coefficient of solute in surfactant solution, respectively; C_{mic} is the equilibrium concentration of surfactant micelles in aqueous phase.

The partition of PAHs into various solid-sorbed non-ionic surfactants was investigated (Zhu and Zhou, 2008), which was evaluated with the solute distribution coefficient of K_{sm} on unit surfactant mole basis. The K_{sm} values for PAHs increased with increasing solid-sorbed surfactant and reached a stable maximum values ($K_{sm}{}^{m}$) with surfactants sorption in saturation state, at which the solid surface was completely covered with the surface micelle (or admicelle) and the sorbed surfactant, on a carbon-normalized basis, is more effective as a sorbent for HOCs than soil organic matter. With various nonionic surfactants, the $K_{sm}{}^{m}$ values for a given PAHs were found to have a good linear relationship with the corresponding K_{mc} values of PAHs to surfactant micelles in solution. A model also was developed to describe and predict the distribution of PAHs in solid-water-surfactant systems with the available and measurable system parameter (Zhu and Zhou, 2008).

As the sorption of surfactants onto soils and the partition of HOCs into the soil-sorbed surfactants, the apparent soil-water distribution coefficient of HOCs with surfactant (K_d^*) would increase or decrease relative to the intrinsic distribution coefficient without surfactant (K_d) (Fig. 2 (a)) (Zhou and Zhu, 2005). Depending on the ratio of K_d^* to K_d, the desorption percentage (R_d^*) of HOCs by surfactants also would be less or greater than the intrinsic desorption percentage by water (R_d) and the condition favorable removal of contaminants from a natural soil requires $R_d^* > R_d$ or $K_d^* < K_d$. A term of relative efficiency coefficient (REC), defined as the ratio of R_d^* to R_d with equation (3) can be used to evaluate the performance of surfactant in enhancing desorption for HOCs relative to water. Only when REC>1 ($R_d^* > R_d$ or $K_d^* < K_d$), surfactants would enhance HOCs desorption relative to water (Zhou and Zhu, 2008). With REC=1 ($R_d^* = R_d$ or $K_d^* = K_d$), the corresponding surfactant concentration can be defined as the critical enhanced desorption concentration (CEDC) with equation (4).

$$\mathrm{REC} = \frac{R_d^*}{R_d} = \frac{K_d + \mathrm{W}}{K_d^* + \mathrm{W}} \times 100\% \tag{3}$$

$$\mathrm{CEDC} = C_{aq} + C_{soil} = \frac{K_s Q_s}{K_d K_{mc}} + \mathrm{CMC} + C_{soil} \tag{4}$$

where W is the volume of solution per unit mass of soil; C_{aq} is the equilibrium concentration of surfactant in aqueous solution (M) ; C_{soil} is the concentration loss of surfactant as sorption onto soils. Depending on the surfactant concentration below or above the corresponding CEDC values, the addition of nonionic surfactant would enhance the retardation of soils or promote the removal of HOCs from soil, respectively (Fig. 2 (b)) (Zhou and Zhu, 2005, 2008).

In other words, for the remediation objective, the surfactant initial concentration must be greater than the CEDC. Thus, the CEDC values of surfactant for HOCs are essential for the selection of surfactant dose in surfactant enhanced desorption for the contaminated soils.

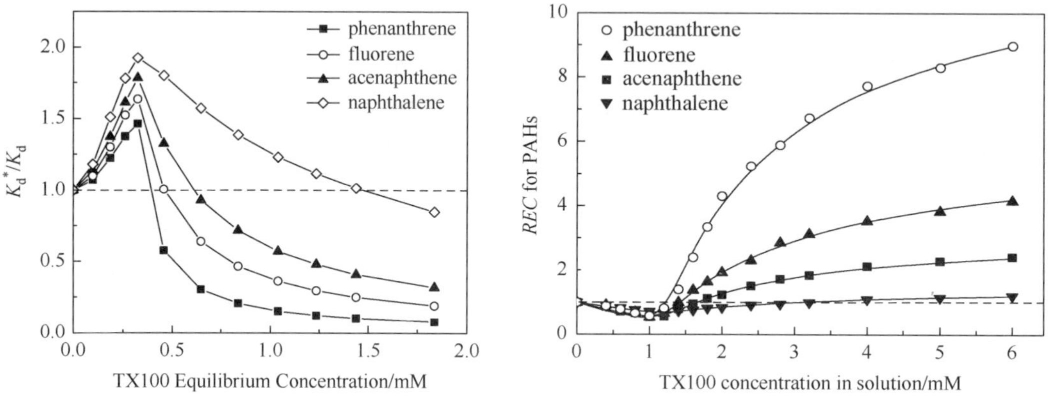

(a) Effect of surfactant on the distribution of PAHs in soil-system

(b) REC values for PAHs as a function of surfactang concentration in solution

Fig. 2 Effect of surfactants sorption on the removal of organic contaminants from contaminated soils

Since the sorption of surfactants onto soils and the corresponding partition of HOCs into soil-sorbed surfactant, the component characteristics of soil-surfactant-HOCs systems would govern the efficiency of surfactant-enhanced desorption in an important way, which may affect the degree of surfactant sorption onto soils, the partition of HOCs into soil-sorbed surfactant, the solubility enhancement of HOCs by surfactant micelles and the distribution of HOCs in soil-water-surfactant system. The effects of soil composition, surfactant structure and PAH properties on the efficiency of surfactant in enhancing PAHs desorption relative to water were investigated (Zhou and Zhu, 2008). It was observed that the sorption capability of soils for nonionic surfactants appeared to depend on the clay content (or surface area) of soils and decreased with an increase in surfactant hydrophilicity (or HLB value). The selection of surfactant for enhancing desorption for contaminated soil should consider not only the solubilization enhancement for HOCs, but also more importantly the sorption of surfactant onto soils and the partition of HOCs into the soil-sorbed surfacatant, both of which are closely related to the surfactant structure. The REC values for the selected PAHs on a given soil appeared to be proportional with the K_{ow} values of compounds. Thus, surfactants are more effective for the desorption of the relative hydrophobic HOCs from the soils with lower clay and higher organic matter content. For the contaminated soils with higher clay content, the relatively hydrophilic surfactants would be more efficient in enhancing HOC desorption, although the solubilization enhancement of such HOCs was relatively weak.

2.2 Enhanced soil flushing of HOCs by anionic-nonionic mixed surfactants

The sorption of surfactant onto soils and the subsequent partitioning of HOCs into the soil-sorbed surfactant will potentially influence the remediation time scales as well as the surfactant addition strategies. An improved strategy to soil washing technology for the remediation of contaminated soils is to reduce the sorption of surfactant onto soils and enhance the removal of HOCs from contaminated soils in order to obtain the optimal remediation efficiency with the minimum surfactant dose.

Mixed surfactants generally exhibit a number of synergistic advantages because of the formation of mixed micelles and could be employed to be more applicable over a wide range of subsurface conditions such as temperature, salinity, and hardness conditions than individual surfactants (Zhou and Zhu, 2004, 2005). The studies on solubilization of PAHs in anionic-nonionic surfactant mixture also showed the higher solubility of PAHs in mixed surfactant systems than those in single surfactant solutions at comparable surfactant concentrations (Zhu and Chiou, 2001; Zhu and Feng, 2003; Zhou and Zhu, 2004, 2005). Thus, the anionic-nonionic mixed surfactant may have a better performance than the individual surfactant in the practical applications of surfactant enhanced washing for contaminated soils.

A new approach using an anionic/nonionic mixed surfactant, sodium dodecyl sulphate (SDS) with Triton X-100 (TX100), was utilized to enhance the desorption and flushing of phenanthrene from contaminated natural soil in an aim to improve the efficiency of surfactant enhanced remediation technology (Zhou and Zhu, 2007a, 2007b). It was observed that the sorption of TX100 onto soil was severely restricted in the presence of SDS in batch and column experiments and decreased with the increasing mole fraction of SDS in mixed surfactant solutions (Fig. 3 (a)) ; meanwhile the apparent solubilities of phenanthrene in mixed surfactant appeared to be greater than that in single TX100 solution. Both resulted in the distribution of phenanthrene in soil-water system with TX100 being reduced with the presence of SDS and the aqueous phase concentration of phenanthrene appeared to be positively related to the mole fraction of SDS in solution. These results can be attributed to the formation of mixed micelles in surfactant solution and the corresponding decrease in the critical micelle concentration of TX100 in mixed solution. Thus, the removal percentages of phenanthrene from the contaminated soil by SDS-TX100 mixed solution were greater than that by individual surfactants and increased with the increasing mole fraction of SDS in mixed surfactant solutions (Fig. 3 (b)). Using anionic-nonionic mixed surfactants (SDS-TX100), the higher flushing efficiency for phenanthrene-contaminated soil can be obtained with the lower surfactant solution volumes in comparison with the individual nonionic surfactants. Thus, anionic/nonionic mixed surfactants may be a better choice for the applications of surfactant remediation technology.

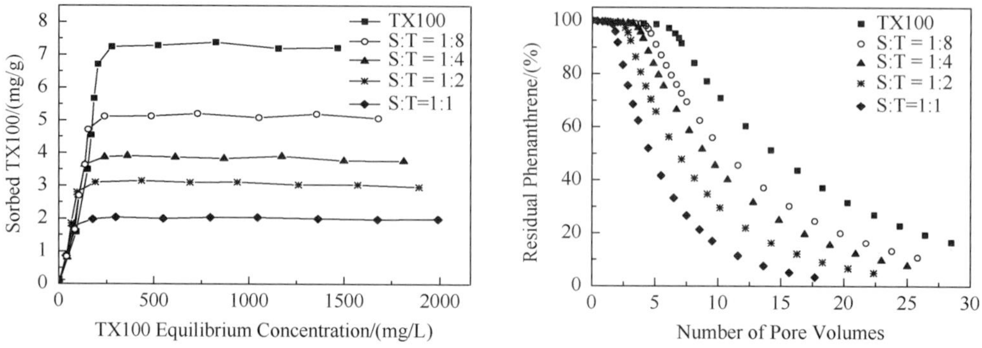

(a) The reduced sorption of TX100 onto soil with the presence of SDS in mixed surfactants

(b) The enhanced removal of phenanthrene from the contaminated soil column by SDS-TX100mixed surfactants

Fig. 3 Enhanced soil flushing of HOCs by anionic-nonionic mixed surfactants

2.3 Recovery of surfactant solution from soil washing solution

During the soil washing process by surfactant solutions, large volumes of waste soil washing solution containing surfactants and HOCs are generated; such a solution pose potential toxic hazards to the environment and must be treated after soil washing. In addition, surfactant-enhanced soil washing is of ten not economically acceptable due to the high cost of the washing agent. These factors are expected to potentially influence the practical implementation of surfactant enhanced soil washing technology. Thus, recovering and then recycling surfactant solutions is one of the most economical and effective strategies for treating waste soil washing solutions and reducing operation costs for surfactant enhanced soil washing technology.

Selective sorption with organobentonite was employed for the removal of polycyclic aromatic hydrocarbons (PAHs) from aqueous surfactant solution as a potential means of recovering surfactant solution after soil washing (Zhou et al., 2013). PAHs can be effectively removed by organo-bentonite from surfactant solutions in a high proportion relative to the sorption loss of surfactant, and surfactant recovery by selective sorption with organobentonite is more feasible for surfactant solution of a lower concentration, higher hydrophilicity, or containing HOCs of higher hydrophobicity. Moreover, organobentonite could be repeatedly used for recovering surfactant solution, and the re-used organobentonite greatly reduced the sorption loss of surfactant and subsequently increased the selectivity for surfactant recovery considerably. The fixed-bed columns with activated carbon can also selectively remove HOCs from a surfactant solution in a high proportion relative to the sorption losses of the surfactant due to the excellent sorption capacity of the solid-sorbed surfactant for HOCs (Zhou et al.,

2015). The high bed heights and low flow rates both are favorable for fixed-bed in selectively removing HOC from surfactant solution by increasing the contact time. Based on the constant-pattern model and the apparent partition coefficient of the HOCs between the solid and aqueous phases, the dynamic sorption process of the surfactant and HOCs in the fixed-bed columns were well described and predicted, which is required for the design of a fixed-bed system in practical application. These results suggested that selective sorption with organobentonite or fixed-bed with activated carbon can be used as effective means of treating washing effluents and recovering surfactant solutions in the surfactant-enhanced flushing of HOC-contaminated soils.

3. Surfactant enhanced bioremediation of organic contaminated soils

Bioremediation including microbial remediation and phytoremediation has attracted great interest for in situ treatment of HOC contaminated soils for economic and environmental reasons. However, HOCs in contaminated soils usually exhibit low bioavailability to both microorganisms and plants due to their strong adsorption to the soil matrix, which limits the bioremediation of contaminated soils with HOCs and is a key problem to solve for improving bioremediation efficiency of contaminated soils with HOCs. Since surfactant can increase the HOCs aqueous-phase concentration via micelle solubilization and the mobilization of HOCs from solid into aqueous phases, the addition of a surfactant to contaminated soil with HOCs might improve the bioavailability of HOCs in soil, facilitating their biodegradation and plant uptake.

3.1 Surfactant enhanced plant uptake of HOCs

Phytoremediation is a potential “green”, cost-effective, and secure remediation technology that uses plants for the in-situ cleanup of contaminated soils. Plant uptake is one of the important pathways for the removal of HOCs from the soil environment. Uptake and accumulation of phenanthrene and pyrene by 12 plant species grown in various treated soils were comparatively investigated (Gao and Zhu, 2004). Root concentrations or root concentration factors (RCFs) of phenanthrene and pyrene were significantly positively correlated to root lipid contents, indicating that root lipid content might be a good predictor of root accumulation of these compounds. For the same soil-plant treatment, shoot concentrations and concentration factors of phenanthrene and pyrene were generally much lower than root.

For predicting and evaluating the plant uptake of contaminants, a partition limited mode was proposed by Chiou et al. in considering the process as essentially a series of partition uptakes (Zhu et al., 2010). In this respect, the concentration of a contaminant (C_{pt}) in a whole plant or a specific part can be expressed as

$$C_{pt}=\alpha_{pt}+C_w\,(f_{pom}K_{pom}+f_{pw}) \qquad (5)$$

where C_w is the contaminant concentration in external water, K_{pom} is the partition coefficient of the contaminant between plant organic matter and water, f_{pom} is the total weight fraction of the plant organic matter, and f_{pw} is the water weight fraction. α_{pt} is a quasi-equilibrium factor.

The performance of the partition-limited model in predicting plant uptake of polycyclic aromatic hydrocarbons (PAHs) by a wide variety of plant species was evaluated using a greenhouse study (Zhu and Gao, 2004; Gao et al., 2005; Yang and Zhu, 2007; Zhang and Zhu, 2009). The model predictions of root or shoot concentrations for tested plant species were all within an order of magnitude of the observed values. Modeled root concentrations appeared to be more accurate than modeled shoot concentrations. The α_{pt} values for various root uptake of PAHs and the root concentration factors (RCFs) both exhibited significantly correlations with root lipid contents (Zhu and Gao, 2004; Gao et al., 2005). However, it was found that the simulated concentrations in shoots generally exhibited a bigger error than those in roots, and the accuracy of simulated shoot concentrations would be greatly enhanced with the modified model, in which the impact of foliar uptake was excluded (Yang and Zhu, 2007). In addition, it also was found that the sorption of PAHs to ryegrass root was actually regulated by both carbohydrates and lipids rather than lipids individually and the predominant weight fraction of carbohydrates in plant root led to the relative higher contribution of carbohydrates than lipids although the affinity of PAHs for carbohydrates was lower than that for lipids (Zhang and Zhu, 2009). The improved sorption model based on the integral roles of carbohydrates and lipids was suggested and showed excellent accuracy for the prediction of sorption of PAHs to plants.

The effect of surfactants on the uptake of PAHs by ryegrass in hydroponic system was studied (Gao et al., 2004; Zhu and Zhang, 2008; Sun and Zhu, 2009). The presence of rhamnolipids or Tween 80 with relatively low concentrations promoted the plant uptake of phenanthrene and pyrene by ryegrass, whereas surfactants with higher concentrations would inhibit the plant uptake of these compounds (Gao et al., 2004; Zhu and Zhang, 2008). The increase of permeability of ryegrass root cells with the increase of surfactant concentration may lead to the initial enhancement of PAH content in ryegrass roots, and the decrease of PAH adsorption onto the root surface with further increase of surfactant led to the decrease of PAH content in ryegrass roots (Zhu and Zhang, 2008). SDBS-TX100 mixed surfactants with certain compositions and concentrations could also enhance the uptake of phenanthrene and pyrene by ryegrass, which could be attributed to the improved uptake capacity of ryegrass roots for phenanthrene and pyrene (Sun and Zhu, 2009). SDBS-TX100 mixed surfactants can enhance the uptake of phenanthrene and pyrene by ryegrass in a wider range of surfactant concentrations in comparison with corresponding single surfactants, and the maximal contents of phenanthrene and pyrene in ryegrass roots were obtained with the concentrations of SDBS-TX100 around the corresponding critical micelle concentrations. SDBS-Tween 80 mixed surfactants can also enhance the plant uptake of PAHs by ryegrass from contaminates soils (Ni et al., 2014). In the concentration range of 60-150 mg/kg, both ryegrass roots and shoots could accumulate 2-3 times the phenanthrene and pyrene with mixed surfactants than with single Tween 80. These results may be explained by the lower sorption loss and reduced interfacial tension of mixed surfactants relative to Tween 80, which enhanced the bioavailability of PAHs in soils.

3.2 Surfactant enhanced microbial degradation of HOCs

Surfactant-enhanced bioremediation (SEBR) has been considered as an effective remediation technique for contaminated soils with HOCs as surfactants can improve the bioavailability of HOCs for microorganisms by increasing their aqueous phase concentration via micelle solubilization and the mobilization of HOCs from solid into aqueous phases (Zhang and Zhu, 2010, 2014). Thus, an improved strategy for SEBR is to enhance the removal of HOCs from contaminated soils and then to improve the biodegradation rate of HOCs in order to obtain optimal remediation efficiency. It has been observed that anionic-nonionic mixed surfactants can promote HOC biodegradation in comparing with single nonionic surfactant as mixed surfactants can reduce the sorption of nonionic surfactants onto soils and then enhance HOC desorption from contaminated soils (Zhao et al., 2005; Yu et al., 2007). However, surfactants may also promote or inhibit the biodegradation of HOCs with microorganisms by affecting the interfacial interaction between HOCs with microorganisms (Zhang and Zhu, 2014). Thus, it is necessary to investigate the interfacial interaction of HOCs with microorganisms and then to reveal the role mechanisms of surfactants in promoting or inhibiting HOC biodegradation, which is essential for evaluating the performance of surfactants in enhancing HOC biodegradation with microorganisms.

The effects of different surfactants on the sorption and biodegradation of pyrene by *Klebsiella oxytoca* PYR-1, as well as their interactions with bacterial cell surface and membrane lipids were investigated (Zhang and Zhu, 2012; Zhang et al., 2013). It was found that surfactants enhanced or inhibited pyrene biodegradation depending on their effects on the sorption of pyrene onto bacterial cell, which occurred mainly through modifying cell surface hydrophobicity or disrupting bacterial membrane, respectively. A relatively high positive correlation was observed between biodegradation promotion and enhancement of sorption coefficients for pyrene in the presence of surfactant, indicating that surfactant-induced sorption played the dominant role during pyrene biodegradation and magnitude of influence depended on their effects on the sorption of pyrene onto bacterial cell. The improved biodegradation of pyrene was mainly due to surfactant-facilitated sorption, and the changes of cell surface hydrophobicity (CSH) suggested that sorbed surfactant take place of hydrophilic sites of bacterial surface and facilitated HOC-biomass contact. Different type surfactants imposed various impacts on the biodegradation of organic pollutants via different mechanisms such as modification of CSH and disrupting bacterial membranes.

In order to better understand how surfactants affect biodegradation of HOCs, the effects of surfactants on the cell surface hydrophobicity (CSH) and microbial activities were investigated (Li and Zhu, 2012). The effects of surfactants on the microbial CSH depend on the interfacial interaction between cell surface and surfactant molecules. Surfactants can enhance lipopolysaccharide (LPS) release by the direct removal of surfactant solubilization for LPS or detergent properties of surfactants. The removal of LPS have exposed the outer membrane to the aqueous phase and then resulted in the increased hydrophobic character of the cell surface, which facilitated the uptake of PAHs by direct contact between hydrophobic cells and hydrophobic HOC molecules. This allowed the penetration of PAH molecules into the intracellular environment by passive diffusion and then catechol 1, 2-dioxygenase (C12) activities were enhanced by the increasing intracellular concentrations of PAHs, meanwhile electron transport system (ETS) activities were also enhanced. Thus, enhanced LPS release probably increased ETS and C12 activities as well as phenanthrene biodegradation by increasing CSH. Therefore, surfactant-enhanced CSH provides a simple, yet effective strategy in the future field applications of surfactant-enhanced bioremediation of HOCs.

The effects of the surfactants on a membrane's fatty acid composition and the transmembrane transport of phenanthrene were also investigated (Li and Zhu, 2014). Surfactants could modify the composition of fatty acids of *Citrobacter* sp. Strain SA01 cells and increase the amount of unsaturated fatty acids. An increased amount of unsaturated fatty acids corresponded to greater membrane fluidity and the increased unsaturated fatty acids promoted phenanthrene to partition from the extracellular matrix to cell debris, which increased reverse partitioning from the cell debris to the cytochylema. As a result, increased membrane fluidity enhanced transmembrane tansport of phenanthrene, which is a limiting step for HOC biodegradation as the transmembrane transport of phenanthrene is less favorable than the phenanthrene partitioning between the aqueous phase and cell debris. As expected, surfactant addition is a simple and effective method to accelerate the rate-limiting step of transmembrane transport of HOCs in bioremediation.

To advance our understanding of surfactant-enhanced biodegradation at the gene level, the expression levels of three genes of *Arthrobacter* sp. SA02 in the biodegradation of phenanthrene were investigated. The Δ9 *fatty acid desaturase* gene codes for Δ9 fatty acid desaturase, which can convert saturated fatty acid to its unsaturated form by introducing double bonds into bacterial fatty acids through the aerobic pathway of fatty acid biosynthesis. The ring-hydroxylating dioxygenase (*RHDase*) and the 1-hydroxyl-2-naphthoate dioxygenase (*1H2Nase*) genes code for the RHDase and 1H2Nase enzymes, respectively, which play a key role in decomposing doubly hydroxylated aromatic compounds. Along with the increasing biodegradation of phenanthrene as surfactant concentrations increased, the expression levels of the Δ9 *fatty acid desaturase*, *RHDase*, and *1H2Nase* genes were significantly increased. The expression level of the Δ9 *fatty acid desaturase* gene presented a good linear correlation with the saturated/unsaturated ratio and was almost perfectly linearly related to that of the *RHDase* and *1H2Nase* genes, which suggested an interaction effect among the three genes. On the basis of the genetic and physiological changes, a pathway was proposed to link the gene expression of the surfactant-enhanced biodegradation of HOCs with the changes in the cell surface hydrophobicity (CSH), membrane fluidity, and intracellular biodegradation. Surfactants can decrease the lipoteichoic acid (LTA) content in the cell membrane, which in turn increases the CSH and membrane fluidity by regulating the expression level of the Δ9 *fatty acid desaturase* gene, which increased the content of the unsaturated fatty acid and in turn promoted the partition and transmembrane transport of phenanthrene, bringing more HOC molecules to the cell and up-regulating the expression levels of degradation genes, *RHDase* and *1H2Hase* genes (Fig. 4).

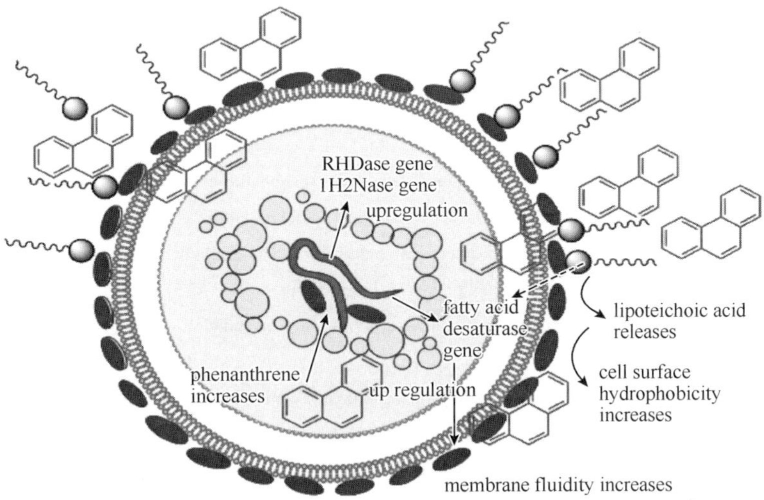

Fig. 4 The gene expression of surfactant-enhanced biodegradation of hydrophobic organic compounds

References

Gao Y, Zhu L. 2004. Plant uptake, accumulation and translocation of phenanthrene and pyrene in soils. Chemosphere, 55 (9): 1169-1178.

Gao Y, Zhu L, Ling W. 2005. Application of the partition-limited model for plant uptake of organic chemicals from soil and water. Science of the Total Environment, 336 (1): 171-182.

Gao Y, Zhu L, Hu C, et al. 2004. Effects of Tween 80 on plant uptake of phenanthrene and pyrene from water. Acta Scientiae Circumstantiae (in Chinese), 24 (4): 713-718.

Jin H, Zhou W, Zhu L. 2013. Utilizing surfactants to control the sorption, desorption, and biodegradation of phenanthrene in soil-water system. Journal of Environmental Sciences, 25 (7): 1355-1361.

Li F, Zhu L. 2012. Effect of surfactant-induced cell surface modifications on electron transport system and catechol 1, 2-dioxygenase activities and phenanthrene biodegradation by *Citrobacter* sp. SA01. Bioresource Technology, 123: 42-48.

Li F, Zhu L. 2014. Surfactant-modified fatty acid composition of *Citrobacter* sp. SA01 and its effect on phenanthrene transmembrane transport. Chemosphere, 107: 58-64.

Li F, Zhu L, Wang L, et al. 2015. Gene expression of an Arthrobacter in surfactant-enhanced biodegradation of a hydrophobic organic compound. Environmental Science & Technology, 49 (6): 3698-3704.

Li F, Zhu L, Zhang D. 2014. Effect of surfactant on phenanthrene metabolic kinetics by *Citrobacter* sp. SA01. Journal of Environmental Sciences, 26 (11): 2298-2306.

Lu L, Zhu L. 2009. Reducing plant uptake of PAHs by cationic surfactant-enhanced soil retention. Environmental Pollution, 157 (6): 1794-1799.

Lu L, Zhu L. 2012. Effect of a cationic surfactant on the volatilization of PAHs from soil. Environmental Science and Pollution Research, 19 (5): 1515-1523.

Lu L, Zhu L. 2012. Effect of soil components on the surfactant-enhanced soil sorption of PAHs. Journal of Soils and Sediments, 12 (2): 161-168.

Ni H, Zhou W, Zhu L. 2014. Enhancing plant-microbe associated bioremediation of phenanthrene and pyrene contaminated soil by SDBS-Tween 80 mixed surfactants. Journal of Environmental Sciences, 26 (5): 1071-1079.

Paria S. 2008. Surfactant-enhanced remediation of organic contaminated soil and water. Advances in Colloid and Interface Science 138, 24-58.

Sun L, Zhu L Z. 2009. Effect of anionic-nonionic mixed surfactant on ryegrass uptake of phenanthrene and pyrene from water. Chinese Science Bulletin, 54 (3): 387-393.

Yang Z, Zhu L. 2007. Performance of the partition-limited model on predicting ryegrass uptake of polycyclic aromatic hydrocarbons. Chemosphere, 67 (2): 402-409.

Yu H, Zhu L, Zhou W. 2007. Enhanced desorption and biodegradation of phenanthrene in soil-water systems with the presence of anionic-nonionic mixed surfactants. Journal of Hazardous Materials, 142 (1): 354-361.

Zhang D, Zhu L. 2014. Controlling microbiological interfacial behaviors of hydrophobic organic compounds by surfactants in biodegradation process. Frontiers of Environmental Science & Engineering, 8 (3): 305-315.

Zhang D, Zhu L. 2012. Effects of Tween 80 on the removal, sorption and biodegradation of pyrene by Klebsiella oxytoca PYR-1. Environmental Pollution, 164: 169-174.

Zhang D, Zhu L, Li F. 2013. Influences and mechanisms of surfactants on pyrene biodegradation based on interactions of surfactant with a Klebsiella oxytoca strain. Bioresource Technology, 142: 454-461.

Zhang M, Zhu L. 2009. Sorption of polycyclic aromatic hydrocarbons to carbohydrates and lipids of ryegrass root and implications for a sorption prediction model. Environmental Science & Technology, 43 (8): 2740-2745.

Zhang M, Zhu L. 2010. Effect of SDBS-Tween 80 mixed surfactants on the distribution of polycyclic aromatic hydrocarbons in soil-water system. Journal of Soils and Sediments, 10 (6): 1123-1130.

Zhao B, Zhu L, Li W, et al. 2005. Solubilization and biodegradation of phenanthrene in mixed anionic-nonionic surfactant solutions. Chemosphere, 58 (1): 33-40.

Zhou W, Wang X, Chen C, et al. 2013. Removal of polycyclic aromatic hydrocarbons from surfactant solutions by selective sorption with organo-bentonite. Chemical Engineering Journal, 233: 251-257.

Zhou W, Yang Q, Chen C, et al. 2015. Fixed-bed study and modeling of selective phenanthrene removal from surfactant solutions. Colloids and Surfaces A: Physicochemical and Engineering Aspects, 470: 100-107.

Zhou W, Zhu L. 2005. Solubilization of polycyclic aromatic hydrocarbons by anionic-nonionic mixed surfactant. Colloids and Surfaces A: Physicochemical and Engineering Aspects, 255 (1): 145-152.

Zhou W, Zhu L. 2007. Efficiency of surfactant-enhanced desorption for contaminated soils depending on the component characteristics of soil-surfactant-PAHs system. Environmental Pollution, 147 (1): 66-73.

Zhou W, Zhu L. 2007. Enhanced desorption of phenanthrene from contaminated soil using anionic/nonionic mixed surfactant. Environmental Pollution, 147 (2): 350-357.

Zhou W, Zhu L. 2008. Enhanced soil flushing of phenanthrene by anionic-nonionic mixed surfactant. Water Research, 42 (1): 101-108.

Zhou W, Zhu L. 2008. Influence of surfactant sorption on the removal of phenanthrene from contaminated soils. Environmental Pollution, 152 (1): 99-105.

Zhu L, Chen B, Tao S. 2004. Sorption behavior of polycyclic aromatic hydrocarbons in soil-water system containing nonionic surfactant. Environmental Engineering Science, 21 (2): 263-272.

Zhu L, Chiou C T. 2001. Water solubility enhancements of pyrene by single and mixed surfactant solutions. Journal of Environmental Sciences, 13 (4): 491-496.

Zhu L, Gao Y. 2004. Prediction of phenanthrene uptake by plants with a partition-limited model. Environmental Pollution, 131 (3): 505-508.

Zhu L, Lu L, Zhang D. 2010. Mitigation and remediation technologies for organic contaminated soils. Frontiers of Environmental Science & Engineering in China, 4 (4): 373-386.

Zhu L, Zhang M. 2008. Effect of rhamnolipids on the uptake of PAHs by ryegrass. Environmental Pollution, 156 (1): 46-52.

Zhou W, Zhu L. 2005. Distribution of polycyclic aromatic hydrocarbons in soil-water system containing a nonionic surfactant. Chemosphere, 60 (9): 1237-1245.

Zhu L, Zhou W. 2008. Partitioning of polycyclic aromatic hydrocarbons to solid-sorbed nonionic surfactants. Environmental Pollution, 152 (1): 130-137.

Phytoremediation of Polychlorinated Biphenyl Contaminated Soil by Transgenic Alfalfa Associated Bioemulsifier AlnA

Hejun REN, Yan WAN, Yongsheng ZHAO

Key Laboratory of Groudwater Resources and Environment of the Ministry of Education, College of Environment and Resources, Jilin University

Abstract

Polychlorinated biphenyls (PCBs) accumulation in the environment poses great risks to human and animal health. the effective strategies for disposal of PCBs are urgently needed. In this study, we generated transgenic alfalfa plants expressing bacteria *bphC* gene and characterized desorption efficiencies of PCBs in soil by bioemulsifier protein AlnA. The remediation efficiency of PCBs-contaminated soils by the combination of AlnA and alfalfa expressing *bphC* was investigated also. Results showed the phytoremediation trials by transgenic alfalfa associated AlnA had considerable application potential in remediation of PCBs pollution.

1. Introduction

Polychlorinated biphenyls (PCBs) are a class of synthetic organic compounds characterized with high hydrophobicity, toxicity, and bioaccumulation. Though production bans, PCBs remain a ubiquitous contaminant worldwide due to extremely slow biodegradation for semi-volatile, low solubility in water, and high tendency of soils or sediments adsorption (Åslund et al., 2007) Therefore, effective strategies are needed for remediationof PCBs in environment.

Comparing with physical and chemical methods, phytoremediation is now emerging as a promising strategy and attracting much attention due to its advantages of being less expensive (the fact that it is carried out in situ), environmentally sustainable, and aesthetically pleasing (Krämer, 2005). Many studies showed alfalfa (*Medicago sativa* L.) is an ideal natural resource for remediation of contaminated soils with a variety of elite characteristics, such as highly productive biomass, drought tolerance, fast growing, and available in large amount during several months of the year (Zhang et al., 2007). Nevertheless, recent study have shown the removal rate of PCBs using wild alfalfa is inadequate and slow (Zeeb et al., 2006). The primary reason is that alfalfa lack the necessary enzymatic machinery involved in bacteria or mam-mals for efficient cleavage of aromatic structure. Alternatively, there is increasing in the efficiency of phytoremediation, which will be greatly enhanced by transgenic plants bearing bacterial genes such as 2, 3-dihydroxybiphenyl-1, 2-dioxygenase (*bphC*), involved in xenobiotic metabolism, leading to a wider application in the field (Novakova, et al., 2010). Whereas, there are scarcely reports about transgenic alfalfa plants designed for phytoremediation of contaminated soil with PCBs.

Generally, PCBs has lower solubility, more adsorptive to soil and more persistent with increasing in the degree of chlorination (Borja et al., 2005), which is also one of the reasons for low phytoremediation efficiency of PCBs in the environmental. Therefore, enhancing soil washing for high chlorinated PCBs by solubilization with surfactants is more important. Unfortunately, the large-scale application of synthetic surfactants is limited by their environmental toxicity, unstable chemical properties, and expensive cost (Xia et al., 2009). The use of biosurfactants is particular interest because of their ecofriendly and biodegradable properties. Among the readily available biosurfactants, the bioemulsifier protein AlnA produced by *Acinetobacter radioresistens* can enhancing the solubility of PAHs through the process of intercalation has been documented (Toren et al., 2002). We have recently found that the recombinant AlnA protein as a solubilizing agent can significantly enhance the extractability of PCBs from contaminated soils (Wang et al., 2014). It seems to be a desired candidate for feasibility of industrial production and large-scale application. However, further characterization of the performance of AlnA in highly-complex soil was necessary.

This report attempted to generate transgenic alfalfa plants expressing bacteria 2, 3-dihydroxybiphenyl-1, 2-dioxygenase gene (*bphC*. B) and focused on the potential feasibility of using AlnA to promote the disposal of PCBs in soil by transgenic alfalfa. Moreover, soil microbial community profiles were characterized to determine the microbial response during the phytoremediation of PCBs.

2. Materials and methods

2.1 Chemicals, plants and soil

PCB standards were purchased from AccuStandard Inc., USA. Alfalfa seeds (*Medicago sativa* L. cv. Gongnong No. 1) were kindly provided by Jilin Academy of Agricultural Sciences, China. Soil samples were collected from a trial plot of Jilin University in Changchun, Jilin Province. Transformer oil was dissolved in acetone and then spiked into the soils. These artificially contaminated soils were stirred vigorously for 30 min to promote the homogeneous distribution of PCBs and aged for 1 month at 25 ℃ before pot experiments.

2.2 Transformation and selection of alfalfa transformants lines

The gene fragment coding *bphC*. B (cloned from a soil metagenomic library) was inserted into a plant binary expression vector pCAMBIA 3301 by replacing the GUS gene between CaMV 35S promoter and terminator The p3301-*bphC*. B and null p3301 plasmids were introduced into *Agrobacterium tumefaciens* EHA105 by freeze-thaw method and subsequently transformed into alfalfa using the leaf-dipping method (Montague et al., 2007). Regenerated shoots from embryos with 3-5 roots were selected on solid MS medium containing Basta (Phosphinothricin) at 2 mg/L for 10 days and then transplanted on solid MS media containing timentin 200 mg/L. Asexual reproductions of transgenic alfalfa lines (T_0) were carried out by the cutting method for continuous passage culture.

2.3 PCBs solubilization and desorption by recombinant protein AlnA

alnA was cloned from *Acinetobacter radioresistens* (ACCC01656) based on the nucleotide acid sequences of OmpA-like protein gene (AY033946.1). The full length 981 bp (minus the signal peptide) open reading frame (ORF) of *alnA* fused to an N-terminal 6-His-tag in pET28a (+) (Clontech, Japan), and the final construct was transformed in *E. coli* BL21 (DE3) plysS (Tiangen, China) using a heat shock method. Recombinant AlnA was purified according to Toren et al (2002). 100 μg of PCB congeners dissolved respectively in hexane were crystallized in the bottom of 1 mL quartz cuvette. The amounts of the solubilized PCBs were determined at 25 ℃ I n a total volume of 1 mL of assay buffer (20 mM Tris, pH 8.0) containing 20 μg of proteins added to the cuvette by a SHIMADZU 2550 spectrophotometer (SHIMADZU, Japan). The desorption of PCBs in the presence of AlnA was determined by combining exactly 1 g of PCBs contaminated soils with 10 mL of an aqueous 0.4 mM AlnA, Tween-80 and cyclodextrin in 100 mL glass vials.

2.4 Pot experiments design

The pot experiments were conducted in an illumination incubator (15/25 ℃, 8 h dark/16 h light). 6 kg (dry wt) of experimental soil was placed in each pot. Three treatments consisting of 10 alfalfas were established: ①soil planted with transgenic alfalfa and irrigated with 200 mL of 10 g/L AlnA protein (TAA), ②soil planted with transgenic alfalfa (TA), and ③soil planted with WT alfalfa (WA). Unplanted soil served as a control (NP). The pots were watered as needed and fertilized every 2 weeks with Hoagland's solution. After 60 days of plant growth, the rhizospheric soils and plants were sampled.

2.5 Quantitative analysis of PCBs

All samples were analyzed in an Agilent GC-MS-6890/5973 equipped with a DB-5MS capillary column (30 m × 0.25 mm × 0.25 μm) for the determination of PCB congener concentrations. Quantification of PCBs was based on external calibration standards containing known concentration of PCB congeners (PCB 3, 5, 28, 52 and 101).

3. Results and discussion

3.1 Combination PCB/2, 4-DCP tolerance in transgenic alfalfa plants

To confirm the present of *bphC*. B gene in transgenic alfalfa plants, total DNA and RNA from the seedlings were extracted, respectively. DNA-PCR and RT-PCR analysis were subsequently performed. As expected, a 0.9-kb DNA product was detected in the Basta-resistant independent transgenic alfalfas (BB3, BB4, BB9, BB11, BB22, and BB27), but not in wild-type. Transgenic line BB11 with healthy growth and high activity of *bphC* enzyme was chosen for removal efficiency of plants to PCBs. Alfalfa plants tolerance to the combination of PCB 5 and 2, 4-DCP was assayed using transgenic plant BB11 planted in glass flasks for 4 weeks. As seen in Fig. 1, no visual evidence of damage was observed for transgenic plant BB11 and it exhibited strong resistance upto 120/50 mg/L of PCB 5/2, 4-DCP complex contaminants (Fig.1 (a)). Transgenic plants thrived well in the coexistence of PCB 5 (80 mg/L) and 2, 4-DCP (25 mg/L), whereas, non-transgenic control plants showed marked symptoms of severe damage possibly induced by the complex pollutants and were almost dead (Fig.1 (b)). As expected, the transgenic alfalfa BB11 expressing *bphC*. B was found to show cross-tolerance to PCBs and 2, 4-DCP in vegetative growth stages.

0/0mg/L 40/10mg/L 80/25mg/L 120/50mg/L
(a) Transgenic alfalfa plant BB11

0/0mg/L 40/10mg/L 80/25mg/L 120/50mg/L
(b) Wild-type alfalfa plants

Fig. 1 Transgenic alfalfa plant tolerance to PCB 5 and 2, 4-DCP (Wang et al., 2015)

Transgenic alfalfa plant BB11 and wild-type alfalfa plants with PCB 5/2, 4-DCP at 0/0, 40/10, 80/25, and 120/50 mg/L, respectively. PCB/2, 4-DCP tolerance in transgenic alfalfa lines was observed as plant growth

3.2 Solubilization and desorption of PCBs by AlnA

The solubility of 7 PCB congeners was examined in 0 and 20 μg/mL of AlnA respectively. In all cases, equilibrium was reached in about 1 h. It was demonstrated that AlnA could greatly enhancethe aqueous solubility of PCBs (Fig. 2 (a)). The solubility of PCBs was different significantly varying from 4.56 (PCB 3) to 32.11 (PCB 52) μg/mL and the ratio of PCBs solubilized by AlnA compared to buffer, from 1.87 (PCB 1) to 6.12 (PCB 52). The highest solubility rate of PCBs could reach 32% (PCB 52) in the present of 20 μg/mL AlnA. In the desorption experiment, AlnA was comparable to Tween-80 (a nonionic surfactant) and cyclodextrin (a biosurfactant). The percent of PCBs desorbed from the soil amended with AlnA was 41%-76%, significantly greater than that with Tween-80 and cyclodextrin at the same concentration of 0.4 mM (Fig.2 (b)). For Tween-80 and cyclodextrin, the percent of PCBs desorbed decreased along with increasing number of chlorine in biphenyl-ring. All of the above results indicated AlnA had higher desorption capacity of higher chlorinated PCBs.

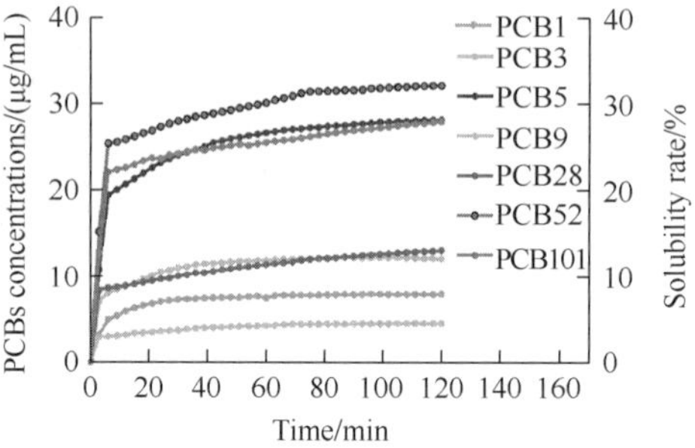

(a) Kinetics of PCBs solubilization in the presence of 20 μg/mL of AlnA. Quartz cuvettes containing 100μg of crystalline PCB congeners was added 1.0 mL of buffer.Data obtained every 3 min were shown

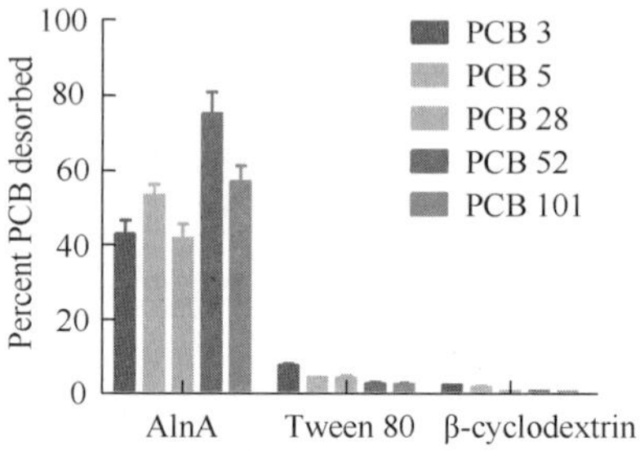

(b) PCBs desorbed from soils in the presence of AlnA, Tween-80 and β-cyclodextrin at a dose of 0.4 mM. Soil/solution ratio was 1:10. The absence of error bars in some data indicates very low standard deviation values.

Fig. 2 Solubilization and desorption of PCBs by AlnA (Wang et al., 2014)

3.3 AlnA combined with transgenic alfalfa remediates PCBs-contaminated soils

Plant biomass (fresh weight), were measured to investigate the feasibility of surfactant-enhanced phytoremediation with transgenic alfalfa and AlnA and the effect of PCBs on plant growth in different treatments. A significant difference in plant physiological traits was observed in the different treatments (Fig. 3 (a)). Quantitative analysis of PCBs in soils and plants was conducted on the basis of the chromatograph peak area. The removal rates of PCB congeners in soils are shown in Fig. 3 (b). Compared with NP treatment, the application of alfalfa in soils significantly increased PCB removal rates with the exception of di-PCBs (PCB4/PCB9). The removal rate of TA treatment was higher than that of WA treatment, especially for di-PCBs (PCB 6), tri-PCBs (PCB 17, 19, and 27), and tetra-PCBs (PCB 49). For most PCB congeners, the highest removal rate (33%-81%) was observed for TAA treatment, followed by TA, WA and NP treatment. In particular, the removal rates of tri-PCBs (PCB 16/PCB 32 and PCB 31/PCB 28) and tetra-PCBs (PCB 49) in TAA treatment were 3.6-, 1.1-, and 2-fold higher than in TA treatment. These results indicate that using the AlnA enhances the transgenic alfalfa remediation efficiency of soils contaminated with PCBs, particularly tri-and tetra-PCBs.

(a) Alfalfa exposed to PCBs for 60 d in pot experiment

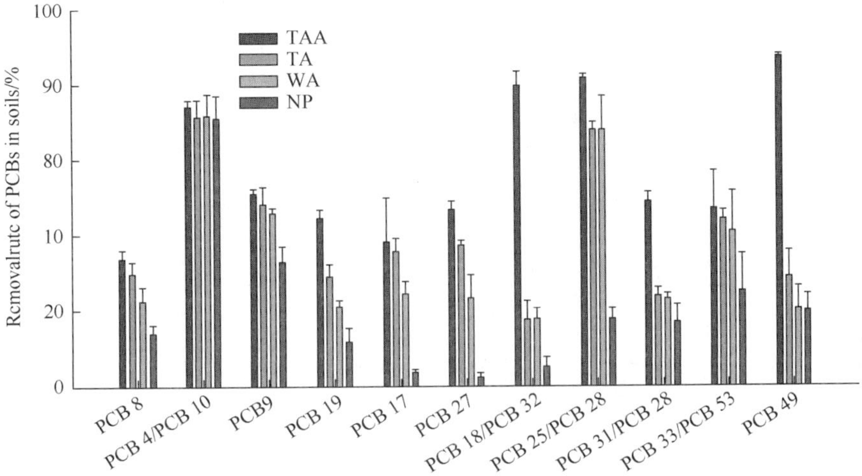

(b) Removal rates of PCBs in soils. TAA, Transgenic alfalfa -AlnA, TA Transgenic alfalfa, WA Wild-type alfalfa, NP Unplanted alfalfa

Fig. 3 Remediation of PCBs-contaminated soils by AlnA combined with transgenic alfalfa

3.4 Microbial response during the phytoremediation of PCBs

Microbial community analysis was performed by Illumina high-throughput sequencing to understand the biological process of PCB degradation by indigenous microbes in phytoremediation. Microbial community composition showed that OTUs were affiliated to 38 known (Bacteria 34 and Archaea 4) and 1 unclassified phyla, and to 124 known (Bacteria 119 and Archaea 5) and 1 unclassified classes. The relative importance of the most abundant phyla and genus is shown in Fig.4. The most abundant phylum was *Proteobacteria* (on average 34.8%-48.5%). *Bafcteroidetes* and *Acidobacteria* were the subdominant groups (Fig. 4 (a)). The abundance of *Proteobacteria* was significantly greater in transgenic TAA treatment (48.5%) than under the other treatments. Within this phylum, the classes *Alpha-, Beta-, Delta-,* and *Gammaproteobacteria* were the four most frequently observed. *Alphaproteobacteria* with the highest abundance accounted for 15.5%-18.4%. At the genus level (Fig.4 (b)), the most abundant genera, including *Kaistobacter, Arthrobacter, Pseudomonas, Rhodoplanes,* and *Sphingomonas,* accounted for approximately 12% of all sequences. This result indicates that the microbes detected were closely related to the biodegradation of refractory organics. *Arthrobacter*, *Pseudomonas,* and *Sphingomonas* with outstanding degradation abilities widespread in long-term PCB-contaminated soils were observed (Field and Sierra Alvarez, 2008). *Pseudomonas, Methylophaga*, and *Flavobacterium* were significantly enriched in TAA treatment but rarely detected in the other treatments. In particular, *Pseudomonas* was 4-fold more abundant in TAA treatment than in NP treatment. Furthermore, PCB removal may be promoted by increasing the abundance of indigenous bacteria that display PCB degradation abilities in soils with AlnA addition.

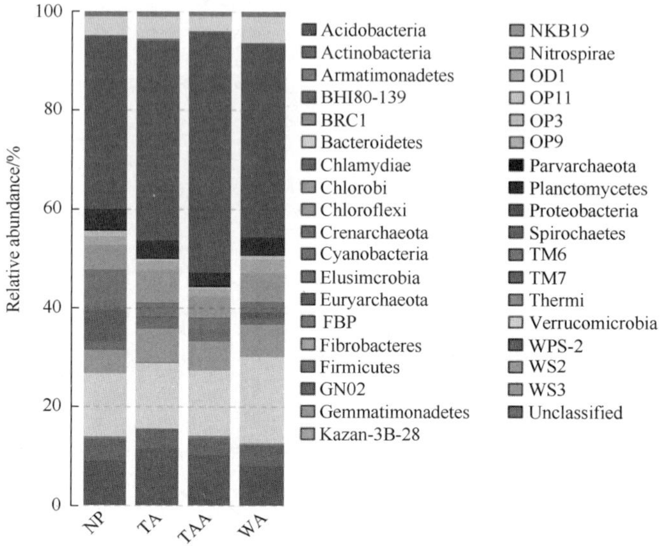

(a) Phylum

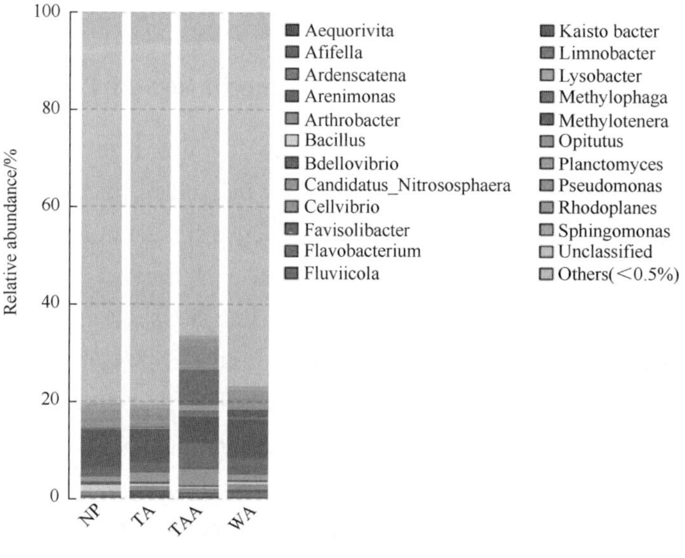

(b) Genus

Fig. 4　Phylum and genus level distribution

NP: Unplanted alfalfa, TA: Transgenic alfalfa, TAA: Transgenic alfalfa-AlnA, WA: Wild-type alfalfa.

4. Conclusions

Alfalfa expressing *bphC* resulted in cross-tolerance to PCBs and 2, 4-DCP in vegetative growth stages. Recombinant protein could significantly enhance solubility of crystal PCBs and easier recover of PCBs from contaminated soils. Phytoremediation trials by transgenic alfalfa associated AlnA promoted PCB contaminated soil remediation. The bacterial guild mainly comprising the genera *Pseudomonas*, *Arthrobacter*, and *Sphingomonas* correlated with the removal rates of PCBs.

References

Åslund M L W, Zeeb B A, Rutter A., et al. 2007. In situ phytoextraction of polychlorinated biphenyl- (PCB) contaminated soil. Science of the Total Environment, 374 (1): 1-12.

Krämer U. 2005. Phytoremediation: Novel approaches to cleaning up polluted soils. Current Opinion in Biotechnology, 16 (2): 133-141.

Zhang J Y, Qiu L M, Jia H, et al. 2007. Occurrence and congeners specific of polychlorinated biphenyls in agricultural soils from Southern Jiangsu, China. Journal of Environmental Sciences, 19 (3): 338-342.

Zeeb B A, Amphlett J S, Rutter A, et al. 2006. Potential for phytoremediation of polychlorinated biphenyl- (PCB) -contaminated soil. International Journal of Phytoremediation, 8 (3): 199-221.

Novakova M, Mackova M, Antosova Z, et al. 2010. Cloning the bacterial *bphC* gene into *Nicotiana tabacum* to improve the efficiency of phytoremediation of polychlorinated biphenyls. Bioengineered Bugs, 1 (6): 419-423.

Borja J, Taleon D M, Auresenia J, et al. 2005. Polychlorinated biphenyls and their biodegradation. Process Biochemistry, 40 (6): 1999-2013.

Xia H L, Chi X Y, Yan Z J, et al. 2009. Enhancing plant uptake of polychlorinated biphenyls and cadmium using tea saponin. Bioresource Technology 100: 4649-4653.

Toren A, Ron E, Bekerman R, et al. 2002. Solubilization of polyaromatic hydrocarbons by recombinant bioemulsifier AlnA. Applied Microbiology and Biotechnology, 59 (4-5): 580-584.

Wang Y, Wang C, Ren H, et al. 2014. Effectiveness of recombinant protein AlnA in enhancing the extractability of polychlorinated biphenyls from contaminated soils. Journal of Hazardous Materials, 279: 67-74.

Montague A, Ziauddin A, Lee R, et al. (2007). High-efficiency phosphinothricin-based selection for alfalfa transformation. Plant Cell, Tissue and Organ Culture, 91 (1): 29-36.

Field J, Sierra Alvarez R.2008. Microbial transformation and degradation of polychlorinated biphenyls. Environmental Pollution, 155: 1-12

Wang Y, Ren H, Pan H, et al. 2015. Enhanced tolerance and remediation to mixed contaminates of PCBs and 2, 4-DCP by transgenic alfalfa plants expressing the 2, 3-dihydroxybiphenyl-1, 2-dioxygenase. Journal of Hazardous Materials, 286: 269-275.

Microbial Degradation of Chemical Pesticides and Bioremediation of Pesticide-Contaminated Sites in China

Jiandong JIANG, Shunpeng LI

Department of Microbiology, Key Lab of Microbiological Engineering of Agricultural Environment, Ministry of Agriculture, College of Life Sciences, Nanjing Agricultural University

Abstract

The intensive use of chemical pesticides has also resulted in serious environmental problems. Biodegradation of chemical pesticides and bioremediation of pesticide-contaminated sites by microorganisms is considered as the most effective, less expensive and non-secondary pollution method. In this article, the achievements in the lab for Environmental Microbiology in Nanjing Agricultural University will be summarized, including isolation and collection of high efficient pesticides-degrading bacteria, elucidation of the catabolic pathway of pesticides in microbes, cloning of the key gene (s) involved in pesticide-degradation, the ecological process during biodegradation of pesticides, and removal of pesticides contamination by microbes.

Keywords: Pesticides, Microbial degradation, Bioremediation

Chemical pesticides as the major compounds to control pests and diseases have been popularly used world widely in modern agricultural systems. However, the intensive use of chemical pesticides has also resulted in serious environmental problems because they are either recalcitrant or biodegraded very slowly. Owing to their highly toxic nature, there is significant concern regarding the large quantities of pesticide residues in the environment. Many methods were thought out for remediation these pesticide residues including incineration, landfills and bioremediation. Due to the limitation of physical-chemical remediation methods, bioremediation constitutes an attractive alternative to physicochemical methods of remediation, and is considered as the most effective, less expensive and non-secondary pollution method by many biological and environmental scientists.

Bioremediation is an innovative technology that has the potential to alleviate numerous pesticide contamination problems and was often chosen for the cleanup of contaminated sites. Bioremediation processes can be broadly categorized into two groups: *ex situ* and *in situ*. *Ex situ* bioremediation technologies include bioreactors, biofilters, land farming and some composting methods. *In situ* bioremediation technologies include bioventing, biosparging, biostimulation, liquid delivery systems and some composting methods. *In situ* treatments tend to be more attractive because they require less equipment, generally have a lower cost and generate less disturbance to the environment.

Microorganisms play key roles in the biodegradation of chemical pesticides. In recent years, there is an increasing interest in the use of microorganisms with versatile biodegradative properties to deal with the pollution of pesticides. This can be done by using indigenous microorganisms or by adding an enriched culture of microorganisms that have specific characteristics to degrade the desired pesticides at a quicker rate. Ideally, bioremediation results in the complete mineralization of chemical pescitides to H_2O and CO_2 without the build up of intermediates.

The lab for Environmental Microbiology in Nanjing Agricultural University has been engaged in the pesticides degradation and remediation by microorganisms for more than 20 years and has made some progress in this field, including isolation and collection of high efficient pesticides-degrading bacteria, elucidation of the catabolic pathway of pesticides in microbes, cloning of the key gene (s) involved in pesticide-degradation, the ecological process during biodegradation of pesticides, removal of pesticides contamination by microbes.

1. Isolation and collection of high efficient pesticides-degrading bacteria

More than 150 bacterial strains capable of degrading various herbicides have been isolated from pesticide-contaminated soil or water samples all over China. These isolates are able to degrade various kinds of pesticides including organochlorine, organophosphorus and organic nitrogenous insecticides, fungicides and herbicides (Table 1 and Table 2). As regard as herbicides, sulfonylurea herbicides, phenylurea herbicides, chloroacetanilide herbicides, aryloxyphenoxy propanoate herbicides, benzonitrile and diphenylether herbicides are all found to be degraded by these isolates. These isolates will provide us with abundant microbial resources for the study the mechanisms involved in the herbicide biodegradation and the application of bioremediation technology for the herbicide-contaminated sites.

Table 1 Microbial isolates capable of degrading various pesticides

Isolates	Pesticides degraded	Reference
Lysinibacillus sp. ZB-1	Diphenylethers Herbicides (Fomesafen, Lactofen, Fluoroglycofen)	Chemosphere. 2009. 77: 1614-1619
Pseudomonas sp. LW3	Sulfonylurea Herbicides (Chlorimuron-Ethyl)	FEMS Microbiology Letters. 2009. 296 (2): 203-209
Brevundimonas sp. LY-2	Diphenylethers Herbicides (Lactofen)	Journal of Agricultural and Food Chemistry. 2010. 58 (17): 9711-9715
Pseudomonas aeruginosa L36	Sulfonylurea Herbicides (Metsulfuron-Methyl)	World Journal of Microbiology and Biotechnology. 2010. 26: 515-521
Sphingomonas sp. Y57	Propanil, Isoproturon	Journal of Hazardous Materials. 2011. 196: 412-419
Ancylobacter sp. XJ-412-1	Sulfonylurea Herbicides (Metsulfuron-Methyl, Thifensulfuron-Methyl, Bensulfuron-Methyl)	Current Microbiology. 2011. 62 (6): 1718-1725

Continue table

Isolates	Pesticides degraded	Reference
Pseudomonas azotoformans QDZ-1	Aryloxyphenoxy Propanoate Herbicide (Cyhalofop-Butyl, Quizalofop-P-Ethyl, Fenoxaprop-P-Ethyl, Diclofop-Methyl, Haloxyfop-P-Methyl)	Journal of Agricultural and Food Chemistry. 2011. 59: 6040-6046
Paracoccus sp. FLY-8	Chloroacetanilide Herbicides (Alachlor, Acetochlor, Propisochlor, Butachlor, Pretilachlor)	Journal of Agricultural and Food Chemistry. 2011. 59 (9): 4614-4621
Pseudomonas zeshuii BY-1	Diphenylethers Herbicides (Fomesafen, Lactofen, Acifluorfen, Fluoroglycofen)	Journal of Agricultural and Food Chemistry. 2012. 60: 7104-7110
Rhodococcus sp. B1	Chloroacetanilide Herbicides (Butachlor, Alachlor, Acetochlor, Pretilachlor)	Journal of Agricultural and Food hemistry. 2012. 60 (50): 12238-12244
Sphingobium sp. YBL2	Phenylurea Herbicides (Isoproturon)	International Biodeterioration and iodegradation. 2012. 70: 8-13
Catellibacterium caeni DCA-1	Chloroacetanilide Herbicides (Butachlor)	International Biodeterioration and Biodegradation. 2012. 73: 16-22
Paracoccus sp. FLN-7	Chlorpropham, Propham, Propanil	Applied and Environmental Microbiology. 2012. 78 (6): 4848-4855
Sphingopyxis sp. OB-3	Bromoxynil (Bromoxynil Octanoate)	Journal of Agricultural and Food Chemistry. 2013. 61 (47): 11550-11559
Comamonas sp. 7D-2	Benzonitriles Herbicides (Bromoxynil)	Molecular Microbiology, 2013, 89 (6): 1121-1139
Sphingobium quisquiliarum DC-2 and *Sphingobium baderi* DE-13	Chloroacetanilide Herbicides (Acetochlor)	Bioresource Technology. 2013. 148: 628-631

It was found that 38 strains are the first time reported from that genus to be capable of degrading specific herbicides, and 20 species were found to be novel species (Table 2), showing the diverse of microbial strains for the degradation of chemical herbicides.

Table 2 Novel species isolated capable of degrading pesticides

Novel species isolated	Pesticide degraded	Reference
Rhodococcus qingshengii sp. nov.	Carbendazim	International Journal of Systematic and Evolutionary Microbiology. 2007. 57: 2754-2757
Candida mengyuniae sp. nov.	Metsulfuron-Methyl	International Journal of Systematic and Evolutionary Microbiology. 2009. 59: 1237-1241
Flavobacterium haoranii sp. nov.	Cypermethrin	International Journal of Systematic and Evolutionary Microbiology. 2010. 60: 2882-2886
Rhodococcus jialingiae sp. nov.	Carbendazim	International Journal of Systematic and Evolutionary Microbiology. 2010. 60: 378-381
Sphingobium faniae sp. nov.	Pyrethroid	International Journal of Systematic and Evolutionary Microbiology. 2010. 60: 408-412
Sphingobium qiguonii sp. nov.	Carbaryl	International Journal of Systematic and Evolutionary Microbiology. 2010. 60: 2724-2728
Sphingobium wenxiniae sp. nov.	Pyrethroid	International Journal of Systematic and Evolutionary Microbiology. 2011. 61: 1776-1780

Continue table

Novel species isolated	Pesticide degraded	Reference
Lysobacter ruishenii sp. nov.	Chlorothalonil	International Journal of Systematic and Evolutionary Microbiology. 2011. 61: 674-679
Comamonas zonglianii sp. nov.	Phenol	International Journal of Systematic and Evolutionary Microbiology. 2011. 61: 255-258
Methylopila jiangsuensis sp. nov.	Acetochlor	International Journal of Systematic and Evolutionary Microbiology. 2011. 61: 1561-1566
Catellibacterium caeni sp. nov.	Butachlor	International Journal of Systematic and Evolutionary Microbiologyc. 2011. 61: 1921-1926
Hansschlegelia zhihuaiae sp. nov.	Sulfonylurea Herbicides	International Journal of Systematic and Evolutionary Microbiology. 2011. 61: 1114-1117
Sphingobium jiangsuense sp. nov.	3-Phenoxybenzoic acid	International Journal of Systematic and Evolutionary Microbiology. 2012. 62: 800-805
Sphingobacterium wenxiniae sp. nov.	Cypermethrin	International Journal of Systematic and Evolutionary Microbiology. 2012. 62: 683-687
Burkholderia zhejiangensis sp. nov.	Methyl Parathion	International Journal of Systematic and Evolutionary Microbiology. 2012. 62, 1337-1341
Catellibacterium nanjingense sp. nov.	Propanil	International Journal of Systematic and Evolutionary Microbiology. 2012. 62: 495-499
Pseudomonas zeshuii sp. nov.	Fomesafen	International Journal of Systematic and Evolutionary Microbiology. 2012. 62: 2608-2612
Dokdonella kunshanensis sp. nov.	Butachlor	International Journal of Systematic and Evolutionary Microbiology. 2013. 63: 1519-1523
Paracoccus huijuniae sp. nov	Chlorpropham, Propham, Propanil	International Journal of Systematic and Evolutionary Microbiology. 2013. 63: 1132-1137

2. Elucidation of the catabolic pathway of pesticides in microbes

Through GC-MS, HPLC-MS, MS/M, the metabolites produced during herbicide-degradation were identified, and 17 metabolic pathways have been elucidated including the pathways for chlorimuron-ethyl, fomesafen, lactofen, propanil, butachlor, metsulfuron-methyl, cyhalofop-butyl, thifensulfuron-methyl, isoproturon, chlorpropham, acetochlor, bromoxynil and bromoxynil octanoate (Table 3). Among these pathways, 8 pathways are the first time reported, indicating the diversity of metabolism of herbicides.

Table 3 The identified metabolic pathways of herbicides in microbes

Pesticides	Degrading strains	Metabolites identified	Reference
Cypermethrin	*Sphingobium wenxiniae* JZ-1	2, 2-dimethyl-3- (2, 2-dichlorovinyl) -cyclopropanecarboxylic acid, 3-phenoxybenzoic acid, catechol	Applied and Environmental Microbiology. 2009. 75 (17): 5496-5500
Triazophos	*Diaphorobacter* sp. TPD-1	1-phenyl-3-hydroxy-1, 2, 4-triazole, 2-phenyl hydrazinecarboxylic acid	Current Microbiology. 2011. 62: 1294-1301
Dimethoate	*Paracoccus* sp. lgjj-3	dimethoate carboxylic acid, O, O, S-trimethyl thiophosphorothioate, phosphorothioic O, O, S-acid, O, O, O-trimethyl phosphoric ester	International Biodeterioration and Biodegradation. 2010. 64: 51-57

Continue table

Pesticides	Degrading strains	Metabolites identified	Reference
Isocarbophos	*Arthrobacter* sp. scl-2	isopropyl salicylate, salicylic acid, gentisate	Journal of Microbiology and Biotechnology. 2009. 19 (11): 1439-1446
Methamidophos	*Hyphomicrobium* sp. MAP-1	O, S-dimethyl hydrogen thiophosphate, S-methyl dihydrogen thiophosphate, methyl dihydrogen phosphate, phosphoric acid	Biodegradation. 2010. 21: 513-523
Chlorothalonil	*Pseudomonas* sp. CTN-3	4-hydroxyl-chlorothalonil	Journal of Bacteriology. 2010. 192 (11): 2737-2745; Current Microbiology. 2010. 61: 226-230
Buprofezin	*Pseudomonas* sp. DFS35-4	2-imino-5-phenyl-3- (propan-2-yl) -1, 3, 5-thiadiazinan-4-one, 2-imino-5-phenyl-1, 3, 5-thiadiazinan-4-one, methyl (phenyl) carbamic acid	Biodegradation. 2011. 22: 1135-1142
Carbofuran	*Novosphingobium* sp. FND-3	2-hydroxy-3-(3-methypropan-2-ol) phenol, 2-hydroxy-3-(3-methylpropan-2-ol) benzene-N-methylcarbamate, 5-hydroxycarbofuran	FEMS Microbiology Letters. 2007. (271): 207-213
Carbendazim	*Rhodococcus jialingiae* djl-6-2	2-aminobenzimidazole, 2-hydroxybenzimidazole	Chemosphere. 2010. 81: 639-644
Chlorimuron-Ethyl	*Pseudomonas* sp. LW3	N-formyl-benzosulfimide, 4-chloro chloromethoxy-pyrimidin-2-amine	FEMS Microbiology Letters. 2009. 296: 203-209
Fomesafen	*Lysinibacillus* sp. ZB-1	N-[4-{4- (trifluoromethyl) phenoxy}-2-methanamidephenyl]acetamide	Chemosphere. 2009. 77: 1614-1619
Lactofen	*Brevundimonas* sp. LY-2	1- (carboxy) ethyl-5- (2-chloro-4- (trifluoromethyl) phenoxy) -2-nitrobenzoate	Journal of Agricultural and Food Chemistry. 2010. 58: 9711-9715
Propanil	*Sphingomonas* sp. Y57	3, 4-dichloroaniline, 4, 5-dichlorocatechol	Journal of Hazardous Materials. 2011. 196: 412-419
Butachlor	*Paracoccus* sp. FLY-8	[2-chloro-N- (2, 6-dimethylphenyl) -N- (methoxymethyl) acetamide]; 2-chloro-N- (2, 6-dimethylphenyl) acetamide; 2, 6-diethylaniline	Journal of Agricultural and Food Chemistry. 2011. 59 (9): 4614-4621
Metsulfuron-Methyl	*Ancylobacter* sp. XJ-412-1	2-[[[[(4-methoxy-6-methyl-1, 3, 5-triazin-2-yl) amino]carbonyl]amino]sulfonyl]benzoic acid (metsulfuron acid)	Current Microbiology, 2011, 62 (6): 1718-1725
Cyhalofop-Butyl	*Pseudomonas azotoformans* QDZ-1	butanol, cyhalofop acid	Journal of Agricultural and Food Chemistry, 2011, 59, 6040-6046
Butachlor	*Rhodococcus* sp. B1	Butoxymethanol; 2-chloro-N- (2, 6-dimethyl phenyl) -acetamide	Journal of Agricultural and Food Chemistry. 2012. 60: 12238-12244
Thifensulfuron-Methyl	*Hansschlegelia zhihuaiae* S113	thifensulfuron acid	Applied and Environmental Microbiology. 2012. 78 (6): 1962-1968
Isoproturon	*Sphingobium* sp. YBL2	MDIPU; DDIPU; 4-IA	International Biodeterioration and Biodegradation. 2012. 70: 8-13
Chlorpropham	*Paracoccus* sp. FLN-7	3-chloroaniline	Applied and Environmental Microbiology. 2012. 78 (6): 4848-4855
Fomesafen	*Pseudomonas zeshuii* BY-1	[5- (4-trifluoromethylphenoxy) -2-nitro-benzoyl]-sulfinamic acid; [5- (2-chloro-4-difluoromethyl-phenoxy) -2-nitro-benzoyl]-sulfinamic acid; N-[4- (difluoromethyl-phenoxyl) -2-formylamino-phenyl]-acetamide.	Journal of Agricultural and Food Chemistry. 2012. 60: 7104-7110

Continue table

Pesticides	Degrading strains	Metabolites identified	Reference
Butachlor	*Catellibacterium caeni* DCA-1	N-hydroxymethyl-2-chloro-N- (2, 6-diethyl-phenyl) -acctamide; 2-chloro-N- (2, 6-diethyl -phenyl) -acetamide; N- (2, 6-diethyl-pheny) -N-hydroxymethyl-formamide; (2, 6-diethyl -phenyl) -ethoxymethyl-carbamic acid	International Biodeterioration and Biodegradation. 2012. 73: 16-22
Acetochlor	*Sphingobium quisquiliarum* DC-2 and *Sphingobium baderi* DE-13	2-chloro-N- (2-methyl-6-ethylphenyl) acetamid; 2-methyl-6-ethylaniline; 2-methyl-6-ethylaminophenol; 2-methyl-6-ethylbenzoquinoneimine	Bioresource Technology. 2013. 148: 628-631
Bromoxynil	*Comamonas* sp. 7D-2	3, 5-dibromo-4-hydroxybenzoate, 3-bromo-4-hydroxybenzoate, p-hydroxybenzonic acid, protocatechuic	Molecular Microbiology. 2013. 89 (6): 1121-1139
Bromoxynil Octanoate	*Sphingopyxis* sp. OB-3	3, 5-dibromo-4-hydroxybenzonitrile, octanoic acid	Journal of Agricultural and Food Chemistry. 2013. 61 (47): 11550-11559
Isoproturon	*Sphingobium* sp. YBL2	(1- (4-isopropylphenyl) -3-methylurea), (1- (4-isopropylphenyl) urea)	Applied and Environmental Microbiology. 2013. 79 (24): 7846-7856

3. Cloning of the key gene (s) involved in pesticide-degradation

During the past two decades, over 20 key genes involved in the pesticide degradation have been cloned, including the genes for the catabolism of organochlorine, organophosphorus and organic nitrogenous insecticides, fungicides and herbicides. Among these genes, 15 genes were the first time reported novel genes and showed relatively low similarities to known genes (Table 4).

Table 4 The key genes cloned that are involved in the degradation of pesticides

Gene	Pesticides degraded	Reference
mpd	Organophosphorus Insecticide	Applied and Environmental Microbiology. 2001
pytH	Permethrin	Applied and Environmental Microbiology. 2009
chd	Chlorothalonil	Journal of Bacteriology. 2010
sulE	Sulfonylurea Herbicides	Applied and Environmental Microbiology. 2012
ampA	Dimethoate	Applied and Environmental Microbiology. 2012
chbH	Cyhalofop	Journal of Agricultural and Food Chemistry. 2011
feH	Fenoxaprop-P-Ethyl	FEMS Microbiology Letters, 2011
pamh	Propanil	Applied Microbiology and Biotechnology. 2012
cmeH	Diclofop-Methyl	Bioresource Technology. 2013
broH	Bromoxynil Octanoate	Journal of Agricultural and Food Chemistry. 2013
bhbAB	Bromoxynil	Molecular Microbiology. 2013
pdmAB	Isoproturon	Applied and Environmental Microbiology. 2013
pbaAB	3-Phenoxybenzoic acid	Applied and Environmental Microbiology. 2014
cndA	Acetochlor	Applied and Environmental Microbiology. 2014
ddhA	Isoproturon	Environmental Microbiology. 2016

3.1 A novel pyrethroid-hydrolyzing carboxylesterase gene from *Sphingobium* sp. strain JZ-1

A novel esterase gene, *pytH*, encoding a pyrethroid-hydrolyzing carboxylesterase was cloned from *Sphingobium* sp. strain JZ-1. The gene contained an open reading frame of 840 bp. Sequence identity searches revealed that the deduced enzyme shared the highest similarity with many α/β-hydrolase fold proteins (20% to 24%identities). *PytH* was expressed in *Escherichia coli* BL21 and purified using Ni-nitrilotriacetic acid affinity chromatography. It was a monomeric structure with a molecular mass of approximately 31 kDa and a pI of 4.85. *PytH* was able to transform p-nitrophenyl esters of short-chain fatty acids and a wide range of pyrethroid pesticides, and isomer selectivity was not observed. No cofactors were required for enzyme activity.

3.2 A novel hydrolytic dehalogenase gene involved in the degradation of the fungicide of chlorothalonil

Dehalogenases play key roles in the detoxification of halogenated aromatics. Interestingly, only one hydrolytic dehalogenase for halogenated aromatics, 4-chlorobenzoyl-CoA dehalogenase, has been reported. Here, we characterize another novel hydrolytic dehalogenase for halogenated aromatic compound from the 2, 4, 5, 6-tetrachloroisophthal-onitrile (chlorothalonil) degrading strain of *Pseudomonas* sp. CTN-3, which we have named Chd. Chd catalyzes a hydroxyl substitution at the 4-chlorine atom of chlorothalonil. The metabolite of the Chddehalogenation, 4-hydroxy-trichloroisophth- alonitrile, was identified by reverse-phase high-performance liquid chromatography (HPLC), tandem mass spectrometry (MS/MS) and nuclear magnetic resonance (NMR). Chd dehalogenates chlorothalonil under anaerobic and aerobic conditions and does not require the presence of cofactors such as coenzyme A (CoA) and ATP. Chd contains a putative conserved domain of the metallo-β-lactamase superfamily and shows the highest identity with several metallohydrolases (24%-29%). Chd is a monomer (36 kilodaltons) and the isoelectric point (pI) of Chd is estimated to be 4.13. Chd has a dissociation constant (*K*m) of 0.112 mmol/L and an overall catalytic rate (*K*cat) of 207 s^{-1} for chlorothalonil. Chd is completely inhibited by 1, 10-phenanthroline, diethyl pyrocarbonate and N-bromosuccinic acid. Site-directed mutagenesis of Chd revealed that histidine 128 and 157, serine 126, aspartate 45, 130 and 184, and tryptophan 241 were essential for the dehalogenase activity. Chd differs from other reported hydrolytic dehalogenases based on the analysis of amino acid sequences and catalytic mechanisms. This study provides an excellent dehalogenase candidate for mechanistic study of hydrolytic dehalogenation of halogenated aromatic compound.

3.3 Novel gene clusters involved in the catabolism of the herbicide bromoxynil

Dehalogenation is the key step in the degradation of halogenated aromatics, while reductive dehalogenation is originally thought to rarely occur in aerobes. An aerobic strain of *Comamonas* sp. 7D-2 was shown to degrade the brominated aromatic herbicide bromoxynil completely and release two equivalents of bromides under aerobic conditions. The enzymes involved in the degradation of bromoxynil to 4-carboxy-2-hydroxymuconate-6-semialdehyde, including nitrilase, reductive dehalogenase (*BhbA*), 4-hydroxybenzoate 3-monooxygenase and protocatechuate 4, 5-dioxygenase, were molecularly characterized. The novel dehalogenase *BhbA* was shown to be a complex of a respiration-linked reductive dehalogenase (RdhA) domain and a NAD (P) H-dependent oxidoreductase domain and to have key features of anaerobic respiratory RdhAs, including two predicted binding motifs for [4Fe-4S] clusters and a close association with a hydrophobic membrane protein (*BhbB*). *BhbB* was confirmed to anchor *BhbA* to the membrane.

BhbA was partially purified and found to use NAD (P) H as electron donors. Full-length *bhbA* homologues were found almost exclusively in marine aerobic proteobacteria, suggesting that reductive dehalogenation occurs extensively in aerobes and that *bhbA* is horizontally transferred from marine microorganisms. The discovery of a functional reductive dehalogenase and ring-cleavage oxygenases in an aerobe opens up possibilities for basic research as well as the potential application for bioremediation.

3.4 Sulfonylurea herbicide de-esterification esterase gene from *Hansschlegelia zhihuaiae* S113

De-esterification is an important degradation or detoxification mechanism of sulfonylurea herbicide in microbes and plants. However, the biochemical and molecular mechanisms of sulfonylurea herbicide de-esterification are still unknown. A novel esterase gene, *sulE*, responsible for sulfonylurea herbicide de-esterification, was cloned from *Hansschlegelia zhihuaiae* S113. The gene contained an open reading frame of 1194 bp, and a putative signal peptide at the N terminal was identified with a predicted cleavage site between Ala37 and Glu38, resulting in a 361-residue mature protein. *SulE* minus the signal peptide was synthesized in *Escherichia coli* BL21 and purified to homogeneity. *SulE* catalyzed the de-esterification of a variety of sulfonylurea herbicides that gave rise to the corresponding herbicidally inactive parent acid and exhibited the highest catalytic efficiency toward thifensulfuron-methyl. *SulE* was a dimer without the requirement of a cofactor. The activity of the enzyme was completely inhibited by Ag^{+}, Cd^{2+}, Zn^{2+}, methamidophos, and sodium dodecyl sulfate. A *sulE*-disrupted mutant strain, *sulE*$^{-}$, was constructed by insertion mutation.

sulE^{-} lost the de-esterification ability and was more sensitive to the herbicides than the wild type of strain S113, suggesting that *sulE* played a vital role in the sulfonylurea herbicide resistance of the strain. The transfer of *sulE* into *Saccharomyces cerevisiae* BY4741 conferred on it the ability to de-esterify sulfonylurea herbicides and increased its resistance to the herbicides. This study has provided an excellent candidate for the mechanistic study of sulfonylurea herbicide metabolism and detoxification through de-esterification, construction of sulfonylurea herbicide-resistant transgenic crops, and bioremediation of sulfonylurea herbicide-contaminated environments.

3.5 A novel arylamidase gene from *Paracoccus* sp. strain FLN-7 that is involved in the hydrolysis of amide pesticides

The bacterial isolate *Paracoccus* sp. strain FLN-7 hydrolyzes amide pesticides such as diflubenzuron, propanil, chlorpropham, and dimethoate through amide bond cleavage. A gene, *ampA*, encoding a novel arylamidase that catalyzes the amide bond cleavage in the amide pesticides was cloned from the strain. *ampA* contains a 1395-bp open reading frame that encodes a 465-aminoacid protein. *AmpA* was expressed in *Escherichia coli* BL21 and homogenously purified using Ni-nitrilotriacetic acid affinity chromatography. *AmpA* is a homodimer with an isoelectric point of 5.4. *AmpA* displays maximum enzymatic activity at 40℃ and a pH of between 7.5 and 8.0, and it is very stable at pHs ranging from 5.5 to 10.0 and at temperatures up to 50℃. *AmpA* efficiently hydrolyzes a variety of secondary amine compounds such as propanil, 4-acetaminophenol, propham, chlorpropham, dimethoate, and omethoate. The most suitable substrate is propanil, with *K*m and *K*cat values of 29.5 μmol/L and 49.2/s, respectively. The benzoylurea insecticides (diflubenzuron and hexaflumuron) are also hydrolyzed but at low efficiencies. No cofactor is needed for the hydrolysis activity. *AmpA* shares low identities with reported arylamidases (less than 23%), forms a distinct lineage from closely related arylamidases in the phylogenetic tree, and has different biochemical characteristics and catalytic kinetics with related arylamidases. The results in the present study suggest that *AmpA* is a good candidate for the study of the mechanism for amide pesticide hydrolysis, genetic engineering of amide herbicide-resistant crops, and bioremediation of amide pesticide-contaminated environments.

3.6 Novel bacterial *N*-demethylase genes *PdmAB* involved in the initial step of *N*, *N*-dimethyl-substituted phenylurea herbicide degradation

The environmental fate of phenylurea herbicides has received considerable attention in recent decades. The microbial metabolism of *N*, *N*-dimethyl-substituted phenylurea herbicides can generally be initiated by mono-*N*-demethylation. In this study, the molecular basis for this process was revealed. The *pdmAB* genes in *Sphingobium* sp. strain YBL2 were shown to be responsible for the initial mono-*N*-demethylation of commonly used *N*, *N*-dimethyl-substituted phenylurea herbicides. *PdmAB* is the oxygenase component of a bacterial Rieske non-heme iron oxygenase (RO) system. The genes *pdmAB*, encoding the α subunit *PdmA* and the β subunit PdmB, are organized in a transposable element flanked by two direct repeats of an insertion element resembling ISRh1. Furthermore, this transposable element is highly conserved among phenylurea herbicide-degrading sphingomonads originating from different areas of the world. However, there was no evidence of a gene for an electron carrier (a ferredoxin or a reductase) located in the immediate vicinity of *pdmAB*. Without its cognate electron transport components, expression of *PdmAB* in *Escherichia coli*, *Pseudomonas putida*, and other sphingomonads resulted in a functional enzyme. Moreover, coexpression of a putative [3Fe-4S]-type ferredoxin from *Sphingomonas* sp. strain RW1 greatly enhanced the catalytic activity of *PdmAB* in *E. coli*. These data suggested that *PdmAB* has a low specificity for electron transport components and that its optimal ferredoxin may be the [3Fe-4S] type. PdmA exhibited low homology to the α subunits of previously characterized ROs (less than 37% identity) and did not cluster with the RO group involved in *O*-or *N*-demethylation reactions, indicating that *PdmAB* is a distinct bacterial RO *N*-demethylase.

3.7 Comparative genomic analysis of isoproturon-mineralizing sphingomonads reveals the isoproturon catabolic genes

The worldwide use of the phenylurea herbicide, isoproturon (IPU), has resulted in considerable concern about its environmental fate. Though many microbial metabolites of IPU are known and IPU-mineralizing bacteria have been isolated, the molecular mechanism of IPU catabolism has not been elucidated yet. In this study, complete genes that encode the conserved IPU catabolic pathway were revealed, based on comparative analysis of the genomes of three IPU-mineralizing sphingomonads and subsequent experimental validation. The complete genes included a novel hydrolase gene *ddhA*, which is responsible for the cleavage of the urea side chain of the IPU demethylated products; a distinct aniline dioxygenase gene cluster *adoQTA1A2BR*, which has a broad substrate range; and an inducible catechol *meta*-cleavage pathway gene cluster *adoXEGKLIJC*. Furthermore, the initial mono-*N*-demethylation genes *pdmAB* were further confirmed to be involved in the successive *N*-demethylation of the IPU mono-*N*-demethylated product. These IPU-catabolic genes were organized into four transcription units and distributed on three plasmids. They were flanked by multiple mobile genetic elements and highly conserved among IPU-mineralizing sphingomonads. The elucidation of the molecular mechanism of IPU catabolism will enhance our understanding of the microbial mineralization of IPU and provide insights into the evolutionary scenario of the conserved IPU-catabolic pathway.

4. Horizontal transfer of the chlorothalonil hydrolytic dehalogenase gene facilitates bacterial adaptation to chlorothalonil-contaminated sites

Microorganisms have evolved multiple mechanisms to adapt to environmental stresses, such as mutation, DNA rearrangement, and horizontal gene transfer. The horizontal transfer of genes plays a key role in the evolution of catabolic genes, thereby facilitating bacterial adaptation to pollutant-contaminated sites. Bacterial dehalogenases catalyze the cleavage of carbon-halogen bonds of many man-made chlorinated compounds, which is a key step in the detoxification of these priority organic pollutants. Notably, many dehalogenase genes are associated with transmissible elements. In our laboratory, many diverse TPN-dechlorinating strains have been isolated, and the novel chlorothalonil hydrolytic dehalogenase (Chd), which catalyzes a hydroxyl substitution at the 4-chlorine atom of TPN to form 2, 4, 5-trichloro-6-hydroxybenzene-1, 3-dicarbonitrile (4-TPN-OH), was also identified. In this study, we discovered a close association between the highly conserved *chd* gene and a novel insertion sequence (IS), IS*Ocsp1*, in diverse TPN-dechlorinating strains. We suggest that the ecological role of horizontal gene transfer is to facilitate bacterial adaptation to TPN-contaminated sites, by allowing these bacteria to rapidly transform toxic TPN to less toxic 4-TPN-OH.

5. Removal of pesticide contamination by microbes

5.1 Degradation of fomesafen in soil by strain *Lysinibacillus* sp. ZB-1

The soil sample was collected from the top 0-20 cm from agricultural field in the campus of Nanjing Agricultural University, Nanjing, China. The soil with a pH value of 6.82 and has never been treated with fomesafen. Soil samples were sterilized by autoclaving at 121℃for 60 min. Subsamples (200 g) of fresh soil and sterile soil were weighed, and the solution of fomesafen was added to obtain a final concentration of 10 mg/kg soil, and mixed well. One set of fresh soil and of sterile soil were inoculated with strain ZB-1 (10^8 cells/g). Another set of uninoculated soil was kept as a control. The inoculum was thoroughly mixed into the soils under sterile conditions, and the moisture was adjusted to 35% (w/w of dry weight of soil). Each soil microcosm was incubated at 30℃in the dark. At intervals of 7 d, 10 g of the soil samples were collected and the concentration of fomesafen was detected. It was found that only 11.1% of the initially added 10 mg/L fomesafen was degraded in uninoculated sterilized soil after five weeks. When strain ZB-1 was inoculated to the sterilized soil, the degradation of fomesafen increased to 67.1% during the same period, which showed obviously strain ZB-1 enhanced the degradation of fomesafen. In fresh soil samples with and without inoculation, the degradation of fomesafen was 77.9% and 14.8% after five weeks respectively. The degradation rate of fomesafen in fresh soil inoculated with strain ZB-1 was clearly more rapid than that of without inoculation, which confirmed that strain ZB-1 could cooperate well with the indigenous microorganisms to degrade fomesafen in soil. The results of fomesafen degradation in soil indicated that strain ZB-1 could be used potentially for the bioremediation of fomesafen contaminated soils.

5.2 Degradation of the chloroacetamide herbicide butachlor by *Catellibacterium caeni* sp. nov DCA-1^T

Three different soil samples (red soil, fluvo-aquic soil and high sandy soil) were collected from Yingtan (Jiangxi Province), Suqian (Jiangsu Province) and Nantong (Jiangsu Province), China, respectively. Soil samples were treated as described previously. Two sets of non-sterile and sterile soils were inoculated with 5% (v/w) cells of strain DCA-1^T, respectively. Soil samples inoculated with heat-killed DCA-1^T cells were kept as the control. Butachlor was added to soil at a concentration of 50 mg/kg dry soil and then adsorbed in the dark overnight. Sterile water was employed to adjust the soil moisture to 40%. Soil samples were collected for butachlor analysis at 1, 3, 5, 10, 20, 30 and 45 d, respectively.

The recovery efficiencies of butachlor from the three soils ranged from 89.8% to 101.5%, indicating that butachlor could be recovered from these soils efficiently. In all three sterile soils inoculated with heat-killed cells, less than 20% of butachlor was degraded in 45 days. However, in the non-sterile soils about 60%-80% of butachlor was degraded over 45 days, indicating that indigenous microorganisms could also degrade butachlor. The inoculation of DCA-1^T cells significantly accelerated the butachlor degradation in both sterile and non-sterile soils, with 57.2%-90.4% of 50 mg/kg butachlor removed in 5 days compared to 5.4%-36% in the controls (inoculated with heat-killed cells), showing that strain DCA-1^T played major roles in the removal of butachlor. In the sterile soils, the inoculation of strain DCA-1^T removed 68.6%-86.0% of 50 mg/k butachlor in 30-45 days. However, in the non-sterile soils, the inoculation of strain DCA-1^T removed nearly the whole 50 mg/kg butachlor in the same time span, indicating that the inoculated exogenous strain DCA-1^T was in good cooperation with indigenous microorganisms. In the sterile soils, the degradation rate of butachlor in the red soil (pH 4.8) by the inoculated strain DCA-1^T cells was 68.6% in 45 days, which was relatively lower than that in the fluvo-aquic soil (82.8%, pH 6.3) and high sandy soil (86%, pH 8.2), showing that soil type significantly affected the butachlor degradation by strain DCA-1^T.

5.3 Elimination of chlorothalonil inhibition by chlorothalonil hydrolytic dehalogenase in alcoholic fermentation

The effect of different kinds of pesticide residues in grapes on alcoholic fermentation by

Saccharomyces cerevisiae was evaluated. Among four types of pesticides added into the grape slurry, omethoate, triadimefon and cyhalothrin did not inhibit the alcoholic fermentation at their proposed spraying concentration of 0.21, 0.10 and 0.10 g/L, respectively, whereas chlorothalonil concentration above 0.03 g/L behaved significantly negative influence on both S. cerevisiae growth and alcoholic fermentation efficiency. When the chlorothalonil concentration was lower than 0.01 g/L, the fermentation proceeded smoothly without any degradation of chlorothalonil. Considering the cumulative toxicity and adverse effect of chlorothalonil on fermentation, chlorothalonil hydrolytic dehalogenase (Chd) extracellularly expressed from the recombinant *Bacillus subtilis* WB800 was used to pretreat the chlorothalonilcontaminated grape slurry. After treatment by the Chd enzyme in an activity of 7.25 mU/L slurry for 60 min, the inhibition effect could be substantially eliminated even at an initial concentration of 0.10 g/L chlorothalonil. This study provides a potential approach for solving the conflict in fermentation industry with pesticides inhibition.

References

Hang B J, Hong Q, Xie X T, et al. 2012. *SulE*, a sulfonylurea herbicide de-esterification esterase from *Hansschlegelia zhihuaiae* S113. Applied and Environmental Microbiology, 78 (6): 1962-1968.

Liang B, Wang G, Zhao Y F, et al. 2011. Facilitation of bacterial adaptation to chlorothalonil-contaminated sites by horizontal transfer of the chlorothalonil hydrolytic dehalogenase gene. Applied and Environmental Microbiology, 77 (12): 4268-4272.

Liang B, Jiang J D, Zhang J, et al. 2012. Horizontal transfer of dehalogenase genes involved in the catalysis of chlorinated compounds: evidence and ecological role. Critical Reviews in Microbiology, 38 (2): 95-110.

Liang B, Li R, Jiang D, et al. 2010. Hydrolytic dechlorination of chlorothalonil by *Ochrobactrum* sp. CTN-11 isolated from a chlorothalonil-contaminated soil. Current Microbiology, 61: 226-233.

Liang B, Lu P, Li H H, et al. 2009. Biodegradation of fomesafen by strain *Lysinibacillus* sp. ZB-1 isolated from soil. Chemosphere. 77: 1614-1619.

Liang B, Zhao Y K, Lu P, et al. 2010. Biotransformation of the diphenyl ether herbicide lactofen and purification of a lactofen esterase from *Brevundimonas* sp. LY-2. Journal of Agricultural and Food Chemistry, 58 (17): 9711-9715

Meng C, He Q, Huang J W, et al. 2015. Degradation of chlorothalonil through a hydrolytic dehalogenase secreted from *Bacillus subtilis* WB800. International Biodeterioration and Biodegradation, 104: 97-104.

Yang C L, Li R, Song Y, et al. 2011. Dentification of the biochemical degradation pathway of triazophos and its intermediate in *Diaphorobacter* sp. TPD-1. Current Microbiology, 62: 1294-1301.

Wang G L, Wang L, Chen H H, et al. 2011. *Lysobacter ruishenii* sp. nov., a chlorothalonil-degrading bacterium isolated from a long-term chlorothalonil-contaminated soil in China. International Journal of Systematic and Evolutionary Microbiology, 61: 674-679.

Wang G L, Li R, Li S P, et al. 2010. A novel hydrolytic dehalogenase for the chlorinated aromatic compound chlorothalonil. Journal of Bacteriology, 192 (11): 2737-2745.

Hong Y F, Zhou J, Hong Q, et al. 2010. Characterization of a fenpropathrin-degrading strain and construction of a genetically engineered microorganism for simultaneous degradation of methyl parathion and fenpropathrin. Journal of Environmental Management, 91: 2295-2300.

Liu H M, Cao L, Lu P, et al. 2012. Biodegradation of butachlor by *Rhodococcus* sp. strain B1 and purification of its hydrolase (ChlH) responsible for N-dealkylation of chloroacetamide herbicides. Journal of Agricultural and Food Chemistry, 60 (50): 12238-12244.

Yang H, Wang X, Zheng J, et al. 2013. Biodegradation of acetamiprid by *Pigmentiphaga* sp. D-2 and the degradation pathway. International Biodeterioration and Biodegradation, 85: 95-102.

Zhang J, Sun J Q, Yuan Q Y, et al. 2011. Characterization of the propanil biodegradation pathway in *Sphingomonas* sp. Y57 and cloning of the propanil hydrolase gene *prpH*. Journal of Hazardous Materials, 196: 412-419.

Zhang J, Hong Q, Li Q, et al. 2012. Characterization of isoproturon biodegradation pathway in *Sphingobium* sp. YBL2. International Biodeterioration and Biodegradation, 70: 8-13.

Jiang J D, Zhang R F, Li R, et al. 2007. Simultaneous biodegradation of methyl parathion and carbofuran by a genetically engineered microorganism constructed by mini-Tn5 transposon. Biodegradation, 18: 403-412.

Zheng J W, Li R, Zhu J C, et al. 2012. Degradation of the chloroacetamide herbicide butachlor by *Catellibacterium caeni* sp. nov DCA-1^T. International Biodeterioration and Biodegradation, 73: 16-22.

Zheng J W, Chen Y G, Zhang J, et al. 2011. Description of *Catellibacterium caeni* sp. nov., reclassification of Rhodobacter changlensis Anil Kumar et al. 2007 as *Catellibacterium changlense* comb. nov. and emended description of the genus Catellibacterium. International Journal of Systematic and Evolutionary Microbiology, 61 (8): 1921-1926.

Ma J P, Wang Z, Lu P, et al. 2009. Biodegradation of the sulfonylurea herbicide chlorimuron-ethyl by the strain *Pseudomonas* sp. LW3. FEMS Microbiology Letters, 296 (2): 203-209.

Zhang J, Yin J G, Hang B J, et al. 2012. Cloning of a novel arylamidase gene from *Paracoccus* sp. strain FLN-7 that hydrolyzes amide pesticides. Applied and Environmental Microbiology, 78 (14): 4848-4855.

Zhang J, Chen S A, Zheng J W, et al. 2012. *Catellibacterium nanjingense* sp. nov., a propanil-degrading bacterium isolated from activated sludge, and emended description of the genus Catellibacterium. International Journal of Systematic and Evolutionary Microbiology, 62 (3): 495-499.

Chen K, Huang L L, Xu C F, et al. 2013. Molecular characterisation of the enzymes involved in the degradation of a brominated aromatic herbicide. Molecular Microbiology, 89 (6): 1121-1139.

Chen K, Liu X M, Li R, et al. 2011. Isolation of a buprofezin co-metabolizing strain of *Pseudomonas* sp. DFS35-4 and identification of the buprofezin transformation pathway. Biodegradation, 22: 1135-1142.

Chen K, Xu X H, Zhang L, et al. 2016. Comparison of four *Comamonas* catabolic plasmids reveals the evolution of pBHB to catabolize haloaromatics. Applied and Environmental Microbiology, 82 (5): 1401-1411

Chen K, Liu Y, Mao D M, et al. 2013. An essential esterase (BroH) for the mineralization of bromoxynil octanoate by a natural consortium of *Sphingopyxis* sp. strain OB-3 and *Comamonas* sp. strain 7D-2. Journal of Agricultural and Food Chemistry. 61 (47): 11550-11559.

Wang L, Wen Y, Guo X Q, et al. 2010. Degradation of methamidophos by *Hyphomicrobium* species MAP-1 and the biochemical degradation pathway. Biodegradation, 21: 513-523.

Wang L, Wang G L, Li S P, et al. 2010. *Luteibacter jiangsuensis* sp. nov.: a methamidophos-degrading bacterium isolated from a methamidophos-manufacturing factory. Current Microbiology, 62: 289-295.

Gu L F, Jiang J D, Li X H, et al. 2007. Biodegradation of ethametsulfuron-methyl by *Pseudomonas* sp. SW4 isolated from contaminated soil. Current microbiology, 55 (5): 420-426.

Sun L N, Zhang J, Chen Q, et al. 2013. *Sphingobacterium caeni* sp. nov., isolated from activated sludge. International Journal of Systematic and Evolutionary Microbiology, 63 (6): 2260-2264.

Sun L N, Zhang J, Kwon S W, et al. 2013. *Paracoccus huijuniae* sp. nov., an amide pesticide-degrading bacterium isolated from activated sludge of a wastewater biotreatment system. International Journal of Systematic and Evolutionary Microbiology, 63 (3): 1132-1137.

Liu H M, Lou X, Ge Z J, et al. 2015. Isolation of an aryloxyphenoxy propanoate (AOPP) herbicide-degrading strain *Rhodococcus* ruber JPL-2 and the cloning of a novel carboxylesterase gene (feh). Brazilian Journal of Microbiology, 46 (2): 425-432.

Cheng M, Chen K, Guo S, et al. 2015. PbaR, an IclR family transcriptional activator for the regulation of the 3-phenoxybenzoate 1', 2'-dioxygenase gene cluster in *Sphingobium wenxiniae* JZ-1T. Applied and Environmental Microbiology, 81 (23): 8084-8092.

Lu P, Jin L, Liang B, et al. 2011. Study of biochemical pathway and enzyme involved in metsulfuron-methyl degradation by *Ancylobacter* sp. XJ-412-1 isolated from soil. Current Microbiology, 62 (6): 1718-1725.

He Q, Huang J, Yang X, et al. 2016. Effect of pesticide residues in grapes on alcoholic fermentation and elimination of chlorothalonil inhibition by chlorothalonil hydrolytic dehalogenase. Food Control, 64: 70-76.

Chen Q, Wang C H, Deng S K, et al. 2014. Novel three-component Rieske non-heme iron oxygenase system catalyzing the N-dealkylation of chloroacetanilide herbicides in Sphingomonads DC-6 and DC-2. Applied and Environmental Microbiology, 80 (16): 5078-5085.

Li R, Zheng J W, Wang R, et al. 2010. Biochemical degradation pathway of dimethoate by *Paracoccus* sp. Lgjj-3 isolated from treatment wastewater. International Biodeterioration and Biodegradation, 64: 51-57.

Li R, Guo X Q, Chen K, et al. 2009. Isolation of an isocarbophos-degrading strain of *Arthrobacter* sp. scl-2 and identification of the degradation pathway. Journal of Microbiology and Biotechnology, 19 (11): 1439-1446.

Li R, Liu Y, Zhang J, et al. 2012. An isofenphos-methyl hydrolase (Imh) capable of hydrolyzing the P-O-Z moiety of organophosphorus pesticides containing an aryl or heterocyclic group. Applied Microbiology and Biotechnology, 94: 1553-1564

Ali S W, Yu F, Li L, et al. 2012. Studies revealing bioremediation potential of the strain *Burkholderia* sp. GB-01 for abamectin contaminated soils. World Journal of Microbiology and Biotechnology, 28 (1): 39-45.

Gu T, Zhou C, Sørensen S R, et al. 2013. The novel bacterial N-demethylase *PdmAB* is responsible for the initial step of N, N-dimethyl-substituted phenylurea herbicide degradation. Applied and Environmental Microbiology, 79 (24): 7846-7856.

Li X, Jiang J, Gu L, et al. 2008. Diversity of chlorpyrifos-degrading bacteria isolated from chlorpyrifos-contaminated samples. International Biodeterioration and Biodegradation, 62 (4): 331-335.

Yan X, Gu T, Yi Z, et al. 2016. Comparative genomic analysis of isoproturon - mineralizing sphingomonads reveals the isoproturon catabolic mechanism. Environmental Microbiology, doi: 10.1111/1462-2920.13413

Huang X, He J, Sun J, et al. 2007. Isolation and characterization of a metsulfuron-methyl degrading bacterium *Methylopila* sp. S113. International Biodeterioration and Biodegradation, 60 (3): 152-158.

Huang X, He J, Sun X F, et al. 2010. Characterization and molecular mechanism of a naturally occurring metsulfuron-methyl resistant strain of *Pseudomonas aeruginosa*. World Journal of Microbiology and Biotechnology, 26 (3): 515-521.

Huang X, Pan J, Liang B, et al. 2009. Isolation, characterization of a strain capable of degrading imazethapyr and its use in degradation of the herbicide in soil. Current microbiology, 59 (4): 363-367.

Li Y, Zhang J, Chen Q, et al. 2013. *Dokdonella kunshanensis* sp. nov., isolated from activated sludge, and emended description of the genus Dokdonella. International Journal of Systematic and Evolutionary Microbiology, 63 (4): 1519-1523.

Li Y, Chen Q, Wang C H, et al. 2013. Degradation of acetochlor by consortium of two bacterial strains and cloning of a novel amidase gene involved in acetochlor-degrading pathway. Bioresource Technology, 148: 628-631.

Zhang J, Zheng J W, Liang B, et al. 2011. Biodegradation of chloroacetamide herbicides by *Paracoccus* sp. FLY-8 in vitro. Journal of Agricultural and Food Chemistry, 59 (9): 4614-4621.

Feng Z, Li Q, Zhang J, et al. 2012. Microbial degradation of fomesafen by a newly isolated strain *Pseudomonas zeshuii* BY-1 and the biochemical degradation pathway. Journal of Agricultural and Food Chemistry, 60 (29): 7104-7110.

Nie Z J, Hang B J, Cai S, et al. 2011. Degradation of cyhalofop-butyl (Cyb) by *Pseudomonas azotoformans* strain qdz-1 and cloning of a novel gene encoding cyb-hydrolyzing esterase. Journal of Agricultural and Food Chemistry, 59 (11): 6040-6046.

Trans-Membrane Transport Theory and Synergistic Remediation Technology for the Bioremediation of Oil Polluted Soil

Hongqi WANG, Mengyuan SU, Shuo DIAO, Jie XU, Xiaoxiong WU

College of Water Sciences, Beijing Normal University

Abstract

The main researches of our team are mainly divided into 3 different part. The first main research is combined with theoretical innovation. We proposed the theory of Trans-membrane transport of petroleum hydrocarbons, and creatively apply the ^{13}C stable isotope and ^{14}C radioactively labeled element to trace the transport of the PAHs in the subcellular scale. We have successfully found the rule of BaP-1 degrading the ^{14}C-fluoranthene, and the optimum condition during the degradation. The second main research is combined with the selection of bacterium which have the ability to degrade the PAHs in low temperature condition (10℃). Furthermore, we have constructed three different communities. The third part of our research is mainly about the technologies and applications of plant-microbial combined remediation of petroleum contaminated soil.

Keywords: trans-membrane transport theory, subcellular image, low temperature degradation, plant-microbial combined remediation

1. Introduction

Polycyclic aromatic hydrocarbons (PAHs) are a concern in the whole world because of their toxic, mutagenic, teratogenic and carcinogenic effects. PAHs are a class of petroleum pollutants and are typical persistent organic pollutants. PAHs contain a group of potential environmental pollutants which can be found in soils, water, air and sediments, which are considered to be the possible hazard to the environment, and their distributions in the environment will cause a significant health risk for human beings.

Bioremediation is widely known as an appropriate remediation technology for hydrocarbon-contaminated soils. However, the successful application of bioremediation depends on appropriate biodegradative microorganisms. It is important to research the mechanism of biodegradation and the method to increase the degradation rate of PAHs in this process. Considering that the degradation system of PAHs in microorganisms is complex, and the key link and rate-controlling steps in the process are not clear, we have studied the uptake and transportation processes of PAHs by microorganisms, which are confirmed to be the key steps in the remediation of PAHs. Besides, some application of bioremediation technology, such as immobilization technology and plant-microbial combined remediation, have been researched in recent years.

2. Soil bacterium absorption uptake and trans-membrane transport of petroleum hydrocarbons

2.1 Isotope tracer technique for trans-membrane transport of petroleum hydrocarbon micro-mechanism analysis of microorganisms

Polycyclic aromatic hydrocarnons are typical persistent organicpollutant, and also a class of petroleum pollutants. They are widely diffused in soil, air, and water. Although PAHs might adsorb by water, evaporate into air and photolysis, the degradation by microbial is still the major process of the whole degradation. And PAHs degradation efficiency mostly depends on the environmental conditions, number and types of the microorganisms, (Haritash et al., 2009). We isolated *Rhodoccossus* sp. BaP-1 from soil at Beijing Chemical Plant (Beijing, China) that has been contaminated by organic compounds and contains high concentration of PAHs. The indigenous biosurfactant-producing bacterium, *Rhodococcus* sp. BaP-1 has high PAHs degradation ability. In the present study, the mechanism of ^{14}C-fluoranthene of *Rhodococcus* sp. BaP-1 was examined, outside and inside the bacteria cell membrane of the ^{14}C were also investigated. The transport of fluoranthene involves concurrent catabolism of ^{14}C that leading to the generation of significant amount of $^{14}CO_2$. A significant influence of environmental factors on uptake and trans-membrane transport process was investigated, in order to screen for the key factors that influence the biodegradation process of PAHs contaminated environments. Then, we used the Box-Behnken design to optimize and enhance the transe-membrane transport process. And the result showed that the max-imum cellular uptake rate of the fluoranthene could achieve 0.0308 μmol/(min·mg) protein (observed) and 0.304 μmol/(min·mg) protein (predicted) when the premier temperature, pH and saltility were set at 20℃, 9% and 1%, respectively (Li et al., 2013).

During growth on mineral salts medium, the bacterium could produce biosurfactants, higher hydrophobic lead to better contact between bacteria and fluoranthene. When concentration of fluoranthene was much higher, it would cause lower biodegradation rate. Associated with Michaelis-Menten equation to describe the uptake and trans-membrane transport kinetics, on the basis of the experimental results, emulsification was probably not the leading reason that occurred simultaneously during the biodegradation process. Fluoranthene uptake was measured with ^{14}C-fluoranthene (ChemDeop, 07-43-060, 99% pure, 60 mCi/mmol). The uptake process of the uptake process of fluoranthene by *Rhodococcus* sp. BaP-1was evaluated further by measuring ^{14}C-fluoranthene mineralization. Mineralization was carried out at concentration of ^{14}C-fluoranthene ranging from 0.2 to 20 μmol/L. After 48 h incubation, mineralization was measured as $^{14}CO_2$ evolution from^{14}C-fluoranthene. The distribution of ^{14}C in supernatant, cell pellet, $^{14}CO_2$ trap, as well as abiotic lost during the experiments, are presentedin Table 1. When cells were incubated with 0.2, 0.4, 1.6, 3.2, 9.6 and 20 μmol/L of ^{14}C-fluoranthene, the proportion of radioactivityfrom $^{14}CO_2$ was 34.58%, 46.23%, 44.42%, 38.18%, 37.49% and 36.83%, respectively. The abiotic loss of ^{14}C-fluoranthene was lowerthan 7%. As an example, for lower initial concentration of 0.4 μmol/L ^{14}C-fluoranthene, 46.23% of the radioactivity lost fromthe aqueous phase (72.1%), was recovered in form of $^{14}CO_2$, whileonly 25.0% was recovered in cell pellet, and the abiotic lost was 1.68%. But in higher initial concentration incubation, mineralization by cells was lower. The results indicated that the transport offluoranthene involves concurrent catabolism of ^{14}C, that leading to the generation of significant amount of $^{14}CO_2$.

Table 1 Distribution of ^{14}C-fluoranthene after incubation in the presence of *Rhodococcus* sp. BaP-1

Totaradion cpm	Supernatant cpm	Cell pellet cpm	Trapped $^{14}CO_2$/cpm	Abiotic lost/cpm
15408	5875	3633	5328	825
30816	8273	7866	14247	518
123264	44553	20395	54749	4305
246528	99818	42036	94125	14517
739584	270423	127980	277237	45490
1540800	548199	327231	567463	97805

Table 2 Variables and test levels for Plackett-Burman design

No.	Variables	Levels			*F*-value	*p*-Value
		-1	0	+1		
X1	Temperature/℃	20	25	30	7.27	0.0308
X2	pH	7	8	9	62.74	<0.0001
X3	Salinity/%	0	2	4	8.38	0.0231
X4	Inoculum level/%	0.5	1.0	1.5	0.22	0.6554
X5	C/N	5	10	15	1.95	0.2049
X6	Oscillation frequency/rpm	100	125	150	0.108	0.9744
X7	Contact time/min	1	3	5	0.25	0.639

2.2 Subcellular imaging of stable isotopically labeled organic compounds in a single cell by NanoSIMS

Stable isotopes (e.g., ^{13}C, ^{15}N, etc.) have the ability to tag wide range of inorganic and organic compounds such as $NaH^{13}CO_3$, $^{13}CH_4$, and have already been used in many ways intending to have a good understanding of the biological process. And at the subcellar scale, secondary iron mass spectrometry (NanoSIMS) have successfully addressed many technical questions associated with premiere low elementary sensitivity, aggressive sample destruction and inadequate spatial resolution (Peta et al., 2007).

In our research, we highlighted the potential of stable isotope combined with NanoSIMS to detect and image the uptake, metabolism, distribution of labeled compounds of ^{13}C-glucose. We cultured *Halomonas* sp. BDG-3 within ^{13}C-glucose for 15 h, then the cultured sampled are chemically fixed, dehydrated and embedded with resin to form the ultra-thin slices for TEM (transmission electron microscope), and NanoSIMS at the same time. We use the TEM to confirm the location of the single cell. And the NanoSIMS examination will show the subcellular distribution of labeled element at subcellular level. Specifically, we have already determined 3 different growth stages during the growth of BDG-3 by different metabolism of ^{13}C signal and ^{14}N signal.

Considering that high resolution image of NanoSIMS couldn't reflect the exact area of cells directly. For choosing the exact examining area of interest, we chose transmission electron microscopy (TEM) imaging to locate single cell before NanoSIMS analysis.

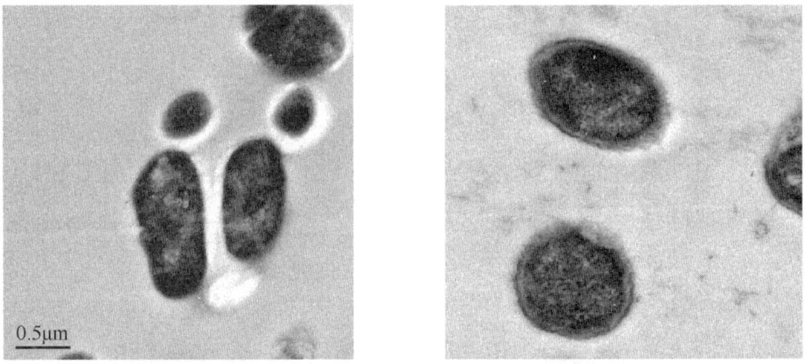

M image of individual BDG-3 cell, demonstrating the structure

(a) BDG-3 was grown in MSM with ^{12}C-glucose　(b) BDG-3 was grown in MSM with ^{13}C-glucose

Fig. 1　TEM image of individual BDG-3 cell, demonstrating the structure

In Fig. 1 we could clearly observe the structure of single cells. And the complicated pre-preparation for NanoSIMS did not destruct the integrity of cells. SIMS analysis was carried out on the CamecaNanoSIMS 50L. Cs^+primary ion was generated by a caesium source and accelerated towards the sample surface. In Fig. 2 showed the signal of $^{13}C^-$ and $^{12}C^{14}N^-$ respectively. In Fig. 2 (a) showed the different intensity of $^{13}C^-$ in the area of the sample. Different areas have obvious difference of intensity. And $^{12}C^{14}N^-$ represents the signal of the N signal. N signal plays the role of locating, depicting, quantity the area of cytomembrane, nucleoid, and organelle. Combining Fig. 2 (a) and Fig. 2 (b), we could easily find that different single cells have the different uptake of ^{13}C-glucose. And ^{13}C-glucose plays as the sole source of the C source we could easily get the different metabolism situation.

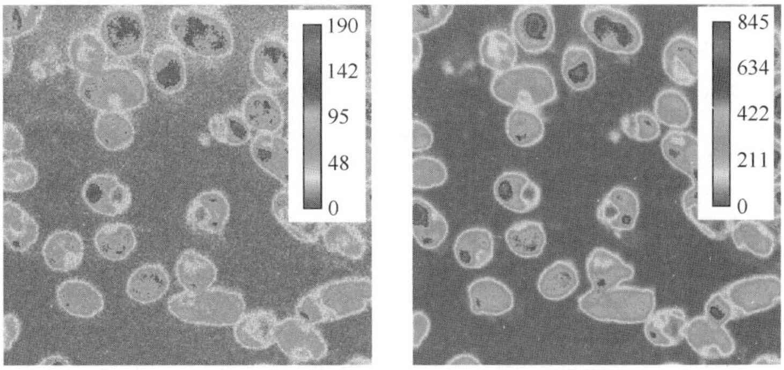

(a) ^{13}C (b) $^{12}C^{14}N^{-}$

A thin section(0.5 μm) was observed in primary Cs^{+} source. Observed field: 10 μm×10 μm.

Fig. 2 High resolution NanoSIMS images of BDG-3 observed after incubation with ^{13}C-glucose

By this high-resolution method, we can define the different stage of uptake and transport process in the different cells. Further we could reveal novelk insights into metabolic activities. By mastering these complicated specimen preparation data analyses of NanoSIMS, we could replace the large and chemically complex tags used for fluorescence microscopy, without altering the chemical and physical properties of the molecule. Through the study of methodology, this is a new and reliable way to analysis the metabolism process in further research about degradation of more complex pollutants.

3. Biodegradation technology and application of oil polluted soil

Many PAHs have highly toxic, mutagenic, or carcinogenic properties (Sato and Aoki, 2002), but are quite recalcitrant to biodegradation. Microbial degradation is an important measure for PAHs-contaminated soil. A study by Greece researchers showed that bacterial strains which obtained by screening from petroleum-contaminated soil may prove to be promising microorganisms for bioremediation to remove PAH-containing pollutants from contaminated sites (Zhang et al., 2004).

3.1 Isolation, classification and identification, and the species diversity of high effectively and cold-resistant PAHs-degrading bacteria in low temperature condition (10℃)

3.1.1 Screen and isolation

Under the conditions of 10℃, benzo[a]pyrene (BaP), benzo[a]anthracene (BaA), benzo[b]fluoranthene (BbF), benzo[k]fluoranthene (BkF), indeno[1, 2, 3-cd]pyrene (IPY) were used as carbon and energy source singly or in combination and 50 strains were isolated by steps including enrichment culture and plate cultivation from the soil at a coking plant in Beijing (Li, 2010).

3.1.2 The improvement of polymerase chain reaction (PCR) and the analysis of species diversity

We deal with bacteria by the methods of PCR-digestion directly without the extracting of DNA. What's more, the relationship between species could be reflected into the library (Hao, 2009).

Pure culture microbial strains which were isolated from coking plant No.3 soil samples were used as templates, biological methods have been improved to make a rapid removal of duplication between strains, using PCR technique (primer 8f, 1492r) and ARDRA patterns which enzyme digestion (RsaI and MspI) of cloned 16S rDNA gene sequences. The results showed that under the conditions of 10℃, 50 strains are clustered into 15 operational taxonomic units (OTUs) (Fig. 3) and belong to genera *Sphingomonas*, *Methylobacterium*, *Burkholderia*, *Rhodococcus*, *Bradyrhizobium*, *Phyllobacterium*, *Chryseobacterium*, *Microbacteriumm*, of which *Sphingomonas* is the dominant species. These 15 sequences were submitted to GenBank and the accession number were GQ249211-GQ249225 (Li, 2010).

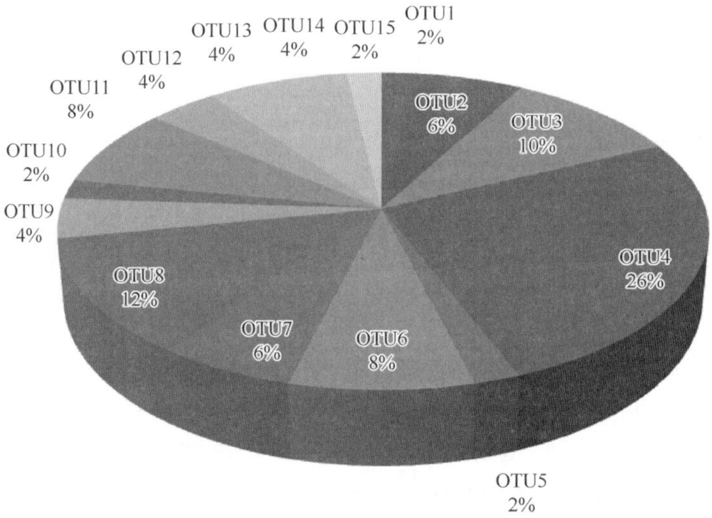

Fig. 3 Microbial community structure of PAHs-contaminated soil

3.2 Degradation properties of high effectively and cold-resistant PAHs-degrading bacteria, and community construction

Degradation property is almost the most significant properties for bacteria, and the degradation properties of sole bacteria was limited. Therefore, the study of community construction for heavy contaminated soil remediation is of profound theoretical and practical significance.

3.2.1 Degradation properties under the conditions of 10℃

In the first experiment: the degradation performance of all the strains under the conditions of 10℃ was tested. In the liquid phase, No. C1, D2, D9, D17, D24 strains were in the best growing conditions after 20-day culture, among which D17 and D24 increased obviously in growth (Fig. 4 (a)). Change of the growth and oil degradation had the same trend of these 5 strains. The degradation rate of D24 was 39.81%, while the degradation rate of D17 was 47.64% (Fig. 4 (b)). Conditions in the soil phase, after 15 days of degradation, TPH in the system with D17 is 86.5 mg (per 7 g soil) less than which not adding strains (empty), with the degradation rate of 15.9%; TPH in the system with D24 is 15.8 mg (per 7 g soil) less than that not adding strains (empty) (Fig. 4 (c)). GC-MS results of D24 told that D24 had little effect on the degradation of saturated hydrocarbon. Possibly because the sample has been partly degraded by the original microbe in the soil, which made further degradation more difficult. However, adding D24 had a significant impact on the aromatic components, in which the degradation level of heavy aromatic components is higher than the light component (Hao, 2009).

In the second experiment: the degradation performance of all the strains was tested with ultrosonic extraction method and HPLC. The highest degradation rates of BaA, BaP, BbF, BkF, IPY were 39.64%, 33.52%, 38.57%, 25.37% and 31.17% respectively when tested with

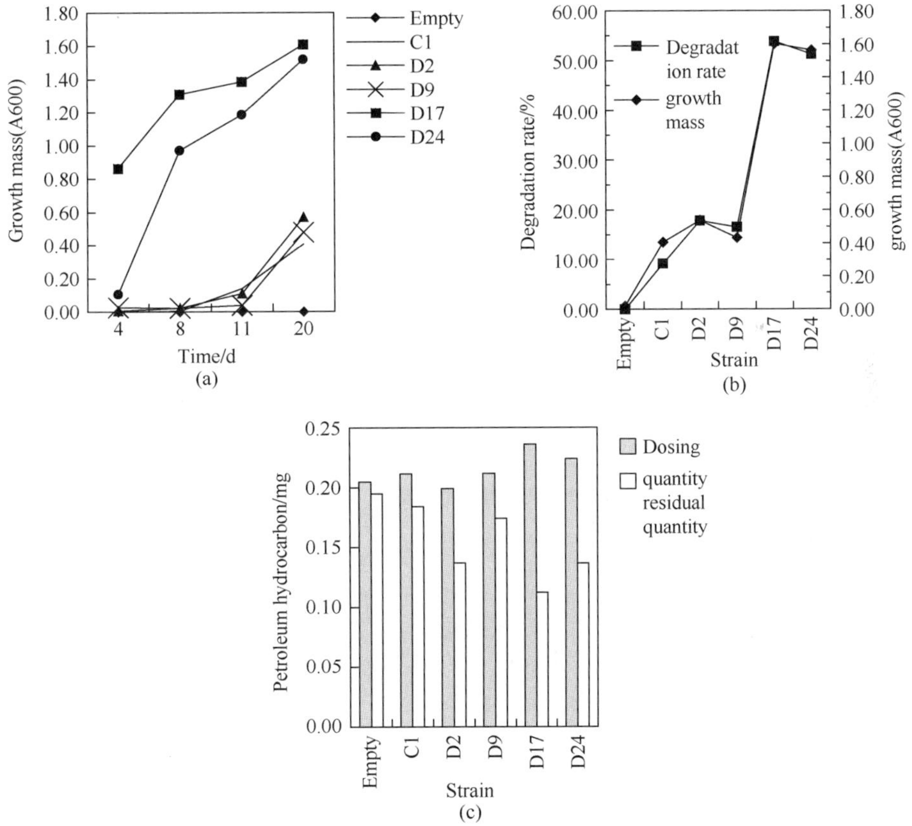

Fig. 4 (a) Growth curve of 5 strains (10℃) ; (b) Growth curve of 5 strains and crude oil-degradation; (c) TPH in the system

strain 3-6-12 which was identified as Burkholderia in 12 days. The degradation rate of *Sphingomonas* was not the highest in the tests of these strains (Li, 2010).

3.2.2 Construction of bacterium community

The limited degradation capability of each signal strain on PAHs, especially the high-molecular weight PAHs, was confirmed. Therefore we construct different bacterial consortias to form better degradation system to enhance its efficiency.

According to the difference of degradation for five PAHs by single strains, we obtained three bacterial consortia. The first bacterial consortia contains 7 strains. The degradation rate of benzene (a) pyrene (BaP), benzo[a]anthracene (BaA), benzo[b]fluoranthene (BbF), benzo[k]flu oranthene (BkF) was the highest by different strain. The second bacterial consortia consists of 5 strains, which has a height advantage on growth in all of the 15 OUTs. The third bacterial consortia consists of the top four strains which have the higher degradation capability of mixed PAHs (Li, 2010).

Using these consortia, the average degradation rates reached 77.22%, 62.14% and 69.04%, respectively in 30 days (Table 3).

Table 3 Degradation of PAHs by construction of bacterial consortium

Bacterial consortium	Strains	PAHs-degradation rates/%				
		BaA	BbF	BkF	BaP	IPY
1	1-1-1、1-1-2、3-2-1、3-2-3、1-3-1、2-4-1、3-4-1	76.30	78.57	77.00	76.86	77.39
2	3-6-5、3-6-10、3-6-15、3-6-25、3-6-30	60.34	64.68	61.82	61.76	62.11
3	3-6-12、3-6-15、3-6-25、3-6-27	68.32	70.35	68.89	68.29	69.37

3.3 Studying of optimum soil environmental conditions for constructed community

The degradation rate of amount of 14 kinds of PAHs was determined after 14 days, 28 days and 52 days (Fig. 5). In order to obtain the optimum soil environmental conditions of biodegradation, this orthogonal experiment selected seven influencing factors: bacterial consortia, amount of bacteria, temperature, soil water content, the ratio of carbon, nitrogen and phosphate, surfactant (TW-80), Fenton reagent and corn oil. The results of range analysis showed the primary factor for influencing the effects of bioremediation was C: N: P during the first 14 days, temperature and bacterial consortia were the primary factor of 28d and 52d respectively (Fig. 6) (Li, 2010).

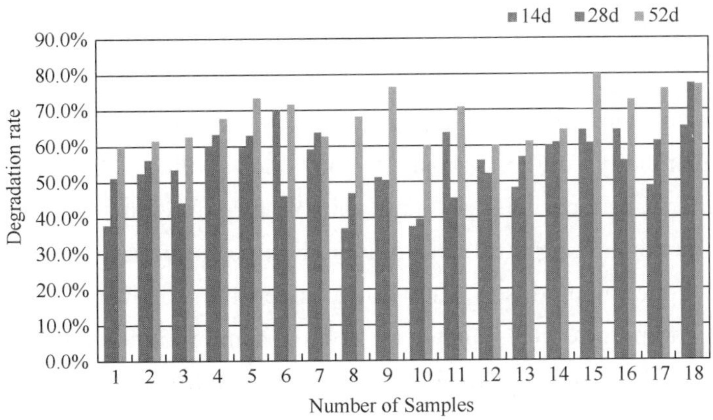

Fig. 5 Degradation rate of the samples at different times

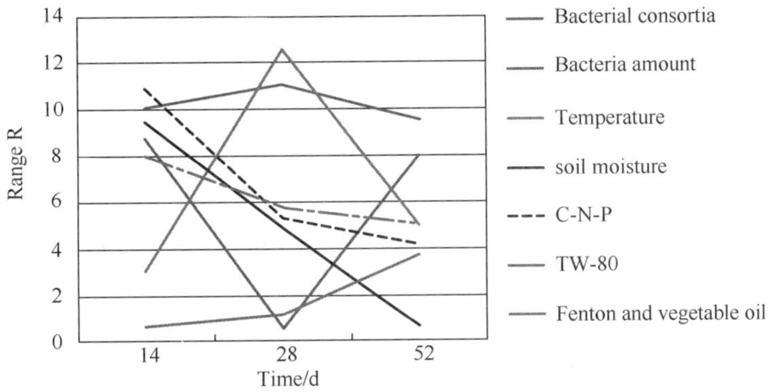

Fig. 6 The importance and optimal levels of the factors

3.4 Study on immobilization technology of cold-resistant petroleum hydrocarbon-degrading bacteria in soil

3.4.1 Factors influencing immobilization process

The optimized results of process parameters are as following: SA-attapulgite clay immobilized microorganism, water temperature 50 ℃, microbial mass 4%, reaction time 18h, multiplication time 60 h; SA-attapulgite clay-$CaCO_3$ immobilized microorganism, water temperature 60℃, microbial mass 2%, reaction time 18 h, multiplication time 48 h. The unified technical parameters offer the potential of large-scale preparation and production of immobilized microorganism.

3.4.2 Comparison between two immobilized microorganism

Comparing the basic performance by microscopic view, scanning electron microscopy (SEM) shows that, with the introduction of attapulgite, the apparent roughness is reinforced; internal drape and pore are increased; SA-attapulgite clay-$CaCO_3$ immobilized microorganism has internal cellular structure and a dense layer of immobilized ball surface, which can effectively prevent internal bacteria outflows, and be good for transfer of substrates, product and dissolved oxygen.

Comparing two immobilized microorganism by petroleum hydrocarbon degradation, experiment in artificially oil-contaminated soil shows that SA-attapulgite clay-$CaCO_3$ immobilized microorganism particles are more suitable for pollution site remediation.

Comparing the degrading petroleum hydrocarbon performance, degradation experiments in artificially oil contaminated soil show that, under the condition of different substrate concentration, SA-attapulgite clay-$CaCO_3$ immobilized microorganism has more advantages of degrading. Under 5% and 10% inoculation quantity condition, the mass transfer ability of SA-attapulgiteclay-$CaCO_3$ immobilized microorganism can provide security for microorganism using petroleum hydrocarbon as carbon source. In short term, the oil degradation rates by SA-attapulgite clay-$CaCO_3$ immobilized microorganism can reach up to 60.41% (Wang, 2011).

4. Technologies and applications of plant-microbial combined remediation of petroleum contaminated soil

The combined pollution of petroleum and heavy metals in unsaturated zones in soil is a common problem in China. Taking the advantage of both phytoremediation and microbial remediation, the ecological effects and remediation of PAHs and heavy metal composite pollution in soils were studied by pot experiments at the indoor controllable conditions. The ecological effects in single or composite conditions were explored, as well as the interactions between pollutions. The mechanism of collaborative remediation of composite polluted soils by plants and microorganism were also studied.

4.1 Pollution of lead and benzo[a]pyrene

One study focused on the combined ecological effects of lead (Pb) and benzo[a]pyrene (BaP) on soil enzymes and physiological characters of plants and their fates in soil-plant systems through greenhouse pots experiments. Studies on a modified joint-action model and its application to describing ecological effects of Pb-BaP combined pollution were also covered in this study.

Seed germination, shoot elongation, root elongation and antioxidant enzymes in shoot were measured and tested to explore the psychological responses of wheat. A uniform design of 6 levels for two factors was employed here to study the effect of the interaction of Pb and BaP on the soil enzymes, including dehydrogenase activities (dha), urease activities (ura) and catalase activities (cat). The effects of rhizophere on the responses of soil enzymes to combined pollution of Pb and BaP were also studied here. The adherence of BaP to root of corn was found to be related with Pb in soil.

A vigorous joint action model was drawn in this study to describe the combined effect of Pb and BaP. This model included the ratio of pollutants as well as the multiplication of pollutants as the interaction forms between pollutants. The model has been successfully applied to prediction of variation of soil enzymes under the combination of Pb and BaP. This model was considered to be very useful in discovering the mechanisms of interaction between pollutants (Wang, 2010).

4.2 Pollution of cadmium and lead

Plant-microbial combined remediation was comprehensively conducted to treat cadmium and lead compound pollution in the soil by applying the knowledge of pollution ecology and biology. The study started with screening strains that had tolerance to heavy mental, took *Solanumnigrum* L. and *Bidens aximowicziana* Oett as tested plants, then designed 16 groups orthogonal experiment with plant, microbial, chelating agents, Cd treatment, and Pb treatment as the influencing factors on the basis of single cadmium and lead pollution to systematic study the growth conditions of the plants, plants in particular, the uptake and accumulation content of heavy metal in the plant, especially the ground plant parts, the heavy-metal content in rhizosphere soil, and rhizosphere controlling experiments including the rhizosphere pH, microbes, soil enzyme activity.

All of these studies revealed the uptake and accumulation characteristics of plants, the activation effect of heavy metal, the accumulation and transport capacity of plants, and rhizosphere controlling of and influencing factors, which could help seek the complex interaction mechanisms between various factors under the combined stress of Cd and Pb pollution (Wang, 2010).

4.3 Pollution of oil

In order to explore the feasibility of the combined remediation of petroleum contaminated soil, one study focused on the combined remediation of petroleum contaminated soil by degrading bacterial and the winter wheat through indoor experiments and a field simulation trial, and to explore the optimum conditions of the combined remediation, the greenhouse pots experiments were conducted (Wang, 2011).

Cold-adapted bacteria using petroleum hydrocarbon as the sole carbon source which isolated from Tianjin Dagang oil contaminated soil were selected, while wheat and alfalfa were chosen as supplied plants, different remediation systems were constructed then. The result of oil degradation was compared and the soil enzyme activity was determined. The diversity of rhizobacteria which included the microorganism function diversity and the genetic diversity as well as the influence of inoculation bacteria was analyzed with the method of molecular biology. The response of micro-ecosystem of the hydrocarbon phytoremediation inoculated with a microbial was studied.

After 70 d remediation, it could be observed that wheat, alfalfa and inoculation bacteria could enhance the petroleum hydrocarbon degradation rate to different degree. The degradation rate was 18.09% in wheat-immobilization inoculation bacteria system which had better remediation potential (Ying, 2011).

5. Conclusions

In the present study, the uptake and trans-membrane transport mechanism of petroleum hydrocarbons by degrading bacteria and immobilization technology of cold-resistant petroleum hydrocarbon-degrading bacteria in soil were both used in actual plant-microbial combined remediation of petroleum contaminated soil, thus a realistic new technique is provided for remediation method for the petroleum hydrocarbons contaminated soil in low temperature condition. Main conclusions of the study included:

(1) The uptake and trans-membrane transport mechanism of petroleum hydrocarbons by degrading by bacteria were examined; the distribution mechanism of petroleum hydrocarbons outside and inside the bacteria cell membrane were also investigated. A significant influence of environmental factors on uptake and trans-membrane transport process were investigated, in order to screen for the key factors that influence the biodegradation process of PAHs and apply some significant implications for the bioremediation of PAHs contaminated environments. Correlating with the assistance of the NanoSISM combined with the stable isotope, we could easily have a better understanding of the subcellular distribution of different process of the uptake and transport of the organic compounds even more complex pollutants.

(2) In low temperature, oil-contaminated soil samples were used and crude oil was used as the sole carbon source to isolate and screen the major functional bacteria. 31 degradation bacteria were isolated under the conditions 10℃. We completed classification and identification, and the analysis of species diversity of high effectively and cold-resistant PAHs-degrading bacteria. What's more, we improve the degradation properties of high effectively and cold-resistant PAHs-degrading bacteria by construct community in different ways.

(3) Based on SA-attapulgite clay, SA-attapulgite clay-$CaCO_3$ composite materials as the carrier and highly efficient petroleum-degrading bacteria screened by laboratory, immobilized microorganism system was constructed. Immobilization technology can not only enhance the resistance to environmental impact, but also promote the growth of microorganism. In addition, the dynamic change of petroleum hydrocarbon degradation of artificial contaminated soil and undisturbed contaminated soil with SA-attapulgite clay-$CaCO_3$ immobilized microorganisms were inspected. The degradation petroleum hydrocarbon experiment in undisturbed contaminated soil shows that, by degradation effects contrast with free bacteria, immobilized microorganism

got more degradation ability enhancement.

(4) Taking the advantages of both phytoremediation and microbial remediation, the ecological effects and remediation of PAHs and heavy-metal composite pollution in soils were studied by pot experiments at the indoor controllable conditions. The ecological effects in single or composite conditions were explored, as well as the interactions between pollutions. The mechanism of collaborative remediation of composite polluted soils by plants and microorganism were also studied. Plants which had high tolerance for coexistence of PAHs and heavy metals in soils were screened. The plants were also found to have excellent ability to enrich heavy metals in shoot. Besides, the mutual promotion and mutual restraint between plants and microbe were further discussed. A detailed description of the interaction between petroleum and heavy metals to plant growth and soil enzyme activities was given under low concentration conditions. Improving joint action module between petroleum and heavy metals were proposed and tested successfully. The distribution and fates of the compound pollutants in soil-plant system, as well as the controlling techniques of rhizosphere conditions, were also discussed. All the obtained results provided useful technical support for remediation of soils polluted by both oil and heavy metals in the future.

References

Clode P L, Stern R A, Marshall A T. 2007. Subcellular imaging of isotopically labeled carbon compounds in a biological sample by ion microprobe (NanoSIMS). Microscopy Research and Technique, 70 (3): 220-229.

Hao X G. 2009. Screening and identification of the predominant cold-resistant bacteria in oil-contaminated soil and studying of their composition and degradation ability. Beijing: Beijing Normal University.

Haritash A K, Kaushik C P. 2009. Biodegradation aspects of polycyclic aromatic hydrocarbons (PAHs): A review. Journal of Hazardous Materials, 169 (1): 1-15.

Li X B. 2010. Isolation of high-molecular weight PAHs-degrading bacteria, community construction and studying ofoptimum soil environmental conditions. Beijing: Beijing Normal University.

Li Y, Wang H, Hua F, et al. 2014. Trans-membrane transport of fluoranthene by *Rhodococcus* sp. BaP-1 and optimization of uptake process. Bioresource Technology, 155: 213-219.

Li Y, Wang H, Hua F. 2013. Uptake modes of fluoranthene by strain *Rhodococcus* Sp. BaP-1. Biotechnology and Biotechnological Equipment, 27 (6): 4256-4262.

Sato H, Aoki Y. 2002. Mutagenesis by environmental pollutants and bio-monitoring of environmental mutagens. Current Drug Metabolism, 3: 311-319.

Su M Y, et al.2016. Subcellular imaging and metabolic activity analysis of isotopically labeled carbon compounds in biological sample by ion microprobe (Nano SIMS). Beijing: Beijing Normal University.

Wang Q. 2011. Cold-adapted bacterium and winter wheat combined remediation of oil-polluted soil. Beijing: Beijing Normal University.

Wang S. 2010. Ecological effects and fates of combined pollution of lead and benzo[a]pyrene in soil-plant system. Beijing: Beijing Normal University.

Wang X. 2010. Plant-microbial combined bioremediation of cadmium, lead compound contaminated soil. Beijing: Beijing Normal University.

Wang X. 2011. Study on immobilization technology of cold-resistant petroleum hydrocarbon-degrading bacteria in soil.

Beijing: Beijing Normal University.

Ying X. 2011. Resarch on remediation of petroleum contaminated soil by plant-inoculation bacteria and the change of rhizosphere micro-ecosystem. Beijing: Beijing Normal University.

Zhang H M, Kallimanis A, KoukkouA, et al. 2004. Isolation and characterization of novel bacteria degrading polycyclic aromatic hydrocarbons from polluted Greek soils. Applied Microbiology and Biotechnology, 65 (1): 124-131.

Site Remediation Technology Research Based on Dynamic Balance Mechanism of Organic Contaminants in Soil/Water/Air Environmental System

Wenhui ZHANG

CECEP DADI Environmental Remediation Co., Ltd.

Abstract

Based on the dynamic balance of organic contaminants in soil/water/air environmental system, soil vapor extraction (SVE), in-situ chemical oxidation (ISCO) and thermal remediation technologies applicable for varied organic composite contaminated sites were developed and integrated, and successfully applied in remediating multiple representative chemical contaminated sites in Jiangsu and Hangzhou. Proprietary owned indirect thermal desorption unit with horizontal double cylinder rotary kiln and new method and complete set of equipment for multi-stage purification for off gas were successfully researched and developed, to treat recalcitrant/persistent organic contaminated soil, which initiated the site scale application of thermal desorption technology in decommissioned sites of pesticide chemical plants. Mobile and module-designed SVE equipment was developed based on dynamic balance mechanism of volatile organic contaminants (VOCs), characterized with easy installation and operation, high efficiency, more evenly extraction effect and reduced extraction blindness. Fenton/sodium persulfate combined remediation technology was researched and developed in the basis of dynamic balance mechanism of organic contaminants desorption-dissolution-oxidation in soil/water environmental system. This developed technology accelerates the degradation and removal process of adsorbed contaminants, improves the treatment effect and shortens the treatment period. First site scale application of ISCO in industrial contaminated site, significantly promotes the application level of ISCO in China.

1. Research background

Relocation of industrial enterprises is an effective measure to rapidly improve the municipal environment and promote the upgrade of enterprises. Recent years, with the implementation of policies such as "suppress the second industry and develop the third industry", "relocation of industrial enterprises from urban area to industrial park/suburbs" and "industry transfer", almost all of the middle-large size cities are facing the shut down and relocation of heavily contaminated industrial enterprises, a large number of abandoned sites are emerging in urban areas. According to the statistics, the shutdown and relocated enterprises were increased from 6.61×10^3 to 2.25×10^4 between 2001 and 2008, with a growth of 1984 per year and a total number of over 100 000 enterprises. These relocated enterprises, characterized with a long production history, relatively out-dated processing equipment and extensive management, were normally heavily contaminated. Direct development and utilization of these contaminated sites without any remediation will put the human health in danger. Industrial contaminated site remediation has been an extremely urgent major event, which also related to people' s livehood and China' s international imagine. Industrial site pollution can be divided into inorganic pollution (mainly refers to heavy metals), organic pollution and complex inorganic and organic pollution, based on the types of contaminants. Organic pollution is featured with more complex and diverse contaminant types and more commonly found, comparing with heavy metal pollution.

Organic contaminants generally exist in dissolved, adsorbed and gaseous form in soil/water/air environmental system, or exist in the form of non aqueous phase liquid (NAPL) when the concentrations are extremely high. Different existing forms of contaminants has varied environmental behaviors, existing form conversion of subsurface contaminants runs through the entire environmental process and forms a dynamic balance. Through artificially induced and development of control technologies, the dynamic balance can be driven to specific direction (for example: degradation of organic matter in contaminated soil can make the adsorbed contaminants convert to dissolved contaminants, and thus enhance the chemical oxidation effect), which can overcome the shortage of traditional remediation technologies and develop an efficient, green and economic technologies for industrial contaminated site remediation. Through the dynamic balance control, SVE, ISCO and thermal desorption technologies were developed and integrated by our company, to solve complex organic contaminated site remediation with different concentrations, pollution types and site characteristics.

2. Research achievement

2.1 Thermal desorption technology

Based on the dynamic balance of recalcitrant/persistent organic contaminants driven by heat in soil/water/air environmental system, proprietary owned indirect thermal desorption unit with horizontal double cylinder rotary kiln and new method and complete set of equipment for multi-stage purification for off gas were successfully researched and developed, to treat recalcitrant/persistent organic contaminated soil. This initiated the site scale application of thermal desorption technology in decommissioned sites of pesticide chemical plants, providing technical supports for localization of thermal desorption equipment, engineering application of thermal desorption technology in industrial contaminated site, and implementation of POPs convention in China.

The developed thermal desorption unit is indirect fired, double cylinder rotary kiln is adopted to enhance the heating effect and realize rapid and efficient treatment of contaminated soil. Contaminated soil does not have direct contact with the heat source, incomplete combustion will not occur and efficiently avoid the production of dioxin and other harmful toxic substances during the desorption process. Meanwhile, the adsorption off gas is relatively simply composed, which decreases its treatment cost and secondary pollution risk to the environment. Heat transfer processes mainly include: heat conduction between soil and inner kiln wall, convection and radiation between soil and carrier gas, convection and radiation between carrier gas and inner kiln wall, soil radiation to inner kiln wall, flue gas convection and radiation to inner and outer kiln wall, conduction between outer kiln wall and insulating layer, convection and radiation between the insulating layer and external environment. Kiln sealing structure was designed and developed, which efficiently solved its sealing difficulty in high temperature and dynamic conditions. The off gas generated from the thermal desorption process will be disposed and purified by non-incineration treatment chain: cyclone dust removal-two stage spraying-drying-activated carbon adsorption/gliding arc degradation, without production of dioxin and other harmful toxic substances.

Diagram of the thermal desorption unit can be seen in the figure below.

Fig. 1 Thermal desorption unit

More than 99% removal rate of organic contaminants such as VOCs, SVOCs, PAHs, PCBs and PBDEs can be reached, by using the thermal desorption unit to treat contaminated soil (see Fig. 2).

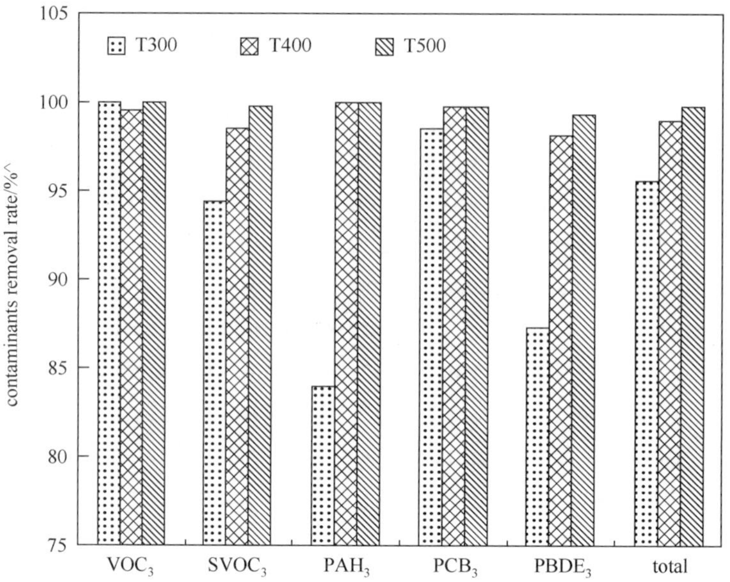

Fig. 2 Contaminants removal rate in different temperature conditions (T300, T400 and T500 respectively represents 300℃, 400℃ and 500℃)

2.2 SVE technology

Mobile and module-designed SVE equipment was developed based on dynamic balance mechanism of VOCs under normal pressure-negative pressure-normal pressure circulation. Extraction, electric, detection and off gas treatment units were integrated into a whole set of system, the SVE system is characterized with easy installation and operation and high efficiency. Air supply to soil is realized by specialized equipment in designated locations, which can ensure that SVE can be conducted in certain negative pressure and air supply can be added evenly to the soil, therefore more evenly extraction effect and reduced extraction blindness can be realized. SVE technology has been successfully adopted in various types of contaminated sites, providing technology and equipment support for economic and efficient remediation of VOCs and safe land use.

The researched and developed SVE technology adopts intermittent treatment process and saves energy consumption. Test was conducted on TPH contaminated soil to study the phase change during SVE process. Contaminant levels in off gas were measured daily when the SVE system starts and shuts down, by using VOCs fast detector (PID). The measurements show that as the extraction time extends, PID readings measured when the system switches on gradually reduces, which explains that concentrations of VOCs are gradually reduced and efficiently removed. Meanwhile, PID readings measured when the system shuts down are obviously lower than those measured when switches on, when the system restarts the next day, the readings are apparently higher than those measured when shuts down last time, this is possibly related with the phase change of contaminants in the soil. Data can be seen in Table 1 and Fig. 3.

Table 1 PID readings measured during the SVE test process

Date	PID readings measured when starts up	PID readings measured when shuts down
1 d	400	100
16 d	248	60
19 d	113	43
48 d	58.1	22
55 d	31.2	20
62 d	16.1	0.6

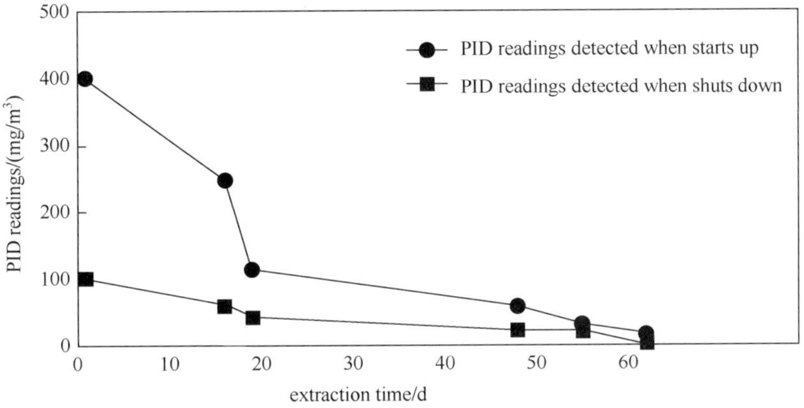

Fig. 3 Variation of PID readings during the SVE test process

Besides, field study shows that with the increase of extraction gas flow, it takes less time to reach the extraction tail, i.e. the required purification time is shorter, the contaminants removal rates are higher, while the removal rate is not in proportion to extraction flow. When the ventilating flow arrives at a certain degree, the removal rate of contaminants will not significantly increase any more.

Developed SVE technology is adopted to treat contaminated soil from a pesticide plant, as Table 2 shows, concentrations of all VOCs achieve the remediation goals after 5 days extraction, with a removal rate between 86.5%-100% and a total VOCs removal rate of 99.7%. These data explains that the developed SVE equipment and technology has good treatment effect to silty soil.

Table 2 SVE treatment effects

No.	Target contaminant	Remediation goals / (mg/kg)	Pre-remediation concentrations/ (mg/kg)	Post remediation concentrations/ (mg/kg)	Removal rate/%
1	benzene	2.9	579	2.71	99.5
2	toluene	183	25100	82.2	99.7
3	ethylbenzene	37	297	0.83	99.7

Continue table

No.	Target contaminant	Remediation goals / (mg/kg)	Pre-remediation concentrations/ (mg/kg)	Post remediation concentrations/ (mg/kg)	Removal rate/%
4	trichloromethane	0.06	88	<0.05	100
5	tetrachloroethylene	0.1	2.7	<0.05	100
6	Total		26066.7	85.74	99.7

2.3 Chemical oxidation technology

Organic contaminants solubilization-oxidation combined remediation technology is researched and developed, based on the dynamic balance mechanism of organic contaminants desorption-dissolution-oxidation in soil/water environmental system. Non-selective oxidants or solubilizer are introduced into the subsoil to consume soil organic matter, which promotes desorption and dissolution of adsorbed phase contaminants in water phase. Afterwards, long-lasting strong oxidizing agent is injected into the subsoil and dissolved organics are further degraded. This combined remediation technology accelerates diffusion rate of groundwater and remediation agent, improves treatment effect and shortens treatment period, has the advantage of oriented remediation in subsoil at a minimum of agent loss and usage. First site scale application of ISCO in industrial contaminated site, significantly promotes the application level of ISCO in China.

ISCO was conducted by using the self-designed in-situ injection system and organic contaminants solubilization-oxidation combined remediation technology, which is researched and developed based on the dynamic balance mechanism of organic contaminants desorption-dissolution-oxidation. Altogether five rounds injection was conducted, PID readings detected at the well head were recorded (see readings in Table 3). The data shows that PID readings are not very high after first and second rounds of oxidant (sodium persulfate) injection, and it shows a trend of decline. While PID readings rise dramatically after the fourth round of injection, which explains that a large number of adsorbed contaminants are desorbed, and generates contaminant vapor by Fenton reaction.

Table 3 PID readings measured at well head

	After 1st round injection	After 2nd round injection	After 4th round injection	After 5th round injection
Well 1	49.1	35.2	15000	34.5
Well 2	50.4	36.1	15000	22.5

Groundwater samples were taken and concentrations were also measured, after each round of injection. Experiment results can be seen in Fig. 4 below.

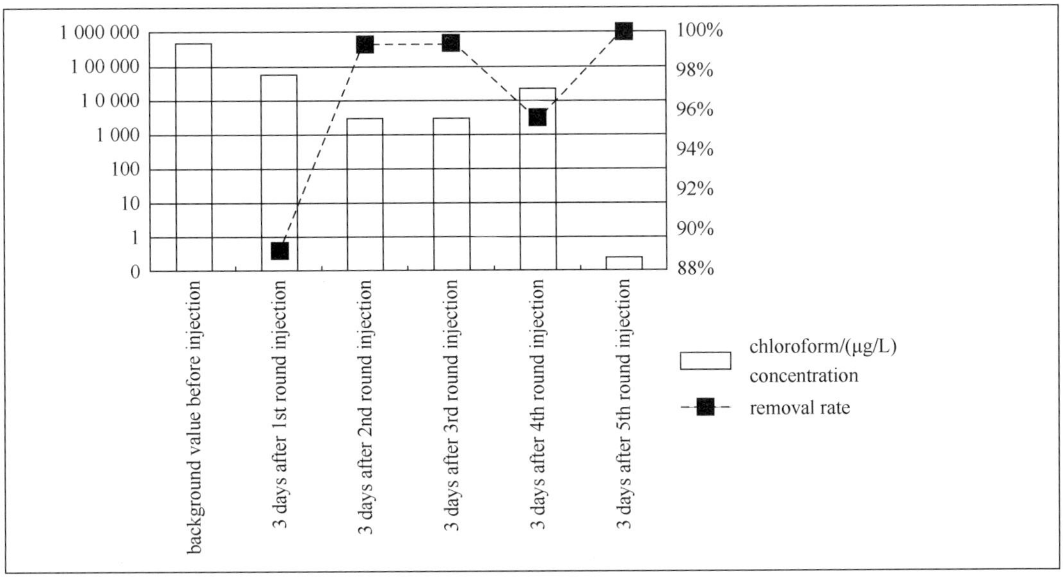

Fig. 4 Pilot test results

Research results show that ISCO remediation technology, based on the dynamic balance mechanism of organic contaminants desorption-dissolution-oxidation in soil/water environmental system, can remediate the organic contaminated soil and groundwater in an efficient and rapid manner, efficiently avoiding remediation rebound and tail.

3. Conclusions

High concentrations and complex contaminant compositions are characteristic of industrial site pollution in China, facing the poor applicability, low treatment efficiency and high operation cost of current remediation technologies, technologies are researched and developed based on the dynamic balance mechanism of contaminants in soil/water/air environmental system. The self-developed technologies include: SVE technology targets VOCs, chemical oxidation technology targets VOCs and SVOCs, thermal desorption technology targets recalcitrant/persistent organic contaminants. In addition, proprietary owned combined remediation technology system and complete set of equipment are successfully developed and integrated, which realizes efficient, green and economic remediation of organic contaminated site. Representative industrial contaminated soil and groundwater remediation projects with high risk and complex pollution were successfully completed in Jiangsu and Zhejiang Province by using these technologies, with a volume of 300, 000 m^3, eliminating the environmental risks and meanwhile contributing to the development of local brownfield site.

Thermal Desorption Technology and Application

Ligang WANG, Yong YUE, Pixue WANG

Jereh Environment Group

1. Introduction

Thermal desorption technology (TDT) is a treating process with contaminants to be volatilized and separated out from soil through direct or indirect heat-exchange which makes contaminants' temperature exceed their respective boiling point, and exhaust gases to be collected and treated. TDT's advantage lies in heating temperature and holding time to be regulated according to different boiling points of contaminants for their targeted elimination. 95% of volatile organic contaminants (VOCs) and semi-VOCs in soil could be eliminated using TDT, even petroleum hydrocarbons, poly-aromatic hydrocarbons (PAHs), and volatile heavy metals could be also treated with good eliminating effect (Araruna et al., 2004; Chang et al., 2006; Aresta et al., 2008; Comuzzi et al., 2011; Falciglia et al., 2011; Yang et al., 2014).

Jereh Environment Group researched and developed a set of TDT equipment, and this technology has applied in remediating soil contaminated by organic chemicals and/or volatile heavy metals. TDT is one of the most competitive and mature technologies used in environmental protection field in Jereh Environment Group.

2. Technology and equipment

2.1 Technical principal

Under anaerobic condition, material is indirectly heated by high-temperature smoke and fire which first heat thermal desorption chamber (TDC) wall conducting heat to materials, target contaminants in material are selectively heated to a certain temperature above their respective boiling point through regulating heating process and material retention time, thus, target contaminants and water are volatilized and separated out to form high-temperature mixed gases. These mixed gases become liquid after aspirated into quenching unit for cooling down, these liquid are further separated in the following procedure. In the end, oil phase separated could be used as fuel with reusable value, separated water could be recycled in treatment system, while non-condensable gas separated are turned into high-temperature compartment in burning unit and combust after purified. The finally discharged include treated solid and exhaust gas which meets national standard for waste gas discharge, the TPH content in treated solid is below 1% which meets national criteria on solid waste discharge.

2.2 Equipment and process flow

TDT consists of Pre-treatment Unit, Thermal Desorption Unit, Water Treatment Unit and Non-condensable gas treatment Unit, the designed process is as follows:

(1) After air-dried, contaminated soil would be treated by the screening and crushing equipment firstly, until all of their diameters could meet the feed dimension. The bulk solid separated out are collected together for the further treatment.

(2) After treated by the pre-treatment unit, the contaminated soil would be transported to the Thermal Desorption Unit (TDU), meanwhile N_2 produced from nitrogen generator is sent into TDU, making the TUD anaerobic. When material is heated to 400-600℃, contaminants volatilize out and form into high-temperature oil-gas mixture.

(3) Then the oil-gas mixture is aspirated into quenching skid, where water is sprayed to cool down the oil-gas to become liquid, subsequent treatment make oil and water mixture separate into water and oil phase respectively. Non-condensable gases are transported into specific device, after treated and purified, sent back to high-temperature compartment in burning chamber for combusting.

(4) Waste water is transported into water treatment system. Treated water is recycled as spraying water or condensation water.

(5) Sludge sourced from water treatment system is treated through filter press, produced solid waste is sent back to pretreatment unit for further treated, waste water is transported back to water treatment system.

(6) Separated oil phase is collected and stored in oil tank.

(7) The end product (solid) would be transported to discharge port after treated by TDU, and then loaded on stacking field after cooling process.

(8) Continuous Emission Monitoring System (CEMS) in TDT consists of gas monitoring subsystem, particulate matter monitoring subsystem, pretreatment system, and data acquisition system. The monitor object could be customized according to different project. It could monitor the concentration of pollutants real-time and alert when the concentration exceeds the limit.

3. Advantages

3.1 Technical advantages

3.1.1 General technical advantages

With the largest processing capacity in single module, no dioxins produced under an anaerobic condition; no chemicals added, no secondary pollution; with perfect pretreatment, optional matching equipment, and extensive adaptability; with skid equipment designed easily to disassemble and transport.

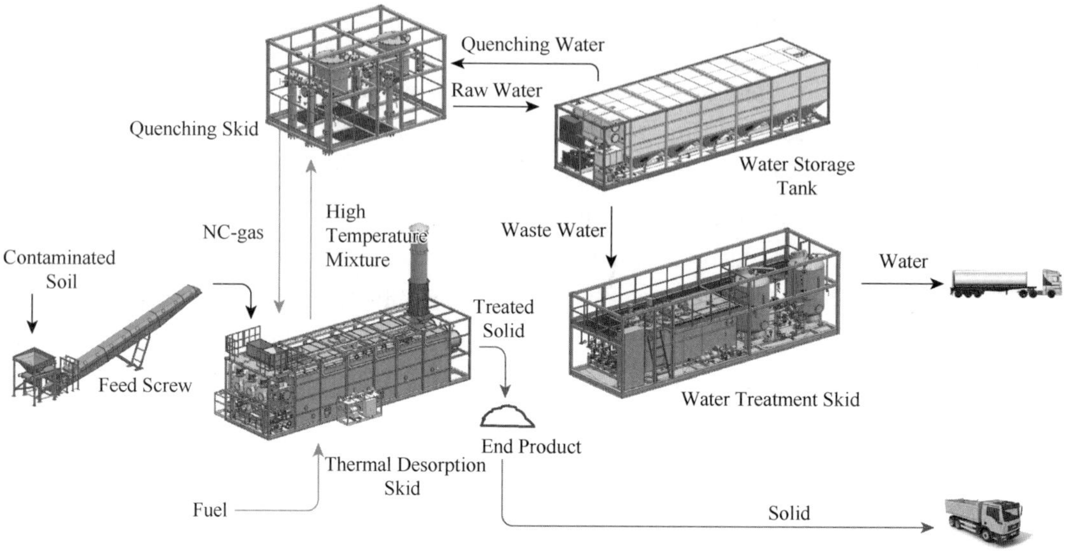

Fig. 1 Equipment and general process flow

3.1.2 Technical achievements

Recycled oil phase contains BS&W$<$1% which is below international standard (BS&W$<$2%), could be used as preparing oil base mud or energy oil.

Treated solid contains TPH$<$0.3% which is below international standard (TPH$<$1%), could be used as paving material, bricking material, and land-filling material.

Discharged gases could meet national standard *Gaseous pollutants emission standard* (GB16297), particulate matter$<$120 mg/m^3, $SO_2$$<$550 mg/m^3, fluoride$<$9.0 mg/m^3, non- condensed gases are entirely recycled, no dioxins produced.

All the remediation targets for contaminated soil are attained. For example, DDT or HCH content is below 0.1 mg/kg in treated soils from Hangzhou pesticide plant.

3.2 Equipment advantages

3.2.1 High-degree automation

Visualization: monitoring system is configured to realize visible operation on whole treatment process.

Abundant data: data bank function includes data collection, storage, and analysis, exporting various curves and report forms.

Network function: handhold data-collecting control system and remote command centre maintenance system are realized through connecting industrial communication network.

Intelligence: intelligence system adopts closed-loop control to realize fully automatic operation through multiple linking protection and early-warning function.

3.2.2 Treatment capacity

The maximum treatment capacity is 8T/H for single set of the equipment, 16-32T/H of treatment capacity could be realized using integrated stack design according to customer's request. Up to 2015, 50, 000 tons of oily wastes had been treated, and 150, 000 tons of contaminated soils had been remediated.

3.2.3 Safety security

Safety is ensured, potential dangers is completely eliminated.

TDC has special feed inlet with sealing structure to isolate from the outside air and ensure an anaerobic condition in the desorption process. Nitrogen generator is configured, high purity nitrogen is filled into TDC to control oxygen content. Oxygen monitoring system is configured to detect the oxygen content in TDC in the process of thermal desorption. Multiple protection systems are configured to prevent over temperature, overpressure, and overload etc., and realize warning function and automatic shutdown protection. Automatic fire-extinguishing system with high-pressure CO_2 is configured, which jointly works with oxygen monitoring system to protect equipment. Fire-stopping valve and safety valve are configured to protect oil gases with over-press and fire-resistance in the process of thermal desorption.

3.3 Economy advantages

Maximum of oil recovery and the highest economic benefit are realized. Exhaust-gas recycling device is configured, recycled exhaust-gas with high temperature is used to preheat air, which decrease energy consumption. Equipment is designed in skid shade to reduce transport and installment cost and time. Universal combustion system is configured, in which recycled oil and non-condensable gas could be used as fuel. Temperature real-time monitoring system is configured to monitor material properties in the process of treatment, which ensures treating effect to be controlled and makes equipment run in the most economic condition.

4. Practical examples

Fig. 2 Client: Daqing oilfield; Pollutant type: oil-based drilling cuttings;
Treatment capacity: 2T/H; Treated quantity: 3000 tons
(August 2014)

Fig. 3 Client: Ningbo thermal power plant; Pollutant type: PAHs;
Treatment capacity: 8T/H; Treated quantity: 10, 000 tons
(September 2015)

Fig. 4 Client: Hangzhou pesticide plant; Pollutant type: organic pesticides;
Treatment capacity: 21T/H; Treated quantity: 70, 000 tons
(December 2014)

Fig. 5 Client: Sichuan oil-gas field; Pollutant type: oil-based mud;
Treatment capacity: 2T/H; Treated quantity: 2000 tons
(November 2014)

References

Araruna J J T, Portes V L O , Soares A P L, et al. 2004. Oil spills debris clean up by thermal desorption. Journal of Hazardous Materials 110: 161-171.

Aresta M , Dibenedetto A , Fragale C , et al. 2008. Thermal desorption of polychlorobiphenyls from contaminated soils and their hydrodechlorination using Pd-and Rh-supported catalysts. Chemosphere 70: 1052-1058.

Chang T C, Yen J H. 2006 On-site mercury-contaminated soils remediation by using thermal desorption technology. Journal of Hazardous Materials 128: 208-217.

Comuzzi C, Lesa B, Aneggi E, et al. 2011. Salt-assisted thermal desorption of mercury from contaminated dredging sludge. Journal of Hazardous Materials, 193: 177-182.

De Percin P R Paul. 1995. Application of thermal desorption technologies to hazardous waste sites. Journal of Hazardous Materials, 40: 203-209.

Falciglia P P , Giustra M G , Vagliasindi F G A. 2011 Low-temperature thermal desorption of diesel polluted soil: Influence of temperature and soil texture on contaminant removal kinetics. Journal of Hazardous Materials. 185: 392-400.

Yang B , Xue N, Ding Q, et al. 2014 Polychlorinated biphenyls removal from contaminated soils using a transportable indirect thermal dryer unit: Implications for emissions. Chemosphere 114: 84-92.

Part V Metal-Organic Combined Pollution and Remediation

Study on Remediation Technologies of Organic and Heavy Metal Contaminated Soils

Shuhai GUO, Fengmei LI, Peijun LI, Sa WANG, Qing ZHAO, Gang LI, Bo WU, Peidong TAI

Institute of Applied Ecology, Chinese Academy of Sciences

Soil pollution is a global problem which has been gained enormous attentions especially in China. Institute of Applied Ecology, Chinese Academy of Sciences is the first research institute to study remediation technology for contaminated soils. In 1990s, this institute carried out the Major Programs of Chinese Academy of Sciences namely "Study of clean technology on contaminated soil" . Since the year 2000, Environmental Engineering Group and Pollution Ecology Group of the Institute of Applied Ecology have carried out much research on the mechanisms, technology, materials and equipment required for the remediation of soils contaminated by organic pollutants and heavy metals. They have undertaken forty-six projects on soil remediation supported by the Chinese government and Chinese Academy of Sciences, such as the National High Technology Research and Development Program of China (863 Program), the Major State Basic Research Development Program of China (973 Program), the National Natural Science Foundation of China and the Major Programs of the Chinese Academy of Sciences. Many significant results have been obtained, including six remediation technology, five remediation equipments, ten remediation materials, and the publication of forty-six patents and sixty-seven papers.

1. Remediation of soils contaminated with organic compounds

Petroleum and polycyclic aromatic hydrocarbons (PAHs) are ubiquitous organic pollutants. As a result of their persistence and their toxic, mutagenic and carcinogenic properties, petroleum and PAHs are significantly risky to both the environment and public health. The importance of preventing environmental contamination by petroleum and PAHs and the need to remove them from the environment has been recognized by a number of government agencies. During the last decade, various strategies have been considered for the remediation of soils contaminated with organic pollutants, including physical, chemical and biological techniques. Microbial degradation is a promising route for the removal of organic pollutants and can safely achieve the complete degradation of organic pollutants at lower costs and with less disturbance to the soil than physicochemical treatments. Environmental Engineering Group and Pollution Ecology Group are currently focusing on developing methods to remove or degrade organic pollutants through biological processes involving microorganisms. Research is being carried out on the mechanisms, technologies and equipment required for both microbial remediation and the enhanced bioremediation of contaminated soils; in addition, soil remediation projects have been set up in an oil field and at a coking site.

1.1 Bioremediation

1.1.1 Isolation and screening of microorganisms for the degradation of petroleum

The microbial degradation of petroleum usually begins with an attack on the alkane or light aromatic fractions, whereas the high molecular weight hydrocarbons, aromatics, resins and asphaltenes are considered to be recalcitrant. The saturated components of petroleum (alkanes and cycloalkanes), particularly the intermediate length *n*-alkanes (C10-C20) are usually degraded first. Shorter chain length alkanes are more toxic to microorganisms, whereas the longer chain length (C20-C40) alkanes are more difficult to degrade as a result of their low solubility in aqueous liquids and low bioavailability. Resins and asphalt are degraded slowly and with low efficiency because of their low aqueous solubility and the high resonance energies of their structures.

Six different strains of fungi were separated from a soil contaminated with petroleum and were screened for their potential to degrade persistent petroleum based on analyses of their lipase activities (Guo et al., 2005). A thick oil containing high concentrations of the resin and asphalt fractions was mixed with fine sand; the concentrations of oil and resin/asphalt were 3.76 and 0.095 g/kg, respectively. The results indicated that six strains of fungi showed a higher rate of degradation of the thick oil with resin/asphalt over a treatment period of 80 days. The degradation rates of the heavy thick oil using *Phanerochaete chrysoporium*, *Cviolus versicolor*, *Trichoderma* sp., *Cunninghamella* sp., *Aspergillus* sp. and *Fusarium* sp. were 20.4%, 19.1%, 22.5%, 15.9%, 16.5% and 15.0%, respectively. However, only *C. versicolor* and *Cunninghamella* sp. were able to degrade the resins and asphalt at rates of 11.5% and 11.6%, respectively. Enrichment culture and plate cultivation methods were used to obtain 11 strains of bacteria and 10 strains of actinomyces from soils contaminated by viscous oil in the Liaohe oil field (Li et al., 2006). These strains all had the ability to degrade resins and asphalt. The highest degradation rate for samples containing both resins and asphalt was 57.6%. These strains may provide a possible resolution to the degradation of persistent petroleum hydrocarbons in soils.

1.1.2 Isolation and screening of PAH-degrading bacteria

We isolated and screened some specialized PAH-degrading microorganisms to accelerate the removal of PAHs from the environment. *Pseudomonas* sp. was isolated from an enriched culture in PAH-contaminated soils in Benxi City, Liaoning Province, northeast China (Yuan et al., 2011). The ability of these strains to degrade phenanthrene, pyrene andbenzo[a]pyrene was determined after 28 days of incubation at 28℃. The results showed degradation rates for phenanthrene, pyrene and benzo[a]pyrene of 88.4%, 54.0%and 68.4%, respectively. The degradation of benzo[a]pyrene in a contaminated soil by a colony of microorganisms and also by individual strains of *Cephalosporium*, *Penicillium*, *Aspergillus*, *Fusarium*, *Bacillus* and *Arthrobacter* was reported by Li et al. (2001). The results showed that the degradation ability of microorganisms was higher in fresh soils than in dried soils. The microorganism colony had a higher ability to degrade benzo[a]pyrene than the individual microorganisms and the fungi were more effective in degradation than the bacteria. *Fusarium* (2209) was shown to be a highly effective strain for the degradation of benzo[a]pyrene. Three strains (*Fusarium* sp., *Penicillium* sp. and *Mucor* sp.) were isolated from an oil-contaminated soil. All the three strains showed the ability to degrade benzo[a]pyrene (Gong et al., 2001). Microbial consortia isolated from aged oil-contaminated soil were used to degrade 16 PAHs (15.72 mg/kg) in the soil and slurry phases (Li et al., 2008). The three microbial consortia (bacteria, fungi and bacteria-fungi complexes) were able to degrade the PAHs and the highest rates of removal of PAHs were found in soils and slurries inoculated with fungi (50.1% and 55.4%, respectively). The biodegradation of PAHs in the slurries was lower than in the soil samples for the treatments using inoculation with bacteria and bacteria-fungi complexes. The degradation of three- to five-ring PAHs treated by microbial consortia was observed in both the soils and the slurries. The highest degradation rates for individual PAHs (anthracene, fluoranthene and benz[a]anthracene) in the soils were 45.9%-75.5%, 62%-83.7% and 64.5%-84.5%, respectively, and 46.0%-75.8%, 50.2%-86.1% and 54.3%-85.7%, respectively, in the slurries. Therefore the inoculation of microbial consortia (bacteria, fungi and bacteria-fungi complexes) isolated from in situ contaminated soils could be considered as a successful method by which to degrade PAHs.

The biodegradation of PAHs (8.15 mg/kg) in an aged contaminated soil by isolated microbial consortium (five fungi and three bacteria) during an incubation period of 64 days was reported by Li et al. (2009). The treatments applied were: ①biodegradation by the addition of the microbial consortium to sterile soils (BM); ②biodegradation by the addition of the microbial consortium to non-sterile soils (BMN); and ③biodegradation by *in situ* "natural" microbes in non-sterile soils (BNN). The fungi in the BM and BMN soils grew rapidly from day 0 to day 4 of the incubation period and then reached a relative equilibrium. In contrast, the fungi in the BNN soil remained at a constant level for the entire incubation period. The bacteria in the BMN soils grew rapidly from day 0 to day 2 of the incubation period and then reached a relative equilibrium, whereas those in the BM and BMN soils grew slowly throughout the 64-day incubation period. After 64 days of incubation, the biodegradation of PAHs was

35.0%, 40.7% and 41.3% in the BNN, BMN and BM soils, respectively. A release of sequestrated PAHs from aged contaminated soils was observed in this experiment, especially in the BM soil. Therefore although the bioaugmentation of introduced microbial consortium increased the biodegradation of PAHs in aged contaminated soils with low PAH concentrations, it may be possible to optimize the microbial consortia to use bioremediation more effectively in the field.

Low molecular weight PAHs with two or three fused aromatic rings are more soluble in water and thus degrade more easily than high molecular weight PAHs with four to six fused aromatic rings, which are more hydrophobic and more difficult to break down. Three strains of bacteria (*Sphingomonas* sp., *Ochrobactrum* sp. and *Achromobacter* sp.) capable of degrading high molecular weight PAHs were isolated from a PAH-contaminated soil around an iron and steel in Benxi (Gao et al., 2016; Li et al., 2016). Their ability to degrade PAHs was studied in a soil contaminated with PAHs and they were shown to have a stronger ability to degrade heavy PAHs. Compared with treatments that inoculated the soils with *Sphingomonas* sp., *Ochrobactrum* sp. and *Achromobacter* sp. individually, the degradation rate in the soil inoculated with a mixture of the bacteria was enhanced by 9.2%, 11.5% and 16.1%, respectively. The degradation rates of four-, five- and six-ring PAHs were enhanced by 12.1%, 13.4% and 14.5% in a bioremediation-electrokinetic experiment than in the bioremediation alone experiment, which indicated that *Sphingomonas* sp., *Ochrobactrum* sp. and *Achromobacter* sp. were able to adapt to an electric field.

1.1.3 Isolation and characterization of biosurfactant-producing bacteria

Biosurfactants are surface-active compounds produced by some microorganisms and can be either totally or partially extracellular. They are polymeric amphipathic molecules containing both polar and non-polar moieties, which allows them to form micelles that accumulate at the interphase between liquids of different polarities, such as water and oil. Microorganisms that degrade hydrocarbons produce biosurfactants with various chemical properties and different molecular sizes that decrease the surface tension of hydrophobic, water-insoluble substrates and thereby enhance their bioavailability and rate of bioremediation. Fifty-two isolates of bacteria were obtained from the oil-mud and sewage from Daqing Oil Field by enrichment culture to use in the bioremediation of oil-polluted soils (Niu et al., 2005). Three isolates (B22, B24 and B25) were selected based on their degreasing activity and the determination of the surface tension experiments. The surface activities of the biosurfactants were both low and stable. It was shown that the main components of the purified biosurfactants were lipopeptides, rhamnolipids, sophorolipids and glyceride compounds. The degradation rate for oils and muds reached an average of 70%, or even higher if fermented liquids of the three isolates were used. Compared with the control samples, the degradation rate increased by about seven to nine times when using the isolates (Fig. 1).

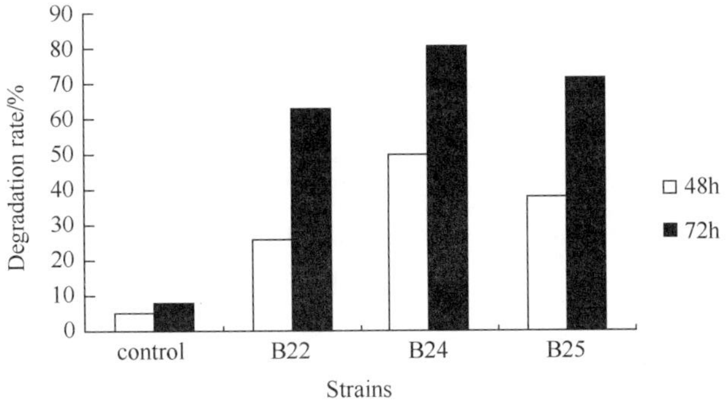

Fig. 1. Effect of biosurfactants on the removal of oil from contaminated soils

1.2 Bioremediation technologies for soils polluted with petroleum and PAHs

1.2.1 Soil composting in windrows

Composting is a technology that utilizes microbes to clean pollutants. A soil from the Liaohe Oil Field contaminated with crude oil was treated using off-site bioremediation technology by composting on a 20 m×10 m prepared bed (Jiang et al., 2001). Eight composting windrow units were set, each 8 m long, 2 m wide and 0.35 m high. When the contamination from total petroleum hydrocarbons (TPHs) was in the range 4.16-7.72 g/100 g soil, the total degradation rate of TPHs was 45.19%-56.74% after 53 days of operation. This suggests a technological basis for the bioremediation of oil-contaminated soils. Micronutrient-enriched chicken manure and rice husks were applied as nutrition and a bulking agent, respectively (Li et al., 2002a). The lipase activities of the indigenous microorganisms were analyzed and three indigenous fungi with high lipase activities was identified. An inoculum consisting of these three indigenous fungi and one introduced (exotic) fungus was applied to four different types of oil-contaminated soil. The inoculum of indigenous fungi increased both the total number of colony-forming units and the rate of degradation of the TPHs in all contaminated soils, but at different rates. In sharp contrast with other studies, the introduction of exotic microorganisms did not improve the remediation, which suggests that the inoculation of oil-contaminated sites with non-indigenous species is likely to be unsuccessful in terms of remediation. However, indigenous genera of microbes were found to be very effective in increasing the rate of degradation of TPHs. The degradation of TPHs was mainly controlled by the composition of the aromatic hydrocarbons, asphaltenes and resins. 38-57% degradation of the crude oils (with densities ranging from 25, 800 to 77, 200 mg/kg dry weight) in contaminated soils was achieved after 53 days of operation. The degradation patterns followed typical first-order reactions. The remediation of contaminated soils by composting mainly relies on two mechanisms: ①adsorption by organic matter and ②degradation by microorganisms. The decomposition of organic pollutants in the soil/compost mixture mostly relies on the microbial activity and therefore suitable growth conditions for the microorganisms must be present during the composting process.

The influence of fertilizers, soil dressings, moisture and pH on the bioremediation of sludge containing petroleum has been studied and it has been shown that the application of a soil dressing was an important factor in the efficiency of treatment (Guo et al., 2005). At natural temperatures, the optimized parameters were: amount of soil dressing, 20%; fertilizer, 5%-10%; solid microbial materials, 5%; moisture, 30%; and pH 7. When the original TPH concentrations were in the range 20.6-22.3 g/kg, the removal rate was>49%. Thus the construction and operation of field-scale composting biopiles in windrows with passive aeration is a cost-effective bioremediation technology.

1.2.2 Two-phase bioremediation of oil-contaminated soil from the Liaohe Oil Field

A prepared bed was set up for the bioremediation of oil-contaminated soils. Soils contaminated with different types of oil were composted in the prepared bed for a treatment period of 210 days divided into two phases (Li et al., 2003). When the concentration of TPHs (consisting of thin oil, high condensate oil, special viscous oil and viscous oil) was in the range 25.8-77.2 g/kg dry soil, the oil removal rate reached 38.4%-56.7% after 53 days. In the second phase, the oil removal rate reached 66.6%-81.0% after 156 days (Fig. 2). These results show that most of the easily degraded hydrocarbons were removed during the first phase and that the efficiency of remediation decreased during the second phase.

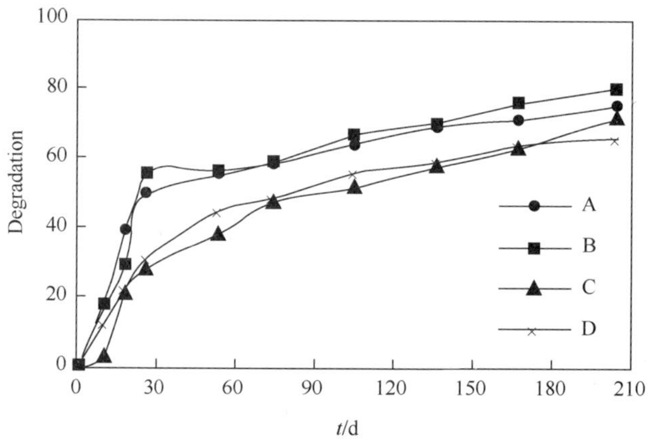

Fig. 2 Degradation rate of total petroleum hydrocarbons during composting

1.2.3 Immobilized microorganism technology

The biodegradation of PAHs in soil using immobilized microorganisms has been extensively studied using the fungus *Fusarium* sp. and the bacteria *Zoogloea* sp., *Mucor* sp. and *Bacillus* sp. (Li et al., 2005a, 2005b; Su et al., 2006). Different carriers were used, e.g. polyvinyl alcohol + sodium alginate + powdered activated carbon or vermiculite (Li et al., 2005b; Su et al., 2006). Fig. 3 shows the procedures used in this technique.

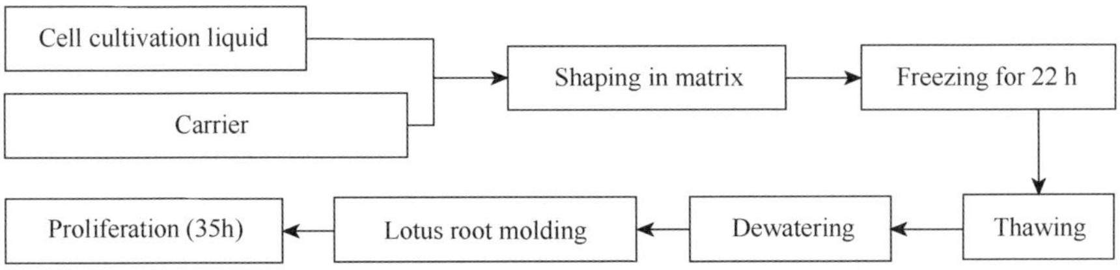

Fig. 3 Flow diagram for the immobilization of microorganisms

1.3 Enhanced bioremediation

1.3.1 Application of biosurfactants in the biodegradation of oil-contaminated soils

Heavy oil contains alkanes, aromatic hydrocarbons, organic acids and resins; the high-number carbon chains and aromatic substances account for the highest proportion and it is therefore difficult to remediate soils contaminated with heavy oils. Two bacteria and four fungi were selected and matched after antagonistic tests (Niu et al., 2005), one of which produced biosurfactants. They were used to remediate a soil contaminated with high condensate oil, viscous oil, thick oil and thin oil. It was shown that the presence of a biosurfactant could increase the rate of biodegradation. The degradation rates of high condensate oil, viscous oil, thick oil and thin oil in the treatments inoculated with the strain producing a biosurfactant were enhanced by 16.1%, 22.1%, 20% and 13.2%, respectively, compared with treatment with the strain that did not produce a biosurfactant (Table 1). The increase in the rate of biodegradation was highest between 90 and 120 days. These results showed that biosurfactants can increase the efficiency of bioremediation.

Table 1. Comparison of degradation rates for oil in a prepared bed

Sample type	Degradation rate of oils/%			
	30 days	90 days	120d ays	150 days
soil contaminated with high condensateoil	8.2	17.1	43.7	43.7
soil contaminated with high condensateoil+biosurfactant	11.0	33.2	41.8	59.8
soil contaminated with viscous oil	1.2	6.1	21.8	27.3
soil contaminated with viscous oil+biosurfactant	6.6	28.1	30.8	49.4
soil contaminated with thick oil+biosurfactant	9.5	23.5	33.1	31.9
soil contaminated with thick oil	13.6	43.9	55.2	51.9
soil contaminated with thin oil+biosurfactant	2.8	17.4	26.2	32.5
soil contaminated with thin oil+biosurfactant	28.4	26.9	39.4	45.7

1.3.2 Bioslurry remediation of soils contaminated with polycyclic aromatic hydrocarbons

The bioremediation of soils contaminated with PAHs was studied using a bioslurry reactor process (Fig. 4). Phenanthrene and pyrene were chosen as the test pollutants (Li et al., 2002b). The results showed that the physical and chemical properties of the pollutants were the key factors affecting the bioremediation of PAHs in soils. PAHs with fewer benzene rings and lower molecular weights were easier to biodegrade. Phenanthrene was easier to remove from soils than pyrene. The temperature, the ratio of water to soil and the airflow were the most important control factors in the bioslurry reactor process (Gong et al., 2001). Good remediation results were achieved when the ratio of water to soil was 2:1, the temperature was 20-25℃ and the aeration flux was 60 L/h. When fungi isolated as pure cultures from the contaminated soil were used to degrade PAHs, 90% of pyrene and 33.3% of benz[a]anthracene were degraded by *Fusarium*, 81.5% of pyrene and 49.2% of benz[a]anthracene were degraded by *Mucor*, and 52% of pyrene and 46% of benz[a]anthracene were degraded by *Penicillium* after 34 days of incubation.

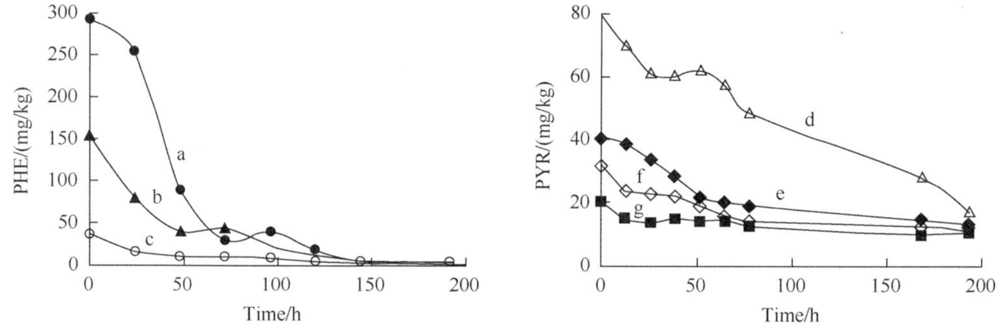

Fig. 4　Degradation of phenanthrene (PHE) and pyrene (PYR) in soils at different levels of contamination using a slurry reactor

1.3.3　Bioremediation combined with chemical oxidation

Two methods of enhancing the degradation of PAHs were compared: degradation by coupling *Penicillium chrysogenum* with $KMnO_4$ and degradation by *P. chrvsogenum* alone. The parameters of degradation in the more efficient method were optimized (Zang et al., 2007a). It was shown that the method of coupling *P. chrysogenum* with $KMnO_4$ gave better results; this was the first method to be used in the degradation of BaP and its metabolites. The metabolite BP 1, 6-quinone was the species most likely to be accumulated in pure cultures and the highest degradation was obtained when the concentration of $KMnO_4$ in the cultures was 0.01% (w/v).

Based on the experimental results, a novel concept with regard to the bioremediation of BaP-contaminated soils was discussed, considering the influence of the accumulated metabolites on environmental toxicity. Hydrogen peroxide-zinc (H_2O_2-Zn), the fungus *Aspergillus niger* and the bacteria *Zoogloea* sp. played important parts in the different phases. The degradation parameters of the system were optimized and it was shown that degradation was the highest when the fungus-bacteria were combined with H_2O_2-Zn. The concentration of BaP in the cultures was 30-120 mg/L, and the initial pH of the cultures was 6.0 (Zang et al., 2007b). However, phenanthrene, as a co-metabolite, significantly inhibited the degradation of BaP. Compared with the conventional method of degradation by domestic fungus only, this combined system enhanced the degradation BaP by > 20% in 12 days. The highest accumulation of BP 1, 6-quinone and 3-OH BP was reduced by nearly 10% in the degradation experiments, which further proved that the combined degradation system was more effective in removing BaP and its metabolites.

1.3.4 Electro-bioremediation

Electro-bioremediation is a promising hybrid technology for the remediation of organic contaminants. The technique consists of the controlled application of a low-intensity direct electric current to contaminated soils between appropriately distributed electrodes. During the electrokinetic treatment of a soil, nutrients and microorganisms are spread throughout the soil and the organic pollutants can be transported to the area containing the microbial populations, where they are then degraded. The electric current also induces redox reactions on the surface of the electrodes, resulting in the destruction of organic compounds. The electric field is also able to stimulate microbial enzyme activity in the soil. In addition, the electrical current can heat the soil, which can help the remediation processes in cold climates. Although electrokinetic remediation has been applied successfully in soils with organic contaminants, its application for the remediation of complex molecular structures, such as petroleum and PAHs, is limited. To enhance the efficiency of electrokinetic remediation in soils, we studied the mechanisms, materials and equipment used inelectro-bioremediation.

1) Mechanisms of electro-bioremediation

Direct currents applied across electrodes inserted in a soil generate an electric field that can mobilize, extract and degrade contaminants by electrokinesis, electrochemical oxidation and biogeochemical modification. For organic pollutants, oxidative degradation is the primary mechanism in treatment by electrokinetics alone. Direct and indirect oxidation are the main types of reaction in the degradation of soil pollutants. In direct oxidation, the contaminants react on the surface of the electrode, which relies on electron transfer between the electrode surface and the organic pollutant. Indirect oxidation occurs via the novel oxidizing substances produced by electrochemical reactions that further oxide the contaminants.

In an experiment to remove oil from a soil, 9.9% of the oil was removed from near to an electrode with an electric intensity of 1.02 V/cm (Fig. 5) (Li et al., 2015). A significant positive correlation was found between the electrical conductivity and the rate of removal of pyrene (R=0.829, $p<0.01$) in an experiment that used an anode-cathode separated system; the rate of removal of pyrene in the anode chamber was higher than that in the cathode chamber, indicating that electrochemical oxidation at the anode was a more effective pathway for the removal of pyrene (Xu et al., 2014). Also, the variations in the rates of removal were related to changes in the current. A rapid rate of removal of pyrene occurred when the current increased; as the current decreased in the later stages of treatment, the removal rate showed a slower increase.

The organic pollutant was also removed in the center of the electric field and oxidation reactions occurred at all the interfaces between the soil particles and the pore water. This is consistent with the microconductor principle, which is another degradation mechanism in electrokinetic remediation. Wet soil may act as a 'diluted' electrochemical solid bed reactor and electrochemical reactions can be induced on the reactor as a result of its electron-conducting properties. The study of Xu et al. (2014) showed that although less pyrene was removed in the cathode chamber than that in anode chamber, at least 20% of the pyrene was degraded. Wang et al. (2013) also reported that the degradation of PAHs reached 19.0% in the central region of the electric field (1 V/cm) altered directional operation, although there was some contribution from the metabolism of indigenous fungi. An indirect oxidation method that combined a treatment technology using the direct electro-Fenton process and bioremediation was established by Xu et al. (2015) and Gan et al. (2015). This technique makes use of the strong oxidant ·OH produced by the electro-Fenton process to effectively degrade and mineralize persistent organic pollutants such as pyrene.

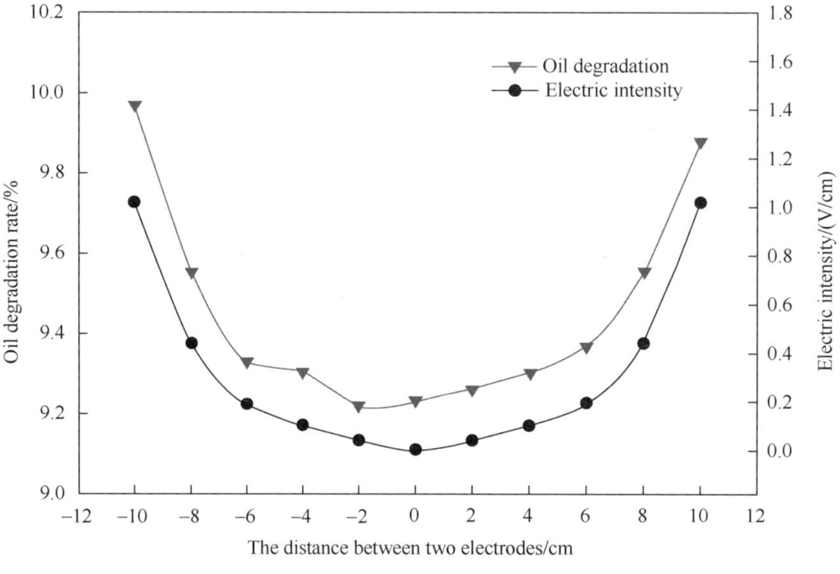

Fig. 5 Spatial distribution of oil removal and electric field between two electrodes in an electrokinetic-polarity reversal experiment

For electrokinetic-enhanced bioremediation, the total remediation effect is the combined action of oxidative degradation and biological metabolism. Electrokinetic-enhanced bioremediation is not simply the sum of the two remediation effects, but reflect a coupling effect. The synergistic effects of bioremediation and electrokinetics is related to both nutrient transport and the metabolic processes of the microorganisms. For an *n*-alkane such as *n*-hexadecane, the degradation rate in electro-bioremediation was 53.7%, an increase of 20.3% compared with bioremediation without an electric field and *n*-hexadecanoic acid was the only intermediate product detected by the monoterminal oxidation pathway (Yuan et al. 2013) (Fig. 6). *n*-Hexadecanoic acid was removed by intracellular β-oxidation and the rate of β-oxidation in electro-bioremediation was 2.2 times that of bioremediation alone, indicating an increase in the rate of β-oxidation of the corresponding alkanoic acid as a result of stimulation by the electric field. Similar data were obtained by Guo et al. (2014) that is, the extent of degradation in electrokinetic-enhanced bioremediation was significantly higher than that in the simulated curve (the sum of the rate of degradation in bioremediation and electrokinetic remediation) after 60 days (Fig. 7). Both the transport of nutrients and the activation of biological metabolic enzymes contributed to the increase in biometabolism and enhanced the efficiency of bioremediation.

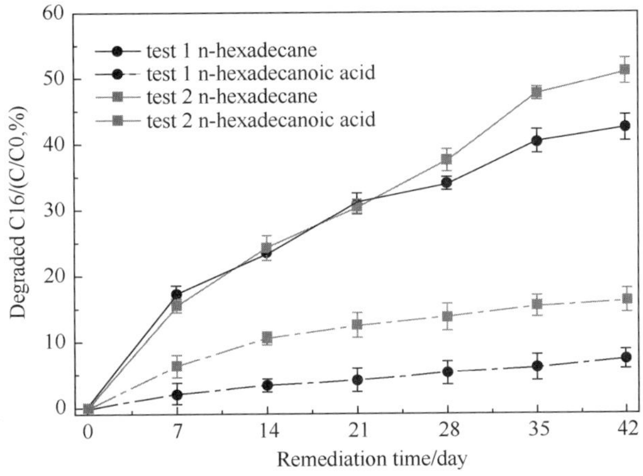

Fig. 6 Degradation rates of n-hexadecane and n-hexadecanoic acid over a 42-day period. Test1, bioremediation alone; test2, electro-bioremediation

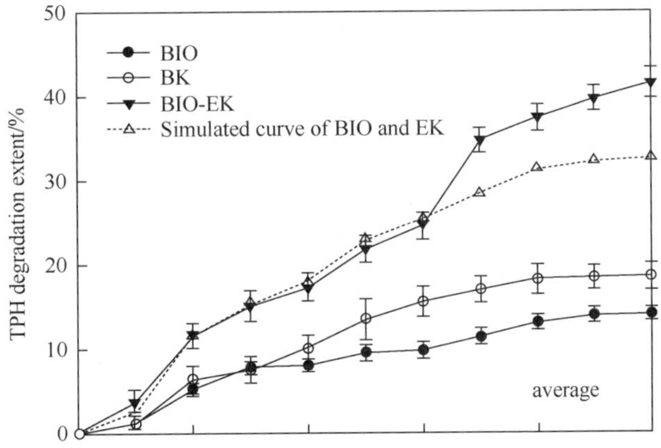

Fig. 7. Average degradation extent of total petroleum hydrocarbons (TPHs) for bioremediation alone (BIO), electrokinetic (EK) and bioremediation-electrokinetic (BIO-EK) tests. The simulated curve calculated from the sum of the extent of degradation in the BIO and EK tests can be used as a comparison with that in the BIO-EK test. Error bars represent one standard error

2) Materials for electro-bioremediation

(1) PAH-degrading bacteria adapted to electrokinetic conditions

Bacteria that degrade PAHs and that were capable of growing under electrokinetic conditions were isolated using an adjusted acclimatization and enrichment procedure based on a soil contaminated with heavy PAHs in the presence of an electric field (Li et al., 2016). Their abilities to degrade heavy PAHs under the influence of an electric field was investigated individually in artificially contaminated soils. The results showed that strains PB4 (*Pseudomonas fluorescens*) and FB6 (*Kocuria* sp.) were the most efficient degraders of heavy PAHs under electrokinetic conditions. They were re-inoculated into a polluted soil from an industrial site with a PAH concentration of 184.95 mg/kg. Compared with the experiments without an electric field, the degradation abilities of *P. fluorescens* and *Kocuria* sp. were enhanced in the industrially polluted soil under electrokinetic conditions. The extent of degradation of the total PAHs was increased by 15.4% and 14.0% in the electrokinetic PB4 and FB6 experiments (PB4+EK and FB6+EK) relative to the PB4 and FB6 experiments without electrokinetic conditions (PB4 and FB6), respectively. These results indicated that *P. fluorescens* and *Kocuria* sp. could efficiently degrade heavy PAHs under electrokinetic conditions and that they have the potential to be used for the electro-bioremediation of PAH-contaminated soils, especially if the soil is contaminated with heavy PAHs.

(2) Composite functional electrodes

The removal of persistent organic pollutants is much lower in the center of soil samples than in the soil near the electrodes because of their poor solubility. Therefore it is important to find a method of enhancing the migration of the organic pollutants through the soil. We developed composite functional electrodes with a cover of polypropylene/oil-absorbent cotton on the electrodes (Li et al., 2011). The oil-absorbent cotton contains a hole with a diameter of 0.5-1.0 cm. Column-shaped graphite electrodes were used to generate the electric field. The composite functional electrodes have the advantages of adsorbing pollutants from the center of the soil sample and increasing the efficiency of removal. It was shown that the removal rate of the pollutants could be improved by 26% using the composite functional electrodes (Table 2).

Table 2. Effect of type of electrode on the removal rate of oil from contaminated soils

Treatment	Electrodes	Removal rate/%
Control sample	No	4
Test 1	Graphite electrodes	24
Test 2	Composite functional electrodes	50

(3) Optimization of the electric field

There is a significant positive correlation between electric intensity and the rate of degradation of the organic substance (e.g. oil) (Fig. 8) in the electric oxidation mechanism

(Li et al., 2010). Therefore it is essential to optimize the electric field conditions, including the electric field intensity and the electric field distribution, to regulate the efficiency of remediation. The effect of the voltage gradient on the degradation of pyrene and the microbial community in a soil was studied during the remediation process (Wei et al., 2015). The electric field conditions were not only crucial in oxidative degradation, but also affected microbial metabolism. A moderate voltage gradient stimulated the synergistic effect of bioremediation and electrokinetics in the remediation of soils contaminated with organic species such as pyrene or petroleum (Wei et al., 2015; Fan et al., 2015). According to the optimized data, the most suitable voltage gradient for the degradation of pyrene and stabilization of the microbial community was 1.2 V/cm (Fig. 9) (Wei et al., 2015). A higher or lower voltage gradient weakened the chemical oxidation or microbial metabolic activity. A constant voltage gradient of 1.0 or 2.0 V/cm was applied to study the effect of a unidirectional electric field on the rate of removal of PAHs and the abundance and diversity of indigenous fungi in a PAH-contaminated soil at the Benxi Iron and Steel Group Corporation, Liaoning Province, Northeast China (Wang et al., 2013). The indigenous fungal 18S rRNA gene copy numbers and community diversity were significantly higher under a voltage gradient of 1 V/cm than under a voltage gradient of 2 V/cm. The mode of application of an electric field was also an important method of improving the remediation effect on the basis of the application of the optimum electric field intensity. Polarity reversal was an efficient method with which to ameliorate the effect of the electric field on the physicochemical properties of the soil, such as the moisture content and soil pH. The properties of soils were less affected under alternating polarity electrokinetics and microorganisms adapted their growth. The distribution of soil moisture remained even and the soil pH remained broadly constant. There was a positive correlation between the extent of degradation of PAHs, the bacterial counts and the moisture content under electrokinetic remediation, as determined by Pearson correlation analysis. According to Li (2012), the extent of degradation of total PAHs under alternating polarity electrokinetics was higher than that under constant polarity, indicating that alternating polarity electrokinetics enhanced the biodegradation of PAHs more effectively than constant polarity electrokinetics. A similar conclusion was reached by Huang et al. (2012) (Fig. 10); the alternating polarity treatment effectively made up for the adverse changes in the soil physicochemical properties induced by electrokinetics and the effect of electrokinetics combined with bioremediation for pyrene was equal to the sum of the extent of degradation from electrokinetics and bioremediation. To make the removal of organic contaminants uniform and to realize the synergistic effect of biological degradation and electrochemical stimulation, a rotatory 2D electric field was designed and applied to the remediation of petroleum-contaminated soil (Li et al., 2010). The 2D electric field resulted in a uniform distribution of the electric intensity, soil moisture content, nutrient content and microorganisms and provided an appropriate micro-environment for bioremediation (moderate acidity/alkalinity, soil temperature and electric field stimulation) (Fig. 11). The composition of the microbial community and the number of bacteria also remained stable with no significant temporal and spatial shifts, which maintained the activity and metabolic capacity of the microorganisms. Guo et al. (2014a) showed that the measured degradation of petroleum exceeded the simulated curve for bioremediation and electrokinetics, indicating the synergistic effect obtained during the later phase of treatment. The optimized electric field guaranteed suitable soil physicochemical properties, extended the effective duration of the electric field, and made full use of the repair functions of both bioremediation and electrokinetics.

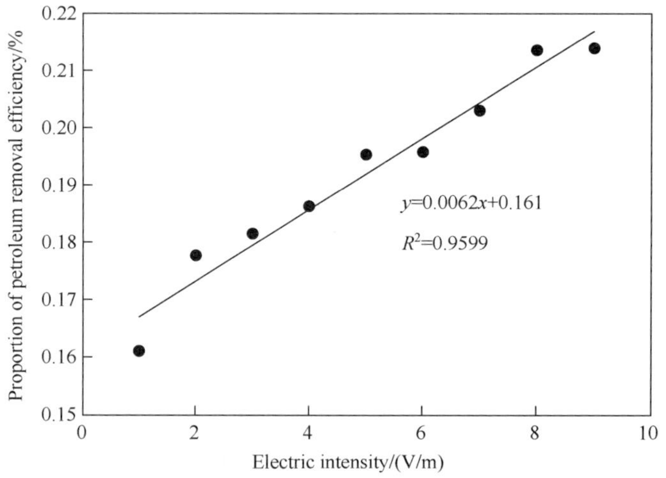

Fig. 8　Correlative analysis of the electric intensity and the efficiency of petroleum removal

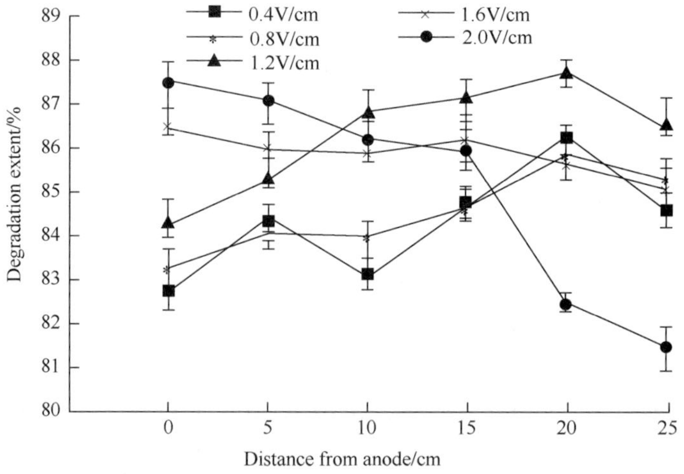

Fig. 9　Extent of degradation of pyrene at each site under different voltage gradients after 35 days of treatment

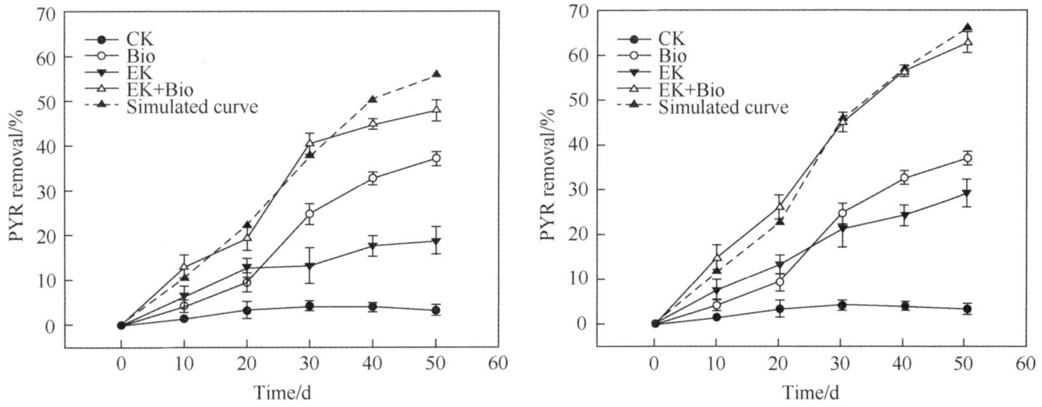

(a) Electrokinetics (EK) without polarity reversal

(b) EK with polarity reversal.

Fig. 10 Changes in removal of pyrene (PYR) by different types of treatment

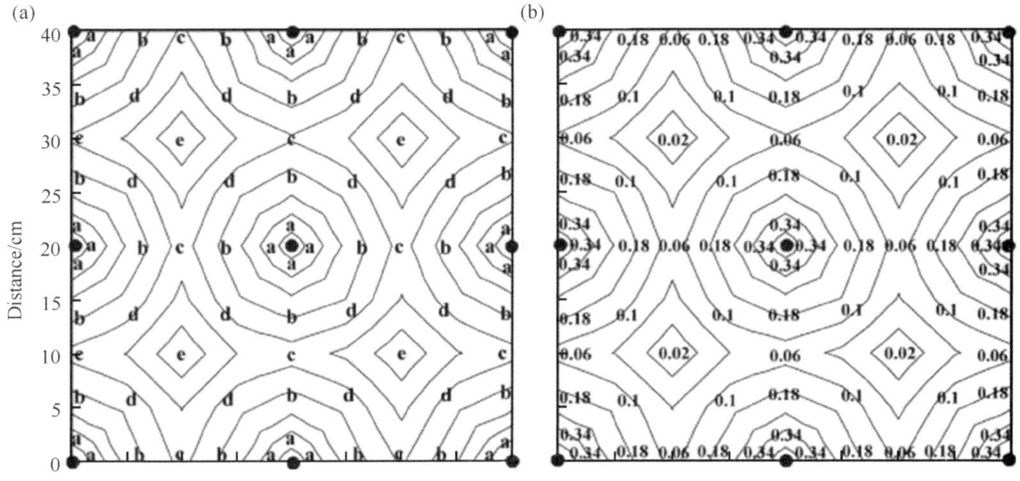

(a) Sampling positions

(b) Electric field intensity /(V/cm)

Fig. 11. Contour plots showing the distribution of sampling positions and electric field intensity/ (V/cm)

(4) Equipment for electro-bioremediation

The equipment used for the electrokinetic remediation of soils contaminated with organic species has been developed on the basis of the principles of remediation and the monitoring requirements. A device for the salt bridge enhanced remediation of pyrene-contaminated soils was developed to realize separation of the anode and cathode (Guo et al., 2013a). Two separate electrokinetic regions were set up in a contaminated soil and the two regions were connected with salt bridges (1 cm diameter, 30 cm long) saturated with agar-KCl (Fig. 12). This experimental set up resulted in an acidic anode region and an alkaline cathode region; a stable acidic environment is conducive to providing the oxidation potential required to remove organic pollutants.

The electric field distribution and intensity are crucial parameters in determining the efficiency of electrokinetic remediation. Guo et al. (2014b) developed an electric field monitoring device to collect and control data on electric fields. The main components included a probe and an electrode polarity detection module, a reference negative pressure calibration module, an A/D module, a field strength conversion module and a control module (Fig. 13). The probes collect data on the voltage and the signal is then transferred to the main control component. The electric field intensity is calculated to show the real-time distribution of the electric field during the electrokinetic remediation process. As a result of the non-uniform electric field supplied by the cylindrical electrodes, it was necessary to reinforce the weak electric field at certain locations using an electrode transposition device equipped with a polarity switching component (Guo et al., 2015a). Total coverage of the treatment regions with an electric field was realized through polarity switching in four-electrode groups, but with a relatively high electric intensity around the region of the electrodes. To enhance the electric intensity in the central region, a supplementary electrode was used to enhance the electric intensity by 50% and to achieve uniform remediation with high-intensity electrokinetic remediation (Fig. 14). Fig. 15 shows another set up in which an electrode matrix of $M \times N$ electrodes was used to obtain a uniform field intensity with a symmetrical electric field structure (Guo et al., 2013b). This system solved the problem of "dead corners" -that is, regions without an electric field-while increasing the percentage of electric field used. In a defined field, any point at an equal distance from an electrode had the same electric field strength.

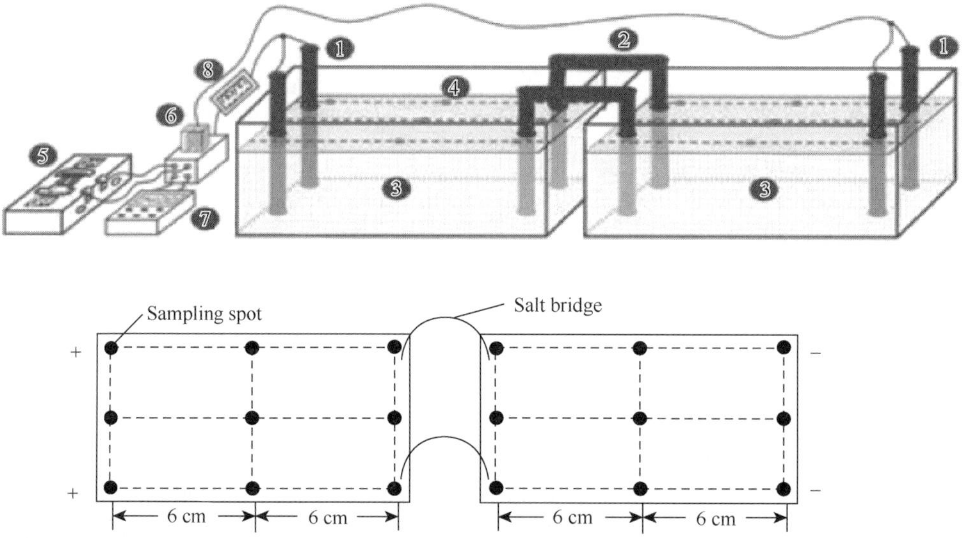

Fig. 12 Schematic representation of the experimental set up and sampling positions for a system using a salt bridge

① electrode; ② salt bridge; ③ soil sample; ④sampling spot; ⑤ power supply; ⑥ relay; ⑦ timer; ⑧ ampere meter

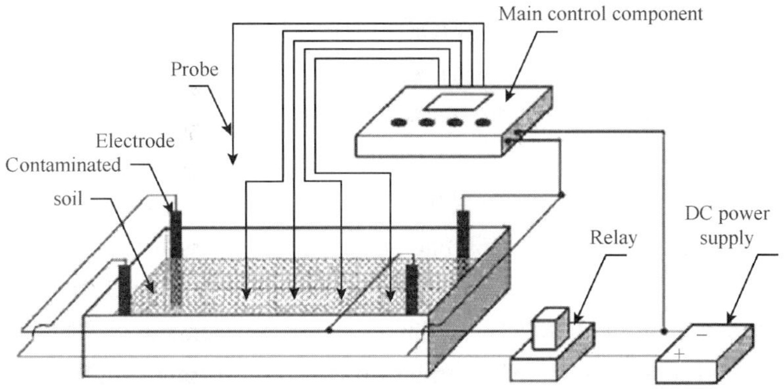

Fig. 13 Electric field monitoring device for use in the electrokinetic remediation of soils

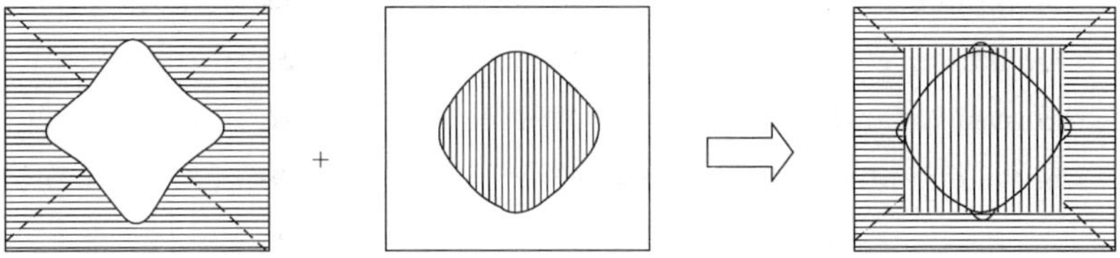

Fig. 14 Electrode transposition device equipped with a polarity switching component

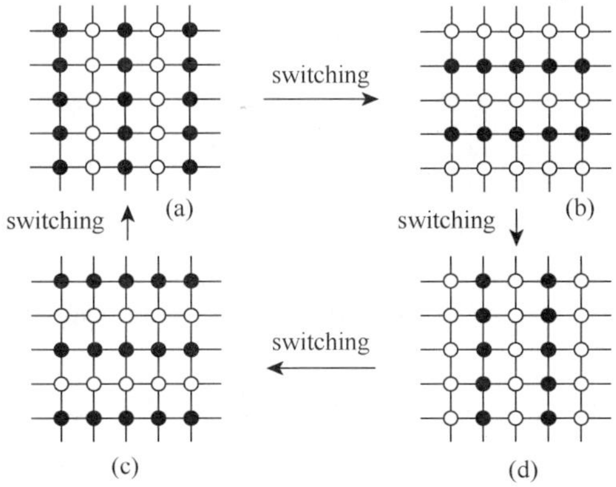

Fig. 15 Design for an electrode matrix with polarity switching to produce the same intensity of electric field at any point of equal distance from the electrodes

2. Remediation of soils contaminated with heavy metals

There are interests in electrokinetic remediation as a new technology to treat soils contaminated with chromium. However, the process of remediation is complex because chromium exists as both cationic and anionic species in soils. Approaching anodes (AAs) was used to decrease the "focusing" effects during the electrokinetic remediation of chromium-contaminated soils (Guo et al., 2012) ; this method enhanced the currents and decreased the pH of the soil. This technique not only improved the removal of Cr (Ⅵ) (92.50%), but also Cr (total) (35.96%). Citric acid was more suitable as a catholyte to reduce the focusing effect and to enhance the rate of removal of Cr (total). After remediation, the four fractions of Cr (total) showed broader changes in distribution. Electrokinetic remediation with approaching anodes decreased the content of BCR2, BCR3 and BCR4 near the cathode (Li et al., 2012a, 2012b). However, it increased the ratio of BCR1 in the sections near the anode, especially with citric acid as the electrolyte. In addition, the method of combining reduction of Cr (VI) with electrokinetic remediation with approaching anodes has been proposed. Sodium bisulfite ($NaHSO_3$) was selected as the reducing agent. When $NaHSO_3$ was added to the soil before electrokinetic treatment, 90.3% of the Cr (VI) was reduced to Cr (III). Electrokinetic experiments showed that the adverse effect of the contrasting migration of Cr (III) and Cr (VI) species, which can limit the practical application of this technique, was eliminated in the presence of the reducing agent (Wei et al., 2016). After remediation, the pH of the soil in the electrokinetic remediation with approaching anodes experiments was between 1.8 and 5.0. The efficiency of removal of total Cr was 64.4% (except in the "focusing" region) when the reduction reaction was combined with the electrokinetic remediation with approaching anodes method.

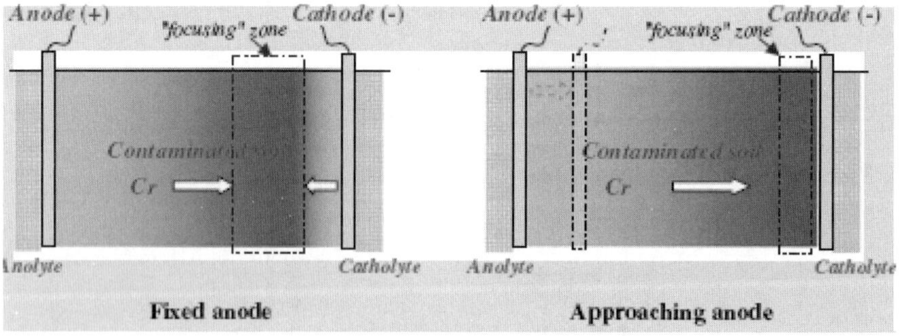

Fig. 16 The "focusing" effect during electrokinetic (EK) remediation of chromium-contaminated soils

References

Fan R, Guo S, Li T, et al. 2015. Contributions of electrokinetics and bioremediation in the treatment of different petroleum components. CLEAN-Soil, Air, Water, 43 (2): 251-259.

Gan X H, Guo S H, Xu W D, et al. 2015. VAnalysis of Factors Influencing Hydroxyl Radical Formation in Electro-Fenton Slurry Reactor, Journal of Agro-Environment Science.34: 44-49.

Gao Y M, Yang X L, Li F M, et al. 2016. Isolation of heavy PAH-degrading bacteria and their characteristics of degradation. Doi: 10.13292/j.1000-4890.201606.008.

Gong Z Q, Li P J, Wang X, et al. 2001. Cometabolic degradation of Benzo[a]pyrene in the soil by the introduced fungi. Research of Environmental Sciences, 14 (6): 36-39.

Gong Z, Li P, Guo S, et al.2001. Bioslurry remediation of soil contaminated with polycyclic aromatic hydrocarbons. Environmental Science, 22 (5): 112-116.

Guo S H, Li F M, Xu S J, et al. 2013a. A salt bridge-enhanced electrokinetics remediation method for the removal of soil organic pollutants.

Guo S H, Li T T, Wang Y H, et al. 2015. System and method for remediation of organic contaminated soil.

Guo S H, Li T T, Wu B. 2013b. A method of electrode transpose through polarity switching and field strength supplement.

Guo S H, Wu B, Zou Z H, et al. 2014b. Electric field monitoring device and method for electrokinetics remediation of polluted soil.

Guo S H, Zhang H R, Zhang C H.2005. Screening for degradation fungi of persistent petroleum hydrocarbon. Journal of Agro-Environment Science, 24 (1): 161-164.

Guo S, Fan R, Li T, et al. 2014a. Synergistic effects of bioremediation and electrokinetics in the remediation of petroleum-contaminated soil. Chemosphere, 109: 226-233.

Guo S, Zhang H, Li F, et al. 2005. Treatment Technology of Composting Piles for Mud Containing Oil. Journal of Agro-Environment Science, 24 (4): 812-815.

Huang D, Guo S, Li T, et al. 2013. Coupling interactions between electrokinetics and bioremediation for pyrene removal from soil under polarity reversal conditions. CLEAN-Soil, Air, Water, 41 (4): 383-389.

Jiang C L, Sun T H, Li P J, et al. 2001. An off site petroleum-contaminated soil bioremediation technology: soil composting in windrow. Chinese Journal of Applied Ecology.

Li F M, Guo S H, Niels Hartog, et al. 2016. Isolation and characterization of heavy polycyclic aromatic hydrocarbon-degrading bacteria adapted to electrokinetic conditions. Biodegradation, 27: 1-13.

Li F M, Guo S H, Niu Z X, et al. 2006. Isolation of viscous-oil degradative microorganism and biodegradation to resin and asphaltene. Chinese Journal of Soil Science, 37 (4): 764-767.

Li F M, Guo S H, Zhang C C, et al. 2016. Isolation of PAHs degrading bacteria and its application to remediation of polluted soil in coking site. Environmental Pollution & Control, 38 (4): 1-5.

Li F, Guo S, Hartog N. 2012. Electrokinetics-enhanced biodegradation of heavy polycyclic aromatic hydrocarbons in soil around iron and steel industries. Electrochimica Acta, 85: 228-234.

Li G, Guo S, Li S, et al. 2012. Comparison of approaching and fixed anodes for avoiding the 'focusing' effect during electrokinetic remediation of chromium-contaminated soil. Chemical Engineering Journal, 203: 231-238.

Li G, Yang X L, Liang L L, et al. 2012. Evaluation of the potential redistribution of chromium fractionation in contaminated soil by citric acid/sodium citrate washing, Arabian Journal of Chemistry. http: //dx.doi.org/10.1016/j.arabjc.2012.10.016.

Li P J, Gong Z Q, Jing X, et al. 2002b. Bioremediation of PAHs contaminated soil using bio-slurry reactor process. Chinese Journal of Applied Ecology, 13 (3), 327-330.

Li P J, Wang X, Stagnitti F, et al. 2005b. Degradation of phenanthrene and pyrene in soil slurry reactors with immobilized bacteria zoogloea sp.. Environmental Engineering Science, 22 (3), 390-399.

Li P J, Xu H X, Zhang C G. 2001. The degradation of B (a) P by microorganismin contaminated soil. Techniques and Equipment for Environmental Pollution Control, 2 (5): 37-40.

Li P, Li H, Stagnitti F, et al. 2005. Biodegradation of pyrene and phenanthrene in soil using immobilized fungi Fusarium sp. Bulletin of Environmental Contamination & Toxicology, 75 (3).

Li P, Sun T, Stagnitti F, et al. 2002. Field-scale bioremediation of soil contaminated with crude oil. Environmental Engineering Science, 19 (5): 277-289.

Li P, Tai P, Guo S, et al.2003. Two phases bioremediation of oil contaminated soil from Liaohe oil field. Environmental Science, 24 (3): 74-78.

Li S C, Li T T, Li G, et al. 2012. Enhanced electrokinetic remediation of chromium contaminated soil using approaching anodes. Front Environ Sci Eng. 6: 869-874.

Li T T, Guo S H, Li F M, et al. 2011. Polypropylene composite functional electrode and its application. China, 201110439187.1.

Li T, Guo S, Wu B, et al. 2010. Effect of electric intensity on the microbial degradation of petroleum pollutants in soil. Journal of Environmental Sciences, 22 (9): 1381-1386.

Li T, Guo S, Wu B, et al. 2015. Effect of polarity - reversal and electrical intensity on the oil removal from soil. Journal of Chemical Technology and Biotechnology, 90 (3): 441-448.

Li X, Li P, Lin X, et al. 2008. Biodegradation of aged polycyclic aromatic hydrocarbons (PAHs) by microbial consortia in soil and slurry phases. Journal of Hazardous Materials, 150 (1): 21-26.

Li X, Lin X, Li P, et al. 2009. Biodegradation of the low concentration of polycyclic aromatic hydrocarbons in soil by microbial consortium during incubation. Journal of Hazardous Materials, 172 (2): 601-605.

Niu M F, Han X R, Guo S H, et al.2005. Application of biosurfactant in the biodegradation of petroleum contaminated soil with Bio-prepared bed. Chinese Journal of Soil Science, 36 (5): 712-715.

Niu M F, Li F M, Han X R, et al. 2005. Isolation of biosurfactant producing microorganisms and their stablility. Chinese Journal of Ecology, 24 (6): 631-634.

Su D, Li P, Frank S, et al. 2006. Biodegradation of benzo[a]pyrene in soil by Mucor sp. SF06 and Bacillus sp. SB02 co-immobilized on vermiculite. Journal of Environmental Sciences, 18 (6): 1204-1209.

Wang J, Li F, Li X, et al. 2013. Effects of electrokinetic operation mode on removal of polycyclic aromatic hydrocarbons (PAHs), and the indigenous fungal community in PAH-contaminated soil. Journal of Environmental Science and Health, Part A, 48 (13): 1677-1684.

Wei W, Li F M, Yang X L, et al. 2015. Influence of voltage on pyrene removal and microbial community in soil during electrokineticsremediation. Chinese Journal of Ecology. 34, 1382-1388.

Wei X, Guo S, Wu B, et al. 2016. Effects of reducing agent and approaching anodes on chromium removal in electrokinetic soil remediation. Frontiers of Environmental Science & Engineering, 10 (2): 253-261.

Xu S J, Guo S H, Wu B, et al. 2014. An assessment of the effectiveness and impact of electrokinetic remediation for pyrene-contaminated soil, J Environ Sci (China), 26, 2290-2297.

Xu W, Guo S, Li G, et al. 2015. Combination of the direct electro-Fenton process and bioremediation for the treatment of pyrene-contaminated soil in a slurry reactor. Frontiers of Environmental Science & Engineering, 9 (6): 1096-1107.

Yuan S, Liang C H, Li M, et al. 2011. Screening of PAHs-degrading bacteria strains and determination of their capability in degrading PAHs. Chinese Journal of Ecology, 30 (2): 315-319.

Yuan Y, Guo S H, Li F M, et al. 2013. Effect of an electric field on n-hexadecane microbial degradation in contaminated soil. International Biodeterioration & Biodegradation, 77: 78-84.

Zang S Y, Li P J, Yu X C, et al. 2007a. Degradation of metabolites of benzo[a]pyrene by coupling Penicillium chrysogenum with KMnO4. Journal of Environmental Sciences, 19, 238-243.

Zang S, Li P, Li W, et al. 2007. Degradation mechanisms of benzo[a]pyrene and its accumulated metabolites by biodegradation combined with chemical oxidation. Chemosphere, 67 (7): 1368-1374.

Biodegradable Chelant-Assisted Phytoextraction

Chunling LUO[1], Xiangdong LI[2], Zhenguo SHEN[3]

[1]Guangzhou Institute of Geochemistry, Chinese Academy of Sciences

[2]Department of Civil and Environmental Engineering, The Hong Kong Polytechnic University

[3]College of Life Sciences, Nanjing Agricultural University

Abstract

The present manuscript reviews the research progresses on the phytoextraction of metals using biodegradable chelants in our research group. It mainly focuses on the theoretical considerations including metal solubilization, root uptake and translocation upwards to the shoots, the effects of chelant application on metal leaching and soil microbes, and the optimization of the chelant-assisted phytoextraction. In addition, the potential of phytoextraction of heavy metals and persistent organic pollutants (POPs) with biodegradable chelant in the heavy metal-POPs co-contaminated is reviewed as well.

Keywords: heavy metal, biodegradable chelant, EDDS, phytoextraction, leaching, POPs

1. Chelant-assisted phytorextraction

The idea of using plants to remediate metal-contaminated soil has attracted a great deal of research in the last two decades. But due to the limited plant species with a high capacity to accumulate metals, especially metals with low bioavailability in soil, such as Pb, and to produce a large amount of biomass, one alternative approach using chelants to improve the uptake of metals by high biomass plants has been proposed, inspired by studies on plant nutrition (Marschner, 1995).

Careful assessment and evaluation is required to determine the biodegradation and toxicity of the chelating agents and their metal complexes in soils. Although EDTA (ethylene diamine tetraacetic acid) was recognized as the most efficient chelant to increase metal uptake by plants, especially for the uptake of Pb, the low biodegradability of the chemical does not make it a good choice for large-scale field applications. In recent years, the focus of research has shifted to some more biodegradable chelants, such as NTA (nitrilotriacetate), EDDS (S, S- ethylene diamine tetraacetic acid), and others. The use of these biodegradable chelants in improving the uptake of metals by plants and in limiting the leaching of metals from soil has become an attractive field of research. Most of this kind of research has been carried out in the form of studies comparing the previous

EDTA results in metal uptake efficiencies with additional data on the biodegradability of chelants and the metal leaching potential from the application of the chemicals. The optimization and application of this technology should be based on the full understanding of important processes involved, such as metal solubilization from the application of chelants, the uptake of metals by the roots of plants, and their transport upwards to the shoots of the plants. To prevent the possible movement of metal-chelants into groundwater and to reduce the impact of the remaining chelant on soil microorganisms, the selection of chelants and the amount and process of their application are important.

2. Theoretical considerations

2.1 Metal solubilization by chelants

In the process of chelant-assisted phytoextraction, chelant is applied to the soils. First, chelant can desorb metals from the soil matrix, and the mobilized metals move to the rhizosphere for uptake by plant roots. The amounts of bioavailable metals in soil solution are mainly determined by the properties of the soil and the chelant which is applied (Huang et al., 1997; Tandy et al., 2004). The efficacy of a chelant in the extraction of metals is usually rated with the stability constants Ks of the chelant-metal complexes. According to Elliott and Brown (1989), the order of magnitude of the Ks can be used to rank different chelants according to their general efficacy, but not to rank the efficacies of a specific chelant toward different metals because the latter is also influenced by the metal speciation in a given soil matrix. We found the effectiveness of a variety of synthetic chelants to induce Pb desorption from soil in decreasing order, was EDTA>DTPA (diethylene triamine pentaacetic acid) >NTA>citric acid (Shen et al., 2002). EDTA is more efficient than EDDS in the extraction of Pb and Cd, but EDDS is more effective in the extraction of Cu and Zn (Luo et al., 2005).

2.2 The sensitiveness of plan species to chelants

The potential of 18 different species/cultivars of plants, including 13 dicotyledons and 5 graminaceous monocotyledons, to be used in the chemically enhanced phytoextraction of Cu, Pb, Zn and Cd was assessed using pot experiments. It found for different plant species, even different cultivars within the same species, there are significant differences in the metal tolerance and uptake ability (Luo et al., 2006a). The dicotyledon species suffered from more severe phytotoxicity than the graminaceous monocotyledon species after the application of EDTA and the enhancement of metal concentrations in the shoots of the plants was more pronounced for the dicotyledon plants than for the graminaceous monocotyledon plants (Luo et al., 2006a). Therefore, it is possible that the roots of dicotyledon species would experience from heavier physiological damages, which could lead to a breakdown of the root exclusion mechanism, and in turn to the indiscriminate uptake of solutes by plants, than the roots of monocotyledon species.

Of all the plants tested, garland chrysanthemum (*Chrysanthemum coronarium* L.) showed the greatest sensitivity of growth to the application of EDTA and the highest enhancement of the concentrations of Cu, Pb, Zn in the shoots. Compared with EDTA, EDDS was more effective in enhancing the concentration of Cu in the shoots of *Chrysanthemum coronarium* L. grown on multi-metal contaminated soils.

2.3 Mechanism of metal-chelant uptake by plants

The predominant theory for metal-chelant uptake is the split-uptakemechanism, bywhich only free metal ions can be absorbed by plant roots (Marschner et al., 1986). Another important theory suggests that some of the purportedly intact metalechelant complexes are taken up by plants (Bell et al., 1991; Salt et al., 1995). A schematic display of this process is shown in Fig. 1 (Leštan et al., 2008). As a typical soil metal contaminant, Pb has been extensively studied. The metal can be absorbed by plant roots and transferred as a Pb-EDTA complex (Vassil et al., 1998; Epstein et al., 1999).

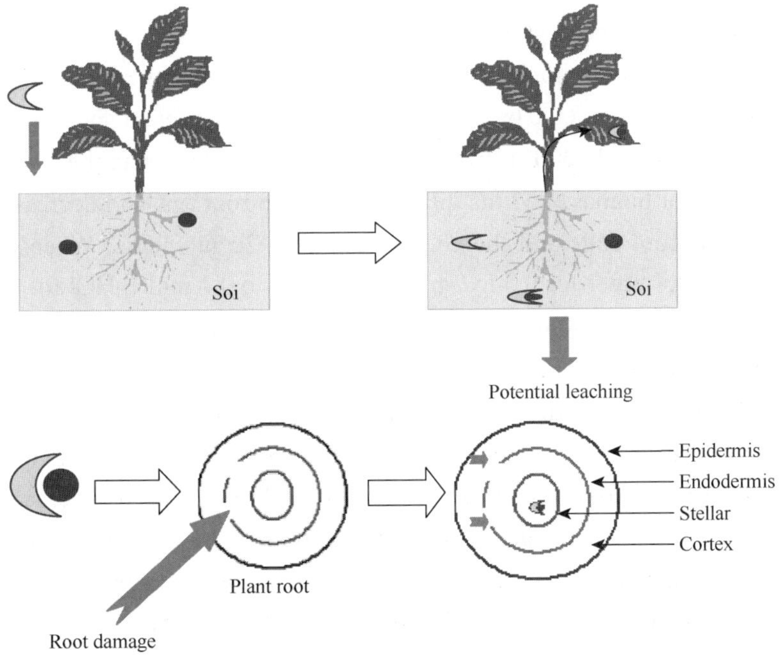

Fig. 1. The schematic representation of the uptake of metal-chelant complexes by plant roots, their translocation upward, and the potential leaching of metals into the surrounding environment in the process of chelant-enhanced phytoextraction (the red circle and yellow moon represent the metals in the soil and the applied chelant, respectively)

We carried out two studies on the mechanism of plant metal uptake with the addition of biodegradable chelants citric acid and EDDS. In one study, we investigated the non-selective apoplastic passage of Cu and Cu-citrate complexes into the root stele of monocotyledonous corn and dicotyledonous soybean using an inorganic-salt-precipitation technique (Fu et al., 2015). Either Cu ions or Cu-citrate complexes were drawn into root through the apoplast from the root growth medium, and $K_4[Fe(CN)_6]$ was subsequently perfused through xylem vessels or the entire root cross section. Based on microscopic identification of the reddish-brown precipitates of copper ferrocyanide in the cell walls of the xylem of corn and soybean roots, Cu^{2+} passed through the endodermal barrier into the xylem of both species. When the solution containing 200 μmol/L $CuSO_4$ and 400 μM sodium citrate (containing 199.98 μM Cu-citrate, 0.02 μmol/L Cu^{2+}) was drawn via differential pressure gradients into the root xylem while being perfused with $K_4[Fe(CN)_6]$ through the entire root cross-section, reddish-brown precipitates were observed in the walls of the stele of soybean, but not corn root. However, when a $CuSO_4$ solution containing 0.02 or 0.2 μmol/L free Cu^{2+} was used, no reddish- brown precipitates were detected in the stele of either of the two plants. Results indicated that endodermis was permeable to Cu-citrate complexes in primary roots of soybean, but not corn. The permeability of the endodermal barrier to the Cu-citrate complex may vary between dicotyledonous and monocotyledonous plants, which has considerable implications for chelant-enhanced phytoextraction.

In our another study, hydroponic cultures were conducted to investigate the effects of the EDDS on the growth and Cu uptake by garland chrysanthemum, a plant species sensitive to soil chelant amendment (Wei et al., 2007). In the presence of 50 μmol/L Cu, the addition of EDDS increased shoot and root biomass and the vitality of cells in root tips and decreased the relative electrolyte leakage of root cells and the concentration of Cu in shoots. When the roots were pretreated with 65℃ hot water for 0.5 to 2 h or with 0.001 to 0.1 mol/L HCl for 24 h before the exposure to 50 μmol/L Cu + 100 μmol/L EDDS, the concentration of Cu in shoots increased considerably compared with the plants without any pretreatment. A statistically significant correlation was found between the Cu concentrations in shoots and the relative electrolyte leakage of root cells. These results indicated that Cu-EDDS might be a less bioavailable form to plants and that some physiological damage to the roots led to enhanced accumulation of Cu in plant shoots. Therefore, some damage to the root may be helpful for the indiscriminate uptake of metal-chelant complex by plant roots. The damage could be caused by the toxicity of metals, chelants and other artificial means (Vassil et al., 1998; Luo et al., 2006a).

3. Evaluation of metal leaching and effects on soil microbes

A novel experimental setup (one pot placed above another) was used to investigate the residual effects of EDDS application on plant growth and metal uptake (Yang et al., 2013).

Two plant species, garland chrysanthemum and ryegrass, were grown in the upper pots (mimicking the upper soil layers) and were harvested 7 days after EDDS application. During this period the upper pots were watered twice. The lower pots (mimicking the subsoil under the upper soil layers) served as leachate collectors. Thereafter, the two pots were separated, and the same plants were grown in the upper and lower pots in two continuous croppings. Results showed that EDDS application restrained the growth of the first crop and resulted in a dramatic enhancement of Cu accumulation in plants grown in the upper pots. However, no negative growth effects were identified for the second and third crops, which were harvested 81 and 204 days after the EDDS application, respectively. In the lower pots, the leachate from the upper pots after EDDS application exhibited the increased total and $CaCl_2$-extractable Cu concentrations in the soil. However, the growth of garland chrysanthemum and ryegrass, and their shoot Cu concentrations were unaffected. These data suggest that the residual risk associated with EDDS application was limited, and that subsoil to which EDDS leachate was applied may exhibit reduced Cu bioavailability for plants due to the biodegradation of EDDS.

A field study was carried out and the metal leaching along the 60 cm depth soil profiles were evaluated as well within 36 days after the biodegradable chelant EDDS was applied (Wang et al., 2012). Results showed EDDS significantly increased soluble Cu in the top 5 cm soil layer 1 day after the application, and the increase of soluble metals was generally limited in the top 20 cm soil. Metal speciation analysis indicated all Cu and Zn were in forms of Cu-EDDS and Zn-EDDS complexes in soil solution, and Ca was the major competitor with trace metals to EDDS. The soluble metals decreased quickly with time, and no significant difference was observed in the extractable Cu between EDDS treatments and the controls 22 days after the EDDS addition. The potential leaching associated with biodegradable EDDS addition may be controlled under field conditions. We also evaluated the effects of EDDS on enzyme activities and microbial community composition in Cu contaminated soils, planted with either corn or beans (Yang et al., 2013). Results showed that the application of EDDS did not affect urease and acid phosphatase activities in the soil, but greatly reduced catalase and saccharase activities, and increased β-glucosidase activity on the seventh day after EDDS application. On the 28th day, no significant difference was observed in the enzyme activities (except for β-glucosidase) of EDDS-treated soils compared to the controls. Analysis of phospholipid fatty acids (PLFAs) showed that the application of 3 mmol/kg EDDS did not cause significant stress to soil microbial communities. However, PCR-denaturing gradient gel electrophoresis (PCR-DGGE) fingerprint revealed that EDDS influenced the bacterial communities in the soils, and the effects on bean soils were more significant than that with corn. In general, the enzyme activities and bacterial communities were influenced by the application of EDDS, but the impact became weaker or even disappeared with the biodegradation of EDDS.

4. Optimizing chelant-assisted phytoextraction

For a given chelant, different methods of application can produce different levels of phytoextraction efficiency. Exploring effective strategies for the application of chelants is useful in optimizing the technology. We found applying chelant in several smaller dosages (versus in one application) can result in the enhanced phytoextraction of Pb (Shen et al., 2002). The combined application of different chemicals can also greatly improve the metal phytoextraction efficiency.

The combined application of different chelants based on the interactions between metals and different chelants was studied, in which the solubility of metals by a chelant can be increased by another chelant through the reduction of competition from other metals in soil. It found hat the combined application of EDTA and EDDS led to a higher level of efficiency (i.e., a synergy effect) in the phytoextraction of Cu, Pb, Zn and Cd than could be obtained by the application of either chelant alone (Luo et al., 2006b). There are two reasons for the result: the fact that EDTA and EDDS have different levels of efficiency in extracting metals from soils; and a decrease in the competitive cations for trace metals with EDTA, such as soil-soluble Ca, due to the addition of EDDS (Tandy et al., 2004).

The utilization of some artificially physiological damage to destroy the plant root structure to facilitate the direct uptake of metal-chelants and their translocation into the shoots was proved efficient in enhancing the phytoextraction. In our study, plant roots were pretreated with MC (methanol:trichloromethane), HCl and hot water, and from treatment with DNP (2, 4- dinitrophenol, an un coupler of oxidative phosphorylation), and dramatic increase in the concentrations of Pb in shoots was achieved with the EDTA treatment (Luo et al., 2006c). Applying similar treatments in a pot experiment, Luo et al. (2006d) found that when chelants were applied as hot solutions at the rate of 1 mmol/kg, the concentrations and total phytoextraction of Cu, Zn and Cd by plant shoots exceeded or at least approximated those in the shoots of plants treated with normal chelants at a rate of 5 mmol/kg (Luo et al., 2006d, 2007, 2008). This result indicated that the amount of chelant applied could be greatly decreased for the given effectiveness of chelants in enhancing the phytoextraction of trace metals from contaminated soils. The soil leaching study demonstrated that there was no significant difference in the soluble metals between the hot and normal chelant applications when the chelant was applied at the same dosage. The decreased dosage of chelant resulted in decreased concentrations of soluble metals in soils, which meant that the hot chelant application did not increase metal leaching compared with the normal chelant application.

5. Application of chelants in metal and organic pollutants co-contaminated soils

Co-contamination by heavy metals and persistent organic pollutants (POPs) is ubiquitous in the environment. Fate of POPs within soil/water-plant system is a significant concern and an area where much uncertainty still exists when plants suffered co-toxicity from POPs and metals. Our study investigated the fate of polybrominated diphenyl ethers (PBDEs) when Cu was present within the soil/water-plant system using pot and hydroponic experiments. The presence of Cu was found to induce damage to the root cell membranes of corn with increasing concentration in both shoots and roots. The PBDE congeners BDE209 and BDE47 in shoots were also enhanced with the increasing electrolytic leakage from root, attributed to Cu damage, and the highest shoot BDE209 and BDE47 levels were observed under the highest Cu dosage (Wang et al., 2016). In addition, positive correlations were observed between the PBDE content of corn shoots and the electrolytic leakage of corn roots. These results indicated that within a defective root system, more PBDEs will penetrate the roots and are acropetally translocated in the shoots.

Based on the knowledge of effects of Cu on the uptake of PBDEs by plants, we evaluated the influence of the biodegradable chelant EDDS on plant uptake of polychlorinated biphenyls (PCBs), PBDEs, and Cu by corn from electronic waste (e-waste) -contaminated soil (Wang et al., 2015). The highest concentration and highest total uptake of Cu in corn were observed in the treatment with 5 mmol/kg EDDS, which resulted in a 4-fold increase of the Cu translocation factor (C_{shoot}/C_{root}) compared to the control. The concentrations of PCBs and PBDEs in shoots and roots increased with increasing application rates of EDDS, and 1.58-fold and 1.32-fold average increases in the concentrations of PCBs and PBDEs, respectively, were observed in shoots in the EDDS treatments. A significant positive correlation was observed between shoot Cu and shoot PCBs and PBDEs. We speculate that PCBs and PBDEs were activated by the EDDS-triggered dissolved organic carbon (DOC) and then indiscriminately taken up by roots and translocated to shoots following damage to the roots mainly by the increased extractable Cu resulting from the EDDS application. In addition, we also investigated the effects of EDDS and the plant-growth-promoting bacterium DGS6 on pollutant uptake by corn from ewaste-contaminated soils (Luo et al., 2015). The highest concentration and total uptake of Cu and Zn in corn shoots were observed in the presence of EDDS and DGS6+EDDS, respectively. The PCB concentrations in shoots ranged from 0.53 to 0.72 ng/g, and the highest PCB concentration was observed in the presence of EDDS. This could be ascribed to the enhanced dissolved organic carbon, increased dissolution and efficient translocation of PCBs from roots to shoots, as well as potential root damage due to increased soluble metal levels in soil solution. In contrast, the highest total uptake of PCBs in shoots was observed in the presence of DGS6, likely due to enhanced shoot biomass and high levels of air deposition.

References

Bell P F, Chaney R L, Angle J S. 1991. Free metal activity and total metal concentrations as indexes of micronutrient availability to barley (*Hordeum vulgare* cv "Klages"). Plant and Soil, 130: 51-62.

Elliott H A, Brown G. 1989. Comparative evaluation of NTA and EDTA for extractive decontamination of Pb-polluted soils. Water, Air and Soil Pollution, 45: 361-369.

Epstein A L, Gussman C D, Blaylock M J, et al. 1999. EDTA and Pb-EDTA accumulation in *Brassica juncea* grown in Pb-amended soil. Plant and Soil, 208: 87-94.

Fu Y Z, Luo C L, Shen Z G. 2015. Permeability of plant young root endodermis to Cu ions and Cu-citrate complexes in corn and soybean. International Journal of Phytoremediation, 17: 822-834.

Huang J W, Chen J J, Berti W R, et al. 1997. Phytoremediation of lead-contaminated soils: role of synthetic chelates in lead phytoextraction. Environmental Science and Technology, 31: 800-805.

Leštan D, Luo C L, Li X D. 2008. The use of chelating agents in the remediation of metal-contaminated soils—a review. Environmental Pollution, 153: 3-13.

Luo C L, Wang S R, Wang Y, et al. 2015. Effects of EDDS and plant-growth-promoting bacteria on plant uptake of trace metals and PCBs from e-waste-contaminated soil. Journal of Hazardous Materials, 286: 379-385.

Luo C L, Shen Z G, Lou L Q, et al. 2006a. EDDS and EDTA-enhanced phytoextraction of metals from artificially contaminated soil and residual effects of chelant compounds. Environmental Pollution, 144: 862-871.

Luo C L, Shen Z G, Li X D, et al. 2006b. Enhanced phytoextraction of Pb and other metals from contaminated soils through the combined application of EDTA and EDDS. Chemosphere, 63: 1773-1784.

Luo C L, Shen Z G, Li X D, et al. 2006c. The role of root damage in the EDTA-enhanced accumulation of lead by Indian mustard plants. International Journal of Phytoremediation, 8: 323-337.

Luo C L, Shen Z G, Baker A J M, et al. 2006d. A novel strategy for chemically enhanced phytoremediation of heavy metal-contaminated soils. Plant and Soil, 285: 67-80.

Luo C L, Shen Z G, Li X D. 2005. Enhanced phytoextraction of Cu, Pb, Zn and Cd with EDTA and EDDS. Chemosphere, 59: 1-11.

Luo C L, Shen Z G, Li X D. 2007. Plant uptake and the leaching of metals during the hot EDDS-enhanced phytoextraction process. International Journal of Phytoremediation, 9: 181-196.

Luo C L, Shen Z G, Li X D. 2008. Hot NTA application enhanced metal phytoextraction from contaminated soil. Water Air and Soil Pollution, 188: 127-137.

Marschner H. 1995. Mineral Nutrition of Higher Plants. London: Academic Press.

Marschner H, Romheld V, Kissel M. 1986. Different strategies in higher plants in mobilization and uptake of iron. Journal of Plant Nutrition, 9: 695-713.

Salt D E, Prince R C, Pickering I J, et al. 1995. Mechanisms of cadmium mobility and accumulation in Indian mustard. Plant Physiology, 109: 1427-1433.

Shen Z G, Li X D, Wang C C, et al. 2002. Lead phytoextraction from contaminated soil with high-biomass plant species. Journal of Environmental Quality, 31: 1893-1900.

Tandy S, Bossart K, Mueller R, et al. 2004. Extraction of heavy metals from soils using biodegradable chelating agents. Environmental Science and Technology, 38: 937-944.

Vassil A D, Kapulnik Y, Raskin I, et al. 1998. The role of EDTA in lead transport and accumulation by Indian mustard. Plant Physiology, 117: 447-453.

Wang A G, Luo C L, Yang R X, et al. 2012. Metal leaching along soil profiles after the EDDS application—a field study. Environmental Pollution, 164: 204-210.

Wang S R, Wang Y, Lei W R, et al. 2015. Simultaneous enhanced removal of Cu, PCBs, and PBDEs by corn from e-waste contaminated soil using the biodegradable chelant EDDS. Environmental Science and Pollution Research, 17: 1-8.

Wang S R, Wang Y, Luo C L, et al. 2016. Could uptake and acropetal translocation of PBDEs by corn be enhanced following Cu exposure?Evidence from a root damage experiment. Environmental Science & Technology, 50: 856-863.

Wei L, Luo C L, Wang C C, et al. 2007. Biodegradable chelating agent ethylenediaminedisuccinic acid reduces uptake of copper through alleviating the physiological damage of copper to the roots of *Chrysanthemum coronarium* L. from hydroponical experiments. Environmental Toxicology and Chemistry, 26: 749-754.

Yang L, Luo C L, Liu Y, et al. 2013. Residual effects of EDDS on plants grown in a mine soil during EDDS-assisted Cu phytoextraction. Science of the Total Environment, 444: 263-270.

Yang L, Wang G P, Cheng Z N, et al. 2013. Influence of the application of chelant EDDS on soil enzymatic activity and microbial community structure. Journal of Hazardous Materials, 262: 561-570.

Speciation, Activity, Transformation and Degradation of Heavy Metals and Organochlorines in Red Soils

Yongtao LI, Wenyan LI, Huijuan XU

College of Natural Resources and Environment, South China Agricultural University

China launched a national soil quality and pollution survey (NSQPS) from 2006 to 2010, and released the nationwide survey of soil pollution bulletin on April 17, 2014. The survey showed that the main pollution type in China was inorganic pollution (82.2%), followed by organic pollution and combined pollution (Ministry of Environmental Protection and Ministry of Land and Resources of P. R. China, 2014). The main inorganic pollutants are heavy metals, including cadmium, arsenic, copper, lead, chromium, zinc, etc. Heavy metal pollution is covert, persistent and irreversible (Wang et al., 2001), may remain in local ecosystem for a long time and pose a potential threat to public health (Fahmi et al., 2014). Organic pollutants are known to persist in the environment, bioaccumulate in biota, and pose hazard to the ecological system and human health. How to clean the heavy metals and organic pollutants from the environment in order to avoid their entrance into the food chain is important for protecting the health of animals and human beings. Therefore, understanding the status of heavy metal and organic pollutions are the basic idea for remediating the pollution from the environment. Guangdong Province, especially the Pearl River Delta (PRD) has undergone a rapid transition from a traditional agriculture-based region to an increasingly industry-and technology-based region. As a consequence, the soils in PRD have been extensively under severe environmental stresses, and considered to be a perfect region for regional researching and evaluation of pollutions (Hu et al., 2013). Due to non-strictly controlled emissions from mining, smelting and industrial activities in PRD over the past several decades, large amounts of wastewater, sludge, e-waste and exhaust, which often contain elevated concentrations of heavy metals, have been released to the environment (Zhou et al., 2007; Zhao et al., 2012). And the agricultural boom in the recent past was accompanied with widespread applications of pesticide as DDTs and HCHs, making the PRD a region with the highest usage of pesticide in China (Hua and Shan, 1996). Therefore, increasing efforts have been made to deal with environmental pollution issues in the PRD region. Our article would compile our research results, which have already been published in professional journals, to try to provide a comprehension of the heavy metal pollution in PRD and microbial remediation of organic pollutant.

1. Speciation, activity, transformation and environmental effects of soil heavy metals

1.1 Ion activity and distribution of heavy metals in acid mine drainage polluted subtropical soils

The mining industry has produced a large amount of metal-rich wastes and caused serious environmental pollution and soil degradation (Sheoran and Sheoran, 2006). The oxidative dissolution of these mine wastes give rise to acidic, metal-enriched mine drainage (AMD) (Johnson and Hallberg, 2005). The degraded land associated with mining account for 3.2 Mha in China in 2004, 80%-90% of which still remain unreclaimed (Li, 2006). However, the nature and extent of this contamination are highly variable, depending on the nature of mine ore body and associated geological strata and climate (Johnson and Hallberg, 2005). Accordingly, it is important to accurately assess the availability and toxicity of heavy metals in specific AMD-polluted soils.

The poly-metallic mine Dabaoshan, as an example, is one of the biggest opencast mines in South China. Mining wastes severely polluted surrounding agricultural fields and have been threatening downriver metropolitan area since 1960s (Li et al., 2005; Lin et al., 2007). Etiologic survey (Liu et al., 2005) showed that 153 villagers died from alimentary tract malignant tumor in a nearby village Shangba from 1984 to 2001, which accounted for 51% of total death toll. The high incidence caused people to suspect the causality with toxicity of soil heavy metals. Our researches chose Dabaoshan as an excellent site to assess the availability and toxicity of heavy metals and their main controlling factor, help to understand the processes affecting the surrounding polluted agricultural fields.

However, it is difficult to accurately measure free metal ion in soil solution equilibrium because the individual chemical species are often present at nano-and picomolar concentration in natural systems (Weng et al., 2001a). The recently developed soil column Donnan membrane technology (SC-DMT) has advantage of allowing simultaneous measurement of multi-metal free ions at low concentrations, with minimized disturbance of soil chemical equilibrium (Weng et al., 2001a, 2001b; Cancès et al., 2003. Additionally, selective sequential extractions (SSEs), despite some limitations, provide operationally defined soil-phase fractionation according to the targeted geochemical compartment (Quantin et al., 2002; Peijnenburg and Jager, 2003). This extraction procedure has been widely accepted for quantitative and comprehensive prediction and identification of metal fractions bound to soil solid phases (Kennedy et al., 1997).

Our researches combined SC-DMT with SSEs to estimate the amount of the most available forms of metal ions. SC-DMT determined the free and complexed ions of heavy metals in soil solution of agricultural fields around Dabaoshan. The average concentrations of total dissolved

Cu, Pb, Zn and Cd in soil solution were 2.5, 2.0, 1.9 and 2.9 times higher than Chinese national standards for irrigation water quality (GB5084—2005), respectively. Paddy soils contained higher concentration of metals in soil solution than non-flooded cultivated fields. In addition, ANOVA ($p<0.001$) shows that concentration of Cd^{2+} ion in solution as a percentage of the total soil Cd content (12.5%) was significantly higher than Cu (1.2%), Pb (0.3%) and Zn (3.3%). Similarly, complexed Cd in solution expressed as a proportion of the total soil Cd content (3.3%) and Zn (2.3%) was remarkably higher than Cu (0.5%) and Pb (0.07%). It indicates Cd and Zn were preferentially leached from the soils and had higher solubility than Cu and Pb. SSEs determined the seven fractions in total contents of four heavy metals. The difference between sum of metals with the seven SSEs steps and total soil metal contents determined independently after acid digestion was less than ±14% of total soil metal contents. The sums of six extractable fractions contributed on average to 61.4% (Cu), 55.1% (Pb), 49.2% (Zn) and 67.4% (Cd) of total metals. Cu and Cd showed extremely high proportion over 85% and over 90% in some sites. Heavy metals bound to Fe, Mn and Al oxides were predominant extractable fractions, which accounted on average 80% or more of non-residual forms of Cu, Pb and Zn, except for Cd (only 30%). In most samples, amorphous Fe and Al oxides were the main bearing phases for heavy metals. In some samples, Pb bound to Mn oxide and Zn bound to crystalline Fe oxides are dominant. Proportion of soluble, exchangeable and organic bound Cu, Pb and Zn was relatively low. Adversely, percentages of soluble and exchangeable Cd were pronouncedly high (30%). SC-DMT and SSEs both showed that despite total contents of the four heavy metals were not extremely high compared to EU standards, solubility and extractability of heavy metals, especially Cd and Zn, were considerably high in the long-term mine drainage polluted soils of Dabaoshan area. In all labile fractions, free ions and metal-ligand complexes in the soil solution are considered as the most active forms for biota uptake (Parker and Pedler, 1996), and provide a good assessment of potential risks and toxicity through heavy metal contamination (Sauvé et al., 2000a, 2002b). Consequently, high metal ion activity likely enhanced intake and accumulation of metals in crops, and eventually posed long-term toxicity to villager health in the area (Li et al., 2009).

1.2 The dominant environment factors controlling metal activity in the long-term AMD-polluted subtropical soils

Heavy metals in the investigated soils may originate from dissolved metals and suspended mineral waste particles in irrigation water. Main mineral components in Dabaoshan mine are sulphidic metal minerals (Li et al., 2005; Lin et al., 2007). The sulphidic ores, once exposed to oxygen and water due to mining, transportation and tillage, can result in accelerated oxidative dissolution and strong acidity of iron pyrite and other sulphidic minerals (Johnson and Hallberg, 2005; Lin et al., 2007). The acidic soil environment further enhanced metal cation release from metal-bearing minerals, and promoted dissolution of dominant soil clay minerals. Then in our soils, Al, Fe and Mn ions may result from the dissolution of gibbsite and kaolinite (Monterroso et al., 1994), of goethite or amorphous $Fe(OH)_3$ (Stumm and Morgan, 1981), and of manganese oxides (Davranche and Bollinger, 2000), respectively. Low redox potential caused by reduction condition both in mine waters and in paddy soils also promoted the release of Fe and Mn into solution and the trace metals which are sorbed on these minerals. The process enhances mobility and bioavailability of metals in the soil. Accordingly, apart from direct irrigation input, oxidative and acidic dissolution of metal-bearing mineral particles may be the main sources of dissolved metals in the soil solution.

In our researches, stepwise regression analyses and PCA analysis were performed to determine soil environmental factors affecting free and complexed metal ions in soil solution. The stepwise regression analyses showed that exchangeable fraction was the most important effect factor that affect free and complexed metal ions distribution, except for complexed Zn. Cu bound to crystalline Fe oxides was specifically related to free and complexed Cu ions, while Pb bound to Mn oxides was significantly associated with corresponding dissolved Pb fractions. Dissolved Al showed marked influence on both Cu and Pb. Adversely, other reactive compartments and environmental factors, such as organic matter and pH, had little effect on active heavy metals. In addition, in spite of remarked spatial variability in metal contents and distribution, the association between dissolved metals and metals binding to soil solid phases was similar in samples from different sites and land uses. However, the PCA analysis showed that pH and soil organic matter unconventional behavior should be clarified. Metal binding to soils is mostly controlled by four reactive compartments: soil organic matter (SOM), clay minerals, iron and manganese oxi-hydroxides. In our researches, we considered that soil samples under the very low pH conditions could be fully protonated and preventing metal binding to clay minerals and iron and manganese oxi-hydroxides. And also since a shift of at least 1 or 2 order of magnitude in the pH value is needed to measure an effect on the metal ion binding, the narrow range of pH values in Dabaoshan area justifies the lack of pH effects for the SOM fraction. To SOM, it was assumed to behave like a Humic Acid (HA) and the corresponding generic parameters for metal and proton binding of the NICA-Donnan model published by Milne et al. (2003) was used for the calculation. The difference between the initial amount of metal ion bound to the SOM (i.e. 100%) and the remaining amount calculated for the HMnO extraction conditions corresponds to the pool of metal ion that was extracted from the SOM and erroneously attributed to HMnO. We considered that average 40% of the Cu and 85%-95% of the Cd, Zn and Pb bound to the SOM is removed by the HMnO extraction step. The calculation does not account for re-adsorption to the other components of the soil namely iron oxides. Therefore, the values probably correspond to a maximal effect. This calculation could explain the free metal ion concentration in the SC-DMT and lack of involvement of SOM in the soil metal. As clarified the dominant environment factors controlling metal activity, we used the NICA-Donnan model to simulate four metal ions in the Donor solution of the SC-DMT, and found the good agreement between model and measurements by SC-DMT experiments, confirmed our researches (Li et al., 2009).

1.3 Microbial biomass, enzyme and mineralization activity in relation to soil organic C, N and P turnover influenced by acid metal stress

The researches on ion activity and distribution of heavy metals in acid mine drainage polluted subtropical soils and their controlling factors indicated that the acid metal-rich wastes might pose serious risks to the soil environment, and it is important to accurately assess ecotoxicity of metals under strongly acidic soil conditions. Soil microorganisms and enzymes are the primary mediators of soil biological processes, including organic matter degradation, mineralization and nutrient recycling. They play an important role in maintaining soil ecosystem quality and functional diversity (Kandeler et al., 1996). A number of general and specific biochemical parameters were used as indicators to estimate soil degradation caused by heavy metal stress (He et al., 2003). However, the limitations of individual biochemical properties or simple indices have been criticized because such ecotoxicological studies remain few and somewhat contradictory (Giller et al., 1998; Vig et al., 2003). Gil-Sotres et al. (2005) pointed out that a promising solution to better reflect the complexity of soil system is to use complex indices expressed by a set of biochemical parameters, such as the cascade of enzyme activities approach proposed by Nannipieri et al. (2002), and to use statistical techniques, such as principal component analysis. We hypothesized that comprehensive set of microbiological indices and their relationships may allow us to elucidate explicitly the combined stress of metal and acidity, and explain the pattern of scattered responses of single parameters involved in specific soil processes. Therefore, we chose soil microbial biomass, enzyme and mineralization activities involved in organic C, N and P turnover, to evaluate the quality of a subtropical agricultural soil affected by long-term acid metal stress. Fractions of C, N and P involved in soil organic matter, microbial biomass and mineralization processes were estimated. Total enzyme activity and eight hydrolase activities (xylanase, amylase, b-glucosidase, invertase, N-acteyl-glucosaminidase, urease, alkaline and acid phosphatases) in different decomposition stages of organic C, N and P were selected to characterize the soil functional diversity.

These biochemical parameters were compared with soil metal variables (total contents and free and ligand-complexed ions of Cu, Pb, Zn, Cd, Al and Mn), using principal component analyses, co-inertia and discriminant analyses. The multiple statistics indicate that the metal variables were significantly related with not only general biological factors, but also respective datasets of biomass, enzyme activities and mineralization rates (all $p<0.001$). In general, metal variables were inversely related to parameters and indices of microbial biomass C, N and P, and

C-related polysaccharidase and heterosidase activities, and P mineralization. As comparison, metal variables exhibited positive relationships with parameters and indices of N-related N-acteyl-glucosaminidase, urease, ammonification, total N mineralization and metabolic quotient, compared with inhibited nitrification. Specifically, free and complexed metal cations showed higher bioavailability than total contents in most cases. Cu, Pb, Al and Mn had different ecotoxicological impacts than Cd and Zn did. Stepwise regression models demonstrated that metal variables are key stress factors, but most of them excluded soil pH. Furthermore, spatial distribution in land uses and of sampling sites clearly separated the soil samples in these models ($p < 0.001$). We conclude that such a statistical analysis of microbiological and biochemical indices can provide a reliable and comprehensive indication of changes in soil quality and organic nutrient cycling, after exposure to long-term acid metal stress (Li et al., 2009).

1.4 The oxidative transformation of sodium arsenite at the interface of α-MnO_2 and water

Cu, Pb, Zn, Cd were most studied in our researches, however metalloids such as arsenic (As) often fall into the heavy metal category due to similarities in chemical properties and environmental behavior (Chen et al., 1999). Arsenic is ubiquitous in the environment and has four oxidation states: 0 (arsenic), -3 (arsine), +3 (arsenite) and +5 (arsenate). The toxicity of arsenic depends strongly on its chemical forms. Two soluble inorganic forms, arsenite [As (III)] and arsenate [As (V)], are toxic to biological systems, with arsenite considered more toxic than arsenate (Schaeffer et al., 2005). As (III) is a neutral species of H_3AsO_3 at neutral pH, while As (V) is the most stable species under aerobic conditions and normally exists as deprotonated oxyanions of arsenic acid $H_2AsO_4^-$ in solution of pH 4-10 (Lakshmipathiraj et al., 2006). Compared with As (V), As (III) has high mobility and weak adsorption in soils, and thereby is more poisonous (Coddington, 1986). Oxidative transformation of arsenite is a complicated process because the reaction occurred both in solution and on surface of oxidants (Wang et al., 2008). Similarly the different valents of arsenic (III and V) exist both in solution and on surface of oxidants, and some of them can even enter the structure of oxidant and be fixed as a part of the oxidant permanently (Lin and Wu, 2001). However, systematic information about the oxidative transformation of arsenite species is yet not enough. Therefore, our aim was to investigate the effect of α-MnO_2, one of the most stable natural manganese oxides, on oxidative transformation of arsenite. Manganese, as the third most abundant transition metal in earth crust, is important to both biological and environmental processes (Duckworth and Sposito, 2005). Manganese oxides are naturally found in soils, aquifers, oceanic and aquatic systems (Nesbitt et al., 1998), and are considered as important natural oxidizing agents, with reductive half potentials of 1.50 and 1.23V for

MnOOH and MnO_2, respectively (Raven et al., 1998; Stone, 1987). Manganese oxides were reported readily oxidizing many reduced organic/inorganic pollutants in aqueous environment, such as chromium (Fendorf and Zasoski, 1992) and arsenite (Stollenwerk et al., 2007).

We investigated the oxidative transformation of arsenite into arsenate with batch experiments under different reaction conditions, based on different initial arsenite concentrations, α-MnO_2 dosages and pHs. The special focus was on the different states of arsenic species that occurred homogeneously in the reaction solution or heterogeneously on the surface and in the structure of α-MnO_2 during arsenite transformation process. The changes in morphologies and structures of α-MnO_2 during reaction were also studied to explore the reaction mechanisms. The results showed arsenite transformation occurred and was accompanied by the adsorption and fixation of both As (III) and As (V) on α-MnO_2. About 90% of sodium arsenite (10 mg/L) were transformed by α-MnO_2 under the conditions of 25℃ and pH 6.0, 36.6% of which was adsorbed and 28.9% fixed by α-MnO_2. Increased α-MnO_2 dosages promoted As (III) transformation rate and adsorption of arsenic species. The transformation rate and adsorption of arsenic species raised with increasing pH values of reaction solution from 4.7 to 8.0. The oxidation rate decreased and adsorbed As (III) and As (V) increased with increasing initial arsenite concentration. The enhancement on oxidative transformation of sodium arsenite may result from abundant active sites of α-MnO_2. Along with adsorption and fixation of arsenic species during the reaction, the crystal structure of α-MnO_2 did not change, but the surface turned petty and loosen. Our results demonstrated that α-MnO_2 has important potential in arsenic transformation and removal as the environmentally friendly natural oxidant in soil and surface water (Li et al., 2010).

2. Cooperative biodegradation of earthworm and microorganism on soil organochlorines

Due to the environmental persistence, bioaccumulation and potential carcinogenicity to humans and wildlife, the degradation and reduction of organic pollutants have been the focus of researchers for decades. Unlike trace metal, organic pollutants are degradable by microbes, which provides a research idea of enhancing the microbial degradation to remediate soil polluted by organic pollutants. Our researches mainly aim to develop environmental friendly soil remediation technique by studying the enhancement of microbial degradation by earthworm and vermicomposting.

2.1 Enhancement effect of earthworms on removal and degradation processes of organochlorine pesticides

Organochlorine pesticides are characteristically fat soluble, have very low solubility in water, and are resistant to metabolism. They have been labeled as environmental hazards for decades. 1, 1, 1-Trichloro-2, 2-bis (4-chlorophenyl) ethane (DDT) and Pentachlorophenol (PCP) are two typical organochlorine pesticides. DDT was the first to be used for large-scale applications in agriculture production in the 1940s (Galiulin et al., 2002). Due to its stable chemical structure, environmental persistence, bioaccumulation and potential carcinogenicity to humans and wildlife, the use of DDT has been banned in most countries since the early 1970s (Aislabie et al., 1997; Mitra et al., 2001). However, it can still be broadly detected in soils, waters, air, plants and animal tissues all over the world (Aislabie et al., 1997; Mitra et al., 2001; Galiulin et al., 2002; Turusov et al., 2002; Binelli and Provini, 2003). Particularly, a considerable amount of DDT residues was found in farmland soils and sediments in China (Guo et al., 2009; Jiang et al., 2009; Yang et al., 2010), which can directly affect soil quality, agricultural product safety, and ecological health (Turusov et al., 2002; Huang et al., 2007). PCP has been widely used as pesticide, herbicide and wood-preserving agent (Czaplicka, 2004). Due to its carcinogenic, low biodegradability, chemical stability teratogenic and mutagenic toxicity, PCP contamination seriously affects soil quality, agricultural product safety and ecological health (Li et al., 2008; Puglisiet al., 2009). As DDT, PCP has also been widely detected in soils, sediments, water, plants and human breast milk (Hong et al., 2005; Gao et al., 2008). Therefore, the removal and remediation of soil DDT and PCP residues are meaningful and essential for environmental safety all over the world.

In the natural environment, soil DDT and PCP can be removal through many processes, such as volatilization, leaching, adsorption, fixing, photolysis, chemical reduction, and biodegradation (Aisllabie et al., 1997; Binelli and Provini, 2003; Stokes et al., 2006; Puglisi et al., 2009; Wong and Bidleman, 2011), among which the biodegradation is the most important soil DDT and PCP removal process. However, natural degradation process of soil DDT and PCP are quite slow (Kelee et al., 1995; Galiulin et al., 2002). Hence, some enhancement technologies were important to accelerate the soil DDT and PCP removal. Several abiotic methods have been explored for enhanced organic pollutant degradation, such as reduction with zero-valent iron and ferrous ion and oxidation by hydrogen peroxide and ferricion (Li et al., 2008; Sayles et al., 1997). However, most of these abiotic methods are unsuitable for large-scale applications because of high costs and potential secondary contamination.

In contrast, bioremediation methods have the potential to degrade soil organic pollutants in a cost-effective and environmentally friendly manner (McAllister et al., 1996; Mitra et al., 2001). In recent years, some microorganisms such as the bacterium *Sphingobacterium* sp. and the fungus *Fusarium solani* have been applied for soil organic pollutant removal (Mitra et al., 2001; Fang et al., 2010). However, these bioremediation technologies still meet some difficulties in the implementation process, such as limitation of nutrient and other available carbon sources, low bioavailability of soil pollutants, insufficiency of oxygen supply for aerobic bioremediation, and competition between the endogenous and indigenous microorganisms

(Benime et al., 2003; Li et al., 2003; Huang et al., 2007; Tyagi et al., 2011). Hence, exploration of a new, functional, biological resource will be helpful in strengthening the bioremediation technology of organic pollutants in soils.

Earthworms represent 60%-80% of the total soil biomass and play an important part in soil ecology. They have strong environmental adaptability and reproductive capacity, and show high tolerance and resistance to the organic pollutants (Langdon et al., 1999; Langdon et al., 2003; Reid and Watson, 2005). Through their mucilaginous secretions and soil organic matter transformation, earthworms can increase microbial activity and nutrient availability of organic matter (Benckiser, 1997; Suthar, 2007). Earthworms' movement and burrowing activity increase soil aeration, improve transport and distribution of microorganisms through bioturbation, and enhance contacting opportunities between microorganisms and reactants (Lavelle and Spain, 2001; Luepromchai et al., 2002; Hickman and Reid, 2008). Through these ecological functions, earthworms can effectively improve soil conditions, and offset the bioremediation limitations.

2.1.1 Enhancement effect of earthworms on removal and degradation processes of DDT

Effects of two ecological earthworm species (epigeic *Eisenia foetida* and endogeic *Amynthas robustus* E. Perrier) with different densities (15 and 30 individuals per kg of soil) on the removal of soil DDT with two pollution levels (2 and 4 mg/kg) were investigated (Lin et al., 2012). Concentrations of DDT and its metabolites, including 1, 1-dichloro-2, 2-bis (4-hlorophenyl) ethane (DDD), 1, 1-dichloro-2, 2-bis (4-chlorophenyl) ethylene (DDE), and 1-chl oro-2, 2-bis (4- chlorophenyl) ethylene (DDMU), were monitored after 60, 180, and 360 days of incubation. The results obtained showed that both earthworm species can significantly enhance degradation of soil DDT to its metabolites. For *E. foetida*, the higher earthworm density showed significantly higher rate of DDT degradation than the lower one. Anaerobic reductive dechlorination was the main degradation pathway over 180 days of incubation, while the aerobic dechlorination process was promoted between 180 and 360 days of incubation. Some earthworm amended treatments showed significantly higher microbial biomass carbon and nitrogen than the control, which suggested that earthworms might enhance the microbial degradation of DDT. Both earthworm species would have the potential to be applied to enhance the remediation of agricultural lands polluted by DDT.

1) Effect of the two earthworm species on soil DDT residues

Earthworms could enhance the microbial degradation of soil DDT. In an incubation of 360 days, soil DDT concentrations in treatments with and without earthworms all gradually decreased. But soil DDT concentrations in earthworm treatments were significantly lower than that in controls. At a lower initial DDT concentration of 2.0 mg/kg, after 360 days of incubation, 46.4 and 37.5% of residual DDT were observed in the presence of both 15 (E1) and 30 (E2) *E. foetida* per kg soil, as well as only 36.2 and 29.2% of DDT in the presence of

15 (A1) and 30 (A2) *Amynthas robustus* E. Perrier, all of which were significantly lower than that (74.5%) in the control. For the higher DDT pollution level of 4.0 mg/kg, t-test analysis displayed similar results as the lower pollution level. Soil residual DDT in the E1 and E2 only accounted for 50.0 and 35.8% after a 360-day incubation, while those in the A1 and A2 were 37.8 and 51.8%, all of which were significantly lower than that in the control (76.3%). The results suggested that both earthworm species could enhance soil DDT removal.

Different earthworms have different appropriate density to enhance the microbial degradation of soil DDT. For *E. foetida*, in both lower and higher DDT pollution level, a relatively high density of *E. foetida* with 30 individuals per kg of soil is more favorable for the enhanced DDT removal than a density with 15 individuals per kg of soil. *E. foetida* falls is within the epigeic category and has a relatively small size (0.10-0.25g), living usually in topsoil and the litter layer. These organisms have a preference for loose topsoil rich in organic matter due to their relatively poor burrowing ability (Schaefer and Filser, 2007; Yadav and Garg, 2011) ; consequently, they can increase the aeration of the topsoil. They crush organic debris into tiny particles, accelerating organic matter decomposition and secreting products of metabolism which may favor reproduction and growth of soil indigenous microorganisms Benckiser, 1997; Zhang et al., 2000). Therefore, *E. foetida* might accelerate the DDT removal by improving soil nutrient availability for microorganisms. Meanwhile, mortality has to be considered when earthworms are to be considered in situ remediation. High survival rates (96.1%-98.9%) were observed in the *E. foetida* treatments at the two DDT pollution levels, and cocoons and juvenile earthworms were discovered in all epigeic treated soils, though the reproduction was not counted in the survival percentage. These results suggested that most *E. foetida* could survive under both 2.0 and 4.0 mg/kg soil DDT pollution levels.

A. robustus E. Perrier belongs to endogeic species, with relatively large size (0.53-0.15 g). The contact opportunity between DDT and soil microorganisms can be enhanced through their bioturbation and movement (Luepromchai et al., 2002; Singer et al., 2001). These organisms have strong burrowing capacity and mostly stay in the burrow systems (Schaefer et al., 2005) ; hence, they provide larger soil porosity and more oxygenation under the soil surface (Benckiser, 1997). They not only feed on the decayed organic matter but also feed on soil directly. Therefore, the DDT adsorbed onto soil particles can be degraded by the intestinal flora and digestive enzymes in gut of an earthworm (Zhang et al., 2000; Langdon et al., 2003; Hickman and Reid, 2008). As a result, *A. robustus* E. Perrier may also enhance the soil DDT removal via increasing the DDT bioavailability, soil aeration and intestinal digestion. However, for the lower pollution level (2 mg/kg soil) with *A. robustus* E. Perrier, both the rates of A1 and A2 were similar, whereas the A1 showed a significantly higher rate than the A2 for the higher pollution level (4 mg/kg soil), in contrast to that observed with *E. foetida*. The survival rates of *A. robustus* E. Perrier were 94.4% and 88.9% at the lower pollution level and 82.8% and 68.9% at the higher pollution level in the A1 and A2 treatments, respectively.

According to Schaefer and Filser (2007), high density incubation with the earthworm species of big size might restrict their growth and reproduction, leading to high mortality. Furthermore, *A. robustus* E. Perrier, mainly feeding on soils, can ingest much more DDT and easily be harmed in the highly polluted soils. Thus, appropriate densities should be considered according to the earthworm species and pollution level for remediation application in the DDT contaminated soils.

2) Effect of earthworms on products and pathways of DDT degradation

Addition of earthworms could significantly increase DDD, DDE and DDMU productions compared to control. After adding DDT, concentration of soil DDD, DDE and DDMU all increased over time with and without earthworms. But soils from treatments with earthworms had significantly higher DDD, DDE and DDMU productions compared to the controls without earthworms at the end of incubation of 360 days in both higher (4.0 mg/kg) and lower (2.0 mg/kg) DDT pollution levels. It was observed that the degradation of DDT to DDD, DDE and DDMU was enhanced by both *E. foetida* and *A. robustus* E. Perrier in the same order of magnitude. In both higher and lower pollution levels, the main metabolites found after 60 days of incubation were DDD and DDE, while DDMU was hardly detected. The DDD concentrations were higher than those of DDE in all earthworm treatments. For the 180-day incubation, the DDD and DDE concentrations increased in all earthworm treatments, and those of DDD were higher than those of DDE. In addition, only small amounts of DDMU were detected. At the end of the test (after 360 days), the concentrations of DDE and DDMU increased considerably, but both of them were still lower than those of DDD. Dechlorination has been shown to be a key step in DDT degradation in many studies (Guo et al., 2009). Through a reductive dechlorination pathway, DDT can be reduced to DDD, 1-chloro-2, 2-bis (4-chlorophenyl) ethane (DDMS) and 2, 2-bis (4-chlorophenyl) ethane (DDNS) step by step under anoxic conditions. Through a dehydrochlorination pathway, DDT can be transformed to DDE, and the product of DDD and DDMS can also be degraded to DDMU and 2, 2-bis (4-chlorophenyl) ethylene (DDNU) under aerobic conditions (Kelee et al., 1995; Galiulin et al., 2002). Our results demonstrated that the reductive dechlorination pathway might dominate in the 0-180 days of incubation, and the dehydrochlorination pathway was strengthened after 180 days. From the 0 to 180 days, DDD may primarily come from the reductive dechlorination of DDT under anaerobic conditions. The gut of an earthworm is suitable for anaerobic microbial activities (Luepromchai et al., 2002). Large amounts of microorganisms are hosted in the gut of an earthworm, which easily improves the DDT anaerobic dechlorination process. Additionally, the excretion of anaerobic microorganisms from the gut of an earthworm increases the diversity of anaerobic microorganisms in soils, which may help to enhance the DDT degradation under the soil surface (Langdon et al., 2003; Suthar, 2007). Zhang et al. (2000) also reported that quantities of water-stable aggregates increased after adding earthworms to soils. Such aggregates have a high water holding capacity and their internal fine pores are easily filled with water which is favorable to form anaerobic conditions. From 180 to 360 days, DDE and DDMU may be originated from the dehydrochlorination of DDT and DDD under aerobic conditions. After the earthworms were applied and adapted to the soil environment, their movement and burrow activity increased, which can aerate the soil, providing more dissolved oxygen for enhanced aerobic DDT degradation (Lavelle, 2001). Based on the above discussion, the earthworms can provide both anaerobic and aerobic conditions for DDT degradation in soils.

3) Effect of earthworms on fate and removal processes of soil DDT

The results showed that the increment of three metabolites in the earthworm treatments is related to the disappearance of DDT residual in both pollution levels. The average undetected partition of DDTs in the controls and earthworm-treatments was 4.67%±1.26% and 8.26%±2.41%, both of which were at a similar level. Therefore, we concluded that degradation was the dominant process for the enhancement effect of earthworms on the soil DDT removal. Many processes, such as volatilization, leakage, aging, bioaccumulation and degradation, can contribute to DDT removal from the soil environment (Aislabie et al., 1997; Galiulin et al., 2002). In our experiment, the minor partition of undetected DDTs in the controls may partially result from the DDT loss through the volatilization, leakage and aging processes. Through burrowing and digestion, earthworms can increase soil porosity, oxygenation (Benckiser, 997), and water retention (Schaefer and Filser, 2007), which may enhance the volatilization and leakage processes for the soil DDT loss. In addition, earthworms can also accumulate the DDTs due to their ingestion of soil (Wågman et al., 1999; Fang et al., 2010; Li et al., 2010; Vermeulen et al., 2010). In the present study, DDT loss via volatilization, leakage and bioaccumulation in the presence of earthworms could be included in the 3.59% difference of average undetected DDTs between the controls and earthworm treatments. Soil DDT degradation involves both abiotic and biotic processes (Galiulin et al., 2002). The abiotic degradation mainly includes photolysis in the topsoil under aerobic conditions and chemical reduction by reductants, such as organic acids and metals, under anaerobic conditions (Turusov et al., 2002; Li et al., 2008). All incubations in this study were covered with hay and kept in the dark, so the photolysis process may contribute little to the DDT degradation. Many low molecular weight organic acids secreted by earthworms may help in chemical reduction of DDT under anaerobic conditions (Zhang et al., 2000). As discussed above, earthworms might enhance soil DDT biodegradation through increasing DDT bioavailability, soil oxygenation for aerobic degradation, biomass and activity of soil microorganisms, and degradation in the gut. For the C1 pollution level, the microbial carbon in the E1 and microbial nitrogen in the A1 were significantly higher than the control. For the C2 pollution level, the microbial carbon in the A1 and A2 and microbial nitrogen in the E1, A1 and A2 were significantly higher than the control. These results suggested that the microbial activities in some earthworm treatments were improved compared with the controls, which support the enhanced biodegradation of soil DDT.

Therefore, degradation played the main role in the enhanced soil DDT removal by earthworms in our study, and earthworms may be useful remediation tools in DDT contaminated soils with appropriate earthworm density. Further experiments should be conducted to elucidate the microorganisms in soils and in the gut of earthworms responsible for DDT biodegradation.

2.1.2 Enhancement effect of earthworms on microbial degradation processes of PCP

The effect of two earthworm species (*Amynthas robustus* Perrier and *Eisenia fetida* Savigny) on the soil microbial degradation of pentachlorophenol (PCP) were investigated (Li et al., 2015). PCP-degrading microbes were identified using DNA-stable isotope probing (SIP). The results showed that adsorption and fixing to soil particles and organic fractions dominated the fate of PCP in soil without any amendments. The inoculation of both earthworm species significantly enhanced soil PCP disappearance and basal respiration. The DNA-SIP results revealed that *Klebsiella*, *Cupriavidus*, *Aeromonas*, and *Burkholderia* spp. were present at higher relative abundances in [^{13}C]-labeled-PCP- amended soil microcosms than [^{12}C]-PCP-amended soil in the presence of *A. robustus*, indicating that these bacterial species were responsible for PCP assimilation. *Cupriavidus* and *Aeromonas* sp. were also detected in the earthworm gut before inoculation, and their relative abundance was affected by earthworms. These results demonstrated that earthworms can introduce functional bacteria into soils and increase the population of PCP-degrading bacteria, thereby accelerating soil PCP degradation.

1) Ethanol-and water-extractable PCP and soil pH in the sterile soil treatments

The ethanol-extractable PCP concentrations of all treatments decreased gradually as time elapsed. No significant difference ($p > 0.05$) was observed between the sterile soil (S_S) and sterile soil amended with sterile compost (S_SC_S) during the incubation period, while the soils to which earthworms were added (S_SC_SA/S_SC_SE) had significantly lower PCP concentrations than S_SC_S after 14 days. At the end of the incubation period (day 42), the ethanol-extractable PCP concentrations were ranked as $S_SC_SA < S_SC_SE < S_SC_S \approx S_S$. At the beginning of the incubation period (day 0), the water-extractable PCP level was 12.1 mg/kg (mass percentage=12.7% of total PCP) in S_S, whereas it was approximately 23.7 mg/kg in the S_SC_S, S_SC_SA, and S_SC_SE treatments. The water-extractable PCP level decreased dramatically after 7 days in S_S, while in S_SC_S, it decreased over time to a similar level as S_S at the end of the experiment. The water-extractable PCP level decreased significantly to 6.20 and 8.79 mg/kg by day 14 in S_SC_SA and S_SC_SE, respectively, but increased to 15.7 mg/kg by day 42. The percentage of the water-extractable PCP declined 58.4%, 85.8%, 35.7% and 32.5% in S_S, S_SC_S, S_SC_SA and S_SC_SE, respectively. Initially, the pH of S_S was detected as 3.86, while the addition of sterile compost increased the soil pH to 5.23 in S_SC_S. The soil pH in the S_S and S_SC_S treatments increased gradually over time and reached 4.82 and 5.70 at the end of incubation, respectively. The soil pH in the S_SC_SA and

S_SC_SE treatments were approximately 5.24 initially and increased significantly to approximately 7.0 at the end of incubation.

2) Ethanol-extractable PCP in non-sterile soil treatments

There was no significant difference ($p > 0.05$) between the non-sterile soil (S_N) and sterile-compost-amended non-sterile soil (S_NC_S) during the incubation period, whereas the treatments including earthworms (S_NC_SA/S_NC_SE) exhibited significantly lower PCP levels than S_NC_S after 28 days. Furthermore, S_NC_SA had a significantly lower PCP level than S_NC_SE. Regarding the non-sterile soil amended with non-sterile compost, the ethanol-extractable PCP concentrations in the S_NC_N, S_NC_NA, and S_NC_NE treatments were significantly lower than that in S_N after 14 days, and the residual ethanol-extractable PCP levels were ranked in the order: $S_NC_NA < S_NC_NE < S_NC_N < S_N$; the differences were significant ($p < 0.05$) at the end of the incubation period.

3) Distributions of different PCP fractions

To evaluate the fate of PCP, the water-extractable PCP, humus-fixed PCP, and earthworm-accumulated PCP were determined after the 42-day incubation. The difference between the ethanol- and water-extractable PCP was designated as soil-bound PCP, and the remaining PCP, excluding the fractions mentioned above, was considered the disappeared PCP fraction. The mass percentages of the PCP fractions in the various treatments were calculated. The water-extractable, soil-bound, humus-fixed and disappeared PCP accounted for 12.7%, 64.2%, 3.89% and 19.2%, respectively, of the total PCP in S_S. The levels of the four PCP fractions in either S_N or S_SC_S were not significantly different to those in S_S ($p > 0.05$), although the disappeared PCP increased to 22.3% in S_N. S_NC_S did not show significantly higher levels of disappeared PCP ($p > 0.05$) than S_N, but S_NC_N exhibited a significant increase (to 57.7%) in disappeared PCP. Both S_SC_SA and S_SC_SE had significantly higher levels of water-extractable, humus-fixed, and disappeared PCP fractions but significantly lower soil-bound PCP levels than S_SC_S. For the treatments of non-sterile soil with sterile/non-sterile compost, the compost-amended soils with earthworms also contained significantly higher disappeared PCP levels than did the soils without earthworms. The disappeared PCP fractions were 30.2%, 79.2%, and 83.8% in soils containing *A. robustus* (i.e., S_SC_SA, S_NC_SA, and S_NC_NA, respectively), which were significantly higher than in soils containing *E. fetida* (i.e., S_SC_SE, S_NC_SE, and S_NC_NE; 25.1%, 45.9%, and 72.9%, respectively). Additionally, the PCP concentrations accumulated by earthworms were too low to be observed. S_SC_SA and S_NC_NA contained significantly higher earthworm-accumulated PCP levels than S_SC_SE and S_NC_NE, respectively.

2.1.3 The identification of functional bacteria using DNA-stable isotope probing

1) Identification of functional bacteria

The functional bacteria in the *A. robustus*-amended treatments (S_NC_NA) were investigated using DNA-SIP. T-RFLP electropherograms revealed that fragments of 152 and 198 bp were present at a significantly higher density in the heavy fractions (1.733 g m/L and 1.730 g m/L) of [^{13}C]-labeled-PCP-amended soil microcosms than in the respective [^{12}C]-unlabeled controls on the 28th day. Two other fragments (215 and 309 bp) were relatively more abundant in the heavy fractions (1.726 g m/L) in the [^{13}C]-PCP-amended microcosms. The peak T-RFLP relative abundance values for the 152-, 198-, 215-, and 309-bp fragments were 89.0%, 1.61%, 3.24%, and 2.37%, respectively, in the heavy fractions. The taxonomic identities of each enriched fragment (cloned 16S rDNA sequence) all belong to the *Proteobacteria* phylum. The 152-, 198-, 215- and 309-bp fragments had high similarities (≥97%) with *Klebsiella*, *Cupriavidus*, *Aeromonas*, and *Burkholderia* spp., respectively.

2) Functional bacteria in the earthworm gut

The presence and relative abundance of individual functional bacteria in the guts of earthworms following the S_SC_SA and S_SC_SE treatments were investigated using T-RFLP. Only the 198- and 215-bp fragments were found in the guts of both earthworms before they were added to the soils. The relative abundances of the 198- and 215-bp fragments in the earthworm guts were initially low (<6%) but increased significantly to 18% on day 14 in the gut of *A. robustus* and then decreased gradually. In contrast, the abundances of these fragments increased significantly after 28 days in the gut of *E. fetida*. Neither the 198- nor 215-bp fragment was initially detected in the sterile-compost-amended sterile soil. However, after being amended with earthworms, the relative abundances of the 198- and 215-bp fragments in the soils increased after 28 days.

2.2 Enhancement effect of vermicomposting on pentachlorophenol degradation

Vermicomposting is an effective and environmentally friendly approach for soil organic contamination clean-up. We investigated the roles and mechanisms of earthworm (*Eisenia foetida*) on soil pentachlorophenol (PCP) degradation with sterile and non-sterile soil-compost treatment (Lin et al., 2016a). Limited soil PCP degradation was observed in the control and sterile compost treatments, whereas the synergetic effects of earthworm and compost contributed to the PCP biodegradation acceleration by significantly improving microbial biomass and activities. Sequence analysis and phylogentic classification of soil bacterial and fungal community structure after 42 days treatment identified the dominancy of indigenous bacterial families *Pseudomonadaceae*, *Sphingobacteriaceae* and *Xanthomonadaceae*, and fungal family *Trichoco-maceae*, which were responsible for PCP biodegradation and stimulated by vermicomposting. Further investigation revealed the dominant roles of sterile compost during PCP biodegradation as the formation of humus-PCP in soil rather than neutralizing soil pH and increasing PCP availability. The mechanisms of vermicomposting include humus-PCP complex degradation, humus consumption and soil pH neutralization. This part provides a comprehensive understanding of the synergetic effect of vermicomposting on microbial community functions and PCP degradation enhancement in soils.

2.2.1 Effects of vermicomposting on soil PCP residues

PCP residues have significantly decreased in the compost and vermicomposting treatments. Residual PCP in treatment with non-sterile compost (C_N), treatment with earthworm (E), treatment with sterile compost and earthworm (EC_S) and treatment with non-sterile compost and earthworm (EC_N) was significantly lower than in control (C_K) and treatment with sterile compost (C_S) ($P<0.05$). After 42 days, only 14.0% and 30.4% of residual PCP were found in EC_N and EC_S, while 39.1% and 37.1% in C_N and E. The PCP half-life in EC_S and EC_N was 15 and 22 days, respectively, lower than C_N (31 d), E (29 d), C_K (87 d) and C_S (88 d). The results suggested that vermicomposting with non-sterile compost enhances soil PCP biodegradation via stimulating local microbial communities. The humus-fixed PCP is the dominant PCP bound residue in soils, significantly decreased after 42 days in E and EC_N, lower than EC_S, C_S and C_K. The humus-fixed PCP contents in C_K and C_N were lower than C_S, while no significant difference was observed between other treatments. Additionally, PCP concentrations accumulated in earthworms were 0.03, 0.03 and 0.04 mg/kg in E, EC_S and EC_N treatments, only accounting for 0.08%-0.10% of initial PCP addition. Little PCP accumulated in earthworms, although they could ingest PCP-contaminated soil. Meanwhile, high *E. foetida* survival rates were observed as 96.67%-100% during the vermicomposting remediation.

2.2.2 Effects of vermicomposting on soil properties and microbial activities

Soil pH in EC_S and EC_N was significantly higher than E, C_S and C_N, all of which were significantly higher than C_K. The compost addition evidently increased the organic matters. TOC in C_S, S_N, EC_S and EC_N were at the same level ($p>0.05$), all significantly higher than E and C_K. Humus was the lowest in E treatment. Huminin C_S was significantly higher than that in C_N and C_K, all significantly higher than that in EC_S and EC_N. Humic acid in C_S was significantly higher than that in EC_S, EC_N, C_N and C_K, and fulvic acid in C_N and EC_N were significantly higher than EC_S, C_S and C_K.

Soil respiration rates showed high variability for all vermicomposting treatments, especially for EC_S and EC_N. After 42 days, EC_S and EC_N had the highest soil respiration rates as 6.87 mg/kg and 7.16 mg/kg, and the lowest in C_K (1.98 mg/kg) and C_S (2.27 mg/kg). The results further demonstrated higher biomass carbon in E, EC_S and EC_N than C_N. Similarly, the biomass nitrogen in EC_S and EC_N were significantly higher than E, C_S and C_N, and C_K was the lowest of all the treatments. The results suggested that vermicomposting showed positive impacts on microbial activity and biomass, consequently resulting in soil PCP biodegradation improvement.

2.2.3 Soil microbial community change in vermicomposting treatment

The difference and succession of microbial community in various vermicomposting treatments was evaluated by PCR-DGGE. The observable changes involved the possible bacterial and fungal PCP degraders in vermicomposting treatments, originally from the initial PCP contaminated soils or the new exogenous species driving from the earthworm and compost.

The bacterial community changed substantially and the dominant bands were unique in different treatment (Fig. 1). The bands $B_B/B_D/B_E/B_F$ were enhanced by either earthworm or compost, while bands B_A/B_C were enhanced only by earthworm and non-sterile compost, respectively. After 14 days, the bands Be/Bf in vermicomposting treatments have higher abundance than those in C_K. Vermicomposting stimulated the bands Ba/Bb/Bc in E, EC_S and EC_N, while C_N enhanced the band Bd. Cluster analysis showed that the bacterial community in C_S and C_N gathered for a class, and then with C_K, while E, EC_S and EC_N had a higher degree of similarity. The bacterial community was the most diverse after 28 days degradation, and cluster analysis indicated similar category within C_S and C_N, followed by C_K, whereas earthworm treatments (E, EC_S and EC_N) had high similarity. Compared to C_K, the similarity order of treatments is $C_S > C_N > (E/EC_S/EC_N)$, suggesting vermicomposting significantly influenced the soil bacterial community. Soil fungal community also had noticeable change with the additive of earthworm or/and compost (Fig. 2). The more significance of relative abundance of specific bands at 42 days indicated the stimulation of vermicomposting treatments. The most significant enhancement occurred in EC_S and EC_N. The enhanced bands $F_a/F_b/F_c/F_d$ at 14 days were also found at 28 and 42 days, marked as $F_A/F_B/F_C/F_D$ and $F_1/F_2/F_3/F_4$, respectively. Cluster analysis showed that the fungi community in earthworm treatments (E, EC_S and EC_N) had higher similarity, whereas compost treatments (C_S and C_N) formed another similar category and C_K in separate one. DGGE results indicated that earthworm and compost significantly affected bacterial and fungal community structure, and their synergic effects stimulated more microbial populations compared to the treatments with only earthworm or compost. The targeted bacterial and fungal bands were sequenced and phylogenetically classified. Compared with C_K, compost stimulated bacterial band B_5 assigned to *Sphingobacteriaceae*. The bands B_2/B_7 have higher abundance in E, EC_S and EC_N with phylogenetic similarity to *Sphingobacteriaceae* and *Xanthomonadaceae*, whereasless in C_S and C_N. The band $B_1(C_N)$ was assigned to *Flavobacteriaceae* and stimulated by earthworm. It is noteworthy that, in EC_S and EC_N treatments, the bands $B_3/B_4/B_6$ were of higher abundance than in C_K and close to TM7, *Pseudomonadaceae* and *Opitutaceae*. Earthworm and compost significantly stimulated the fungal bands $F_1/F_2/F_3$, assigned to *Mucoraceae*, *Tremellaceae* and *Trichocomaceae*, respectively. The addition of non-sterile compost promoted *Hypocreaceae* (F_4).

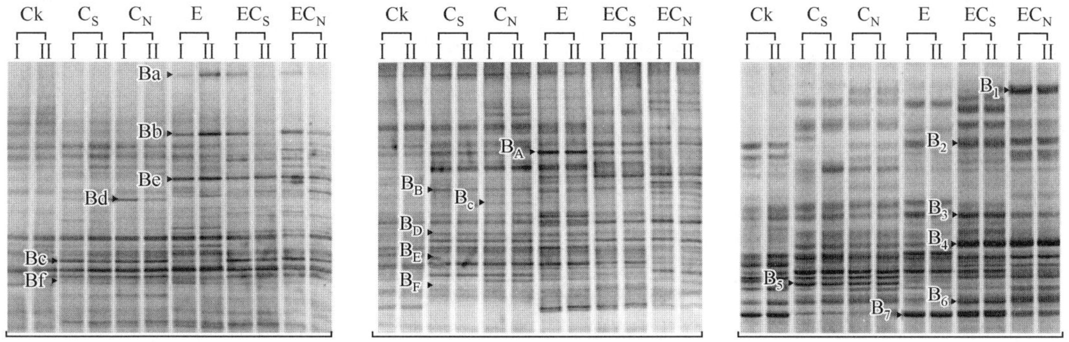

(a) 14d (b) 28d (c) 42d

Fig. 1 Cluster analysis of bacterial (16S rRNA) community structure in different treatments by DGGE (Lin et al., 2016a)

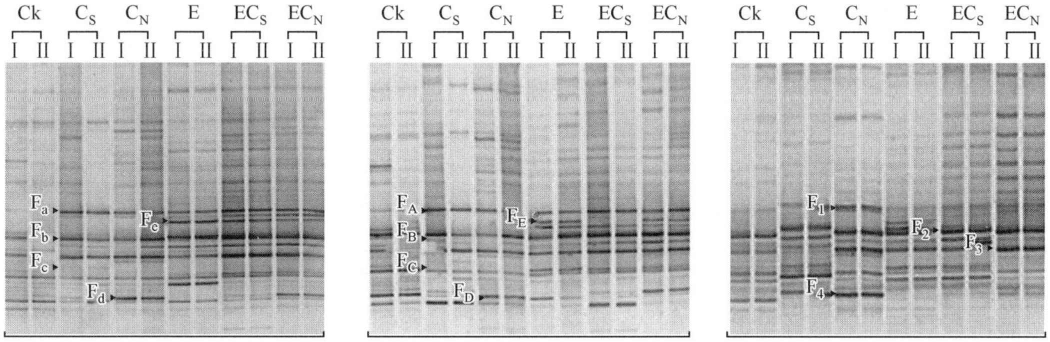

(a) 14d (b) 28d (c) 42d

Fig. 2 Cluster analysis of fungal (18S rRNA) community structure in different treatments by DGGE (Lin et al., 2016a)

This part demonstrated that earthworm or non-sterile compost had unique roles in enhancing soil PCP removal. Vermicomposting contributed to humus consumption and soil pH neutralization, releasing humus-PCP complex and increasing PCP availability. Sterile compost slowed PCP mineralization by increasing humus-fixed PCP in soil. Soil bacterial and fungal community structure was also significantly affected by vermicomposting, and the phylogenetic classification uncovered some indigenous microorganisms mineralizing PCP, including bacterial *Pseudomonadaceae*, *Sphingobacteriaceae* and *Xanthomonadaceae* and fungal *Trichocomaceae*. Compost decomposition also provided microbial available substrates to stimulate PCP degradation, supported by the activities of bacterial family *Flavobacteriaceae* in non-sterile compost, and fungal families *Mucoraceaea* and *Tremellaceae*. Vermicomposting has the potential to enhance the bioremediation of PCP contaminated soil.

2.3 The impact on the soil microbial community and enzyme activity of two earthworm species during the bioremediation of pentachlorophenol-contaminated soils

The ecological effect of earthworms on the fate of soil pentachlorophenol (PCP) differs with species. This study addressed the roles and mechanisms by which two earthworm species (epigeic *Eisenia fetida* and endogeic *Amynthas robustus* E. Perrier) affect the soil microbial community and enzyme activity during the bioremediation of PCP-contaminated soils (Lin et al., 2016b). *A. robustus* removed more soil PCP than *E. foetida* did. *A. robustus* improved nitrogen utilisation efficiency and soil oxidation more than *E. foetida* did, whereas the latter promoted the organic matter cycle in the soil. Both earthworm species significantly increased the amount of cultivable bacteria and actinomyces in soils, enhancing the utilisation rate of the carbon source (i.e. carbohydrates, carboxyl acids, and amino acids) and improving the richness and evenness of the soil microbial community. Additionally, earthworm treatment optimized the soil microbial community and increased the amount of the PCP-4-monooxygenase gene. Phylogenic classification revealed stimulation of indigenous PCP bacterial degraders, as assigned to the families *Flavobacteriaceae*, *Pseudomonadaceae* and *Sphingobacteriacea*, by both earthworms. *A. robustus* and *E. foetida* specifically promoted *Comamonadaceae* and *Moraxellaceae* PCP degraders, respectively.

2.3.1 PCP biodegradation process

Extractable PCP residues in all treatments gradually decreased during the 42 days of bioremediation. There was no difference ($p > 0.05$) amongst treatments at 0, 7 and 14 d. However, by the end of the incubation period (42 days), CE1 (treated by endogeic *A. robustus* earthworm and 4.5% sterile compost) removed more PCP (83.4%) than CE2 (treated by epigeic *E. fetida* earthworm and 4.5% sterile compost, 49.2%), and these were both higher than CK (neither earthworm or compost, 28.3%) and C (treated by 4.5% sterile compost, 28.1%). The removal of soil-bound PCP residues was similar; the highest in C (1.66 mg/kg), followed by CK (1.34 mg/kg), CE2 (1.45 mg/kg) and CE1 (1.00 mg/kg). PCP accumulation was small in both epigeic (0.04 mg/kg) and endogeic (0.05 mg/kg) earthworms, only accounting for 0.10% and 0.14% of initial PCP concentrations. The survival rate of epigeic *E. foetida* and endogeic *A. robustus* was 100% and 87%, respectively. Earthworms, especially the endogeic species (*A. robustus*), significantly improved the removal efficiency of soil extractable and bound residual PCP.

2.3.2 Soil chemical properties

After 42 days of bioremediation, the chemical properties of the soil differed depending on the treatments. The soil pH in CE1 and CE2 was higher than in C and CK; the latter had the lowest pH. The addition of sterile compost increased the content of organic matter in C, CE1 and CE2. Although total nitrogen was equal between treatments ($p>0.05$), the amount of NH_4^+-N and NO_3^--N in earthworm treatments (CE1 and CE2) was higher than in CK and C. There was no difference in fulvic acid between treatments, whereas humic acid was higher in C and CE1 and humin was higher in CK and C.

2.3.3 Cultivable soil microorganisms

There were more cultivable bacteria in the earthworm treatments (CE1 and CE2) than in C and CK. After 42 days of bioremediation, the number of cultivable bacteria per treatment was 29.0×10^5 CFU/g (CE1), 25.6×10^5 CFU/g (CE2), 19.6×10^5 CFU/g (C) and 14.9×10^5 CFU/g (CK). Cultivable fungi showed a dramatic increase from 0 to 14 d, with no difference between treatments throughout the bioremediation process. The final amount of cultivable fungi in CE1, CE2, C and CK treatments was 5.0×10^4 CFU/g, 4.8×10^4 CFU/g, 5.3×10^4 CFU/gand 4.5×10^4 CFU/g, respectively. Endogeic *A. robustus* stimulated cultivable actinomyces and after 42 days of bioremediation, they were higher (5.2×10^5 CFU/g) than for the other treatments (CE2, 2.9×10^5 CFU/g; C, 1.5×10^5 CFU/g; and CK, 1.3×10^5 CFU/g). The contribution of endogeic and epigeic earthworms to cultivable actinomyces and bacteria was identified, indicating their roles in the microbial community and their function in PCP degradation.

2.3.4 Soil enzyme activity

Soil enzyme activity represents the capacity to metabolise carbon and nitrogen in soils. Earthworm treatments (CE1 and CE2) enhanced urease activity from 0.13-0.14 mg/g to 0.32-0.46 mg/g, which did not change in C and CK. Similarly, catalase activity did not change in C and CK, whereas in CE1 and CE2 it increased from 0.22-0.23 mg/g at 0 d to 0.76-0.95 mg/g at 28 d, and then stabilised. It shows an increase in invertase activity after 42 days of bioremediation in C, CE1 and CE2, of 7.7 mg/g, 5.5 mg/g and 6.0 mg/g, respectively. Earthworm treatments remarkably enhanced soil cellulase activity, 2.2 mg/g in CE1 and 2.7 mg/g in CE2, both over twice as high as initial cellulase activities in C and CK. Soil enzyme activity was significantly improved by earthworms.

2.3.5 Total microbial activity and biomass

Soil respiration rates varied throughout the bioremediation process. The highest rates occurred from 7 to 28 d and were 1.2, 3.6, 7.1 and 4.7 mg CO_2/kg soil for CK, C, CE1 and CE2, respectively. The biomass carbon in CE1 (447.2 mg/kg) and CE2 (367.3 mg/kg) was much higher than in CK (146.1 mg/kg) and C (205.2 mg/kg), which was similar to biomass nitrogen, in which CE1 (99.9 mg/kg) and CE2 (96.3 mg/kg) were higher than C (62.0 mg/kg) and CK (40.1 mg/kg). Both microbial activity and biomass were enhanced by earthworm treatment, indicating an effect of earthworms on the microbial community.

2.3.6 Soil bacterial functional diversity

AWCD was used to represent the carbon utilization functional diversity of soil bacteria. The two earthworm treatments CE1 and CE2 (1.87±0.39 and 1.72±0.39) had similar AWCD values, which were higher than the non-earthworm treatments C and CK (1.34±0.73 and 1.31±0.58, $p<0.05$). Compared to the behaviour of PCP bioremediation, the results indicated that earthworms increased the carbon utilization capacities of indigenous microbes and therefore improved their PCP degradation performance. The higher richness (S) in CE1 (endogeic *A. robustus*) than CE2 (epigeic *E. fetida*), C and CK, suggested a significantly stronger effect of endogeic earthworms on microbial carbon utilization. The carbon source utilisation efficiency data did not reveal any impacts on miscellaneous or polymer metabolism. The indigenous microbes in earthworm treatments (CE1 and CE2) used carbohydrates, carboxyl acids, amino acids andamines more efficiently, indicating that earthworms can stimulate the use of small molecule carbon sources, attributing to the enhanced carbon utilization capacities.

2.3.7 Analysis of microbial community structure

There was no difference in soil microbial community diversity (16S rRNA DGGE) within replicates (Ⅰ, Ⅱ and Ⅲ for each treatment), whereas the dominant bands between treatments (with or without earthworms) were different. Cluster analysis showed more similarity between the two earthworm treatments (CE1 and CE2), than between these and the cluster formed by the control (CK) and sterile compost (C). Thus, earthworm addition had a significantly greater impact on the soil microbial community than compost addition. Some indigenous microbes (B1-B9) were stimulated by earthworms, significantly different from the bands (A1-A7) in the uncontaminated treatments. It is possible that they might have PCP-degrading capacities or associate with compost organic matter metabolism to accelerate PCP degradation. Both endogeic and epigeic earthworms stimulated the bacterial populations represented by band B1 (KM284689), B2 (KM284690), B4 (KJ137169) and B9 (KM284692), with high sequence similarity to the families *Flavobacteriaceae* and *Pseudomonadaceae*. Two strains (KM284691 and KM284693) were only promoted by endogeic earthworm treatment and were close to the families *Comamonadaceae* and *Sphingobacteriaceae*. Epigeic earthworm addition stimulated the PCP-degrading or organic matter metabolism microorganisms (KM284694, KJ137165 and KJ137166) with high sequence similarity to the families *Moraxellaceae*, *Flavobacteriaceae* and *Sphingobacteriaceae*. Compared to our previous research of the effect of earthworms in PCP-contaminated and sterile soils, these species are not related to the PCP degraders in the earthworm intestinal microbiota (*Cupriavidus* and *Aeromonas*), and they are therefore the indigenous microbes responsible for PCP degradation or organic matter metabolism to accelerate PCP degradation.

2.3.8 PCP-4-monooxygenase gene (*pcpB*)

Using the qPCR calibration curve, the copy numbers of 16S rRNA and the PCP-4-monooxygenase gene (*pcpB*) were quantified for each treatment. There was no significant difference ($p>0.05$) in total bacterial 16S rRNA between CK ($3.27\pm0.15\times10^{10}$ copies/g), C [$(3.22\pm0.09)\times10^{10}$ copies/g], CE1 [$(2.37\pm0.17)\times10^{10}$ copies/g] and CE2 [$(3.09\pm0.01)\times10^{10}$ copies/g] treatment. CE1 treatment had the highest *pcpB* gene abundance [$(5.97\pm0.23)\times10^{6}$ copies/g], followed by CE2 treatment [$(4.16\pm0.01)\times10^{6}$ copies/g]. They were both significantly higher (approximately 1.73 to 2.16 times) than the C and CK treatments [$(2.41\pm0.01)\times10^{6}$ copies/g and $(2.22\pm0.01)\times10^{6}$ copies/g, respectively]. Because the PCP-4-monooxygenase *pcpB* gene was found in the family *Sphingobacteriaceae* (Beaulieu et al., 2000), the results provide additional evidence that the PCP degrader *Sphingobacteriaceae* was enriched in the CE1 and CE2 systems, similar to our conclusion from the microbial community structure.

Both endogeic *A. robustus* and epigeic *E. foetida* earthworms accelerated PCP degradation. Cultivable bacteria and actinomyces were promoted, increasing their ability to use small molecule carbon sources. The soil microbial community was also changed by the earthworm treatments. Endogeic *A. robustus* had a better PCP degradation performance than epigeic *E. foetida*, due to its strong bioturbation and humin consumption, which released bound residual PCP absorbed by sterile composts and soils. Because of their differing effects on soil enzyme activity, endogeic *A. robustus* and epigeic *E. foetida* preferentially enhanced the soil nitrogen and carbon cycle, respectively. The change in the soil bacterial community and PCP degradation genes in this study therefore suggested that the dominance and activities of indigenous PCP bacterial degraders were stimulated by both the rich organic matter from compost additives and the direct soil microenvironment improvement by earthworms. The dominant functional microbes were assigned to the families *Flavobacteriaceae*, *Pseudomonadaceae* and *Sphingobacteriacea*. In particular, endogeic *A. robustus* stimulated *Comamonadaceae*, whereas epigeic *E. foetida* stimulated *Moraxellaceae*. Consequently, these two types of earthworms have different ecological roles in the PCP bioremediation process.

3. Conclusions

Based on our research, special attention should be paid to potential risks of heavy metals to human health in PRD. The risks of heavy metals are mainly from speciation, activity of heavy metals and their effect factors in soil. The researches based on poly-metallic mine Dabaoshan, from which mining wastes severely polluted surrounding agricultural fields, showed Cd and Zn, were considerably high in the long-term mine drainage polluted soils. And the high metal ion activity, which might be controlled by several environmental factors, likely enhanced intake and accumulation of metals in crops, and eventually posed long-term toxicity to villager health in the area. Furthermore, based on a statistical analysis of microbiological and biochemical combined with soil metal, provided a reliable and comprehensive indication of changes in soil quality and organic nutrient cycling, after exposure to long-term acid metal stress.

For soil organic pollutants, our research mainly explored the possible measures and mechanisms to accelerate its degradation rate. We found that both earthworms and vermicomposting have the potential to enhance the bioremediation of organochlorine-pesticides contaminated soil (DDT and PCP). Results of DNA-SIP further indicated that earthworms can introduce functional bacteria into soils and increase the abundance of PCP-degrading bacteria, thus accelerating soil PCP degradation. Vermicomposting accelerate PCP degradation via increasing PCP availability, affecting soil bacterial and fungal community structure, and providing microbial available substrates.

References

Aislabie J M, Richards N K, Boul H L. 1997. Microbial degradation of DDT and its residues—a review. New Zealand Journal of Agricultural Research, 40 (2): 269-282.

Beaulieu M, Becaert V, Deschenes L, et al. 2000. Evolution of bacterial diversity during enrichment of PCP-degrading activated soils. Microbial Ecology, 40 (4): 345-355.

Benimeli C S, Amoroso M J, Chaile A P, et al. 2003. Isolation of four aquatic streptomycetes strains capable of growth on organochlorine pesticides. Bioresource Technology, 89 (2): 133-138.

Binelli A, Provini A. 2003. DDT is still a problem in developed countries: the heavy pollution of Lake Maggiore. Chemosphere, 52 (4): 717-23.

Coddington K. 1986. A review of arsenicals in biology. Toxicological & Environmental Chemistry, 11 (4): 281-290.

Cances B, Ponthieu M, Castrec-Rouelle M, et al. 2003. Metal ions speciation in a soil and its solution: experimental data and model results. Geoderma, 113 (3): 341-355.

Czaplicka, M., 2004. Sources and transformations of chlorophenols in the natural environment. Science of the Total Environment, 322: 21-39.

Davranche M, Bollinger J C. 2000. Heavy metals desorption from synthesized and natural iron and manganese oxyhydroxides: effect of reductive conditions. Journal of Colloid and Interface Science, 227 (2): 531-539.

Dong W Q Y, Cui Y, Liu X. 2001. Instances of soil and crop heavy metal contamination in China. Soil and Sediment Contamination, 10 (5): 497-510.

Duckworth O W, Sposito G. 2005. Siderophore-manganese (III) interactions II. Manganite dissolution promoted by desferrioxamine B. Environmental Science & Technology, 39 (16): 6045-6051.

Eijsackers H, Van Gestel C A M, De Jonge S, et al. 2001. Polycyclic aromatic hydrocarbon-polluted dredged peat sediments and earthworms: a mutual interference. Ecotoxicology, 10 (1): 35-50.

Fendorf S E, Zasoski R J. 1992. Chromium (III) oxidation by. delta.-manganese oxide (MnO2). 1. Characterization. Environmental Science & Technology, 26 (1): 79-85.

Fang H, Chu X, Wang X, et al. 2010. Using matrix solid-phase microextraction (matrix-SPME) to estimate bioavailability of DDTs in soil to both earthworm and vegetables. Archives of Environmental Contamination and Toxicology, 58 (1): 62-70.

Fang H, Dong B, Yan H, et al. 2010. Characterization of a bacterial strain capable of degrading DDT congeners and its use in bioremediation of contaminated soil. Journal of Hazardous Materials, 184 (1): 281-289.

Fredj F B, Wali A, Khadhraoui M, et al. 2014. Risk assessment of heavy metal toxicity of soil irrigated with treated wastewater using heat shock proteins stress responses: case of El Hajeb, Sfax, Tunisia. Environmental Science and Pollution Research,

21 (6): 4716-4726.

Greig-Smith P W, Becker H, Edwards P J, et al. 1992. Ecotoxicology of earthworms. Intercept.

Giller K E, Witter E, Mcgrath S P. 1998. Toxicity of heavy metals to microorganisms and microbial processes in agricultural soils: a review. Soil Biology and Biochemistry, 30 (10): 1389-1414.

Galiulin R V, Bashkin V N, Galiulina R A. 2002. Review: Behavior of persistent organic pollutants in the air-plant-soil system. Water, Air, and Soil Pollution, 137 (1-4): 179-191.

Gil-Sotres F, Trasar-Cepeda C, Leirós M C, et al. 2005. Different approaches to evaluating soil quality using biochemical properties. Soil Biology and Biochemistry, 37 (5): 877-887.

Gao J J, Liu L H, Liu X R, et al. 2008. Levels and spatial distribution of chlorophenols 2, 4-dichlorophenol, 2, 4, 6-trichlorophenol, and pentachlorophenol in surface water of China. Chemosphere, 71: 1181-1187.

Guo Y, Yu H Y, Zeng E Y. 2009. Occurrence, source diagnosis, and biological effect assessment of DDT and its metabolites in various environmental compartments of the Pearl River Delta, South China: a review. Environmental Pollution, 157 (6): 1753-1763.

Hua X M, Shan Z J. 1996. The production and application of pesticides and factor analysis of their pollution in environment in China. Adv Environ Sci, 4 (2): 33-45.

Huamain C, Chunrong Z, Cong T U, et al. 1999. Heavy metal pollution in soils in China: status and countermeasures. Ambio: 130-134.

He Z L, Yang X E, Baligar V C, et al. 2003. Microbiological and biochemical indexing systems for assessing quality of acid soils. Advances in Agronomy, 78: 89-138.

Hong H C, Zhou H Y, Luan TG, et al. 2005. Residue of pentachlorophenol in freshwater sediments and human breast milk collected from the Pearl River Delta, China. Environment International, 31: 643-649.

Huang Y, Zhao X, Luan S. 2007. Uptake and biodegradation of DDT by 4 ectomycorrhizal fungi. Science of the Total Environment, 385 (1): 235-241.

Hickman Z A, Reid B J. 2008. Earthworm assisted bioremediation of organic contaminants. Environment International, 34 (7): 1072-1081.

Hu Y, Liu X, Bai J, et al. 2013. Assessing heavy metal pollution in the surface soils of a region that had undergone three decades of intense industrialization and urbanization. Environmental Science and Pollution Research, 20 (9): 6150-6159.

Johnson D B, Hallberg K B. 2005. Acid mine drainage remediation options: a review. Science of the total environment, 338 (1): 3-14.

Jian-Min Z, Zhi D, Mei-Fang C A I, et al. 2007. Soil heavy metal pollution around the Dabaoshan mine, Guangdong province, China. Pedosphere, 17 (5): 588-594.

Jiang Y F, Wang X T, Jia Y, et al. 2009. Occurrence, distribution and possible sources of organochlorine pesticides in agricultural soil of Shanghai, China. Journal of Hazardous Materials, 170 (2): 989-997.

Keeney D R, Nelson D W. 1982. Nitrogen-Inorganic forms. nitrate leaching and yield on a coarse-textured soil. J. Environ.

Kelce W R, Stone C R, Laws S C, et al. 1995. Persistent DDT metabolite p, p' DDE is a potent androgen receptor antagonist.

Kandeler F, Kampichler C, Horak O. 1996. Influence of heavy metals on the functional diversity of soil microbial communities. Biology and Fertility of Soils, 23 (3): 299-306.

Kennedy V H, Sanchez A L, Oughton D H, et al. 1997. Use of single and sequential chemical extractants to assess radionuclide and heavy metal availability from soils for root uptake. Analyst, 122 (8): 89R-100R.

Langdon C J, Piearce T G, Black S, et al. 1999. 3Resistance to arsenic-toxicity in a population of the earthworm Lumbricus rubellus. Soil Biology and Biochemistry, 1 (14): 1963-1967.

Lavelle P, Spain A V. 2001. Soil ecology. Springer Science & Business Media.

Lin T F, Wu J K. 2001. Adsorption of arsenite and arsenate within activated alumina grains: equilibrium and kinetics. Water Research, 35 (8): 2049-2057.

Luepromchai E, Singer A C, Yang C H, et al. 2002. Interactions of earthworms with indigenous and bioaugmented PCB-degrading bacteria. FEMS Microbiology Ecology, 41 (3): 191-197.

Langdon C J, Piearce T G, Meharg A A, et al. 2003. Interactions between earthworms and arsenic in the soil environment: a review. Environmental Pollution, 124 (3): 361-373.

Li Y T, Becquer T, Quantin C, et al. 2005. Microbial activity indices: sensitive soil quality indicators for trace metal stress. Pedosphere , 15 (4): 409-416.

Liu Y S, Gao Y, Wang K W, et al. 2005. Etiologic study on alimentary tract malignant tumor in villages of high occurrence. China Tropical Medicine, 5 (5): 1139-1141.

Lakshmipathiraj P, Narasimhan B R V, Prabhakar S, et al. 2006. Adsorption studies of arsenic on Mn-substituted iron oxyhydroxide. Journal of Colloid and Interface Science, 304 (2): 317-322.

Li M S. 2006. Ecological restoration of mineland with particular reference to the metalliferous mine wasteland in China: A review of research and practice. Science of the Total Environment, 357 (1): 38-53.

Lin C, Wu Y, Lu W, et al. 2007. Water chemistry and ecotoxicity of an acid mine drainage-affected stream in subtropical China during a major flood event. Journal of Hazardous Materials, 142 (1): 199-207.

Li F B, Wang X G, Liu C S, et al. 2008. Reductive transformation of pentachlorophenol on the interface of subtropical soil colloids and water. Geoderma, 148: 70-78.

Li F, Wang X, Li Y, et al. 2008. Enhancement of the reductive transformation of pentachlorophenol by polycarboxylic acids at the iron oxide-water interface. Journal of Colloid and Interface Science, 321 (2): 332-341.

Li Y T, Rouland C, Benedetti M, et al. 2009. Microbial biomass, enzyme and mineralization activity in relation to soil organic C, N and P turnover influenced by acid metal stress. Soil Biology and Biochemistry, 41 (5): 969-977.

Li Y T, Becquer T, Dai J, et al. 2009. Ion activity and distribution of heavy metals in acid mine drainage polluted subtropical soils. Environmental Pollution, 157 (4): 1249-1257.

Li H, Gu Y, Zhao Y, et al. 2010. Leachate treatment using a demonstration aged refuse biofilter. Journal of Environmental Sciences, 22 (7): 1116-1122.

Li X H, Wang X Z, Wang W, et al. 2010. Profiles of Organochlorine pesticides in earthworms from urban leisure areas of Beijing, China. Bulletin of Environmental Contamination and Toxicology, 84 (4): 473-476.

Li X, Liu C, Li F, et al. 2010. The oxidative transformation of sodium arsenite at the interface of α-MnO2 and water. Journal of Hazardous Materials, 173 (1): 675-681.

Lin Z, Li X M, Li Y T, et al. 2012. Enhancement effect of two ecological earthworm species (Eisenia foetida and Amynthas robustus E. Perrier) on removal and degradation processes of soil DDT. Journal of Environmental Monitoring, 14: 1551-1558.

Li X M, Lin Z, Luo C L, et al. 2015. Enhanced microbial degradation of pentachlorophenol from soil in the presence of earthworms: Evidence of functional bacteria using DNA-stable isotope probing. Soil Biology & Biochemistry, 81: 168-177.

Lin Z, Bai J, Zhen Z, et al., 2016a. Enhancing pentachlorophenol degradation by vermicomposting associated bioremediation. Ecological Engineering, 87: 288-294.

Lin Z, Zhen Z, Wu Z H, et al. 2016b. The impact on the soil bacterial community and enzyme activity of two earthworm species during the bioremediation of pentachlorophenol-contaminated soils. Journal of Hazardous Materials, 301: 35-45.

Monterroso C, Alvarez E, Macías F. 1994. Speciation and solubility control of Al and Fe in minesoil solutions. Science of the Total Environment, 158: 31-43.

McAllister K A., Lee H, Trevors J T. 1996. Microbial degradation of pentachlorophenol. Biodegradation, 7: 1-40.

Mitra J, Mukherjee P K, Kale S P, et al. 2001. Bioremediation of DDT in soil by genetically improved strains of soil fungus Fusarium solani. Biodegradation, 12 (4): 235-245.

Milne C J, Kinniburgh D G, Van Riemsdijk W H, et al. 2003. Generic NICA-Donnan model parameters for metal-ion binding by humic substances. Environmental Science & Technology, 37 (5): 958-971.

Nesbitt H W, Canning G W, Bancroft G M. 1998, XPS study of reductive dissolution of 7Å-birnessite by H3AsO3, with constraints on reaction mechanism. Geochimica et Cosmochimica Acta, 62 (12): 2097-2110.

Nannipieri P, Kandeler E, Ruggiero P. 2002. Enzyme activities and microbiological and biochemical processes in soil. Enzymes in the environment. Marcel Dekker, New York: 1-33.

Parker D R, Pedler J F. 1997. Reevaluating the free-ion activity model of trace metal availability to higher plants//Plant Nutrition for Sustainable Food Production and Environment. Springer Netherlands: 107-112.

Peijnenburg W, Jager T. 2003. Monitoring approaches to assess bioaccessibility and bioavailability of metals: matrix issues. Ecotoxicology and Environmental Safety, 56 (1): 63-77.

Puglisi E, Vernile P, Bari G, et al. 2009. Bioaccessibility, bioavailability and ecotoxicity of pentachlorophenol in compost amended soils. Chemosphere, 77: 80-86.

Quantin C, Becquer T, Rouiller J H, et al. 2002. Redistribution of metals in a New Caledonia Ferralsol after microbial weathering. Soil Science Society of America Journal, 66 (6): 1797-1804.

Raven K P, Jain A, Loeppert R H. 1998. Arsenite and arsenate adsorption on ferrihydrite: kinetics, equilibrium, and adsorption envelopes. Environmental Science & Technology, 32 (3): 344-349.

Reid B J, Watson R. 2005. Lead tolerance in Aporrectodea rosea earthworms from a clay pigeon shooting site. Soil Biology and Biochemistry, 37 (3): 609-612.

Stumm W, Morgan J J. 1970. Aquatic chemistry; an introduction emphasizing chemical equilibria in natural waters.

Stone A T. 1987. Reductive dissolution of manganese (Ⅲ/Ⅳ) oxides by substituted phenols. Environmental Science & Technology, 21 (10): 979-988.

Sayles G D, You G, Wang M, et al. 1997. DDT, DDD, and DDE dechlorination by zero-valent iron. Environmental Science & Technology, 31 (12): 3448-3454.

Sauve S, Hendershot W, Allen H E. 2000. Solid-solution partitioning of metals in contaminated soils: dependence on pH, total metal burden, and organic matter. Environmental Science & Technology, 34 (7): 1125-1131.

Singer A C, Jury W, 2001. Luepromchai E, et al. Contribution of earthworms to PCB bioremediation. Soil Biology and Biochemistry, 33 (6): 765-776.

Turusov V, Rakitsky V, Tomatis L. 2002. Dichlorodiphenyltrichloroethane (DDT): ubiquity, persistence, and risks. Environmental Health Perspectives, 110 (2): 125.

Schaefer M, Petersen S O, Filser J. 2005. Effects of Lumbricus terrestris, Allolobophora chlorotica and *Eisenia fetida* on microbial community dynamics in oil-contaminated soil. Soil Biology and Biochemistry, 37 (11): 2065-2076.

Schaeffer R, Soeroes C, Ipolyi I, et al. 2005. Determination of arsenic species in seafood samples from the Aegean Sea by liquid chromatography- (photo-oxidation) -hydride generation-atomic fluorescence spectrometry. Analytica Chimica Acta, 547 (1): 109-118.

Sheoran A S, Sheoran V. 2006. Heavy metal removal mechanism of acid mine drainage in wetlands: a critical review. Minerals Engineering, 19 (2): 105-116.

Stokes J D, Paton G I, Semple K T. 2006. Behaviour and assessment of bioavailability of organic contaminants in soil: relevance for risk assessment and remediation. Soil Use and Management, 21: 475-486.

Suthar S. 2007. Nutrient changes and biodynamics of epigeic earthworm Perionyx excavatus (Perrier) during recycling of some agriculture wastes. Bioresource Technology, 98 (8): 1608-1614.

Schaefer M, Juliane F. 2007. The influence of earthworms and organic additives on the biodegradation of oil contaminated soil. Applied Soil Ecology, 36 (1): 53-62.

Stollenwerk K G, Breit G N, Welch A H, et al. 2007. Arsenic attenuation by oxidized aquifer sediments in Bangladesh. Science of the Total Environment, 379 (2): 133-150.

Tyagi M, da Fonseca M M R, de Carvalho C C C R. 2011. Bioaugmentation and biostimulation strategies to improve the effectiveness of bioremediation processes. Biodegradation, 22 (2): 231-241.

Vance E D, Brookes P C, Jenkinson D S. 1987. An extraction method for measuring soil microbial biomass C. Soil biology and Biochemistry, 19 (6): 703-707.

Vig K, Megharaj M, Sethunathan N, et al. 2003. Bioavailability and toxicity of cadmium to microorganisms and their activities in soil: a review. Advances in Environmental Research, 8 (1): 121-135.

Vermeulen F, Covaci A, D' Havé H, et al. 2010. Accumulation of background levels of persistent organochlorine and organobromine pollutants through the soil-earthworm-hedgehog food chain. Environment International, 36 (7): 721-727.

Wågman N, Strandberg B, van Bavel B, et al. 1999. Organochlorine pesticides and polychlorinated biphenyls in household composts and earthworms (Eisenia foetida). Environmental Toxicology and Chemistry, 18 (6): 1157-1163.

Weng L, Temminghoff E J M, Van Riemsdijk W H. 2001. Contribution of individual sorbents to the control of heavy metal activity in sandy soil. Environmental Science & Technology, 35 (22): 4436-4443.

Weng L, Temminghoff E J M, Van Riemsdijk W H. 2001. Determination of the free ion concentration of trace metals in soil solution using a soil column Donnan membrane technique. European Journal of Soil Science, 52 (4): 629-637.

Wang Y, Morin G, Ona-Nguema G, et al. 2008. Arsenite sorption at the magnetite-water interface during aqueous precipitation of magnetite: EXAFS evidence for a new arsenite surface complex. Geochimica et Cosmochimica Acta, 72 (11): 2573-2586.

Wong F, Bidleman T F. 2011. Aging of organochlorine pesticides and polychlorinated biphenyls in muck soil, volatilization, bioaccessibility, and degradation. Environment Science & Technology, 45: 958-963.

Yang L, Xia X, Liu S, et al. 2010. Distribution and sources of DDTs in urban soils with six types of land use in Beijing, China. Journal of Hazardous Materials, 174 (1): 100-107.

Yadav A, Garg V K. 2011. Recycling of organic wastes by employing *Eisenia fetida*. Bioresource technology, 102 (3): 2874-2880.

Zhao H, Xia B, Fan C, et al. 2012. Human health risk from soil heavy metal contamination under different land uses near Dabaoshan Mine, Southern China. Science of the Total Environment, 417: 45-54.

Zhang B G, Li G T, Shen T S, et al. 2000. Changes in microbial biomass C, N, and P and enzyme activities in soil incubated with the earthworms *Metaphire guillelmi* or *Eisenia fetida*. Soil Biology and Biochemistry, 32 (14): 2055-2062.

Remediation of Contaminated Soils with Biochar

Hailong WANG[1, 2, 3], Xing YANG[1], Lizhi HE[1], Kouping LU[1], Karin MÜLLER[4], Kim MCGROUTHER[5], Song XU[2], Xiaokai ZHANG[1], Jianwu LI[1], Huagang HUANG[6], Guodong YUAN[3], Guotao HU[1], Xingyuan LIU[3]

[1] Key Laboratory of Soil Contamination Bioremediation of Zhejiang Province, School of Environmental and Resource Sciences, Zhejiang A & F University

[2] School of Environmental and Chemical Engineering, Foshan University

[3] Guangdong Dazhong Agriculture Science Co., Ltd.

[4] The New Zealand Institute for Plant & Food Research Limited, Ruakura Research Centre

[5] Scion

[6] Yancao Production Technology Center, Bijie Yancao Company of Guizhou Province

1. Introduction

The rapid development of the Chinese economy in the last few decades, has been accompanied by some environmental problems. Increasingly more soils have been found to be contaminated with organic and inorganic toxins due to waste emissions from industrial production, mining activities, organic waste applications, wastewater irrigation, and inadequate management of pesticides and chemicals in agricultural production systems (Zhang et al., 2013). Many studies have focused on remediation of soils contaminated with heavy metals and organic pollutants through chemical, bioremediation as well as integrated methods (Park et al., 2011; Zhang et al., 2013; Sun et al., 2015; Hu et al., 2016a, 2016b).

Biochar is a carbonaceous solid derived from residual organic materials produced by agriculture and forestry (Wu et al., 2012; Yang et al., 2016). Biochar typically has a microporous structure, a high surface alkalinity and adsorption capacity, and can be used for reducing greenhouse gas emissions (Liu et al., 2011; Dong et al., 2013), enhancing soil productivity (Deng et al., 2016; Guo et al., 2016) and environmental quality (Wang et al., 2010; Zhang et al., 2010; Xu et al., 2013; He et al., 2015).

In the past few years, a range of studies were conducted at the Key Laboratory of Soil Contamination Bioremediation of Zhejiang Province, China, to investigate the effect of biochar application on the adsorption and desorption, mobility and bioavailability of heavy metals and organic pollutants (e.g., phthalic acid esters or PAEs) in soil (e.g., Lu et al., 2014, 2016; Zhang et al., 2014, 2016; He et al., 2015, 2016; Yang et al., 2016). The aim of this chapter is to provide an overview of the findings obtained from these studies.

2. Using biochar for the remediation of soil contaminated by heavy metals

Incubation (Yang et al., 2016) and pot experiments (Lu et al., 2014, 2016) were carried out with a naturally co-contaminated paddy soil. The soil contained 1.4 mg/kg cadmium (Cd), 693 mg/kg copper (Cu), 527 mg/kg lead (Pb) and 1471 mg/kg Zinc (Zn), and was sampled in the southwest of Hangzhou City, Zhejiang Province, China. For each experiment, three rates (0%, 1% and 5%) and two particle sizes (<0.25 and <1mm) of rice straw and bamboo biochar application were investigated. In the incubation experiment, the physico-chemical soil properties, as well as extractable heavy metals, available phosphorus and enzyme activities were analyzed in the soil samples after one year. In the pot experiment, a metal-tolerant plant, *S. plumbizaincicola*, was chosen as an indicator plant. After three months of growth, the mobility and chemical fractions of the heavy metals in the soil and their concentrations in plant shoots were analyzed.

Another pot experiment was conducted to explore the influence of different application rates (0%, 1%, 2.5% and 5%) of tobacco stalk biochar and dead pig biochar on the bioavailability, and redistribution of heavy metals as well as the growth of tobacco plants (*Nicotiana tabacum* L.) in a naturally contaminated soil (Yang et al., 2015). The soil was co-contaminated with 3 mg/kg Cd and 429 mg/kg Zn and was collected from a paddy field in the southeast of Wenzhou City, Zhejiang Province, China. After 80 days of growth, the plants were harvested and the aboveground biomass and the roots were separated. In addition, the soil samples were retrieved from each pot and mixed thoroughly before air drying.

2.1 Effect of biochar on soil properties

In the incubation experiment, Yang et al. (2016) found that the soil pH increased with the increasing application rates of rice straw biochar, and that the fine rice straw biochar was more effective at increasing the soil pH than the coarse straw biochar. However, the application of bamboo biochar did not significantly alter the soil's pH. After the 1-year incubation period, the soil's electrical conductivity (EC) of the two fine rice straw treatments at 1% and 5% application rates increased approximately by twofold compared to the control. The same trend was found in the pot experiment carried out by Lu et al. (2014). The soil pH increased significantly ($P<0.05$) with the biochar application and the highest soil pH corresponded to the 5% fine rice straw biochar treatment. Moreover, the rice straw biochar was more effective than the bamboo biochar in increasing soil pH. The results also indicated that the 5% rice straw biochar treatments increased the EC by 350%. However, there was no significant difference after the addition of bamboo biochar. In the pot experiment with tobacco plants, Yang et al. (2015) observed that tobacco stalk biochar and dead pig biochar significantly ($p<0.05$) increased the pH, and the tobacco stalk biochar was more effective than the dead pig biochar, but had no significant effect on the EC of the soil.

Yang et al. (2016) also determined the water soluble organic carbon (WSOC), cation exchange capacity (CEC) and the concentration of available phosphorus (P) at the end of their incubation experiment. Only the coarse straw biochar treatment at the 5% amendment rate significantly ($p<0.05$) increased the WSOC content in soil. This result might be due to the WSOC, released from biochar, adsorbing to the soil. The CEC increased with increasing application rates of rice straw biochar, and overall, the fine particle size treatments were more effective than adding the coarser materials (Yang et al., 2016). Addition of rice straw biochar, with its higher pH and CEC was more effective than bamboo biochar in enhancing the availability of P (Yang et al., 2016). Therefore, out of the ten treatments the high application rate (5%) of fine sized (<0.25 mm) rice straw biochar was the most effective in increasing soil P availability.

In addition, activities of urease, catalase and acid phosphatase were determined in this incubation experiment (Yang et al., 2016). The urease activity increased by 143% and 107% after the application of 5% coarse and fine rice straw biochars, respectively. Both bamboo and rice straw biochars significantly ($p<0.05$) increased the catalase activity but had no significant impact on the activity of acid phosphatase (Fig. 1).

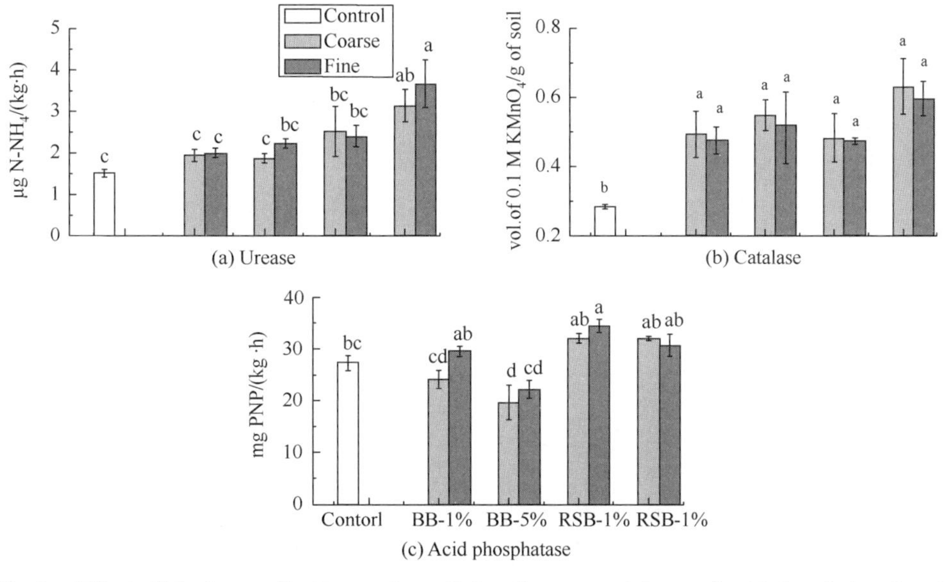

Fig. 1 Effect of biochar application on the activity of urease, catalase and acid phosphatase in soil
Treatments: control, 1% and 5% bamboo biochar (BB) and rice straw biochar (RSB) with two particle sizes (coarse and fine). Error bars are standard error of the means (n=4). Different letters above the column indicate significant differences between treatments at $p<0.05$ level (from Yang et al., 2016)

2.2 Effect of biochar on the mobility and extractability of heavy metals in co-contaminated soil

The effect of biochar application on the extractability of heavy metals was investigated by

Yang et al. (2016). The highest reductions in extractable Cu (97.3%) and Zn (62.2%) were achieved by applying 5% of fine rice straw biochar (Fig. 2). The coarse rice straw biochar treatment applied at the same rate reduced the Cu and Zn extractability by 94.8% and 52.9%, respectively. The two treatments reduced the extractable Cd by 17.7% to 25.8% and Pb by 81.9% to 97.9%. These results are in line with the results from the pot experiment conducted by Lu et al. (2014) who found that the higher application rate and finer particle size were similarly more effective in reducing the concentrations of extractable heavy metals in the soil. This was explained by the larger surface area of the finer particles. The presence of both, bamboo biochar and rice straw biochar was more effective in reducing the concentrations of extractable Cu and Pb than in reducing the concentrations of extractable Cd and Zn in soil, and rice straw biochar was more effective than bamboo biochar (Lu et al., 2016; Yang et al., 2016). The pot experiment with tobacco and dead pig-based biochars also demonstrated that the concentrations of $CaCl_2$-extractable Cd and Zn in soil decreased significantly ($p<0.05$) with increasing application rates of both types of biochar. The tobacco stalk biochar was more effective than the dead pig biochar in reducing the concentrations of extractable Cd and Zn.

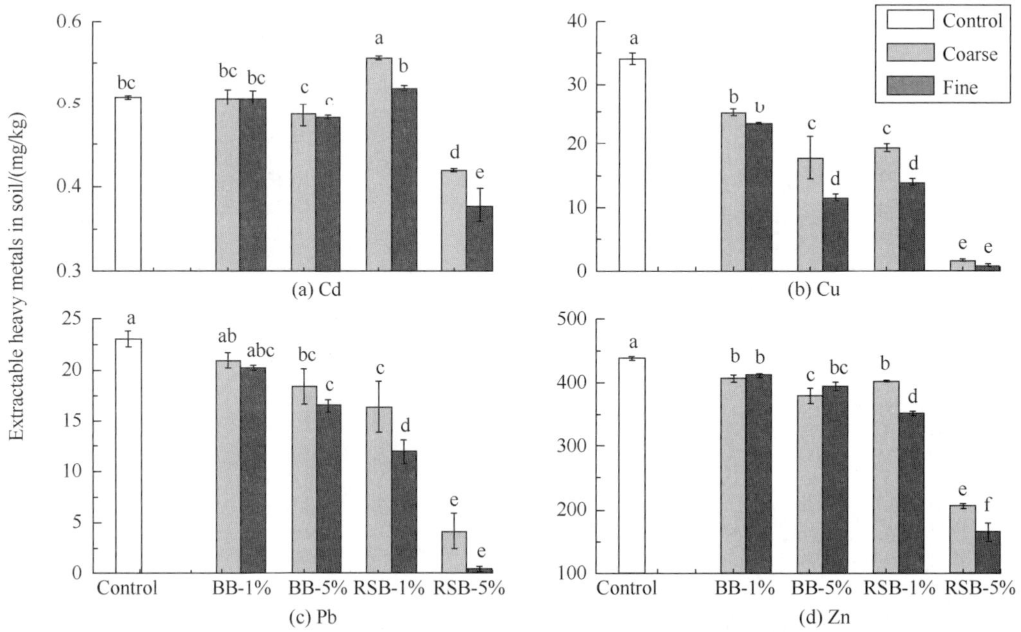

Fig. 2 Effect of biochars addition on the concentration of extractable heavy metals in soil (Yang et al., 2016)
Treatments: control, 1% and 5% bamboo biochar (BB) and rice straw biochar (RSB) with two particle sizes (coarse and fine). Error bars are standard error of the means (n=4). Different letters above the columns indicate significant differences between treatments at $p<0.05$ level

2.3 Effect of biochar on the redistribution of heavy metals in soil

The sequential extraction procedure from the European Union Bureau of Reference

(EUBCR) (Zemberyová et al., 2006) was applied in order to determine the redistribution and extractability of heavy metals in soil. In the incubation experiment (Liu et al., 2015) (similar with the experiment conducted by Yang et al., 2016), the acid-soluble fractions of Cd and Cu in the soil were transformed to reducible and oxidizable fractions in the fine (0.25 mm) rice straw biochar treatment. However, the acid-soluble fraction of Zn was mainly transformed to reducible Zn after the application of biochar. Adding coarse (1 mm) bamboo biochar to soil enhanced the transformation of Cu and Zn from the acid soluble fraction to reducible, oxidizable and residual fractions. Lu et al. (2016) also observed that biochar application significantly ($p<0.05$) increased the metal fractions bound to organic matter, but had no effect on the residual metal fractions. Higher biochar application rates resulted in larger fractions of acid extractable and Fe/Mn oxide metals with a corresponding decrease in the more stable organic matter bound metal fractions, indicating that biochar applications are likely to immobilize heavy metals in soil. In the tobacco pot experiment (Yang et al., 2015), a 5-step sequential extraction was performed following the method suggested by Tessier et al. (1979) to identify changes in the bonding forms and redistribution of Cd and Zn in soil. Both tobacco stalk and dead pig biochars significantly ($p<0.05$) increased Cd and Zn immobilization in soil, and the tobacco stalk biochar was more effective than the dead pig biochar. Biochar application may affect heavy metal immobilization by direct and indirect mechanisms including adsorption, precipitation, inner-sphere complexation and cation exchange (Park et al., 2011; Zhang et al., 2013). For instance, the application of biochar may immobilize heavy metals through surface adsorption owing to the large surface area and highly porous structure of biochars (Yang et al., 2015). Adsorption may also be attributed to the presence of O-containing functional groups such as-OH, -COOH and C=C on the biochar surface (Park et al., 2011). Additionally, the functional groups and mineral oxides may also contribute to the formation of inner-sphere complexation (Zhang et al., 2013).

2.4 Effect of biochar on the plant uptake of heavy metals

The type of biochar source material and grain size had a significant effect on the uptake of Cd, Cu, Pb and Zn by *S. plumbizincicola* (Fig. 3) (Lu et al., 2014). The concentration of Cd in the shoots significantly ($p<0.05$) decreased with bamboo biochar as well as rice straw biochar applications. However, there was no significant change in the Cd concentration in plant shoots with increasing biochar application rates. Addition of biochars significantly reduced the plant uptake of Cu and Pb with the exception of the 1% bamboo biochar treatment. The 5% rice straw biochar treatment significantly ($p<0.05$) reduced the uptake of Cu and Pb in plant shoots. Compared to the control, biochar addition also significantly ($p<0.05$) reduced the uptake of Zn. The bamboo biochar appeared to be more effective than the rice straw biochar in reducing Cd uptake by *S. plumbizincicola*, whereas the rice straw biochar was more effective in reducing Cu and Pb uptake by the plant. This highlights that the type of biochar influences the heavy metal uptake by plants. Also, differences were noted depending on the heavy metal tested.

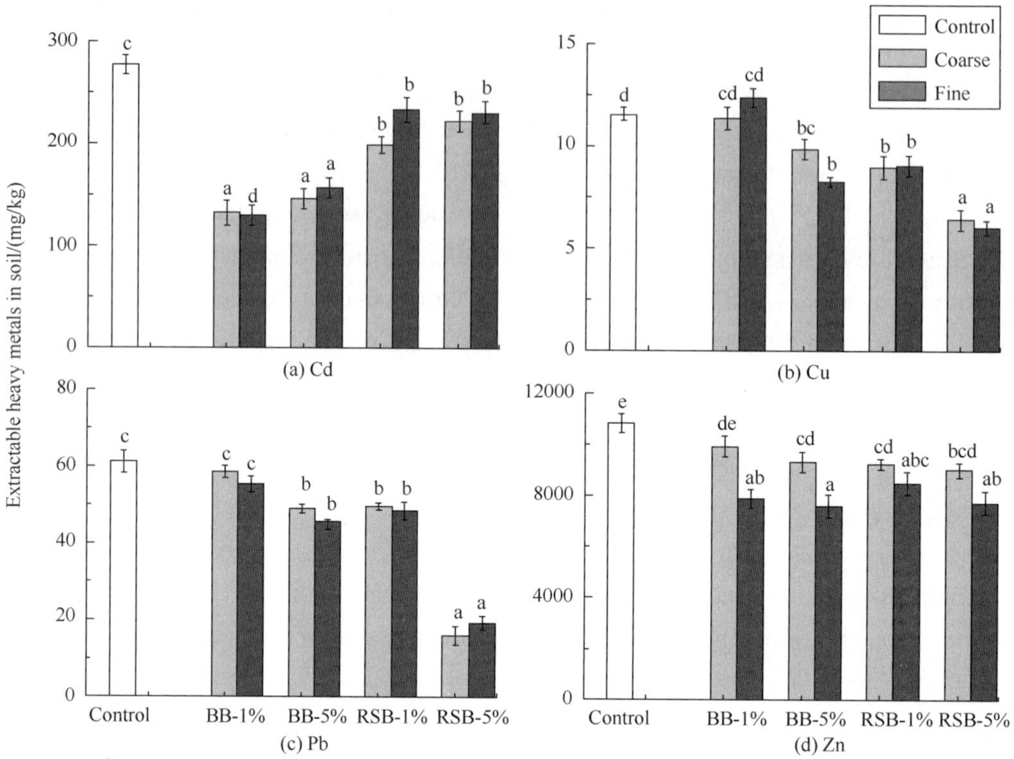

Fig. 3 Effect of bamboo and rice straw biochars on concentrations of Cd, Cu, Pb and Zn in the shoot of *Sedum plumbizincicola*. Treatments: control, 1% and5% bamboo biochar (BB), rice straw biochar (RSB) with two particle sizes (coarse and fine). Error bars are standard deviation of the means (n=4). Different letters above the columns indicate significant ($p<0.05$) difference between treatments (from Lu et al., 2014)

The concentrations of Cd and Zn in the shoots and roots of tobacco plants were determined in the pot experiment introduced in section 2 by Yang et al. (2015). Tobacco stalk biochar and dead pig biochar significantly ($p<0.05$) reduced Cd and Zn accumulation in tobacco roots and shoots. The tobacco stalk biochar was more effective in reducing the accumulation of Cd, whereas dead pig biochar was more effective for reducing the accumulation of Zn (Fig. 4). Overall, the concentration of Cd was higher in plant shoots than in the roots, indicating that Cd was not translocated in the plant from the roots to the shoots. Furthermore, Yang et al. (2015) also found that the soil pH was positively correlated ($p<0.05$) with both shoot and root biomass, and negatively ($p<0.01$) correlated with Cd and Zn concentration in the shoots and roots of the tobacco plants. Furthermore, there was a positive correlation ($p<0.01$) between the concentration of $CaCl_2$-extractable heavy metals and the accumulation of heavy metals in the tobacco plants.

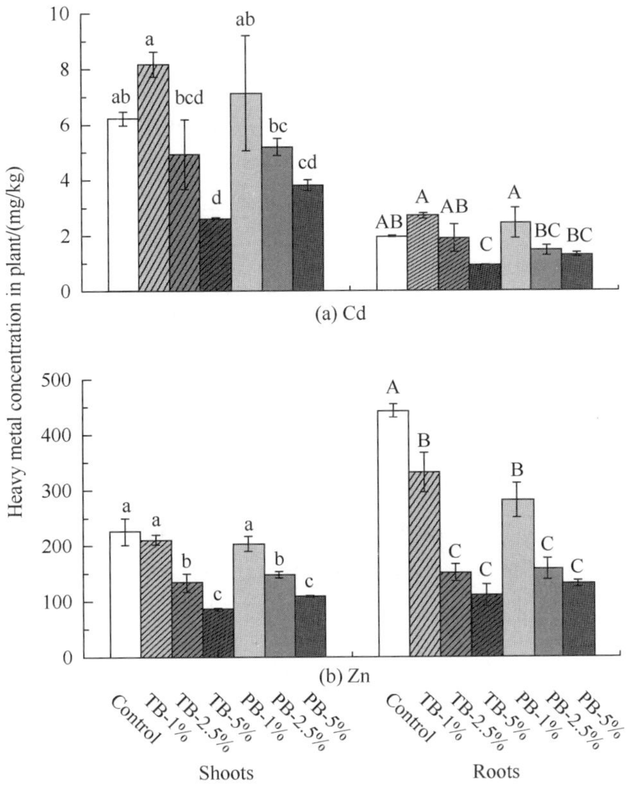

Fig. 4 Effect of biochar addition on Cd and Zn concentrations in shoot and root of tobacco plant
Treatments: control, 1%, 2.5%, and 5% tobacco stalk biochar (TB) and dead pig biochar (PB). Error bars are standard error of the means (n=3). Different letters above the columns indicate significant differences between treatments at $p<0.05$ level (Yang et al., 2015)

2.5 Mechanisms of remediating heavy metal-contaminated soils with biochar

2.5.1 Effect of the type of biochar

Our results showed that rice straw biochar appeared to be more effective in immobilizing and redistributing metals than bamboo biochar (Lu et al., 2014; Liu et al., 2015; Yang et al, 2016). The result is not unexpected as the rice straw biochar had a higher pH, ash content and surface alkalinity, a higher content of Si and K and more O-containing functional groups than the bamboo biochar. The bamboo biochar appeared to be more effective than the rice straw biochar in reducing Cd uptake by *S. plumbizincicola*, whereas the rice straw biochar was more effective in reducing Cu and Pb uptake by the plant. Similarly, it seemed that higher levels of O-containing functional groups, organic carbon content, and a higher CEC made the tobacco stalk biochar a more effective soil amendment than the dead pig biochar. The tobacco stalk biochar was more effective for immobilizing Cd, whereas, the dead pig biochar was more effective for Zn immobilization. According to the literature, an increased pH could reduce heavy metal extractability and mobility in soil by promoting the formation of precipitates (Lu et al.,

2014). The presence of P, Si, Al, O, carbonate and chloride in soil may also contribute to the formation of stable $CdCO_3$, $PbCO_3$ and pyromorphite-like phases (Yang et al., 2016). In addition, functional groups such as phenolic OH, -COOH and C=N have been shown to form stable complexes with heavy metals, thereby increasing the specific adsorption of metals by soils amended with biochar (Huang et al., 2011).

2.5.2 Effect of the biochar application rate

Overall, the concentration of extractable heavy metals decreased significantly with increasing application rates of biochar. While biochar may directly reduce the extractable and mobile heavy metal fraction in soils through adsorption or precipitation (Zhang et al., 2013), the addition of biochar may also enhance the pH, EC and affect other soil properties. This in turn may indirectly lead to decreased heavy metal extractability and mobility (Lu et al., 2014). However, there was no significant change in Cd concentration with increasing biochar application rates. The associated mechanism is yet to be explored. In the case of tobacco plant growth, dead pig biochar applied at 2.5% was more effective than applied at 1% or 5%.

2.5.3 Effect of the particle size of biochar

In the studies introduced above, the particle size of biochar only occasionally showed a significant effect on the mobility of heavy metals. In those cases, the finer sized biochar treatments were more effective in decreasing the pore water concentrations and plant availability concentrations of Cd, Pb and Zn than the coarser sized biochar. This may be attributed to a larger specific surface area of fine biochar than coarse biochar. The particle size of biochar had no significant impact on decreasing the fraction bound to organic matter and residual metals. The following findings in the literature corroborate our observations. Cao et al. (2003) found that phosphorus concentrations in biochar played an important role for immobilizing Cd, Pb and Zn. Zheng et al. (2012) found that water-extractable phosphorus levels were higher in fine biochar ($<$0.18 mm) than in coarse biochar (0.5-2 mm). They also noted that P reduced the metal translocation from roots to shoots by the formation or co-precipitation of insoluble metal phosphate in the roots.

3. Using biochar for the remediation of soils contaminated by phthalic acid esters (PAEs)

3.1 Effect of biochar on the adsorption of PAEs in soils

Diethyl phthalate (DEP) is one of the most widely used PAEs, which mainly originates from the leaching and volatilization from plastic greenhouses and plastic mulch and the discharge of wastewater (Stales et al., 1997). In China, increasingly more soils used for vegetable production have been contaminated by PAEs (e.g., DEP) which were released from the plastic greenhouse roofs and film mulch. When DEP is released to soils, it can be taken up by plants which may eventually enter the human food chain (Zhang et al., 2014). Therefore, a laboratory study was conducted by Zhang et al. (2014, 2015) to evaluate the effect of biochar on soil adsorption and desorption of DEP, and to assess the effect of different aging processes on the adsorption capacity of DEP to soil amended with biochar.

3.1.1 Effect of soil organic carbon content on diethyl phthalate adsorption

Zhang et al. (2014) found that soil organic carbon (OC) contents not only affected the ability of soils to adsorb DEP but also influenced the effectiveness of biochars to adsorb DEP in contaminated soils. A soil with a higher OC content (22 mg/kg) adsorbed DEP more efficiently than a soil with a lower OC content (3.5 mg/kg). In addition, the adsorption capacities of the low OC soil treated with 1% and 0.5% (*w*/*w*) bamboo biochar were about 1.5 to 15 times, respectively, higher than those of the high OC soil under the same treatments. This was explained by the blocking of pores by soil-derived dissolved organic matter or by surface competition between co-adsorbing dissolved organic matter molecules and biochar particles (Zhang et al., 2014).

3.1.2 Effect of pyrolysis temperature of biochar on diethyl phthalate adsorption

Generally, the properties of biochar may strongly affect its adsorption capacity for organic pollutants. Pyrolysis temperature has been identified as a crucial factor for the adsorption capacity of biochar (Zhang et al., 2013). The biochar adsorption capacity for DEP was higher for rice straw biochar (RSB) produced at 350 ℃ than for RSB produced at 650 ℃ (Fig. 5) (Zhang et al., 2014). Lower biochar pyrolysis temperature increased the adsorption capacities for DEP. This phenomenon was also observed by others and explained by the larger specific surface area (SSA) of biochar produced at lower temperatures (Bornemann et al., 2007). However, Zhang et al. (2014) found that the adsorption capacity of the biochar decreased with the increase of the specific surface area of the biochar. Thus, there were other underlying mechanisms. A Fourier transform infrared (FTIR) analysis revealed that functional groups such as O-H stretching (3, 200-3, 600/cm), aromatic C=C and C=O (1, 400-1, 600/cm) of the rice straw biochar disappeared as the pyrolysis temperature reached 650°C. This indicates that the sorption mechanism of DEP to the straw biochar may be attributed to the coordination of molecules to C=C and C=O (π-electrons) bonds and intermolecular hydrogen bonding (Zhang, et al., 2014).

Contrasting results were observed for the bamboo biochar (BB). The adsorption of DEP was higher in the soil treated with BB pyrolyzed at 650℃ than in the soil treated with biochar pyrolyzed at 350℃ (Fig. 5) (Zhang et al., 2014). Moreover, the SSA also followed the same order, with the surface area of the biochar pyrolyzed at 650℃ being much greater than that of the biochar pyrolyzed at 350℃.

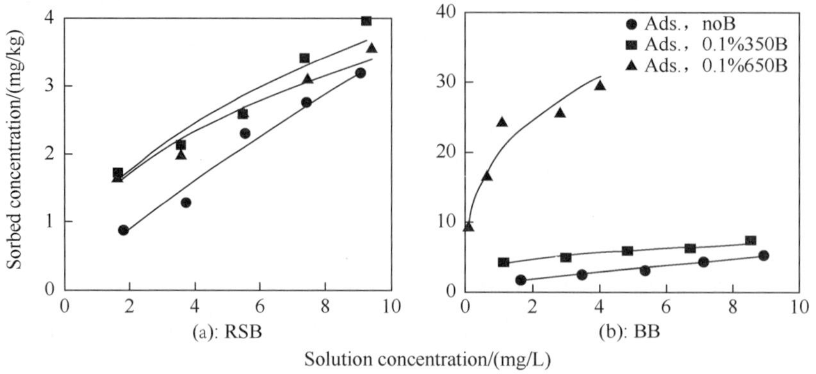

Fig. 5　Effect of pyrolysis temperature of biochar (RSB: rice straw biochar, BB: bamboo biochar, 0.1% is the application rate), on the adsorption (Ads.) of diethyl phthalate in a low organic carbon content (3.5 mg/kg) soil. Symbols are measured data and the solid lines are Freundlich model fits (from Zhang et al., 2014)

3.1.3　Effect of feedstock material of biochar on diethyl phthalate adsorption

Furthermore, the SSA, pore volume, and microporosity of biochar are dependent on the feedstock material (Cesarino et al., 2012). In the experiments described above, the bamboo biochar application was more effective in adsorbing DEP than the straw biochar application. For example, the adsorption capacity of the low OC soil treated with 0.5% BB pyrolyzed at 650℃ was thousands times greater than that of the soil treated with RSB pyrolyzed at 650℃ (Zhang et al., 2014). Haghseresht et al. (1999) found that the abundance of sorption-active edge sites may be increased by high amounts of amorphous carbon. In other words, higher amount of amorphous carbon may lead to a higher adsorption capacity of biochar. Usually, the aromatic carbon content in plant tissue is mainly determined by its lignin content (Bornemann et al., 2007; Cesarino et al., 2012). Studies showed that the lignin content of bamboo (37%) is considerably higher than that of the straw (Binod et al., 2010). Therefore, it is conceivable that lignin-derived biochar is more likely to have a pronounced effect on adsorbing DEP.

3.1.4　Effect of aging of biochar on diethyl phthalate adsorption

When biochar is applied to soil, a range of biogeochemical interactions take place, a process commonly referred to as "aging". Biochar properties may change during the process of aging (Kookana, 2010). This in turn may eventually also change the adsorption capacity of biochar for organic contaminants. Zhang et al. (2015) studied the effect of alternating wet and dry aging conditions on the adsorption capacity of biochar. The results showed that the sorption capacity for DEP varied between aging treatments (Fig. 6) (Zhang et al., 2015). One of the most important reasons that reduced the adsorption capacity of biochar is the blocking of sorption sites of biochar by dissolved organic carbon (DOC). Both treatments, alternating wet and dry and constantly moist aging conditions can enhance DOC concentrations in soil (Zhang et al., 2010). Yemadje et al. (2015) showed that dry-wet cycles increased moderate C mineralization compared to a continuously moist soil. Zhang et al. (2015) also found that the alternating wet and dry aging condition was more effective for soil organic carbon mineralization, and thus, the DOC concentrations were increased more than under the constant moist aging condition.

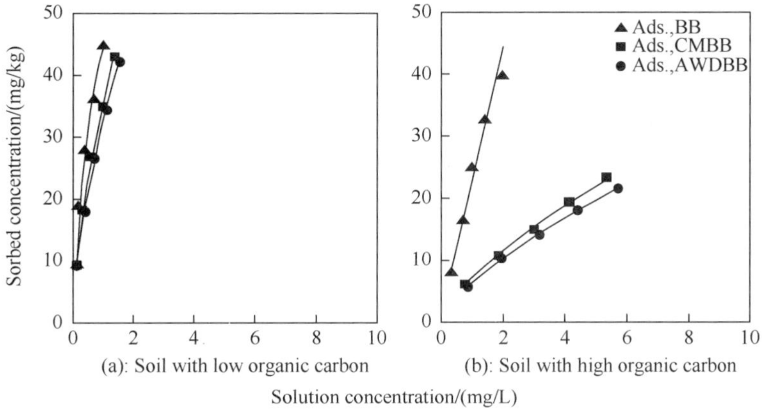

Fig. 6 Effect of aging processes (AWD: alternating wet and dry aging; CM: constantly moist aging) on the adsorption (Ads.) of diethyl phthalate on bamboo biochar (BB) in soils. Symbols are measured data and the lines are Freundlich model fits (from Zhang et al., 2015)

3.1.5 Effect of biochar on diethyl phthalate desorption

Compared with the untreated control, the soils amended with biochar indicated that desorption of DEP was affected by sorption hysteresis in all treatments. DEP adsorbed to soils treated with biochar was less likely to be desorbed (Zhang et al., 2014).

All aging treatments displayed a tendency to more easily desorb DEP. Zhang et al. (2015) observed that the degree of desorption hysteresis was quite different between the two aging conditions described above, and the constantly moist treatments showed a higher desorption hysteresis than the alternating wet and dry treatments. However, the underlying mechanism of this difference is not understood and needs further investigation. Also, whether and under which conditions sorbed DEP will release into the soil requires further investigation.

In conclusion, the application of biochars to soil can increase the DEP adsorption of soils. Pyrolysis temperature, feedstock material and the biochar aging condition will strongly affect the adsorption capacity of biochar for organic pollutants (Zhang et al., 2014, 2015).

3.2 Effect of biochar on the degradation and bioavailability of PAEs (DEHP) in soils

Di-(2-ethylhexyl) phthalate (DEHP) is one of the most commonly used PAEs in plastic products, such as polyethylene, polystyrene, polyethylene terephthalate and polyvinyl chloride. The low water solubility and degradation rate of DEHP contribute to its accumulation in soils (Katayama et al., 2010). In most regions of China, the concentrations of DEHP in soils exceed the recommended value for soil cleanup guidelines set by the United States Environmental Protection Agency. It has been shown that DEHP disrupts endocrine functions in humans (He et al., 2015). It is important to better understand the effect of soil properties and organic amendments on the degradation dynamics and bioavailability of DEHP in soil amended with biochar. Thus, an incubation experiment was conducted to determine the biodegradation of DEHP in soils in the presence of biochar and composted manure. In addition, a pot experiment was conducted to evaluate the effect of biochar on the bioavailability of DEHP in two soils, one low in organic carbon (LOC) and one high (HOC), using *Brassica chinensis* L. as an indicator plant (He et al., 2016).

3.2.1 Effect of biochar on the degradation rate of PAEs (DEHP) in soil

3.2.1.1 Effect of soil properties on the degradation rate of DEHP in soil

Degradation was faster in the HOC soil, where almost 90% of the DEHP were removed at the end of a 112-day incubation experiment. In comparison, 50% of the DEHP remained in the LOC soil at the end of the experiment. Consequently, the half-lives of DEHP were longer in all LOC treatments than in HOC treatments. The higher pH of the HOC soil had a higher alkaline hydrolysis leading to a faster DEHP rate (Stales et al., 1997). In addition, the high OC content may also have affected the degradation of DEHP through increasing the soil's biological activity (Xu et al., 2008).

3.2.1.2 Effect of organic amendments on the degradation rate of DEHP in soil

When the LOC soil was treated with organic amendments the DEHP half-lives were significantly ($p < 0.05$) lower compared to the LOC control soil. In contrast, the amendments of the HOC soil had no significant effect on the degradation of DEHP. This might be explained by the fact that soil organic carbon played an important role in covering reactive surfaces and pores of biochar (Pignatello and Xing, 1995). The actual capacity of the two types of biochar measured in isolation is thus masked (He et al., 2016) when biochar is added to soil.

The half-lives of the DEHP in the HOC treatments were not significantly different across the whole incubation period. The addition of biochar and biochar plus composted manure to the LOC soil increased the DEHP degradation rates except for the bamboo biochar (BB) treatment. While in the LOC treatments, the DEHP half-life was lower when amended with pig biochar, this may due to the biochar properties. The larger specific surface area of pig biochar (PB) could provide more shelters for microorganisms. Meanwhile, the mineral contents and pH of the pig biochar were also higher than those of the BB.

In this study, we also studied the effects of biochar plus compost manure on the fate of

DEHP in contaminated soils. The result showed an increasing DEHP degradation (Tabl is widely known that biochar can provide shelter for microorganisms and composted m can provide nutrients for microorganisms, thus, enhancing the microbial density degradation rate of organic pollutants (Xu et al., 2015). In addition, the soil pH incre when amended with biochar plus composted manure because they are both alkaline orga amendments. This affected the solubility of organic carbon and the microbial bioma (Andersson et al., 2000).

Table 1 Parameters of the first-order kinetic equation for DEHP degradation in two soils amended with biochar and biochar plus composted manure (unpublished data)

Treatments	Kinetic equation	Rate constant/day	Half-life/days	R^2
LOC	$y=-0.0063x+4.6057$	0.0063	110	0.94
LOCPB	$y=-0.0119x+4.7732$	0.0119	58	0.96
LOCBB	$y=-0.0063x+4.6555$	0.0063	110	0.88
LOCPB-M	$y=-0.0096x+4.6257$	0.0096	72	0.78
LOCBB	$y=-0.0137x+4.6374$	0.0137	51	0.95
HOC	$y=-0.0196x+4.3872$	0.0196	35	0.93
HOCPB	$y=-0.0173x+4.411$	0.0173	40	0.94
HOCBB	$y=-0.0176x+4.472$	0.0176	39	0.93
HOCPB-M	$y=-0.0185x+4.3902$	0.0185	38	0.96
HOCBB-M	$y=-0.0172x+4.2919$	0.0172	40	0.96

Note: LOC: low organic carbon content (0.35%) soil without amendments; LOCPB: low organic carbon content soil amended with 1.0% pig biochar; LOCBB: low organic carbon content soil amended with 1.0% bamboo biochar; LOCPB-M: low organic carbon content soil amended with 1.0% pig biochar and 1.0% composted manure; LOCBB-M: low organic carbon content soil amended with 1.0% bamboo biochar and 1.0% composted manure; HOC: high organic carbon content (2.2%) soil without amendments; HOCPB: high organic carbon content soil amended with 1.0% pig biochar; HOCBB: high organic carbon content soil amended with 1.0% bamboo biochar; HOCPB-M: high organic carbon content soil amended with 1.0% pig biochar and 1.0% composted manure; HOCBB-M means high organic carbon content soil amended with 1.0% bamboo biochar and 1.0% composted manure.

3.2.2 Effect of biochar on the bioavailability of PAEs (DEHP) in soils

3.2.2.1 Effect of biochar on DEHP residues in soils

The residual concentrations of DEHP tended to be higher in the soils treated with biochar than in the control soils at the end of the 50-day pot experiment. They were lower ($p<0.05$) in the high organic carbon content soil (HOC; 2.2% C) than in the low organic carbon content soil (LOC; 0.35% C) (Fig. 7 (a)) (He et al., 2016). In the LOC treatments, the rice straw biochar treatment retained significantly more DEHP compared to the bamboo biochar treatment, which was at least partly explained by its larger specific surface area. However, the opposite result was found for the HOC soil treatments (He et al., 2016). This may be due to the soil's higher OC content. Organic carbon plays an important role in covering reactive surfaces and blocking pores of biochar, and/or competing for biochar surface areas with organic contaminants (Pignatello and Xing, 1995), and thereby, masking the actual sorption capacity of the two types of biochar measured in isolation. The DEHP concentrations in plant shoots grown in the HOC soils were lower than those in the LOC soils ($p<0.05$) (He et al., 2016).

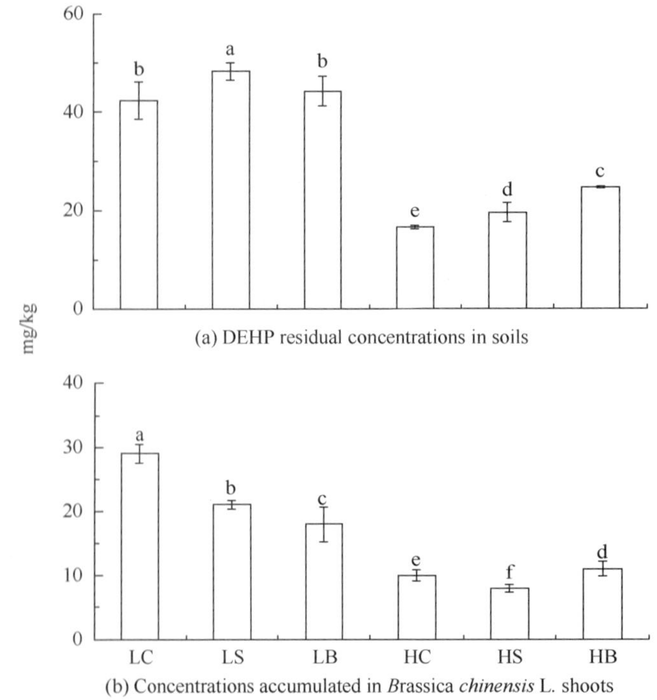

Fig. 7 DEHP residual concentrations in soils and concentrations accumulated in *Brassica chinensis* L. shoots. LC: Shoots from plants grown in low organic carbon content (LOC: 0.35%) soil without biochar (Control) ; LS: Shoots from plants grown in LOC soil with 2% straw biochar; LB: Shoots from plants grown in LOC soil with 2% bamboo biochar; HC: Shoots from plants grown in high organic carbon content (HOC: 2.2%) soil without biochar (Control) ; HS: Shoots from plants grown in HOC soil with 2% straw biochar; HB: Shoots from plants grown in HOC soil with 2% bamboo biochar. Error bars are standard errors of the means (n=3). The lower case letters above the column indicate significant difference between treatments according to Duncan's test ($p<0.05$) (from He et al., 2016)

3.2.2.2 Effect of biochar on DEHP residues in plant shoots

In the LOC soil treatments, the DEHP concentrations in the plants of the two biochar treatments were significantly lower than those in the control soils ($p<0.05$) (He et al., 2016). The result was similar to previous research, which had shown that soil amendments with biochar reduced the accumulation of organic pollutants in plants. However, in the HOC soil treatments, the effect of adding biochar was not. This indicates that the interactions of a soil's OC content with biochar can affect the uptake of organic pollutants by plants to some extent, meaning that the actual sorption capacity of biochar cannot be used as indicator for its effect on the bioavailability of DEHP (He et al., 2016).

In addition, the bio-concentration factors (BCFs) values of the biochar treatments were significantly ($p < 0.05$) lower than those of the controls. These results indicate that the addition of biochar to soils reduced the accumulation of DEHP in *B. chinensis* L., which decreases the potential risk of human uptake of DEHP (He et al., 2016).

The results above suggest that soil OC contents and biochar amendments of soils are both important factors for the immobilization of DEHP. They affected the environmental fate of DEHP in soils. They both reduced the bioavailability of DEHP, while the effect of biochar in soils depended on soil properties, such as the OC content (He et al., 2016).

3.3 Effect of biochar on the diversity of microorganisms in dibutyl phthalate-contaminated soil

It is commonly accepted that biochar amendments can increase the porosity, water hold capability and nutrient content of soils, as well as offer shelter for microorganisms, which will increase the microbial activity and the diversity of biological community (Lehmann et al., 2011). However, there is little information on the effect of biochar on the diversity of microorganisms in PAEs contaminated soil. Thus, a 56-day incubation experiment was conducted by Fan et al. (2016) to evaluate the effects of biochar type (bamboo and rice straw biochars) and application rate (0, 0.5 and 2.0%) on the diversity of the microbial community structure in two soils (HOC: 2.2% and LOC: 0.35%), which was contaminated by another kind of PAEs-dibutyl phthalate (DBP). The diversity of the microbial community structure was analyzed by the phospholipid fatty acids (PLFA) method.

3.3.1 Effect of biochar on bacterial PLFA and fungal PLFA in dibutyl phthalate-contaminated soil

The bacterial PLFA, fungal PLFA and microbial community's total PLFA of the HOC soil was significantly ($p < 0.05$) higher than those of the LOC soil (Fig. 8) (Fan et al., 2016). This highlights that organic carbon controls the circulation of energy and nutrients in soils. Bacterial PLFA, fungal PLFA and microbial community's total PLFA of the HOC soil significantly ($p < 0.05$) increased in the 2% straw biochar treatment, whereas a significant ($p < 0.05$) reduction in the PLFA values was observed for the bamboo biochar treatments. This was due to the properties of bamboo biochar, compared to rice straw biochar, their larger specific surface area and higher porosity provided better shelter for the microorganisms in soil (Fan et al., 2016).

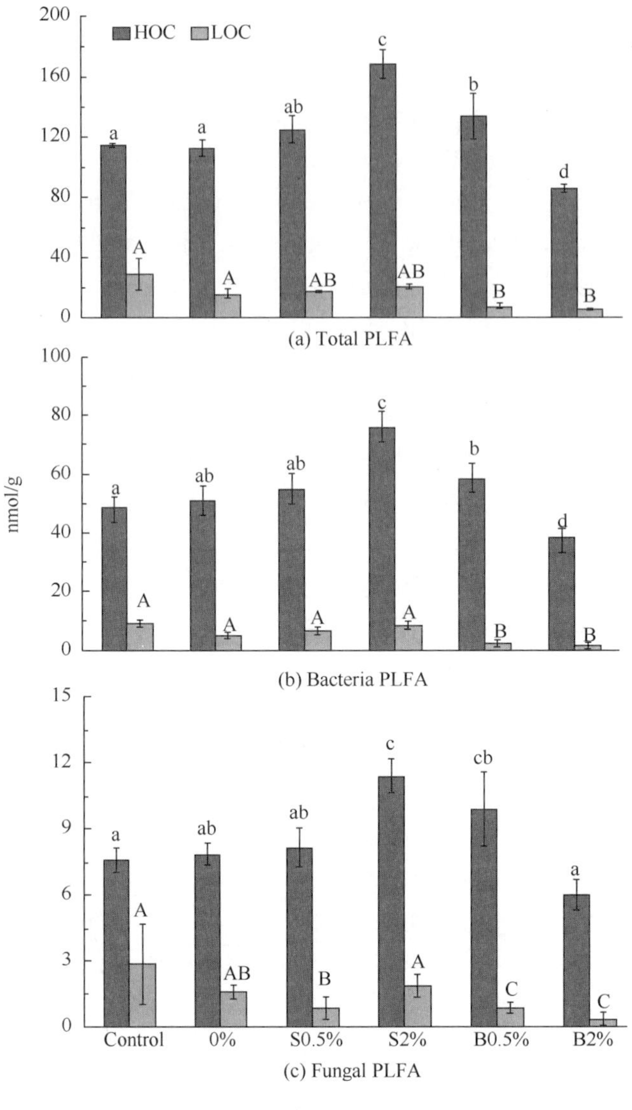

Fig. 8　Effect of biochar treatments on the bacterial PLFA, fungal PLFA and total quantity of microbial PLFA in soil. Error bars are standard deviation of the means (*n*=3). Different letters above the columns indicate significant ($p < 0.05$) difference between treatments (from Fan et al., 2016)

Note: Control: soils without any amendments and DBP; 0%: soils without any amendments; S0.5%: DBP contaminated soil amended with 0.5% straw biochar; S2%: DBP contaminated soil amended with 2.0% straw biochar; B0.5%: DBP contaminated soil amended with 0.5% bamboo biochar; B2%: DBP contaminated soil amended with 2.0% bamboo biochar; HOC: high organic carbon soil (2.2% OC); LOC: low organic carbon soil (0.35%)

3.3.2　Effect of biochar on Shannon Index in dibutyl phthalate-contaminated soil

The Shannon Index of HOC soils was higher than that of the LOC soils. This means that the diversity of the microbial community was higher in the HOC than in the LOC soil. In the LOC soil, the 2.0% bamboo biochar treatment reduced the soil microbial community Shannon Index the most (Fan et al., 2016).

In addition, the ratio of Gram-negative bacteria/Gram-positive bacteria reached the highest level in the treatment with 2.0% rice straw biochar addition, whereas the biochar type and application rate had no significant influence on the ratios of fungi/bacteria and Gram-negative bacteria/Gram-positive bacteria in the soil with high organic matter content. The greatest reduction in the soil microbial stress level was observed in the DBP-contaminated soil treated with 2% straw biochar (Fan et al., 2016). The differences were due to the content of soil organic matter, dosage and types of biochar. Parameters, such as the pH, SSA influenced the soil microbial community structure (Fan et al., 2016).

4. Outlook

Biochar can potentially be used as soil amendment for remediating contaminated soils through adsorption and immobilization of heavy metals and organic contaminants. The impact of biochar treatment on immobilization of contaminants varied with the type, application rate and particle size of the biochar as well as the contaminant types. Application of biochar to soil can also improve soil properties that enhance the activities of soil microorganisms. However, many aspects still need to be investigated and process understanding needs to be improved. Some potential future research directions and knowledge gaps are listed below.

(1) Conduct large-scale field trials with different soil types and climatic conditions before operational scale remediation projects can be implemented.

(2) Identify the effect of biochar on soil properties and especially on microbial communities in long-term field experiments.

(3) Investigate and better understand the long-term environmental fate of the sequestered contaminants.

(4) Explore how pyrolysis conditions affect biochar properties as soil amendment.

(5) Find an abundant, low-cost biomass feedstock for large-scale biochar production.

(6) Characterize different biomass materials and pyrolysis conditions and optimize production systems to produce designer biochar products to be used effectively for specific remediation work.

(7) Explore new technologies by combining biochar applications with other methods such as phytoremediation and soil washing.

Acknowledgements

These studies were funded by the National Natural Science Foundation of China (21577131; 4171337), the Key Program of the Zhejiang Provincial Natural Science Foundation, China (LZ15D010001), Zhejiang A & F University Research and Development Fund (2010FR097 and 2012FK031), the Science and Technology Foundation of the Guizhou Province, China ([2013]2193), the Scientific Research and Technology Development Foundation of Bijie Yancao Company of Guizhou Province, China (BJYC-201308), and the Special Funding for the Introduced Innovative R&D Team of Dongguan (2014607101003).

References

Andersson S, Nilsson S I, Saetre P. 2000. Leaching of dissolved organic carbon (DOC) and dissolved organic nitrogen (DON) in mor humus as affected by temperature and pH. Soil Biology & Biochemistry, 32 (1): 1-10.

Binod P, Sindhu R, Singhania R R, et al. 2010. Bioethanol production from rice straw: an overview. Bioresource Technology, 101 (13): 4767-4774.

Bornemann L C, Kookana R S, Welp G. 2007. Differential sorption behaviour of aromatic hydrocarbons on charcoals prepared at different temperatures from grass and wood. Chemosphere, 67 (5): 1033-1042.

Cao X, Ma L Q, Shiralipour A. 2003. Effects of compost and phosphate amendments on arsenic mobility in soils and arsenic uptake by the hyperaccumulator, *Pteris vittata* L. Environmental Pollution, 126 (2): 157-167.

Cesarino I, Araújo P, Domingues Júnior A P, et al. 2012. An overview of lignin metabolism and its effect on biomass recalcitrance. Brazilian Journal of Botany, 35 (4): 303-311.

Deng W, Van Zwieten L, Lin Z, et al. 2016. Sugarcane bagasse biochars impact respiration and greenhouse gas emissions from a latosol. Journal of Soils and Sediments (in press).

Dong D, Yang M, Wang C, et al. 2013. Responses of methane emissions and rice yield to applications of biochar and straw in a paddy field. Journal of Soils and Sediments, 13 (8): 1450-1460.

Fan S, He L, Qin H, et al. 2016. Effect of biochar on diversity of microbial community in soils contaminated with dibutyl phthalate (in Chinese). Acta Scientiae Circumstantiae, 36 (5): 1800-1809.

Guo X, Lu K, Hu G, et al. 2016. Effect of dead pig and bamboo biochars on greenhouse soil properties and vegetable yield. Journal of Zhejiang A & F University (in press).

Haghseresht F, Lu G Q, Whittaker A K. 1999. Carbon structure and porosity of carbonaceous adsorbents in relation to their adsorption properties. Carbon, 37 (9): 1491-1497.

He L, Fan S, Müller K, et al. 2016. Biochar reduces the bioavailability of di- (2-ethylhexyl) phthalate in soil. Chemosphere, 142: 24-27.

He L, Gielen G, Bolan N S, et al. 2015. Contamination and remediation of phthalic acid esters in agricultural soils in China: a review. Agronomy for Sustainable Development, 35 (2): 519-534.

Hu G, Yang X, Chen X, Lu K, et al. 2016a. Physiological responses of bamboo-willow plants to heavy metal stress (in Chinese). Acta Scientiae Circumstantiae (in press), DOI: 10.13671/j.hjkxxb.2016.0030.

Hu, G, Yu Y, Yang X, et al. 2016b. Uptake, Accumulation and Translocation of Cadmium in Bamboo-willow Plant (in Chinese). Acta Scientiae Circumstantiae, 36 (4): 1508-1514.

Huang J H, Hsu S H, Wang S L. 2011. Effects of rice straw ash amendment on Cu solubility and distribution in flooded rice paddy soils. Journal of Hazardous Materials, 186 (2): 1801-1807.

Katayama A, Bhula R, Burns G R, et al. 2010. Bioavailability of xenobiotics in the soil environment//Reviews of environmental contamination and toxicology. Springer New York: 1-86.

Kookana R S. 2010. The role of biochar in modifying the environmental fate, bioavailability, and efficacy of pesticides in soils: a review. Soil Research, 48 (7): 627-637.

Lehmann J, Rillig M C, Thies J, et al. 2011. Biochar effects on soil biota-a review. Soil Biology and Biochemistry, 43 (9): 1812-1836.

Liu B R. 2010. Changes in soil microbial biomass carbon and nitrogen under typical plant communies along an altitudinal gradient in east side of Helan Mountain. Ecology and Environmental Sciences, 19 (4): 883-888.

Liu J, Yang X, Lu K, et al. 2015. Effect of bamboo and rice straw biochars on the transformation and bioavailability of heavy metals in soil (in Chinese). Acta Scientiae Circumstantiae, 35: 3679-3687.

Liu Y, Yang M, Wu Y, et al. 2011. Reducing CH_4 and CO_2 emissions from waterlogged paddy soil with biochar. Journal of Soils and Sediments, 11 (6): 930-939.

Lu K, Yang X, Gielen G, et al. 2016. Effect of bamboo and rice straw biochars on the mobility and redistribution of heavy metals (Cd, Cu, Pb and Zn) in contaminated soil. Journal of Environmental Management (in press).

Lu K, Yang X, Shen J, et al. 2014. Effect of bamboo and rice straw biochars on the bioavailability of Cd, Cu, Pb and Zn to *Sedum plumbizincicola*. Agriculture, Ecosystems & Environment, 191: 124-132.

Park J H, Choppala G K, Bolan N S, et al. 2011. Biochar reduces the bioavailability and phytotoxicity of heavy metals. Plant and Soil, 348 (1-2): 439-451.

Pignatello J J, Xing B. 1995. Mechanisms of slow sorption of organic chemicals to natural particles. Environmental Science & Technology, 30 (1): 1-11.

Smernik R J. 2009. Biochar and sorption of organic compounds. Biochar for Environmental Management: Science and Technology: 289-300.

Staples C A, Peterson D R, Parkerton T F, et al. 1997. The environmental fate of phthalate esters: a literature review. Chemosphere, 35 (4): 667-749.

Sun T, Mao X, Lu K, Wang H, 2015. Removal of heavy metals from co-contaminated soil by washing with citric acid (in Chinese). Acta Scientiae Circumstantiae, 35 (8): 2573-2581.

Tessier A, Campbell P G C, Bisson M. 1979. Sequential extraction procedure for the speciation of particulate trace metals. Analytical Chemistry, 51 (7): 844-851.

Wang H, Lin K, Hou Z, et al. 2010. Sorption of the herbicide terbuthylazine in two New Zealand forest soils amended with biosolids and biochars. Journal of Soils and Sediments, 10 (2): 283-289.

Wu W, Yang M, Feng Q, et al. 2012. Chemical characterization of rice straw-derived biochar for soil amendment. Biomass and Bioenergy, 47: 268-276.

Xu G, Li F, Wang Q. 2008. Occurrence and degradation characteristics of dibutyl phthalate (DBP) and di- (2-ethylhexyl) phthalate (DEHP) in typical agricultural soils of China. Science of the Total Environment, 393 (2): 333-340.

Xu Q, Gu G, Zhang M, 2015. Promoting antibiotics degradation via application of organic fertilizers organicfertilizers. Acta Agriculturae Zhejiangensis, 27: 417-422.

Xu X, Cao X, Zhao L, et al. 2013. Removal of Cu, Zn, and Cd from aqueous solutions by the dairy manure-derived biochar. Environmental Science and Pollution Research, 20 (1): 358-368.

Yang X, Liu J, McGrouther K, et al. 2016. Effect of biochar on the extractability of heavy metals (Cd, Cu, Pb, and Zn) and enzyme activity in soil. Environmental Science and Pollution Research, 23 (2): 974-984.

Yang X, Lu K, McGrouther K, et al. 2015. Bioavailability of Cd and Zn in soils treated with biochars derived from tobacco stalk and dead pigs. Journal of Soils and Sediments (in press).

Yemadje PL, Guibert H, Bernoux M, Deleporte P, Chevallier T, 2015. Dry-wet cycles affect carbon mineralization of soil. International Conference Agroecology for Africa-Afa.

Žemberyová M, Bartekova J, Hagarova I. 2006. The utilization of modified BCR three-step sequential extraction procedure for the fractionation of Cd, Cr, Cu, Ni, Pb and Zn in soil reference materials of different origins. Talanta, 70 (5): 973-978.

Zhang H, Lin K, Wang H, et al. 2010. Effect of Pinus radiata derived biochars on soil sorption and desorption of phenanthrene. Environmental Pollution, 158 (9): 2821-2825.

Zhang X, He L, Sarmah A K, et al. 2014. Retention and release of diethyl phthalate in biochar-amended vegetable garden soils. Journal of Soils and Sediments, 14 (11): 1790-1799.

Zhang X, Sarmah A K, Bolan N S, et al. 2016. Effect of aging process on adsorption of diethyl phthalate in soils amended with bamboo biochar. Chemosphere, 142: 28-34.

Zhang X, Wang H, He L, et al. 2013. Using biochar for remediation of soils contaminated with heavy metals and organic pollutants. Environmental Science and Pollution Research, 20 (12): 8472-8483.

Zheng R L, Cai C, Liang J H, et al. 2012. The effects of biochars from rice residue on the formation of iron plaque and the accumulation of Cd, Zn, Pb, As in rice (Oryza sativa L.) seedlings. Chemosphere, 89 (7): 856-862.

The Research and Development of Technology for Contaminated Site Remediation

Xiaoyong LIAO*, Xiulan YAN, Dong MA, Dan ZHAO, Lu SUN, You LI, Yang FEI, Peng LI, Longyong LIN, Huan TAO

Beijing Key Laboratory of Environmental Damage Assessment and Remediation, Institute of Geographic Sciences and Natural Resources, Chinese Academy of Science

*Corresponding author

Abstract

The higher values of soil and groundwater pollution index of site contamination presented a significant threat of negative impact on human health. Contamination of industrial lands has become a new environmental problem in China during the process of industrial and urbanization development. Key technologies, such as in situ chemical oxidation (ISCO), soil vapor extraction (SVE), soil washing, solidification/stabilization (S/S), bioremediation, due to their site contamination remediation were researched and innovated due to their higher efficiency, better equipment, and lower cost. We expect those technologies to be used to solve the problems of site contamination coursed by human industrial activities in China.

Keywords: remediation, contaminated site, in situ chemical oxidation, soil vapor extraction, soil washing, solidification/stabilization, bioremediation

1. Introduction

The problem of site contamination has become a focus in the environment protection field in the recent years. Contamination of urban industrial lands is a new environmental problem in China during the process of upgrade of industrial structure and adjustment of urban layout (Liao et al., 2011). There is a huge market need for contaminated site remediation in China. However, our country is faced with more barriers to solve site contamination, such as a lack of technologies, funds and policy frameworks.

According to our special environmental circumstances, our team began to engage in the research of site remediation technology since 2007. We studied in situ chemical oxidation, soil vapor extraction, soil washing, solidification/stabilization and bioremediation technologies with intellectual property rights. We focus on new remediation methods, remediation process and mechanism; in addition, new equipments and relevant medicaments were also developed. Our group also originally designed a vehicle mounted site remediation equipment, a double-tank chemical oxidation remediation equipment and complete sets of soil vapor extraction equipment, which costs were far lower than analogous products in the world. An accurate assessment prior to the contaminated site remediation is essential and pre-conditional to carry out a site remediation project (Liao et al., 2014). We established a number of domestic influential contamination remediation pilot test sites in Beijing, Shanxi, Hunan and Guangdong province.

2. Research of key remediation technology

2.1 Chemical oxidation

Coking plant is one of the most common class of facilities with contamination in China, where polycyclic aromatic hydrocarbons (PAHs) and benzene, toluene, ethyl benzene, and xylene (BTEX) are target indicator compounds. In situ chemical oxidation (ISCO) is a remediation technology where a chemical oxidation is injected into the subsurface to degrade organic contaminants in place, which can be used to treat a wide variety of contaminants of concern, including chlorinated solvents, pesticides, PAHs, BTEX, explosives and some other organics. Different oxidants were evaluated for removal of PAHs and BTEX from contaminated soils at coking to screen the optimal reagents (Zhao et al., 2011a, 2011b).Permanganate and activated persulfate showed the highest PAHs removal (Zhao et al., 2011a), while Fenton reagent and modified Fenton reagent were the optimum oxidants for removing BTEX (Zhao et al., 2011b).

A wide range of naturally occurring reactants other than the target contaminant (s) also react with oxidant, which increased the total oxidant demand (TOD) during ISCO. The concentrations of permanganate decreased rapidly during the first 8 d of the reaction with PAHs, and the reaction follows first order kinetics, with rate constant ranging from 0.01 to 0.3/h (Liao et al., 2011). The variation of TOD with depth in PAHs contaminated site was also determined, and the typical TOD is about 50 g MnO_4^-/kg soil for clayey silt fill, 20-40 g MnO_4^-/kg soil for silt, silty clay and 1-7 g MnO_4^-/kg soil for fine sand (Liao et al., 2011). Statistical analysis revealed that TOD was positively correlated with total organic carbon (TOC) content, clay content and PAHs concentrations, besides sand content (Liao et al., 2011).

Persulfate was a newly emerging oxidant used for ISCO, and it can be catalyzed with various reactants to form radicals, which were more powerful oxidants. The effect and mechanism of persulfate activated by different methods for PAHs removal in soil were revealed (Zhao et al., 2013). Compared with citrate chelated iron activated persulfate, alkaline activated persulfate and a hydrogen peroxide-persulfate binary mixture, thermal activation showed the highest removal of PAHs (99.1%) (Zhao et al., 2013). Electron spin resonance spectra indicated the presence of hydroxyl radicals in thermally activated systems and weak hydroxyl radical activity in the H_2O_2-persulfate system, while superoxide radicals were predominant in alkaline activated systems (Fig. 1).

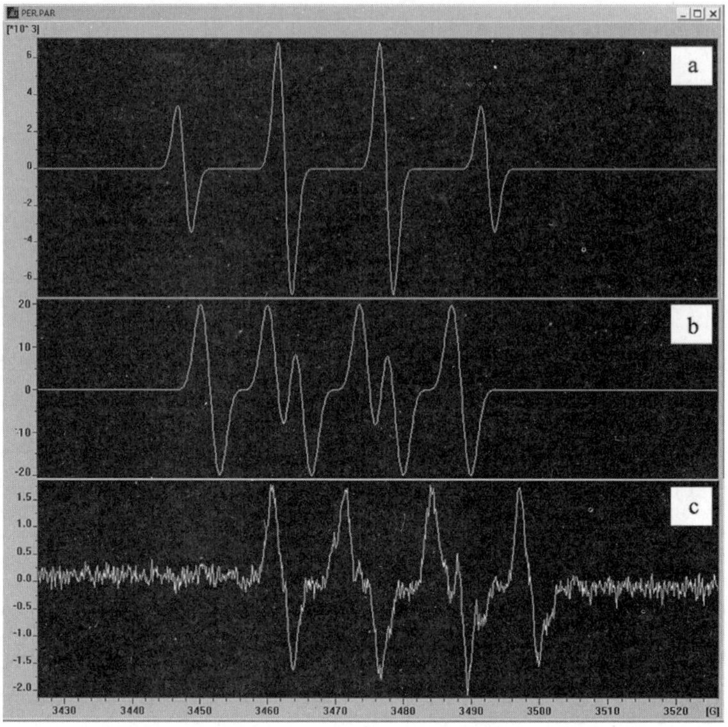

Fig. 1　Electron spin resonance characteristic spectra of the DMPO-radical adducts for (a) the hydroxyl radical, (b) the sulfate radical, (c) the superoxide radical (Zhao et al., 2013)

Reaction byproducts during chemical oxidation of polycyclic aromatic hydrocarbons were a key issue, which should be taken into consideration. Fenton oxidation of PAHs produce a mixture of ketones, quinones, aldehydes, and car-boxylic acids (Lee et al., 2011), permanganate oxidation mainly generate three types of products, i.e., aromatic ketones, aromatic quinones, and aliphatic or aromatic acids (Forsey et al., 2011), and a large number of short-chain alkanes, aldehydes, and monocarboxylic acids formed during ozone oxidation (Liang et al., 2009). 1H-phenalen-1-one, 9H-fluoren-9-one, and 1, 8-naphthalic anhydride were produced during persulfate oxidation of PAHs, and can be further oxidized (Liao et al., 2014). Persulfate oxidation imposed limited environmental risk related to the reaction byproducts.

Injection wells (with or without recirculation) was frequently used for ISCO, and specifically designed injection equipment may be beneficial for oxidant delivery. Vehicle mounted and fixed oxidant injection equipments for in situ chemical oxidation were firstly developed by our research team and patented (Fig. 2). The equipments were applied in several remediation cases and successfully improved the efficiency, which was a milestone of China's contaminated sites remediation.

Fig. 2 Vehicle mounted and fixed oxidant injection equipment and its operating interface

2.2 Soil vapor extraction

Volatile organic compounds (VOCs) such as BTEX are common organic pollutants in industrial contaminated sites. Qualitative and quantitative detection of VOCs pollutants is a prerequisite for accurate assessment of site environmental risks and the development of effective remediation schemes. Static headspace and purge-and-trap combined gas chromatography were useful to determine BTEX in the soils (Ma et al., 2011). In recent years soil vapor extraction (SVE) and bioventing (BV) have been used extensively to remove volatile organic compounds (VOCs) from the vadose zone (USEPA, 2007). Removal efficiency and influencing factors of benzene by thermal enhanced soil vapor extraction technology were studied (Li et al., 2014). A series of column experiments were conducted to study volatile contaminant (benzene) removal and process principles using SVE under different thermal conditions. The removal efficiency of benzene with the electrical resistance heating method was 96.5%, which was significantly higher than that with the steam injection method (90.5%, p=0.002) or with the normal SVE method (85.2%, p=0.001). In addition, SVE with electrical resistance heating showed a more stable heating effect than SVE with steam injection in homogeneous sandy soil. Based on existing equilibrium SVE models, we established a one-dimensional non-equilibrium mass transfer model for estimating volatile contaminant removal under thermal enhancement. The improved model can accurately reflect benzene content variations using SVE with electrical resistance heating and provides a theoretical basis for preliminary site remediation assessments using SVE under continuous thermal treatment.

Li et al. (2014) investigated the effect of thermal enhanced soil vapor extraction on benzene removal from sand, loam and clay and the mechanism. Compared to the routine control treatment, the benzene removal rates were improved by 13.1% and 12.3% and the remediation periods were reduced by 75% and 14%, from sand and loam respectively using thermal enhanced SVE. Thermal enhancement decreased the moisture content and increased the soil permeability of clay. On the surface of clay particles, absorption peaks of carboxyl and ethyl disappeared and the content of soil organic substances decreased significantly. Compared to the conventional SVE, the benzene removal rate was improved by 34% in clay soil treated by thermal enhanced SVE. For sand and loam, thermal enhancement could increase the removal rate by promoting the diffusion of benzene in the soil and achieve substantial removal of pollutants in a relatively short period of time. For clay, it could enhance the effect of SVE by reducing the absorption capacity between soil particle surface and contaminant and improving the performance of the gas diffusion in soil by decreasing the moisture content and increasing the soil permeability.

Bioventing has emerged as one of the most cost-effective in situ technologies available to address petroleum hydrocarbon spills (Magalhães et al., 2010). Our research selected a diesel oil contaminated site to conduct microorganism enhanced bioventing for petroleum hydrocarbon pollutants remediation. After thirty days' SVE remediation, the mean concentration of diesel hydrocarbon in soils was lower than National Soil Environmental Standard Ⅱ, with the mean removal rate of 82.12%. SVE showed high ability to remove VOCs. The concentrations of VOCs of the soils reduced from several hundred mg/L to tens of mg/L or even lower after vapor extraction of each day (6h). The pollutants concentration prone to the phenomenon of rebound after a decline due to diffusion and redistribution of VOCs. The remediation efficiency to soil contaminated by diesel was 81.47% after twenty five days of the following bioventing. The mean concentration of diesel in soil decreased to 152.44 mg/kg, a little higher than the value of National Soil Environmental Standard Ⅰ.

On the basis of the key technology parameters and process, we have developed the SVE/BV integrated equipment, which can monitor vacuum degree, temperature, the concentration of pollutants in real-time and automatically record remediation data. Supported by the National 863 Project of "Equipment development and demonstration of soil vapor extraction and remediation technology for volatile organic compounds contaminated sites" undertaken by our group joint Tianjin University and Peking University, a solvent absorption device for exhaust gas has been developed based on the principle of chemical industry. These SVE remediation equipments, that have a wide range of applications and good market development prospects, are helpful for solving the problem of VOCs contaminated sites in China.

2.3 Soil washing

Soil washing is one of the most time-efficient and versatile treatment alternatives for the elimination of organic pollutants and heavy metals from contaminated soils, which can be applied to pilot/full-scale field remediation. A series of studies on soil washing have been conducted, including the screening for the type of eluent and adsorbents, the development of complex reagents, the optimization of the washing process conditions, the establishment of the complete technological system and development of repairing equipment etc.

Washing agent such as acids, surfactants, and chelating agents are often added into eluents to solubilize contaminants. The unique molecular structure of surfactant allows to enhance the water solubility of soil contaminants, especially for the hydrophobic organic compounds. Desorption of polycyclic aromatic hydrocarbons (PAHs) from an abandoned manufactured gas plant (MGP) soil was evaluated using four surfactants. The results showed that only 8 g/L TX100 could remove all types of the sixteen PAHs partly in the MGP soil, and the removal efficiencies of different PAHs ranged from 13% to 77.8% (Chong et al., 2014). A pilot study confirmed that a new type of agent (a mixture of 1.0% rhamnolipids in 2.0 $mol{\cdot}L^{-1}$ HCl) was suitable for treating the heavy metals and arsenic co-contaminated soil from an abandoned mine, with the removal efficiencies of 79.1% Pb, 79.6% Cd, 63.7% Zn and 38.5% As (Liao et al., 2016). Besides, the best washing effect of As is achieved by 10% KH_2PO_4 (data not published).

The remediation of heavy metal-contaminated soil by sieving combined with soil washing was proven to concentrate the pollutants into smaller volumes of soil and could effectively remove metals. Washing mechanism was primarily analyzed through SEM-EDS observations and correlation analysis; the results showed that the leaching regularity of the heavy metals and arsenic was found to be closely related to Fe, Mn, and Ca contents of the soil fractions. Besides, practical remediation parameters were also recommended (Liao et al., 2016). Further, the effects of pH, temperature and washing times on removal for heavy metals and arsenic were evaluate for different particle size fractions. The optimal washing conditions achieved in the study could be acted as a guide for the technology parameters of the remediation project (Li et al., 2016).

One of the environmentalists' pursuits is to develop the new, highly effective and economic technology applied to leachate. In order to avoid secondary pollution on the environment, the selective adsorption of five organic wastewater treatment materials (activated carbon, diatomite, resin, bentonite and zeolite) was compared for recovering and reusing surfactants from leachate. As a result, the selectivity value of activated carbon was the highest in the five adsorbing materials, with the Selective adsorption coefficient of 109. In the view point of environmental protection, pure activated carbon was suggested to apply to the organic wastewater remediation project (data not published). In order to select the best adsorbent for arsenic, the arsenic adsorption efficiency and the water absorption character of eight materials (iron powder, iron oxide, activated carbon, bentonite and their mixtures) were studied. The results showed that iron oxide and iron oxide+activated carbon (1:1) had the best arsenic removal effects, and their removal rate all reached 100%. While the iron oxide+activated carbon (1:1) was recommended as adsorbent when we took both effect and cost into account (Wang et al., 2013). It is very interesting that we have synthesized a new adsorbent from straw extracting which presents high-efficiency, low cost and environmentally friendly. The adsorption efficiency of heavy metals can reach as high as 95%.

A complete technological system of ex-situ soil washing has been built which can provide new methods for soil remediation. What's more, a repairing method and device of arsenic-polluted soil and waste was established (Fig. 3). In order to turn the sci-tech fruits into real productivities, we have developed the first soil/tailings washing equipment, which was carrying out the experiments on soil remediation in Hunan province, with a daily contaminated soil/tailings capacity of 3 tons. The equipment has high degree of autonomy and modularized, consisting of sieving module, washing module, wastewater treatment module, sludge disposal module and electrical control system. The equipment can achieve a quick and relatively thorough method of dealing with soil pollution and simultaneously provide the theory and scientific parameters for practical engineering.

Fig. 3　The remediation equipment for soil washing

2.4　Solidification/stabilization

Solidification/stabilization (S/S) is a widely-used remediation technology for treatment of contaminated soils at home and abroad, which is a process of blending treatment reagents into contaminated soils to impart physical and/or chemical changes that result in reduced environmental impact of the contaminants to groundwater and/or surface water (USEPA, 2011). Our team carried out lots of studies on the development of novel, economic, efficient, risk-limited S/S agents for industrial sites and farmlands, and clarification of the remediation effects and mechanism mainly on heavy metal contaminated soils, but also including heavy metals and organic pollutants co-contaminated soils. Moreover, a series of site S/S remediation demostration projects were carried out to identify its effectiveness in the field.

(1) Solidification/stabilization agents and its effectiveness

Several novel S/S agents were developed for the remediation of heavy metal contaminated soils, such as FMBO, FC, which were mainly iron-based materials. FMBO could be an effective stabilization agent to remediate arsenic and heavy metal contaminated soils. In the adsorption tests, the results showed FMBO was more effective for both As (Ⅴ) and As (Ⅲ) removal, particularly for As (Ⅲ) because of its oxidation. The maximum adsorption capacities calculated by Langmuir model were 80.6 mg/g for As (Ⅴ) and 129.9 mg/g for As (Ⅲ) on FMBO, much better than Fe_2O_3 and Nano-Fe_2O_3 (Fei et al., 2015). In the culture experiments, the results showed that FMBO could significantly stabilize As and Pb in the soils, which could decrease 95.2%-100% of TCLP-extractable As and 95.5%-97.5% of TCLP-extractable Pb after adding FMBO under 5% mass percentage dosage (Fig. 4). Meanwhile, it did not lead to the activation of other heavy metals, such as cadmium, zinc and copper (Fei et al., 2016). In the

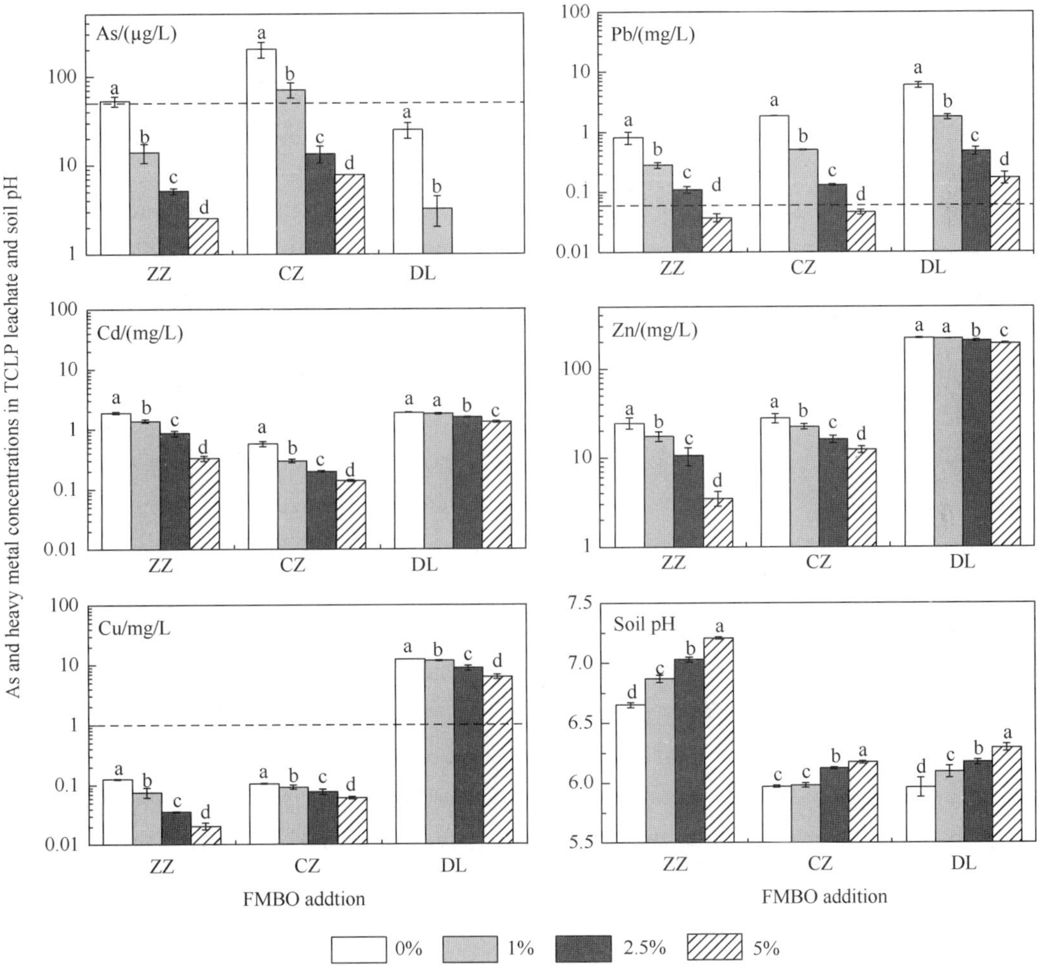

Fig 4. As and heavy metal concentrations in TCLP leachate and soil pH change after adding FMBO in different soils

sequential extraction tests, the results showed that FMBO could transform non-specially sorbed and specially sorbed As mainly to amorphous and poorly-crystalline ones (Fei et al., 2015), while acid exchangeable Pb mainly to reducible fractions (Fei et al., 2016). FC was one of the best agents to stabilize As-contaminated soils from Dalian, Chenzhou and Wenshan, which could decrease TCLP and available As concentrations of soils by 84.07%-98.26% and 12.30%-31.78%, respectively. The results also suggested that FC showed small impact on soil pH and significantly reduced the soil available P concentration (still higher than the high yield fertility level standard 25 mg/kg) (data not published).

(2) Solidification/stabilization mechanism of transformation and transport

Microscopic characteristics of reaction products and transport processes were investigated after/or during S/S treatment. From the analysis of XRD, FTIR, FMBO stabilized As mainly through adsorption by surface hydroxyl groups, resulting in the formation of inner-sphere complexes (Fei et al., 2015, 2016; Komarek et al., 2013). While to Pb, Cd, Zn, Cu, the stabilization achieved through multiple processes, including adsorption and precipitation (Fei et al., 2016). To FC, the FTIR and XPS analysis revealed that Fe-OH groups were the main binding sites for As adsorption, and imbedding of Ce made the surface hydroxyl groups rise to 45.86% (data not published).

Soil pore is one of the most important locations for the reaction between iron-based materials and As. In our recent research, a silicon-based pillar array microfluidic observation platform was established to study the migration mechanism of the FC colloidal particles inside the porous media. The results showed that the FC colloidal particles' migration and the sedimentation inside the microfluidic chamber were significantly affected by the pore size. The FC colloidal particles could migrate into about 57.02% surface area in the large porosity region, only 23.27% in the small porosity region. Benefit from the LIBS technology's in-situ capabilities, the similar distributions of Fe and As had been detected in the large porosity region.

(3) S/S remediation at a heavy metal contaminated industrial site in Gansu Province

A series of site remediation demonstration projects were carried out to identify S/S effectiveness in the field in Hunan, Guangdong, Gansu Province and so on, which were seriously polluted by various heavy metals. For example, with the support of National Natural Science Foundation of China (No. 2013AA06A206), one S/S remediation of heavy metals contaminated industrial site was implemented in Gansu Province (Fig. 5). The soil was polluted heavily by lead, zinc, cadmium mainly, due to the relatively backward technology of mining, dressing and smelting in history. P-based materials and S-based materials were used, and in-situ and ex-situ technologies were applied in the project for comparison. The S/S effects were good, for most soil samples tested by leaching method of GB5085.3—2007 reached the standards of heavy metals.

Besides the remediation projects, we explored the re-use methods of polluted soils after adding S/S agents to form cement, baking-free bricks, roadbed fillers, ceramsite, etc. In our recent work, we successfully developed the production technologies of manufacture, test, and application to make baking-free bricks and ceramsite.

Fig. 5 Remediation project of heavy metal contaminated industrial site in Gansu Province

2.5 Phytoremediation

(1) Assisted phytoremediation by Arsenate Reducing Bacteria

Pteris vitatta L., the first known and the most important arsenic hyperaccumulator, has been widely recognized as the effective material for arsenic contaminated soil remediation. Generally, rate of plant biomass production and high arsenic accumulation in the fronds are two key aspects for successful application of phytoextraction. In addition to the remarkable ability of *P. vittata* to tolerate high internal arsenic, its extraction of low levels of arsenate from soil is extraordinary, considering that arsenate mobility in soil is limited. It is necessary to develop phytoremediation strategies of promoting the growth of *P. vittata* and increasing its ability to remove As from contaminated soil.

It is interesting to apply plants combined to some microorganisms to increase the efficiency of contaminants extraction. The metabolism of microorganisms influences the biogeo-chemical cycle of As, affecting both its speciation and toxicity, and may facilitate plant growth either indirectly or directly, thus affecting the capacity of phytoextraction by plants. Recently, efforts have been made to isolate rhizobacterial associated with *P. vittata*, specifically, arsenic solubilizing bacteria, which can potentially enhance phytoremediation through improving metal removal rate and increasing harvested biomass. Yang et al. (2012) identified and purified a group of the As-reducing bacteria associated with *P. vittata*, including *Streptomyces lividans PSQ*, *Comamonas* sp. *Ts37*, *Agrobacterium* sp. *C13*, *Streptomyces lividans shf2*, etc. A pot experiment was carried out to demonstrate that these isolated arsenic-reducing rhizobacterium significantly enhanced fern growth and soil arsenic phytoremediation efficiency with arsenic accumulation of fern increased by 88%-206% compared to the non-bacterial inoculation treatment (Zhao et al., 2010). Soil rhizosphere characteristics of species of *Streptomyces lividans PSQ* As transformation ability to improve plant growth and As uptake via solubilizing As from soil was further determined (Wen et al., 2013). PSQ can effectively promote reduction of As (V) to As (III), which significantly reduced the content of arsenic in residual fraction from 48.15 mg/kg to 28.75 mg/kg. Inoculation with PSQ significantly influenced the rhizosphere environment. Specifically, pH and DOC concentrations increased, and the transformation of arsenic species were promoted to enhance the bioavailability. The results permitted some insight into the mechanisms by which the arsenate reducing bacteria enhanced the effectiveness of *P. vittata* to remove As from the polluted soil.

(2) Remediation potential for As-PAHs co-contamination by *Pteris vittata* L. and PAH-degrading bacteria combination

The simultaneous accumulation of arsenic and polycyclic aromatic hydrocarbons (PAHs) are frequently found and evaluated in several types of anthropogenic industry contaminated sites, such as coking & chemical industry site, mining and metallurgy industry site, lumber and wood production site, etc. (Zhu et al., 2012). However, methods for remediation of arsenic and PAHs co-contamination are scarce to report.

It is proposed that the hyperaccumulation of arsenic, together with the enhanced PAHs dissipation in rhizosphere to a certain extent, is supposed to be an ideal remediation approach for arsenic and PAHs co-contamination. Thus, firstly, whether *P. vitatta* would be a potential remediation material for arsenic and PAHs co-contamination is deserved to discuss. According to the industry field survey, Sun et al. (2014) demonstrated that endemic plant species of *P. vitatta* growing at the industrial sites can colonize and survive in soils with multiple heavy metals and PAHs. Although spatial distributions of the heavy metal and PAHs varied greatly among industrial sites (arsenic and PAHs reaching 228 and 34 mg/kg, respectively), it is possible to conclude that the species *P. vitatta* have potential in co-exposure tolerating and accumulating arsenic (1276 mg/kg dry biomass) and PAHs. The study of interactions of arsenic and phenanthrene on plant uptake and antioxidative response of *P. vitatta* were further in pot experiment to evaluate the key process. The co-existence of arsenic and phenanthrene had little impact on plant arsenic accumulation, although synergistic effect on antioxidants and arsenic transformation in plant was observed. *P. vitatta* was identified with the high ability to tolerate arsenic and PAHs co-exposure, accumulate considerable arsenic and enhance PAHs dissipation, suggesting the application potential of *P. vitatta* in phytoremediation of arsenic and PAHs co-contamination (Sun et al., 2011).

In order to alleviate the inhibition of co-exposure to *P. vitatta* growth and improve the remediation efficiency of PAHs, PAHs-degrading bacterial with the ability of facilitating the low water-solubility of PAHs degradation was inoculated in *P. vitatta* rhizosphere to evaluate the remediation efficiency by the Liao's group. The selected PAHs-degrading bacterial was isolated from high level of arsenic and PAHs co-contaminated soil and tolerant to the arsenic content of 4000 mg/L. The interaction between *P. vitatta* ecotypes and PAHs degrading bacteria in plant tolerance capacity and arsenic and PAHs speciation and bioavailability in the rhizosphere was also discussed (data not published). It was found that inoculation of bacterial significantly promoted the rhizosphere metal oxide bound arsenic desorption, and stimulated the plant growth and antioxidative activity for *P. vitatta* of Guangxi ecotypes, thus significantly enhanced the plant arsenic accumulation capacity; However, significantly reduced the arsenic accumulation in *P. vitatta* of Hunan ecotypes, which may be ascribed to the detoxification mechanisms of the bacteria to reduce the arsenic exposure for plant. In addition, inoculation of bacteria significantly stimulated the phenanthrene dissipation in plant rhizosphere, especially in the treatment of Guangxi ecotypes of *P. vitatta*. The results of the research would enrich the basic information for arsenic and PAHs co-contamination and provide theoretical significance for the co-contamination remediation technology development.

3. Conclusions

Our teams have developed a series of key remediation technologies and equipment. Besides, a decision support system (DSS) for remediation of contaminated sites was designed to select the optimal remediation technologies and formulating economic and efficient remediation plans based on site investigation and risk assessment. However, considering the complexity of the characteristics of site pollution, we begin to explore the integration of multiple technologies. We look forward to the remediation technologies can be more promoted and applied in industrial contaminated sites in China.

Acknowledgements

This study was financed by the Natural Science Foundation of China (41271339, 40771184, 41571309, 40801205), the National 863 Program (2009AA063102, 2013AA06A206, 2012AA06A201), Beijing Science and Technology Program (D08040900360803, D161100004716001, Z141100000914011, Z121109003112077), National Science and Technology Support Program (2012BAC10B04).

References

Chong Z Y, Liao X Y, Yan X L, et al. 2014. Enhanced desorption of PAHs from manufactured gas plant soils using different types of surfactants. Pedosphere, 24 (2): 209-219.

Fei Y, Yan X L, Liao X Y, et al. 2015. Effects of iron-based oxides on arsenic stabilization in soils of different water contents. Acta Scientiae Circumstantiae. 35 (10): 3252-3260. (In Chinese)

Fei Y, Yan X L, Liao X Y, et al. 2016. Stabilization effects and mechanisms of Fe-Mn binary oxide on arsenic and heavy metal co-contaminated Soils. Acta Scientiae Circumstantiae, in press. (In Chinese)

Forsey S P, Thomson N R,Barker J F.S.P. Forsey, N.R. Thomson, J.F. Barker, 2010. Oxidation kinetics of polycyclic aromatic hydrocarbons by permanganate. Chemosphere, 79 (6): 628-636.

Lee B D, Iso M, Hosomi M.B.D. Lee, M. Iso, M. Hosomi. 2001. Prediction of Fenton oxidation positions in poly-cyclic aromatic hydrocarbons by frontier electron density. Chemosphere, 42 (4): 431-435.

Li P, Liao X Y, Yan X L, et al. 2014. Effect of thermal enhanced soil vapor extraction on benzene removal in different soil textures. Environmental Science, 35 (10): 3888-3895. (In Chinese)

Li P, Liao X Y, Yan X L, et al. 2014. Mass transfer model of soil vapor extraction under thermal treatment for removing a volatile contaminant. Environmental Engineering and Management Journal, In press.

Li Y, Liao X Y, Yan X L, et al. 2015. The study of optimization technological conditions of washing remediation for heavy metal contaminated soil with rhamnolipid. Journal of Agro-environment Science, 34 (7): 1287-1292. (In Chinese)

Liang Y T, J.D.V. Nostrand, Wang J, et al. 2009. Microarray-based functional gene analysis of soil microbial communities during ozonation and biodegradation of crude oil. Chemosphere, 75 (2): 193-199.

Liao X Y, Chen T B, Xiao X Y, et al. 2007. Selecting appropriate forms of nitrogen fertilizer to enhance soil arsenic removal by *Pteris vittata*: a A new approach in phytoremediation. International journal of phytoremediation, 9 (4): 269-280.

Liao X Y, Chen T B, Xie H, et al. 2005. Soil As contamination and its risk assessment in areas near the industrial districts of Chenzhou city, Southern China. Environment International, 31 (6): 791-798.

Liao X Y, Chen T B, Yan X L, et al. 2008. Effects of different forms of P fertilizers on phytoremediation for As-contaminated soils using As-hyperaccumulator *Pteris vittata* L. Environmental Science, 29 (10): 2906-2911. (In Chinese)

Liao X Y, Chong Z Y, Yan X L, et al. 2011. Urban industrial contaminated sites: a new issue in the field of environmental remediation in China. Environmental Science, 32 (3): 784-794. (In Chinese)

Liao X Y, Li Y, Yan X L. 2016. Removal of heavy metals and arsenic from a co-contaminated soil by sieving combined with washing process. Journal of Environmental Sciences, 41 (3): 202-210.

Liao X Y, Ma D, Yan X L, et al. 2013. Distribution pattern of polycyclic aromatic hydrocarbons in particle-size fractions of coking plant soils from different depth. Environmental Geochemistry and Health, 35 (3): 271-282.

Liao X Y, Tao H, Yan X L, et al. 2014. Discussion on several key points of decision support system for remediation of contaminated sites. Environmental Science, 35 (4): 1576-1585. (In Chinese)

Liao X Y, Yan X L, Wang Y Z, et al. 2012. Environmental Risk Presented by Arsenic Contamination of Building and Facility Surfaces in a Coking Plant. Bulletin of Environmental Contamination and Toxicology, 88 (6): 915-921.

Liao X Y, Zhao D, Yan X L, et al. 2011. Determination of potassium permanganate demand variation with depth for oxidation-remediation of soils from a PAHs-contaminated coking plant. Journal of Hazardous Materials, 193 (20): 164-170.

Liao X Y, Zhao D, Yan X L, et al. 2014. Identification of persulfate oxidation products of polycyclic aromatic hydrocarbon during remediation of contaminated soil. Journal of Hazardous Materials, 276 (9): 26-34.

Liu Q X, Yan X L, Liao X Y, et al. 2015. Effects of Soil Moisture on Phytoremediation of As-contaminated Soils Using As-Hyperaccumulator *Pteris vittata* L. Environmental Science, 36 (8): 3056-3061. (In Chinese)

Magalhães S M C, Ferreira Jorge R M, Castro P M L. 2009. Investigations into the application of a combination of bioventing and biotrickling filter technologies for soil decontamination processes—a transition regime between bioventing and soil vapor extraction, Journal of Hazardous Materials. 170 (2-3): 711-715.

Sun L, Liao X Y, Yan X L, et al. 2014. Evaluation of heavy metal and polycyclic aromatic hydrocarbons accumulation in plants from typical industrial sites: potential candidate in phytoremediation for co-contamination. Environmental Science and Pollution Research, 21 (21): 12494-12504.

Sun L, Yan X L, Liao X Y, et al. Ferreira Jorge, R M,Castro P M L.S.M.C. Magalhães, R.M. Ferreira Jorge, P.M.L. Castro phenanthrene on their uptake and antioxidative response in *Pteris vittata* L. Environmental Pollution, 159 (12): 3398-3405.

Tao H, Liao X Y, Yan X L, et al. 2014. Uncertainty analysis and pollution volumetric calculation of soil BaP contents in a contaminated site. Geographical Research. 33 (10): 1857-1865. (In Chinese)

USEPA. 2007. Treatment technologies for site cleanup: annual status report (Twelfth Edition) (EPA-542-R-07-012). Washington DC: Office of Solid Waste and Emergency Response.

Wang J Y, Zhang Y Z, Yan X L, et al. 2013. Removal efficiencies of different adsorption materials to arsenic in soil leachate. Hunan Agricultural Sciences. 21 (09): 46-49. (In Chinese)

Wen Y, Liao X Y, Yan X L. 2013. Arsenic-resistance of *Streptomyces* sp. and its effects on arsenic enrichment of *Pteris vittata* L. Asian Journal of Ecotoxicology, 8 (2): 186-193.

Yan X L, Lin L Y, Liao X Y, et al. 2012. Arsenic accumulation and resistance mechanism in Panax notoginseng, a traditional rare medicinal herb. Chemosphere, 87 (1): 31-36.

Yan X L, Lin L Y, Liao X Y, et al. 2013. Arsenic stabilization by zero-valent iron, bauxite residue, and zeolite at a contaminated site planting Panax notoginseng. Chemosphere, 93: 661-667.

Yan X L, Zhang M, Liao X Y, et al. 2012. Influence of amendments on soil arsenic fractionation and phytoavailability by *Pteris vittata* L. Chemosphere, 88 (2): 240-244.

Yang Q, Tu S X, Wang G, et al. 2012. Effectiveness of applying arsenate reducing bacteria to enhance arsenic removal from polluted soils by *Pteris vittata* L. International Journal of Phytoremediation, 14 (1): 89-99.

Zhai L M, Liao X Y, Chen T B, et al. 2008. Regional assessment of cadmium pollution in agricultural lands and the potential health risk related to intensive mining activities: A case study in Chenzhou city, China. Journal of Environmental Sciences, 20 (6): 696-703.

Zhao D, Liao X Y, Yan X L, et al. 2011. Chemical oxidants for remediation of soils contaminated with polycyclic aromatic hydrocarbons at a coking site. Environmental Science, 32 (3): 857-863. (In Chinese)

Zhao D, Liao X Y, Yan X L, et al. 2013. Effect and mechanism of persulfate activated by different methods for PAHs removal. Journal of Hazardous Materials, 254-255: 228-235.

Zhao D, Yan X L, Liao X Y, et al. 2011. Chemical Oxidants for Remediation of BTEX-contaminated Soils at Coking Sites. Environmental Science, 32 (3): 849-856. (In Chinese)

Zhao G C, Liao X Y, Yan X L, et al.2010. Enhancement of As-accumulation by *Pteris vittata* L. affected by microorganisms. Environmental Science, 31 (2): 431-436. (In Chinese)

Zhu G H, Sun L, Liao X Y, et al.2012. Combined pollution of heavy metals and PAHs and its risk assessment in industrial sites of Chenzhou city. Geographical Research, 31 (5): 831-839. (In Chinese)

Application and Case Study of Barrier Technology in Soil and Groundwater Remediation

Li WEI, Zhengyong LV, Shucai LI, Guojie FENG, Jingwen LI, Kun SHEN, Zhu MIAO, Hudi ZHU

Beijing GeoEnviron Engineering and Technology

1. Barrier Technology

1.1 Introduction

Barrier technology is a remediation method which use impermeable barriers to isolate contaminated soil and surrounding environment. Transportation and diffusion of pollutants to the surroundings can be cut off through impermeable landfills or impermeable barrier layers, which could further avoid human body and surrounding environment injury.

Barrier technology includes: ①Vertical Barrier System, ②Bottom Barrier System, and ③ Surface Barrier System.

1.1.1 Vertical barrier system

The vertical barrier system could prevent the horizontal migration of pollutants and their leachate, therefore isolate contaminate matter from external surface water and groundwater infiltration/inflow.

Hydraulic control has used the barrier system with Construction Engineering of Common Underground Continuous Wall technology (such as deep foundation pit, water conservancy project curtain grouting, concrete continuous wall, mud wall, steel pile, etc.).

However, there are still some problems with the construction of physical enclosure technology. For instance, the defects of connection between wall units and the existence of construction joints/deformation joints, which could lead to a discontinuity, low deformation adaptability of site settlement, high permeability coefficient, and low resistance to chemical corrosion. All of these defects limit the application of geotechnical membrane in environmental engineering field.

High density polyethylene (HDPE) vertical barrier technology can be a suitable solution to the above problems. As a flexible polymer composite material, HDPE membrane has a high tensile strength, elongation rate (>700%), and tear resistance, and extremely low penetration coefficient ($k<1\times10^{-13}$ cm/s) (Zhen, 2012).

HDPE membrane is non-toxic and harmless to the environment, with no pollution to water bodies and other contact medium. It is resistance to most of acidic, alkaline, salty, and oil chemical corrosion. HDPE membrane also has the ability to endure long-term bioerosion and sunniness without aging. The lifetime of HDPE membranes can be as much as 100 years.

The vertical barrier system of HDPE membrane, which is 2-7 m in width, is a homogeneous and continuous barrier system which installed underground. Two adjacent membranes are connected by a connection lock, and are installed by professional installation equipment. According to the engineering geological conditions of the site, the system can be installed by penetration installation (Gund wall) or slotted installation (curtain wall).

1.1.2 Bottom barrier system

In the lower layer of the pollution source, a natural and continuous low permeable layer (or weak permeability layer) is used as the bottom barrier system to prevent the diffusion of contaminate to the environment. This low permeable layer can be a natural clay layer or a weak weathered rock layer. If there is no low permeable layer or weak permeable layer, the construction of artificial barrier layer can be reached by laying the impervious material.

The artificial barrier layer could be constructed by the laying of HDPE membrane, clay, impermeable concrete materials, etc.

1.1.3 Surface barrier/cover system

In order to lower the environmental risk, prevent the migration of pollution to the ground surface, surface barrier system is constructed on the ground surface of the contaminated region.

The system is generally composed of ecological vegetable layer, ecological soil layer, ecological support layer, rainwater collection and drainage system, horizontal barrier layer, gas collection and guide system, and biological treatment system, etc.

According to the different materials used in the barrier layer, the surface barrier system can be divided into clay barrier/cover system, HDPE film barrier/cover system, bentonite blanket barrier/cover system.

1.2 Construction method

1.2.1 Penetration installation (Gund Wall)

Vibration hammer (Hammer) and auxiliary rigid board are mostly used during penetration installation. Based on the designed depth of the barrier system and other requirements, membranes are inserted directly into the required low permeable aquifer by the force of the vibration hammer and driven force of the rigid board. Two adjacent membranes are connected by the patented male and female wedge connection lock.

Penetration installation is suitable for soft soil and loose sand soil site conditions, the depth of penetration could generally reach 15 meters below ground surface.

1.2.2 Slotted installation (Curtain Wall)

Slotted installation is an installation technology of HDPE membrane wall in deep groove which has been excavated.

Slotted installation equipment generally consists of a slot machine, diaphragm support frame, lifting equipment and trench backfill equipment. With slurry wall and other protective measures, the equipment can slot the designed depth and reach the impermeable layer; membranes are fixed on the support frame of a certain rigidity and sent to the bottom of the groove by lifting equipment. Two adjacent membranes are connected by patented multi tooth interlocking device. Sand, clay, bentonite or concrete can be used in accordance with the engineering requirements.

Slotted systems and their installation techniques are applicable to any soil conditions, including clay, sand, gravel, and rock. Depth of slotted system can be up to 100 meters below ground surface.

2. Case study-Barrier system

2.1 Barrier system for Zijin Mining leachate leaking accident

Due to the continuous heavy rainfall, rainwater accumulated in Zijin Mining Area, Shanghang County, Fujian Province, China, in July 2010. With the increasing water pressure, the impermeable membrane of the leachate pond broke, resulting huge amount of leachate water leaked into the Tingjiang River. The acidic wastewater containing copper caused serious pollution to the surrounding soil, surface water and groundwater.

After comprehensive consideration of the local hydrological environment characteristic, the pollution situation, the production process of wet smelting system and the safety factors of the production process, a three-dimensional barrier system which combined bottom barrier and vertical barrier was used to protect the surrounding environment.

2.1.1 Horizontal flexible barrier system

1) Wet heap leaching field-Control from pollution source

A horizontal flexible barrier system was constructed in flat and firm ground with selected geotextile materials such as HDPE membrane and clay. The system can effectively control the wet heap leaching of copper ore, avoid the copper-contained acid leakage during mining process and improve the efficiency of copper recovery.

2) Solution pool-Control of transportation process

HDPE membrane and other geotechnical materials were used to build a horizontal flexible barrier to ensure safe delivery and storage of metal liquid, to reduce the leak of metal liquid and contamination of soil and groundwater.

3) Flood control pool-Emergency control

To improve flood control ability, a large flood control pool with horizontal flexible barrier was constructed using HDPE membrane. In case of emergency, the copper containing solution can be temporarily stored in the flood control pool, so as to avoid the production system for overload operation and to avoid possible leakage of the acid liquid.

2.1.2 Vertical barrier system

In order to control the sewage and possible emergent situation, two vertical flexible barriers were set ①between the copper heap leaching field and the flood control regulation pool; ②between the flood control pool and Dingjiang river.

HDPE membrane was the main barrier material for the vertical flexible barrier. The vertical barrier system was constructed into the underground until micro weathered rock layer (rock is complete, hard, permeability, low permeability coefficient $k \leqslant 10^{-7}$ cm/s). The scientific located epoxy concrete (high strength acid and corrosion resistance) was used to connect the vertical barrier and rock layers. With the use of HDPE membrane, rock layer, and epoxy concrete, the environmental friendly vertical barrier system was built to control the leaking and diffusion of pollutants and acid waste water to the environment (Zheng et al., 2014).

After the soil and groundwater contamination is effectively controlled, soil washing technique using chemical agent is chosen based on the local rainfall characteristics.

To achieve the effective pollution control and circulation utilization, natural precipitation was used to clean the contaminated soil and reduce the concentration of copper ion in soil. Contaminated groundwater was collected by extraction/monitoring wells and treated afterward. Valuable metals copper was collected during this process.

(a) Vertical barrier construction

(b) Zijin Mining Region

Fig. 1 Remediation of Zijin Mining pollution area using vertical barrier system

2.2 Groundwater barrier system for a slagheap site in Huize, Yunnan

The slagheap of this project belongs to the Huize branch of Yunnan Chihong Xinzhe Company. The site is heavy metal polluted, which greatly influence the local soil, water and gas environment.

To minimize the impact of the surrounding environment, barrier technology was selected as the remediation approach according to the local hydro-geological condition. A 212 meter long groundwater interception zone with 3 flexible vertical barrier and 3 hydraulic vertical wells were constructed on the south side of the slag pile (downstream). Surface intercepting ditch was built with a total length of about 180 m on the north side of the slagheap. On the east side of the slag heap (downstream), the flexible vertical wall of groundwater was built with a total length of about 243 m. The landform of the slag heap was reconstructed and soil was covered for planting.

Fig. 2 Vertical barrier system construction

With the implementation of the barrier system, the stockpile can be recycled with minimized impact to local environment and reduce soil erosion at the same time. The implementation of this project fully reflects the principle of green and sustainable remediation.

3. Case study-Solidification and stabilization technology

3.1 Heavy metal contamination remediation project (Phase one) in Xiawan river in Zhuzhou, Hunan province, China

This project was listed into the first batch of "The National Special Planning Projects of Heavy Metal Contamination Remediation in Xiangjiang River Basin".

The Xiawan river is located in Qingshuitang Industrial Zone in Zhuzhou. The wastewater produced by factories beside the river was unregulated discharged for ages, with non-properly disposed industrial solid wastes. The ambient soil, surface water and groundwater had been heavily polluted by the cadmium, lead, mercury, arsenic and other heavy metals in these wastes. The industrial wastewater was mainly discharged into Xiangjiang River through Xiawan river. The concentration of mercury, cadmium, lead, arsenic and other heavy metals in the sediment exceeded the national applicable standards severely after about sixty years of accumulation and precipitation. As the drinking water security was being threatened, corresponding treatment and remediation were desperately needed.

The remediation process included environmental dredging, solidification & stabilization, barrier isolation and landfill. A total of 38, 360 cubic meter of heavy metal polluted sediment and 11, 720 cubic meter of contaminated shore-side soil had been remediated (Feng, 2013). The implementation of this project was of great significance to improve the water quality of Xiawan river, reduce the heavy metal pollution loaded into Xiangjiang River, improve the water quality of Xiangjiang River from Zhuzhou to Changsha reach, and ensure the drinking water safety in Changsha, Zhuzhou and Xiangtan region.

(a) S/S process (b) Ecological remediation of shore-side contaminated soil

Fig. 3 Heavy metal pollution control and ecological restoration project of Xiawan river

3.2 The integrated treatment project of heavy metal contaminated waste in Qingshuitang industrial zone, in Zhuzhou of Hunan province

This project was listed into first batch of "The National Special Planning Projects of Heavy Metal Contamination Remediation in Xiangjiang River Basin" .

Qingshuitang Industrial Zone (QIZ) was established in 1950 s, with over 130 enterprises operated inside. The major business of QIZ was metallurgy, chemistry and energy industry. Millions tons of industrial waste sediments were generated each year. The ambient environment was contaminated in different degrees by the waste sediments due to historical disorderly disposal. The heavy metal contamination was most critical, which seriously threatened the safety of surrounding ecological environment.

The industrial waste treated in this project were in various type and had complicated composition, in which major pollutants were cadmium, lead, arsenic, mercury, zinc, acid and alkali, etc. The contaminated should be treated separated according to the different pollutants and their corresponding contamination degree, which lead to different usage of reagents. In-situ solidification & stabilization was combined with ex-situ solidification & stabilization and landfill to remediate the contaminated and groundwater and shallow perched water at the same time. About 2 million cubic meter of industrial waste and contaminated soil were properly treated.

This project effectively treated the heavy metal contamination in the industrial zone, which guaranteed the ecological safety, lowered the environmental risk and provided the foundation of transition from old industrial zone to ecological new city.

(a) Contaminated waste sampling

(b) Current situation after treated

Fig. 4 Comprehensive control project of heavy metal contaminated waste in Qingshuitang Industrial Zone

3.3 The Cr polluted soil remediation demonstration project in Yun xian County, Hubei province

In order to ensure the water source safety on the central route of the South-to-North Water

Diversion Project, about 0.16 million cubic meter of heavy metal contaminated soil were treated in this project in Yunxian county in Hubei province. A-level standard of "Standard of Soil Quality Assessment for Exhibition Land" (HJ350-2007) was applied during contaminated soil excavation process, and IV-class water standard of "Environmental Quality Standards for Surface Water" (GB3838—2002) was adopted as that of heavy metal leachate concentration and pH of contaminated soil after remediation. Ex-situ solidification/stabilization technology was used to treat contaminated soil. Sulfur-based and iron-based solidification/stabilization reagents were applied, which were independently developed by the remediation company of this project. About 0.2 million cubic meter remediated soil were transported to the safety landfill constructed for this project for final disposal (Li et al., 2016). The successful implementation of this project provided an important guarantee for the water quality and safety of the reservoir area of the South-to-North Water Diversion Project.

(a) The temporary storage land for the contaminated soil (under construction)

(b) The S/S treatment factory

(c) S/S process

(d) The current situation of after-treated site

Fig. 5 Demonstration project of heavy metal contaminated soil in Hubei Province

References

Feng G J, Wei L, Li S C, et al. 2013. Case study on treatment and remediation of heavy-metal contaminated sediment in a canal in hunan province. Environmental Engineering. 31 (6): 130-133.

Li S, Feng G J, Kong X B, et al. 2016. Case project of solidification/stabilization of heavy metal contaminated soils in Hubei province. Environmental Protection Science. 42 (2): 103-107.

Zhen S L. 2012. Introduction of the technology of river pollution control and ecological restoration. 2012 national river management and ecological restoration technology summary: 168-171.

Zheng Z H, Zhu T G, Luo B, et al. 2014. Application of vertical ecological barrier system in the prevention and control of groundwater pollution in mine. The academic conference of Chinese Society for Environmental Sciences (2014): 4385-4392.

Practice of Green and Sustainable Remediation, Risk-Based Megasite Management in China

Hongzhen ZHANG, Jingqi DONG, Ning SUN, Jinnan WANG, Shunze WU

Department of Environmental Engineering, Chinese Academy for Environmental Planning

1. Integrated risk-based management of river basin megasites

1.1 Investigation and preliminary source tracing analyses for a small contaminated river basin caused by non-ferrous metal mining activities

Non-ferrous metal mining areas are often seriously contaminated in a long history, large areas, and multi-environmental media involved (Moore et al., 1990). To understand the current status of heavy metal contamination and the sources, we investigated in a small river basin where mining activities kept for a long time. The results show that the mining activities have caused metal contaminations in river courses, farmlands, surface water and groundwater, within which As, Pb, Sb and Zn are the dominant pollutants. There are nine potential metal pollution sources, among which the Changchengling mine, groundwater and spring around the junction of Qingtoujiang river and Yuxi river, and the tailings and contaminated soils around the Xialian dressing factories are the three most important explicit sources. Sources of heavy metal pollution in the river basin are complex, including many point and non-point sources, such as tailings, abandoned ores, mining water, groundwater and spring. The areas of Qingtoujiang and Xialian contributed about 60%-75% and 15%-20% to the Sb concentration in the downstream water separately. Since the historical non-point pollution is serious in the region and it is difficult to quantify the Sb contributions of sediments and tailings covered the area to the downstream surface water, the emergency responding measures focusing on the surface water may not be able to solve the problem of increasing Sb concentration thoroughly.

1.2 The construction of a heavy metal pollution alert system in the scales of river basins

Currently, heavy metal pollution events in the scale of river basins occur frequently in China, the watershed heavy metal pollution monitoring and warning capabilities lag behind and high ecological and health risks to the surrounding people are widely posed (Gross et al., 2007). Targeted to the Chinese situation and referring to the heavy metal pollution emergency response experiences of developed countries, a heavy metal pollution alert system in the scales of river

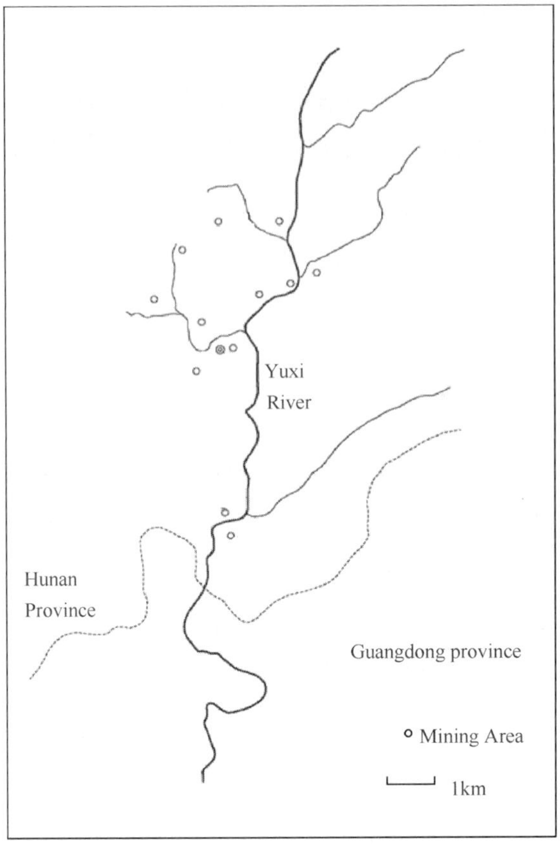

Fig. 1 Contribution of different regions to increasing Sb concentration in the river basin

basins (METALert) was developed. Based on the five key industries defined in the 12th Five-Year Plan of Heavy Metal Pollution Prevention, the system aggregated the methods of hydrological and meteorological monitoring, geographical information system and hydrological and hydraulic modeling to realize the functions of heavy metal pollution source tracking, modeling and forecast, emergency response, spatial visualization and information report. The system acts as a decision support tool for Chinese heavy metal pollution alert and emergency response at watershed scales and could help to improve and complete the current environmental emergency response and warning system in China.

1.3 The nexus between environmental risk-damage assessment of non-ferrous metal mining mega-site and integrated pollution control principle

At present, China is in short of systemic investigation and effective pollution control experiences on typical non-ferrous metal mining mega-site, and there is little research in source identification, inversion and modeling, risk quantification, regional hazardous characterization, integrated regulation and control mechanism regarding meta-sites (Gustavson et al, 2007;

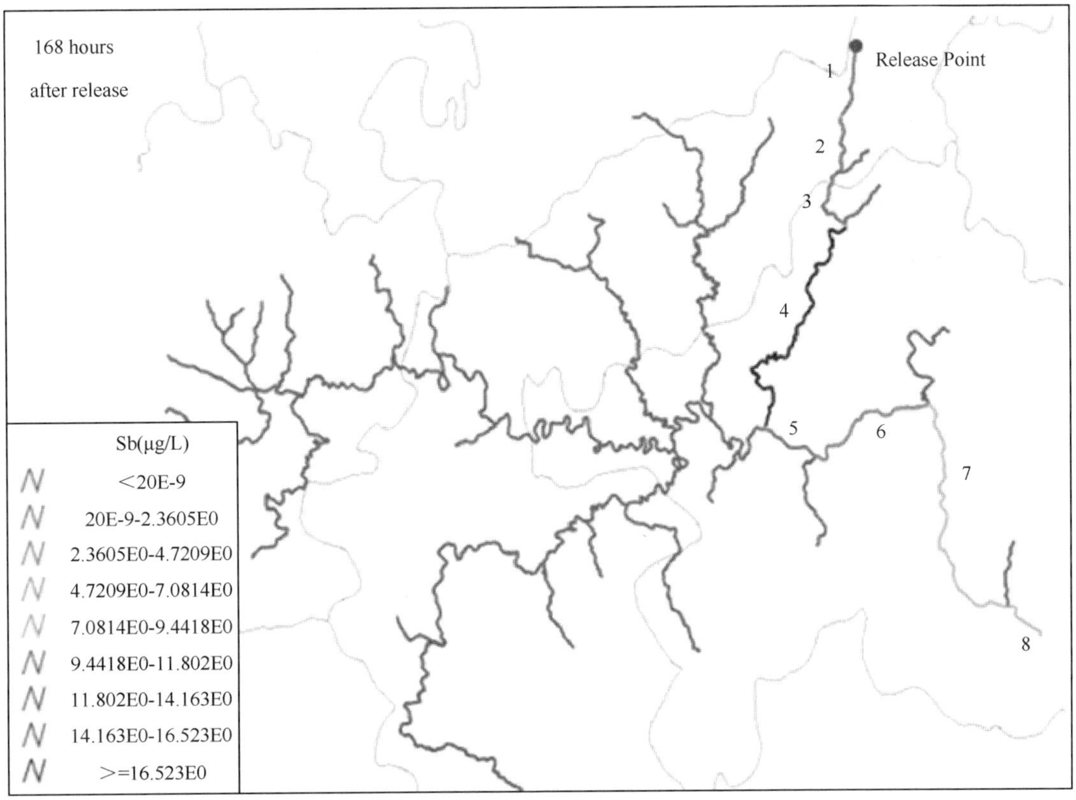

Fig. 2 Demonstration of case study using METALert at a river basin (Drown by VITO, Belgium)

Pascoe et al., 1994). This study focuses on severe environmental risk and damage caused by mega-sites, which often have complex causes, result in wide ranges of investigation, involve multi environmental media, and include secondary pollution sources. Based on detail investigation on the metal contents and speciation of typical mega-sites' soil, groundwater, surface-water, sediment, crops and other media, we will analyze the main sources and their contributions to the environment downstream, simulate contaminants transformation and migration progresses, forecast the temporal and spatial evolution processes of the mega-sites' contaminants. Risk transmission and injury trigger mechanism of contaminants at mega-sites will be analyzed, an integrated conceptual model will be established based on the analyses of nexus among human health risk assessment, ecological risk assessment and environmental damage assessment, and a coordinated quantification technique will also be developed. The purpose of this study is to illuminate the integrated environmental control mechanism which is based on the site remediation and ecological restoration oriented concept using the net environmental benefit analysis method, and to pursue key factors that affecting mega-sites' risk reduction and the effects of pollution control and remediation measurements. It's expected that the results could provide important theoretical basis for the metal-contaminated site control and management at regional levels or in watersheds (Moore and Langner, 2012).

2. Green and sustainable remediation practice in China

2.1 Optimization model for redevelopment of contaminated soils (brownfields) in pilot cities at early planning stage

Targeting the types and the spatial distributions of potential contaminated sites/soils in pilot cities, following the general city planning directions, the factors including contaminated sites/soils remedial technology feasibilities, investment modes, public society concerns and the requirements of City general sustainable development are comprehensively considered (Diamond et al., 1999). The land use scenarios and the sequences of urban brownfields redevelopment are optimized rationally, to guarantee the economic, environmental and social benefit and the urban sustainable developments (Zhang et al., 2010). Optimizing feasible redevelopment scenarios for the brownfields contaminated soils (including land use modes, land redevelopment priorities and launch time). The remediation and redevelopment of contaminated soils in central downtown areas, suburban areas and rural and mining areas are included, mainly focusing on the recognized (potential) contaminated sites or sites could be recognized based on current information and drawn attentions from governments and societies, that may cause environmental problems to sensitive receptors or human health risks/damages. Acontaminated sites inventory, and land redevelopment and reuse mode alternatives will be proposed, by focusing on the key objects, areas and categories. A qualitative method (levels of high, medium and low) is adopted to ranking the treatment/remediation of contaminated sites/soils, taking contaminated sites/soils and the urban redevelopment planning and implementation spots as the single evaluation object, and considering core elements (damage degrees, technical feasibilities of remediation and funding sources) (Foley et al., 2010; Geisler et al., 2014)

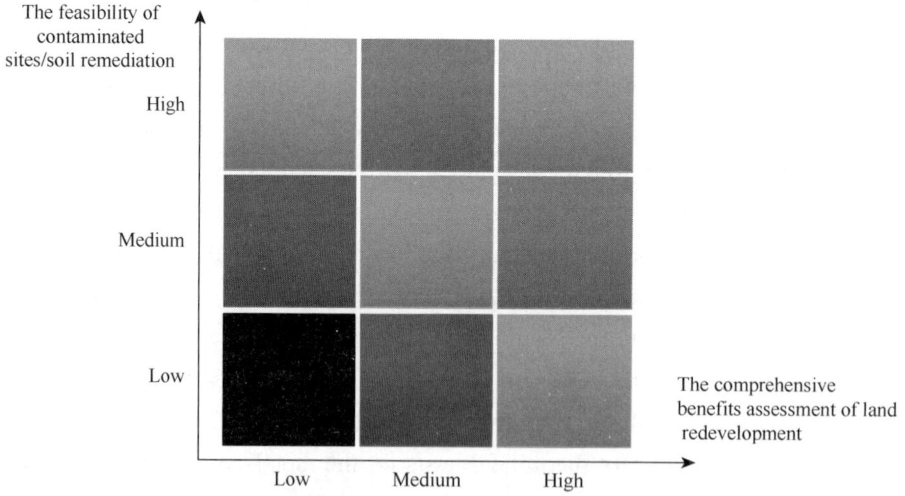

Fig. 3 Concept of ptimization model for redevelopment of contaminated soils

2.2 Cost-benefit analysis model for contaminated sites/soils remediation

For prioritizing the city contaminated sites remediation in the progresses of land redevelopment and providing decision supports for land planning. The Cost-Benefit Analysis (CBA) is applied on typical gathering zones of contaminated sites in pilot cities (same or similar contaminated industrial parks, industrial accumulation zones, etc.) ; alternatively, on the contaminated sites as urban redevelopment patches in spatial scale (Garcao et al., 2015). By listing related necessary information and references for each alternative (remediation + redevelopment land use types), the items of cost, benefit and others (which includes the relative important items but difficult to be monetized) are sort out. Then, all the Sub-COST and Sub-Benefit items are determined and the relative importance is assessed, following with the calculations of monetization or normalization by setting a suitable discount rate. At last, the references to decision support are provided after analyzing the uncertainties of the assessment results. The above factors are monetized by appropriate methods, keeping consistence of the units and picking rational discount rate, to calculate the overall costs and benefits after monetization. By combining with the qualitative evaluation results of non-monetary factors, the uncertainty of CBA assessment results could be analyzed.

2.3 Study on coupling method of Life Cycle Impact Assessment (LCIA) and common risk-based control for contaminated site management

Problems of contaminated sites are becoming prominent over the world. With the fact that contaminated sites remediation methods in China are primarily ex-situ remediation with high energy consumption and adverse effects to the environment, sustainable management strategies are highly deficient (Kennedy and Cheong, 2013). The LCA-based researches are hot topics at present, i.e. the selection of remediation alternatives for contaminated sites, the impact assessment on remedial activities and the standard nalyses on the impacts of remediation technologies, being used to assess comprehensive effect of the risk-based controls. However, problems exist during LCIA, the most important phase of LCA, such as lack of unified assessment standards, generalized assessing models and high uncertainties of assessment results. With the assistance of statistical data, on-site tracking investigations, visiting and interviews, questionnaire surveys and monitoring, the coupling model based on regular risk management approaches, such as the pollution dispersion and transportation, risk assessment and site remediation scenario simulations, combining with site remediation LCIA is attempted to be established. The sensitivities of key factors and parameters of the LCIA coupling model's primary and secondary impacts are evaluated using Fourier amplitude sensitivity test. The approach using Markov-Chain Monte Carlo Method to quantify the uncertainties of LCIA results is proposed. A Cr contaminated site which is currently in the progress of pilot remediation, is selected as an example applied with the LCA method. The research is supposed to contribute practically and methodologically to the quantitative analysis on environment and health impacts by

2.4 Environmental damage assessment case study of a cooking plant contaminated site

Soil and groundwater contamination is becoming worse in China, it's of emergency to establish environmental damage assessment (EDA) methods based on equivalency analysis and site remediation cost (Gastineau and Taugourdeau, 2014; Söderqvist et al., 2015). Taking a cooking plant site contaminated by Polycyclic Aromatic Hydrocarbons (PAHs), benzene as example, soil and groundwater contamination damage assessment was carried out using Resources equivalency analysis (REA) methods, and uncertainty of results were also analyzed. The results show that, there is significant difference between the damage assessment results, which is 2.71 billion RMB, and actual site remediation cost, which is 2.05 billion RMB. The former is evidently larger than later. Social discount rate influences the soil contamination damage assessment result most, and for groundwater, the most important factors are thickness of contaminated groundwater layer, porosity of aquifer, social discount rate and remediation cost per unit groundwater. ERA can be practically used to assess soil and groundwater contamination damage at contaminated industrial site, with a strong theoretically and technically basis. But we need Refinement of legislation and specific technical guideline to standard the EDA content, process and method.

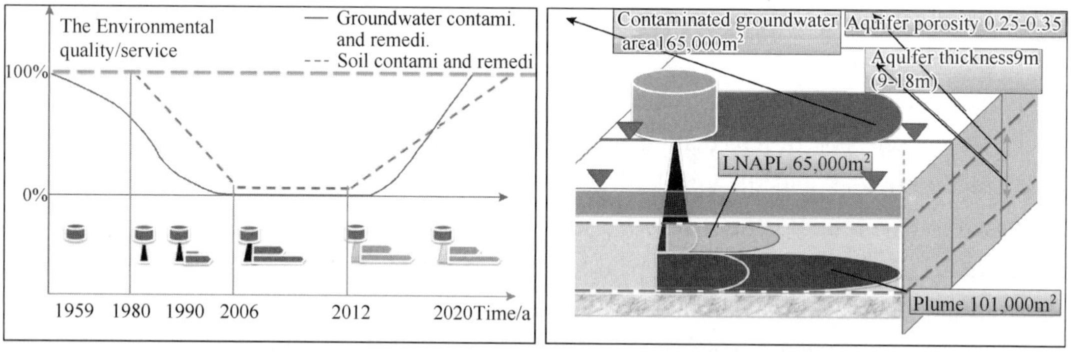

(a) The contamination process　　(b) The scheme of groundwater contamination

Fig. 4　Contaminated site concept model of the cooking plant case

References

Diamond M L, Page C A, Campbell M, et al. 1999. Life-cycle framework for assessment of site remediation options: method and generic survey. Environmental Toxicology and Chemistry, 18 (4): 788-800.

Foley J M, Rozendal R A, Hertle C K, et al. 2010. Life cycle assessment of high-rate anaerobic treatment, microbial fuel cells, and microbial electrolysis cells. Environmental Science & Technology, 44 (9): 3629-3637.

Garcao R. 2015. Assessment of alternatives of urban brownfield redevelopment-Application of the SCORE tool in early planning stages. Master's Thesis, Chalmers University of Technology, Sweden.

Gastineau P, Taugourdeau E. 2014. Compensating for environmental damages. Ecological Economics, 97: 150-161.

Geisler G, Hellweg S, Liechti S, et al. 2004. Variability assessment of groundwater exposure to pesticides and its consideration in life-cycle assessment. Environmental Science & Technology, 38 (16): 4457-4464.

Gross M. 2007. Communicating ignorance and the development of post-mining landscapes. Science Communication, 29 (2): 264-270.

Gustavson K E, Barnthouse L W, Brierley C L, et al. 2007. Superfund and mining megasites. Environmental Science & Technology, 41 (8): 2667-2672.

Kennedy C J, Cheong S M. 2013. Lost ecosystem services as a measure of oil spill damages: a conceptual analysis of the importance of baselines. Journal of Environmental Management, 128: 43-51.

Moore J N, Langner H W. 2012. Can a river heal itself?Natural attenuation of metal contamination in river sediment. Environmental Science & Technology, 46 (5): 2616-2623.

Moore J N, Luoma S N. 1990. Hazardous wastes from large-scale metal extraction. A case study. Environmental Science & Technology, 24 (9): 1278-1285.

Pascoe G A, Blanchet R J, Linder G, et al. 1994. Characterization of ecological risks at the milltown reservoir - clark fork river sediments superfund site, montana. Environmental Toxicology and Chemistry, 13 (12): 2043-2058.

Söderqvist T, Brinkhoff P, Norberg T, et al. 2015. Cost-benefit analysis as a part of sustainability assessment of remediation alternatives for contaminated land. Journal of Environmental Management, 157: 267-278.

Zhang Y I, Baral A, Bakshi B R. 2010. Accounting for ecosystem services in life cycle assessment, part II: toward an ecologically based LCA. Environmental Science & Technology, 44 (7): 2624-2631.

Zhang Y, Singh S, Bakshi B R. 2010. Accounting for ecosystem services in life cycle assessment, Part I: a critical review. Environmental Science & Technology, 44 (7): 2232-2242.